THE UNIVERSITY OF ARIZONA SPACE SCIENCE SERIES
Richard P. Binzel, General Editor

The Solar System Beyond Neptune
M. Antonietta Barucci, Hermann Boehnhardt, Dale P. Cruikshank, and Alessandro Morbidelli, editors, 2008, 592 pages

Protostars and Planets V
Bo Reipurth, David Jewitt, and Klaus Keil, editors, 2007, 951 pages

Meteorites and the Early Solar System II
D. S. Lauretta and H. Y. McSween, editors, 2006, 943 pages

Comets II
M. C. Festou, H. U. Keller, and H. A. Weaver, editors, 2004, 745 pages

Asteroids III
William F. Bottke Jr., Alberto Cellino, Paolo Paolicchi, and Richard P. Binzel, editors, 2002, 785 pages

Tom Gehrels, General Editor

Origin of the Earth and Moon
R. M. Canup and K. Righter, editors, 2000, 555 pages

Protostars and Planets IV
Vincent Mannings, Alan P. Boss, and Sara S. Russell, editors, 2000, 1422 pages

Pluto and Charon
S. Alan Stern and David J. Tholen, editors, 1997, 728 pages

Venus II—Geology, Geophysics, Atmosphere, and Solar Wind Environment
S. W. Bougher, D. M. Hunten, and R. J. Phillips, editors, 1997, 1376 pages

Cosmic Winds and the Heliosphere
J. R. Jokipii, C. P. Sonett, and M. S. Giampapa, editors, 1997, 1013 pages

Neptune and Triton
Dale P. Cruikshank, editor, 1995, 1249 pages

Hazards Due to Comets and Asteroids
Tom Gehrels, editor, 1994, 1300 pages

Resources of Near-Earth Space
John S. Lewis, Mildred S. Matthews, and Mary L. Guerrieri, editors, 1993, 977 pages

Protostars and Planets III
Eugene H. Levy and Jonathan I. Lunine, editors, 1993, 1596 pages

Mars
Hugh H. Kieffer, Bruce M. Jakosky, Conway W. Snyder, and Mildred S. Matthews, editors, 1992, 1498 pages

Solar Interior and Atmosphere
A. N. Cox, W. C. Livingston, and M. S. Matthews, editors, 1991, 1416 pages

The Sun in Time
C. P. Sonett, M. S. Giampapa, and M. S. Matthews, editors, 1991, 990 pages

Uranus
Jay T. Bergstralh, Ellis D. Miner, and Mildred S. Matthews, editors, 1991, 1076 pages

Asteroids II
Richard P. Binzel, Tom Gehrels, and Mildred S. Matthews, editors, 1989, 1258 pages

Origin and Evolution of Planetary and Satellite Atmospheres
S. K. Atreya, J. B. Pollack, and Mildred S. Matthews, editors, 1989, 1269 pages

Mercury
Faith Vilas, Clark R. Chapman, and Mildred S. Matthews, editors, 1988, 794 pages

Meteorites and the Early Solar System
John F. Kerridge and Mildred S. Matthews, editors, 1988, 1269 pages

The Galaxy and the Solar System
Roman Smoluchowski, John N. Bahcall, and Mildred S. Matthews, editors, 1986, 483 pages

Satellites
Joseph A. Burns and Mildred S. Matthews, editors, 1986, 1021 pages

Protostars and Planets II
David C. Black and Mildred S. Matthews, editors, 1985, 1293 pages

Planetary Rings
Richard Greenberg and André Brahic, editors, 1984, 784 pages

Saturn
Tom Gehrels and Mildred S. Matthews, editors, 1984, 968 pages

Venus
D. M. Hunten, L. Colin, T. M. Donahue, and V. I. Moroz, editors, 1983, 1143 pages

Satellites of Jupiter
David Morrison, editor, 1982, 972 pages

Comets
Laurel L. Wilkening, editor, 1982, 766 pages

Asteroids
Tom Gehrels, editor, 1979, 1181 pages

Protostars and Planets
Tom Gehrels, editor, 1978, 756 pages

Planetary Satellites
Joseph A. Burns, editor, 1977, 598 pages

Jupiter
Tom Gehrels, editor, 1976, 1254 pages

Planets, Stars and Nebulae, Studied with Photopolarimetry
Tom Gehrels, editor, 1974, 1133 pages

The Solar System Beyond Neptune

The Solar System Beyond Neptune

Edited by

M. Antonietta Barucci
Hermann Boehnhardt
Dale P. Cruikshank
Alessandro Morbidelli

With the assistance of

Renée Dotson

With 101 collaborating authors

THE UNIVERSITY OF ARIZONA PRESS
Tucson

in collaboration with

LUNAR AND PLANETARY INSTITUTE
Houston

About the front cover:

View from a satellite looking at a primary Kuiper belt object. The Sun and zodiacal light are in the background. The shadow cast by the parent object is sharp because the Sun is nearly a point source, <1 minute of arc across. *Painting by William K. Hartmann, Planetary Science Institute, Tucson, Arizona.*

About the back cover:

Discovery of the first TNO: 1992 QB_1. D. Jewitt, University of Hawaii, and J. Luu, University of California at Berkeley, reported in IAU Circular 5611 (September 14, 1992) on the discovery of a very faint object with very slow (3"/hour) retrograde near-opposition motion, detected in CCD images obtained with the University of Hawaii's 2.2-m telescope at Mauna Kea. The object appears stellar in 0".8 seeing, with an apparent Mould magnitude R = 22.8 ± 0.2 measured in a 1".5-radius aperture and a broadband color index V–R = +0.7 ± 0.2. *Image courtesy of David Jewitt, University of Hawaii.*

The University of Arizona Press
in collaboration with the Lunar and Planetary Institute

♾ This book is printed on acid-free, archival-quality paper.
Manufactured in the United States of America

13 12 11 10 09 08 6 5 4 3 2 1

Library of Congress Cataloging-in-Publication Data

The solar system beyond Neptune / M.A. Barucci ...
[et al.], with the assistance of Renee Dotson ; with
101 collaborating authors ; foreword by David C. Jewitt.
p. cm. — (The University of Arizona space science series)
Includes bibliographical references and index.
ISBN 978-0-8165-2755-7 (hardcover : alk. paper)
1. Trans-Neptunian objects. I. Barucci, M. A. (M.
Antonietta)
QB694.S65 2008
523.48—dc22 2007047068

Contents

PART IX: LABORATORY

PART X: PERSPECTIVES

List of Contributing Authors

Scientific Organizing Committee

Acknowledgment of Reviewers

The editors gratefully acknowledge the following reviewers for their time and effort in reviewing chapters for this volume:

Erik Asphaug
Irina Belskaya
Max Bernstein
Gary M. Bernstein
Richard P. Binzel
Michael E. Brown
John Robert Brucato
Alberto Cellino
Sébastien Charnoz
Joshua E. Colwell
Jacques Crovisier
Catherine de Bergh
Audrey Delsanti
Alain Doressoundiram
Martin Duncan
James Elliot
Vacheslav V. Emel'yanenko
Julio A. Fernández
Rodney Gomes
Tommy Grav
Keith Grogan
Joe Hahn
Olivier Hainaut
Alan W. Harris (DLR)
William K. Hartmann
Daniel Hestroffer
Matthew J. Holman
Kevin R. Housen
Robert R. Howell
Reggie Hudson
Željko Ivezić
Peter Jenniskens
Scott J. Kenyon
Sun-Kun King
Zoran Knežević
Leif Kahl Kristensen
Monica Lazzarin
Daniela Lazzaro
Emmanuel Lellouch
Javier Licandro
J.-C. Liou
Amaya Moro-Martín
Robert M. Nelson
Stanton J. Peale
Nuno Peixinho
Jean-Marc Petit
Dina Prialnik
Hans Rickman
Ted L. Roush
Philippe Rousselot
Darrell F. Strobel
Mario Trieloff
Chadwick A. Trujillo
Sebastian Wolf

Foreword

Only rarely are we, as scientists and as people, able to witness a whole new research tree grow and blossom from essentially nothing within just a few years, yet this is exactly what has happened with the Kuiper belt since 1992. Those of us whose research has "helped the tree grow," a group that includes most of the authors in this book, share a real and justifiable sense that we are involved in something big, even something historic in the exploration of the solar system. A milestone has been passed. We have at last moved decisively beyond the domains of the rocky planets and the giant planets into the domain of the comets. Almost everything we find in this third domain is new and surprising: Theory has so far been a rather unhelpful guide. But the shock of the new only shows that we are *really* exploring and adds considerably to the allure of the subject. Even better, most of the key work so far has been done by individuals or very small groups, lending a strong personal dimension to the endeavor. No team meetings or group-think sessions for us! As the field develops this will surely change. But, for now, what a privilege and what a thrill it is to be able to do science this way.

The Kuiper belt is amazing in its ability to link together research areas that previously seemed unlinked. Different populations of small bodies used to be described separately and given different labels, like strange animals in an exotic zoo. Now, we see their connections more clearly. The Centaurs are probably escaped Kuiper belt objects (KBOs). The Jupiter-family comets are Centaurs that have run the gauntlet of the giant planets and survived (the more typical fate of such bodies is to be batted out of the solar system, never to be seen again). Some Earth-crossing "asteroids" are dead or dormant comets recently arrived from the Kuiper belt. Even the Trojans and the irregular satellites of the giant planets could have originated in the Kuiper belt, although convincing evidence for (or against) this possibility is lacking. The context provided by the Kuiper belt shows that the whole system is beautifully interconnected: It has become meaningless to study the different parts in splendid isolation, as we did before. Planetary science will never be the same again.

The Kuiper belt is also a transitional structure that helps us to relate our planetary system to others. We observe dusty disks around main-sequence stars in which the dust lifetime is short compared to the age of the central star. Dust in these "debris disks" cannot be primordial and must be continually supplied from a hidden source if steady-state is to be maintained. Extrapolating from our own system, collisions between unseen extrasolar KBOs constitute a likely source of dust for the debris disks. If so, it is entirely appropriate to think of the Kuiper belt as the Sun's own debris disk. Perhaps in the future we can use the Kuiper belt to help interpret the debris disks, and the debris disks to set the Kuiper belt in a proper stellar context. Planetary scientists and stellar astronomers have something to talk about.

So, what do we really know? A foreword like this is not the place to discuss details but I think it is worthwhile to at least present key results from the first ~15 years of exploration of the Kuiper belt, in the form of a timeline (see Fig. 1). There I have indicated the dates of major observational findings that have changed the way in which we think about the outer solar system. The scale on the right shows the steady rise in the number of objects known.

The broad inferences from these findings are well known. The resonant KBOs, including the abundant 3:2 mean-motion-resonance Plutinos, were probably trapped as Neptune migrated outward, perhaps toward the end of its growth phase. The obvious implication is that the KBOs, including big ones like Pluto and Eris, had formed before Neptune settled in place. While Neptune is not gas-dominated like Jupiter and Saturn, it does contain a few Earth masses of hydrogen and helium that must have been accreted from the rapidly

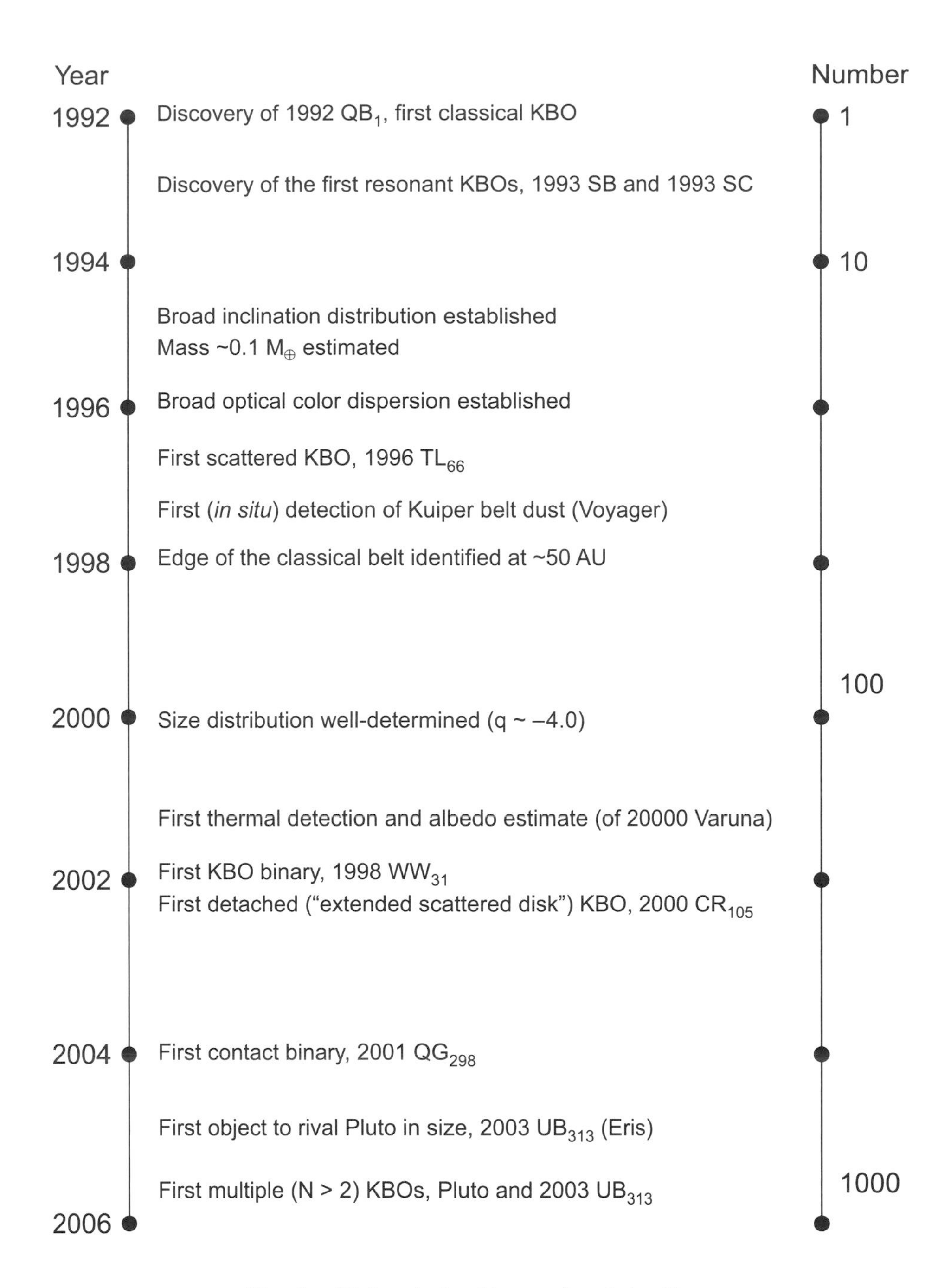

Fig. 1. Kuiper belt: Observational timeline.

dissipating gas nebula. Observations of disks around other stars show that gas survives only a few (at most ten) million years, and so this sets an upper limit to the time available to grow Neptune and, probably, the KBOs. However, the small (0.1 $M_\oplus$) mass and large extent of the belt imply low disk surface densities (~0.005 kg m^{-2}, roughly 10^4 times smaller than at the orbit of Jupiter) and unreasonably long (e.g., billions of years) binary accretion growth times for the KBOs, if they formed in place. A possible solution is to form them closer to the Sun where the protoplanetary disk densities were higher, and growth times shorter, and then to transport them out to their current locations (but how?). Another possibility is that growth rates in the Kuiper belt were much higher than is believed because of local concentration of solids ("particle pile-up") from the action of aerodynamic or other nongravitational forces, but this has yet to be adequately explored. The most direct way to achieve growth times short compared to the age of the solar system is by increasing the density in the Kuiper belt by increasing the mass. A mass augmentation factor of ~100 to ~1000 times is needed, meaning that the primordial belt mass would have been ~10 $M_\oplus$ to 100 $M_\oplus$ (i.e., in the range of a Uranus or a Neptune, perhaps as high as a Saturn mass). It is interesting that high initial densities are also needed to account for the observed high abundance of Kuiper belt binaries. Specifically, none of the major models proposed for binary formation can operate with the necessary efficiency unless the density of objects started much higher than now.

If these high initial masses can be believed, then the real puzzle (and I think still the most important puzzle) for the Kuiper belt is how 99% or 99.9% of the mass was lost. Collisional grinding and subsequent mass loss by radiation drag seem attractive, because they remind us of processes likely to operate in the dusty "debris disks" of nearby stars. However, the KBO size distribution (a differential power law proportional to diameter^{-4}) places a lot of the mass in KBOs too large to have ever been collisionally destroyed, so I suspect this model cannot be the right one. The alternative is dynamical ejection, in which most of the mass is scattered away by a passing star, or careening planets or violent shaking of the solar system in response to resonant interactions between giants, or some other way. Whatever happened left us with a delicately and remarkably structured Kuiper belt, including populations of resonant, scattered, and detached ("extended scattered disk") objects in addition to the classical belt. The unexpectedly sharp outer edge of the classical belt at ~48 AU may also reflect some destructive process at the end of the clearing phase: There are many suggested explanations for the edge, all of them rather *ad hoc*. The clearing process probably also pumped up the velocity dispersion in the belt to its current value of ~1.7 km s^{-1}. Collisions between KBOs today are erosive, not agglomerative. The Kuiper belt is dynamically very far from the thin accretion disk suggested in some early models.

Beyond the models that try to fit specific features of the Kuiper belt, we are beginning to see attempts to propose "global" models that account for the overall structure of the belt *and* the rest of the solar system. In one such model, Uranus and Neptune are survivors of a set of approximately five "oligarchs" whose gravitational interaction ejected the others and simultaneously cleared the Kuiper belt. The efficacy of this model appears to be strongly sensitive to the assumed surface density of the protoplanetary disk, working best for very low values. In another model, no doubt described elsewhere in this book, the key destabilizing event is instead assumed to be caused by Jupiter and Saturn crossing the 2:1 mean-motion resonance. If it happened, resonance crossing would shake the solar system, perhaps hurling Neptune into the Kuiper belt where it would dislodge most of the mass and shower the planetary region with liberated projectiles.

This model further seeks to fit the late heavy bombardment (LHB) that is inferred from lunar data to have occurred 3.8 G.y. ago. Migration into the resonance hence must be delayed by ~700 m.y. after planet formation. In the model, this delay is achieved by making judicious assumptions about the initial orbits of the planets and the magnitude of the torque from the massive primordial Kuiper belt. An everyday analogy for this delayed instability is a pencil balanced on its tip on a flat table top. If the initial conditions (the orientation and initial angular momentum of the pencil about the tip) and external perturbations (vibrations, drag from air currents) are controlled with sufficient precision, the fall of the pencil could be delayed, in principle, for a very long time. (Try this yourself, to get a feel for the fine-tuning required.) But it's not easy. Indeed, honest experts disagree about whether or not the LHB occurred at all because of uncertainties inherent in the interpretation of limited and biased samples of lunar rock. There might have been no LHB "spike" at all, in which case the model, somewhat perversely, would offer us a beautiful explanation of something that did not happen!

So, is figuring out the history of the solar system a hopeless game of guesswork? It sometimes looks like that, but the answer is a definite "no." The flood of data from the first 15 years of the Kuiper belt has unarguably focused our attention on important processes (like migration and planet-planet interactions) that were previously more or less ignored. Detector advances (Moore's Law) and better telescopes bring us greater observational power every year. Observations that were at the limits with cameras of 4 Megapixels in 1992 are trivial with 400-Megapixel cameras now. The next step is Pan STARRS in Hawai'i (1.4 Gigapixels and a 7 deg^2 field of view), a precursor to the Large Synoptic Survey Telescope project. Better observations offer us a real chance to finally figure out what's in our solar system. Better models, when used to cast observationally testable predictions, will eventually show us why it is the way it is.

David Jewitt
Honolulu, Hawai'i
January 29, 2007

Preface

You hold in your hands the gateway to the outer solar system. In just over a decade, what was previously our solar system's *regio incognita* has become the new frontier. *The Solar System Beyond Neptune* provides a benchmark for this nascent exploration with the goal of providing a foundation upon which scientific advancement over the next decade can be built. Herein, more than 100 authors encapsulate our current knowledge on the nature and evolution of the Kuiper belt, the scattered disk, their related populations, and their environment as it extends to the outermost boundary of our solar system. Of greatest importance is the inclusion of our most difficult current challenges and perspectives for how major new advancements may be made. Our target audience is new students and researchers drawn to the mysteries of the outer solar system as well as current practitioners seeking to advance and broaden their understanding.

Historical perspective suggests we are on the threshold of an astonishing new era in our understanding of the outer solar system. Driving our advancement is the rapid pace of discovery and physical characterization enabled by modern technology. Over a century passed between the discovery of the first main-belt asteroid and the thousandth; that milestone required just over a decade for the Kuiper belt. New surveys currently planned may succeed in making the *known* outer solar system population exceed (in number) all others. We are on the threshold of our first *in situ* exploration of the transneptunian region with the New Horizons mission on its way to the Pluto system and beyond into the Kuiper belt. Additional driving forces include our advancing theoretical understanding of the formation and evolution of the transneptunian region as well as the chemistry and thermophysical state of its constituents. It is the interplay of checks and balances between theory, observations, and laboratory work that weave the tapestry of our current and future understanding. Tomorrow's masters are today's students and we challenge new students picking up this volume for the first time to take hold of that fabric and pull, stretch, tear, shred, and re-weave it as needed to create the future.

The authors and editors of *The Solar System Beyond Neptune* deserve our thanks and admiration for the foundation they have built and the future they are enabling through the creation of this volume. While they may seem to be the lucky few who have been the early pioneers, their story is no exception within the history of exploration. Hard work and perseverance are the key ingredients to finding luck and success in any age. We are particularly grateful to Renée Dotson and co-workers of the Lunar and Planetary Institute (LPI) who brought this volume to reality, literally page by page. The unflagging support and dedication for the ongoing success of the Space Science Series by LPI Director Stephen Mackwell and the staff of the University of Arizona Press is key for creating these volumes as gateways to the present and future of space science. Welcome inside!

Richard P. Binzel
Space Science Series General Editor
Lexington, Massachusetts
November 2007

Part I: Introduction

The Solar System Beyond Neptune: Overview and Perspectives

M. Antonietta Barucci
Observatoire de Paris

Hermann Boehnhardt
Max-Planck-Institut für Astronomie Lindau

D. P. Cruikshank
NASA Ames Research Center

Alessandro Morbidelli
Observatoire de la Cote d'Azur

1. INTRODUCTION

We are in a new era in which the frontiers of our solar system have been completely redefined, thanks to the discoveries of Centaurs and transneptunian objects (TNOs). As of 2007, 15 years after the first discovery, more than 1200 new icy bodies have been detected and observed at increasingly greater distances from the Sun. The discovery of the TNOs resulted in the immediate realization that Pluto is a member of a much larger population. A resolution of the International Astronomical Union (August 24, 2006) defined a new category of objects, the "dwarf planets," and Pluto was recognized as the prototype of this group.

Fifteen years of discoveries and advanced studies give today a completely new view of the solar system beyond Neptune, which has allowed us to develop new models of the formation and evolution of our planetary system. These icy bodies can be considered the remnants of the external planetesimal swarms and they can provide essential information and con-straints on the processes that dominated the evolution of the early solar nebula, as well as of other planetary systems around young stars.

As can be seen in the various chapters of this book, different terms are used in reference to these icy bodies. Many authors use "Kuiper belt objects," as this was the historical terminology used immediately after the first discoveries and is still very common in the literature. Although other names were used, for example, "Edgeworth-Kuiper objects," we prefer and we suggest the use of the more neutral name TNOs to avoid the controversy over who first hypothesized the existence of this population, as described in the "historical" chapter by Davies et al.

The discovery of Pluto by Tombaugh in 1930 triggered early ideas concerning solar system objects beyond the orbit of Neptune, at a time where neither the Kuiper belt nor the Oort cloud of comets was known (although many comets in long-periodic orbits had been observed, e.g., the work of Edgeworth and Kuiper in the 1940s and 1950s). Later, in 1982, a more conclusive study by Fernández and Ip argued for the existence of a source of short-periodic comets close to the ecliptic and beyond the known planetary orbits. About 10 years later, Jewitt and Luu discovered the object 1992 QB_1, now numbered 15670, the first body in a near-circular orbit beyond Neptune. This was an epochal astronomical discovery, since it triggered within a few months the detections of further asteroid-like objects in the outskirts of the planetary system. It did not take longer than two to three years to find the first ~100 distant bodies, representative of a remnant entity from the formation period of the planetary system, i.e., the TNO population. However, the discovery story was — and most likely still is — not yet over, leading to the recognition of a "zoo of transneptunian objects" with distinct orbital and physical properties.

The science of the solar system beyond Neptune is continuously and rapidly evolving. The understanding of this region is one of the most active research fields in planetary science at the present, and many new discoveries can be expected in the coming years. The study of this region and the objects it contains will contribute to the understanding of the still puzzling formation age of the solar system.

2. THE TRANSNEPTUNIAN OBJECT POPULATION

Why do we study the transneptunian population? This population carries the scars of the accretional and evolutionary processes that sculpted the current form of the outer solar system. To understand the history of a rock, the radioactive elements are the most useful, even if they are often a negligible fraction of the total rock mass. Likewise, in our quest to understand the evolution of the solar system, the small bodies appear to provide the richest information, even if their total mass is negligible with respect to that of the planets.

In addition to the physical properties of TNOs, which give us information on the thermal and chemical processes in the outer protoplanetary disk — when and where the objects formed — there are two broad characteristics of the TNO population that give us fundamental clues to unveil the history of the solar system: the size distribution and the orbital distribution.

Determining the orbits of TNOs is made difficult by their faintness and the associated complications in following them for several months. The specific difficulties and problem solutions for the orbit determination of TNOs are presented in the chapter by Virtanen et al. These objects, by definition (and unlike most other solar system bodies), are observed only over very short orbit arcs. Here one should remember that Pluto has moved only 155° in true anomaly since its discovery, i.e., less than 45% of its orbit around the Sun. Hence, specific methods for orbit determination are developed, allowing statistical predictions of ephemeris uncertainties that can help improve the orbital parameters by new measurements. Nonetheless, a substantial fraction of the discovered bodies are not reobserved early enough to ensure future secure recovery, and many objects are still lost.

The size distribution of TNOs is reviewed in the chapter by Petit et al. It is now certain that the size distribution of TNOs is very steep at the large size end. The exact value of the exponent of the differential distribution is still debated, but should be between –5 and –4. Pluto and its companions of comparable size fit this single-slope power law. Therefore they appear not to be a special category of objects, but rather the largest statistical members of the TNO population. The steep size distribution cannot extend indefinitely to small sizes, otherwise the total mass of the population would be infinite. Therefore, the size distribution has to "roll over" toward a power law with a shallower slope. The size at which the change in the power-law exponent occurs is a subject of debate. Previous work, using published results and new HST observations that showed a deficit of objects at apparent magnitude ~26, claimed that the rollover is at a diameter range of about 100–300 km. The authors of the Petit et al. chapter challenge this conclusion, presenting new observations that show a unique power law distribution up to magnitude 25–25.5. Settling this controversy requires a larger statistical dataset that will become available with the observations enabled by a new generation of instruments (see the chapter by Trujillo et al.).

The current uncertainty in the size distribution of TNOs does not allow a precise assessment of the total mass of the population. Current estimates range from 0.01 to a few times 0.1 $M_\oplus$. Upper estimates on the total mass also come from the absence of detected perturbations on the motion of Neptune and of Halley-type comets. Whatever the real value, it appears low (by 2 to 3 orders of magnitude) with respect to the primordial mass in the transneptunian region inferred from a radial extrapolation of the solid mass contained in the giant planets. In terms of mass deficit, therefore, the transneptunian region is similar to the asteroid belt.

The steep size distribution at large sizes is usually interpreted as a signature of the accretion process, whereas the shallower slope at small sizes is expected to be the consequence of collisional erosion. The accretion/erosion process is reviewed in the chapter by Kenyon et al. Their chapter explains that the two processes occur contemporaneously. While the larger bodies are still growing, they excite the orbital eccentricities and inclinations of the small bodies, whose mutual collisions start to become disruptive. Because the dispersion velocity of the small bodies is on the order of the escape velocity from the largest bodies, the system is always on the edge of an instability. If some processes (collisional damping, gas drag, weakened solar radiation due to the low optical depth of dust population) reduce somewhat the dispersion velocity of the small bodies or the evacuation rate of the dust, then accretion wins and a substantial fraction of the total mass is incorporated in large, unbreakable bodies. If, conversely, some external perturbation (from a fully grown planet or close stellar passages) enhances the velocity dispersion, then the accretion stalls, and most of the mass remains in small bodies, is eventually ground down to dust size, and is then evacuated by radiation effects. The simulations presented in the Kenyon et al. chapter suggest that in the transneptunian region most of the mass remained in small bodies and that, consequently, the mass deficit of the TNO population was caused primarily by collisional grinding. These same simulations predict the formation of a few Pluto-sized bodies as the largest members of the TNO population. However, several lines of evidence argue in favor of the past existence of 100–1000 Pluto-sized bodies in the planetesimal disk. Strictly speaking, these lines of evidence apply to the planetesimal disk in the region of the giant planets, and not in the transneptunian region. If this is really the case (many large bodies in the inner part of the disk and most of the mass in small bodies in the outer part), the origin of this drastic change of size distribution with heliocentric distance remains to be understood. It should also be noted that current coagulation models fail to produce the cores of the giant planets.

The complex orbital structure has prompted astronomers to divide the transneptunian population into subclasses, often defined slightly differently and given different names in different papers. The chapter by Gladman et al. aims to bring some order and consensus to the classification of TNOs. The authors of the chapters in this book have tried to adopt the Gladman et al. nomenclature and definitions, although this has not always been possible. The idea behind the approach of Gladman et al. is to have a classification that reflects what we see today, making an abstraction of formation and orbital sculpting models that might be largely trusted today but will be discarded in the future. Of course, this is a difficult exercise, because what we actually see when looking at the orbital distribution is inevitably influenced by what we think generated that distribution.

The chapter by Gladman et al. is logically paired with the chapter by Kavelaars et al. on the orbital distribution of TNOs. A crucial issue addressed in that chapter is the understanding of how the orbital structure of the TNOs discovered so far is influenced by observational biases. Some biases are easy to model, in principle; they depend on the pointing history and limiting magnitude of the surveys. Unfortunately, this information is not available for many surveys, in particular those from the early years of TNO discoveries. Other biases are more subtle. For instance, objects that receive incorrect provisional orbits might not be recovered later because of the wrong ephemeris, and therefore are lost. The objects that are not lost are those for which the provisional orbits are correct. In such a circumstance, the resulting catalog of transneptunian orbits looks extremely similar to what the person assigning provisional orbits had in mind! Luckily the situation is not that paradoxical. Modern (and future) surveys give special attention to careful followup of the discovered objects in order to be able to determine the orbits accurately, without the need of *a priori* assumptions. Thus, today some features in the picture of the orbital structure of the transneptunian population appear secure and not due to biases. We will return to these features in section 4.

3. TRANSNEPTUNIAN OBJECT PHYSICAL PROPERTIES

Knowledge of the composition and properties of these icy bodies forming the TNO population would help in better understanding the processes that shaped the solar nebula at large heliocentric distances and that determined the formation and evolution of the planets.

The availability of very large groundbased telescopes (8 and 10 m) and telescopes in space (the Hubble and Spitzer telescopes), equipped with modern sensors in many wavelength regions, has enabled observational studies of the physical properties of a significant number of TNOs. In addition to information about surface composition, the observations have given us fundamental information about the bulk properties of several TNOs. Those properties include the size, shape, presence of satellites, and in some instances the bulk density and porosity. Knowledge of bulk properties is critical to an understanding of the origin and evolution of TNOs.

Photometric observations over time can give insight into the shape of a solar system body that is too small or distant to be seen with any spatial resolution, and nearly a century of electronic photometry with telescopes has taught astronomers how to do this well. Seen from Earth, an object of spherical shape and uniform surface brightness will give a steady brightness as it rotates, while an irregular shape (or nonuniform surface brightness) will result in a lightcurve of variable brightness over time. Experience has shown that the largest bodies have near-spherical shapes and that their variable lightcurves result from an irregular surface brightness (e.g., Pluto). Somewhat smaller bodies have irregular shapes, and those shapes, rotation periods, and orientation of the spin axes can be determined from precision photometric measurements made over time. In their chapter, Sheppard et al. have analyzed the rotation periods and photometric ranges of a number of TNOs, Centaurs, and main-belt asteroids, and show that some objects are spinning so fast that their equilibrium shapes are considerably elongated. The measured rotation periods of TNOs larger than about 50 km in radius range from about 3 to 18 h, with the peak in the frequency distribution at 8–9 h; there is very little information on smaller objects in the TNO population. Objects less than about 50 km in radius are expected to be collisional remnants, with irregular shapes generated at the time of disruption of the original body.

The size of a KBO can be estimated from a measurement of the brightness of the sunlight reflected from it, but only if the reflectance (albedo) of its surface is known. In cases where both the reflected light and the thermal radiation at long wavelengths can be measured, it is possible to calculate both the dimensions of the object and its surface albedo. This "radiometric technique" has been extensively used in asteroid studies, and is well calibrated from many independent measurements of the sizes of asteroids. Transneptunian objects are very cold (~30 to 50 K) because of their great distance from the Sun, and consequently their thermal radiation is very weak and reaches its blackbody peak at a wavelength near 100 μm. Long-wavelength thermal emission can be measured in a few wavelength bands from Earth-based telescopes, but the most sensitive telescope used in this work is the Spitzer Space Telescope, with capabilities to detect exceedingly weak radiation at 24 and 70 μm. For approximately 40 Centaurs and transneptunian objects, the thermal emission at one or both of these wavelengths has been measured by Spitzer, and the sizes have been derived from these and groundbased measurements of the visible radiation (reflected sunlight). Their dimensions are thus known with precision on the order of 10% or 20% (see chapter by Stansberry et al.), and their surface albedos are seen to range widely from about 3% to 85%, with most objects having low values.

Patterns have begun to emerge from statistical studies of albedos and colors, such that the redder TNOs and Centaurs have higher albedos than those with more neutral spectral reflectances. Additionally, albedo appears to be correlated with an object's mean heliocentric distance, diameter, and spectral reflectance. Several possible trends of colors and orbital elements are described in the chapters by Doressoundiram et al. and Tegler et al. In particular, there is a well-known correlation between color and orbital inclination. Highly inclined classical objects have diverse colors ranging from gray to red, while low-inclination classical objects are mostly very red. The significance of these relationships in terms of the origin, evolution, and space environment of TNOs and Centaurs has only just begun to be explored. A taxonomic scheme based on multivariate statistics (see chapter by Fulchignoni et al.) is proposed to distinguish groups of TNOs having the same colors. The differences among these groups could provide some evidence on the evolution processes affecting the TNO population. The effect of the phase angle on the photometric and polarimetric data to analyze the properties, such as grain sizes and albedo as well as porosity of the surface material, is reviewed in the chapter by Belskaya et al.

Satellites have been detected for several TNOs and a few Centaurs using optical methods of high-resolution imaging. Most TNO satellites have been found with the Hubble Space Telescope (see the chapter by Noll et al.), while adaptive optics with groundbased telescopes (e.g., Keck) has revealed others. When the orbital period and distance of the satellites can be determined, the mass of the primary body can be calculated from Kepler's third law, and the mass of the satellite can be estimated. Nearly 10% of the TNOs studied at high spatial resolution from groundbased telescopes have one or more known satellites. Pluto has three known.

In a few cases, TNOs having satellites are also sufficiently large to be detected at thermal wavelengths, as with the Spitzer Space Telescope. In those special cases, it is possible to combine the mass determinations with the size of the body in order to calculate its mean density. The mean density is reflective of the internal composition, particularly the relative fractions of ices, rocky material, and metals, as well as the porosity. Similar information on mean density and porosity has become available for a few asteroids and comets, making it possible to compare small bodies originating in both the outer and inner regions of the solar system. The calculated densities of TNOs have surprised investigators by their wide range, from ~0.5 to nearly 3 g/cm^3. Pluto's density is 2.03 g/cm^3, corresponding to an internal mix of rock and ice. The TNOs with densities less than 1 g/cm^3 are presumed to be porous to varying degrees. In some cases, e.g., (47171) 1999 TC_{36}, the low density requires that some 50–75% of the interior consists of void space.

Many of the bulk physical properties of TNOs carry important implications for their origin and evolution. The occurrence of binaries, for example, cannot be explained by close encounters or collisions in the TNO population that presently exists. Instead, it appears to be a remnant of the early, larger, population in which multiple encounters and mutual collisions were far more frequent than is possible today. The wide range in mean density of TNOs and Centaurs challenges us to explore scenarios of formation, collisional history, and internal thermal processing. Concurrent discoveries about the physical properties and compositions of comets, presumed to originate from the Kuiper belt, have given surprising results on the heterogeneity of these bodies, which include large fractions of high-temperature minerals from regions in the solar nebula closer to the Sun than Mercury, as well as materials representative of condensation at large heliocentric distances.

Observational studies of the compositions of TNOs, Centaurs, and the comets that came from the Kuiper belt depend on high-quality observational data, mostly obtained with groundbased telescopes. The most diagnostic spectroscopic information occurs in the near-infrared, with the spectral region at 1.0–2.5 μm carrying much of the information about ices and some minerals in the surface layers of these objects. Fortunately, this spectral region is readily detectable with groundbased telescopes. However, remote sensing observations are limited to probing only the "optical" surfaces of TNOs, and the subsurface composition must mostly be inferred rather than measured directly. Larger telescopes and improving spectrometers continue to expand the range of objects that can be observed, but a fundamental limitation outside the observatory is the paucity of laboratory spectroscopic data on candidate materials on the surfaces of TNOs, Centaurs, and comets.

Recently surveys of TNOs have begun to reveal objects of comparable and even larger size than Pluto. Most of the large TNOs are sufficiently bright for detailed physical study and most of the TNOs, like Pluto, have unique dynamical and physical histories (see chapter by Brown). As a whole, the largest TNOs appear to be more diverse in surface composition, presence of satellites, and density. It is probable that about three more TNOs of large size await discovery, but perhaps tens to hundreds more can exist in the distant region where Sedna resides during its 11,000-year orbit. Among the large objects detected, Sedna appears dynamically distinct from the entire transneptunian population. It has a perihelion beyond the main concentration (more than twice the semimajor axis of Neptune) and an extreme eccentric orbit with an aphelion at 927 AU.

Although the discovery of such objects presages a large population in the distant region, no surveys for fainter objects have yet succeeded in detecting such distant objects. The four largest known TNOs (Eris, Pluto, Sedna, and 2005 FY_9) have surfaces spectrally dominated by frozen methane, but the surface characteristics differ on each body. Eris is currently the largest known object of the population, with a remarkably high albedo of about 0.87, and a satellite, named Dysnomia.

The most striking difference between the largest TNOs and the remainder of the population is the presence of volatiles in the spectra of the large objects compared to relatively featureless spectra of the smaller ones. All the large objects show the presence of some ices on their surfaces.

Investigations of the surface compositions of TNOs and Centaurs consist of measurements of color, i.e., the shape and slope of the spectral energy distribution of reflected sunlight, and spectroscopic observations aimed at the detection of specific molecules and minerals (see chapter by Barucci et al.). The enabling laboratory data for ices, minerals, and refractory organic materials consist of a miscellany of spectra and optical constants (complex refractive indices) obtained at various spectral resolutions over various wavelength intervals. In many cases the resolution and wavelength regions are inadequate or inappropriate for the observational data at hand, and since the observations cannot be made at any arbitrary resolution or spectral region, it is essential that the laboratory data be taken under the appropriate conditions and in the appropriate ways. The chapter by de Bergh et al. gives an extensive review of the available laboratory data for candidate materials found or expected on the surfaces of TNOs and Centaurs, and identifies the gaps in available information. As they note, the complex refractive indices of ices, organic materials, and minerals are of special importance because they are used in radiative transfer model calculations of synthetic spectra to match the observational data. These indices are often very difficult to measure, but their importance is underscored by the successes that have been achieved in modeling TNO and Centaur spectra, sometimes with as many as five different materials.

In their chapter, de Bergh et al. emphasize the nature and importance of complex refractory organic solids, because for some outer solar system bodies with especially red colors, only these organics have been able to match the spectral characteristics obtained at the telescope. Imperfect and imprecisely diagnostic as they are, a class of refractory organics called tholins has proven to be the organic material of choice in modeling the surfaces of outer solar system bodies, in part because they are the only materials for which reliable optical constants are readily available. Whatever their limitations, the tholins are found to account for the colors of a great many bodies in the solar system, including planetary satellites, certain asteroids, Centaurs, and TNOs.

4. PHYSICAL PROCESSES

The TNO population embraces the most pristine objects in the solar system, but over the course of 4.5 G.y. they have suffered various weathering processes, including damage from cosmic rays and ultraviolet radiation, sputtering and erosion, and mutual collisions. An understanding of all these processes is critically important to the interpretation of the surface compositions and the internal compositions and structures of the TNOs.

The surface structure and chemistry of a TNO or Centaur is affected by the aggressive attack by energetic phenomena in the space environment, with the result that molecular complexes are structurally changed and the molecular compositions of ice and minerals are altered over time. Laboratory experiments on a variety of appropriate materials, irradiated by energetic particles that simulate the space environment, are described in the chapter by Hudson et al., who show how molecular transformations lead to the production of different components when (primarily) ices are irradiated. This topic is in its infancy, and as it develops, it is certain to produce results important and relevant to the study of TNOs and Centaurs.

Collisions are very important for the evolution of TNOs, as is evident from the occurrence of interplanetary dust particles and the size distribution and total mass of the TNOs. In their chapter, Leinhardt et al. analyze the physical effect of collisions, describing recent advances on the effects of collisions by laboratory experiments and new numerical simulations. Their chapter discusses possible relevant consequences of collisions as they pertain to the alteration of the surface properties and the modification of the internal structure of the target. Collisions are relevant both to small and large objects. The moons of Pluto are modeled to have been formed from a disk of debris ejected during the collision of Pluto with a projectile of almost equal size, in a process similar to that of the formation of Earth's Moon. Very recently, the first collisional family has been discovered in the transneptunian population, associated with the nearly Pluto-sized body (136108) 2003 EL_{61}. This object is an anomalously fast rotator, has an unusually high density, and has two satellites. These features alone suggested that 2003 EL_{61} was originally a body with a high-density rocky core and a low-density icy mantle that suffered a giant collision that spun it up and ejected a large fraction of the mantle into space. Now, five other objects, with diameters ranging from 150 to 400 km, have been identified to be tightly clustered in orbital space around 2003 EL_{61}. These objects share the same physical properties: an icy surface whose spectrum shows deep-water absorption bands and a "gray" color. No other object in the TNO population has both these properties (see chapters by Brown and Barucci et al.). Altogether, these aspects make this collisional family a very compelling case: a unique "beast" in the solar system, given that no family in the asteroid belt contains objects of comparable sizes.

Very little is known about the internal structure of TNOs although they are presumed to display great diversity. The interplay between the effects of solar heating and internal heating in bodies with different initial conditions can result in completely different configurations. In their chapter, Coradini et al. investigate the thermal evolution of these bodies, and how and when it can procede to internal differentiation. By estimating the presumed surface expressions of differentiation and evolution of a TNO, the authors try to link the surface properties with those of the interiors. In particular, they investigate the link with the comets, and they conclude that the great variety observed in comets can either reflect their initial compositions or can be related to the collisional disruption of previously differentiated bodies, thus giving rise to objects with different volatile content.

The presence of some ices on the surfaces of all large TNOs indicates that space weathering and collisional resurfacing are not the only mechanisms that can affect the surface properties, and probably some internal geological activity such as cryovolcanism should be considered. The detection of surface volatiles and high albedos on some TNOs indicates the possible existence of atmospheres, at least as transient phenomena on other TNOs in addition to Pluto. The structure of a TNO atmosphere will depend on the distance from the Sun, the atmospheric composition, internal radiative transfer, and many other physical processes. As described in the chapter by Stern and Trafton, the principal requirement for atmospheric formation on transneptunian objects is the presence of gases or sublimating/evaporating materials on or near the surface. In the case of low-gravity bodies such as TNOs, which have prodigious atmospheric escape rates, it implies some resupply mechanism to the surface, such as internal activity, or the import or excavation of volatiles by impactors.

Outgassing from the interior is another important process, not only because of the high volatile content of TNOs, but also because they may be unusually porous (see the chapters by McKinnon et al. and Coradini et al.). Such porosity increases the conductivity of volatiles to the surface and therefore the effective size of the reservoir that supports an escaping atmosphere or coma. The importance of internal release, from the near surface (as for the geysers of Triton and Enceladus) or from deeper inside, should not be underestimated.

5. FORMATION AND EVOLUTION

Beyond Neptune, but inside the Oort cloud, we note there are two major populations, each containing a comparable number of objects. One group, called the scattered disk, occupies unstable orbits scattered by Neptune; another group has orbits that are stable, at least on timescales of several billions of years. The latter can be subdivided into a resonant population (inhabiting major mean-motion resonances with Neptune) and a classical population (not affected by any notable resonance). To make things more complex, there is also a stable population of objects whose distribution is remarkably similar to that of the scattered disk. This population is called "detached" (sometimes "fossilized" or "extended scattered"). Whereas the scattered disk and the detached population span all values of semimajor axis (with larger eccentricities for larger semimajor axes), the classical population is sharply bounded at 48–50 AU. Something happens there; either there is a sharp drop in the density of the population (an outer "edge"), or a sharp change in the size distribution (all objects beyond this limit being too small — or, less likely, too dark — to be detected by current surveys).

The eccentricities of the resonant population can be as large as 0.3 or 0.4. In fact, resonant objects are stable even if they are Neptune crossers (within some limits), thanks to the stabilizing effect of resonant dynamics. Conversely, the eccentricities in the classical population are bounded to the range 0.1–0.2 (depending on semimajor axis). In fact, with larger eccentricity values, the objects would be scattered by Neptune. A remarkable feature is that the eccentricity distribution in the classical population does not peak at zero.

For semimajor axes up to 45 AU, the eccentricity distribution is rather flat in the allowed stability range. Between 45 AU and the edge of the classical belt, there is even a deficit of low-eccentricity objects! Finally, the inclination distribution in the classical belt is bimodal. About half the population has a peaked inclination distribution in the range 0°–4°, and the remainder has a flat inclination distribution ranging up to 35°. These two subpopulations with small and large inclinations are called "cold" and "hot," respectively. Curiously, these subpopulations seem to have different physical properties. Their colors are different (see the chapter by Doressoundiram et al.), as are their size distributions (see the chapter by Petit et al.).

The theoretical models of the origin of the orbital structure of the transneptunian population are reviewed in the chapters by Gomes et al. and Morbidelli et al. The former focuses on the scattered disk and the detached population. It reviews the dynamics of the scattered population, characterized by several episodes of trapping into mean-motion resonances with Neptune of even larger order. It demonstrates that the current scattered disk should be the remnant of a much more populated structure, formed when Neptune scattered away the planetesimals from its neighboring region. Gomes et al. also show that the scattered disk, not the classical or the resonant populations, is the source of Centaurs and Jupiter-family comets, and they also discuss a possible connection between scattered disk and Halley-type comets. Finally, they address the origin of the detached population during a primordial phase in which Neptune's orbit was still evolving (expanding and with possible large changes in eccentricity) due to the interactions with the other planets and with the neighboring planetesimals.

The chapter by Morbidelli et al. focuses on the classical and the resonant populations. First, it reviews the now "classical" model of Neptune's migration and the origin of the resonant populations, and then addresses the issue of formation of the outer edge of the classical belt, presenting the various models proposed so far. Finally, Morbidelli et al. come to the problem of the mass deficit of the transneptunian population, arguing in favor of an alternative solution to the collisional grinding scenario presented by Kenyon et al. They propose that the original planetesimal disk had an outer edge somewhere around 30 AU (which helped stabilize Neptune at its current location). The objects that we see today in the classical population would have formed within 30 AU and implanted into their current region during a chaotic phase of the evolution of the giant planets. If this view is right, the current low mass of the transneptunian population is not due to the elimination of mass from the 40–50-AU region, but to the low efficiency of the implantation process from the inner part of the disk.

What emerges from the chapters by Gomes et al. and Morbidelli et al. is the ambition to explain the current structure of the transneptunian population in the framework of a unitary model of evolution of the solar system. The model that is presented — called the Nice model — aims to simultaneously explain the current orbits of the giant planets, the capture of their irregular satellites, the late heavy bombardment of the Moon, the properties of the transneptunian population, and the origin of the Trojans of Jupiter and Neptune. Time will tell whether this model will stand as a template for our view of solar system history, or will eventually fail in the light of some discovery that it cannot explain.

Somewhat related to the Gomes et al. chapter is that by Duncan et al., devoted to the origin of the Oort cloud. The Oort cloud is in some sense the distant end of the scattered disk, perturbed by passing stars and the galactic tides. The Duncan et al. chapter reviews the reference formation model and the remaining open problems. The discovery of Sedna, with a semimajor axis of roughly 500 AU and a perihelion distance of ~80 AU, brought about a real revolution in our view of the Oort cloud structure and of its formation. Sedna could be considered as a member of the detached population. However (as shown in the Gomes et al. chapter), the models that successfully reproduce the distribution of the detached population at smaller semimajor axes fail to explain the origin of the orbit of Sedna. The fact that no objects have ever been discovered with a perihelion distance comparable to that of Sedna but with a much smaller semimajor axis (despite the more favorable observational biases) suggests that Sedna is actually at the edge of the inner Oort cloud (see also the nomenclature chapter by Gladman et al.). The extension of the inner Oort cloud down to 500 AU in semimajor axis places an important constraint on the galactic environment in which the solar system formed. This feature is successfully reproduced by postulating that the Sun formed in a cluster with a central density of gas and stars on the order of 10^4–10^5 $M_\odot$ per cubic parsec. These densities compare well with those observed in young stellar associations that are a few million years old, and are slightly lower than that of the Trapezium region in Orion. Thus, the Sun should have formed in a quite typical environment and not as a rare, isolated star.

6. LINKS WITH THE OTHER POPULATIONS, BOUNDARIES, AND COMPARISON WITH OTHER STELLAR SYSTEMS

Although this book is devoted primarily to the transneptunian population, we also explore the interrelationships between TNOs and other populations inside Neptune's orbit, such as the Centaurs, Jupiter Trojans, etc. We also must consider the question of "What is the boundary of our solar system?" and the effects on TNOs (including comets) caused by the interactions between our solar system's boundary and interstellar space. Of particular fascination is how our solar system appears from afar, and therefore, how might we recognize "Kuiper belts" or "Oort clouds" around other stars that are similar to or different from our own.

In a "traditional" view, where the jovian Trojans formed around Jupiter's orbit and the irregular satellites were captured from the regions in the vicinity of the giant planets, the similarities among Trojans, satellites, and TNOs tell us that the primordial planetesimal disk had quite uniform physical properties with respect to heliocentric distance, at least beyond the so-called snowline (the distance at which water vapor condenses as ice). The differences just remind us that the disk could not be *totally* uniform. In the view of the Nice model, Trojans, satellites, and TNOs are all captured into their current residence regions from an annulus of planetesimals roughly between 20 and 30 AU.

The Oort cloud would have been assembled in two stages. In the first stage, occurring immediately after giant planet formation when the Sun was presumably in a dense galactic environment, the inner Oort cloud formed from the planetesimals within 20 AU. Sedna should have been emplaced onto its current orbit during this stage. The second stage occurred when the 20–30-AU planetesimal annulus was dispersed at the time of the late heavy bombardment, and resulted in the formation of the outer Oort cloud (see chapters by Duncan et al. and Gomes et al.).

Thus, through the interrelationships in their formation scenarios, the resulting similarities are obvious: these objects are just brothers and sisters. The physical differences need to be explained by the different evolutionary paths that the three categories of objects followed to reach their current orbits. For instance, Trojans do not have extremely red colors, unlike TNOs. But the nuclei of Jupiter-family comets also do not show extremely red colors, although we are confident that they come from the Centaurs (characterized by a bimodal color distribution), which in turn come from the scattered disk (characterized by a mixed variety of colors). In conclusion, studying similarities and differences between TNOs and other populations inside Neptune's orbit is interesting, although the interpretation is model-dependent. In principle, a deep understanding of differences and similarities might help us in discriminating among different views of the origins of these populations.

The chapter by Nicholson et al. focuses on the irregular satellites of the giant planets, reviewing the observational efforts that made their discovery possible, the satellites' orbital distribution, and their physical properties. Finally, it discusses merits and drawbacks of the various capture mechanisms proposed so far, including the Nice model.

The chapter by Dotto et al. reviews the orbital and physical properties of the Trojans of Jupiter, noting similarities and differences with both primitive asteroids and TNOs. While the actual relationship among irregular satellites, Trojans, and TNOs is still debated, the Jupiter-family comets (see the chapter by Lowry et al.) are believed with confidence to be representatives of the *in situ* unobservable population of kilometer-sized TNOs. These comets escaped from the transneptunian region and were transported into the inner solar system by the combined scattering actions of the four giant planets. Hence, the physical properties of the Jupiter-family comets can help to illuminate and understand those of the more distant companions remaining in the transneptunian region, in particular since the former are easier to observe from Earth and have even spacecraft measurements available.

An interesting (although for the moment still debatable) scenario for links with other solar system bodies is illustrated in the chapter by Gounelle et al., i.e., meteorites from the outer solar system. Here, in particular the CI1 chondrites are suspected to originate from the population of Jupiter-family comets, which, in turn, as we have seen, are dynamically linked to the transneptunian population. Hence, CI1 chondrites can provide information on the physico-chemical properties and constitution of matter in the outskirts of the planetary system. An intriguing finding from the CI1 chondrites is the indication of hydrothermal alteration of the material, implying that their parent bodies experienced physical conditions leading to the presence of liquid water in their interiors.

Last, but not least, the collision history of the Kuiper belt calls for the existence of a dust cloud at larger heliocentric distances, a debris disk entity that can also be observed around some other stars. With a production of ~10^{15} g/yr of dust (measured from the two Pioneer spacecraft), the transneptunian population is probably the main generator of small particles in the solar system, slightly exceeding the combined roles of the asteroid belt and of sublimating comets in the inner solar system. The evolution of the dust generated in the transneptunian region is reviewed in the chapter by Liou and Kaufmann. The dust particles that are sufficiently large not to be blown away by radiation pressure spiral inward under the Poynting-Robertson drag effect. Like comets, most of the dust is eventually scattered into hyperbolic orbits during close encounters with the giant planets. However, a fraction of the dust population manages to pass through the giant planet system in a relatively unperturbed manner, and penetrates into the inner solar system with orbits characterized by moderate eccentricities and inclinations, quite typical of the dust particles produced in the main asteroid belt. Liou and Kaufmann estimate that the dust produced in the transneptunian region contributes up to 5% of the dust flux at 1 AU.

Beyond Neptune, the steady-state distribution of the transneptunian dust does not have a cylindrical symmetry. It has specific azimuthal structures that would reveal the presence of Neptune to a putative observer from outside the solar system. The connection between the transneptunian dust and the debris disks observed around other stars is discussed in the chapter by Moro-Martín et al. Of course, the debris disks that we observe are much more massive than the current transneptunian dust disk, and probably correspond to the phase when the transneptunian planetesimal population was much more numerous than now. Frequently they show structures in the disk light distribution — such as spirals, rings, and clumps — that are interpreted to originate from collision events. The detection of inner cavities and brightness asymmetries and other features in disks might be a signature for the presence of massive planets shaping the disk geometry through their gravitational interaction. Many disks do not show solid-state features, indicating that the particles therein are larger than 10 μm in size. A few disks, however, display strong silicate emission that might be due to the release of small grains ejected by collision events of major planetesimals in the disk.

Finally, our view of the solar system beyond Neptune must consider where our solar system ends. The outermost nongravitational influence of our solar system on the galaxy comes from the Sun's magnetic field and the outflow of the solar wind. In essence, these create a "bubble" in interstellar space, within which our solar system resides. Outside this boundary, called the heliopause, one finds the interstellar medium. The chapter by Richardson and Schwadron examines our current knowledge of the heliopause and its probing and eventual crossing by the (hopefully) still transmitting Voyager spacecraft. At that time we will know directly that for the first time an object mady by human beings has left our solar system and entered interstellar space.

When Edgeworth and Kuiper conjectured the existence of a belt of small bodies beyond Neptune, they certainly were imagining a disk of planetesimals preserving the pristine conditions of the protoplanetary disk (e.g., extremely small orbital eccentricities and inclinations). But, since the first discoveries of transneptunian objects, astronomers have realized that this picture is not correct; the disk has been affected by a number of processes that have given the population a very complex structure. As we anticipated above, this structure can potentially help us to unveil what actually happened out there, and in turn understand how the giant planets of the solar system formed and evolved.

7. PERSPECTIVES

The last section of the book provides perspectives for the future and new directions for the exploration of the transneptunian region. Unlike the surveys performed up to now, future survey projects (see the chapter by Trujillo) aim at a near-complete coverage of the transneptunian region that is within the reach of 2–8-m-class groundbased telescopes. Not only will the instrumentation enable the discovery of a tremendous number of new objects, but it will also provide high-quality orbit determinations (enabling proper dynamical analysis), and it will facilitate measurements of their physical properties. A putative population of small and very distant objects that are invisible to conventional Earth-based observations because they are too faint might be detectable through occultations of background objects, through a new observing option described in the chapter by Roques et al. In fact, this method could enable the investigation the subkilometer-sized population of the TNO size distribution, as well as the outer regions of the Kuiper belt.

Stellar occultations observed from Earth are also probably the most powerful tool available for the detection of TNO atmospheres. Although occultations are capable of revealing atmospheres down to microbar pressure levels, stellar occultations by TNOs are rare because of their tiny angular sizes and uncertainties in the knowledge of their orbits. Further in the future, the American-European-Chilean-Atacama Large Millimeter Array (ALMA), the Cornell Caltech Atacama Telescope (CCAT), and the joint ESA/NASA Herschel mission could be used for a large survey of 100-km-class-sized TNOs.

And finally, NASA's *New Horizons* spacecraft is already on its way to Pluto-Charon. In 2015 its instruments (see the chapter by Weaver et al.) will completely open new insights into the physics and chemistry of these two Kuiper belt representatives, no doubt disclosing many of the secrets of these bodies that only an *in situ* investigation can reveal. And with some luck *New Horizons*, after its visit at Pluto-Charon, will have the opportunity to fly by one or more so-far-undetected Kuiper belt objects farther along the road to the outskirts of our planetary system.

The Early Development of Ideas Concerning the Transneptunian Region

John K. Davies
UK Astronomy Technology Centre

John McFarland and Mark E. Bailey
Armagh Observatory

Brian G. Marsden
Harvard-Smithsonian Center for Astrophysics

Wing-Huen Ip
National Central University, Taiwan

We review the history of the prediction of, and searches for, a population of comets and transneptunian planetesimals. Starting with initial speculations before and after the discovery of Pluto, we examine various predictions by Edgeworth, Kuiper, and others on the existence of such a population and review the increasingly sophisticated theoretical efforts that eventually showed that the number of short-period comets requires that an ecliptic transneptunian population exists. We then recount various search programs that culminated in the discovery of the first few transneptunian objects and led to the realization that this region is dynamically much more complicated than first suspected and has important links both to Centaurs and the dense inner core of the Oort cloud.

1. REACTIONS TO THE DISCOVERY OF PLUTO

"In the little cluster of orbs which scampers across the sidereal abyss under the name of the solar system there are, be it known, nine instead of a mere eight, worlds." Datelined Flagstaff, Arizona, March 13, 1930, the seventy-fifth anniversary of the birth of its founder, this announcement from the Lowell Observatory via the Associated Press was brilliantly concocted. It could hardly fail to attract the attention of the educated population of the third of those worlds. It did not matter that the subhead in the next day's *New York Times* to the effect that "The Sphere, Possibly Larger than Jupiter and 4,000,000,000 Miles Away, Meets Predictions" represented a gross misinterpretation of the truth. The Lowell Observatory had announced that the solar system now had nine planets, and there could be no argument about that. No matter that most astronomical textbooks written a decade before Lowell was born had stated that there were then already 11 known planets.

It was the addition of Neptune in 1846 and the growing number of discoveries of small bodies between Mars and Jupiter that prompted the astronomical community to count just eight bodies as "planets" and to relegate the lesser bodies to the status of "minor planets" (kleine Planeten, petites planètes, etc.) or "asteroids." Although the term asteroid ("star like") had been coined by William Herschel, it had rarely been used outside the United States, perhaps because, as the discoverer of the substantially larger planet Uranus, Herschel had deliberately intended to convey a somewhat derogatory meaning.

Although in 1930 few astronomers doubted that the young Clyde Tombaugh (who at the time received very little credit for his single-handed and tremendously laborious search) had come across a particularly interesting object, it did not help that the Lowell Observatory provided nothing in terms of quantitative information about the new body apart from a rough estimate of its sky position on the day before their grandiose announcement. The first real evidence came from George(s) van Biesbroeck, whose measurements from photographs obtained at the Yerkes Observatory three and four days later, and published March 20, 1930, on Harvard Announcement Card 112, suggested — but did not by themselves prove — that the new object might be located beyond Neptune.

Confirmation came from the orbital computations by Ernest C. Bower and Fred L. Whipple, graduate students of the University of California at Berkeley and members of what was then the world's leading school for the computation of orbits. On the basis of Van Biesbroeck's data and a three-week series of observations obtained at the Lick Observatory by F. W. Meyer, Bower and Whipple showed (Harvard Announcement Card 118, April 7, 1930) that the new body was some 41 AU from the Earth in an orbit inclined at 17° to the ecliptic with a well-defined nodal direction. Since the orbital eccentricity was completely indeterminate, whether the body was or was not a bona fide transneptunian body was idle speculation, although Bower and Whipple were able definitively to state that, even if the orbit were parabolic, the perihelion distance could not be less than 17 AU.

Finally, a week after the publication of the Bower-Whipple conclusions, the Lowell group published its own orbital calculation on Harvard Announcement Card 121 (April 14, 1930). This was based on just three observations at monthly intervals and although there was a warning that "considerable revision . . . is not unexpected," the Lowell report indicated specific values of 0.909 for the orbital eccentricity and more than 3000 yr for the orbital period.

Since the press had been eagerly awaiting another statement from the Lowell Observatory, they asked Armin O. Leuschner, director of the "Students' Observatory" that hosted Bower and Whipple, for a comment. Leuschner had been impatient over the Lowell group's persistent failure to support its claims and irritated by its public relations success. In the *New York Times* of April 14, 1930, under the subhead "Lowell Observatory Estimates Put Trans-Neptunian Object in Asteroid or Comet Class," Leuschner made the most of his opportunity to speak out: "The Lowell result confirms the possible high eccentricity announced by us on April 5. Among the possibilities are a large asteroid greatly disturbed in its orbit by close approach to a major planet such as Jupiter, or it may be one of many long-period planetary objects yet to be discovered, or a bright cometary object." Then came his coup de grâce: "I have frequently referred to the close orbital and physical relationship of minor planets and comets. High eccentricity and small mass would seem to eliminate object as being planet X predicted by Lowell, and singly an unexpected discovery, nevertheless of highest astronomical importance and interest on account of the great distance of the object in the solar system at discovery."

In fact, authorities such as *Campbell* (e.g., 1916, 1919), *Aitken* (e.g., 1926), and *Leuschner* (e.g., 1927) had been speculating for many years about the possibility of transneptunian planets and the orbital distribution of small bodies in the outer planetary system (cf. *Leuschner,* 1932; *Öpik,* 1932), and had frequently considered comets in general to represent material that had been left over on the outskirts of the solar system beyond the orbit of Neptune. This suggests that these authors may have contemplated the existence of an entity similar to that which is nowadays variously called the Kuiper belt, the Edgeworth-Kuiper belt, or the transneptunian belt. Indeed, *Beekman* (1999) has argued that Leuschner ". . . suggested that Pluto 'could be the first of a large group of such objects'," continuing " . . . in view of its size, Pluto as a comet would of course be exceptional, but in the asteroid belt — between the orbits of Mars and Jupiter — did one not find among the large ensemble of dwarfs also a few giants, such as Ceres, Pallas and Vesta?" However, in playing down the idea that the object soon to be known as Pluto was the "transneptunian planet" predicted by Percival Lowell, it is clear that Leuschner was merely stressing that there are many long-period planetary or cometary objects yet to be discovered. Thus, especially in view of his remarks about high orbital eccentricity, it seems more likely that he was envisaging a much more extended distribution of transneptunian objects.

In this context, a popular article published by Frederick C. Leonard soon after the discovery of Pluto seems more à propos (*Marsden,* 2000). By the middle of May 1930, recognition of a likely prediscovery observation of Pluto from three years earlier had allowed Andrew C. D. Crommelin (*Circ. Brit. Astr. Assoc., 93*) to conclude that the new object had an orbital eccentricity rather less than 0.3 and a perihelion point just inside the orbit of Neptune, results that were confirmed during the following weeks as further old images were located. By August that year, *Leonard* (1930) could therefore write with some confidence: ". . . Now that a body of the evident dimensions and mass of Pluto has been revealed, is there any reason to suppose that there are not other, probably similarly constituted, members revolving around the Sun outside the orbit of Neptune? . . . As a matter of fact, astronomers have recognized for more than a century that this system is composed successively of the families of the terrestrial planets, the minor planets, and the giant planets. Is it not likely that in Pluto there has come to light the *first* of a *series* of ultra-Neptunian bodies, the remaining members of which still await discovery but which are destined eventually to be detected?"

2. FIRST QUANTITATIVE APPROACHES

2.1. Edgeworth

A more comprehensive approach to the problem was made by the independent Irish astronomer Kenneth E. Edgeworth (*McFarland,* 1996) during the 1930s. After a successful military and civilian career, Edgeworth retired to his family home in Ireland and began developing his ideas on the cosmogony of the solar system. This work, "The Evolution of the Solar System," culminated in a manuscript submitted for publication in 1938 (*McFarland,* 2004), which essentially developed the very old idea [dating back, at least, to Kant's (1755) *Universal Natural History and Theory of the Heavens*] that the formation of planets could be understood as a consequence of the accumulation of numerous smaller bodies, or condensations, in a protoplanetary disk that extended far beyond the known planetary orbits. Edgeworth's manuscript lay in the hands of several publishing houses (e.g., George Allen and Unwin Ltd., Methuen and Co. Ltd.) as early as the spring of 1938. It also reached several leading astronomers of the day. For example, at the suggestion of R. A. Lyttleton, a copy was sent by F. J. M. Stratton to W. J. Luyten, who commented favorably upon Edgeworth's approach to the problem in a personal communication to the latter (*Luyten,* 1938).

His published work (*Edgeworth,* 1943, 1949) appears to have been the first quantitative investigation into the possible existence of a vast number of potential comets in an ecliptic annulus beyond the orbits of Neptune and Pluto. Postulating a primordial disk of gas and small particles orbiting around an already well-developed Sun, he proposed, in what was a very early discussion of the effects of viscous and tidal forces on the dissipation of angular momentum in the protoplanetary disk, that if the system was sufficiently dense to cause it to condense into various subregions, then these would coalesce to form the major planets.

On the outskirts of the system, however, beyond Neptune and Pluto, the density of the disk would be lower and the condensation processes that formed the major planets would have insufficient time to operate fully and form large single planets. Thus, again following ideas that can be traced to Kant's cosmogony, Edgeworth noted that owing to the decrease of density in the outskirts of the nebula and the lower velocities of condensations in this region, the rate of growth of individual bodies would decrease rapidly with increasing heliocentric distance (cf. *Bailey,* 1994).

In this way, Edgeworth calculated that at great distances the condensation processes would produce a system comprising a very large number of relatively small "heaps of gravel" that would survive to the present day. He felt that if these bodies were seen at close quarters they would appear as partially condensed clusters composed of a small nucleus with a concomitant Saturn-like disk (*Edgeworth,* 1961). These bodies would become visible as observable comets if perturbed on to Sun-approaching orbits.

In his unpublished manuscript (*Edgeworth,* 1938), he also made order-of-magnitude calculations of the approximate number and sizes of the potential comets beyond Neptune, first for a total mass in the annulus of 0.33 $M_\oplus$ and then for 0.1 $M_\oplus$. These calculations yielded figures of 200 million and 2000 million objects with individual masses of about 2×10^{-9} $M_\oplus$ and 5×10^{-11} $M_\oplus$, respectively, i.e., they would be smaller and more numerous than most of the then-known minor planets in the main asteroid belt. The annulus, Edgeworth reasoned, extended from about 65 AU to perhaps over 260 AU and he felt that these numbers and sizes matched those required to replenish the continual loss of comets (*Edgeworth,* 1938).

From his calculations, Edgeworth concluded that Neptune represented the limiting case for the formation of a single large planet in the outer solar system. Unless there was considerably more mass than seemed reasonable in the transneptunian disk, it would be impossible to form a single large transneptunian planet. The status of Pluto, in Edgeworth's mind, appeared to alternate between that of a planet and that of an escaped satellite of Neptune. Of Pluto, he wrote: "Pluto, the latest addition to our list of members of the solar system, is too small to be classed as a major planet, in spite of its position; it has been suggested that it is an escaped satellite of Neptune's and we shall find in due course that there are good reasons for placing it in that category" (*Edgeworth,* 1938). In making this remark he was presumably referring to the paper of *Lyttleton* (1936) on a possible origin for Pluto. Later, in his book (*Edgeworth,* 1961), he sometimes ranks it among the planets.

Overall, Edgeworth had a remarkably interesting and productive life and many of his astronomical ideas anticipated future developments. Given his "amateur" position, it is difficult to know the extent to which his quantitative analysis would have influenced other key workers in the field, which at the time was in a highly fluid state. Nevertheless, it is clear that he had a firm grasp of the problem and a variety of independent views, and it has been argued (e.g., *Brück,* 1996; *McFarland,* 1996, 2004; *Green,* 1999, 2004) that his work should be given greater credit.

2.2. Kuiper

A second significant contribution to the study of the origin of the solar system came from Gerard P. Kuiper (for a biography, see *Cruikshank,* 1993) in a paper published in a symposium to mark the progress of astrophysics during the half-century since the establishment of the Yerkes Observatory (*Kuiper,* 1951a). Although *Kuiper* (1951b) states that this symposium paper had been submitted for publication in November 1949 and was given limited circulation in February 1950, he evidently had time to include discussion of both *Oort*'s (1950) and *Whipple*'s (1950a,b) seminal papers, published in the first quarter of 1950. In his section entitled "Comets and Unknown Planets," Kuiper considered the fate of a belt of nebular material beyond Neptune and extending as far as Pluto's aphelion distance (i.e., from approximately 38 AU to 50 AU). He assumed that the temperature in this relatively stable region was low enough for water vapor, methane, and ammonia to condense first to form "snowflakes" and then objects a few tens of centimeters across (see also *Kuiper,* 1956). He stated that these "snowballs" would continue to combine even long after the dissipation of the solar nebula, so that after a gigayear, the average size of the bodies would be in the region of 1 km across, with the largest ones perhaps up to 100 km across. If the belt of material had a mass of 5×10^{24} kg, Kuiper estimated that this would agree with *Oort*'s (1950) estimate of $\approx 10^{11}$ members of total mass 10^{24} kg in his giant spheroidal comet reservoir.

Kuiper's work resonated with Whipple's icy conglomerate picture for the cometary nucleus (*Whipple,* 1950a,b), although it was developed apparently quite independently of Whipple's work. Kuiper felt that comets had probably not been formed between Mars and Jupiter, as Oort had speculatively suggested, but postulated instead that many of these "snowballs" could be delivered by Pluto's perturbations first toward Neptune and then by further planetary perturbations, including those of Jupiter, into Oort's "comet trap" (cf. *Öpik,* 1932). This mechanism required Pluto to have a mass in the range 0.1–1.0 $M_\oplus$, which, although later disproved by the discovery of Charon (*Christy and Harrington,* 1978), was widely believed at this time. Kuiper concluded that the comets we see today were sent from the giant cometary cloud into the inner solar system by Oort's mechanism of random perturbations by passing stars, which had resulted in their isotropic distribution of directions of approach. Beyond Pluto's aphelion distance of 50 AU, where its dynamical sweeping would be negligible, Kuiper reintroduced the important idea, dating from the previous generation, that a primordial belt of residual nebular material may still exist, and be populated by comets. Kuiper also considered that the fragility of comets and their tendency to disintegrate into small meteoroids was in accord with this scenario.

3. COMET BELT

3.1. Whipple and a Comet Ring

Although Pluto's intrinsic faintness and measurements by Kuiper of its angular size suggested an object having no more than half the diameter of Earth, attempts to determine its mass from its perturbations on other bodies in the outer planetary system persisted in giving figures as large as 0.9 $M_\oplus$ (*Brouwer,* 1951), even into the 1960s. Concerned that the resulting density was impossibly large, *Whipple* (1964a,b) considered that the perturbations might instead come from a ring of icy cometary bodies, of which Pluto would merely be one member. He found that a ring of material having 10–20 $M_\oplus$ at a solar distance of 40–50 AU was one of a number of nonunique solutions that might fit the observations, and he urged that this be tested by better determinations of the orbits of Uranus, Neptune, and Pluto. Supposing that the comet ring consisted of objects of diameter more than 1 km and albedo 0.07 in a disk 2° thick at heliocentric distance 40 AU, Whipple calculated that, even with a total mass of 100 $M_\oplus$, the surface brightness of the disk would be no brighter than 7th magnitude per square degree and therefore undetectable against the glow of the zodiacal light and the gegenschein. He also remarked that, with an apparent magnitude of 22, an individual body as large as 100 km across would still not be detectable with the instrumentation available at the time.

3.2. Observational Constraints

In an attempt to place more exacting demands on the mass of the Whipple comet ring, *Hamid et al.* (1968) computed the effect of the secular perturbations of such a ring on the orbits of seven known periodic comets with aphelia greater than 30 AU. They found that the strongest test would be provided by Comet 1P/Halley, and that their calculations did not support the existence of a comet belt of more than 0.5 $M_\oplus$ to a distance of 40 AU and of more than 1.3 $M_\oplus$ to 50 AU. Although the computation of cometary orbits is complicated by the effects of nongravitational forces, there was some credence to a result in terms of perturbations of the cometary orbital planes, because these are not obviously affected by such nongravitational effects.

Nevertheless, the apparent existence of unexplained perturbations on the orbital planes of Neptune and Uranus continued to be a worry, and it caused others to conclude that moderately massive unknown planets, as well as comets, remained to be discovered within 100 AU of the Sun (cf. *Brady,* 1972; *Goldreich and Ward,* 1972; *Seidelmann et al.,* 1972), and various suggestions were made to detect such hypothetical material (e.g., *Whipple,* 1975; *Bailey et al.,* 1984). *Bailey* (1976) appears to have been the first to consider the role of stellar occultations as a possible probe of these "invisible" outer solar system bodies, and in later work (*Bailey,* 1983a,b, 1986) noted that a suitable density distribution of comets in a spheroidal distribution could be a source of the unmodeled forces previously attributed to "Planet X" as well as a potential additional source for short-period comets. We note the recent detection of apparent "shadows" caused by distant subkilometer objects occulting the compact X-ray source Scorpius X-1 (*Chang et al.,* 2006; cf. *Jones et al.,* 2006), and similarly, the apparent detection by *Roques et al.* (2006) of distant subkilometer objects at visual wavelengths using the high-speed ULTRACAM camera mounted on the 4.2-m William Herschel Telescope.

Another approach was taken by *Jackson and Killen* (1988). They considered that the far-infrared flux emitted by dust produced during the grinding down of bodies through mutual collisions might be detectable. Although they admitted that the number of free parameters made drawing any conclusions from their models difficult, and no such detection of solar system dust was ever made in data taken by IRAS or COBE, submillimeter observations of cool dust disks around other nearby stars have recently spawned a lively area of research.

Thus, during the 1960s through the mid-1980s many authors had begun to consider different models for a trans-neptunian cometary density distribution (e.g., *Cameron,* 1962; *Whipple,* 1964b; *Safronov,* 1969, 1977; *Mendis,* 1973; *Öpik,* 1973; *Biermann and Michel,* 1978; *Hills,* 1981), and thoughtful reviews of the position up to about 1990 were provided by *Hogg et al.* (1991) and *Tremaine* (1990). Soon after, however, from a careful analysis of data from the Voyager mission, *Standish* (1993) appeared finally to lay Lowell's Planet X to rest. He concluded that there was no evidence for any significant unobserved mass in the outer solar system if correct values were used for the masses and orbital elements of the known planets.

3.3. Jupiter-Family Comets

The problem of the origin of the majority of short-period comets — those with periods less than about 20 yr and often described as "Jupiter-family" comets — had confounded, for a century or more, theoretical predictions based on the classical capture of comets from the near parabolic flux. The key difficulty lay in the efficiency of the capture process, i.e., how many short-period comets would be produced from the observed long-period flux. Analytic work (e.g., *Newton,* 1878) had demonstrated that it was impossible to produce the observed number of short-period comets as a result of single close approaches of objects in nearly parabolic orbits to Jupiter. The introduction of powerful new computational tools during the 1970s, however, increasingly focused attention on the process of gravitational capture of comets into short-period orbits by a more gradual random-walk evolution: either "diffusion" of orbital energy (e.g., *Everhart,* 1972) or a more complex process. The latter would involve the exchange of an object's perihelion and aphelion distances as a result of exceptionally close planetary approaches (*Strömgren,* 1947), leading to the "handing down" of comets in the outer solar system from one planet to another (e.g., *Kazimirchak-Polonskaya,* 1972, 1976; *Vaghi,* 1973; *Everhart,* 1976, 1977).

Everhart's work (e.g., *Everhart,* 1972) had highlighted the important role of the so-called "capture zone" in the dynamical evolution of nearly parabolic orbits to short-period, Jupiter-family types. This showed that the majority of captured short-period comets appeared to originate from a rather narrow region of phase space, i.e., from originally nearly parabolic orbits with initial perihelion distances, q, in the range 4–6 AU and initially low ($i < 9°$) inclinations, the capture probability from all other parts of the (q, i) plane being much smaller. According to Everhart's detailed investigations, the gravitational influence of Jupiter, and to a lesser extent that of Saturn, resulted in the capture to short-period orbits of 0.7% of the original near-parabolic flux within this region by the time they had orbited the Sun 2000 times.

Although Everhart had been careful to state that this was not the only evolutionary picture (and the issue of the number of orbits before dynamical capture had occurred was also an important consideration), an influential paper by *Joss* (1973) provided a rather damning counterargument. Given the low efficiency of the perturbative process demonstrated by Everhart, and the fact that inclinations less than 9° account for only a very small fraction (some 0.6%) of the observed isotropic near-parabolic flux, Joss showed that the predicted steady-state number of short-period comets was still too small. Thus, neither "diffusion" nor capture by a single close approach to Jupiter seemed capable of explaining the observed number of Jupiter-family comets, at least from the observed near-parabolic flux. He concluded simply (and correctly!) that the origin of short-period comets was not then understood.

Another approach was highlighted by *Fernández* (1980). He showed that if the observed Jupiter-family comets originated from a steady-state isotropic nearly parabolic flux, the process was so highly inefficient that it should have led to the loss from the Oort cloud (and the planetary system) of more than 10^{12} long-period comets over the age of the solar system. This was many times more than the total number of comets thought to have been originally present. This led him to consider a new source for the short-period comets, namely the transneptunian belt introduced by Whipple and others, tacitly placing the ring of small icy bodies (comets and planetesimals) between 35 AU and 50 AU from the Sun.

The second key innovation made by Fernández was to estimate the rate of orbital diffusion as a result of random gravitational encounters between the comets and planetesimals. The actual efficiency for scattering the bodies on to Neptune-crossing orbits, so that they could in turn be injected on to short-period orbits by the sequential "handing down" process mentioned above, depends on the mass ($M_{max} \sim 10^{21}–10^{22}$ kg) of the largest member of the distribution and the differential mass-distribution index ($\alpha \sim 1.5–1.9$). As we have now learned (*Torbett,* 1989; *Torbett and Smoluchowski,* 1990; *Duncan et al.,* 1995), the orbital evolution of these transneptunian objects is driven both by such close approaches and the long-term chaotic gravitational effects of the outer planets, for example, the e–i excitation mechanisms associated with mean-motion resonances in the outer planetary region. Nevertheless, by postulating the existence of Pluto-sized objects in the transneptunian disk, Fernández made a bold suggestion that has since stood the test of time.

After this pioneering work, Fernández began a series of collaborative projects with W.-H. Ip on the orbital evolution of icy planetesimals in the outer planetary accretion zones. Making use of the statistical method of orbital calculation invented by *Öpik* (1951) and *Arnold* (1965), they explored the injection of such icy planetesimals into the Oort cloud and their subsequent return to the inner solar system as near-parabolic comets (*Fernández and Ip,* 1981, 1983). An unexpected result from their numerical modeling effort concerned the outward migration of Saturn, Uranus, and Neptune, accompanied by the inward migration of Jupiter, during the accretion phase of the two outer planets (*Fernández and Ip,* 1984). This process is driven by the extensive exchange of orbital energy and angular momentum of the widely scattered planetesimals, which have total masses comparable to that of the major planets. As discussed below, such an orbital migration process has formed the theoretical basis (*Malhotra,* 1995) for the trapping mechanism of Pluto and other transneptunian objects in the 2:3 mean-motion resonance with Neptune (the so-called "Plutinos").

3.4. Kuiper Belt

A major departure came not just with the potential to integrate the orbits of thousands of comets for timescales comparable to the age of the solar system, but with the focus on a new question, namely the distribution of the *inclinations* of the short-period comets. Noting that the process of gravitational capture should roughly conserve the orbital inclinations of the captured comets, at least in a statistical sense, *Duncan et al.* (1988) found that capture from an initial nearly isotropic parabolic flux would tend to produce short-period comets with a much broader spread of inclinations than are observed. Setting aside the question of how many orbits would be required for the dynamical capture from long-period orbits to take place (the process would generally take longer for high-inclination retrograde orbits than for low-inclination direct types), they concluded that the generally low inclinations of the majority of "short-period" comets with periods less than 200 yr required a flattened distribution of source orbits. This was contrary to the results of Everhart, who had focused on comets with orbital periods less than a dozen years. In particular, they proposed that the observed short-period comets must be fed from a low-inclination cometary reservoir close to the orbit of Neptune. They proposed naming the region the "Kuiper belt," but Tremaine has since noted that when the paper was written they were unfamiliar with the work of Edgeworth. For a review of the later discussion surrounding the name "Kuiper belt," see *Davies* (2001) and *Fernández* (2005).

In order to reduce the amount of computer time required for these direct integrations of orbital evolution, *Duncan et al.* (1988) had increased the masses of the giant planets by a factor $\mu = 40$ in some cases, arguing that this should not significantly affect the relative proportions of objects captured from initially low vs. high inclinations. Although their results failed to conform with those derived from standard "diffusion" theory (e.g., *Stagg and Bailey,* 1989), subsequent work using the rather smaller planetary mass-enhancement factor $\mu = 10$ (*Quinn et al.,* 1990), as well as complementary simulations based on the Öpik-Arnold computational scheme (*Ip and Fernández,* 1991; but cf. *Bailey,* 1992), appeared to confirm the validity of the approximation. Thus, in spite of later investigations (e.g., *Manara and Valsecchi,* 1992; *Valsecchi and Manara,* 1997) to the effect that even $\mu = 10$ would significantly affect the frequency distribution of orbital energy changes per revolution and so distort the long-term dynamical evolution (cf. *Everhart,* 1979), *Duncan et al.*'s (1988) key result — the need for a flattened initial source distribution to explain the observed low-inclination Jupiter-family comets — became firmly established.

3.5. Prediction of Icy Planetoids

In the wake of these dynamical investigations, and specifically following the suggestion by *Fernández* (1980) that there may exist a significant population of massive trans-neptunian planetesimals or "planetoids" with masses up to that on the order of Pluto, and the earlier suggestions to the same effect by *Drobyshevski* (e.g., 1978, 1981), the threads were finally drawn together in an influential work by *Stern* (1991). Here, he hypothesized the existence of a population of 1000-km-sized ice dwarfs located in an extended disk-like distribution at heliocentric distances ranging from approximately 30–500 AU. Stern based this proposal on the high axial tilts of Uranus and Neptune (suggestive of collisions), the existence of Neptune's large, retrograde satellite Triton (suggestive of a capture event), and the improbability of forming the Pluto-Charon binary (cf. *McKinnon,* 1984). Stern argued that these characteristics of the outer solar system implied that there was once a large population of 1000-km-sized bodies between approximately 20 and 50 AU and that these objects should have been scattered into what he called the "Kuiper disk" and the Oort cloud (e.g., *Stern,* 1998, 2003). He pointed out that optical and infrared sky surveys offered the capability of detecting, or severely constraining, the presence of such objects out to distances of at least 100 AU.

4. EARLY SEARCHES AND DISCOVERY

4.1. Search Programs

A systematic search for distant minor planets was carried out by Charles Kowal between December 1976 and February 1985. Kowal used the 48-in Schmidt telescope at the Palomar Observatory to record 6400 deg^2 of sky to a limiting magnitude of approximately $m_V = 21$ (*Kowal,* 1989). The plates were searched by blinking in the manner of the Pluto search by Clyde Tombaugh. Due to trailing losses, etc., Kowal estimated that slow-moving objects in his survey were detectable to a limiting magnitude of about 20. Although this survey did result in the discovery of the first Centaur, (2060) Chiron (*Kowal,* 1977, 1979; *Kowal et al.,* 1979), plus several comets and Apollo-Amor planet-crossing asteroids, he did not detect any transneptunian objects (TNOs). Due to the nonuniformity of the survey coverage, which was a function of seasonal weather effects, no detailed statistical analysis of the results was considered feasible.

Another early search for distant slow-moving objects (defined as having an apparent motion less than 10 arcsec h^{-1}) was made by Jane Luu and David Jewitt in 1987 (*Luu and Jewitt,* 1988). They used both the 0.6/0.9-m twin Schmidt telescopes at Kitt Peak National Observatory (KPNO) and Cerro Tololo Inter-American Observatory (CTIO) and the McGraw-Hill 1.3-m telescope at KPNO fitted with a 390 × 584 CCD camera. They searched each of 11 5.5-deg^2 Schmidt fields, covering a total of 297 deg^2 to a limiting magnitude m_V of approximately 20, plus a 0.338-deg^2 field with the CCD camera to a limit of $m_R \simeq 24$. No distant objects were found, but in their analysis they noted that the empirical limits set by the existing surveys were too weak to contradict the hypothesis that the Oort cloud might extend into the planetary region.

In April 1989, *Levison and Duncan* (1990) used the U.S. Naval Observatory 1-m telescope at Flagstaff, Arizona, to image 4.88 deg^2 of sky with a 2048 × 2048 CCD. They then used an automated search program to search for moving objects and visually examined any promising candidates reported by the software. They searched for objects with reflex motions that would place them beyond about 25 AU, but were unable to discover any slow-moving objects to a completeness limit of $m_V = 22.5$.

Another unsuccessful search was made by *Tyson et al.* (1992). They imaged a 40 arcmin^2 area repeatedly over several nights with the CTIO 4-m telescope using relatively short exposures to minimize trailing losses. After normal flat-fielding and cosmic-ray removal, they assembled a single deep exposure by summing all the images and then removed this from each of the individual frames to give a set of residual images. They then co-added these residual images in a grid of reference frames centered on potential outer solar system objects with apparent motions between 1" and 4" per hour. Although their methodology was sound, they failed to detect any objects of $m_R < 25$. Published only as an AAS abstract, these results were being written up in more detail by P. Guhathakurta et al. but the paper was never published, being preempted by the announcement of the discovery of 1992 QB_1.

About the same time Anita and William Cochran carried out a survey using the imaging grism instrument mounted on the 2.7-m telescope of the McDonald Observatory. They observed on part or all of 22 nights between November 14, 1990, and March 25, 1993. *Cochran et al.* (1991) claimed that they would be able to detect what they referred to as "giant comets" at 50 AU if they existed, but no discoveries were ever reported from this program.

4.2. First Discoveries

The first object having an orbit that is completely transneptunian was recorded by Jewitt and Luu using the 2.2-m University of Hawaii telescope on Mauna Kea, Hawaii, on August 30–September 1, 1992. Designated 1992 QB_1, it was reported by *Jewitt and Luu* (1992) on September 14. The same circular presented a calculation by Marsden showing that, as had been the case when Pluto was announced, the orbit was completely indeterminate, the current distance from the Earth being anywhere between 37 AU (for a direct parabolic orbit) and 59 AU (for a retrograde parabolic orbit). The assumption of a direct circular orbit yielded a radius of 41 AU and inclination 2°.

Observations over a four-month arc rendered it likely that 1992 QB_1 was indeed the first discovery of an object in a low-eccentricity orbit entirely well beyond Neptune (*Marsden,* 1992), and with the availability of observations at the 1993 opposition the orbital shape and size could be refined to perihelion distance 41 AU and mean distance 44 AU. The discovery was described in detail by *Jewitt and Luu* (1993a). The object is now numbered (15760).

Another object, 1993 FW, located on the opposite side of the sky but that turned out to have a rather similar orbit (perihelion and mean distances 42 AU and 44 AU, inclination 8°) was reported by *Luu and Jewitt* (1993) on March 29.

There were four further discoveries of distant objects during September 1993. 1993 RO and 1993 RP were found by *Jewitt and Luu* (1993b,c) from Hawaii and 1993 SB and 1993 SC by *Williams et al.* (1993) with the 2.5-m Isaac Newton Telescope at La Palma (see also *Williams et al.,* 1995). Commenting on possible orbits, Marsden remarked in particular that direct circular solutions for all four had radii of 32–36 AU, i.e., much closer to Neptune's distance than had been the case for 1992 QB_1 and 1993 FW. The true nature of the orbits of these last four objects will be discussed in the next section.

5. NEW CLASSES OF TRANSNEPTUNIAN OBJECTS

5.1. Plutinos

Although the existence of objects some 60° from Neptune and for which the assumption of direct, circular orbits placed them only slightly beyond Neptune might have hinted that they were Neptune "Trojans," librating in 1:1 orbital resonance with Neptune, it seemed at least as likely that they were instead relatively near the perihelion points of orbits in the 2:3 resonance, which has a much larger phase space. After all, Pluto itself librates in 2:3 resonance with Neptune, a possibility apparently not even suggested until it was firmly established in 1964 (*Cohen and Hubbard,* 1965).

In May 1994, the availability of followup observations of 1993 SC finally provided an opportunity for the publication (*Marsden,* 1994) of the result that the assumption that this object was near perihelion and in the Neptune 2:3 resonance (mean distance 39 AU) could ensure that it would always be more than 14 AU from Neptune. This was at a time when perihelic orbits having mean distances ranging from 34 AU to more than 44 AU would also reasonably fit the observations. The same possibility was also found to be viable for the three less-well-observed discoveries from September 1993. With the availability of observations from later oppositions, the nature of the orbits of 1993 RO, 1993 SB, and 1993 SC could be confirmed as near-perihelion "Plutinos" — a term introduced by *Jewitt and Luu* (1995, 1996) — and the two *Williams et al.* (1993) objects were numbered (15788) and (15789). For a more detailed account, see *Marsden* (1996).

5.2. Other Resonant Types

More often than not, as the pace of TNO discoveries from 1994 onward increased, the initial assumption of perihelic 2:3 Neptune-resonant motion in appropriate cases turned out to be valid. However, in February 1995 another Luu-Jewitt object, now known as (15836) 1995 DA_2, was found for which a circular orbit solution indicated a radius of 34 AU, but which had a longitude almost 180° from Neptune. Since the perihelic 2:3 assumption in this case would yield a close approach to Neptune when the object was at the same longitude, a perihelic orbit in 1:2 resonance (mean distance 48 AU) was initially assumed instead (*Marsden,* 1995a). When the observations extended for more than a month, however, it seemed that a perihelic 3:4 Neptune-resonant orbit would be more viable (*Marsden,* 1995b), and the correctness of this assumption was proven when observations were made at the next opposition. Also in early 1995, another borderline case having an initial circular solution with radius 37 AU was shown at the second opposition to avoid Neptune by virtue of its being in the 3:5 resonance (*Marsden,* 1995c). This object, (15809) 1994 JS, has perihelion and mean distances of 33 AU and 42 AU.

On the theoretical front, it should be noted that various authors were developing the idea that the structure of the transneptunian region might be rather complicated. *Levison and Duncan* (1993) had carried out integrations of test particles over billion-year timescales and found that these led to complex structures in a process that they described as "gravitational sculpting" of the region. Similarly, *Morbidelli et al.* (1995) had begun to explore the resonant structure of the region and *Malhotra* (1993) was considering that, following the ideas of *Fernández and Ip* (1984) concerning planetary migration, the capture of Pluto into the 2:3 resonance was a consequence of early planetary migration and subsequent dynamical evolution of the outer solar system. *Malhotra* (1995) suggested that, in addition to the 2:3 resonance, TNOs should be found librating in the 1:2 resonance with Neptune, and maybe others such as 3:4, 3:5, etc. She further developed these ideas in *Malhotra* (1996). After a few false alarms, the first confirmed cases of 1:2 libration, namely 1997 SZ_{10} and (20161) 1996 TR_{66}, were recognized in December 1998 (*Marsden,* 1998a,b).

In more recent years, both theory (e.g., *Nesvorný and Roig,* 2000, 2001; *Nesvorný and Dones,* 2002; *Lykawka and Mukai,* 2006) and observations have advanced rapidly, with objects now progressively confirmed to be librating in the 4:7, 4:5, 1:1, 5:9, 2:5, 3:7, and 1:3 resonances with Neptune (see chapter by Gomes et al., but cf. chapter by Gladman et al.). The mean distance (44 AU) corresponding to the 4:7 resonance is close to those of the first modern TNO discoveries 1992 QB_1 and 1993 FW, but the librating cases tend to have significantly smaller perihelion distances (as low as 33 AU) than their nonlibrating neighbors. The 5:9 resonance is nearby, although the single 5:9 librator so far confirmed has a perihelion distance of 40 AU (*Chiang et al.,* 2003).

The 42 AU mean distance of the 3:5 librator (15809) lies near the inner edge of a rather extensive population of objects in low-eccentricity orbits. These are variously termed "classical Kuiper belt objects" (classical KBOs) or (as a word-play on 1992 QB_1) "Cubewanos," first introduced by *Marsden* (1997). Similarly, the mean distance (≃48 AU) of the 1:2 mean-motion resonance lies near the outer edge of this "core" TNO population. Of course, these boundaries for the low-eccentricity population may well owe their origin to secular resonances with Neptune, such as the ν_8. In any case, with the discovery of numerous Cubewanos and a growing number of confirmed mean-motion librators, as well as other objects in a wide range of more-eccentric, higher-inclination orbits, it soon became clear that the whole transneptunian region comprises a complex ensemble of different dynamical types, with some subclasses of orbit possibly having quite different dynamical histories from others.

5.3. Classical Kuiper Belt Objects

The objects comprising the core of the transneptunian population, in what effectively covers the range of the Whipple-Fernández comet belt, represent what some authors have called the "Classical Kuiper" or "Edgeworth-Kuiper" belt. These objects were originally thought to be the principal reservoir for short-period comets, with a total population, for semimajor axes in the range 30–50 AU and diameters greater than 100 km, on the order of 10^5 objects. However, with the provision of increasingly accurate orbits for some objects with multiple oppositions and smaller perihelion distances (and so with the potential to undergo close approaches to Neptune), many were found to be Neptune librators. Thus, at least over relatively short timescales, they are protected against close approaches to Neptune and cannot readily evolve onto short-period orbits.

At the time of this writing, it is not known whether these resonant objects are permanently protected against close approaches with Neptune, or whether they might still represent a significant source of short-period comets. What has become clear, however, is that the core nonresonant population mostly comprises orbits with relatively large perihelion distances and very long dynamical lifetimes (*Duncan et al.,* 1995), and so cannot be the dominant source of short-period comets (e.g., *Emel'yanenko et al.,* 2005), as was originally proposed.

5.4. Scattered Disk Objects and Centaurs

The term *scattered disk* was applied by *Torbett* (1989) to postulated icy objects generally in highly eccentric and substantially inclined orbits with perihelion distances beyond Neptune. These were the planetesimals supposedly scattered by Uranus and Neptune toward the Oort cloud, and Torbett remarked that the survival of this hypothetical disk would depend on the efficiency of stellar perturbations in raising their perihelion distances. Following earlier work by *Fernández and Ip* (1981) and *Duncan et al.* (1987), *Torbett* (1989) concentrated on objects with modest inclinations and perihelia in the range 30–45 AU, with a view to finding unstable regions for the subsequent production of short-period comets as the result of successive interactions with all four giant planets.

Whereas *Levison and Duncan* (1997) had focused on the classical Kuiper belt as the principal source of short-period comets, *Duncan and Levison* (1997) had alternatively shown that the same short-period comet population could equally (or perhaps more plausibly) arise as a result of the evolution of a population of Uranus- and Neptune-scattered planetesimals, which they called "scattered icy objects." These comets, as they were assumed to be, would have been formed close to the major planets coevally with the planetesimals that had led to the formation of these same planets, and were identified simply as residual primordial objects from this region, scattered outward by Uranus and Neptune during the final phases of planetary migration and accretion.

With the discovery of the object (15874) 1996 TL_{66} (*Luu et al.,* 1997), having an orbit with perihelion and mean distances of 35 AU and 83 AU and an inclination of 24°, there was an appreciation that this could be representative of the scattered bodies discussed by *Torbett* (1989) and *Duncan and Levison* (1997). 1996 TL_{66} therefore became recognized as the prototype for a new class of known *scattered* KBOs, or later — following the terminology of *Levison and Duncan* (1997) — the first *scattered Kuiper belt* or *scattered disk objects* (SDOs). The second SDO to be recognized was 1998 XY_{95} (*Marsden,* 1999), with perihelion and mean distances 37 AU and 64 AU, respectively, and inclination 7°.

The term SDO is a little unfortunate, because it has never been clear whether it is also applicable to high-eccentricity objects with perihelia that are *inside* the orbit of Neptune — the planet-crossing Centaurs — or indeed whether some of the objects with large perihelion distances and/or large eccentricity are scattered at all. If the term is applied to the Centaurs, should there be an arbitrary lower limit of 25 AU on the perihelion distance of SDOs (cf. *Morbidelli et al.,* 2004a), or should orbits with substantially smaller perihelion distances also be included, for example, (29981) 1999 TD_{10} [perihelion and mean distances 12 AU and 95 AU and inclination 6°; see *Williams* (1999)], and the exceptional object (127546) 2002 XU_{93}, with its record inclination of 78°?

In principle, as noted by *Horner et al.* (2003) and *Emel'yanenko et al.* (2004), there is no clear dynamical distinction between Centaurs and SDOs, as they overlap in space and on long timescales their respective orbital elements can vary over a wide range, evolving from one class to the other. Thus, the Minor Planet Center has for some years presented the orbits of both classes of object on the same web page. In general, a purely dynamical classification scheme should avoid, if possible, drawing arbitrary lines between objects that are on dynamically similar orbits, and implying (or depending upon) information about an individual object's origin, as in the phrase "scattered disk object," until the origin and evolution of the whole solar system is much better understood.

In current parlance, Centaurs are usually regarded as planet-crossing objects moving within the range of the giant planets and SDOs as objects on Neptune-approaching orbits with much longer orbital periods. However, even the second Centaur to be discovered (in 1992), namely (5145) Pholus, with perihelion and mean distances of 9 AU and 20 AU respectively, has an aphelion beyond Neptune and so periodically enters the transneptunian region. As soon as one includes the results of long-term numerical integrations of their orbits (e.g., *Hahn and Bailey,* 1990; *Horner et al.,* 2003, 2004) the essential equivalence of the two classes of object can hardly be avoided, demonstrating that Centaurs and SDOs merely represent different phases of evolution of the same underlying "Centaur" population.

A further important consideration is the maximum perihelion distance within which a TNO may be considered as "scattered" by Neptune or coming under its dynamical influence. Although *Torbett* (1989) had suggested 45 AU, the scattered disk produced in *Duncan and Levison*'s (1997) 4-G.y. integrations had perihelia out to about 40 AU. More recent computations (e.g., *Emel'yanenko et al.,* 2003) have suggested that 38 AU may be a more realistic figure, but in practice there will not be a perfectly sharp boundary: The effects of mean-motion and secular resonances, not to mention the long-term effects of small galactic perturbations, especially on orbits of relatively large semimajor axes, all have significant effects on the objects' long-term evolution (*Emel'yanenko,* 1999; *Mazeeva and Emel'yanenko,* 2002).

Transneptunian objects with perihelia somewhat above 38 AU that otherwise have similar dynamical characteristics to SDOs are sometimes termed *outer* TNOs or *extended* SDOs (*Gladman et al.,* 2002; *Emel'yanenko et al.,* 2003; *Morbidelli et al.,* 2004b), and represent another new population of objects. Their intrinsic number is at least 10 times as many as that in both the classical Kuiper belt and the scattered disk (*Gladman,* 2005). The largest such object so far known (albeit possibly a "borderline" SDO) is (136199) Eris (formerly 2003 UB_{313}) (*Brown et al.,* 2005). It was discovered near the aphelion point of its orbit, and has perihelion and mean distances of 38 AU and 68 AU, and inclination 44°, and is notable also for being somewhat larger than Pluto. Other objects in this new class of outer TNOs (*Gladman et al.,* 2002; *Emel'yanenko et al.,* 2003), which represent either an extended "disk" population or a class of objects merging into the inner Oort cloud, are 2000 CR_{105}, with perihelion distance 44 AU, and (90377) Sedna = 2003 VB_{12}, which is widely regarded as occupying part of the inner Oort cloud (*Brown et al.,* 2004). It has a perihelion distance of 76 AU, an orbital period of around 12,000 yr, and an inclination of 12°.

5.5. More Recent Discoveries

In recent years new discoveries have added further complexity to this already complicated picture. For example, although 90% of the currently known classical Kuiper belt objects have orbital inclinations under 15°, a few objects, including the largest known member, (136472) 2005 FY_9 (*Brown et al.,* 2005), have inclinations around 30°, and there is one (2004 DG_{77}) confirmed at inclination 48°. Similarly, with a perihelion distance of 35 AU, (136108) 2003 EL_{61}, the third of the large TNOs reported in mid-2005 (*Brown et al.,* 2005), qualifies more as an outer TNO or as a borderline Centaur/SDO than as a classical Kuiper belt object. Its inclination (28°) is again somewhat high for the "classical" Kuiper belt, although its mean distance of 43 AU and relatively low eccentricity is arguably in the classical range.

The inclinations of the Plutinos, now numbering approximately 100 objects from among the ~650 TNOs with reliable orbit determinations, also range up to around 30° (with 2005 TV_{189} having the largest value, 34°). However, the Plutinos have a much broader inclination distribution, with less than 25% of the objects having inclinations less than 15°.

The 4:5 resonance is of course somewhat inward of the 3:4 resonance and involves low-eccentricity orbits with semimajor axes around 35 AU. It seems rather unlikely that objects will be found at the 5:6 and higher first-order resonances, as the sequence reaches its limit with the much stronger 1:1 "Trojan" case, of which the first definite example found was 2001 QR_{322} (*Marsden,* 2003; *Chiang et al.,* 2003). At the time of this writing, a total of four Neptune Trojans have been identified, all librating about the leading Lagrangian point L4 (*Sheppard and Trujillo,* 2006).

The most recently recognized populated mean-motion resonances with Neptune, namely the 3:7, 2:5, and 1:3, are beyond the 1:2 resonance, lying near mean distances 53, 55, and 62 AU. Since these librators have perihelion distances around 32–33 AU, they are at first glance indistinguishable from Centaurs, especially as a principal dynamical characteristic of Centaurs is their tendency between episodes of strong planetary perturbations to have semimajor axes close to one or another mean-motion resonance with a controlling planet. Although there is only one confirmed librator at each of the 3:7 and 1:3 resonances — (95625) 2002 GX_{32} = 1994 JV and (136120) 2003 LG_7 (*Wasserman and Marsden,* 2004; *Marsden,* 2005) — the 2:5 resonance (*Chiang et al.,* 2003) is remarkable in that it has more than a dozen well-established librators. Also well beyond the 1:2 resonance there is the exceptional low-eccentricity object 2004 XR_{190}, with perihelion and mean distances 51 AU and 57 AU, close to the 3:8 resonance, and inclination as high as 47° (*Allen et al.,* 2006).

6. CONCLUSIONS

This review of the development of ideas concerning the transneptunian region has brought us from a discussion of the discovery of Pluto through early theories of the structure and evolution of the protoplanetary disk and the origin of short-period comets, to the dynamical links between genuinely *trans*neptunian objects and Centaurs on the one hand, and comets in both the inner and outer Oort cloud on the other. The subject has important implications not just for the origin of the solar system and the formation of the outer planets, but also for understanding the evolution of the inner solar system and the development of life on Earth, for example, through the effects of large comets that can evolve onto Earth-crossing orbits from this outer region (*Bailey et al.,* 1994).

It has taken 75 years since the discovery of Pluto for astronomers to come to grips with some of the dynamical complexity of this outer part of the solar system. This is a period comparable to the century between the discovery of the first few main-belt minor planets and the development of a coherent understanding of the whole region between Mars and Jupiter. In an interesting historical "resonance" with our understanding of the main-belt population, the 2006 International Astronomical Union (IAU) General Assembly in Prague attempted to construct a formal definition of a planet. After much discussion, there was overwhelming support from those present for a motion to restrict the definition of the solar system's "planets" to the eight bodies known both to be in hydrostatic equilibrium and dynamically to dominate their regions of space. Since these eight are not the only solar system bodies in hydrostatic equilibrium, the IAU also agreed to classify as "dwarf planets" other such bodies that do not dynamically dominate their local regions. These "dwarf planets" are numbered in the general catalog of noncometary "small solar system bodies" that have the most reliable orbit determinations and currently include (1) Ceres, the largest of the bodies in the region between Mars and Jupiter, as well as (134340) Pluto and (136199) Eris.

Acknowledgments. We thank many colleagues for discussions. We thank the two referees for their comments, which were carefully considered. Many of their suggestions have been incorporated into the final manuscript. Astronomy at Armagh Observatory is supported by the Northern Ireland Department of Culture, Arts and Leisure. The work of W.-H.I. presented here was supported by grant NSC 94-2112- M-008-033.

REFERENCES

Aitken R. G. (1926) The solar system — some unsolved problems. *Publ. Astron. Soc. Pac., 38,* 277–295.

Allen R. L., Gladman B., Kavelaars J. J., Petit J.-M., Parker J. W., and Nicholson P. (2006) Discovery of a low-eccentricity, high-inclination Kuiper belt object at 58 AU. *Astrophys. J. Lett., 640,* L83–L86.

Arnold J. R. (1965) The origin of meteorities as small bodies. II. The model. *Astrophys. J., 141,* 1536–1547.

Bailey M. E. (1976) Can 'invisible' bodies be observed in the solar system? *Nature, 259,* 290–291.

Bailey M. E. (1983a) Is there a dense primordial cloud of comets just beyond Pluto? In *Asteroids, Comets, Meteors* (C.-I. Lagerkvist and H. Rickman, eds.), pp. 383–386. Uppsala Observatory, Uppsala, Sweden.

Bailey M. E. (1983b) Comets, Planet X, and the orbit of Neptune. *Nature, 302,* 399–400.

Bailey M. E. (1986) The near-parabolic flux and the origin of short-period comets. *Nature, 324,* 350–352.

Bailey M. E. (1992) Origin of short-period comets. *Cel. Mech. Dyn. Astron., 54,* 49–61.

Bailey M. E. (1994) Formation of outer solar system bodies: Comets and planetesimals. In *Asteroids Comets Meteors 1993* (A. Milani et al., eds.), pp. 443–459. IAU Symposium 160, Kluwer, Dordrecht.

Bailey M. E., McBreen B., and Ray T. P. (1984) Constraints on cometary origins from isotropy of the microwave background and other measurements. *Mon. Not. R. Astron. Soc., 209,* 881–888.

Bailey M. E., Clube S. V. M., Hahn G., Napier W. M., and Valsecchi G. B. (1994) Hazards due to giant comets: Climate and short-term catastrophism. In *Hazards Due to Comets and Asteroids* (T. Gehrels., ed.), pp. 479–533. Univ. of Arizona, Tucson.

Beekman G. (1999) Wie voorspelde het bestaam van de Kuipergordel. *Zenit October,* 432–433.

Biermann L. and Michel K. W. (1978) The origin of cometary nuclei in the presolar nebula. *Earth Moon Planets, 18,* 447–464.

Brady J. L. (1972) The effect of a trans-plutonian planet on Halley's comet. *Publ. Astron. Soc. Pac., 84,* 314–322.

Brouwer D. (1951) Comparison of numerical integration of the motions of Uranus and Neptune with Newcomb's tables. *Astron. J., 56,* 35.
Brown M. E., Trujillo C., and Rabinowitz D. (2004) Discovery of a candidate inner Oort cloud planetoid. *Astrophys. J., 617,* 645–649.
Brown M. E., Trujillo C. A., and Rabinowitz D. (2005) 2003 EL_{61}, 2003 UB_{313}, and 2005 FY_9. *IAU Circular 8577.*
Brück M. (1996) The Edgeworth-Kuiper belt. *Irish Astron. J., 23,* 3–6.
Cameron A. G. W. (1962) The formation of the Sun and planets. *Icarus, 1,* 13–69.
Campbell W. W. (1916) The solar system. *Publ. Astron. Soc. Pac., 28,* 222–246.
Campbell W. W. (1919) What we know about comets. *The Adolfo Stahl Lectures in Astronomy.* First delivered December 8, 1916. Stanford Univ., San Francisco.
Chang H.-K., King S.-K., Liang J.-S., Wu P.-S., Lin L.C.-C., and Chiu J.-L. (2006) Occultation of X-rays from Scorpius X-1 by small trans-neptunian objects. *Nature, 442,* 660–663.
Chiang E. I., Lovering J. R., Millis R. L., Buie M. W., Wasserman L. H., and Meech K. J. (2003) Resonant and secular families of the Kuiper Belt. *Earth Moon Planets, 92,* 49–62.
Christy J. W. and Harrington R. S. (1978) The satellite of Pluto. *Astron. J., 83,* 1005–1008.
Cochran A. L., Cochran W. D., and Torbett M. V. (1991) A deep imaging search for the Kuiper disk of comets. *Bull. Am. Astron. Soc., 23,* 1314.
Cohen C. J. and Hubbard E. C. (1965) Libration of the close approaches of Pluto to Neptune. *Astron. J., 70,* 10–13.
Cruikshank D. (1993) Gerard Peter Kuiper. *Biograph. Mem. Natl. Acad. Sci., 62,* 259–295.
Davies J. K. (2001) *Beyond Pluto.* Cambridge Univ., Cambridge. 233 pp.
Drobyshevski E. M. (1978) The origin of the solar system: Implications for transneptunian planets and the nature of the long-period comets. *Moon and Planets, 18,* 145–194.
Drobyshevski E. M. (1981) The history of Titan, of Saturn's rings and magnetic field, and the nature of short-period comets. *Moon and Planets, 24,* 13–45.
Duncan M. J. and Levison H. F. (1997) A disk of scattered icy objects and the origin of Jupiter-family comets. *Science, 276,* 1670–1672.
Duncan M., Quinn T., and Tremaine S. (1987) The formation and extent of the solar system comet cloud. *Astron. J., 94,* 1330–1338.
Duncan M., Quinn T., and Tremaine S. (1988) The origin of short-period comets. *Astrophys. J. Lett., 328,* L69–L73.
Duncan M., Levison H. F., and Budd S. M. (1995) The dynamical structure of the Kuiper belt. *Astron. J., 110,* 3073–3081.
Edgeworth K. E. (1938) The evolution of the solar system. Unpublished manuscript, Trustees of the National Library of Ireland, Dublin. *Manuscript Nos. 16869/47 and /48.* 226 pp.
Edgeworth K. E. (1943) The evolution of our planetary system. *J. Brit. Astr. Assoc., 53,* 181–188.
Edgeworth K. E. (1949) The origin and evolution of the solar system. *Mon. Not. R. Astron. Soc., 109,* 600–609.
Edgeworth K. E. (1961) *The Earth, the Planets and the Stars: Their Birth and Evolution.* Macmillan, New York.
Emel'yanenko V. V. (1999) From the solar system comet cloud to near-Earth space. In *Evolution and Source Regions of Asteroids and Comets* (J. Svoreñ et al., eds.), pp. 339–344. IAU Colloquium No. 173, Astron. Inst. Slovak Acad. Sci., Tatranská Lomnica.
Emel'yanenko V. V., Asher D. J., and Bailey M. E. (2003) A new class of trans-Neptunian objects in high-eccentricity orbits. *Mon. Not. R. Astron. Soc., 338,* 443–451.
Emel'yanenko V. V., Asher D. J., and Bailey M. E. (2004) High eccentricity trans-Neptunian objects as a source of Jupiter family comets. *Mon. Not. R. Astron. Soc., 350,* 161–166.
Emel'yanenko V. V., Asher D. J., and Bailey M. E. (2005) Centaurs from the Oort cloud and the origin of Jupiter-family comets. *Mon. Not. R. Astron. Soc., 361,* 1345–1351.
Everhart E. (1972) The origin of short-period comets. *Astrophys. Lett., 10,* 131–135.
Everhart E. (1976) The evolution of cometary orbits. In *The Study of Comets: Part I* (B. Donn et al., eds.), pp. 445–461. IAU Colloquium 25, NASA SP-393, Washington, DC.
Everhart E. (1977) Evolution of comet orbits as perturbed by Uranus and Neptune. In *Comets Asteroids Meteorites: Interrelations, Evolution and Origins* (A. H. Delsemme, ed.), pp. 99–104. IAU Colloquium 39, Univ. of Toledo, Toledo, Ohio.
Everhart E. (1979) Past and future orbit of 1977 UB, object Chiron. *Astron. J., 84,* 134–139.
Fernández J. A. (1980) On the existence of a comet belt beyond Neptune. *Mon. Not. R. Astron. Soc., 192,* 481–491.
Fernández J. A. (2005) *Comets: Nature, Dynamics, Origin, and Their Cosmogonical Relevance.* Springer, Dordrecht. 383 pp.
Fernández J. A. and Ip W.-H. (1981) Dynamical evolution of a cometary swarm in the outer planetary region. *Icarus, 47,* 470–479.
Fernández J. A. and Ip W.-H. (1983) On the time evolution of the cometary influx in the region of the terrestrial planets. *Icarus, 54,* 377–387.
Fernández J. A. and Ip W.-H. (1984) Some dynamical aspects of the accretion of Uranus and Neptune: The exchange of orbital angular momentum with planetesimals. *Icarus, 58,* 109–120.
Gladman B. (2005) The Kuiper belt and the solar system's comet disk. *Science, 307,* 71–75.
Gladman B., Holman M., Grav T., Kavelaars J., Nicholson P., Aksnes K., and Petit J.-M. (2002) Evidence for an extended scattered disk. *Icarus, 157,* 269–279.
Goldreich P. and Ward W. R. (1972) The case against Planet X. *Publ. Astron. Soc. Pac., 84,* 737–742.
Green D. (1999) Book review. *Intl. Comet Quarterly, 21,* 44–46.
Green D. (2004) Fred Lawrence Whipple (1906–2004). *Intl. Comet Quarterly, 26,* 115–118.
Hahn G. and Bailey M. E. (1990) Rapid dynamical evolution of giant comet Chiron. *Nature, 348,* 132–136.
Hamid S. E., Marsden B. G., and Whipple F. L. (1968) Influence of a comet belt beyond Neptune on the motions of periodic comets. *Astron. J., 73,* 727–729.
Hills J. G. (1981) Comet showers and the steady-state infall of comets from the Oort cloud. *Astron. J., 86,* 1730–1740.
Hogg D. W., Quinlan G. D., and Tremaine S. (1991) Dynamical limits on dark mass in the solar system. *Astron. J., 101,* 2274–2286.
Horner J., Evans N. W., Bailey M. E., and Asher D. J. (2003) The populations of comet-like bodies in the solar system. *Mon. Not. R. Astron. Soc., 343,* 1057–1066.
Horner J., Evans N. W., and Bailey M. E. (2004) Simulations of

the population of Centaurs — I. The bulk statistics. *Mon. Not. R. Astron. Soc., 354,* 798–810.

Ip W.-H. and Fernández J. A. (1991) Steady-state injection of short-period comets from the trans-Neptunian cometary belt. *Icarus, 92,* 185–193.

Jackson A. A. and Killen R. M. (1988) Infrared brightness of a comet belt beyond Neptune. *Earth Moon Planets, 42,* 41–47.

Jewitt D. C. and Luu J. X. (1992) 1992 QB_1. *IAU Circular 5611.*

Jewitt D. C. and Luu J. X. (1993a) Discovery of the candidate Kuiper belt object 1992 QB_1. *Nature, 362,* 730–732.

Jewitt D. and Luu J. (1993b) 1993 RO. *IAU Circular 5865.*

Jewitt D. and Luu J. (1993c) 1993 RP. *IAU Circular 5867.*

Jewitt D. C. and Luu J. X. (1995) The solar system beyond Neptune. *Astron. J., 109,* 1867–1876.

Jewitt D. C. and Luu J. X. (1996) The Plutinos. In *Completing the Inventory of the Solar System* (T. W. Rettig and J. M. Hahn, eds.), pp. 255–258. ASP Conf. Series 107, San Francisco.

Jones T. A., Levine A. M., Morgan E. H., and Rappaport S. (2006) Millisecond dips in Sco X-1 are likely the result of high-energy particle events. Available online at *arXiv.org/abs/astro-ph/0612129.*

Joss P. C. (1973) On the origin of short-period comets. *Astron. Astrophys., 25,* 271–273.

Kazimirchak-Polonskaya E. I. (1972) The major planets as powerful transformers of cometary orbits. In *The Motion, Evolution of Orbits, and Origin of Comets* (G. A. Chebotarev et al., eds.), pp. 373–397. IAU Symposium 45, Kluwer, Dordrecht.

Kazimirchak-Polonskaya E. I. (1976) Review of investigations performed in the U.S.S.R. on close approaches of comets to Jupiter and the evolution of cometary orbits. In *The Study of Comets: Part I* (B. Donn et al., eds.), pp. 490–536. IAU Colloquium 25, NASA SP-393, Washington, DC.

Kowal C. T. (1977) Slow-moving object Kowal. *IAU Circular 3129.*

Kowal C. T. (1979) Chiron. In *Asteroids* (T. Gehrels, ed.), pp. 436–439. Univ. of Arizona, Tucson.

Kowal C. T. (1989) A solar system survey. *Icarus, 77,* 118–123.

Kowal C. T., Liller W., and Marsden B. G. (1979) The discovery and orbit of (2060) Chiron. In *Dynamics of the Solar System* (R. L. Duncanbe, ed.), pp. 245–250. IAU Symposium 81, Reidel, Dordrecht.

Kuiper G. P. (1951a) On the origin of the solar system. In *Astrophysics: A Topical Symposium* (J. A. Hynek et al., eds.), pp. 357–424. McGraw-Hill, New York.

Kuiper G. P. (1951b) On the origin of the solar system. *Proc. Natl. Acad. Sci., 37,* 1–14.

Kuiper G. P. (1956) The formation of the planets. Part III. *J. R. Astron. Soc. Canada, 50,* 158–176 (Part I: *J. R. Astron. Soc. Canada, 50,* 57–68; Part II: *J. R. Astron. Soc. Canada, 50,* 105–121).

Leonard F. C. (1930) The new planet Pluto. *Leaflet Astron. Soc. Pac. August,* 121–124.

Leuschner A. O. (1927) The Pons-Winnecke comet. *Publ. Astron. Soc. Pac., 39,* 275–294.

Leuschner A. O. (1932) The astronomical romance of Pluto. *Publ. Astron. Soc. Pac., 44,* 197–214.

Levison H. F. and Duncan M. J. (1990) A search for proto-comets in the outer regions of the solar system. *Astron. J., 100,* 1669–1675.

Levison H. F. and Duncan M. J. (1993) The gravitational sculpting of the Kuiper belt. *Astrophys. J. Lett., 406,* L35–L38.

Levison H. F. and Duncan M. J. (1997) From the Kuiper belt to Jupiter-family comets: The spatial distribution of ecliptic comets. *Icarus, 127,* 13–32.

Luu J. X. and Jewitt D. C. (1988) A two-part search for slow moving objects. *Astron. J., 95,* 1256–1262.

Luu J. and Jewitt D. (1993) 1993 FW. *IAU Circular 5730.*

Luu J. X., Marsden B. G., Jewitt D., Trujillo C. A., Hergenrother C. W., Chen J., and Offutt W. B. (1997) A new dynamical class of object in the outer solar system. *Nature, 387,* 573–575.

Luyten W. J. (1938) Letter to K. E. Edgeworth, dated June 6th, 1938. Trustees of the National Library of Ireland, Dublin. *Manuscript No. 16869/45.* 2 pp.

Lykawka P. S. and Mukai T. (2006) Long term dynamical evolution and classification of classical TNOs. *Earth Moon Planets, 97,* 107–126.

Lyttleton R. A. (1936) On the possible results of an encounter of Pluto with the Neptunian system. *Mon. Not. R. Astron. Soc., 97,* 108–115.

Malhotra R. (1993) The origin of Pluto's peculiar orbit. *Nature, 365,* 819–821.

Malhotra R. (1995) The origin of Pluto's orbit: Implications for the solar system beyond Neptune. *Astron. J., 110,* 420–429.

Malhotra R. (1996) The phase space structure near Neptune resonances in the Kuiper belt. *Astron. J., 111,* 504–516.

Manara A. and Valsecchi G. B. (1992) Cometary orbital evolution in the outer planetary region. In *Asteroids, Comets, Meteors 1991* (A. W. Harris and E. Bowell, eds.), pp. 381–384. Lunar and Planetary Institute, Houston.

Marsden B. G. (1992) 1992 QB_1. *IAU Circular 5684.*

Marsden B. G. (1994) 1993 RO, 1993 RP, 1993 SB, 1993 SC. *IAU Circular 5983.*

Marsden B. G. (1995a) 1995 DA_2. *Minor Planet Electronic Circular 1995-E05.*

Marsden B. G. (1995b) 1995 DA_2. *Minor Planet Electronic Circular 1995-G04.*

Marsden B. G. (1995c) 1994 JS. *Minor Planet Electronic Circular 1995-K04.*

Marsden B. G. (1996) Searches for planets and comets. In *Completing the Inventory of the Solar System* (T. W. Rettig and J. M. Hahn, eds.), pp. 193–207. ASP Conf. Series 107, San Francisco.

Marsden B. G. (1997) Distant minor planets (1997 Aug. 30.0 TT). *Minor Planet Electronic Circular 1997-P12.*

Marsden B. G. (1998a) 1997 SZ_{10}. *Minor Planet Electronic Circular 1998-Y09.*

Marsden B. G. (1998b) 1996 TR_{66}. *Minor Planet Electronic Circular 1998-Y28.*

Marsden B. G. (1999) 1998 XY_{95}. *Minor Planet Electronic Circular 1999-X33.*

Marsden B. G. (2000) Leuschner and the Kuiper belt. *Distant EKOs: The Kuiper Belt Electronic Newsletter, Issue 10,* pp. 3–5.

Marsden B. G. (2003) 2001 QR_{322}. *Minor Planet Electronic Circular 2003-A55.*

Marsden B. G. (2005) Daily orbit update (2005 May 24 UT). *Minor Planet Electronic Circular 2005-K33.*

Mazeeva O. A. and Emel'yanenko V. V. (2002) Variations of the Oort cloud comet flux in the planetary region. In *Asteroids, Comets, Meteors — ACM 2002* (B. Warmbein, ed.), pp. 445–448. ESA SP-500, Noordwijk, The Netherlands.

McFarland J. (1996) Kenneth Essex Edgeworth — Victorian polymath and founder of the Kuiper belt? *Vistas Astron., 40,* 343–354.

McFarland J. (2004) K. E. Edgeworth and TNOs. In *Proceedings of Ceres 2001 Workshop,* pp. 19–23. Institut de Mécanique Céleste et de Calcul des Ephémérides, Paris, France.

McKinnon W. B. (1984) On the origin of Triton and Pluto. *Nature, 311,* 355–358.

Mendis D. A. (1973) The comet-meteor stream complex. *Astrophys. Space Sci., 20,* 165–176.

Morbidelli A., Thomas F., and Moons M. (1995) The resonant structure of the Kuiper belt and the dynamics of the first five trans-Neptunian objects. *Icarus, 118,* 322–340.

Morbidelli A., Emel'yanenko V. V., and Levison H. F. (2004a) Origin and orbital distribution of the trans-Neptunian scattered disc. *Mon. Not. R. Astron. Soc., 355,* 935–940.

Morbidelli A., Brown M. E., and Levison H. F. (2004b) The Kuiper belt and its primordial sculpting. *Earth Moon Planets, 92,* 1–27.

Nesvorný D. and Dones L. (2002) How long-lived are the hypothetical Trojan populations of Saturn, Uranus, and Neptune? *Icarus, 160,* 271–288.

Nesvorný D. and Roig F. (2000) Mean motion resonances in the trans-Neptunian region. I. The 2:3 resonance with Neptune. *Icarus, 148,* 282–300.

Nesvorný D. and Roig F. (2001) Mean motion resonances in the trans-Neptunian region. II. The 1:2, 3:4 and weaker resonances. *Icarus, 150,* 104–123.

Newton H. A. (1878) On the origin of comets. *Am. J. Sci. Arts (3rd Ser.), 16,* 165–179.

Oort J. H. (1950) The structure of the cloud of comets surrounding the solar system, and a hypothesis concerning its origin. *Bull. Astron. Inst. Netherlands, 11,* 91–110.

Öpik E. J. (1932) Note on stellar perturbations of nearly parabolic orbits. *Proc. Am. Acad. Arts Sci., 67,* 169–183.

Öpik E. J. (1951) Collision probabilities with the planets and the distribution of interplanetary matter. *Proc. R. Irish Acad. Sect. A, 54,* 165–199.

Öpik E. J. (1973) Comets and the formation of planets. *Astrophys. Space Sci., 21,* 307–398.

Quinn T., Tremaine S., and Duncan M. (1990) Planetary perturbations and the origins of short-period comets. *Astrophys. J., 355,* 667–679.

Roques F., Doressoundiram A., Dhillon V., Bickerton S., Kavelaars J. J., Moncuquet M., Auvergne M., Belskaya I., Chevreton M., Colas F., Fernandez A., Fitzsimmons A., Lecacheux J., Mousis O., Pau S., Peixinho N., and Tozzi G. P. (2006) Exploration of the Kuiper belt by high-precision photometric stellar occultations: First results. *Astron. J., 132,* 819–822.

Safronov V. S. (1969) *Evolution of the Protoplanetary Cloud and Formation of the Earth and the Planets.* Nauka, Moscow. Translated by Israel Program for Scientific Translations, Jerusalem, 1972.

Safronov V. S. (1977) Oort's cometary cloud in the light of modern cosmogony. In *Comets Asteroids Meteorites: Interrelations, Evolution and Origins* (A. H. Delsemme, ed.), pp. 483–484. IAU Colloquium 39, Univ. of Toledo, Toledo, Ohio.

Sheppard S. S. and Trujillo C. A. (2006) A thick cloud of Neptune Trojans and their colors. *Science, 313,* 511–514.

Seidelmann P. K., Marsden B. G., and Giclas H. L. (1972) Note on Brady's hypothetical trans-Plutonian planet. *Publ. Astron. Soc. Pac., 84,* 858–864.

Stagg C. R. and Bailey M. E. (1989) The stochastic capture of short-period comets. *Mon. Not. R. Astron. Soc., 241,* 507–541.

Standish E. M. (1993) Planet X: No dynamical evidence in the optical observations. *Astron. J., 105,* 2000–2006.

Stern S. A. (1991) On the number of planets in the outer solar system: Evidence of a substantial population of 1000-km bodies. *Icarus, 90,* 271–281.

Stern S. A. (1998) Pluto and the Kuiper disk. In *Solar System Ices* (B. Schmitt et al., eds.), pp. 685–709. Kluwer, Dordrecht.

Stern S. A. (2003) The evolution of comets in the Oort cloud and Kuiper belt. *Nature, 424,* 639–642.

Strömgren E. (1947) The short-periodic comets and the hypothesis of their capture by the major planets. *Publ. Copenhagen Obs. No. 144; Kgl. Danske Vid. Selskab, Mat.-Fys. Medd., XXIV, No. 5,* 3–11.

Torbett M. V. (1989) Chaotic motion in a comet disk beyond Neptune — The delivery of short-period comets. *Astron. J., 98,* 1477–1481.

Torbett M. V. and Smoluchowski R. (1990) Chaotic motion in a primordial comet disk beyond Neptune and comet influx to the solar system. *Nature, 345,* 49–51.

Tremaine S. (1990) Dark matter in the solar system. In *Baryonic Dark Matter* (D. Lynden-Bell and G. Gilmore, eds.), pp. 37–65. Kluwer, Dordrecht.

Tyson J. A., Guhathakurta P., Bernstein G. M., and Hut P. (1992) Limits on the surface density of faint Kuiper Belt objects. *Bull. Am. Astron. Soc., 24,* 1127.

Vaghi S. (1973) The origin of Jupiter's family of comets. *Astron. Astrophys., 24,* 107–110.

Valsecchi G. B. and Manara A. (1997) Dynamics of comets in the outer planetary region II. Enhanced planetary masses and orbital evolutionary paths. *Astron. Astrophys., 323,* 986–998.

Wasserman L. H. and Marsden B. G. (2004) 2002 GX_{32} = 1994 JV. *Minor Planet Electronic Circular 2004-N31.*

Whipple F. L. (1950a) On tests of the icy conglomerate model for comets. *Astron. J., 55,* 83.

Whipple F. L. (1950b) A comet model. I. The acceleration of Comet Encke. *Astrophys. J., 111,* 375–394.

Whipple F. L. (1964a) The history of the solar system. *Proc. Natl. Acad. Sci., 32,* 565–594.

Whipple F. L. (1964b) Evidence for a comet belt beyond Neptune. *Proc. Natl. Acad. Sci., 32,* 711–718.

Whipple F. L. (1975) Do comets play a role in galactic chemistry and γ-ray bursts? *Astron. J., 80,* 525–531.

Williams G. V. (1999) 1999 TD_{10}. *Minor Planet Electronic Circular 1999-V07.*

Williams I. P., Fitzsimmons A., and O'Ceallaigh D. (1993) 1993 SB and 1993 SC. *IAU Circular 5869.*

Williams I. P., O'Ceallaigh D. P., Fitzsimmons A., and Marsden B. G. (1995) The slow-moving objects 1993 SB and 1993 SC. *Icarus, 116,* 180–185.

Transneptunian Orbit Computation

Jenni Virtanen
Finnish Geodetic Institute

Gonzalo Tancredi
University of Uruguay

G. M. Bernstein
University of Pennsylvania

Timothy Spahr
Smithsonian Astrophysical Observatory

Karri Muinonen
University of Helsinki

We review the orbit computation problem for the transneptunian population. For these distant objects, the problem is characterized by their short observed orbital arcs, which are known to be coupled with large uncertainties in orbital elements. Currently, the observations of even the best observed objects, such as the first-ever transneptunian object (TNO), Pluto, cover only a fraction of their revolution. Furthermore, of the some 1200 objects discovered since 1992, roughly half have observations from only one opposition. To ensure realistic analyses of the population, e.g., in the derivation of unbiased orbital distributions or correlations between orbital and physical properties, realistic estimation of orbital uncertainties is important. We describe the inverse problem of orbit computation, emphasizing the short-arc problem and its statistical treatment. The complete solution to the problem can be given in terms of the orbital-element probability density function (p.d.f.), which then serves as a starting point for any further analysis, where knowledge of orbital uncertainties is required. We give an overview of the variety of computational techniques developed for TNO orbital uncertainty estimation in the recent years. After presenting the current orbital distribution, we demonstrate their application to several prediction problems, such as classification, ephemeris prediction, and dynamical analysis of objects. We conclude with some future prospects for TNO orbit computation concerning the forthcoming next-generation surveys, including the anticipated evolution of TNO orbital uncertainties over the coming decades.

1. INTRODUCTION

After over a decade of dedicated observations, the transneptunian population of small bodies has recently passed the landmark of 1000 known objects. The observational data for individual objects extend over several decades, but as a population the transneptunian objects (TNOs) are still characterized by their short observed orbital arcs.

Due to their distance from the observer as well as their long orbital periods, observing the population can be both costly and tedious. Both discovery and followup observations have mostly been acquired in dedicated surveys that have been adapted to the slow motions and faint magnitudes of the targets. While many of the first discoveries were made in the pencil-beam surveys (e.g., *Gladman et al.,* 1998), a large portion of observing efforts have subsequently been carried out by the Deep Ecliptic Survey (*Millis et al.,* 2002; *Elliot et al.,* 2005). While the number of well-observed objects has been steadily increasing, so has the number of unfollowed discoveries: In the 2000s, roughly half the TNOs discovered each year have remained as one-apparition objects, which is also the fraction of one-apparition objects in the current population (as of May 2006). It is worth noting that one of the earliest TNO discoveries, 1993 RP, remains lost some 13 years after discovery.

In the chain of efforts to analyze the observed population, orbit computation is in practice the first task to be accomplished. It is also a critical one, since all followup questions depend, in one way or the other, on the orbital elements, which need to be derived already from the often-limited discovery data. For the individual objects, ephemeris predictions for followup observations require reasonable estimates for the orbital elements as well as their uncertainties. For the population, realistic evaluation of uncertainties from astrometric data offer the starting point for the derivation of unbiased orbital distributions (see the chapter by Kavelaars et al.) and for analyzing the dynamical structure of the transneptunian region (chapter by Gladman et al., hereafter G07).

The short-arc orbit computation problem has been extensively studied for the inner solar system, particularly in the recent years in the quest for Earth-approaching asteroids. A great number of numerical techniques have been developed to derive orbits for asteroids since the discovery of asteroid Ceres in 1801. We refer to *Bowell et al.* (2002) and *Milani et al.* (2002) for general reviews of the progress made in the past decade and restrict ourselves to those advances particularly relevant for TNOs. The most important development has been the change in viewpoint from determination of orbits to estimation of orbital uncertainty. Many of the advances are related to nonlinear uncertainty estimation, which is essential for short arcs, or to specific use of the linear approximation, particularly for TNOs.

This active research has resulted in new tools for the users of the end product, the orbital elements, either by means of software or orbital databases, both of which we describe in the current chapter. The orbits of TNOs are nowadays made available, in essence, through four well-established web-based services: at the Minor Planet Center (MPC), at Lowell Observatory, at the University of Pisa ("AstDys"), and at the Jet Propulsion Laboratory (JPL). The MPC maintains a database of TNO orbits together with all their astrometric data. Careful checking of incoming observations and putative identifications by MPC staff limits the number of incorrect observations or identifications for these objects, and as a result the MPC database is probably >99% accurate in terms of matching the proper observations and orbits. It is worth noting that with such sparsely observed objects, bad linkages can occur at fairly high rates, and these linkages can often result in an acceptable differential correction but also a spurious orbit. Of the four services, the AstDys and JPL databases provide also formal orbital uncertainties for the objects in terms of covariance matrices. In addition, AstDys also offers proper elements for outer-solar-system objects. Ephemeris prediction service TNOEPH (*Granvik et al.,* 2003) is on the other hand the first web-based realization of nonlinear uncertainty propagation.

The chapter is organized as follows. We first describe the inverse problem of orbit computation, emphasizing the short-arc problem and its statistical treatment. We demonstrate the use of various computational techniques developed for TNOs (section 2) and evaluate the current orbital uncertainties for the observed population (section 3). In section 4, we discuss the orbital distributions in connection with several prediction problems, such as classification and ephemeris prediction. In section 5, we conclude with some future prospects for TNO orbit computation.

2. ORBITAL INVERSION AND UNCERTAINTY ESTIMATION

In orbital inversion, we want to derive information on the orbital parameters from astrometric observations. Through the awareness of the existing uncertainty in observed asteroid positions, orbits are today treated in a probabilistic fashion. A fundamental question in orbital inversion is the accuracy of the parameters derived. The orbital elements are often not used as such, rather one wants to have an estimate for their *uncertainty* that can then be propagated to another epoch in time or to different parameter space (e.g., to sky-plane coordinates). According to statistical inverse theory (e.g., *Lehtinen,* 1988; *Menke,* 1989), the complete solution to the inversion is given in terms of an orbital-element probability density function (p.d.f.). While the evaluation of the p.d.f. can be a complicated task, the advantage is that, once derived, there is no need for additional error analysis, which is sometimes a major obstacle in traditional, deterministic approaches. Access to the full p.d.f. is useful, if not crucial, in many applications. This is the case when evaluating collision probabilities, which rely on knowledge of the orbital-element p.d.f., or when analyzing the orbital characteristics of small bodies as a population, as in the present paper.

2.1. Statistical Inverse Problem

In the following, we will summarize the formulation of orbit computation in the framework of statistical inverse theory following *Muinonen and Bowell* (1993). The principal idea is that all the available information involved in the inversion can be presented with the *a posteriori* probability density for the orbital elements.

There are several choices for the parameterization of orbits; we denote the osculating orbital elements of an asteroid at a given epoch t_0 by the six-vector $\mathbf{P}$. The Cartesian elements $\mathbf{P} = (X, Y, Z, \dot{X}, \dot{Y}, \dot{Z})^T$ are well-suited for numerical computations; in a given Cartesian reference frame, the coordinates $(X, Y, Z)^T$ denote the position and the coordinates $(\dot{X}, \dot{Y}, \dot{Z})^T$ the velocity (T is transpose). For illustrative purposes, we use Keplerian elements, $\mathbf{P} = (a, e, i, \Omega, \omega, M_0)^T$ and the elements are, respectively, the semimajor axis, eccentricity, inclination, longitude of ascending node, argument of perihelion, and mean anomaly. The three angular elements i, Ω, and ω are currently referred to the ecliptic at equinox J2000.0. The equations describing the inverse problem are the *observation equations*

$$\psi = \Psi(\mathbf{P},\mathbf{t}) + \varepsilon \qquad (1)$$

which relate the astrometric observations $\psi = (\alpha_1, \delta_1; \ldots;$

α_N, $\delta_N)^T$ made at times $\mathbf{t} = (t_1, \ldots, t_N)^T$ to the theory, $\Psi(\mathbf{P}, \mathbf{t})$, and where $\varepsilon = (\varepsilon_1^{(\alpha)}, \varepsilon_1^{(\delta)}; \ldots; \varepsilon_N^{(\alpha)}, \varepsilon_N^{(\delta)})^T$ describe the (random) observational errors.

According to the statistical inverse theory, the solution of these equations is the *a posteriori* probability density, p_p, for the orbital elements

$$p_p(\mathbf{P}) \propto p_{pr}(\mathbf{P})p(\psi|\mathbf{P}) \quad (2)$$

The ingredients needed for its construction are (1) *a priori* probability density p_{pr} for the unknown parameters $\mathbf{P}$; (2) the conditional density of the observations $p(\psi|\mathbf{P})$ expressed with the observational error p.d.f., p_ε. Or in practice, $p_\varepsilon(\Delta\psi)$, as evaluated for the O–C ("observed minus computed") residuals $\Delta\psi$ (see equation (1)).

The equation above is also one form of Bayes' theorem, which enables the user to bring *a priori* information to the inversion process. Despite its controversial nature, the Bayesian approach has in practice turned out to be very useful in many applications. When the solved-for parameters have large uncertainties, *a priori* information can sometimes be used to constrain wide *a posteriori* p.d.f.s. Furthermore, *Muinonen et al.* (2001) pointed out that when solving for p_p in nonlinear inversion, the invariance of the inverse solution under parameter transformations can only be guaranteed by introducing a regularizing *a priori* p.d.f. (e.g., *Mosegaard and Tarantola,* 2002). They suggested the use of Jeffreys's prior (*Jeffreys,* 1946; see also *Box and Tiao,* 1973)

$$p_{pr}(\mathbf{P}) \propto \sqrt{\det\Sigma^{-1}(\mathbf{P})} \quad (3)$$

where Σ is the information matrix, or inverse covariance matrix, for the orbital elements. [The information matrix for the orbital elements follows from the inverse covariance matrix for the observations (taken to be known) and the partial derivatives between the parameters and the observed coordinates (computed analytically or numerically).] *Virtanen and Muinonen* (2006) demonstrated that securing the invariance can be crucial when deriving probability measures, such as impact probability, for very short-arc objects. For well-observed objects, the regularizing prior can typically be assumed constant (section 2.2).

The second ingredient in equation (2) is the error p.d.f. Ideally, the observational data should include a complete error model, but this is rarely the case (cf. future surveys such as Gaia). In fact, in the current format of distribution of astrometric observations, no information on measurement errors is included, thus the orbit computer needs to make an educated guess for the distribution of errors. A practical assumption is a multivariate Gaussian noise probability density with zero mean

$$p(\varepsilon;\Lambda) = \frac{1}{(2\pi)^{2N}\sqrt{\det\Lambda}} \exp\left[-\frac{1}{2}\varepsilon^T\Lambda^{-1}\varepsilon\right] \quad (4)$$

where Λ is the joint $2N \times 2N$ error covariance matrix of the angular observations. Typically, the observed angular coordinates are taken to be independent, although the above form for the error model allows the inclusion of correlation information when available.

Finally, we can write the *a posteriori* orbital-element probability density as

$$p_p(\mathbf{P}) \propto \sqrt{\det\Sigma^{-1}(\mathbf{P})} \exp\left[-\frac{1}{2}\chi^2(\mathbf{P})\right] \quad (5)$$

where $\chi^2 = \Delta\psi^T\Lambda^{-1}\Delta\psi$ is evaluated for the elements $\mathbf{P}$.

The Gaussian assumption, although widely used, has also been shown to fail. Again, in the ideal case, the measurement errors should be more or less randomly distributed. However, the observations can include (unknown, i.e., not corrected) systematic errors, and their distribution can have non-Gaussian features (*Carpino et al.,* 2003). Inversion results can be highly sensitive to the noise assumption (e.g., *Virtanen and Muinonen,* 2006), and thus the handling of observational errors is an important feature of any inversion technique. Transneptunian object astrometry already typically has well-confined measurement errors, at the 0.1-arcsec level. It is expected that astrometric error modeling in general will be improving in the future, due to improving measurement accuracy of the coming surveys (see section 3.1) and improving documentation (the pending new format for astrometric data). However, as the measurement accuracies improve to the milliarcsec level, the modeling demands will also be increasing as the angular difference between photocenter and barycenter will become detectable (*Kaasalainen et al.,* 2005).

How should we characterize the p.d.f.? Numerical techniques for solving the p.d.f. can roughly be divided into two categories: (1) point estimates and (2) p.d.f. sampling via Monte Carlo (MC) means. Point estimates — obtained through optimization methods — are typically sets of parameters describing the p.d.f., such as the least-squares solution, which consists of the maximum likelihood (ML) orbital elements and their covariance matrix (section 2.2).

Point estimates typically work in cases with well-confined p.d.f.s, where the parameter uncertainties are relatively small. For p.d.f.s spreading over a wide range of values in the parameter space, point estimates can be misleading and the only way to reliably characterize the p.d.f. is via MC sampling. While the use of point-estimate methods may be tempting, e.g., due to their speed, many MC methods are today equally feasible thanks to improving computer resources.

Since the observation equations are nonlinear, an important factor when considering the variety of available techniques is their degree of linearity. To highlight the need for the validity consideration, we describe the different methods in two parts: first, techniques relying on the linear approximation and, second, fully nonlinear techniques.

2.2. Linearization in Uncertainty Estimation

In the linear approximation, the relationship between the observations and the parameters in equation (1) can be linearized as

$$\Psi(\mathbf{P},t) = \Psi(\mathbf{P}_{ls},t) + \sum_{j=1}^{6} \Delta P_j \frac{\partial \Psi}{\partial P_j}(\mathbf{P}_{ls},t)$$

which is valid in the vicinity of the reference orbit, $\mathbf{P}_{ls}$, denoting the least-squares orbital elements, and where $\Delta\mathbf{P} = \mathbf{P} - \mathbf{P}_{ls}$. Introducing the approximation through equations (1) and (5), the resulting orbital-element p.d.f. is Gaussian

$$p_p(\mathbf{P}) \propto \sqrt{\det\Sigma^{-1}(\mathbf{P}_{ls})} \cdot \exp\left[-\frac{1}{2}\Delta\mathbf{P}^T\Sigma^{-1}(\mathbf{P}_{ls})\Delta\mathbf{P}\right] \quad (6)$$

Here assuming constant *a priori* p.d.f. is acceptable, $\det\Sigma^{-1}(\mathbf{P}) \approx \det\Sigma^{-1}(\mathbf{P}_{ls})$, because the exponential part of equation (6) makes the p.d.f. well-confined. The least-squares orbital elements $\mathbf{P}_{ls}$ are obtained through a differential correction procedure, and the covariance matrix follows from linearized propagation of uncertainties

$$\Sigma(\mathbf{P}_{ls}) = \left(\frac{\partial \mathbf{P}}{\partial \psi}\right)_{ls}^{T} \Sigma(\psi) \left(\frac{\partial \mathbf{P}}{\partial \psi}\right)_{ls} \quad (7)$$

where $\Sigma(\psi) = \Lambda$. $\mathbf{P}_{ls}$ coincides with the ML estimate, and together with the covariance matrix Σ, constitutes the full, concise solution to the inverse problem [LSL; nonlinear least-squares with linearized covariances; see, e.g., *Muinonen et al.* (2006) for details].

The covariance matrix defines a six-dimensional uncertainty region in the orbital-element space, centered on the ls orbit

$$\Delta\chi^2 = (\mathbf{P} - \mathbf{P}_{ls})^T\Sigma^{-1}(\mathbf{P}_{ls})(\mathbf{P} - \mathbf{P}_{ls}) \quad (8)$$

For practical purposes, the element uncertainty can be described using the projections of this six-dimensional error hyperellipsoid. By replacing $\psi \rightarrow \mathbf{P}$ and $\mathbf{P} \rightarrow \mathbf{F}$, equation (7) is the general law of error propagation and can be used to transform the orbital uncertainties into new set of parameters, $\mathbf{F}$, sky-plane uncertainties, for example.

When describing the method of Bernstein and Khushalani, *Bernstein and Khushalani* (2000, hereafter BK) note that the TNO population presents several simplifications over the general orbit-fitting problem that greatly extend the utility of linearized fitting. For the TNOs, it is possible to choose a basis for the orbital elements such that degeneracies or uncertainties large enough to invalidate the linearization are confined to one of the six elements for arcs of ≈1 week or longer, and are well understood even for single-night arcs.

BK choose a Cartesian coordinate system with z axis along the Earth–TNO vector at the first observation epoch. If we expand the apparent motion in powers of the inverse distance $\gamma \equiv (1\ \text{AU})/z$, then the fact that $\gamma \leq 10^{-1.5}$ for TNOs means that only the first few terms are important, until the orbit becomes very well determined — at which point a linearized solution is accurate for any choice of orbital elements. For example, the transverse gravitational acceleration of the TNO is γ^3 smaller than that of Earth, and can hence be largely ignored until the orbit solution reaches high accuracy.

The BK orbital element set is $\{x/z, y/z, \dot{x}/z, \dot{y}/z, \gamma = 1/z, \dot{z}/z\}$. The first two elements are simply the angular position at the first observation, hence are well constrained immediately. A second observation determines the angular velocity $\dot{x}/z - \gamma\dot{x}_E$, where the second term is the reflex from Earth's motion. Hence the third, fourth, and fifth elements are degenerately constrained by a pair of observations. This degeneracy is broken by an arc long enough to determine the angular acceleration, which is $\approx\gamma\ddot{x}_E$. From the ground, an arc of ≈1 week suffices to determine γ, except for targets near opposition; from Hubble Space Telescope (HST), a single orbit often suffices to determine distance, as the reflex of the spacecraft orbit gives a detectable parallax.

The last BK element, $\dot{z}/z$, generally requires multiopposition arcs for precise constraint. It is intuitive that the line-of-sight velocity is the most difficult Cartesian element to discern from imaging data. The counterpoint is that ephemerides are quite insensitive to this orbital element (as long as the orbit is bound), and hence a determination of the first five BK elements can provide ephemerides (and uncertainties) that are sufficiently accurate for scheduling recovery observations several years into the future. Thus while a poor constraint on the sixth BK element precludes the determination of Keplerian elements and hence assignment to a TNO dynamical class, the BK system does allow linearized fitting to make full use of the information available in a short arc. Once the arc is long enough to constrain all six elements, then the error ellipse in the six BK elements is straightforwardly transformed into Cartesian or Keplerian elements via equation (7).

Kristensen (2004) derives a linearization of the TNO equations of motion that is more accurate than the BK version, yielding arcsecond precision for typical orbits over one-year arcs. Linearizations of some potential prior distributions are also presented by Kristensen. It is hence possible to work with orbit parameterizations that are trivially solved and in which the p.d.f.s are strictly Gaussian, assuming Gaussian errors for the observations.

The nearly inertial motion of TNOs eliminates the multiple branches of short-arc solutions that sometimes arise for asteroids, and the BK fitting process is very fast and robust. This simplicity, plus the ability to produce ephemerides for recovery (with error ellipses), have made the method popular for observers. The BK code becomes less convenient for objects interior to Jupiter's orbit, as the expansion in γ converges poorly. BK is poorly suited to investigations in which the likelihood distribution along poorly constrained elements is important, e.g., attempting to infer the

Keplerian-element distribution for populations of short-arc objects. In such cases a fully nonlinear, MC characterization of the likelihood, including relevant priors, is required.

2.3. Nonlinear Sampling of Uncertainty

Linear methods typically work well when the orbital uncertainties are well constrained, i.e., relatively small. This is not the case when we are dealing with objects with small amounts of data, either by number or by their orbital arc covered; such data are known to be coupled with large uncertainties. Problems with linear covariance mapping can also arise from other sources than the data. The uncertainties often need to be propagated to another epoch or different parameter space, and this mapping can be strongly nonlinear.

For TNOs, the most obvious problem is connected with their short arcs. While for inner-solar-system objects nonlinear p.d.f.s are well known to result for new discoveries — corresponding to observed arcs from a few hours up to weeks — due to their long orbital periods, the nonlinearities for TNOs can be dominant much longer, up to several years, that is, over several apparitions. Second, long propagations needed, e.g., in solar-system integrations, which may entail additional nonlinearities in terms of deep close approaches with massive bodies, can cause the failure of the linear approximation.

Luckily, there are several ways to solve these problems. For the uncertainty propagation, one solution is to introduce semilinear approximations (see section 2.4). However, when the element uncertainties/nonlinearities are large, they can in practice only be estimated by sampling the uncertainty region with a discrete map. This also solves the propagation problem since the mapping points can be propagated to the desired epoch, or coordinates, individually and nonlinearly. Below, we describe perhaps the most general approach to nonlinear sampling, by Monte Carlo means; we reserve section 2.4 for other advances in the subject.

The MC techniques are based on random sampling in some selected coordinates to generate large numbers of sample orbits mapping the p.d.f. Orbital elements (**P**) can be sampled directly if the p.d.f. is already well-confined, e.g., Gaussian (*Chodas and Yeomans,* 1996; *Muinonen and Bowell,* 1993), or close to being bell-shaped (see VoV technique below). Sampling in observation space (R.A. and Dec.) is practical if the orbital-element p.d.f. is too extended to allow direct sampling (e.g., in BK). If there are very few observations, including the object's unknown topocentric distance in the sampling parameters ("extended" observation space; R.A., Dec., rho) reduces the minimum number of observations needed in the inversion to two (see Ranging below).

In contrast to the linear case, large uncertainties typically imply strongly non-Gaussian p.d.f.s. Methods based on orbital or observational MC typically make use of the linear approximation, in the form of least-squares fitting, while sampling in the extended observation space is in practice the only way to handle strongly nonlinear cases. Below, we describe two nonlinear techniques, the first of which uses linearization as a intermediate step, but both of which result in fully nonlinear p.d.f. sampling.

2.3.1. Volume-of-variations (VoV). *Virtanen et al.* (2003) pointed out that, for TNOs, the linear approximation can break down even for the numbered objects. As a solution, they suggested six-dimensional sampling of the uncertainty region using least-squares covariances (see also *Muinonen and Bowell,* 1993), a generalization of the so-called line-of-variation methods (*Milani,* 1999) (section 2.4). *Muinonen et al.* (2006) devised a new technique that samples a six-dimensional volume in the phase space instead of a one-dimensional curve. In the initialization part of VoV, starting from the global LSL solution, local linear approximations are computed for a reduced number of orbital elements as a function of one or more mapping elements. Currently, the semimajor axis or the Cartesian X-coordinate is chosen as the single mapping parameter, and by correcting for the remaining five orbital elements, a discrete set of points corresponding to the local p.d.f. maxima is computed. In the second part, guided by the local linear approximations, MC sampling is introduced in the six-dimensional phase-space volume for a fully nonlinear treatment.

The degree of nonlinearity can be increased in the initialization part by selecting several mapping parameters (e.g., the longitude of the ascending node in addition to semimajor axis), but the technique may quickly become computationally excessive for more numerous mapping parameters. In this paper, the VoV technique has been systematically applied to a large number of TNOs and, at the same time, the technique was fully automated.

2.3.2. Statistical orbital ranging (Ranging). When the orbital uncertainties are large, sampling of the p.d.f. in orbital elements is no longer efficient nor possible. *Virtanen et al.* (2001) and *Muinonen et al.* (2001) established the idea of sampling the orbital-element uncertainty in topocentric spherical coordinates. In Ranging, two observation dates are chosen from the complete observation set and the corresponding topocentric distances (or ranges), as well as the R.A. and Dec. angles, are MC sampled using predefined intervals, which are subject to iteration. The two Cartesian positions lead to an unambiguous set of orbital elements based on well-established techniques in celestial mechanics. The elements then qualify as a sample orbit if the fit to the observation set is acceptable. *Virtanen et al.* (2003) applied the automated Ranging to TNOs, presenting the first systematic analysis of the orbital uncertainties of the then-known population.

2.4. Other Advances

Another approach to dealing with the nonlinearities in orbit computation was put forward in *Milani* (1999). They have developed one-dimensional techniques based on the line of variations (LOV), which is aligned with the principle axis of the covariance ellipsoid (cf. equation (8)). To overcome the problems faced with linear propagation, they have proposed semilinear approximations for mapping and propagating orbital uncertainties starting from the LSL (*Milani and Valsecchi,* 1999). The concept of multiple solutions along the LOV has been used for better mapping the orbital uncertainty in many applications such as identification or impact monitoring (*Milani et al.,* 2002, 2005).

For what they term "very short arcs," *Milani et al.* (2004) have taken a geometric approach to uncertainty estimation. Through least-squares fitting in the observation space (two angles and angular rates), they use a set of virtual asteroids to describe the uncertainty region by triangulation of the unknown range-range-rate plane. This approach is also suitable for making ephemeris predictions for Centaurs (*Milani and Knežević,* 2005) but the need for the estimation of angular rates may become problematic for more distant objects due to their slow relative motions (*Milani et al.,* 2006).

Closely resembling the initialization part of VoV, *Chesley* (2005) introduces a two-dimensional plane-of-variation technique that utilizes the range and range rate as the orbital elements to be systematically varied and uses it to estimate the encounter probability for a short-arc asteroid. However, the four remaining dimensions of the inverse problem are not treated.

Techniques similar to Ranging, based on the variation of topocentric coordinates, have been described by others (see, e.g., *Bowell et al.,* 2002, and references therein), but the one by *Goldader and Alcock* (2003) is specifically tailored for TNOs. The idea therein is identical to Ranging as far as the sampling procedure is considered. However, they do not provide solid probabilistic interpretation of the uncertainty mapping, although the two methods produce similar results for the *extent* of the uncertainty region.

The discovery of several multiple systems in the Kuiper belt has spurred research on dynamics of binary objects. Although the particulars of the physical problem differ from the one discussed in the present paper, and are thus omitted here, there are some links between the two problems that we consider worth discussing. Multiple systems are observationally demanding, and optimal datasets for orbit computation can be difficult to obtain. The shortage of data is even more obvious for distant and faint systems such as TNO binaries. Recently, *Hestroffer et al.* (2005) introduce an MC-based sampling to Thiele-Innes method for evaluating the uncertainties in binary orbits in general. As followup observations for TNO binaries are especially costly, Grundy et al. (in preparation) modify Ranging for binary systems to optimize the scheduling of Hubble Space Telescope (HST) observations. Both methods are well-suited for TNO binaries with very sparse data, and can be used to impose realistic uncertainties for the derivable physical parameters, such as the total mass and density of the objects.

3. TRANSNEPTUNIAN OBJECT ORBITAL UNCERTAINTIES

To illustrate our current understanding of the orbital uncertainties for the TNO population, we have carried out TNO orbital inversion systematically for the known objects using the orbit computation software package Orb developed at the University of Helsinki Observatory. The package currently provides tools for full, automated orbital inversion of solar system objects, as well as tools for many prediction problems. Below, we demonstrate the use of the statistical techniques just described, in the order of increasing degree of nonlinearity: (nonlinear) least-squares estimation, VoV technique, and Ranging. As a comparison, uncertainties are also evaluated using the BK method, which has been widely used, e.g., for TNO ephemeris prediction.

Some of the questions we would like to discuss here include determining the optimal application regions for the different methods. In particular, as the amount of observational data increases, when do point estimates (vs. sample orbits) become useful? Reflecting the expected improvement in astrometric accuracies, what is the role of observational accuracy in TNO orbital inversion?

We have included in the analysis all outer-solar-system objects listed as transneptunian objects, scattered-disk objects, or Centaurs in the MPC database. As of May 2006, there were 1159 objects, of which 148 were numbered. [In addition, Pluto, numbered (134340) in August 2006, was included for completeness.] For the unnumbered objects, TNO orbital-element p.d.f.s are represented with 2000 sample orbits, which all satisfy the observations with predefined accuracy (requiring both the weight in equation (5) and the O–C residuals to be acceptable). On a modern workstation, the computations for each object took on average two minutes, the typical number of trial orbits needed being some tens of thousands. For numbered objects, least-squares orbital elements and covariance matrices are used. For the observational error p.d.f. in equation (4), we assume uncorrelated R.A. and Dec. p.d.f.s with zero mean and standard deviation of 0.5 arcsec. There was need to adjust this noise assumption for several objects, as discussed below in section 3.1. No Bayesian *a priori* information is imposed in the computations except for the shortest arcs, for which a practical constraint in semimajor axis is used; the maximum allowed value for a was 1000 AU (see *Virtanen and Muinonen,* 2006). Computations were mostly carried out using the two-body dynamical model, but the planetary perturbations were included for the numbered objects. The validity of the two-body approximation was also discussed.

To quantify the results of the orbital inversion, the orbital uncertainties are illustrated in Fig. 1 using two quality metrics (*Muinonen and Bowell,* 1993): the standard deviations (σ) of the semimajor axis and inclination p.d.f.s. Based on these metrics, as proposed in *Virtanen et al.* (2003), the population can be divided into three categories: one-apparition objects (513), two-apparition objects (96), and multiapparition objects (550). The first category consists of objects that have been observed up to six months and is characterized by very large uncertainties. Semimajor axis, together with eccentricity, is typically poorly determined for short arcs (σ_a >~ 10 AU), while inclination can on occasion be determined rather well (σ_i ~ 0.001°) even with such limited data. The two-apparition objects have been observed in the subsequent opposition after discov-

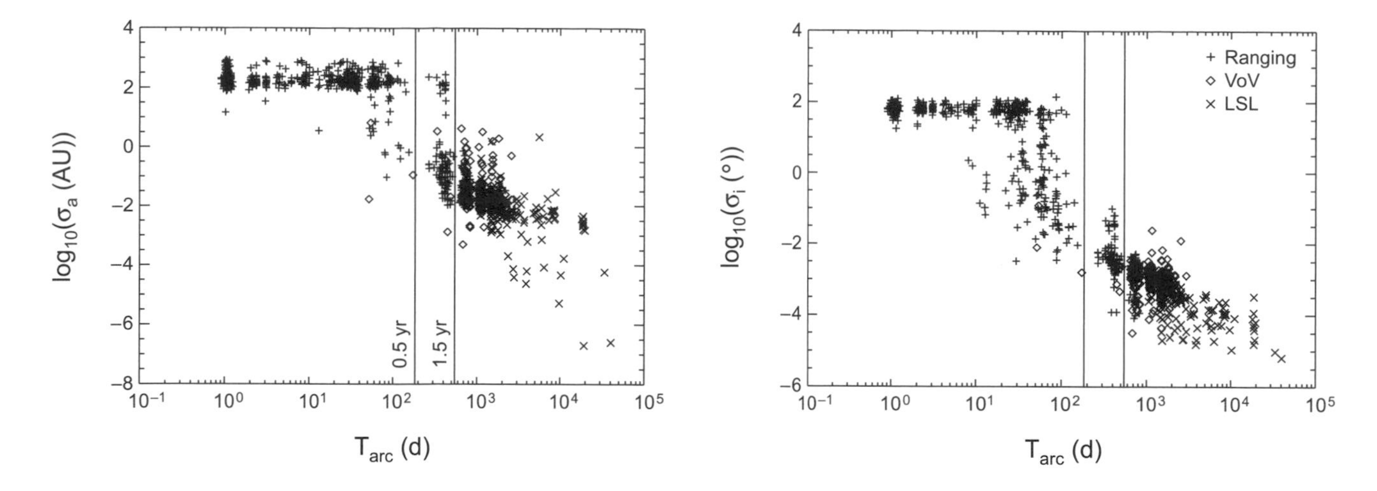

Fig. 1. Phase transition in the orbital uncertainties as a function of the length of the observed time arc. Based on the semimajor-axis (left) and inclination (right) standard deviations, objects can be divided into three categories: one-, two-, and multi-apparition. The application of different inverse techniques is also illustrated; Ranging (pluses), VoV (diamonds), LSL (crosses).

ery ($180 < T_{arc} < 540$ d). The accuracy of the orbit varies greatly, and the data can fix the orbital elements rather well, or leave some of the elements very poorly constrained. Multiapparition objects include all the objects with observational arcs longer than 1.5 yr that already have well-confined uncertainties for all the elements.

Figure 2 shows the distribution of TNO orbits using the joint orbital-element p.d.f.s obtained by summing the p.d.f.s for the individual objects, which have been normalized (all objects have equal weights). The histograms for only objects observed on several oppositions (solid curves) display the known characteristics of the region; for semimajor axis, the two largest groupings of objects are the classical belt and the 3:2 resonance population. The inclination p.d.f. shows the twofold structure with low-inclination (<5°, "cold") and high-inclination (10°–30°, "hot") populations. However, the short-arc objects (included in dashed curves) introduce several biases in the joint distributions, which do not show up for longer arcs. Most notably there is an abundance of high-eccentricity orbits that corresponds to the peak at small values for the perihelion distance (q). An *a priori* constraint from the well-observed orbits can be used to smooth away this behavior for short-arc objects.

Since the TNO population includes many very-short-arc objects with small numbers of observations, the applicability of the statistical interpretation of the orbit computation results should be discussed, as pointed out by *Bowell et al.* (2002). The effect of Jeffreys' regularizing *a priori* p.d.f. used to maintain the invariance was monitored for the shortest arcs ($T_{arc} < 180$ d), as suggested by *Virtanen and Muinonen* (2006). The joint p.d.f.s of orbital elements were found to be insensitive to the choice of the prior (Jeffreys' or constant prior). Furthermore, we point out that, when the analysis is based on only a few data points, the orbital element p.d.f. itself may lose its meaning but the extent of the orbital uncertainty can still be mapped. One of the objects analyzed, 2004 PR_{107}, had only two observations, on distinct nights. In such a case, the orbital-element p.d.f. can be assumed uniform, i.e., in Ranging, the sample orbits are accepted simply if they produce acceptable residuals at the observation dates.

3.1. Phase Transitions

As the orbital uncertainties gradually decrease with increasing observational data, there typically exists a time interval over which the accuracy of the orbit quickly improves. This nonlinear collapse in the uncertainties has been termed the "phase transition" (e.g., *Muinonen and Virtanen,* 2002) and can also be seen in Fig. 1 for the TNO uncertainties. The exact location of the phase transition has been shown to vary from object to object (*Muinonen et al.,* 2006) — depending on the observational data and dynamical class of the object — and it is also different for the different elements. For Keplerian elements, this typically happens first for the inclination and last for semimajor axis and eccentricity. For Cartesian elements, phase transition occurs when the observational data begin to constrain elements to a region that is entirely contained by our expectation of a bound orbit or other prior constraints. The line-of-sight velocity is the orbital element, which takes the longest to be meaningfully constrained by the data.

Figure 1 presents an average or "statistical" view of the phase transition for the whole population. The transition regime coincides with the two-apparition category of objects, among which the scatter in the semimajor-axis uncertainty is the largest. Note that the smallest uncertainties across the time axis typically correspond to the closest Centaur orbits, while the isolated object close to [5000 d, 1 AU] (cross symbol) is (90377) Sedna, one of the most distant TNOs (a_{ML} = 529 AU).

The phase transition can help to study when the different techniques described in the previous section can and should be applied. *Virtanen et al.* (2003) already showed that the

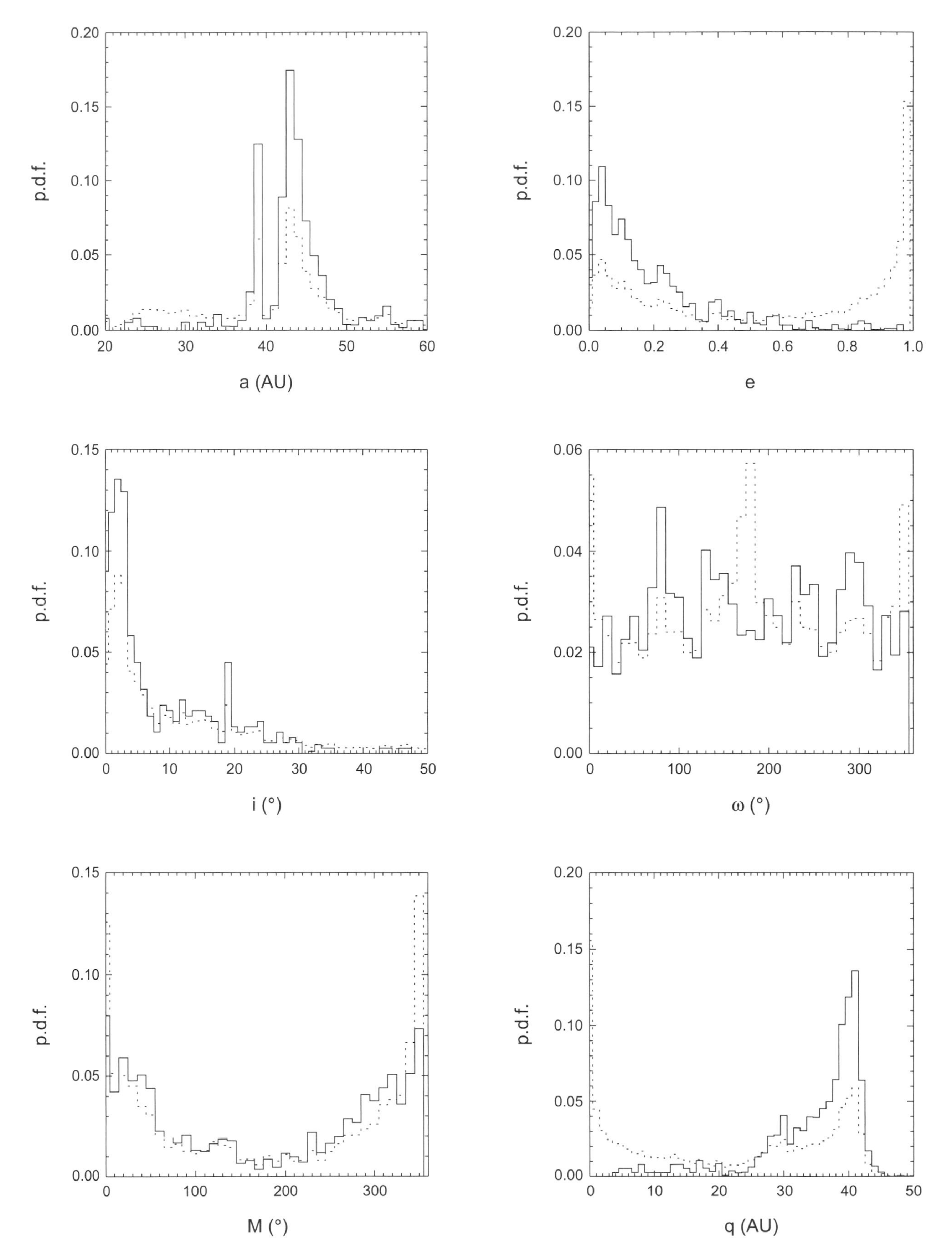

Fig. 2. Joint orbital-element p.d.f.s for all known objects (dashed line) and for objects observed longer than $T_{arc} > 180$ d (solid line). The prominent characteristics of the region are clearly visible, but the short-arc objects introduce several biases in the joint distributions, which do not show up for longer arcs.

Ranging technique could be applied to 90% of the population known in 2001, including multiapparition TNOs. Since the fraction of one-apparition objects has been steady at 50%, it was *a priori* clear that most of the computations for unnumbered objects can be carried out using one method. Figure 1 shows Ranging solutions for 736 objects, longest arc extending up to six years of observations (2218 d), which is well beyond the transition regime.

However, the observational data for the known objects has also been steadily growing — numbered objects from 19 (February 2001) to 148 (May 2006), multiapparition objects from 93 to 550 — highlighting the need for efficient methods for longer arcs. Methods such as BK and VoV have been developed for this purpose. The BK method has been put to practice in followup work (see section 4.1). Here we have tested VoV extensively for multiapparition TNOs (272 objects in Fig. 1).

To continue the comparison of the different techniques, we plot the semimajor-axis standard deviations computed using all four methods for a selection of objects (Fig. 3). The uncertainty estimates mostly agree, indicating that the choice of method is not very crucial in estimating modest to small uncertainties. The most notable discrepancies appear when the uncertainties are large, but there are several cases of modest uncertainties where almost order-of-magnitude differences can occur. A defect of the present comparison is that it does not tell about the nonlinear correlations between the elements that may exist and can only be mapped with nonlinear techniques.

As discussed in section 2.1, modeling of observational errors is a significant factor in the orbital inversion process. Once the phase transition occurs, the uncertainties in the orbital elements will scale linearly with the uncertainties in the astrometric data. More precise astrometric data can also shorten the observed time arc required to reach the transition. Since, however, the measurement uncertainty in line-of-sight velocity typically shrinks faster than linearly with arc length, the time of phase transition advances more slowly than linearly with improved astrometric errors.

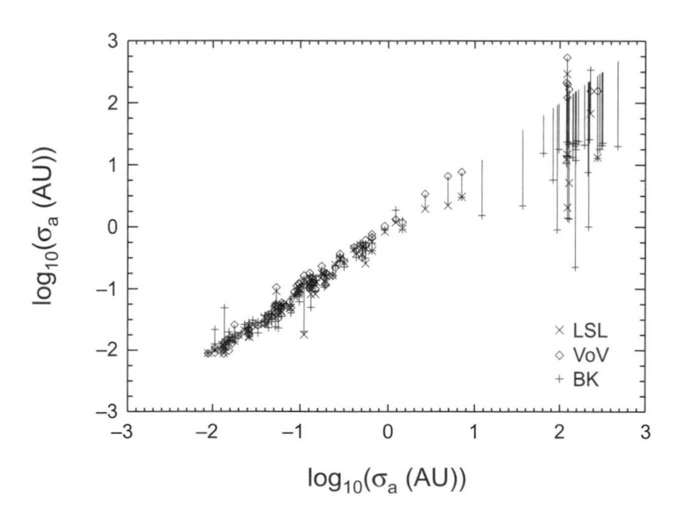

Fig. 3. Comparison between orbital uncertainties evaluated using different inversion techniques. Standard deviations of semimajor axis from Ranging in the x-axis and from the other techniques in the y-axis. Note that the largest VoV uncertainties (upper right corner) result from incomplete p.d.f. sampling (constrained by unphysical solutions).

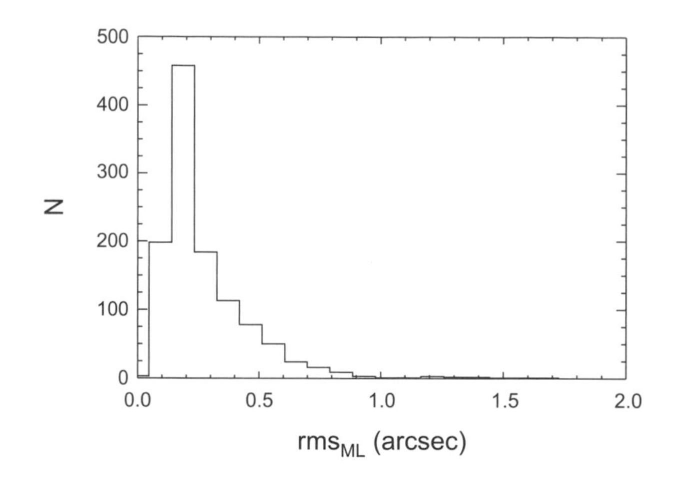

Fig. 4. Distribution of the rms values for the maximum likelihood (ML) orbits. For most of the objects, observational data is accurate to ~0.2 arcsec, but there are 14 objects with rms_{ML} > 1.0 arcsec.

The current accuracy of TNO astrometry can be estimated from the distribution of O–C residuals. Figure 4 shows the distribution of rms values for the ML orbits. For majority of TNOs, the noise assumption of 0.5 arcsec turned out to be a conservative one; the rms distribution suggests that most of the observational data is accurate to ~0.2 arcsec. However, it is evident that the quality of data varies a lot, as the best-fitting orbits can have rms values from 0.1 to close to 2.0 arcsec. For some 50 multiapparition objects, the large fractions of apparent outliers in their datasets were only reduced by including the planetary perturbations, indicating the failure of two-body approximation. We note that including the perturbations will also reduce the residuals for some objects with large rms. Furthermore, for some 20 objects, acceptable orbit solution could only be found if the assumed R.A. and Dec. standard deviations were increased to 1.0 arcsec. This set of objects included five objects that are TNO binary candidates (see the chapter by Noll et al.). In fact, as the binary systems are rather numerous among known TNOs, residual distributions could be studied to find signatures for unknown binary systems among the known objects.

In astrometry, there are two sources of errors. First is the uncertainty in determination of the TNO position relative to astrometric field stars in its vicinity. Modern groundbased techniques and good PSF fitting, including accurate timing, should reduce this well below 0.1 arcsec, to 10-milliarcsec level for high-S/N objects, and even below for spacebased observations. The second source of ephemeris errors is inaccuracy in the placement of the astrometric standard stars on the global inertial coordinate system. The quality of global astrometric standard systems is improving rapidly, with the ~0.2-arcsec errors of the USNO-B catalog to be reduced substantially in the next decade by several astrometric surveys (e.g., *Zacharias,* 2004). These include, in the expected "first-light" order, the USNO CCD Astrograph Catalog (UCAC), Panoramic Survey Telescope and Rapid Response System (Pan-STARRS), ESA's Gaia mission, and Large Synoptic Survey Telescope (LSST).

While ESA's astrometric space observatory Gaia can only observe a handful of the brightest TNOs due to its limiting magnitude of ~20 (F. Mignard, personal communication), it will do so with accuracies in the 10-microarcsec level at best. On the other hand, Pan-STARRS is expected to discover large numbers of objects, but with a more modest accuracy of some 10 milliarcsec (*Jedicke et al.,* 2007). To study the influence of the improving accuracy on orbital uncertainties, we created simulated sets of observations for an object on a distant orbit by assuming different astrometric accuracies. The orbit and observing dates correspond to TNO (87269), which has a semimajor axis of 550 AU and eccentricity of 0.96. The original dataset was broken down night by night, and the observations were simulated with accuracies of 0.5, 0.005, and 0.00005 arcsec. Figure 5 shows the phase transition in the observational accuracy. For the longer arc lengths, there is a linear trend: an order-of-magnitude improvement in the observational accuracy results in a similar improvement in the orbital accuracy. On the other hand, to reach a given orbital accuracy, an improvement of 2 orders of magnitude in the astrometric accuracy implies a decrease of 1 order in the length of observational arc required.

3.2. Filtering Unstable Orbits

Up to now we have not made any specific considerations about the dynamical properties of the orbit solutions computed. The transneptunian region is supposed to be a reservoir of objects that has lasted since the origin of the solar system, with objects henceforth slowly leaking away or entering into planet-crossing orbits (e.g., the Centaurs). The lifetime during the planet-crossing state is several orders of magnitudes shorter than the time spent in the transneptunian region. Therefore, it could be expected that most of the objects were in stable nonencountering orbits, and only a small proportion of objects might have unstable orbits allowing encounters with the planets at present.

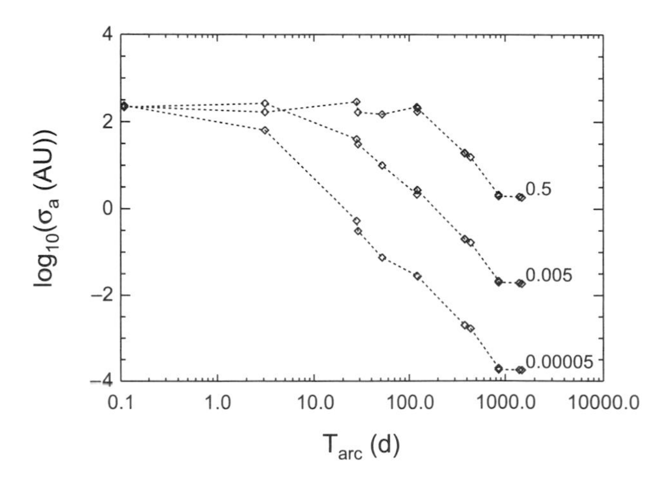

Fig. 5. Time evolution of semimajor-axis uncertainty for improving observational accuracy of a simulated dataset. More precise astrometric data can shorten the time required to reach the phase transition. In the case of (87269), the present orbital accuracy would have been reached in ~3 months with the expected accuracy of Pan-STARRS and in ~10 d with Gaia accuracy.

In order to check the orbital stability, one may integrate the orbits for a few millions of years and check for possible encounters. For the MC sampled orbital-element p.d.f.s, this would be an extremely time-consuming procedure. We therefore implement a stability criterion based just on the computed orbital elements. The filtering process is described in detail in *Virtanen et al.* (2003); we hereby present a summary of it. We perform an analysis of the minimum orbital intersection distances (MOIDs) of all the orbits with respect to the four giant planets, namely Jupiter, Saturn, Uranus, and Neptune. We have defined a possible close planetary approach as MOID being less than 3 Hill radii (r_H) from the planet. A low MOID would generally imply a close approach to the planet, unless the object is protected by a mean-motion resonance with the corresponding planet. We only check for mean-motion resonance with Neptune, because a capture into a resonance by the other giant planets in the Centaur region would generally be short-term in nature. To confirm a resonance capture, the orbit needs again to be numerically integrated for several million years to check whether the resonance argument of the expected resonance is librating or circulating (as done in G07). Instead, to obtain a preliminary estimate, we have adopted a simpler criterion based on proximity to the theoretical location of the libration zone. If in the a–e plane the orbit falls inside any region of possible libration with respect to Neptune, the orbit is tentatively considered to be in the resonance. The libration zones are taken from Fig. 1 of *Morbidelli et al.* (1995) up to the 1:6 resonance, assuming a diamond shape in the a–e plane and a maximum width of 2 AU. Since we have seen that there are objects in higher-order resonances, we extend their figure to include resonances from 1:7 to 1:21 using the above parameters for the libration zones. For multiapparition objects, the fraction of nonresonant objects with low MOIDs is just 4%. For one-apparition objects, 99% have at least one orbit fulfilling these criteria, corresponding to 50% of the total number of orbits. We thus conclude that the huge orbital uncertainties present for the one-apparition objects actually imply a large variety of possible dynamical evolutions for the objects, not all of which are realistic considering our current understanding of the dynamics of the outer solar system.

For some of the objects, more than 95% of the computed orbits are nonresonant and have low MOIDs. These are objects that are very likely experiencing close encounters with the giant planets. A detailed analysis of these cases is out of scope for this chapter. Nevertheless, we would like to mention two examples with low MOID with respect to Neptune: a one-apparition object 2002 PQ_{152} (Ranging

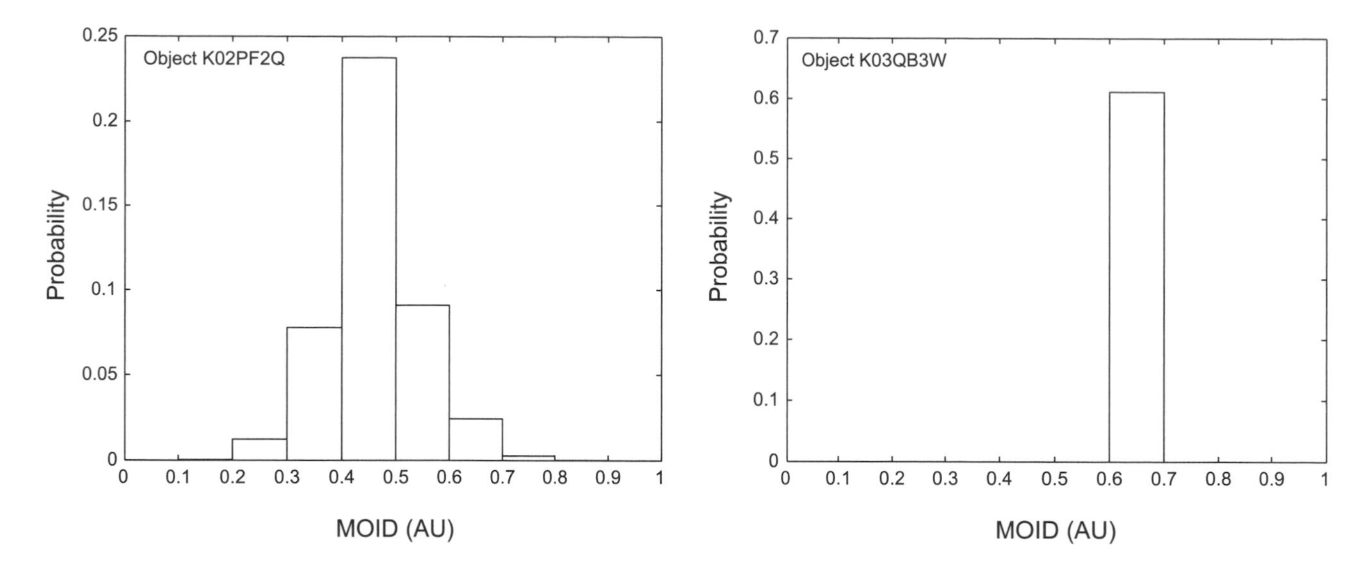

Fig. 6. Distribution of MOID values with respect to Neptune for two example cases, 2002 PQ_{152} and 2003 QW_{113}.

solutions) and a two-apparition object 2003 QW_{113} (VoV solutions). In Fig. 6 we present their MOID p.d.f.s with respect to Neptune. Note the spread in the MOID values for 2002 PQ_{152} around a mean value of ~0.45 AU. Nevertheless, there are no solutions at very short distances that could imply a possible collision with the planet. The distribution of MOID values for 2003 QW_{113} is well defined around a value of 0.65 AU.

The chances of a collision with Neptune among all the orbit solutions can be estimated to be very low but nonzero. We found 346 orbits among the total of approximately 1 million solutions for one- and two-apparition objects having MOIDs with respect to Neptune less than 3 planetary radius. A procedure to estimate the collision probability to any of the planets of a newly discovered object would be worth implementing.

3.3. Orbital Database

The current computations form the basis for a TNO orbital-element database. The orbital-element p.d.f.s are summarized in one of the following four formats: (1) ML orbital elements and standard deviations (e.g., least-squares solution), (2) a set of — all or best-fitting — sample orbits (e.g., from Ranging), (3) a set of orbital elements mapping the boundaries of the uncertainty region (from the VoV technique). In addition, orbit solutions after the filtering procedure can be obtained. The database is available from the authors (contact J. Virtanen).

The second and third format can be useful in more detailed dynamical studies, where a set of sample orbits mapping the orbital uncertainty can be used as initial values for numerical integrations. This kind of approach has been used for analyzing resonance occupation (*Chiang et al.,* 2003; *Elliot et al.,* 2005; see also G07). Another application of the database is in ephemeris prediction. The difficulty there is that the mapping of the uncertainty region from orbital-element space to sky-plane coordinates is nonlinear. While in the linear approximation, LSL covariance can be directly transformed into the sky-plane coordinates (see equation (7)), in the nonlinear case the full orbital-element p.d.f. needs to be propagated. Another solution for this nonlinear mapping problem has been given by *Milani* (1999) in terms of semilinear confidence boundaries and it has been widely used for asteroid recovery. The AstDys database offers orbital elements and covariances also for TNOs, but does not currently provide information on the nonlinearities that are dominant for the majority of objects in this short-arc population.

4. PREDICTION PROBLEMS

4.1. Ephemeris Uncertainty

Because of the nearly linear motion of TNOs, the projection of the element p.d.f. to the sky plane typically results in fairly linear distribution even if the original p.d.f. is highly nonlinear. This behavior is in contrast to the near-Earth objects (NEOs), for which the sky-plane uncertainties based on short-arc orbits quickly extend all over the celestial sphere. While the NEO community includes a large group of follow-up astrometrists, this is not the case for TNOs. Most of this is due to the large size of the telescope required for routine followup, as well as the long integration times required, which makes the efficiency of TNO observing rather low ("time spent per object" is high).

The BK method was designed for the practical purpose of helping observers to evaluate sky-plane uncertainties for the costly observations of TNOs. It has turned out as a useful tool for followup in several surveys; the method has been applied to objects discovered in the Deep Ecliptic Survey (*Millis et al.,* 2002; *Elliot et al.,* 2005) as well as in the Very Wide component of the Canada-France-Hawaii Telescope Legacy Survey (*Jones et al.,* 2006).

To illustrate the current ephemeris uncertainty for the population, the orbital-element p.d.f.s described in section 3

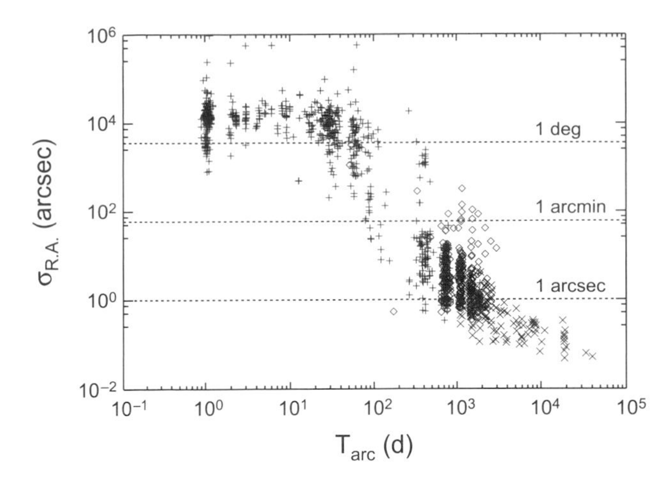

Fig. 7. Current ephemeris uncertainties for the known TNOs in terms of the standard deviation of the R.A. marginal p.d.f.s (July 4, 2006). A large majority of the one-apparition objects can be considered lost, and for several numbered objects (stars) the current uncertainty in their position is already close to 10 arcsec. Symbols as in Fig. 1.

have been propagated to a joint epoch and transformed into spherical coordinates. Figure 7 illustrates ephemeris uncertainty with the help of the standard deviations of the R.A. marginal p.d.f.s. The relevance of recovery attempts depends on the size of the sky-plane uncertainty region to be searched. Assuming that, in practice, 1σ uncertainty in R.A. (or Dec.) over 1° means that an object is lost, nearly half of the known population is beyond routine recovery.

At discovery, the MPC usually attempts to place each object in one of a few narrow dynamical classes, such as "plutino," "cubewano," or "scattered-disk object." While this may aid recovery of the object in the future if the guess is correct, it can also result in the loss of an object. Because of the large uncertainties involved for short-arc objects, reclassifications of TNOs in MPC listings are frequent (listed in the web-based *Distant EKOs Newsletter*). The probabilistic approach to orbit computation makes dedicated recovery attempts possible, however. With the help of Bayesian *a priori* information, wide ephemeris uncertainties can be constrained and an observing strategy can be planned (*Virtanen et al.,* 2001). As *a priori* knowledge, one can use dynamical filtering based on the TNO classification scheme (see section 4.2) or a full *a priori* p.d.f. based on, preferably unbiased, orbital-element distribution of known TNOs.

In an ideal search program, the followup of discovered objects is carried out as part of the survey. While the observing concept of future all-sky surveys will ensure followup more or less automatically, this has not been the case with past and current surveys. The natural goal of astrometric followup is to secure TNO orbits, as a result of which they are eventually assigned a number. As with asteroids, the condition for numbering is that the orbital uncertainties are considered small enough to enable routine followup observations for a reasonable time into the future. The ephemeris uncertainties for the known objects were propagated 10 years ahead from the last observation for each object. For the numbered objects, the 1σ uncertainties are mostly in the arcsecond level but, for several cases, the uncertainty grows to tens of arcseconds, or even arcminutes, if followup observations were to cease for 10 years. This highlights the need for continuous astrometric efforts for these distant objects.

While the question of optimized observing strategy may become to a large degree obsolete in the next few years, it will still be important for so-called science alerts, which are typically new discoveries with some property that makes them require immediate attention. Depending on the observing concept of the survey in question, there may be need for external observing efforts before the routine recovery by the survey itself.

4.2. Dynamical Classification

One of the first questions one wants to pose once a discovery of a new solar system object has been made is whether this object can be classified in one of the known dynamical groupings. While finding the answer to this question might not be as crucial and urgent for TNOs as it is for Earth-approaching objects, it is nevertheless needed in order to form a dynamical picture of our planetary system. Statistical methods provide firsthand tools for dynamical classification.

The final classification algorithm requires detailed dynamical studies using numerical integrations, and for the multiapparition objects, that has been the subject of G07, where the BK method is applied. Here we want to address the question of short-arc classification, and since this is by no means a static problem, we study the evolution of the classifications with the increasing observed time arc.

There has not been a clear consensus among the planetary research community about dynamical classification of the variety of solar system objects. In particular, the classification of TNOs has been the subject of many proposals, e.g., *Gladman et al.* (1998), G07, and *Virtanen et al.* (2003). Although we do not entirely agree with the classification scheme proposed by G07, we have adopted their scheme for consistency but introduced some modifications to speed up the short-arc classification process. As in section 3.2, to avoid heavy numerical integrations, we use some simpler criteria based on the orbital elements. We follow the flowchart of G07 presented in their Fig. 1 (left), but the resonance occupation is decided with a plot of the location of libration zones in the a–e plane, as explained in section 3.2. The other criteria based on the numerical integrations in G07's scheme is the classification of a scattered disk object (SDO). An SDO is by definition an object that has experienced close approaches to Neptune, and therefore the orbit must come close to that of the planet. We have adopted the criterion that SDOs should have a perihelion distance less than the semimajor axis of Neptune plus 3 r_H of the planet (i.e., 2.3 AU), although it is not necessary that the MOID with respect to Neptune is that low at present. As we shall see, a large fraction of the short-arc orbits are retrograde. Retrograde orbits would have a Tisserand parameter with respect to Jupiter less than 3, and in many cases a negative value. In G07's scheme a retrograde orbit with small q would be incorrectly classified as a Jupiter-family comet. Therefore, we introduce a new criterion in the flow chart: In the second box from top to bottom, we ask if the orbit is retrograde (inclination greater than 90°).

We then have the following groups: (1) not the subject of this book (NotInc), (2) retrograde (Retro), (3) resonance (Res), (4) Jupiter-family comets (JFC), (5) Centaurs (Centa), (6) Inner Oort (InOort), (7) scattered disk objects (Scatt), (8) detached (Detac), (9) outer belt (Outer), and (10) classical belt (Class).

Using the orbital-element p.d.f.s computed either with Ranging, VoV, or least-squares estimation (section 3), we apply the classification scheme to the TNO population. Figures 8–10 show the phase-space structure of the transneptunian region. For one-apparition objects, the oversup-

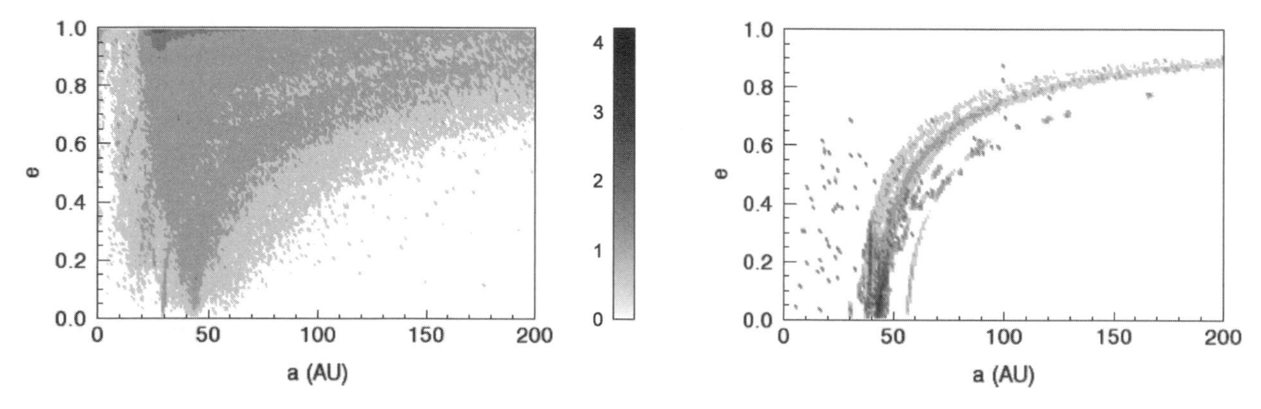

Fig. 8. Contour plot in the a–e plane illustrating the phase-space structure of the transneptunian region, left for $T_{arc} < 180$ and right for $T_{arc} > 180$ d. The intensity of the grayscale shading is proportional to $\sqrt{(N)}$, where N is the average number of objects in each bin (bin size 1 AU × 1 AU). The distribution is very flat for short-arc objects, illustrating the difficulty in their classification.

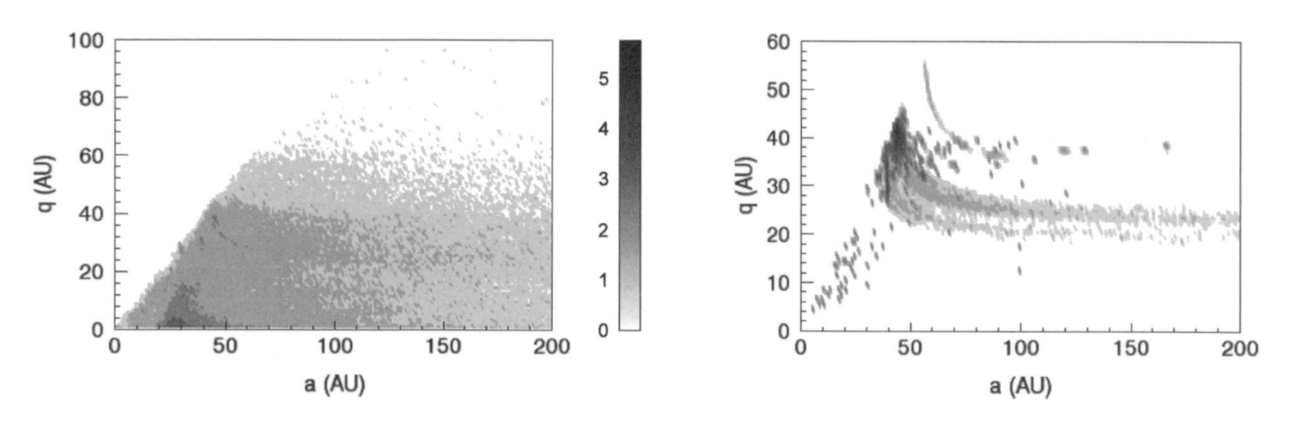

Fig. 9. As in Fig. 8 but in the a–q plane (bin size 1 AU × 1 AU).

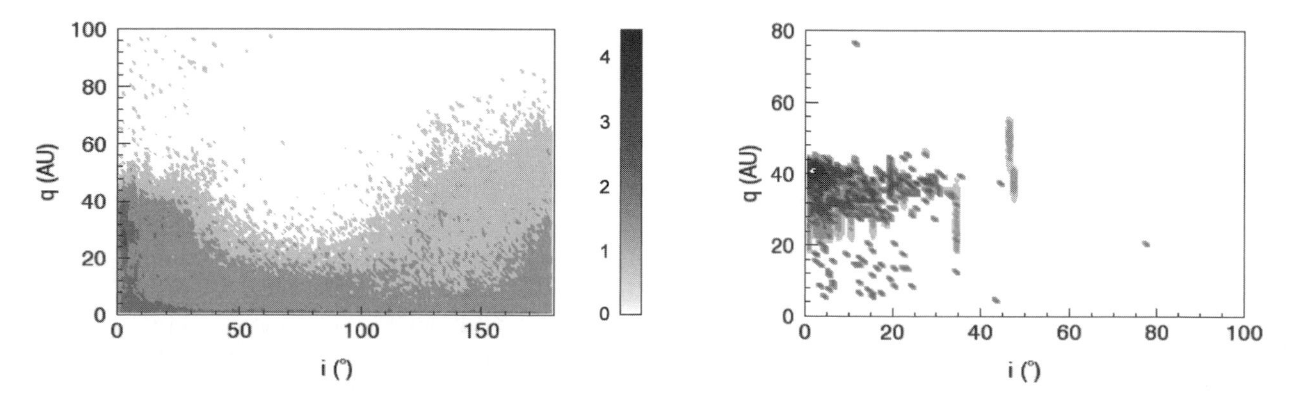

Fig. 10. As in Fig. 8 but in the i–q plane (bin size 1° × 1 AU). Note that the few vertically elongated clouds in the right plot correspond to objects with rather well-defined i but large uncertainties in q (i.e., in a and e).

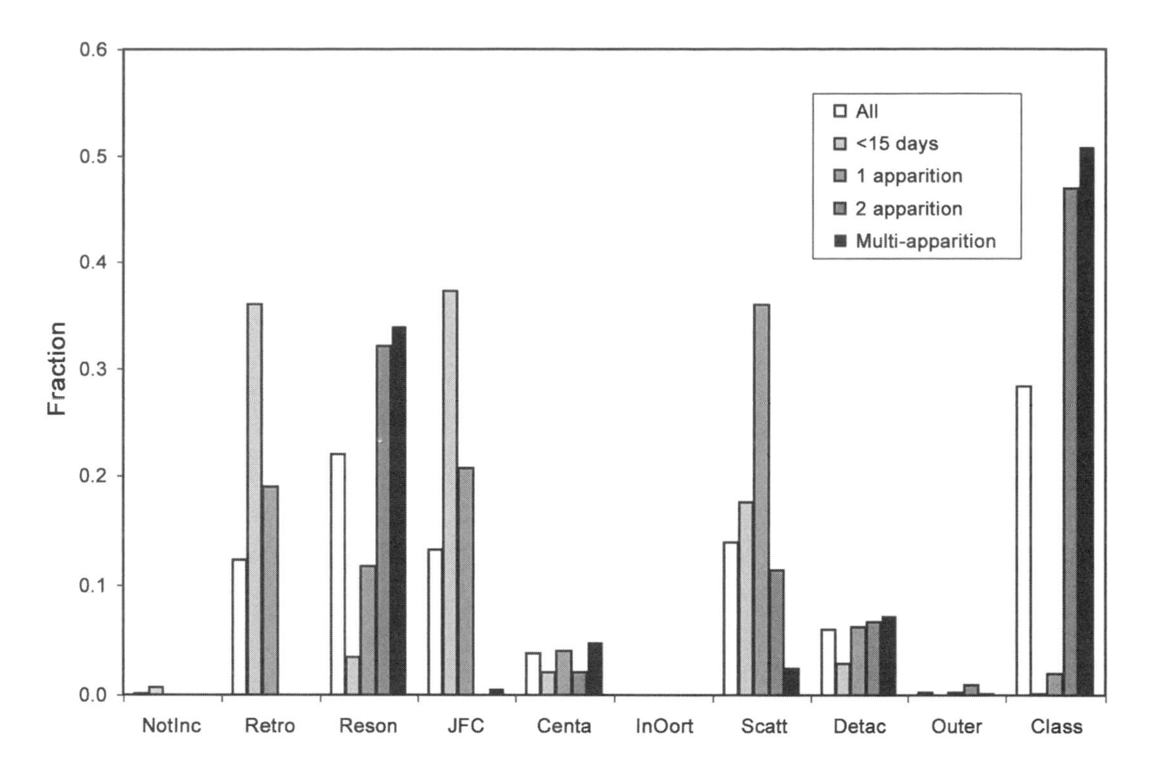

Fig. 11. Dynamical classification and its evolution as a function of increasing observational arc. The definitions for the different groupings are described in the text.

ply of high-eccentricity small-perihelion-distance orbits is again evident. The detailed dynamical structure of the region becomes visible, when only objects having observational arc T_{arc} > 180 d are included. The difficulty in the classification of objects becomes apparent from Fig. 11, which shows the probabilities for the adopted dynamical groupings for increasing observed time arcs. Short-arc data is well represented with cometary-type and even retrograde orbits. Thus even the TNO membership of these very-short-arc objects is uncertain. The percentages of two- and multi-apparition objects are very similar in all the groups. Therefore, for a secure dynamical classification we require an observed arc longer than ~2 yr.

We would like to point out some potentially interesting cases. There has been discussion about the existence of the outer edge for the belt. Based on our classification, there are two confirmed objects, which are outside the 2:1 resonance and in low-e orbits. 2003 UY_{291} is very close to the classical population with $(a, e, i)_{ML}$ = (49.7, 0.18, 3.5), while 2004 XR_{190} is clearly on the outside: $(a, e, i)_{ML}$ = (57.4, 0.11, 46.7) (see also *Allen et al.,* 2006). From Fig. 8 for the one-apparition objects, there is evidence for new candidates for the Neptune Trojans, and in fact, during the writing of this chapter, second-apparition data became available for two objects (2005 TN_{53} and 2005 TO_{74}), which strengthen their Trojan classification. The existence of other short-arc Trojans needs to be studied with integrations.

A small fraction of objects exist that could not be classified in any of the proposed groupings. There are in particular two short-arc objects [2004 PR_{107} (two observations, one day) and 2001 DU_{108} (four observations, three days)] that have a high probability of being inner-solar-system objects. With their current datasets, their classification is extremely uncertain and NEO classification cannot be overruled.

5. CONCLUSION

We have reviewed the orbit computation problem in the transneptunian region. While the TNO population continues to be a poorly observed one after over a decade of dedicated observations, the recent advances in orbit computation methods help us to examine the TNO orbital uncertainties even from exiguous data. Careful assessment of orbital uncertainties of the population is in place for several reasons. First of all, to keep track of the discovered objects, their sky-plane uncertainties need to be estimated. The orbital information is the starting point for the attempts to define the dynamical structure of the transneptunian region as well as for deriving unbiased orbital distributions. Also, the correlation between TNO physical and dynamical properties are of interest. These are topics that are discussed in other chapters of this volume.

We have described the numerical methods available for TNO orbit computation. Nearly half the population have been observed during one apparition only, and thus their large uncertainties need to be estimated with a fully nonlinear inversion method. Ranging efficiently unveils the extended orbital-element p.d.f.s and strongly nonlinear correlations between elements. BK and VoV methods have been developed and shown to successfully handle cases of more modest orbital uncertainties. While the Monte Carlo methods are today feasible to use due to computer resources available, methods using linearization are attractive due to their speed whenever the approximation can be justifiably adopted. In particular, linearized methods suffice in most cases for ephemeris prediction due to the nearly linear TNO motions.

An important application for the orbital information is in classification of the discovered objects. Due to the large uncertainties involved, dynamical classification based on one-apparition data is highly uncertain. Any detailed dynamical analysis of the population should be based on multiapparition data. In the statistical approach for orbit computation, tentative classification can still be made by assigning an object probabilities for being in the different groupings. Such information can be used, for example, as constraining *a priori* information in recovery attempts.

We note that the dynamical model adopted in TNO orbit computation should be studied further. Previously, it has been general practice to consider perturbative forces of the nine planets in orbit integrations, but recently our view of the solar system has been revisited. It would be interesting to study in practice what the effect of (134340) Pluto is in TNO orbit computation. In a similar manner, the perturbative effects of the other large objects, such as (136199) Eris, as well as the effect of mutual close encounters between TNOs on orbit computation will be a subject for further studies.

Should studies of TNO collisions within timescales of tens of thousands of years be of interest, the orbital-element p.d.f.s allow detailed monitoring and evaluation of mutual close approaches among TNOs as well as TNO close approaches to giant planets. It is evident from the p.d.f.s of TNOs with exiguous observational data that, within the orbital uncertainties, a wide spectrum of dynamical evolutions be possible for individual TNOs. Furthermore, TNO orbital-element p.d.f.s can mimic cometary orbital-element p.d.f.s and thereby offer test cases for the assessment of terrestrial collision probabilities for long-period comets. The advent of next-generation astrometric surveys will have a large impact on TNO orbits but they will also introduce new challenges for orbit computation. Deep all-sky observations will accumulate huge databases of observations for solar system objects, which need to be correctly linked and identified with known objects (e.g., *Granvik and Muinonen,* 2005; *Milani et al.,* 2005). For TNOs, Ranging has already been used to automatically verify linkages between observation sets from different apparitions (*Virtanen et al.,* 2003), and the approach by *Granvik and Muinonen* (2005) is straightforward to implement for the search of linkages between short-arc observation sets of TNOs. On the other hand, improving astrometric accuracies will unveil the coupling of physical and dynamical properties of the objects and make it necessary to solve for them simultaneously. With much of the work already underway, we can in a decade anticipate having an improved TNO orbit database with an increase by a factor of several in the number of objects included, with a similar improvement in their estimated orbital accuracies.

Acknowledgments. We thank the two reviewers for constructive comments. J.V. wishes to thank M. Granvik for software development for the Orb package as well as helpful discussions on orbits in general. The research of J.V. was, in part, supported by the Magnus Ehrnrooth Foundation.

REFERENCES

Allen R. L., Gladman B., Kavelaars J., Petit J-M., Parker J., and Nicholson P. (2006) Discovery of a low-eccentricity, high inclination Kuiper belt object at 58 AU. *Astron. J., 640,* 83–86.

Bernstein G. M. and Khushalani B. (2000) Orbit fitting and uncertainties for Kuiper belt objects. *Astron. J., 120,* 3323–3332.

Bowell E., Virtanen J., Muinonen K., and Boattini A. (2002) Asteroid orbit computation. In *Asteroids III* (W. F. Bottke Jr. et al., eds.), pp. 27–43. Univ. of Arizona, Tucson.

Box G. E. P. and Tiao G. C. (1973) *Bayesian Inference in Statistical Analysis.* Addison-Wesley, Reading, Massachusetts.

Carpino M., Milani A., and Chesley S. R. (2003) Error statistics of asteroid optical astrometric observations. *Icarus, 166(2),* 248–270.

Chesley S. R. (2005) Very short arc orbit determination: The case of 2004 FU_{162}. In *Dynamics of Populations of Planetary Systems* (Z. Knežević and A. Milani, eds.), pp. 255–258. Cambridge Univ., Cambridge.

Chiang E. I. and 10 colleagues (2003) Resonance occupation in the Kuiper Belt: Case examples of the 5:2 and Trojan resonances. *Astron. J., 126,* 430–443.

Chodas, P. W. and Yeomans D. K. (1996) The orbital motion and impact circumstances of comet Shoemaker-Levy 9. In *The Collision of Comet Shoemaker-Levy 9 and Jupiter* (K. S. Noll et al., eds.), pp. 1–30. Cambridge Univ., Cambridge.

Elliot J. L., Kern S. D., Clancy K. B., Gulbis A. A. S., Millis R. L., et al. (2005) The Deep Ecliptic Survey: A search for Kuiper belt objects and Centaurs. II. Dynamical classification, the Kuiper belt plane, and the core population. *Astron. J., 129,* 1117–1162.

Gladman B., Kavelaars J. J., Nicholson P. D., Loredo T. J., and Burns J. A. (1998) Pencil-beam surveys for faint trans-neptunian objects. *Astron. J., 116,* 2042–2054.

Goldader J. D. and Alcock C. (2003) Constraining recovery observations for trans-neptunian objects with poorly known orbits. *Publ. Astron. Soc. Pac., 115,* 1330–1339.

Granvik M. and Muinonen K. (2005) Asteroid identification at discovery. *Icarus, 179,* 109–127.

Granvik M., Virtanen J., Muinonen K., Bowell E., Koehn B., and Tancredi G. (2003) Transneptunian object ephemeris service (TNOEPH). In *First Decadal Review of the Edgeworth-Kuiper Belt* (J. K. Davies and L. H. Barrera, eds.), pp. 73–78. Kluwer, Dordrecht [reprinted from *Earth Moon Planets, 92(1–4)*].

Hestroffer D., Vachier F., and Balat B. (2005) Orbit determination of binary asteroids. *Earth Moon Planets, 97,* 245–260.

Jedicke R., Magnier E. A., Kaiser N., and Chambers K. C. (2007) The next decade of solar system discovery with Pan-STARRS. In *Near Earth Objects, Our Celestial Neighbors: Opportunity and Risk* (A. Milani et al., eds.), pp. 341–352. IAU Symposium No. 236, Cambridge Univ., Cambridge.

Jeffreys H. (1946) An invariant form for the prior probability in estimation problems. *Proc. R. Statistical Soc. London, A186,* 453–461.

Jones R. L., Gladman B., Petit J.-M., Rousselot P., Mousis O., et al. (2006) The CFEPS Kuiper belt survey: Strategy and presurvey results. *Icarus, 185,* 508–522.

Kaasalainen M., Hestroffer D., and Tanga P. (2005) Physical models and refined orbits for asteroids from Gaia photo- and astrometry. In *The Three Dimensional Universe with Gaia*, p. 301. ESA SP 576, European Space Agency, Noordwijk.

Kristensen L. K. (2004) Initial orbit determination for distant objects. *Astron. J., 127,* 2424–2435.

Lehtinen M. S. (1988) On statistical inversion theory. In *Theory and Applications of Inverse Problems* (H. Haario, ed.), pp. 46–57. Pitman Research Notes in Mathematical Series 167.

Menke W. (1989) *Geophysical Data Analysis: Discrete Inverse Theory.* Academic, New York.

Milani A. (1999) The asteroid identification problem I: Recovery of lost asteroids. *Icarus, 137,* 269–292.

Milani A. and Knežević Z. (2005) From astrometry to celestial mechanics: Orbit determination with very short arcs. *Celest. Mech. Dyn. Astron., 92,* 1–18.

Milani A. and Valsecchi G. B. (1999) The asteroid identification problem II: Target plane confidence boundaries. *Icarus, 140,* 408–423.

Milani A., Chesley S. R., Chodas P. W., and Valsecchi G. B. (2002) Asteroid close approaches: Analysis and potential impact detection. In *Asteroids III* (W. F. Bottke Jr. et al., eds.), pp. 55–70. Univ. of Arizona, Tucson.

Milani A., Gronchi G. F., Michieli Vitturi M., and Knežević Z. (2004) Orbit determination with very short arcs. I Admissible regions. *Celest. Mech. Dyn. Astron., 90,* 57–85.

Milani A., Gronchi G. F., Knežević Z., Sansaturio M. E., and Arratia O. (2005) Orbit determination with very short arcs. II Identifications. *Icarus, 179,* 350–374.

Milani A., Gronchi G. F., Knežević Z., Sansaturio M. E., Arratia O., Denneau L., Grav T., Heasley J., Jedicke R., and Kubica J. (2006) Unbiased orbit determination for the next generation asteroid/comet surveys. In *Asteroids Comets Meteors 2005* (D. Lazzaro et al., eds.), pp. 367–380. Cambridge Univ., Cambridge.

Millis R. L., Buie M. W., Wasserman L. H., Elliot J. L., Kern S. D., and Wagner R. M. (2002) The Deep Ecliptic Survey: A search for Kuiper belt objects and Centaurs. I. Description of methods and initial results. *Astron. J., 123,* 2083–2109.

Morbidelli A., Thomas F., and Moons M. (1995) The resonant structure of the Kuiper belt and the dynamics of the first five trans-Neptunian objects. *Icarus, 118,* 322–340.

Mosegaard K. and Tarantola A. (2002) Probabilistic approach to inverse problems. In *International Handbook of Earthquake and Engineering Seismology (Part A)* (W. H. K. Lee et al., eds.), pp. 237–265. Academic, New York.

Muinonen K. and Bowell E. (1993) Asteroid orbit determination using Bayesian probabilities. *Icarus, 104,* 255–279.

Muinonen K. and Virtanen J. (2002) Computation of orbits for near-Earth objects from high-precision astrometry. In *Proc. of the International Workshop on Collaboration and Coordination Among NEO Observers and Orbit Computers* (Kurashiki, October 23–26, 2001, Organized by Japan Spaceguard Association), pp. 105–113.

Muinonen K., Virtanen J., and Bowell E. (2001) Collision probability for Earth-crossing asteroids using orbital ranging. *Celest. Mech. Dyn. Astron., 81,* 93–101.

Muinonen K., Virtanen J., Granvik M., and Laakso T. (2006) Asteroid orbits using phase-space volumes of variation. *Mon. Not. R. Astron. Soc., 368,* 809–818.

Virtanen J. and Muinonen K. (2006) Time evolution of orbital uncertainties for the impactor candidate 2004 AS_1. *Icarus, 184,* 289–301.

Virtanen J., Muinonen K., and Bowell E. (2001) Statistical ranging of asteroid orbits. *Icarus, 154,* 412–431.

Virtanen J., Muinonen K., Tancredi G., and Bowell E. (2003) Orbit computation for transneptunian objects. *Icarus, 161,* 419–430.

Zacharias N. (2004) Astrometric reference stars: From UCAC to URAC. *Astron. Nachr. 325(6),* 631–635.

Part II:

Transneptunian Object Populations

Nomenclature in the Outer Solar System

Brett Gladman
University of British Columbia

Brian G. Marsden
Harvard-Smithsonian Center for Astrophysics

Christa VanLaerhoven
University of British Columbia

We define a nomenclature for the dynamical classification of objects in the outer solar system, mostly targeted at the Kuiper belt. We classify all 584 reasonable-quality orbits, as of May 2006. Our nomenclature uses moderate (10 m.y.) numerical integrations to help classify the *current* dynamical state of Kuiper belt objects as resonant or nonresonant, with the latter class then being subdivided according to stability and orbital parameters. The classification scheme has shown that a large fraction of objects in the "scattered disk" are actually resonant, many in previously unrecognized high-order resonances.

1. INTRODUCTION

Dynamical nomenclature in the outer solar system is complicated by the reality that we are dealing with populations of objects that may have orbital stability times that are either moderately short (millions of years or less), appreciable fractions of the age of the solar system, or extremely stable (longer than the age of the solar system). While the "classical belt" is loosely thought of as what early searchers were looking for (the leftover belt of planetesimals beyond Neptune), the need for a more precise and complete classification is forced by the bewildering variety we have found in the outer solar system.

1.1. Philosophy

The inner solar system is somewhat analogous, and from it we take some inspiration. In the main asteroid belt the recognized subpopulations are generically demarcated by resonances, be they mean-motion (the 2:1 is often taken as the outer "edge" of the main belt; the Hildas are in or near the 3:2) or secular resonances (which separate the Hungarias from the rest of the belt, for example). In contrast, once "out of the main belt," the near-Earth objects (NEOs) are separated from each other by rather arbitrary cuts in orbital element space; the semimajor axis a = 1 AU separation between Aten and Apollo has no real dynamical significance for these unstable orbits, but this well-accepted division makes discussion easier since, for example, Atens are a heavily evolved (both dynamically and potentially physically) component of the NEA population while Apollos will be on average much younger.

1.2. Classification Outline

For small-a comets, historical divisions are rather arbitrary (e.g., based on orbital period), although recent classifications take relative stability into account by using the Tisserand parameter (*Levison,* 1996) to separate the rapidly depleted Jupiter-family comets (JFCs) from the longer-lived Encke and Chiron-like (Centaur-like) orbits.

In the transneptunian region, for historical reasons the issue of stability has been important due to arguments about the primordial nature of various populations and the possibility that the so-called "Kuiper belt" is the source of the Jupiter-family comets. Already-known transneptunian objects (TNOs) exhibit the whole range of of stabilities from strongly planet-coupled to stable for >4.5 G.y. Because TNOs (like NEOs) might change class in the near or distant future, we adopt the fundamental philosophy that *the classification of a TNO must be based upon its current short-term dynamics* rather than a belief about either where it will go in the future or what its past history was. In addition, we accept the *fait accompli* that there will necessarily be a level of arbitrariness in some of the definitions. We have attempted to find a balance among historical intent, recent usage, and the need to tighten the nomenclature. We have liberally used ideas from the literature, with a goal of developing a scheme that has practical utility, while keeping an eye toward the intention that stability should be part of the nomenclature.

The classification scheme is a process of elimination (outlined in Fig. 1) based on either the object's current orbital elements and/or the results of a 10-m.y. numerical integration into the future. Using this flowchart, we have clas-

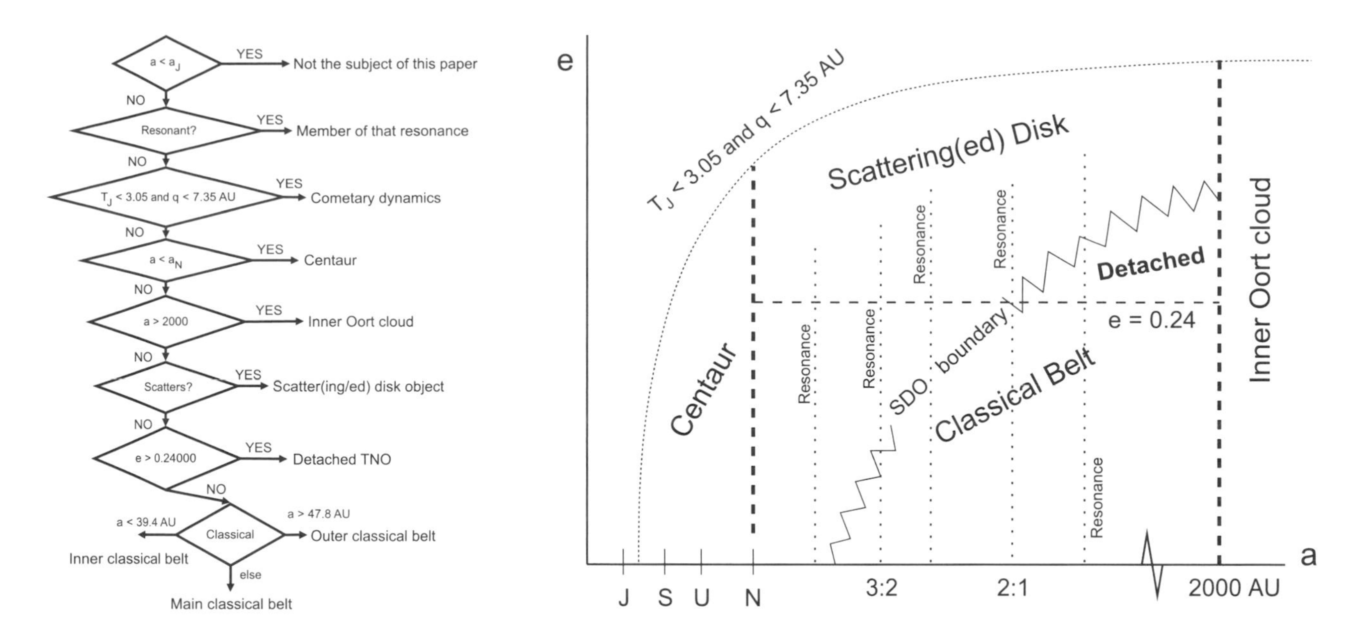

Fig. 1. *Left:* Flowchart for the outer solar system nomenclature. When orbital elements are involved they should be interpreted as the osculating barycentric elements. *Right:* A cartoon of the nomenclature scheme (not to scale). The boundaries between the Centaurs, JFCs, scattered disk, and inner Oort cloud are based on current orbital elements; the boundaries *are not* perihelion distance curves. Resonance inhabitance and the "fuzzy" SDO boundary are determined by 10-m.y. numerical integrations. The classical belt/detached TNO split is an arbitrary division.

sified the entire three-opposition (or longer) sample present in the IAU's Minor Planet Center (MPC) as of May 2006. The tables provide our SSBN07 classification for this *Solar System Beyond Neptune* book. Transneptunian objects that are now numbered also have their original provisional designation to aid their identification in previously published literature.

We have found that in order to make a sensible TNO dynamical classification, we were forced to define what is *not* a TNO; we thus begin with the regions that bound TNO semimajor axes (Centaurs and the Oort cloud) and eccentricities (the JFCs).

2. CENTAURS AND COMETARY OBJECTS

Historically, periodic comets were classified according to their orbital period P, with short-period comets having P < 200 yr and long-period comets with P > 200 yr. While there existed at one point a classification system that assigned the short-period comets to planetary "families" according to which of the giant planets was the closest to their heliocentric distances at aphelion, it became evident that there was little dynamical significance to such a classification, except in the case of the Jupiter family, which (by virtue of the typical orbital eccentricities involved) has tended to apply to comets having P < 20–30 years. This suggested it would be reasonable to cement this classification with the use of the Tisserand parameter with respect to Jupiter (first attempted by *Kresák,* 1972), defined by the Tisserand parameter T_J with respect to Jupiter

$$T_J \equiv \frac{a_J}{a} + 2\sqrt{\frac{a}{a_J}(1 - e^2)} \cos i \qquad (1)$$

where a, e, and i are the orbital semimajor axis, eccentricity, and inclination of a comet and a_J is the semimajor axis of Jupiter (about 5.2 AU). A circular orbit in the reference plane (approximately the ecliptic but more correctly Jupiter's orbital plane) with $a = a_J$ yields $T_J = 3.0$. Exterior coplanar orbits with $q = a_J = a(1 - e)$ (i.e., perihelion at Jupiter) have T_J just below 3, and thus as long as i is small the condition $T_J < 3$ is nearly the same as having q interior to Jupiter. However, if the inclination in increased or q pushed considerably below Jupiter, T_J drops well below 3 and can even become negative for retrograde orbits.

Because of this, *Carusi et al.* (1987) and *Levison* (1996) suggested that $T_J = 2.0$ provided a convenient lower boundary for a Jupiter-family comet (JFC). Comets with $T_J < 2$ include retrograde and other high-i, high-e comets, while high-e but low-i orbits can remain in the range $T_J = 2$–3. *Carusi et al.* (1987) considered a $T_J = 1.5$ lower boundary, which has the merit of including comets 96P and 8P as JFCs; the most notable comet that might be on the "wrong" side is then 27P/Crommelin ($T_J = 1.48$), although with P ≃ 27 yr this object could appropriately be relegated to the comet group variously categorized as of "Halley type" (HT) or of "intermediate period" (Comet 1P/Halley itself has $T_J = -0.61$). Since it is not directly relevant to our Kuiper belt nomenclature, we drop the issue of the lower T_J boundary.

TABLE 1. Centaurs and Jupiter-coupled objects (SSBN07 classification).

Jupiter-Coupled			
60558 = 2000EC$_{98}$ = Echeclus	52872 = 1998SG$_{35}$ = Okyrhoe		
Centaurs			
02060 = 1977UB = Chiron	05145 = 1992AD = Pholus	07066 = 1993HA$_2$ = Nessus	08405 = 1995GO = Asbolus
10199 = 1997CU$_{26}$ = Chariklo	10370 = 1995DW$_2$ = Hylonome	31824 = 1999UG$_5$ = Elatus	32532 = 2001PT$_{13}$ = Thereus
49036 = 1998QM$_{107}$ = Pelion	54598 = 2000QC$_{243}$ = Bienor	52975 = 1998TF$_{35}$ = Cyllarus	55576 = 2002GB$_{10}$ = Amycus
63252 = 2001BL$_{41}$	83982 = 2002GO$_9$ = Crantor	88269 = 2001KF$_{77}$	95626 = 2002GZ$_{32}$
119315 = 2001SQ$_{73}$	119976 = 2002VR$_{130}$	120061 = 2003CO$_1$	121725 = 1999XX$_{143}$
J94T00A = 1994TA	K00CA4O = 2000CO$_{104}$	K00F53Z = 2000FZ$_{53}$	K01XP5A = 2001XA$_{255}$
K02D05H = 2002DH$_5$	K03W07L = 2003WL$_7$	K05Uh8J = 2005UJ$_{438}$	

So what is the *upper* limit of T_J for a JFC, beyond which are Centaurs and scattering objects? *Levison* (1996) used an upper limit of $T_J = 3.0$. For a circular jovian orbit, objects with $T_J > 3$ do not cross Jupiter's orbit, but Jupiter's e is sufficiently large that this approximation is not good enough to prevent complications. Not only are there several comets with T_J slightly greater than 3.0 that according to any reasonable definition should be called JFCs, but in some cases T_J oscillates about 3.0 within a matter of decades. Comets like 39P/Oterma, which over a quarter-century interval moved from an osculating orbit entirely outside Jupiter to one entirely inside Jupiter *and back* (with T_J remaining in the range 3.00–3.04), lead us to solve the problem by allowing JFCs to have T_J up to 3.05. We note in passing that the Centaurs (60558) Echeclus (q = 5.8 AU, $T_J = 3.03$) and (52782) Okyrhoe (q = 5.8 AU, $T_J = 2.95$), which already bear the names of Centaurs, are reclassified as Jupiter coupled; our numerical integration confirms these objects to be rapidly perturbed by Jupiter.

Finally, there is a terrible generic problem with the Tisserand invariant (leading us below to reject its use in the main Kuiper belt); orbits with perihelia far outside Jupiter but sufficiently high i eventually have $T_J < 3.05$ since the second term of equation (1) becomes small. The TNO 127546 = 2002 XU$_{93}$ has a/q/i = 66.5/21.0/77.9° and T_J = 1.2 but clearly is not remotely coupled to Jupiter. We thus feel that a pericenter qualifier must be added to the $T_J < 3.05$ condition to keep large-i outer solar system objects out of the dynamical comet classes. In analogy with the upper Aten q boundary (about halfway to Mars), we simply use q < 7.35 AU (halfway to Saturn) as an additional qualifier. With this definition, the combined T_J and q condition (Fig. 1) tells us what is beyond Jupiter's reach and in need of classification (Table 1) for our present purposes.

This brings us to the Centaurs, whose perihelia are sufficiently high that they are not JFCs. The prototype (2060) = 95P/Chiron (q = 8.5 AU, a = 13.7 AU, i = 6.9°, and T_J = 3.36) is a planet-crossing object, as is (5145) Pholus (q = 8.7 AU, a = 20.4 AU, i = 24.7°, T_J = 3.21). Indeed, while a Centaur has historically been broadly defined as an object of low i and low-to-moderate e in the distance range of the giant planets, the historical intent was that its evolution was not currently controlled by Jupiter. Since the JFC definition essentially takes care of this latter condition, we are left with the question of where the Centaurs stop. While there were historical definitions that involved aphelion distance Q > 11 AU, it is useful to have an outer bound on Q so that Centaurs retain their identity as objects mostly *between* the giant planets. We do this by using $a < a_N$ (Neptune's semimajor axis) as the boundary; the resonant (see below) Neptune Trojans fall on the boundary between the Centaurs and the SDOs.

3. THE INNER OORT CLOUD

We will not spend a great deal of time dealing with Oort cloud nomenclature, but feel obligated to put an outer bound on the scattering disk. Although the production mechanism of the Oort cloud and the past galactic environment of the Sun are unclear, since we are basing our definitions on the *current dynamics* we ask the question: Where does the current dynamics of a distant object become dominated by external influences? *Dones et al.* (2004) show that a very evident transition in the dynamics begins at a = 2000 AU for TNOs scattered out by the giant planets; for a > 2000 AU the galactic tidal field and passing stars cause appreciable alteration of the perihelia and inclinations. We thus adopt a = 2000 AU as the formal (somewhat arbitrary) beginning of the inner Oort cloud (and thus end of the Kuiper belt). Objects with a > 2000 AU but with $T_J < 3.05$ and q < 7.35 AU would be considered JFCs since their evolution is dominated by Jupiter (see chapter by Duncan et al.).

Note that the definitions above give, for the first time, a formal sharp demarcation of the Kuiper belt, which is bounded on the inner and outer "a" boundaries by the Centaurs and Oort cloud, and above in eccentricity by the JFC population (defined by the Tisserand parameter). This definition makes SDOs (see below) part of the Kuiper belt.

4. RESONANT OBJECTS

While the Centaurs and JFCs are rapidly evolving, the resonant TNO populations may be critical to our understanding of the region's history. We adopt the convention that the p:q resonance denotes the resonance of p orbital periods of the inner object (usually Neptune) to q periods of the TNO (and thus external resonance have p > q). The "order" of the resonance is p–q, with high-order resonances being weak unless e is large. After the suggestion in 1994 that there were TNOs other than Pluto in the 3:2 resonance with Neptune, a host of other resonances have been shown to be populated (although in this chapter we search only for mean-motion, rather than secular, resonances). These resonances are important to the structure of the belt because (1) they allow large-e orbits to survive for 4.5 G.y despite approach or even crossing of Neptune's orbit; (2) the chaotic nature of the resonance borders allow both (a) the temporary trapping of SDOs near the border of the resonance (*Duncan and Levison,* 1997) or (b) nearly resonant objects to escape into the Neptune-coupled regime (*Morbidelli,* 1997); and (3) the relative population of the resonances may be a diagnostic of the amount and/or rate of planet migration (*Chiang and Jordan,* 2002; *Hahn and Malhotra,* 2005).

Resonant occupation (or proximity) can really only be addressed by a direct numerical calculation of the orbital evolution, because simply having a TNO with corresponding period near a rational ratio of Neptune's (one-half for the 2:1 resonance, for example) is not at sufficient condition to be in the resonance. The angular orbital elements must also be appropriately arranged so that a resonant argument, for example, of the form

$$\phi_{94} = 9\lambda_N - 4\lambda - 5\varpi \tag{2}$$

oscillates ("librates") around some value, rather than progressing nearly uniformly ("circulating"). Equation (2) gives a resonant argument of the 9:4 mean-motion resonance (the "plutinos" are found in the 3:2); λ_N and λ are the mean longitudes ($\Omega + \omega + M$) of Neptune and the TNO, and ϖ is the longitude of perihelion ($\Omega + \omega$) of the TNO. It is the geometrical relation between the angles embodied by the resonant angle that prevents the resonant TNOs from approaching Neptune even if they have high e. (Other 9:4 resonant arguments than ϕ_{94} exist; these involve ϖ_N, Ω, and Ω_N. However, these tend to be weaker because $e_N \ll e$ and inclinations are usually small.)

4.1. Classification Algorithm

Because we are involving a numerical integration in a dynamical classification, it is worth pausing to discuss possible weaknesses of this approach. One could say that instead of classifying objects, one is actually classifying nominal orbits, since the object's orbital elements are necessarily only known to some finite precision. With a nomenclature involving an arbitrary cut in orbital parameter space, even the orbit-fitting method influences the potential classification. We believe that TNO dynamical classifications are only reliable for objects with observations in (at least) *each and every* of three or more oppositions; two-opposition orbits may still have a-uncertainties of an AU or more (and two-year arcs with no observations one year after discovery are also insufficient). Since a numerical integration is involved in resonance classification, even a reasonably well-observed object with an orbit near the boundary of a resonance could have enough uncertainty in its orbital elements to straddle the border. In principle the numerical integration algorithm could influence the result, especially for these objects on the border. We view these practical difficulties as unavoidable problems whose shortfalls are far outweighed by the need to establish a classification based on stability.

Our resonance identification algorithm is similar to that proposed by *Chiang et al.* (2003) and *Elliot et al.* (2005), but with modifications related to both the definitions of the scattering populations (to be discussed later) and to the evaluation of uncertainty in the TNO orbital parameters. For each three-opposition object, the best-fit orbit is integrated, along with two other nominal orbits that are the extremes of the orbital uncertainty in semimajor axis; the details of this procedure are discussed in the following subsection. If in a 10-m.y. integration a particular resonance argument librates for all three initial conditions, the object has a "secure" resonant classification. If two of these initial conditions show the resonant behavior the object will be classified as "probably" in the resonance; if only one of the arguments is resonant, we classify the object according to the behavior of the other two trajectories, noting the vicinity of the resonance. This is important for objects near resonance borders or near the portions of those resonances where the semimajor axis range of that resonance is small (at low-e, for example), as this will flag the object as being especially in need of further observation despite having a reasonably high-quality orbit.

As already concluded by *Chiang et al.* (2003), diagnosing the maximal semimajor axis variation is the most important element for determining resonance occupancy since resonances are confined to small ranges of a. Currently only the eccentricity-type mean-motion resonances between TNOs and Neptune are securely known to be populated, ranging from the 1:1 Trojan resonance with Neptune (*Chiang et al.,* 2003; *Marsden,* 2003) to the 3:1 at a = 63.0 AU (*Marsden,* 2005). In principle there is no limit to the order of the resonance that could be occupied (and thus classified). In particular, the a > 50 AU population is often discovered near perihelion and thus on orbits of large eccentricity, which allows higher-order resonances (like the second-order 3:1 or the third-order 5:2) to be occupied. We have examined all eccentricity-type resonant arguments up to sixth order routinely, and searched to much higher order for objects with perihelia q < 38 AU that show stable behavior in the 10-m.y. numerical integration, as we find their relative stability is often due to resonance occupation.

Note that the 10-m.y. future window used here does not require the object to be resonant over the age of the solar system, but only for a small number of resonant librations. This is in keeping with the philosophy that it is the *current* dynamics of the TNOs that we are classifying. For example, it is immaterial whether an object in the 2:1 resonance with $e \simeq 0.3$ arrived there by (1) eccentricity pumping after trapping in resonances during an outward migration of Neptune (*Malhotra,* 1993), (2) being trapped into the 2:1 from a scattered orbit (e.g., *Duncan and Levison,* 1997; *Gomes,* 2003; *Hahn and Malhotra,* 2005), or (3) diffusing up to $e \simeq$

0.3 from an e ≃ 0 orbit due to slow dynamical diffusion over the age of the solar system. Despite these rather different orbital histories (which will of course be impossible to discriminate among for a given object) the nomenclature classifies the object as 2:1 resonant because its *current* dynamical state is resonant.

4.2. Numerical Method

For a given object, we begin with the astrometric observations from each and every one of three oppositions (or more), and perform a best fit *barycentric* orbit solution using the method of *Bernstein and Khushalani* (2000). That is, the position and velocity vectors at the time of the first observation are computed, giving osculating elements relative to the center of mass of the giant planets and the Sun. The orbital elements are thus determined to a fractional precision of $\sim 10^{-5}$, several orders of magnitude more precise than the uncertainty in the orbital elements.

Our method then asks the question: What is the set of possible orbits that are consistent with the orbit solution, as judged by the residual quality of the best-fit orbit? *Chiang et al.* (2003) approached this question by diagonalizing the covariance matrix around the best fit and using the diagonal elements to generate a set of new orbits that are on the 3σ surface, assuming that all astrometric observations have the same error. Inspection of the residuals from the best-fit orbits for our MPC object sample shows enormous variance in the astrometric quality of the observations. While there are objects like K03QB3X (= 2003 QB_{103} = CFEPS L3q03) with maximum residuals of 0.22" over the entire three-year arc, most TNOs have many residuals of 0.6–0.9", and others have observations with >2" residuals and RMS residuals signficantly more than 0.5". In our algorithm, classification certainty is based on the actual orbit quality of the object in question, as shown by its internal consistency. We therefore search in parameter space for other orbits (1) that have no residuals >1.5× the worst residual of the best fit and (2) whose RMS residual is <1.5× the residual RMS of the best fit.

We accomplish this search with a set of numerical subroutines, some of which are taken from the latest release of the *Bernstein and Khushalani* (2000) package. Using the best-fit orbit expressed in their (α, β, γ) coordinate system, corresponding to two on-sky axes and one radial axis through the observation (along with three velocity axes), the diagonal elements of the covariance matrix in these axes are used as the step size in our Monte Carlo parameter search. A search for the largest semimajor axis orbit is then begun. A random displacement in all six axes is made, with a variation in each of those coordinates of up to ±3σ (as determined by the covariance matrix), yielding a candidate orbit. If this candidate orbit passes the consistency checks based on its residuals relative to the observations, and if the test orbit's a is larger than the current highest-a orbit (or smaller than the current lowest-a) orbit, it becomes the new high-a orbit and the search is repeated using this new orbit as the starting point. The maximum-a orbit is the last one for which 10^6 trials yield no consistent orbit with a still-larger a. At this point a search is begun for the lowest-a orbit starting from either the best-fit orbit or a lower-a orbit that might have already been discovered in the high-a search. Figure 2 shows the result of the process for the TNO 2001 KG_{76}; the parameter-space search shows that this low-i TNO has an a known to ≃±0.1 AU (about 2%) with an e accuracy of about 0.0015. As is usual for TNOs, the orbital inclination and ascending node are known extremely precisely, while the argument of pericenter ω and time of pericenter passage T (or mean anomaly) are known to a fractional accuracy several times worse.

Once the extremal orbits are found, we generate heliocentric position and velocity vectors for the best fit and two extremal orbits at the instant corresponding to the first observation, along with the planetary position and velocity vectors for the giant planets for that epoch (we add the mass of the terrestrial planets to the Sun). We then numerically integrate forward in time using the swift-rmvs3 integrator (*Duncan and Levison,* 1994) based on the mixed-variable symplectic algorithm of *Wisdom and Holman* (1991). The three orbits are followed for 10 m.y. and mean-motion resonances with Neptune up to at least sixth order are routinely examined. If all three test particles librate in a resonance for 10 m.y., the object is classified securely in that resonance (see Table 2). If two of the three test particles librate then the resonance identification is insecure; this is indicated by the classification list being in "( )". Insecure TNOs require further observation to reduce the orbital uncertainty. (As Table 5 shows, most insecure resonant classifications are associated with high-order resonances, which are "thin" in semimajor axis.) If only one of three test particles librates the object is classed as nonresonant; we do not here record all such cases of potential resonance, which will only be identified with the help of further observation. In our philosophy, insecure classifications mean *more observations are needed* to remove the uncertainty; this is a simple matter for these objects, as their positions are known to the level of a few arcseconds in almost all cases.

If we had used only 3-m.y. integrations, 5 of the 584 TNOs would have been classified slightly differently, although this is not disturbing since all 5 were insecure under the 10-m.y. analysis. Four resonant TNOs had a high-a or low-a clone become nonresonant in the 3–10-m.y. period (becoming insecurely resonant instead of securely), and one insecurely scattering TNO became insecurely resonant. These objects are not ready to be securely classified and the solution is to acquire more observations.

For Deep Ecliptic Survey TNOs classified as resonant by *Elliot et al.* (2005) using a very similar method, we find the same resonant classification. However, with longer arcs available, we now recognize some other objects (e.g., 2002 PA_{149}) as low-order resonant. Since we search to much higher order when we see what is clearly resonant behavior in our inspection of the orbital history, we have also found a suite of objects in resonances of order >4 (e.g., the

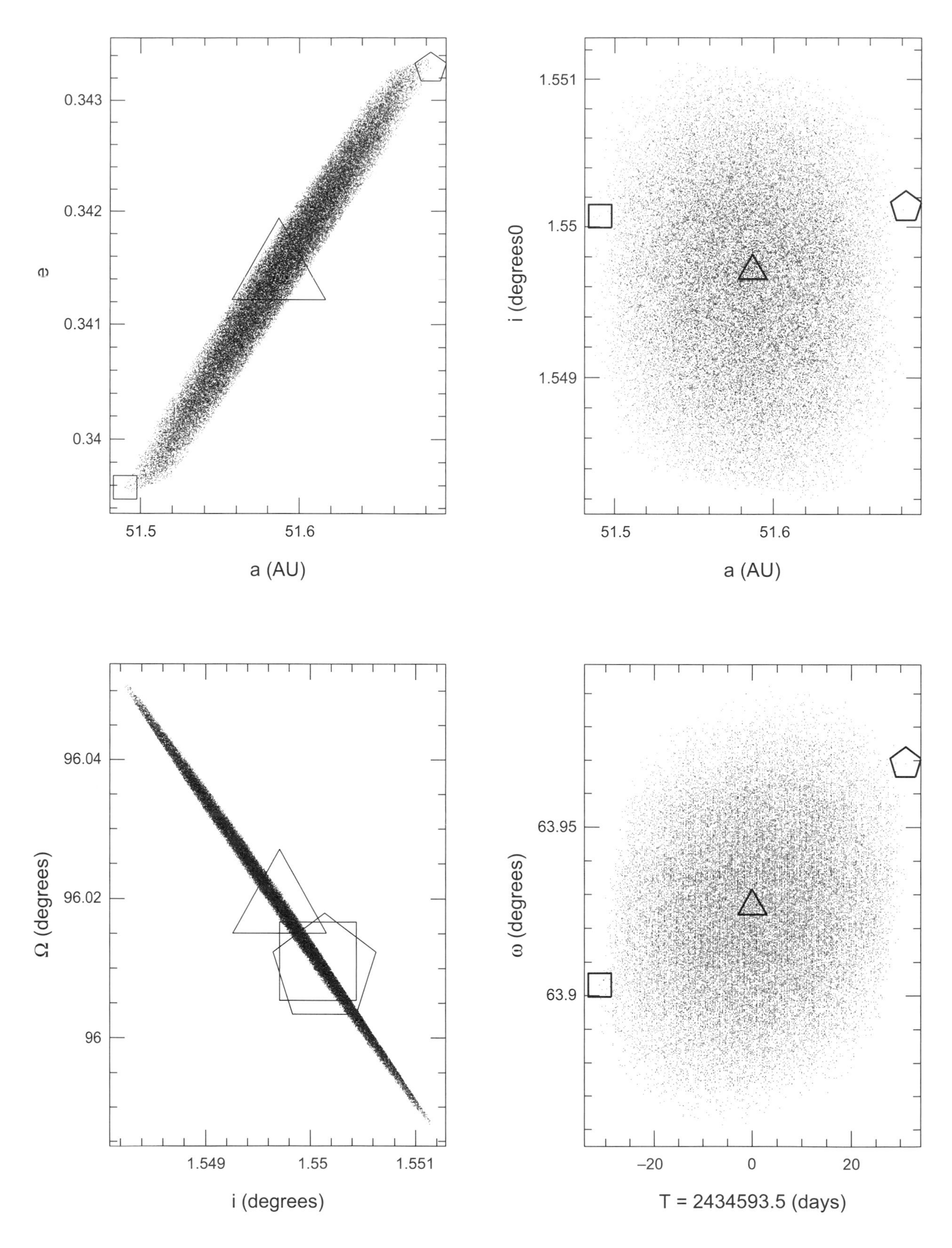

Fig. 2. Example of the determination of the bounds on the orbital elements for an object before classification (here 2001 KG_{76}). Starting at the best-fit orbit (large triangle), we search for the highest (pentagon) and lowest (square) semimajor axis orbits (using the procedure described in the text) that give acceptable orbit fits (upper left), where dots show all orbits that were discovered. As is often the case, the extremal orbits also have nearly the maximal variation in e. The values of the other orbital parameters for each viable as well as the best-fit and extremal orbits are in the other panels.

TABLE 2. Resonant objects (SSBN07 classification).

1:1	K01QW2R = 2001QR$_{322}$			
5:4	79969 = 1999CP$_{133}$	127871 = 2003FC$_{128}$	131697 = 2001XH$_{255}$	
4:3	15836 = 1995DA$_{2}$	J98U43U = 1998UU$_{43}$	J99RL5W = 1999RW$_{215}$	(K00CA4Q = 2000CQ$_{104}$)
	K03SV7S = 2003SS$_{317}$			
11:8	(131695 = 2001XS$_{254}$)			
3:2	15788 = 1993SB	15789 = 1993SC	15810 = 1994JR$_{1}$	15820 = 1994TB
	15875 = 1996TP$_{66}$	19299 = 1996SZ$_{4}$	20108 = 1995QZ$_{9}$	24952 = 1997QJ$_{4}$
	28978 = 2001KX$_{76}$ = Ixion	32929 = 1995QY$_{9}$	33340 = 1998VG$_{44}$	38628 = 2000EB$_{173}$ = Huya
	47171 = 1999TC$_{36}$	47932 = 2000GN$_{171}$	55638 = 2002VE$_{95}$	69986 = 1998WW$_{24}$
	69990 = 1998WU$_{31}$	84719 = 2002VR$_{128}$	84922 = 2003VS$_{2}$	90482 = 2004DW = Orcus
	91133 = 1998HK$_{151}$	91205 = 1998US$_{43}$	118228 = 1996TQ$_{66}$	(119069 = 2001KN$_{77}$)
	119473 = 2001UO$_{18}$	120216 = 2004EW$_{95}$	126155 = 2001YJ$_{140}$	129746 = 1999CE$_{119}$
	131318 = 2001FL$_{194}$	133067 = 2003FB$_{128}$	J93R00O = 1993RO	J95H05M = 1995HM$_{5}$
	J96R20R = 1996RR$_{20}$	J98HF1H = 1998HH$_{151}$	J98HF1Q = 1998HQ$_{151}$	J98U43R = 1998UR$_{43}$
	J98W31S = 1998WS$_{31}$	J98W31V = 1998WV$_{31}$	J98W31Z = 1998WZ$_{31}$	J99CF8M = 1999CM$_{158}$
	J99RL5K = 1999RK$_{215}$	K00CA5K = 2000CK$_{105}$	K00F53V = 2000FV$_{53}$	K00GE7E = 2000GE$_{147}$
	K00Y02H = 2000YH$_{2}$	K01FH2U = 2001FU$_{172}$	K01FI5R = 2001FR$_{185}$	K01K76Y = 2001KY$_{76}$
	K01K77B = 2001KB$_{77}$	K01K77D = 2001KD$_{77}$	K01K77Q = 2001KQ$_{77}$	K01QT8F = 2001QF$_{298}$
	K01QT8G = 2001QG$_{298}$	K01QT8H = 2001QH$_{298}$	K01RE3U = 2001RU$_{143}$	K01V71N = 2001VN$_{71}$
	K02CP1E = 2002CE$_{251}$	K02CM4W = 2002CW$_{224}$	K02G32F = 2002GF$_{32}$	K02G32L = 2002GL$_{32}$
	K02G32V = 2002GV$_{32}$	K02G31W = 2002GW$_{31}$	K02G32Y = 2002GY$_{32}$	K02VD0U = 2002VU$_{130}$
	K02VD0X = 2002VX$_{130}$	K02X93V = 2002XV$_{93}$	K03A84Z = 2003AZ$_{84}$	K03FC7L = 2003FL$_{127}$
	K03H57A = 2003HA$_{57}$	K03H57D = 2003HD$_{57}$	K03Q91B = 2003QB$_{91}$	K03Q91H = 2003QH$_{91}$
	K03SV7O = 2003SO$_{317}$	K03SV7R = 2003SR$_{317}$	K03T58H = 2003TH$_{58}$	K03UT2T = 2003UT$_{96}$
	K03UT2V = 2003UV$_{96}$	K03WJ1A = 2003WA$_{191}$	K04E96H = 2004EH$_{96}$	K04FG4W = 2004FW$_{164}$
	K05TI9V = 2005TV$_{189}$		134340 = Pluto	
5:3	15809 = 1994JS	126154 = 2001YH$_{140}$	J99CD1X = 1999CX$_{131}$	K00P30L = 2000PL$_{30}$
	K00QP1N = 2000QN$_{251}$	K01XP4P = 2001XP$_{254}$	(K02G32S = 2002GS$_{32}$)	K02VD1A = 2002VA$_{131}$
	K02VD0V = 2002VV$_{130}$	K03UT2S = 2003US$_{96}$	K03YH9W = 2003YW$_{179}$	
7:4	60620	118378 = 1999HT$_{11}$	118698 = 2000OY$_{51}$	(119066 = 2001KJ$_{76}$)
	(119067 = 2001KP$_{76}$)	119070 = 2001KP$_{77}$	119956 = 2002PA$_{149}$	134568 = 1999RH$_{215}$
	135024 = 2001KO$_{76}$	(135742 = 2002PB$_{171}$)	(J99CF8D = 1999CD$_{158}$)	J99H12G = 1999HG$_{12}$
	J99K18R = 1999KR$_{18}$	(K00F53X = 2000FX$_{53}$)	K00O67P = 2000OP$_{67}$	(K00Y01U = 2000YU$_{1}$)
	K01QT8E = 2001QE$_{298}$	K01K76N = 2001KN$_{76}$	(K03QB1W = 2003QW$_{111}$)	
9:5	K01K76L = 2001KL$_{76}$	K02G32D = 2002GD$_{32}$		
11:6	(K01K76U = 2001KU$_{76}$)			
2:1	20161 = 1996TR$_{66}$	26308 = 1998SM$_{165}$	119979 = 2002WC$_{19}$	130391 = 2000JG$_{81}$
	J97S10Z = 1997SZ$_{10}$	J99RL5B = 1999RB$_{215}$	J99RL6B = 1999RB$_{216}$	K00QP1L = 2000QL$_{251}$
	K01FI5Q = 2001FQ$_{185}$	K01U18P = 2001UP$_{18}$	K02PH0U = 2002PU$_{170}$	
19:9	(K03QB3X = 2003QX$_{113}$)			
9:4	42301 = 2001UR$_{163}$	K01K76G = 2001KG$_{76}$		
7:3	131696 = 2001XT$_{254}$	(95625 = 2002GX$_{32}$)	(J99CB8V = 1999CV$_{118}$)	
12:5	(79978 = 1999CC$_{158}$)	119878 = 2002CY$_{224}$		
5:2	26375 = 1999DE$_{9}$	38084 = 1999HB$_{12}$	60621 = 2000FE$_{8}$	69988 = 1998WA$_{31}$
	(84522 = 2002TC$_{302}$)	119068 = 2001KC$_{77}$	135571 = 2002GG$_{32}$	K00SX1R = 2000SR$_{331}$
	K01XP4Q = 2001XQ$_{254}$	K02G32P = 2002GP$_{32}$	K03UB7Y = 2003UY$_{117}$	
8:3	82075 = 2000YW$_{134}$			
3:1	136120 = 2003LG$_{7}$			
7:2	(K01K76V = 2001KV$_{76}$)			
11:3	(126619 = 2002CX$_{154}$)			
11:2	(26181 = 1996GQ$_{21}$)			
27:4	(K04PB2B = 2004PB$_{112}$)			

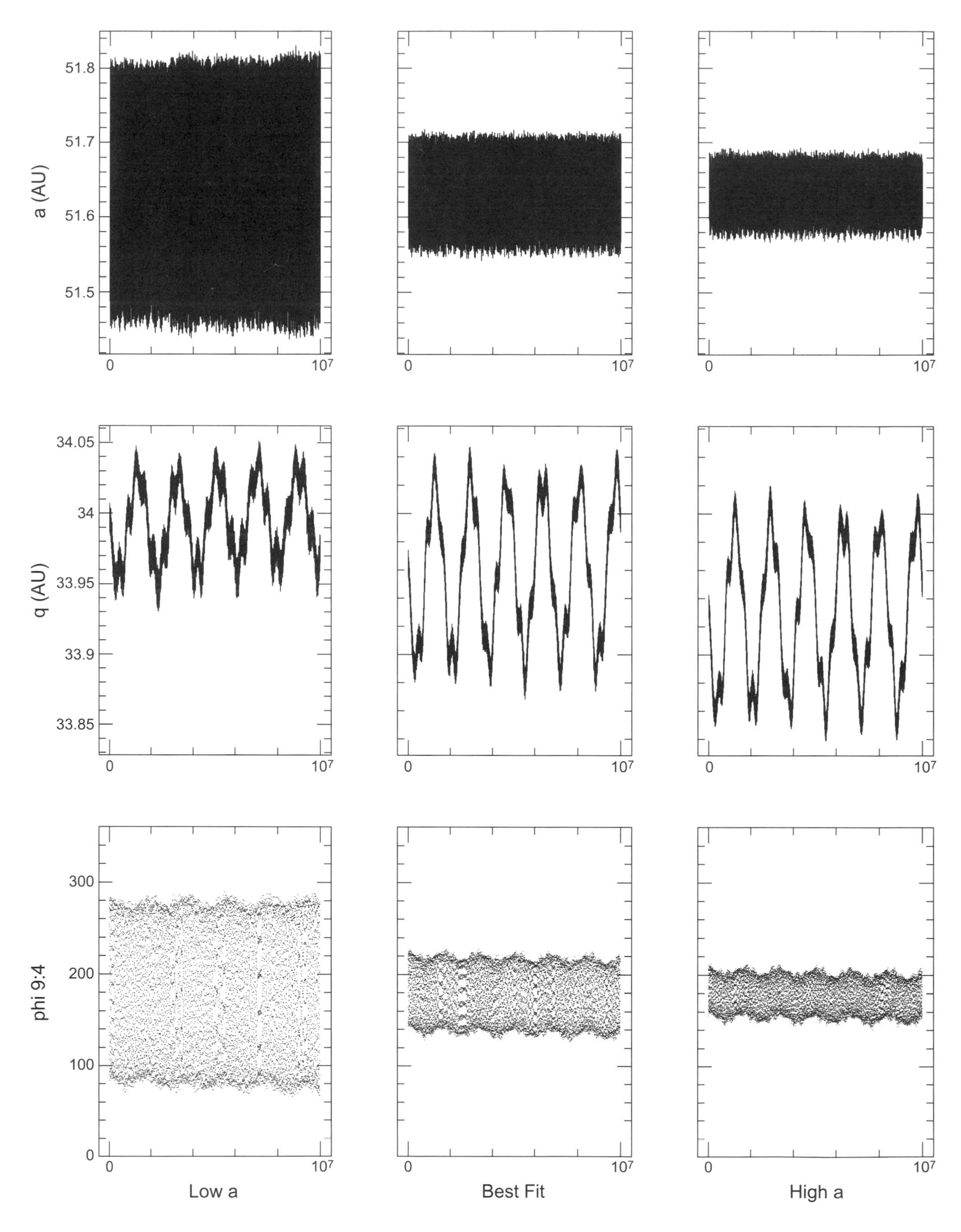

Fig. 3. The time evolution of the low-a (*left column*), best-fit (*center*), and high-a (*right column*) orbits for 2001 KG_{76} = K01K76G. Horizontal axis is time in years. The likelihood of a high-order resonance is indicated by the stability of a and periodic oscillations in q, despite $q < 35$. Indeed, all three integrated initial conditions librate in the fifth-order 9:4 mean-motion resonance, and this is thus a secure classification. Clearly the amplitude of the resonant argument is still quite uncertain and requires further observations.

seventh-order 12:5 libration of 119878 = 2002 CY_{224}). Figure 3 shows the occupancy of the fifth-order 9:4 resonance by 2001 KG_{76}.

Our study has revealed a surprisingly high number of a > 48.4 AU objects that appear to be resonant. Since many of these have been previously classified as SDOs, these are discussed in the following section. Figure 4 shows all the resonant TNO identifications inside a = 73 AU, along with the remaining classifications discussed below.

5. NONRESONANT OBJECTS

The TNOs remaining to be classified are those that the numerical integration shows are nonresonant. The problem is the desire to involve orbital stability in the nomenclature for historical and cosmogonic reasons, due to the great interest in knowing the source regions of comets. Unlike boundaries between two stable or two unstable populations, arbitrary cuts in phase space (e.g., simple cuts in perihelion

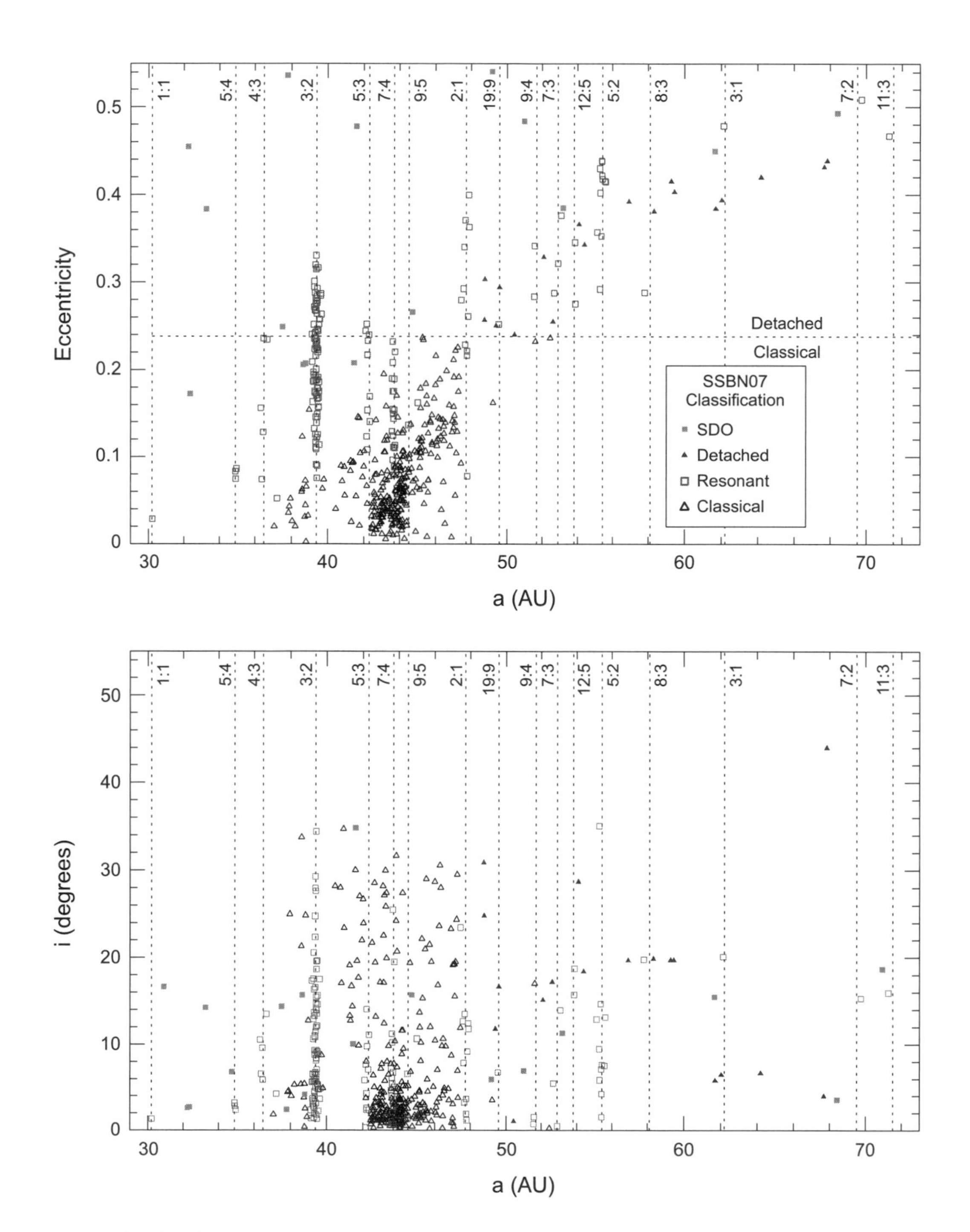

Fig. 4. The SSBN07 classification of the TNO region for a = 30–73 AU. Resonant semimajor axes are labeled and indicated by dotted lines. The horizontal dashed line gives the arbitrary division between the detached TNOs and classical belt. Beyond a = 73 AU all objects are detached or SDOs with the exception of two insecure resonant classifications (see tables).

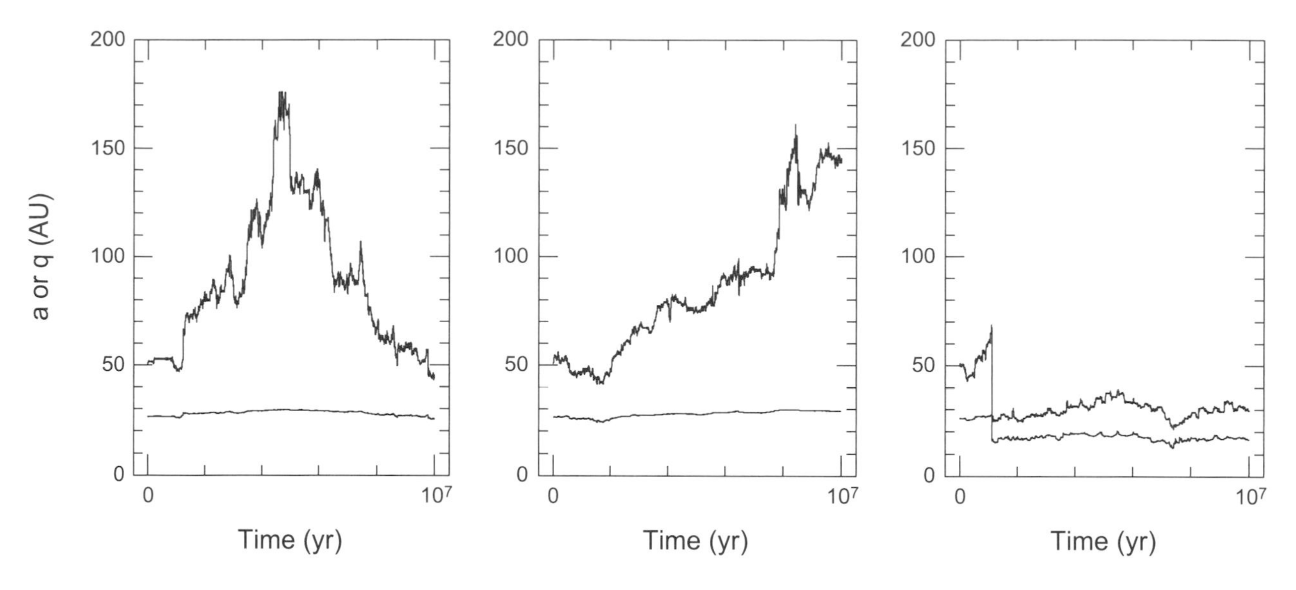

Fig. 5. The nonresonant (secure) classification of scattering(ed) disk object 2003 QW_{113} = K03QB3W, for the low-a (*left*), best-fit (*center*), and high-a (*right*) orbit. The lower history shown in each panel is q = a(1 – e), the upper is a.

distance q) do not provide a satisfactory separation between the classical belt and the scattered disk.

Although a separation based on the Tisserand parameter with respect to Neptune (analogous to the JFC definition) is attractive (*Elliot et al.,* 2005), we have abandoned this. Unlike the JFC/Centaur and JFC/SDO boundaries that we use (which are between two unstable populations), the SDO/classical belt boundary is *not* well modeled by such a simple division because the border between the two populations is extremely complex, involving all orbital parameters. *Duncan et al.* (1995) show the very intricate nature of the boundary, which cannot be modeled as a constant q cut (especially since the inner q boundary varies with i). Since physical studies of TNOs would want to cleanly separate the unstable SDOs away from the classical belt objects in this region, we decided to exploit the integrations that had already occurred for all objects to decide if the objects are actually currently heavily interacting with Neptune.

5.1. Scatter(ing/ed) Disk Objects

The term "scattered disk" was originally intended for TNOs scattered to large-e orbits with q near Neptune. While not stable, because of the long orbital periods and sometimes near-resonant behavior with Neptune, some SDOs can survive ~4.5 G.y. (*Duncan and Levison,* 1997). Given that there is also a well-populated "extended scattered"/detached disk (section 5.2), it is not entirely clear how the scattered disk was produced. For example, it may be possible that a passing star, rogue planet(s), or sweeping resonances have emplaced objects in this region, rather than direct scattering by Neptune. Therefore, our philosophy is that the SDOs are those objects that are *currently scattering actively off Neptune*, rather than ascribing to this population any specific ideas about their origin.

Fortunately, the 10-m.y. numerical integrations (already executed to look for resonance occupancy) cleanly identifies SDOs due to their rapid variation in semimajor axis. Figure 5 shows a prototype example of the orbit evolution of an SDO. We adapt a criterion similar to *Morbidelli et al.* (2004): An excursion in a of ≥1.5 AU during the 10-m.y. integration classifies the object as an SDO. We find that the exact value used (1–2-AU variation) makes little difference, as SDOs suffer large-a changes in short times. Although in principle SDOs can be "on the edge" of showing significant a mobility, we rarely find any confusion. Thus, SDOs in this definition (Table 3) are "scatter*ing*" objects rather than "scatter*ed*," even though we acknowledge that the latter term is entrenched in the literature.

We have found many cases of objects currently classified as SDOs that are in fact resonant (cf. *Hahn and Malhotra,* 2005; *Lykawka and Mukai,* 2007). For example, the TNO

TABLE 3. Scatter(ing/ed) disk objects (SSBN07 classification).

15874 = $1996TL_{66}$	29981 = $1999TD_{10}$	33128 = $1998BU_{48}$	42355 = $2002CR_{46}$
44594 = $1999OX_3$	54520 = $2000PJ_{30}$	59358 = $1999CL_{158}$	60608 = $2000EE_{173}$
65489 = $2003FX_{128}$	73480 = $2002PN_{34}$	78799 = $2002XW_{93}$	82155 = $2001FZ_{173}$
82158 = $2001FP_{185}$	87269 = $2000OO_{67}$	87555 = $2000QB_{243}$	91554 = $1999RZ_{215}$
120181 = $2003UR_{96}$	127546 = $2002XU_{93}$	J99CB8Y = $1999CY_{118}$	(K00QP1M = $2000QM_{251}$)
(K01K77G = $2001KG_{77}$)	(K01OA9M = $2001OM_{109}$)	K02G32B = $2002GB_{32}$	K02G32E = $2002GE_{32}$
(K03FC9H = $2003FH_{129}$)	K03H57B = $2003HB_{57}$	K03Q91Z = $2003QZ_{91}$	K03QB3W = $2003QW_{113}$
K03WH2U = $2003WU_{172}$	(K04D71J = $2004DJ_{71}$)		

in Figs. 2 and 3 is currently classified as an SDO in the MPC lists, but is really in a fifth-order resonance. We find that the orbital evolution of a resonant TNO with q < 38 AU exhibits a much-muted semimajor variation compared to a nonresonant object, and upon hunting we usually identify a high-order resonance. Figure 6 shows one of the few boundary cases we have found, where the TNO might either be (1) resonant in the 5:1 if a drops slightly given further observations, (2) a detached object (section 5.2) that migrates only slowly due to the 21° inclination, or (3) conceivably an object that has "stuck" temporarily to the border of the mean-motion resonance (see *Duncan and Levison,* 1997, for a discussion).

Owing to the use of the numerical integration, in this nomenclature SDOs exist over a large a range and are not confined to a > 50 AU as has often been done in the literature; instead the SDO population extends down to a = 30 AU where the Centaurs begin. There is essentially an SDO upper-e limit where coupling to Jupiter occurs. At very large a, where external influences become important, the inner Oort cloud begins.

5.2. The Detached Transneptunian Objects

After the recognition that there must be a large population of objects in the outer Kuiper belt with pericenters decoupled from Neptune (*Gladman et al.,* 2002), the boundaries of this region have expanded as more large-a TNOs are discovered. We have dropped the term "extended scattered" because it is unclear if this population was emplaced by scattering. In any case the term "detached" (adopted from *Delsanti and Jewitt,* 2006) can be understood in the present tense and keeps with our philosophy of using an TNO's current dynamical behavior for the classification. We have elected not to adopt the Tisserand value with respect to Neptune (*Elliot et al.,* 2005) as part of a definition, since the prevelance of high-i TNOs here and in the classical belt makes for a very messy mix (where large i forces $T_N < 3$ for orbits with essentially no dynamical coupling to Neptune). The numerical integration of each object separates the SDOs from the detached TNOs. But we are left with the thorny problem of where the detached population should end at low eccentricity. While in principle one could call all nonresonant, nonscattering TNOs "classical," having 2000 CR_{105} or Sedna lumped in with a circular orbit at 44 AU is both not useful and not in line with the recent literature. *Elliot et al.* (2005) proposed using the arbitrary lower bound of e = 0.2 on the population; we amend this to e = 0.24 because at moderate inclination (10°–20°) there are stable orbits interior to the 2:1 resonance (*Duncan et al.,* 1995) that are more comfortably thought of as classical belt objects than detached objects (see Fig. 4). The e = 0.24 division thus gives the symmetry that stable TNOs with identical e but on either side of the 2:1 resonance will both be considered classical belt objects.

This definition results in the detached TNOs (Table 4) being those nonscattering TNOs with large eccentricities (e > 0.24) and not so far away that external influences are important to their current dynamics (a > 2000 AU). The mechanisms that emplaced the detached TNOs on these orbits are undergoing active current research (see chapters by Gomes et al., Levison et al., Kenyon et al., and Duncan et al.). We hypothesize that as future work extends the observed arcs of some of these detached TNOs, more of them will be securely identified as being in high-order mean-motion resonances, with interesting cosmogonic implications.

5.3. The Classical Belt

By the process of elimination (Fig. 1), those objects that are left belong to the "classical belt," whose previous definitions have usually had no clear outer or inner boundaries. Here, the classical belt is not confined between the 3:2 and 2:1 resonance, to which it was sometimes limited to. Rather, it also extends inward to the dynamically stable low-e region inside the 3:2 (*Gladman,* 2002; *Petit and Gladman,* 2003), and out to the lightly populated low-e orbits outside the 2:1 resonance. There may be a popular misconception that the stable classical orbits with a < 39.4 AU (i.e., interior to the 3:2 resonance) are somehow "disconnected" from the stable phase space of the a = 42–48 AU region; the "gap" present in the often-used low-i stability diagram of *Duncan et al.* (1995) is only present at low-i and is (as those authors showed) due to the ν_8 secular resonance. The ν_8 is located at a ≃ 41 AU for low-i, but then moves to lower a at i ≃ 10°, thus stablizing the a = 39.4–42-AU region for moderate and high inclinations.

Although formally we take all the nonresonant low-e TNOs to be in the classical belt (removing any SDOs, of course), it may be useful terminology to divide the classical belt into an *inner* classical belt (a < 39.4 AU, nonresonant), an *outer* classical belt (a > 48.4 AU, nonresonant, and e < 0.24), and a *main* classical belt (sometimes called cubewanos). There is a continuous region of stable phase space connecting the inner classical belt to the main classical belt, and only the 2:1 resonance separates the main belt from the outer classical belt. The utility of these terms is thus simply descriptive, but has the practical advantage of giving an adjective to isolate these cosmogonically interesting semimajor axis regions. While these subclasses serve no strong nomenclature purpose, as there is little current *dynamical* difference between these regions, we flag these objects as such in Table 5.

The outer classical belt is currently inhabited only by the high-quality-orbit TNOs 2003 UY_{291} and 2001 QW_{297}, which the discoverers classified as detached (*Elliot et al.,* 2005), and 48639 = 1995 TL_8, classified as detached by *Gladman et al.* (2002). They will soon be joined by 2004 XR_{190} [aka Buffy (*Allen et al.,* 2006)] with a ≃ 57.5, e ≃ 0.10. While TL_8 and UY_{291} have inclinations below 5°, one may be uncomfortable with i = 17° or 47° TNOs being in the classical belt; there are objects with similarly high i in the main classical belt (e.g., 2004 DG77 with i = 47.6°), and thus i cannot be a guide for membership in the classical belt even if the dynamically hot state of the classical belt is somewhat of a surprise. It may not be generally realized

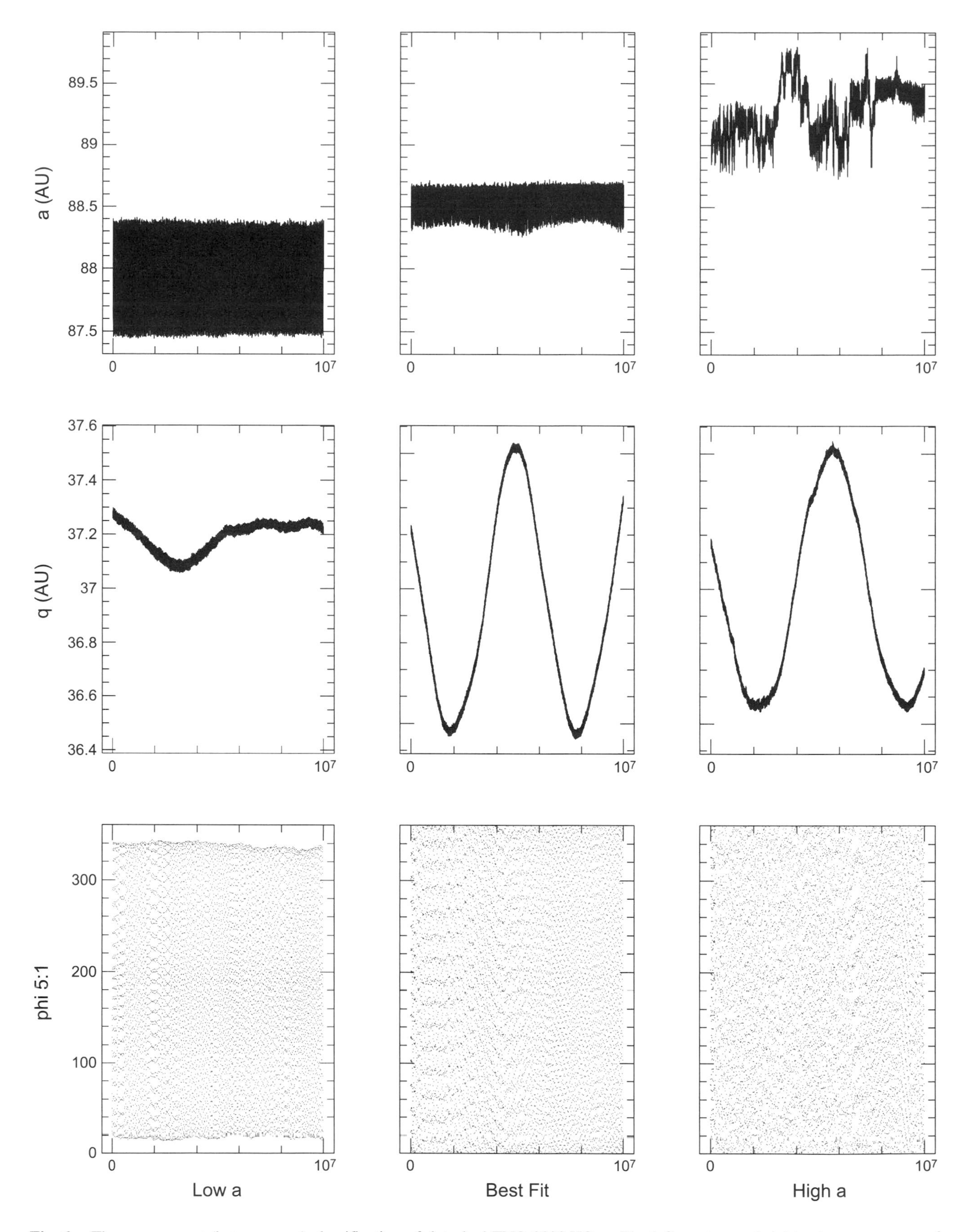

Fig. 6. The nonresonant (but unsecure) classification of detached TNO 2003 YQ_{179}. The left, center, and right columns correspond to the orbital histories from the numerical integration of the lowest-a, best-fit, and highest-a orbits from the orbit uncertainty calculation. The best-fit and highest-a orbits show nonresonant behavior, and the nominal classification is "detached." However, with the roughly ±0.4 AU semimajor axis uncertainty, we find that the lowest-a orbit exhibits large-amplitude libration in the 5:1 mean-motion resonance. Further observations are needed to ensure that the true orbit is not in the 5:1.

TABLE 4. Detached TNOs (SSBN07 classification).

(40314 = 1999KR$_{16}$)	60458 = 2000CM$_{114}$	90377 = 2003VB$_{12}$ = Sedna	118702 = 2000OM$_{67}$
120132 = 2003FY$_{128}$	134210 = 2005PQ$_{21}$	136199 = 2003UB$_{313}$ = Eris	(J98X95Y = 1998XY$_{95}$)
(J99CB8Z = 1999CZ$_{118}$)	J99CB9F = 1999CF$_{119}$	(J99CB9G = 1999CG$_{119}$)	J99H11W = 1999HW$_{11}$
J99RL4Z = 1999RZ$_{214}$	(J99RL5D = 1999RD$_{215}$)	(J99RL5J = 1999RJ$_{215}$)	K00AP5F = 2000AF$_{255}$
K00CA5Q = 2000CQ$_{105}$	K00CA5R = 2000CR$_{105}$	K00P30E = 2000PE$_{30}$	K00P30F = 2000PF$_{30}$
(K00P30H = 2000PH$_{30}$)	(K00Y02C = 2000YC$_{2}$)	(K01FJ4M = 2001FM$_{194}$)	(K02G31Z = 2002GZ$_{31}$)
(K02G32A = 2002GA$_{32}$)	K03FC9Z = 2003FZ$_{129}$	(K03Q91K = 2003QK$_{91}$)	(K03YH9Q = 2003YQ$_{179}$)

TABLE 5. Classical objects (SSBN07 classification; italics indicate inner belt, bold indicates outer belt).

15760 = 1992QB$_{1}$	15807 = 1994GV$_{9}$	15883 = 1997CR$_{29}$	16684 = 1994JQ$_{1}$
19255 = 1994VK$_{8}$	19308 = 1996TO$_{66}$	19521 = 1998WH$_{24}$ = Chaos	20000 = 2000WR$_{106}$ = Varuna
24835 = 1995SM$_{55}$	24978 = 1998HJ$_{151}$	33001 = 1997CU$_{29}$	*35671 = 1998SN$_{165}$*
(38083 = 1999HX$_{11}$ = Rhadamanthus)		45802 = 2000PV$_{29}$	**48639 = 1995TL$_{8}$**
49673 = 1999RA$_{215}$	50000 = 2002LM$_{60}$ = Quaoar	52747 = 1998HM$_{151}$	53311 = 1999HU$_{11}$ = Deucalion
55565 = 2002AW$_{197}$	55636 = 2002TX$_{300}$	55637 = 2002UX$_{25}$	58534 = 1997CQ$_{29}$ = Logos
60454 = 2000CH$_{105}$	66452 = 1999OF$_{4}$	66652 = 1999RZ$_{252}$	69987 = 1998WZ$_{25}$
(76803 = 2000PK$_{30}$)	79360 = 1997CS$_{29}$	79983 = 1999DF$_{9}$	80806 = 2000CM$_{105}$
82157 = 2001FM$_{185}$	85627 = 1998HP$_{151}$	85633 = 1998KR$_{65}$	86047 = 1999OY$_{3}$
86177 = 1999RY$_{215}$	88267 = 2001KE$_{76}$	88268 = 2001KK$_{76}$	88611 = 2001QT$_{297}$
90568 = 2004GV$_{9}$	118379 = 1999HC$_{12}$	*119951 = 2002KX$_{14}$*	120178 = 2003OP$_{32}$
120347 = 2004SB$_{60}$	*120348 = 2004TY$_{14}$*	123509 = 2000WK$_{183}$	126719 = 2002CC$_{249}$
(129772 = 1999HR$_{11}$)			
J93F00W = 1993FW	J94E02S = 1994ES$_{2}$	J94E03V = 1994EV$_{3}$	J95D02B = 1995DB$_{2}$
J95D02C = 1995DC$_{2}$	J95W02Y = 1995WY$_{2}$	J96K01V = 1996KV$_{1}$	J96R20Q = 1996RQ$_{20}$
J96T66K = 1996TK$_{66}$	J96T66S = 1996TS$_{66}$	J97C29T = 1997CT$_{29}$	(J97C29V = 1997CV$_{29}$)
J97Q04H = 1997QH$_{4}$	J97R05T = 1997RT$_{5}$	J97R09X = 1997RX$_{9}$	
J98FE4S = 1998FS$_{144}$	J98HF1L = 1998HL$_{151}$	*J98HF1N = 1998HN$_{151}$*	J98HF1O = 1998HO$_{151}$
J98K61Y = 1998KY$_{61}$	J98K62G = 1998KG$_{62}$	J98K65S = 1998KS$_{65}$	J98W24G = 1998WG$_{24}$
J98W24V = 1998WV$_{24}$	J98W24X = 1998WX$_{24}$	J98W24Y = 1998WY$_{24}$	J98W31T = 1998WT$_{31}$
J98W31W = 1998WW$_{31}$	J98W31X = 1998WX$_{31}$	J98W31Y = 1998WY$_{31}$	
J99CB9B = 1999CB$_{119}$	J99CB9C = 1999CC$_{119}$	J99CB9H = 1999CH$_{119}$	J99CB9J = 1999CJ$_{119}$
J99CB9L = 1999CL$_{119}$	J99CB9N = 1999CN$_{119}$	J99CD3Q = 1999CQ$_{133}$	J99CF3M = 1999CM$_{153}$
(J99CF3O = 1999CO$_{153}$)	J99CF3U = 1999CU$_{153}$	J99CF4H = 1999CH$_{154}$	J99CF8K = 1999CK$_{158}$
J99D00A = 1999DA	J99D08H = 1999DH$_{8}$	(J99G46S = 1999GS$_{46}$)	J99H11S = 1999HS$_{11}$
J99H11V = 1999HV$_{11}$	J99H12H = 1999HH$_{12}$	J99H12J = 1999HJ$_{12}$	J99O03Z = 1999OZ$_{3}$
J99O04A = 1999OA$_{4}$	J99O04C = 1999OC$_{4}$	J99O04D = 1999OD$_{4}$	J99O04E = 1999OE$_{4}$
J99O04G = 1999OG$_{4}$	J99O04H = 1999OH$_{4}$	*J99O04J = 1999OJ$_{4}$*	J99O04K = 1999OK$_{4}$
J99O04M = 1999OM$_{4}$	J99O04N = 1999ON$_{4}$	J99RL4T = 1999RT$_{214}$	J99RL4Y = 1999RY$_{214}$
J99RL5E = 1999RE$_{215}$	J99RL5G = 1999RG$_{215}$	J99RL5N = 1999RN$_{215}$	J99RL5U = 1999RU$_{215}$
J99RL5X = 1999RX$_{215}$	J99RL6A = 1999RA$_{216}$	J99XE3Y = 1999XY$_{143}$	
K00CA4L = 2000CL$_{104}$	K00CA4P = 2000CP$_{104}$	K00CA5E = 2000CE$_{105}$	K00CA5F = 2000CF$_{105}$
K00CA5G = 2000CG$_{105}$	K00CA5J = 2000CJ$_{105}$	K00CA5L = 2000CL$_{105}$	(K00CA5N = 2000CN$_{105}$)
K00CA5O = 2000CO$_{105}$	K00CB4N = 2000CN$_{114}$	K00CB4Q = 2000CQ$_{114}$	K00F08A = 2000FA$_{8}$
K00F08C = 2000FC$_{8}$	K00F08F = 2000FF$_{8}$	K00F08G = 2000FG$_{8}$	K00F08H = 2000FH$_{8}$
(K00F53R = 2000FR$_{53}$)	K00F53S = 2000FS$_{53}$	K00F53T = 2000FT$_{53}$	K00GE6V = 2000GV$_{146}$
K00GE6X = 2000GX$_{146}$	K00GE6Y = 2000GY$_{146}$	K00GI3P = 2000GP$_{183}$	K00J81F = 2000JF$_{81}$
K00K04K = 2000KK$_{4}$	*K00K04L = 2000KL$_{4}$*	*K00O51B = 2000OB$_{51}$*	K00O67H = 2000OH$_{67}$
K00O67J = 2000OJ$_{67}$	K00O67K = 2000OK$_{67}$	K00O67L = 2000OL$_{67}$	K00O67N = 2000ON$_{67}$
K00O69U = 2000OU$_{69}$	K00P29U = 2000PU$_{29}$	K00P29W = 2000PW$_{29}$	K00P29X = 2000PX$_{29}$
K00P29Y = 2000PY$_{29}$	K00P30A = 2000PA$_{30}$	K00P30C = 2000PC$_{30}$	K00P30D = 2000PD$_{30}$
(K00P30G = 2000PG$_{30}$)	K00P30M = 2000PM$_{30}$	K00P30N = 2000PN$_{30}$	K00QM6C = 2000QC$_{226}$
(K00Sb0Y = 2000SY$_{370}$)	K00W12V = 2000WV$_{12}$	K00WG9T = 2000WT$_{169}$	K00WI3O = 2000WO$_{183}$
K00Y01V = 2000YV$_{1}$	K00Y01X = 2000YX$_{1}$	K00Y02A = 2000YA$_{2}$	*K00Y02B = 2000YB$_{2}$*
K00Y02E = 2000YE$_{2}$	K00Y02F = 2000YF$_{2}$		

TABLE 5. (continued).

K01C31Z = 2001CZ$_{31}$	K01DA6B = 2001DB$_{106}$	K01DA6D = 2001DD$_{106}$	K01FI5K = 2001FK$_{185}$
K01FI5L = 2001FL$_{185}$	K01FI5N = 2001FN$_{185}$	K01FI5O = 2001FO$_{185}$	K01FI5T = 2001FT$_{185}$
K01FJ3K = 2001FK$_{193}$	(K01H65Y = 2001HY$_{65}$)	K01K76F = 2001KF$_{76}$	K01K76H = 2001KH$_{76}$
(K01K76T = 2001KT$_{76}$)	(K01K76W = 2001KW$_{76}$)	K01K77A = 2001KA$_{77}$	(K01K77E = 2001KE$_{77}$)
K01K77O = 2001KO$_{77}$	K01OA8K = 2001OK$_{108}$	K01OA8Q = 2001OQ$_{108}$	K01OA8Z = 2001OZ$_{108}$
K01OA9G = 2001OG$_{109}$	*K01P47K = 2001PK*$_{47}$	K01QT7O = 2001QO$_{297}$	K01QT7P = 2001QP$_{297}$
K01QT7R = 2001QR$_{297}$	**(K01QT7W = 2001QW$_{297}$)**	K01QT7X = 2001QX$_{297}$	K01QT7Y = 2001QY$_{297}$
K01QT7Z = 2001QZ$_{297}$	K01QT8A = 2001QA$_{298}$	K01QT8B = 2001QB$_{298}$	K01QT8C = 2001QC$_{298}$
K01QT8D = 2001QD$_{298}$	K01QT8J = 2001QJ$_{298}$	K01QW2Q = 2001QQ$_{322}$	K01QW2S = 2001QS$_{322}$
K01QW2T = 2001QT$_{322}$	K01QW2W = 2001QW$_{322}$	K01RE3W = 2001RW$_{143}$	K01RE3Z = 2001RZ$_{143}$
K01U18N = 2001UN$_{18}$	K01U18Q = 2001UQ$_{18}$	K01XP4R = 2001XR$_{254}$	K01XP4U = 2001XU$_{254}$
K02CF4S = 2002CS$_{154}$	K02CF4T = 2002CT$_{154}$	K02CM4X = 2002CX$_{224}$	K02CM5B = 2002CB$_{225}$
K02CO8Y = 2002CY$_{248}$	K02CP1D = 2002CD$_{251}$	K02F06U = 2002FU$_{6}$	K02F06V = 2002FV$_{6}$
K02F36W = 2002FW$_{36}$	K02F36X = 2002FX$_{36}$	K02G32H = 2002GH$_{32}$	K02G32J = 2002GJ$_{32}$
K02K14W = 2002KW$_{14}$	(K02M04S = 2002MS$_{4}$)	K02PE5Q = 2002PQ$_{145}$	K02PE9D = 2002PD$_{149}$
K02PE9O = 2002PO$_{149}$	K02PE9P = 2002PP$_{149}$	K02PF5D = 2002PD$_{155}$	K02PH0T = 2002PT$_{170}$
K02PH0V = 2002PV$_{170}$	K02PH0W = 2002PW$_{170}$	K02PH0X = 2002PX$_{170}$	K02PH0Y = 2002PY$_{170}$
K02PH1A = 2002PA$_{171}$	K02PH1C = 2002PC$_{171}$	K02VD0F = 2002VF$_{130}$	(K02VD0S = 2002VS$_{130}$)
K02VD0T = 2002VT$_{130}$	K02VD1B = 2002VB$_{131}$	K02VD1D = 2002VD$_{131}$	K02W21L = 2002WL$_{21}$
K02X91H = 2002XH$_{91}$			
(K03E61L = 2003EL$_{61}$)	K03FC7K = 2003FK$_{127}$	*K03FC8D = 2003FD*$_{128}$	K03FD0A = 2003FA$_{130}$
K03G55F = 2003GF$_{55}$	(K03H56X = 2003HX$_{56}$)	K03H56Y = 2003HY$_{56}$	K03H56Z = 2003HZ$_{56}$
K03H57C = 2003HC$_{57}$	K03H57E = 2003HE$_{57}$	(K03H57G = 2003HG$_{57}$)	K03H57H = 2003HH$_{57}$
(K03K20O = 2003KO$_{20}$)	K03L09D = 2003LD$_{9}$	(K03M12W = 2003MW$_{12}$)	K03Q90W = 2003QW$_{90}$
K03Q90X = 2003QX$_{90}$	K03Q90Y = 2003QY$_{90}$	*K03Q91Q = 2003QQ*$_{91}$	K03QB3F = 2003QF$_{113}$
K03SV7N = 2003SN$_{317}$	(K03SV7P = 2003SP$_{317}$)	K03SV7Q = 2003SQ$_{317}$	(K03T58G = 2003TG$_{58}$)
(K03T58K = 2003TK$_{58}$)	K03T58L = 2003TL$_{58}$	K03UB7Z = 2003UZ$_{117}$	**K03UT1Y = 2003UY$_{291}$**
K03UT2B = 2003UB$_{96}$	(K03YH9J = 2003YJ$_{179}$)	K03YH9K = 2003YK$_{179}$	*K03YH9L = 2003YL*$_{179}$
K03YH9M = 2003YM$_{179}$	K03YH9N = 2003YN$_{179}$	K03YH9O = 2003YO$_{179}$	K03YH9P = 2003YP$_{179}$
K03YH9R = 2003YR$_{179}$	K03YH9S = 2003YS$_{179}$	K03YH9T = 2003YT$_{179}$	K03YH9U = 2003YU$_{179}$
K03YH9V = 2003YV$_{179}$	K03YH9X = 2003YX$_{179}$		
K04XJ0X = 2004XX$_{190}$	K05F09Y = 2005FY$_{9}$		

that discovery of 2003 UY$_{291}$ has provided the first low-i, e < 0.2 TNO beyond the 2:1 resonance. The reason for the sparse population beyond the 2:1 is still an area of active research.

This brings us to a potential division of the classical belt into "hot" and "cold" components based on orbital inclination. While there are compelling arguments for interesting structure in the i-distribution (*Doressoundiram et al.,* 2002; *Gulbis et al.,* 2006), we do not feel the situation is yet sufficiently explored to draw an arbitrary division, especially since the plane with which to reference the inclinations is unclear (this choice of plane will move many objects in and out of the category). Although a cut near i_{cut} = 5° into hot and cold populations may eventually be useful, this cut reflects no dynamical separation. Although the high abundance of low-i TNOs is partially a selection effect of surveys being largely confined to the ecliptic (*Trujillo et al.,* 2001; *Brown,* 2001), an additional "cold" population does seem to be required, but strangely only in the a ≃ 42–45-AU region (see chapter by Kavelaars et al.); this cold component does not seem to be present in the inner or outer classical belt, or any of the resonant populations.

Elliot et al. (2005) essentially performed the hot/cold cut via their Tisserand parameter (with respect to Neptune) definition, which puts almost all TNOs with i > 15° into the SDO or detached populations. We are uncomfortable with calling extremely stable objects (e.g., a ~ 46, e ~ 0.1, i ~ 20°) SDOs, and thus propose that if a hot/cold division becomes enshrined, it be applied only to the classical belt and not SDOs.

6. CONCLUSION

The SSBN07 nomenclature algorithm defined herein separates outer solar system objects into unique groups with no gray areas to produce future problems. The term transneptunian region and the Kuiper belt become the same and the *transneptunian region* becomes defined as the union of the classical belt, SDO/detached populations, and the resonant objects exterior to the Neptune Trojans. There is a very large fraction of the a > 48 TNOs that are resonant, and further observations are required to hone their orbits.

Acknowledgments. We thank R. L. Jones, M. Ćuk, L. Dones, M. Duncan, J. Kavelaars, P. Nicholson, J-M. Petit, J. Rottler, and P. Weissman for helpful discussions. Note: After submission we became aware of the recent paper of *Lykawka and Mukai* (2007) with many similar results.

REFERENCES

Allen R. L., Gladman B., Kavelaars J., Petit J-M., Parker J., and Nicholson P. (2006) Discovery of a low-eccentricity, high-inclination Kuiper belt object at 58 AU. *Astron. J., 640,* 83–86.

Bernstein G. and Khushalani B. (2000) Orbit fitting and uncertainties for Kuiper belt objects. *Astron. J., 120,* 3323–3332.

Brown M. (2001) The inclination distribution of the Kuiper belt. *Astron. J., 121,* 2804–2814.

Carusi A., Kresák L., Perozzi E., and Valsecchi G. B. (1987) High-order librations of Halley-type comets. *Astron. Astrophys., 187,* 899–905.

Chiang E. and Jordan A. (2002) On the Plutinos and Twotinos of the Kuiper belt. *Astron. J., 124,* 3430–3444.

Chiang E. I. and 10 colleagues (2003) Resonance occupation in the Kuiper Belt: Case examples of the 5:2 and Trojan resonances. *Astron. J., 126,* 430–443.

Delsanti A. and Jewitt D. (2006) The solar system beyond the planets. In *Solar System Update* (Ph. Blondel and J. Mason, eds.), pp. 267–294. Springer-Praxis, Germany.

Doressoundiram A., Peixinho N., de Bergh C., Fornasier S., Thébault P., Barucci M. A., and Veillet C. (2002) The color distribution in the Edgeworth-Kuiper belt. *Astron. J., 124,* 2279–2296.

Dones L., Weissman P., Levison H., and Duncan M. (2004) Oort cloud formation and dynamics. In *Comets II* (M. C. Festou et al., ed.), pp. 153–174. Univ. of Arizona, Tucson.

Duncan M. and Levison H. (1994) The long-term dynamical behavior of short-period comets. *Icarus, 108,* 18–36.

Duncan M. and Levison H. (1997) A scattered comet disk and the origin of Jupiter family comets. *Science, 276,* 1670–1672.

Duncan M., Levison H. F., and Budd S. M. (1995) The dynamical structure of the Kuiper belt. *Astron. J., 110,* 3073–3081.

Elliot J. and 10 colleagues (2005) The Deep Ecliptic Survey: A search for Kuiper belt objects and Centaurs. II. Dynamical classification, the Kuiper belt plane, and the core population. *Astron. J., 129,* 1117–1162.

Gladman B. (2002) Nomenclature in the Kuiper belt. In *Highlights of Astronomy, Vol. 12* (H. Rickman, ed.), pp. 193–198. Astronomical Society of the Pacific, San Francisco.

Gladman B., Holman M., Grav T., Kavelaars J., Nicholson P., Aksnes K., and Petit J-M. (2002) Evidence for an extended scattered disk. *Icarus, 157,* 269–279.

Gomes R. (2003) The origin of the Kuiper belt high-inclination population. *Icarus, 161,* 404–418.

Gulbis A. A. S., Elliot J. L., and Kane J. F. (2006) The color of the Kuiper belt core. *Icarus, 183,* 168–178.

Hahn J. and Malhotra R. (2005) Neptune's migration into a stirred-up Kuiper belt: A detailed comparison of simulations to observations. *Astron. J., 130,* 2392–2414.

Kresák L. (1972) Jacobian integral as a classification and evolutionary parameter of interplanetary bodies. *Bull. Astron. Inst. Czechoslovakia, 23,* 1.

Levison H. (1996) Comet taxonomy. In *Completing the Inventory of the Solar System* (T. W. Rettig and J. M. Hahn, eds.), pp. 173–191. ASP Conf. Series 107, San Francisco.

Lykawka P. and Mukai T. (2007) Origin of scattered disk resonant TNOs: Evidence for an ancient excited Kuiper belt of 50 AU radius. *Icarus, 186,* 331–341.

Malhotra R. (1993) The origin of Pluto's peculiar orbit. *Nature, 365,* 819–821.

Marsden B. G. (2003) 2001 QR_{322}. *Minor Planet Electronic Circular 2003-A55.*

Marsden B. G. (2005) 2003 LG_7. *Minor Planet Electronic Circular 2005-L33.*

Morbidelli A. (1997) Chaotic diffusion and the origin of comets from the 2/3 resonance in the Kuiper belt. *Icarus, 127,* 1–12.

Morbidelli A., Emel'yanenko V. V., and Levison H. F. (2004) Origin and orbital distribution of the trans-Neptunian scattered disc. *Mon. Not. R. Astron. Soc., 355,* 935–940.

Petit J-M. and Gladman B. (2003) Discovery and securing TNOs: The CFHTLS ecliptic survey. *C. R. Phys., 4,* 743–753.

Trujillo C., Jewitt D., and Luu J. (2001) Properties of the trans-Neptunian belt: Statistics from the Canada-France-Hawaii Telescope Survey. *Astron. J., 122,* 457–473.

Wisdom J. and Holman M. (1991) Symplectic maps for the n-body problem. *Astron. J., 102,* 1528–1538.

The Orbital and Spatial Distribution of the Kuiper Belt

JJ Kavelaars
Herzberg Institute of Astrophysics

Lynne Jones
University of Washington

Brett Gladman
University of British Columbia

Joel Wm. Parker
Southwest Research Institute

Jean-Marc Petit
Observatoire de Besançon

Models of the evolution of Neptune's migration and the dynamical processes at work during the formation of the outer solar system can be constrained by measuring the orbital distribution of the remnant planetesimals in the Kuiper belt. Determining the true orbit distribution is not simple because the detection and tracking of Kuiper belt objects (KBOs) is a highly biased process. In this chapter we examine the various biases that are present in any survey of the Kuiper belt. We then present observational and analysis strategies that can help to minimize the effects of these biases on the inferred orbital distributions. We find that material currently classified as the classical Kuiper belt is well represented by two subpopulations: a high-inclination component that spans and uniformly fills the stable phase space between 30 and 47 AU combined with a low-inclination, low-eccentricity population enhancement between 42 and 45 AU. The low-i, low-e component may be that which has long been called the "Kuiper belt." We also find weaker evidence that the high-i component of the classical Kuiper belt may extend beyond the 2:1 mean-motion resonance with Neptune. The scattering/detached disk appears to extend to larger semimajor axis with no evidence for a falloff steeper than r^{-1}. This population is likely at least as large as the classical Kuiper belt population and has an i/e distribution much like that of the hot classical Kuiper belt. We also find that the fraction of objects in the 3:2 resonance is likely around 20% and previous estimates that place this population at ~5% are inconsistent with present observations. Additionally, high-order mean-motion resonances play a substantial role in the structure of the Kuiper belt.

1. OVERVIEW OF THE HISTORY OF KUIPER BELT SURVEYS

In 1949 K. Edgeworth, and in 1951 G. Kuiper, postulated the existence of a debris disk beyond the orbit of Neptune based on the hypothesis that material in this zone had likely not formed into large planets (*Irwin et al.,* 1995). The 1930 discovery of Pluto (*Tombaugh,* 1961) was the result of a search for an object that would explain the (incorrectly) measured motions of Neptune. Although, at the time, Pluto was not recognized as a large member of an ensemble of material, many researchers recognized that a search of the outer solar system could prove fruitful (see chapter by Davies et al. for a thorough historical review). In particular, a measure of the mass and orbital distribution of material in this zone of the solar system could explain the source of short-period comets.

Initial surveys of the Kuiper belt were conducted with the assumption that objects discovered beyond Pluto would be undisturbed, pristine relics of planet formation. Dynamically cold, circular, and low-inclination orbits were expected. The first two Kuiper belt objects discovered after Pluto, 1993 QB_1 (*Jewitt et al.,* 1992) and 1993 FW (*Luu et al.,* 1993), fit the expectation. These discoveries were followed by the discovery of objects at distances of just ~35 AU, much closer than 1992 QB_1 or 1993 FW. Initially the astrometric positions of these new objects [1993 RO (*Jewitt et al.,* 1993), 1993 RP (*Liller,* 1993), and 1993 SB and 1993 SC (*Williams et al.,* 1993)] were fit with orbits assumed to be circular (a necessary constraint due to the short discovery arcs available).

In *IAU Circular 5869,* Williams et al. noted ". . . there is rapidly developing a very severe problem of securing adequate astrometric follow-up, which is absolutely essential for any understanding of this exciting development in the outer solar system" (*Williams et al.,* 1993).

Soon after discovery, however, followup observations of 1993 SC were reported to the Minor Planet Center by *Tholen et al.* (1994) and demonstrated that the measured positions were better matched by a Pluto-like orbit with eccentricity of ~0.2. As a result of these observations the orbits of 1993 RO, 1993 RP, and 1992 SB were also assumed to be Pluto-like and new ephemerides were computed. At this time, B. Marsden released the following statement in *IAU Circular 5985*: "If the true orbits differ significantly from those on IAUC 5983, the ephemerides could be substantially in error . . ." (*Marsden,* 1994).

Even with these early warnings the avalanche of Kuiper belt object (KBO) discoveries quickly overwhelmed the followup efforts available, creating a distorted view based on ephemeris bias (defined below). Selection effects such as those described below also influenced the theoretical view of the Kuiper belt that has resulted from these surveys. *Jewitt et al.* (1996) provided one of the first attempts to quantify the size and shape of the Kuiper belt. In this early work they considered the biases against detections of high-inclination objects and determined that the intrinsic width of the Kuiper belt is likely around ~30° FWHM, much broader than anticipated. As surveys of the Kuiper belt have continued, observers are becoming more aware that a correct assessment of all observational bias is critical if one is to correctly measure the distribution of material in the distant solar system.

2. BIASES IN THE OBSERVED ORBIT DISTRIBUTION

2.1. Flux Bias

The most obvious bias in any optical imaging survey is "flux bias": Objects that are brighter are easier to detect and thus make up a disproportionate fraction of the detected population. Kuiper belt objects are discovered in the optical via reflected solar light, thus

$$\text{flux} \propto \frac{D^2}{\Delta^2 R^2} \quad (1)$$

where D is the object's diameter, Δ is the distance between Earth and the KBO, and R is the distance between the Sun and the KBO ($R \approx \Delta$). This results in objects at 30 AU being ~8× (or 2.3 mag) brighter than the same-sized objects at 50 AU. In addition, an object with a diameter of 1000 km is ~100× (or 5 mag) brighter than a 100-km-diameter object, even if there is no difference in albedo between them. Determining the true population of the Kuiper belt requires accurate knowledge of the flux limits of the survey.

A result of this flux bias is that Plutinos are a large fraction of the observed population, as they can spend some fraction of their orbit interior to Neptune, making them easier to detect than an object on a circular orbit beyond 40 AU.

The immediate impact is that there is nearly no solid information about the population of small ($D < 10$ km) objects and only limited knowledge of the distant ($\Delta > 50$ AU) KBO population.

2.2. Pointing Bias

A second bias depends on the ecliptic latitude and longitude of the survey; each region of the sky will contain a different fraction of each orbital class than another region.

Most obviously, a survey of fields near ecliptic latitude 10° will not detect any objects with inclination below this value, regardless of the fraction of objects with low inclinations. On the other hand, conducting a survey directly in the ecliptic plane is more efficient at detecting low-i than high-i objects. Objects with higher inclinations, which spend only a small fraction of their orbits near the ecliptic plane, will be poorly represented in an "on ecliptic" survey (see Fig. 11 in *Trujillo et al.,* 2001b). With typical KBO sample sizes of a few hundred objects at most, this results in poor statistical sampling of high-i KBOs.

Furthermore, Plutinos come to pericenter at solar longitudes that are far from Neptune and anti-Neptune. Thus, a flux-limited survey will be more sensitive to nearby objects and detect a higher fraction of Plutinos (vs. classical KBOs) when observing near quadrature with Neptune than when observing near Neptune or anti-Neptune, where the Plutinos are near aphelion. More generally, each mean-motion resonance has specific longitude ranges (relative to Neptune) where objects come to perihelion.

Knowledge of the pointing history of a survey must be available if one hopes to disentangle the true underlying population of the Kuiper belt from the biased observed one.

2.3. Ephemeris Bias

Ephemeris bias is particularly insidious. During a nominal survey of solar system objects, the initial discovery observation provides a first estimate of the orbit of the body. For objects in the asteroid belt, an observational arc of a week is sufficient to provide a reasonably robust measure of the asteroid's orbit, such that the object can be found again at some later date.

For the Kuiper belt, an arc of a few days does little more than to estimate the current distance and the orbital inclination of the object to within ~10–30%. Beyond these first-order measures, little orbital information is available. To predict the location of the object at a later date, strong orbital assumptions must be made, the most common being to assume that the object is on a circular orbit. The assumption made is almost certainly incorrect but does allow a first order estimate of the location of the object (accurate to ~60") a few months after discovery. If, however, the newly discovered KBO is not observed again within a few months of discovery, then only those objects for which the orbital assumptions are correct will be recovered one year later. Unfortunately, those objects for which incorrect assumptions are made will be lost and the part of the orbit parameter space they represented may be lost with them (see Fig. 1).

Directed followup more than a few months after initial discovery tends to cause surveys to "leak" those objects that are most likely to have been indicating some new part of parameter space. Ephemeris bias keeps the orbit distribution looking like the assumed distribution that went into the ephemeris estimate, which may not look anything like the underlying orbit distribution.

2.4. Detection Bias

Observational surveys for KBOs proceed via comparing images (i.e., blinking), or comparing source catalogs constructed at different epochs. However, the sky density of asteroids in any given survey field is much higher than that of KBOs and so the time between epochs is kept short enough (~1 h) that asteroid confusion is minimized. Kuiper belt objects at 75 AU only move at ~2"/h; in a typical survey they will move only two seeing disks between epochs. More distant objects have even smaller sky motions and their detection is even more problematic. Combining the difficulty in aligning images with the image quality of typical groundbased large-area surveys, it is little wonder that objects beyond 75 AU are found so rarely. A careful examination of detection efficiency as a function of source distance must be conducted in order to determine the true radial limit of a survey.

Given the flux, pointing, ephemeris, and detection biases, it is little wonder that 500-km-diameter objects on high-inclination orbits with pericenters outside 50 AU are seldom detected.

2.5. Survey Design

Completely removing all observer biases from a given survey is impossible. Flux, pointing, and detection biases reflect the intrinsic problem of observing solar system objects and cannot be eliminated by clever observing strategy. Ephemeris bias, however, can be greatly reduced by a carefully planned observing program.

Ephemeris bias is maximized when observers rely completely on ephemeris predictions based on poorly constrained orbits. For KBOs the uncertainty in the ephemeris based on observational arcs of only a few hours to days exceed the field of view (~10 arcmin) of most facilities where recovery observations are attempted in the following opposition (see Fig. 1). Although this problem was realized quite early in Kuiper belt surveys, the insidious nature of the bias ensured that the full impact could not be realized because "you don't know what you're missing."

A straightforward approach to eliminating ephemeris bias is, simply, to follow the targets more frequently during the discovery opposition and to avoid targeted followup observations in favor of repeatedly observing the same large area of sky. The slow motion of KBOs ensures that they can be reobserved within a few degrees of their discovery location over the course of many months and even years. Thus, a survey that worked in patches larger than a few square degrees could repeatedly image the patches in order to obtain "serendipitous" followup that would then be free of ephemeris bias. This notion of blind followup is precisely what has been employed by the Canada-France-Ecliptic Plane Survey (CFEPS) (*Jones et al.,* 2006) and is the planned strategy for surveys to be run by the LSST and Pan-Starrs projects.

Reliable ephemeris predictions, which ensure successful pointed recovery of a given object, are achievable once the object has been tracked through the second apparition using a frequent followup strategy. Figure 2 presents the evolution of the orbital and ephemeris uncertainty for an object tracked by CFEPS. This figure demonstrates the rapid growth in the ephemeris uncertainty and indicates that, without a ~6–8-week arc of observation in the discovery opposition, an object ephemeris uncertainty is >2000 arcsec. Additionally, until observations in the third opposition are available, precise classification is not feasible.

3. DEBIASING THE SURVEYS

The past decade and a half has seen a large number of surveys of KBOs. Initially these surveys focused on the detection of objects, proving the existence of the Kuiper belt. During this initial "discovery" phase, any new detection provided new insight into structure of the Kuiper belt. The discovery of 1992 QB_1 proved the existence of the Kuiper belt. The realization that there are a large number of objects in 2:3 resonance with Neptune made clear the importance of resonances. The subsequent discovery of 1999 TL_{66} (*Luu et al.,* 1997) and other objects on "scattering" orbits introduced a further complication to the dynamics of this region (*Duncan and Levison,* 1997). The existence of "detached" objects (*Gladman et al.,* 2002) on a variety of large-a and large-pericenter orbits, such as 2000 CR_{105} (*Elliot et al.,* 2005), Sedna (*Brown et al.,* 2004), Eris (*Brown et al.,* 2005), and 2004 XR_{190} (aka Buffy) (*Allen et al.,* 2006), is further challenging models of the formation of the Kuiper belt.

Table 1 lists Kuiper belt surveys that are currently in the literature along with the basic parameters of each survey. Some of the surveys listed in Table 1 are ongoing and quantities listed in the table provide a snapshot of the current detection statistics.

With these discoveries also come questions: What is the total population of the material in the "Kuiper belt"? What fraction of the population is in resonance? How can these objects come into resonance? What is the relative size of the "scattered" component? How stable is this component? How large a population do objects like 2001 CR_{105} and

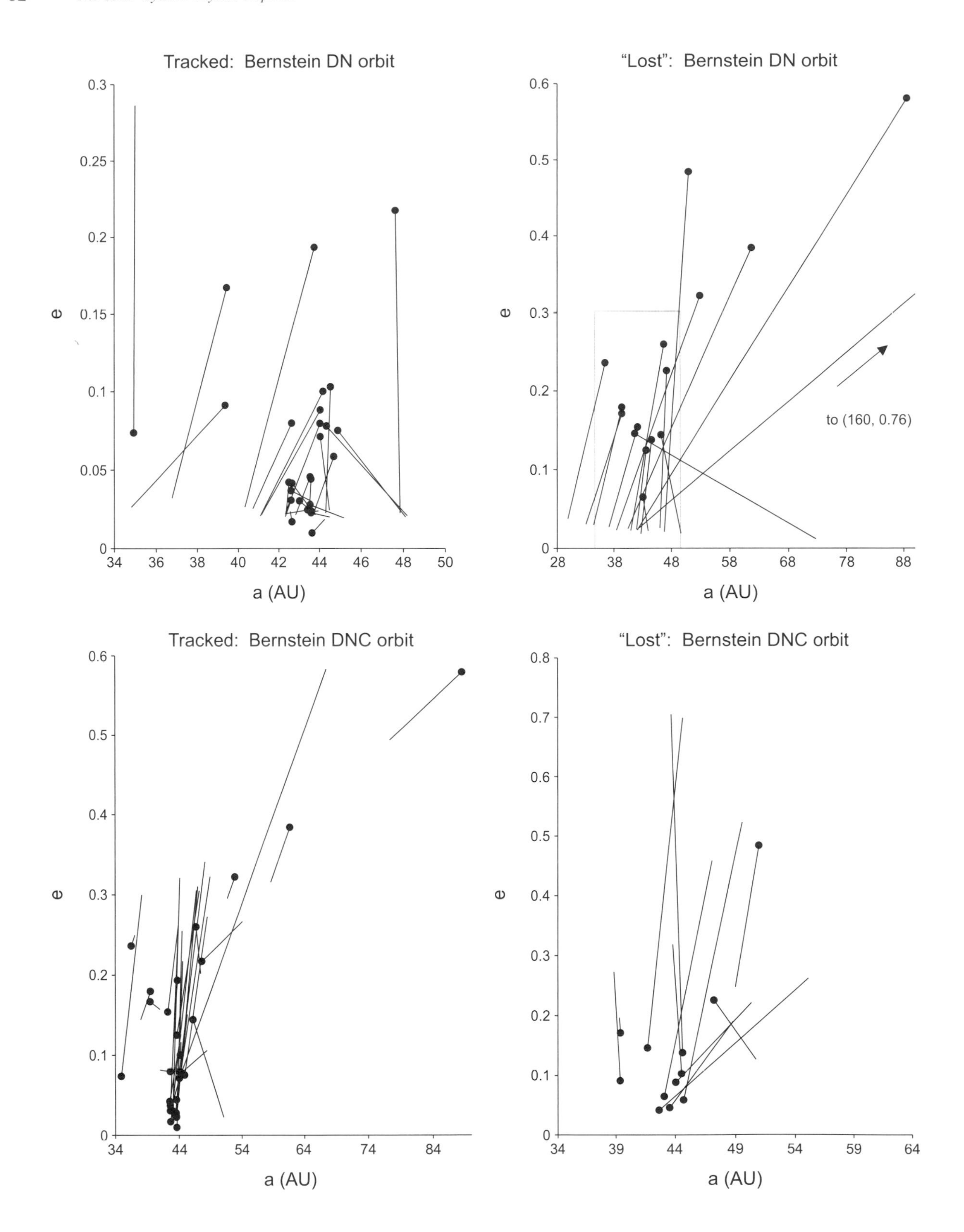

Fig. 1. An indication of the effects of ephemeris biases. The top left panel shows the a/e change from discovery-arc-based orbits (start of lines) to final orbits based on multi-opposition tracking (points at end of line) for the CFEPS "L3" release. The right panel presents the same information for those CFEPS L3 objects that, one year after discovery, were more than 10 arcmin from the ephemeris based on their discovery-arc orbits. Those objects in the right panel would have been "lost" by recovery attempts at telescopes with FOVs of 10 arcmin or less. The figure demonstrates that those objects whose orbits are outside the classical belt region are more likely to be lost, due to ephemeris bias. The bottom set of panels demonstrates that the loss rate is much lower when a 60-d arc in the discovery opposition is available before small FOV recovery is attempted.

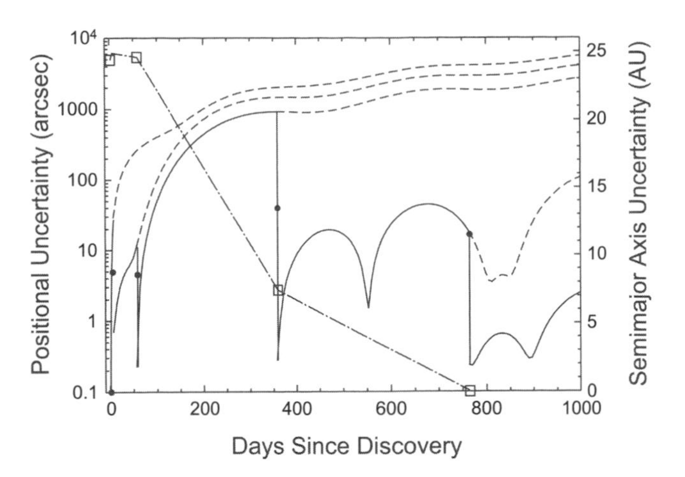

Fig. 2. Time evolution of positional and semimajor axis uncertainty for a classical-belt object. The solid line indicates the evolution of the predicted ephemeris uncertainties (based on orbfit from *Bernstein and Khushalani,* 2000) as additional observations are acquired; solid dots indicate the observed ephemeris error at each recovery. The dashed lines indicate the ephemeris uncertainty growth without the additional observations at each epoch. The dash-dot lines indicate the semimajor axis uncertainty as additional observations are acquired. Clearly tracking with a large (>10') field of view facility is required for the second opposition recovery observations. The semimajor axis uncertainty is too large for classification until after recovery observations in the third opposition.

Sedna represent? How did they come to be in their current orbits? What fraction of objects are like 2004 XR_{190}? Is there a vast reserve of distant large bodies like Eris? Clearly surveys of this region of the solar system continue to provide challenging observations.

3.1. Analytical Approaches to Debiasing

The sources of biases introduced into the observed representation of the underlying Kuiper belt are all well understood (see previous section), making the bulk properties of the underlying population determinable based on the analytic modeling of these biases. Such analytic modeling approaches can even be useful for deriving bulk information from surveys with little to no characterization information available.

3.1.1. The inclination distribution. Unfortunately, one cannot invert the observed inclination distribution and from that determine the intrinsic distribution function. *Jewitt et al.* (1996) first noted that the inclination distribution of material in the Kuiper belt must be quite broad. This conclusion was based on their detection of a number of high-inclination (>10°) KBOs even though their survey was confined to be near the plane of the ecliptic. Jewitt et al. found that either a "box" distribution [i.e., N(i) ~ *constant*] or a single broad Gaussian provided reasonable representations of this

TABLE 1. List of Kuiper belt surveys.

Survey	Facility	Area	Depth	Secure	Detections	Reference
Irregular Sat.	CFHT/12K	11.85	24.0	0	66	*Petit et al.* (2006a)
DES	KPNO/CTIO	550	22.0	217	486	*Elliot et al.* (2005)
ACS	HST	0.02	28.3	0	3	*Bernstein et al.* (2004)
Caltech	Pal 0.6 m	19389	20.5	54	71	*Trujillo and Brown* (2003)*
Allen1	KPNO	1.5	25.5	6	24	*Allen et al.* (2001)
Allen2	CTIO	1.4	24.8	0	10	*Allen et al.* (2002)
SSDS	Sloan	100	21.5	0	1	*Ivezić et al.* (2001)
KPNO-Large	KPNO0.9/Mosaic	164	21.1	4	4	*Trujillo et al.* (2001a)
CFH/12	CFHT 3.6	0.31	25.93	2	17	*Gladman et al.* (2001)
CFHT	CFHT 3.6	73	23.7	59	86	*Trujillo et al.* (2001b)
Spacewatch	KPNO 0.9 m	1483.5	21.5	36	39	*Larsen et al.* (2001)
sKBO	CFHT/12K	20.2	23.6	0	3	*Trujillo et al.* (2000)
sKBO	UH2.2/8K	51.5	22.5	0	1	*Trujillo et al.* (2000)
Baker-Nunn	APT-0.5 m	1428	18.8	1	1	*Sheppard et al.* (2000)
CB99	KECK/LRIS	0.01	27.0	0	2	*Chiang and Brown* (1999)
SSO	Siding Spring	12	21.0	0	1	*Brown and Webster* (1998)
JL Deep	Keck	0.3	26.3	1	6	*Luu and Jewitt* (1998)
G98a	CFHT/UH8k	0.35	24.6	1	1	*Gladman et al.* (1998)
G98b	Pal 5 m	0.049	25.6	1	4	*Gladman et al.* (1998)
G98c	Pal 5 m	0.075	25.0	0	0	*Gladman et al.* (1998)
JLT	CFHT/UH8k	51.5	23.4	12	13	*Jewitt et al.* (1998)
Pluto-Express	CFH12K	2.2	23.5	3	4	*Trujillo and Jewitt* (1998)
MKCT	UH2.2 m	3.9	24.2	10	14	*Jewitt et al.* (1996)
MKCT	CTIO 1.5 m	4.4	23.2	1	3	*Jewitt et al.* (1996)
ITZ	WHT	0.7	23.5	2	2	*Irwin et al.* (1995)

*See also the chapter by Brown.

very early dataset. *Brown* (2001) developed a procedure that allows the comparison of analytical models and observations of the inclination distribution of KBOs. Brown suggested, based on an examination of the observed inclination distribution, that a reasonable analytical approximation for the inclination distribution is a superposition of two Gaussian distributions such that the intrinsic inclination distribution has the form

$$f_t = \sin(i)\left[ae^{-\frac{i^2}{2\sigma_1^2}} + (1-a)e^{-\frac{i^2}{2\sigma_2^2}}\right] \quad (2)$$

and the observed ecliptic distribution takes the form $f_e(i) = f_t(i)/\sin(i)$ (see *Brown,* 2001).

We use this same approach to examine the inclination distribution of different KBO subpopulations using the classification list compiled in the chapter by Gladman et al. (see Table 2).

We find that equation (2) provides a reasonable representation of the inclinations distributions for the four subpopulations of the belt examined (see Table 2). However, as previously found in *Brown* (2001), the range of parametric values allowed for the scattering, detached, and Plutino populations are all consistent with a nonexistent "low-inclination" component. Using a similar approach to debiasing their survey, although more carefully tuned to their observations, *Elliot et al.* (2005) match three different model distributions to their observed inclination distribution. They find that the global KBO population is best represented as either a double Gaussian (equation (2)) or as a single narrow Gaussian plus a broad Lorentzian. The main conclusion from their analysis is that the bulk inclination distribution is double peaked. They conclude, however, that the narrow-inclination component is dominated by the classical belt population and find that the "scattered" disk objects show little concentration toward low inclinations.

In effect, the double-Gaussian fit to the classical population tells us that a low-inclination (cold) component must exist for this subpopulation of the Kuiper belt. The other components of the belt, however, are well represented as having just a high-inclination (hot) component with a Gaussian width that is similar to the "hot" component of the classical Kuiper belt.

TABLE 2. Parametric representation of the inclination distribution of various subpopulations of the Kuiper belt for the inclination model in equation (2).

Population	σ_1	σ_2	*	$D\sqrt{N}$
Classical	1.5 ± 0.4	13 ± 3	0.95 ± 0.02	0.56
Scattered	1.6 ± 1.6	13 ± 5	$0.4^{0.8}_{0.0}$	0.72
Detached	1.1 ± 0.4	18 ± 6	$0.8^{1.0}_{0.0}$	1.26
Plutinos	1.7 ± 0.4	13 ± 5	$0.4^{0.8}_{0.0}$	0.7

*Values of $D\sqrt{N} > 1.5$ rejected at the ~1σ level (see *Brown,* 2001). None of the $D\sqrt{N}$ values for the models fits reported here reject the underlying model distribution. The parameters determined in this work are consistent with those reported in *Brown* (2001).

3.1.2. The distance distribution. *Trujillo and Brown* (2001) established that the radial distribution of material in the Kuiper belt can be determined using the apparent radial distribution via the equation

$$f(R) = \frac{\beta(R)f_{app}(R)}{\Gamma'(m_V)} \qquad \beta \equiv \left[\frac{R^2 - R}{R_0^2 - R_0}\right]^{q-1} \quad (3)$$

where $f_{app}(R)$ is the observed (apparent) radial distribution, R_0 is the inner edge of the Kuiper belt (taken to be R_0 ~ 42 AU for the Kuiper belt as whole and R_0 ~ 35 AU for the scattering component when treated separately), β(R) is the bias correction factor, and $\Gamma'(m_V)$ is the normalization constant, which depends only on the flux from the detected objects and does not affect the radial distribution (see *Trujillo and Brown,* 2001, for a complete derivation). Trujillo and Brown required that the following assumptions be met for their debiasing approach: (1) all KBOs follow the same size distribution, described by a differential power law; (2) observations are conducted at opposition allowing the transformation $\Delta = R - 1$; (3) the albedo is not a function of size or heliocentric distance R.

Evidence of a "break" in the luminosity function of KBOs has been detected (*Bernstein et al.,* 2004) [see also the chapter on the luminosity function (LF) by Petit et al.] and the albedos of KBOs are now recognized to have some dependence on object size (see chapter by Stansberry et al.). Even in light of these changes in circumstance, one can still use the above formalism, since the break in the LF is faintward of the majority of objects in the MPC database used in our analysis and albedo only appears to vary for the largest KBOs. Additionally, as shown in *Trujillo and Brown* (2001), the debiasing of the radial distribution is only weakly dependent on albedo.

Figure 3 presents the debiased radial distribution for three subsets of the classified KBOs with H > 3.5. The figure clearly demonstrates that the "cold classical" KBOs radial distribution is peaked near ~44 AU while the "hot classical" component of the population is more broadly distributed. Interestingly, the "hot" and "cold" components of the belt have identical radial distributions beyond ~46 AU. The scattering/detached KBOs exhibit more extended radial distributions with no obvious preferred distance.

The radial distribution of the classified populations has an unknown amount of ephemeris bias, since we don't know which objects from the originally detected population were lost due to being misclassified, nor if there was a strong distance/orbit correlation among those losses.

Although Fig. 3 provides a compelling picture of a truncated disk, the debiasing approach described by Trujillo and Brown is only effective in those parts of the solar system where we have actual detections, *unless we know a priori that the surveys present in the MPC were sensitive to objects beyond the apparent limit of the classical belt,* which seems likely to be true only out to distances of ~75 AU. Based on the available detections and this analytic debiasing approach, we conclude that the radial falloff of the classical

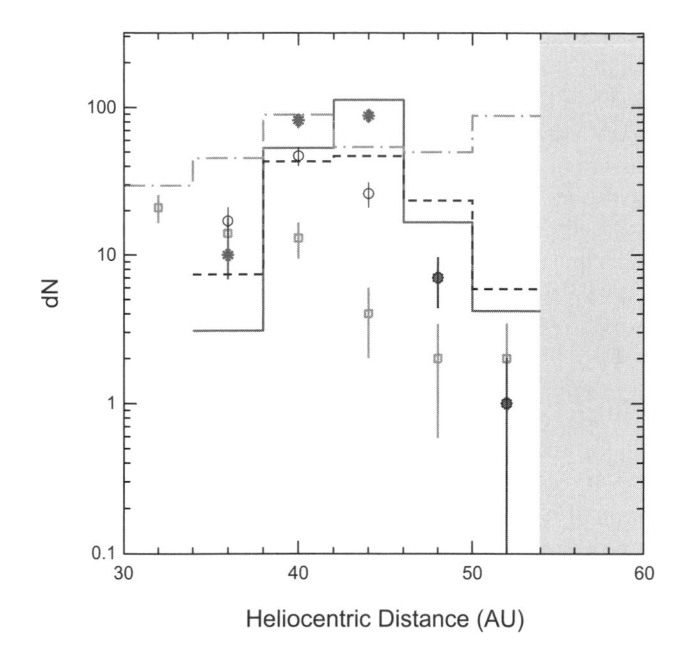

Fig. 3. The radial distribution at detection of classified KBOs with H > 3.5. The points represent the distribution of detected KBOs (asterisks: classical KBOs with inclinations <5°; black open circles: classical KBOs with inclinations >5°; gray boxes: scattering and detached KBOs). The lines (solid line: "cold" classical; black dashed: "hot" classical; gray dot-dashed: scattering/detached) represent the radial distribution after debiasing based on the method in *Trujillo and Brown* (2001). The "cold classical" belt object density must decay very rapidly with distance beyond ~45 AU; the "hot classical" appears less "peaked" than the "cold" classical component. Remarkably, the cold and hot radial distributions are identical beyond 45 AU, indicative of a single, underlying, population. The scattering/detached belt object density does not appear to be a strong function of distance. The shaded area of the plot represents that part of the solar system where no dynamically classified objects with H > 3.5 have been detected. Vertical scale indicates relative population at each distance.

belt must be quite steep and the "hot" classical belt objects extend inward to slightly lower heliocentric distances where their larger inclinations help ensure their stability. The scattering and detached objects do not appear to share this distribution.

3.2. Survey Simulator Debiasing

The complex interaction between the various biases outlined above and the orbital distribution results in a situation where detailed comparisons between model and observations cannot be achieved using analytic debiasing approaches. Like the construction of the models themselves, detailed comparison requires the use of a simulator. Essentially the survey simulator entails mapping the model space into the observation space. Such approaches have been detailed in some previous Kuiper belt surveys (see, e.g., *Jewitt et al.,* 1998; *Trujillo et al.,* 2001b; *Jones et al.,* 2006).

A survey simulator provides a programmatic method of introducing into the model those same biases that are present in the observational data. By correctly emulating the processes of surveying, a meaningful comparison can be achieved. For this method to be successful the user must look to as many constraining surveys as possible for testing their model orbital distribution.

We provide here a rough outline of the operation of the CFEPS survey simulator:

1. A randomly selected model particle is assigned a size, based on some externally calibrated understanding of the size distribution.
2. The sky location and brightness are determined based on the model particle's orbit.
3. The observability of the object is compared to the characterization of the survey and detected objects are kept.
4. The above steps are repeated until a sample with the same size as the survey is achieved,
5. The orbital elements of the simulated survey are compared to the observed Kuiper belt.
6. The model is tuned to better represent reality and the above steps are repeated.

When a model that provides a compatible simulation of the observations is found, this model can be accepted as a reasonable (although possibly not unique) representation of the true underlying population. Various bulk orbital properties and other details can be determined by examining the model distributions.

3.2.1. Survey characterization. A full and correct characterization of the survey's detection efficiency is critical to the successful use of a survey simulator approach. For each of the biases present in the survey, the information needed to model this bias must be determined from the survey observations. For each field of the survey the ecliptic longitude and latitude must be reported along with a determination of the flux-based detection efficiency as a function of the sky rate and angle of motion of KBOs in the field.

Providing the details required to allow accurate survey simulations is a great burden on the survey observer. Without this information, however, the modeler is left with only the rough analytic comparisons detailed above and determining the true underlying orbit distribution may never be possible.

3.2.2. Model uniqueness. A strong caveat must be made for both the analytic and survey simulator approaches to comparing model distribution with the observed population. A model, when convolved through a survey simulator, may look like the observed population, or the observed population when analytically debiased may resemble the model; however, this does not ensure any uniqueness of the model distribution. For an accurate picture to emerge modelers must produce a range of possible scenarios for the dynamical state of the belt and then through piecewise comparison various unfavorable scenarios can be eliminated with no guarantee that the true underlying population will ever be found.

Most of the survey simulator results come in the form of constraints that can be placed on those orbital elements that require longer observing periods before they can be determined accurately, such as a and e. In addition, survey simulators are an excellent approach to determining the relative strengths of the various orbital populations. For each of the orbital distributions discussed in the analytic section, we present a short review of similar results that have been derived using survey simulator approaches.

We conclude each section by describing the results we obtain using the CFEPS survey simulator and the CFEPS L3 (*Petit et al.,* 2006b) and Pre-Survey (Jones et al. 2006) data releases. Figure 4 presents the a/e/i distributions for the CFEPS + Pre-Survey detections, our base model, and the "simulated detections." This model is discussed further in the following sections.

3.2.3. The inclination distribution. Based on early evidence of a broad (in latitude) Kuiper disk, *Trujillo et al.* (2001b) conducted observations at a variety of ecliptic latitudes and then, using their knowledge of the pointing history of the survey, matched the observed latitude density distribution to the underlying inclination distribution. They found that the latitude distribution of material in the Kuiper belt could be reproduced by assuming that the underlying inclination distribution followed the form of a single Gaussian of width $\sigma \sim 20°$. Trujillo et al. also found that the functional form in equation (2) or a simple "constant" distribution like that proposed in *Jewitt et al.* (1996) did an equally good job of matching their observations (an example of survey nonuniqueness). Inclination distributions that are nonzero at i = 0 while nonphysical appeared to provide a reasonable match to the data.

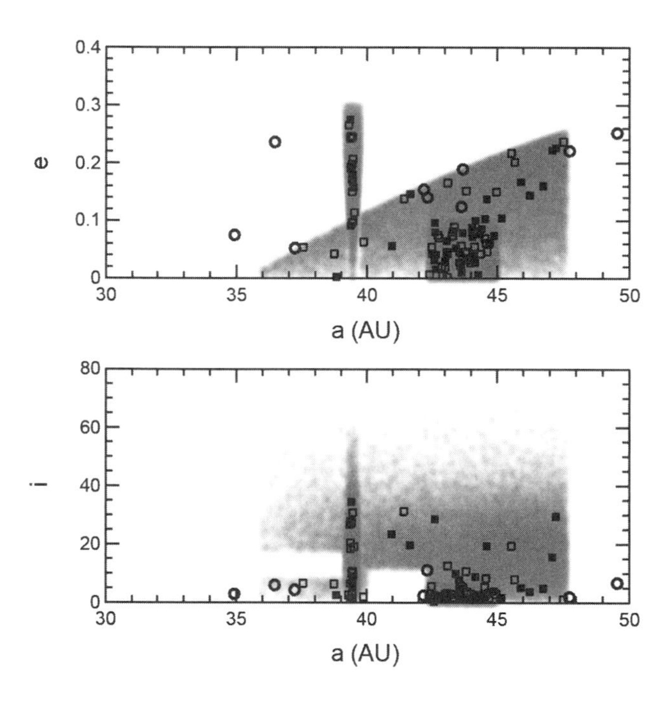

Fig. 4. The CFEPS L3 + Pre-Survey base model of the a/e distribution of material in the classical Kuiper belt plus the Plutino population (taken from *Kavelaars et al.,* 2006). Dots represent the model parent population, open squares represent the simulated observed population, filled squares are the orbital elements of the actual L3 + Pre-Survey detections, open circles represent the orbital parameters of L3 sources that were found to be in various Neptune mean-motion resonances and are not represented by the base model. The processes of "sampling" the base population through the survey simulator is highly stochastic and the sampling shown here is merely representative of one of many realizations performed to evaluate the validity of the model.

Based on the CFEPS survey simulator we find that the inclination distribution of the L3 + Pre-Survey sample, taken as a whole, is well represented by a either a double Gaussian, as in section 3.1.1, or by a flat inclination distribution modified by sin(i); this is very similar to the result found by *Jewitt et al.* (1996) and *Trujillo et al.* (2001b).

When we consider only those KBOs that are found to be part of the "classical" Kuiper belt between 40 < a < 47.5 AU, we find that models with only a single broad inclination distribution are rejected, much as found by *Elliot et al.* (2005) and *Brown* (2001) and in section 3.1.1. We conclude that *Jewitt et al.* (1996) and *Trujillo et al.* (2001b) did not require substructure in the inclination distribution because this structure is only present in the classical Kuiper belt populations.

Using the CFEPS survey simulator we explored the range of "double Gaussian" models allowed for the 40 < a < 47.5 AU zone and find that the results are basically identical to those found in section 3.1.1. Also, as determined via analytic debiasing in section 3.1.1, we find that there is no evidence of a double-component inclination distribution in the Plutino population.

3.2.4. The radial extent of the classical Kuiper belt. The initial explosion of discoveries of KBOs soon pointed to an apparent lack of objects on low-eccentricity orbits with semimajor axis beyond ~50 AU. Initial work on this problem by *Dones* (1997) indicated that, with only a handful of KBOs known at the time and the assumption of a flat luminosity function, a nonzero number of KBOs should have been detected with a > 50 AU if such objects existed. Further Monte Carlo or "survey simulator" analysis in *Jewitt et al.* (1998) indicated that a "classical" Kuiper belt smoothly extending to a > 50 AU was rejected by the available observations. A deep survey by *Gladman et al.* (2001) found a number of objects beyond the 50-AU limit, consistent with a model of radially decaying density. Since none of their detected KBOs beyond 50 AU were on circular orbits, Gladman et al. concluded that the lack of a population of KBOs on circular orbits beyond 50 AU was not inconsistent with their survey results. Further observational work by *Allen et al.* (2001) demonstrated that if the classical belt did extend beyond 50 AU, then the surface density in that region must be exceptionally low. Additional compelling evidence for an edge at ~50 AU has come from survey-simulator-style analysis from *Trujillo et al.* (2001b). More recently, *Hahn and Malhotra* (2005) attempted to construct a pseudo-survey-simulator analysis of the KBO orbits in the Minor Planet Center database, and they remarked that the lack of known KBOs on circular orbits with a > 50 appears to be consistent with a belt that is truncated at a ~ 45 AU.

Using our survey simulator approach we have found that the semimajor axis distribution in the zone between 40 and 47 AU cannot be modeled using a single functional form. We are forced to separate the semimajor axis distributions for the hot and cold components and discuss the radial extent of the Kuiper belt in this context.

3.2.5. The "hot" and "cold" classical Kuiper belt. Guided by the previous analytic and survey simulator efforts as well as plots of the a/e/i distribution of the classifiable KBOs (see chapter by Gladman et al.), we used the CFEPS survey simulator to examine the semimajor axis distribution of the nonresonant KBOs with 30 < a < 50 AU and e < 0.2. Our best-fit model for the "cold" population has semimajor axis distributed between 42 and 45 AU with e uniformly distributed between 0 and 0.1. We also find that the semimajor axis distribution of the "cold" population (i.e., those objects with inclinations drawn from a Gaussian of width 1.5°) is inconsistent with a semimajor axis distribution that extends to a > 45 AU; the falloff in density is very steep beyond a ~ 45. We find the remarkable result that this low-inclination (cold) component appears to be tightly confined to a region between 42.5 < a < 44.5 AU with eccentricities randomly distributed between 0 and 0.1. Those "classical" KBOs in the 42.5–44.5-AU zone with large inclinations, as well as all "classical" KBOs outside this narrow range of semimajor axis, appear to be drawn from a single high-inclination component with eccentricities drawn from a uniform phase-space distribution, P(e) ∝ e. Based on these models, we find that ~35% of the "classical" Kuiper belt resides in a cold low-e/low-i component in the 42.5 < a < 44.5-AU zone, with the remaining ~65% in the "hot" component of the inclination/eccentricity distribution. Thus, we find that there is an enhancement of low-inclination orbits in the 42–45-AU region superposed on a uniform hot population covering all stable phase space from 30 to 47.5 AU and perhaps even to ~60 AU if the detached KBOs are considered to be the large-a extension of the "hot" classical belt.

3.2.6. The Plutino fraction. Nearly contemporaneously with the discovery of the first "Plutinos" *Malhotra* (1993) explained Pluto's "peculiar" orbit as being the result of a resonance trapping caused by the outward migration of Neptune. A mapping of structure and size of the Plutino population provides a strong constraint on this migration model.

Initial estimates of the fractional sizes of various subpopulations of the Kuiper belt were made in *Jewitt et al.* (1998), who reported on the detection of 13 KBOs in a survey of ~51 deg^2 and then compared them to the complete set of observed Kuiper belt objects with Monte Carlo realizations of various models of the belt populations. Using the ≈50 KBOs then known, Jewitt et al. found that 10–30% of the Kuiper belt is made up of 3:2 resonant objects; the lack of 2:1 resonators (at that time) was in mild conflict with models of Neptune's migration. However, Jewitt et al. cautioned that these conclusions were not firmly established by the available observations.

Trujillo et al. (2001b) determine a Plutino fraction by debiasing their detected population using a previously determined scaling relation (*Jewitt et al.,* 1998). *Trujillo et al.* (2001b) determined that the intrinsic ratio of Plutinos to classical belt objects is ~5%. The bias estimate in *Jewitt et al.* (1998), however, assumed that the Plutinos have an inclination distribution like that of the classical belt. In addition, Jewitt et al. had determined the bias fraction for the selection of fields that they observed, and the application of this same bias factor to the fields observed in *Trujillo et al.* (2001b) is not correct.

Hahn and Malhotra (2005) use their pseudo-survey-simulator approach to estimate the intrinsic Plutino fraction based on orbits in the Minor Planet Center database. They find that the observed Plutino fraction is quite low, near just 4% for the faintest Plutinos, and conclude that the intrinsic fraction is far below that predicted by resonance capture models alone.

Using the CFEPS L3 survey simulator we find that an eccentricity distribution of $P(e) \propto \exp((e - e_o)^2/\sigma_e)$ with a cut at e > 0.3 does a satisfactory job in reproducing the observed eccentricity distribution of the L3 + Pre-Survey sample when $e_o \sim 0.15$ and $\sigma_e \sim 0.1$. This eccentricity distribution, combined with an inclination distribution of a single component Gaussian ($P(i) \propto \sin(i)e^{-i^2/2\sigma^2}$) of width like that of the "hot" classical KBOs and a flat distribution of libration amplitudes between 0° and that of Pluto, does an excellent job of reproducing the L3 + Pre-Survey Plutino detections. Based on our Plutino model and comparing with our "hot" + "cold" Kuiper belt model, we find that the Plutino population is ~20 ± 10% the size of the "hot" + "cold" components of the classical Kuiper belt. This larger size for the Plutino population is much more consistent with models of migration of Neptune that normally result in very efficient capture into the 3:2 resonance and agrees well with estimates from *Jewitt et al.* (1998) but starkly contrasts the population estimates in *Trujillo et al.* (2001b) and *Hahn and Malhotra* (2005), who find that only a few percent of the intrinsic population is trapped in the 3:2 resonance with Neptune. We find that models with Plutino fractions as low as 4% are rejected at the 95% confidence level.

The *Trujillo et al.* (2001b) lower estimate of the Plutino fraction is on based their comparison of the entire population of objects they declared as Plutinos with the entire population of objects they declared as classical KBOs without recognizing that the latitude distribution of Plutinos is different from that of the low-inclination classical KBOs. The situation is further complicated: Of the seven objects that Trujillo et al. report as Plutinos, Gladman et al. (see chapter in this book) find that only five are robustly classifiable; two have been lost. Of the five classifiable Plutino candidates two, 1999 CP_{133} and 1999 RW_{215}, are in the 5:4 and 4:3 mean-motion resonance, not the 3:2 as determined from their discovery orbits, thus the sample has shrunk to just three Plutinos. Of the remaining three robustly classified Plutinos, one was found in the "high" ecliptic fields while two classical KBOs were found in those fields. Thus, the observed Plutino fraction in these fields, where the latitude distribution of Plutinos and classical KBOs is more similar than on the ecliptic, is 50% the size of the "hot" classical population, thus the intrinsic Plutino fraction may be much higher than the estimated 5% value.

3.2.7. The 2:1 and other resonances. *Chiang et al.* (2003a,b), *Hahn and Malhotra* (2005), *Lykawka and Mukai* (2007), and the Gladman et al. chapter present methods of searching for the resonance characteristics of KBOs with well-determined orbits. Based on these classification works, an understanding of the importance of high-order resonances is emerging. Many objects that were thought to be scattering off Neptune are trapped in high-order resonance with that planet, perhaps indicating that Neptune's mean-motion resonance swept through an already excited Kuiper belt population (*Hahn and Malhotra,* 2005).

Murray-Clay and Chiang (2005) find that the structure of the 2:1 resonance may play a critical role in determining the timescales for Neptune's migration and conclude that the current population of 2:1 resonators excludes migration timescales shorter than ~10^6 yr. A careful mapping of the structure and populations of the various Kuiper belt resonances will provide many clues to the history of planet formation and evolution in this region. At this time, unfortunately, no strong constraints on the size and distribution of the 2:1 resonance is available. This is an area where future observational work would be fruitful. Unfortunately, the CFEPS projects L3 + Pre-Survey sample contains only one (or perhaps two) 2:1 resonant objects and very few constraints can be placed on this population at this time.

3.2.8. The scattering and detached objects. *Trujillo et al.* (2000) provide an estimate of the size of the "scattering" disk population. Based on their four initial detections, Trujillo et al. conclude that the underlying scattering population must be substantial, likely as large as the classical Kuiper belt population. This conclusion is based on the straightforward observation that scattered disk objects (SDOs) spend only a small fraction of their orbital period in the 30–50-AU zone where they can be detected, and therefore the discovery of four such objects indicates a relatively large parent population. One of the four SDOs reported by Tru-jillo et al. has subsequently been found to be in resonance (see chapter by Gladman et al.), reducing the original estimate of *Trujillo et al.* (2000) by 25%.

Gladman et al.'s chapter reports that 59 of the 541 classifiable KBOs are in either the "scattering" or "detached" orbital class, indicating a rather substantial population of such objects. We do not report here a measure of this interesting population of objects using the two detached and two scattering objects in the CFEPS L3 sample, as our estimates are extremely model dependent. We do note, however, that a large population (at least as large as that of classical belt) of SDOs with pericenters between 36 and 38 AU and semimajor axes extending out to hundreds of AU is consistent with all currently available observations.

4. SUMMARY

Many of the bulk properties of the Kuiper belt are now coming fully to light. Figure 4 provides a good visual representation of the emerging view of the Kuiper belt:

1. The "cold classical" Kuiper belt is a low-i, low-e component tucked into the 42.5 < a < 44.5-AU zone. This very "cold" component represents about 35% of the classical Kuiper belt population. Models where the cold belt extends beyond 45 AU are rejected.

2. The "hot classical" Kuiper belt contains a population of objects weighted toward large e ($P(e) \propto e$) and drawn from a broad inclination distribution that is well-represented by a Gaussian of width ~15°. This "hot" belt appears to uniformly fill the stable orbital phase-space between 35 and 47 AU and the detached objects *may* be the smooth extension of this population to larger a. The "hot" population represents about 65% of the classical Kuiper belt.

3. Many KBOs are trapped in (very) high-order mean-motion resonances with Neptune. Some of these high-order resonators have been mistakenly thought of as "scattered disk" objects. Clearly these high-order resonators provide an important clue to the evolution of the outer solar system. About 10% (open circles in Fig. 4) of the observed CFEPS sample is in this class.

4. The scattering and detached objects, although losing members to more accurate classification, continue to indicate that a substantial population of KBOs must reside in these populations. The debiased distance distribution of scattered objects is flat, indicating a shallow radial dependence (~r^{-1}) in this population. The detached population *may* be seen as a continuation of the hot classical KBOs beyond a ~ 50 AU.

We have demonstrated the strength of utilizing the CFEPS survey simulator to interpret models of the Kuiper belt's underlying populations. Future Kuiper belt surveys that provide such simulators will provide greatly enhanced constraints on the detailed structure of this region of the solar system.

REFERENCES

Allen R. L., Bernstein G. M., and Malhotra R. (2001) The edge of the solar system. *Astrophys. J. Lett., 549,* L241–L244.

Allen R. L., Bernstein G. M., and Malhotra R. (2002) Observational limits on a distant cold Kuiper belt. *Astron. J., 124,* 2949–2954.

Allen R. L., Gladman B., Kavelaars J. J., Petit J.-M., Parker J. W., and Nicholson P. (2006) Discovery of a low-eccentricity, high-inclination Kuiper belt object at 58 AU. *Astrophys. J. Lett., 640,* L83 L86.

Bernstein G. and Khushalani B. (2000) Orbit fitting and uncertainties for Kuiper belt objects. *Astron. J., 120,* 3323–3332.

Bernstein G. M., Trilling D. E., Allen R. L., Brown M. E., Holman M., and Malhotra R. (2004) The size distribution of trans-neptunian bodies. *Astron. J., 128,* 1364–1390.

Brown M. E. (2001) The inclination distribution of the Kuiper belt. *Astron. J., 121,* 2804–2814.

Brown M. E., Trujillo C., and Rabinowitz D. (2004) Discovery of a candidate inner Oort cloud planetoid. *Astrophys. J., 617,* 645–649.

Brown M. E., Trujillo C. A., and Rabinowitz D. L. (2005) Discovery of a planetary-sized object in the scattered Kuiper belt. *Astrophys. J. Lett., 635,* L97–L100.

Brown M. J. I. and Webster R. L. (1998) A search for bright Kuiper Belt objects. *Publ. Astron. Soc. Australia, 15,* 176–178.

Chiang E. I. and Brown M. E. (1999) Keck pencil-beam survey for faint Kuiper belt objects. *Astron. J., 118,* 1411–1422.

Chiang E. I., Lovering J. R., Millis R. L., Buie M. W., Wasserman L. H., and Meech K. J. (2003a) Resonant and secular families of the Kuiper belt. *Earth Moon Planets, 92,* 49–62.

Chiang E. I., Jordan A. B., Millis R. L., Buie M. W., Wasserman L. H., Elliot J. L., Kern S. D., Trilling D. E., Meech K. J., and Wagner R. M. (2003b) Resonance occupation in the Kuiper belt: Case examples of the 5:2 and Trojan resonances. *Astron. J., 126,* 430–443.

Dones L. (1997) Origin and evolution of the Kuiper belt. In *From Stardust to Planetesimals* (Y. J. Pendleton and A. G. G. M. Tielens, eds.), p. 347. ASP Conf. Series 122, San Francisco.

Duncan M. J. and Levison H. F. (1997) A scattered comet disk and the origin of Jupiter family comets. *Science, 276,* 1670–1672.

Elliot J. L. and 10 colleagues (2005) The Deep Ecliptic Survey: A search for Kuiper belt objects and Centaurs. II. Dynamical classification, the Kuiper belt plane, and the core population. *Astron. J., 129,* 1117–1162.

Gladman B., Kavelaars J. J., Nicholson P. D., Loredo T. J., and Burns J. A. (1998) Pencil-beam surveys for faint trans-neptunian objects. *Astron. J., 116,* 2042–2054.

Gladman B., Kavelaars J. J., Petit J.-M., Morbidelli A., Holman M. J., and Loredo T. (2001) The structure of the Kuiper belt: Size distribution and radial extent. *Astron. J., 122,* 1051–1066.

Gladman B., Holman M., Grav T., Kavelaars J., Nicholson P., Aksnes K., and Petit J.-M. (2002) Evidence for an extended scattered disk. *Icarus, 157,* 269–279.

Hahn J. M. and Malhotra R. (2005) Neptune's migration into a stirred-up Kuiper belt: A detailed comparison of simulations to observations. *Astron. J., 130,* 2392–2414.

Irwin M., Tremaine S., and Zytkow A. N. (1995) A search for slow-moving objects and the luminosity function of the Kuiper belt. *Astron. J., 110,* 3082.

Ivezić Ž. and 32 colleagues (2001) Solar system objects observed in the Sloan Digital Sky Survey commissioning data. *Astron. J., 122,* 2749–2784.

Jewitt D., Luu J., and Marsden B. G. (1992) 1992 QB1. *IAU Circular 5611,* 1.

Jewitt D., Luu J., and Marsden B. G. (1993) 1993 RO. *IAU Circular 5865,* 1.

Jewitt D., Luu J., and Chen J. (1996) The Mauna Kea-Cerro-Tololo (MKCT) Kuiper belt and Centaur survey. *Astron. J., 112,* 1225.

Jewitt D., Luu J., and Trujillo C. (1998) Large Kuiper belt objects: The Mauna Kea 8K CCD survey. *Astron. J., 115,* 2125–2135.

Jones R. L. and 16 colleagues (2006) The CFEPS Kuiper belt survey: Strategy and presurvey results. *Icarus, 185,* 508–522.

Kavelaars J. J., Jones L., Gladman B., Petit J., Parker J., and the CFEPS Team (2006) CFEPS [Canada-France Ecliptic Plane Survey]: The details. *AAS/Division for Planetary Sciences Meeting Abstracts, 38,* #44.02.

Larsen J. A., Gleason A. E., Danzl N. M., Descour A. S., McMillan R. S., Gehrels T., Jedicke R., Montani J. L., and Scotti J. V. (2001) The Spacewatch wide-area survey for bright Centaurs and trans-neptunian objects. *Astron. J., 121,* 562–579.

Liller W. (1993) Possible nova in Lupus. *IAU Circular 5867,* 2.

Luu J. X. and Jewitt D. C. (1998) Deep imaging of the Kuiper belt with the Keck 10 meter telescope. *Astrophys. J. Lett., 502,* L91.

Luu J., Jewitt D., and Marsden B. G. (1993) 1993 FW. *IAU Circular 5730,* 1.

Luu J.-X., Marsden B. G., Jewitt D., Truijillo C. A., Hergenrother C. W., Chen J., and Offutt W. B. (1997) A new dynamical class in the trans-neptunian solar system. *Nature, 387,* 573–575.

Lykawka P. S. and Mukai T. (2007) Origin of scattered disk resonant TNOs: Evidence for an ancient excited Kuiper belt of 50 AU radius. *Icarus, 186,* 331–341.

Malhotra R. (1993) The origin of Pluto's peculiar orbit. *Nature, 365,* 819.

Marsden B. G. (1994) 1993 RO, 1993 RP, 1993 SB, 1993 SC. *IAU Circular 5985,* 1.

Murray-Clay R. A. and Chiang E. I. (2005) A signature of planetary migration: The origin of asymmetric capture in the 2:1 resonance. *Astrophys. J., 619,* 623–638.

Petit J.-M., Holman M. J., Gladman B. J., Kavelaars J. J., Scholl H., and Loredo T. J. (2006a) The Kuiper belt luminosity function from $m_R = 22$ to 25. *Mon. Not. R. Astron. Soc., 365,* 429–438.

Petit J.-M., Gladman B., Kavelaars J. J., Jones R. L., Parker J., and Bieryla A. (2006b) The Canada-France Ecliptic Plane Survey: First (L3) data release. *AAS/Division for Planetary Sciences Meeting Abstracts, 38,* #44.01.

Sheppard S. S., Jewitt D. C., Trujillo C. A., Brown M. J. I., and Ashley M. C. B. (2000) A wide-field CCD survey for Centaurs and Kuiper belt objects. *Astron. J., 120,* 2687–2694.

Tholen D. J., Senay M., Hainaut O., and Marsden B. G. (1994) 1993 RO, 1993 RP, 1993 SB, 1993 SC. *IAU Circular 5983,* 1.

Tombaugh C. W. (1961) The trans-neptunian planet search. In *Planets and Satellites* (G. Kuiper and B. Middlehurst, eds.), p. 12. Univ. of Chicago, Chicago.

Trujillo C. A. and Brown M. E. (2001) The radial distribution of the Kuiper belt. *Astrophys. J. Lett., 554,* L95–L98.

Trujillo C. A. and Brown M. E. (2003) The Caltech wide area sky survey. *Earth Moon Planets, 92,* 99–112.

Trujillo C. and Jewitt D. (1998) A semiautomated sky survey for slow-moving objects suitable for a Pluto Express mission encounter. *Astron. J., 115,* 1680–1687.

Trujillo C. A., Jewitt D. C., and Luu J. X. (2000) Population of the scattered Kuiper belt. *Astrophys. J. Lett., 529,* L103–L106.

Trujillo C. A., Luu J. X., Bosh A. S., and Elliot J. L. (2001a) Large bodies in the Kuiper belt. *Astron. J., 122,* 2740–2748.

Trujillo C. A., Jewitt D. C., and Luu J. X. (2001b) Properties of the trans-neptunian belt: Statistics from the Canada-France-Hawaii Telescope survey. *Astron. J., 122,* 457–473.

Williams I. P., Fitzsimmons A., and O'Ceallaigh D. (1993) 1993 SB and 1993 SC. *IAU Circular 5869,* 1.

Size Distribution of Multikilometer Transneptunian Objects

J-M. Petit
Observatoire de Besançon

JJ Kavelaars
Herzberg Institute of Astrophysics

B. Gladman
University of British Columbia

T. Loredo
Cornell University

There are two main goals in studying the size distribution of the transneptunian objects (TNOs). The first is the quest to determine the mass in the transneptunian region. The second is to understand the competition between accretion and collisional erosion. The size distribution of the largest bodies is controlled by the accretion process, while that of bodies smaller than 50–100 km in diameter is believed to result from collisional evolution. An accessible way to determine the size distribution of TNOs is to determine their luminosity function (LF) and then try to convert magnitude to size. Interpreting the survey data to determine the correct LF, with confidence region for the parameters, is a subtle problem and is only beginning to be properly understood. Converting the LF into a size distribution is very complex and involves modeling, both dynamical and of physical surface properties of the TNOs. Several papers have been published that address this question, yielding LF slope of 0.3 to 0.9, and 1 object per deg^2 brighter than R magnitude 23–23.5. The exponent of the size distribution is most likely on the order of 4–5 for bodies larger than a few tens of kilometers, and the number of objects with diameter larger than 100 km is on the order of a few 10^4. However, this subject is still in its infancy, and much observational and modeling work needs to be done before we claim to know the size distribution of the various populations of TNOs.

1. INTRODUCTION

In this chapter, we consider another aspect of transneptunian object (TNO) discovery called the luminosity function (LF), which we then relate to the size distribution of these small bodies. Discovering TNOs goes beyond the simple fact of finding yet other small bodies in the solar system. Our ultimate goal is to understand, through the knowledge of the current small-body populations, the formation and evolution processes of our solar system, and potentially of other planetary systems. Much of the motivation for observational and cosmochemical studies of small bodies stems from the desire to use the results to constrain or otherwise illuminate the physical and chemical conditions in the early solar system, in the hope of learning more about the processes that led to the formation of our planetary system. As will be seen in the chapter by Kenyon et al., the size distribution of the TNO population holds clues to the process of giant-planet and small-body formation and the collisional evolution of the latter.

Large bodies are most likely immune to collisional disruption over the age of the solar system in the current Kuiper belt environment (*Farinella and Davis,* 1996), and their size distribution is therefore directly linked to the accretion processes. Even in the denser collisional environment of the early solar system, these bodies must have been safe, as the mass depletion of the belt must have resulted from dynamical erosion rather than collisional erosion (*Petit and Mousis,* 2004; *Morbidelli and Brown,* 2004). Smaller than 50–100 km in diameter, the TNOs should have suffered strong collisional evolution, and their current size distribution is connected to their physical properties and collisional environment (*Davis and Farinella,* 1997).

2. HISTORICAL REVIEW

The long history of asteroid observations designed to explore the asteroid main-belt size distribution in order to study collisional physics extends naturally to the Kuiper belt. Even after more than a decade of Kuiper belt exploration, the value of the fundamental property of the belt's mass still varies in the literature. The asteroid belt's size distribution is decently approximated by power laws over certain diameter ranges. Assuming this also holds for the

Kuiper belt, astronomers have tried to estimate the mass in objects of a given (observable) size, then use the slope of the apparent magnitude distribution to estimate the slope of the diameter distribution and finally estimate the Kuiper belt mass by integrating the power law from the reference point to upper and lower limits.

Firm determination of a reference object size requires either resolving objects in the optical (thus directly measuring diameter) or obtaining both optical and thermal infrared observations to use thermal modeling to estimate a diameter. Only three of the largest TNOs (Pluto, Quaoar, and Eris, formerly 2003 UB_{313}) have been resolved, all using the Hubble Space Telescope (HST) (*Albrecht et al.,* 1994; *Brown and Trujillo,* 2004, 2006); the diameter estimates for Eris thus obtained are still only accurate at the 5% level. Pluto and Eris are shown to have albedos more than an order of magnitude larger than the value of 4% often used in the literature. The promising avenue of detection of the thermal IR flux from large TNOs has yielded mixed results, as several of the known large TNOs have not been detected by the Spitzer Space Telescope (see chapter by Stansberry et al.), presumably also because most TNOs have albedos far above the 4% figure. Only the most nearby (and thus warmest) large TNOs have yielded diameter estimates from their IR emission.

That determining the size distribution was necessary for measuring the Kuiper Belt mass was recognized at the time of the discovery of 1992 QB_1 (*Jewitt and Luu,* 1993). With additional discoveries, the measurement of the so-called luminosity function, or LF, has become an important goal of observational Kuiper belt surveys. The LF simply gives either the cumulative or incremental number of TNOs brighter than a given apparent magnitude; it is often given relative to a surveyed area of 1 deg^2. The usual functional form used is an exponential for the cumulative LF like

$$\Sigma(m) = 10^{\alpha(m - m_o)} \quad (1)$$

with α being the slope and m_o the magnitude at which one would expect to have 1 object per deg^2 of sky. Conversion of the LF to a size distribution requires certain assumptions (discussed below), which can lead to a power law. But even the measurement of the apparent magnitude distribution requires careful analysis of sky surveys.

Determining the slope of the apparent magnitude distribution requires a reasonable number of TNOs to be discovered in a survey of known area and known sensitivity. In order to be of any use for further modeling and/or comparison with other works, *the surveys need to publish their areal coverage, the TNO magnitudes (with errors), and (very importantly) the characterization of their detection efficiency as a function of magnitude* (at least; giving it as a function of other observing parameters like the rate of motion can be very useful too) *for each portion of their discovery fields.* Actually, the publication of the full information necessary for the reader to be able to redo the work is mandatory in a scientific publication. Otherwise, we are no longer using a scientific approach, but are merely relying on faith.

In the following we list the works that explicitly addressed the question of LF determination. We separate the surveys between those that satisfied the above requirements (Table 1) and those that did not (Table 2). In the first category, we have the work by *Jewitt and Luu* (1995), *Irwin et al.* (1995), *Jewitt et al.* (1998), *Gladman et al.* (1998, 2001), *Chiang and Brown* (1999), *Trujillo et al.* (2001a,b), *Bernstein et al.* (2004), *Petit et al.* (2006), and Fraser et al. (personal communication, 2007), whose main characteristics are summarized in Table 1. We divided them into two categories: surveys in which the objects are visible on each individual frames (wide-area surveys), and surveys in which several images were stacked together after shifting to reveal the objects (small-area deep surveys).

For the sake of completeness, we also list the surveys that addressed the LF determination, but did not meet the above requirements: *Jewitt et al.* (1996), *Luu and Jewitt* (1998), *Sheppard et al.* (2000), *Larsen et al.* (2001), and *Elliot et al.* (2005). The first two works did not publish their efficiency function, while the last three only sparsely sampled the efficiency function on a few frames and/or did not provide the information necessary to match efficiency functions to specific sky coverage. Table 2 gives the characteristics of this second set of surveys.

Other KBO surveys have been performed over the years, but were intended to simply find objects, and/or determine dynamical information, not LF, and are thus not described in this chapter. However, the most important ones, which were used as constraints in the works presented here, are *Tombaugh* (1961) (T61), *Luu and Jewitt* (1988) (LJ88), *Kowal* (1989) (K89), *Levison and Duncan* (1990) (LD90), and *Cochran et al.* (1995) (C95). T61 and K89 were photographic plate surveys, LJ88 and LD90 were groundbased CCD surveys, and C95 was a space-based, HST survey.

2.1. Wide-Area Surveys of Transneptunian Objects

Large-scale surveys typically cover from several square degrees (deg^2) up to a few thousand, reaching a limiting m_R magnitude of 24 or brighter. The goal is to detect a large number of objects in each of a small number of CCD images taken of the same sky region at 1- to 24-h spacing. They generally use detections on single images and search for objects whose measured position changes from frame to frame at rates consistent with outer solar system targets. The relatively bright targets detected are then suitable for tracking over the several-year baseline needed to determine an orbit; however, under certain assumptions, knowledge of the orbit is not required to determine the LF. Such survey can essentially provide an estimate of the "zeropoint" (at which magnitude m_o there is one object per deg^2 brighter than m_o) and the slope.

A major potential complication of such an approach is that there are good reasons to expect that the on-sky surface density will vary with ecliptic latitude and longitude. A change in m_o (to fainter magnitude) is expected as one departs from the plane of the solar system as the spatial den-

TABLE 1. List of past characterized luminosity function (LF) surveys.

Reference	Abbrev.	Ω* (deg²)	N†	η_{max}‡	m_{50}§	R.A.¶ (2000)	l** (deg)	Comments
Wide-Area Surveys								
Jewitt and Luu (1995)	JL95	1.2	7	1.	24.8	21:30–01:10	0–5	Fall
						10:00–15:10	0–5	Spring
Irwin et al. (1995)	ITZ95	0.7	2	1.	23.5	—	0–10	
Jewitt et al. (1998)	JLT98	51.5	13	0.91	22.5	23:50–02:10	0–5	Oct. 1996
						07:30–10:40	0–5	Feb. 1997
Trujillo et al. (2001a)	TJL01	73	86	0.83	23.7	08:00–14:00	−10, 0, 10	1999
						21:20–01:00	0, 20	Mar. 2000
Trujillo et al. (2001b)	T01	164	4	0.85	21.1	22:18–01:25	0–12	
						09:00–12:05	0–5	
Petit et al. (2006)	P06	5.97	39	0.90	24.6	21:08–21:17	0–1.9	Uranus
		5.88	26	0.90	24.2	20:17–20:26	0–1.7	Neptune
Deep Surveys								
Gladman et al. (1998)	G98	0.25	2	1.	24.6	11:50	0	CFHT/8K
		0.175	3	1.	25.6	23:00, 00:10	0, 4.5	5-m Hale
Chiang and Brown (1999)	CB99	0.009	2	1.	27.0	22:55	0.5	
Gladman et al. (2001)	G01	0.27	17	1.	25.9	09:32	2.6	CFHT/12K
		0.012	0	1.	26.7	19:24	1.0	VLT/FORS1
Bernstein et al. (2004)	B04	0.019	3	1.	28.7	14:08	1.5	
Fraser et al. (2007)	F07	0.64	6	0.96	25.4	21:40	−0.7	CFHT/12K
		0.85	19	0.97	25.7	22:24	−0.8	MEGAPrime
		0.76	14	0.92	25.4	20:39	1.3	CTIO/Blanco

*Actual search area of the survey.
†Number of TNOs used for LF determination in that work.
‡Maximum efficiency of the survey.
§R magnitude at which efficiency drops to 50% of its maximum value.
¶Range of right ascension.
**Range of ecliptic latitude.

TABLE 2. List of past luminosity function (LF) surveys (not meeting our requirements).

Reference	Abbrev.	Ω* (deg²)	N†	η_{max}‡	m_{50}§	R.A.¶ (2000)	l** (deg)	Comments
Jewitt et al. (1996)	JLC96	4.4	3	—	23.2	12:15–16:00	0–20	CTIO 1.5-m
		3.9	12	—	24.2	08:30–00:40	0–5	UH 2.2-m
Jewitt and Luu (1998)	JL98	0.28	5	—	26.1	—	—	Keck wide
		0.028	1	—	26.6	—	—	Keck deep
Sheppard et al. (2000)	S00	1428	0	0.92	18.8	07:00–12:00	0–20	0.5-m APT
Larsen et al. (2001)	L01	550.1††	8	0.97	21.5	00:00–24:00‡‡	0–5	SpaceWatch
Elliot et al. (2005)	E05	~500§§	512¶¶	0.96§§	22.0§§	00:00–24:00††	0–5	

*Actual search area of the survey.
†Number of TNOs used for LF determination in that work.
‡Maximum efficiency of the survey.
§R magnitude at which efficiency drops to 50% of its maximum value.
¶Range of right ascension.
**Range of ecliptic latitude.
††Effective area on the ecliptic, correcting for density decrease at large ecliptic latitudes; see L01 for details.
‡‡Regions close to the galactic plane were not included in this survey.
§§Values estimated from Fig. 15 of E05; magnitude refers to the V–R filter.
¶¶TNOs only, no Centaurs or objects closer than 30 AU.

sity of the thin belt drops off. Since most surveys have been near the ecliptic plane, this effect might be thought to be small (but see below). Similarly, the existence of resonant populations means that not all longitudes are equal in a flux-limited survey. Certain longitudes relative to Neptune (which dominates the resonant structure) are the preferred pericenter locations of each mean-motion resonance; e.g., longitudes 90° ahead of and behind Neptune are the preferred pericenter locations for the 3:2 mean-motion resonance (*Malhotra,* 1996; *Chiang and Jordan,* 2002). Thus, surveys directed at these locations will discover more TNOs, since the much more abundant small objects from the size distribution become plentiful in the survey volume. Therefore, the interpretation of large-area surveys is very complex.

2.2. Deep Small-Area Surveys of Transneptunian Objects

Deep surveys cover only a fraction of a square degree of sky, and reach an R magnitude fainter than 24.5. They combine a large number of frames of a confined region of sky, shifting them according to the typical rate of motion of TNOs in the sky in order to discover objects with low signal-to-noise in a given frame. In the combined image, the signal from objects at the assumed rate and sky direction adds constructively to give detectable signal with a confined PSF, while stars and other fixed objects will trail. This technique is often called *pencil-beam*, analogous to extragalactic studies, since it is capable of probing to objects at great distance.

2.3. Size Distribution Determination

Two methods have been used through the years to determine the size distribution. The first relies on determining the LF, which is certainly more directly accessible and requires little, if any, modeling. The second was a direct modeling of size distribution and comparison with observations. Three of the works mentioned above restricted themselves to LF determination (ITZ95, G98, B04) only.

ITZ95 first assumed a differential power-law size distribution

$$n(r) \propto r^{-q} \qquad (2)$$

and fixed albedo and showed that the absolute magnitude distribution would follow a form given by equation (1). Next, with proper assumptions (not explicitly given, but hinting at a power-law dependence) on the heliocentric distance dependence of the number density of TNOs, they linked it to an exponential LF. The correspondence between the indices of these functional forms is

$$\alpha = (q - 1)/5 \qquad (3)$$

After this, they determined the LF of apparent magnitude, and converted it into an absolute magnitude LF for the purpose of comparison with previous works and other populations. No further mention was made of the size distribution.

G01 showed that a simple assumption of power-law behavior of the distance distribution of the object is sufficient to derive equation (3). In this way, the increment in number of objects when reaching 1 mag fainter is independent of the distance considered in a flux-limited survey. This holds as long as the survey is not wide enough that it samples the size where there is only one object, nor deep enough that it reaches small enough objects for which one would expect to have a different index in equation (2) (see below). This direct connection between luminosity and size distribution provides a tangible connection between the observation and physical property being sought and at only the cost that the LF must be a uniform single exponential.

CB99, G01, P06, and F07 used this relation to give a size distribution mostly for comparison purposes or as a mean of deriving other quantities of interest such as the mass of the belt. CB99 also estimated the number of 1–10-km-sized comet progenitors in the Kuiper belt and found it compatible with the estimate from *Levison and Duncan* (1997) to supply the rate of Jupiter-family comets.

JL95 estimated the size distribution using a Monte Carlo simulation of their survey. For this, they used a very simple model for the Kuiper belt, assuming power-law size and heliocentric distributions and estimated a limiting value on q from comparing a graph of the expected detection to the actual ones. JLT98 further refined this method to determine the LF of their survey, either alone or together with previous ones, in parallel. For the size distribution determination, they used a two-population model, with classical KBOs (CKBOs) and Plutinos (bodies in the 3:2 mean-motion resonance with Neptune). According to a rough description of their survey, they selected the objects from the model that would have been observed. The intrinsic population of Plutinos was adjusted to reproduce the apparent fraction of Plutinos in their survey ($\simeq$35%). Finally, they compared the binned differential LF to the observed one. As expected, their best-fit index of the size distribution was roughly related to the LF slope by the relation derived by ITZ95. This work was extended to larger surveys by JLT01 and T01.

2.4. Mass of the Belt, Distant Belt, Largest Body, etc.

In many cases, the size distribution was only a step toward determining other quantities of interest like the mass of the belt, the existence of an outer edge, or the largest body one should find. However, these generally require some extra assumption to be derived. For example, as showed by G01, estimating the mass of the belt requires knowledge of the radial extent of the belt and the size at which the size distribution becomes shallower (see below). JLT98 determined the mass of the belt from bodies larger than 50 km in radius between 30 and 50 AU, excluding the scattered TNOs, to be ~0.1 $M_\oplus$. Interestingly, TJL01, using the same parameters and size-distribution slope, found the mass of the belt to be ~0.06 $M_\oplus$. T01 estimated that the mass due to bodies larger than 500 km is ~1/5 of the previous value. CB99 determined the mass of the belt inside 48 AU, from bodies brighter than $m_R = 27$, to be ~0.2 $M_\oplus$. S00 did the same for Centaurs larger than 50 km and found a mass of ~10^{-4} $M_\oplus$. G01 gave the mass of the belt as a function of the mass at which q becomes smaller than 4 and quote a value of ~0.1 $M_\oplus$. Selecting a smaller population of TNOs, namely the *classical Kuiper belt* with inclination $i < 5°$, B04 gave a smaller mass of ~0.01 $M_\oplus$.

The size distribution was also used to estimate the fraction of objects that one should detect further out than 50 AU. But this again requires some assumptions on the plausible distance distribution. G98 found that the lack of detection of distant objects in their survey was to be expected, independent of the presence of an edge of the Kuiper belt. Later works (G01, TJL01, and B04) showed, however, that the lack of detections at large distances was consistent with an edge of the large-body belt, only allowing a significance mass in bodies smaller than ~40 km outside 50 AU. G01, however, raised the problem of the lack of detection of scattered disk objects, which are known to be present in that region.

P06 used equation (3) to assess the reality of the depletion of distant objects from their LF with the *Trujillo and Brown* (2001) method.

3. SIZE DISTRIBUTION VERSUS (APPARENT) MAGNITUDE DISTRIBUTION

As mentioned before, we are interested in the size distribution of this small-body population rather than just the LF. The LF is simply an initial proxy to the size distribution.

3.1. Converting from Magnitude to Size

We first review the different factors that connect the size of an object to its apparent (or measured) magnitude. The apparent magnitude of a TNO can be represented as

$$m = m_\odot - 2.5 \log \left[\frac{\nu r^2 \phi(\gamma) f(t)}{2.25 \times 10^{16} R^2 \Delta^2} \right] \quad (4)$$

where $m_\odot$ is the apparent magnitude of the Sun in the filter used for observations [–26.92 in the AB system, for Bessel R or KPNO R filter, *www.ucolick.org/cnaw/sun.html*, from *Bruzual and Charlot* (2003) and *Fukugita et al.* (1995)]; ν the geometric albedo in the same filter; r the radius of the object (expressed in km); γ the phase angle, i.e., the angle Sun-object-observer; $\phi(\gamma)$ the phase function (equal to 1 for $\gamma = 0$); f(t) the rotational lightcurve function; and R and Δ are the heliocentric and geocentric distances (expressed in AU). R, Δ, and γ depend only on the geometry of the observation and are due to the orbits of the object and Earth around the Sun. ν, r, $\phi(\gamma)$, and f(t) depend on the physical and chemical properties of the object itself.

f(t) is typically a periodic function of time with a rather short period (few hours to few tens of hours) and moderate amplitude variations (for large TNOs) with mean value of 1. For asteroids, the amplitude tends to be larger for smaller objects presumably due to greater relative departure from sphericity. For TNOs, the trend will probably be the same, although with possible large departures from the general trend (for example, 2003 EL_{61} has a lightcurve amplitude of 0.3, while some smaller objects have no lightcurve). According to *Lacerda and Luu* (2006), 30% of the objects have a lightcurve amplitude $\Delta m \geq 0.15$ and 10% with amplitude $\Delta m \geq 0.4$. So the typical lightcurve amplitude is on the order of the uncertainty on magnitude estimates of the corresponding TNOs for large-area surveys. The lightcurve is due to the rotation of the object and either or both an elongated shape and surface features and albedo variations. $\phi(\gamma)$ is a function of the physical and chemical properties of the surface of the TNO. It often manifests itself by an opposition surge, i.e., a nonlinear increase in surface brightness that occurs as the phase angle decreases to zero. Two causes to give rise to the opposition effect are usually considered: (1) shadow-hiding and (2) interference-enhancement, often called coherent-backscatter. Some general regolith property-dependent characteristics of each mechanism are understood, and some papers are devoted to a discussion on the relative contribution of both mechanisms (*Drossart,* 1993; *Helfenstein et al.,* 1997; *Hapke et al.,* 1998; *Shkuratov and Helfenstein,* 2001). The width and amplitude of the opposition surge depends on the dominant mechanism, shadow-hiding giving a narrower and brighter opposition surge. One can check for the effect of coherent backscatter and/or shadow-hiding by studying the influence of wavelength dependence on the opposition brightening.

To determine the size distribution from the reflected light from the source we must first remove the geometrical effects. One first computes the absolute magnitude: the apparent magnitude the object would have at a heliocentric and geocentric distances of 1 AU, neglecting phase corrections and assuming the object is visible only by reflected sunlight. The absolute magnitude (*Bowell et al.,* 1989) corresponds to $\Delta = R = 1$, $\gamma = 0$, i.e., $\phi(\gamma) = 1$, and averaging over one rotational period

$$\begin{aligned} H &= \langle m \rangle + 2.5 \log \left[\frac{\phi(\gamma)}{R^2 \Delta^2} \right] \\ &= m_\odot - 2.5 \log \left[\frac{\nu r^2}{2.25 \times 10^{16}} \right] \end{aligned} \quad (5)$$

The heliocentric and geocentric distances are easily determined with an accuracy of about 10% at discovery, even with an arc of just 1 day. With a few follow-up observations at 2 months, 1 year and 2 years after discovery, the distance can be estimated with a precission of 1% (2-month arc) or less than 0.1% (1-year arc or more). The uncertainty regarding the distance gives an error of 0.4 mag on HR at discovery time, which is then easily reduced by a factor of 10 or 100.

Accounting for the rotational lightcurve and the phase effect requires many more observations. The object must be observed during one or more full rotational periods, at a given phase angle, to determine the rotationally averaged magnitude at that phase angle $\langle m \rangle$. Although a few bright objects have had their rotational periods determined in this way, such observations are impractical for large-scale surveys and for objects at the limit of detection of these surveys. Hence this effect is often omitted altogether, or modeled with a simple fixed-amplitude periodic function.

The next phenomenon to account for is the phase effect. We need to know the variation of $\phi(\gamma)$ between zero phase angle and the actual observation angle. Modeling this variation is still in its infancy, and one usually resorts to empirical formulae that were developed for asteroids [H–G formalism from *Bowell et al.* (1989)], or simple linear approximations (*Shaefer and Rabinowitz,* 2002; *Sheppard and Jewitt,*

2002), both of which fail to reproduce the strong and narrow opposition surge at very small phase angle that has been detected for several TNOs and Centaurs (*Rousselot et al.,* 2006). Linear and H–G formalisms tend to underestimate the magnitude at zero phase angle by up to 0.1–0.2 mag.

The last parameter needed is the geometric albedo of the object. The only model-independent method to determine the albedo is to actually directly measure the size and brightness of the object. Since this is not possible, the next best thing to do is to try and measure the brightness of the object both in the visible and in the thermal infrared, and use some thermal modeling of the object. Knowing the visual band brightness and distance of the object gives a one-parameter family of solutions for the size, parameterized by the albedo. The thermal infrared flux gives another, independent, family of solutions. The intersection of the two families gives an estimate of the size and the albedo. The resulting estimate is only as good as the thermal model used to derive it. The uncertainty is at least a factor of 2 in surface area if pole position is unknown (pole-on vs. equatorial) (see chapter by Stansberry et al.). Even measuring the thermal flux, however, is very difficult and possible only for the biggest objects, requiring the use of the largest and most sensitive instruments available. Hence several of the biggest objects have been assigned different size and albedo from groundbased observations and from Spitzer [1999 TC_{36} (*Altenhoff et al.,* 2004; *Stansberry et al.,* 2006); 2002 AW_{197} (*Cruikshank et al.,* 2005)].

Because of all these difficulties, not all authors have dared to convert from apparent magnitude distributions to size distributions and those who do must use many assumptions and simplifications. For example, in all the works presented before, the rotational lightcurve has been completely neglected. This can be partly justified as the largest number of objects detected in a given survey is usually close to the limiting magnitude of that survey's detection. In fact, half of the objects are generally within the 0.5–1 mag at the faint end of the survey. For these faint objects, the uncertainty of the magnitude measurement is on the order of or even larger than the expected amplitude of the rotational lightcurve (*Lacerda and Luu,* 2006). Note, however, that this can introduce a bias if the large-amplitude objects are detected only at their brightest rotation phase.

3.2. The Limits of Power-Law Distributions

All the previous works have used exponential functions for their LF and power-law functions for their size distributions. Classically, scientists look for scale-free, self-similar functions to represent physical phenomena that do not have an obvious scale. This is particularly true when those phenomena extend over several decades of the governing parameter, such as the size or mass of a small body, distances, or stellar masses. Another driver in choosing the functional representation in modeling is the need to combine functions from different parts of the models while still having an easy-to-use function. Both power-law and exponential functions satisfy this requirement of keeping the same form when combined. Finally, in several instances in astrophysics, plotting data on double logarithmic graphs results in aligned points. One is then tempted to represent such data as a straight line, yielding a power-law function in the original variables.

In the case of the size distribution of small bodies, the work of *Dohnanyi* (1969) has been responsible for the widespread use of power-law distributions. Dohnanyi has shown that under very strong assumptions regarding the effects of hypervelocity collisions [the main assumption, that the collisional process is scale-independent, has been proven to be wrong, i.e., *Benz and Asphaug* (1999)], a quasi-steady-state distribution is reached. The final distribution is the product of a slowly decreasing function of time and a power-law of index 11/6 for the differential mass distribution [$n(M) \propto M^{-11/6}$, M being the mass of the asteroid], corresponding to an index $q = 3(11/6) - 2 = 7/2$ for the differential size distribution. The resulting "equilibrium power-law slope" is mostly due to the adoption of a power-law functional form to model the outcomes of hypervelocity collision (*Gault et al.,* 1963). Since the size of fragments in fragmentation experiments span several orders of magnitudes, one tends to show the logarithm of the number of fragments vs. the logarithm of their size. Fragmentation processes being random in nature, this is usually a rather scattered plot, with some kind of trend in it (see, e.g., *Giblin et al.,* 1994, 1998; *Ryan et al.,* 1999). Using a power law here is a very rough approximation.

Likewise, the observed LF of the asteroid belt show a wave-like structure superposed over an approximately power-law trend. Although the general trend of the size distribution of fragments and the LF of the asteroid belt can be roughly approximated by a power law and an exponential, the details may depart noticeably from these models.

At the small end of the size spectrum, a problem arises depending on the value of the index q. The mass of objects with size in the range $r_{min} < r < r_{max}$ is

$$\begin{aligned}\mathcal{M}(r_{min}, r_{max}) &= \int_{r_{min}}^{r_{max}} n(r)M(r)dr \\ &= \frac{4\pi\rho_v A}{3}\int_{r_{min}}^{r_{max}} r^{3-q}dr \\ &= \frac{4\pi\rho_v A}{3(4-q)}\left[r_{max}^{4-q} - r_{min}^{4-q}\right]\end{aligned} \qquad (6)$$

where A is the normalizing constant of the differential size distribution, M(r) the mass of an object of radius r, and ρ_v its volume density, $q \neq 4$. When $q > 4$, the total mass diverges at small sizes (G01). Most of the surveys presented above have found that q is on the order of, but likely larger than, 4, hence there clearly needs to be a limit to the power-law size distribution at some small size r_k beyond which a lower size index is required. This change in q was proposed as soon as the surveys suggested a rather large value of q, since astronomers expected to have q = 3.5 at small sizes, where they assumed a collisional equilibrium would have been reached.

As the B04 survey began to reach very faint objects, they started to see a departure from the uniform exponential LF, which they attributed to the expected change in size distribution shape. They first modeled this change using a rolling exponential

$$\Sigma(m) = \Sigma_{23} 10^{\alpha(m-23) + \alpha'(m-23)^2} \quad (7)$$

with Σ_{23} being the sky density of objects at magnitude 23. They also investigated a double exponential fit, as the harmonic mean of two exponentials

$$\Sigma(m) = (1 + c)\Sigma_{23}[10^{-\alpha_1(m-23)} + c10^{-\alpha_2(m-23)}]^{-1} \quad (8)$$

$$c \equiv 10^{(\alpha_2 - \alpha_1)(m_{eq} - 23)} \quad (9)$$

The asymptotic behavior is an exponential of indices α_1 at one end of the size spectrum and α_2 at the other end, with the two exponentials contributing equally at m_{eq}.

Equation (3) was widely used in all those works, but relied on a constant albedo for all objects. JLT98 noticed that there seems to be a variation of albedo with size, ranging from 0.04 for the small bodies to 0.13 for 2060 Chiron to 0.6 for Pluto. Regardless, they used a fixed albedo in their derivation of the size distribution. F07 explore the effect of a varying albedo on equation (3) by examining the effect of a power-law albedo $\nu \propto r^{-\beta}$ as a toy model (here again, a power law is used for computational conveniency). Equation (3) then becomes

$$q = 5(\alpha - \beta/2) + 1 \quad (10)$$

The albedo of Pluto is 0.6 for a size of r ~ 1000 km (*Albrecht et al.,* 1994), while smaller objects (r ~ 100 km or smaller) seems to be $\nu \sim 0.06$ with large fluctuations (*Grundy et al.,* 2005; chapter by Stansberry et al.), implying $\beta < 0$, possibly down to $\beta \sim -1$. From Figure 3 of the chapter by Stansberry et al., the situation can be even more complex, with an albedo almost independent of size for diameters smaller than 200–300 km, and a very steep rise of the albedo for objects larger than 1000 km. In any case, an estimate of q that assumes $\beta = 0$ potentially underestimates the steepness of the size distribution by up to 1 or 2.

4. HOW TO BEST ESTIMATE THE POPULATION CHARACTERISTICS OF TRANSNEPTUNIAN OBJECTS

A survey may be characterized by the angular region surveyed, the survey efficiency function, and, for each detected object, an estimate of its apparent magnitude and uncertainty. We want to use these data to infer properties of the TNO population as a whole. Most directly, we may seek to estimate the LF. Less directly, but of more direct physical interest, we may seek to estimate the size distribution of TNOs, or the distribution of orbital elements.

To infer properties of the TNO population using data from multiple surveys is nontrivial for several reasons. For a particular survey, the analysis must account for selection effects (which cause some parts of the population to be over- or underrepresented in the sample with respect to others) and measurement error (which distorts the magnitude distribution shape as objects "scatter" in magnitude). To combine surveys, the analysis must consider systematic differences between surveys (e.g., due to use of different bandpasses). These complications cannot be removed in a model-independent way; there is no such thing as a unique "debiased" survey summary. The nature and amount of the distortions depends on the true population distribution, so biases cannot be accounted for without making assumptions about the distribution. The analyst must make some modeling assumptions, explicitly or implicitly, and account for the distortions in tuning the model.

These challenges are hardly unique to TNO studies. They arise and have received significant attention in analyses of the magnitude or flux distributions of stars, optical and radio galaxies, X-ray sources, γ-ray bursts, and AGN, to name a few notable examples (*Loredo and Wasserman,* 1995; *Drell et al.,* 2000; *Loredo and Lamb,* 2002). Similar challenges are widely and deeply studied in the statistics literature, in the field of survey sampling. In each discipline, early analyses rely on intuitively appealing but fundamentally flawed methods such as least-squares analysis of binned or cumulative counts (possibly weighted) or adjusted sample moments. As sample sizes grow, the need for greater care is gradually recognized, and methodology matures. In all the fields mentioned, attention has gradually converged on likelihood-based methods (at least for analysis with parametric models).

The basic idea behind likelihood-based methods is that hypotheses that make the observed data more probable should be preferred. Thus the central quantity of interest is the *likelihood function,* the probability for the observed *data,* considered as a function of the hypotheses under consideration.

We need to use the likelihood function to quantify our uncertainty about the *hypotheses,* and there are rival approaches for creating confidence statements about hypotheses using the likelihood function. Perhaps the best-known approach, in the case of parameter estimation (where the hypotheses are indexed by values of continuous parameters), is to draw likelihood contours at levels chosen to givea desired frequency of coverage of the true parameter values ("confidence level") in repeated sampling. Unfortunately, except in simple settings, accurate coverage can only be guaranteed asymptotically (i.e., in the limit of large sample size), a significant drawback when surveys may have few or even no detected TNOs. In addition, accounting for measurement error within this "frequentist" framework is problematic and a topic of current research in statistics; the best-developed solutions are also only asymptotically valid. Finally, we often need to summarize the implications of the data for a subset of the parameters (e.g., for the slope of the LF, or the location of a break or cutoff). Properly accounting for the uncertainty in the uninteresting "nuisance parameters" in such summaries remains an open problem in frequentist statistics despite decades of study.

These are some of the reasons recent works have adopted a Bayesian approach for TNO population inference. In this approach, one calculates a *posterior probability density* for the parameters, interpreted more abstractly as indicating the degree to which the data and model assumptions imply that the true parameter values lie in various regions. Adopting this more abstract goal for inference carries with it many benefits. One can straightforwardly calculate probabilities for parameter regions (now called "credible regions") that are accurate for any sample size. Measurement error is easily handled, and nuisance parameters are easily dealt with (*Gull,* 1989). Furthermore, statisticians have shown that parametric Bayesian procedures have good performance when viewed as frequentist procedures, with asymptotic accuracy as good as and sometimes superior to that of common frequentist procedures. A possible drawback is that the posterior distribution is found by multiplying the likelihood by a *prior density* for the parameters, expressing an initial state of uncertainty (before considering the data). When data are sparse, one's conclusions can depend on the prior, although the dependence is explicit and can be used to quantitatively probe the degree to which the data are informative. A more challenging drawback is the absence of straightforward "goodness-of-fit" (GoF) tests in the Bayesian approach. See *Sivia and Skilling* (2006) for a tutorial on Bayesian methods.

ITZ95 first introduced likelihood methods for TNO population studies. They adopted a frequentist approach, and did not consider complications due to measurement error or detection efficiency. Several works in the last decade (G98, G01, B04, P06, F07) adopted the Bayesian approach, but have not presented the full correct details in the TNO literature. We derive the posterior distribution based on TNO survey data in three steps. First, we use the Poisson distribution and the product rule from probability theory to construct the likelihood function for an idealized survey reporting precise magnitude measurements above some hard threshold. Next, we use the law of total probability to modify the likelihood to account for measurement error. Finally, we use Bayes's theorem to get the posterior from the likelihood and a prior. We focus here on inferring the LF; in principle it is straightforward to generalize such an analysis to other TNO population descriptions, e.g., to explicitly model the size distribution, or, more ambitiously, the distribution of sizes and orbital elements. In practice, existing surveys are not sufficiently well characterized to permit accurate calculation of the resulting likelihoods. Thus we are presently reduced to inferring the size distribution indirectly via LF estimation, and to inferring orbital element distributions approximately via simple scatter plots and histograms. Future large surveys should ameliorate these issues and allow more careful inferences; we will describe the generalized methodology in future work.

We consider a Poisson point process model M of the LF with parameters $\mathcal{P}$, specified by the differential magnitude distribution, $\sigma(m)$, defined so that $\sigma(m)dmd\Omega$ is the probability for there being a TNO of magnitude in $[m,m + dm]$ in a small patch of the sky of solid angle $d\Omega$ [so $\Sigma(m)$ is its integral]. For simplicity we assume no direction dependence in the surveyed patch. For idealized data, we imagine the m_i values spread out on the magnitude axis. We divide the axis into empty intervals indexed by ε with sizes Δ_ε, between small intervals of size δm containing the N detected values m_i. The expected number of TNOs in empty interval ε for a suvery covering solid angle Ω is

$$\mu_\varepsilon = \Omega \int_{\Delta_\varepsilon} dm\sigma(m) \qquad (11)$$

The expected number in the interval δm associated with detected TNO i is $\mu_i = \Omega\delta m\sigma(m)$, where we take δm small enough so the integral over δm is well approximated by this product. The probability for seeing no TNOs in empty interval ε is the Poisson probability for no events when μ_ε are expected, given by $e^{-\mu_\varepsilon}$. The probability for seeing a TNO of magnitude m_i in δm is the Poisson probability for one event when μ_i are expected, given by $\mu_i e^{-\mu_i}$. Multiplying these probabilities gives the likelihood for the parameters, $\mathcal{P}$, of model M, specifying $\sigma(m)$. The expected values in the exponents sum to give the integral of $\sigma(m)$ over all accessible m values, so the likelihood can be written

$$\mathcal{L}(\mathcal{P}) = (\Omega\delta m)^N \exp\left[-\Omega \int dm\Theta(m_{th} - m)\sigma(m)\right] \times \prod_{i=1}^{N} \sigma(m_i) \qquad (12)$$

where $\Theta(m_{th} - m)$ is a Heaviside function restricting the integral to m values smaller than the threshold, m_{th}. The factor in front is a constant that will drop out of Bayes' theorem and can henceforth be ignored.

Now we consider actual survey data, which differs from idealized data in two ways: the presence of a survey efficiency rather than a sharp threshold, and the presence of magnitude uncertainties. We immediately run into difficulty with a point process model because we cannot make the previous construction, since we do not know the precise values of the TNO magnitudes. To cope with this, we make use of the law of total probability, which states that to calculate a desired probability $p(A|I)$, we may introduce a set of auxiliary propositions $\{B_i\}$ that are independent and exhaustive (so $\Sigma_i p(B_i|I) = 1$). Then

$$p(A|I) = \sum_i p(A,B_i|I) = \sum_i p(B_i|I)p(A|B_i,I) \qquad (13)$$

Here we will take A to be the data, and B_i to specify the unknown true magnitudes for the detected TNOs.

To facilitate the calculation, we need to introduce some notation. When occurring as an argument in a probability, let m_i denote the proposition that there is a TNO of magnitude m_i in an interval δm at m_i. We divide the data, D, into two parts: the data from the detected objects, $\{d_i\}$, and the proposition, $\mathcal{N}$, asserting that no other objects were detected. Then, using the law of total probability and the product rule, the likelihood can be written

$$\begin{aligned}\mathcal{L}(\mathcal{P}) &= p(D|\mathcal{P},M) \\ &= \int \{dm_i\} p(\{m_i\},\mathcal{N}|\mathcal{P},M) \times \\ &\qquad p(\{d_i\}|\{m_i\},\mathcal{N},\mathcal{P},M)\end{aligned} \tag{14}$$

The first factor in the integrand can be calculated using a construction similar to that used for the idealized likelihood above, with one important difference: The presence of the $\mathcal{N}$ proposition means that we cannot assume that no TNO is present in a Δ_ε interval, but rather that no TNO was *detected*. Thus these probabilities are Poisson probabilities for no events when μ_ε are expected, with

$$\mu_\varepsilon = \Omega \int dm \eta(m) \sigma(m) \tag{15}$$

the *detectable* number of TNOs in the interval, not the total number. Thus the first factor in the integrand in equation (14) resembles the righthand side of equation (12), but with $\eta(m)$ replacing the Heaviside function in the integral in the exponent.

The second factor in the integrand in equation (14) is the probability for the data from the detected objects, given their magnitudes. If we know the magnitude for a particular source, we can easily calculate the probability for its detection data (independent of the value of other data, or of the LF model parameters). So this probability factors as a product of factors $p(d_i|m_i)$, the probability for the data d_i from TNO i presuming it has magnitude m_i. We call this the *source likelihood function,* $\ell_i(m_i) \equiv p(d_i|m_i)$. It will often be adequately summarized by a Gaussian function specified by the best-fit (maximum likelihood) magnitude for the TNO and its uncertainty (more rigorously, if one is doing photometry using some kind of χ^2 method, it would be proportional to exp $(-\chi^2(m_i)/2)$). With this understanding, we can now calculate equation (14) as

$$\begin{aligned}\mathcal{L}(\mathcal{P}) = \exp\left[-\Omega \int dm \eta(m) \sigma(m)\right] \times \\ \prod_i \int dm \ell_i(m) \sigma(m)\end{aligned} \tag{16}$$

where we have simplified the notation by dropping the indices from the m_i variables in the integrals, since they are just integration variables for independent integrals. We can easily evaluate these integrals using a Gauss-Hermite quadrature rule.

Finally, Bayes' theorem indicates that the posterior density for the parameters is proportional to the product of the likelihood and a prior density

$$p(\mathcal{P}|D,M) = \frac{1}{C} p(\mathcal{P}|M) \mathcal{L}(\mathcal{P}) \tag{17}$$

where C is a normalization constant equal to the integral of the product of the prior, $p(\mathcal{P}|M)$, and the likelihood. As a conventional expression of prior ignorance, we use flat priors (uniform probability) for most parameters, and log-flat priors [uniform probability for log(x) of parameter x, or probability proportional to 1/x] for nonzero scale parameters [e.g., a multiplicative amplitude parameter for $\sigma(m)$ (*Jeffreys,* 1961)]. The Bayesian "answer" to the parameter estimation problem is the full, multivariate posterior distribution. But it is useful to summarize it. The posterior mode (the value of $\mathcal{P}$ that maximizes the posterior) provides a convenient "best-fit" summary. Credible regions are found by integrating the posterior within contours of equal density; this can be done with simple quadrature rules in one or two dimensions, or via Monte Carlo methods in higher dimensions. Finally, to summarize implications for a subset of the parameters (e.g., just the LF slope), we integrate out ("marginalize") the uninteresting parameters.

Two surprising aspects of the likelihood in equation (16) are worth highlighting (for further discussion, see *Loredo and Lamb,* 2002; *Loredo,* 2004). First is the absence of an $\eta(m)$ factor in the source integrals. Adding it produces inferences that are significantly biased, as is readily shown with simulation studies. To understand its absence, recall that the definition of $\ell_i(m)$ as $p(d_i|m)$, the probability for the data from a detected source. Separate d_i into two parts: $\mathcal{D}_i$, which stands for *"the raw data for candidate TNO i indicate a detection,"* and n_i, which stands for all the data we measured on that object (e.g., counts in pixels). Then $\ell_i(m) = p(\mathcal{D}_i,n_i|m)$, which we may factor two ways

$$\ell_i(m) = p(n_i|m)p(\mathcal{D}_i|n_i,m) \tag{18}$$

$$= p(\mathcal{D}_i|m)p(n_i|\mathcal{D}_i,m) \tag{19}$$

Now note that the detection criterion is a "yes/no" criterion decided by the data values. Thus the second factor in equation (18) — the probability that object i is detected, given its raw data values — is unity (for detected objects). The first factor summarizes the photometry, so $\ell_i(m)$ is just as described above. The second line gives an alternative expression, whose first factor is the probability for detecting an object of magnitude m, i.e., $\eta(m)$. But if we factor $\ell_i(m)$ this way, the second factor must be properly calculated. It requires finding the probability for the raw data, *given that we know the data satisfy the detection criteria.* This is not the usual likelihood or χ^2 photometry calculation; in fact, one can show it corresponds to the photometry likelihood function *divided by* $\eta(m)$, thus canceling the first factor.

The second surprise is the importance of the $\ell_i(m)$ integrals. Intuition may suggest that if uncertainties are small, the integrals can be eliminated [the $\ell_i(m)$ functions can be approximated as δ functions at the best-fit magnitudes]. Intuition may also suggest that this should be especially valid with large samples, i.e., that the uncertainties should "average out." Statisticians have known for half a century that such intuitions are invalid. Measurement error, if ignored, not only biases estimates, but makes them formally *inconsistent* (estimates converge to incorrect values even for infinite sample size). One way to understand this is to realize that, fundamentally, each new object brings with it a new parameter (its m value) that the analyst must estimate (implicitly or explicitly). The number of (latent) parameters in the problem thus grows proportional to the sample size, invalidating our intuition from fixed-parameter problems. The surprising consequence is that it becomes *more* important to properly account for measurement error as sample size increases. Current TNO survey sizes appear to be right on the border of where measurement error bias becomes important, so it is imperative that future analyses properly account for it.

So far we have discussed modeling a single survey. For a group of surveys with consistent calibrations, surveying nearby regions, the joint likelihood function based on all data is just the product of the likelihoods from the individual surveys. But in reality, calibration errors vary from survey to survey, and different groups may use different bandpasses requiring color-dependent photometric conversions to a common bandpass, introducing systematic offsets. Also, even when surveyed regions are small enough that the TNO density is nearly constant across each region, the anisotropy of the TNO density can lead to significant differences in the amplitude of the LF accessible to each survey if the surveyed regions are not all near to each other. These issues were not accounted for in analyses prior to P06 and F07, although CB99 noted the inconsistency of combining different surveys. P06 noticed that LF estimates from individual surveys had similar slopes but different amplitudes, so they only used their own survey. F07 presents the first analysis of multiple surveys explicitly accounting for these complications.

We can handle these complications quantitatively by suitably expanding the model. Photometric zero-point offsets can be handled by introducing an offset parameter for each survey, denoted m_s for survey s. In the likelihood function for survey s, we replace $\sigma(m)$ everywhere by $\sigma(m - m_s)$. Without constraints on m_s, the data may allow unrealistically large shifts, particularly for surveys with few or no detections, so it is important to quantify the systematic errors, and include this information in the analysis via a prior for m_s for each survey [e.g., via a Gaussian whose mean is the estimated offset between the survey magnitude scale and that adopted in the $\sigma(m)$ model, and with standard deviation reflecting calibration uncertainties]. This prior is essentially the likelihood function for m_s from analysis of auxiliary calibration data.

A natural way to parameterize anisotropy effects is to introduce direction explicitly (indicated by the unit vector, **n**) and write the anisotropic LF as $\sigma(\mathbf{n},m) = Af(\mathbf{n})\sigma(m)$, the product of an amplitude parameter, A, a direction dependence $f(\mathbf{n})$, and a normalized LF shape function, $\rho(m)$ (with $\int dm\rho(m) = 1$). Hence we separate the LF parameters, $\mathcal{P}$, into shape parameters, $\mathcal{S}$, that specify $\rho(m)$, direction parameters, $\mathcal{O}$, that specify $f(\mathbf{n})$, and the amplitude. For most surveys, anisotropy within the surveyed region will be negligible. Denoting the centers of the surveyed regions by $\mathbf{n}_s$, the likelihood for survey s can be approximated by substituting $Af(\mathbf{n}_s)\rho(m)$ for $\sigma(m)$. The direction parameters $\mathcal{O}$ then enter the likelihood solely via $f(\mathbf{n}_s)$. As a rough approximation, one could use the values of $f(\mathbf{n})$ at the survey centers directly as direction parameters. Equivalently, one can replace the product $Af(\mathbf{n}_s)$ with a survey-dependent amplitude parameter, $A_s \equiv Af(\mathbf{n}_s)$. The apparent simplicity of this parameterization is somewhat illusory; e.g., one should somehow enforce that the sky density falls away from the invariable plane. Thus it is probably best to introduce some simple parameterization of $f(\mathbf{n})$, provided enough surveys are available to allow meaningful estimates of the parameters.

For the LF model of equation (1), there is a possible identifiability problem with these parameterizations: For a single survey, m_0 (which plays the role of A) and m_s are conflated. The m_s priors thus play a crucial role, making the amplitude and magnitude shift parameters identifiable. A similar issue arises if we parameterize anisotropy via the amplitude at the field center; this, too, is conflated with m_0. It is not clear how to handle this via priors, again arguing for explicit parameterization of anisotropy.

Finally, we note that the normalization constant, $C = \int d\mathcal{P}p(\mathcal{P}|M)\mathcal{L}(\mathcal{P})$, although playing a minor role in parameter estimation, plays a crucial role in comparison of rival parameterized models, where it acts as the likelihood of the model *as a whole*. This constant is called the *marginal likelihood* for the model. One of the main issues in LF determination is whether the TNO magnitude distribution has curvature. A uniquely Bayesian way to address this is to compare a *flat* (i.e., exponential) model for $\sigma(m)$ with one that has curvature (like the ones proposed by B04). We can compare models by looking at the marginal likelihood ratio in favor of one over the other, called the *Bayes factor*, $B_{12} = C_1/C_2$, where C_1 and C_2 are the normalization constants for the rival models (*Wasserman,* 1997; *Sivia and Skilling,* 2006). These are the odds favoring model M_1 over model M_2 if one were to consider the two models equally plausible *a priori*. The convention for interpreting Bayes factors is that values less than 3 or so do not indicate significant evidence for the stronger model. Values from 3 to 20 indicate positive but not compelling evidence. Values over 20 or so (i.e., a probability >0.95 for the favored model) indicate strong evidence. An appealing and important aspect of Bayesian inference is that Bayes factors implement an automatic and objective "Ockham's razor" (*Jefferys and Berger,* 1992; *Gregory,* 2005), penalizing more complex models for their extra parameters. This happens because C is an *average* of

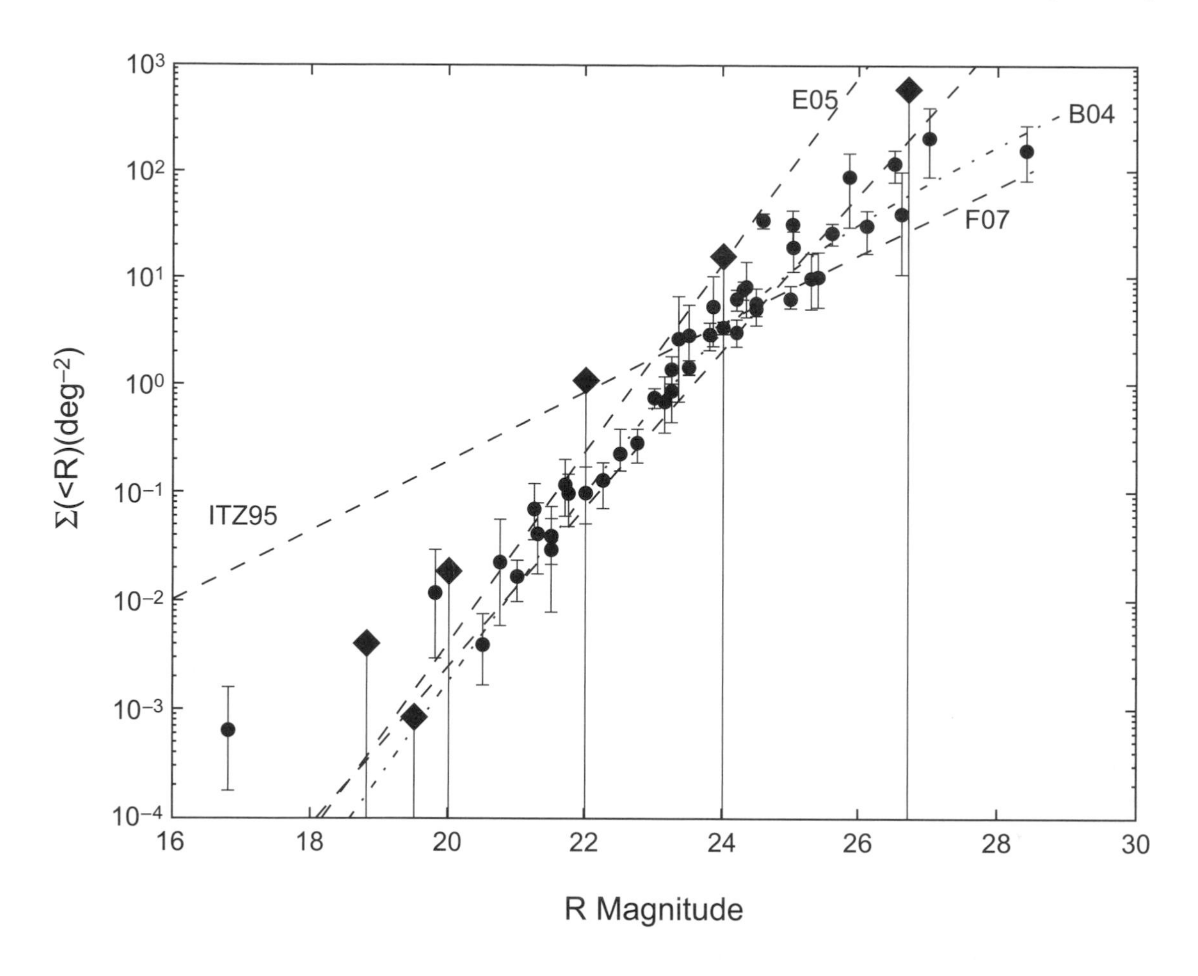

Fig. 1. Cumulative surface density of TNOs brighter than a given R magnitude. The solid circles represent the values derived from the works by T61, JLC96, LJ98, L01, JL95, ITZ95, JLT98, G98, CB99, TJL01, G01, T01, B04, P06, and F07. The error bars correspond to a 68.3% probability. The solid diamonds represent the upper limit at 99.7% probability for nondetection from K89, LJ88, LD90, S00, and G01. The dashed lines show the shallowest (ITZ95), the most recent (F07), and the steepest (E05, $\alpha = 0.88$) LF given in Table 3. The dash-dotted line shows the LF of B04, integrated to give a cumulative density. Note that such a plot should never be used to "best-fit" the LF.

the likelihood function. A more complex model may have a larger maximum likelihood, but by having a larger parameter space to average over, its marginal likelihood can be smaller than that of a simpler rival.

5. CURRENT SIZE DISTRIBUTION ESTIMATES FROM DIFFERENT SURVEYS

A measure of the size distribution of the Kuiper belt is a fundamental property that must be determined if we are to understand the processes of planet formation. Given this importance, a number of authors have provided estimates of the size distribution, as derived from the LF. In this section we discuss the differences between the results of these different authors, in an attempt to provide a clearer picture of the true distribution of material in the Kuiper belt. As a visual aid, Fig. 1 presents the surface densities (or upper limits in case of nondetection) provided by the surveys to date. We also added a few of the proposed LF. Note that this is given only from an illustrative point of view. One should not use this kind of representation to derive the LF (although this was used in earlier work, but no longer in the recent works), but rather resort to Bayesian likelihood methods, as explained in section 4.

Table 3 gives a summary of the LF and/or size distributions proposed by the works listed in Table 1. Below, we now comment on those values, their meaning, and also mention some results from works listed in Table 2.

Initial published estimates (JL95, Monte Carlo approach; ITZ95, maximum likelihood frequentist approach) found the size distribution of material to be well represented by a power law $N(>D) \propto D^{(1-q)}$, with $q < 3$, and D being the diameter of the TNO. Already at this early stage, ITZ95 realized that a break in the exponential was needed to account for the lack of detection in former shallow surveys. The slope proposed for their detections, together with those of JL95, was very shallow, $\alpha = 0.32$, much shallower than any other population of small bodies in the solar system. Such a shallow slope was inconsistent with the lack of detection by previous surveys by LD90 (4 objects predicted) and LJ88

TABLE 3. Luminosity function (LF) and size distributions of past surveys.

Survey	LF slope (α)	Normalization Constant*	Size Distribution Slope (q)	$N(D \geq 100$ km)	Comments List of Data Used†
Wide-Area Surveys					
JL95		$C_{25} = 6 \pm 2.3$	$q < 3$	~35,000	$30 \leq R \leq 50$ AU region
					T61, K89, LJ88, LD90, JL95
ITZ95	$0.32^{+0.10}_{-0.08}$	$C_{25} = 7.9^{+2.9}_{-2.3}$			JL95, ITZ95
	~0.6				everything but JL95
	~0.13‡		$C_{25} = 5.6$		T61, K89, LJ88, LD90, JL95, ITZ95
JLT98	0.58 ± 0.05	$m_0 = 23.27 \pm 0.11$	4.0 ± 0.5		$20 \leq m_R \leq 25$
					Published data with $m_R > 23$
TJL01	0.63 ± 0.06	$m_0 = 23.04^{+0.8}_{-0.9}$	$4.0^{+0.6}_{-0.5}$		Maximum likelihood§
					Own data only
T01	0.66 ± 0.06	$m_0 = 23.32 \pm 0.09$	$4.2^{+0.4}_{-0.3}$	$4.7^{+1.6}_{-1.0} \times 10^4$	TJL01, T01
P06	0.76 ± 0.1	$m_0 = 23.3^{+0.2}_{-0.25}$	$q \simeq 4.8$		Own data only
Deep Surveys					
G98	$0.76^{+0.10}_{-0.11}$	$m_0 = 23.4^{+0.20}_{-0.18}$			LJ88, LD90, C95, ITZ95, JLT98, G98
CB99	0.52 ± 0.02	$m_0 = 23.5 \pm 0.06$	3.6 ± 0.5		All surveys and upper limits
	0.66 ± 0.04		4.3 ± 0.5		ITZ95, JLT98, G98, CB99
G01	0.69 ± 0.07	$m_0 = 23.5 \pm 0.03$	$q \gtrsim 4$		LJ88, LD90, C95, ITZ95, JLT98, G98, CB99, G01
B04¶	$\alpha_1 = 0.88$	$\Sigma_{23} = 1.08$			$R_{eq} = 23.6$
	$\alpha_2 = 0.32$				CB99, L01, TJL01, G01, ABM02, TB03, B04
F07	0.6 ± 0.15	$m_0 = 23.25 \pm 0.5$			Own data only
	0.73 ± 0.06**	$m_0 = 23.55 \pm 0.15$**	4.6 ± 0.3**		F07, unpublished data by Kavelaars et al.

*C_x: number of object brighter than R magnitude x per deg²; m_0: R magnitude at which a cumulative density of 1 object per deg² is reached; Σ_x: Number of object per deg², per unit magnitude at R magnitude x.

†A reference to A on the line of B means that data acquired in work A are used in work B, not all the data used by A to derive theLF.

‡Two exponentials, break at $m_0 = 22.4$, bright end slope = 1.5.

§Fitting the differential LF with a frequentist approach, $\alpha = 0.64^{+0.11}_{-0.10}$ and $m_0 = 23.23^{+0.15}_{-0.20}$.

¶Double exponential differential LF $\Sigma(R) = (1 + c)\Sigma_{23}[10^{-\alpha_1(R-23)} + c10^{-\alpha_2(R-23)}]^{-1}$ with $c = 10^{(\alpha_2 - \alpha_1)(R_{eq} - 23)}$.

**3σ uncertainties.

(41 objects predicted). So they added a cutoff at bright magnitudes in the form of a steeper slope for magnitudes brighter than a fitted value m_0 and a shallower slope for fainter objects. They fixed the value of the bright end slope at α_b = 1.5 and found a best fit slope for the faint end $\alpha_f = 0.13$ and a break at $m_0 = 22.4$.

These initial, groundbreaking attempts were based on small samples of a few to tens of objects. The large number of parameters that come into play in producing a full-up model of the Kuiper belt (radial, inclination, and size distributions, which perhaps differ between objects in different orbital classes) ensures that attempting to constrain the problem with only a dozen or so detections will very likely lead to false conclusions.

Later works gradually steepened the slope of the size distribution. Using a Monte Carlo approach, JLC96 claimed q ~ 4, while JLT98 showed that q ~ 4, depending only mildly on the choice of radial distribution. With a larger (73 deg²) and fainter ($m_R \sim 23.7$) survey that detected 86 TNOs, TJL01 concluded that $q = 4.0 \pm 0.5$. To reach this conclusion, they used only their 74 detections close to the ecliptic plane. They also directly estimated the LF and found $\alpha = 0.64^{+0.11}_{-0.10}$, but did not use this to derive the size distribution (this would yield $q = 4.2^{+0.55}_{-0.5}$), although they quote transformation equation (3). G98 introduced a Baysian weighted maximum likelihood approach to fit an exponential LF and then translated this to a size distribution of q ~ 4.5 using equation (3).

CB99 extended the work toward the faint end with their two-discovery survey, including the faintest TNO detected at that time ($m_R = 26.9 \pm 0.2$). Combining a range of previous surveys, they find that $0.5 < \alpha < 0.7$, and that the LFs determined using different combinations of surveys result in exponential slopes that are formally inconsistent at the ~5σ level. This indicates that lumping all observations together does not provide a very consistent picture of the LF of the belt. The reason for this was first demonstrated by P06, and an attempt to solve the problem was proposed by F07, along the lines mentioned in section 4. G01 also find that only a subsample of the available surveys of the Kuiper belt can be combined in a self-consistent manner and, using just that subsample, find α ~ 0.7 and conclude $q \gtrsim 4.0$.

Using the HST Advanced Camera for Surveys, B04 searched 0.019 deg² of sky, to a flux limit of $m_R = 28.5$, more than 1.5 mag fainter than any previous groundbased survey. They discovered 3 objects, while a linear extrapolation of the LF from G01 predicted the detection of ~46 objects. Clearly the deviation from a uniform exponential has been reached.

B04 modeled the observed LF as a combination of two exponential functions and also as a function whose power "rolled over" from one slope onto another to account for the observed lack of objects faintward of $m_R = 26$. In order to constrain the bright end of the LF, B04 combined all previous TNO-LF surveys together. Recall that previous authors had already found that this often results in variations of the exponential slope that are outside the range allowed by the uncertainty measures of the individual projects. Based on this combined fitting B04 determined that the bright end slope is actually steeper than previous estimates $\alpha \sim 0.8$, while faintward of $m_R \sim 24$ the LF starts to be dominated by a flatter slope of $\alpha \sim 0.3$. Because the Kuiper belt has an extent of about ~20 AU, the rollover occupies a fairly large range of magnitude and, although starting at $m_R \sim 23$, the shallower slope is not dominant until $m_R \sim 25.5$. Because the LF is no longer a simple exponential, there is not a complete direct connection between the size distribution power q and the LF slope α. However, the asymptotic behavior of B04's LF imply a size distribution power of $q \sim 2.5$, much shallower than the 3.5 value expected for Donanyhi-like distribution.

Other surveys have subsequently attempted to determine the LF of the Kuiper belt in the $22 < m_R < 26$ region, thus more firmly establishing the steep component of the LF and better constraining planetesimal accretion models. E05 employ a novel survey efficiency model, but flawed method in order to determine the slope of the size distribution based on analysis of their Deep Ecliptic Survey observations. They did not effectively measure their detection efficiency, so they tried to parameterize it and solve for these new parameters together with the LF parameters. Unfortunately, this is strongly degenerate and a slight change in the efficiency function has strong implications for the derived slope for the LF. They had to fix the efficiency parameters in several cases in order to get reasonable slopes. Finally, E05 find that $q \sim 4.7$ for objects in the classical Kuiper belt. P06 present an analysis of the LF of Kuiper belt objects detected in their survey for irregular satellites of Uranus and Neptune and find that $q \simeq 4.8$. They also show that this slope is similar to that of G01, but with an offset in the sky density, which they relate to change in the direction of the surveys (see Fig. 2 and their Fig. 8). For these surveys we see that there appears to be consensus on steepness of the slope of the size distribution of Kuiper belt material in the $22 < m_R < 25$. These steep values for the slope of q stand in stark contrast to the original estimates of $q \sim 2.5$.

F07 present a more complete approach to combining data from multiple surveys than has previously been attempted. F07 observe that the value of m_o (the point at which a survey sees 1 object per deg^2) in the LF of observed TNOs at different ecliptic longitudes and latitudes need not be identical. This is due to the narrow width of the Kuiper belt and the effects of resonance libration angles causing the sky density of Plutinos to vary from month to month. Additionally, F07 note that the transformation between filter sets and the absolute calibrations of various surveys cannot be perfect. Therefore, F07 modified the single exponential LF by allowing m_o to vary between surveys (which all attain different depths) and thus account for these unknown variations in m_o. This change has the effect of removing the variations in α seen by CB99 and allows F07 to provide a very robust measure of the slope between $21 < m_R < 25.7$. F07 find that values of $q < 4$ for this magnitude range are now formally rejected with greater than 5σ confidence. In the same time, F07 rejects a change to shallower slope occurring brightward of $m_R \sim 24.3$ at more than 3σ, and propose that this change occurs around $m_R \sim 25.5$.

It is now demonstrated that for the whole ensemble of TNOs, the size distribution slope is steeper than 4, most likely around $q \sim 4.5$, brightward of $m_R \sim 25$.

To our knowledge, all published works on TNO accretion assumed an unperturbed accretion phase that produce a cumulative size distribution slope $q \sim 3$ (and always $q < 4$) up to the largest bodies. They also tend to produce disks that are much more massive than the current Kuiper belt mass. Steeper slopes for the largest bodies are actually achieved during the accretion phase, and retaining them requires some perturbations (external or endogenic) to halt or change the collisional accretion of large bodies at some early time (chapter by Kenyon et al.). The slope reached for large bodies, and the size at which the transition to a shallower slope for small bodies occurs, are strong constraints on the time at which these perturbations happened. The slope for smaller bodies depends on the relative velocity distribution (eccentricity and inclination) and the strength of the bodies. The main perturbation mechanisms are a close flyby of a star and Neptune steering. Flybys tend to leave too much mass in the dust grains, while Neptune steering can produce the correct slope and low dust mass implied by the observations (chapter by Kenyon et al.). The stopping of coagulation accretion (slope at large sizes) and the dynamical removal of most of the remaining material (transition size between steep and shallow slopes) require that Neptune steering occurred before 100–200 m.y. This seems to present a serious timing problem for the Nice model (chapters by Kenyon et al. and Morbidelli et al.).

6. SIZE DISTRIBUTION OF THE DIFFERENT DYNAMICAL POPULATIONS

The different dynamical populations of small bodies of the outer solar system described in the chapter by Gladman et al. are thought to result from different mechanisms. They could very well present various size distributions, coming from different places, and having suffered different accretion and collisional evolution.

In the early ages of size distribution determination, the small number of objects made it difficult, in fact almost impossible to search for different size distributions for each

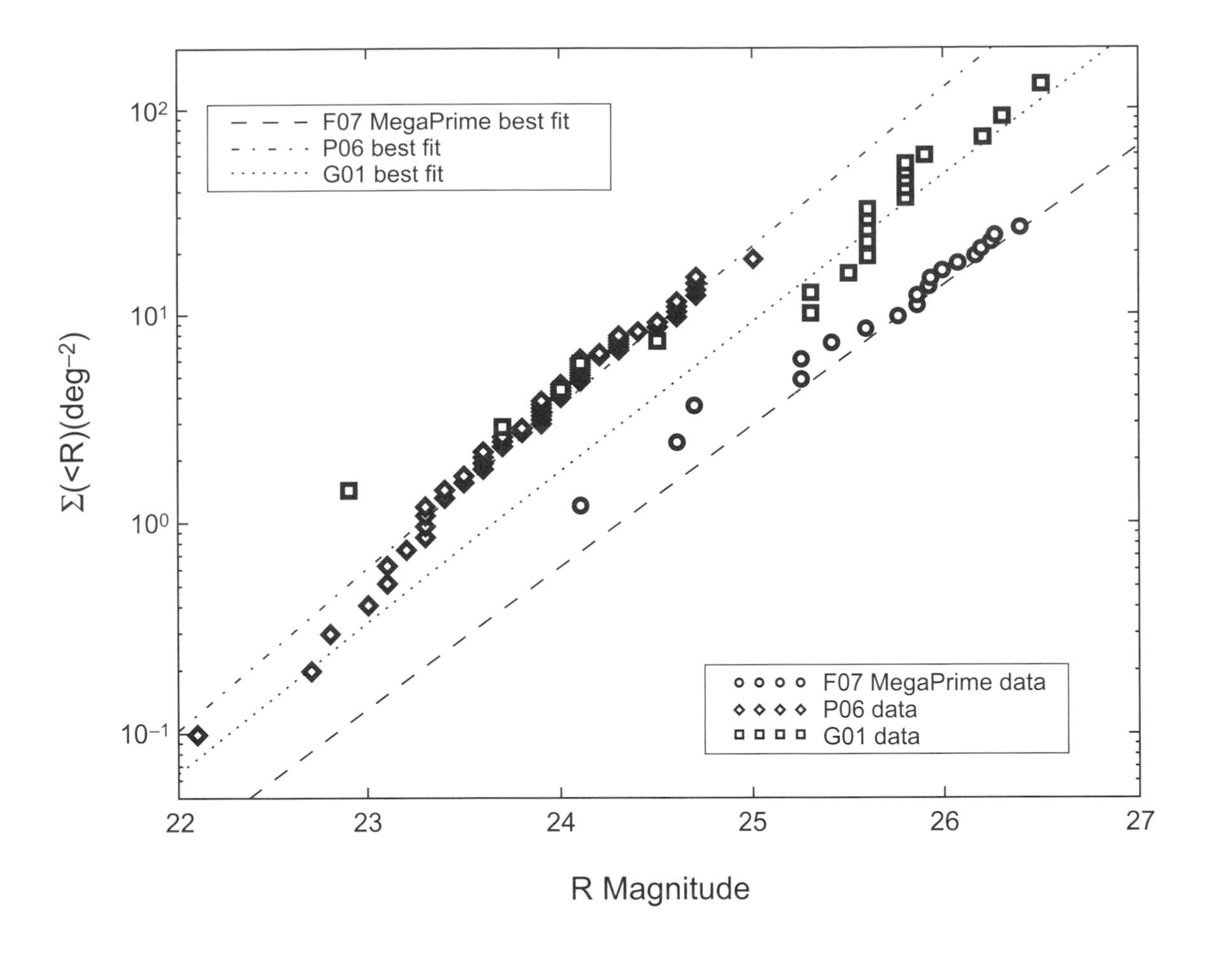

Fig. 2. Debiased cumulative surface density of TNOs for G01 (open squares), P0 (open diamonds), and MEGAPrime data of F07 (open circles). The data have been debiased using the published efficiency functions. The different lines represent the best-fit single-exponential LF for G01 (dotted line), P06 (dash-dotted line), and F07 (dashed line). All three surveys, taken independently, are well approximated by a single exponential function with index ranging from 0.7 to 0.8. Clearly, one cannot simply combine them as was done up to and including B04 (although CB99 noted that combining leads to inconsistent results). One must account for an offset in surface density due to longitude and latitude variation, magnitude difference in different passband filters, and calibration uncertainties. An attempt to apply these corrections was done by F07, but this requires external determination of these offset in order to avoid unreasonable fit values for m_0 for surveys with small number of detections (F07).

dynamical class. Thus, few people tried to address this question, and most of those who did assumed a unique size distribution for all classes, with only a change in the normalizing factor.

JLT98 had to assume some orbital distribution of objects to solve for the size distribution. They used a two-population model composed of a classical Kuiper belt and Plutinos. In doing so, they used the same size distribution form for both populations, and assumed a simplified model of the semimajor axis and eccentricity distributions of the classical belt, as well as the eccentricity distribution of the Plutinos. In doing so, they estimated that the *apparent* Plutino fraction of ~38% found in their data corresponds to a 10–20% fraction of the TNOs in the 30–50-AU region. Furthermore, they computed the bias in their survey against finding objects in the 2:1 resonance with Neptune and concluded that the nondetection of those could not formally disprove the hypothesis that the 2:1 and 3:2 resonances are equally populated (*Malhotra,* 1995). Note that the classification used rather short arcs, typically 1 year or less for many objects in the survey, which may lead to misclassification. TJL01 applied the same approach to their new survey, for which they attempted to obtain better orbital determination by longer tracking effort. Using the bias toward finding 3:2 resonant objects computed by JLT98, they converted their observed 8% Plutino fraction to a 3–4% real fraction, much smaller than the previous estimate, and more in line with the findings of *Petit and Gladman* (2003). As for the 2:1 population, their conclusions were similar to those of JLT98.

L00 used also a single slope for their LF for all TNOs and give their sky density of classical belt and of scattering (scattered in their paper) disk and find the latter to be about six times less populated than the former. They also considered what they called Centaurs (not a population of TNOs *per se*), of which they discovered four, and derived a shallower slope than for TNOs, and a much lower sky density of 0.017 ± 0.011 Centaurs brighter than $m_R = 21.5$. However, according to the new classification scheme KBB07 described in the chapter by Gladman et al., only two objects are Centaurs, one (1998 SG_{35} = Okyrhoe) is a JFC, and the last one (33128=1998 BU_{48}) is a scattered disk object. We note that the objects used in that work typically had a decently determined orbit, and the change in classification is due to the fuzzy definitions used at that time. T01 confirmed the sky density of Centaurs at the bright end, finding one such object (using the same definition as L00) in their survey.

B04 were the first to try and fit completely different LFs to different dynamical classes. They define three different classes that span one or more classes of KBB07. The TNO sample contains all objects with heliocentric distance R > 25 AU. The CKBO sample roughly correspond to the classical belt, with 38 AU < R < 55 AU and inclination i < 5°. The excited sample is the complement of the CKBOs in the TNOs. They find that the differential LF of the CKBO sample is well described by a very steep slope ($\alpha \sim 1.4$) at the bright end, and a shallower, but still rising, slope ($\alpha \sim 0.4$) at the faint end, while the differential LF of the excited sample would have a bright end slope similar to that derived by other works to the whole belt ($\alpha \sim 0.65$), and would be decreasing at the faint end (slope $\alpha \sim -0.5$). This result is to be taken with great caution as the classification used was purely practical and has no connection to any common (or lack of) origin of the bodies.

E05 had the best survey at the time to address the question of size distribution vs. dynamical population, having the largest number of objects discovered in a single survey. They were also able to determine precisely the orbit of a fair fraction of their objects. They consider three major classes of TNOs — mainly the classical belt, the resonant objects, and the scattered objects — with boundaries similar to those of KBB07. They find that the slope of the LF varies significantly from the classical belt (0.72), to the resonant objects (0.60), to the scattered objects (1.29). Interestingly, the slope of their overall sample is steeper (0.88) than for any subsample but the scattered one, which make up only a small fraction of the objects. However, a caveat is in order here, since E05 did not actually measure their efficiency at discovery, but rather used a heuristic approach to estimate it. Also, only about half of their objects had an orbit precise enough to allow for classification.

7. FUTURE WORK AND LINK WITH FORMATION AND COLLISIONAL EVOLUTION MODELS

The two major outstanding issues on this topic concern the size distribution of the various dynamical classes and the size(s) at which the shape of the distribution changes.

The work of B04 has shown that, going to the faint end of the LF, there is a knee in the size distribution, which they estimate to be around $m_R \sim 24$–25. But P06 and F07 have shown the straight exponential to extend to at least m_R = 26. To settle this question, we will need a deg² survey down to R magnitude 28 or so, beyond the expected knee in the size distribution. At the other end of the LF, earlier works (ITZ95, and to a lesser extend JLT98) have reported the need for a steeper slope of the LF for objects brighter than $m_R \sim 22$–23. ITZ95 were the only ones to give an estimate of the needed slope, at 1.5. All subsequent works assumed a single slope brightward of $m_R \sim 24$–25. However, these works were dealing mostly with LF faintward of $m_R \sim 22$, either explicitly or implicitly. Deciding if there is really a change in the LF slope brightward of $m_R \sim 22$–23 requires surveying the ~1500 deg² around the ecliptic plane where the classical Kuiper belt resides, down to R magnitude 22–23.

The challenge for determining the size distribution of the different dynamical classes is that we need to have a fully characterized survey finding a decent number of TNOs in each class, and follow them all (or at least a large fraction of them, without orbital bias) to firmly establish their class. This requires a follow-up program for at least 2 years after discovery. The only current survey of this kind is the Very Wide component of the CFHT Legacy Survey, which discovers TNOs as faint as $m_R \simeq 23.5$–24 depending on the seeing conditions. It has covered 300 deg² in its discovery phase and is still chasing the objects it has discovered. The forthcoming Pan-STARRS survey will cover the entire sky to 0.5–1 mag fainter than the current CFHTLS-VW survey. Both can address the question of the size distribution for the large-sized bodies, larger than the transition knee detected by B04. In the 5 to 10 years to come, we should have a good knowledge of the size distribution of the large bodies (larger than 100 km in radius) in each of the main dynamical classes, allowing useful comparison with early accretion models.

The size distribution of TNOs smaller than the knee for the CKBO class (the most abundant one in most of the surveys) will also be determined during the same period, thanks to large efforts on 8-m-class telescopes. Determining the shape of the size distribution of small TNOs in each dynamical class is the step after determining that of the TNOs as a whole. This will require a large sample in each dynamical class, hence surveying several deg² of sky to be observed to magnitude ~28. All objects thus discovered will have to be followed until their orbit has been firmly established. This will occur only when we have deg² cameras on either Extremely Large Telescopes or on the new generation of Space Telescopes.

Acknowledgments. We thank M. Holman, S. Kenyon, A. Morbidelli, and an anonymous reviewer for comments that considerably improved the text. J-M.P. acknowledges support from CNRS and the Programme National de Planétologie.

REFERENCES

Albrecht R., Barbieri C., Adorf H.-M., Corrain G., Gemmo A., Greenfield P., Hainaut O., Hook R. N., Tholen D. J., Blades J. C., and Sparks W. B. (1994) High-resolution imaging of the Pluto-Charon system with the Faint Object Camera of the Hubble Space Telescope. *Astrophys. J. Lett., 435,* L75–L78.

Altenhoff W. J., Bertoldi F., and Menten K. M. (2004) Size estimates of some optically bright KBOs. *Astron. Astrophys., 415,* 771–775.

Benz W. and Asphaug E. (1999) Catastrophic disruptions revisited. *Icarus, 142,* 5–20.

Bernstein G. M., Trilling D. E., Allen R. L., Brown M. E., Holman M., and Malhotra R. (2004) The size distribution of trans-neptunian bodies. *Astron. J., 128,* 1364–1390.

Bowell E., Hapke B., Domingue D., et al. (1989) Application of photometric models to asteroids. In *Asteroids II* (R. P. Binzel et al., eds.), pp. 524–556. Univ. of Arizona, Tucson.

Brown M. E. and Trujillo C. A. (2004) Direct measurement of the size of the large Kuiper Belt object (50000) Quaoar. *Astron. J., 127,* 2413–2417.

Brown M. E., Schaller E. L., Roe H. G., Rabinowitz D. L., and Trujillo C. A. (2006) Direct measurement of the size of 2003 UB313 from the Hubble Space Telescope. *Astrophys. J. Lett., 643,* L61–L63.

Bruzual G. and Charlot S. (2003) Stellar population synthesis at the resolution of 2003. *Mon. Not. R. Astron. Soc., 344,* 1000–1028.

Chiang E. I. and Brown M. E. (1999) Keck pencil-beam survey for faint Kuiper belt objects. *Astron. J., 118,* 1411–1422.

Chiang E. I. and Jordan A. B. (2002) On the Plutinos and twotinos of the Kuiper belt. *Astron. J., 124,* 3430–3444.

Cochran A. L., Levison H. F., Stern S. A., and Duncan M. J. (1995) The discovery of Halley-sized Kuiper belt objects using the Hubble Space Telescope. *Astrophys. J., 455,* 342–346.

Cruikshank D. P., Stansberry J. A., Emery J. P., Fernández Y. R., Werner M. W., Trilling D. E., and Rieke G. H. (2005) The high-albedo Kuiper belt object (55565) 2002 AW_{197}. *Astrophys. J. Lett., 624,* L53–L56.

Davis D. R. and Farinella P. (1997) Collisional evolution of Edgeworth-Kuiper belt objects. *Icarus, 125,* 50–60.

Dohnanyi J. S. (1969) Collisional model of asteroids and their debris. *J. Geophys. Res., 74,* 2531–2554.

Drell P. S., Loredo T. J., and Wasserman I. (2000) Type Ia supernovae, evolution, and the cosmological constant. *Astrophys. J., 530,* 593–617.

Drossart P. (1993) Optics on a fractal surface and the photometry of the regoliths. *Planet. Space Sci., 41(5),* 381–393.

Elliot J. L., Kern S. D., Clancy K. B., Gulbis A. A. S., Millis R. L., Buie M. W., Wasserman L. H., Chang E. I., Jordan A. B., Trilling D. E., and Meech K. J. (2005) The Deep Ecliptic Survey: A search for Kuiper belt objects and Centaurs. II. Dynamical classification, the Kuiper belt plane, and the core population. *Astron. J., 129,* 1117–1162.

Farinella P. and Davis D. R. (1996) Short-period comets: Primordial bodies or collisional fragments? *Science, 273,* 938–941.

Fukugita M., Shimasaku K., and Ichikawa T. (1995) Galaxy colors in various photometric band systems. *Publ. Astron. Soc. Pac., 107,* 945.

Gault D. E., Shoemaker E. M., and Moore H. J. (1963) *Spray Ejected from the Lunar Surface by Meteoroid Impact.* NASA Rept. TND-1167, U.S. Govt. Printing Office, Washington, DC.

Giblin I., Martelli G., Smith P. N., Cellino A., DiMartino M., Zappalà V., Farinella P., and Paolocchi P. (1994) Field fragmentation of macroscopic targets simulating asteroidal catastrophic collisions. *Icarus, 110,* 203–224.

Giblin I., Martelli G., Farinella P., Paolicchi P., Di Martino M., and Smith P. N. (1998) The properties of fragments from catastrophic disruption events. *Icarus, 134,* 77–112.

Gladman B., Kavelaars J. J., Nicholson P. D., Loredo T. J., and Burns J. A. (1998) Pencil-beam surveys for faint trans-neptunian objects. *Astron. J., 116,* 2042–2054.

Gladman B., Kavelaars J. J., Petit J.-M., Morbidelli A., Holman M. J., and Loredo T. (2001) The structure of the Kuiper belt: Size distribution and radial extent. *Astron. J., 122,* 1051–1066.

Gregory P. G. (2005) *Bayesian Logical Data Analysis for the Physical Sciences: A Comparative Approach with Mathematica Support.* Cambridge Univ., Cambridge.

Grundy W. M., Noll K. S., and Stephens D. C. (2005) Diverse albedos of small trans-neptunian objects. *Icarus, 176,* 184–191.

Gull S. L. (1989) Bayesian data analysis — Straight line fitting. In *Maximum Entropy and Bayesian Methods* (J. Skilling, ed.), pp. 511–518. Kluwer, Dordrecht.

Hapke B., Nelson R., and Smythe W. (1998) The opposition effect of the Moon: Coherent backscatter and shadow hiding. *Icarus, 133,* 89–97.

Helfenstein P., Veverka J., and Hillier J. (1997) The lunar opposition effect: A test of alternative models. *Icarus, 128,* 2–14.

Irwin M., Tremaine S., and Zytkow A. N. (1995) A search for slow-moving objects and the luminosity function of the Kuiper belt. *Astron. J., 110,* 3082–3092.

Jeffreys H. (1961) *Theory of Probability.* Oxford Univ., Oxford.

Jefferys W. H. and Berger J. O. (1992) Ockham's Razor and Bayesian Analysis. *Am. Sci., 80,* 64–72.

Jewitt D. and Luu J. (1993) Discovery of the candidate Kuiper belt object 1992 QB_1. *Nature, 362,* 730–732.

Jewitt D. and Luu J. X. (1995) The solar system beyond Neptune. *Astron. J., 109,* 1867–1876.

Jewitt D., Luu J., and Chen J. (1996) The Mauna Kea-Cerro-Tololo (MKCT) Kuiper belt and Centaur survey. *Astron. J., 112,* 1225–1238.

Jewitt D. C., Luu J. X., and Trujillo C. (1998) Large Kuiper belt objects: The Mauna Kea 8k CCD survey. *Astron. J., 115,* 2125–2135.

Kowal C. (1989) A solar system survey. *Icarus, 77,* 118–123.

Lacerda P. and Luu J. (2006) Analysis of the rotational properties of Kuiper belt objects. *Astron. J., 131,* 2314–2326.

Larsen J. A., Gleason A. E., Danzi N. M., Descour A. S., McMillan R. S., Gehrels T., Jedicke R., Montani J. L., and Scotti J. V. (2001) The Spacewatch wide-area survey for bright Centaurs and trans-neptunian objects. *Astron. J., 121,* 562–579.

Levison H. F. and Duncan M. J. (1990) A search for proto-comets in the outer regions of the solar system. *Astron. J., 100,* 1669–1675.

Levison H. F. and Duncan M. J. (1997) From the Kuiper belt to Jupiter-family comets: The spatial distribution of ecliptic comets. *Icarus, 127,* 13–32.

Loredo T. J. (2004) Accounting for source uncertainties in analyses of astronomical survey data. In *Bayesian Inference and Maximum Entropy Methods in Science and Engineering* (R. Fischer et al., eds.), pp. 195–206. AIP Conference Proceedings 735, American Institute of Physics, New York.

Loredo T. J. and Lamb D. Q. (2002) Bayesian analysis of neutrinos observed from supernova SN 1987A. *Phys. Rev. D, 65,* 063002.

Loredo T. J. and Wasserman I. M. (1995) Inferring the spatial and energy distribution of gamma-ray burst sources. I. Methodology. *Astrophys. J. Suppl., 96,* 261–301.

Luu J. X. and Jewitt D. (1988) A two-part search for slow-moving objects. *Astron. J., 95,* 1256–1262.

Luu J. X. and Jewitt D. (1998) Deep imaging of the Kuiper belt with the Keck 10 meter telescope. *Astrophys. J. Lett., 502,* L91–L94.

Malhotra R. (1995) The origin of Pluto's orbit: Implications for the solar system beyond Neptune. *Astron. J., 110,* 420–429.

Malhotra R. (1996) The phase space structure near Neptune resonances in the Kuiper belt. *Astron. J., 111,* 504–516.
Morbidelli A. and Brown M. E. (2004) The Kuiper belt and the primordial evolution of the solar system. In *Comets II* (M. C. Festou et al., eds.), pp. 175–191. Univ. of Arizona, Tucson.
Petit J.-M. and Gladman B. (2003) Discovering and securing TNOs: The CFHTLS ecliptic survey. *C. R. Phys., 4,* 743–753.
Petit J.-M. and Mousis O. (2004) KBO binaries: How numerous were they? *Icarus, 168,* 409–419.
Petit J.-M., Holman M. J., Gladman B. J., Kavelaars J. J., Scholl H., and Loredo T. J. (2006) The Kuiper belt's luminosity function from m_R = 22–25. *Mon. Not. R. Astron. Soc., 365,* 429–438.
Rousselot P., Belskaya I. N., and Petit J.-M. (2006) Do the phase curves of KBOs present any correlation with their physical and orbital parameters? *Bull. Am. Astron. Soc., 38,* 44.08.
Ryan E. V., Davis D. R., and Giblin I. (1999) A laboratory impact study of simulated Edgeworth-Kuiper belt objects. *Icarus, 142,* 56–62.
Schaefer B. E. and Rabinowitz D. L. (2002) Photometric light curve for the Kuiper belt object 2000 EB173 on 78 nights. *Icarus, 160,* 52–58.
Sheppard S. S. and Jewitt D. C. (2002) Time-resolved photometry of Kuiper belt objects: Rotations, shapes, and phase functions. *Astron. J., 124,* 1757–1775.
Sheppard S. S., Jewitt D. C., Trujillo C. A., Brown M. J. I., and Ashley M. B. C. (2000) A wide-field CCD survey for Centaurs and Kuiper belt objects. *Astron. J., 120,* 2687–2694.
Shkuratov Y. G. and Helfenstein P. (2001) The opposition effect and the quasi-fractal structure of regolith: I. Theory. *Icarus, 152,* 96–116.
Sivia D. and Skilling J. (2006) *Data Analysis: A Bayesian Tutorial,* second edition. Oxford Univ., New York.
Stansberry J. A., Grundy W. M., Margot J. L., Cruikshank D. P., Emery J. P., Rieke G. H., and Trilling D. E. (2006) The albedo, size, and density of binary Kuiper belt object (47171) 1999 TC_{36}. *Astrophys. J., 643,* 556–566.
Tombaugh C. W. (1961) The trans-neptunian planet search. In *Planets and Satellites* (G. P. Kuiper and B. M. Middlehurst, eds), pp. 12–30. Univ. of Chicago, Chicago.
Trujillo C. A. and Brown M. E. (2001) The radial distribution of the Kuiper belt. *Astrophys. J. Lett., 554,* L95–L98.
Trujillo C. A., Jewitt D. C., and Luu J. X. (2001a) Properties of the trans-neptunian belt: Statistics from the Canada-France-Hawaii Telescope survey. *Astron. J., 122,* 457–473.
Trujillo C. A., Luu J. X., Bosh A. S., and Elliot J. L. (2001b) Large bodies in the Kuiper belt. *Astron. J., 122,* 2740–2748.
Wasserman L. (1997) *Carnegie Mellon University Dept. of Statistics Technical Report 666.* Carnegie Mellon Univ., Pittsburgh.

Part III:
Bulk Properties

Color Properties and Trends of the Transneptunian Objects

A. Doressoundiram
Observatoire de Paris

Hermann Boehnhardt
Max-Planck Institute for Solar System Research

Stephen C. Tegler
Northern Arizona University

Chad Trujillo
Gemini Observatory

The color of transneptunian objects (TNOs) is the first and basic information that can be easily obtained to study the surface properties of these faint and icy primitive bodies of the outer solar system. Multicolor broadband photometry is the only tool at the moment that allows characterization of the entire population that is relevant for statistical work. Using the colors available for more than 170 objects it is possible to get a first glance at the color distribution in the Edgeworth-Kuiper belt. First, results show that a wide color diversity characterizes the outer solar system objects. Transneptunian objects have surfaces showing dramatically different colors and spectral reflectances, from neutral to very red. At least one cluster of objects with similar color and dynamical properties (the red, dynamically cold classical TNOs beyond 40 AU) could be identified. Furthermore, evidence for correlations between colors and orbital parameters for certain objects have been found at a high significance level. Both color diversity and anisotropy are important because they are diagnostic of some physical effects processing the surfaces of TNOs and/or some possible composition diversity. In this paper, we will review the current knowledge of the color properties of TNOs, describe the observed color distribution and trends within the Edgeworth-Kuiper belt, and address the problem of their possible origin.

1. INTRODUCTION

It is now clear that Pluto, Charon, and the ≈1200 known (as of January 2007) transneptunian objects (TNOs) are what remain of an ancient disk of icy debris that did not accrete into a giant, jovian-like planet beyond the orbit of Neptune. By studying TNOs, it is possible to examine the preserved building blocks of a planet, and thereby shed some light on the process of planet building in our solar system as well as extrasolar planetary systems.

A photometric survey is a natural first step in the investigation of TNOs because it is the only kind of survey capable of providing a global view of the transneptunian belt. Specifically, it is possible to obtain accurate magnitudes and colors for all of the known TNOs using 2-m- to 10-m-class telescopes. Accurate magnitudes and colors for a large number of TNOs provide a starting point and context for more in-depth techniques such as spectroscopy and spacecraft flybys. It is also possible to look for statistically significant correlations between colors and other properties of TNOs (e.g., orbital properties). Such correlations may yield insight into the important formation and evolution processes in the outer solar system.

An early expectation was that all TNOs should exhibit the same red surface color. But why were outer-main-belt asteroids and jovian trojans known to exhibit red colors (e.g., see *Degewij and van Houten,* 1979)? In addition, *Gradie and Veverka* (1980) suggested that organic materials — molecules defined by C–H bonds — should readily form at large heliocentric distances and cause redder colors at large heliocentric distances. *Gradie and Tedesco* (1982) discovered a correlation between color and semimajor axis of asteroids, implying that asteroids preserve a record of solar system formative and evolutionary processes at various heliocentric distances, rather than being a chaotic mixture. The extraordinary red color of the early Centaur [5145 Pholus (*Mueller et al.,* 1992)] and TNO [1992 QB1 (*Jewitt and Luu,* 1993)] discoveries seemed to confirm the idea that objects at large heliocentric distances exhibited red surface colors. Transneptunian objects were thought to form over a small range of distance from the Sun where the temperature in the young solar nebula was the same. The similar temperature suggested that TNOs formed out of the solar nebula with the same mixture of molecular ices and the same ratio of dust to icy material. In addition, their similar formation distance from the Sun suggested that TNOs experienced a similar evolution. For example, the irradiation of surface CH_4 ice by cosmic rays, solar ultraviolet light, and solar wind particles could convert some surface CH_4 ice into red, complex, organic molecules. CH_4 ice was known to exist on the surface of Pluto (*Cruikshank et al.,* 1976). Laboratory simulations supported the conversion of CH_4 ice into complex organic material (e.g., *Moore et al.,* 1983; *Strazzulla et al.,* 1984; *Andronico et al.,* 1987). By their nature, the complex organic molecules were expected to absorb more incident blue sunlight than red sunlight. Therefore, the light reflected from the surfaces of TNOs, and observed on Earth, was expected to consist of a larger ratio of red to blue light than the incident sunlight. It was a surprise, therefore, to discover that TNOs exhibited a wide range of colors.

Dynamically, the Edgeworth-Kuiper belt is strongly structured, and three main populations are usually distinguished within this region. The resonant TNOs are trapped in mean-motion resonances with Neptune, in particular the 2:3 at 39.4 AU (the so-called "Plutinos"), and are usually on highly eccentric orbits. The less-excited classical TNOs, also called "classicals," populate the region between the 2:3 and the 1:2 (say 40 AU < a < 50 AU) resonances. The "scattered disk objects" (SDOs) make up a less-clearly-defined population and are mainly objects with high eccentricity e and perihelion near Neptune that were presumably placed in these extreme orbits by a weak interaction with Neptune. The "Centaurs," finally, represent a dynamical family of objects in unstable orbits with semimajor axes between Jupiter and Neptune. The classification given here is historically the first one adopted by the community, but the frontiers between each group are not easy to define. Moreover, the classicals are a diverse population consisting of objects ranging from low to very high degrees of dynamical excitation. Obviously this population needs to be dissected carefully. A most complete and rigorous classification is discussed in the chapter by Gladman et al.

How do the color properties compare within each dynamical group of TNOs? Answering this question will provide important clues to understand how these objects have been formed and how their surface properties have evolved. Indeed the different orbital properties characterizing each dynamical family of TNO should have a distinct impact on the color properties of the surfaces of TNOs. For example, those objects in extreme orbits (high i and e) experience more energetic collisions than those in circular orbits. Such energetic collisions in turn have a significant impact on resurfacing, thus on the color of the surface.

In the sections that follow, we discuss observational strategies, data reduction techniques, and analysis methods of TNO photometric surveys (section 2), observed optical and near-infrared colors (section 3), correlations between colors and orbital elements (section 4), links between colors and chemical and mineralogical properties of TNOs (section 5), and possible processes responsible for the observed colors of TNOs (section 6).

2. PRINCIPLE AND TOOLS OF TRANSNEPTUNIAN OBJECT PHOTOMETRY

2.1. Observational Strategy

Due to the faintness of the TNOs and Centaurs, it is a very difficult and telescope-time-consuming task to obtain spectroscopy of a statistically representative sample that could be used for in-depth studies of surface properties and their correlations with other physical, dynamical, formation, and evolutionary parameters of the objects. However, a first and much faster (in terms of telescope time) assessment of intrinsic surface properties of these objects is possible through photometric characterization of their reflectance spectra by determining broadband filter colors and spectral gradients derived from the filter measurements. It is assumed that sunlight is reflected at the surfaces of the objects only, and no atmosphere or dust coma is present. Activity induced by intrinsic processes (gas evaporation or cometary activity) or external processes (impacts) is presumed to play a role in the resurfacing of TNOs and Centaurs. In rare cases, such as Pluto (with a tenuous atmosphere) and Chiron (showing episodic cometary activity), the presence of an atmosphere and/or a dust coma may affect the interpretation of the measured colors of the objects.

Since they are well characterized and available at many observatories, astronomical broadband filters are used for the photometric measurements of TNOs and Centaurs, i.e., BVRI for the visible wavelength range and JHK for the near-IR. Due to the faintness of the objects in the respective wavelength range, U and L and M filters can only be used for the very brightest objects (e.g., Pluto-Charon). For the visible, the vast majority of data are available in the Bessel filter set, while only a few measurements are obtained using the Johnson filter set (*Sterken and Manfroid,* 1992, and references therein), which differs from the former one in the transmission ranges of the R and I filters. For the near-IR filter set, the differences in the J_s and K_s transmission ranges compared to standard J and K filters (*Glass,* 1997, and references therein) is of minor importance for color studies of the objects.

Transneptunian objects are challenging to observe because they are faint, move relative to the "fixed" background of stars and galaxies, and exhibit brightness variations (lightcurves). In this section, we discuss these challenges and techniques to overcome them. Note that these problems are reminiscent of the first works performed on main-belt asteroids in the early 1960s and 1970s when available technology at that time did not allow the use of more sophisticated techniques, i.e., spectroscopy (see *McCord and Chapman,* 1975; *Bowell and Lumme,* 1979, and references therein). Then the same logic was applied as observers progressed from the main belt to Hildas, Trojans (*Hartmann et al.,* 1982), and finally Centaurs and TNOs (*Luu and Jewitt,* 1996).

2.1.1. Faintness. First, the most challenging problem is that the TNO population represents some of the faintest objects in the solar system. The typical apparent visual magnitude of a TNO is about 23 mag, although a few objects brighter than 20 mag are known. Another criterion is the minimum signal-to-noise ratio (SNR) of the photometry: Typically, SNR ~30 (corresponding to an uncertainty of 0.03–0.04 mag) is required to accomplish the spectral classification. Indeed, the differences among the known types of minor bodies (which to first order have the color of the Sun) are much more subtle than for stellar types, so considerably higher photometric accuracy is necessary. The more limited the range of wavelengths, the higher the accuracy needed to distinguish objects of different type. However, to reach such a SNR goal, one needs to overcome two problems: (1) the sky contribution, which could be important and critical for faint objects not only in the infrared, but also for visible observations; and (2) the contamination by unseen background sources such as field stars and galaxies. For instance, the error introduced by a 26-mag background source superimposed on a 23-mag object is as large as 0.07 mag. One solution to alleviate these problems is the use of a smaller aperture for the flux measurement of the object. This method is described in the next section.

2.1.2. Motion relative to background. Transneptunian objects, while at very remote heliocentric distance, still have noticeable motion that restricts the exposures to relatively short integration times. At opposition, the motion of TNOs at 30, 40, and 50 AU are, respectively, about 4.2, 3.2, and 2.63/h, thus producing a trail of ~1.03 in 15 min exposure time in the worst case. The trailing of an object during an exposure has devastating effects on the SNR, since the flux is diluted over a larger area of background sky. This dilution, in turn, introduces higher noise. Thus, increasing exposure time will not necessarily improve the SNR for TNO photometry. For practical purposes, the exposure time is chosen such that the trailing due to the object's motion does not exceed the size of the seeing disk. The observer then takes and combines a number of untrailed exposures to improve the SNR. An alternative is to follow the object at its proper motion. In this case, however, one faces another problem: The point-spread function of the object (PSF) would be obviously different from that of the field stars. Moreover, the so-called aperture correction technique used very frequently for accurate TNO photometry would become very difficult since this method requires the PSF calibration of nearby field stars (see next section).

2.1.3. Lightcurves. The color of a TNO or Centaur is ideally measured simultaneously for the two or more filters in order to guarantee that the same surface area of the object is imaged at the same time. However, the majority of color measurements are obtained by exposing the object through different filters sequentially. In order to avoid color changes induced by albedo and/or shape variations of the likely rotating body, it is important to measure the object flux through the various filters quasi-simultaneously, i.e., within a time interval that is short compared to the rotation period of the object. For instance, for an object with a 10-h rotation period, the phase angle of the surface changes by 10° in 30 min. Hence, very fast rotators can mimic strange surface colors or unexpected color variations [see, e.g., 1994 EV_3 (*Boehnhardt et al.,* 2002)]. Since for many objects the exact rotation period is unknown, it is thus advisable to monitor changes in the object brightness due to rotation by repeated exposures in a single filter (e.g., V filter) intermittent during the complete sequence of filter exposures needed for the color measurement of the object. If the rotation period is known, one can phase the filter exposures to appropriately image the same surface area under the same illumination conditions.

2.2. Data Reduction Techniques

The data reduction consists of the basic reduction steps applicable for the visible and near-IR photometric datasets, i.e., bias and flatfield corrections, cosmic-ray removal, alignment and co-addition of the jittered images, and flux calibration through standard stars.

The brightness of the object is frequently measured through the so-called aperture correction technique (*Howell,* 1989; *Tegler and Romanishin,* 1997; *Doressoundiram et al.,* 2001) as justified by the faintness of the TNOs. The basis of this method is that the photometric measurement is performed by using a small aperture on the order of the size of the seeing disk. Consequently, the uncertainty in the measurement is reduced because less noise from the sky background is included in the aperture. However, by doing so, one loses light from the object. Thus, to determine how much light is thrown away, the so-called "aperture effect" is calibrated using a large number of nearby field stars. This is reasonable as long as the motion of TNOs during each exposure is smaller than the seeing, and hence the TNOs' point-spread functions (PSFs) are comparable to those of field stars. For all the photometry, the sky value can be computed as the median of a sky annulus surrounding the object. The advantages in the use of a small aperture are (1) a decrease of the contribution of the sky, which could be important and critical for faint objects; and (2) a reduction of the risk of contamination of the photometry by unseen background sources.

2.3. Analysis Methods

Many useful physical parameters of TNOs can be derived from broadband photometry. These parameters include color, spectral gradient, absolute magnitude, and size.

2.3.1. Color indices. The color indices used here (such as U–V, B–V, etc.) are the differences between the magnitudes obtained in two filters. In other words, the F_1–F_2 color index (where F_1 and F_2 are any of the UBV, etc., filters) is the ratio of the surface reflectance approximately valid for the central wavelengths λ_1 and λ_2 of the corresponding filters.

2.3.2. Reflectance spectrum. The information contained in the color indices can be converted into a very-low-resolution reflectance spectrum R_F using

$$R_F = 10^{-0.4\,(M_F - M_F\mathrm{Sun})}$$

where M_F and M_FSun are the magnitudes in filter F of the object and of the Sun, respectively (M_VSun = –26.76). Normalizing the reflectance to 1 at a given wavelength (conventionally, the V central wavelength is used), we have

$$R_{F,V} = 10^{-0.4\,[(M_F - M_V) - (M_F - M_V)\mathrm{Sun}]}$$

The filter magnitudes and colors of the Sun for the filters commonly used can be found in *Hardorp* (1980) and *Degewij et al.* (1980). Values of the main colors for the Sun used here are B–V = 0.67, V–R = 0.36, V–I = 0.69, V–J = 1.03, V–H = 1.36, and V–K = 1.42. Figure 1 illustrates the coarse reflectance spectrum using broadband filter photometry compared to spectroscopy of a TNO.

2.3.3. Spectral gradient. The spectral gradient S is a measure of the reddening of the reflectance spectrum between two wavelengths. It is expressed in percent of reddening per 100 nm

$$S(\lambda_2 > \lambda_1) = (R_{F2,V} - R_{F1,V})/(\lambda_2 - \lambda_1)$$

where λ_1 and λ_2 are the central wavelengths of the F_1 and F_2 filters, respectively. S, given in percent of reddening per 100 nm wavelength difference, is intimately related to the constitution of the surface of the object, although at present it is not possible to draw any detailed conclusions from spectral gradients (or colors) on specific surface properties. By definition, spectral gradient values are a direct measure for the intrinsic reddening of the object produced by the surface properties of the object, i.e., the solar colors are "removed"; hence, S = 0 means exactly solar colors. If several filters are measured over a wider wavelength range (e.g., BVRI to cover the visible spectrum), the spectral gradients can be averaged over the main adjacent color indices (e.g., B–V, V–R, R–I) in order to obtain the overallslope of the reflectance spectrum between the two limiting wavelengths.

2.3.4. Absolute magnitude. The absolute magnitude of a TNO is the magnitude at zero phase angle and at unit heliocentric and geocentric distances. Geometrical effects are removed by reducing the M_F visual magnitude (in the F filter) to the absolute magnitude $M_F(1,1,0)$ using

$$M_F(1,1,0) = M_F - 5\log(r\Delta) - \alpha\beta$$

where r, Δ, and α are respectively the heliocentric distance (AU), the geocentric distance (AU), and the phase angle (deg). The term $\alpha\beta$ is the correction for the phase brightening effect (*Belskaya and Shevchenko,* 2000; see also the chapter by Belskaya et al.). However, the phase function is

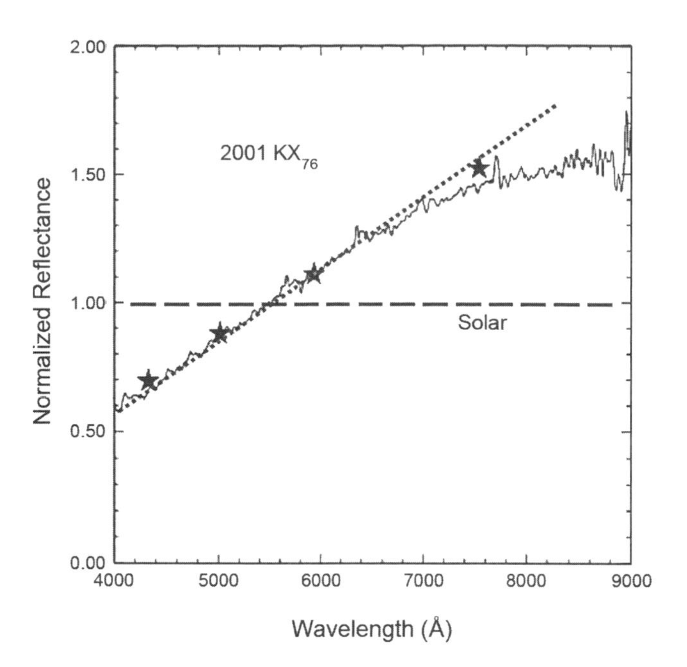

Fig. 1. Reflectance spectrum, broadband filter brightness, and spectral gradient of the TNO 2001 KX_{76} or (28978) Ixion. The reflectance is shown normalized to unity at 550 nm. The BVRI broadband filter brightness is plotted by "star" symbols (from short to long wavelength) over the central wavelength of the respective filter bandpass. The spectral gradient fitted to the spectroscopy and photometry of the object is shown as a dotted line. Neutral/solar reflectance is represented by the dashed line.

completely unknown for most of the TNOs. Based on a first-phase-curve study of a few TNOs, *Sheppard and Jewitt* (2002) have shown almost linear and fairly steep phase curves in the range of phase angles from 0.2° to 2°. They found an average β of 0.15 mag/deg. A similar steep slope was found for Centaurs (*Bauer et al.,* 2002). This implies a possible considerable error in H calculations disregarding the phase correction. Further studies have also been made measuring phase curves in parallel with polarimetry (*Bagnulo et al.,* 2006; *Boehnhardt et al.,* 2004).

2.3.5. Size. Size is the most basic parameter defining a solid body. Unfortunately, sizes (and albedos) are difficult to determine as they mostly require thermal measurements. Hence objects with accurate size determination are few (see chapter by Stansberry et al.). Consequently, for most objects, a canonical geometric albedo p is assumed and the absolute magnitude $M_F(1,1,0)$ can be converted into the radius R of the object (km) using the formula by *Russell* (1916)

$$pR^2 = 2.235 \times 10^{16}\, 10^{0.4(M_F\mathrm{Sun} - M_F(1,1,0))}$$

where M_FSun is the magnitude of the Sun in the filter F (M_VSun = –26.76). Owing to the lack of available albedo measurements, it has become the convention to assume an albedo of 0.04, common for dark objects and cometary nuclei. However, one should be aware of the fact that the sizes are purely indicative and are largely uncertain. For instance, if we would use, instead, an albedo of 0.14 (i.e., the albedo of the Centaur 2060 Chiron), the results for size estimates would have to be divided by a factor of about 2.

Having these limitations in mind (unconstrained phase function and albedo), in all the following analysis and figures that make use of absolute magnitude and size, we have taken the above values for the phase correction and an average albedo of 0.09 for the geometric albedo (R band).

3. COLOR DIVERSITY AND SPECTRAL GRADIENT

O. Hainaut from the European Southern Observatory in Chile has compiled and maintains a database of color measurements of TNOs, Centaurs, and short-period comets. The database can be accessed at *www.sc.eso.org/~ohainaut/MBOSS*. As of late September 2006, this MBOSS database contains values for 209 objects (Plutinos 32; classicals 95; SDOs 32; Centaurs 30, comets 20) at 1077 epochs covering the UBVRIJHKLM filter set (not all the objects have the full filter set). It also provides references to the values stored. (Note that Tegler and Romanishin have also put colors and magnitudes from their survey for ~120 TNOs and Centaurs on the web at *www.physics.nau.edu/~tegler/research/survey.htm*.)

Both colors and spectral gradients are "integral" characterization parameters of surface properties of the objects: "integral" over the illuminated and visible part of the surface, which for TNOs and distant Centaurs is almost equivalent to a complete hemisphere since the phase angles are small, and "integral" over the wavelength range of the filters used.

3.1. Color-Color Diagrams of Transneptunian Objects

Color-color diagrams show two measured color indices of the object sample vs. each other, e.g., V–R color vs. B–V color as measured for TNOs and/or Centaurs. Solar colors provide the reference for zero intrinsic reddening. Objects with "higher" color indices than solar are called "redder" than the Sun, those with "smaller" values are "bluer" than the Sun. The "redder"/"bluer" colors are equivalent to a positive/negative spectral gradient compared to the solar spectrum. For guidance, the reddening trend line can be drawn in the color-color diagrams: This line connects locations in the color-color diagram with increasing values of constant spectral reflectance slope. Deviations from the reddening trend line indicate a nonlinear behavior of the intrinsic reddening over the spectral range covered by the filters of the color-color diagram. Examples for color-color diagrams of TNOs and Centaurs are shown in Fig. 2.

Color changes over the body's surface combined with rotation of the object can result — in principle — in variations of the object's colors depending on the rotation phase when the measurements are performed (even for simultaneous or quasi-simultaneous filter exposures). For Centaurs in eccentric orbits, changes in the aspect angle due to orbital motion can also result in color variations for objects with nonhomogenous surface colors. (Due to the long orbital periods of TNOs, changes of the aspect angle along the orbit may happen over timescales of decades only, a time interval much longer than the time span during which filter observations of TNOs have been performed up to now.) Similar statements were also made in the 1960s to 1980s when interpreting rotational effects and albedos and colors (see, e.g., *Bowell and Lumme,* 1979; *French and Binzel,* 1989). Rotation phase resolved color measurements of only a few objects are published so far (see chapters by Sheppard et al. and Belskaya et al.). In order to achieve a complete characterization of the color properties of a body, full coverage of the rotation (aspect) phase range is required (but rarely done). In cases where several color measurements are available, averaging of the results can be performed to obtain a mean value for the colors and spectral gradients.

Color-color diagrams of TNOs (as well as of Centaurs and cometary nuclei; see Fig. 2) in the visible show a wide distribution from neutral and even slightly "bluish" to very red values. Figure 2 also shows the various dynamical classes of the TNOs (i.e., classical objects, Plutinos, scattered disk objects) and the Centaurs. All subclasses are color undistinguishable, which is consistent with a common origin for these objects.

For the overall TNO population (of more than 150 objects with BVRI color measurements) clustering is not obvious although originally proposed based upon much smaller samples (*Tegler and Romanishin,* 1998). However, clustering in color-color diagrams is found when considering different populations of TNOs and Centaurs (see discussion below). The trend line overplotted in Fig. 2 represent the locus of object with a linear reflectance spectrum. The "diagonal orientation" of the TNO cloud in the V–R vs. B–V plot in Fig. 2, following the trend line of increased reddening, suggests constant surface reddening from B to R for the majority of the objects. Instead, the respective R–I vs. B–V color-color plot in Fig. 2 clearly indicates systematic deviations from the trend line toward smaller reddening at the long wavelength end of the visible spectrum. In near-IR color-color diagrams (H–K vs. J–H in Fig. 3), a strong clustering of TNOs around solar colors is found that clearly indicates flat and close to neutral surface colors for the majority of the objects. Very few objects show significant reddish colors (*Delsanti et al.,* 2006). Outliers toward bluer than solar colors exist and may be due to absorption bands of surface ices (e.g., H_2O, CH_4) with influence on broadband colors (in particular in H and K filters). In this respect "bluish" near-IR colors may be used to identify potential objects with ice absorption spectra — although the photometric indices are certainly not qualified to replace spectroscopic analysis of the objects for ice signature detections. The changeover from the red color in visible to the more neutral near-IR spectrum in the individual objects happens between about 750 and 1400 nm and thus affects mostly the I and J filter reflectance. However, it also explains the systematic deviations from the trend lines in R–I vs. B–V color-color diagrams. In other words, visible colors are mutually correlated, while they appear to be unrelated to infrared colors (*Doressoundiram et al.,* 2002; *Delsanti et al.,* 2006).

3.2. Spectral Gradient Histograms of Transneptunian Objects and Centaurs

The full range of spectral gradients of the object sample should be covered by a dense set of [S_{min}, S_{max}] pairs, the lower and upper limit for the spectral gradient of the objects to be counted in the respective histogram bin. Examples for spectral gradient histograms of TNOs and Centaurs are shown in Fig. 4. In the following analysis spectral gradients among TNOs and Centaurs have been obtained from BVRI broadband photometry extracted from the MBOSS database.

The surface characterization by spectral gradients somehow implies and is best used for objects with rather featureless reflectance spectra (i.e., no absorption due to surface ices). This requirement may reasonably be fulfilled in the visible wavelength range where only weak absorption bands are found in a few TNOs (see chapter by Barucci et al.) without noticeable effects on the filter magnitudes. The near-IR wavelength region contains various strong absorption bands of H_2O and CH_4 ices, which have been found in some TNOs. Hence, effects on the surface colors are possible and, indeed, the "bluish" colors J–H and/or J–K may be due to these ice absorptions (see below).

The range of spectral gradients of TNOs and Centaurs in the visible wavelength region ranges from –5%/100 nm to 45%/100 nm, with the exception of some very red Centaurs. The spectral gradient distributions for the four major dynamical classes (classicals, SDOs, Plutinos, and Centaurs) display evident differences (Fig. 4): Classicals cluster between 25 and 35%/100 nm and are abundant also at somewhat smaller reddening slopes (i.e., between 15 and 25%/100 nm). Plutinos and SDOs peak at moderately red levels, while Centaurs show a double peak distribution with a high number at low reddening (5–15%/100 nm, comparable to the range of the distribution maximum for the Plutinos and SDOs) and a shallower peak with very red slopes (35–

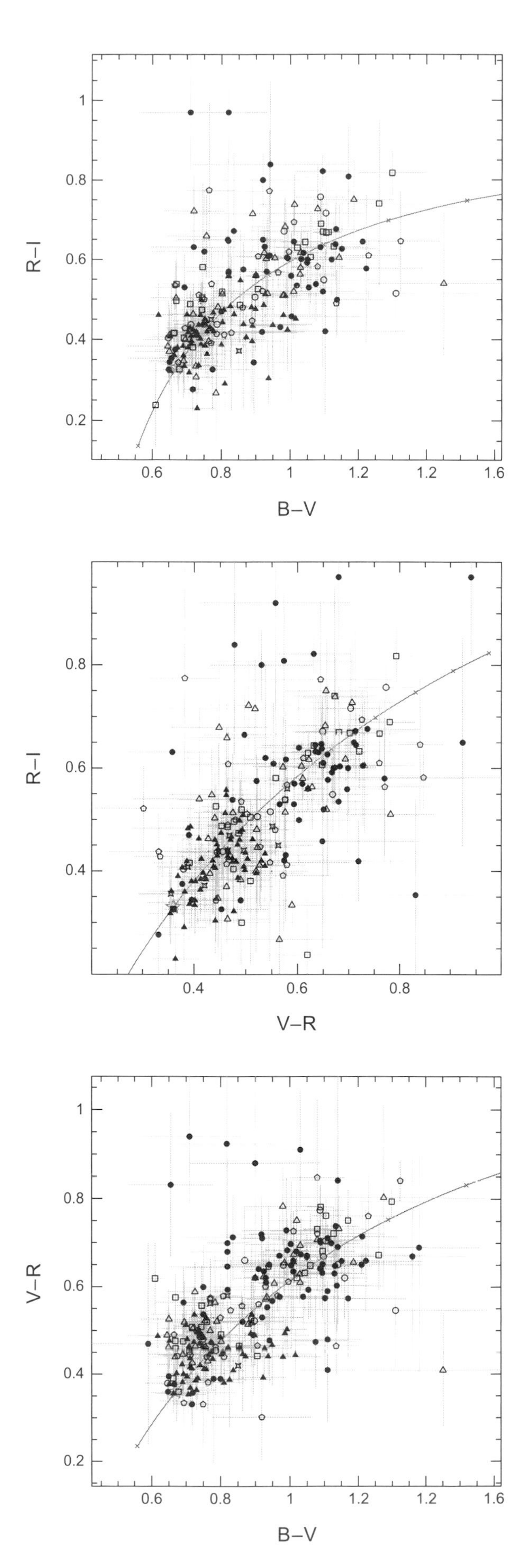

Fig. 2. Color-color diagrams of TNOs (Plutinos = open triangles, classicals = filled circles, SDOs = open pentagons, Centaurs = open squares, Trojans = filled triangles, comets = open four-spike star). The diagrams apply for broadband filter results in BVRI. Solar colors are indicated by an open five-spike star symbol. The curve is the locus of object with a linear reflectivity spectrum. The Sun, as a reference, has a linear reflectivity spectrum with a null slope. A tick mark is placed every 10%/100 nm; the curve is graduated in units of spectral slope from 0 (solar) to 70%/100 nm (very red). Plots are from the web page of O. Hainaut (*www.ls.eso.org/~ohainaut/MBOSS*), who compiles photometric results of minor bodies in the outer solar system.

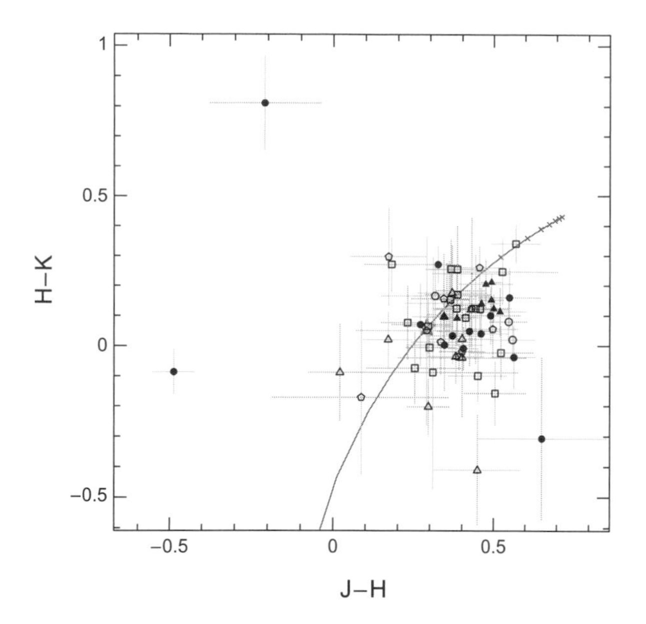

Fig. 3. J–H vs. H–K color-color diagram of TNOs and Centaurs. Symbols are Plutinos = open triangles, classicals = filled circles, SDOs = open pentagons, Centaurs = open squares, Trojans = filled triangles, comets = open four-spike star, Sun = open five-spike star, trend line of increased reddening = solid line. This line is the locus of objects with a reflectivity spectrum of increasing linear slope (see also Fig. 2). Most of the points lie below the reddening curve. This is an indication for a change and decrease of the spectral slope over the JHK range. The two outliers are (19308) 1966 TO_{66} (top left) and (24835) 1995 SM_{55} (bottom left).

45%/100 nm, similar to the main maximum in the classicals histogram). The structured histograms suggest that further subpopulations exist for classicals and Centaurs, which are described and analyzed by more sophisticated statistical methods below. The overall reddening distributions of the TNOs and Centaurs need an interpretation based upon surface processing scenarios and modeling. The spectral gradient in the near-IR wavelength region tend to flatten off (close to solar-type colors are mostly measured) and is not very instrumental for population classification except for the identification and analysis of outliers.

Such analysis is reminiscent of early work done with asteroids in the 1970s and 1980s. In particular, a relation between color (or spectral gradient) and ice abundance in outer solar system objects, as caused by spectral absorption bands, was the background of *Hartmann et al.*'s (1982) work and a subsequent series of papers. These authors showed a direct correlation between position in a V–J vs. J–K diagram, and the measured albedos and ice contents of well-observed satellites and asteroids.

4. COLOR DISTRIBUTION OF THE TRANSNEPTUNIAN OBJECTS: STRONGLY SHAPED

The main purpose is to see if color is a tracer for any physical process. While any physical processes are of interest, we have very little detailed knowledge of physical mechanisms. The physical processes that are easiest to test

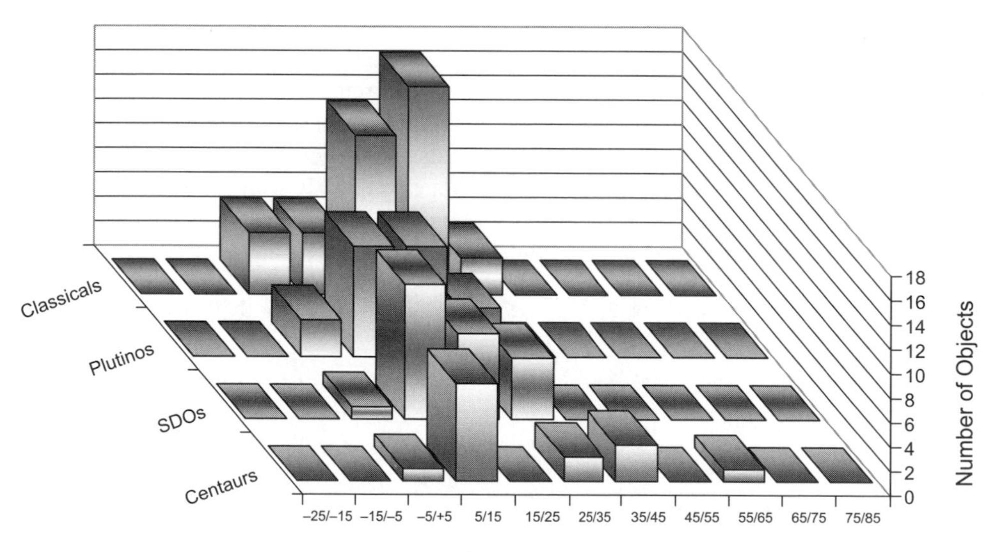

	–25/–15	–15/–5	–5/+5	5/15	15/25	25/35	35/45	45/55	55/65	65/75	75/85
Centaurs	0	0	1	8	0	2	3	0	1	0	0
SDOs	0	0	1	11	7	5	0	0	0	0	0
Plutinos	0	0	3	9	9	4	0	0	0	0	0
Classicals	0	0	5	5	13	17	3	0	0	0	0

Fig. 4. Spectral gradient histogram of TNOs and Centaurs (after *Doressoundiram and Boehnhardt,* 2003).

with such limited knowledge consist of comparisons of color with combinations of orbital physical parameters.

4.1. Database of Colors and Color Indices

For this analysis, we restrict our sample to those collated in the MBOSS database used in the previous section. Here we focus on the visible colors of the TNO surfaces. The analysis is made simpler by the fact that visible colors are mutually correlated throughout the VRI regime (*Boehnhardt et al.,* 2001; *Doressoundiram et al.,* 2002).

This is the basic premise behind the supposition that the coloring agent could be tholin-like materials (*Jewitt and Luu,* 2001; *Cruikshank et al.,* 2005), which tend to produce a uniform color gradient throughout the visible region. This fact allows the transformation of most color values in the MBOSS database to a common reddening scale, the percent of reddening per 100 nm, S (see section 2.3) implemented for the MBOSS database by *Hainaut and Delsanti* (2002). For objects with multiple color values, all are combined into this metric with formal error propagation.

A plot of true V–R vs. the S metric appears in Fig. 5, illustrating that this transformation is valid for objects with multiple color data. The MBOSS database allows us to look at the VRI color characteristics among 176 TNOs and Centaurs listed in the Minor Planet Center database.

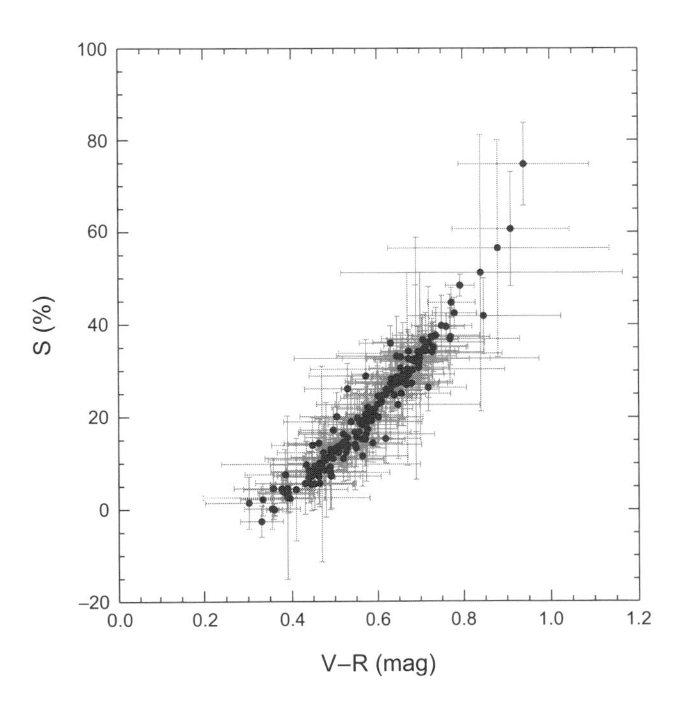

Fig. 5. This plot shows a strong correlation between the reddening factor S (computed in the visible wavelength) and the V–R color of objects. Thus, our transformation of all color data to the common S scale is consistent with measured values of the most common color filter difference, V–R. The use of the S reddening factor allows combinations of all visible colors to be used in correlation analysis and reduces error bars by significant amounts by combining data throughout the VRI wavelength region.

4.2. Correlations

The main purpose of collecting a wide range of color data is to consider whether TNO surface color is related to dynamical factors and can give an indication of how solar system wide factors may impact TNO composition. Plates 1 and 2 are an overview of the color distribution of TNOs obtained within the Meudon Multicolor Survey (2MS) of *Doressoundiram et al.* (2005). The advantage of this view is that it offers to the eye global patterns between color and orbital elements. Although the search for correlations between colors and dynamical information such as orbital elements is open ended, we here limit the search to basic orbital element data and a few derived quantities. The most basic and robust statistical test searching for correlations is the Spearman rank correlation test, which does not presuppose a functional form for the correlation, as do many other statistical tests (*Spearman,* 1904).

4.2.1. Criteria for statistical significance. We compare our color metric S against the most basic of dynamical parameters, orbital elements. Since we are looking for a correlation between color and many potential parameters, we must use a higher criterion for statistical significance than the standard 3σ used throughout astronomy. Indeed, simple simulations show that if a set of 176 uniformly distributed numbers (representing our 176 color values) are compared against n = 6 other uniformly distributed datasets (e.g., representing the 6 basic orbital parameters), a > 3σ spurious "correlation" occurs with the Spearman test 1.35% of the time, a factor (n – 1) = 5 in excess of the 0.27% rate expected and seen for comparison to a single orbital parameter.

Therefore, to properly compare color correlations among the orbital parameters, we must raise our threshold to $\approx 3.5\sigma$ for six orbital parameters to realize the 0.27% spurious correlation rate for a 3σ event for a single orbital parameter. Since the probability for spurious correlation increases roughly linearly with the number of extra parameters tested, and Gaussian probabilities are a strong function of σ values, even a moderate increase to 3.8σ is appropriate for considering correlations of up to n = 20 parameters as we do here. We adopt this 3.8σ value here for the remainder of our discussion.

4.2.2. Populations considered. It is quite clear that neither the Plutinos (in the 2:3 mean-motion resonance at 39.4 AU) nor the SDOs (q near Neptune) have any significant correlation between the MBOSS colors and any orbital parameter, and no researchers have reported such correlations. For completeness, we do test the Plutinos and SDOs here. It is also quite clear that Centaurs (q > 5.2 AU and a < 30.1 AU) have a bimodal color distribution of a strength not seen in any TNO population (*Peixinho et al.,* 2003; *Tegler et al.,* 2003; chapter by Tegler et al.). Discussion of the Centaur population is not included in this work as they represent a population that is much more dynamically unstable than the TNOs and is potentially affected by recent cometary activity. Thus, there remain only two populations in the Edgeworth-Kuiper belt that could have any correlation between color and orbital parameters: the classical TNOs and the "other TNOs," TNOs with semimajor axes in the range (40 AU < a < 50 AU) but with eccentricities too high to be considered classicals. We define this population as the "near-scattered TNOs," since they have similar perihelion properties to the SDOs, but are close enough that the separation between them and the classical TNOs is not easily defined.

It is not easy to determine where the classical TNOs end and the near-scattered TNOs begin. Observers traditionally arbitrarily choose a minimum perihelion value to distinguish between the classical and near-scattered TNOs somewhere between 36 AU and 41 AU. However, dynamical studies have shown that the 40 AU to 50 AU region has many narrow resonances that can mask objects that are true classical TNOs. The *lower limit on the perihelion for discriminating the classical TNO population from the near-scattered TNOs is critical for detecting a color/orbital element correlation* with either population. Evaluating exactly where the perihelion criterion should be set to define classical TNOs is difficult at best and beyond the scope of this work as it is primarily a dynamical argument (see chapter by Gladman et al.). Instead we treat all these TNOs as one class, included in the single denomination of *classical* TNOs, defined as 40 AU < a < 50 AU whatever their eccentricity.

4.2.3. Summary of significant correlations. A statistically significant clustering of very red classical TNOs with low eccentricity (e < 0.05) and low inclination (i < 5°) orbits beyond ~40 AU from the Sun, as first suggested by *Tegler and Romanishin* (2000) from a much smaller dataset, has been confirmed by *Doressoundiram et al.* (2002) and *Trujillo and Brown* (2002). *Doressoundiram et al.* (2002) have shown that this cluster of low-i classical TNOs is statistically distinct from the others classical TNOs (higher i, diverse colors) at the 99% significance level. This red cluster is clearly seen in Plate 1, where objects with perihelion distances around and beyond 40 AU are mostly very red. The cluster members have similar dynamical and surface properties, and they may represent the first taxonomic family in the Edgeworth-Kuiper belt.

Many studies have been performed to search for correlations between color and physical parameters (e.g., size). So far, no significant trend has been found between color and size, or absolute magnitude. This points out that the coloring process (whatever it is) is not sensitive to size, at least in the size range accessible by groundbased observations.

We tested the following four populations: all TNOs (q > 25 AU), the Plutinos (38.5 AU < a < 40 AU), the SDOs (q near Neptune), and the classical TNOs (40 AU < a < 50 AU). These were tested against the following orbital parameters: inclination, eccentricity, perihelion, and the root-mean-square orbit velocity, V_{rms}, as discussed by *Stern* (2002). Indeed, collisional resurfacing could also provide a possible explanation for the orbital element dependency of the color indices: One would naturally expect highly excited objects to suffer more energetic impacts. No other orbital parameters were tested, because no correlations have been reported for other parameters. The results of the correlation tests are shown in Table 1. It is *clear that the entire TNO population shows a statistically significant correlation between color and both rms velocity and inclination*, this trend dominated by the TNOs with 40 AU < a < 50 AU (see Fig. 6), which also show a statistically significant correlation with inclination and rms velocity. There is a moderate color/perihelion correlation for the entire TNO population, but it is at the threshold of significance considering the number of orbital parameters we are testing and is not significant for any subpopulation. Thus, we consider both

TABLE 1. Color correlations.

Population	Correlation Color vs. . . .	N	Significance (Gaussian)	Significance (1 – Conf.)	Significant?
All TNOs (q > 25 AU)	rms velocity	141	5.7σ	1×10^{-8}	Yes
All TNOs (q > 25 AU)	Inclination	141	5.5σ	4×10^{-8}	Yes
All TNOs (q > 25 AU)	Perihelion	141	3.8σ	2×10^{-4}	Maybe
All TNOs (q > 25 AU)	Eccentricity	141	3.1σ	2×10^{-3}	No
Classical TNOs (40 < a < 50 AU)	rms velocity	78	4.5σ	7×10^{-6}	Yes
Classical TNOs (40 < a < 50 AU)	Inclination	78	4.9σ	8×10^{-7}	Yes
Classical TNOs (40 < a < 50 AU)	Perihelion	78	2.5σ	1×10^{-2}	No
Classical TNOs (40 < a < 50 AU)	Eccentricity	78	1.3σ	0.2	No
Scattered (a > 50 AU)	rms velocity	26	1.2σ	0.2	No
Scattered (a > 50 AU)	Inclination	26	1.2σ	0.2	No
Scattered (a > 50 AU)	Perihelion	26	0.6σ	0.5	No
Scattered (a > 50 AU)	Eccentricity	26	1.3σ	0.8	No
Plutinos (38.5 < a < 40 AU)	rms velocity	32	0.0σ	1.0	No
Plutinos (38.5 < a < 40 AU)	Inclination	32	0.0σ	1.0	No
Plutinos (38.5 < a < 40 AU)	Perihelion	32	0.2σ	0.8	No
Plutinos (38.5 < a < 40 AU)	Eccentricity	32	0.1σ	0.9	No

The criterion for statistical significance is set to 3.8σ in this investigation as we tested for correlations between color and 16 different parameter/subpopulation combinations, as explained in the text.

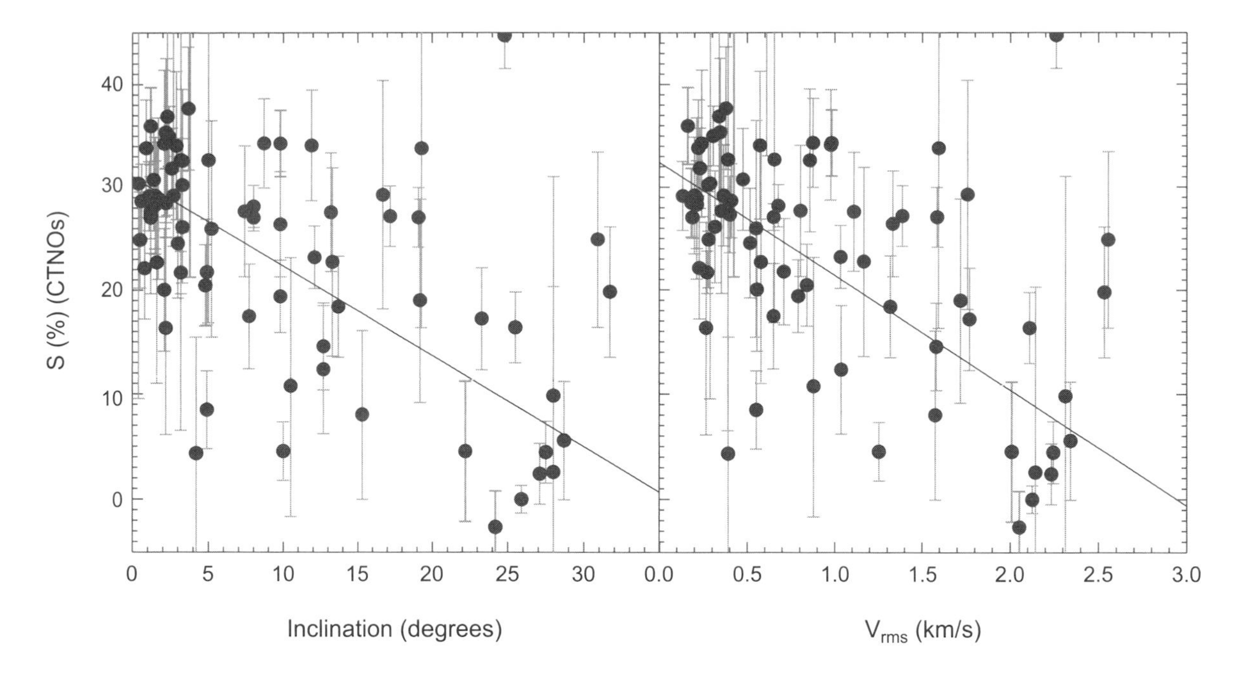

Fig. 6. Color metric S for the classical TNOs vs. inclination and rms orbit velocity V_{rms}. A least-squares linear slope has been fitted to the data to guide the eye. Both of these correlations are significant at the 4σ level using the Spearman test, which does not assume a functional form to the correlation.

the color/inclination and color/rms velocity correlations to be statistically significant.

Although the Spearman rank correlation test excels at determining whether a correlation is statistically significant, it is not useful as a test to determine which correlation is more significant, the inclination or the rms velocity correlation. Thus, it remains to be seen whether a specific testable mechanism can be found that produces either an inclination/color correlation or a rms velocity/color correlation.

5. COLORS AND SURFACE PROPERTIES

5.1. Colors and Chemical Properties

Is there any correlation between the presence of a particular ice or mineralogical absorption band in the spectra of TNOs and their colors? For example, do TNOs with CH_4-ice bands in their spectra display redder surfaces than TNOs with H_2O-ice bands in their spectra? This is a reasonable correlation to study because Pluto exhibits CH_4 bands and it has a redder color than Charon, which exhibits H_2O-ice bands.

The difficulty with looking for such a correlation among TNOs is that there are only a handful of objects that are known to exhibit absorption bands in their spectra. In particular, H_2O-ice bands are seen in spectra of Charon, 1996 TO_{66} (*Brown et al.,* 1999), Varuna (*Licandro et al.,* 2001), Quaoar (*Jewitt and Luu,* 2004), Orcus (*Fornasier et al.,* 2004a), and 2003 EL_{61} and its satellite (*Barkume et al.,* 2006). CH_4-ice bands are seen in spectra of Pluto (*Cruikshank et al.,* 1976); Neptune's satellite Triton, which may be a captured TNO (*Cruikshank et al.,* 1993); Eris (*Brown et al.,* 2005); and 2005 FY_9 (*Licandro et al.,* 2006). Mineralogical bands are seen in the spectra of 2003 AZ_{84} (*Fornasier et al.,* 2004b), Huya, and 2000 GN_{171} (*Lazzarin et al.,* 2003).

Despite the small sample size, is there any correlation between the presence of CH_4, H_2O bands, and surface color? The seven TNOs exhibiting H_2O bands display colors ranging from B–V = 0.63 ± 0.03 for 2003 EL_{61} (*Rabinowitz et al.,* 2006) to B–V = 0.96 ± 0.08 for Varuna. For reference, Pluto and Charon have B–V = 0.87 ± 0.01 and B–V = 0.70 ± 0.01 (*Binzel,* 1988). The four TNOs exhibiting CH_4-ice bands and the three TNOs exhibiting mineralogical bands display a similar wide range of B–V colors. So far, there is not any obvious correlation between the presence of an ice or mineralogical band and the B–V colors of TNOs. However, larger samples may allow a better test.

5.2. Modeling of Surface Colors

What can modeling tell us about the surface colors? By comparing broadband photometric and spectroscopic observations with radiative transfer models (*Hapke,* 1981; *Douté and Schmitt,* 1998), it is possible to constrain the chemical composition of TNO surfaces. In particular, radiative transfer models make it possible to transform laboratory optical constants for candidate surface materials such as organic material (tholins and kerogen), dark material (amorphous carbon), ices, and silicates (tremolite) into reflectance spectra for comparison with observations. By inputting different percentages of candidate materials in a radiative transfer model and finding the best match between the theoretical spectra and the observed spectra, it is possible to constrain the chemical composition and grain sizes on the surfaces of TNOs.

What sort of compositions do the models suggest for objects at the extreme ends of the TNO color range, i.e., a red vs. a gray TNO? It appears the observed red colors of TNOs are consistent with the presence of a tholin-like material (*Cruikshank et al.,* 1998; see also chapter by de Bergh et al.). Tholins are synthetic macromolecular compounds produced from irradiated gaseous or solid mixtures of simple hydrocarbons, water, or nitrogen (*Sagan and Khare,* 1979). Although their optical properties depend on the original mixture and conditions of irradiation, all tholins show a common characteristic, i.e., colors ranging from yellow to red. Evidently, the numerous overlapping electronic bands of these complex organic molecules produce a red pseudo-continuum at ultraviolet and optical wavelengths. *Doressoundiram et al.* (2003) find the red color, V–J = 2.44 ± 0.06, of 26181 (1996 GQ_{21}) is consistent with a mixture of 15% Titan tholin (*Khare et al.,* 1984), 35% ice tholin I (*Khare et al.,* 1993), and 50% amorphous carbon. It is important to note that such a model is only a suggestion of what might be the nature of the surface material. The solution is not unique, given the many unknown parameters in the models.

At the other extreme of color, *de Bergh et al.* (2005) found that the visible and infrared colors and spectra of the gray TNO Orcus, V–J = 1.08 ± 0.04, is consistent with a Hapke-type model that has 5% H_2O-ice at 40 K, 42% kerogen, and 53% amorphous carbon. Terrestrial kerogens are the product of decay of organic matter, but interstellar dust and some carbonaceous meteorites contain materials of similar structure of nonbiological origin. Kerogens have been used to model the spectra of some asteroids. All components in the model of Orcus had a grain size of 10 μm. For reference, the Sun has V–J = 1.08. The chapters of de Bergh et al. and Barucci et al. in this book give a more complete overview on spectral modeling and links between color properties and composition.

6. ORIGIN OF THE COLOR ANISOTROPY AND PROPOSED RESURFACING SCENARIOS

6.1. A Primordial Origin

A dynamical simulation (*Gomes,* 2003; see also chapter by Morbidelli et al.) and some thoughts on methane chemistry in the outer solar system provide a way of interpreting the colors of TNOs. The dynamical simulation predicts that as Neptune migrated outward, it scattered objects originally 25 AU from the Sun into the orbits of the present-day SDOs, high-i classical TNOs (i > 5°), and high-i Plutinos. In contrast, classical TNOs in low-e and low-i orbits (the red cluster of classical TNOs discussed in section 4) remained far enough away from Neptune that they were never perturbed much by the planet. Perhaps around 40 AU from the Sun, methane went from condensing in a water-ice rich clathrate to condensing as pure methane (*Lewis,* 1972). Whereas the loss of methane from a clathrate surface would result in a lag made up of colorless water-ice crust, a pure methane-ice crust would provide much material for alteration into red organic compounds, even if there were a substantial amount of methane sublimation. If the dynamical simulation and this idea about methane surface chemistry are correct, SDOs, high-i classical TNOs, and high-i Plutinos should exhibit gray surface colors and low-i classical TNOs should exhibit red surface colors. The SDOs are largely gray and the low-i classical TNOs are red; however, additional observations are necessary to see if the high-i classical TNOs and the high-i Plutinos are largely gray.

6.2. Evolutionary Processes

The red colors of TNOs and Centaurs are usually attributed to the effects of surface aging and darkening due to high-energy radiation and particle bombardment in interplanetary space, also called space weathering (see chapters by Hudson et al. and de Bergh et al.). Blue surface colors can be produced by major collisions with other objects through deposits of fresh icy material from the body interior or from the impactor. Smaller impacts could also refresh the surface through so-called impact gardening. Both resurfacing processes (irradiations and collisions) have very long timescales (on the order of millions of years). A complicating situation may result from a reduction of the material reddening for very intense and/or very long irradiation times. This is indicated by laboratory experiments performed on asphaltite material (*Moroz et al.,* 2003). Resurfacing on a much shorter timescale could happen due to ice recondensation and/or dust deposition from a temporary atmosphere produced by outgassing due to solar heating, and from dust coma activity. N_2 and CO ices may be able to sublimate at distances of 40 AU and beyond (*Delsemme,* 1982). This ice sublimation process works quite efficiently for Pluto, as well as possibly also for Charon (*Yelle and Elliot,* 1997) and Chiron (*Meech et al.,* 1997).

The first numerical simulation results of TNO colors were performed using simplified reddening and impact resurfacing models (*Luu and Jewitt,* 1996). The results show that the range of colors observed in TNOs can be well reproduced. Typically, red objects are exposed to high-energy radiation for a long time (on the order of 10^6 yr) without resurfacing by collision. In turn, resurfacing by collision events produces "sharp" short-term color changes to neutral or at least less red values in the objects. The modeling, and more recent work by *Delsanti et al.* (2004), did not constrain conclusively the importance of collisions in the Edgeworth-Kuiper belt. The resurfacing by ice deposits from outgassing and from high-level irradiations is viewed as an additional "randomizing" complication of the modeling. Size- and population-dependent color modeling has not yet been performed.

7. CONCLUSION

Broadband photometry represents the simplest observing technique to study physical properties of TNOs and Centaurs. By assuming canonical albedo values together with magnitude estimates in the visible, one can get an initial idea on the typical sizes of these primitive bodies in the solar system. Color data and results derived from spectral gradient estimations provide an accurate measure of the global surface reflectance of the objects. There is no doubt that a large number of TNOs and Centaurs have very red surface colors, in fact much redder than other solar system bodies and in particular those in the inner solar system. Nevertheless, objects with neutral colors are also found.

With the large and high-quality color datasets available, strong and significant results have been found. Statistical analyses point to correlations between optical colors and orbital inclination and orbital rms velocity for the classical Edgeworth-Kuiper belt. On the other hand, no clear trend is obvious for Plutinos, scattered objects, or Centaurs, and no trend can be drawn regarding correlation of colors with size or heliocentric distance. These strong results can be summarized as follows: (1) The classical low-inclination TNOs are redder than the other populations; (2) the colors in the visible appear to be consistent throughout the visible region, but are generally unrelated to the colors in the near-IR; (3) the TNOs show strong color diversity when compared to other solar system bodies; and (4) there are color-orbital inclination and color-rms velocity correlations for the classical TNOs ($40 < a < 50$ AU).

At present, no fully convincing mechanism exists that explains the anisotropic color distribution within the Edgeworth-Kuiper disk. We do not know whether the color diversity is the result of true compositional diversity or whether it is the result of some evolutionary processes, e.g., collisions or surface irradiation processes. On the other hand, with the computational models performed so far coupled with the observational facts, we can conclude that collision alone cannot explain all, but it certainly plays a role in color distribution seen at present time in the TNOs. Not only collisions, but other evolutionary processes as well (e.g., outgassing, complex space weathering, micrometeorites, surface gardening, atmospheric interaction, etc.), have to be modeled. More observational data, numerical simulations, and overall laboratory experiments are required to unveil the origin of the color properties and trends in the present Edgeworth-Kuiper belt.

REFERENCES

Andronico G., Baratta G. A., Spinella F., and Strazzulla G. (1987) Optical evolution of laboratory-produced organics: Applications to Phoebe, Iapetus, outer belt asteroids and comet nuclei. *Astron. Astrophys., 184,* 333–336.

Bagnulo S., Boehnhardt H., Muinonen K., Kolokolova L., Belskaya I., and Barucci M. A. (2006) Exploring the surface properties of trans-neptunian objects and Centaurs with polarimetric FORS1/VLT observations. *Astron. Astrophys., 450,* 1239–1248.

Barkume K. M., Brown M. E., and Schaller E. L. (2006) Water ice on the satellite of Kuiper belt object 2003 EL_{61}. *Astrophys. J. Lett., 640,* L87–L89.

Bauer J. M., Meech K. J., Fernández Y. R., Farnham T. L., and Roush T. L. (2002) Observations of the Centaur 1999 UG_5: Evidence of a unique outer solar system surface. *Publ. Astron. Soc. Pac., 114,* 1309–1321.

Belskaya I. N. and Shevchenko V. G. (2000) Opposition effect of asteroids. *Icarus, 147,* 94–105.

Binzel R. P. (1988) Hemispherical color differences on Pluto and Charon. *Science, 241,* 1070–1072.

Boehnhardt H., Tozzi G. P., Birkle K., Hainaut O., Sekiguchi T., Vlair M., Watanabe J., Rupprech G., and the Fors Instrument Team (2001) Visible and near-IR observations of trans-neptunian objects. Results from ESO and Calar Alto telescopes. *Astron. Astrophys., 378,* 653–667.

Boehnhardt H., Delsanti A., Barucci A., Hainaut O., Doressoundiram A., Lazzarin M., Barrera L., de Bergh C., Birkle K., Dotto E., Meech K. J., Ortiz J. L., Romon J., Sekiguchi T., Thomas N., Tozzi G. P., Watanabe J., and West R. M. (2002) ESO large program on physical studies of trans-neptunian objects and Centaurs: Visible photometry — first results. *Astron. Astrophys., 395,* 297–303.

Boehnhardt H., Bagnulo S., Muinonen K., Barucci M. A., Kolokolova L., Dotto E., and Tozzi G. P. (2004) Surface characterization of 28978 Ixion (2001 KX_{76}). *Astron. Astrophys., 415,* L21–L25.

Bowell E. and Lumme K. (1979) Colorimetry and magnitudes of asteroids. In *Asteroids* (T. Gehrels, ed.), pp. 132–169. Univ. of Arizona, Tucson.

Brown R. H., Cruikshank D. P., and Pendleton Y. (1999) Water ice on Kuiper belt object 1996 TO_{66}. *Astrophys. J. Lett., 519,* L101–L104.

Brown M. E., Trujillo C. A., and Rabinowitz D. L. (2005) Discovery of a planetary-sized object in the scattered Kuiper belt. *Astrophys. J. Lett., 635,* L97–L100.

Cruikshank D. P., Pilcher C. B., and Morrison D. (1976) Pluto — Evidence for methane frost. *Science, 194,* 835–837.

Cruikshank D. P., Roush T. L., Owen T. C., Geballe T. R., de Bergh C., Schmitt B., Brown R. H., and Bartholomew M. J. (1993) Ices on the surface of Triton. *Science, 261,* 742–745.

Cruikshank D. P. and 14 colleagues (1998) The composition of Centaur 5145 Pholus. *Icarus, 135,* 389–407.

Cruikshank D. P., Imanaka H., and Dalle Ore C. M. (2005) Tholins as coloring agents on outer solar system bodies. *Adv. Space Res., 36,* 178–183.

De Bergh C., Delsanti A., Tozzi G. P., Dotto E., Doressoundiram A., and Barucci M. A. (2005) The surface of the transneptunain object 90482 Orcus. *Astron. Astrophys., 437,* 1115–1120.

Delsanti A., Hainaut O., Jourdeuil E., Meech K. J., Boehnhardt H., and Barrera L. (2004) Simultaneous visible and near-IR photometric study of Kuiper belt object surfaces with the ESO/Very Large Telescopes. *Astron. Astrophys, 417,* 1145–1158.

Delsanti A., Peixinho N., Boehnhardt H., Barucci A., Merlin F., Doressoundiram A., and Davies J. K. (2006) Near-infrared color properties of Kuiper belt objects and Centaurs: Final results from the ESO Large Program. *Astrophys. J., 131,* 1851–1863.

Degewij J. and van Houten C. J. (1979) Distant asteroids and outer jovian satellites. In *Asteroids* (T. Gehrels, ed.), pp. 417–435. Univ. of Arizona, Tucson.

Degewij J., Cruikshank D., and Hartmann W. (1980) Near-infrared colorimetry of J7 Himalia and S9 Phoebe: A Summary of 0.3 to 2. m reflectances. *Icarus, 44,* 541–547.

Delsemme A. H. (1982) Chemical composition of cometary nuclei. In *Comets* (L. L. Wilkening, ed.), pp. 85–130. Univ. of Arizona, Tucson.

Doressoundiram A. and Boehnhardt H. (2003) Multicolour photometry of trans-neptunian objects: Surface properties and structures. *Compt. Rend. Phys., 7,* 755–766.

Doressoundiram A., Barucci M. A., Romon J., and Veillet C. (2001) Multicolor photometry of trans-neptunian objects. *Icarus, 154,* 277–286.

Doressoundiram A., Peixinho N., De Bergh C., Fornasier S. Thebault Ph., Barucci M. A., and Veillet C. (2002) The color distribution of the Kuiper belt. *Astron. J., 124,* 2279–2296.

Doressoundiram A., Tozzi G. P., Barucci M. A., Boehnhardt H., Fornasier S., and Roman J. (2003) ESO large programme on trans-neptunian objects and Centaurs: Spectroscopic investigation of Centaur 2001 BL_{41} and TNOs (26181) 1996 GQ_{21} and (26375) 1999 DE_9. *Astron. J., 125,* 2721–2727.

Doressoundiram A., Peixinho N., Doucet C., Mousis O., Barucci M. A., Petit J. M., and Veillet C. (2005) The Meudon multicolor survey (2MS) of Centaurs and trans-neptunian objects: Extended dataset and status on the correlations reported. *Icarus, 174,* 90–104.

Douté S. and Schmitt B. (1998) A multilayer bi-directional reflectance model for the analysis of planetary surface hyperspectral images at visible and near-infrared wavelengths. *J. Geophys. Res., 103,* 31367–31390.

Fornasier S., Dotto E., Barucci M. A., and Barbieri C. (2004a) Water ice on the surface of the large TNO 2004 DW. *Astron. Astrophys., 422,* L43–L46.

Fornasier S., Doressoundiram A., Tozzi G. P., Barucci M. A., Boehnhardt H., de Bergh C., Delsanti A., Davies J., and Dotto E. (2004b) ESO large program on physical studies of trans-neptunian objects and Centaurs: Final results of the visible spectrophotometric observations. *Astron. Astrophys., 421,* 353–363.

French L. M. and Binzel R. P. (1989) CCD photometry of asteroids. In *Asteroids II* (R. P. Binzel et al., eds), pp. 54–65. Univ. of Arizona, Tucson.

Glass I. S. (1997) *Handbook of Infrared Astronomy.* Cambridge Univ., Cambridge.

Gomes R. S. (2003) The origin of the Kuiper belt high-inclination population. *Icarus, 161,* 404–418.

Gradie J. and Tedesco E. (1982) Compositional structure of the asteroid belt. *Science, 216,* 1405–1407.

Gradie J. and Veverka J. (1980) The composition of the Trojan asteroids. *Nature, 283,* 840–842.

Hainaut O. R. and Delsanti A. C. (2002) Colors of minor bodies in the outer solar system. *Astron. Astrophys., 389,* 641–664.

Hapke B. (1981) Bidirectional reflectance spectroscopy. 1. Theory. *J. Geophys. Res., 86,* 3039–3054.

Hardorp J. (1980) The sun among the stars. III — Energy distribution of 16 northern G-type stars and solar flux calibration. *Astron. Astrophys., 91,* 221–232.

Hartmann W. K., Cruikshank D. P., and Degewij. J. (1982) Remote comets and related bodies — VJHK colorimetry and surface materials. *Icarus, 52,* 377–409.

Howell S. B. (1989) Two-dimensional aperture photometry — Signal-to-noise ratio of point-source observations and optimal data-extraction techniques. *Publ. Astron. Soc. Pac., 101,* 616–622.

Jewitt D. and Luu J. (1993) Discovery of the candidate Kuiper belt object 1992 QB_1. *Nature, 362,* 730–732.

Jewitt D. and Luu J. X. (2001) Colors and spectra of Kuiper belt objects. *Astron. J., 122,* 2099–2114.

Jewitt D. C. and Luu J. (2004) Crystalline water ice on the Kuiper belt object (50000) Quaoar. *Nature, 432,* 731–733.

Khare B. N., Sagan C., Arakawa E. T., Suits F., Callcott T. A., and Williams M. W. (1984) Optical constants of organic tholins produced in a simulated titanian atmosphere — from soft X-ray to microwave frequencies. *Icarus, 60,* 127–137.

Khare B. N., Thompson W. R., Cheng L., Chyba C., Sagan C., Arakawa E. T., Meisse C., and Tuminello P. S. (1993) Production and optical constants of ice tholin from charged particle irradiation of (1:6) C_2H_6/H_2O at 77K. *Icarus, 103,* 290–300.

Lazzarin M., Barucci M. A., Boehnhardt H., Tozzi G. P., de Bergh C., and Dotto E. (2003) ESO large programme on physical studies of trans-neptunian objects and Centaurs: Visible spectroscopy. *Astron. J., 125,* 1554–1558.

Lewis J. S. (1972) Low temperature condensation from the solar nebula. *Icarus, 16,* 241.

Licandro J., Oliva E., and Di Martino M. (2001) NICS-TNG infrared spectroscopy of trans-neptunian objects 2000 EB_{173} and 2000 WR_{106}. *Astron. Astrophys., 373,* L29–L32.

Licandro J., Pinilla-Alonso N., Pedani M., Oliva E., Tozzi G. P., and Grundy W. M. (2006) The methane ice rich surface of large TNO 2005 FY9: A Pluto-twin in the trans-neptunian belt? *Astron. Astrophys., 445,* L35–L38.

Luu J. and Jewitt D. (1996) Color diversity among the Centaurs and Kuiper belt objects. *Astrophys. J., 112,* 2310.

McCord T. B. and Chapman C. R. (1975) Asteroids — Spectral reflectance and color characteristics. *Astrophys. J., 195,* 553–562.

Meech K. J., Buie M. W., Samarasinha N. H., Mueller B. E. A., and Belton M. J. S. (1997) Observations of structures in the inner coma of Chiron with the HST Planetary Camera. *Astrophys. J., 113,* 844–862.

Moore M. H., Donn B., Khanna R., and A'Hearn M. F. (1983) Studies of proton-irradiated cometary-type ice mixtures. *Icarus, 54,* 388–405.

Moroz L. V., Baratta G., Distefano E., Strazzulla G., Starukhina L. V., Dotto E., and Barucci M. A. (2003) Ion irradiation of asphaltite: Optical effects and implications for trans-neptunian objects and Centaurs. *Earth Moon Planets, 92,* 279–289.

Mueller B. E. A., Tholen D. J., Hartmann W. K., and Cruikshank D. P. (1992) Extraordinary colors of asteroidal object (5145) 1992 AD. *Icarus, 97,* 150–154.

Peixinho N., Doressoundiram A., Delsanti A., Boehnhardt H., Barucci M. A., and Belskaya I. (2003) Reopening the TNOs color controversy: Centaurs bimodality and TNOs unimodality. *Astron. Astrophys., 410,* L29–L32.

Rabinowitz D. L., Barkume K., Brown M. E., Roe H. S. M., Tourtellotte S., and Trujillo C. (2006) Photometric observations constraining the size, shape, and albedo of 2003 EL_{61}, a rapidly rotating, Pluto-sized object in the Kuiper belt. *Astrophys. J., 639,* 1238–1251.

Russell H. N. (1916) On the albedo of the planets and their satellites. *Astron. J., 43,* 173.

Sagan C. and Khare B. N. (1979) Tholins — Organic chemistry of interstellar grains and gas. *Nature, 277,* 102–107.

Sheppard S. S. and Jewitt D. C. (2002) Time-resolved photometry of Kuiper belt objects: Rotations, shapes and phase functions. *Astron. J., 124,* 1757–1775.

Spearman C. (1904) The proof and measurement of association between two things. *Am. J. Psych., 57,* 72–101.

Sterken C. and Manfroid J. (1992) *Astronomical Photometry — A Guide.* Astrophys. Space Sci. Library, Kluwer, Dordrecht.

Stern S. A. (2002) Evidence for a collisional mechanism affecting Kuiper belt object colors. *Astron. J., 124,* 2297–2299.

Strazzulla G., Cataliotti R. S., Calcagno L., and Foti G. (1984) The IR spectrum of laboratory synthesized polymeric residues. *Astron. Astrophys., 133,* 77–79.

Tegler S. C. and Romanishin W. (1997) The extraordinary colors of trans-neptunian objects 1994 TB and 1993 SC. *Icarus, 126,* 212–217.

Tegler S. C. and Romanishin W. (1998) Two distinct populations of Kuiper-belt objects. *Nature, 392,* 49.

Tegler S. C. and Romanishin W. (2000) Extremely red Kuiper-belt objects in near-circular orbits beyond 40 AU. *Nature, 407,* 979–981.

Tegler S. C., Romanishin W., and Consolmagno G. J. (2003) Color patterns in the Kuiper belt: A possible primordial origin. *Astrophys. J. Lett., 599,* L49–L52.

Trujillo C. A. and Brown M. E. (2002) A correlation between inclination and color in the classical Kuiper belt. *Astrophys. J. Lett., 566,* L125–L128.

Yelle R. V. and Elliot J. L. (1997) Atmospheric structure and composition: Pluto and Charon. In *Pluto and Charon* (S. A. Stern and D. J. Tholen, eds.), p. 347. Univ. of Arizona, Tucson.

Colors of Centaurs

Stephen C. Tegler
Northern Arizona University

James M. Bauer
Jet Propulsion Laboratory

William Romanishin
University of Oklahoma

Nuno Peixinho
Grupo de Astrofisica da Universidade de Coimbra

Minor planets on outer planet-crossing orbits, called Centaur objects, are important members of the solar system in that they dynamically link Kuiper belt objects to Jupiter-family comets. In addition, perhaps 6% of near-Earth objects have histories as Centaur objects. The total mass of Centaurs (10^{-4} $M_\oplus$) is significant, about one-tenth of the mass of the asteroid belt. Centaur objects exhibit a physical property not seen among any other objects in the solar system, their B–R colors divide into two distinct populations: a gray and a red population. Application of the dip test to B–R colors in the literature indicates there is a 99.5% probability that Centaurs exhibit a bimodal color distribution. Although there are hints that gray and red Centaurs exhibit different orbital elements, application of the Wilcoxon rank sum test finds no statistically significant difference between the orbital elements of the two color groups. On the other hand, gray and red Centaurs exhibit a statistically significant difference in albedo, with the gray Centaurs having a lower median albedo than the red Centaurs. Further observational and dynamical work is necessary to determine whether the two color populations are the result of (1) evolutionary processes such as radiation-reddening, collisions, and sublimation or (2) a primordial, temperature-induced, composition gradient.

1. INTRODUCTION

October 18, 1977, marks the discovery of the first minor planet with a perihelion distance far beyond the orbit of Jupiter (*Kowal,* 1979). Because the minor planet was on an orbit largely lying between Saturn and Uranus, Kowal named his discovery after the Centaur Chiron, son of Kronos (Saturn) and grandson of Uranus. Fifteen years would elapse before the discovery of the next Centaur (1992 AD; 5145 Pholus) by the Spacewatch asteroid search project (*Scotti,* 1992). There are now several dozen Centaurs known.

There is no generally agreed upon definition of the term Centaur in the literature. The term is frequently and loosely defined as a minor planet on an outer-planet-crossing orbit. Here, we use a precisely constrained definition for Centaur given by the Minor Planet Center, an object on an orbit with semimajor axis, a, less than Neptune's orbit at 30.1 AU and a perihelion distance, q, larger than Jupiter's orbit at 5.2 AU. As of September 30, 2006, there are 39 objects in the Lowell Observatory Deep Ecliptic Survey database and 62 objects in the Minor Planet Center database with $q > 5.2$ AU and $a < 30.1$ AU.

Planetary perturbations and mutual collisions in the Kuiper belt are probably responsible for the ejection of objects from the Kuiper belt onto Centaur orbits. The Kuiper belt dynamical classes (e.g., Plutinos, classical objects, scattered disk objects) that are the sources of Centaurs are unknown. Numerical simulations of Neptune's gravitational influence on Kuiper belt objects (KBOs) indicate that more objects are perturbed onto orbits with larger semimajor axes, while fewer objects are perturbed onto orbits with smaller semimajor axes, i.e., Centaur orbits (*Levison and Duncan,* 1997). Because Centaurs cross the orbits of the outer planets, they are dynamically unstable, with lifetimes $\sim 10^6$ yr (*Tiscareno and Malhotra,* 2003; *Horner et al.,* 2004). Some Centaurs evolve into Jupiter-family comets, others are ejected from the solar system, and yet others impact the giant planets. In addition, some Jupiter-family comets evolve back into Centaurs (*Hahn and Bailey,* 1990; *Horner et al.,* 2004). Unfortunately, it is impossible to determine from numerical simulations alone whether a given Centaur was a Jupiter-family comet in the past.

Centaurs probably contribute to the near-Earth object (NEO) population. One study finds approximately 6% of the NEO population ultimately comes from the Kuiper belt (*Morbidelli et al.,* 2002). Another study finds a Centaur becomes an Earth-crossing object for the first time about every 880 yr (*Horner et al.,* 2004). Although the percentage of NEOs with a history as a Centaur is small, their potential as hazards is important. Specifically, most NEOs are collision fragments and are significantly smaller than 10 km in diameter. If a large Centaur, like Chiron or Pholus, were to cross Earth's orbit, the debris and dust from a fragmentation could create problems in the space near Earth (*Hahn and Bailey,* 1990).

The number of Centaurs and their total mass is a significant component of the solar system. An analysis of the number of Centaur discoveries in a wide-field optical survey suggests there are about 10^7 Centaurs with diameters larger than 2 km and about 100 Centaurs with diameters larger than 100 km (*Sheppard et al.,* 2000). The same analysis estimates the total mass of Centaurs at about 10^{-4} $M_\oplus$. For comparison, the total mass of main belt asteroids is about 10^{-3} $M_\oplus$ (*Davis et al.,* 2002).

A considerable amount of what we know about the physical and chemical properties of Centaurs comes from photometry. An analysis of groundbased optical photometry and Spitzer Space Telescope infrared photometry for a sample of the known Centaurs yields diameters ranging from 32 to 259 km and albedos ranging from 2% to 18% (see the chapter by Stansberry et al.). Optical lightcurves yield Centaur periods of rotation ranging from 4.15 to 13.41 hours (see the chapter by Sheppard et al.). The evolution of Pholus' lightcurve over a decade suggests it has a highly nonspherical shape, i.e., it has axial ratios of 1.9:1:0.9 (*Farnham,* 2001; *Tegler et al.,* 2005). If Pholus is a strengthless rubble pile and its nonspherical shape is due to rotational distortion, then its axial ratios and period of rotation (9.980 ± 0.002 h) indicate it has a density of 0.5 g cm^{-3}, suggestive of an ice-rich and porous interior (*Tegler et al.,* 2005).

The volatile-rich nature of Centaurs is evident in near-infrared spectra of four Centaurs. In particular, H_2O-ice and possibly CH_3OH-ice bands are seen in the spectrum of Pholus (*Cruikshank et al.,* 1998). In addition, H_2O-ice bands are seen in the spectra of (10199) Chariklo (*Brown et al.,* 1998), (32522) Thereus (*Licandro and Pinilla-Alons,* 2005), and (83982) Crantor (*Doressoundiram et al.,* 2005a).

Further evidence for the volatile-rich nature of Centaurs comes from images of four additional Centaurs. 2060 Chiron (*Meech and Belton,* 1990), 166P/NEAT (*Bauer et al.,* 2003a), 167P/CINEOS (*Romanishin et al.,* 2005), and (60558) Echeclus (*Choi et al.,* 2006) all exhibit comae. Some Centaurs (by our dynamical definition) with comae have comet names (e.g., 167P/CINEOS) and others have both a Centaur and a comet name (e.g., 60558 Echeclus is also known as 174P/Echeclus). Echeclus is on both the Centaur and short-period-comet lists of the Minor Planet Center. At r = 12.9 AU, Echeclus displayed an extraordinarily large and complex coma (Fig. 1). At such large distances, it's too cold for sublimation of H_2O-ice to drive the coma formation. It is possible much more volatile molecular ices such as CO-ice and CO_2-ice are driving the coma formation in Centaurs. Alternatively, *Blake et al.* (1991) suggest the formation of a CH_3OH clathrate from a mixture of amorphous H_2O and CH_3OH can result in the exhalation of excess CH_3OH in a burst of activity at large heliocentric distances.

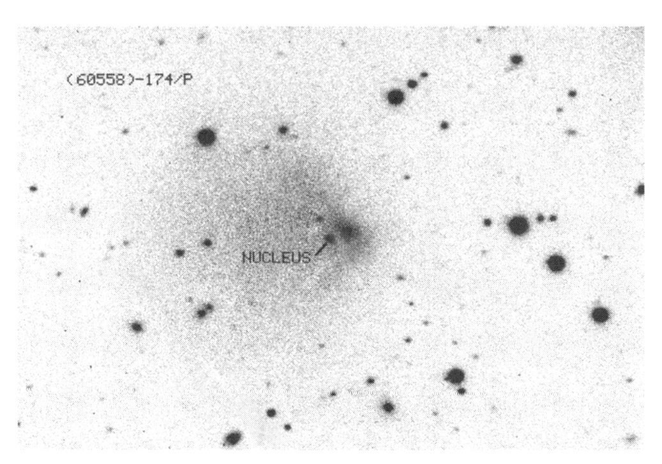

Fig. 1. A 360-s R-band image of Echeclus and a complex coma structure. The image was obtained on 2006 April 2.3 UT with the Vatican Advanced Technology 1.8-m telescope on Mt. Graham, Arizona. It has dimensions of 285 arcsec by 195 arcsec. North is toward the top and east is toward the left. Echeclus is at the position marked "nucleus," with an R magnitude of 20.1. The complex coma structure includes a low-surface brightness coma of diameter 2 arcmin that is centered 1 arcmin east of Echeclus, R = 16, and a higher-surface brightness condensation about 12 arcsec in diameter centered about 7 arcsec west of Echeclus, R = 17.9. Echeclus was at r = 12.9 AU. If the complex coma structure is due to a second object at the position of the higher-surface brightness condensation, the relationship between the two objects is not clear, nor is why the second object appears to exhibit a coma and Echeclus does not exhibit a coma.

Perhaps the most surprising discovery about Centaurs concerns their surface colors. Centaurs and KBOs should exhibit a range of surface colors because (1) solar and cosmic radiation should redden the surfaces, and (2) occasional impacts by smaller objects should puncture the radiation reddened crusts and expose interior, pristine, gray ices (*Luu and Jewitt,* 1996). Surprisingly, measurements of the optical B–R colors of Centaurs divide into two distinct color populations (*Peixinho et al.,* 2003, *Tegler et al.,* 2003).

This chapter describes the two color populations and their statistical significance. Then, we use the two populations to try and constrain important formation and evolution processes in the outer solar system.

2. COLOR MEASUREMENTS

2.1. Objects with Coma

Besides the four Centaurs with coma mentioned above (2060 Chiron, 166P/NEAT, 167P/CINEOS, and 60558 Echeclus), there are five additional objects that are often classified as comets because they exhibit coma: the orbits of 29P/Schwassmann-Wachmann 1, 39P/Oterma, 165P/LINEAR, C/2001 M10 (NEAT), and P/2004 A1 (LONEOS) are consistent with our definition of a Centaur, q > 5.2 AU and a < 30.1 AU. Indeed, Oterma's close encounter with Jupiter in 1963 moved its perihelion distance from that of a Jupiter-family comet at 3.4 AU to that of a Centaur at 5.47 AU (see discussion by *Bauer et al.,* 2003a).

It is quite difficult to measure the surface colors of active Centaurs as it requires observations during unpredictable windows of inactivity when the surfaces are not embedded in coma gas and dust. In general, Centaurs with activity appear to exhibit surfaces that absorb sunlight with nearly equal efficiency between wavelengths of 5500 and 6500 Å. In other words, Centaurs with activity display surfaces with nearly solar colors, $(V-R)_\odot = 0.36$ (*Bauer et al.,* 2003a). A notable exception is 166P/NEAT, which had an unusually red color during a recent period of activity, $(V-R) = 0.95 \pm 0.02$ (*Bauer et al.,* 2003a). Regardless of whether the red color is due to the surface or coma, 166P/NEAT is one of the reddest Centaurs known. Measurements of additional active Centaurs are essential to determine the effect of coma activity on surface colors and evolution.

2.2. Objects Without Coma

2.2.1. B–R colors. Two teams independently and simultaneously discovered that Centaurs exhibit two different slopes in their reflectance spectra between 4500 and 6500 Å, i.e., they exhibit two distinct B–R color populations (*Peixinho et al.,* 2003; *Tegler et al.,* 2003). An examination of their samples finds 15 objects in common. The mean difference between the 15 pairs of measurements is 0.02 mag, i.e., the two teams find essentially the same colors for the same objects. The excellent agreement says two things. First, the two color measurements for each Centaur were obtained at random rotational phases, so it appears Centaurs do not in general exhibit large color variations over their surfaces. Second, there are no apparent systematic effects in the photometry of either team that could yield the observed bimodality in the B–R color.

The excellent agreement between the two teams suggests it is reasonable to combine the two samples into a single sample of 26 Centaurs (Table 1). For each of the 15 overlap objects, an average value is given in Table 1 where the two individual measurements are weighted by the inverse square of their corresponding uncertainties. The orbital elements come from the Deep Ecliptic Survey (*www.lowell.edu/users/buie/kbo/kbofollowup.html*). Objects in the table are ordered from grayer, i.e., solar-type colors (B–R = 1.03 on the Kron-Cousins system), to redder colors. Colors for Chiron and Echeclus come at times when they had little or no coma contamination.

Figure 2 shows a histogram of the B–R colors in Table 1. The gray population has $1.0 < B-R < 1.4$, and the red population has $1.7 < B-R < 2.1$. Notice there are no objects with $1.4 < B-R < 1.7$ among the sample of 26 Centaurs.

2.2.2. V–R and R–I colors. A V–R and R–I color survey of 24 Centaurs does not exhibit two color populations; see Fig. 1 of *Bauer et al.* (2003b). The color dichotomy

TABLE 1. Centaur colors and orbital elements.

Name	Number	Prov Des	B–R	Source*	a	Q	q	i	e	H
	95626	2002 GZ_{32}	1.03 ± 0.04	TRC	23.03	28.02	18.03	15.02	0.217	6.84
Chiron	2060	1977 UB	1.04 ± 0.05	Pei	13.49	18.57	8.40	6.99	0.377	6.16
		2002 DH_5	1.05 ± 0.07	Pei	22.19	30.41	13.96	22.46	0.371	10.20
Bienor	54598	2000 QC_{243}	1.12 ± 0.03	avg	16.48	19.76	13.20	20.73	0.199	7.52
	119315	2001 SQ_{73}	1.13 ± 0.02	TRC	17.51	20.60	14.42	17.42	0.176	9.57
Hylonome	10370	1995 DW_2	1.15 ± 0.06	avg	25.11	31.36	18.86	4.14	0.249	8.93
		2000 FZ_{53}	1.17 ± 0.05	TRC	23.67	34.99	12.34	34.90	0.478	11.41
Thereus	32532	2001 PT_{13}	1.18 ± 0.01	TRC	10.71	12.86	8.55	20.34	0.202	8.67
	63252	2001 BL_{41}	1.20 ± 0.03	avg	9.79	12.71	6.87	12.47	0.298	11.51
Okyrhoe	52872	1998 SG_{35}	1.21 ± 0.02	avg	8.41	10.99	5.84	15.62	0.306	10.93
		2003 WL_7	1.23 ± 0.04	TRC	20.14	25.35	14.94	11.17	0.258	8.98
Asbolus	8405	1995 GO	1.23 ± 0.05	avg	18.16	29.46	6.86	17.61	0.622	9.07
	120061	2003 CO_1	1.24 ± 0.04	TRC	20.87	30.79	10.94	19.73	0.476	8.81
Pelion	49036	1998 QM_{107}	1.25 ± 0.04	avg	20.14	22.96	17.32	9.36	0.140	10.37
Chariklo	10199	1997 CU_{26}	1.26 ± 0.04	avg	15.82	18.50	13.13	23.38	0.170	6.40
Echeclus	60558	2000 EC_{98}	1.38 ± 0.04	avg	10.71	15.63	5.80	4.35	0.458	9.50
Elatus	31824	1999 UG_5	1.70 ± 0.02	avg	12.74	18.02	7.46	5.59	0.414	9.88
Amycus	55576	2002 GB_{10}	1.79 ± 0.03	avg	25.01	34.81	15.21	13.35	0.392	7.45
	88269	2001 KF_{77}	1.81 ± 0.04	TRC	25.87	31.96	19.77	4.36	0.236	9.49
Crantor	83982	2002 GO_9	1.85 ± 0.02	avg	19.34	24.65	14.04	12.78	0.274	8.60
Cyllarus	52975	1998 TF_{35}	1.86 ± 0.05	avg	26.41	35.56	16.26	12.62	0.384	9.25
	121725	1999 XX_{143}	1.86 ± 0.07	Pei	17.98	26.30	9.66	6.77	0.463	8.53
Nessus	7066	1993 HA_2	1.88 ± 0.06	Pei	24.83	37.85	11.81	15.63	0.524	9.54
		2001 XZ_{255}	1.92 ± 0.07	TRC	16.03	16.59	15.47	2.61	0.035	11.13
		1994 TA	1.92 ± 0.06	avg	16.76	21.84	11.67	5.40	0.303	11.43
Pholus	5145	1992 AD	2.04 ± 0.07	avg	20.25	31.81	8.69	24.71	0.571	6.89

*Source: TRC = *Tegler et al.* (2003); Pei = *Peixinho et al.* (2003). Avg = weighted average of colors from TRC and Pei.

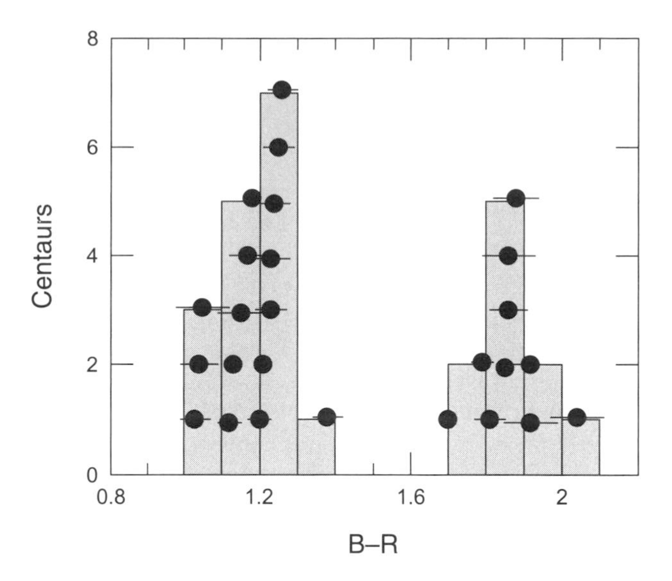

Fig. 2. Histogram of Centaur B–R colors in Table 1. Centaurs appear to divide into two distinct color populations, one with solar to slightly red colors, 1.0 < B–R < 1.4, and the other with red colors, 1.7 < B–R < 2.1. For reference, the Sun has B–R = 1.03.

appears largest in optical surveys that include short-wavelength B-band measurements.

2.2.3. H–K colors. In a near-infrared color survey of 17 Centaurs, *Delsanti et al.* (2006) found two interesting results. First, they found that the objects appear to divide into two color groups on a H–K vs. B–R color-color plot (see their Fig. 3), suggesting a correlation between the absorbers at B-band and H- or K-band.

Second, they found Centaurs with the reddest B–R colors display H–K colors bluer than the Sun (H–K = 0.06). Delsanti et al. suggest that an absorption feature between 1.7 and 2.2 μm is responsible for the bluer than solar colors. The H_2O-ice band at 2.2 μm is a prime candidate as three of the objects exhibit the band in spectroscopic observations. However, H_2O-ice does not absorb near 4500 Å and so it cannot be responsible for the B–R colors.

3. STATISTICAL TEST OF B–R AND H–K BIMODALITY

3.1. Discussion of Dip Test

A visual inspection of Fig. 2 in this chapter or Fig. 3 in *Delsanti et al.* (2006) suggest that Centaurs divide into two B–R and two H–K color populations; however, it is essential to quantify the statistical significance of the apparent divisions. So, what's a good statistical test to apply to Centaur colors?

Present knowledge of Centaur colors is insufficient to justify the assumption of a particular probability distribution, e.g., a normal distribution. Therefore, it is safest to use nonparametric tests that do not assume a particular probability distribution. Furthermore, tests based on bins (e.g., *Jewitt and Luu,* 2001) and Monte Carlo simulations (e.g., *Hainaut and Delsanti,* 2002; *Tegler and Romanishin,* 2003) are very dependent on the way the bins are chosen.

The dip test (*Hartigan and Hartigan,* 1985) is a nonparametric test that does not assume a particular probability distribution and does not require binning data. It is designed to test for one population (unimodality) vs. two populations (bimodality) using monotone regression.

The dip test finds the best fitting unimodal function and measures the maximum difference (*dip*) between that function and the empirical distribution of the sample. This *dip* approaches zero for samples from a unimodal distribution and approaches a positive number for a multimodal distribution. The probability that such a *dip* is not due to pure chance may be obtained from the tables published by *Hartigan and Hartigan* (1985). The dip test and tables for significance levels are also available in the R statistical package (*www.r-project.org*). The dip test is similar to the Kolmogorov Test (*Kolmogorov,* 1933).

3.2. Result of Dip Test on B–R Colors

Application to the dip test in the R package to the sample of N = 26 B–R colors in Table 1 and Fig. 2 results in *dip* = 0.11597, implying that the B–R color distribution is bimodal at a confidence level of CL = 99.5%. There are 10 objects in the red group and 16 in the gray group.

3.3. Result of the Dip Test on H–K Colors

An application of the dip test to 17 H–K Centaur colors indicates there is only a 32% probability that the H–K color distribution is bimodal (*Delsanti et al.,* 2006). On the other hand, application of the Kolmogorov test to the H–K and B–R colors in their Fig. 3 suggests there is a 96.4% probability that there are two H–K and B–R populations. It will be interesting to see what increasing the size of the Delsanti et al. H–K sample does to the statistical probability of two H–K color populations.

4. DIFFERENCES BETWEEN TWO COLOR POPULATIONS

Since the dip test finds the B–R color distribution is bimodal at the 99.5% level, it is natural to look for additional differences between the two color populations in the hope that they will lead to a physical explanation for the two color populations. The Wilcoxon rank sum test is a nonparametric test that does not require data binning (*www.r-project.org*). It provides a way to test the statistical significance of any differences.

4.1. Orbital Elements and Absolute Magnitude

The three left panels in Fig. 3 display semimajor axis, a, aphelion distance, Q, and perihelion distance, q, vs. B–R color for the 26 Centaurs. As there is no apparent correlation between color and a, Q, or q within a color population, the median value of each orbital element within a color

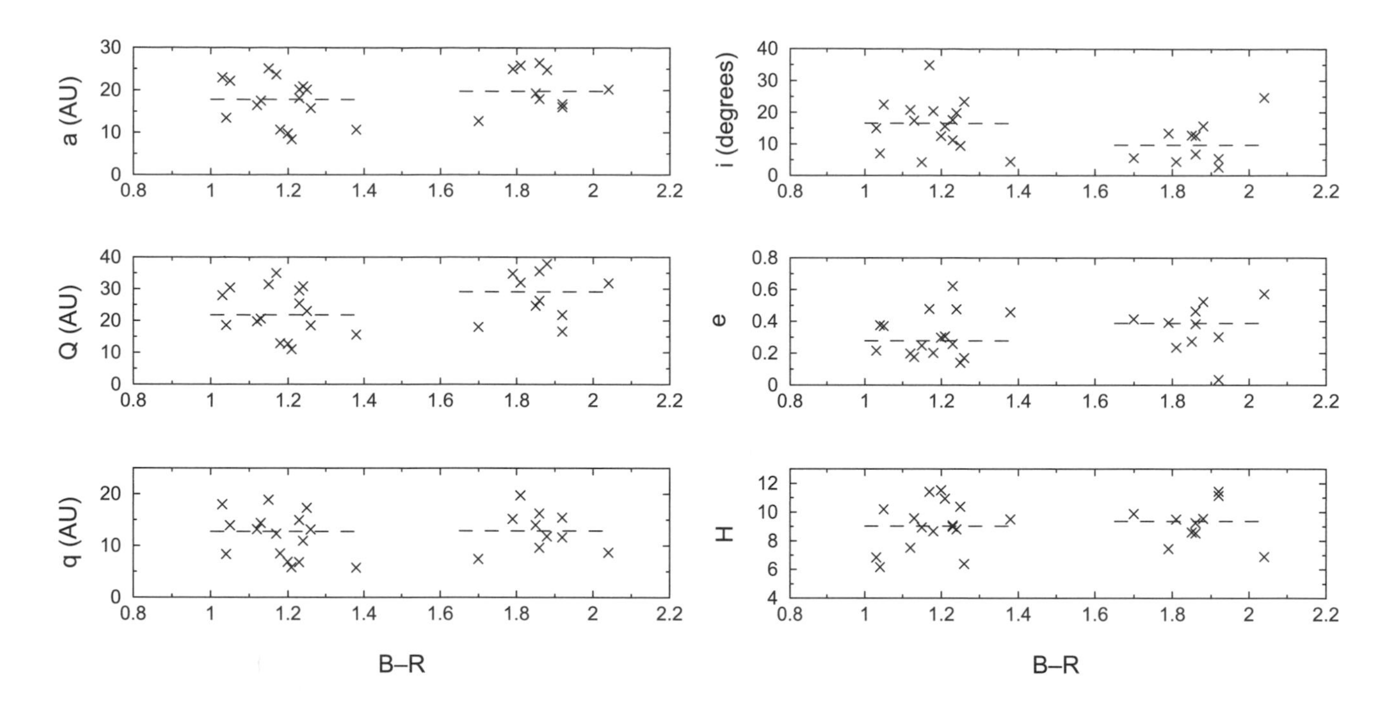

Fig. 3. There are no major differences between the gray and red Centaur values of a, Q, q, i, and e as well as absolute magnitude H. The dashed lines in each panel indicate the median value of a, Q, q, i, e, or H for the gray and red populations. The greatest probability of a difference between the two populations occurs for inclination angle. A Wilcoxon rank sum test says the probability that the gray and red Centaurs have the same inclination angle distribution is 10%.

population is marked with a dashed line extending across the color range of the population. It appears red Centaurs have slightly larger orbits than gray Centaurs (median red a = 19.8 AU, median gray a = 17.8 AU). Interestingly, V–R and R–I color measurements suggest that redder Centaurs exhibit slightly larger semimajor axis values (see Fig. 6 in *Bauer et al.,* 2003b). It also appears that redder Centaurs extend farther from the Sun than gray Centaurs (median red Q = 29.1 AU, median gray Q = 21.8). The red and gray Centaur populations appear to have virtually identical values for q.

The three right panels in Fig. 3 display inclination angle, i, eccentricity, e, and absolute visual magnitude, H, vs. B–R color. The i distribution is most interesting. All but one of the red Centaurs lie at an inclination angle lower than the median inclination angle of the gray population, suggesting that red Centaurs have significantly lower inclination angles than gray Centaurs. Red Centaurs appear to have slightly larger eccentricities than gray Centaurs. V–R and R–I color measurements appear to exhibit the same eccentricity pattern (*Bauer et al.,* 2003b). The absolute magnitudes of the red and gray populations appear virtually identical.

Despite the apparent patterns visible to the eye in Fig. 3, application of the Wilcoxon rank sum test finds no statistically significant difference between the gray and red Centaur values of a, Q, q, i, e, and H. The parameter exhibiting the most statistically significant difference between the gray and red Centaurs is inclination angle. The Wilcoxon test indicates the probability of red and gray Centaurs having the same inclination angle distribution is 10%.

4.2. Albedo

Figure 4 displays albedo measurements from Spitzer Space Telescope and groundbased optical photometry (see the chapter by Stansberry et al.) vs. B–R color for 15 Centaurs (Table 1). It appears gray Centaurs exhibit lower albedos than red Centaurs. The Wilcoxon rank sum test indicates that the probability of gray and red Centaurs having the same albedo distribution is only 1%. In other words, the median albedos of gray and red Centaurs exhibit a statistically significant difference. Stansberry et al. apply the Spearman test to albedos and slopes of optical spectra (colors) and find the likelihood of a correlation is 98%.

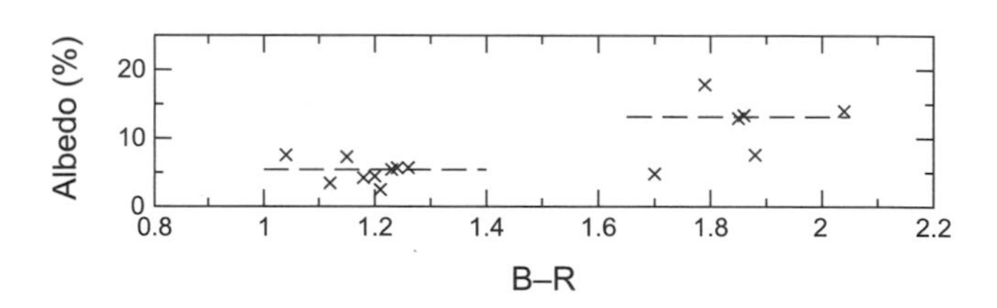

Fig. 4. Gray Centaurs exhibit smaller albedos than red Centaurs. The two dashed lines indicate the median value of albedo for the gray and red populations. A Wilcoxon rank sum test says the probability that the gray and red Centaurs have the same albedo distribution is only 1%. The median albedos of gray and red Centaurs exhibit a statistically significant difference. Albedos are from the chapter by Stansberry et al. and B–R colors come from Table 1.

5. COLORS OF OTHER SOLAR SYSTEM OBJECTS

Some or all of the solar system objects below may have a dynamical link to Centaur objects. Therefore, it is natural to compare their colors to Centaur colors in the hope the comparison will result in a reason for the two Centaur B–R color populations.

5.1. Comet Nuclei

Dynamical simulations suggest some Centaurs evolve into Jupiter-family comets and some Jupiter-family comets evolve into Centaurs (see section 1). B–R colors exist for 12 Jupiter-family comet nuclei (*Jewitt and Luu,* 1990; *Luu,* 1993; *Meech et al.,* 1997; *Delahodde et al.,* 2001; *Jewitt,* 2002; *Li et al.,* 2006) and two Oort cloud comet nuclei (*Jewitt,* 2002; *Abell et al.,* 2005). Observing bare comet nuclei requires observations at large heliocentric distances so that the nuclei are not shrouded in sublimating gas and dust. However, at large heliocentric distances the nuclei are faint and difficult to observe. From a comparison of Figs. 5a and 5b, it is apparent comet nuclei only overlap the gray Centaur population. None of the 14 nuclei exhibit colors similar to the red Centaur population.

5.2. Jupiter Trojans

Recent dynamical simulations suggest Trojan asteroids (located 60° ahead and behind Jupiter at the L4 and L5 Lagrangian points) may have formed well beyond Jupiter and were subsequently captured by Jupiter early in the history of the solar system (*Morbidelli et al.,* 2005). (See the chapter by Dotto et al. for a review of current ideas on the formation of Jupiter Trojans.) From a comparison of Figs. 5a and 5c, it is apparent Trojan asteroids overlap the gray Centaur population. None of the 26 Trojans in Fig. 5c (*Fornasier et al.,* 2004) exhibit colors similar to the red Centaur population.

5.3. Irregular Satellites

Irregular satellites of the jovian planets have larger inclination angles, eccentricities, and semimajor axes than regular satellites. These orbital characteristics suggest irregular satellites were captured by their parent planets. Figure 5d is a B–R histogram of 1 neptunian (*Schaefer and Schaefer,* 2000), 8 saturnian (*Grav et al.,* 2003), and 12 jovian satellites. Again, the colors of these irregular satellites overlap the gray Centaur population. B–R colors for the uranian satellites (Caliban and Sycorax) do not appear in Fig. 5d because colors in the literature are inconsistent (*Maris et al.,* 2001; *Rettig et al.,* 2001; *Romon et al.,* 2001).

5.4. Neptune Trojans

Neptune Trojans were likely captured in the L4 and L5 Lagrangian regions during or shortly after Neptune's formation (see discussion by *Sheppard and Trujillo,* 2006). Four Neptune Trojans are now known and they have essentially gray colors (Fig. 5e) (Sheppard and Trujillo).

5.5. Kuiper Belt Objects

The chapter by Doressoundiram et al. contains a thorough discussion of KBO colors. Here the focus is on KBO dynamical classes that are possible sources of Centaurs and their B–R colors for comparison to Centaur B–R colors.

5.5.1. Plutinos. The presence of objects at a = 39.6 AU in the 2:3 mean-motion resonance with Neptune, Plutinos, may be the result of sweeping resonance capture of the migrating planets (*Hahn and Malhotra,* 2005). *Romanishin and Tegler* (2007) have combined their B–R colors with those in the literature to produce a B–R histogram with 41 Plutinos (Fig. 5f). Plutino colors span the range of the two Centaur color populations, but their distribution appears continuous rather than dividing into two populations. Application of the dip test yields only a 70% probability of two color populations.

5.5.2. Cold classical Kuiper belt objects. Dynamically cold classical KBOs are on orbits with q > 40 AU, 42 < a < 45 AU, e < 0.1, and i < 10°. They may have formed close to or at their current location. They are dominated by extremely red surface colors (*Tegler and Romanishin,* 2000; *Jewitt and Luu,* 2001; *Tegler et al.,* 2003; *Peixinho et al.,* 2004; *Doressoundiram et al.,* 2005b). A histogram of B–R colors for a sample of 25 cold classical KBOs is given in Fig. 5g. The B–R color measurements in the figure come from 17 papers in the literature. Only 1 out of 25 objects overlaps the color range of gray Centaurs.

5.5.3. Scattered disk objects. Scattered disk objects (SDOs) are thought to have been scattered by Neptune, or another giant planet, onto orbits with large inclinations, i > 15°, large eccentricities, e > 0.3, and large semimajor axes, a > 45 AU. A histogram of 17 SDOs in Fig. 5h exhibits a lack of red surface colors. The B–R color measurements in the figure come from 22 papers in the literature.

6. POSSIBLE EXPLANATIONS

The possible processes responsible for two distinct B–R color populations of Centaurs and the colors of objects with dynamical links to the Centaurs divide into two groups: evolutionary and primordial.

6.1. Evolutionary

Evolutionary models carry the implicit assumption that KBO subsurface material is pristine whereas surface material reddens, darkens, and becomes refractory as a result of continual bombardment by solar radiation (e.g., ultraviolet photons and solar wind particles).

In addition, evolutionary models assume random and occasional collisions puncture radiation-reddened crusts and expose pristine, volatile, subsurface material. Such a model explains the range of colors seen among KBOs irrespective of dynamical class (*Luu and Jewitt,* 1996). Surface tempera-

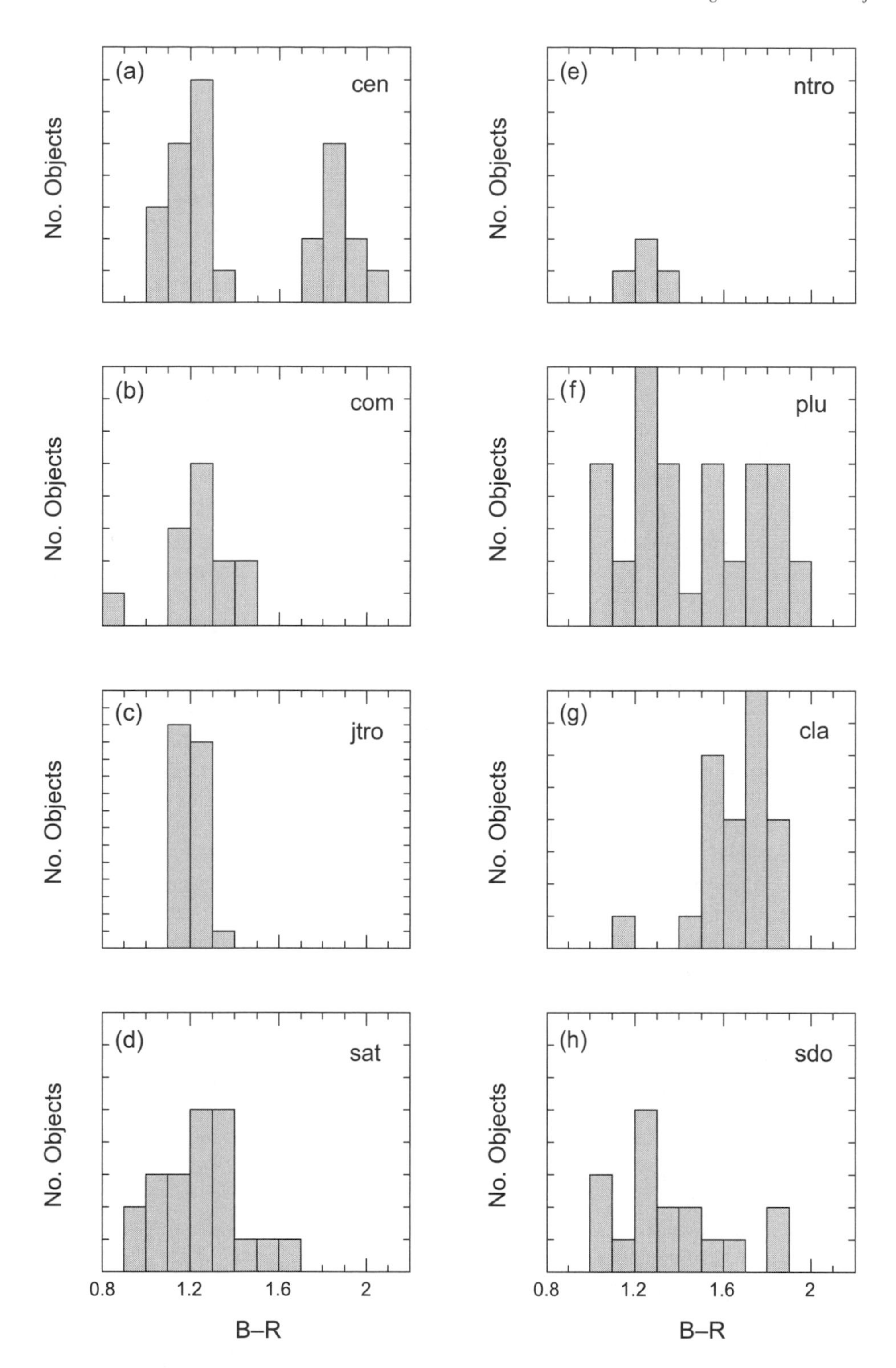

Fig. 5. Histograms of B–R colors for outer solar system objects. The objects in panels **(b)**–**(h)** are ordered in increasing semimajor axis: **(a)** Centaur objects; **(b)** comet nuclei; **(c)** jovian Trojans; **(d)** irregular satellites of Jupiter, Saturn, and Neptune; **(e)** neptunian Trojans; **(f)** Plutinos; **(g)** cold classical KBOs; **(h)** scattered disk objects. Notice that Centaurs are the only class of objects interior to the Kuiper belt that exhibit B–R > 1.7. Each tick mark on the y-axis corresponds to one object.

tures are probably too low in the Kuiper belt for most molecular ices to sublimate (25–50 K).

As some KBOs make their way onto Centaur orbits and thereby get closer to the Sun, their evolution is probably dominated by radiation-reddening and sublimation. Because Centaurs reside in a less-densely populated region than the Kuiper belt, collisions are probably not as important as they are for KBOs.

In evolutionary models, red KBOs with thick, global, reddish crusts become red Centaurs that are able to resist solar heating of subsurface material. In other words, red Centaurs have ancient surfaces. Gray KBOs with surfaces recently reworked by impacts become gray Centaurs. The red crusts of these objects are largely destroyed, making it easy for them to sublimate volatile icy material (e.g., CO and N_2 ices). *Doressoundiram et al.* (2005b) suggest there are no Centaurs with intermediate colors because when KBOs with small patches of exposed interior material and intermediate colors become Centaurs, they quickly sublimate and either coat their surfaces with gray debris or destroy more red crust and so give these objects a more globally gray color.

Burial of red crusts by sublimation debris is a reasonable explanation of the lack of red surface colors among comet nuclei, Trojan asteroids, and irregular satellites (Fig. 5). *Jewitt* (2002) first put forth such a mechanism to explain the lack of red colors among comet nuclei.

It is possible to test the sublimation mechanism as an explanation of Centaur colors. Specifically, observations of KBOs with intermediate colors should exhibit B–R color variations as they rotate.

6.2. Primordial

It is possible that KBOs and Centaurs retain some signature of their primordial colors. For example, perhaps at a heliocentric distance slightly smaller than 40 AU CH_4 went from condensing in a H_2O-ice rich clathrate to condensing as pure CH_4 (*Lewis,* 1972). Whereas the loss of CH_4 from from a clathrate surface results in a lag made up of colorless H_2O-ice crust, a pure CH_4-ice crust provides much material for alteration into red organic compounds, even if there was a substantial amount of CH_4 sublimation. So perhaps objects that formed less than 40 AU from the Sun originally had gray surface colors and objects that formed beyond 40 AU had red surface colors. CH_4-ice bands are seen in spectra of (134340) Pluto (*Cruikshank et al.,* 1976; *Fink et al.,* 1980; *Grundy and Fink,* 1996), Neptune's satellite Triton, which may be a captured KBO (*Cruikshank et al.,* 1993), (136199) Eris (*Brown et al.,* 2005), and (136472) 2005 FY9 (*Licandro et al.,* 2006; *Tegler et al.,* 2007). It is also possible that CH_3OH could provide the carbon and hydrogen for radiation-induced reddening events.

A complication to looking for such a primordial color signature is that all KBOs and Centaur objects (with the possible exception of low-inclination classical KBOs and a few resonant objects) have been scattered by Neptune from their original orbits. For example, a dynamical simulation predicts that as Neptune migrated outward it scattered objects originally 25 AU from the Sun onto orbits of the present-day SDOs, high-inclination classical KBOs, and high-inclination Plutinos. In contrast, low-inclination classical KBOs remained far enough away from Neptune that they weren't perturbed much by the planet (*Gomes,* 2003).

If the CH_4 chemistry idea and the dynamical simulation are correct, SDOs, high-inclination classical KBOs, and high-inclination Plutinos formed less than 40 AU from the Sun and they should exhibit gray surface colors, whereas low-inclination KBOs formed more than 40 AU from the Sun and they should exhibit red surface colors. In general, observations support these patterns (*Tegler et al.,* 2003).

How would such a mechanism explain two distinct color populations of Centaurs? Centaurs have short orbital lifetimes, and so their color distribution is dominated by recent Neptune scatterings of KBOs. Perhaps Neptune is now sufficiently far enough from the Sun and hence close enough to the low-inclination classical belt that it now scatters some red low-inclination classical KBOs onto Centaur orbits. Remember, there is a hint in Fig. 3 that the red Centaurs have lower orbital inclinations than the gray Centaurs. Although it may be difficult or impossible to use current day orbital parameters of Centaurs to say much about orbital parameters before their scatterings, inclination angle may be the one orbital parameter that retains some memory (*Levison and Duncan,* 1997). So, perhaps the two color populations of Centaurs are pointing to two separate source regions in the Kuiper belt (e.g., gray Centaurs come from SDOs and red Centaurs come from low-inclination classical belt objects).

7. CONCLUSIONS

It is clear that Centaur objects exhibit two distinct B–R color populations. Although there are ideas as to why there are two color populations, further theoretical and observational work is necessary to determine whether the dichotomy is primordial or due to evolutionary processes.

Acknowledgments. S.C.T. and W.R. acknowledge support from NASA Planetary Astronomy grant NNG 06GI38G. N.P. acknowledges support from the European Social Fund and the Portuguese Foundation for Science and Technology (SFRH/BPD/18729/2004).

REFERENCES

Abell P. A. and 10 colleagues (2005) Physical characteristics of comet nucleus C/2001 OG108 (LONEOS). *Icarus, 179,* 174–194.

Bauer J. M., Fernández Y. R., and Meech K. J. (2003a) An optical survey of the active Centaur C/NEAT (2001 T4). *Publ. Astron. Soc. Pac., 115,* 981–989.

Bauer J. M., Meech K. J., Fernández Y. R., Pittichova J., Hainaut O. R., Boehnhardt H., and Delsanti A. C. (2003b) Physical survey of 24 Centaurs with visible photometry. *Icarus, 166,* 195–211.

Blake D., Allamandola L, Sandford S., Hudgins D., and Friedemann F. (1991) Clathrate hydrate formation in amorphous cometary ice analogs in vacuo. *Science, 254,* 548–551.

Brown M. E., Trujillo C. A., and Rabinowitz D. L. (2005) Discovery of a planetary-sized object in the scattered Kuiper belt. *Astrophys. J. Lett., 635,* L97–L100.

Brown R. H., Cruikshank D. P., Pendleton Y., and Veeder G. J. (1998) Identification of water ice on the Centaur 1997 CU26. *Science, 280,* 1430–1432.

Choi Y. J., Weissman P. R., and Polishook D. (2006) (60558) 2000 EC_{98}. *IAU Circular 8656.*

Cruikshank D. P., Pilcher C. B., and Morrison D. (1976) Pluto — Evidence for methane frost. *Science, 194,* 835–837.

Cruikshank D. P., Roush T. L, Owen T. C., Geballe T. R., de Bergh C., Schmitt B., Brown R. H., and Bartholomew M. J. (1993) Ices on the surface of Triton. *Science, 261,* 742–745.

Cruikshank D. P. and 14 colleagues (1998) The composition of Centaur 5145 Pholus. *Icarus, 135,* 389–407.

Davis D. R., Durda D. D., Marzari F., Bagatin A. C., and Gill-Hutton R. (2002) Collisional evolution of small body populations. In *Asteroids III* (W. F. Bottke Jr. et al., eds.), pp. 545–558. Univ. of Arizona, Tucson.

Delahodde C. E., Meech K. J., Hainaut O. R., and Dotto E. (2001) Detailed phase function of comet 28P/Neujmin 1. *Astron. Astrophys., 376,* 672–685.

Delsanti A., Peixinho N., Boehnhardt H., Barucci A., Merlin F., Doressoundiram A., and Davies J. K. (2006) Near-infrared properties of Kuiper belt objects and Centaurs: Final results from the ESO large program. *Astron J., 131,* 1851–1863.

Doressoundiram A., Barucci M. A., Tozzi G. P., Poulet F., Boehnhardt H., de Bergh C., and Peixinho N. (2005a) Spectral characteristics of the trans-Neptunian object (55565) 2002 AW197 and the Centaurs (55576) 2002 GB10 and (83982) 2002 GO9: ESO large program on TNOs and Centaurs. *Planet. Space Sci., 53,* 1501–1509.

Doressoundiram A., Peixinho N., Doucet C., Mousis O., Barucci M. A., Petit J. M., and Veillet C. (2005b) The Meudon multicolor survey (2MS) of Centaurs and trans-Neptunian objects: Extended dataset and status on the correlations reported. *Icarus, 174,* 90–104.

Farnham T. L. (2001) The rotation axis of Centaur 5145 Pholus. *Icarus, 152,* 238–245.

Fink U., Smith B. A., Johnson J. R., Reitsema H. J., Benner D. C., and Westphal J. A. (1980) Detection of a CH_4 atmosphere on Pluto. *Icarus, 44,* 62–71.

Fornasier S., Dotto E., Marzari F., Barucci M. A., Boehnhardt H., Hainaut O., and de Bergh C. (2004) Visible spectroscopic and photometric survey of L5 Trojans: Investigation of dynamical families. *Icarus, 172,* 221–232.

Gomes R. S. (2003) The origin of the Kuiper belt high-inclination population. *Icarus, 161,* 404–418.

Grav T., Holman M. J., Gladman B. J., and Aksnes K. (2003) Photometric survey of irregular satellites. *Icarus, 166,* 33–45.

Grundy W. M. and Fink U. (1996) Synoptic CCD spectrophotometry of Pluto over the past 15 years. *Icarus, 124,* 329–343.

Hahn G. and Bailey M. E. (1990) Rapid dynamical evolution of the giant comet Chiron. *Nature, 348,* 132–136.

Hahn J. M. and Malhotra R. (2005) Neptune's migration into a stirred-up Kuiper belt: A detailed comparison of simulations to observations. *Astron J., 130,* 2392–2414.

Hainaut O. R. and Delsanti A. C. (2002) Colors of minor bodies in the outer solar system. A statistical analysis. *Astron. Astrophys., 389,* 641–664.

Hartigan J. A. and Hartigan P. M. (1985) The dip test of unimodality. *Ann. Stat., 13,* 70–84.

Horner J., Evans N. W., and Bailey M. E. (2004) Simulations of the population of Centaurs — I. The bulk statistics. *Mon. Not. R. Astron. Soc., 354,* 798–810.

Jewitt D. C. (2002) From Kuiper belt object to cometary nucleus: The missing ultrared matter. *Astron. J., 123,* 1039–1049.

Jewitt D. C. and Luu J. X. (1990) CCD spectra of asteroids. II — The Trojans as spectral analogs of cometary nuclei. *Astron. J., 100,* 933–944.

Jewitt D. C. and Luu J. X. (2001) Colors and spectra of Kuiper belt objects. *Astron. J., 122,* 2099–2114.

Kolmogorov A. N. (1933) About the empirical determination of a distribution law. *Giornale dell' Istituto Italiano degli Attuari, 4,* 83–91.

Kowal C. T. (1979) Chiron. In *Asteroids* (T. Gehrels, ed.), pp. 436–439. Univ. of Arizona, Tucson.

Levison H. F. and Duncan M. J. (1997) From the Kuiper belt to Jupiter-family comets: The spatial distribution of ecliptic comets. *Icarus, 127,* 13–32.

Lewis J. S. (1972) Low temperature condensation from the solar nebula. *Icarus, 16,* 241–252.

Li J.-Y., A'Hearn M. F., McFadden L. A., Sunshine J. M., Crockett C. J., Farnham T. L., Lisse C. M., Thomas P. C., and the Deep Impact Science Team (2006) Deep Impact photometry of the nucleus of Comet 9P/Tempel 1. In *Lunar and Planetary Science XXXVII,* Abstract #1839. Lunar and Planetary Institute, Houston (CD-ROM).

Licandro J. and Pinilla-Alonso N. (2005) The inhomogeneous surface of Centaur 32522 Thereus (2001 PT13). *Astrophys. J. Lett., 630,* L93–L96.

Licandro J., Pinilla-Alonso N., Pedani M., Oliva E., Tozzi G. P., and Grundy W. M. (2006) The methane ice rich surface of large TNO 2005 FY9: A Pluto-twin in the trans-Neptunian belt? *Astron. Astrophys., 445,* L35–L38.

Luu J. X. (1993) Spectral diversity among the nuclei of comets. *Icarus, 104,* 138–148.

Luu J. and Jewitt D. (1996) Color diversity among the Centaurs and Kuiper belt objects. *Astron. J., 112,* 2310–2318.

Maris M., Carraro G., Cremonese G., and Fulle M. (2001) Multicolor photometry of the Uranus irregular satellites Sycorax and Caliban. *Astron. J., 121,* 2800–2803.

Meech K. J. and Belton M. J. S. (1990) The atmosphere of Chiron. *Astron. J., 100,* 1323–1338.

Meech K. J., Bauer J. M., and Hainaut O. R. (1997) Rotation of comet 46P/Wirtanen. *Astron. Astrophys., 326,* 1268–1276.

Morbidelli A., Bottke W. F., Froeschle C., and Michel P. (2002) Origin and evolution of near Earth objects. In *Asteroids III* (W. F. Bottke Jr. et al., eds.), pp. 409–422. Univ. of Arizona, Tucson.

Morbidelli A., Levison H. F., Tsiganis K., and Gomes R. (2005) Chaotic capture of Jupiter's Trojan asteroids in the early solar system. *Nature, 435,* 462–465.

Peixinho N., Doressoundiram A., Delsanti A., Boehnhardt H., Barucci M. A., and Belskaya I. (2003) Reopening the TNO color controversy: Centaurs bimodality and TNOs unimodality. *Astron. Astrophys., 410,* L29–L32.

Peixinho N., Boehnhardt H., Belskaya I., Doressoundiram A., Barucci M. A., and Delsanti A. (2004) ESO large program on Centaurs and TNOs: Visible colors — final results. *Icarus, 170,* 153–166.

Rettig T. W., Walsh K., and Consolmagno G. (2001) Implied evolutionary differences of the jovian irregular satellites from a BVR color survey. *Icarus, 154,* 313–320.

Romon J., de Bergh C., Barucci M. A., Doressoundiram A., Cuby J. G., Le Bras A., Doute S., and Schmitt B. (2001) Photometric and spectroscopic observations of Sycorax, satellite of Uranus. *Astron. Astrophys., 376,* 310–315.

Romanishin W. and Tegler S. C. (2007) Statistics and optical colors of KBOs and Centaurs. In *Statistical Challenges in Modern Astronomy IV* (E. Feigelson and G. J. Babu, eds.), in press. Astronomical Society of the Pacific, San Francisco.

Romanishin W., Tegler S. C., Boattini A., de Luise F., and di Paola A. (2005) Comet C/2004 PY_{42}. *IAU Circular 8545.*

Schaefer B. E. and Schaefer M. W. (2000) Nereid has complex large-amplitude photometric variability. *Icarus, 146,* 541–555.

Scotti J. V. (1992) 1992 AD. *IAU Circular 5434.*

Sheppard S. S. and Trujillo C. A. (2006) A thick cloud of Neptune Trojans and their colors. *Science, 313,* 511–514.

Sheppard S. S., Jewitt D. C., Trujillo C. A., Brown M. J. I., and Ashley M. C. B. (2000) A wide-field CCD survey for Centaurs and Kuiper belt objects. *Astron. J., 120,* 2687–2694.

Tegler S. C. and Romanishin W. (2000) Extremely red Kuiper belt objects in near-circular orbits beyond 40 AU. *Nature, 407,* 979.

Tegler S. C. and Romanishin W. (2003) Resolution of the Kuiper belt color controversy: Two distinct color populations. *Icarus, 161,* 181–191.

Tegler S. C., Romanishin W., and Consolmagno G. (2003) Colors patterns in the Kuiper belt: A possible primordial origin. *Astrophys. J. Lett., 599,* L49–L52.

Tegler S. C., Romanishin W., Consolmagno G. J., Rall J., Worhatch R., Nelson M., and Weidenschilling S. (2005) The period of rotation, shape, density, and homogeneous surface color of the Centaur 5145 Pholus. *Icarus, 175,* 390–396.

Tegler S. C., Grundy W. M., Romanishin W., Consolmagno G. J., Mogren K., and Vilas F. (2007) Optical spectroscopy of the large Kuiper belt objects 136472 (2005 FY9) and 136108 (2003 EL61). *Astron. J., 133,* 526–530.

Tiscareno M. S. and Malhotra R. (2003) The dynamics of known Centaurs. *Astron. J., 126,* 3122–3131.

Surface Properties of Kuiper Belt Objects and Centaurs from Photometry and Polarimetry

Irina N. Belskaya
Kharkiv National University

Anny-Chantal Levasseur-Regourd
Université Pierre et Marie Curie (Paris 6)/Aéronomie

Yuriy G. Shkuratov
Kharkiv National University

Karri Muinonen
University of Helsinki

Physical properties of Kuiper belt objects (KBOs) can be assessed by studying their photometric and polarimetric phase effects close to opposition, i.e., the brightness opposition effect and the negative polarization branch. The first phase-curve observations and their promising preliminary interpretations are reviewed for KBOs and Centaurs. Despite the limited range of phase angles accessible from groundbased observations, distinct features have been discovered both in brightness and polarization. Recent results of relevant numerical and laboratory simulations of phase-angle effects are reviewed. The possibility of constraining the geometric albedos of the surfaces based on photometric and polarimetric observations is discussed.

1. INTRODUCTION

Probing the surface properties of solar system objects by photometric and polarimetric observations at different phase angles (angle between the Sun and the Earth as seen from the object) is a traditional technique in solar system remote sensing. Scattered light measured at varying illumination and observation geometries contains information about the physical properties of the topmost surface layer such as particle size, heterogeneity, complex refractive index, porosity, and surface roughness. The physical parameters define such observable parameters as the geometric albedo of the surface. The intricate inverse problem to constrain surface properties from phase-angle-resolved photometry and polarimetry is almost the only way to assess the microscopic properties of the surface from remote observations.

In the particular case of Kuiper belt objects (KBOs), the geometry of groundbased observations is limited to small phase angles, with the maximum accessible phase angle being α_{max} = arcsin (1/r), where r is the heliocentric distance in AU. For transneptunian objects (TNOs), the phase-angle range is typically less than 2°, increasing to 7°–8° for Centaurs. These distant objects are thus observable only close to the backscattering direction.

At small phase angles, two intriguing phenomena are typically observed for solid atmosphereless solar system bodies: the so-called opposition effect in brightness and the negative branch in the degree of linear polarization. The opposition effect manifests itself as a considerable nonlinear increase of surface brightness as the phase angle decreases to zero. Negative polarization is a peculiar case of partially linearly polarized scattered light where the electric field vector component parallel to the scattering plane predominates over the perpendicular component. Observations of the abovementioned opposition phenomena for various kinds of solar system bodies have a long history, whereas the understanding of their physical nature has progressed considerably only during the last two decades (for reviews, see, e.g., *Muinonen et al.,* 2002; *Shkuratov et al.,* 2004). There are several physical mechanisms that may contribute to the opposition phenomena depending on the properties of the surfaces. Both opposition phenomena are considered to have, at least partly, similar physical causes and their joint analysis can result in constraints on the physical properties of the surfaces under consideration.

At present, the observational data on the photometric and polarimetric phase effects of KBOs are still scarce for any statistical conclusions. Polarimetric observations of KBOs have just begun and many of the photometric observations are of insufficient accuracy for the derivation of firm conclusions.

In the present chapter, we focus on the study of the opposition phenomena for KBOs and Centaurs and on what constraints can be derived for the surface properties on the basis of such a study. In section 2, we review the photometric and polarimetric observations of KBOs and Centaurs, with emphasis on the phase effects. Polarimetry is here described in more detail, whereas photometry is also discussed in the chapter by Sheppard et al. Section 3 gives a brief description of the physical mechanisms for the opposition phenomena and describes recent results of relevant numerical and laboratory simulations. In section 4, we compare the KBO observations to those of other solar system bodies and discuss which surface properties can be constrained by the observations. Finally, we outline the future prospects of the photometric and polarimetric studies.

2. OBSERVATIONS

2.1. Advances in Photometry

The first tentative derivations of the magnitude dependences on the phase angle for a few KBOs were obtained as a byproduct of the observational programs devoted to the determination of the rotational properties of KBOs. *Sheppard and Jewitt* (2002) reported phase curves of seven KBOs in the R band, with a small number (2–4) of observations within phase angles less than 2°. The authors emphasized the similarity of the phase coefficients (slopes in the phase curves) for all KBOs studied, with the mean phase coefficient being large at $\beta = 0.15$ mag/deg, suggesting comparative uniformity of the surface properties. Independently, *Schaefer and Rabinowitz* (2002) carried out numerous photometric observations for the KBO (38628) Huya and derived a comparable phase coefficient of 0.125 ± 0.009 mag/deg in the R band. They concluded that the observed magnitude variations of (38628) Huya vs. the phase angle (ranging 0.3°–2.0°) are consistent with a linear dependence. Additional phase curves were published by *Rousselot et al.* (2003, 2005a), *Boehnhardt et al.* (2004), and *Bagnulo et al.* (2006).

Magnitude dependences were also observed for a few Centaurs in a larger range of phase angles up to 7° in the R band (*Bauer et al.*, 2002, 2003; *Rousselot et al.*, 2005a; *Bagnulo et al.*, 2006) and for the Centaur (10199) Chariklo in the J band (*McBride et al.*, 1999). The phase curves were fitted according to the H, G magnitude system with G values ranging from –0.39 to 0.18 for the R band, considering them to be consistent with low-albedo surfaces (*Bauer et al.*, 2003). Note that the negative G values have no physical meaning and indicate problems with the H, G system. Generally, the phase curves obtained for Centaurs are mutually quite compatible within the uncertainties of the observations.

Belskaya et al. (2003) indicated a possible existence of a very narrow opposition surge starting at phase angles below 0.1°–0.2°. For all available observations at such extremely small phase angles, the brightness was found to be systematically higher than the brightness extrapolated from a linear phase function. The same result was obtained for the TNO (55637) 2002 UX_{25} observed at a phase angle of 0.02° by *Rousselot et al.* (2005a).

Photometric observations with an emphasis on extremely small phase angles were made by *Hicks et al.* (2005) for the TNO (20000) Varuna. They found an opposition surge of about 0.1 mag at phase angles less than 0.1°. Further observations of Varuna confirmed the pronounced opposition surge, allowing its more precise estimation (*Belskaya et al.*, 2006). The magnitude-phase dependence for Varuna is shown in Fig. 1.

Rabinowitz et al. (2007) presented a large survey dedicated to the measurements of phase curves for distant solar system objects, including 18 TNOs and 7 Centaurs. They found a wide range of phase coefficients from almost zero to 0.4 mag/deg and reported a significant wavelength dependence (from B band to I band) of the coefficients for some objects. Their conclusions are based on long-term observing campaigns with typically very few observations for an object per night. Because of the difficulties in distinguishing between the effects from rotation and varying phase angle (see discussion below), such an approach can give only rough estimates for phase coefficients.

Recalling that, in general, the phase-curve behavior near opposition is nonlinear, a linear function used for phase-curve fitting should be treated as a rough approximation for a limited phase-angle range. Thus, the values of phase coefficients depend on the phase-angle range used in the linear fitting. This is well illustrated by the phase coefficient of Varuna varying from 0.33 to 0.11 mag/deg (*Belskaya et al.*, 2006).

The values of the most accurate phase coefficients for KBOs and Centaurs are compiled in Table 1, which also gives the ranges of phase angles covered by the observations. Although the uncertainties of the phase coefficients are rather large, the observational data indicate two main features in the photometric phase curves of KBOs and Centaurs: (1) a steep linear phase dependence for most objects; (2) an opposition surge at extremely small phase angles below 0.1°.

2.1.1. Pluto. The photometric phase curve of Pluto has been observed repeatedly (*Binzel and Mulholland,* 1984; *Tholen and Tedesco,* 1994; *Buie et al.,* 1997; *Buratti et al.,* 2003). All observations show a small phase coefficient in the

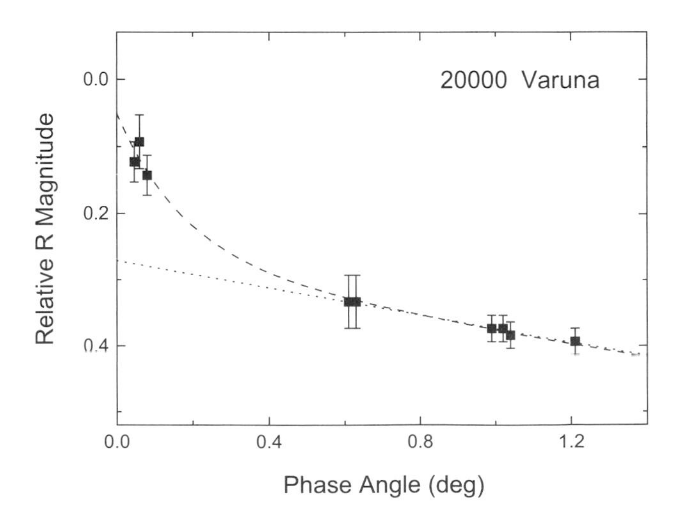

Fig. 1. Magnitude phase curve of Varuna fitted by a linear-exponential model (for details see *Belskaya et al.*, 2006).

TABLE 1. Phase coefficients measured for KBOs and Centaurs.

Object	Phase-angle range (deg)	Phase coefficient in the R band (mag/deg)	σ (mag/deg)	Reference
(2060) Chiron	1.4–4.2	0.045	0.023	*Bagnulo et al.* (2006)
(5145) Pholus	0.7–4.0	0.075	0.006	*Buie and Bus* (1992)
(10370) Hylonome	1.3–3.0	0.058	0.045	*Bauer et al.* (2003)
(20000) Varuna	0.06–1.23	0.11*	0.03	*Belskaya et al.* (2006); *Jewitt and Sheppard* (2002)
(28978) Ixion	0.25–1.34	0.20	0.04	*Boehnhardt et al.* (2004)
(29981) 1999 TD_{10}	0.77–3.5	0.12	0.06	*Rousselot et al.* (2005b)
(31824) Elatus	1.3–7.1	0.082	0.006	*Bauer et al.* (2002)
(32532) Thereus	0.2–5.6	0.072	0.004	*Rabinowitz et al.* (2007)
(38628) Huya	0.5–1.9	0.14	0.02	*Sheppard and Jewitt* (2002)
(38628) Huya	0.28–2.0	0.125	0.02	*Shaefer and Rabinowitz* (2002)
(40314) 1999 KR_{16}	0.31–1.3	0.14	0.02	*Sheppard and Jewitt* 2002
(47932) 2000 GN_{171}	0.02–2.0	0.143 (V)	0.030	*Rabinowitz et al.* (2007)
(50000) Quaoar	0.25–1.2	0.16	0.06	*Bagnulo et al.* (2006)
(50000) Quaoar	0.17–1.3	0.159 (V)	0.027	*Rabinowitz et al.* (2007)
(54598) Bienor	0.3–3.0	0.095 (V)	0.016	*Rabinowitz et al.* (2007)
(60558) Echeclus	0.1–3.9	0.18	0.02	*Bauer et al.* (2003); *Rousselot et al.* (2005a)
(90482) Orcus	0.4–1.2	0.114 (V)	0.030	*Rabinowitz et al.* (2007)
(136108) 2003 EL_{61}	0.5–1.1	0.091 (V)	0.025	*Rabinowitz et al.* (2006)
(134340) Pluto	0.6–2.0	0.029 (V)	0.001	*Buie et al.* (1997)

*Opposition surge observed at $\alpha < 0.1°$ was excluded from calculations of phase slope.

V band of 0.03–0.04 mag/deg. A nonlinear opposition surge has not been observed so far, but its existence cannot be excluded at phase angles below 0.4°, not covered by the observations. *Buratti et al.* (2003) noticed a weak wavelength dependence for the phase coefficient of Pluto ranging from 0.037 mag/deg in B to 0.032 mag/deg in R. However, most of the observations are related to the Pluto-Charon system. An attempt to constrain the individual photometric properties of these objects was made by *Buie et al.* (1997). They obtained a smaller phase coefficient for Pluto (0.029 mag/deg) and the first estimation of the phase coefficient for Charon (0.087 mag/deg) in the phase-angle range of 0.6°–2°.

Small phase coefficients, comparable to that of Pluto, were reported by *Rabinowitz et al.* (2006, 2007) for the large TNOs (136199) Eris, (136472) 2005 FY_9, and (136108) 2003 EL_{61}. They concluded that phase coefficients below 0.10 mag/deg are a salient feature for Pluto-scale TNOs with neutral colors, high albedos, and icy surfaces.

2.1.2. Uncertainties of phase coefficients. Most phase-curve observations were made with an accuracy of about 0.1 mag. Taking into account the narrow range of phase angles available, phase coefficient estimates based on such observations are not reliable. The small errors of the phase coefficients given in some papers were derived without taking into account the individual observational uncertainties pertaining to each photometric point. Using proper weights in the linear least-squares analyses, the uncertainties of the phase coefficients become considerably larger.

Another problem in deriving phase coefficients is connected to difficulties in assessing the brightness variations due to rotation. Lightcurve amplitudes were taken into account only for a few objects. The values of the phase coefficients derived by omitting the lightcurve variations should be treated with great caution. For precise derivation of photometric phase curves, observations made at different phase angles should be reduced to a standard geometry by using a proper spin and shape model for the object. Such models are, however, rarely available for TNOs. In practice, assuming a linear phase-curve dependence, the phase coefficient and the rotation period are simultaneously derived from minimization of lightcurve scatter. Such an approach provides only rough estimates for phase coefficients. The subsequent step comprises a recomputation of the phase curve by using well-defined lightcurves. However, when a composite lightcurve is based on short-term observations at different phase angles, it is hard to distinguish between the effects from rotation and varying phase angle. In that case, the recalculated phase coefficient is always very close to its rough estimate used for the period determination and remains uncertain.

2.1.3. Small-phase-angle effects in lightcurves. The photometric observations available for KBOs have demonstrated rather large increases in brightness toward the zero phase angle, with rapidly increasing steepness of the phase curve. When studying the lightcurve variations due to rotation, the phase effects need to be taken into account (see chapter by Sheppard et al.). Usually, a linear magnitude phase dependence is assumed, which can be used only as a rough approximation at phase angles far from the opposition surge. When observations are available at phase angles $\alpha < 0.2°$, great caution is in place when including them in the construction of composite lightcurves. As in the case of Varuna (*Belskaya et al.,* 2006), the opposition surge can reach the amplitude of 0.2 mag relative to the extrapolation from the linear fit of the phase curve at $\alpha > 0.2°$. Omission of the opposition surge will result in an overestimation of the lightcurve amplitude, when observations at extremely small phase angles are included in lightcurve analyses.

Variations of scattering properties over the TNO surfaces can also influence the lightcurve shape and amplitude. Differences both in the amplitude and the positions of lightcurve extrema measured at extremely small (0.1°) and somewhat larger (1°) phase angles were found for Varuna (*Belskaya et al.,* 2006). A plausible explanation for the lightcurve changes is the variation of the light-scattering properties over the surface, giving rise to differing opposition surges, e.g., between two hemispheres of the object. In this case, lightcurves measured at phase angles beyond the opposition surge can be different from those obtained at extremely small phase angles. Generally, the phase effect results in an increase of the lightcurve amplitude toward the zero phase angle, and may give rise to misinterpretations of the rotational properties of KBOs. On the other hand, the study of the amplitude-phase relationship can give additional information on the physical characteristics of the surface.

2.2. Advances in Polarimetry

2.2.1. Principles. In general, the polarization state of light can be fully described by the Stokes vector S (I, Q, U, V) where I is the intensity, Q and U characterize the linear polarization state, and V describes the circular polarization state. Sunlight scattered by solid planetary surfaces (or dust clouds) becomes partially linearly polarized. Two quantities are used to describe the linear polarization state, the total degree of linear polarization P and the position angle θ. The total linear-polarization degree P is equal to $(I_{max} - I_{min})/(I_{max} + I_{min})$, where I_{max} and I_{min} are respectively the maximum and minimum intensities of the polarized light. The position angle θ corresponds to the plane of I_{max}, and is measured with respect to the celestial coordinate system from the direction of the north celestial meridian. These quantities are related to the Stokes parameters

$$P = \sqrt{q^2 + u^2}, \quad \theta = \frac{1}{2}\arctan\frac{u}{q}$$

where u = U/I and q = Q/I are the normalized Stokes parameters.

For sunlight scattered by planetary surfaces, the polarization plane position is usually either normal or parallel to the scattering plane (the Sun-object-Earth plane). Therefore the polarization degree can be expressed through $P_r = P\cos(2\theta_r)$, where θ_r is the angle between the measured position angle of polarization θ and the normal to the scattering plane: $\theta_r = \theta - (\varphi \pm 90°)$. The angle φ is the position angle of the scattering plane and the sign inside the bracket is chosen to ensure the condition $0° \leq (\varphi \pm 90°) \leq 180°$. The degree of linear polarization P_r may also be expressed with the help of the intensities scattered perpendicular $I_\perp$ or parallel $I_\parallel$ to the scattering plane

$$P_r = (I_\perp - I_\parallel)/(I_\perp + I_\parallel)$$

It means that the sign of the polarization degree can be positive or negative depending on which component predominates. Negative linear polarization was first discovered by *Lyot* (1929) for the Moon and later found to be a ubiquitous phenomenon for planetary surfaces and interplanetary dust at small phase angles. The degree of polarization varies with the phase angle of observations (the angle between the incident and emergent light) and with the wavelength of the observations, characterizing the properties of the surface, mainly its albedo and texture.

Typical polarization phase curves $P(\alpha)$ of atmosphereless bodies have a negative polarization branch reaching $P_{min} \sim -0.2$ to 2% at phase angles $\alpha_{min} \sim 7°$–10° turning to positive values at the inversion angle $\alpha_{inv} \sim 20°$–25° (see, e.g., *Levasseur-Regourd and Hadamcik,* 2003). Most of the polarization phase curves observed have rather symmetric negative branches with $\alpha_{min} \sim \alpha_{inv}/2$ and P near zero at phase angles reaching 0°. For very bright surfaces, an asymmetric peak of negative polarization of about 0.4% has been found at small phase angles of less than 1°–2° (*Rosenbush et al.,* 1997, 2005).

2.2.2. Instrumentation and data reduction. At small phase angles (as observable from the Earth), TNOs are expected to present a relatively small polarization degree. To measure it with an accuracy better than 0.1%, the signal-to-noise ratio needs to be better than 500, which can be achieved only by using large telescopes for these faint objects. At present, all observations available for TNOs and one Centaur have been made with the 8.2-m Very Large Telescope (VLT) in Cerro Paranal Observatory (Chile) using the Focal Reducer/Low Dispersion Spectrograph (FORS1) instrument in its imaging polarimetry mode. Polarimetric observations and data reduction are discussed in detail by *Bagnulo et al.* (2006). A thorough error analysis for dual-beam optical linear polarimetry is given by *Patat and Romaniello* (2006).

2.2.3. Results. The first polarimetric observations for a TNO (except Pluto) were carried out in 2002 by *Boehnhardt et al.* (2004). Observations of the Plutino (28978) Ixion have shown rather pronounced negative polarization noticeably changing as a function of the phase angle in spite of the small range of phase angles covered (0.2°–1.3°). These observations have demonstrated the capability of the polarimetric technique to study distant objects even if they are observable only at very small phase angles, below α_{min}. Later, three additional objects have been observed: (2060) Chiron and (50000) Quaoar (*Bagnulo et al.,* 2006) and 29981 (1999 TD_{10}) (*Rousselot et al.,* 2005b).

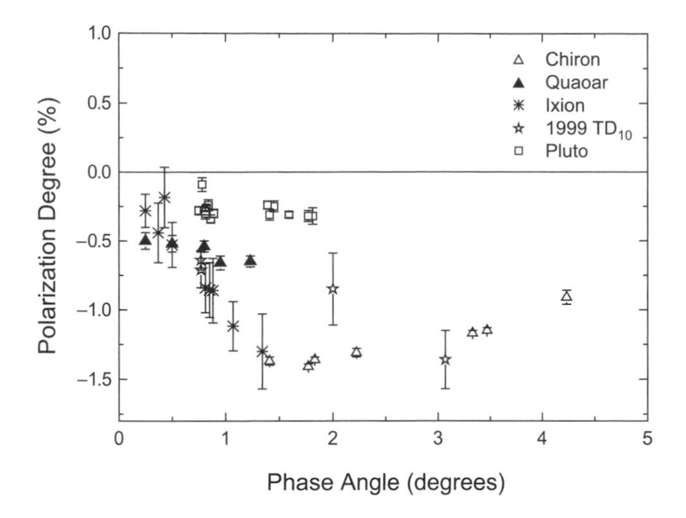

Fig. 2. Polarization phase curves in the R band of KBOs, Centaur Chiron, and for Pluto in the V band (see text for references).

The polarimetric observations available for three KBOs and one Centaur are shown in Fig. 2, which also includes polarimetric data for Pluto, obtained by a number of authors (*Fix and Kelsey,* 1973; *Breger and Cochran,* 1982; *Avramchuk et al.,* 1992). Pluto's polarization curve exhibits a behavior characterized by shallow and nearly constant negative polarization. Although only a few KBOs or Centaurs have been observed up to now, polarimetric observations provide unequivocal evidence for their differing phase curves. Table 2 presents the maximal absolute values of polarization degree $|P_R|$ measured for each object, their geometric albedos p in the R band, and the range of phase angles covered by the observations. The observational data available can be summarized as (1) a noticeable negative polarization is inherent in the surfaces of all distant objects observed, varying from –0.3% to –1.4%; (2) two different polarimetric trends are seen at small phase angles: slow changes in the negative polarization degree for the largest objects (Pluto and Quaoar) and a rapid increase (in absolute terms) in the negative polarization of ~1%/deg in the phase-angle range of 0.3°–1.3° (Ixion); (3) the minimum of the negative polarization branch of Chiron occurs at small phase angles less than 2°; and (4) no wavelength dependence of the polarization degree (BVR bands) is observed for Chiron; a color effect (VR bands) is suspected for 1999 TD_{10}, but its confirmation would require more observations.

3. SIMULATIONS OF PHOTOMETRIC AND POLARIMETRIC OPPOSITION PHENOMENA

3.1. Physical Mechanisms

When considering the physical mechanisms causing the photometric and polarimetric phase dependences, we assume that the TNOs have no noticeable atmospheres that could influence such dependences. This may not always be true. For instance, unusual photopolarimetric properties of Pluto might be partially attributed to its thin atmosphere. Nevertheless, we here ignore possible atmospheric contributions, since there is little data on the atmospheres of TNOs.

There are several physical mechanisms relevant for particulate media that may potentially contribute to the opposition effect. Their contribution depends on the physical parameters of the surface that scatters light, e.g., particle size or porosity. For the surfaces of atmosphereless solar system objects, two main mechanisms are usually considered to be relevant: the shadow-hiding (SM) and coherent-backscattering mechanisms (CBM). Moreover, single-particle scattering also contributes to the opposition phenomena, providing broad backscattering peaks and negative polarization branches (e.g., *Shkuratov et al.,* 2002, 2006; *Muinonen et al.,* 2007). However, single-particle scattering can be envisaged to have a minor direct contribution at small phase angles. In the case of KBOs, SM and CBM are the most important physical mechanisms for the phenomena. Note that, on one hand, SM contributes to the photometric phase dependencies only. On the other hand, CBM contributes both to photometric and polarimetric phase dependencies. Below, we briefly describe both mechanisms. More details can be found in the reviews by *Muinonen* (1994) and *Shkuratov et al.* (2002).

3.1.1. Coherent backscattering mechanism. In order to describe the mechanism, let us consider an incident electromagnetic plane wave (wavelength λ and wave number $k = 2\pi/\lambda$) propagating in the negative direction of the z-axis and interacting with two end scatterers A and B, which are on

TABLE 2. Polarization degree measured for KBOs and a Centaur.

Object	p_R*	Phase-angle Range (deg)	$\|P_R\|$ (%)	Reference
(2060) Chiron	0.08	1.41–4.23	1.40	*Bagnulo et al.* (2006)
(28978) Ixion	0.20	0.25–1.34	1.30	*Boehnhardt et al.* (2004)
(29981) 1999 TD_{10}	0.05	0.77–3.10	1.35	*Rousselot et al.* (2005b)
(50000) Quaoar	0.26	0.25–1.23	0.65	*Bagnulo et al.* (2006)
(134340) Pluto	0.61†	0.75–1.81	0.32†	*Breger and Cochran* (1982); *Fix and Kelsey* (1973)

*Albedo is given according to Stansberry et al. (this volume).
†Data are given for the V band.

the order of one to hundreds of wavelengths apart, via an arbitrary number of intermediate scatterers in between them. Two scattered wave components due to the two opposite propagation directions between scatterers A and B always interfere constructively in the backward direction, whereas in other directions, the interference characteristics vary. Three-dimensional averaging over scatterer locations results in a backscattering enhancement with decreasing angular width for increasing number of intermediate scatterers, because the average distance between the end scatterers is larger for higher numbers of intermediate scatterers.

In order to explain the CBM for the negative degree of linear polarization for unpolarized incident light, the derivation and proper averaging of the Stokes vectors corresponding to the scattered electromagnetic fields are required for two orthogonal linear polarization states of the incident plane wave. Consider incident polarizations parallel and perpendicular to the scattering plane (here yz-plane) defined by the light source, object, and the observer. For simplicity, consider two scatterers A and B at a distance d from one another aligned either on the x-axis or the y-axis, while the observer is in the yz-plane. Since first-order scattering is typically positively polarized (e.g., Rayleigh scattering and Fresnel reflection), the scatterers sufficiently far away from each other ($kd = 2\pi d/\lambda \gg 1$) interact predominantly with the electric field vector perpendicular to the plane defined by the source and the two scatterers, while interaction with the electric field vector parallel to that plane is suppressed. The observer in the yz-plane will detect negative polarization from the geometry with the scatterers aligned along the x-axis and the incident polarization along the y-axis, and positive polarization from the geometry with the scatterers aligned along the y-axis and the incident polarization along the x-axis. However, the positive polarization suffers from the phase difference $kd \sin\alpha$, whereas the phase difference for the negative polarization is zero for all phase angles. Averaging over scatterer locations will result in negative polarization near the backward direction. Scattering orders higher than the second, which experience similar preferential interaction geometries, can also contribute to negative polarization. As above for the opposition effect, the contributions from increasing orders of scattering manifest themselves at decreasing phase angles.

3.1.2. Shadowing mechanism. On the surfaces of TNOs, SM is bound to be relevant for length scales significantly larger than the wavelength of incident light. It is the most prominent first-order scattering mechanism. In principle, we can distinguish between shadowing by the rough interface between the regolith and the free space and shadowing by the internal geometric structure of the regolith, whereas in practice, for disk-integrated photometric data, it can be difficult if not impossible to discriminate between the two shadowing contributions. In both cases, SM is due to the fact that a ray of light penetrating into the scattering medium and incident on a certain particle can always emerge back along the path of incidence, whereas in other directions, the emerging ray can be blocked by other particles. The internal SM depends mainly on the volume density of the scattering medium, whereas the interfacial SM depends mainly on surface roughness (here standard deviation of the interfacial slopes). Recent studies of the internal SM and interfacial SM have been carried out by, e.g., *Muinonen et al.* (2001), *Shkuratov et al.* (2005), and *Parviainen and Muinonen* (2007).

3.1.3. Analytical models. There are no rigorous electromagnetic solutions for light scattering by random rough surfaces, whereas in specific cases, analytical approximations are available for, e.g., the interpretation of photometric observations. The most popular model is the photometric model proposed by *Hapke* (1986). It accounts for (1) single-particle scattering, (2) shadowing due to particulate surfaces and their large-scale topography, and (3) multiple scattering between the particles. The model was later modified to incorporate coherent backscattering and the influence of anisotropic single scatterers on multiple scattering, with seven free parameters in the latest version (*Hapke,* 2002). Unfortunately, the theory does not allow an unambiguous interpretation of observed or measured phase curves. Even for asteroids observed at large ranges of phase angles, the Hapke model does not give reliable estimates for photometric parameters (see *Helfenstein and Veverka,* 1989). Besides, the model incorporates the so-called diffusion approximation to describe coherent backscattering, which cannot be applied in the case of particulate surfaces with low and intermediate albedo. Moreover, any coherent backscattering model has to take into account polarization in order to correctly describe the photometric spike (*Mishchenko,* 1991); the Hapke model is a scalar one and does not consider polarization at all.

Mishchenko et al. (2000) made use of the analytical theories by *Ozrin* (1992) and *Amic et al.* (1997) to compute coherent backscattering by nonabsorbing media of Rayleigh scatterers. They were able to offer reference results, i.e., accurate predictions for the values of the amplitude and width of the opposition effect and the shape and depth of the negative polarization surge. One of the main predictions is that the photometric opposition effect is accompanied by a polarization opposition effect of the same angular width (e.g., *Mishchenko and Dlugach,* 1993). There are, however, several shortcomings in the approach by *Mishchenko et al.* (2000): (1) absorbing scatterers are not treated; (2) the results are limited to Rayleigh scatterers; and (3) the shadowing effects are excluded.

3.2. Numerical Modeling

The backscattering phenomena can be studied using Monte Carlo (MC) simulations of radiative transfer and coherent backscattering for particulate surfaces of spherical scatterers (*Shkuratov et al.,* 2002; *Muinonen,* 2004). The application of MC modeling to the polarimetric and photometric observations is bound to require certain simplifications. For example, the single scatterers are assumed to be Rayleigh-scatterers (e.g., *Muinonen et al.,* 2002), allowing simultaneous modeling of the photometric and polarimetric phase effects using a minimum number of physical parameters, i.e., the single-scattering albedo and mean free path. The MC method for coherent backscattering by absorbing and scattering media mimics radiative transfer but, for each multiple-scattering event, it additionally computes the coherent-backscattering contribution with the help of the reciprocity relation in electromagnetic scattering. Application of the Rayleigh-scatterer model to the observations of KBOs and one Centaur showed that a possible way to explain their polarization properties is to assume two-component surface media consisting of Rayleigh scatterers with small and large single-scattering albedos. Details of model fits are given in *Boehnhardt et al.* (2004), *Bagnulo et al.* (2006), and *Rousselot et al.* (2005b). Note that the fits are by no means unambiguous; rather, they demonstrate the capability of the two-component model to explain the observations. Moreover, the model considers the geometric albedo as a parameter known *a priori*, e.g., from thermal infrared observations. Since the present KBO albedos are poorly known, potential changes in them can imply changes in the parameters of the two-component model [cf. Ixion (*Bagnulo et al.*, 2006)].

Polarimetric observations of bright objects like Pluto are difficult to explain using a Rayleigh-scatterer model. This is in agreement with the findings in *Muinonen et al.* (2002) for bright asteroids. However, the Rayleigh-scatterer model that includes a coherent-backscattering component with a long mean free path is capable of explaining the observations. Clearly, a more sophisticated scatterer model in coherent backscattering could allow more significant CBM contributions to the opposition effect.

There are some intriguing features predicted by the modeling of the observations available for KBOs. Figure 3 shows model phase curves of brightness and polarization for Chiron, Ixion, and Pluto. The modeling suggests a narrow spike in both polarimetry and photometry for all objects that can be checked with future observations. We emphasize that a sharp surge in brightness and polarization at phase angles less than 0.2° is also suggested for Pluto. Further discussion will be given in section 4. Here we would like to emphasize that detailed phase curves for both brightness and polarization are a necessity for further progress in modeling and that observations at phase angles very close to zero are crucial for verifying the coherent-backscattering contributions.

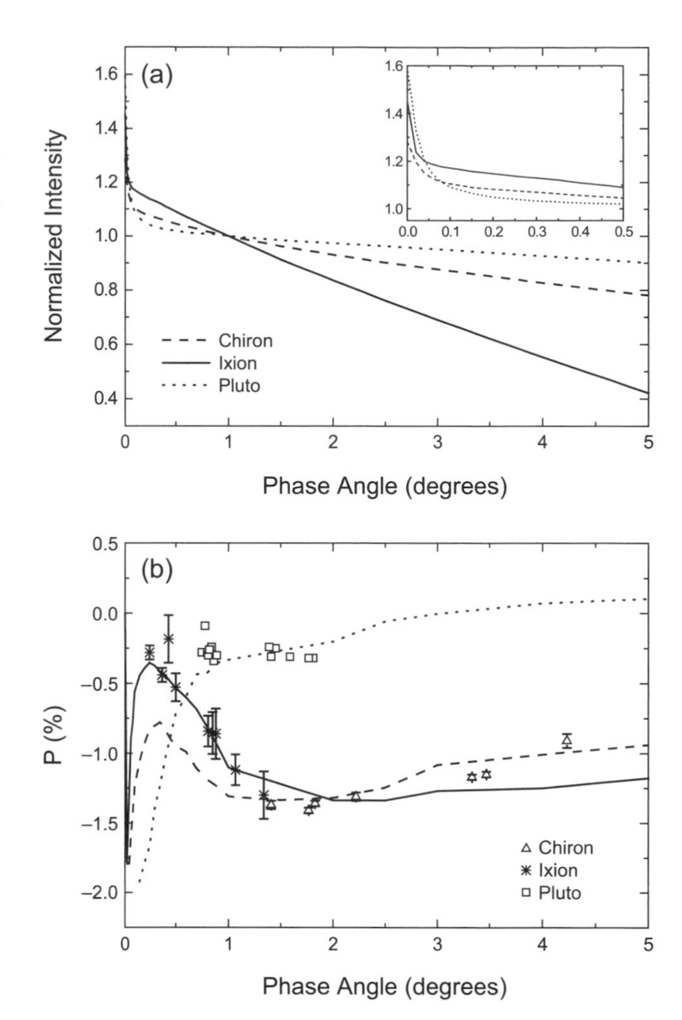

Fig. 3. Examples of **(a)** model intensity and **(b)** polarization fits for Chiron, Ixion, and Pluto. The inset in the upper corner of **(a)** shows the small phase angle region in greater detail. The available polarimetric observations of these objects are also plotted (see text for references).

3.3. Laboratory Simulations

3.3.1. Instruments for experimental simulations. In order to be applicable to the interpretation of KBO observations, laboratory measurements should be carried out down to very small phase angles. Measurements at such phase angles are technically difficult and require special instruments. For that very purpose, a new laboratory photometer was designed at the Institute of Astronomy of Kharkiv National University to perform measurements in the phase-angle range of 0.008°–1.5°. Such small phase angles become feasible due to the small linear apertures of the light source (unpolarized laser) and the receiver, and the large distances from the scattering sample surface to the light source and the detector (*Psarev et al.*, 2007). The new device complements the other Kharkiv laboratory photopolarimeter, which works in the phase-angle range of 0.2°–17° and uses a lamp as the light source (*Shkuratov et al.*, 2002). Numerous small-phase-angle measurements were also made at the Jet Propulsion Laboratory (JPL) using the long-arm goniometric photopolarimeter with a laser as the light source, covering the phase-angle range of 0.05°–5° (*Nelson et al.*, 2000). The Kharkiv and JPL instruments were intercalibrated by measuring the same sample. A laboratory instrument suitable for photometric measurements down to the phase angle of 0.2° was used at the University of Helsinki Observatory with an unpolarized laser (*Kaasalainen et al.*, 2002). The laboratory experiments were also made to simulate polarimetric properties of both regoliths and dust clouds under microgravity conditions (e.g., *Levasseur-Regourd and Hadamcik*, 2003).

3.3.2. Brightness opposition spike. A key question to be answered by the laboratory measurements is: Which surfaces can produce narrow and sharp brightness opposition spikes? The opposition spike is found to be more prominent with increasing surface albedo (e.g., *Shkuratov et al.,* 2002; *Psarev et al.,* 2007). Results from photometric measurements in the phase-angle range of 0.008°–1.5° for two samples with substantially differing albedos, a very bright sample of smoked MgO and a very dark sample of carbon soot, are shown in Fig. 4. For both samples, thick layers of smoked materials were used, and both surfaces were very fluffy. Results shown in Fig. 4 allow one to compare data obtained with the Kharkiv lamp and laser photometer at the equivalent wavelength of 0.63 μm. The curves for the same sample measured by different instruments are in good agreement. The bright and dark samples have albedos of 99% and 2.5%, respectively, determined at a 2° phase angle relative to the photometric standard Halon. The MgO sample has a very prominent opposition spike at phase angles $<0.8^\circ$. The dark sample of carbon soot does not show any significant opposition features with an almost linear phase dependence in the range of 0.008°–2°. However, if we consider a wider phase-angle range up to 15°, the nonlinearity of the carbon-soot phase curve can be clearly seen (Fig. 4). The laboratory data suggests that surfaces giving rise to the opposition spike have inevitably rather high albedos necessary to produce the coherent-backscattering enhancement. Further measurements of mixtures of MgO and carbon soot samples in various proportions (albedos 9%, 19%, 47%) have shown that the opposition spike becomes noticeable for albedos greater than 19% (*Psarev et al.,* 2006). It was also shown that the width of the spike depends on the surface texture, mainly its packing density.

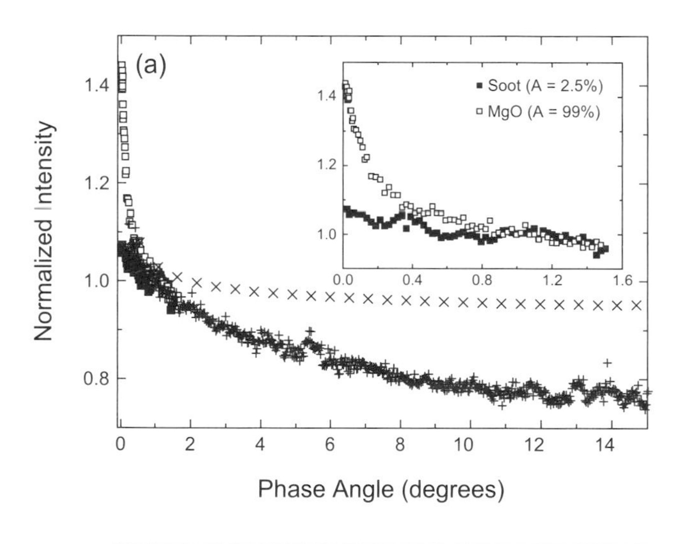

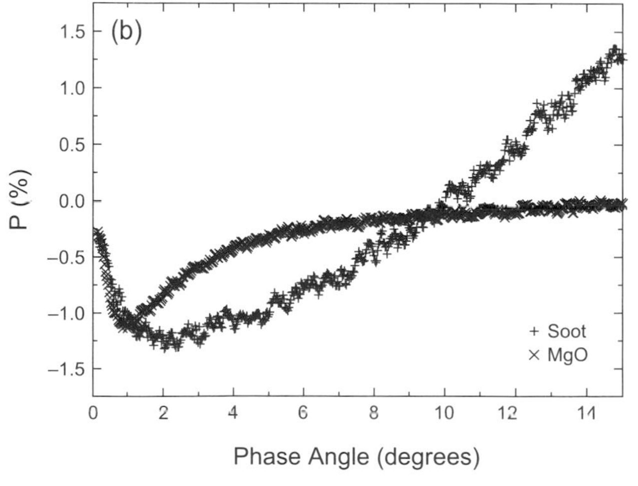

Fig. 4. (**a**) Photometric and (**b**) polarimetric phase curves for MgO measured with the laser laboratory photometer (open and filled squares) and the lamp photopolarimeter (crosses and pluses). The inset in the upper corner of (**a**) shows the small phase angle region in greater detail.

Thus, laboratory measurements suggest that observations of the opposition spike can put constraints on the surface albedo and texture. In particular, a very narrow and sharp opposition spike observed for Varuna and probably for other KBOs could indicate rather high surface albedos and a fluffy surface texture.

3.3.3. Negative polarization. The first polarimetric observations of three KBOs and one Centaur raise two main questions, insofar as laboratory measurements are concerned: (1) Which surface properties influence the position of the minimum polarization, and (2) which surfaces can produce a significant negative polarization at small phase angles?

Numerous laboratory polarimetric measurements have been carried out to study negative polarization (for surveys see *Muinonen et al.,* 2002; *Shkuratov and Ovcharenko,* 2002). The depth and shape of the negative-polarization branch was found to depend strongly on such parameters as complex refractive index, particle size, packing density, and microscopic optical heterogeneity of the laboratory samples, all affecting the albedo of the surface. An attempt to distinguish the effects of individual surface parameters on the photometric and polarimetric phase curves was made by *Shkuratov et al.* (2002) based on the laboratory measurements of diverse samples in the phase-angle range of 0.2°–4°. Here we briefly summarize the current understanding of the roles of the abovementioned parameters in the formation of the negative-polarization branch.

3.3.4. Albedo. Although the geometric albedo of the surface follows from the physical parameters, it is valuable to summarize certain empirical relations including the albedo as one of the parameters. The inverse correlation between the albedo and the maximum of the negative polarization degree (in absolute terms) is usually observed for pulverized silicates and meteorites (e.g., *Zellner et al.,* 1977). As a rule, the lower the albedo, the deeper the negative polarization branch. The correlation is successfully applied in the estimation of geometric albedos for asteroids. It is not yet known whether it can be applied to KBOs and Centaurs. At small phase angles, the correlation is destroyed by the interplay of the physical parameters of the surfaces (*Shkuratov et al.,* 2002).

3.3.5. Refractive index. The real part of the refractive index affects the shape of the negative-polarization branch. Increasing the real part usually makes the negative polarization branch more prominent. The imaginary part of the refractive index affects primarily the albedo of the surface (see above).

3.3.6. Particle size. An increase in the depth of negative polarization is observed for powdered dielectric surfaces with decreasing particle size down to the wavelength of incident light. It is well illustrated by laboratory measurements of particle-size separates of bright powders of Al_2O_3 made by *Shkuratov et al.* (2002). For grain sizes larger than 1 μm, the depth of the negative-polarization branch does not effectively exceed 0.25%. However, the smallest particle fractions of 0.1 μm and 0.5 μm show astonishingly similar negative-polarization branches with the minimum of about 0.8% situated at $\alpha \sim 1.6°$. Qualitatively, the measurements reproduce the polarimetric features of KBOs suggesting that their surfaces include a large portion of submicrometer- to micrometer-sized particles. In fact, a pronounced negative polarization is typical of granular structures in 1-μm size scales and of complex small-scale topographies. Measurements of the fine SiO_2 powder with fractal-like structures, where tiny particles (~10 nm) form aggregates and the aggregates form larger aggregates, etc., shows an extremely narrow negative polarization branch (Fig. 5).

3.3.7. Packing density. Measurements of laboratory samples of fine dispersed particles before and after compression clearly show differing phase-angle dependences (Fig. 5). Fluffy samples of MgO powder (~1 μm) and superfine SiO_2 powder (~10 nm) produce asymmetric negative polarization branches with minima at small phase angles. After compressing, the negative-polarization branch became wider. Details of the measurements are given in *Shkuratov et al.* (2002).

3.3.8. Microscale optical heterogeneity. Optically contrasting media give rise to pronounced negative-polarization branches. A significant increase of negative polarization (in absolute terms) is found when fine powders with highly different albedos are mixed together (*Shkuratov,* 1987). Even small amounts of a bright powder added to a dark powder (e.g., a mixture of 5% of MgO and 95% of carbon soot) noticeably increase the negative polarization as compared to that of the individual mixture components. Laboratory measurements reproducing phase curves of cometary dust also require mixtures of bright and dark fluffy materials to obtain relevant negative-polarization branches (*Hadamcik et al.,* 2006).

3.3.9. Interrelations between the negative-polarization branch and the opposition effect. In many cases, the parameters of the opposition effect and the negative-polarization branch are closely correlated. Generally, the opposition effect becomes more prominent with increasing surface albedo whereas the negative-polarization branch is neutralized. However, there are a number of exceptions to the typically observed relationships between the opposition effect and the negative-polarization branch (for details see *Shkuratov et al.,* 2002). This conclusion is well-illustrated in Figs. 4–5, which show both brightness and polarization measurements for carbon soot, MgO, and SiO_2. One can see that there is no

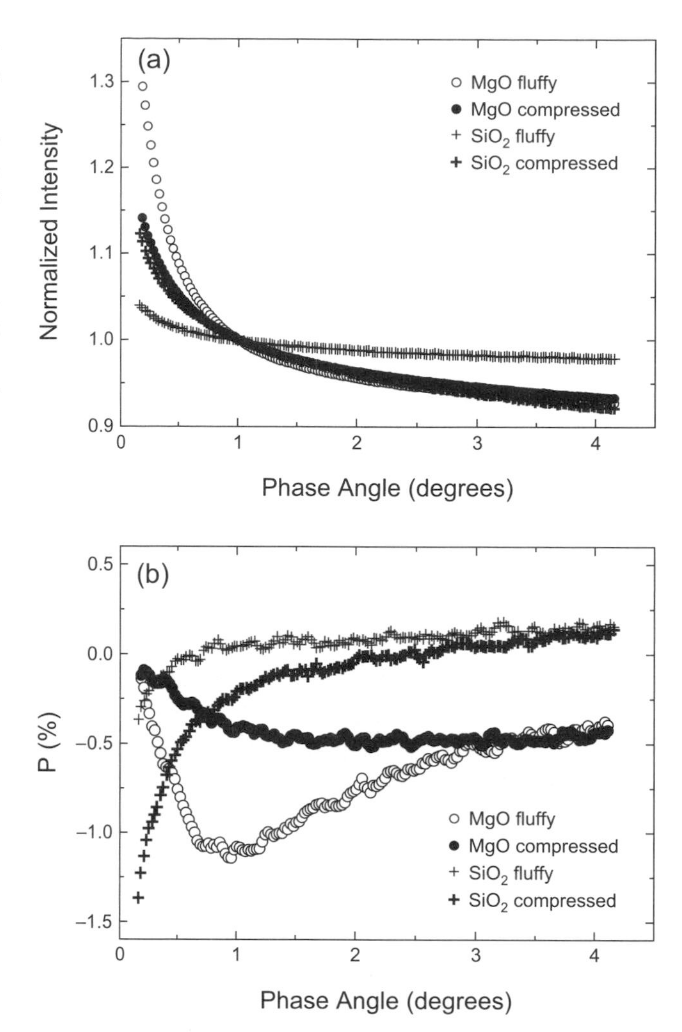

Fig. 5. **(a)** Photometric and **(b)** polarimetric phase curves for MgO and SiO_2 before and after compressing with alcohol drying at λ = 0.63 μm (*Shkuratov et al.,* 2002).

strict correlation between the photometric and polarimetric phase dependences.

4. DISCUSSION AND CONCLUSIONS

The first observations of the phase-angle effects in brightness and polarization for KBOs and Centaurs underscore their significance in constraining the albedos and surface textures of these objects. In spite of the limited range of phase angles accessible from groundbased observations, distinct features have been observed both in brightness and polarization. Their interpretation based on numerical and laboratory simulations may suggest the following surface properties: (1) fluffy surfaces with a large portion of submicrometer- or micrometer-sized particles; (2) inhomogeneous surface matter (scatterers with small and large single-scattering albedos); (3) moderate to high albedos of surfaces showing sharp opposition spikes. Note that fluffy surfaces are possible but not necessary, e.g., similar phase effects can arise when shadowing and coherent backscattering take

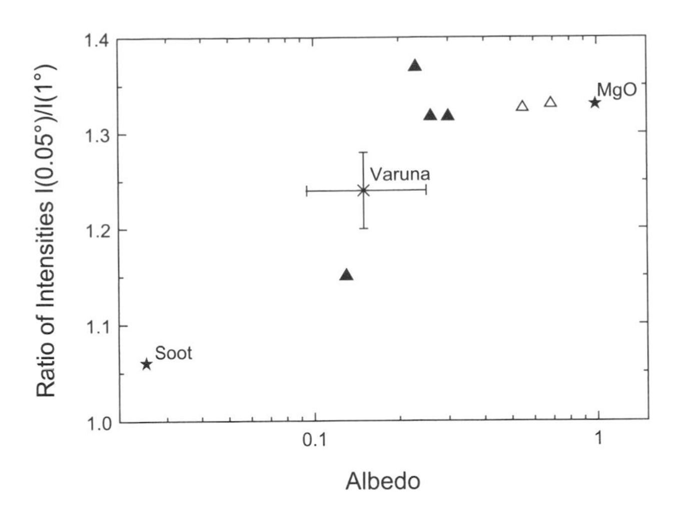

Fig. 6. Ratio of intensities at the phase angle of 0.05° to that at the phase angle of 1° for uranian satellites (black triangles), saturnian satellites (open triangles), TNO Varuna, and the laboratory samples.

place among scatterers embedded in an otherwise optically homogeneous host material.

Additional constraints on the surface properties can be derived from the comparison of the observational features obtained for KBOs and Centaurs with the corresponding features for other solar system objects. Phase curves near opposition have been measured for a variety of solar system objects. In spite of the different compositions and origins, most of the objects give rise to rather similar photometric and polarimetric characteristics mainly depending on surface albedo. The comparison suggests a way to constrain the surface albedo by using small-phase-angle observations. Figure 6 shows the ratio of intensities measured at $\alpha = 0.05°$ to that at $\alpha = 1°$ for both laboratory samples and planetary surfaces. Data of uranian and saturnian satellites were taken according to *Buratti et al.* (1992) and *Verbiscer et al.* (2005), respectively. Note that we are using only data obtained at these small phase angles avoiding any model-dependent extrapolations. Observations of main-belt asteroids cannot be used in such a comparison due to the influence of the finite angular size of the solar disk. The data of the uranian and saturnian satellites were taken from *Buratti et al.* (1992) and *Verbiscer et al.* (2005), respectively. The ratio of I(0.05°)/I(1°) tends to increase with surface albedo and seems to saturate for albedos of 30–40%, when further increasing of albedo does not change the ratio. This is likely to be connected to the usage of 0.05° as the limiting phase angle for the ratio estimations. Good coincidence of the laboratory data and available telescopic observations suggests a similarity of the surface textures of the smoked samples to those of the satellites and the TNO Varuna. The ratio of I(0.05°)/I(1°) observed for Varuna is typical for moderate-albedo surfaces.

The phase coefficients of atmosphereless bodies estimated at larger phase angles usually exhibit a simple relation to albedo: The higher the phase coefficient, the lower the surface albedo (e.g., *Belskaya and Shevchenko,* 2000). It can be explained by the dominant influence of the shadowing mechanism on the phase curves at phase angles beyond the opposition surge. A similar trend could exist for the TNO and Centaur phase coefficients as pointed out by *Rousselot et al.* (2006) and *Rabinowitz et al.* (2007). Further observations are needed to confirm the trend.

Many observers have mentioned the large phase coefficients measured for Centaurs and TNOs. An important question is whether they really differ from those of other solar system objects. Unfortunately, uncertaintics in the phase-coefficient estimation in the limited phase-angle range of TNOs are so large that no conclusions can be currently made. However, the phase-coefficient determinations for Centaurs up to the phase angles of 4°–7° provide irrefutable evidence of a larger steepness of their phase curves as compared to those of other solar system objects. As one can see from Fig. 7, the ratios of intensities I(1°)/I(4°) measured for Centaurs are considerably larger than those of other objects. A possible correlation of the phase coefficient with the albedo needs to be verified with additional observations.

A pronounced branch of negative polarization deeper than 1% at small phase angles close to 1° is also unique among solar system bodies observed so far. At such small phase angles, the polarization degree typically does not fall below –0.6% for a variety of minor bodies (asteroids, cometary dust, satellites of major planets), being almost independent of their surface albedos (Fig. 8). The observations available for the KBOs and the one Centaur do not yet allow to conclude whether we have observed a "normal" negative branch of polarization or an additional narrow peak predicted by theoretical modeling of coherent backscattering (e.g., *Mishchenko et al.,* 2000). *Rosenbush et al.* (1997, 2005) claimed a detection of such a peak in the observations

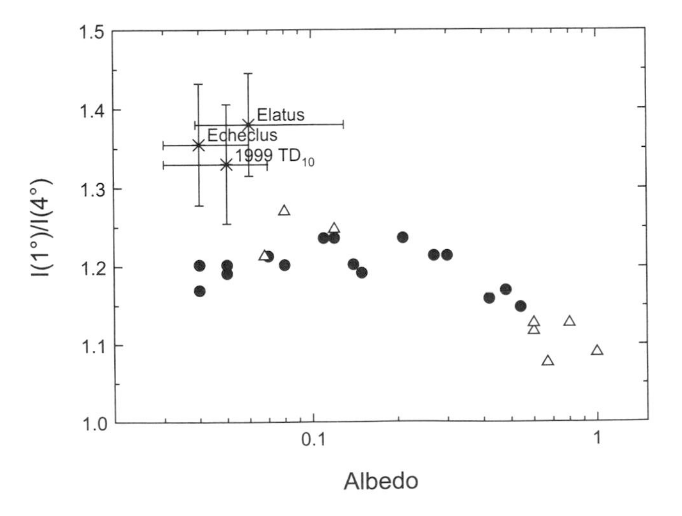

Fig. 7. Phase coefficient vs. albedo for asteroids, satellites (triangles), and Centaurs. For references for asteroid and satellite data, see *Belskaya and Shevchenko* (2000) and *Bauer et al.* (2006). Albedos of Centaurs were taken from Stansberry et al. (this volume).

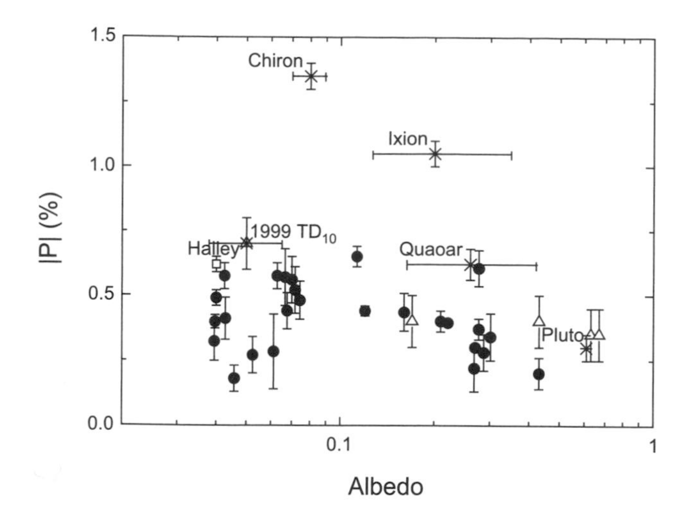

Fig. 8. Polarization degree at $\alpha = 1°$ vs. albedo for asteroids (circles) from *Cellino et al.* (2005) and *Fornasier et al.* (2006), satellites (triangles) from *Rosenbush et al.* (1997), a comet from *Kiselev et al.* (2005), and TNOs (for references see text).

of bright satellites and asteroids but its small depth (typically less than 0.4%) together with the large scatter of data raise doubts about its existence. Observations of moderate- and low-albedo asteroids did not show any particular features at small phase angles (e.g., *Cellino et al.,* 2005). Numerical modeling (see section 3.2) has suggested a secondary peak of polarization for Chiron, Ixion, and Pluto, which are important to verify with future observations.

An alternative explanation is that the polarization phase dependences of KBOs and Centaurs are extremely asymmetric and, thus, the phase angle, at which P_{min} is observed, is close to 1°–2°. This may explain the trend seen between the albedo and the negative polarization degree for the KBOs and Centaur observed (Fig. 8), resembling that for asteroids. Further observations of objects with known albedos are needed to study whether the trend really exists. If confirmed, it may allow an independent estimation of surface albedos based on polarimetric observations.

Note that the negative branches of polarization measured for a few comets (*Kiselev et al.,* 2005) are, in general, quite similar to those of low-albedo asteroids while they are more symmetric. The phase angle of the minimum polarization for S-type asteroids is systematically smaller than that of cometary dust (see, e.g., *Levasseur-Regourd,* 2004).

Although reliable observational data on phase effects in photometry and polarimetry are very scarce for KBOs and Centaurs, the existing data give a first look into the microscopic properties of the surface layers and suggest different surface properties as compared to less distant small solar system objects.

The study of the photometric and polarimetric phase effects for KBOs and Centaurs is just beginning. The available observations have revealed remarkable features in the opposition brightening and negative polarization that can be used to constrain surface properties on these distant objects. Further photometric and polarimetric observations are needed to understand how common the features are and to probe the differences in surface properties of objects having different dynamical histories. Special observation programs of objects with well-determined albedos can help to establish possible empirical correlations between the photometric and polarimetric parameters and the geometric albedo of the surface. Storage of the original photometric observations of KBOs and Centaurs is of great importance since it allows the compilation of all data from different observers and the subsequent new interpretations of the phase curves. Such a database was initiated by Rousselot et al. (*www.obs-besancon.fr/bdp/*). Finally, it is important to develop a new empirical system to compute absolute magnitudes for KBOs: the H, G magnitude system fails to fit the photometric phase curves.

Further progress in remote sensing of the surface properties requires further developments in theoretical and numerical modeling, and laboratory simulations of light scattering. Numerical modeling is able to give predictions that need to be checked by further observations. Detailed observations of polarimetric and photometric phase curves down to extremely small phase angles are of great value to improve the theoretical models. Only close coordination of observations and modeling can provide further progress in obtaining meaningful results on surface textural properties from remote-sensing data. Moreover, an extension of the spectral range of observations and using other types of measurements for joint analyses (e.g., spectroscopy, radiometry) is a promising way to improve our knowledge on the surface properties of KBOs and Centaurs.

Acknowledgments. We thank A. Verbiscer and A. Ovcharenko for providing us with important unpublished data and A. Cellino and Yu. Krugly for valuable comments. I.B. and Yu.S. are grateful to Université Pierre et Marie Curie for financing their stay in France. K.M. is grateful to the Academy of Finland for partial funding of his research.

REFERENCES

Amic E., Luck J. M., and Nieuwenhuizen Th. M. (1997) Multiple Rayleigh scattering of electromagnetic waves. *J. Phys. I, 7,* 445–483.

Avramchuk V. V., Rakhimov V. I., Chernova G. P., and Shavlovskii N. M. (1992) Photometry and polarimetry of Pluto near its perihelion. *Kinemat. Fiz. Nebesn. Tel, 8(4),* 37–45 (in Russian).

Bagnulo S., Boehnhardt H., Muinonen K., Kolokolova L., Belskaya I., and Barucci M. A. (2006) Exploring the surface properties of Transneptunian Objects and Centaurs with polarimetric FORS1/VLT observations. *Astron. Astrophys., 450,* 1239–1248.

Bauer J. M., Meech K. J., Fernandez Ya. R., Farnham T. L., and Roush T. L. (2002) Observations of the Centaur 1999 UG5: Evidence of a unique outer solar system surface. *Publ. Astron. Soc. Pac., 114,* 1309–1321.

Bauer J., Meech K., Fernandez Y., Pittichova J., Hainaut O., Boehnhardt H., and Delsanti A. (2003) Physical survey of 24 Centaurs with visible photometry. *Icarus, 166,* 195–211.

Bauer J. M., Grav T., Buratti B. J., and Hicks M. D. (2006) The phase curve survey of the irregular saturnian satellites: A possible method of physical classification. *Icarus, 184,* 181–197.

Belskaya I. N. and Shevchenko V. G. (2000). Opposition effect of asteroids. *Icarus, 147,* 94–105.

Belskaya I. N., Barucci M. A., and Shkuratov Yu. G. (2003) Opposition effect of Kuiper belt objects: Preliminary estimations. *Earth Moon Planets, 92,* 201–206.

Belskaya I. N., Ortiz J. L., Rousselot P., Ivanova V., Borisov G., Shevchenko V. G., and Peixinho N. (2006) Low phase angle effects in photometry of trans-neptunian objects: 20000 Varuna and 1996 TO_{66}. *Icarus, 184,* 277–284.

Binzel R. P. and Mulholland J. D. (1984) Photometry of Pluto during the 1983 opposition — A new determination of the phase coefficient. *Astron. J., 89,* 1759–1761.

Boehnhardt H., Bagnulo S., Muinonen K., Barucci M. A., Kolokolova L., Dotto E., and Tozzi G. P. (2004) Surface characterization of 28978 Ixion (2001 KX_{76}). *Astron. Astrophys., 415,* L21–L25.

Breger M. and Cochran W. D. (1982) Polarimetry of Pluto. *Icarus, 49,* 120–124.

Buie M. W. and Bus S. J. (1992) Physical observations of (5145) Pholus. *Icarus, 100,* 288–294.

Buie M., Tholen D., and Wasserman L. (1997) Separate lightcurves of Pluto and Charon. *Icarus, 125,* 233–244.

Buratti B. J., Gibson J., and Mosher J. A. (1992) CCD Photometry of the uranian satellites. *Astron. J., 104,* 1618–1622.

Buratti B. J., Hillier J. K., Heinze A., Hicks M. D., Tryka K. A., Mosher J. A., Ward J., Garske M., Young J., and Atienza-Rosel J. (2003) Photometry of Pluto in the last decade and before: Evidence for volatile transport? *Icarus, 162,* 171–182.

Cellino A., Gil Hutton R., Di Martino M., Bendjoya Ph., Belskaya I. N., and Tedesco E. F. (2005) Asteroid polarimetric observations using the Torino UBVRI photopolarimeter. *Icarus, 179,* 304–324.

Fix L. A. and Kelsey J. D. (1973) Polarimetry of Pluto. *Astrophys. J., 184,* 633–636.

Fornasier S., Belskaya I., Shkuratov Yu. G., Pernechele C., Barbieri C., Giro E., and Navasardyan H. (2006) Polarimetric survey of asteroids with the Asiago telescope. *Astron. Astrophys., 455,* 371–376.

Hadamcik E., Renard J. B., Levasseur-Regourd A. C., and Lasue J. (2006) Light scattering by fluffy particles with the $PROGRA^2$ experiment: Mixtures of materials. *J. Quant. Spectros. Rad. Transfer, 100,* 143–156.

Hapke B. (1986) Bidirectional reflectance spectroscopy. 4. The extinction coefficient and the opposition effect. *Icarus, 67,* 264–280.

Hapke B. (2002) Bidirectional reflectance spectroscopy. 5. The coherent backscatter opposition effect and anisotropic scattering. *Icarus, 157,* 523–534.

Helfenstein P. and Veverka J. (1989) Physical characterization of asteroid surfaces from photometric analysis. In *Asteroids II* (R. Binzel et al., eds.), pp. 557–593. Univ. of Arizona, Tucson.

Hicks M. D., Simonelli D. P., and Buratti B. J. (2005) Photometric behavior of 20000 Varuna at very small phase angles. *Icarus, 176,* 492–498.

Jewitt D. and Sheppard S. (2002) Physical properties of trans-Neptunian object (20000) Varuna. *Astron. J., 123,* 2110–2120.

Kaasalainen S., Piironen J., Muinonen K., Karttunen H., and Peltoniemi J. I. (2002) Laboratory experiments on backscattering from regolith samples. *Appl. Opt., 41,* 4416–4420.

Kiselev N., Rosenbush V., Jockers K., Velichko S., and Kikuchi S. (2005) Database of comet polarimetry: Analysis and some results. *Earth Moon Planets, 97,* 365–378.

Levasseur-Regourd A. C. (2004) Polarimetry of dust in the solar system: Remote observations, in-situ measurements and experimental simulations. In *Photopolarimetry in Remote Sensing* (G. Videen et al., eds.), pp. 393–410. NATO Sci Series, Kluwer, London.

Levasseur-Regourd A. C. and Hadamcik E. (2003) Light scattering by irregular dust particles in the solar system: Observations and interpretation by laboratory measurements. *J. Quant. Spectros. Rad. Transfer, 79-80,* 903–910.

Lyot B. (1929) Recherches sur la polarisation de la lumière des planètes et de quelques substances terrestres. *Ann. Obs. Paris, 8,* 1-161.

McBride N., Davies J. K., Green S. F., and Foster M. J. (1999) Optical and infrared observations of the Centaur 1997 CU_{26}. *Mon. Not. R. Astron. Soc., 306,* 799–805.

Mishchenko M. I. (1991) Polarization effects in weak localization of light: Calculation of the copolarized and depolarized backscattering enhancement factors. *Phys. Rev., B44,* 12597–12600.

Mishchenko M. I. and Dlugach J. M. (1993) Coherent backscatter and the opposition effect for E-type asteroids. *Planet. Space Sci., 41,* 173–181.

Mishchenko M. I., Luck J.-M., and Nieuwenhuizen T. M. (2000) Full angular profile of the coherent polarization opposition effect. *J. Opt. Soc. Am. A, 17,* 888–891.

Muinonen K. (1994) Coherent backscattering by solar system dust particles. In *Asteroids, Comets, Meteors 1993* (A. Milani et al., eds.), pp. 271–296. IAU Symposium No. 160, Kluwer, Dordrecht.

Muinonen K. (2004) Coherent backscattering of light by complex random media of spherical scatterers: Numerical solution. *Waves Random Media, 14(3),* 365–388.

Muinonen K., Stankevich D., Shkuratov Yu. G., Kaasalainen S., and Piironen J. (2001) Shadowing effect in clusters of opaque spherical particles. *J. Quant. Spectros. Rad. Transfer, 70,* 787–810.

Muinonen K., Piironen J., Shkuratov Yu. G., Ovcharenko A. A., and Clark B. E. (2002) Asteroid photometric and polarimetric phase effects. In *Asteroids III* (W. F. Bottke Jr. et al., eds.), pp. 123–138. Univ. of Arizona, Tucson.

Muinonen K., Zubko E., Tyynelä J., Shkuratov Yu. G., and Videen G. (2007) Light scattering by Gaussian random particles with discrete-dipole approximation. *J. Quant. Spectros. Rad. Transfer, 106,* 360–377.

Nelson R., Hapke B., Smythe W., and Spilker L. (2000) The opposition effect in simulated planetary regolith. Reflectance and circular polarization ratio change at small phase angle. *Icarus, 147,* 545–558.

Ozrin V. D. (1992) Exact solution for coherent backscattering of polarized light from a random medium of Rayleigh scatterers. *Waves Random Media, 2,* 141–164.

Parviainen H. and Muinonen K. (2007) Rough-surface shadowing for self-affine random rough surfaces. *J. Quant. Spectrosc. Radiat. Transfer, 106,* 389–397.

Patat F. and Romaniello M. (2006) Error analysis for dual-beam optical linear polarimetry. *Publ. Astron. Soc. Pac., 118,* 146–161.

Psarev V., Ovcharenko A., Shkuratov Yu., Belskaya I., and Videen G. (2007) Photometry of surfaces with complicated structure at extremely small phase angles. *J. Quant. Spectrosc. Radiat. Transfer, 106,* 455–463.

Rabinowitz D., Barkume K., Brown M., et al. (2006) Photometric observations constraining the size, shape and albedo of 2003 EL_{61}, a rapidly rotating, Pluto-sized object in the Kuiper belt. *Astrophys. J., 639,* 1238–1251.

Rabinowitz D., Schaefer B., and Tourtellote S. (2007) The diverse solar phase curves of distant icy bodies. Part 1: Photometric observations of 18 trans-neptunian objects, 7 Centaurs, and Nereid. *Astron. J., 133,* 26–43.

Rosenbush V. K., Avramchuk V. V., Rosenbush A. E., and Mishchenko M. I. (1997) Polarization properties of the Galilean satellites of Jupiter: Observations and preliminary analysis. *Astrophys. J., 487,* 402–414.

Rosenbush V. K., Kiselev N. N., Shevchenko V. G., Jockers K., Shakhovskoy N. M., and Efimov Yu. S. (2005) Polarization and brightness opposition effects for the E-type asteroid 64 Angelina. *Icarus, 178,* 222–234.

Rousselot P., Petit J.-M., Poulet F., Lacerda P., and Ortiz J. (2003) Photometry of the Kuiper-Belt object 1999 TD_{10} at different phase angles. *Astron. Astrophys., 407,* 1139–1147.

Rousselot P., Petit J.-M., Poulet F., and Sergeev A. (2005a) Photometric study of Centaur (60558) 2000 EC_{98} and trans-neptunian object (55637) 2002 UX_{25} at different phase angles. *Icarus, 176,* 478–491.

Rousselot P., Levasseur-Regourd A. C., Muinonen K., and Petit J.-M. (2005b) Polarimetric and photometric phase effects observed on transneptunian object (29981) 1999 TD_{10}. *Earth Moon Planets, 97,* 353–364.

Rousselot P., Belskaya I. N., and Petit J.-M. (2006) Do the phase curves of KBOs present any correlation with their physical and orbital parameters? *Bull. Am. Astron. Soc., 38(3),* 565.

Schaefer B. E. and Rabinowitz D. L. (2002) Photometric light curve for the Kuiper belt object 2000 EB_{173} on 78 nights. *Icarus, 160,* 52–58.

Sheppard S. S. and Jewitt D. C. (2002) Time-resolved photometry of Kuiper belt objects: Rotations, shapes, and phase functions. *Astron. J., 124,* 1757–1775.

Shkuratov Yu. G. (1987) Negative polarization of sunlight scattered from celestial bodies: Interpretation of the wavelength dependence. *Sov. Astron. Lett., 13,* 182–183.

Shkuratov Yu. and Ovcharenko A. (2002) Experimental modeling of opposition effect and negative polarization of regolith-like surfaces. In *Optics of Cosmic Dust* (G. Videen and M. Kocifaj, eds.), pp. 225–238. NATO Sci Series, Kluwer, London.

Shkuratov Yu., Ovcharenko A., Zubko E., Miloslavskaya O., Nelson R., Smythe W., Muinonen K., Piironen J., Rosenbush V., and Helfenstein P. (2002) The opposition effect and negative polarization of structurally simulated planetary regoliths. *Icarus, 159,* 396–416.

Shkuratov Yu., Videen G., Kreslavsky M. A., Belskaya I. N., Ovcharenko A., Kaydash V. G., Omelchenko V. V., Opanasenko N., and Zubko E. (2004) Scattering properties of planetary regoliths near opposition. In *Photopolarimetry in Remote Sensing* (G. Videen et al., eds.), pp. 191–208. NATO Sci Series, Kluwer, London.

Shkuratov Yu. G., Stankevich D. G., Petrov D. V., Pinet P. C., Cord Au. M., and Daydou Y. H. (2005) Interpreting photometry of regolith-like surfaces with different topographies: Shadowing and multiple scatter. *Icarus, 173,* 3–15.

Shkuratov Yu., Bondarenko S., Ovcharenko A., Videen G., Pieters C., Hiroi T., Volten H., Hovenier J., and Munos O. (2006) Comparative studies of the reflectance and degree of linear polarization of particulate surfaces and independently scattering particles. *J. Quant. Spectrosc. Radiat. Transfer, 100,* 340–358.

Tholen D. J. and Tedesco E. F. (1994) Pluto's lightcurve: Results from four oppositions. *Icarus, 108,* 200–208.

Verbiscer A. J., French R. G., and Helfenstein P. (2005) Saturn's inner satellites at true opposition: Observations during a central transit of the Earth across the solar disk. *Bull. Am. Astron. Soc., 37,* 702.

Zellner B., Leake M., Lebertre T., Duseaux M., and Dollfus A. (1977) The asteroid albedo scale. I. Laboratory polarimetry of meteorites. *Proc. Lunar Sci. Conf. 8th,* pp. 1091–1110.

Photometric Lightcurves of Transneptunian Objects and Centaurs: Rotations, Shapes, and Densities

Scott S. Sheppard
Carnegie Institution of Washington

Pedro Lacerda
Grupo de Astrofisica da Universidade de Coimbra

Jose L. Ortiz
Instituto de Astrofisica de Andalucia

We discuss the transneptunian objects and Centaur rotations, shapes, and densities as determined through analyzing observations of their short-term photometric lightcurves. The lightcurves are found to be produced by various different mechanisms including rotational albedo variations, elongation from extremely high angular momentum, as well as possible eclipsing or contact binaries. The known rotations are from a few hours to several days with the vast majority having periods around 8.5 h, which appears to be significantly slower than the main-belt asteroids of similar size. The photometric ranges have been found to be near zero to over 1.1 mag. Assuming the elongated, high-angular-momentum objects are relatively strengthless, we find most Kuiper belt objects appear to have very low densities (<1000 kg m^{-3}) indicating high volatile content with significant porosity. The smaller objects appear to be more elongated, which is evidence of material strength becoming more important than self-compression. The large amount of angular momentum observed in the Kuiper belt suggests a much more numerous population of large objects in the distant past. In addition we review the various methods for determining periods from lightcurve datasets, including phase dispersion minimization (PDM), the Lomb periodogram, the Window CLEAN algorithm, the String method, and the Harris Fourier analysis method.

1. INTRODUCTION

The transneptunian objects (TNOs) are a remnant from the original protoplanetary disk. Even though TNOs may be relatively primitive, their spins, shapes, and sizes from the accretion epoch have been collisionally altered over the age of the solar system. The rotational distribution of the TNOs is likely a function of their size. In the current Kuiper belt the smallest TNOs (radii r < 50 km) are susceptible to erosion and are probably collisionally produced fragments (*Farinella and Davis,* 1996; *Davis and Farinella,* 1997; *Bernstein et al.,* 2004). These fragments may have been disrupted several times over the age of the solar system, which would have highly modified their rotational states (*Catullo et al.,* 1984). Intermediate-sized TNOs have probably been gravitationally stable to catastrophic breakup but are likely to have had their primordial rotations highly influenced through relatively recent collisions. The larger TNOs (r > 100 km) have disruption lifetimes longer than the age of the solar system and probably have angular momentum and thus spins that were imparted during the formation era of the Kuiper belt. Thus the largest TNOs may show the primordial distribution of angular momenta obtained through the accretion process while the smaller objects may allow us to understand collisional breakup of TNOs through their rotations and shapes.

Like the rotations, the shape distribution of TNOs is probably also a function of their size. The largest TNOs should be dominated by their gravity with shapes near their hydrostatic equilibrium point. The smaller TNOs are probably collisional fragments with self-gravity being less important, allowing elongation of the objects to dominate their lightcurves.

The main technique in determining the rotations and shapes of TNOs is through observing their photometric variability (*Sheppard and Jewitt,* 2002; *Ortiz et al.,* 2006; *Lacerda and Luu,* 2006). For the largest TNOs most photometric variations with rotation can be explained by slightly nonuniform surfaces. Two other distinct types of lightcurves stand out in rotation period and photometric range space for the largest TNOs (r > 100 km). The first type of lightcurve [examples are (20000) Varuna and (136108) 2003 EL_{61}] have large amplitudes and short periods that are indicative of rotationally elongated objects near hydrostatic equilibrium (*Jewitt and Sheppard,* 2002; *Rabinowitz et al.,* 2006). The second type, of which only 2001 QG_{298} is a member to date, show extremely large amplitude and slow rotations and are best described as contact binaries with similar-sized components (*Sheppard and Jewitt,* 2004).

Two objects, (19308) 1996 TO_{66} and (24835) 1995 SM_{55}, may have variable amplitude lightcurves, which may result from complex rotation, a satellite, cometary effects, a recent collision, or most likely phase-angle effects (*Hainaut et al.,* 2000; *Sheppard and Jewitt,* 2003; *Belskaya et al.,* 2006; see also chapter by Belskaya et al.).

This chapter is organized as follows: In section 2 we discuss how rotation periods are determined from lightcurve observations and the possible biases involved. In section 3 the possible causes of the detected lightcurves are considered. In section 4 we mention what the measured lightcurves may tell us about the density and composition of the TNOs. In section 5 we look at what the shape distribution of the TNOs looks like when assuming that elongation is the reason for the larger lightcurve amplitudes and double-peaked rotation periods. In section 6 we discuss what the observed angular momentum of the ensemble of TNOs may tell us about the Kuiper belt's past environment. Finally, section 7 examines possible correlations between spin periods and amplitudes, and the dynamical and physical properties of the TNOs.

2. ANALYZING LIGHTCURVES

2.1. Period-Detection Techniques

There are currently several period-detection techniques such as phase dispersion minimization (PDM) (*Stellingwerf,* 1978), the Lomb periodogram (*Lomb,* 1976), the Window CLEAN algorithm (*Belton and Gandhi,* 1988), the String method (*Dworetsky,* 1983) and a Fourier analysis method (*Harris et al.,* 1989) that can be efficiently used to fit asteroid lightcurves. All these techniques are suitable to data that are irregularly spaced in time. Although the photometry data of TNOs are usually unevenly sampled, the sampling times are not random. This results in what is usually called "aliasing problems" (see section 2.3).

The PDM method (*Stellingwerf,* 1978) is especially suited to detect periodic signals regardless of the lightcurve shape. The PDM method searches for the best period that minimizes a specific parameter Θ. This parameter measures the dispersion (variance) of the data phased to a specific period divided by the variance of the unphased data. Therefore, the best period is the one that minimizes the dispersion of the phased lightcurve.

$$\Theta = s^2/\sigma^2 \tag{1}$$

where

$$\sigma^2 = \sum_{i=1}^{N} (x_i - \bar{x})^2/(N-1)$$

and

$$s^2 = \frac{\sum_{j=1}^{M} (n_j - 1)s_j^2}{\sum_{j=1}^{M} (n_j - M)}$$

N is the number of observations, x_i are the measurements, $\bar{x}$ is the mean of the measurements and s_j are the variances of M distinct samples. The samples are taken such that all the members have a similar ϕ_i, where ϕ_i is the phase corresponding to a trial period. Usually the phase interval (0,1) is divided into bins of fixed size, but the samples can be chosen in any other way that satisfies the criterion mentioned above.

The Lomb method (*Lomb,* 1976) is essentially a modified version of the well-known Fourier spectral analysis, but the Lomb technique takes into account the fact that the data are unevenly spaced and therefore the spectral power is "normalized" so that it weights the data in a "per point" basis instead of on a "per time" interval basis. The Lomb-normalized spectral power as a function of frequency $P_N(\omega)$ is

$$P_N(\omega) = \frac{1}{2\sigma^2}\frac{\left[\sum_j (h_j - \bar{h})\cos\omega(t_j - \tau)\right]^2}{\sum_j \cos^2\omega(t_j - \tau)} + \frac{\left[\sum_j (h_j - \bar{h})\sin\omega(t_j - \tau)\right]^2}{\sum_j \sin^2\omega(t_j - \tau)} \tag{2}$$

where ω is angular frequency (2πf), σ^2 is the variance of the data, $\bar{h}$ is the mean of the measurements, h_j and t_j are the measurements and their times, and τ is a kind of offset that makes $P_N(\omega)$ independent of shifting all the t_i by any constant. Quantitatively, τ is such that

$$\tan(2\omega\tau) = \sum_j \sin 2\omega\tau_j / \sum_j \cos 2\omega\tau_j$$

In this method, the best period is the one that maximizes the Lomb-normalized spectral power.

The String method (e.g., *Dworetsky,* 1983) finds the best period by searching for the period that minimizes a parameter that can be regarded as a length

$$L = \sum_{i=1}^{n} [(m_i - m_{i-1})^2 + (\Phi_i - \Phi_{i-1})^2]^{1/2} + [(m_i - m_n)^2 + (\Phi_i - \Phi_n + 1)^2]^{1/2} \tag{3}$$

where Φ_i are the phases from a trial period, m_i are the experimental values, and n is the number of observations.

The Window CLEAN algorithm (*Belton and Gandhi,* 1988) is a special application of the CLEAN algorithm (well known to radio astronomers, as it is widely used to synthesize images when dealing with synthetic aperture data). In the case of time series analysis, its application is made by computing a window function (which takes into account the observing times having zero value at the times when no data were taken and 1 for the rest). This window function is used to deconvolve the true signal by means of the CLEAN algorithm. In other words, the window function is used to generate sort of a filter in the frequency domain to be applied to the regular Fourier spectrum in order to smooth out the spectral power of the signals that arise from the sampling pattern. Therefore the true periodic signal shows up more clearly in the corrected spectrum.

The Harris method (*Harris et al.,* 1989) was specifically developed for asteroid lightcurve studies and is basically a fit of the data to a Fourier series, which can be chosen to be of any degree

$$H(\alpha,t) = \bar{H}(\alpha) + \sum_{l=1}^{m} A_l \sin\frac{2\pi l}{P}(t - t_0) + B_l \cos\frac{2\pi l}{P}(t - t_0) \quad (4)$$

where $H(\alpha,t)$ is the computed magnitude at solar phase angle α and time t, $\bar{H}(\alpha)$ is the mean magnitude at phase angle α and A_l, B_l are Fourier coefficients. For a given period P, the fit is carried out by finding the minimum of a bias-corrected variance

$$s^2 = \frac{1}{n-k}\sum_{i=1}^{n}\left(\frac{\delta_i}{\varepsilon_i}\right)^2$$

where $\delta_i = V_i(\alpha_j) - H(\alpha_j,t_i)$ is the deviation from the observations to the model (with α_j the phase angle of night j) and ε_i are *a priori* error estimates of the measurements. On the other hand, $k = 2m + p + 1$, where m is the degree of the Fourier series and p is the total number of days of data.

The minimum value of s^2 corresponds to the best solution. If increasing the degree of the solution by one fails to decrease s^2 by $s^2/(n - k)$, the new highest-order harmonic is taken as nonsignificant.

2.2. Computation of Significance Levels

After a period is identified by means of a suitable method one usually wants to know how confident that determination is. One of the nice features of the Lomb periodogram method is the fact that it readily gives a confidence level in the form $P(>z) = 1 - (1 - \exp - z)^M$ (*Scargle,* 1982) where z is the maximum spectral power and M is the number of independent frequencies, which can easily be estimated. In the case of the PDM method, the lower the value of Θ the higher the significance level. Although there is no formal expression for the significance in PDM a Θ value less than about 0.2 is desired in order for the found period to be considered highly significant. Other methods also give parameters that are associated to a significance level or give criteria to accept/reject periods, but in all cases it is assumed that the errors follow a Gaussian distribution.

The best approach to analyze the confidence levels may be by using Monte Carlo simulations. One can run Monte Carlo simulations in which one generates random photometry values within the range observed at exactly the same times as each data point was taken. The simulated datasets can be analyzed with the particular technique that the author has chosen and one can generate a distribution of values for, e.g., maximum Lomb-normalized spectral power or a distribution of values of minimum Θ or a distribution of the output of a given method. This distribution can be compared with the value from the analysis of the actual data and in that way one can assess a probability by comparing how many times a value larger/lower than a given threshold appeared in the simulation, divided by the number of simulations performed.

Unfortunately, the photometry errors are not completely random. Sometimes there are clear systematic errors that can be corrected for, but other times the systematic errors are not very evident. For instance, it is not unusual that photometry datasets from two different nights may have a small shift due to differential extinction (not adequately accounted for), or perhaps one night's data are noisier than another one because of larger seeing or extinction, or background source contamination may vary over a few hours. These systematic errors can also be simulated in Monte Carlo methods and therefore a more reliable confidence level can be computed.

When systematic errors are simulated, the confidence levels obtained always drop substantially compared to the case in which pure random errors were assumed. As an example, Fig. 1 depicts confidence levels for a real case using (1) Monte Carlo simulations in which only random Gaussian errors are assumed and (2) Monte Carlo simulations in which night-to-night random offsets of 0.03 variance are added to the otherwise random data.

Therefore, the inclusion of small 0.03-mag shifts in the data from different nights reveals that the spectral power can increase. In some cases these kind of systematic errors can give rise to periodicities that would be identified as signifi-

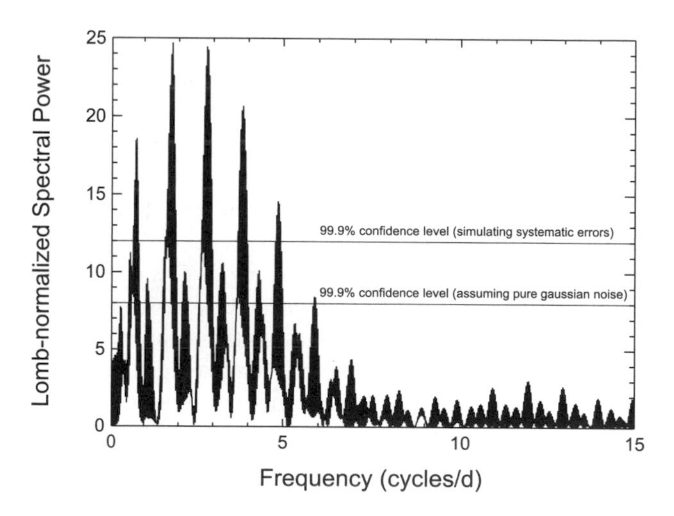

Fig. 1. Lomb Periodogram of Centaur (31824) Elatus (data from *Gutierrez et al.,* 2001). The main peak corresponds to the rotation period, whereas the other peaks are aliases at nearly 1 cycle/day spacing. Significance levels are indicated by the two horizontal lines (assuming two different error models).

cant by a regular Lomb periodogram in which the significance level was estimated by the equation $1 - (1 - \exp(-z))^M$.

2.3. Problems of Aliasing

As stated before, the data taken from a groundbased telescope are not randomly spaced in time, because there are more or less regular gaps in data acquisition sequences. For this reason there are inherent frequencies in the data that interfere with the true periodic variability of the object, giving rise to aliases. The main aliases are associated with the night-to-night observing gap and are such that

$$P_{alias}^{-1} = 1.0027 \text{ d} \pm P_{true}^{-1}$$

where 1.00027^{-1} is the length of the sidereal day.

Other minor aliases are seen at

$$P_{alias}^{-1} = k \times 1.0027 \text{ d} \pm P_{true}^{-1}$$

where k is an integer. See Fig. 1 as an example of a Lomb periodogram from a real case. The aliases are easy to identify when the plot is shown as a function of frequency instead of period, because of the ~1 spacing. These aliases usually have decreasing power as k increases. This is also illustrated in Fig. 1.

One can readily see that two waves $\exp(2\pi i f_1 t)$ and $\exp(2\pi i f_2 t)$ give the same samples at an interval Δ if and only if f_1 and f_2 differ by a multiple of $1/\Delta$.

When data are scarce, it is sometimes difficult to distinguish between the true period and an alias. Visual inspection of the lightcurve phased to one period or the other can sometimes help, but in some cases the ambiguity cannot be entirely resolved until more data are taken.

2.4. Selection Effects and Biases

Most of the TNO photometry reported is based on observations carried out in few-day observing runs, which implies that only short-term rotation variability can be detected. In addition, the photometry has some noise associated with it. Therefore, the current data sample may have some biases in the sense that short-term periods and large amplitudes are favored.

Unfortunately, long-term monitoring of TNOs (to try to debias the sample) is difficult to schedule in most telescopes as it requires many observing nights with medium-large telescopes. Besides, long-term monitoring requires careful absolute calibrations, which implies more observing time and photometric conditions that are not always met. Therefore, only a few cases have been studied. International collaboration to use different telescope resources would improve the situation and would also allow a better study of phase effects, which are important and sometimes might have caused misinterpretation of rotation periods due to opposition surges (see, e.g., *Belskaya et al.*, 2006). Also, the creation of a database where all TNO lightcurves could be accessible would allow to mitigate this problem. *Rousselot et al.* (2005a) have already created the infrastructure for that at the website *www.obs-besancon.fr/bdp/*.

3. CAUSES OF BRIGHTNESS VARIATIONS

Observed time-resolved photometric brightness variations of TNOs can be caused by several processes that may be periodic or variable (Fig. 2). A TNO's apparent magnitude is determined by its geometrical position relative to Earth and the Sun as well as its physical attributes and can be calculated from

$$m_R = m_{\odot} - 2.5\log[p_R r^2 \phi(\alpha)/2.25 \times 10^{16} R^2 \Delta^2)] \quad (5)$$

in which r [km] is the radius of the TNO, R [AU] is the heliocentric distance, Δ [AU] is the geocentric distance, $m_{\odot}$ is the apparent red magnitude of the Sun (–27.1), m_R is the apparent red magnitude of the TNO, p_R is the geometric albedo in the R band, and $\phi(\alpha)$ is the phase function, normalized in such a way that when $\alpha = 0°$ at opposition, $\phi(0) = 1$. The phase function depends on the surface properties of the object. For TNO lightcurve studies the rough approxi-

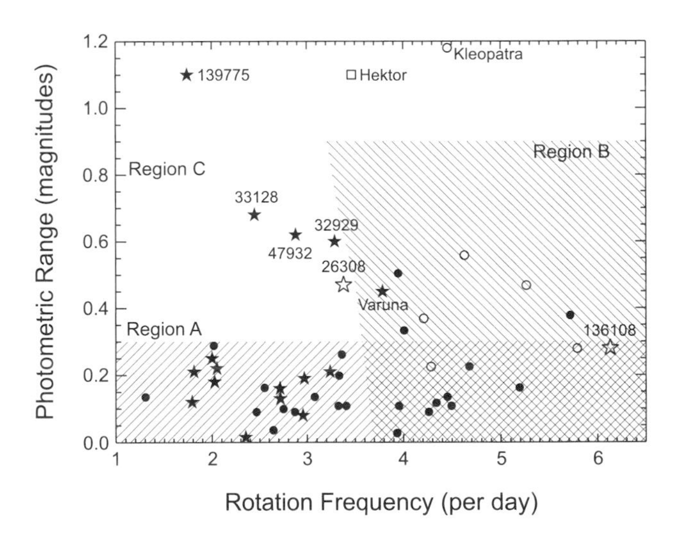

Fig. 2. Shown here are the rotation periods and photometric ranges of known TNO lightcurves and the main-belt asteroids with r > 100 km. Region A: The range of the lightcurve could be equally caused by albedo, elongation or binarity. Region B: The lightcurve range is most likely caused by rotational elongation. Region C: The lightcurve range is most likely caused by binarity of the object. Stars denote KBOs, circles denote main-belt asteroids (radii ≥ 100 km), and squares denote the Trojan 624 Hektor and the main-belt asteroid 216 Kleopatra, which are mentioned in the text. Open symbols signify known binary objects. Objects just to the left of Region B would have densities significantly less than 1000 kg m^{-3} in order to be elongated from rotational angular momentum. Binary objects are not expected to have photometric ranges above 1.2 mag. The TNOs that have photometric ranges below the photometric uncertainties (~0.1 mag) have not been plotted since their periods are unknown. These objects would all fall into Region A. The asteroids have been plotted at their expected mean projected viewing angle of 60° in order to more directly compare to the TNOs of unknown projection angle. Modified from *Sheppard and Jewitt* (2004).

mation $\phi(\alpha) = 10^{-0.4\beta\alpha}$ describes the linear phase dependence in magnitudes where β is the linear phase coefficient.

The brightness variations caused by the differing positional geometry of a TNO in equation (5) can usually be easily removed by simply using the absolute magnitude, defined as $m_R(1,1,\alpha) = m_R - 5\log(R\Delta) - \beta\alpha$, instead of the apparent magnitude. The absolute magnitude is the magnitude of the TNO if it were 1 AU from Earth and the Sun and at zero phase angle. The effects of differing heliocentric and geocentric distances are well understood, but the phase coefficient is not since it is dependent on the surface characteristics of the TNO (see chapter by Belskaya et al.).

3.1. Rotational Surface Variations

If a TNO has varying albedo or topography across its surface, the markings will cause a photometric lightcurve correlated with the object's rotation rate. For atmosphereless bodies, these surface variations usually do not create large-amplitude lightcurves and have been empirically found to be less than about 10–20% on most atmosphereless bodies (*Degewij et al.,* 1979; *Magnusson,* 1991). To date, no significant color variations with rotation have been observed from atmosphereless TNOs, which, if observed, may be an indication of compositional differences across the surface. The majority of low-amplitude lightcurves of large objects are believed to be caused by surface variations. Large objects able to substain atmospheres, such as Pluto, may obtain slightly higher albedo variations on their surfaces since dark areas will efficiently absorb light, creating warmth that may release volatiles into the atmosphere, which can then condense as bright spots on cooler surfaces (*Spencer et al.,* 1997).

3.2. Elongation from Material Strength

A TNO that is elongated will display photometric variations with rotation caused by changes in its projected cross-section. A rotationally elongated object will show a double-peaked lightcurve since each of two long and short axes will be observed during one full rotation. Assuming the lightcurve is produced from elongation of the object, we can use the maximum and minimum flux to determine the projection of the bodies' elongation into the plane of the sky through (*Binzel et al.,* 1989)

$$\Delta m = 2.5\log\left(\frac{a}{b}\right) - 1.25\log\left(\frac{a^2\cos^2\theta + c^2\sin^2\theta}{b^2\cos^2\theta + c^2\sin^2\theta}\right) \quad (6)$$

where $a \geq b \geq c$ are the semiaxes of a triaxial object with rotation about the c axis, Δm is the difference between the maximum and minimum flux expressed in magnitudes, and θ is the angle at which the rotation (c) axis is inclined to the line of sight (an object with $\theta = 90°$ is viewed equatorially).

Lightcurves of both asteroids and planetary satellites show that for the most part objects with radii ≥50 to 75 km have shapes mostly dominated by self-gravity and not by material strength (*Farinella,* 1987; *Farinella and Zappala,* 1997). This is attributed to the bodies having weak structures from fragmentation accrued in past harsh collisional environments (e.g., *Davis and Farinella,* 1997). Unlike the largest TNOs, the smaller TNOs are expected to not be dominated by self-gravity and thus may be structurally elongated. To date few small TNOs have been observed for rotational data, but they appear to have larger-amplitude lightcurves (*Trilling and Bernstein,* 2006).

It is assumed that the smaller TNOs have random pole orientations due to collisions, yet it remains to be seen if this is true for the largest TNOs. For a random distribution, the probability of viewing an object within the angle range θ to $\theta + d\theta$ is proportional to $\sin(\theta)d\theta$. Since the average viewing angle is one radian ($\theta \approx 60°$) the average sky-plane ratio of the axes of an elongated body is smaller than the actual ratio by a factor $\sin(60) \approx 0.87$. Collisionally produced fragments on average have axis ratios $2:2^{1/2}:1$ (*Fujiwara et al.,* 1978; *Capaccioni et al.,* 1984). When viewed equatorially, such fragments will have $\Delta m = 0.38$ mag. At the mean viewing angle $\theta = 60°$ we obtain $\Delta m = 0.20$ mag.

3.3. Rotational Elongation from High Angular Momentum

An object will fly apart if it reaches the critical rotational period, P_{crit}, when the centrifugal acceleration of a rotating body equals the gravitational acceleration. This occurs at

$$P_{crit} = \left(\frac{3\pi}{G\rho}\right)^{1/2} \quad (7)$$

where G is the gravitational constant and ρ is the density of the object. Even with longer rotation periods an object will be rotationally deformed. In the main-belt asteroids, only the smallest (~0.1 km-sized) asteroids have the tensile strength to resist rotational deformation (*Pravec et al.,* 2003). The amount of deformation depends on the structure and strength of the body. Strengthless rubble-pile-type structures become triaxial "Jacobi" ellipsoids at rotations just above the critical rotation point (*Chandrasekhar,* 1969; *Weidenschilling,* 1981). Since we can estimate the shape and specific angular momentum from the amplitude and rotation period of an object we can estimate their bulk densities (see section 4.1). Both lightcurves of the TNOs (20000) Varuna (*Jewitt and Sheppard,* 2002) and (136108) 2003 EL_{61} (*Rabinowitz et al.,* 2006) have been explained through rotational deformation, while several large main-belt asteroids have similar characteristics (*Farinella et al.,* 1981).

3.4. Eclipsing or Contact Binaries

Transneptunian object lightcurves could be generated from eclipsing or contact binaries. The wider the separation, the more distinctive or "notched" the expected lightcurve, unlike the more sinusoidal lightcurves caused by most other rotational effects. The axis ratio of 2:1, which is a contact binary consisting of two equally sized spheres, corresponds to a peak-to-peak lightcurve range $\Delta m = 0.75$ mag, as seen from the rotational equator. Viewed at the average viewing angle of $\theta = 60°$ would give a lightcurve range of $\Delta m =$ 0.45 mag. Very close binary components should be elongated by mutual tidal forces, giving a larger lightcurve range (*Leone et al.,* 1984). Leone et al. find that the maximum range for a tidally distorted nearly contact binary is about 1.2 mag. The contact binary hypothesis is the likely explanation of TNO 2001 QG_{298}'s lightcurve (*Sheppard and Jewitt,* 2004) as well as Jupiter Trojan (624) Hektor's lightcurve (*Hartmann and Cruikshank,* 1978; *Weidenschilling,* 1980; *Leone et al.,* 1984), and could also explain the lightcurve of main-belt asteroid (216) Kleopatra (*Leone et al.,* 1984; *Ostro et al.,* 2000; *Hestroffer et al.,* 2002).

3.5. Variable

Nonperiodic time-resolved photometric variations could be caused by a recent impact on the TNO's surface, a complex rotational state, a binary object that has different rotation rates for each component, or even cometary activity (*Hainaut et al.,* 2000). Impacts are thought to be exceedingly rare in the outer solar system and the probability that we would witness such an event is very small.

The timescale for a complex rotation state to damp to principal axis rotation (*Burns and Safronov,* 1973; *Harris,* 1994) is $T_{damp} \sim \mu Q/(\rho K_3^2 r^2 \omega^3)$ where μ is the rigidity of the material composing the asteroid, Q is the ratio of the energy contained in the oscillation to the energy lost per cycle, ρ is the bulk density of the object, ω is the angular frequency of the rotation, r is the mean radius of the object, and K_3^2 is the irregularity of the object in which a spherical object has $K_3^2 \sim 0.01$ while a highly elongated object has $K_3^2 \sim 0.1$. All TNOs and Centaurs observed to date are relatively large, and with reasonable assumptions about the other parameters, one finds all are expected to be in principal-axis rotation because the damping time from a complex rotational state is much less than the age of the solar system.

Because of the very long orbital periods for the TNOs (>200 yr) we don't expect the pole orientation to our line of sight to change significantly over the course of several years and thus we should not expect any significant lightcurve changes from differing pole orientations from year to year for TNOs. Centaurs may have shorter orbital periods and thus their pole orientation to our line of sight may significantly change over just a few years. A few attempts have been made at determining possible pole orientations for Centaurs from their varying lightcurve amplitudes (*Farnham,* 2001; *Tegler et al.,* 2005).

Cometary activity is not expected at such extreme distances from the Sun, although several attempts have been made to observe such processes with no obvious activity reported to date for objects beyond Neptune's orbit. Two TNOs have been reported to have possible variability, (19308) 1996 TO_{66} (*Hainaut et al.,* 2000) and (24835) 1995 SM_{55} (*Sheppard and Jewitt,* 2003). The reported variability of 1996 TO_{66} seems to have been caused by several observations obtained at very different phase angles (*Belskaya et al.,* 2006). The variability of 1995 SM_{55} has been suggested but not confirmed.

4. SPIN STATISTICS

4.1. Densities

It is widely believed that mutual collisions have significantly affected the inner structure of TNOs. Objects with radii r > 200 km have presumably never been disrupted by impacts, but have probably been increasingly fractured and gradually converted into gravitationally bound aggregates of smaller pieces (*Davis and Farinella,* 1997; Lacerda et al., in preparation). Whenever shaken by subsequent collisions, such pieces will progressively rearrange themselves into energetically more stable configurations, and the overall shape of the objects will relax to an equilibrium between gravitational and inertial accelerations (due to rotation). In contrast, smaller TNOs (r < 100 km) have probably been produced in catastrophic collisions between intermediate-sized (r ~ 100–200 km) bodies, and may be coherent fragments whose shapes depend mostly on material strength, or reaccumulations of impact ejecta (see, e.g., *Leinhardt et al.,* 2000). Indirect evidence for the latter comes from, e.g., the tidal disruption of Comet Shoemaker-Levy 9 (*Asphaug and Benz,* 1996, and references therein), which is believed to have originated in the transneptunian region (*Fernandez,* 1980; *Duncan et al.,* 1988). The very largest TNOs, which constitute a large fraction of the sample considered here, have also probably relaxed to equilibrium shapes as a consequence of high internal pressures (*Rabinowitz et al.,* 2006, and references therein).

If we assume that TNOs have relaxed to equilibrium shapes then their rotation states can be used to set limits on their densities: The centripetal acceleration due to self-gravity (bulk density) must be sufficient to hold the material together against the inertial acceleration due to rotation (spin period). In the extreme case of fluids, the balance of the two accelerations restricts the shapes of the rotating objects to certain well-studied figures of equilibrium (*Chandrasekhar,* 1969). Although TNOs are composed of solid material, their presumed fragmentary structure (or sheer size and internal pressure in the case of the largest bodies) validates the fluid approximation as a limiting case, which we will assume as valid in the remainder of this section [i.e., we ignore any friction that may change their shapes slightly (*Holsapple,* 2001, 2004)].

The figures of equilibrium that produce lightcurves are the Jacobi ellipsoids since they are triaxial. Assuming an equator-on geometry, we can use equation (6) with the measured peak-to-peak amplitude of a lightcurve to calculate a lower limit for the a/b axis ratio of the Jacobi ellipsoid that best approximates the shape of the TNO. Chandrasekhar's formalism relates the shape and spin period of the TNO to its density [see *Chandrasekhar* (1969) for an in-depth analysis of the simple density relation shown in equation (7)]. Since the estimated a/b is a lower limit, due to the unknown geometry, the derived density will also be a lower limit. An upper limit can also be obtained from the fact that ellipsoids with a/b > 2.31 are unstable to rotational fission (*Jeans,* 1919).

In Fig. 3 we plot density ranges, calculated as described above, vs. $m_R(1,1,0)$ for TNOs brighter than absolute magnitude 6 (r > 100 km assuming moderate albedos), more likely to be figures of equilibrium (safely in the gravity dominated regime, with clear double-peaked lightcurves, and spin period P < 10 h). Also shown are TNOs Pluto, Charon, and 1999 TC_{36}, which belong to multiple systems and are thus suitable for density measurement (see chapter by Noll et al.). A trend of larger (brighter) objects being denser is apparent. This trend can be attributed to porosity (volume fraction of void space), to rock/ice mass fraction, or to a combination of both. Indeed, bodies with density lower than water must have some internal porosity, even more so if they carry significant rock/ice mass fraction (*Jewitt and Sheppard,* 2002). Asteroid densities, calculated assuming they are equilibrium figures (*Sheppard and Jewitt,* 2002), are also plotted in Fig. 3 for five bodies that are probably rotationally deformed rubble piles (*Farinella et al.,* 1981). Asteroids are believed to have high refractory content, which explains why they have densities higher than similar-sized TNOs. Furthermore, their densities are lower than that of solid rock, which also implies internal porosity (*Yeomans et al.,* 1997). Two TNOs, 2003 EL_{61} ($\rho \sim 2500$ kg m^{-3}) and 2001 CZ_{31} ($\rho \sim 2000$ kg m^{-3}), have rotational properties that require comparatively higher densities. Although this may be an indication of slightly higher rock/ice mass fractions or lower porosities in the case of these TNOs, the small numbers do not permit us to rule out any scenario. We note that substantial porosity may exist for even the largest icy TNOs if no significant heating has taken place in the body over the age of the solar system (*Durham et al.,* 2005). This may not be true for more rocky-type objects.

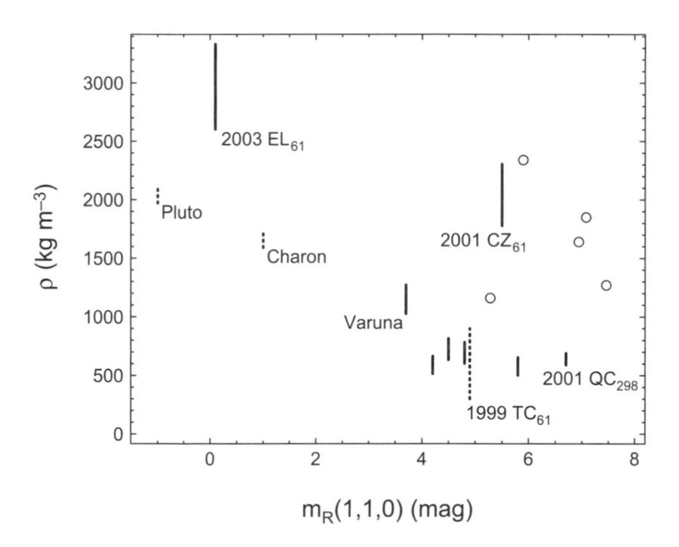

Fig. 3. Estimated density ranges as a function of absolute magnitude $m_R(1,1,0)$. Dashed lines correspond to densities of transneptunian binaries, estimated based on the satellite orbital properties. Solid lines indicate density ranges found for rotationally elongated TNOs assumed to be in hydrostatic equilibrium. Circles correspond to asteroids 15 Eunomia, 87 Sylvia, 16 Psyche, 107 Camilla, and 45 Eugenia, also thought to be rotationally elongated and in hydrostatic equilibrium (see *Sheppard and Jewitt,* 2002; *Farinella et al.,* 1981).

4.2. Spin Rate Versus Size

The number of TNOs with a well-measured spin period is about 40 as of July 2007 (Table 1). Besides being small, this sample is certainly biased. For instance, brighter objects with larger brightness variation are overrepresented, as are those with rotation periods P < 24 h. It is nevertheless interesting to compare the distribution of spin rates of TNOs with that of main-belt asteroids (MBAs). *Pravec et al.* (2003) presents a detailed study of asteroid rotation rates, using a sample of nearly 1000 lightcurves (*cfa-www.harvard.edu/iau/lists/LightcurveDat.html*). These lightcurves sample a range of sizes still mostly inaccessible in transneptunian studies, starting at bodies a few tens of meters in radius; only a few lightcurves have been reported for r ~ 10 km TNOs (*Trilling and Bernstein,* 2006) and Centaurs. Below we present a more indepth comparison between TNOs and MBAs. One TNO, 2003 EL_{61}, is outstanding in that it has the fastest rotation (P = 3.9154 ± 0.0002 h) measured for a solar system object larger than 100 km (*Rabinowitz et al.,* 2006).

Figure 4 shows the distributions of TNO and main-belt-asteroid spin periods. To minimize the effects of the aforementioned biases, only objects brighter than $m_R(1,1,0)$ = 6.5 mag, with at least Δm = 0.15 mag brightness variation range, and spinning faster than P = 20 h per full rotation were considered. In this range TNOs seem to spin slower, on average, than asteroids, with mean periods P_{TNO} = 8.4 h and P_{MBA} = 6.0 h (*Sheppard and Jewitt,* 2002; *Lacerda and Luu,* 2006). We used two nonparametric tests to test the null hypothesis that these samples are drawn from the same distribution, the Mann-Whitney U test and the Kolgomorov-Smirnov (K-S) test. The respective probabilities are 1.4% and 4.1%, which indicates that the parent distributions are likely different, but not unequivocally so. The different collisional environments as well as compositions within the main asteroid belt and Kuiper belt should account for any differences between the rotational distributions of the two populations. A similar plot of the TNOs and main-belt asteroids amplitude distributions shows no obvious differences (Fig. 5).

In Fig. 6 we plot spin period vs. absolute magnitude for the TNO data. To look for possible trends of spin rate with size, we follow *Pravec et al.* (2003) and plot a running mean

TABLE 1. Well-observed TNOs for variability.

Name			H* (mag)	Δm_R† (mag)	Single‡ (h)	Double§ (h)	Reference¶
(136199)	Eris	2003 UB$_{313}$¶	–1.2	<0.01	—	—	S
(134340)	Pluto¶		–1.0	0.33	153.2		B
(136472)		2005 FY$_{9}$	–0.4	<0.05	—	—	RST
(136108)		2003 EL$_{61}$¶	0.1	0.28 ± 0.04	—	3.9154 ± 0.0002	RB
(90377)	Sedna	2003 VB$_{12}$	1.6	0.02	10.273	—	G
(90482)	Orcus	2004 DW¶	2.3	0.04 ± 0.02	10.08 ± 0.01	—	OGS
				<0.03	—	—	S
(50000)	Quaoar	2002 LM$_{60}$¶	2.6	0.13 ± 0.03	—	17.6788 ± 0.0004	OG
(28978)	Ixion	2001 KX$_{76}$	3.2	<0.05	—	—	SS,OGC
(55565)		2002 AW$_{197}$	3.3	0.08 ± 0.07	8.86 ± 0.01	—	OGS
				<0.03	—	—	S
(55636)		2002 TX$_{300}$	3.3	0.08 ± 0.02	8.0 or 12.1	16.0 or 24.2	SS,OS
(55637)		2002 UX$_{25}$¶	3.6	<0.06	—	—	SS
				0.21 ± 0.06	7.2 or 8.4	—	RP
(20000)	Varuna	2000 WR$_{106}$	3.7	0.42 ± 0.03	—	6.34 ± 0.01	JS,OGC
		2003 AZ$_{84}$	3.9	0.12 ± 0.02	6.72	13.44	SS,OGS
(90568)		2004 GV$_{9}$	4.0	<0.08	—	—	S
(42301)		2001 UR$_{163}$	4.2	<0.08	—	—	SS
(84922)		2003 VS$_{2}$	4.2	0.21 ± 0.02	—	7.41	S,OGS
(19308)		1996 TO$_{66}$	4.5	0.25 ± 0.05	5.9	11.8	OH,SS,BO,SB
(120348)		2004 TY$_{364}$	4.5	0.22 ± 0.02	5.85 ± 0.01	11.70 ± 0.01	S
		2001 QF$_{298}$	4.7	<0.12	—	—	SS
(26375)		1999 DE$_{9}$	4.7	<0.10	>12?	—	SJ
(38628)	Huya	2000 EB$_{173}$	4.7	<0.06	—	—	SJ,SR,OGC,LL
(24835)		1995 SM$_{55}$	4.8	0.19 ± 0.05	4.04 ± 0.03	8.08 ± 0.03	SS
(19521)	Chaos	1998 WH$_{24}$	4.9	<0.10	—	—	SJ,LL
(47171)		1999 TC$_{36}$¶	4.9	<0.05	—	—	SS,OGC,LL
(82075)		2000 YW$_{134}$	5.0	<0.1	—	—	SS
(120132)		2003 FY$_{128}$	5.0	<0.08	—	—	S
(79360)		1997 CS$_{29}$	5.1	<0.08	—	—	SJ
(119979)		2002 WC$_{19}$¶	5.1	<0.05	—	—	S
(26181)		1996 GQ$_{21}$	5.2	<0.10	—	—	SJ
(55638)		2002 VE$_{95}$	5.3	<0.06	—	—	SS
				0.08 ± 0.04	6.76,7.36,9.47	—	OGS
(126154)		2001 YH$_{140}$	5.4	0.21 ± 0.04	13.25 ± 0.2	—	S,OGS
(15874)		1996 TL$_{66}$	5.4	<0.12	—	—	RT,LJ,OGS
(148780)		2001 UQ$_{18}$	5.4	<0.3	—	—	S
(88611)		2001 QT$_{297}$ 1¶	5.5	<0.1	—	—	OKE
		2001 QT$_{297}$ 2¶		0.6	4.75	—	OKE
(150642)		2001 CZ$_{31}$	5.7	<0.20	—	—	SJ
				0.21 ± 0.02	4.71	—	LL
		2001 KD$_{77}$	5.8	<0.07	—	—	SS
(26308)		1998 SM$_{165}$**	5.8	0.45	7.1		R,SS
(40314)		1999 KR$_{16}$	5.8	0.18 ± 0.04	5.840 or 5.929	11.680 or 11.858	SJ
(35671)		1998 SN$_{165}$	5.8	0.16 ± 0.01	8.84	—	LL
(66652)		1999 RZ$_{253}$	5.9	<0.05	—	—	LL
(47932)		2000 GN$_{171}$	6.0	0.61 ± 0.03	—	8.329 ± 0.005	SJ
(82158)		2001 FP$_{185}$	6.1	<0.06	—	—	SS
(79983)		1999 DF$_{9}$	6.1	0.40 ± 0.02	6.65	—	LL
(82155)		2001 FZ$_{173}$	6.2	<0.06	—	—	SJ
(80806)		2000 CM$_{105}$	6.2	<0.14	—	—	LL
		2003 QY$_{90}$ 1¶	6.3	0.34 ± 0.12	3.4 ± 1.1	—	KE
		2003 QY$_{90}$ 2¶		0.90 ± 0.36	7.1 ± 2.9	—	KE
		1996 TS$_{66}$	6.4	<0.15	—	—	LL
(33340)		1998 VG$_{44}$	6.5	<0.10	—	—	SJ
(139775)		2001 QG$_{298}$	6.7	1.14 ± 0.04	—	13.7744 ± 0.0004	SSJ
(15875)		1996 TP$_{66}$	6.8	<0.15	—	—	RT,CB
(15789)		1993 SC	6.9	<0.15	—	—	RT,D
(15820)		1994 TB	7.1	<0.15	—	—	SS
(33128)		1998 BU$_{48}$	7.2	0.68 ± 0.04	4.9 or 6.3	9.8 or 12.6	SJ

TABLE 1. (continued).

Name			H* (mag)	Δm_R† (mag)	Single‡ (h)	Double§ (h)	Reference¶
(42355)	Typhon	2002 CR_{46}¶	7.2	<0.05	—	—	SS,OGC
		1997 CV_{29}	7.4	0.4	—	16	CK
(32929)		1995 QY_9	7.5	0.60	7.3	—	RT,SS
(91133)		1998 HK_{151}	7.6	<0.15	—	—	SS
		2000 FV_{53}	8.2	0.07	3.79 or 7.5	—	TB
		2003 BG_{91}	10.7	0.18	4.2	—	TB
		2003 BF_{91}	11.7	1.09	7.3 or 9.1	—	TB
		2003 BH_{91}	11.9	0.42	?	—	TB
Centaurs							
(10199)	Chariklo	1997 CU26	6.4	?	?	?	PLO
(2060)	Chiron	1977 UB**	6.5	0.09 to 0.45	—	5.917813	BBH,L,MB
(5145)	Pholus	1992 AD	7.0	0.15 to 0.6	—	9.98	BB,H,F,TRC
(54598)	Bienor	2000 QC_{243}	7.6	0.75 ± 0.09	4.57 ± 0.02	—	OBG
(29981)		1999 TD_{10}	8.8	0.65 ± 0.05	7.71 ± 0.02	—	OGC,RPP,MH
(73480)		2002 PN_{34}	8.2	0.18 ± 0.04	4.23 or 5.11	—	OGC
(120061)		2003 CO_1	8.9	0.10 ± 0.05	4.99	—	OGS
(8405)	Asbolus	1995 GO	9.0	0.55		8.93	BL,DN
(32532)	Thereus	2001 PT_{13}	9.0	0.16 ± 0.02	4.1546 ± 0.0001	—	OBG
(83982)	Crantor	2002 GO_9	9.1	0.14 ± 0.04	6.97 or 9.67	—	OGC
(60558)		2000 EC_{98}	9.5	0.24 ± 0.06	13.401	—	RP
(31824)	Elatus	1999 UG_5	10.1	0.102 to 0.24	13.25 or 13.41	—	BM,GO
(52872)	Okyrhoe	1998 SG_{35}	11.3	0.2	8.3	—	BMF

*Absolute magnitude of the object.
†The peak to peak range of the lightcurve.
‡The lightcurve period if there is one maximum per period. If not shown the uncertainties are at the last significant digit.
§The lightcurve period if there is two maximum per period. If not shown the uncertainties are at the last significant digit.
¶A known binary TNO with both components lightcurves 1 and 2 known if labeled.
**Centaurs observed to have coma.

References: BBH = *Bus et al.* (1989); L = *Luu and Jewitt* (1990); BB = *Buie and Bus* (1992); H = *Hoffmann et al.* (1993); MB = *Marcialis and Buratti* (1993); BL = *Brown and Luu* (1997); B = *Buie et al.* (1997); D = *Davies et al.* (1997); DN = *Davies et al.* (1998); LJ = *Luu and Jewitt* (1998); CB = *Collander-Brown et al.* (1999); RT = *Romanishin and Tegler* (1999); OH = *Hainaut et al.* (2000); F = *Farnham* (2001); GO = *Gutierrez et al.* (2001); R = *Romanishin et al.* (2001); PLO = *Peixinho et al.* (2001); JS = *Jewitt and Sheppard* (2002); SJ = *Sheppard and Jewitt* (2002); SR = *Schaefer and Rabinowitz* (2002); BM = *Bauer et al.* (2002); OBG = *Ortiz et al.* (2002); SB = *Sekiguchi et al.* (2002); SS = *Sheppard and Jewitt* (2003); OGC = *Ortiz et al.* (2003a); OG = *Ortiz et al.* (2003b); OKE = *Osip et al.* (2003); BMF = *Bauer et al.* (2003); RPP = *Rousselot et al.* (2003); SSJ = *Sheppard and Jewitt* (2004); OS = *Ortiz et al.* (2004); CK = *Chorney and Kavelaars* (2004); MH = *Mueller et al.* (2004); G = *Gaudi et al.* (2005); TRC = *Tegler et al.* (2005); TB = *Trilling and Bernstein* (2006); RP = *Rousselot et al.* (2005b); OGS = *Ortiz et al.* (2006); RB = *Rabinowitz et al.* (2006); LL = *Lacerda and Luu* (2006); KE = *Kern and Elliot* (2006); BO = *Belskaya et al.* (2006); RST = *Rabinowitz et al.* (2007); S = *Sheppard* (2007).

of sets of four consecutive data points (jagged solid line). We also plot a running median for the same box size (jagged dashed line). Although the data may seem very scattered, the running mean hints at a trend of smaller TNOs spinning slightly faster. To test this hypothesis we employ the runs test for randomness (*Wall and Jenkins,* 2003), using as binary statistic the position of the measured periods relative to the median of all the measurements: Each measurement is either above or below the median, with probability 1/2. This test determines if successive (sorted by object absolute magnitude) spin period measurements are independent by checking if the number of runs (sequences of periods above and below the median) is sufficiently close to the expected value given the sample size. For the data plotted in Fig. 6, 11.52 ± 2.40 runs are expected and 12 are found. The data are thus perfectly consistent with consecutive measurements being independent.

4.3. Amplitude Versus Size

The available TNO lightcurve ranges (Δm) are plotted vs. object absolute magnitude in Fig. 7. The data suggest aslight tendency of higher variability for smaller TNOs. Except for the high-angular momentum TNOs 2003 EL_{61} and (20000) Varuna, all other objects intrinsically brighter than $m_R(1,1,0) \sim 6.0$ mag ($r \sim 110$ km assuming a 10% albedo) have relatively low variability ($\Delta m \lesssim 0.25$ mag). The same is not true for smaller (fainter) TNOs for which a much larger spread in lightcurve range exists. We used both the Mann-Whitney and K-S tests to find the absolute magni-

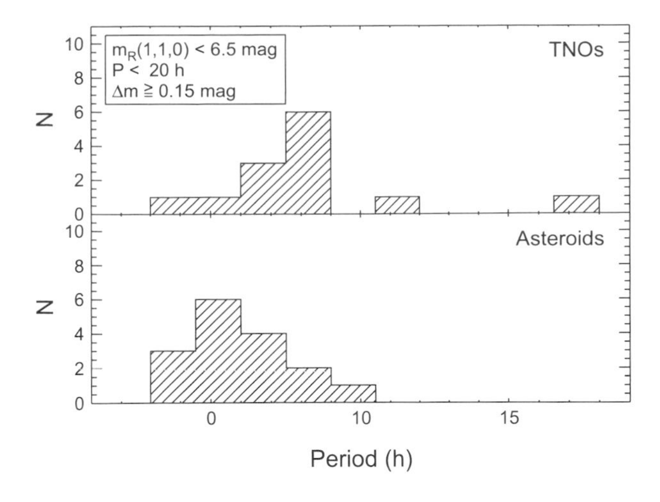

Fig. 4. Histogram of spin period for TNOs and asteroids. To minimize biases, only objects brighter than $m_R(1,1,0) = 6.5$ mag, spinning faster than P = 20 h, and with lightcurve ranges larger than Δm = 0.15 mag have been considered. As discussed in the text, the TNOs have statistically longer periods ($\bar{P}_{TNO} \sim 8.4$ h) than the main-belt asteroids ($\bar{P}_{MBA} \sim 6.0$ h) of similar size.

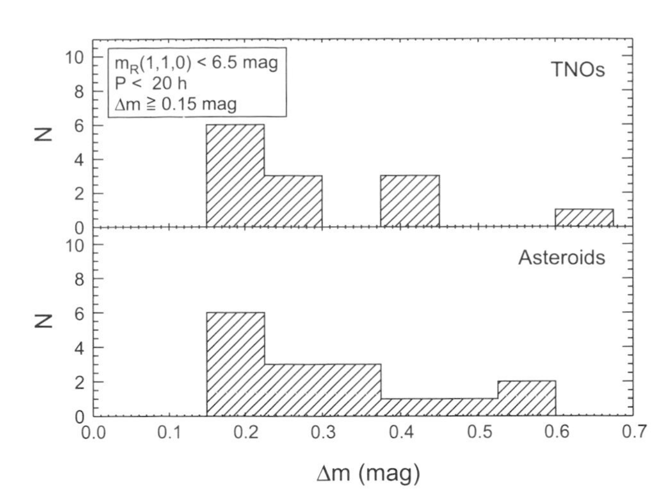

Fig. 5. Histogram of lightcurve range Δm for TNOs and asteroids. To minimize biases, only objects brighter than $m_R(1,1,0) = 6.5$ mag, spinning faster than P = 20 h, and with lightcurve ranges larger than Δm = 0.15 mag have been considered.

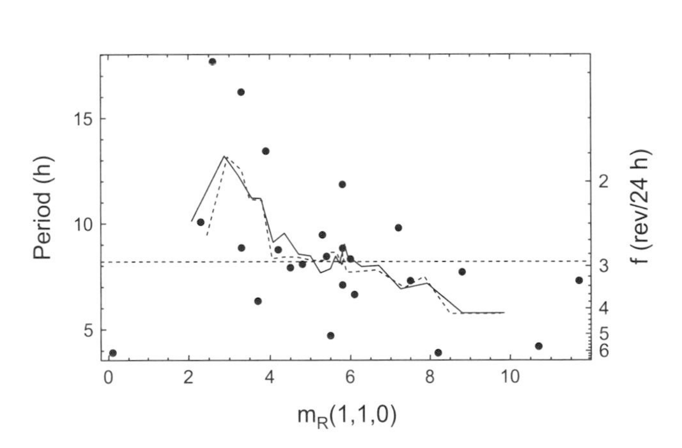

Fig. 6. Absolute magnitude vs. spin periods of the TNOs with well-measured lightcurves. As explained in the text, it appears that smaller TNOs may spin slightly faster than the larger TNOs. Pluto has not been plotted.

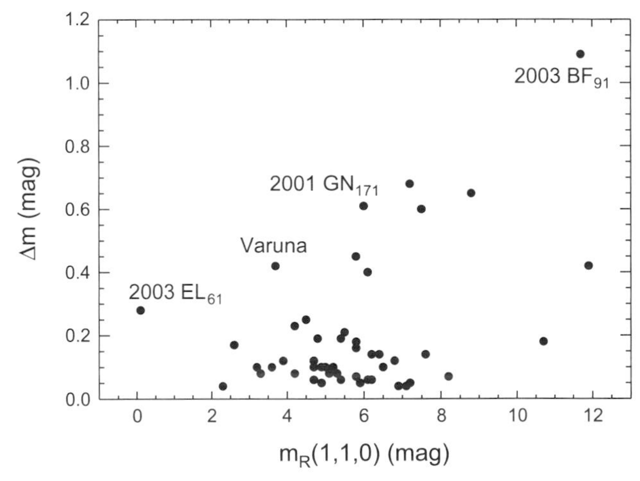

Fig. 7. Total lightcurve range plotted as a function of absolute magnitude for some of the TNOs showing rotational lightcurves in Table 1. High specific angular momentum TNOs 2003 EL_{61}, (20000) Varuna, 2001 GN_{171}, and 2003 BF_{91} are labeled. The likely contact binary, 2001 QG_{298}, has not been plotted. A statistically obvious trend of smaller TNOs having larger-amplitude lightcurves can be seen.

tude boundary that maximizes the difference in the peak-to-peak amplitude distributions of larger and smaller objects. We split TNOs at values of absolute magnitude from 3.0 to 8.5 in steps of 0.5, and calculated the probabilities $p_{U\text{-test}}$ and $p_{K\text{-}S}$ that the two populations were drawn from the same parent distribution. Both tests indicate $m_R(1,1,0) = 5.5$ (r ~ 150 km assuming 10% albedo) as maximum difference boundary, with $p_{U\text{-test}} = 0.16$ and $p_{K\text{-}S} = 0.19$. Choosing $m_R(1,1,0) = 6.5$ (r ~ 90 km assuming 10% albedo) yields comparable probabilities, $p_{U\text{-test}} = 0.17$ and $p_{K\text{-}S} = 0.20$. In conclusion, the data suggest that smaller TNOs could have larger lightcurve variability. This appears consistent with the idea that the smaller objects are more irregular in shape and collisionally evolved.

5. SHAPE DISTRIBUTION OF TRANSNEPTUNIAN OBJECTS

Due to their minute angular size, the shapes of most individual TNOs cannot be measured directly. However, under the assumption that the periodic brightness variations in a TNO lightcurve are caused by the object's nonspherical shape, we can statistically investigate the TNO shape distribution from the distribution of lightcurve peak-to-peak amplitudes (*Sheppard and Jewitt,* 2002, *Lacerda and Luu,* 2003, *Luu and Lacerda,* 2003). In simple terms, a population of elongated objects will typically produce large brightness variations, whereas a population of nearly spherical objects will predominantly cause nearly flat lightcurves.

As a simplification we will consider TNOs to be prolate ellipsoids with semiaxis $a > b = c$ and use $\tilde{a}$ to represent the shape of a given TNO. The shape distribution can be approximated by function $f(\tilde{a})$, which when multiplied by the element $d\tilde{a}$ gives the probability of finding a TNO with axis-ratio between $\tilde{a}$ and $\tilde{a} + d\tilde{a}$. As mentioned in section 3.2, the aspect angle θ, defined as the smallest angular distance between the line of sight and the TNO's spin axis, also influences the range of brightness variation. Since the distribution of spin orientations of TNOs is unknown, the most reasonable *a priori* assumption is that the orientations are random. Using these presumptions, Lacerda and Luu (2003) have shown that if TNOs have a shape distribution $f(\tilde{a})$ then the probability of finding a TNO with a lightcurve range $\Delta m \geq 0.15$ mag is given by

$$p(\Delta m \geq 0.15) \approx \int_{\sqrt{K}}^{\infty} f(\tilde{a}) \sqrt{\frac{\tilde{a}^2 - K}{(\tilde{a}^2 - 1)K}} \, d\tilde{a} \qquad (8)$$

where $\sqrt{K} = 10^{0.4 \times 0.15}$ is the axis ratio $\tilde{a}$ at which $\Delta m =$ 0.15 mag for an object viewed equatorially (see *Lacerda and Luu,* 2003). This equation can be used to constrain the shape distribution $f(\tilde{a})$.

The best estimate of $p(\Delta m \geq 0.15)$ is the fraction of TNOs that show brightness variations larger than 0.15 mag. From Table 1 we see that about 38% of the listed TNOs have lightcurve ranges $\Delta m \geq 0.15$ mag. Following previous authors, we adopt $\Delta m = 0.15$ mag as a threshold for variability detection because most ranges below this value are uncertain and usually taken as upper limits. Table 1 also shows that there is a significant fraction of objects with large peak-to-peak brightness variations: 16% have $\Delta m > 0.40$ mag. These observational constraints seem to indicate that any candidate shape distribution must allow a large fraction of nearly round objects, but also a significant amount of very elongated objects. A power-law type distribution of the form $f(\tilde{a}) \propto \tilde{a}^{-q}$ has been shown to fit best the available data (*Sheppard and Jewitt,* 2002; *Luu and Lacerda,* 2003). The best-fit slope, calculated using the method described in *Lacerda and Luu* (2003) and the data from Table 1, is $q = 4.8^{+1.2}_{-1.0}$.

More recently it has been shown that larger and smaller TNOs may have different shape distributions (*Lacerda and Luu,* 2006). This is to be expected because the material strength in smaller TNOs is likely sufficient to maintain irregular shapes, while the larger TNOs should have rounder shapes as a result of their gravity. Figure 7 shows lightcurve ranges plotted against absolute magnitude for the TNOs in Table 1. Albedo measurements exist only for a few of the listed objects. For this reason we choose to sort these TNOs by absolute magnitude, as proxy for size. We place the line between *larger* and *smaller* TNOs at $m_R(1,1,0) =$ 5.5 mag. Assuming a 0.10 (0.04) albedo, this corresponds to about 360 km (570 km) diameter. If we apply the procedure described above to TNOs brighter and fainter than 5.5 mag separately, we find that the power-law shape distributions that best fit each of the two groups are considerably different: for larger ($m_R(1,1,0) \leq 5.5$ mag) TNOs we find $q = 6.0^{+2.2}_{-1.7}$, while for smaller objects the best slope is $q = 3.8^{+1.6}_{-1.3}$. Although the significance is low (~1.5σ), the data indeed show the trend of more elongated (or irregular) shapes for smaller objects.

Inspection of Fig. 7 suggests that a simple relation between size and shape may not exist, because the latter will certainly depend on other factors such as the collisional history of individual objects. The cluster of objects with $\Delta m \leq$ 0.3 mag variability may have had a milder collisional evolution than larger specific angular momentum TNOs such as (136108) 2003 EL_{61}, (20000) Varuna, (47932) 2000 GN_{171}, and 2003 BF_{91}.

6. ANGULAR MOMENTUM OF TRANSNEPTUNIAN OBJECTS

The extremely fast rotations observed for several large TNOs [e.g., (20000) Varuna and (136108) 2003 EL_{61}] as well as relatively small satellites that are known around large TNOs [such as Pluto, 2003 EL_{61}, and Eris (2003 UB_{313}); see chapter by Noll et al.] show that many of the Kuiper belt objects (KBOs) have high amounts of specific angular momentum (Fig. 2). It is probably safe to assume this high angular momentum was imparted through collisions. In the current Kuiper belt the collision timescale to significantly modify the angular momentum of the largest TNOs is about 10^{12} yr (*Jewitt and Sheppard,* 2002). Thus the collisions likely occurred in an earlier Kuiper belt that had over 100× more large KBOs than we see there today.

The likely outcome of a large collision on a self-gravitating body is a fractured, rubble-pile-type structure (*Asphaug et al.,* 1998). Once formed, rubble-pile structures can insulate the object from disruption from further collisions by absorbing the energy of impact efficiently. In addition, the porous ices probably found in KBOs may be efficient at dissipating impact energy (*Arakawa et al.,* 2002; *Giblin et al.,* 2004). The true outcome of an impact depends on several parameters including the size of the impactor, target, and angle of impact. A glancing low-velocity collision will substantially alter the spin of a target body and may create some of the satellites and fast-spinning KBOs (*Leinhardt et al.,* 2000; *Durda et al.,* 2004).

7. CORRELATIONS WITH ROTATIONAL CHARACTERISTICS

It seems plausible that the evolutionary path that resulted in an object having a specific orbit might have affected the rotation state of that particular object. Therefore, the study of rotational properties as a function of orbital and other physical parameters might yield useful information concerning evolutionary paths. A similar reasoning has been fol-

TABLE 2. Correlation (>95%) of rotational characteristics with physical and orbital parameters.

Param*	Spear ρ[†]	Error	Sig (%)[‡]	N[§]
Δm vs. H	0.335	0.005	99.45	73
Δm vs. Q	–0.292	0.016	97.12	58
Δm vs. e	–0.279	0.022	96.19	58
Δm vs. H	0.285	0.026	96.13	58
P vs. i	0.614	0.034	95.98	13
Δm vs. Q	–0.240	0.044	95.48	73
P vs. H	–0.347	0.055	94.99	36

*Orbital parameters where Δm is the lightcurve amplitude, H is absolute magnitude, Q is aphelion distance, e is eccentricity, P is the rotation period, and i is inclination of the orbit.

[†]Spearman.

[‡]Significance level.

[§]Number of objects used. 73 means all TNOs and Centaurs used in the correlation, 58 means only the TNOs were used, 36 means only TNOs with well-measured lightcurves used, and 13 means only Plutinos were used.

lowed for colors (see chapter by Doressoundiram et al.). Unfortunately, the number of TNOs whose photometric variability has been measured is small, but the sample size is now large enough to start some analysis. *Santos-Sanz et al.* (2006) studied possible correlations of rotation periods, P, and amplitudes, Δm, vs. orbital and physical parameters using 73 TNOs and Centaurs; this includes the 30 or so with determined periods and amplitudes as well as the more numerous objects that have been well observed but have no detectable variability (see Table 1).

The strongest significant correlation found (>99%) was discussed in section 4.3, in which the rotational amplitude is correlated with the absolute magnitude (size) of the object. A few weaker and less evident correlations were found (>95%) but more data are needed to confirm them as significant (see Table 2). The next highest correlation is that of the amplitude vs. aphelion distance, Q (in this case anticorrelation), which is difficult to explain in terms of plausible physical processes that would decrease the amplitude of the lightcurve (and presumably the degree of irregularity) for the objects with larger aphelion distances. Maybe objects further out collide less often or sublimate less material over the age of the solar system.

Binary objects may also affect these statistics since any satellite (which maybe unknown) may influence the rotation, although in most cases any companions are expected to have negligible effects. It may also be interesting to analyze the large and small bodies separately as both families have clearly different photometric amplitudes but the current sample size is too small.

8. CONCLUDING REMARKS

Our knowledge on the rotational information of the largest TNOs is still a work in progress. The recent discovery of several large TNOs has shown that the lightcurve measurements of such objects are still in their infancy. Future lightcurve measurements of large objects are highly desirable using small- and medium-class telescopes. Once we increase the lightcurve inventory and their time bases to decades, we will start to be able to determine the pole orientations of TNOs.

To date there is little or no information about the rotations of small TNOs (r < 50 km). Future observations of these TNOs would be beneficial to determine if their rotation periods and amplitudes are similar to the larger objects observed to date. A transition between gravitational to mechanical structural domination should be observed for objects with radii between 50 and 100 km. The smaller objects (r < 50 km) should show a significantly different distribution of rotation periods and amplitudes than the larger objects (r > 100 km). TNOs with radii smaller than about 50 km are probably just collisional shards with shapes and rotations presumably set by the partitioning of kinetic energy delivered by the projectile responsible for breakup. Unlike the larger TNOs their rotation states are much more influenced from recent collisional events. These smaller TNOs would be much fainter than the larger objects and thus would require a number of nights on large-class telescopes (6–10 m) to obtain the signal-to-noise needed to detect their lightcurves.

In addition, it would be beneficial to obtain lightcurve information on the binary TNOs to determine their angular momentum and orbital rotational states.

Acknowledgments. We thank P. Santos-Sanz for sharing results prior to publication. S.S.S. was supported for this work by NASA through Hubble Fellowship grant #HF-01178.01-A awarded by the Space Telescope Science Institute, which is operated by the Association of Universities for Research in Astronomy, Inc., for NASA, under contract NAS 5-26555. P.L. is grateful to Fundação para a Ciência e a Tecnologia (BPD/SPFH/18828/2004) for financial support. J.L.O. acknowledges support from AYA-2005-07808-C01-03.

REFERENCES

Arakawa M., Leliwa-Kopystynski J., and Maeno N. (2002) Impact experiments on porous icy-silicate cylindrical blocks and the implication for disruption and accumulation of small icy bodies. *Icarus, 158,* 516–531.

Asphaug D. and Benz W. (1996) Size, density, and structure of Comet Shoemaker-Levy 9 inferred from the physics of tidal breakup. *Icarus, 121,* 225–248.

Asphaug D., Ostro S., Hudson R., Scheeres D., and Benz W. (1998) Disruption of kilometre-sized asteroids by energetic collisions. *Nature, 393,* 437.

Bauer J., Meech K., Fernandez Y., Farnham T., and Roush T. (2002) Observations of the Centaur 1999 UG5: Evidence of a unique outer solar system surface. *Publ. Astron. Soc. Pac., 114,* 1309–1321.

Bauer J., Meech K., Fernandez Y., Pittichova J., Hainaut O., Boehnhardt H., and Delsanti A. (2003) Physical survey of 24 Centaurs with visible photometry. *Icarus, 166,* 195–211.

Belskaya I., Ortiz J., Rousselot P., Ivanova V., Vorisov G., Shevchenko V., and Peixinho N. (2006) Low phase angle effects in photometry of trans-neptunian objects: 20000 Varuna and 19308 (1996 TO_{66}). *Icarus, 184,* 277–284.

Belton J. and Gandhi A. (1988) Application of the CLEAN algorithm to cometary light curves. *Bull. Am. Astron. Soc., 20,* 836.

Bernstein G., Trilling D., Allen R., Brown M., Holman M., and Malhotra R. (2004) The size distribution of trans-Neptunian bodies. *Astron. J., 128,* 1364–1390.

Binzel R., Farinella P., Zappala V., and Cellino A. (1989) Asteroid rotation rates: Distributions and statistics. In *Asteroids II* (R. P. Binzel et al., eds.), pp. 415–441. Univ. of Arizona, Tucson.

Brown W. and Luu J. (1997) CCD photometry of the Centaur 1995 GO. *Icarus, 126,* 218–224.

Buie M. and Bus S. (1992) Physical observations of (5145) Pholus. *Icarus, 100,* 288–294.

Buie M., Tholen D., and Wasserman L. (1997) Separate lightcurves of Pluto and Charon. *Icarus, 125,* 233–244.

Burns J. and Safronov V. (1973) Asteroid nutation angles. *Mon. Not. R. Astron. Soc., 165,* 403–411.

Bus S., Bowell E., Harris A., and Hewitt A. (1989) 2060 Chiron — CCD and electronographic photometry. *Icarus, 77,* 223–238.

Capaccioni F., Cerroni P., Coradini M., Farinella P., Flamini E., et al. (1984) Shapes of asteroids compared with fragments from hypervelocity impact experiments. *Nature, 308,* 832–834.

Catullo V., Zappala V., Farinella P., and Paolicchi P. (1984) Analysis of the shape distribution of asteroids. *Astron. Astrophys., 138,* 464–468.

Chandrasekhar S. (1969) *Ellipsoidal Figures of Equilibrium.* Yale Univ., New Haven.

Chorney N. and Kavelaars J. (2004) A rotational light curve for the Kuiper belt object 1997 CV29. *Icarus, 167,* 220–224.

Collander-Brown S., Fitzsimmons A., Fletcher E., Irwin M., and Williams I. (1999) Light curves of the trans-Neptunian objects 1996 TP66 and 1994 VK8. *Mon. Not. R. Astron. Soc., 308,* 588–592.

Davies J., McBride N., and Green S. (1997) Optical and infrared photometry of Kuiper belt object 1993SC. *Icarus, 125,* 61–66.

Davies J., McBride N., Ellison S., Green S., and Ballantyne D. (1998) Visible and infrared photometry of six Centaurs. *Icarus, 134,* 213–227.

Davis D. and Farinella P. (1997) Collisional evolution of Edgeworth-Kuiper belt objects. *Icarus, 125,* 50–60.

Degewij J., Tedesco E., and Zellner B. (1979) Albedo and color contrasts on asteroid surfaces. *Icarus, 40,* 364–374.

Duncan M., Quinn T., and Tremaine S. (1988) The origin of short-period comets. *Astrophys. J. Lett., 328,* L69–L73.

Durda D., Bottke W., Enke B., Merline W., Asphaug E., Richardson D., and Leinhardt Z. (2004) The formation of asteroid satellites in large impacts: Results from numerical simulations. *Icarus, 170,* 243–257.

Durham W., McKinnon W., and Stern L. (2005) Cold compaction of water ice. *Geophys. Res. Lett., 32,* L18202.

Dworetsky M. (1983) A period-finding method for sparse randomly spaced observations, or How long is a piece of string? *Mon. Not. R. Astron. Soc., 203,* 917–924.

Farinella P. (1987) Small satellites. In *The Evolution of the Small Bodies of the Solar System* (M. Fulchignoni and L. Kresak, eds.), p. 276. North-Holland, Amsterdam.

Farinella P. and Davis D. (1996) Short-period comets: Primordial bodies or collisional fragments? *Science, 273,* 938–941.

Farinella P. and Zappala V. (1997) The shapes of the asteroids. *Adv. Space Res., 19,* 181–186.

Farinella P., Paolicchi P., Tedesco E., and Zappala V. (1981) Triaxial equilibrium ellipsoids among the asteroids. *Icarus, 46,* 114–123.

Farnham T. (2001) The rotation axis of Centaur 5145 Pholus. *Icarus, 152,* 238–245.

Fernandez J. A. (1980) On the existence of a comet belt beyond Neptune. *Mon. Not. R. Astron. Soc., 192,* 481–491.

Fujiwara A., Kamimoto G., and Tsukamoto A. (1978) Expected shape distribution of asteroids obtained from laboratory impact experiments. *Nature, 272,* 602–603.

Gaudi B. S., Stanek K. Z., Hartman J. D., Holman, M. J., and McLeod B. A. (2005) On the rotation period of (90377) Sedna. *Astrophys. J., 629,* L49–L52.

Giblin I., Davis D., and Ryan E. (2004) On the collisional disruption of porous icy targets simulating Kuiper belt objects. *Icarus, 171,* 487–505.

Gutierrez P., Oritz J., Alexandrino E., Roos-Serote M., and Doressoundiram A. (2001) Short term variability of Centaur 1999 UG5. *Astron. Astrophys., 371,* L1–L4.

Hainaut O., Delahodde C., Boehnhardt H., Dott E., Barucci M. A., et al. (2000) Physical properties of TNO 1996 TO66. Lightcurves and possible cometary activity. *Astron. Astrophys., 356,* 1076–1088.

Harris A. (1994) Tumbling asteroids. *Icarus, 107,* 209–211.

Harris A., Young J., Bowell E., Martin L., Millis R., Poutanen M., Scaltriti F., Zappala V., Schober H., Debehogne H., and Zeigler K. (1989) Photoelectric observations of asteroids 3, 24, 60, 261, and 863. *Icarus, 77,* 171–186.

Hartmann W. and Cruikshank D. (1978) The nature of Trojan asteroid 624 Hektor. *Icarus, 36,* 353–366.

Hestroffer D., Marchis F., Fusco T., and Berthier J. (2002) Adaptive optics observations of asteroid (216) Kleopatra. *Astron. Astrophys., 394,* 339–343.

Hoffmann M., Fink U., Grundy W., and Hicks M. (1993) Photometric and spectroscopic observations of 5145 Pholus. *J. Geophys. Res., 98,* 7403–7407.

Holsapple K. (2001) Equilibrium configurations of solid cohesionless bodies. *Icarus, 154,* 432–448.

Holsapple K. (2004) Equilibrium figures of spinning bodies with self-gravity. *Icarus, 172,* 272–303.

Jeans J. (1919) *Problems of Cosmogony and Stellar Dynamics.* Cambridge Univ., London/New York.

Jewitt D. and Sheppard S. (2002) Physical properties of trans-Neptunian object (20000) Varuna. *Astron. J., 123,* 2110–2120.

Kern S. and Elliot J. (2006) Discovery and characteristics of the Kuiper belt binary 2003QY90. *Icarus, 183,* 179–185.

Lacerda P. and Luu J. (2003) On the detectability of lightcurves of Kuiper belt objects. *Icarus, 161,* 174–180.

Lacerda P. and Luu J. (2006) Analysis of the rotational properties of Kuiper Belt objects. *Astron. J., 131,* 2314–2326.

Leinhardt Z., Richardson D., and Quinn T. (2000) Direct N-body simulations of rubble pile collisions. *Icarus, 146,* 133–151.

Leone G., Farinella P., Paolicchi P., and Zappala V. (1984) Equilibrium models of binary asteroids. *Astron. Astrophys., 140,* 265–272.

Lomb N. (1976) Least-squares frequency analysis of unequally spaced data. *Astrophys. Space Sci., 39,* 447–462.

Luu J. and Jewitt D. (1990) Cometary activity in 2060 Chiron. *Astron. J., 100,* 913–932.

Luu J. and Jewitt D. (1998) Optical and infrared reflectance spectrum of Kuiper belt object 1996 TL66. *Astrophys. J., 494,* L117–L120.

Luu J. and Lacerda P. (2003) The shape distribution of Kuiper belt objects. *Earth Moon Planets, 92,* 221–232.

Magnusson P. (1991) Analysis of asteroid lightcurves. III — Albedo variegation. *Astron. Astrophys., 243,* 512–520.

Marcialis R. and Buratti B. (1993) CCD photometry of 2060 Chiron in 1985 and 1991. *Icarus, 104,* 234–243.

Mueller B., Hergenrother C., Samarasinha N., Campins H., and McCarthy D. (2004) Simultaneous visible and near-infrared time resolved observations of the outer solar system object (29981) 1999 TD10. *Icarus, 171,* 506–515.

Ortiz J., Baumont S., Gutierrez P., and Roos-Serote M. (2002) Lightcurves of Centaurs 2000 QC243 and 2001 PT13. *Astron. Astrophys., 388,* 661–666.

Ortiz J., Gutierrez P., Casanova V., and Sota A. (2003a) A study of short term variability in TNOs and Centaurs from Sierra Nevada observatory. *Astron. Astrophys., 407,* 1149–1155.

Ortiz J., Gutierrez P., Sota A., Casanova V., and Teixeira V. (2003b) Rotational brightness variations in trans-Neptunian object 50000 Quaoar. *Astron. Astrophys., 409,* L13–L16.

Ortiz J., Sota A., Moreno R., Lellouch E., Biver N., et al. (2004) A study of trans-Neptunian object 55636 (2002 TX300). *Astron. Astrophys., 420,* 383–388.

Ortiz J., Gutierrez P., Santos-Sanz P., Casanova V., and Sota A. (2006) Short-term rotational variability of eight KBOs from Sierra Nevada observatory. *Astron. Astrophys., 447,* 1131–1144.

Osip D., Kern S., and Elliot J. (2003) Physical characterization of the binary Edgeworth-Kuiper belt object 2001 QT297. *Earth Moon Planets, 92,* 409–421.

Ostro S., Hudson R. S., Nolan M. C., Margot J.-L., Scheeres D. J., et al. (2000) Radar observations of asteroid 216 Kleopatra. *Science, 288,* 836–839.

Pravec P., Harris A., and Michalowski T. (2003) Asteroid rotations. In *Asteroids III* (W. F. Bottke Jr. et al., eds.), pp. 113–122. Univ. of Arizona, Tucson.

Peixinho N., Lacerda P., Ortiz J., Doressoundiram A., Roos-Serote M., and Gutiérrez P. (2001) Photometric study of Centaurs 10199 Chariklo (1997 CU_{26}) and 1999 UG_5. *Astron. Astrophys., 371,* 753–759.

Rabinowitz D., Barkume K., Brown M., Roe H., Schwartz M., et al. (2006) Photometric observations constraining the size, shape and albedo of 2003 EL61, a rapidly rotating, Pluto-sized object in the Kuiper belt. *Astrophys. J., 639,* 1238–1251.

Rabinowitz D., Schaefer B., and Tourtellote S. (2007) The diverse solar phase curves of distant icy bodies. Part 1: Photometric observations of 18 trans-Neptunian objects, 7 Centaurs, and Nereid. *Astron. J., 133,* 26–43.

Romanishin W. and Tegler S. (1999) Rotation rates of Kuiper-belt objects from their light curves. *Nature, 398,* 129–132.

Romanishin W., Tegler S., Rettig T., Consolmagno G., and Botthof B. (2001) *Proc. Natl. Acad. Sci., 98,* 11863.

Rousselot P., Petit J., Poulet F., Lacerda P., and Ortiz J. (2003) *Astron. Astrophys., 407,* 1139–1147.

Rousselot P., Petit J., and Belskaya I. (2005a) Besancon photometric database for Kuiper-belt objects and Centaurs. Abstract presented at Asteroids, Comets, Meteors 2005, August 7–12, 2005, Rio de Janeiro, Brazil.

Rousselot P., Petit J., Poulet F., and Sergeev A. (2005b) Photometric study of Centaur (60558) 2000 EC98 and trans-Neptunian object (55637) 2002 UX25 at different phase angles. *Icarus, 176,* 478–491.

Scargle J. D. (1982) Studies in astronomical time series analysis. II — Statistical aspects of spectral analysis of unevenly spaced data. *Astrophys. J., 263,* 835–853.

Schaefer B. and Rabinowitz D. (2002) Photometric light curve for the Kuiper belt object 2000 EB173 on 78 nights. *Icarus, 160,* 52–58.

Santos-Sanz P., Ortiz J., and Gutiérrez P. (2006) Rotational properties of TNOs and Centaurs. Abstract presented at the International Workshop on Trans-Neptunian Objects: Dynamical and Physical Properties, July 3–7, 2006, Catania, Italy.

Sekiguchi T., Boehnhardt H., Hainaut O., and Delahodde C. (2002) Bicolour lightcurve of TNO 1996 TO66 with the ESO-VLT. *Astron. Astrophys., 385,* 281–288.

Sheppard S. and Jewitt D. (2002) Time-resolved photometry of Kuiper belt objects: Rotations, shapes, and phase functions. *Astron. J., 124,* 1757–1775.

Sheppard S. and Jewitt D. (2003) Hawaii Kuiper belt variability project: An update. *Earth Moon Planets, 92,* 207–219.

Sheppard S. and Jewitt D. (2004) Extreme Kuiper belt object 2001 QG298 and the fraction of contact binaries. *Astron. J., 127,* 3023–3033.

Sheppard S. (2007) Light curves of dwarf plutonian planets and other large Kuiper belt objects: Their rotations, phase functions, and absolute magnitudes. *Astron. J., 134,* 787–798.

Spencer J., Stansberry J., Trafton L., Young E., Binzel R., and Croft S. (1997) Volatile transport, seasonal cycles, and atmospheric dynamics on Pluto. In *Pluto and Charon* (S. Stern and D. Tholen, eds.), p. 435. Univ. of Arizona, Tucson.

Stellingwerf R. (1978) Period determination using phase dispersion minimization. *Astron. J., 224,* 953–960.

Tegler S., Romanishin W., Consolmagno G., Rall J., Worhatch R., Nelson M., and Weidenschilling S. (2005) The period of rotation, shape, density, and homogeneous surface color of the Centaur 5145 Pholus. *Icarus, 175,* 390–396.

Trilling D. and Bernstein G. (2006) Light curves of 20–100 km Kuiper belt objects using the Hubble space telescope. *Astron. J., 131,* 1149–1162.

Wall J. and Jenkins C. (2003) *Practical Statistics for Astronomers.* Cambridge Univ., Cambridge.

Weidenschilling S. (1980) Hektor — Nature and origin of a binary asteroid. *Icarus, 44,* 807–809.

Weidenschilling S. (1981) How fast can an asteroid spin. *Icarus, 46,* 124–126.

Yeomans D., Barriot J., Dunham D., Farquhar R. W., Giorgini J. D., et al. (1997) Estimating the mass of asteroid 253 Mathilde from tracking data during the NEAR flyby. *Science, 278,* 2106–2109.

Composition and Surface Properties of Transneptunian Objects and Centaurs

M. Antonietta Barucci
Observatoire de Paris

Michael E. Brown
California Institute of Technology

Joshua P. Emery
SETI Institute and NASA Ames Research Center

Frederic Merlin
Observatoire de Paris

Centaurs and transneptunian objects are among the most primitive bodies of the solar system and investigation of their surface composition provides constraints on the evolution of our planetary system. An overview of the surface properties based on space- and groundbased observations is presented. These objects have surfaces showing a very wide range of colors and spectral reflectances. Some objects show no diagnostic spectral bands, while others have spectra showing signatures of various ices (such as water, methane, methanol, and nitrogen). The diversity in the spectra suggests that these objects represent a substantial range of original bulk compositions, including ices, silicates, and organic solids. The methods to model surface compositions are presented and possible causes of the spectral diversity are discussed.

1. INTRODUCTION

The investigation of the surface composition of transneptunian objects (TNOs) and Centaurs provides essential information on the conditions in the early solar system at large distances from the Sun. The transneptunian and asteroid belts can be considered as the "archeological sites" where the nature of planet-building material may be examined. The investigation of the properties of these icy bodies, as remnants of the external planetesimal swarms, is essential to understanding the formation and the evolution of the population. The knowledge of the compositional nature of the whole population can provide constraints on the processes that dominated the evolution of the early solar nebula as well as of other planetary systems around young stars. Even though space weathering due to solar radiation, cosmic rays, and interplanetary dust can affect the uppermost surface layer of these bodies (see chapter by Hudson et al.), and energetic collisions could have played an important role (see chapter by Leinhardt et al.), TNOs represent the most pristine material available for groundbased investigation.

Studies of the physical properties of these objects are still limited by their faintness, and many open questions remain concerning their surface composition. Compositional determination remains a technically challenging practice for these astronomical targets. Several irregular satellites (see chapter by Nicholson et al.) seem to have a Kuiper belt origin. Some of these have been well studied, but their distinct histories preclude direct interpretation in terms of the transneptunian region. Pluto and its satellite Charon remain the best observed TNOs (see chapters by Stern and Trafton and Weaver et al.), although the system formation remains a puzzling question.

2. OBSERVATIONAL TECHNIQUES

Photometry has been the most extensively used technique to investigate the surface properties of these remote objects, since most of them are extremely faint. Many different photometric observations have been performed, particularly in the visible region, providing data for a large number of objects.

Photometric surveys have observed more than 130 objects and have revealed a very surprising color diversity. Various statistical analyses have been applied and a wide range of possible correlations between optical colors and physical and orbital parameters have been investigated (see chapters by Tegler et al. and Doressoundiram et al.).

Phase functions and polarimetry provide additional information on surface properties. The behavior of polarization phase angle depends on properties of the upper surface layer, such as albedo, particle size distribution, porosity, heterogeneity, etc. These characteristics can be constrained through numerical modeling of light scattering by the surface material, taking into account the chemical and mineralogical composition (see chapter by Belskaya et al.). Measuring thermal fluxes in the far-infrared is also a fundamental technique for albedo determination (see chapter by Stansberry et al.).

However, these techniques can provide only limited constraints on the surface composition of the population. For instance, colors can be influenced not only by composition, but also by scattering effects in particulate regoliths and by viewing geometry. Colors cannot, in general, be used to determine composition, but they can be used to classify objects into groups. A new taxonomy based on color indices (B–V, V–R, V–I, V–J, V–H, and V–K) has been derived that identifies four groups: BB, BR, IR, and RR. The BB group contains objects with neutral colors, the RR group contains those with very red colors (the reddest among the solar system objects), and the other two groups have intermediate behaviors (*Barucci et al.,* 2005b; chapter by Fulchignoni et al.). The physical significance of color diversity is still unclear, although it is reasonable to assume that the different colors reflect intrinsically different composition and/or different evolutional history.

The most detailed information on the compositions of TNOs can be acquired only from spectroscopic observations. The wavelength range between 0.4 and 2.5 μm provides the most sensitive technique available from the ground to characterize the major mineral phases and ices present on TNOs. Diagnostic spectral features of silicate minerals, feldspar, carbonaceous assemblages, organics, and water-bearing minerals are present in the visible (V) and near-infrared (NIR) spectral regions. At the near-infrared wavelengths there are also signatures from ices and hydrocarbons. Cometary activity has been detected on several Centaurs (*Luu and Jewitt,* 1990; *Pravdo et al.,* 2001; *Fernandez et al.,* 2001; *Choi and Weissman,* 2006). Weakly active Centaurs or TNOs could also show fluorescent gaseous emission bands.

Most of the known TNOs and Centaurs are too faint for spectroscopic observations, even with the world's largest telescopes. As a result, only the brightest bodies have been observed spectroscopically. The exposure times required are generally long, and as the objects rotate around their maximum inertia principal axis, the resulting spectra often contain signals from both sides of the object.

2.1. Major Spectroscopy Ground Surveys

The brightest Centaurs can be observed with small telescopes, particularly for the V range, but fainter objects have required the use of 8–10-m-class telescopes. *Luu and Jewitt* (1996) were the first to observe these distant objects. They used the Keck 1 telescope, and have continued their program since that time with observations at the Keck and Subaru telescopes on Mauna Kea.

2.1.1. European Southern Observatory (ESO) survey. As soon as VLT-ESO began operating, *Barucci et al.* (2000) and the associated team started an observational campaign in the visible and near-infrared at unit 1 Antu, unit 3 Melipan, and unit 4 Yepun. To date, about 20 objects have been observed at VLT. For most of these objects, simultaneous V + NIR spectra were measured.

2.1.2. California Institute of Technology (Caltech) survey. *Brown* (2000) and collaborators started observations of TNOs and Centaurs with the low-resolution infrared spectrograph at the Keck Observatory. To date, about 30 objects have been observed at Keck in H + K band.

The observational strategies vary depending on the telescope and instrumentation. In general, a sequence of observations includes several spectra of the object, solar analogs, and a series of calibration that include bias frames, flat field, and a lamp for wavelength calibration. Careful removal of the dominant sky background (atmospheric emission bands) in the infrared and the choice of good solar analogs are essential steps to ensure high-quality data. The solar analog has to be observed during the same night at the same air mass as the target. Using large telescopes, the known good solar analogs generally are too bright and cannot be used as they could saturate the instruments. Simultaneous measures of the absolute calibration are essential to adjust the different spectral pass-bands and provide reliable final spectra.

2.2. Space Surveys

Although discovery and characterization of TNOs and Centaurs are dominated by groundbased measurements, several programs with both the Hubble and Spitzer Space Telescopes are also relevant. The Hubble programs span the visible and near-infrared, detecting sunlight reflected off the surfaces. Several authors have observed with Hubble; in particular, *Noll et al.* (2000) used NICMOS to measure broadband reflectances in the near-infrared ($\lambda < 2.5$ μm) of four TNOs. A near-infrared reflectance spectrum (1–2 μm) of the Centaur 8405 Asbolus was also measured using NICMOS on Hubble (*Kern et al.,* 2000).

The Spitzer Space Telescope (*Werner et al.,* 2004) allows low- and moderate-spectral-resolution spectroscopy from 5.2 to 38 μm. Broadband imaging photometry is also possible in nine bands from 3.6 to 160 μm. The lower end of this wavelength range is sensitive to reflected sunlight from the distant (cold) TNOs and Centaurs. At longer wavelengths, however, thermal emission radiated by the bodies themselves is detected. The crossover between reflected and emitted radiation depends on surface temperature and can occur anywhere from near 6 μm for some Centaurs to ~15 μm for colder TNOs. Thermal emission measurements with Spitzer can also be used to derive sizes and albedo (see chapter by Stansberry et al.). Furthermore, thermal emission spectra offer the opportunity to detect emissivity features, which are diagnostic of surface composition. Unfortunately, spectral measurements are less sensitive than broadband measurements, so only the thermally brightest TNOs and Centaurs can be usefully detected with the spectrograph on Spitzer.

Spitzer has a relatively short lifetime due to finite supplies of cryogen to cool the detectors, but several current and pending programs are taking advantage of Spitzer for TNO studies while it lasts.

3. SPECTROSCOPY RESULTS

3.1. The Visible Spectra

Visible spectra, generally obtained with a low-resolution grism, are mostly featureless with a large variation in the spectral gradient from neutral to very red, confirming the diversity seen in broadband colors. The visible wavelength range provides important constrains on surface composition, particularly for reddest objects, whose reflectance increases rapidly with wavelength. Such ultrared slopes are usually interpreted to indicate the presence of organic material on the surface. The measured spectral slopes range between –1%/10³ Å and ~55%/10³ Å, with the Centaurs Pholus and Nessus being the reddest objects known up to now in the solar system. The visible range is also important for detecting aqueously altered minerals such as phyllosilicates. Three objects, all Plutinos (Fig. 1), have had reports of broad absorptions present in their visible spectra. These features are very similar to those due to aqueously altered minerals found in spectra of some main-belt asteroids, irregular satellites, and meteorites (*Vilas and Gaffey,* 1989, and subsequent papers). In the case of the three Plutinos, however, all attempts to confirm these absorptions have shown only featureless spectra. While spectral variability due to rotational modulation cannot be excluded, these detections must remain uncertain until the observations are confirmed. The presence of phyllosilicates has also been suggested by *Jewitt and Luu* (2001), who reported absorption bands (around 1.4 and 1.9 μm) in the spectrum of the Centaur 26375 1999 DE_9. All these features are rather weak, and the reality of these bands also requires confirmation.

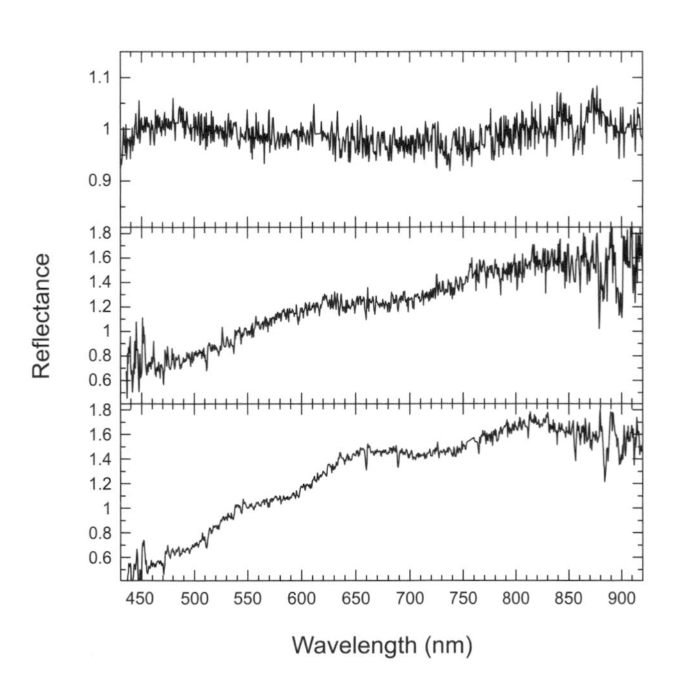

Fig. 1. Visible spectra of three TNOs showing aqueous alteration bands. The top spectrum of 2003 AZ_{84} was obtained by *Fornasier et al.* (2004b), and a continuum computed with a linear least-squares fit to the observed spectrum has been removed. The broad absorption band has been identified centered at about 700 nm. The other two spectra were obtained by *Lazzarin et al.* (2003). The 47932 (GN171) spectrum exhibits a feature around 725 nm, while 38628 Huya presents two absorption bands centered respectively at about 600 and 730 nm. On asteroids and meteorites these bands have been attributed to an $Fe^{2+} \rightarrow Fe^{3+}$ charge transfer in iron oxides in phyllosilicates (*Vilas and Gaffey,* 1989).

How aqueous alteration process could have occurred far from the Sun is not well understood, but formation of hydrated minerals directly in the early solar nebula cannot be excluded. Finding aqueously altered materials in TNOs would not be too surprising (*de Bergh et al.,* 2004), since hydrous materials seem to be present in comets, and hydrous silicates are detected in interplanetary dust particles (IDPs) and in micrometeorites.

3.2. Near-Infrared Spectra

The near-infrared wavelength range (1–2.5 μm) is the most diagnostic region for determining the presence of ices. Signatures of water ice are present at 1.5, 1.65, 2.0 μm, and signatures of other ices include those due to CH_4 around 1.7 and 2.3, CH_3OH at 2.27 μm, and NH_3 at 2 and 2.25 μm, as well as solid C-N bearing material at 2.2 μm. The first observations in this wavelength range were carried out on the Centaurs 2060 Chiron and 5145 Pholus (see *Barucci et al.,* 2002b, for a review on Centaurs) while the first spectrum of a TNO, 15789 (1993 SC), was obtained by *Luu and Jewitt* (1996) in the visible and by *Brown et al.* (1997) in the near-infrared. These early data showed a very noisy red dish spectrum with some features that they attributed to hydrocarbon ice, but that did not appear in higher-quality observations obtained later (*Jewitt and Luu,* 2001). In the near-infrared region some Centaur and TNO spectra are featureless, while others show signatures of ices. Reflectance spectra from 1.4 to 2.4 μm of four representative TNOs observed at Keck with various signal precision are reported in Fig. 2.

3.3. Results from Groundbased Spectroscopy

More than 40 objects have been observed spectroscopically to date, but only a few have been well studied in both the visible and near-infrared and rigorously modeled. These objects are faint and even observations with long exposure time and with the largest telescopes (Keck, Gemini, Subaru, and VLT) often do not yield high-quality spectra. All the objects observed spectroscopically in the near-infrared and available in literature have been listed in Table 1.

In Fig. 3 the visible and the near-infrared spectra of some Centaurs and TNOs observed at VLT are shown along with the best-fit spectral model. In general, TNOs and Centaurs

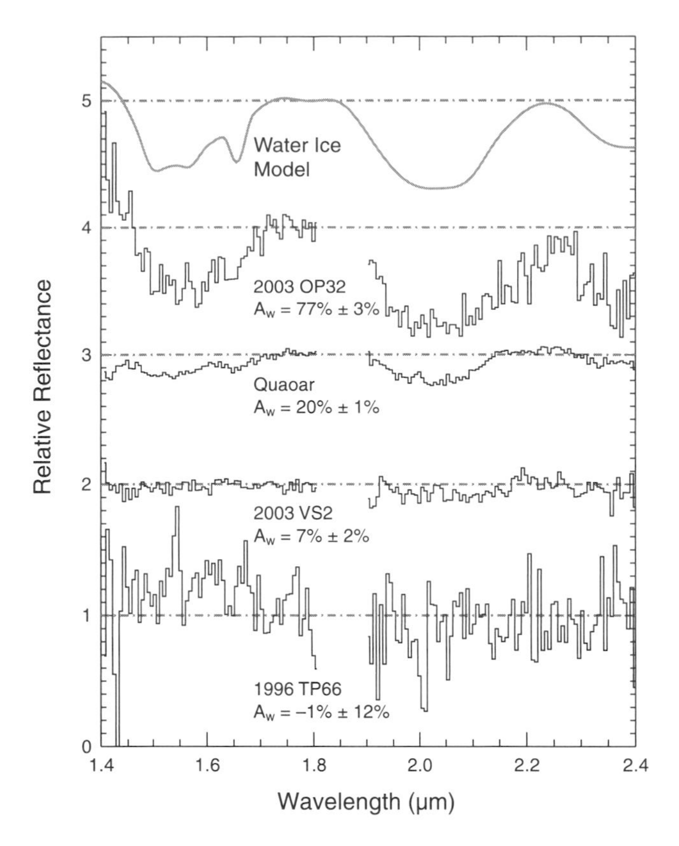

Fig. 2. The infrared reflectance spectra of four representative KBOs obtained with the Near Infrared Camera at the Keck Observatory. The spectra are normalized by the reflectance at 1.7 μm and are shifted by units of 1.0. The gray dashed lines mark the 0% and 100% relative reflectance for each spectrum. A model spectrum of pure water ice (smooth, gray line) created using a Hapke model and an ice temperature of 40 K with a grain size of 50 μm, is presented for comparison. The model shows the broad water ice absorptions at 1.5 and 2.0 μm that are also seen in the spectra of 2003 OP_{32} and Quaoar. The feature at 1.65 μm due to crystalline water ice feature is also seen in these spectra. The quantity A_w, quantifying the amount of water ice absorption at 2.0 μm, is reported next to each spectrum (see text for definition).

have similar spectral behavior. Both populations show a large diversity in spectra, which presumably indicates a similarly large diversity in surface characteristics. It is necessary to combine the visible and near-infrared spectra in order to place strong constraints on surface composition. However, in some cases this is not a sufficient step.

Radiative transfer models have been used to interpret the V + NIR spectra using intimate or geographical mixtures of organics, silicate minerals, carbonaceous assemblages, ices, and/or hydrocarbons. The red slopes are in general well reproduced by organic compounds like tholins or kerogen on the surface. The fractional abundance of these materials depends on the steepness of the spectral slope. Flat slopes and low-albedo objects can be well modeled with a large fraction of amorphous carbon. Silicates also seem to be present on some surface. For example, olivine on the surface of 5145 Pholus (*Cruikshank et al.*, 1998) significantly improves the model fit when included. Ices are easily detected when high-quality spectra are available (see Table 1). Other objects, however, show completely featureless spectra, as in the case of Asbolus. *Romon-Martin et al.* (2002) reported detailed near-infrared observations for Asbolus over the complete rotational period with high signal precision and they did not detect any features or any variability with rotation. They model the surface composition with a mixture of Triton and Titan tholins, ice tholin, and amorphous carbon. For some objects, even when observed with large telescopes, the measured spectra are incomplete or of fairly low quality.

With a small number of exceptions described below, all the spectra are either featureless to the level of the noise or show varying levels of the 1.5- and 2.0-μm absorptions due to water ice. *Barkume et al.* (2007) quantify the water-ice absorption depth and classify each of the spectra by a simple proxy for the amount of water-ice absorption. They define A_w as the fractional difference between the reflectance at 1.7 μm, in the water ice continuum, and at 2.0 μm, in the middle of the water-ice absorption band. This proxy, along with its associated error, gives a good indication of not only the shape of the spectrum, but also the quality. Barkume et al. (in preparation) found that 65% of their observed TNOs show no definitive absorptions due to water ice. Some objects with poorer signal precision, like 1996 TP_{66} (Fig. 2), may have small signatures. A plot of visible spectral gradient vs. water-ice absorption depth is shown in Fig. 4. *Brown et al.* (2007) identified a cluster of objects with absorption depths greater than 43% associated with a collisional family, whereas the remaining TNOs have water absorption depth less than 25% or no absorption. No correlation between the presence/amount of water-ice absorption and slope gradient appears when the hypothesized dynamical family has been removed.

Only a small number of objects show near-infrared absorption features attributable to any species other than water ice. Most prominently, the largest TNOs — Eris, Pluto, and 2005 FY_9 — each have complex spectra dominated by absorption bands due to methane (see chapter by Brown). The Centaur Pholus shows strong absorption bands at 2.04 and at 2.27 μm. *Cruikshank et al.* (1998) modeled the complete spectrum of Pholus from 0.4 to 2.5 μm interpreting the surface composition with the presence of the silicate olivine, Titan tholins, water ice, methanol, and carbon. They suggested that Pholus could be a primitive object, a large comet nucleus that has never been active. Two TNOs also show the 2.27-μm absorption, which has been interpreted as methanol. 55638 2002 VE_{95} was observed at VLT-ESO by *Barucci et al.* (2006) in the visible and near-infrared. The spectra show clear absorption at 1.5, 2.0, and 2.27 μm, implying the presence of H_2O and CH_3OH (and/or a photolytic product of methanol). The spectrum is similar to that of Pholus, though 2002 VE_{95} is less red. 1996 GQ_{21} also has a spectrum similar to that of Pholus (*Brown,* 2003; Barkume et al., in preparation). These detections are potentially important, suggesting a chemically primitive surface, since heating and other processes remove the light hydrocarbons in favor of macromolecular carbon.

3.4. Far Infrared: Spitzer Space Telescope Observations

As a technique for characterization of solid surfaces, thermal emission spectroscopy has been somewhat underexplored, not only for TNOs and Centaurs, but for minor bodies as a whole. The main reason is that strong absorption bands and rapid variability in Earth's atmosphere make such observations very difficult from the ground, even for the brightest asteroids. The ISO satellite demonstrated the utility of this technique for small bodies, detecting spectral emissivity features on several bright asteroids, particularly several belonging to primitive taxonomic classes (*Barucci et al.,* 2002c; *Dotto et al.,* 2002, 2004). Spitzer's significantly higher sensitivity as compared to ISO and groundbased observations opens the door for thermal emission spectroscopy of TNOs and Centaurs.

The mid-IR (5–38 μm) is well-suited to investigating the composition and structure of the surfaces of these objects. This spectral region contains the Si-O stretch and bend fundamental molecular vibration bands (typically in the ranges 9–12 and 14–25 μm, respectively). Interplay between surface and volume scattering around these bands creates complex patterns of emissivity highs and lows that are very

TABLE 1. Diameter (in kilometers), ices observed, taxonomic group (*Barucci et al.,* 2005b), and dynamical group (C: Centaurs, Cl: classical, R: resonant, S: scattered, and D: detached objects), as defined in the chapter by Gladman et al., of the TNOs and Centaurs available in the literature.

Object	Ices (Ref)	Diameter (Ref)	Tax. G	Dyn. G.	H_2O depth
2060 Chiron (1977 UB)	H_2O, var. (Lu00, Fo99, Ro03)	150 ± 10 (Fe02)	BB	C	—
5145 Pholus (1992 AD)	H_2O, CH_3OH (Cr98)	155 ± 44 (St)	RR	C	12 ± 3
8405 Asbolus (1995 GO)	none (Ke00*, Ba00, Br00a, Ro02)	85 ± 12 (Fe02, St)	BR	C	—
10199 Chariklo (1997 CU_{26})	H_2O (BrK98, Dot03c)	262 ± 18 (St)	BR	C	11 ± 3
31824 Elatus (1999 UG_5)	H_2O, var. (Bau02)	48 ± 14 (St)	RR	C	—
32532 Thereus (2001 PT_{13})	H_2O, var. (Ba02a, Me05, Li05)	61 ± 13 (St)	BR	C	—
52872 Okyrhoe (1998 SG_{35})	H_2O? (Dot03a)	52 ± 10 (St)	BR	C§	—
55576 Amycus (2002 GB_{10})	H_2O? (Dor05)	77 ± 12(St)	RR	C	—
63252 (2001 BL_{41})	none (Dor03)	20‡	BR	C	—
83982 Crantor (2002 GO_9)	H_2O (Dor05)	67 ± 19 (St)	RR	C	—
15789 (1993 SC)	none (Je01)	330 ± 66 (Th00)	RR	R	0 ± 10
15874 (1996 TL_{66})	none (Lu98)	630 ± 90 (Th00)	BB	S	0 ± 8
15875 (1996 TP_{66})	none (Bark07)	171 ± 49 (St)	RR	R	–1 ± 11
19308 (1996 TO_{66})	H_2O, var. (Br99)	<900 (Al04)	BB	Cl	65 ± 5
19521 Chaos (1998 WH_{24})	none (Bark07)	450‡	IR	Cl	–6 ± 4
20000 Varuna (2000 WR_{106})	H_2O? (Li01), (Bark07)	624 (–140, +179) (St)	IR	Cl	6 ± 3
24835 (1995 SM_{55})	H_2O (Bark07)	140‡	BB	Cl	43 ± 13
26181 (1996 GQ_{21})	none (Dor03) H_20,CH_3OH (Bark07)	400‡	RR	R	9 ± 3
26375 (1999 DE_9)	H_2O (Je01, Bark07)	461 ± 45 (St)	IR	R	18 ± 5
28978 Ixion (2001 KX_{76})	H_2O? (Li02, Boe04, Bark07)	574 ± 140 (St)	IR/RR	R	9 ± 4
29981 (1999 TD_{10})	H_2O (Bark07)	104 ± 13(St)	BR	S	13 ± 4
38628 Huya (2000 EB_{173})	H_2O? (Br00b, Je01, dB04)	540±40 (St)	IR	R	6 ± 3
42301 (2001 UR_{163})	none (Bark07)	620‡	RR	R	–7 ± 10
47171 (1999 TC_{36})	H_2O (Dot03a, Me05)	414 ± 38(St)	RR	R	14 ± 4
47932 (2000 GN_{171})	none (dB04, Bark07	400 (Sh02), 322 ± 54 (St)	IR	R	–8 ± 6
50000 Quaoar (2002 LM_{60})	H_2O (Je04)	1280 ± 190 (Br04)	RR?	Cl	20 ± 1
54598 Bienor (2000 QC_{243})	H_2O (Dot03a)	206 ± 30 (St)	BR	S	—
55565 (2002 AW_{197})	H_2O (Dor05), none (Bark07)	700±50 (Cr05)	IR	Cl	0 ± 3
55636 (2002 TX_{300})	H_2O (Li006b, Bark07)	643±210	BB	Cl	65 ±4
55637 (2002 UX_{25})	none (Bark07)	682 ± 114	IR	Cl	3 ± 3
55638 (2002 VE_{95})	H_2O, CH_3OH (Bar06)	400‡	RR	R	20 ± 4
65489 Ceto (2003 FX_{128})	H_2O (Bark07)	229 ± 18	—	D	14 ± 11
66652 (1999 RZ_{253})	none (Bark07)	280‡	RR	Cl	–13 ± 13
79360 (1997 CS_{29})	none (Gr05)	400‡	RR	Cl	0 ± 10
84522 (2002 TC_{302})	none (Bark07)	1149 ± 325	—	S	–1 ± 7
84922 (2003 VS_2)	H_2O? (Bark07)	729 ± 188(St)	—	R	7 ± 2
90377 Sedna (2003 VB_{12})	CH_4, N_2 (Ba05a, Tr05†)	2000‡	RR	D	—
90482 Orcus (2004 DW)	H_2O (Fo04a, dB05, Tr05)	951±74 (St)	BB	R	22 ± 4
119951 (2002 KX_{14})	none (Bark07)	565 ± 182	—	C	1 ± 17
120178 (2003 OP_{32})	H_2O (Bark07)	850‡	BB?	Cl	77 ± 4
134340 Pluto	CH_4, CO, N_2 (Ow93)	2350 ± 60 (Mi93, Th89)	BR	R	—
136108 (2003 EL_{61})	H_2O (Tr07)	2000 ± 500 (Ra) 1342±133 (St)	BB	Cl	55 ± 1
136199 Eris (2003 UB_{313})	CH_4, N_2? (Br05, Mer06)	3000±300 (Be06) 2450 ± (Br06)	BB	D	—
136472(2005 FY_9)	CH_4 (Li06a)	1905 ± 100 (St)	BR	Cl	—

TABLE 1. (continued).

Object	Ices (Ref)	Diameter (Ref)	Tax. G	Dyn. G.	H_2O depth
2003 AZ_{84}	H_2O (Bark07)	688 ± 99(St)	BB	Cl	22 ± 7
2005 RN_{43}	none (Bark07)	220‡	—		0±3
2005 RR_{43}	H_2O (Bark07)	200‡	—		60±7
Charon	HO_2O, NH_3 (Br00c)	1208 ± 15 (Si06, Gu06)	BB?	~	58 ± 3
S/2005 (2003 EL_{61}) 1	H_2O (Bark06)		BB?		87 ± 11

**Kern et al.* (2000) claimed to have detected water ice on one side, but the results have been contradicted by later observations.
†*Trujillo et al.* (2005) obtained a largely featureless spectrum.
‡If we assume an albedo of 0.10.
§Now classified as JFC (see chapter by Gladman et al.).

In the last column the H_2O depth has been reported as computed by Barkume et al. (in preparation); var = reported for objects whose spectra show variation in ice content.

References: Al04 = *Altenhoff et al.* (2004); Ba00 = *Barucci et al.* (2000); Ba02 = *Barucci et al.* (2002a); Ba05a = *Barucci et al.* (2005a); Br06 = *Brown et al.* (2006); Bark06 = *Barkume et al.* (2006); Bark07 = Barkume et al. (in preparation); Bau02 = *Bauer et al.* (2002); Be06 = *Bertoldi et al.* (2006); Boe04 = *Boehnhardt et al.* (2004); Br99 = *Brown et al.* (1999); Br00a = *Brown* (2000); Br00b = *Brown et al.* (2000); Br00c = *Brown and Calvin* (2000); Br04 = *Brown and Trujillo* (2004); Br05 = *Brown et al.* (2005); BrK98 = *Brown and Koresko* (1998); Bark = *Barkume et al.* (2006); Cr98 = *Cruikshank et al.* (1998); Cr05 = *Cruikshank et al.* (2005); dB04 = *de Bergh et al.* (2004); dB05 = *de Bergh et al.* (2005); Dor03 = *Doressoundiram et al.* (2003); Dor05 = *Doressoundiram et al.* (2005); Dot03a = *Dotto et al.* (2003a); Dot03c = *Dotto et al.* (2003b); Fe02 = *Fernandez et al.* (2002); Fo04a = *Fornasier et al.* (2004a); Fo99 = *Foster et al.* (1999); Gr96 = *Grundy and Fink* (1996); Gr05 = *Grundy et al.* (2005); Gu06 = *Gulbis et al.* (2006); Je01 = *Jewitt and Luu* (2001); Je04 = *Jewitt and Luu* (2004); Ke00 = *Kern et al.* (2000); Li01 = *Licandro et al.* (2001); Li02 = *Licandro et al.* (2002); Li05 = *Licandro and Pinilla-Alonso* (2005); Li06a = *Licandro et al.* (2006a); Li06b = *Licandro et al.* (2006b); Lu98 = *Luu and Jewitt* (1998); Lu00 = *Luu et al.* (2000); Me05 = *Merlin et al.* (2005); Mi93 = *Millis et al.* (1993); Ow93 = *Owen et al.* (1993); Ra = *Rabinowitz et al.* (2006); Ro02 = *Romon-Martin et al.* (2002); Ro03 = *Romon et al.* (2003; Sh02 = *Sheppard and Jewitt* (2002; Si06 = *Sicardy et al.* (2006; St = chapter by Stansberry et al.; Th89 = *Tholen and Buie* (1989); Th00 = *Thomas et al.* (2000); Tr05 = *Trujillo et al.* (2005); Tr07 = *Trujillo et al.* (2007).

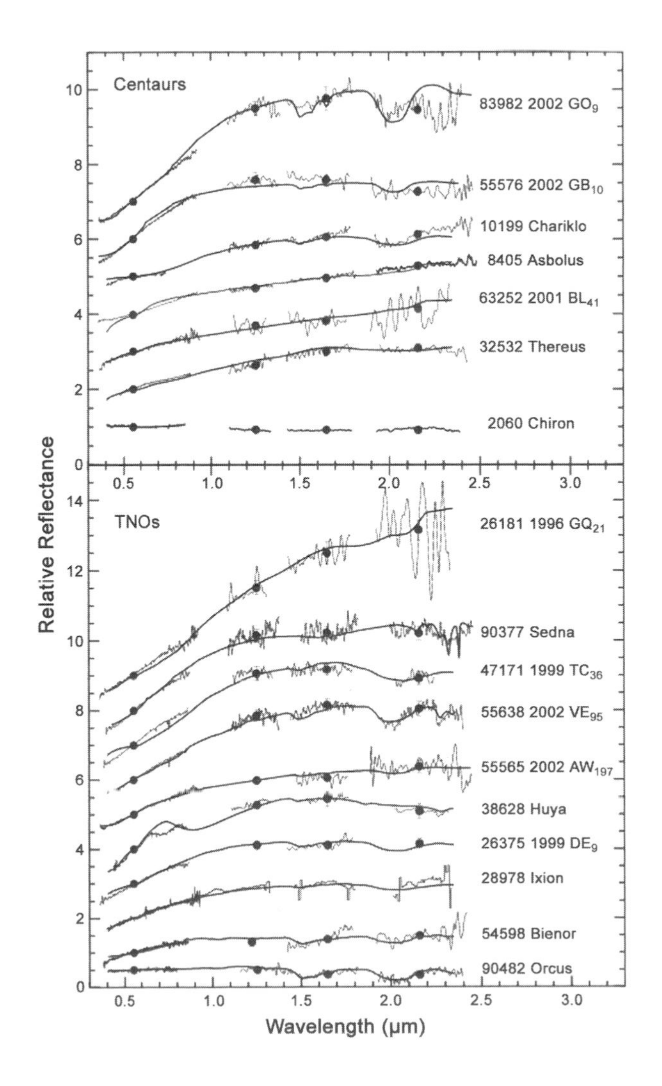

Fig. 3. Visible and near-infrared spectra with broadband photometric points at V, J, H, and K of some Centaurs and TNOs observed at VLT (ESO). The continuous lines superimposed on the spectra are the best-fit spectral models (see Table 1 for references). The spectra are normalized at 0.55 μm. Spectra are shifted by one unit for clarity.

sensitive to, and therefore diagnostic of, silicate mineralogy (e.g., *Salisbury et al.,* 1992; *Hapke,* 1996; *Witteborn and Roush,* 2000; *Cooper et al.,* 2002). Relevant ices (e.g., H_2O, CH_3OH) also exhibit features in this wavelength range. These bands, particularly the Si-O fundamentals, are also very sensitive to grain size and surface structure.

The Infrared Spectrograph (IRS) (*Houck et al.,* 2004) measures the thermally emitted flux density (e.g., units of Jy) as a function of wavelength of TNOs and Centaurs. In this thermal flux spectrum, the compositionally diagnostic emissivity features are superposed on the thermal continuum. The thermal continuum depends on many factors, including the object's size, albedo, thermal inertia, distance from the Sun, and surface roughness. Emissivity spectra are derived by dividing the measured flux spectrum by a model of the thermal continuum. Allowing the radius and albedo to vary in the model in order to find the best thermal continuum fit to the SED results in estimates of these parameters. This is the same method as described for the MIPS radiometry in the chapter by Stansberry et al., but with a different dataset. The absolute calibration of IRS has an uncertainty of ~10%, which propagates to uncertainties of ~5% in the size estimate (see *Emery et al.,* 2006, for further discussion).

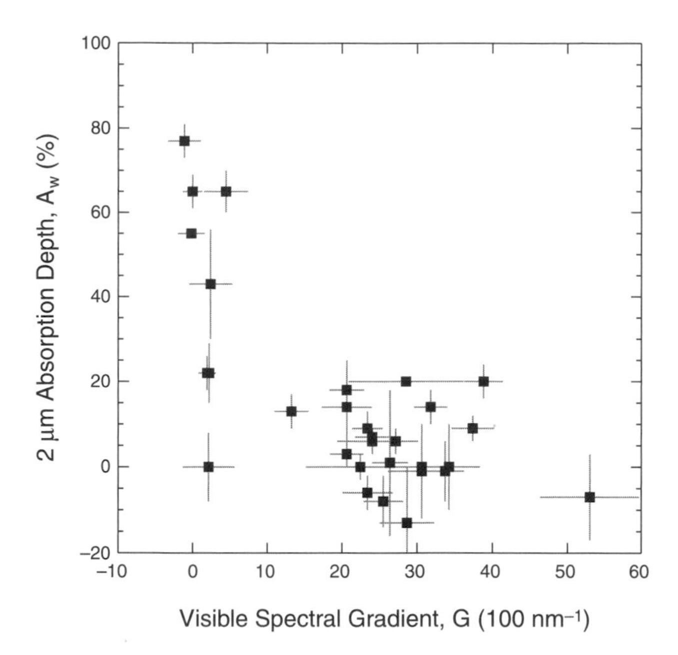

Fig. 4. The 2-µm absorption depth vs. visible color gradient for all TNOs (see Table 1) with available near-infrared spectra and colors. The cluster of objects with no spectral gradient and high water ice absorption are all associated with the 2003 EL_{61} collisional family (*Brown et al.,* 2007).

The large heliocentric distances (and therefore cold surfaces) of TNOs and Centaurs preclude observations over the full range of IRS. Current Spitzer/IRS programs include two Centaurs that are observed from 7.5 to 38 µm, four Centaurs that cover 14.2–38 µm, and 11 TNOs and Centaurs only in the longest wavelength module (20–38 µm). Five of these objects were not detected, presumably because their albedos are higher than expected and therefore their thermal fluxes were lower than expected. Initial results in terms of surface characteristics are summarized below.

The Centaur 8405 Asbolus was bright enough to be observed from 7.5–38 µm (Fig. 4). The emissivity spectrum of Asbolus exhibits emission bands with strong spectral contrast, similar to those of three Trojan asteroids (see chapter by Dotto et al.). An emissivity high is evident near 10 µm. This is presumably a plateau like in the Trojan spectra, although the Asbolus spectrum cuts off on the short wavelength side before the emissivity comes back down. There is also a broad rise from about 18 to 28 µm with an apparent double peak near 19 and 24 µm. The strong emissivity plateau near 10 µm and the broader emissivity high near 20–25 µm are interpreted as fine-grained silicates. Large grain sizes behave very differently, with emissivity lows at these locations (e.g., *Christensen et al.,* 2000). However, the emissivity bands measured by IRS in Asbolus (and the Trojans) also do not exactly match those expected for regolith surfaces; the 10-µm plateau is narrower and the spectra do not rise as rapidly near 15 µm. Analysis of these differences is in progress. Surface structure is very likely playing an important role. In particular, having fine-grained silicate particles embedded in a matrix that is fairly transparent at these wavelengths (e.g., macromolecular organic solids) may help resolve some of these differences (*Emery et al.,* 2006).

The V-NIR reflectance spectrum (0.3–2.5 µm) of Asbolus is moderately red with no detectable absorption features. The V-NIR colors place it in the BR spectral class (see chapter by Fulchignoni et al.), and the albedo is very low. These characteristics are also very similar to Trojan asteroids and are consistent with silicate surface compositions (*Cruikshank et al.,* 2001; *Emery and Brown,* 2004).

Most of the other Centaurs and TNOs observed with IRS are much fainter than Asbolus and the Trojans, and consequently the emissivity spectra obtained for them are somewhat poorer in quality (lower signal precision). Spectra of four objects are shown along with Hektor and Asbolus in Fig. 5. These have been binned to lower spectral resolution to improve the signal precision. The emissivity spectrum of Elatus is similar to that of Asbolus, with a broad high from ~17 to 27 µm, although possible double peaks are at slightly different wavelengths. *Bauer et al.* (2002) report

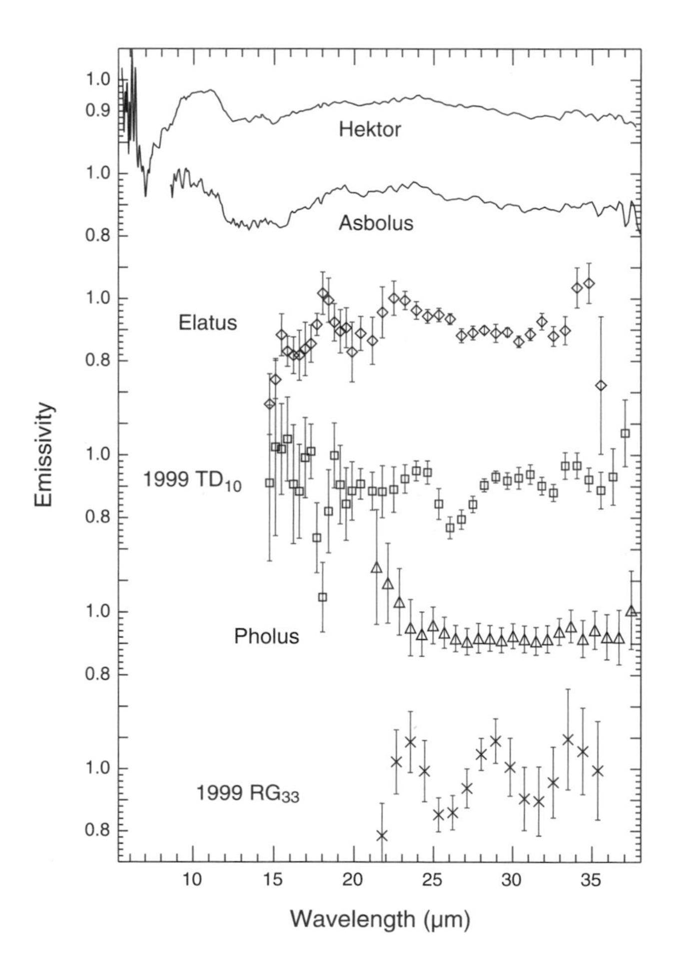

Fig. 5. Emissivity of several objects from the Spitzer Space Telescope. The spectra of 31824 Elatus (diamonds), 29981 1999 TD_{10} (squares), and 5145 Pholus (triangles) are Gaussian binned with a FWHM of eight spectral channels, and the spectrum of 15504 1999 RG_{33} (× symbols) is similarly binned with a FWHM of 10 channels. 624 Hektor is a Trojan asteroid, shown for comparison.

rotational variation in the spectrum of Elatus, which varies in spectral class from BR and featureless to IR with H_2O-ice bands. The BR and featureless spectrum is consistent with the similarity of the emissivity spectrum to that of Asbolus. The emissivity spectra of 1999 TD_{10} and 1999 RG_{33} both contain an emissivity low near 26 μm. The origin of this feature is uncertain at this point, but may be diagnostic of silicate mineralogy or ices on the surface (although it does not seem to match any known H_2O bands). It is interesting to note that although both of these objects are in highly eccentric orbits (0.77 and 0.87 for RG_{33} and TD_{10}, respectively), their perihelion distances are significantly different (2.16 and 12.28 AU). No near-infrared spectrum has been published for either of these two objects. The emissivity spectrum of Pholus displays no diagnostic bands, despite having a very active near-infrared reflectance spectrum.

4. INTERPRETATION AND SPECTRAL MODELING

4.1. Scattering Theories

Several methods have been developed to calculate synthetic reflectance spectra for comparison with the observational data for planetary surfaces. The goal of such modeling is to constrain grain sizes and mixing characteristics of the various ices, minerals, and organic materials from which those surfaces are composed. Slightly different approximations of radiative transfer theory are used to analyze the diffuse scattering of light from the surface of a semi-infinite particulate medium.

The radiative transfer model is well established for light scattering from dispersed particulate media (*Chandrasekhar,* 1960). However, *Hapke* (1981) was one of the first to formulate a theory of light scattering for surfaces in which the particles are close together. The methods described here, those of *Hapke* (1981) and *Shkuratov et al.* (1999), use the radiative transfer equation within a small elementary volume of the scattering medium, which is characterized by the albedo of single scattering. Both formulations are derived in the geometrical-optics approximation.

4.1.1. Hapke theory. The "bidirectional reflectance theory" of Hapke is based on the equations of radiative transfer for an optically thick, plane-parallel particulate medium. These equations are relatively simple, and can be used even though real planetary surfaces are extremely complex because we do not (and cannot) know many of the physical parameters that would enter into a more complex theory. This simplification permits a fast and accurate numerical evaluation. Moreover, the differences between the approximations of of the Hapke theory (*Hapke,* 1981) and the exact solution of Chandrasekhar are small, especially for low-albedo bodies, which is usually the case in the TNO population (see section 3.3). Because of its simplicity and accuracy, this theory has become the most widely used in the planetary community. We explain it, briefly, in this section.

The bidirectional reflectance of a medium is defined as the ratio of the scattered radiance at the detector to the incident irradiance. The reflectance $r(\mu_0, \mu, g)$ of a surface consisting of particles of arbitrary shape in close proximity depends on incident light angle i ($\mu_0 = \cos i$), emergent light angle e ($\mu = \cos e$), and phase angle (g), and can be computed by

$$r(\mu_0, \mu, g) = \frac{w\mu_0}{4\pi\mu_0 + \mu}[(1 + B(g))P(g) + H(\mu_0)H(\mu) - 1] \quad (1)$$

Equation (1) considers single-scattering with w, the single-scattering albedo, and multiple-scattering with $H(\mu)$ and $H(\mu_0)$, the multiple-scattering functions. P(g) is the phase function due to the singly scattered radiation, which can be isotropic or not, and B(g) describes the opposition effect.

The average single-scattering albedo w is the ratio of the average scattering coefficient of the medium to the average extinction coefficient of the medium and can be expressed as

$$w = S_e + (1 - S_e)\frac{1 - S_i}{1 - S_i\Theta}\Theta \quad (2)$$

This equation (2) depends on the external (S_e) and internal (S_i) surface-scattering coefficients and the absorption coefficient (Θ), which are related to the microscopic quantities n, k (the real and imaginary part of the complex index of refraction respectively), and D (the average diameter). The absorption coefficient, for a given wavelength (λ), is equal to

$$\Theta = e^{\frac{2\alpha D}{3}} \text{ with } \alpha = \frac{4\pi nk}{\lambda}$$

The two multiple-scattering functions are approximated by the same following form

$$H(\mu) = \frac{1 + 2\mu}{1 + 2\mu\sqrt{(1 - w)}} \text{ and } H(\mu_0) = \frac{1 + 2\mu_0}{1 + 2\mu_0\sqrt{(1 - w)}}$$

for low, fixed w, we note that $H(\mu)$ or $H(\mu_0)$ show little variations. A simple approximation for the phase function of nonisotropic scatters is a first-order Legendre polynomial

$$P(g) = 1 + b\cos(g) \quad (2a)$$

In the case of isotropic scattering, P(g) = 1 and in the cases of highly anisotropic scatterers, P(g) = 1 + cos(g) or 1 – cos(g) for backscattering or forward-scattering, respectively. An alternative expression that is sometimes used for the phase function is a Henyey-Greenstein function (1941), which depends on the asymmetry parameter ξ

$$P(g) = \frac{(1 - \xi^2)}{(1 - 2\xi\cos(g) + \xi^2)^{3/2}} \quad (2b)$$

This parameter determines whether the particle is backscattering ($\xi < 0$) or forward-scattering ($\xi > 0$). The first-order expansion (equation (2b)) is adequate if the phase function is single-lobed, but limited in the case of real phase function. Nevertheless, the Henyey-Greenstein function allows the description of a wide range of phase functions and can be used with a high level of confidence.

The opposition effect refers to the tendency of surface reflectance to increase dramatically at very small phase angles. An approximation of the theory of shadow-hiding in a medium, revisited by *Hapke* (1986), is

$$B(g) = \frac{B_0}{1 + \tan(g/2)/h} \tag{3}$$

In equation (3), B(g) describes completely the opposition effect. B_0, an empirical factor comprised between 0 and 1, is the ratio of the near-surface contribution to the total particle scattering at zero phase angle (*Hapke,* 1986). When the particles are opaque then all the scattered light comes from the surface of the particle: B_0 maximizes and becomes equal to 1. The backscatter parameter h depends on the surface properties [it is essentially equal to the ratio of the mean size of the openings between soil particles $\langle a_e \rangle$ to the extinction mean free path (1/E)]

$$h = \frac{E}{2}\langle a_e \rangle$$

Typical values of h are close to 0.4 for uncompacted lunar soil (*Hapke,* 1966) and ~0.1–0.2 for cobalt glass powders (*Hapke and Wells,* 1981).

Several subsequent improvements take into account rough surfaces (*Hapke,* 1984). For TNO observations, which involve low phase angles and relatively low albedos, the correction factors appear to be negligible. We note that the model is reasonably well understood and available for scattering and emission of radiation from regoliths with grain sizes that are large compared to the wavelength. *Hapke et al.* (1997) gives a formula, based on laboratory studies, for grain sizes on the order of the wavelength in the aim of completing his theory for a plane-parallel particulate medium.

Equation (1) can be used to compute the bidirectional reflectance of a medium composed of spheroidal, closely packed particles of a single component. To relate reflectance TNO spectra to synthetic spectra of a mixture of compounds, we have two choices. The first one is to assume an areal mixture. All reflectances of independent components i are linearly combined to represent the total reflectance r_T of the multicomponent surface

$$r_T = \sum_{i=1}^{N} c_i r_i$$

where c_i is the surface proportion of each component of bidirectional reflectance r_i.

On the other hand, we can assume a homogeneous mixture (intimate). The bidirectional reflectance is governed by w_T and if the particles are greater than the wavelength and are close together, w in equation (1) becomes w_T

$$w_T = \frac{\sum_{i=1}^{N} \frac{M_i}{\rho_i D_i} w_i}{\sum_{i=1}^{N} \frac{M_i}{\rho_i D_i}}$$

where w_T is the mean single-scattering albedo of all the particles, M_i is the bulk density of type i particles with solid density ρ_i, single-scattering albedo w_i, and diameter D_i.

4.1.2. Shkuratov method. In the model by *Shkuratov et al.* (1999), multiple reflections in a particle are considered as multiple-scattering in a one-dimensional medium with the same effective reflection coefficients. This approach replaces scattering in a system of particles by scattering in an equivalent system of plates, with the assumption that the calculated one-dimensional reflectance is equal to the reflectance of a three-dimensional medium at small phase angles.

At the boundary of the medium, light propagation is a random branching process and each ray is characterized by its reflection and transmission (R and T respectively), which depend on the complex refractive index n and k (optical constants) of the material and on the local angle of incidence on the particle interface. Transmission and reflection take place when the ray of light enter or leave the particle (see Fig. 6), inducing several types of transmission, and reflection (T_e, T_i, $T_{i'}$...) and (R_e, R_i) respectively.

For a homogeneous particle, there are simple relations between transmissions and reflections. We note that $T_e = 1 - R_e$ and $T_i = 1 - R_i$ and each term is computed using empirical approximations and the usual Fresnel coefficients (r_0)

$$r_0 = \frac{(n-1)^2}{(n+1)^2},\ R_e \approx r_0 + 0.05,\ R_b \approx (0.28 \cdot n - 0.20)R_e$$

$$R_i \approx 1.04 - 1/n^2, \text{ and } R_f = R_e - R_b$$

where R_b and R_f are the average backward and forward reflectance coefficients, respectively.

In this model, the single-scattering albedo of a particle is the sum of the fractions of the fluxes scattered by a particle into the backward (r_b) and forward (r_f) hemispheres (see equations (4a) and (4b)) and depends on the optical density between the branching ($\tau = 4\pi kD/\lambda$). The values (r_b) and (r_f) represent an evaluation of the probabilities of the beam emerging backward or forward at a given scattering number (see *Shkuratov et al.,* 1999, for more details about the evaluation of each of the probabilities of T_i, $T_{i'}$, etc.).

$$r_0 = R_b + \frac{T_e T_i R_i e^{(-2\tau)}}{2(1 - R_i e^{(-\tau)})} \tag{4a}$$

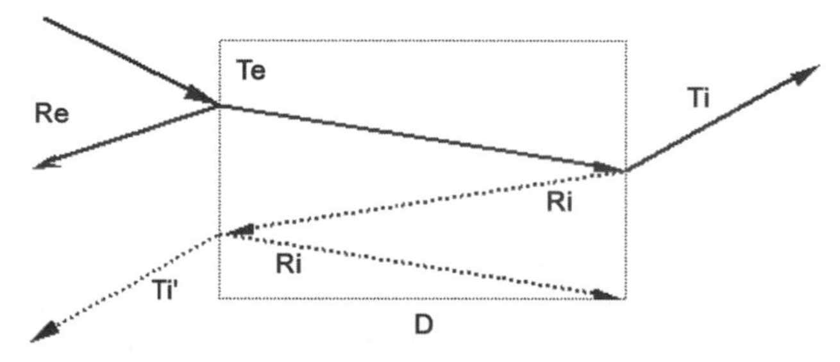

Fig. 6. Schematic diagram of the light propagation in a medium of length D used in the Shkuratov method.

$$r_f = R_f + T_eT_ie^{(-\tau)}\frac{T_eT_iR_ie^{(-2\tau)}}{2(1 - R_ie^{(-\tau)})} \quad (4b)$$

Up to this point, the theories of Hapke and Shkuratov are relatively similar in terms of computing the single-scattering albedo of a homogeneous medium. More significant differences arise in the computation of the reflectance of a particulate surface. Indeed, contrary to Hapke's model, the Shkuratov model takes into account the porosity p of the medium. The reflectance r is given by

$$r = \frac{1 + \beta^2 - \varphi^2}{2\beta} - \sqrt{\left(\frac{1 + \beta^2 - \varphi^2}{2\beta}\right)^2 - 1} \quad (5)$$

where $\beta = (1 - p)r_b$ and $\varphi = (1 - p)r_f + p$.

Moreover, the phase function asymmetry is not a free parameter but a dependent variable that is determined by the grain radius and the composition (see *Poulet et al.,* 2002). Similar to Hapke theory, the different mixtures of ices and mineral can be areal mixtures, or intimate mixtures with or without fine absorbing and independent inclusions. In the case of no inclusions, the relations to compute the reflectance (equation (5)) of a surface composed of several components of concentration c_i are

$$\beta = (1 - p)\sum_i(c_i \cdot r_{bi}) \text{ and } \varphi = (1 - p)\sum_i(c_i \cdot r_{fi}) + p$$

The ability to incorporate inclusions allows us to use components with a particle size that is very small compared to the wavelength λ in the bulk or near the surfaces of coarse particles. These inclusions, which are very small, are approximated to be spherical. Mie theory is applied, and the contribution of absorbing inclusions to the imaginary part of refractive index, k, which can be represented as changed as $k = k + k_a$, were

$$k_a = 3/2c \cdot n \cdot \text{Im}\frac{\varepsilon - 1}{\varepsilon + 2}$$

In this equation, c is the volume concentration of the inclusion in the matrix and e the ratio of the complex dielectric constant of the inclusion to that of the matrix. Note that n is the real part of the matrix refractive index. The dielectric function is related to optical constants through the constitutive relations

$$\varepsilon = \varepsilon' + i\varepsilon'' \text{ with } \varepsilon' = n^2 + k^2 \text{ and } \varepsilon'' = 2nk$$

When the inclusions are formed by space weathering (e.g., solar wind, heating by impact processes, or high-energy flux from the interstellar medium), fine inclusions frequently occur near the surface in a layer of a thickness $d \ll \lambda$. If such inclusions are present on particles, the light beam passes the absorbing layer twice per each scattering, increasing the optical density. For $c \ll 1$ the Mie theory can again be applied, and k becomes

$$k = k + \frac{2d}{2D}k_a$$

One of the most significant differences between the models of Hapke and Shkuratov is the treatment of scattering asymmetry. In the Hapke model, the particle phase function itself is an input parameter to the model (a first-order Legendre polynomial scattering function or a Henyey-Greenstein function), as is the asymmetry parameter, for whichever phase function is used. Recall that the asymmetry parameter describes the relative importance of back- or forward-scattering. In the Shkuratov model, on the other hand, the particle phase function asymmetry is directly determined by the grain size and the composition, and can be characterized by the ratio rf – rb/rf + rb. Scattering asymmetry affects the dark and bright parts of a given spectrum in a nonproportional manner (*Poulet et al.,* 2002). Other small differences are that Shkuratov specifically includes porosity (although it is possible to approximate porosity effects in to Hapke model in an *ad hoc* manner) and the fact that observational geometry is not mentioned in the Shkuratov model, contrary to that of Hapke.

Finally, we mention the radiative transfer model of *Doute and Schmitt* (1998), developed for the calculation of bidirectional reflectance by a parallel plane that is absorbing, scattering, and slightly stratified. This model follows the approach of Hapke but improves the realism of the single- and double-scattering contribution with an unrestricted phase function; the form of the phase function equation (2a) is only used for the higher-order scattering. Concerning stratification, a simple adding algorithm based on the principle of invariance is used (*Doute and Schmitt,* 1998), which improves the results more (e.g., *Doute et al.,* 1999, for Pluto) compared to areal and intimate models. This model allows the analysis of hyperspectral images, even if this aspect is not useful for groundbased observations.

4.2. Limits and Dependences

Although these models are generally able to describe the spectral properties of the analyzed objects quite well, it is important to note that they are only indicative. Mixtures with different grain sizes or different mixing ratios of the constituents can give very similar fits to the data. Therefore, an absolute determination of the surface composition is difficult. Moreover, many of the parameters are not well established or completely unknown. Limits and dependences about the models' parameters are presented below.

4.2.1. Grain size and concentration. The grain size and the abundance of particles directly affect the optical depth of the light. The presence and strength of absorption bands can inform us not only of the terrain composition, but also of its texture. In the Hapke or Shkuratov models, the concentration and grain size are generally linked, particularly for ices, and the determination of both values is difficult. Very similar model spectra can often be obtained with larger grain size and weak concentration or small grain size and large concentration. However, very deep absorption bands usually imply a large grain size and imply that the terrain texture is a compact solid surface rather than a granular surface, especially when the models require centimeter-sized grains. Both models require grains with a diameter several times larger than the wavelength — the entire development of Hapke or Shkuratov models are based on this assumption — and the use of smaller grain sizes violates the principle of geometrical optics. Nevertheless, *Piateck et al.* (2003) has shown by laboratory experiments that the scattering properties of small particles, for reasons not yet understood, disagree strongly with predictions that assume the particles scatter independently, and work is still in progress to understand the detailed scattering properties of very small particles.

4.2.2. Compounds and optical constants. Optical constants are the basis of the intimate mixture modeling. Each optical constant is derived from laboratory measurements for a given molecule. These molecules can be pure or sometimes diluted. Optical constants are most often derived from measured transmittance using the Kramers-Kronig relation. The reflectance of a medium measured in the laboratory is relevant only for the particle size of the given experiment, while optical constants allow the generation of model spectra for different particle sizes and the reproduction of not only geographical mixtures but also intimate mixtures. Optical constants also depend on the properties of the particles, such as the state of the molecules or the degree of crystallization at a given temperature. Obviously, it is essential to have a complete set of optical constants, including a variety of components (ices, organic compounds, minerals, carbons, etc.) at several temperatures (see chapter by de Bergh et al.).

4.2.3. Temperature and dilution. The temperature and dilution of molecules in a matrix are important factors, the variation of which can induce spectral variations (e.g., wavelength shifts and band depths). For example, in the case of crystalline water ice, the absorption bands are deeper at lower temperatures (e.g., the 1.31 and 1.65 μm bands) (*Grundy and Schmitt,* 1998) and also show slight shifts in wavelength. For methane ice diluted in a nitrogen ice medium, the wavelength shift of the methane absorption bands are directly related to the concentration of this ice (*Quirico and Schmitt,* 1999). We note that the pure or diluted state of certain molecules can be observed spectroscopically. For example, not only do the wavelengths of the methane bands shift in dilution, but the 1.69-μm band appears only when methane is pure. Once again, the need to have a large variety of optical constant of numerous molecules at different temperatures and dilution states is mandatory.

4.2.4. Asymmetry parameter. The asymmetry factor is often unknown and its assumed value is based on comparison with other small bodies of the solar system located more or less in the same region. For icy satellites, this value is determined using photometric parameters computed by fitting the Hapke model to remote sensing data. However, the backscattering behavior generally used on icy satellites (corresponding to a negative asymmetry factor) may be a bias introduced by the Hapke model. Indeed, for sparsely distributed, independently scattering grains, the calculated asymmetry parameters are always positive or may be negative only in few cases for densely packed grains (*Mishchenko,* 1994). The under- or overestimation of this factor increases or decreases the entire reflectance in the same way and induces little variations of compound concentration or particle size in the models.

4.2.5. Albedo. The albedo is a measure of the light reflected by the body, and while in the strictest quantitative sense the albedo is dependent on geometric and other factors, here we refer simply to the ratio of reflected light to the incident sunlight, and express this quantity in percent. The knowledge of this value is fundamental to constraining the concentration of ices, carbons, and organics, which can have widely differing reflectivities. For bright bodies, like Pluto or Eris, the high albedo, close to 60–86%, indicates the presence of a large quantity of ices on the surface. This is contrary to dark objects, for which the albedo can be <5%, indicating the presence of dark carbons or organics on the surface.

4.3. Surface Modeling Application

The use of a radiative transfer model is the most realistic and reliable way to interpret the observed spectra and to investigate the surface composition of planetary objects. For an appropriate application of the method, a complete spectrum from the visible and near-infrared is necessary, as is the albedo. In cases where different spectral segments were obtained separately, photometric data are also needed to properly adjust these segments relative to one another.

To illustrate the application of the models described above, we give here an example of the investigation of the surface composition of the Plutino 55638 (2002 VE_{95}), considering areal and intimate mixtures of different materials. Several minerals, ices, and different organic complexes (e.g., tholins and kerogens) at different grain sizes have been considered. The best-fit model is obtained by varying the critical parameters — composition, abundance, and grain size — until the difference between the data and the model spectrum is minimized. In this case, the Levenberg-Marquard minimization algorithm (*Press et al.,* 1986) was used. The albedo of 2002 VE_{95} is unknown, and we assume it equals 0.10 at 0.55 μm, a mean value for TNOs (see chapter by Stansberry et al.). Intimate mixtures give the better matches (as measured by χ^2) than areal mixtures over the spectral range between 0.4 and 2.4 μm. The synthetic spectra obtained with the two models (Hapke and Shkuratov) represent similar quality fits of the spectrum and give similar results but with different percentages of components (Fig. 7).

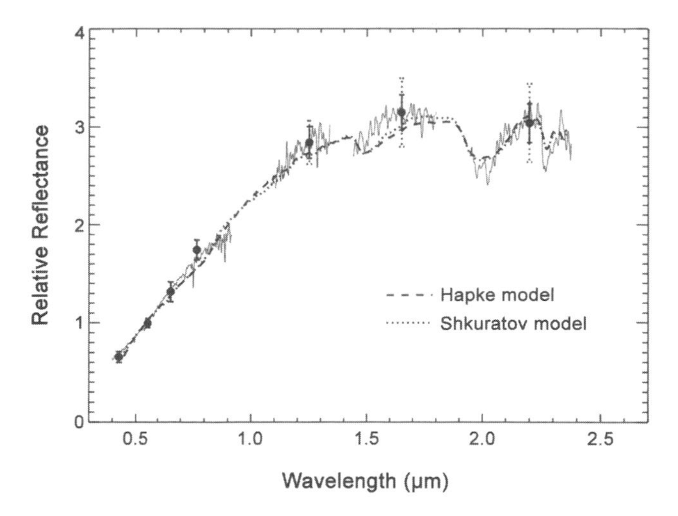

Fig. 7. Spectral reflectances of 55638 (2002 VE_{95}) obtained in V, J, H, and K bands (full lines) and adjusted using photometric colors, which are represented by dots with the relative errors (*Barucci et al.,* 2006). The two best-fit intimate models have also been reported.

The best-fit Hapke model is composed of 12% amorphous H_2O ice (with grain size of 30 µm), 10% CH_3OH ice (40 µm), 47% Titan tholin (4 µm), 14% Triton tholin (4 µm), and 17% amorphous carbon (10 µm). For the Shkuratov intimate model, the best fit is obtained adding some olivine, and it is composed of 13% amorphous H_2O ice (30 µm), 12% CH_3OH ice (40 µm), 20% Titan tholin (4 µm), 29% Triton tholin (4 µm), 19% amorphous carbon (10 µm), and 11% olivine (100 µm). The photometric uncertainty in the H band (Fig. 7) does not allow a more accurate model.

For detailed models and discussions on surface composition obtained for each object, as available in literature, see references reported in Table 1.

5. SURFACE COMPOSITION (STATE OF THE ART)

As discussed above, much of the information obtained from spectral modeling is nonunique, especially if the necessary conditions, as discussed above, are not met (albedo, high signal precision signatures, significant wavelength coverage, etc.). Derived abundances of different materials and even the presence of these materials are easily influenced by choices of model parameters. Nevertheless, the modeling can be used to pick out broad variations in TNO spectra.

Recently, several large objects with strong ice signatures in their spectra have been discovered. This is in contrast with the previous data, in which the majority of the objects showed spectra with faint or no ice signature (*Barucci and Peixinho,* 2006), and often with relatively low signal precision. Analyzing all the available spectra, objects are found to fall into several broad categories following those first identified by *Cruikshank et al.* (2007).

5.1. The Four Spectral Types

The aim of grouping is not to propose the following as taxonomy, but to note the basic surface characteristics of these groups. To date, four spectral groups can be identified:

1. *Methane-dominated spectra.* Most of the very largest transneptunian objects — Eris, Pluto, Sedna, and 2005 FY_9 — have infrared spectra dominated by CH_4 absorption. Some objects show spectra containing absorption bands of both CH_4 and N_2, and the CH_4 is often dissolved in the N_2 ice. Pluto and probably Sedna share these properties, together with Neptune's satellite Triton (generally thought to be a large TNO captured by Neptune). Recent modeling of Eris also seems to favor the presence of N_2 on the surface of Eris (*Merlin et al.,* 2006; *Dumas et al.,* 2007). Some other objects show spectra with absorption bands of CH_4, with clear indication that the CH_4 ice is pure and not dissolved in another ice. This is the case of 2005 FY_9. 2003 EL_{61} is the largest transneptunian object that does not have CH_4 detected on the surface. Detailed discussion of the spectra of these large TNOs can be found in the chapter by Brown.

2. *Water-ice-dominated spectra.* A number of objects have spectra that have moderate to deep absorptions due to water ice. These water ice objects include the largest objects after the methane-dominated objects, but also include smaller objects. The very deepest water-ice absorptions are found exclusively in objects dynamically similar to 2003 EL_{61}, which also has very deep water-ice absorptions. *Brown et al.* (2007) show that this dynamical cluster is consistent with being a family of collisional shards formed from a giant impact on 2003 EL_{61}. Other than this dynamical subclass for the deepest absorptions, no correlation between the presence or amount of water-ice absorption and any other factor is apparent. In particular, if the variation in water-ice absorption is interpreted as variation in the surface exposure of pure water ice with the remainder of the surface covered with red featureless tholins, we would expect to see a correlation between the color index and the water-ice absorption. Once the 2003 EL_{61}-related objects are removed, no correlation is seen. The lack of correlation between color and water-ice absorption is most likely an indication that water ice on the surfaces of these TNOs is intimately mixed with the coloring agent. The physical reason for the variation in color or the variation in water-ice absorption is nonetheless unknown.

All TNOs with spectra showing water-ice absorptions measured with sufficiently high signal precision have been shown to have the 1.65-µm feature due to the presence of crystalline water ice. This is the case for Quaoar, 2003 EL_{61}, 2002 TX_{300}, and 2003 OP_{32}. While crystalline water ice is neither expected at these temperatures nor should be stable against cosmic-ray and UV bombardment, crystalline water ice appears to be ubiquitous in the outer solar system, from the icy satellites of the giant planets, to large TNOs, to even moderate-sized satellites of TNOs (*Barkume et al.,* 2006).

The presence of crystalline water ice implies that ice has been heated to temperatures above 100/110 K. This heating could have occurred in only the uppermost layers by impacts or in the deep interiors. As discussed by *Jewitt and Luu* (2004), the presence of crystalline water could imply cryovolcanic outgassing. Complex geology, as observed on some icy satellites, indicates that solid-state convection could bring warm ices toward the surface, but the presence of crystalline water ice on even relatively small TNOs makes this suggestion appear unlikely. The quality of most of the spectra is too poor to distinguish between amorphous or crystalline water ice. Several other objects, such as Orcus (*de Bergh et al.,* 2005) and 1996 TO_{66} (*Brown et al.*, 1999), have spectra that favor the presence of crystalline water ice, although the data have relatively poor signal to noise precision. The widespread presence of crystalline water ice suggests that it is likely that some more mundane process is at work. While some objects in the outer solar system have been suggested to have a mixture of crystalline and amorphous ices (i.e., the Galilean satellites) (*Hansen and McCord,* 2004), at this point the discovery of an object with water-ice absorption that could be convincingly shown to have no crystalline water ice would be a much greater surprise.

3. *Water-ice spectra with the presence of methanol-like features.* A small number of objects show spectra with the 2.27-μm band characteristic of CH_3OH (and/or a phololytic product of methanol). This detection suggests a chemically primitive surface, since heating and other processes remove the light hydrocarbons in favor of macromolecular carbon. The absorption band has been detected so far only on the surfaces of the TNOs 55638 2002 VE_{95} and 1998 GQ_{21} and the Centaur Pholus. Many objects have a low signal precision, especially in the K band longward of 2.2 μm, which could mask this signature. However, many objects show a decreasing slope after 2.2 μm, implying the possible presence of CH_3OH (or a similar molecule). Nonetheless, the high signal-to-noise in many spectra firmly rules out the presence of this 2.27-μm absorption at the levels seen on these three objects. The presence of methanol on the surface of some Centaurs/TNOs is taken to indicate the chemically primitive nature of these bodies. Methanol is an abundant component of many active comets and of the interstellar medium. No correlation can be found between of the other properties of the three methanol-like objects and the presence of the methanol.

4. *Featureless spectra.* Many objects have featureless spectra in the near-infrared, but a wide range of colors. Some objects are essentially neutral (gray) over this spectral interval, while others are among the reddest objects in the solar system. Geometric albedos appear to lie in the range 0.03–0.2 (see chapter by Stansberry et al.). Many spectra that appear featureless when observed at low signal precision may reveal diagnostic absorption features that place them in other groups once better spectra are available. Nonetheless, many spectra have sufficient signal-to-noise to rule out absorption due to water ice at a very low level. The spectra of this group resemble those of the dead comets and the jovian Trojans (see chapter by Dotto et al.). These bodies could have surface mantles rich in organic material or carbon (see chapter by de Bergh et al.) that mask interior ice.

5.2. Variegated Surfaces: Heterogeneity and Homogeneity

Observations of several icy satellites have shown heterogeneous surfaces due to crater formation or volcanism, which reveal fresh and bright icy components. Photometric and spectroscopic observations of several TNOs and Centaurs have recently shown rotational variability. Photometric variations as represented by the rotational lightcurve of an object depend on the object's shape, rotational properties, and surface variation. Variation of color indices can also reveal heterogeneous surface properties of an object. Spectroscopy is always the best tool to access the surface properties, and the near-infrared, where we can see vibrational harmonics of ices, is the best spectral range to observe such variation. From the visible range, two objects (2000 GN_{171} and 2000 EB_{173}) show visible variable spectra associated with detection of hydrated silicates or featureless materials. This variability is not confirmed and other observations are required (*Fornasier et al.,* 2004). *Barucci et al.* (2002a) obtained two spectra of the Centaur Thereus in the near-infrared region. Although the first one shows absorption features of H_2O ice, the second one is featureless, revealing the heterogeneous nature of its surface. This trend was confirmed by *Licandro and Pinilla-Alonso* (2005) and *Merlin et al.* (2005). Moreover, *Merlin et al.* (2005), using the rotational period of this object, showed the relation between the rotation phase and the presence of water ice in new near-infrared spectra. They assumed that only part of this Centaur is covered by water ice. Available observations of additional Centaurs and TNOs shows that several objects present variations of their surface properties. These include 31824 Elatus (*Bauer et al.,* 2002) and 19308 1996 TO_{66} (*Brown et al.,* 1999). In some cases, observed surface variations may be attributed to different viewing geometry, and possibly to the low signal precision of some of the observational data, in which case it is mandatory to check with higher-quality data. The partial resurfacing by nondisruptive collisions and/or activity would be the most likely cause of the heterogeneity of these objects.

5.3. Causes of the Spectral Types

To investigate the presence of ices on the whole population and the possible connection with the evolution of the TNOs, *Barucci et al.* (2006) analyzed all the NIR spectra available in the literature for which a surface compositional model was performed. No correlation was found between the taxonomic groups and the presence of ices on the surfaces. All four groups show the presence of ices. The majority of nonicy bodies seem concentrated on the reddest group (RR) where organic compounds (causing red slopes) could hide ices present on the surface. The majority of neutral spectra (BB group) bodies have objects for which the content of ice seems generally higher than in the other groups. Similar results have been found by Barkume et al. (in preparation), analyzing the H_2O absorption depth vs. the visible gradient.

Barucci et al. (2006) investigated whether there is a connection between the presence of ice and the size of the objects. They reported that all large objects have ices detected on their surface, while smaller objects can have no ice detection (mantled bodies) or limited ice detection (see Fig. 8).

The presence of methane and other volatiles on the largest TNOs can be understood as a consequence of atmospheric retention by these bodies (see chapters by Brown and by Stern and Trafton). The vast majority of the objects in the Kuiper belt are far too small to have sufficient gravity to retain any volatile species on their surfaces, so we should expect nothing but water ice or involatile long-chain hydrocarbons or tholins on the surface.

While this general expectation is met (with the exception of the methanol-like spectra), the essentially stochastic variation in optical color and in water ice absorption is inexplicable. As a first guess, the spectra of TNOs would be expected to resemble the dark-red featureless spectra of extinct comets and primitive asteroids. While some spectra do indeed resemble these other types of objects, many do not. Some look much more like the icy satellites of the giant planets instead.

While no general explanation for the spectral variation is yet forthcoming, only a small number of possibilities appear to exist. The variations must be due to either variation in the initial composition of the objects or in variation in the subsequent history. While the Kuiper belt itself occupies too small a range in relative heliocentric distance ($\Delta a/a \sim 0.15$) to imagine large temperature or chemical gradients across it, the objects that are now in the Kuiper belt are thought to have come from a large swath of the outer solar system (see chapter by Morbidelli et al.). Indeed, the most homogeneous appearing population within the Kuiper belt, the uniformly red cold classical objects (*Morbidelli and Brown,* 2004), are potentially the only population that is relatively unperturbed. The main chemical variations that could be imagined to affect the final spectrum would be the inclusion or exclusion of hydrocarbons, which would eventually become dark and red. In general, however, the absence of any correlation between water-ice absorption and color does not strongly support this idea.

Variations in the history of the objects could also lead to spectral variations. The stochastic-appearing variations in the spectral properties of the objects make a stochastic process such as impacts an appealing possibility. Early suggestions that recent impacts might excavate fresh ices and cover red tholin-like material leading to color variations appear not to be born out. In particular, such a process would again lead to a general correlation between water-ice absorption depth and color, which is not seen, in addition to frequent rotational color and spectral variation, which has rarely been observed. While none of the explanations for the color and spectral variation of TNOs is particularly satisfying, few other general ideas seem plausible.

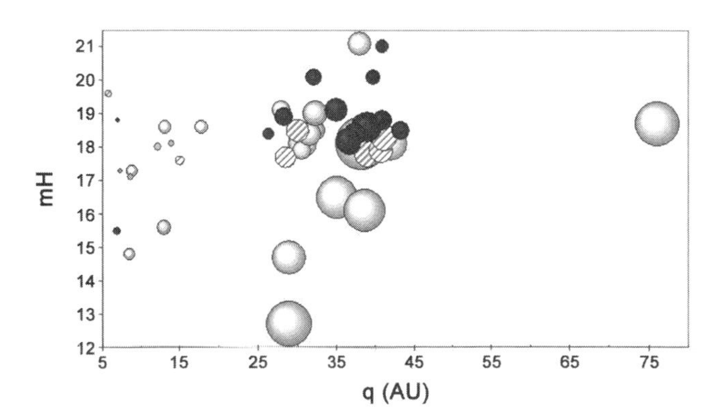

Fig. 8. Distribution of the icy (white) and nonicy (black) bodies (see Table 1) on a plot of the apparent H-band magnitude (mH) at the time of the spectral observation vs. perihelion distance. The objects with no firm detection of ices are reported with dashed lines. The size of the points is proportional to the estimated diameter of the bodies. The smaller points represent the objects with estimated diameters of ~50 km, while the biggest is for Eris, ~2500 km.

6. CONCLUSIONS

In this chapter we presented the knowledge of the composition of the TNO and Centaur population deduced from the different ground- and space-based spectroscopic surveys and the methods commonly used to analyze the resulting data. Surface structure plays an important role on the interpretation of V-NIR and far-infrared data. The knowledge of the surface properties doubtlessly provides constraints on the formation and evolution processes that have affected the population. The spectral behavior of objects within this population shows large differences from one another. Interpretation of these spectra is in general difficult because models of the spectra depend on the choice of many parameters (see section 4). Nevertheless, this is the best way we have to investigate their surface compositions. No correlation has been found between the presence of ice, absorption depth, taxonomic classes, and/or visible spectral gradient. The BB group seems to have objects showing more ice content on their surface. This trend may be influenced by the identification by *Brown et al.* (2007) of a possible family for which they note a correlation between strong water content and a dynamical cluster. They proposed that a giant impact on 2003 EL_{61} could have ejected a large fraction of its original mantle and formed its multiple satellite system and could also have ejected fragments of the original mantle in a clustered dynamical zone. Barkume et al. (in preparation) have shown that 65% of TNOs in their observed sample (which was not restricted to the dynamical cluster) show no definitive absorptions due to water ice. This is represented in Fig. 2 for the objects 2003 VS_2 and 1996 TP_{66}, in which the poorer signal precision may hide small signatures. No correlations have been found between ice abundance and perihelion distance.

The results presented herein also show that all large objects have the presence of some ices on their surface. This indicates that space weathering and collisional resurfacing are not the only mechanisms that can explain the surface properties and composition. It is clear that the largest TNOs can retain materials of higher albedo (ice or frost), but they can also have an outgassing from the interior. A thin atmosphere could be formed around some of these objects, as is the case for Pluto, with seasonal sublimation and recondensation of the most volatile species that would lead to surface modifications.

The variety of surface compositions among TNOs has often been attributed to different collisional evolution states and/or different degrees of surface alteration due to space weathering. In fact, it is known from laboratory experiments (chapter by Hudson et al.; *Brunetto et al.,* 2006) that bombardment by high-energy radiation of ices produces an "irradiation mantle" that is hydrogen-poor and carbon-rich, reddens the spectra, and destroys the ices' bands. If space weathering is the agent of the formation of a mantle, which hides the presence of ices on the surface, this process does not appear to work on the larger objects and other mechanisms have to be investigated.

The lack of correlation between color and water-ice absorption is most likely an indication that water ice on the surfaces of these TNOs is intimately mixed with the coloring agent. The physical reason for the variation in color or the variation in water-ice absorption is nonetheless unknown. Some methanol-like spectra exist, and this implies a chemically primitive nature for these objects, although it is not clear why only these three specific objects show methanol. Energetic impacts have certainly played an important role on the history of this population and may be the cause of at least some of the compositional surface variations.

This field benefits from continuous new discoveries. The investigation of the surface properties of TNOs is still relatively recent and its evolution will depend on the telescopes and instrumentation. A new large program is just started at VLT-ESO to observe an extended sample of objects with higher spectral quality. To investigate the surface of this faint population, albedo remains one of the fundamental parameters to be determined with accuracy. More accurate optical constants are also necessary.

In the future, the James Webb Space Telescope promises to further the exploration of TNOs and Centaurs from space (see chapter by Stansberry et al.). A possible extended mission for New Horizons (see chapter by Weaver et al.), after its encounter with Pluto in 2015, is to fly by one or more TNOs. Such an extended mission would be an exciting opportunity for an up-close look at samples of this population, providing ground truth for the results summarized in this chapter.

REFERENCES

Altenhoff W. J., Bertoldi F., and Menten K. M. (2004) Size estimates of some optically bright KBOs. *Astron. Astrophys., 415,* 771–775.

Barkume K. M., Brown M. E., and Shaller E. L. (2006) Water ice on the satellite of Kuiper belt object 2003 EL61. *Astrophys. J. Lett., 640,* L87–L89.

Barucci M. A. and Peixinho N. (2006) Trans-Neptunian objects' surface properties. In *Asteroids, Comets, and Meteors* (D. Lazzaro et al., eds.), pp. 171–190. IAU Symposium No. 229.

Barucci M. A., de Bergh C., Cuby J. G., Le Bras A., Schmitt B., and Romon J. (2000) Infrared spectroscopy of the Centaurs 8405 Asbolus: First observation at ESO-VLT. *Astron. Astrophys., 357,* L53–L56.

Barucci M. A., Boehnhardt H., Dotto E., Doressoundiram A., Romon J., Lazzarin M., Fornasier S., de Bergh C., Tozzi G. P., Delsanti A., Hainaut O., Barrera L., Birkle K., Meech K., Ortiz J. L., Sekiguchi T., Thomas N., Watanabe J., West R. M., and Davies J. K. (2002a) Visible and near-infrared spectroscopy of the Centaur 32532 (2001 PT_{13}). ESO Large Programme on Trans-Neptunian Objects and Centaurs: First spectroscopy results. *Astron. Astrophys., 392,* 335–339.

Barucci M. A., Cruikshank D. P., Mottola S., and Lazzarin M. (2002b) Physical properties of Trojan and Centaur asteroids. In *Asteroids III* (W. F. Bottke Jr. et al., eds.), p. 273. Univ. of Arizona, Tucson.

Barucci M. A., Dotto E., Brucato J. R., Muller T. G., Morris P., Doressoundiram A., Fulchignoni M., De Sanctis M. C., Owen T., Crovisier J., Le Bras A., Colangeli L., and Mennella B. (2002c) 10 Hygiea: ISO infrared observations. *Icarus, 156,* 202–210.

Barucci M. A., Cruikshank D. P., Dotto E., Merlin F., Poulet F., Dalle Ore C., Fornasier S., and de Bergh C. (2005a) Is Sedna another Triton. *Astron. Astrophys., 439,* L1–L4.

Barucci M. A., Belskaya I. N., Fulchignoni M., and Birlan M. (2005b) Taxonomy of Centaurs and Trans-Neptunian objects. *Astron. J., 130,* 1291–1298.

Barucci M. A., Merlin F., Dotto E., Doressoundiram A., and de Bergh C. (2006) TNO surface ices. Observations of the TNO 55638 (2002 V395) and analysis of the population's spectral properties. *Astron. Astrophys., 455,* 725–730.

Bauer J. M., Meech K. J., Fernández Y. R., Farnham T. L., and Roush T. L. (2002) Observations of the Centaur 1999 UG_5: Evidence of a unique outer solar system surface. *Publ. Astron. Soc. Pac., 114,* 1309–1321.

Bertoldi F., Altenhoff W., Weiss A., Menten K. M., and Albrecht M. (2006) The trans-neptunian object UB313 is larger than Pluto. *Nature, 439,* 563–564.

Boehnhardt H., Bagnulo S., Muinonen K., Barucci M. A., Kolokolova L., et al. (2004) Surface characterization of 28978 Ixion (2001 KX76). *Astron. Astrophys., 415,* L21–L25.

Brown M. E. (2000) Near-Infrared spectroscopy of Centaurs and irregular satellites. *Astron. J., 119,* 977–983.

Brown M. E. (2003) The composition of Kuiper belt objects. *Bull. Am. Astron. Soc., 35,* 969.

Brown M. E. and Calvin W. M. (2000) Evidence for crystalline water and ammonia ices on Pluto's satellite Charon. *Science, 287,* 107–109.

Brown M. E. and Koresko C. C. (1998) Detection of water ice on the Centaur 1997 CU26. *Astrophys. J. Lett., 505,* L65–L67.

Brown M. E. and Trujillo C. (2004) Direct measurement of the size of the large Kuiper belt object (50000) Quaoar. *Astrophys. J., 127,* 2413–2417.

Brown M. E., Blake G. A., and Kessler J. E. (2000) Near-infrared spectroscopy of the bright Kuiper belt object 2000 EB173. *Astrophys. J. Lett., 543,* L163–L165.

Brown M. E., Trujillo C., and Rabinowitz D. L. (2005) Discovery of a planetary-sized object in the scattered Kuiper belt. *Astrophys. J. Lett., 635,* L97–L100.

Brown M. E., Schaller E. L., Roe H. G., Rabinovitz D. L., and Trujillo C. A. (2006) Direct measurement of the size of 2003 UB313 from the Hubble Space telescope. *Astrophys. J. Lett., 643,* L61–L63.

Brown M. E., Barkume K. M., Ragozzine D., and Schaller E. L. (2007) A collisional family of icy objects in the Kuiper belt. *Nature, 446,* 294–296.

Brown R. H., Cruikshank D. P., and Pendleton Y. (1997) Surface composition of Kuiper belt object 1993 SC. *Science, 276,* 973–939.

Brown R. H., Cruikshank D. P., and Pendleton Y. (1999) Water ice on Kuiper belt object 1996 TO66. *Astron. J. Lett., 519,* L101–L104.

Brunetto R., Barucci M. A., Dotto E., and Strazulla G. (2006) Ion Irradiation of frozen methanol, methane and benzene: Linking to the colors of Centaurs and Trans-Neptunian objects. *Astrophys. J., 644,* 646.

Chandrasekhar S. (1960) *Radiative Transfer.* Dover, New York.

Choi Y. J. and Weissman P. R. (2006) *IAU Circular No. 8656.*

Christensen P. R., Bandfield J. L., Hamilton V. E., Howard D. A., Lane M. D., Piatek J. L., Ruff S. W., and Stefanov W. L. (2000) A thermal emission spectral library of rock forming minerals. *J. Geophys. Res., 105(E4),* 9735–9739.

Cooper B. L., Salisbury J. W., Killen R. M., and Potter A. E. (2002) Mid-infrared spectral features of rocks and their powders. *J. Geophys. Res., 107(E4),* 1–12.

Cruikshank D. P., Roush T. L., Bartholomew M. J., Geballe T. R., Pendleton Y. J., White S. M., Bell J. F. III, Davies J. K., Owen T. C., de Bergh C., Tholes D. I., Bernstein M. P., Brown R. H., Tryka K. A., and Dalle Ore C. M. (1998) The composition of Centaur 5145 Pholus. *Icarus, 135,* 389–407.

Cruikshank D. P., Dalle Ore C. M., Roush T. L., Geballe T. R., Owen T. C., de Bergh C., Cash M. D., and Hartmann W. K. (2001) Constraints on the composition of Trojan asteroid 624 Hektor. *Icarus, 153,* 348–360.

Cruikshank D. P., Stansberry J. A., Emery J. P., Fernández Y. R., Werner M. W., et al. (2005) The high albedo Kuiper object (55565) 2002 AW197. *Astrophys. J. Lett., 624,* L53–L56.

Cruikshank D. P., Barucci M. A., Emery J. P., Fernández Y. R., Grundy W. M., Noll K. S., and Stansberry J. A. (2007) Physical properties of trans-Neptunian objects. In *Protostars and Planets V* (B. Reipurth et al., eds.), pp. 879–893. Univ. of Arizona, Tucson.

de Bergh C., Boehnhardt H., Barucci M. A., Lazzarin M., Fornasier S., Romon-Martin J., Tozzi G. P., Doressoundiram A., and Dotto E. (2004) Aqueous altered silicates in the surface of two Plutinos? *Astron. Astrophys., 416,* 791–798.

de Bergh C., Delsanti A., Tozzi G. P., Dotto E., Doressoundiram A., and Barucci M. A. (2005) The surface of the trans-Neptunian object 90482 Orcus. *Astron. Astrophys., 437,* 1115–1120.

Doressoundiram A., Tozzi G. P., Barucci M. A., Boehnhardt H., Fornasier S., Romon J. (2003) ESO Large Programme on Trans-Neptunian Objects and Centaurs: Spectroscopic investigation of Centaur 2001 BL41 and TNOs (26181) 1996 GQ21 and 26375 1999 DE9. *Astron. J., 125,* 2721–2727.

Doressoundiram A., Barucci M. A., Tozzi G. P., Poulet F., Boehnhardt H., de Bergh C., and Peixinho N. (2005) Spectral characteristics and modelling of the trans-Neptunian object (55565) 2002 AW197 and the Centaurs (55576) 2002 GB10 and (83982) 2002 GO9. *Planet. Space Sci., 53,* 1501–1509.

Dotto E., Barucci M. A., Müller T. G., Brucato J. R., Fulchignoni M., Mennella V., and Colangeli L. (2002) ISO observations of low and moderate albedo asteroids: PHT-P and PHT-S results. *Astron. Astrophys., 393,* 1065–1072.

Dotto E., Barucci M. A., Boehnhardt H., Romon J., Doressoundiram A., Peixinho N., de Bergh C., and Lazzarin M. (2003a) ESO Large Programme on Trans-Neptunian Objects and Centaurs: Searching for water ice on 47171 1999 TC36, 1998 SG35, and 2000 QC243. *Icarus, 162,* 408–414.

Dotto E., Barucci M. A., Leyrat C., Romon J., de Bergh C., and Licandro J. (2003b) Unveiling the nature of 10199 Chariklo: Near-infrared observations and modelling. *Icarus, 164,* 122–126.

Dotto E., Barucci M. A., Brucato J. R., Müller T. G., and Carvano J. (2004) Polyxo: ISO-SWS spectrum up to 26 micron. *Astron. Astrophys., 427,* 1081–1084.

Doute S. and Shmitt B. (1998) A multilayer bidirectional reflectance model for the analysis of planetary surface hyperspectral images at visible and near-infrared wavelengths. *J. Geophys. Res., 103(E13),* 31367–31390.

Doute S., Schmitt B., Quirico E., Owen T. C., Cruikshank D. P., de Bergh C., Geballe T. R., and Roush T. L. (1999) Evidence for methane segregation at the surface of Pluto. *Icarus, 142,* 421–444.

Dumas C., Hainault O., Merlin F., Barucci M. A., de Bergh C., Guilbert A., Vernazza P., and Doressoundiram A. (2007) Surface composition of the largest Dwarf Planet 136199 Eris (2003 UB313). *Astron. Astrophys., 471,* 331–334.

Emery J. P. and Brown R. H. (2004) The surface composition of Trojan asteroids: Constraints set by scattering theory. *Icarus, 170,* 131–152.

Emery J. P., Cruikshank D. P., and Van Cleve J. (2006) Thermal emission spectroscopy (5.2–38 μm) of three Trojan asteroids with the Spitzer Space Telescope: Detection of fine-grained silicates. *Icarus, 182,* 496–512.

Fernandez Y. R., and 9 colleagues (2001) *IAU Circular No. 7689,* 3.

Fernandez Y. R., Jewitt D. C., and Sheppard S. S. (2002) Thermal properties of Centaurs Asbolus and Chiron. *Astron. J., 123,* 1050–1055.

Fornasier S., Dotto E., Barucci M. A., and Barbieri C. (2004a) Water ice on the surface of the large TNO 2004 DW. *Astron. Astrophys., 422,* L43–L46.

Fornasier S., Doressoundiram A., Tozzi G. P., Barucci M. A., Boehnhardt H., de Bergh C., Delsanti A., Davies J., and Dotto E. (2004b) ESO Large Programme on Trans-Neptunian Objects and Centaurs: Final results of the visible spectrophotometric observations. *Astron. Astrophys., 421,* 353–363.

Foster M. J., Green S. F., McBride N., and Davies J. K. (1999)

Detection of water ice on 2060 Chiron. *Icarus, 141,* 408–410.

Grundy W. M. and Fink U. (1996) Synoptic CCD spectrophotometry of Pluto over the past 15 years. *Icarus, 124,* 329–343.

Grundy W. M. and Schmitt B. (1998) The temperature-dependent near-infrared absorption spectrum of hexagonal H_2O ice. *J. Geophys. Res., 103(E11),* 25809–25822.

Grundy W. M., Buie M. W., and Spencer J. R. (2005) Near-infrared spectrum of low-inclination classical Kuiper belt object (79360) 1997 CS_{29}. *Astron. J., 130,* 1299–1301.

Gulbis A. A. S., Elliot J. L., Person M. J., Adams E. R., Babcock B. A., et al. (2006) Charon's radius and atmospheric constraints from observations of a stellar occultation. *Nature, 439,* 48–51.

Hansen G. B. and McCord T. B. (2004) Amorphous and crystalline ice on the Galilean satellites: A balance between thermal and radiolytic processes. *J. Geophys. Res., 109,* E01012, 1–19.

Hapke B. (1966) An improved lunar theoretical photometric function. *Astron. J., 71,* 333–339.

Hapke B. (1981) Bidirectional reflectance spectroscopy I, Theory. *J. Geophys. Res., 86(B4),* 3039–3054.

Hapke B. (1984) Bidirectional reflectance spectroscopy III, Correction for macroscopic roughness. *Icarus, 59,* 41–59.

Hapke B. (1986) Bidirectional reflectance spectroscopy IV, The extinction coefficient and the opposition effect. *Icarus, 67,* 264–280.

Hapke B. (1996) A model of radiative and conductive energy transfer in planetary regoliths. *J. Geophys. Res., 101(E7),* 16817–16831.

Hapke B. and Wells E. (1981) Bidirectional reflectance spectroscopy II, Experiments and observations. *J. Geophys. Res., 86(B4),* 3055–3060.

Hapke B., Dimucci D., Nelson R., and Smythe W. (1997) Bidirectional reflectances of regoliths with grain sizes of the order of the wavelength (abstract). In *Lunar and Planetary Science XXVIII,* pp. 513–514.

Henyey C. and Greenstein J. (1941) Diffuse radiation in the galaxy. *Astron. J., 93,* 70–83.

Houck J. and 34 colleagues (2004) The Infrared Spectrograph on the Spitzer Space Telescope. *Astrophys. J. Suppl. Ser., 154,* 18–24.

Jewitt D. and Luu J. (2001) Colors and spectra of Kuiper belt objects. *Astron. J., 122,* 2099–2114.

Jewitt D. and Luu J. (2004) Crystalline water ice on the Kuiper belt object (50000) Quaoar. *Nature, 432,* 731–733.

Kern S. D., McCarthy D. W., Buie M. W., Brown R. H., Campins H., and Rieke M. R. (2000) Compositional variation on the surface of Centaur 8405 Asbolus. *Astrophys. J., 542,* L155–L159.

Lazzarin M., Barucci M. A., Boehnhardt H., Tozzi G. P., de Bergh C., and Dotto E. (2003) ESO Large Programme on Physical Studies of Trans-Neptunian Objects and Centaurs: Visible spectroscopy. *Astron. J., 125,* 1554–1558.

Licandro J. and Pinilla-Alonso N. (2005) The inhomogeneous surface of Centaur 32522 Thereus (2001 PT_{13}). *Astrophys. J., 630,* L93–L96.

Licandro J., Oliva E., and Di Martino M. (2001) NICS-TNG infrared spectroscopy of trans-neptunian 2000 EB173 and 2000 WR106. *Astron. Astrophys., 373,* L29–L32.

Licandro J., Ghinassi F., and Testi L. (2002) Infrared spectroscopy of the largest known trans-Neptunian object 2001 KX_{75}. *Astron. Astrophys., 388,* L9–L12.

Licandro J., Pinilla-Alonso N., Pedani M., Oliva G., Tozzi G., and Grundy W. (2006a) Methane ice rich surface of large TNO 2005 FY9. *Astron. Astrophys., 445,* L35–L38.

Licandro J., di Fabrizio L., Pinilla-Alonso N., de Leon J., and Oliva E. (2006b) Trans-Neptunian object (55636) 2002 TX300, a fresh icy surface in the outer solar system. *Astron. Astrophys., 457,* 329–333.

Luu J. X. and Jewitt D. C. (1990) Cometary activity in 2001 Chiron. *Astron. J., 100,* 913–932.

Luu J. and Jewitt D. (1996) Color diversity among the Centaurs and Kuiper belt objects. *Astron. J., 112,* 2310.

Luu J. and Jewitt D. (1998) Optical and infrared reflectance spectrum of Kuiper belt object 1996 TL66. *Astrophys. J. Lett., 494,* L117–L120.

Luu J., Jewitt D., and Trujillo C. (2000) Water ice in 2060 Chiron and its implications for Centaurs and Kuiper belt objects. *Astrophys. J. Lett., 531,* L151–L154.

Merlin F., Barucci M. A., Dotto E., de Bergh C., and Lo Curto G. (2005) Search for surface variations on TNO 47171 and Centaur 32532. *Astron. Astrophys., 444,* 977–982.

Merlin F., Dumas C., Barucci M. A., Hainaut O., de Bergh C., and Guilbert A. (2006) Spectroscopic analysis of the bright TNOs 2003 UB_{313} and 2003 EL_{61}. *Bull. Am. Astron. Soc., 38,* 556.

Millis R. L., Wasserman L. H., Franz O. G., Nye R. A., Elliot J. L., Dunham E. W., Bosh A. S., Young L. A., Silvan S. M., and Gilmore A. C. (1993) Pluto's radius and atmosphere — Results from the entire 9 June 1988 occultation data set. *Icarus, 105,* 282–297.

Mishchenko M. I. (1994) Asymmetry parameters of the phase function for densely packed scattering grains. *J. Quant. Spectros. Rad. Trans., 52,* 95–110.

Morbidelli A. and Brown M. (2004) The Kuiper belt and the primordial evolution of the solar system. In *Comets II* (M. C. Festou et al., eds.), pp. 175–192. Univ. of Arizona, Tucson.

Noll K. S., Luu J., and Gilmore D. (2000). Spectrophotometry of four Kuiper belt objects with NICMOS. *Astron. J., 119,* 970–976.

Owen T. C., Roush T. L., Cruikshnak D. P., Elliot J. L., Young L. A., de Bergh C., Schmitt B., Geballe T. R., Brown R. H., and Bartholomew M. J. (1993) Surface ices and the atmospheric composition of Pluto. *Science, 261,* 745–748.

Piatek J. L., Hapke B., Nelson R. M., Hale A. S., and Smythe W. D. (2003) Size-dependent measurements of the scattering properties of planetary regolith analogs: A challenge to theory (abstract). In *Lunar and Planetary Science XXXIV,* Abstract #1440.

Poulet F., Cuzzi J. N., Cruikshank D. P., Roush T. and Dalle Ore C. M. (2002) Comparison between the Shkuratov and Hapke scattering theories for solid planetary surfaces: Application to the surface composition of two Centaurs. *Icarus, 160,* 313–324.

Pravdo S., Helim E. F., Hicks M., and Lawrence K. (2001) *IAU Circular No. 7738.*

Press W. H., Flannery B. P., Teulolsky S. A., and Vetterling W. T. (1986) *Numerical Recipes.* Cambridge Univ., Cambridge.

Quirico E., Doute S., Schmitt B., de Bergh C., Cruikshank D. P, Owen T. C., Geballe T. R., and Roush T. L. (1999) Composition state and distribution of ices at the surface of Triton. *Icarus, 139,* 159–178.

Rabinowitz D. L., Barkume K., Brown M. E., Roe H., Schawartz M., Tourtellotte S., and Trujillo C. (2006) Photometric observations constraining the size, shape, and albedo of 2003 EL61, a rapidly rotating, Pluto sized object in the Kuiper belt. *Astrophys. J., 639,* 1238–1251.

Romon-Martin J., Barucci M. A., de Bergh C., Doressoundiram A., Peixinho N., and Poulet F. (2002) Observations of Centaur 8405 Asbolus: Searching for water ice. *Icarus, 160,* 59–65.

Romon-Martin J., Delahodde C., Barucci M. A., de Bergh C., and Peixinho N. (2003) Photometric and spectroscopic observations of (2060) Chiron at the ESO Very Large Telescope. *Astron. Astrophys., 400,* 369–373.

Salisbury J. W., Walter L. S., Vergo N., and D'Aria D. M. (1992). *Mid-Infrared (2.1–25 μm) Spectra of Minerals.* Johns Hopkins Univ., Baltimore.

Sheppard S. S. and Jewitt D. C. (2002) Time-resolved photometry of Kuiper belt objects: Rotations, shapes, and phase functions. *Astron. J., 124,* 1757–1775.

Shkuratov Y., Starukhina L., Hoffmann H., and Arnold G. (1999) A model of spectral albedo of particulate surface: Implications for optcical properties of the moon. *Icarus, 137,* 235–246.

Sicardy B., Bellucci A., Gendro E., Lacombe F., Lacour S., et al. (2006) Charon's size and an upper limit on its atmosphere from a stellar occultation. *Nature, 439,* 52–54.

Tholen D. J. and Buie M. W. (1989) Further analysis of Pluto-Charon mutual event observations. *Bull. Am. Astron. Soc., 21,* 981.

Thomas N., Eggers S., Ip W.-H., Lichtenberg G., Fitzsimmons A., et al. (2000) Observations of the transneptunian objects 1993 SC and 1996 TL66 with the Infrared Space Observatory. *Astrophys. J., 534,* 446–455.

Trujillo C. A., Brown M. E., Rabinowitz D. L., and Geballe T. R. (2005) Near infrared surface properties of the two intrinsically brightest minor planets: (90377) Sedna and (90482) Orcus. *Astrophys. J., 627,* 1057–1065.

Trujillo C. A., Brown M. E., Barkume K. M., Schaller E. L., and Rabinowitz D. L. (2007) The surface of 2003 EL61 in the near infared. *Astrophys. J., 655,* 1172–1178.

Vilas F. and Gaffey M. J. (1989) Phyllosilicate absorption features in main-belt and outer-belt asteroid reflectance spectra. *Science, 246,* 790–792.

Werner M. W. and 25 colleagues (2004) The Spitzer Space Telescope mission. *Astrophys. J. Suppl. Ser., 154,* 1–9.

Witteborn F. C. and Roush T. L. (2000) Thermal emission spectroscopy of 1 Ceres: Evidence for olivine. In *Thermal Emission Spectroscopy and Analysis of Dust, Disks, and Regoliths* (M. L. Sitko et al., eds.), pp. 197–203. ASP Conference Series 196, Astronomical Society of the Pacific, San Francisco.

Physical Properties of Kuiper Belt and Centaur Objects: Constraints from the Spitzer Space Telescope

John Stansberry
University of Arizona

Will Grundy
Lowell Observatory

Mike Brown
California Institute of Technology

Dale Cruikshank
NASA Ames Research Center

John Spencer
Southwest Research Institute

David Trilling
University of Arizona

Jean-Luc Margot
Cornell University

Detecting heat from minor planets in the outer solar system is challenging, yet it is the most efficient means for constraining the albedos and sizes of Kuiper belt objects (KBOs) and their progeny, the Centaur objects. These physical parameters are critical, e.g., for interpreting spectroscopic data, deriving densities from the masses of binary systems, and predicting occultation tracks. Here we summarize Spitzer Space Telescope observations of 47 KBOs and Centaurs at wavelengths near 24 and 70 μm. We interpret the measurements using a variation of the standard thermal model (STM) to derive the physical properties (albedo and diameter) of the targets. We also summarize the results of other efforts to measure the albedos and sizes of KBOs and Centaurs. The three or four largest KBOs appear to constitute a distinct class in terms of their albedos. From our Spitzer results, we find that the geometric albedo of KBOs and Centaurs is correlated with perihelion distance (darker objects having smaller perihelia), and that the albedos of KBOs (but not Centaurs) are correlated with size (larger KBOs having higher albedos). We also find hints that albedo may be correlated with visible color (for Centaurs). Interestingly, if the color correlation is real, redder Centaurs appear to have higher albedos. Finally, we briefly discuss the prospects for future thermal observations of these primitive outer solar system objects.

1. INTRODUCTION

The physical properties of Kuiper belt objects (KBOs) remain poorly known nearly 15 years after the discovery of (15760) 1992 QB_1 (*Jewitt and Luu,* 1993). While KBOs can be discovered, their orbits determined, and their visible-light colors measured (to some extent) using modest telescopes, learning about fundamental properties such as size, mass, albedo, and density remains challenging. Determining these properties for a representative sample of transneptunian objects (TNOs) is important for several reasons. Estimating the total mass of material in the transneptunian region and relating visible magnitude frequency distributions to size- and mass-frequency is uncertain, at best. Quantitative interpretation of visible and infrared spectra is impossible without knowledge of the albedo in those wavelength ranges. Size estimates, when coupled with masses determined for binary KBOs (see the chapter by Noll et al.), constrain the density, and hence internal composition and structure, of these objects. All these objectives have important implications for physical and chemical conditions in the outer protoplanetary nebula, for the accretion of solid objects in the outer solar system, and for the collisional evolution of KBOs themselves. Of course, there is a relative wealth of information about Pluto and Charon, the two longest known KBOs, and we do not address their properties further here.

The Centaur objects, with orbits that cross those of one or more of the giant planets, are thought to be the dynamical progeny of KBOs (e.g., *Levison and Duncan,* 1997). The Centaurs are particularly interesting both because of their direct relation to KBOs, and also because their orbits bring them closer to the Sun and to observers, where, for a given size, they are brighter at any wavelength than their more distant relatives. Because of their planet-crossing orbits, the dynamical lifetimes of Centaurs are relatively short, typically a few million years (e.g., *Horner et al.,* 2004).

The sizes of some KBOs and Centaurs have been determined by a variety of methods. Using the Hubble Space Telescope (HST), *Brown and Trujillo* (2004) resolved the KBO 50000 Quaoar, placed an upper limit on the size of Sedna (*Brown et al.,* 2004), and resolved Eris (*Brown et al.,* 2006). Recently *Rabinowitz et al.* (2005) placed constraints on the size and albedo of 136199 (2003 EL_{61}) based on its short rotation period (3.9 h) and an analysis of the stability of a rapidly rotating ellipsoid. *Trilling and Bernstein* (2006) performed a similar analysis of the lightcurves of a number of small KBOs, obtaining constraints on their sizes and albedos. Advances in the sensitivity of far-IR and submillimeter observatories have recently allowed the detection of thermal emission from a sample of outer solar system objects, providing constraints on their sizes and albedos. *Jewitt et al.* (2001), *Lellouch et al.* (2002), *Margot et al.* (2002, 2004), *Altenhoff et al.* (2004), and *Bertoldi et al.* (2006) have reported submillimeter to millimeter observations of thermal emission from KBOs. *Sykes et al.* (1991, 1999) analyze Infrared Astronomical Satellite (IRAS) thermal detections of 2060 Chiron and the Pluto-Charon system, determining their sizes and albedos. Far-infrared data from the Infrared Space Observatory (ISO) were used to determine the albedos and diameters of KBOs 15789 (1993 SC), 15874 (1996 TL_{66}) (*Thomas et al.,* 2000), and 2060 Chiron (*Groussin et al.,* 2004). *Lellouch et al.* (2000) studied the thermal state of Pluto's surface in detail using ISO. *Grundy et al.* (2005) provide a thorough review of most of the above, and include a sample of binary KBO systems with known masses, to constrain the sizes and albedos of 20 KBOs.

Spitzer Space Telescope (Spitzer hereafter) thermal observations of KBOs and Centaurs have previously been reported by *Stansberry et al.* (2004) (29P/Schwassmann-Wachmann 1), *Cruikshank et al.* (2005) (55565 2002 AW_{197}), *Stansberry et al.* (2006) (47171 1999 TC_{36}), *Cruikshank et al.* (2006), *Grundy et al.* (2007) (65489 2003 FX_{128}), and Grundy et al. (in preparation) (42355 Typhon). Here we summarize results from several Spitzer programs to measure the thermal emission from 47 KBOs and Centaurs. These observations place secure constraints on the sizes and albedos of 42 objects, some overlapping with determinations based on other approaches mentioned above. We present initial conclusions regarding the relationship between albedo and orbital and physical properties of the targets, and discuss future prospects for progress in this area.

2. THERMAL MODELING

Measurements of thermal emission can be used to constrain the sizes, and thereby albedos, of unresolved targets. *Tedesco et al.* (1992; 2002) used IRAS thermal detections of asteroids to build a catalog of albedos and diameters. Visible observations of the brightness of an unresolved object are inadequate to determine its size, because that brightness is proportional to the product of the visible geometric albedo, p_V, and the cross-sectional area of the target. Similarly, the brightness in the thermal IR is proportional to the area, and is also a function of the temperature of the surface, which in turn depends on the albedo. Thus, measurements of both the visible and thermal brightness can be combined to solve for both the size of the target and its albedo. Formally the method requires the simultaneous solution of the following two equations

$$F_{vis} = \frac{F_{\odot,vis}}{(r/1\ AU)^2} R^2 p_V \frac{\Phi_{vis}}{\Delta^2} \tag{1a}$$

$$F_{ir} = \frac{R^2\Phi_{ir}}{\pi\Delta^2} \varepsilon \int B_\lambda(T(\theta,\phi)) \sin\theta d\theta d\phi \tag{1b}$$

where F is the measured flux density of the object at a wavelength in the visible ("vis") or thermal-infrared ("ir"); $F_{\odot,vis}$ is the visible-wavelength flux density of the Sun at 1 AU; r and Δ are the object's heliocentric and geocentric distances, respectively; R is the radius of the body (assumed to be spherical); p_V is the geometric albedo in the visible; Φ is the phase function in each regime; B_λ is the Planck function; and ε is the infrared bolometric emissivity. $T = T(p_V q, \eta, \varepsilon, \theta, \phi)$ is the temperature, which is a function of p_V; ε; the "beaming parameter," η; surface planetographic coordinates θ and ϕ; and the (dimensionless) phase integral, q (see below for discussions of η and q).

In practice, the thermal flux depends sensitively on the temperature distribution across the surface of the target, and uncertainties about that temperature distribution typically dominate the uncertainties in the derived albedos and sizes (see Fig. 1). Given knowledge of the rotation vector, shape, and the distribution of albedo and thermal inertia, it is in principle possible to compute the temperature distribution. Not surprisingly, none of these things are known for a typical object where we seek to use the radiometric method to measure the size and albedo. The usual approach is to use a simplified model to compute the temperature distribution based on little or no information about the object's rotation axis or even rotation period.

2.1. Standard Thermal Model

The most commonly employed model for surface temperature on asteroidal objects is the standard thermal model (STM) (cf. *Lebofsky and Spencer,* 1989, and references therein). The STM assumes a nonrotating (or equivalently, zero thermal inertia) spherical object, and represents the "hot" end member to the suite of possible temperature dis-

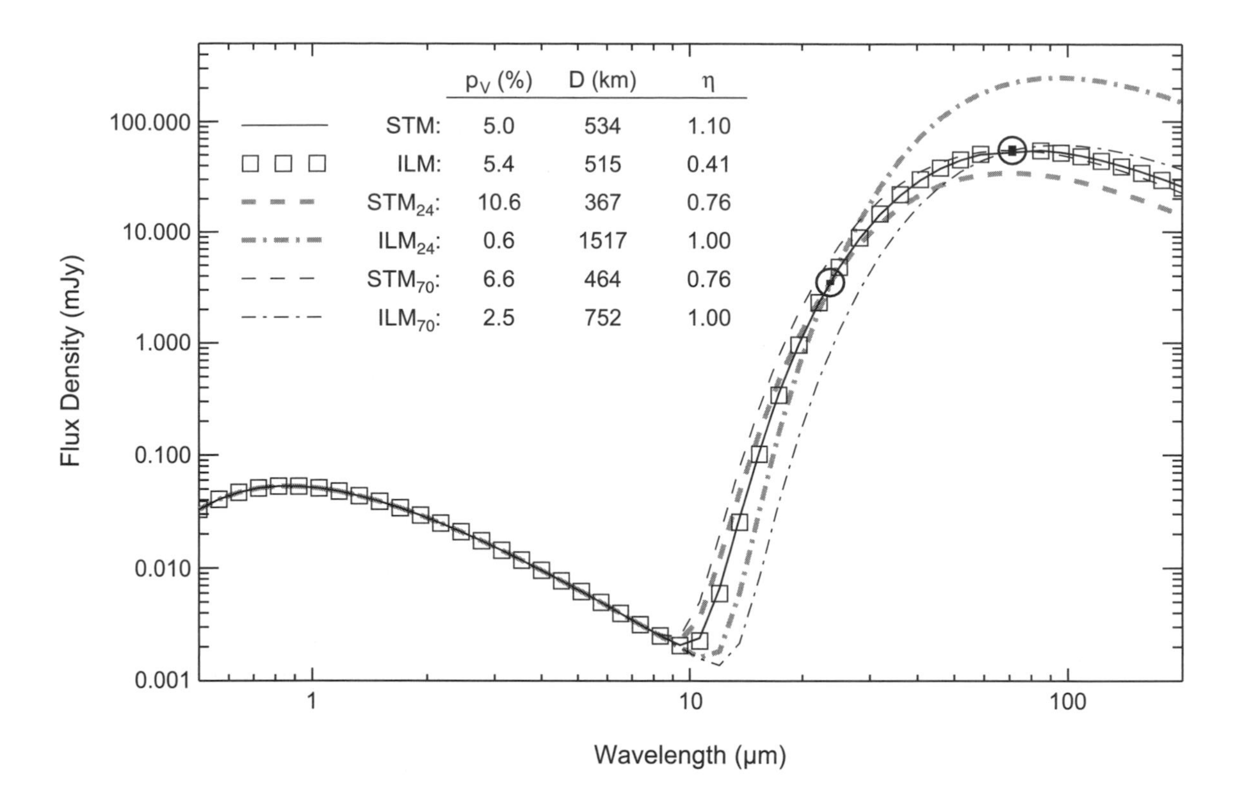

Fig. 1. Thermal models for KBO 38628 Huya (2000 EB_{173}). Spitzer Space Telescope 24- and 70-μm data are shown as circles, with vertical error bars within them indicating the measurement uncertainties. Six models are fit to the data, with the resulting model albedos, diameters, and beaming parameters summarized in the legend. From top to bottom the models are: (1) hybrid STM fit to 24- and 70-μm data, with η as a free parameter (the therm model used here); (2) hybrid ILM fit to 24- and 70-μm data; (3) canonical STM (η = 0.756) fit to the 24-μm data; (4) canonical ILM (η = 1.0) fit to the 24-μm data; (5) canonical STM fit to the 70-μm data; (6) canonical ILM fit to the 70-μm data. Note the close agreement of the albedos and sizes for models 1 and 2. Fits to data from one band, using the canonical asteroid values for η, result in much larger uncertainties in the derived parameters, particularly the fits to the 24-μm data.

tributions. Under STM assumptions, the dayside temperature depends only on the angular distance from the subsolar point, θ: $T(\theta) = T_0\cos^{1/4}\theta$, and the temperature is zero on the nightside. The subsolar point temperature $T_0 = [(1 - A)S/(\eta\varepsilon\sigma)]^{1/4}$. Here $A = qp_V$ is the bolometric albedo, S is the solar constant at the distance of the object, and σ is the Stefan-Boltzmann constant. Even though the STM represents the hottest reasonable distribution of surface temperatures for an object in radiative equilibrium with sunlight, early studies of the emission from asteroids showed that their emission was even hotter than predicted by the STM (*Jones and Morrison,* 1974; *Morrison and Lebofsky,* 1979). That led to the introduction of the beaming parameter, η, which allows for localized temperature enhancements on the dayside, e.g., in the bottoms of craters or other rough features, and the tendency of such warm regions to radiate preferentially in the sunward (and, for outer solar system objects, observerward) direction (i.e., to beam). (Note that while η appears analogously to the emissivity, ε, in the expression for the surface temperature, η does not appear explicitly in the expression for the thermal emission; equation (1b).) *Lebofsky et al.* (1986) derived a value of η = 0.756 based on 10-μm observations of Ceres and Pallas. We refer to the STM with η set to 0.756 as the *canonical STM.*

2.2. Isothermal Latitude Model

The cold end member of the suite of plausible temperature distributions for an object in radiative equilibrium with sunlight is the isothermal latitude model (ILM; also known as the fast-rotator model). The ILM assumes a spherical object illuminated at the equator and rotating very quickly (or equivalently, a slowly rotating object with infinite thermal inertia). The resulting temperature distribution depends only on latitude, ϕ: $T(\phi) = T_0\cos^{1/4}\phi$, where in this case the subsolar point temperature is given by $T_0 = [(1 - A)S/(\pi\eta\varepsilon\sigma)]^{1/4}$. The factor of π in this expression reduces the subsolar point temperature by 33% relative to the STM. Because the ILM is characterized by infinite thermal inertia, local temperature variations, and therefore beaming, are precluded; thus the *canonical ILM* assumes η = 1.

2.3. Hybrid Thermal Model

Figure 1 illustrates the problems inherent in using either the STM or the ILM to measure the sizes and albedos of KBOs. In particular, none of the four canonical STM or ILM models fit to either the 24- or 70-μm data (four lower elements in the figure legend) match the observed 24:70-μm color. As a result, the systematic uncertainties on the albedos and diameters, depending only on whether the STM or ILM is used, are large: p_V is uncertain by a factor >2.5 for the fits to the 70-μm data, and is uncertain by a factor of >17 for the fits to the 24-μm data. (Note, however, that the relative efficacy of these two wavelengths depends on the temperature of the target: If the thermal spectrum peaks near the 24-μm band, observations at that wavelength will be considerably more effective at constraining the physical properties of the target than indicated by this particular example.) However, if the beaming parameter, η, is allowed to be a free parameter of the fit (top two elements in the figure legend), both the color of the thermal emission and its intensity can be matched. More importantly, both the STM and ILM give nearly the same diameters and albedos with η as a free parameter. The basic reason for this is that the 24- and 70-μm data provide a direct determination of the temperature of the thermal emission from the object; equating that color temperature to the effective temperature gives a direct estimate of the size of the target, independent of the details of an assumed temperature distribution (and independent of the visual brightness as well).

While the beaming parameter was introduced to model enhanced localized dayside temperatures and infrared beaming, it can also mimic the effects of other influences on the temperature distribution, such as pole orientation (note that the emission from a pole-on ILM is indistinguishable from the STM), and intermediate rotation rates and thermal inertias. For example, a rotating body with nonzero thermal inertia will have lower dayside temperatures than predicted by the STM, but an STM with a value of η larger than would be supposed based on its surface roughness will have a similar color temperature. Likewise, a quickly rotating body with a low thermal inertia will have higher dayside temperatures than predicted by the ILM, an effect that can be mimicked by an ILM with $\eta < 1$.

Returning to the top two models in the legend of Fig. 1, the STM fit results in $\eta = 1.09$, suggesting that the temperature distribution on the target (the KBO 36828 Huya) is cooler than predicted by the canonical STM with $\eta = 0.756$. Likewise, for the ILM $\eta = 0.41$, suggesting that the surface is significantly hotter than would be predicted by the canonical ILM with $\eta = 1$.

2.4. Thermal Model: Application

In the following we adopt a thermal model in which the beaming parameter, along with size and albedo, are free parameters that we use to simultaneously fit observed flux densities at two thermal wavelengths, and the constraint imposed by the visual brightness of the object. Because such models have temperature distributions intermediate between the canonical STM and ILM, they can be thought of as a hybrid between the two. Furthermore, because the systematic uncertainties in the model albedos and diameters associated with the choice of hybrid STM or hybrid ILM are fairly small relative to the uncertainties in the measured flux densities and other model assumptions, we simply adopt the hybrid STM as our model of choice. (The error bars on p_V and D stemming from the choice of STM or ILM hybrid model in Fig. 1 are ≤4% and ≤2%.) We note that a number of studies have employed a similar approach with variable η (e.g., *Harris,* 1998; *Delbo et al.,* 2003; *Fernández et al.,* 2003).

In order to use the STM, we must make some assumptions regarding the nature of the thermal emission and visible scattering. We assume a gray emissivity, $\varepsilon = 0.9$. The infrared phase function, $\Phi_{ir} = 0.01$ mag/deg, depends only weakly on the emission angle. For our observations, emission angles for all but five targets (29P, Asbolus, Elatus, Thereus, and Okyrhoe) were <5°. Because the effects are small relative to other uncertainties in the models and data, we have neglected the IR phase effect for all the results presented here. We assume standard scattering behavior for the the objects in the visible, i.e., a scattering assymetry parameter, $G = 0.15$, leading to a phase integral $q = 0.39$ (*Bowell et al.,* 1989). This assumption also allows us to directly relate the geometric albedo p_V, the diameter D, and the absolute visual magnitude, H_V via $D = 1346\, p_V^{1/2} 10^{-H_V/5}$, where D is in km (*Bowell et al.,* 1989; *Harris,* 1998). By utilizing the absolute visual magnitude in this way, the scattering phase function, Φ_{vis}, apparently drops out; however, if the actual scattering behavior differs from the assumption above, our albedos and diameters will still be affected because the scattering behavior determines the value of q. We note, also, the results of *Romanishin and Tegler* (2005), who found that absolute magnitudes available through the IAU Minor Planet Center and through the Horizons service at the Jet Propulsion Laboratory have are biased downward (brighter) by 0.3 mag. The H_V values shown in Table 1 are culled from the photometric literature, and should be fairly reliable.

For low-albedo objects, the albedos we derive depend only weakly on the assumed value of q, while for high-albedo objects the value of q exerts a strong influence (see expressions for T_0 in sections 2.1 and 2.2). For the example of 38628 Huya (Fig. 1), changing to $q = 0.8$ makes only a ≤1% difference in the albedo. However, if we use $q = 0.39$ to model the data for the four largest objects in the sample — 90377 Sedna, 136199 Eris, 136108 (2003 EL_{61}), and 136472 (2005 FY_9) — we obtain geometric albedos that exceed a value of 2. While not (necessarily) unphysical, such high values for the geometric albedo are unprecedented. Pluto's phase integral $q = 0.8$, so for these four objects (only) we adopt that value instead.

2.5. Thermophysical Models

More sophisticated extensions to the STM and ILM include the effects of surface roughness and (nonzero, noninfinite) thermal inertia (*Spencer,* 1990), and viewing geometries that depart significantly from zero phase (*Harris,* 1998). However, for the purpose of determining KBO albedos and diameters from their thermal emission, the hybrid

TABLE 1. Orbital and photometric properties.

Number*	Designation*	Name*	a (AU)†	e†	i†	H_V‡	S‡	σ_S‡	TNO?§	Class¶
29P	Schwassmann-Wachmann 1		5.986	0.04	9.39	11.10	15.75	1.10	N	CENTR
2060	1977 UB	Chiron	13.690	0.38	6.93	6.58	1.85	1.18	N	CENTR
5145	1992 AD	Pholus	20.426	0.57	24.68	7.63	50.72	2.44	N	CENTR
7066	1993 HA_2	Nessus	24.634	0.52	15.65	9.7	34.03	9.25	N	CENTR
8405	1995 GO	Asbolus	17.986	0.62	17.64	9.15	19.88	8.58	N	CENTR
10199	1997 CU_{26}	Chariklo	15.865	0.18	23.38	6.66	12.95	1.38	N	CENTR
10370	1995 DW_2	Hylonome	25.202	0.25	4.14	9.41	9.29	2.28	N	CENTR
15820	1994 TB		39.288	0.31	12.14	8.00	40.92	2.87	Y	RESNT
15874	1996 TL_{66}		82.756	0.58	24.02	5.46	0.13	2.24	Y	SCTNR
15875	1996 TP_{66}		39.197	0.33	5.69	7.42	26.52	6.80	Y	RESNT
20000	2000 WR_{106}	Varuna	42.921	0.05	17.20	3.99	23.91	1.25	Y	CLSCL
26308	1998 SM_{165}		47.468	0.37	13.52	6.38	27.77	1.91	Y	RESNT
26375	1999 DE_9		55.783	0.42	7.62	5.21	20.24	3.46	Y	RESNT
28978	2001 KX_{76}	Ixion	39.648	0.24	19.59	3.84	22.90	1.60	Y	RESNT
29981	1999 TD_{10}		95.040	0.87	5.96	8.93	10.37	1.88	Y	CENTR
31824	1999 UG_5	Elatus	11.778	0.38	5.25	10.52	27.75	0.97	N	CENTR
32532	2001 PT_{13}	Thereus	10.617	0.20	20.38	9.32	10.79	0.96	N	CENTR
35671	1998 SN_{165}		37.781	0.04	4.62	5.72	5.05	1.95	Y	CLSCL
38628	2000 EB_{173}	Huya	39.773	0.28	15.46	5.23	22.20	4.80	Y	RESNT
42355	2002 CR_{46}	Typhon	38.112	0.54	2.43	7.65	15.87	1.93	Y	CENTR
47171	1999 TC_{36}		39.256	0.22	8.42	5.39	35.24	2.82	Y	RESNT
47932	2000 GN_{171}		39.720	0.29	10.80	6.2	24.78	3.41	Y	RESNT
50000	2002 LM_{60}	Quaoar	43.572	0.04	7.98	2.74	28.15	1.81	Y	CLSCL
52872	1998 SG_{35}	Okyrhoe	8.386	0.31	15.64	11.04	11.72	5.08	N	CENTR
52975	1998 TF_{35}	Cyllarus	26.089	0.38	12.66	9.01	36.20	2.42	N	CENTR
54598	2000 QC_{243}	Bienor	16.472	0.20	20.76	7.70	6.86	3.17	N	CENTR
55565	2002 AW_{197}		47.349	0.13	24.39	3.61	22.00	2.21	Y	SCTNR
55576	2002 GB_{10}	Amycus	25.267	0.40	13.34	8.07	32.13	4.35	N	CENTR
55636	2002 TX_{300}		43.105	0.12	25.87	3.49	-0.96	1.20	Y	SCTNR
55637	2002 UX_{25}		42.524	0.14	19.48	3.8	26.61	10.90	Y	SCTNR
60558	2000 EC_{98}	174P/Echeclus	10.771	0.46	4.33	9.55	10.43	4.83	N	CENTR
63252	2001 BL_{41}		9.767	0.29	12.45	11.47	14.37	2.75	N	CENTR
65489	2003 FX_{128}	Ceto	102.876	0.83	22.27	6.60	20.72	2.84	Y	CENTR
73480	2002 PN_{34}		30.966	0.57	16.64	8.66	16.21	1.90	Y	CENTR
83982	2002 GO_9	Crantor	19.537	0.28	12.77	9.16	42.19	4.43	N	CENTR
84522	2002 TC_{302}		55.027	0.29	35.12	4.1			Y	SCTNR
84922	2003 VS_2		39.273	0.07	14.79	4.4			Y	RESNT
90377	2003 VB_{12}	Sedna	489.619	0.84	11.93	1.8	33.84	3.62	Y	SCEXT
90482	2004 DW	Orcus	39.363	0.22	20.59	2.5	1.06	1.05	Y	RESNT
90568	2004 GV_9		42.241	0.08	21.95	4.2			Y	SCTNR
119951	2002 KX_{14}		39.012	0.04	0.40	4.6			Y	CLSCL
120061	2003 CO_1		20.955	0.48	19.73	9.29	12.93	1.90	N	CENTR
136108	2003 EL_{61}		43.329	0.19	28.21	0.5	−1.23	0.67	Y	SCTNR
136199	2003 UB_{313}	Eris	67.728	0.44	43.97	−1.1	4.48	4.63	Y	SCTNR
136472	2005 FY_9		45.678	0.16	29.00	0.0	10.19	2.25	Y	RESNT
	2002 MS_4		41.560	0.15	17.72	4.0			Y	SCTNR
	2003 AZ_{84}		39.714	0.17	13.52	3.71	1.48	1.01	Y	RESNT

*Small-body number, provisional designation, and proper name for the target sample.

†Orbital semimajor axis (a), eccentricity (e), and inclination (i).

‡Absolute visual magnitude (H_V) and spectral slope and uncertainty (S and σ_S, in % per 100 nm relative to V band), from the photometric literature.

§Orbital semimajor axis > that of Neptune (30.066 AU).

¶Deep Ecliptic Survey dynamical classification (*Elliot et al.*, 2005): CENTR = Centaur, CLSCL = classical, RESNT = resonant, SCTNR = scattered near, SCEXT = scattered extended.

STM gives results and uncertainties that are very similar to those obtained through application of such thermophysical models (e.g., *Stansberry et al.,* 2006). Because the hybrid STM is much simpler, and it produces results comparable to thermophysical models, we employ only the hybrid STM. (We note that thermophysical models are of significant interest for objects where the pole orientation and rotational period of the target are known, because such models can then constrain the thermal inertia, which is of interest in its own right.)

3. SPITZER OBSERVATIONS

Roughly 310 hours of time on the Spitzer have been allocated to attempts to detect thermal emission from KBOs and Centaurs, with the goal of measuring their albedos and diameters. Spitzer has a complement of three instruments, providing imaging capability from 3.6–160 μm, and low-resolution spectroscopy from 5 to 100 μm (*Werner et al.,* 2004). The long-wavelength imager, MIPS (Multiband Imaging Photometer for Spitzer) (*Rieke et al.,* 2004), has 24-, 70-, and 160-μm channels. Because of the placement of these channels, and the sensitivity of the arrays (which are at least 10× more sensitive than previous far-infrared satellites such as IRAS and ISO), MIPS is well suited to studying the thermal emission from KBOs.

3.1. The Sample

Spitzer has targeted over 70 KBOs and Centaurs with MIPS. About two-thirds of the observations have been successful at detecting the thermal emission of the target, although in some of those cases the detections have a low signal-to-noise ratio (SNR). Here we describe observations of 47 KBOs and Centaurs made during the first 3 years of the mission, focusing on observations of the intrinsically brightest objects (i.e., those with the smallest absolute magnitudes, H_V) and the Centaur objects. Table 1 summarizes the orbital and photometric properties of the sample.

The distribution of the objects in terms of dynamical class is also given, in two forms. The second to last column, labeled "TNO?," indicates whether the orbital semimajor axis is larger than Neptune's. By that measure, 31 of the objects are TNOs, and 17 are what might classically be called Centaur objects; that classification is nominally in agreement with the classification scheme proposed in the chapter by Gladman et al., although they classify Okyrhoe and Echeclus as Jupiter-family comets, rather than Centaurs. Another classification scheme has been proposed by *Elliot et al.* (2005) as a part of the Deep Ecliptic Survey (DES) study, and the target classification there under appears as the last column in Table 1. According to the DES classification, 21 of the targets in the Spitzer sample are Centaurs.

Thus, about 30–40% of the sample we discuss here are Centaurs, and the rest are KBOs. Among the KBOs, only 4 objects are classical, while 12 are in mean-motion resonances with Neptune, 9 are in the scattered disk, and one (90377 Sedna) is in the extended scattered disk: Classical objects are underrepresented. Because classical objects do not approach the Sun as closely as the resonant and scattered disk objects, and because they have somewhat fainter absolute magnitudes, the classical objects are at the edge of Spitzer capabilities. One Spitzer program has specifically targeted 15 of the classical objects, but data analysis is ongoing.

The visible photometric properties of the sample are diverse, and generally span the range of observed variation except in terms of the absolute magnitudes, which for the KBOs are generally $H_V \leq 7$. The spectral properties of KBOs and Centaurs are reviewed in the chapters by Barucci et al., Tegler et al., and Doressoundiram et al.: Here we summarize those characteristics as regards our sample. The visible colors, given in Table 1 as the spectral slope (measured relative to V), cover the range from neutral to very red (Pholus). Visible absorption features have been reported in the 0.6–0.75-μm region for 47932 (2000 GN_{171}), 38628 Huya, and (2003 AZ_{84}) (*Lazzarin et al.,* 2003; *de Bergh et al.,* 2004; *Fornasier et al.,* 2004). Several of the targets exhibit near-IR spectral features, with water and methane ices being the dominant absorbers identified. Water ice detections have been made for 10199 Chariklo, 83982 (2002 GO_9), 47171 (1999 TC_{36}), 47932 (2000 GN_{171}), 90482 Orcus, 50000 Quaoar, and 136108 (2003 EL_{61}). 55638 (2002 VE_{95}) exhibits methanol absorption, as does 5145 Pholus, along with its strong water ice absorption. Methane ice is clearly present on Eris (*Brown et al.,* 2005), and 136472 (2005 FY_9) (*Licandro et al.,* 2006a; *Tegler et al.,* 2007; *Brown et al.,* 2007), and Eris may also have N_2 ice (*Licandro et al.,* 2006b). Two of the objects exhibit surface heterogeneity: 31824 Elatus (*Bauer et al.,* 2003) and 32532 Thereus (*Barucci et al.,* 2002; *Merlin et al.,* 2005).

3.2. The Observations

Most of the targets in the sample presented here were observed in both the 24- and 70-μm channels of MIPS. In a few cases, when the target was predicted to be too faint to observe in the second channel, only one channel was used. Integration times vary significantly, ranging from 200–4000 s. As Spitzer observations of KBOs and Centaurs proceeded, it became clear that they were significantly harder to detect than had been predicted prior to the launch. The difficulty was due to a combination of worse-than-predicted sensitivity for the 70-μm array (by a factor of about 2), and the fact that KBOs are colder and smaller than assumed. As these realities made themselves evident, later observing programs implemented more aggressive observing strategies, and have generally been more successful than the early observations.

In some cases the same target was observed more than once. These observations fall into three categories: (1) repeat observations seeking to achieve higher sensitivity [e.g., 15875 (1996 TP_{66}) and 28978 Ixion], (2) multiple visits to characterize lightcurves [20000 Varuna and 47932 (2000 GN_{171}) are the only cases, and neither observation produced a measurable lightcurve], and (3) multiple visits to allow for the subtraction of background objects (so-called "shadow

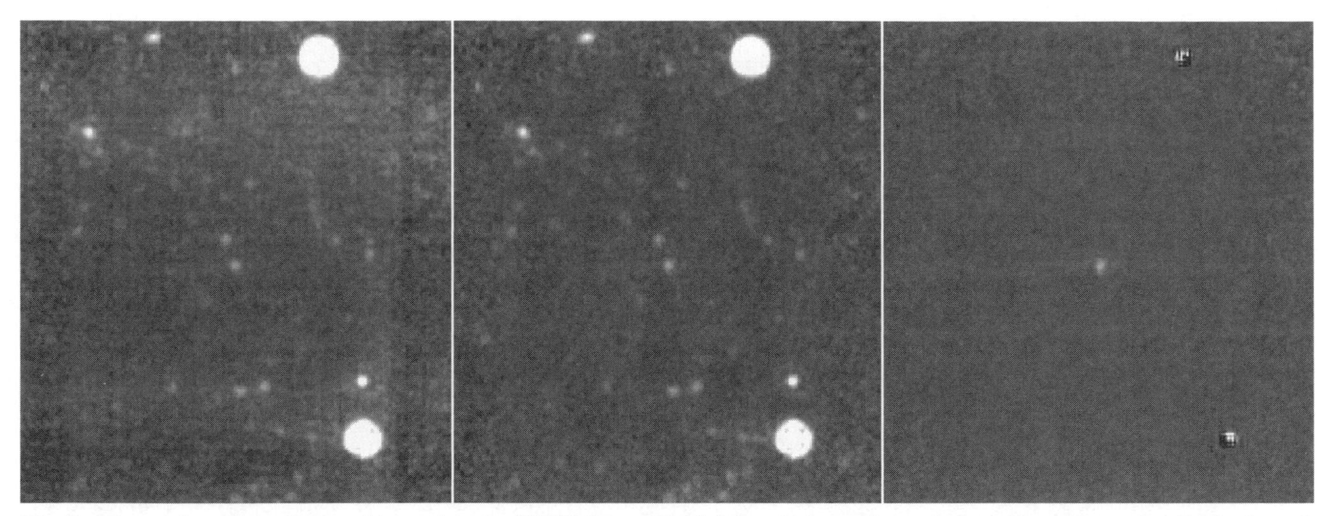

Fig. 2. Processing of the Spitzer 24-μm data for 90482 Orcus. The left panel shows the typical quality of image available through the data pipeline, with scattered light and dark latent artifacts still present. The center panel shows the improvements that can be made by correcting the aforementioned artifacts, and reflects the quality of the data we analyzed for targets that were imaged only once. The right panel shows further improvement due to the subtraction of a shadow observation, and reflects the quality of data we analyzed for targets that were imaged two or more times.

observations"). The basic idea of a shadow observation is to observe the target, wait for it to move out of the way, then re-observe the field. By subtracting the two images, the emission from stationary sources is removed. Figure 2 illustrates the shadow method, as well as some of the extra processing we apply to the Spitzer 24-μm data to improve its quality.

3.3. Photometry

Flux densities were measured using aperture photometry, as described in *Cruikshank et al.* (2005) and *Stansberry et al.* (2006). The apertures used encompassed the core of the PSF, out to about the first Airy minimum (their angular radii were 10" and 15" at 24 and 70 μm). Small apertures were used to maximize the signal-to-noise ratio (SNR) of the measurements. Sky measurements were made in the standard way, with an annulus surrounding the object aperture, and also by placing multiple circular apertures in the region around the target when the presence of background sources or cirrus structure dictated. The photometry was aperture corrected as described in *Engelbracht et al.* (2007) and *Gordon et al.* (2007). Finally, we apply color corrections to our measurements as described in *Stansberry et al.* (2007), resulting in monochromatic flux densities at the effective wavelengths of the 24- and 70-μm filters (23.68 and 71.42 μm, respectively). The MIPS calibration is defined in such a way that the color corrections for stellar spectra are unity. Even though our targets are much colder (typically at or below 80 K), the ≈20% passbands of the MIPS filters result in color corrections that are typically less than 10%.

Uncertainties on the absolute calibration of MIPS are 4% and 5% at 24 and 70 μm, respectively (*Engelbracht et al.*, 2007; *Gordon et al.*, 2007). In our photometry of KBOs and Centaurs we adopt systematic uncertainties of 5% and 10%, to account for the absolute calibration uncertainty and additional uncertainties that may be present, e.g., in our aperture and color corrections. At 70 μm our adopted systematic uncertainty includes significant margin to account for degraded repeatability for faint sources. Additional uncertainty comes from the finite SNR of the detections themselves, which is estimated from the statistics of values falling in the sky annulus and/or sky apertures. We root-sum-square combine the systematic uncertainty with the measurement uncertainty determined from the images to estimate the final error bars on our measurements, and use those total uncertainties in estimating the physical parameters we report. The SNR values we tabulate below reflect the errors estimated from the images, and so reflect the statistical significance of each detection.

4. SPITZER RESULTS

Our flux density measurements, and the albedos and diameters we derive from them, are given in Tables 2 and 3. Table 2 gives our results for those objects observed in both the 24- and 70-μm channel. When only an upper limit on the flux density was achieved, the results in Table 2 bound the albedo and diameter of the target. Table 3 gives the results for those objects observed at only one wavelength, and gives a second interpretation of the data for those objects in Table 2 that were only *detected* at one wavelength.

In both Tables 2 and 3 we give the color corrected flux density of each target, the SNR of the detections, and the temperature we used to perform the color corrections. Where we did not detect the source, we give the 3σ upper limit on the flux density, and the SNR column is blank. When an object was not observed in one of the bands (Table 3 only), the flux and SNR columns are blank. In both Table 2 and 3, albedos (p_V), diameters (D), and beaming parameters (η) follow the fluxes and temperatures.

TABLE 2. Two-band thermal model results.

Number*	Name (Designation)*	AORKEY†	$R_{\odot}$‡	Δ‡	F_{24}§	SNR_{24}§	F_{70}§	SNR_{70}§	$T_{24:70}$¶	p_V (%)**	D**	η**
29P	Schwassmann-Wachmann 1	7864064	5.734	5.561	253.783	48.0	96.1	18.6	164.7¶	$4.61^{+5.22}_{-1.90}$	$37.3^{-11.8}_{+11.3}$	$0.26^{-0.18}_{+0.28}$
2060	Chiron (1977 UB)	9033216	13.462	13.239	54.410	99.0	145.2	23.4	98.1	$7.57^{+1.03}_{-0.87}$	$233.3^{-14.4}_{+14.7}$	$1.13^{-0.13}_{+0.14}$
5145	Pholus (1992 AD)	9040896	18.614	18.152	3.080	66.0	<19.8		>80.2	$>6.56^{+6.38}_{-2.53}$	$<154.5^{-44.5}_{+42.6}$	$<1.37^{-0.48}_{+0.46}$
5145	Pholus (1992 AD)	12661760	19.827	19.768	0.962	18.8	<10.1		>72.9	$>8.12^{+7.93}_{-3.17}$	$<138.9^{-40.1}_{+38.9}$	$<1.78^{-0.60}_{+0.57}$
8405	Asbolus (1995 GO)	9039360	7.743	7.240	202.394	99.0	155.7	23.6	141.8¶	$5.30^{+1.91}_{-1.25}$	$85.4^{-12.2}_{+12.2}$	$0.66^{-0.20}_{+0.23}$
8405	Asbolus (1995 GO)	12660480	8.748	8.388	73.814	99.0	82.7	11.9	127.4	$5.59^{+1.69}_{-1.17}$	$83.2^{-10.3}_{+10.4}$	$0.93^{-0.22}_{+0.25}$
10199	Chariklo (1997 CU_{26})	8806144	13.075	12.684	78.700	99.0	202.5	24.6	99.1	$5.63^{+0.76}_{-0.65}$	$260.9^{-16.0}_{+16.4}$	$1.17^{-0.13}_{+0.14}$
10199	Chariklo (1997 CU_{26})	9038592	13.165	12.890	61.509	99.0	177.0	40.4	96.3	$5.81^{+0.62}_{-0.55}$	$256.8^{-12.8}_{+13.2}$	$1.29^{-0.12}_{+0.13}$
10370	Hylonome (1995 DW_2)	9038080	19.963	19.824	0.503	14.9	<10.2		>65.0	$>1.07^{+1.04}_{-0.42}$	$<168.4^{-48.5}_{+47.3}$	$<2.89^{-0.84}_{+0.80}$
15820	(1994 TB)	9042688	28.562	28.320	<0.062		<11.1		48.2	$>0.55^{+0.64}_{-0.26}$	$<451.3^{-145.4}_{+176.1}$	$4.87^{-1.47}_{+1.90}$
15874	(1996 TL_{66})	9035776	35.125	34.604	0.380	13.5	22.0	4.4	55.6	$3.50^{+1.96}_{-1.07}$	$575.0^{-114.6}_{+115.5}$	$1.76^{-0.33}_{+0.33}$
15875	(1996 TP_{66})	8805632	26.491	26.250	0.689	17.9	<17.6		>62.7	$>1.97^{+1.88}_{-0.76}$	$<310.9^{-88.7}_{+86.1}$	$<1.89^{-0.53}_{+0.50}$
15875	(1996 TP_{66})	12659456	26.629	26.113	0.426	14.6	<6.9		>67.5	$>6.49^{+6.34}_{-2.54}$	$<171.2^{-49.4}_{+48.3}$	$<1.36^{-0.43}_{+0.41}$
20000	Varuna (2000 WR_{106})	9045760	43.209	42.830	<0.086		11.0	4.9	<50.1	$<11.60^{+7.66}_{-4.59}$	$>621.2^{-139.1}_{+178.1}$	$>1.73^{-0.46}_{+0.63}$
26308	(1998 SM_{165})	14402560	36.417	36.087	0.105	15.9	5.2	9.4	56.8	$6.33^{+1.53}_{-1.16}$	$279.8^{-28.6}_{+29.7}$	$1.48^{-0.17}_{+0.17}$
26375	(1999 DE_9)	9047552	34.980	34.468	0.905	38.2	22.6	9.3	62.9	$6.85^{+1.58}_{-1.19}$	$461.0^{-45.3}_{+46.1}$	$1.05^{-0.12}_{+0.12}$
28978	Ixion (2001 KX_{76})	9033472	42.731	42.448	0.584	16.6	19.6	3.5	60.1	$15.65^{+12.00}_{-5.53}$	$573.1^{-141.9}_{+139.7}$	$0.82^{-0.22}_{+0.21}$
28978	Ixion (2001 KX_{76})	12659712	42.510	42.058	0.290	7.9	<18.4		>54.9	$>12.03^{+12.08}_{-4.89}$	$<653.6^{-191.9}_{+194.6}$	$<1.22^{-0.37}_{+0.36}$
29981	(1999 TD_{10})	8805376	14.137	13.945	4.629	31.6	19.5	7.2	87.9	$4.40^{+1.42}_{-0.96}$	$103.7^{-13.5}_{+13.6}$	$1.64^{-0.31}_{+0.32}$
31824	Elatus (1999 UG_5)	9043200	10.333	9.998	6.015	69.8	<12.4		>105.2	$>4.86^{+5.17}_{-1.95}$	$<47.4^{-14.4}_{+13.8}$	$<1.46^{-0.66}_{+0.68}$
31824	Elatus (1999 UG_5)	12661248	11.125	10.826	8.596	99.0	<8.9		>118.3¶	$>9.41^{+11.57}_{-3.97}$	$<34.1^{-11.3}_{+10.8}$	$<0.50^{-0.29}_{+0.33}$
32532	Thereus (2001 PT_{13})	9044480	9.813	9.357	25.938	99.0	32.7	4.8	122.3	$8.93^{+5.35}_{-2.79}$	$60.8^{-12.7}_{+12.5}$	$0.86^{-0.32}_{+0.35}$
32532	Thereus (2001 PT_{13})	12660224	9.963	9.685	23.722	99.0	46.8	10.3	106.5	$4.28^{+1.09}_{-0.80}$	$87.8^{-9.4}_{+9.5}$	$1.50^{-0.28}_{+0.30}$
38628	Huya (2000 EB_{173})	8808192	29.326	29.250	3.630	69.4	57.2	10.9	67.9	$4.78^{+0.94}_{-0.74}$	$546.5^{-47.1}_{+47.8}$	$1.10^{-0.11}_{+0.12}$
38628	Huya (2000 EB_{173})	8937216	29.325	29.210	3.400	69.0	52.9	28.4	68.0	$5.22^{+0.47}_{-0.43}$	$523.1^{-21.9}_{+22.7}$	$1.09^{-0.07}_{+0.07}$
47171	(1999 TC_{36})	9039104	31.098	30.944	1.233	56.4	25.3	10.0	64.9	$7.18^{+1.53}_{-1.17}$	$414.6^{-38.2}_{+38.8}$	$1.17^{-0.12}_{+0.13}$
47932	(2000 GN_{171})	9027840	28.504	28.009	0.258	8.2	11.9	5.6	57.4	$5.68^{+2.54}_{-1.59}$	$321.0^{-54.2}_{+57.4}$	$2.32^{-0.43}_{+0.46}$
50000	Quaoar (2002 LM_{60})	10676480	43.345	42.974	0.279	5.5	24.6	4.2	52.5	$19.86^{+13.17}_{-7.04}$	$844.4^{-189.6}_{+206.7}$	$1.37^{-0.36}_{+0.39}$
52872	Okyrhoe (1998 SG_{35})	8807424	7.793	7.405	28.767	99.0	37.4	9.1	121.0	$2.49^{+0.81}_{-0.55}$	$52.1^{-6.9}_{+6.9}$	$1.46^{-0.35}_{+0.39}$
54598	Bienor (2000 QC_{243})	9041920	18.816	18.350	3.528	78.0	29.7	6.1	76.0	$3.44^{+1.27}_{-0.82}$	$206.7^{-30.1}_{+30.1}$	$1.69^{-0.30}_{+0.30}$
55565	(2002 AW_{197})	9043712	47.131	46.701	0.155	7.7	15.0	6.7	51.9	$11.77^{+4.42}_{-3.00}$	$734.6^{-108.3}_{+116.4}$	$1.26^{-0.20}_{+0.22}$
55576	Amycus (2002 GB_{10})	17766144	15.589	15.155	6.367	86.1	13.6	5.8	99.9¶	$17.96^{+7.77}_{-4.70}$	$76.3^{-12.5}_{+12.5}$	$0.64^{-0.18}_{+0.19}$
55636	(2002 TX_{300})	10676992	40.979	40.729	<0.065		<11.1		48.4	$>17.26^{+20.33}_{-8.33}$	$<641.2^{-206.7}_{+250.3}$	$2.16^{-0.78}_{+0.95}$
55637	(2002 UX_{25})	10677504	42.368	42.413	0.486	15.0	23.0	5.3	57.2	$11.50^{+5.09}_{-3.09}$	$681.2^{-114.0}_{+115.6}$	$1.04^{-0.18}_{+0.18}$
60558	Echeclus (2000 EC_{98})	8808960	14.141	13.736	4.901	84.7	15.5	5.0	94.0	$3.83^{+1.89}_{-1.08}$	$83.6^{-15.2}_{+15.0}$	$1.25^{-0.32}_{+0.33}$
65489	Ceto (2003 FX_{128})	17763840	27.991	27.674	1.463	71.5	14.6	12.2	73.6	$7.67^{+1.38}_{-1.10}$	$229.7^{-18.2}_{+18.6}$	$0.86^{-0.09}_{+0.10}$
73480	(2002 PN_{34})	17762816	14.608	14.153	10.368	99.0	31.0	12.6	95.3	$4.25^{+0.83}_{-0.65}$	$119.5^{-10.2}_{+10.3}$	$1.10^{-0.15}_{+0.16}$

TABLE 2. (continued).

Number*	Name (Designation)*	AORKEY†	$R_\odot$‡	Δ‡	F_{24}§	SNR_{24}§	F_{70}§	SNR_{70}§	$T_{24:70}$¶	p_V (%)**	D**	η**
83982	Crantor (2002 GO_9)	9044224	14.319	13.824	2.276	58.6	<8.7		>89.8	$>8.60^{+8.62}_{-3.36}$	$<66.7^{-19.6}_{+18.7}$	$<1.44^{-0.57}_{+0.56}$
84522	(2002 TC_{302})	13126912	47.741	47.654	0.054	6.5	18.0	3.1	44.8	$3.08^{+2.93}_{-1.24}$	$1145.4^{-325.0}_{+337.4}$	$2.33^{-0.54}_{+0.53}$
84922	(2003 VS_2)	10680064	36.430	36.527	0.304	6.0	25.7	3.5	52.8	$5.84^{+4.78}_{-2.24}$	$725.2^{-187.6}_{+199.0}$	$2.00^{-0.51}_{+0.54}$
90482	Orcus (2004 DW)	13000448	47.677	47.442	0.329	32.4	26.6	12.5	53.1	$19.72^{+3.40}_{-2.76}$	$946.3^{-72.3}_{+74.1}$	$1.08^{-0.09}_{+0.10}$
90568	(2004 GV_9)	13000960	38.992	39.007	0.166	18.2	17.5	9.2	51.4	$8.05^{+1.94}_{-1.46}$	$677.2^{-69.3}_{+71.3}$	$1.94^{-0.20}_{+0.20}$
119951	(2002 KX_{14})	10678016	39.585	39.197	<0.109		<11.7		51.2	$>8.09^{+9.58}_{-3.91}$	$<561.6^{-181.5}_{+219.9}$	$1.91^{-0.66}_{+0.84}$
120061	(2003 CO_1)	17764864	10.927	10.917	21.722	99.0	33.4	11.3	114.7	$5.74^{+1.49}_{-1.09}$	$76.9^{-8.4}_{+8.5}$	$0.91^{-0.18}_{+0.20}$
136108	(2003 EL_{61})††	13803008	51.244	50.920	<0.022		7.8	5.3	<44.6	?‡‡	?‡‡	?‡‡
136199	Eris (2003 UB_{313})††	15909632	96.907	96.411	<0.014		2.7	4.0	<40.1	?‡‡	?‡‡	?‡‡
136472	(2005 FY_9)††	13803776	51.884	51.879	0.296	21.1	14.6	9.4	54.8	?‡‡	?‡‡	?‡‡
	(2002 MS_4)	10678528	47.402	47.488	0.391	20.5	20.0	5.1	56.6	$8.41^{+3.78}_{-2.26}$	$726.2^{-122.9}_{+123.2}$	$0.88^{-0.15}_{+0.14}$
	(2003 AZ_{84})	10679040	45.669	45.218	0.291	12.4	17.8	6.7	55.2	$12.32^{+4.31}_{-2.91}$	$685.8^{-95.5}_{+98.8}$	$1.04^{-0.16}_{+0.16}$

*Small-body number, provisional designation, and proper name for the target sample.

†Unique key identifying the data in the Spitzer data archive.

‡Target distance from the Sun and Spitzer, in AU.

§Color-corrected flux densities (mJy) at 23.68 μm and 71.42 μm. Upper limits are 3σ. SNR is signal-to-noise ratio in the images (see text).

¶The temperature of the blackbody spectrum used to compute the color correction. In most cases this is the 24:70-μm color temperature, but for the four denoted targets, the subsolar blackbody temperature was lower than the color temperature, and we used that instead.

**The visible geometric albedo (p_V, percentage), diameter (D, km), and beaming parameter (η) from hybrid STM fits. Fits to upper limits provide a quantitative interpretation of the constraints they place on p_V and D.

††Results for 136199 Eris, 136199 (2003 EL_{61}), and 136472 (2005 FY_9) assumed a phase integral of 0.8, typical of Pluto.

‡‡No STM with plausible albedo and beaming parameter can simultaneously fit the 24- and 70-μm data. For 136472, models with two albedo terrains can fit the data, and give D ≃ 1500 km.

TABLE 3. Single-band thermal model results.

										KBO-Tuned STM		p_V‡ (%)	D‡
Number*	Name (Designation)*	AORKEY*	$R_\odot$*	Δ*	F_{24}*	SNR_{24}*	F_{70}*	SNR_{70}*	T_{SS}†	p_V (%)‡	D‡	STM_0 ILM_0	STM_0 ILM_0
5145	Pholus (1992 AD)	9040896	18.614	18.152	3.119	66.0	<19.6		91.4	$8.16^{+6.16}_{-?}$	$138.6^{-34.0}_{+?}$	17.07–?	95.8–?
5145	Pholus (1992 AD)	12661760	19.827	19.768	0.987	18.8	<10.0		88.6	$16.18^{+11.55}_{-5.88}$	$98.4^{-23.2}_{+25.0}$	32.74–?	69.2–?
7066	Nessus (1993 HA_2)	9033984	19.501	19.219	0.440	12.4			89.3	$6.53^{+5.14}_{-2.46}$	$59.7^{-15.1}_{+15.9}$	14.02–1.44	40.8–127.4
10370	Hylonome (1995 DW_2)	9038080	19.963	19.824	0.530	14.9	<10.0		88.3	$6.12^{+4.91}_{-2.33}$	$70.5^{-18.0}_{+19.1}$	13.28–1.32	47.9–152.0
10370	Hylonome (1995 DW_2)	12659968	20.333	20.390	0.451	16.0			87.5	$6.33^{+5.12}_{-2.42}$	$69.3^{-17.8}_{+18.9}$	13.80–1.34	46.9–150.5
15875	(1996 TP_{66})	8805632	26.491	26.250	0.720	17.9	<17.3		76.6	$5.17^{+4.98}_{-2.19}$	$191.8^{-54.9}_{+60.9}$	12.54–?	123.1–?
15875	(1996 TP_{66})	12659456	26.629	26.113	0.437	14.6	<6.8		76.4	$8.21^{+7.61}_{-?}$	$152.2^{-42.6}_{+?}$	19.37–?	99.1–?
20000	Varuna (2000 WR_{106})	9045760	43.209	42.830	<0.094		10.9	4.9	60.0	?	?	?–8.09	?–744.1
20000	Varuna (2000 WR_{106})	9031680	43.261	43.030			10.0	5.6	60.0	$17.77^{+6.17}_{-3.79}$	$502.0^{-69.5}_{+64.0}$	26.34–8.68	412.3–718.2
28978	Ixion (2001 KX_{76})	12659712	42.510	42.058	0.303	7.9	<18.3		60.5	25.81§	446.3§	32.28–?	399.1–?
31824	Elatus (1999 UG_5)	9043200	10.333	9.998	5.990	69.8	<12.3		122.7	$6.41^{+3.52}_{-?}$	$41.3^{-8.1}_{+?}$	11.41–?	31.0–?
31824	Elatus (1999 UG_5)	12661248	11.125	10.826	8.596	99.0	<8.9		118.3	?	?	?–?	?–?
35671	(1998 SN_{165})	9040384	37.967	37.542			14.7	6.3	64.0	$4.33^{+1.50}_{-0.91}$	$458.2^{-63.1}_{+57.1}$	6.42–2.17	376.4–648.1
42355	Typhon (2002 CR_{46})	9029120	17.581	17.675			31.4	8.6	94.1	$5.09^{+1.24}_{-0.80}$	$173.8^{-18.0}_{+15.6}$	6.81–3.13	150.3–221.7
52975	Cyllarus (1998 TF_{35})	9046528	21.277	21.001	0.274	8.7			85.5	$11.46^{+8.96}_{-4.36}$	$61.9^{-15.5}_{+16.8}$	24.43–2.42	42.4–134.9
63252	(2001 BL_{41})	9032960	9.856	9.850	4.864	95.6			125.7	$3.90^{+2.12}_{-1.14}$	$34.2^{-6.7}_{+6.5}$	6.93–1.34	25.6–58.3
83982	Crantor (2002 GO_9)	9044224	14.319	13.824	2.310	58.6	<8.6		104.2	$11.18^{+7.09}_{-?}$	$58.5^{-12.7}_{+?}$	21.28–?	42.4–?
90377	Sedna (2003 VB_{12})¶,**	8804608	89.527	89.291			<2.4		41.7	$>20.91^{+8.71}_{-5.29}$	$<1268.8^{-202.7}_{+199.4}$	32.93–8.17	1010.9–2029.0
136108	(2003 EL_{61})¶	13803008	51.244	50.920	<0.025		7.7	5.3	55.1	$84.11^{+9.48}_{-8.10}$	$1151.0^{-59.9}_{+59.8}$	96.41–59.12	1075.1–1372.9
136199	Eris (2003 UB_{313})¶	15909632	96.907	96.411	<0.014		2.7	4.0	40.1	$68.91^{+12.24}_{-9.98}$	$2657.0^{-208.6}_{+216.1}$	84.90–39.17	2393.7–3523.9
136472	(2005 FY_9)¶	13803776	51.884	51.879	††		14.6	9.4	54.8	$78.20^{+10.30}_{-8.55}$	$1502.9^{-90.2}_{+89.6}$	91.63–52.55	1388.3–1833.3
136472	(2005 FY_9)¶	13803776	51.884	51.879	0.296	21.1	††		54.8	$35.99^{+17.56}_{-12.25}$	$2215.2^{-399.2}_{+512.4}$	59.34–6.27	1725.3–5307.0

*The first nine columns are identical to those in Table 2. Flux densities that are blank indicate no data exist.

†The subsolar temperature of a blackbody at the distance of the target. Color corrections are made using a blackbody spectrum with this temperature.

‡The range of visible geometric albedos (given as a percentage) and diameter (in km) derived from fitting the KBO-tuned STM (i.e., $\eta = 1.2 \pm 0.35$) and the canonical STM and ILM. "?" indicates the model emission violates a flux limit.

§Only the KBO-tuned STM using $\eta = 0.85$ did not violate the 70-μm flux limit for this observation of Ixion.

¶Results for 90377 Sedna, 136108 (2003 EL_{61}), Eris, and 136472 (2005 FY_9) assumed a phase integral of 0.8, typical of Pluto.

**Fit to the 70-μm upper limit: lower bound on p_V, upper bound on D.

††Fits to the individual bands for 136472 (2005 FY_9) are shown; it is not possible to simultaneously fit both bands with a single thermal model.

4.1. Two-Wavelength Results

As discussed earlier and demonstrated in Fig. 1, the model-dependent uncertainties in the albedo and diameter we derive for targets detected at both 24 and 70 μm are much smaller than those uncertainties for objects detected in only one of those bands, and in particular are usually very much smaller than for objects detected only at 24 μm. For this reason, we focus first on the targets we either detected at both wavelengths, or for which we have constraints on the flux density at both. We use these results to inform our models for targets with single-band detections and limits.

We apply the hybrid STM to the observed flux densities as follows. For targets *detected* in both bands (Table 2), we fit the observed flux densities and the 1σ error bars, deriving albedo and diameter values and 1σ uncertainties on them. For those objects with an *upper limit* in one band and a detection in the other, we fit the detection and the the upper limit in order to quantitatively interpret the constraints the limit implies for the albedo and diameter. For this second class of observation, we also perform a single-wavelength analysis (see Table 3) in order to derive independent constraints on these properties. While the results given in Table 2 include values of the beaming parameter, η, those values only reflect the departures of the measured emission from the assumptions of the STM; had we chosen to model the data with the ILM, the fitted values for η would be entirely different (even though p_V and D would be very similar). Results from observations made at very similar epochs are averaged. An exception to that rule is the two observations of 38628 Huya. Those data were analyzed independently to provide a check on the repeatability of our overall data analysis and modeling methods for a "bright" KBO, and show agreement at the 4% level for p_V and at the 2% level for D.

The average behavior of the targets is of particular interest for interpreting single-wavelength observations, where we have no independent means for constraining η. Restricting our attention to those targets detected at SNR ≥ 5 at both 24 and 70 μm, and excluding the highest- and lowest-albedo object from each class, we find that for outer solar system objects the average beaming parameter is $\eta = 1.2 \pm 0.35$. We reexamine the average properties of the sample later.

4.2. Single-Wavelength Results

Because we are primarily interested in the albedos and sizes of our targets, we fit our single-wavelength observations with the STM, setting the beaming parameter to the average value determined above: We term this model the "KBO-tuned" STM. We also apply the canonical STM and ILM (i.e., with $\eta = 0.756$ and 1.0, respectively) to the single-wavelength data, to interpret the data in the context of these endmember models and assess the resulting uncertainties in model parameters.

Table 3 gives the results for the single-wavelength sample, including those objects in Table 2 with a detection at one wavelength and an upper-limit at the other. Where a model violates a flux limit, the corresponding albedo and diameter entries appear as a "?". The albedos and diameters we derive using the average beaming parameter from the two-wavelength sample are in the columns labeled "KBO-Tuned STM"; the range of albedos and diameters resulting from application of the canonical STM and ILM are labeled "STM_0" and "ILM_0". Note that the flux densities for objects in both Table 2 and 3 are sometimes slightly different, because in Table 3 the color correction is based on the blackbody temperature at the object's distance, rather than on the 24:70-μm color temperature.

4.3. Spitzer Albedos and Diameters

The results presented above include low SNR detections, nondetections, and multiple results for some targets. In the top portion of Table 4 we present results for the 39 targets that were detected at SNR ≥ 5 at one or both wavelengths. The results for targets that were visited multiple times are averaged unless one observation shows some indication of a problem. Targets with an upper limit in either band appear in both Tables 2 and 3; in the top portion of Table 4 we give values that are representative of all the earlier models. The top portion of the table contains 39 objects: 26 detected at both 24 and 70 μm, 9 at 24 μm only, and 4 at 70 μm only. Seventeen of the objects have orbital semimajor axes inside Neptune, and 21 exterior to Neptune's orbit. Where other albedo and diameter determinations exist, Table 4 summarizes the result, the basis of the determination, and the publication.

4.4. Other Constraints on p_V and D

The albedos and sizes of about 20 TNOs and several Centaurs have been determined by other groups using various methods; the lower portion of Table 4 presents those results not given in the top portion of the table, and the constraints that can be derived from Spitzer data, when those exist (although the SNR for all five cases is low, and for 90377 Sedna only a 70-μm limit is available).

In general our results and those of other groups agree at the $\leq 2\sigma$ level [e.g., 10199 Chariklo, 26308 (1998 SM_{165}), 47171 (1999 TC_{36}), 55565 (2002 AW_{197}), 136199 Eris, 136108 (2003 EL_{62})]. In a few cases there are discrepancies. For example, our results for 20000 Varuna are inconsistent with the millimeter results of *Jewitt et al.* (2001) and *Lellouch et al.* (2002), which suggest a significantly larger size and lower albedo. While our detection at 70 μm nominally satisfied the 5σ threshold for Table 4, the background showed significant structure and the SNR of the detection in the individual visits was actually quite low. Combined with the fact that we were not able to directly fit the beaming parameter, we are inclined to favor the submillimeter results for this object over those from Spitzer. While there is some tendency for the Spitzer diameters to be smaller and albedos higher, there is generally good agreement be-

TABLE 4. Adopted physical properties.

Number*	Name (Designation)*	Physical Properties from Spitzer Data				TNO?*	Other Methods			Reference§
		p_V*	D*	η*	λ_{detect}†		Method‡	p_V*	D*	
29P	Schwassmann-Wachmann	$14.61^{+5.22}_{-1.90}$	$37.3^{-11.8}_{+11.3}$	$0.26^{-0.18}_{+0.28}$	both	Cen	mIR	13 ± 4	40 ± 5	Cr83
2060	Chiron (1977 UB)	$7.57^{+1.03}_{-0.87}$	$233.3^{-14.4}_{+14.7}$	$1.13^{-0.13}_{+0.14}$	both	Cen	mIR	17 ± 2	144 ± 8	Fe02
							ISO	11 ± 2	142 ± 10	Gn04
							mIR	14 ± 5	180	Ca94
5145	Pholus (1992 AD)	8.0^{+7}_{-3}	140^{-40}_{+40}	$1.3^{-0.4}_{+0.4}$	24	Cen	mIR	4.4 ± 1.3	189 ± 26	Da93
							IRS	7.2 ± 2	148 ± 25	Cr06
7066	Nessus (1993 HA_2)	$6.5^{+5.3}_{-2.5}$	60^{-16}_{+16}	$1.2^{-0.35}_{+0.35}$	24	Cen				
8405	Asbolus (1995 GO)	$5.46^{+1.27}_{-0.86}$	$84.2^{-7.8}_{+7.8}$	$0.80^{-0.16}_{+0.17}$	both	Cen	mIR	12 ± 3	66 ± 4	Fe02
							IRS	4.3 ± 1.4	95 ± 7	Cr06
10199	Chariklo (1997 CU_{26})	$5.73^{+0.49}_{-0.42}$	$258.6^{-10.3}_{+10.3}$	$1.23^{-0.09}_{+0.10}$	both	Cen	mm	5.5 ± 0.5	275	Al02
							mIR/mm	7 ± 1	246 ± 12	Gn04
10370	Hylonome (1995 DW_2)	6.2^{+5}_{-3}	70^{-20}_{+20}	$1.2^{-0.35}_{+0.35}$	24	Cen				
15875	(1996 TP_{66})	7.4^{+7}_{-3}	160^{-45}_{+45}	$1.2^{-0.35}_{+0.35}$	24	TNO				
20000	Varuna (2000 WR_{106})	16^{+10}_{-8}	500^{-100}_{+100}	$1.2^{-0.35}_{+0.35}$	70	TNO	submm	6 ± 2	1016 ± 156	Je01, Al04
							mm	7 ± 3	914 ± 156	Le02, Al04
26308	(1998 SM_{165})	$6.33^{+1.53}_{-1.16}$	$279.8^{-28.6}_{+29.7}$	$1.48^{-0.17}_{+0.17}$	both	TNO	mm/bin	9.1 ± 4	238 ± 55	Ma04, Gy05
26375	(1999 DE_9)	$6.85^{+1.58}_{-1.19}$	$461.0^{-45.3}_{+46.1}$	$1.05^{-0.12}_{+0.12}$	both	TNO				
28978	Ixion (2001 KX_{76})	12^{+14}_{-6}	650^{-220}_{+260}	$0.8^{-0.2}_{+0.2}$	24	TNO	mm	>15	<804	Al04
29981	(1999 TD_{10})	$4.40^{+1.42}_{-0.96}$	$103.7^{-13.5}_{+13.6}$	$1.64^{-0.31}_{+0.32}$	both	TNO	IRS	6.5	98	Cr06
31824	Elatus (1999 UG_5)	10^{+4}_{-3}	30^{-8}_{+8}	$1.2^{-0.35}_{+0.35}$	24	Cen	IRS	5.7 ± 2	36 ± 8	Cr06
32532	Thereus (2001 PT_{13})	$4.28^{+1.09}_{-0.80}$	$87.8^{-9.4}_{+9.5}$	$1.50^{-0.28}_{+0.30}$	both	Cen				
35671	(1998 SN_{165})	$4.3^{+1.8}_{-1.2}$	460^{-80}_{+60}	$1.2^{-0.35}_{+0.35}$	70	TNO				
38628	Huya (2000 EB_{173})	$5.04^{+0.50}_{-0.41}$	$532.6^{-24.4}_{+25.1}$	$1.09^{-0.06}_{+0.07}$	both	TNO	mm	>8	<540	Al04
42355	Typhon (2002 CR_{46})	$5.1^{+1.3}_{-0.9}$	175^{-20}_{+17}	$1.2^{-0.35}_{+0.35}$	70	TNO				
47171	(1999 TC_{36})	$7.18^{+1.53}_{-1.17}$	$414.6^{-38.2}_{+38.8}$	$1.17^{-0.12}_{+0.13}$	both	TNO	mm	5 ± 1	609 ± 70	Al04
							mm/bin	14 ± 6	302 ± 70	Ma04, Gy05
47932	(2000 GN_{171})	$5.68^{+2.54}_{-1.59}$	$321.0^{-54.2}_{+57.4}$	$2.32^{-0.43}_{+0.46}$	both	TNO				
50000	Quaoar (2002 LM_{60})	$19.9^{+13.2}_{-7}$	844^{-190}_{+207}	$1.4^{-0.4}_{+0.4}$	both	TNO	image	9 ± 3	1260 ± 190	Br04
52872	Okyrhoe (1998 SG_{35})	$2.49^{+0.81}_{-0.55}$	$52.1^{-6.9}_{+6.9}$	$1.46^{-0.35}_{+0.39}$	both	Cen				
52975	Cyllarus (1998 TF_{35})	11.5^{+9}_{-5}	62^{-18}_{+20}	$1.2^{-0.35}_{+0.35}$	24	Cen				
54598	Bienor (2000 QC_{243})	$3.44^{+1.27}_{-0.82}$	$206.7^{-30.1}_{+30.1}$	$1.69^{-0.30}_{+0.30}$	both	Cen				
55565	(2002 AW_{197})	$11.77^{+4.42}_{-3.00}$	$734.6^{-108.3}_{+116.4}$	$1.26^{-0.20}_{+0.22}$	both	TNO	mm	9 ± 2	977 ± 130	Ma02
55576	Amycus (2002 GB_{10})	$17.96^{+7.77}_{-4.70}$	$76.3^{-12.5}_{+12.5}$	$0.64^{-0.18}_{+0.19}$	both	Cen				
55637	(2002 UX_{25})	$11.50^{+5.09}_{-3.09}$	$681.2^{-114.0}_{+115.6}$	$1.04^{-0.18}_{+0.18}$	both	TNO				
60558	Echeclus (2000 EC_{98})	$3.83^{+1.89}_{-1.08}$	$83.6^{-15.2}_{+15.0}$	$1.25^{-0.32}_{+0.33}$	both	Cen				
63252	(2001 BL_{41})	$3.9^{+2.5}_{-1.3}$	35^{-8}_{+7}	$1.2^{-0.35}_{+0.35}$	24	Cen				

TABLE 4. (continued).

		Physical Properties from Spitzer Data					Other Methods			
Number*	Name (Designation)*	p_V*	D*	η*	λ_{detect}†	TNO?*	Method‡	p_V*	D*	Reference§
65489	Ceto (2003 FX_{128})	$7.67^{+1.38}_{-1.10}$	$229.7^{-18.2}_{+18.6}$	$0.86^{-0.09}_{+0.10}$	both	TNO				
73480	(2002 PN_{34})	$4.25^{+0.83}_{-0.65}$	$119.5^{-10.2}_{+10.3}$	$1.1^{-0.15}_{+0.16}$	both	TNO				
83982	Crantor (2002 GO_9)	11^{+7}_{-4}	60^{-13}_{+15}	$1.20^{-0.35}_{+0.35}$	24	Cen				
90482	Orcus (2004 DW)	$19.72^{+3.40}_{-2.76}$	$946.3^{-72.3}_{+74.1}$	$1.08^{-0.09}_{+0.10}$	both	TNO				
90568	(2004 GV_9)	$8.05^{+1.94}_{-1.46}$	$677.2^{-69.3}_{+71.3}$	$1.94^{-0.20}_{+0.20}$	both	TNO				
120061	(2003 CO_1)	$5.74^{+1.49}_{-1.09}$	$76.9^{-8.4}_{+8.5}$	$0.91^{-0.18}_{+0.20}$	both	Cen				
136108	(2003 EL_{61})	$84.^{+10.}_{-20.}$	$1150.^{-100}_{+250}$		70	TNO	Lcurve	65 ± 6	1350 ± 100	Ra05
136472	(2005 FY_9)	$80.^{+10.}_{-20.}$	$1500.^{-200}_{+400}$		both	TNO				
	(2002 MS_4)	$8.41^{+3.78}_{-2.26}$	$726.2^{-122.9}_{+123.2}$	$0.88^{-0.15}_{+0.14}$	both	TNO				
	(2003 AZ_{84})	$12.32^{+4.31}_{-2.91}$	$685.8^{-95.5}_{+98.8}$	$1.04^{-0.16}_{+0.16}$	both	TNO				
15789	(1993 SC)					TNO	ISO	3.5 ± 1.4	298 ± 140	Th00
15874	(1996 TL_{66})	$3.5^{+2.0}_{-1.1}$	575^{-115}_{+116}	$1.8^{-0.3}_{+0.3}$	both	TNO	ISO	>1.8	<958	Th00
19308	(1996 TO_{66})					TNO	mm	>3.3	<902	Al04, Gy05
19521	Chaos (1998 WH_{24})					TNO	mm	>5.8	<747	Al04, Gy05
24835	(1995 SM_{55})					TNO	mm	>6.7	<704	Al04, Gy05
55636	(2002 TX_{300})	>10	<800		limit	TNO	mm	>19	<709	Or04, Gy05
58534	(1997 CQ_{29})					TNO	bin	39 ± 17	77 ± 18	Ma04, No04, Gy05
66652	(1999 RZ_{253})					TNO	bin	29 ± 12	170 ± 39	No04, Gy05
84522	(2002 TC_{302})	$3.1^{+2.9}_{-1.2}$	1150^{-325}_{+337}	$2.3^{-0.5}_{+0.5}$	both	TNO	mm	>5.1	<1211	Al04, Gy05
88611	(2001 QT_{297})					TNO	bin	10 ± 4	168 ± 38	Os03, Gy05
90377	Sedna (2003 VB_{12})	>16.	<1600.		limit	TNO	image	>8.5	<1800	Br04a
136199	Eris (2003 UB_{313})	$70.^{+15.}_{-20.}$	$2600.^{-200}_{+400}$		70	TNO	mm	60 ± 8	3000 ± 200	Be06
							image	86 ± 7	2400 ± 100	Br06
	(1998 WW_{31})					TNO	bin	6 ± 2.6	152 ± 35	Ve02, Gy05
	(2001 QC_{298})					TNO	bin	2.5 ± 1.1	244 ± 55	Ma04, Gy05

*Columns 1–5 and 7 are as defined in Table 2.

†Wavelengths where the objects were detected at SNR > 5 (above horizontal line) or have lower quality Spitzer data (below line).

‡Method by which the diameter was measured. bin = binary mass plus density assumption; image = HST upper limit; IRS = Spitzer mid-IR spectra; ISO = Infrared Space Observatory; Lcurve = lightcurve + rotation dynamics; mIR = groundbased 10–20 μm; mm = typically 1.2-mm groundbased data; submm = typically 850-μm groundbased data.

§References: Al02 = *Altenhoff et al.* (2002); Al04 = *Altenhoff et al.* (2004); Be06 = *Bertoldi et al.* (2006); Br04 = *Brown and Trujillo* (2004); Br04a = *Brown et al.* (2004); Br06 = *Brown et al.* (2006); Ca94 = *Campins et al.* (1994); Cr83 = *Cruikshank and Brown* (1983); Cr06 = *Cruikshank et al.* (2006); Da93 = *Davies et al.* (1993); Fe02 = *Fernandez et al.* (2002); Gn04 = *Groussin et al.* (2004); Gy05 = *Grundy et al.* (2005); Je01 = *Jewitt et al.* (2001); Le02 = *Lellouch et al.* (2002); Ma02 = *Margot et al.* (2002); Ma04 = *Margot et al.* (2004); No04 = *Noll et al.* (2004); Or04 = *Ortiz et al.* (2004); Os03 = *Osip et al.* (2003); Ra05 = *Rabinowitz et al.* (2005); Th00 = *Thomas et al.* (2000); Ve02 = *Veillet et al.* (2002).

Results above the horizontal line have Spitzer detections with SNR > 5; those below the line have SNR < 5, or no Spitzer data.

tween our Spitzer results and those from other groups and methods.

5. ALBEDO STATISTICS AND CORRELATIONS

The Kuiper belt is full of complexity, in terms of the dynamical history and the spectral character of its inhabitants. It is natural to look for relationships between the albedos of KBOs and their orbital and other physical parameters. Figure 3 shows the Spitzer albedos for detections with SNR ≥ 5 (top portion of Table 4) as a function of orbital semimajor axis, a, perihelion distance, $q_\odot$, object diameter, D, and visible spectral slope, S. Because of their significant intrinsic interest, the data for 136199 (2003 EL_{61}) and 90377 Sedna are also plotted. Immediately apparent in all these plots is the marked distinction between the largest objects [136199 Eris, 136108 (2003 EL_{61}), and 136472 (2005 FY_9)] and the rest of the objects. 90377 Sedna probably also belongs to this class, although our data only place a lower bound on its albedo. Eris and 136472 (2005 FY_9) both have abundant CH_4 ice on their surfaces, and so are expected to have very high albedos. 90377 Sedna's near-IR spectrum also shows evidence for CH_4 and N_2 ices (*Barucci et al.,* 2005; *Emery et al.,* 2007), and *Schaller and Brown* (2007) show that those ices should not be depleted by Jean's escape: It seems likely that 90377 Sedna's albedo is quite high. The surface of 136199 (2003 EL_{61}) is dominated by water-ice absorptions, with no evidence for CH_4 or N_2, yet also has a very high albedo. Charon, which has a similar spectrum, has $p_V \simeq 37\%$, but some saturnian satellites (notably Enceladus and Tethys) have albedos ≥ 80% (*Morrison et al.,* 1986).

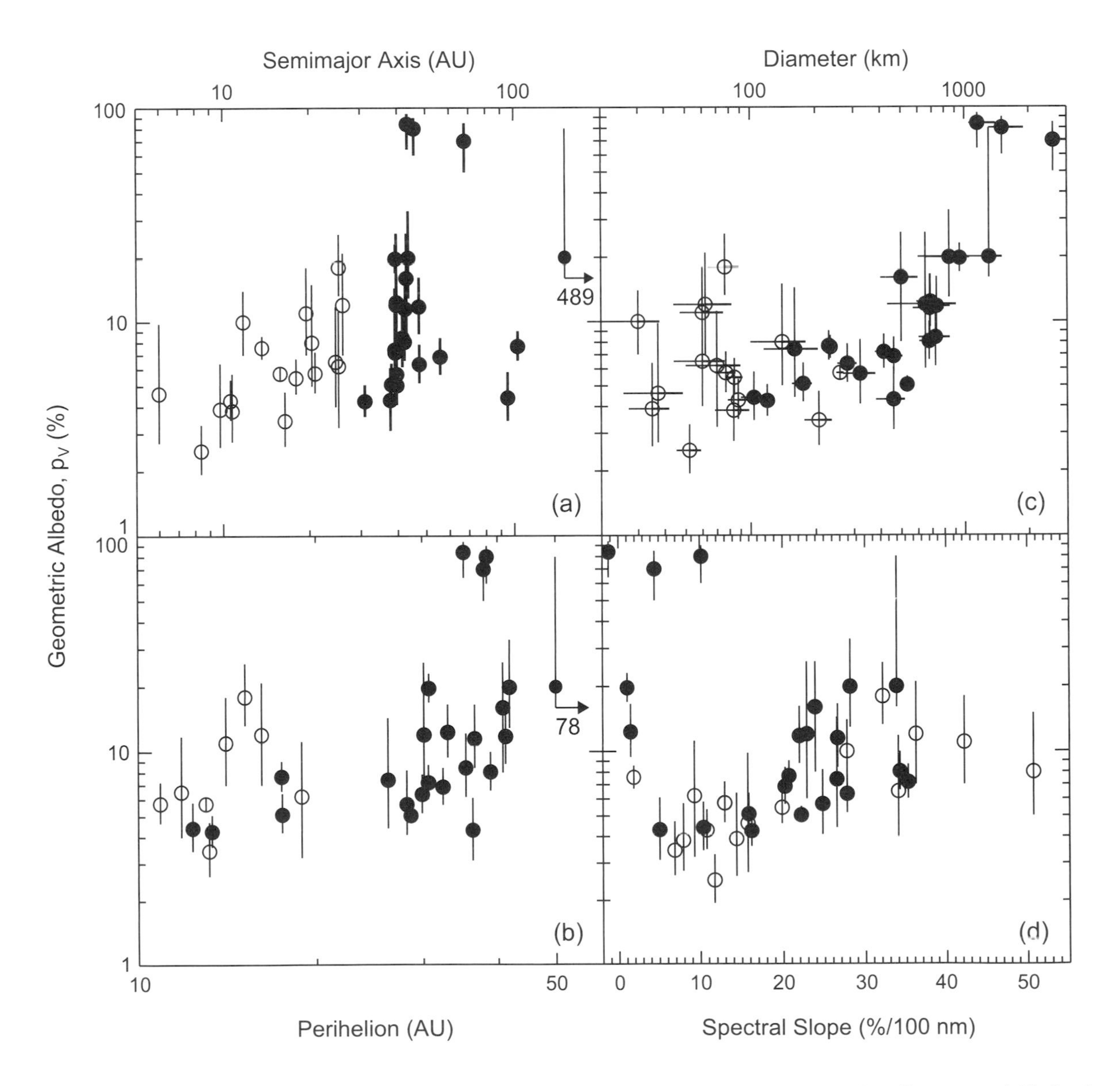

Fig. 3. Geometric albedo plotted vs. **(a)** orbital semimajor axis, **(b)** orbital perihelion distance, **(c)** object diameter, and **(d)** the slope of the object's visible spectrum (i.e., color). Open circles are for Centaur objects (a < 30.066 AU); filled circles are for TNOs. In **(a)** the point for 90377 Sedna has been plotted at a = 150 AU rather than at its true semimajor axis of 489 AU. In **(b)** the point for 90377 Sedna has been plotted at q = 50 AU, rather than at its true perihelion distance of 78 AU.

TABLE 5. Geometric albedo.

	All		MPC Classification*				DES Classification†			
			Centaurs		KBOs		Centaurs		KBOs	
Quantity	Statistics									
Average	8.01		6.55		8.87		6.30		9.88	
Median	6.85		5.74		7.67		5.73		8.41	
σ‡	4.07		2.68		4.22		2.50		4.23	
# Obj.§	35		15		18		19		14	
	Correlations									
Parameter	ρ¶	χ**	ρ¶	χ**	ρ¶	χ**	ρ¶	χ**	ρ¶	χ**
a	0.46	2.74	0.70	2.78	0.24	1.03	0.41	1.81	0.16	0.60
$q_{\odot}$	0.58	3.49	0.58	2.32	0.65	2.83	0.43	1.94	0.53	2.04
D	0.45	2.70	–0.08	0.32	0.77	3.37	–0.07	0.31	0.72	2.80
S	0.40	2.40	0.64	2.58	0.13	0.56	0.66	2.94	–0.08	0.32

*Centaurs classified as objects having orbital semimajor axes <30.066 AU.
†Centaurs classified by dynamical simulations [Deep Ecliptic Survey (*Elliot et al.,* 2005)].
‡Standard deviation of the albedo values.
§Number of Spitzer albedos used (from Table 4). The highest and lowest values were excluded.
¶Spearman rank correlation coefficient between albedo and the parameter in the left column.
**Significance of the correlation, in standard deviations relative to the null hypothesis.

The dichotomy between 136199 Eris, 136108 (2003 EL_{61}), 136472 (2005 FY_9), Pluto (and probably 90377 Sedna), and the rest of the KBOs and the Centaurs, in terms of their albedos and spectral characteristics, suggest that they are members of a unique physical class within the Kuiper belt population (see the chapter by Brown). We will refer to these objects as "planetoids" in the following, and generally exclude them from our discussion of albedo statistics and correlations because of their obviously unique character.

5.1. Albedo Statistics

Table 5 summarizes the statistics of the Spitzer-derived albedos, and the correlations between albedo and other parameters. Because there is no clearly preferred way to differentiate Centaurs from KBOs, we give results for two definitions: $a < 30.066$ AU (which we term the MPC Definition, referring to the Minor Planet Center classification) (see the chapter by Gladman et al.), and the DES Definition (referring to the Deep Ecliptic Survey classification) (*Elliot et al.,* 2005).

Typical geometric albedos for all of the KBOs and Centaurs are in the range 6.9–8.0%, depending on whether the mean or median is used, with a dispersion of about 4.1%. Regardless of which Centaur classification one chooses, it appears that Centaurs may have slightly lower albedos than KBOs, although the differences are not statistically significant relative to the dispersion of the albedos within the classes. The Kuiper variant of the Kolmogorov-Smirnov (K-S) test gives no evidence that the albedos of the KBOs and Centaurs are drawn from different parent populations, regardless of whether the MPC or DES definition of Centaur is used. Typical values for the beaming parameter (excluding results based on an assumed beaming parameter) are in the range 1.1–1.20, with a dispersion of about 0.4. This is in good agreement with the value of 1.2 ± 0.35 we adopted for the "KBO-tuned STM" used to construct Table 3. There does not appear to be any significant difference in the beaming parameter between KBOs and Centaurs.

5.2. Albedo Correlations

Because our errors are nonsymmetric and probably non-Gaussian, we apply the Spearman rank-correlation test to assess the significance of any correlations between albedo and other parameters. Table 5 gives the correlation coefficients (ρ) and their significance (χ) in standard deviations from the noncorrelated case. The albedos for the four planetoids mentioned above are not included in these calculations.

Figures 3a and 3b show p_V as a function of the orbital properties a and $q_{\odot}$. There is an upward trend of p_V vs. a, with the objects at $a < 20$ AU clustering at $p_V \simeq 5\%$, while at larger distances there is significantly more scatter in p_V. As shown in Table 5, the correlation between p_V and a for the entire sample is significant at the $\chi = 2.7\sigma$ level (99.4% likelihood). It appears that most of the correlation is due to the Centaurs, but the significance of the Centaur correlation depends considerably on which definition of Centaur is used. (Note that because the number of objects in the KBO and Centaur subsamples is about half that of the full sample, the significance of the correlations for the subsamples is typically lower than that for the full sample.) Because the significance of the p_V vs. a correlation is below 3σ, it is tentative. Another reason to treat the correlation with some skepticism is that it could reflect biases in the parameter space for KBO discoveries: Low-albedo objects will be harder to detect at visible wavelengths, and the difficulty increases significantly with distance. Because our sample is drawn from optically discovered objects, one might expect a trend such as seen in Fig. 3a even if there is no real correlation between p_V and a.

Figure 3b reveals a similar correlation between p_V and $q_\odot$, and Table 5 suggests that in this case the correlation is significant at the $\chi = 3.5\sigma$ level (99.95% likelihood). This correlation holds up fairly well for both Centaurs and KBOs, regardless of which classification is used. It is possible that this correlation could also be due to the discovery bias mentioned above. However, if it reflects an actual relationship between p_V and q, there may be a fairly simple explanation. Objects closer to the Sun will tend to experience higher temperatures, depleting their surfaces of volatile molecules (which typically have high visible reflectances). Likewise, UV photolysis and solar wind radiolysis will also proceed more quickly closer to the Sun, and could darken those surfaces (although radiolysis by cosmic rays probably dominates beyond about 45 AU).

Figures 3c and 3d show p_V as a function of intrinsic properties of the objects: the diameter, D, and the visible spectral slope (color), S. Figure 3c shows an apparent correlation between p_V and D, and particularly so for the KBOs. This correlation is apparently confirmed in Table 5, where for the MPC classification the p_V vs. D correlation is significant at the $\chi = 3.4\sigma$ level (99.9% likelihood). However, for the DES classification the significance is only $\chi = 2.8\sigma$ (99.5% likelihood), so the correlation is not robust against small changes in which objects are considered as KBOs. Including the planetoids in the correlation calculation increases the significance of the correlation to well above 3σ, but doing so results in a (probably) false impression that the albedos of *all* KBOs are well correlated with diameter. At this time it is difficult to conclude that any such correlation exists at a statistically significant level.

Figure 3d shows an apparent correlation between p_V and S, particularly for the Centaurs. Table 5 shows that this correlation is the second most significant for a subclass, with $2.6 \leq \chi \leq 2.9$ (depending on the classification chosen), second only to the p_V vs. D correlation for KBOs. Here, the Kuiper variant K-S test does indicate a high likelihood (99.95%) that the albedos of red KBOs and Centaurs (with $S > 0.2$) are drawn from a different parent population than the gray ones, a similar result to that found based on the Centaur colors alone (see chapter by Tegler et al.). A natural assumption might be that the color diversity of KBOs and Centaurs results from mixing between icy (bright, spectrally neutral) and organic (dark, red) components. However, this correlation suggests that red objects systematically have higher albedos than the gray ones. On the basis of spectral mixing models between spectrally neutral dark materials (such as charcoal) and red material (represented by Titan tholin), *Grundy and Stansberry* (2003) suggested that just such a correlation between red color and higher albedo might exist. Why the Centaurs might embody this effect more strongly than the KBOs is still a mystery. Interestingly, the three most spectrally neutral objects defy the color-albedo trend, having rather high albedos: There may be at least two mechanisms underlying the observed color diversity. Those objects are 2060 Chiron, 90482 Orcus, and (2003 AZ_{84}), and their unique position in the albedo-color plane may indicate that they share some unique surface character.

6. FUTURE PROSPECTS

At present, Spitzer/MIPS provides the most sensitive method available for measuring thermal fluxes from typical KBOs, but several upcoming observatories and instruments will provide substantially improved sensitivity. The joint ESA/NASA Herschel mission will have at least a factor of 2 better sensitivity at 75 μm (compared to the 70-μm sensitivity of MIPS), and will additionally have a number of photometry channels in the range 70–500 μm. Since cold KBOs have their thermal emission peaks in the range 60–100 μm, observations in the Herschel band passes will map the peak of a KBO SED. Herschel is scheduled for launch in late 2008. The Large Millimeter Telescope (LMT) in central Mexico will have sufficient sensitivity at 1 mm with the SPEED instrument to detect thermal flux from the Rayleigh-Jeans tail of cold KBOs. First light for the LMT is expected in 2008.

Farther in the future, the American-European-Chilean Atacama Large Millimeter Array (ALMA) will provide sufficient sensitivity from 0.35 to 3 mm to detect typical KBOs; first light for ALMA might be as soon as 2012. The Cornell Caltech Atacama Telescope (CCAT) will operate at 200 μm to 1 mm, and its sensitivity at 350 μm will surpass that of ALMA; first light could also be in 2012. Any of Herschel, ALMA, and CCAT (the case is less convincing for the LMT) could be used for a large survey of many moderate-sized (100-km-class) KBOs. Such a program would expand the number of KBOs with good thermal measurements (and therefore radii and albedos) from tens to hundreds.

All these next-generation capabilities operate at wavelengths either near the emission peak of KBOs, or well out on the Rayleigh-Jeans part of their spectra. While albedos and diameters derived from such observations are less model-dependent than those based on single-wavelength observations taken shortward of the emission peak, there are still significant uncertainties. For example, canonical STM and ILM fits to an 850-μm flux density produce albedos that differ by about 30%; if the KBO-tuned STM is used (including its uncertainty on η), that uncertainty is cut almost in half. If the validity of the KBO-tuned STM is born out by further Spitzer observations of KBOs, it can be used to significantly refine the albedos and diameters derived from submillimeter KBO detections.

7. SUMMARY

Efforts to characterize the physical properties of KBOs and Centaurs with Spitzer are beginning to pay off. Considerable improvements have been made in the first three years of the mission in terms of predicting the necessary integration times, developing aggressive and successful observing strategies, and data processing. We present our 24- and 70-μm observations for 47 targets (31 with orbital semimajor axes larger than that of Neptune, 16 inside Neptune's orbit), and apply a modified version of the standard thermal model to derive albedos and diameters for them. Thirty-nine of the targets were detected at signal-to-noise ratios ≥5 at one or both wavelengths. We use that sample to look for relationships between albedo and the orbital and physical parameters of the objects. The most marked such relationship is the distinct discontinuity in albedo at a diameter of about 1000 km, with objects larger than that having albedos in excess of 60%, and those smaller than that having albedos below about 25%. We suggest that these large, very-high-albedo objects [90377 Sedna, 136108 (2003 EL_{61}), 136199 Eris, and 136472 (2005 FY_9)] constitute a distinct class in terms of their physical properties.

The data suggest possible correlations of albedo with orbital distance, and with size and color, but the statistical significance of the correlations is marginal. Two correlations, those of albedo with perihelion distance (for KBOs and Centaurs) and with diameter (for KBOs), are nominally significant at more than the 3σ level. Perhaps the most interesting trend (albeit significant at only about the 2.8σ level) is for distinctly red Centaurs to have higher albedos than those that are more gray, contrary to what might intuitively be expected.

Prospects for improving on and expanding these results are relatively good. Spitzer will be operational into 2009, and more KBO observations will probably be approved. New ground- and spacebased observatories will also contribute significantly, and at wavelengths that are complementary to those used here. In particular, submillimeter to millimeter studies of KBOs should be relatively easy with facilities such as ALMA, CCAT, and LMT. The Herschel mission should also be very productive at far-IR to submillimeter wavelengths.

Acknowledgments. Many of the results presented here are based on observations made with the Spitzer Space Telescope, which is operated by the Jet Propulsion Laboratory, California Institute of Technology, under contract with NASA. Support for this work was provided through multiple awards issued by JPL. Online resources offered by the Solar System Dynamics group at JPL and the Minor Planet Center of the International Astronomical Union were extremely valuable.

REFERENCES

Altenhoff W. J., Menten K. M. and Bertoldi F. (2002) Size determination of the Centaur Chariklo from millimeter-wavelength bolometer observations. *Astron. Astrophys., 366,* L9–12.

Altenhoff W. J., Bertoldi F., and Menten K. M. (2004) Size estimates of some optically bright KBOs. *Astron. Astrophys., 415,* 771–775.

Barucci M. A., Boehnhardt H., Dotto E., Doressoundiram A., Romon J., et al. (2002) Visible and near-infrared spectroscopy of the Centaur 32532 (2001 PT_{13}). ESO Large Program on TNOs and Centaurs: First spectroscopy results. *Astron. Astrophys., 392,* 335–339.

Barucci M. A., Cruikshank D.P., Dotto E., Merlin F., Poulet F., et al. (2005) Is Sedna another Triton? *Astron. Astrophys., 439,* L1–L4.

Bauer J. M., Meech K. J., Fernandez Y. R., Pittichova J., Hainaut O. R., Boehnhardt H., and Delsanti A. (2003) Physical survey of 24 Centaurs with visible photometry. *Icarus, 166,* 195–211.

Bertoldi F., Altenhoff W., Weiss A., Menten K. M., and Thum C. (2006) The trans-Neptunian object UB_{313} is larger than Pluto. *Nature, 439,* 563–564.

Bowell E., Hapke B., Domingue D., Lumme K., Peltoniemi J., and Harris A. (1989) Application of photometric models to asteroids. In *Asteroids II* (R. P. Binzel et al., eds.), pp. 524–556. Univ. of Arizona, Tucson.

Brown M. E. and Trujillo C. A. (2004) Direct measurement of the size of the large Kuiper belt object (50000) Quaoar. *Astron. J., 127,* 2413–2417.

Brown M. E., Trujillo C. A., Rabinowitz D., Stansberry J., Bertoldi F., and Koresko C. D. (2004) A Sedna update: Source, size, spectrum, surface, spin, satellite. *Bull. Am. Astron. Soc., 36,* 1068.

Brown M. E., Trujillo C. A., and Rabinowitz D. L. (2005) Discovery of a planetary-sized object in the scattered Kuiper belt. *Astrophys. J. Lett., 635,* L97–L100.

Brown M. E., Schaller E. L., Roe H. G., Rabinowitz D. L., and Trujillo C. A. (2006) Direct measurement of the size of 2003 UB_{313} from the Hubble Space Telescope. *Astrophys. J. Lett., 643,* L61–L63.

Brown M. E., Barkume K. M., Blake G. A., Schaller E. L., Rabinowitz D. L., Roe H. G., and Trujillo C. A. (2007) Methane and ethane on the bright Kuiper belt object 2005 FY_9. *Astron. J., 133,* 284–289.

Brown R. H., Cruikshank D. P., and Pendleton Y. (1999) Water ice on Kuiper belt object 1996 TO_{66}. *Astrophys. J. Lett., 519,* L101–L104.

Campins H., Telesco C. M., Osip D. J., Rieke G. H., Rieke M. J., and Shulz B. (1994) The color temperature of (2060) Chiron: A warm and small nucleus. *Astron. J., 108,* 2318–2322.

Cruikshank D. P. and Brown R. H. (1983) The nucleus of comet P/Schwassmann-Wachmann 1. *Icarus, 56,* 377–380.

Cruikshank D. P., Stansberry J. A., Emery J. P., Fernández Y. R., Werner M. W., Trilling D. E., and Rieke G. H. (2005) The high-albedo Kuiper belt object (55565) 2002 AW_{197}. *Astrophys J., 624,* L53–L56.

Cruikshank D. P., Barucci M. A., Emery J. P., Fernández Y. R., Grundy W. M., Noll K. S., and Stansberry J. A. (2006) Physical properties of transneptunian objects. In *Protostars and Planets V* (B. Reipurth et al., eds.), pp. 879–893. Univ. of Arizona, Tucson.

Davies J., Spencer J., Sykes M., Tholen D., and Green S. (1993) (5145) Pholus. *IAU Circular 5698.*

de Bergh C., Boehnhardt H., Barucci M. A., Lazzarin M., Fornasier S., et al. (2004) Aqueous altered silicates at the surface of two Plutinos? *Astron. Astrophys., 416,* 791–798.

Delbo M., Harris A. W., Binzel R. P., Pravec P., and Davies J. K. (2003) Keck observations of near-Earth asteroids in the thermal infrared. *Icarus, 166,* 116–130.

Elliot J. L., Kern S. D., Clancy K. B., Gulbis A. A. S., Millis R. L., et al. (2005) The Deep Ecliptic Survey: A search for Kuiper belt objects and Centaurs. II. Dynamical classification, the Kuiper belt plane, and the core population. *Astron. J., 129,* 1117–1162.

Emery J. P., Dalle Ore C. M., Cruikshank D. P., Fernández Y. R., Trilling D. E., and Stansberry J. A. (2007) Ices on (90377) Sedna: Confirmation and compositional constraints. *Astron. Astrophys., 466,* 395–398.

Engelbracht C. W., Blaylock M., Su K. Y. L., Rho J., Rieke G. H., et al. (2007) Absolute calibration and characterization of the Multiband Imaging Photometer for Spitzer. I. The stellar calibrator sample and the 24 micron calibration. *Publ. Astron. Soc. Pac.,* in press.

Fernández Y. R., Jewitt D. C., and Sheppard S. S. (2002) Thermal properties of Centaurs Asbolus and Chiron. *Astron. J., 123,* 1050–1055.

Fernández Y. R., Sheppard S. S., and Jewitt D. C. (2003) The albedo distribution of jovian Trojan asteroids. *Astron. J., 126,* 1563–1574.

Fornasier S., Doressoundiram A., Tozzi G. P., Barucci M. A., Boehnhardt H., et al. (2004) ESO Large Program on Physical Studies of Trans-Neptunian Objects and Centaurs: Final results of the visible spectrophotometric observations. *Astron. Astrophys., 421,* 353–363.

Gordon K. G., Engelbracht C. W., Fadda D., Stansberry J. A., Wacther S., et al. (2007) Absolute calibration and characterization of the Multiband Imaging Photometer for Spitzer II. 70 micron imaging. *Publ. Astron. Soc. Pac.,* in press.

Groussin O., Lamy P., and Jorda L. (2004) Properties of the nuclei of Centaurs Chiron and Chariklo. *Astron. Astrophys., 413,* 1163–1175.

Grundy W. M. and Stansberry J. A. (2003) Mixing models, colors, and thermal emissions. *Earth Moon Planets, 92,* 331–336.

Grundy W. M., Noll K. S., and Stephens D. C. (2005) Diverse albedos of small trans-neptunian objects. *Icarus, 176,* 184–191.

Grundy W. M., Stansberry J. A., Noll K. S., Stephens D. C., Trilling D. E., Kern S. D., Spencer J. R., Cruikshank D. P., and Levison H. F. (2007) The orbit, mass, size, albedo, and density of (65489) Ceto-Phorcys: A tidally-evolved binary Centaur. *Icarus,* in press.

Harris A. W. (1998) A thermal model for near-Earth asteroids. *Icarus, 131,* 291–301.

Horner J., Evans N. W., and Bailey M. E. (2004) Simulations of the population of Centaurs — I. The bulk statistics. *Mon. Not. R. Astron. Soc., 354,* 798–810.

Jewitt D. C. and Luu J. X. (1993) Discovery of the candidate Kuiper belt object 1992 QB_1. *Nature, 362,* 730–732.

Jewitt D. C., Aussel H., and Evans A. (2001) The size and albedo of the Kuiper-belt object (20000) Varuna. *Nature, 411,* 446–447.

Jones T. J. and Morrison D. (1974) Recalibration of the photometric/radiometric method of determining asteroid sizes. *Astron. J., 79,* 892–895.

Lazzarin M., Barucci M. A., Boehnhardt H., Tozzi G. P., de Bergh C., and Dotto E. (2003) ESO Large Programme on Physical Studies of Trans-Neptunian Objects and Centaurs: Visible spectroscopy. *Astrophys. J., 125,* 1554–1558.

Lebofsky L. A. and Spencer J. R. (1989) Radiometry and thermal modeling of asteroids. In *Asteroids II* (R. P. Binzel et al., eds.), pp. 128–147. Univ. of Arizona, Tucson.

Lebofsky L. A., Sykes M. V., Tedesco E. F., Veeder G. J., Matson D. L., et al. (1986) A refined 'standard' thermal model for asteroids based on observations of 1 Ceres and 2 Pallas. *Icarus, 68,* 239–251.

Lellouch E., Laureijs R., Schmitt B., Quirico E., de Bergh C., et al. (2000) Pluto's non-isothermal surface. *Icarus, 147,* 220–250.

Lellouch E., Moreno R., Ortiz J. L., Paubert G., Doressoundiram A., and Peixinho N. (2002) Coordinated thermal and optical observations of trans-neptunian object (20000) Varuna from Sierra Nevada. *Astron. Astrophys., 391,* 1133–1139.

Levison H. F. and Duncan M. J (1997) From the Kuiper belt to Jupiter-family comets: The spatial distribution of ecliptic comets. *Icarus, 127,* 13–32.

Licandro J., Pinilla-Alonso N., Pedani M., Oliva E., Tozzi G. P., and Grundy W. M. (2006a) The methane ice rich surface of large TNO 2005 FY_9: A Pluto-twin in the trans-neptunian belt? *Astron. Astrophys., 445,* L35–L38.

Licandro J., Grundy W. M., Pinilla-Alonso N., and Leysi P. (2006b) Visible spectroscopy of TNO 2003 UB_{313}: Evidence for N_2 ice on the surface of the largest TNO? *Astron. Astrophys., 458,* L5–L8.

Margot J. L., Brown M. E., Trujillo C. A., and Sari R. (2002) The size and albedo of KBO 2002 AW_{197} (abstract). *Bull. Am. Astron. Soc., 34,* 871.

Margot J. L., Brown M. E., Trujillo C. A., and Sari R. (2004) HST observations of Kuiper belt binaries (abstract). *Bull. Am. Astron. Soc., 36,* 1081.

Merlin F., Barucci M. A., Dotto E., de Bergh C., and Lo Curto G. (2005) Search for surface variations on TNO 47171 and Centaur 32532. *Astron. Astrophys., 444,* 977–982.

Morrison D. and Lebofsky L. A. (1979) Radiometry of asteroids. In *Asteroids* (T. Gehrels, ed.), pp. 184–205. Univ. of Arizona, Tucson.

Morrison D., Owen T., and Soderblom L. A. (1986) The satellites of Saturn. In *Satellites* (J. A. Burns and M. S. Matthews, eds.), pp. 764–801. Univ. of Arizona, Tucson.

Noll K. S., Stephens D. C., Grundy W. M., and Griffin I. (2004) The orbit, mass, and albedo of trans-neptunian binary (66652) 1999 RZ_{253}. *Icarus, 172,* 402–407.

Ortiz J. L., Sota A., Moreno R., Lellouch E., Biver N., et al. (2004) A study of trans-neptunian object 55636 (2002 TX_{300}). *Astron. Astrophys., 420,* 383–388.

Osip D. J., Kern S. D., and Elliot J. L. (2003) Physical characterization of the binary Edgeworth-Kuiper belt object 2001 QT_{297}. *Earth Moon Planets, 92,* 409–421.

Rabinowitz D. L., Barkume K., Brown M. E., Roe H., Schwartz M., et al. (2005) Photometric observations constraining the size, shape, and albedo of 2003 EL_{61}, a rapidly rotating, Pluto-sized object in the Kuiper belt. *Astrophys. J., 639,* 1238–1251.

Rieke G. H., Young E. T., Engelbracht C. W., Kelly D. M., Low F. J., et al. (2004) The Multiband Imaging Photometer for Spitzer (MIPS). *Astrophys. J. Suppl., 154,* 25–29.

Romanishin W. and Tegler S. C. (2005) Accurate absolute magnitudes for Kuiper belt objects and Centaurs. *Icarus, 179,* 523–526.

Schaller E. L. and Brown M. E. (2007) Volatile loss and retention on Kuiper belt objects and the depletion of nitrogen on 2005 FY_9. *Astrophys. J. Lett., 659,* L61–L64.

Spencer J. R. (1990) A rough-surface thermophysical model for airless planets. *Icarus, 83,* 27–38.

Stansberry J. A., Van Cleve J., Reach W. T., Cruikshank D. P., Emery J. P., et al. (2004) Spitzer observations of the dust coma and nucleus of 29P/Schwassmann-Wachmann 1. *Astrophys. J. Suppl., 154,* 463–468.

Stansberry J. A., Grundy W. M., Margot J. L., Cruikshank D. P., Emery J. P., et al. (2006) The albedo, size, and density of binary Kuiper belt object (47171) 1999 TC_{36}. *Astrophys. J., 643,* 556–566.

Stansberry J. A., Gordon K. D., Bhattacharya B., Engelbracht

C. W., Rieke G. H., et al. (2007) Absolute calibration and characterization of the Multiband Imaging Photometer for Spitzer III. An asteroid-based calibration at 160 microns. *Publ. Astron. Soc. Pac.,* in press.

Sykes M. V. (1999) IRAS survey-mode observations of Pluto-Charon. *Icarus, 142,* 155–159.

Sykes M. V. and Walker R. G. (1991) Constraints on the diameter and albedo of Chiron. *Science, 251,* 777–780.

Tedesco E. F., Veeder G. J., Fowler J. W., and Chillemi J. R. (1992) *The IRAS Minor Planet Survey.* Phillips Lab. Tech. Report PL-TR-92-2049, Hanscom AFB, Massachussettes.

Tedesco E. F., Noah P. V., Noah M., and Price S. D. (2002) The supplemental IRAS Minor Planet Survey. *Astron. J., 123,* 1056–1085.

Tegler S. C., Grundy W. M., Romanishin W., Consolmagno G. J., Mogren K., and Vilas F. (2007) Optical spectroscopy of the large Kuiper belt objects 136472 (2005 FY_9) and 136108 (2003 EL_{61}). *Astron. J., 133,* 526–530.

Thomas N., Eggers S., Ip W.-H., Lichtenberg G., Fitzsimmons A., Jorda L., et al. (2000) Observations of the trans-neptunian objects 1993 SC and 1996 TL_{66} with the Infrared Space Observatory. *Astrophys. J., 534,* 446–455.

Trilling D. E. and Bernstein G. M. (2006) Light curves of 20–100 km Kuiper belt objects using the Hubble Space Telescope. *Astron. J., 131,* 1149–1162.

Veillet C., Parker J. W., Griffin I., Marsden B., Doressoundiram A., et al. (2002) The binary Kuiper-belt object 1998 WW_{31}. *Nature, 416,* 711–713.

Werner M. W., Roellig T. L., Low F. J., Rieke G. H., Rieke M. J., et al. (2004) The Spitzer Space Telescope mission. *Astrophys. J. Suppl., 154,* 1–9.

Transneptunian Object Taxonomy

Marcello Fulchignoni
LESIA, Observatoire de Paris

Irina Belskaya
Kharkiv National University

Maria Antonietta Barucci
LESIA, Observatoire de Paris

Maria Cristina De Sanctis
IASF-INAF, Rome

Alain Doressoundiram
LESIA, Observatoire de Paris

A taxonomic scheme based on multivariate statistics is proposed to distinguish groups of TNOs having the same behavior concerning their BVRIJ colors. As in the case of asteroids, the broadband spectrophotometry provides a first hint about the bulk compositional properties of the TNOs' surfaces. Principal components (PC) analysis shows that most of the TNOs' color variability can be accounted for by a single component (i.e., a linear combination of the colors): All the studied objects are distributed along a quasicontinuous trend spanning from "gray" (neutral color with respect to those of the Sun) to very "red" (showing a spectacular increase in the reflectance of the I and J bands). A finer structure is superimposed to this trend and four homogeneous "compositional" classes emerge clearly, and independently from the PC analysis, if the TNO sample is analyzed with a grouping technique (the G-mode statistics). The first class (designed as BB) contains the objects that are neutral in color with respect to the Sun, while the RR class contains the very red ones. Two intermediate classes are separated by the G mode: the BR and the IR, which are clearly distinguished by the reflectance relative increases in the R and I bands. Some characteristics of the classes are deduced that extend to all the objects of a given class the properties that are common to those members of the class for which more detailed data are available (observed activity, full spectra, albedo). The distributions of the classes with respect to the distance from the Sun and to the orbital inclination give some hints on the chemico-physical structure of the inner part of the Kuiper belt. An interpretation of the average broadband spectra of the four classes as the result of modifying processes (collisions, space weathering, degassing, etc.), allows us to read the proposed taxonomy in terms of the evolution of TNOs.

1. INTRODUCTION

Taxonomy (from the Greek verb τασσεὶν or *tassein* = "to classify" and νόμος or *nomos* = "law, science") was once only the science of classifying living organisms (alpha taxonomy), but later the word was applied in a wider sense, and may also refer to either a classification of things, or the principles underlying the classification. Almost anything — animate objects, inanimate objects, places, and events — may be classified according to some taxonomic scheme. Such an approach to studying physical properties became an efficient tool in asteroid investigations, which enabled the expansion of our knowledge of physical properties of well-studied asteroids within a taxon to other members belonging to the same taxon.

Multivariate and canonical (parametrical and/or nonparametrical) analysis of the color distribution of transneptunian objects (TNOs) and associated families and their orbital parameters, together with recently developed cluster analysis methods, could be used in order to set the basis of taxonomy. The objectives are to demonstrate, rule out, or constrain the several dynamical, thermal, and surface evolution models and characterize the several dynamical subclasses of TNOs by their surface and orbital properties, defining or constraining their boundaries.

Wood and Kuiper (1963) suggested, on the basis of the distribution of about 40 asteroids in the U–B and B–V plot, the existence of two different compositional classes clustering around the color indices of the Moon and the Sun. These groups were the ancestors of the currently used asteroid S and C classes, respectively.

Tholen and Barucci (1989) summarized the results of their asteroid taxonomies for about 400 asteroids obtained by means of independent multivariate statistical techniques [the principal components (PC) analysis based on eight-color data from *Zellner et al.* (1985) and the G-mode analysis based on the same eight colors database to which the IRAS albedo data from *Tedesco et al.* (1992) were added]. *Tedesco et al.* (1989) obtained three-parameter taxonomy analyzing more than 350 asteroids described by two colors indices (U–V and v–x) and the visual albedo. All these methods provided a similar classification scheme separating the asteroid population into a dozen compositionally homogeneous groups, as outlined by *Barucci and Fulchignoni* (1990).

Analyzing the B–V and V–R colors of 13 TNOs and Centaurs, *Tegler and Romanishin* (1998) obtained two groups, one very red and the other quite neutral with respect to the Sun. *Barucci et al.* (2001) applied both techniques used in classifying the asteroids to a sample of 22 TNOs and Centaurs characterized by four colors (B–V, V–R, V–I, and V–J). The results indicate a clear compositional trend within the examined sample and suggest the possible existence of four homogeneous groups. The increase of the available data allowed *Fulchignoni et al.* (2003) and *Barucci et al.* (2005a) to confirm this result at a higher significance level, analyzing samples of 34 and 51 objects respectively.

An interesting attempt to characterize the TNOs on the basis of their orbital elements and some physical properties has been recently carried out by *De Sanctis et al.* (2006), who used the G mode to analyze 81 TNOs chosen among those with well-known dynamical parameters, described by the color indices B–V and V–R, the absolute magnitude H, the orbital inclination i, orbital eccentricity e, and semimajor axis a. The G-mode analysis separates the 81 objects of the sample in five groups, well separated for the dynamical parameters and less separated in colors. The obtained groups are close to the well-known dynamical ones (classical, Plutinos, scattered, and detached), but the classification firmly identifies two different types of classical objects, corresponding to the "so-called" dynamically cold and hot classical population. Moreover, they found trends and correlations within the members of the different groups.

To distinguish groups of objects with similar properties we need to have a statistically representative dataset. However, the knowledge of physical properties of TNOs is still very limited: To date, the available information concerning TNOs include (1) orbital parameters (known for all the objects with different degrees of precision) and (2) broadband B, V, R, I photometry for about 130 objects; for approximately 70 of them J data are also available, while H and K photometry has been obtained for only about 55 of them.

Hereafter we present the first taxonomic scheme used to obtain a classification of the TNOs. In fact, when dealing with a large number of objects, it is very important to distinguish groups of objects with similar properties, preferably based on measured common characteristics.

2. DATA AND STATISTICAL METHODS

We successively analyzed two homogeneous TNO samples: (1) 67 objects described by four color indices (B–V, V–R, V–I, V–J); (2) 55 objects described by six color indices (B–V, V–R, V–I, V–J, V–H, V–K). The list of objects and their colors together with the references to the original data are given in Table 1.

When multiple observations of an object were available, for the colors we adopted their mean values weighted with the inverse of the error of individual measurement and the standard deviation was assumed as the error. In the case of a single measurement we restricted our consideration to those objects for which colors were determined with an error less than 0.1 mag for BVRIJ colors. The selected data represent a homogeneous dataset in B, V, R, I, J, H, and K bands in general obtained by the same observer during the same run, or intercalibrated through the V measurements. We used nonsimultaneous V and the JHK set of magnitudes in a few cases when it was possible to recalculate V magnitude to the epoch of JHK measurements taking into account the geometry of observations.

The analysis has been carried out using both principal component analysis (PCA) (*Reyment and Joreskog,* 1993) and G-mode analysis (*Coradini et al.,* 1977; *Fulchignoni et al.,* 2000).

The principal components (PC) are linear combinations of the original variables whose coefficients reflect the relative importance of each variable (color) within each principal component. These coefficients are the eigenvectors of the variance-covariance matrix of the colors. The sum of the eigenvalues of this matrix (which is equal to its trace) accounts for the total variance of each sample. Each eigenvalue reflects the percentage of the total variance contributed by each principal component.

We analyzed the same samples with the G-mode multivariate statistics (*Coradini et al.,* 1977), which allowed us to investigate the existence of a finer structure of the samples. G-mode statistics were applied to our samples of 67 and 55 objects described by four and six variables, respectively. The total number of degrees of freedom (268 and 330 respectively) allow us to use this type of statistics. The goal of the analysis is to find groups of objects that have a homogeneous behavior, if any, in terms of their physical characteristics (variables) under consideration. The method provides a quantitative estimation of the weight of each variable in separating the groups. We refer the reader to the quoted literature for details on both PC and G-mode statistics.

TABLE 1. Average colors of the selected sample objects.

Object/Type*	B–V	V–R	V–I	V–J	V–H	V–K	References
Sun	0.67	0.36	0.69	1.08	1.37	1.43	1,2
2060 Chiron/C	0.63 ± 0.02	0.35 ± 0.01	0.70 ± 0.04	1.13 ± 0.01	1.43 ± 0.01	1.50 ± 0.03	14,15,16,17,46
5145 Pholus/C	1.25 ± 0.03	0.77 ± 0.01	1.58 ± 0.01	2.57 ± 0.03	2.94 ± 0.04	2.93 ± 0.04	8,15,16,17,46
7066 Nessus/C	1.09 ± 0.01	0.79 ± 0.01	1.47 ± 0.03	2.29 ± 0.01	2.57 ± 0.10	2.57 ± 0.10	14,15,16,17
8405 Asbolus/C	0.75 ± 0.01	0.47 ± 0.02	0.98 ± 0.01	1.65 ± 0.02	2.06 ± 0.04	2.22 ± 0.08	15,16,17,18,19
10199 Chariklo/C	0.80 ± 0.02	0.48 ± 0.01	1.01 ± 0.01	1.73 ± 0.03	2.14 ± 0.03	2.21 ± 0.03	3,5,16,17,42
10370 Hylonome/C	0.69 ± 0.06	0.43 ± 0.02	0.96 ± 0.03	1.32 ± 0.01	1.50 ± 0.08	1.77 ± 0.09	5,9,15,16,17,42
15788 1993 SB/R	0.80 ± 0.02	0.47 ± 0.01	1.01 ± 0.01	1.43 ± 0.11			3,4,6,20,22,23
15789 1993 SC/R	1.08 ± 0.08	0.70 ± 0.06	1.49 ± 0.04	2.42 ± 0.07	2.82 ± 0.21	2.78 ± 0.20	3,4,7,8,14,
15820 1994 TB/R	1.10 ± 0.02	0.69 ± 0.02	1.43 ± 0.03	2.37 ± 0.09	2.78 ± 0.09	2.93 ± 0.09	3,4,8,9,11,12,20,46
15874 1996 TL_{66}/S	0.73 ± 0.03	0.37 ± 0.02	0.72 ± 0.01	1.46 ± 0.10	1.81 ± 0.17	1.77 ± 0.15	3,4,5,7,10,46
15875 1996 TP_{66}/R	1.05 ± 0.06	0.66 ± 0.02	1.31 ± 0.07	2.26 ± 0.08	2.42 ± 0.08	2.44 ± 0.08	3,4,5,7,9,10,46
19299 1996 SZ_4/R	0.75 ± 0.08	0.52 ± 0.03	0.97 ± 0.14	1.87 ± 0.13			3,4,22,24
19308 1996 TO_{66}/Cl	0.67 ± 0.03	0.40 ± 0.02	0.75 ± 0.02	1.00 ± 0.10	0.79 ± 0.20	1.60 ± 0.18	3,5,7,10,25,26,27
19521 Chaos/Cl	0.94 ± 0.03	0.62 ± 0.01	1.19 ± 0.05	1.89 ± 0.03	2.29 ± 0.03	2.32 ± 0.04	4,9,10,11,12,22,46
20000 Varuna/Cl	0.88 ± 0.02	0.61 ± 0.02	1.24 ± 0.02	1.99 ± 0.01	2.55 ± 0.07	2.52 ± 0.08	4,9,13,28,46
24835 1995 SM_{55}/Cl	0.65 ± 0.01	0.38 ± 0.02	0.71 ± 0.02	1.07 ± 0.05	0.59 ± 0.06	0.49 ± 0.05	4,6,9,10,11,12,28,46
24952 1997 QJ_4/R	0.76 ± 0.04	0.43 ± 0.06	0.81 ± 0.05	1.23 ± 0.31			3,4,6,20,24
26181 1996 GQ_{21}/R	1.01 ± 0.01	0.71 ± 0.01	1.42 ± 0.01	2.39 ± 0.04	2.88 ± 0.04	3.03 ± 0.08	11,13,24,29,46
26308 1998 SM_{165}/R	0.98 ± 0.02	0.65 ± 0.04	1.30 ± 0.01	2.36 ± 0.01	2.91 ± 0.02	2.96 ± 0.07	4,11,12,22,46
26375 1999 DE_9/S	0.97 ± 0.03	0.58 ± 0.01	1.15 ± 0.01	1.84 ± 0.04	2.17 ± 0.05	2.19 ± 0.05	3,9,11,20,29,30,46
28978 Ixion/R	1.03 ± 0.03	0.61 ± 0.03	1.19 ± 0.04	1.88 ± 0.09	2.18 ± 0.11	1.97 ± 0.09	46
29981 1999 TD_{10}/S	0.75 ± 0.02	0.49 ± 0.02	1.02 ± 0.03	1.88 ± 0.07	2.31 ± 0.10	2.43 ± 0.10	4,20,28,46
31824 Elatus/C	1.03 ± 0.03	0.66 ± 0.01	1.28 ± 0.01	2.09 ± 0.07	2.48 ± 0.09	2.51 ± 0.09	9,12,17,28, 46
32532 Thereus/C	0.75 ± 0.01	0.49 ± 0.02	0.94 ± 0.01	1.69 ± 0.05	2.14 ± 0.07	2.30 ± 0.05	17,31,32,46
32929 1995 QY_9/R	0.70 ± 0.02	0.51 ± 0.04	0.86 ± 0.06	2.02 ± 0.01			4,6,9,14
33128 1998 BU_{48}/S	0.95 ± 0.08	0.64 ± 0.02	1.18 ± 0.01	2.27 ± 0.05	2.79 ± 0.11	2.60 ± 0.11	9,11,12,13,17
33340 1998 VG_{44}/R	0.90 ± 0.01	0.59 ± 0.01	1.18 ± 0.08	1.81 ± 0.01	2.21 ± 0.01	2.23 ± 0.01	4,9,10,46
35671 1998 SN_{165}/Cl	0.71 ± 0.06	0.42 ± 0.03	0.82 ± 0.01	1.27 ± 0.05			3,4,6,9,20
38628 Huya/R	0.96 ± 0.02	0.57 ± 0.02	1.20 ± 0.02	1.95 ± 0.02	2.27 ± 0.05	2.37 ± 0.06	3,4,9,24,28,33,34,46
40314 1999 KR_{16}/D	1.06 ± 0.03	0.76 ± 0.01	1.50 ± 0.03	2.37 ± 0.10	2.95 ± 0.12	2.97 ± 0.12	3,11,13,36
42301 2001 UR_{163}/R	1.30 ± 0.11	0.84 ± 0.01	1.46 ± 0.11	2.37 ± 0.06	2.86 ± 0.08		43,46
42355 2002 CR_{46}/S	0.83 ± 0.07	0.55 ± 0.05	0.99 ± 0.06	1.83 ± 0.09	2.18 ± 0.12	2.34 ± 0.09	46
44594 1999 OX_3/S	1.15 ± 0.02	0.69 ± 0.02	1.39 ± 0.02	2.20 ± 0.05	2.63 ± 0.05	2.63 ± 0.10	4,9,17,22,35,43,46
47171 1999 TC_{36}/R	1.03 ± 0.02	0.69 ± 0.01	1.33 ± 0.02	2.32 ± 0.01	2.70 ± 0.03	2.70 ± 0.02	4,9,10,11,12,30,41,46
47932 2000 GN_{171}/R	0.92 ± 0.01	0.62 ± 0.01	1.22 ± 0.02	1.84 ± 0.08	2.21 ± 0.14	2.38 ± 0.14	4,13,24,46
48639 1995 TL_8/Cl	1.04 ± 0.01	0.69 ± 0.01	1.33 ± 0.01	2.42 ± 0.05	2.82 ± 0.09	2.80 ± 0.09	11,20
52872 Okyrhoe/JFC	0.75 ± 0.04	0.47 ± 0.02	0.97 ± 0.02	1.93 ± 0.10	2.40 ± 0.12	2.52 ± 0.10	9,17,20,28,41,46
52975 Cyllarus/C	1.13 ± 0.04	0.69 ± 0.01	1.36 ± 0.03	2.42 ± 0.07	2.87 ± 0.11	2.77 ± 0.11	9,10,11,12,17
54598 Bienor/S	0.69 ± 0.02	0.47 ± 0.02	0.92 ± 0.05	1.74 ± 0.03	2.14 ± 0.05	2.27 ± 0.11	9,11,12,17,30,41,46
55565 2002 AW_{197}/Cl	0.90 ± 0.03	0.62 ± 0.03	1.18 ± 0.03	1.82 ± 0.06	2.15 ± 0.08	2.38 ± 0.10	39

TABLE 1. (continued).

Object/Type*	B–V	V–R	V–I	V–J	V–H	V–K	References
55576 Amycus/C	1.11 ± 0.01	0.71 ± 0.01	1.38 ± 0.05	2.14 ± 0.03	2.44 ± 0.03	2.37 ± 0.03	39,46
55637 2002 UX_{25}/Cl	0.94 ± 0.06	0.54 ± 0.06	1.13 ± 0.05	1.82 ± 0.09	2.22 ± 0.10	2.22 ± 0.11	46
58534 Logos/Cl	0.99 ± 0.01	0.73 ± 0.06	1.29 ± 0.03	1.84 ± 0.37			3,6,9,10,25
60558 Echeclus/JFC	0.85 ± 0.08	0.47 ± 0.01	0.94 ± 0.02	1.49 ± 0.10	2.05 ± 0.12	2.39 ± 0.11	12,17,24
63252 2001 BL_{41}/C	0.72 ± 0.05	0.48 ± 0.03	1.06 ± 0.03	1.65 ± 0.07	2.02 ± 0.09	2.26 ± 0.09	17,29,30,35
66652 1999 RZ_{253}/Cl	0.82 ± 0.17	0.65 ± 0.06	1.30 ± 0.06	2.01 ± 0.07	2.49 ± 0.11	2.59 ± 0.09	4,11,20,37
79360 1997 CS_{29}/Cl	1.08 ± 0.03	0.66 ± 0.04	1.25 ± 0.03	2.06 ± 0.03	2.44 ± 0.08	2.48 ± 0.09	3,4,5,9,10,11
82075 2000 YW_{134}/R	0.87 ± 0.03	0.39 ± 0.02	1.07 ± 0.03	1.68 ± 0.12	2.02 ± 0.16	2.18 ± 0.16	46
82155 2001 FZ_{173}/S	0.87 ± 0.06	0.58 ± 0.04	1.07 ± 0.05	1.92 ± 0.12	2.09 ± 0.17	2.39 ± 0.16	46
83982 Crantor/C	1.10 ± 0.04	0.76 ± 0.01	1.44 ± 0.01	2.46 ± 0.02	2.83 ± 0.02	2.79 ± 0.02	39,46
87555 2000 QB_{243}/S	0.77 ± 0.05	0.38 ± 0.05	1.36 ± 0.08	1.48 ± 0.07	2.09 ± 0.11		46
90377 Sedna/D	1.23 ± 0.09	0.76 ± 0.09	1.37 ± 0.09	2.32 ± 0.06	2.61 ± 0.06	2.66 ± 0.07	21
90482 Orcus/R	0.68 ± 0.04	0.37 ± 0.04	0.74 ± 0.04	1.08 ± 0.04	1.21 ± 0.04	1.25 ± 0.04	40
91133 1998 HK_{151}/R	0.72 ± 0.05	0.49 ± 0.03	0.88 ± 0.01	1.57 ± 0.09			4,9,10
95626 2002 GZ_{32}/C	0.73 ± 0.06	0.51 ± 0.04	0.92 ± 0.05	1.63 ± 0.08	2.08 ± 0.10	2.19 ± 0.10	46
118228 1996 TQ_{66} / R	1.19 ± 0.02	0.66 ± 0.03	1.44 ± 0.14	2.41 ± 0.08			3,4,5,6
134860 2000 OJ_{67}/Cl	1.05 ± 0.06	0.67 ± 0.05	1.27 ± 0.07	1.98 ± 0.10	2.26 ± 0.14	2.33 ± 0.16	9,11
136108 2003 EL_{61}/Cl	0.63 ± 0.03	0.34 ± 0.02	0.68 ± 0.02	1.05 ± 0.02	1.01 ± 0.04	0.94 ± 0.05	45
136199 Eris/D	0.71 ± 0.02	0.45 ± 0.02	0.78 ± 0.02	1.01 ± 0.02	0.72 ± 0.04	0.32 ± 0.05	44
1996 TS_{66}/Cl	1.06 ± 0.03	0.73 ± 0.03	1.31 ± 0.08	1.87 ± 0.03	2.52 ± 0.08		3,4,5,7
1998 WU_{24}/?	0.78 ± 0.03	0.53 ± 0.04	0.99 ± 0.03	1.67 ± 0.04			38
1999 CD_{158}/R	0.86 ± 0.01	0.52 ± 0.02	1.10 ± 0.04	1.86 ± 0.07	2.30 ± 0.07	2.33 ± 0.08	11,12
2000 OK_{67}/Cl	0.82 ± 0.06	0.60 ± 0.05	1.22 ± 0.08	2.42 ± 0.08	2.88 ± 0.11	2.92 ± 0.12	9,11,12
2000 PE_{30}/D	0.75 ± 0.04	0.33 ± 0.06	0.75 ± 0.08	1.65 ± 0.08	2.09 ± 0.08	2.32 ± 0.11	9,11,12,46
2001 CZ_{31}/Cl	0.60 ± 0.15	0.50 ± 0.10	0.80 ± 0.15	1.53 ± 0.10	2.08 ± 0.14	2.24 ± 0.14	11,13
2001 QF_{298}/R	0.65 ± 0.02	0.36 ± 0.02	0.76 ± 0.06	1.30 ± 0.10	1.53 ± 0.14	1.69 ± 0.15	11,43,46
2003 AZ_{84}/R	0.61 ± 0.08	0.45 ± 0.07	0.80 ± 0.07	1.46 ± 0.10	1.48 ± 0.14	1.39 ± 0.16	46

*Type (from the chapter by Gladman et al.): C = Centaurs, Cl = classical, R = resonant, S = scattered, D = detached, JFC = Jupiter-family comet orbit, ? = unusual, Halley-family comet.

References: [1] *Hardorp* (1980); [2] *Hartmann et al.* (1982); [3] *Jewitt and Luu* (2001); [4] *McBride et al.* (2003); [5] *Tegler and Romanishin* (1998); [6] *Gil-Hutton and Licandro* (2001); [7] *Jewitt and Luu* (1993); [8] *Tegler and Romanishin* (1997); [9] *Doressoundiram et al.* (2002); [10] *Boehnhardt et al.* (2001); [11] *Delsanti et al.* (2006); [12] *Delsanti et al.* (2004); [13] *Sheppard and Jewitt* (2002); [14] *Luu and Jewitt* (1996); [15] *Davies et al.* (1998); [16] *Davies et al.* (2000); [17] *Bauer et al.* (2003); [18] *Romanishin et al.* (1997); [19] *Romon-Martin et al.* (2002); [20] *Delsanti et al.* (2001); [21] *Barucci et al.* (2005b); [22] *Tegler and Romanishin* (2000); [23] *Davies et al.* (1997); [24] *Boehnhardt et al.* (2002); [25] *Davies* (2000); [26] *Hainaut et al.* (2000); [27] *Barucci et al.* (1999); [28] *Tegler and Romanishin* (2003); [29] *Doressoundiram et al.* (2003); [30] *Tegler et al.* (2003); [31] *Barucci et al.* (2002); [32] *Farnham and Davies* (2003); [33] *Ferrin et al.* (2001); [34] *Schaefer and Rabinowitz* (2002); [35] *Peixinho et al.* (2004); [36] *Trujillo and Brown* (2002); [37] *Doressoundiram et al.* (2001); [38] *Davies et al.* (2001); [39] *Doressoundiram et al.* (2005a); [40] *De Bergh et al.* (2005); [41] *Dotto et al.* (2003); [42] *McBride et al.* (1999); [43] *Doressoundiram et al.* (2005b); [44] *Brown et al.* (2005); [45] *Trujillo et al.* (2007); [46] *Doressoundiram et al.* (2007).

TABLE 2. Eigenvectors, eigenvalues, and percentage of total variance contributed by each eigenvalue from the PC analysis of the samples of 67 and 55 TNOs.

			Sample of 67 TNOs			
Variable	1	2	3	4		
B–V	0.311	0.508	0.674	–0.436		
V–R	0.235	0.241	0.290	0.896		
V–I	0.450	0.579	–0.678	–0.054		
V–J	0.804	–0.591	0.034	–0.063		
Eigenvalues	0.269	0.015	0.004	0.002		
Percentage of total variance	92.892	5.220	1.269	0.619		
			Sample of 55 TNOs			
Variable	1	2	3	4	5	6
B–V	0.147	0.427	0.332	–0.406	0.721	0.035
V–R	0.116	0.280	0.167	–0.006	–0.314	0.884
V–I	0.234	0.477	0.405	–0.134	–0.569	–0.461
V–J	0.440	0.413	–0.287	0.713	0.202	–0.057
V–H	0.605	–0.076	–0.578	–0.530	–0.111	–0.011
V–K	0.592	–0.579	0.530	0.166	0.070	0.032
Eigenvalues	0.880	0.044	0.011	0.004	0.003	0.001
Percentage of total variance	93.312	4.664	1.199	0.409	0.304	0.112

3. RESULTS

We applied PCA successively to the dataset of 67 TNOs described by 4 variables (B–V, V–R, V–I, and V–J color indices) and to the subset of 55 TNOs described by 6 variables (B–V, V–R, V–I, V–J, V–H, and V–K color indices). The eigenvectors, percentages of total variance contributed by each eigenvector, and eigenvalues of the variance-covariance matrix of the two analyzed samples are reported in Table 2.

These data show that in both cases (the results of the PC analysis carried out on 67 and 55 samples respectively) the first principal component (PC1) accounts for most of the variance of the sample (93%); the second principal component (PC2) adds only 4–5% to the total variance.

The first and the second principal components account for about 98% of the total variance, therefore the PC1 vs. PC2 plane contains practically all the information on the variance of the variables characterizing the considered sample. This analysis establishes the existence of a quasicontinuous trend from neutral to very red objects along the PC1 axis. It is also possible to infer from these results that the degree of reddening is the main distinctive character of the TNO population. The predominance of PC1 (i.e., of V–I and V–J colors in the first case and V–H and V–K colors in the second one) in characterizing TNO behavior is shown by the PC1 scores, which span three times more than PC2. The objects having a neutral color with respect to the Sun have the lower values of the PC1 scores and fall in the left part of the histogram shown in Fig. 1; for larger PC1 scores the objects are redder and redder. The object number density along the PC1 is not homogeneous, indicating the presence of some grouping that overlaps the continuous trend. In Fig. 1 the histogram represents the number of the objects projected onto the PC1, clearly showing four peaks. These groups of objects constitute a finer structure overlapping the general trend from neutral to very red spectra resulting from the PC analysis.

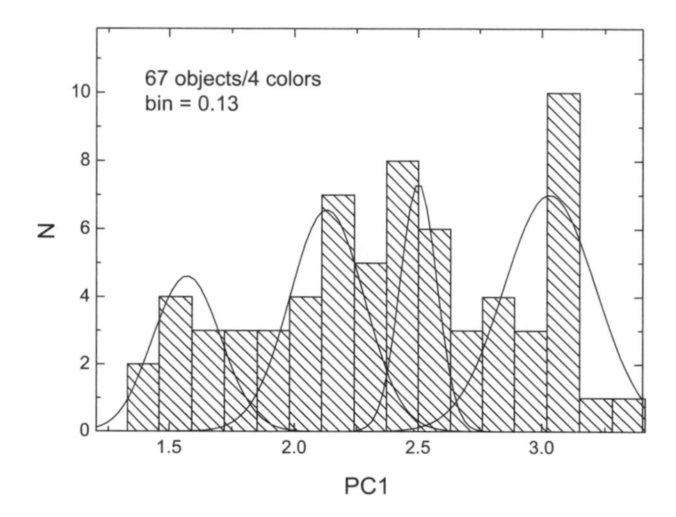

Fig. 1. Histogram showing the distribution of TNOs vs. the PC1 scores. The four superimposed curves represent the four classes obtained by the G mode. The Gaussians described by the parameters characterizing each class (average, standard deviation, and number of objects as reported in Table 3) are projected on the PC1 axis.

The relationship between the variables used is probably nonlinear, so the PC analysis does not allow us to discriminate among the intrinsic structure of these groups. To recognize the structure of the number density distribution on the PC1 axis we used the G-mode method.

When a sample of 67 objects was considered and four colors (B–V, V–R, V–I, and V–J) were taken as variables, for a total of 67 × 4 = 268 degrees of freedom, four groups were recognized at a >99% confidence level. The weight of each variable in separating these groups is 31% for B–V color, 26% for V–I, 22% for V–R, and 21% for V–J, indicating that the B–V variable weight was one-third more than the others in discriminating the classes. One object (2000 QB_{243}) was not attributed to any class, forming a "*single object group*" as in the Tholen asteroid taxonomy: 4 Vesta, 1862 Apollo, and 349 Dembowska formed the V, Q, and R classes (*Tholen and Barucci,* 1989). Today, many small asteroids populate the V class, and some new objects have been added to the R and Q classes (*Binzel and Xu,* 1993; *Bus and Binzel,* 2002).

4. TAXONOMY BASED ON COLOR INDICES

A two-letter designation for the identified groups is introduced to distinguish TNO taxonomy from asteroid taxonomy. Objects having neutral colors with respect to the Sun are classified as the BB ("blue") group; those having a very high red color are classified as RR ("red"). The BR group consists of objects with an intermediate blue-red color, while the IR group includes moderately red objects. For a detailed discussion about the spectral characteristics/composition of objects belonging to the different classes, see the chapter by Barucci et al.

The BB group contains objects having neutral reflectance spectra. Typical objects of the group are 2060 Chiron, 19308 (1996 TO_{66}), 15874 (1996 TL_{66}), 90482 Orcus, 136108 (2003 EL_{61}), and 136199 Eris. The typical spectra are flat, somewhat bluish in the near-infrared. The ice absorption bands seem generally stronger than in the other groups, although the H_2O ice presence in the Chiron spectrum seems connected to temporal/orbital variations, and the spectrum of 1996 TL_{66} is the only one, at present, that is completely flat. A group of these BB objects, which have been found to belong to the same dynamical family as 136108 (2003 EL_{61}), show deeper H_2O ice absorption bands (see the chapter by Brown), which has been interpreted as a consequence of the collisional fragmentation of the family parent body.

The BR group is an intermediate group between BB and IR, even though its color is closer to the behavior of the IR group. Typical members of this taxon are 8405 Asbolus, 10199 Chariklo, 54598 Bienor, and 32532 Thereous. A small percentage of H_2O ice is present on the surface of these objects.

The IR group is less red than the RR group. Typical members of this taxon are 19521 Chaos, 20000 Varuna, 38628 Huya, 47932 (2000 GN_{171}), 26375 (1999 DE_9), and 555565 (2002 AW_{197}). Three of these objects seem to contain hydrous silicates on the surface.

The RR group contains the reddest objects in the solar system, showing a small percentage of ice on the surface. Some well-observed objects are members of this group, such as 47171 (1999 TC_{36}), 55576 (2002 GB_{10}), 83982 (2002 GO_9), and 90377 Sedna. 5145 Pholus, 26181 (1996 GQ_{21}), and 55638 (2002 VE_{95}) contain methanol on their surfaces, and this could imply a primitive chemical nature (see chapter by Barucci et al.).

Unfortunately, we cannot associate an albedo range with each taxonomic group because of the lack of albedo data. The albedo of TNOs is a fundamental property that is still largely undersampled; some albedo measurements have been obtained by the Spitzer Space Telescope (see the chapter by Stansberry et al.), and in the future, ALMA will provide direct measurements of the sizes of TNOs and consequently their albedos. Further developments are also expected from ESA Herschel Space Observatory, which will cover the far-infrared to submillimeter portions of the spectrum (from 60 to 670 μm).

The G-mode method has been applied to analyze a set of 55 objects for which the B–V, V–R, V–I, V–J, V–H, and V–K colors were taken as variables (55 × 6 = 330 d.o.f.). The four groups are also well separated at a confidence level of 99%. The weight of each variable in separating these groups is 21% for the V–I color, 17% for V–J, 16% for V–R, V–K, and V–H, and 14% for B–V, showing that all the variables play the same role in discriminating the taxons. The number of samples, the color average value, and the relative standard deviation for each group obtained by G-mode analysis for both analyzed samples are given in Table 3.

Note that the G mode applied to the sample described by six variables gives always four groups, where the average and the standard deviation values for the first four colors are the same obtained by running the analysis with only four variables. Moreover, these average colors for the taxonomic groups practically coincide with the values obtained from the G-mode analysis of 51 objects (*Barucci et al.,* 2005a). Thus, increasing the number of objects and expanding the set of variables does not result in any significant variations in the identified taxons, clearly indicating the stability of the proposed taxonomy.

The G mode has been extended (*Fulchignoni et al.,* 2000) to assign to one of the previously defined taxons any object for which the same set of variables become available. Moreover, even if a subset of the variables used in the initial development of the taxonomy is known for an object, the algorithm allows us to give at least a preliminary indication of its relevance to a given group. The lack of information on a variable is reflected by the fact that an object could be assigned to two different taxons when the missing variable is the one that discriminates between these taxons.

We applied this algorithm to each of the 66 other TNOs for which B–V, V–R, and V–I colors are available. More-

TABLE 3. The relative number of samples, average colors, and relative standard deviation for the taxons obtained by the G mode.

Sample of 67 TNOs and Centaurs							
Class	N	B–V	V–R	V–I	V–J		
BB	13	0.68 ± 0.05	0.39 ± 0.04	0.75 ± 0.05	1.23 ± 0.21		
BR	18	0.75 ± 0.05	0.49 ± 0.04	0.96 ± 0.05	1.69 ± 0.19		
IR	11	0.92 ± 0.05	0.59 ± 0.04	1.17 ± 0.05	1.87 ± 0.06		
RR	24	1.07 ± 0.07	0.70 ± 0.04	1.36 ± 0.09	2.25 ± 0.19		
Sample of 55 TNOs and Centaurs							
Class	N	B–V	V–R	V–I	V–J	V–H	V–K
BB	10	0.67 ± 0.05	0.38 ± 0.04	0.74 ± 0.04	1.22 ± 0.23	1.27 ± 0.49	1.33 ± 0.60
BR	12	0.75 ± 0.07	0.47 ± 0.03	0.97 ± 0.08	1.69 ± 0.13	2.13 ± 0.12	2.29 ± 0.11
IR	10	0.91 ± 0.04	0.58 ± 0.04	1.14 ± 0.07	1.86 ± 0.06	2.21 ± 0.09	2.31 ± 0.09
RR	23	1.05 ± 0.11	0.69 ± 0.05	1.35 ± 0.10	2.26 ± 0.18	2.66 ± 0.22	2.66 ± 0.25

TABLE 4. Proposed classification on the base of the G-mode classification results; for each TNO and Centaur the respective taxon and the number of colors (N) used in classifying it are reported.

Object	Class	N	Object	Class	N
2060 Chiron	BB	6	40314 1999 KR_{16}	RR	6
5145 Pholus	RR	6	42301 2001 UR_{163}	RR	5
7066 Nessus	RR	6	42355 2002 CR_{46}	BR	6
8405 Asbolus	BR	6	44594 1999 OX_3	RR	6
10199 Chariklo	BR	6	47171 1999 TC_{36}	RR	6
10370 Hylonome	BR	6	47932 2000 GN_{171}	IR	6
15788 1993 SB	BR	4	48639 1995 TL_8	RR	6
15760 1992 QB_1	RR	3	49036 Pelion	BR	3
15788 1993 SB	BR	3	52975 Cyllarus	RR	6
15789 1993 SC	RR	6	54598 Bienor	BR	6
15820 1994 TB	RR	6	52872 Okyrhoe	BR	6
15874 1996 TL_{66}	BB	6	55565 2002 AW_{197}	IR	6
15875 1996 TP_{66}	RR	6	55576 Amycus	RR	6
15883 1997 CR_{29}	BR	3	55636 2002 TX_{300}	BB	3
16684 1994 JQ_1	RR	3	55637 2002 UX_{25}	IR	6
19255 1994 VK_8	RR	3	55638 2002 VE_{95}	RR	6
19299 1996 SZ_4	BR	4	58534 Logos	RR	4
19308 1996 TO_{66}	BB	6	59358 1999 CL_{158}	BR	3
19521 Chaos	IR	6	60454 2000 CH_{105}	RR	3
20000 Varuna	IR	6	60558 Echeclus	BR	6
24835 1995 SM_{55}	BB	6	60608 2000 EE_{173}	BR	3
24952 1997 QJ_4	BR-BB	3	60620 2000 FD_8	RR	3
26181 1996 GQ_{21}	RR	6	60621 2000 FE_8	BR	3
26308 1998 SM_{165}	RR	6	63252 2001 BL_{41}	BR	6
26375 1999 DE_9	IR	6	66452 1999 OF_4	RR	3
28978 Ixion	IR	6	66652 1999 RZ_{253}	RR	6
29981 1999 TD_{10}	BR	6	69986 1998 WW_{24}	BR	3
33001 1997 CU_{29}	RR	3	69988 1998 WA_{31}	BR	3
31824 Elatus	RR	6	69990 1998 WU_{31}	RR	3
32532 Thereus	BR	6	79360 1997 CS_{29}	RR	6
32929 1995 QY_9	BR	4	79978 1999 CC_{158}	IR	3
33128 1998 BU_{48}	RR	6	79983 1999 DF_9	RR	3
33340 1998 VG_{44}	IR	6	82075 2000 YW_{134}	BR	6
35671 1998 SN_{165}	BB	4	82155 2001 FZ_{173}	IR	6
38083 Rhadamanthus	BR	3	82158 2001 FP_{185}	IR	3
38084 1999 HB_{12}	BR	3	83982 Crantor	RR	6
38628 Huya	IR	6	85633 1998 KR_{65}	RR	3

TABLE 4. (continued).

Object	Class	N	Object	Class	N
86177 1999 RY_{215}	BR	3	1998 WV_{31}	BR	3
87555 2000 QB_{243}	U	5	1998 WZ_{31}	BB	3
90377 Sedna	RR	6	1998 XY_{95}	RR	3
90482 Orcus	BB	6	1999 CD_{158}	BR	6
91133 1998 HK_{151}	BR	4	1999 CB_{119}	RR	3
91205 1998 US_{43}	BR	3	1999 CF_{119}	BR	3
91554 1999 RZ_{215}	IR-RR	3	1999 CX_{131}	RR	3
95626 2002 GZ_{32}	BR	6	1999 HS_{11}	RR	3
118228 1996 TQ_{66}	RR	4	1999 OE_4	RR	3
118379 1999 HC_{12}	BR	3	1999 OJ_4	RR	3
119070 2001 KP_{77}	IR	3	1999 OM_4	RR	3
121725 1999 XX_{143}	RR	3	1999 RB_{216}	IR-BR	3
129772 1999 HR_{11}	RR-IR	3	1999 RX_{214}	RR	3
134340 Pluto	BR	3	1999 RE_{215}	RR	3
134860 2000 OJ_{67}	RR	6	1999 RY_{214}	BR	3
136108 2003 EL_{61}	BB	6	2000 CL_{104}	RR	3
136199 Eris	BB	6	2000 GP_{183}	BR	3
1993 FW	IR	3	2000 OK_{67}	RR	6
1993 RO	IR	3	2000 PE_{30}	BB	6
1995 HM_5	BR	3	2001 CZ_{31}	BB-	6
1995 WY_2	RR-IR	3	2001 KA_{77}	RR	3
1996 RQ_{20}	IR-RR	3	2001 KD_{77}	RR	3
1996 RR_{20}	RR	3	2001 QF_{298}	BB	6
1996 TK_{66}	RR	3	2001 QY_{297}	BR	3
1996 TS_{66}	RR	5	2001 UQ_{18}	RR	3
1997 QH_4	RR	3	2002 DH_5	BR	3
1998 KG_{62}	RR-IR	3	2002 GF_{32}	RR	3
1998 UR_{43}	BR	3	2002 GJ_{32}	RR	3
1998 WT_{31}	BB	3	2002 GV_{32}	RR	3
1998 WU_{24}	BR	4	2003 AZ_{84}	BB	6
1998 WS_{31}	BR	3			

over, we classified with the same algorithm 134340 Pluto and 55638 (2002 VE_{95}). In Table 4 we report the resulting classification for 135 objects with the indication of the number of variables used in classifying each object. Some uncertainties (double or nonassignment) remain for only eight objects, but the appurtenance to a given group has to always be considered with some caution because it is only an indication when it is obtained with an incomplete dataset. One of these objects is not classified at all. This fact suggests that other groups could be found if the number of objects analyzed will increase and therefore the proposed taxonomy can still be refined.

The average values of six broadband colors obtained for each class and the relative error bars are represented in Fig. 2 as reflectances normalized to the Sun and to the V colors

$$R_{c_\lambda} = 10^{\pm 0.4(c_\lambda - c_{\lambda_0})}$$

where c_λ and c_{λ_0} are the λ–V colors of the object and of the Sun respectively.

To analyze the behavior of the four TNO taxons with respect to the orbital elements, we calculate the distributions of the TNO taxons with respect to (1) their semimajor axis and (2) their inclination. In Fig. 3 the percentage of objects belonging to a given taxon is reported vs. the semimajor axis (Fig. 3a) and the inclination of their orbits (Fig. 3b) respectively.

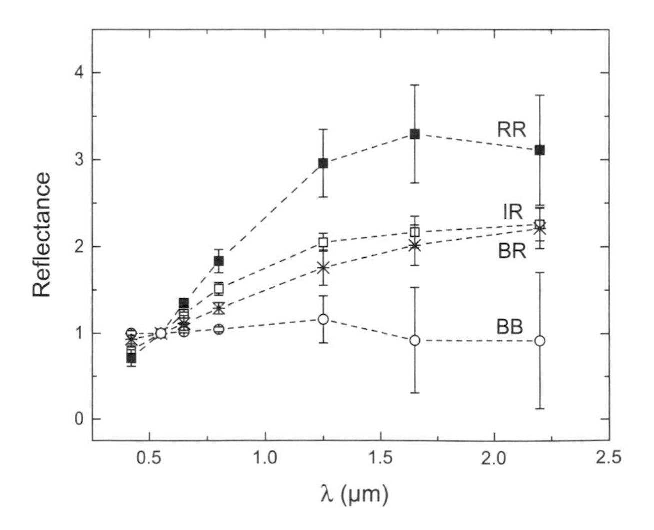

Fig. 2. Average values of the reflectance of the six broadband colors obtained for each taxon normalized to the Sun and to the V colors.

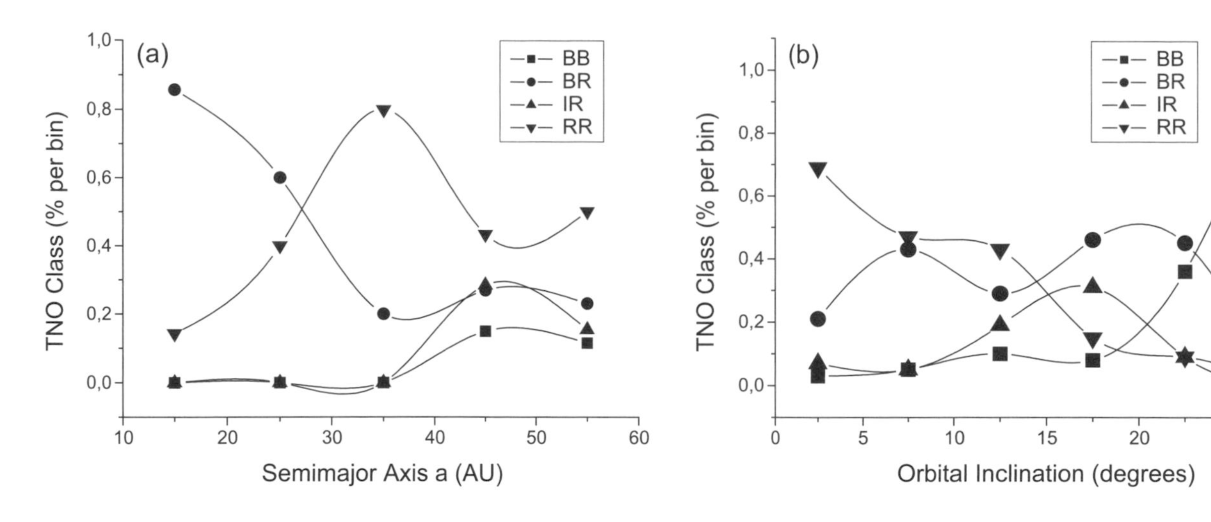

Fig. 3. Percent of objects belonging to a given class vs. **(a)** the semimajor axis and **(b)** the inclination of their orbits.

The following indications can be extracted by these distributions: (1) Centaurs belong to the BR group (the larger number) or to the RR taxons, confirming the known color bimodality of that object (*Tegler and Romanishin,* 1998, *Peixinho et al.,* 2003); the TNOs are distributed in all the taxons, but a larger number of classical objects belongs to the RR taxon. (2) The more distant objects are quite equally subdivided in all the four taxons. (3) The orbital inclinations of the RR TNOs are in general low, confirming the fact that these objects are primordial objects (cold population) with low-inclination orbits, while the high-inclination objects belong to the BB taxon, which seems to be the representative of the "hot" population (*Levison and Stern,* 2001; *Brown,* 2001; *Doressoundiram et al.,* 2002; *Gomes,* 2003; chapter by Doressoundiram et al.). (4) The taxon BR spans the entire range of inclinations, while the taxon IR is characterized by intermediate inclinations (15°–20°).

Moreover, in Fig. 4 the distribution of the different taxons within each TNO dynamical class is shown. The bimodality of the Centaurs is evident, the objects classified as IR seem to be concentrate in the resonant and classical dynamical classes, while the RR objects dominate the classical dynamical class.

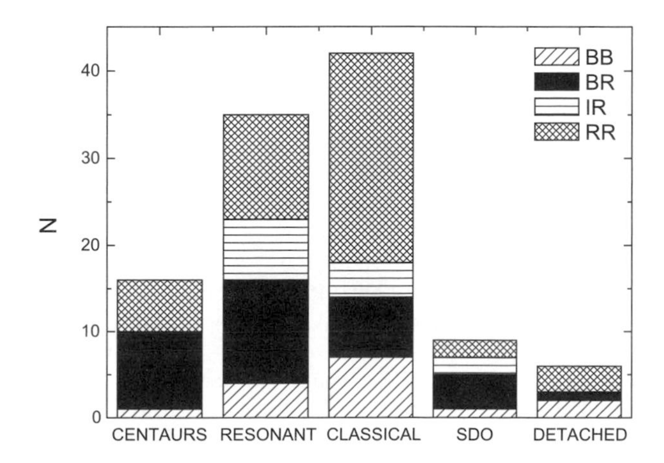

Fig. 4. Distribution of the different taxons within each TNOs' dynamical class.

5. CONCLUSIONS

We applied two multivariate statistical methods (PC analysis and G-mode analysis) to a sample of 67 TNOs and Centaurs for which a homogeneous set of four colors (B–V, V–R, V–I, and V–J) are available. The results provided a quasi-continuous trend from neutral to very red spectra. The results of the G-mode analysis allowed us to distinguish a finer structure superimposed on this trend, separating four groups of homogeneous objects and one single-class object, confirming the results obtained by *Barucci et al.* (2005a), which analyzed 51 objects described by the same variables.

Analyzing a subsample of 55 objects for which a fifth and a sixth variable (the V–H and V–K colors) are available, the same four groups are confirmed. The significance level (99%) of this grouping is larger that (93%) obtained by *Barucci et al.* (1987) in classifying a sample of 438 asteroids with the same statistical techniques. This represents a strong indication that colors reveal real differences in the surface nature of the TNOs, probably having originated through their physico-chemical evolution.

A classification of TNOs and Centaurs based on their broadband colors has been proposed by *Barucci et al.* (2005a) using a two-letter designation for each taxon (BB, BR, IR, and RR). Such classification gives a good indication about the resemblances and/or differences among objects.

Using the extended version of the G-mode analysisfor those objects with only three colors (B–V, V–R, and V–I), we obtain a preliminary classification of another 66 objects (for a total of 135 objects). Among these, one object remains unclassified, and seven have double classification. The availability of at least a fourth variable (V–J or V–H or V–K) appears to be essential for good statistical applications.

The behavior of each group with respect to the orbital elements shows that the RR object inclinations are low and those of the BB objects are high, while BR and IR taxons have inclinations spanning the entire inclination range. The Centaurs exhibit a clear bimodality: Most of them belong to the BR or RR taxon (see chapter by Tegler et al.).

We can conclude that the multivariate analysis of broadband spectrophotometric data of TNOs provides indications for differences in the surface nature of these objects. The quasicontinuous trend, demonstrated by the principal component analysis, is probably a witness to the possible sequence of the alteration processes undergone by the surface of each object, while the different taxons, obtained with G-mode analysis, indicate the present physico-chemical state of the analyzed objects.

It is possible to interpret these results both in terms of evolution of the TNO population and/or in terms of original differences in the composition of the objects. Several scenarios are possible and the lack of fundamental data (albedo, size, mass and size distributions, etc.) make it very difficult to develop a quantitative model describing the history of the TNO population. Here we propose two simple scenarios to provide some clues for interpreting the behavior of the TNOs.

In terms of evolution, we can consider that (1) the position of an object along the trend connecting the BB taxon to the RR taxon would indicate the time during which it has been exposed to the different alteration processes (e.g., activity, collisions, cratering, energetic particle bombardment), starting from a given initial state (original or consequence of a resetting event); (2) each taxon represents a different stage in the evolution of the population; (3) the relative number of objects in each group would account for how long that stage has lasted; and (4) the presence of single-object groups may imply the existence of different (cyclic?) evolution paths.

In terms of original compositional differences, we can conclude, following *Gomes* (2003), that the "hot" population is composed by the objects having quite high orbital inclinations and belonging to the BB taxon, while the "cold" indigenous population is that represented by the RR objects, some of which seem to contain very pristine compounds, such as methanol ice. The intermediate BR and IR taxons would represent the results of the aging of the surfaces of the BB and RR objects respectively, due to the exposure to the environmental modifying processes. The energetic particle bombardment of "dirty" icy surfaces induces the formation of dark, redder hydrocarbon coating films, which could be considered to be responsible for the BB $\rightarrow$ BR transition. A "rain" of micrometeorites over long time intervals may cause the loss of volatiles on the surface of the RR objects and change the characteristics of the overlying ices, which would result in the objects being less red than they were previously: RR $\rightarrow$ IR. Rejuvenation processes (energetic collisions, degassing activity) can reset most of the surface of an affected object to its initial condition, activating a cyclic change in the surface color properties and consequently their classification.

REFERENCES

Barucci M.A. and Fulchignoni M. (1990) Unified asteroid taxonomy. In *Asteroids, Comets and Meteors III* (C-I. Lagerkvist et al., eds.), pp. 7–10. Uppsala Universitet, Uppsala.

Barucci M. A., Capria M. A., Coradini A., and Fulchignoni M. (1987) Classification of asteroids using G-mode analysis. *Icarus, 72,* 304–324.

Barucci M. A., Doressoundiram A., Tholen D., Fulchignoni M., and Lazzarin M. (1999) Spectrophotometric observations of Edgeworth-Kuiper belt objects. *Icarus, 142,* 476–481.

Barucci M. A., Fulchignoni M., Birlan M., Doressoundiram A., Romon J., and Boehnhardt H. (2001) Analysis of trans-neptunian and Centaur colours: Continuous trend or grouping? *Astron. Astrophys., 371,* 1150–1154.

Barucci M. A., Boehnhardt H., Dotto E., Doressoundiram A., Romon J., Lazzarin M., Fornasier S., de Bergh C., Tozzi G. P., Delsanti A., Hainaut O., Barrera L., Birkle K., Meech K., Ortiz J. L., Sekiguchi T., Thomas N., Watanabe J., West R. M., and Davies J. K. (2002) Visible and near-infrared spectroscopy of the Centaur 32532 (2001 PT_{13}). ESO Large Program on TNOs and Centaurs: First spectroscopy results. *Astron. Astrophys., 392,* 335.

Barucci M. A., Belskaya I. N., Fulchignoni M., and Birlan M. (2005a) Taxonomy of Centaurs and trans-neptunian objects. *Astron. J., 130,* 1291–1298.

Barucci M. A., Cruikshank D. P., Dotto E., Merlin F., Poulet F., Dalle Ore C., Fornasier S., and de Bergh C. (2005b) Is Sedna another Triton? *Astron. Astrophys., 439,* L1–L4.

Bauer J. M., Meech K. J., Fernández Y. R., Pittichova J., Hainaut O. R., Boehnhardt H., and Delsanti A. C. (2003) Physical survey of 24 Centaurs with visible photometry. *Icarus, 166,* 195–211.

Binzel R. P. and Xu S.(1993) Chips off of asteroid 4 Vesta — Evidence for the parent body of basaltic achondrite meteorites. *Science, 260,* 186–191.

Boehnhardt H., Tozzi G. P., Birkle K., Hainaut O., Sekiguchi T., Vair M., Watanabe J., Rupprecht G., and the FORS Instrument Team (2001) Visible and near-IR observations of transneptunian objects. Results from ESO and Calar Alto Telescopes. *Astron. Astrophys., 378,* 653–667.

Boehnhardt H., Delsanti A., Barucci A., Hainaut O., Doressoundiram A., Lazzarin M., Barrera L., de Bergh C., Birkle K., Dotto E., Meech K., Ortiz J. E., Romon J., Sekiguchi T., Thomas N., Tozzi G. P., Watanabe J., and West R. M. (2002) *Astron. Astrophys., 395,* 297–303.

Brown M. E. (2001) The inclination distribution of the Kuiper belt. *Astron. J., 212,* 2804–2814.

Brown M. E., Trujillo C. A., and Rabinowitz D. L. (2005) Discovery of a planetary-sized object in the scattered Kuiper belt. *Astrophys. J. Lett., 635,* L97–L100.

Bus S. J. and Binzel R. P. (2002) Phase II of the small main-belt asteroid spectroscopic survey. A feature-based taxonomy. *Icarus, 158,* 146–177.

Coradini A., Fulchignoni M., Fanucci O., and Gavrishin A. I. (1977) A Fortran V program for a new statistical technique: The G mode central method. *Computers and Geosciences, 3,* 85–105.

Davies J. K. (2000) Physical characteristics of trans-neptunian objects and Centaurs. In *Proceedings of the ESO Workshop on Minor Bodies in the Outer Solar System* (A. Fitzsimmons et al., eds.), p. 9. Springer-Verlag, Berlin.

Davies J. K., McBride N., and Green S. F. (1997) Optical and infrared photometry of Kuiper belt object 1993SC. *Icarus, 125,* 61–66.

Davies J. K., McBride N., Ellison S. L., Green S. F., and Ballantyne D. R. (1998) Visible and infrared photometry of six Centaurs. *Icarus, 134,* 213–227.

Davies J. K., Green S., McBride N., Muzzerall E., Tholen D. J., Whiteley R. J., Foster M. J., and Hillier J. K. (2000) Visible and infrared photometry of fourteen Kuiper belt objects. *Icarus, 146,* 253–262.

Davies J. K., Tholen D. J., Whiteley R. J., Green S. F., Hillier J. K., Foster M. J., McBride N., Kerr T. H., and Muzzerall E. (2001) The lightcurve and colours of unusual minor planet 1998 WU_{24}. *Icarus, 150,* 69–77.

De Bergh C., Delsanti A., Tozzi G. P., Dotto E., Doressoundiram A., and Barucci M. A. (2005) The surface of the transneptunian object 90482 Orcus. *Astron. Astrophys., 437,* 1115–1120.

Delsanti A. C., Boehnhardt H., Barrera L., Meech K. J., Sekiguchi T., and Hainaut O. R. (2001) BVRI photometry of 27 Kuiper belt objects with ESO/Very Large Telescope. *Astron. Astrophys., 380,* 347–358.

Delsanti A., Hainaut O., Jourdeuil E., Meech K., Boehnhardt H., and Barrera L. (2004) Simultaneous visible-near IR photometric study of Kuiper belt object surfaces with the ESO/Very Large Telescopes. *Astron. Astrophys., 417,* 1145–1158.

Delsanti A., Peixinho N., Boehnhardt H., Barucci M. A., Merlin F., Doressoundiram A., and Davies J. K. (2006) Near-infrared colour properties of Kuiper belt objects and Centaurs: Final results from the ESO large program. *Astron. J., 131,* 1851–1863.

De Sanctis M. C., Coradini A., and Gavrishin A. (2006) G-mode classification of trans-neptunian objects (abstract). In *Lunar and Planetary Science XXXVII,* Abstract #1109. Lunar and Planetary Institute, Houston (CD-ROM).

Doressoundiram A., Barucci M. A., Romon J., and Veillet C. (2001) Multicolour photometry of trans-neptunian objects. *Icarus, 154,* 277–286.

Doressoundiram A., Peixinho N., de Bergh C., Fornasier S., Thébault P., Barucci M. A., and Veillet C. (2002) The colour distribution in the Edgeworth-Kuiper belt. *Astron. J., 124,* 2279.

Doressoundiram A., Tozzi G. P., Barucci M. A., Boehnhardt H., Fornasier S., and Romon J. (2003) ESO large programme on trans-neptunian objects and Centaurs: Spectroscopic investigation of Centaur 2001 BL_{41} and TNOs (26181) 1996 GQ_{21} and (26375) 1999 DE_9. *Astron. J., 125,* 2721–2727.

Doressoundiram A., Barucci M. A., Boehnhardt H., Tozzi G. P., Poulet F., de Bergh C., and Peixinho N. (2005a) Spectral characteristics and modeling of the trans-neptunian object (55565) 2002 AW_{197} and the Centaurs (55576) 2002 GB_{10} and (83982) 2002 GO_9: ESO large program on TNOs and Centaurs. *Planet. Space Sci., 53,* 1501–1509.

Doressoundiram A., Peixinho N., Doucet C., Mousis O., Barucci M. A., Petit J. M., and Veillet C. (2005b) The Meudon Multicolour Survey (2MS) of Centaurs and trans-neptunian objects: Extended dataset and status on the correlations reported. *Icarus, 174,* 90–104.

Doressoundiram A., Peixinho N., Moullet A., Fornasier S., Barucci M. A., Beuzit J. L., and Veillet C. (2007) The Meudon Multicolour Survey (2MS) of Centaurs and transneptunian objects: From visible to infrared colours. *Astron. J.,* in press.

Dotto E., Barucci M. A., Boehnhardt H., Romon J., Doressoundiram A., Peixinho N., de Bergh C., and Lazzarin M. (2003) Searching for water ice on 47171 1999 TC_{36}, 1998 SG_{35}, and 2000 QC_{243}: ESO large program on TNOs and centaurs. *Icarus, 162,* 408–414.

Farnham T. L. and Davies J. K. (2003) The rotational and physical properties of the Centaur (32532) 2001 PT_{13}. *Icarus, 164,* 418–427.

Ferrin Ignacio, Rabinowitz D., Schaefer B., Snyder J., Ellman N., Vicente B., Rengstorf A., Depoy D., Salim S., Andrews P., Bailyn C., Baltay C., Briceno C., Coppi P., Deng M., Emmet W., Oemler A., Sabbey C., Shin J., Sofia S., van Altena W., Vivas K., Abad C., Bongiovanni A., Bruzual G., Della Prugna F., Herrera D., Magris G., Mateu J., Pacheco R., Sánchez Ge., Sánchez Gu., Schenner H., Stock J., Vieira K., Fuenmayor F., Hernandez J., Naranjo O.; Rosenzweig P., Secco C., Spavieri G., Gebhard M., Honeycutt K., Mufson S., Musser J., Pravdo S., Helin E., and Lawrence K. (2001) Discovery of the bright trans-neptunian object 2000 EB_{173}. *Astrophys. J. Lett., 548,* L243–L247.

Fulchignoni M., Birlan M., and Barucci M. A. (2000) The extension of the G-mode asteroid taxonomy. *Icarus, 146,* 204–212.

Fulchignoni M., Delsanti A., Barucci M. A., and Birlan M. (2003) Toward a taxonomy of the Edgeworth-Kuiper objects: A multivariate approach. *Earth Moon Planets, 92,* 243–250.

Gil-Hutton R. and Licandro J. (2001) VR photometry of sixteen Kuiper belt objects. *Icarus, 152,* 246–250.

Gomes R. (2003) Planetary science: Conveyed to the Kuiper belt. *Nature, 426,* 393–395.

Hardorp J. (1980) The Sun among the stars. III — Energy distributions of 16 northern G-type stars and the solar flux calibration. *Astron. Astrophys., 91,* 221–232.

Hartmann W. K., Cruikshank D. P., and Degewij J. (1982) Remote comets and related bodies — VJHK colourimetry and surface materials. *Icarus, 52,* 377–408.

Hainaut O. R., Delahodde C. E., Boehnhardt H., Dotto E., Barucci M. A., Meech K. J., Bauer J. M., West R. M., and Doressoundiram A. (2000) Physical properties of TNO 1996 TO_{66}. Lightcurves and possible cometary activity. *Astron. Astrophys., 356,* 1076–1088.

Jewitt D. C. and Luu J. X. (1998) Optical-infrared spectral diversity in the Kuiper belt. *Astron. J., 115,* 1667–1670.

Jewitt D. C. and Luu J. X. (2001) Colours and spectra of Kuiper belt objects. *Astron. J., 122,* 2099–2114.

Levison H. F. and Stern S. A. (2001) On the size dependence of the inclination distribution of the main Kuiper belt. *Astron. J., 121,* 1730–1735.

Luu J. X. and Jewitt D. C. (1996) Colour diversity among the Centaurs and Kuiper belt objects. *Astron. J., 112,* 2310.

McBride N., Davies J. K., Green S. F., and Foster M. J. (1999) Optical and infrared observations of the Centaur 1997 CU_{26}. *Mon. Not. R. Astron. Soc., 306,* 799–805.

McBride N., Green S. F., Davies J. K., Tholen D. J., Sheppard S. S., Whiteley R. J., and Hillier J. K. (2003) Visible and infrared photometry of Kuiper belt objects: Searching for evidence of trends. *Icarus, 161,* 501–510.

Peixinho N., Doressoundiram A., Delsanti A., Boehnhardt H., Barucci M. A., and Belskaya I. (2003) Reopening the TNOs colour controversy: Centaurs bimodality and TNOs unimodality. *Astron. Astrophys., 410,* L29–L32.

Peixinho N., Boehnhardt H., Belskaya I., Doressoundiram A., Barucci M. A., and Delsanti A. (2004) ESO large program on Centaurs and TNOs: Visible colours — final results. *Icarus, 170,* 153–166.

Reyment R. and Joreskog K. G. (1993) *Applied Factor Analysis in Natural Sciences.* Cambridge Univ., Cambridge.

Romanishin W., Tegler S. C., Levine J., and Butler N. (1997) BVR photometry of Centaur objects 1995 GO, 1993 HA2, and 5145 Pholus. *Astron. J., 113,* 1893–1898.

Romon-Martin J., Barucci M. A., de Bergh C., Doressoundiram A., Peixinho N., and Poulet F. (2002) Observations of Centaur 8405 Asbolus: Searching for water ice. *Icarus, 160,* 59–65.

Schaefer B. E. and Rabinowitz D. L. (2002) Photometric light curve for the Kuiper belt object 2000 EB_{173} on 78 nights. *Icarus, 160,* 52–58.

Sheppard S. S. and Jewitt D. C. (2002) Time-resolved photometry of Kuiper belt objects: Rotations, shapes, and phase functions. *Astron. J., 124,* 1757–1775.

Tedesco E. F., Williams J. G., Matson D. L., Weeder G. J., Gradie J. C., and Lebofsky L. A. (1989) A three-parameter asteroid taxonomy. *Astron. J., 97,* 580–606.

Tedesco E. F., Veeder G. J., Fowler J. W., and Chillemi J. R. (1992) *The IRAS Minor Planets Survey.* Philips Laboratory Report PL-TR-92-2049, Hanscom Air Force Base, Massachusetts.

Tegler S. C. and Romanishin W. (1997) The extraordinary colours of trans-neptunian objects 1994 TB and 1993 SC. *Icarus, 126,* 212–217.

Tegler S. C. and Romanishin W. (1998) Two distinct populations of Kuiper belt objects. *Nature, 392,* 49.

Tegler S. C. and Romanishin W. (2000) Extremely red Kuiper belt objects in near-circular orbits beyond 40 AU. *Nature, 407,* 979–981.

Tegler S. C. and Romanishin W. (2003) Resolution of the Kuiper belt object colour controversy: Two distinct colour populations. *Icarus, 161,* 181–191.

Tegler S. C., Romanishin W., and Consolmagno S. J. (2003) Colour patterns in the Kuiper belt: A possible primordial origin. *Astrophys. J. Lett., 599,* L49–L52.

Tholen D. J and Barucci M. A. (1989) Asteroid taxonomy. In *Asteroids II* (R. Binzel et al., eds.), pp. 298–315. Univ. of Arizona, Tucson.

Trujillo C. A. and Brown M. E. (2002) A correlation between inclination and colour in the classical Kuiper belt. *Astrophys. J. Lett., 566,* L125–L128.

Trujillo C. A., Brown M. E., Barcume K. M., Schaller E. L., and Rabinowitz D. L. (2007) The surface of 2003 EL_{61} in the near-infrared. *Astron. J., 665,* 1172.

Wood H. J. and Kuiper G. P. (1963) Photometric studies of asteroids. *Astron. J., 137,* 1279.

Zellner B., Tholen D. J., and Tedesco E. F. (1985) The eight-colour asteroid survey — Results for 589 minor planets. *Icarus, 61,* 355–416.

Part IV:

Physical Processes

Physical Effects of Collisions in the Kuiper Belt

Zoë M. Leinhardt and Sarah T. Stewart
Harvard University

Peter H. Schultz
Brown University

Collisions are a major modification process over the history of the Kuiper belt. Recent work illuminates the complex array of possible outcomes of individual collisions onto porous, volatile bodies. The cumulative effects of such collisions on the surface features, composition, and internal structure of Kuiper belt objects (KBOs) are not yet known. In this chapter, we present the current state of knowledge of the physics of cratering and disruptive collisions in KBO analog materials. We summarize the evidence for a rich collisional history in the Kuiper belt and present the range of possible physical modifications on individual objects. The question of how well present-day bodies represent primordial planetesimals can be addressed through future studies of the coupled physical and collisional evolution of KBOs.

1. INTRODUCTION

The Kuiper belt contains some of the least-modified material in the solar system. Some Kuiper belt objects (KBOs) may be similar to the planetesimals that accreted into the larger bodies in the outer solar system. However, KBOs have suffered various modification processes over the lifetime of the solar system, including damage from cosmic rays and ultraviolet radiation, sputtering and erosion by gas and dust in the interstellar medium, and mutual collisions (e.g., *Stern,* 2003). Robust interpretations of the surfaces and internal structures of KBOs require improved insight into the relative weight of each of these processes.

The present understanding of the importance of collisions on the physical evolution of KBOs is limited by the state of knowledge in three fundamental areas: (1) the dynamical history of the different populations within the transneptunian region (see chapter by Morbidelli et al.); (2) the physical properties of KBOs (chapters by Brown and by Stansberry et al.); and (3) how the physical properties of KBOs (expected to be icy and porous) affect the outcome of collisions (this chapter). The dynamical history of a population defines the evolution of mean impact parameters (velocity, angle, mass ratio of the projectile and the target) within and between KBO populations. The impact parameters and the material properties of the colliding bodies determine the outcome of an individual impact event. Finally, the cumulative effects of collisions are determined by the coupled physical and dynamical evolution of KBOs.

Variable progress has been made in these three areas. Over the past decade, great improvements in observations and models have illuminated the rich dynamical history of the Kuiper belt. At present, there is a sparse but growing body of data on the physical properties of KBOs (e.g., size, density, composition, and internal structure). Although a significant body of work has been devoted to collisions between icy, porous bodies, our understanding of the governing physics is still incomplete. The collisional evolution of KBOs is a particularly interesting and challenging problem because of the range of possible outcomes that depend on the changing dynamical structure of the Kuiper belt.

In this chapter, we present a summary of the work to date that can be applied to the physical effects of collisions in the Kuiper belt. We begin with observational evidence for significant past and present-day collisions in the Kuiper belt (section 2). We then present a range of possible outcomes from collisions between KBOs (section 3) and discuss the principal discriminating factors (section 4). Based on the expected physical properties of KBOs, we summarize the results of laboratory and numerical experiments that have been conducted to determine how material properties, such as composition, porosity, and impact conditions, including velocity and mass ratio, affect collision outcomes (section 5). Finally, we discuss several open questions and future research directions for studying collisions in the Kuiper belt (section 6).

2. EVIDENCE FOR A RICH COLLISIONAL HISTORY IN THE KUIPER BELT

In this section, we summarize four observations that support a significant collisional history within the Kuiper belt. First, we discuss observations of interplanetary dust particles (IDPs) by the Pioneer and Voyager spacecraft (section 2.1). Analyses of the orbits of IDPs conclude that the Kuiper belt must be one of the dust source regions. Second, the size distribution of KBOs has at least one break from a simple power law around diameters of tens of kilometers, which is consistent with models of collisional equilibrium among the smaller bodies (section 2.2). Third, the discovery of a possible dynamical family of objects in the Kuiper belt implies conditions that produced at least one near-catastrophic collision of one of the largest KBOs (section 2.3). Finally, models of the accretion of the largest KBOs demonstrate that the mass in the ancient Kuiper belt must have been much larger than observed today. The total mass loss of >90%, and perhaps as much as 99.9%, was driven by a combination of dynamical perturbations and collisional grinding (section 2.4).

In addition to the observable features discussed below, collisions within a small body population will also affect rotation rates, surface colors, and the formation of binaries. The rotation rates of bodies in collisional equilibrium will reflect the angular momentum transfer from typical impact conditions (see, e.g., *Love and Ahrens,* 1997; *Paolicchi et al.,* 2002, for asteroid rotations). The formation of binary KBOs is still a matter of debate. Some binaries seem to have formed via collisions, while others have too much angular momentum for a collision origin (*Margot,* 2002). The observed color diversity in the Kuiper belt is also controversial and not correlated directly with collision energy (see chapter by Doressoundiram et al.). However, the range of outcomes from collisions depends on material properties as well as the impact parameters. The growing data on rotation rates, colors, and binaries will provide in the future additional constraints on the collisional evolution in the Kuiper belt.

2.1. Interplanetary Dust Particles

Dust and small meteoroids were detected in the outer solar system by the Pioneer 10 and 11 and Voyager 1 and 2 missions (*Humes,* 1980; *Gurnett et al.,* 1997). Pioneer 10 and 11 measured the concentration and orbital properties of dust from 1 to 18 AU. The dust impacts detected by Pioneer 11 between 4 and 5 AU were determined to have either high inclination or eccentricity or both. In other words, the IDPs were either not on circular orbits and/or not on near-planar orbits. Hence, the observed increase in particle flux at Jupiter could not be explained by gravitational focusing, which is inefficient for highly inclined and eccentric orbits, and *Humes* (1980) suggested that the dust had a cometary origin.

In a reanalysis of the Pioneer data, *Landgraf et al.* (2002) found that the dust flux was relatively constant at distances exterior to Jupiter's orbit. To produce a constant dust flux from drag forces, the dust must originate from a source beyond the detection locations by the spacecraft. *Landgraf et al.* (2002) modeled the radial dust contribution using three source reservoirs, dust from evaporating Oort cloud and Jupiter-family comets and dust from collisions between KBOs, and argue that the amount of dust observed by Pioneer 10 and 11 can only be explained by a combination of all three reservoirs. They find that comets can account for the material detected inside Saturn's orbit but an additional reservoir is necessary for the dust observed outside Saturn's orbit.

Although Voyager 1 and 2 did not carry specialized detectors for dust, *Gurnett et al.* (1997) found that the plasma wave instruments could detect impacts from small particles with masses $\geq 10^{-11}$ g (2–3 orders of magnitude below the mass detection limit by Pioneer 10 and 11). From data collected between 6 and 60 AU, *Gurnett et al.* (1997) found a severe dropoff in dust detection events after 51 AU and 33 AU for Voyager 1 and 2, respectively. As a result, the authors conclude that the source of the dust cannot be interstellar. Furthermore, the small latitudinal gradient decreases the likelihood that the source objects are planets, moons, or asteroids (if the dust did originate from such objects, one would expect a strong latitudinal gradient since the planets, moons, and asteroids are effectively all in the same plane). The Voyager IDP observations are consistent with a dust source from the Kuiper belt (*Gurnett et al.,* 1997; *Jewitt and Luu,* 2000).

In summary, the radial distribution and orbital properties of outer solar system IDPs cannot be explained by source material solely from Jupiter-family comets and Oort cloud comets and indicate the need for an additional active source of dust in the outer solar system. The IDP observations are well matched by models of dust produced during the collisional evolution of the Kuiper belt (e.g., *Jewitt and Luu,* 2000, and section 2.4). Dust derived from mutual collisions in the present day Kuiper belt is analogous to the zodiacal dust from the asteroid belt (*Müller et al.,* 2005) and observations of rings of dust around other main-sequence stars (see chapters by Moro-Martín et al. and Liou and Kaufmann). Because the removal time of dust is much shorter than the age of the solar system (see chapter by Kenyon et al.), the dust must be replenished by collisions occurring throughout the history of the solar system.

2.2. Size Distribution of Kuiper Belt Objects

Formation models indicate that KBOs accreted within a thin disk with low relative velocities and inclinations (see chapter by Morbidelli et al.). However, the present velocity dispersion (~1 km s^{-1}) and the inclination distribution (about 20° half-width) of KBOs are both much higher than expected during the coagulation stage (*Trujillo et al.,* 2001). The large relative velocities and the large inclination distribution of the KBOs point to significant dynamical interactions with Neptune, which resulted in a rich collisional history (*Davis and Farinella,* 1997).

If the bodies in the Kuiper belt were fully collisionally evolved and collision outcomes were independent of size, the differential size distribution (dN ~ r^{-q}dr, where N is number of bodies in the size bin of radius r) would be described by a self-similar collisional cascade and fit by a single power-law index of q = 3.5 (*Dohnanyi,* 1969; *Williams and Wetherill,* 1994). If the population is only partially collisionally evolved and/or the disruption criteria is dependent on scale, the size distribution will deviate from a single power law. For example, the size distribution in the asteroid belt deviates from a simple power law in part because of strength effects (*O'Brien and Greenberg,* 2005) and recent collisions, such as dynamical family-forming events (*dell'Oro et al.,* 2001).

Recent observations indicate that the size distribution of KBOs has a break at diameters of tens of kilometers, with fewer smaller bodies than expected from extrapolation from bodies hundreds of kilometers in diameter (*Bernstein et al.,* 2004; *Chen et al.,* 2006; *Roques et al.,* 2006). The slope of the differential size distribution of large KBOs (<25 mag, >100-km diameter) is well established, with a slope in the range of 4 to 4.8 (*Trujillo et al.,* 2001; *Petit et al.,* 2006; chapter by Petit et al.). *Bernstein et al.* (2004) also suggested that the classical KBOs have a different size distribution from the other dynamical populations (for KBO population classifications, see chapter by Gladman et al.).

Over the past decade, several groups have made significant progress in modeling the collisional evolution of the Kuiper belt (*Davis and Farinella,* 1997; *Stern and Colwell,* 1997; *Kenyon and Bromley,* 2004; *Pan and Sari,* 2005; *Kenyon and Luu,* 1999; chapter by Kenyon et al.). Their work provides a theoretical basis for a break in the KBO size distribution around tens of kilometers. *Davis and Farinella* (1997) first demonstrated that few of the largest bodies in the Kuiper belt experience catastrophic disruption events in which 50% of the mass is permanently removed. In other words, most of the largest KBOs are primordial; they have persisted since the end of the coagulation stage, although some may have suffered shattering collisions.

Collision evolution models indicate that the break in the size distribution corresponds to the upper size limit of the collisionally evolved population (*Davis and Farinella,* 1997; *Kenyon and Bromley,* 2004; *Pan and Sari,* 2005). Over time, collisions preferentially disrupt smaller objects because of their lower critical disruption energies and their higher number densities (and hence higher collision probabilities) compared to larger bodies. When the disruption criteria is size-dependent, the collisionally evolved size distribution deviates from $q = 3.5$ (see *O'Brien and Greenberg,* 2003). For example, *Pan and Sari* (2005) utilize a disruption criteria proportional to the gravitational binding energy of the body, and the equilibrium power law has $q = 3$. Numerical evolution simulations by *Kenyon and Bromley* (2004) and analytical work by *Pan and Sari* (2005) are in good agreement with the observation by *Bernstein et al.* (2004) of the number of bodies tens of kilometers in size. Note that the location of the break in size between the collisionally evolved and primordial populations increases with time and depends on the dynamical evolution of the system.

The size distribution of KBOs is likely to have a second break in slope at significantly smaller sizes. The second break corresponds to the transition between the collisionally evolved strength-dominated bodies and collisionally evolved gravity-dominated bodies. As the strength of KBOs is essentially unknown at present, the location of the strength to gravity transition based on catastrophic disruption models, e.g., tens to hundreds of meters (section 5.2), is highly uncertain.

In summary, the observed size distribution of KBOs departs from a single power law, which is consistent with the existence of both a collisionally evolved population and a primordial population (*Davis and Farinella,* 1997; *Kenyon and Bromley,* 2004; *Pan and Sari,* 2005; *Stern and Colwell,* 1997). In this scenario, the primordial populations should have experienced primarily surface modification processes through impact cratering events (section 3, Figs. 1 and 2a,b). The population of largest bodies probably has a subpopulation with differentiated internal structures (*Merk and Prialnik,* 2006) and a subpopulation with rubble-pile structures (*Davis and Farinella,* 1997). Both the collisionally evolved strength and gravity-dominated populations would have

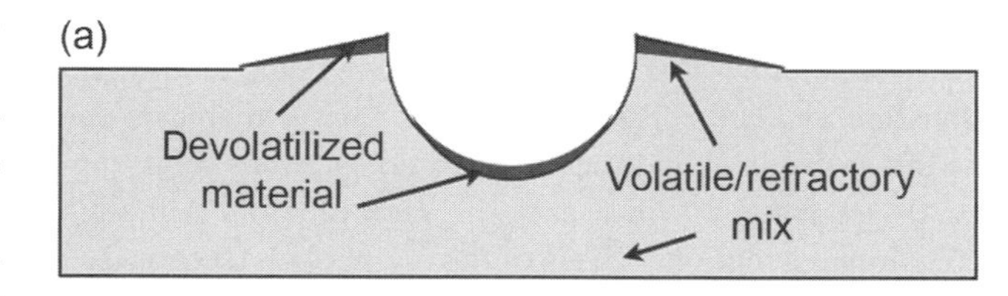

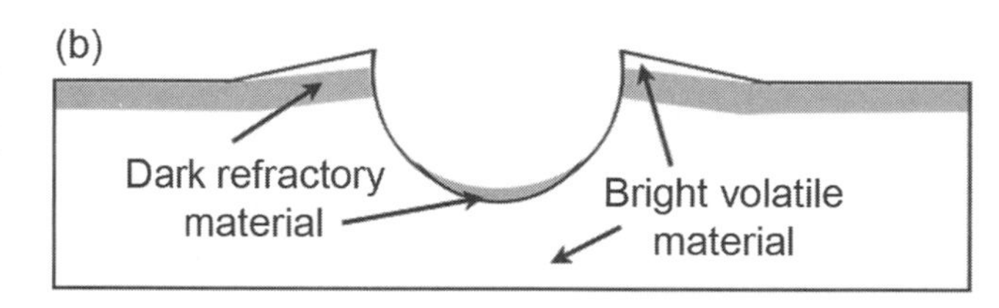

Fig. 1. **(a)** Schematic of an impact crater on a target made of a mixture of volatile and refractory material. The energy of the impact produces melting and devolatization at the base of the crater and in the ejecta. **(b)** Schematic of an impact crater on a target made mostly of volatile material. The surface of the target is covered with a "crust" of darker refractory material. The impact is large enough to excavate fresh volatiles from depth, creating a bright ejecta blanket. Collapse of the crater wall creates a darker region at the bottom of the crater.

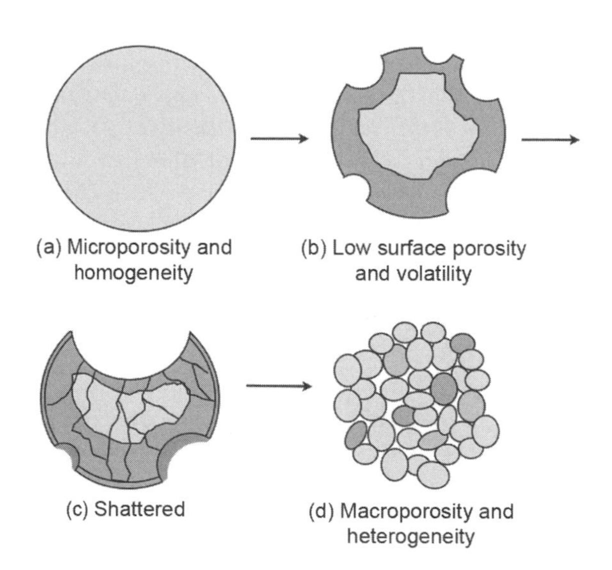

Fig. 2. Schematic showing a possible evolutionary track for a KBO. **(a)** A cross section through a primordial planetesimal with low bulk density and high microporosity. **(b)** After many small impacts, the bulk density has increased and volatile composition decreased at the surface. **(c)** After a large subcatastrophic collision, the body is shattered and the surface is covered with ejecta. **(d)** A catastrophic impact event disrupts the body, creating a rubble pile with high macroporosity and heterogeneous internal composition.

experienced the full range of collisional outcomes, including catastrophic disruption and changes in internal structure and composition (section 3, Fig. 2c).

2.3. Kuiper Belt Family (2003 EL_{61})

At present, a few tens of dynamical families have been identified in the asteroid belt (*Bendjoya and Zappalà,* 2002). These objects are grouped together in proper element space and have similar spectral features where detailed observations are available. The orbits of the family members can be integrated back in time to a common starting point, suggesting formation via a catastrophic collision. Although collisional evolution models of the Kuiper belt (section 2.2) indicate that very few KBOs larger than about 100 km have experienced catastrophic disruption events, *Brown et al.* (2007b) have observed what seems to be a collisional family in the Kuiper belt (see also chapter by Brown).

2003 EL_{61} has two known satellites and five proposed family members. All these objects have similar proper elements, colors, and a deep H_2O spectral feature (*Brown et al.,* 2007b). Although the detection of H_2O ice on the surfaces of KBOs is not unique to these objects, the significant depth of the spectral feature is characteristic of the proposed family, suggesting either more recent or more abundant exposure of surface ice compared to other KBOs. 2003 EL_{61} has a double-peaked rotational light curve with a period of 3.9 h (*Rabinowitz et al.,* 2006). Assuming that the lightcurve is solely due to the equilibrium shape of a rotating, homogeneous, fluid ellipsoid, *Rabinowitz et al.* (2006) and *Lacerda and Jewitt* (2007) derive the size (~1500 km diameter) and density (~2.6 g cm^{-3}). However, the derived size and density are highly uncertain as 2003 EL_{61} is likely to possess nonzero shear strength (*Holsapple,* 2007).

These combined observations suggest that 2003 EL_{61} suffered a significant but subcatastrophic impact event (*Brown et al.,* 2007b). If the modeled bulk density of 2.6 g cm^{-3} is correct and the precollision density of 2003 EL_{61} was comparable to other large KBOs (~2 g cm^{-3}; see chapter by Brown), then about 20% of the original mass was lost. In this model, the dispersed material was preferentially H_2O ice, presumably derived from an ice-rich mantle, producing the shared water spectral feature of the proposed family members.

Further investigation of the proposed 2003 EL_{61} family and search for other dynamical families would provide useful constraints on the collisional history of the Kuiper belt.

2.4. Total Mass of the Kuiper Belt

The total mass in the modern Kuiper belt is depleted from a smooth surface density extrapolation from the giant planet region of the solar system. Based on the observed size distribution of bodies between 30 and 50 AU, the total mass is only about 0.01 $M_\oplus$ (less than 5 $M_♇$) (*Bernstein et al.,* 2004). However, *Stern* (1996) and *Stern and Colwell* (1997) demonstrate that the Kuiper belt must have been more massive in the past for the largest KBOs (100 to 1000 km) to accrete via mutual collisions. At least 90% of the mass in the Kuiper belt was lost through collisions and ejections induced by the stirring and migrating of Neptune (*Stern and Colwell,* 1997; *Hahn and Malhotra,* 1999).

3. POSSIBLE COLLISION OUTCOMES IN THE KUIPER BELT

The observations summarized in the previous section indicate that collisions are an important factor in the evolution of the Kuiper belt. In studying the physical effects of collisions between KBOs, we are guided by the mature studies of collisions in the asteroid belt (*Asphaug et al.,* 2002; *Holsapple et al.,* 2002). However, the possible outcomes of collisions between KBOs are more diverse compared to asteroids because of the dynamical state of the system and the range of physical properties of individual KBOs. The important observations that inform the range of possible collision outcomes are as follows. The dynamical state of the Kuiper belt has changed dramatically with time (see chapter by Morbidelli et al.); hence the mutual collision velocities between KBOs also varied with time. In the modern Kuiper belt, the mutual collision velocities are around 1 km s^{-1} (*Trujillo et al.,* 2001) for classical KBOs and slightly higher for the other populations (see chapter by Gladman et al.). For these impact velocities, most bodies smaller than hundreds of kilometers in size have experienced a catastrophic disruption event, while most of the larger bodies have survived. All bodies should have suffered the production of a significant density of surface impact craters.

In addition to the dynamical impact conditions, the physical properties of KBOs are important. While the present observations are limited, the range of bulk densities of KBOs is <1 to ~2.6 g cm^{-3} (see chapter by Stansberry et al.) and the largest KBOs have a range of surface volatile compositions (H_2O, CH_4, etc.) in addition to a refractory (rock and organic) component (see chapter by Barucci et al.). From these observations and studies of short-period comets, believed to be fragments from KBOs, a typical KBO has a significant (but unknown) fraction of volatiles and high porosity.

Given the possible range of material properties and impact conditions, we outline the potential array of dramatically different outcomes from collisions between KBOs:

1. *Surface impact features.* As a result of the rich collisional history of the Kuiper belt, impact craters are expected to be common on the surfaces of KBOs. However, the morphologies and size distributions are sensitive to the surface and internal structure of the body. Some highly porous bodies survive the formation of multiple, large craters comparable to the radius of the object, as on the low-density asteroid Mathilde (*Veverka et al.,* 1997). Alternatively, only small craters may be observed on rubble piles formed by catastrophic disruption, such as asteroid Itokawa (*Fujiwara et al.,* 2006).

2. *Surface composition and color.* Impact craters and catastrophic disruption events may darken the surface by removing volatiles via heating from the energy of the im-

pact or brighten the surface of a body by excavating fresh ices (Fig. 1).

3. *Density of surface materials.* Laboratory impact craters in highly porous and compressible materials compact the impact site, creating density heterogeneity on the surface (*Housen et al.,* 1999). Over time, cumulative small impacts on a microporous surface may increase the bulk density and decrease the bulk porosity (Fig. 2a,b). In contrast, modeling results indicate that nearly all ejecta from a crater in a macroporous body may reach escape velocities, leaving the bulk density unchanged (*Asphaug et al.,* 1998).

4. *Internal structure and composition.* A subcatastrophic impact may shatter a body (and create a large crater) but leave the original internal material relationships intact (Fig. 2c), while a catastrophic impact both shatters and disperses a body such that the gravitationally reaccumulated remnants are rubble piles with high macroporosity and mixed composition (Fig. 2d).

In the next section, we describe the factors that control the outcome from individual collision events.

4. FACTORS THAT CONTROL OUTCOMES OF COLLISIONS

Recent advances in the understanding of the physics of collisional processes between icy, porous bodies provide new fuel to the study of the role of collisions in the Kuiper belt. The outcome of collision events are governed by the impact conditions (velocity, angle, and mass of each body) and the physical properties of the colliding bodies (strength, composition, and internal structure). Both the impact conditions and physical properties affect the efficiency with which the energy of the impact is coupled to the target. In this section, we summarize four overarching factors that control the outcome of an impact event between KBOs. In the following section (section 5), we will describe laboratory and numerical experiments on KBO analog materials that investigate these controlling factors.

First, the composition and internal structure of the bodies determines the critical velocity required to enter the strong shock regime, where the deformation and coupling of energy and momentum can be described through the Rankine-Hugoniot conservations equations (section 4.1). Slower impact events, where plastic deformation dominates, require more detailed knowledge of the physical properties (particularly strength) of the bodies compared to the strong shock regime. Collisions between KBOs are likely to span the range of plastic and shock deformation.

Second, the final outcome is a balance between the forces of strength and gravity (section 4.2). Scaling laws have been developed for cratering and catastrophic disruption in each regime, but a large transition region exists. Because of the low gravity and expected low strength of KBOs, many collisions may fall in the transition region.

Third, the internal structure and composition of the colliding bodies may significantly affect collision outcomes (section 4.3). Some of the impact energy will be partitioned into phase changes when highly volatile materials are present. High levels of porosity also alter the energy coupling by acting as a shock absorber and localizing shock deformation. The momentum coupling with high porosity changes the excavation flow in the cratering regime and the dispersal of fragments in the disruption regime compared to collisions between solid bodies.

Fourth, collision outcomes are sensitive to the mass ratio of the projectile and target. At the same kinetic energy, a larger projectile is more efficient at removing mass than a smaller projectile (section 4.4).

4.1. Shock Deformation

In most high-energy impact events, the deformation is driven by a shock wave. The energy from the shock controls the physical deformation from the collision, such as fragmentation, pore collapse, heating, and phase changes. The shock also determines the deposition of momentum that leads to crater excavation or dispersal of fragments following a catastrophic disruption event. The amount of deformation can be estimated by considering the volume of material shocked to a given peak pressure.

A strong shock wave is produced in a *hypervelocity* impact event, where the impact velocity exceeds the bulk sound speed (c_b) in both the target and projectile. However, mutual collision velocities between KBOs are likely to span the range from subsonic to supersonic (hypervelocity) collisions (see below). When collisions are comparable to the sound speed, plastic deformation dominates, rather than strong shock deformation. Under subsonic conditions, collisions are simply elastic and governed by the coefficient of restitution.

In this section, we present a summary of the shock physics that determines the outcome in the hypervelocity regime. The peak shock pressure is deduced from the conservation equations and material equation of state, describing the pressure-volume-temperature (P–V–T) states. A shock wave satisfies the Rankine-Hugoniot (R-H) mass, momentum, and energy conservation equations (*Rice,* 1958)

$$u_i - u_0 = U_S\left(1 - \frac{V_i}{V_0}\right) \tag{1}$$

$$P_i - P_0 = \left(\frac{U_S}{V_0}\right)(u_i - u_0) \tag{2}$$

$$E_i - E_0 = \frac{1}{2}(P_i - P_0)(V_0 - V_i) \tag{3}$$

In the above formulae u is particle velocity, U_S is shock velocity, V is specific volume (= $1/\rho$, where ρ is density), P is pressure, and E is specific internal energy. The initial unshocked state is subscripted $_0$ and the final shocked state is subscripted $_i$.

The shock Hugoniot is the curve that describes the locus of possible P–V shock states for a given initial P–V–T state.

For a given impact scenario, the shock pressure is calculated using equation (2), the impact velocity, and the equations of state of the target and projectile. Many materials may be described using a simple linear U_S–u shock equation of state of the form (*Ruoff,* 1967)

$$U_S = c + su \tag{4}$$

where c and s are material constants (for their relationship to finite strain theory, see *Jeanloz,* 1989). The linear shock equation of state is simply a representation of the P–V shock Hugoniot translated into U_S–u space using the R-H equations.

In the planar impact approximation (also called the impedance match solution; see derivation in *Melosh,* 1989), the particle velocities induced by the shock wave reduces the projectile's velocity and mobilizes the target such that continuity at the projectile-target interface is achieved and

$$u_t = v - u_p \tag{5}$$

where v is the impact velocity and subscripts $_t$ and $_p$ refer to the target and projectile, respectively. The shock pressure is derived by solving for u_t using the equality of equation (2) in the target and projectile and substituting equations (4) and (5)

$$\rho_{0,t}(c_t + s_t u_t)u_t = \rho_{0,p}(c_p + s_p(v - u_t))(v - u_t) \tag{6}$$

The quadratic function for u_t is readily solved. In the case of identical shock equations of state in the target and projectile, the particle velocity is equal to v/2, and the peak pressure is given by $\rho_0(c + sv/2)(v/2)$. Because of strength and phase changes, the U_S–u shock equations of state for natural materials are usually fit with multiple linear segments in u, corresponding to different pressure ranges on the shock Hugoniot. The shock equations of state for many rocks and minerals are compiled in *Ahrens and Johnson* (1995a,b), and the equations for nonporous and porous H_2O ice are given in *Stewart and Ahrens* (2004, 2005).

As the shock wave propagates into the target, the peak pressure, derived from the planar impact approximation, decays from rarefaction waves on the free surfaces. The size of the region at peak pressure (known as the isobaric core) and the decay exponent depend on the impact velocity and material properties (e.g., equation of state and porosity) (*Ahrens and O'Keefe,* 1987; *Pierazzo et al.,* 1997). In general, the pressure decay is steeper for high velocities because of energy partitioning into phase changes. The occurrence of impact-induced phase changes can be estimated by considering the critical shock pressures required for melting and vaporization. When the shock pressure is above a critical value, the material is melted/vaporized after passage of the shock wave and return to ambient pressure conditions.

The present mean mutual collision velocity between classical KBOs (~1 km s^{-1}) is lower than the bulk sound speed of full density silicates and ices. Nonporous H_2O ice has a c_b of 3.0 km s^{-1} at 100 K (*Petrenko and Whitworth,* 1999; *Stewart and Ahrens,* 2005). Silicate rocks have larger c_b, typically around 5 km s^{-1} (*Poirier,* 2000). Sound speeds of laboratory preparations of nonporous ice-silicate mixtures, with up to 30 wt% sand, are similar to pure H_2O ice (*Lange and Ahrens,* 1983). Pure porous H_2O ice, on the other hand, can have much lower sound speeds, from 0.1 to 1.0 km s^{-1} for bulk densities of 0.2 to 0.5 g cm^{-3} (*Mellor,* 1975; *Furnish and Remo,* 1997). Silica aerogels with densities of about 0.2 g cm^{-3} have sound speeds of about 200 m s^{-1} (*Gross et al.,* 1988), and 35% porous sand has a sound speed of 130 m s^{-1}. If KBOs are volatile rich and porous, then mean present-day collisions may be supersonic.

During the collisional evolution of the Kuiper belt, collisions span the subsonic to supersonic regimes. Understanding the controlling physics in the subsonic regime, where plastic deformation dominates, will require focused studies on analogs for the range of mechanical structures in the Kuiper belt (section 5). In the strong shock regime, crater scaling relationships and catastrophic disruption theory are applicable, as described in the next section.

4.2. Strength and Gravity

The final outcome of a collision, e.g., crater size or dispersed mass, depends on the balance between strength and gravitational forces. In the case of impact cratering, the relationship between the size and velocity distribution of the impacting population and the observed crater population can provide insight into the collisional history of a system [as has been done for the terrestrial planets (*Strom et al.,* 2005) and asteroids (*O'Brien and Greenberg,* 2005)]. Backing out the impactor properties requires the application of the appropriate crater scaling relationships, which depend on both the impact conditions and material properties. In the case of catastrophic disruption, knowledge of the properties of the populations of disrupted and primordial bodies provide strong constraints on the collisional evolution of the Kuiper belt. Here, we present the crater size and catastrophic disruption scaling laws in the strength and gravity regimes.

4.2.1. Crater scaling theory. Development and validation of the appropriate scaling relationships is crucial for the next generation of collisional evolution models of KBOs that include consideration of physical deformation effects. Because of their low gravity and likely low strength, the outcome of collisions between KBOs is near the transition between the strength and gravity regimes. In this section, we discuss the strength to gravity transition and summarize the equations and material parameters for crater scaling for comparison to laboratory craters in ice and porous targets in section 5.1.

In a cratering event, the shock-driven excavation flow produces a roughly hemispherical cavity, called the transient crater. Assuming that material strength can be represented by a single parameter, Y, the transition size between the strength and gravity-controlled cratering regimes is proportional to $Y/\rho g$. Y is the dominant strength measure that controls crater size (e.g., shear strength); ρ is the bulk density of the target; and g is the gravity of the target (*Melosh,* 1977; *Melosh and McKinnon,* 1978). As the impact energy increases, the outcome of collisions will transition from a cratering regime to a total body disruption regime. The criteria for catastrophic disruption, Q^*_D, is defined as the specific energy (kinetic energy of the projectile divided by the mass of the target) required to disrupt and gravitationally disperse half the mass of the target (*Melosh and Ryan,* 1997). Note that, unlike the cratering regime, disruption is governed by the bulk tensile strength of the body, which is typically an order of magnitude lower than the compressive strength of brittle solids (see sections 4.2.2 and 4.4).

The theory for crater size scaling based on impact parameters and material properties is summarized by *Holsapple* (1993). A common approach utilizes π-scaling, with empirical constants derived from impact and explosion cratering experiments under Earth's gravity and high gravity. Predicting the final crater volume and shape requires two steps: (1) calculating the volume of the transient crater cavity using the π-scaling laws and (2) calculating the amount of collapse of the transient crater to the final crater volume and shape. The first step is better understood than the second.

In π-scaling, the cratering efficiency, π_V, is defined as the ratio of the mass of material ejected and displaced from the transient crater cavity to the mass of the projectile

$$\pi_V = \frac{\rho V}{m_p} = \frac{M_c}{m_p} \tag{7}$$

where V is the volume of displaced and ejected material, m_p is the mass of the projectile, and $M_c = \rho V$. For strength-dominated craters, the cratering efficiency depends on the ratio of a measure of the shear strength of the target, $\bar{Y}$, to the initial dynamic pressure, given by

$$\pi_{\bar{Y}} = \frac{\bar{Y}}{\rho v_\perp^2} \tag{8}$$

where, $v_\perp = v \sin\theta$, v is the impact velocity, and θ is the impact angle from the horizontal. In the gravity-dominated regime, the cratering efficiency depends on the ratio of the lithostatic pressure at a characteristic depth of one projectile radii, r_p, to the normal component of the initial dynamic pressure (the inverse Froude number)

$$\pi_2 = \frac{g r_p}{v_\perp^2} \tag{9}$$

where g is the gravitational acceleration.

Impact experiments demonstrate that the transition from strength- to gravity-dominated cratering spans about two decades in π_2. Following *Holsapple* (1993) and *Holsapple and Housen* (2004), the cratering efficiency can be defined by an empirical, smoothed function of the form

$$\pi_V = K\left(\pi_2 + \pi_{\bar{Y}}^{\beta/\alpha}\right)^{-\alpha} \tag{10}$$

where the exponents are related to a single coupling exponent, μ, by $\alpha = 3\mu/(2 + \mu)$ and $\beta = 3\mu/2$. The coupling exponent μ is bounded by two cratering regimes: momentum scaling (where $\mu = 1/3$) and energy scaling ($\mu = 2/3$) (*Holsapple and Schmidt,* 1987; *Holsapple,* 1987). Note that equation (10) assumes that the target and projectile have the same density. The transition from strength to gravity dominated regimes occurs when $\bar{Y} \sim \rho g r_p$.

In Plate 3, cratering efficiencies are presented for liquid water ($K = 0.98$, $\mu = 0.55$, $\bar{Y} = 0$ MPa), dry sand ($K = 0.132$, $\mu = 0.41$, $\bar{Y} = 0$ MPa, 35% porosity) and weak rocks ($K = 0.095$, $\mu = 0.55$, $\bar{Y} = 3$ MPa) (values from *Holsapple and Housen,* 2004). Dry sand is a noncrushable porous material and weak rock is a reasonable analog for nonporous H_2O ice. Cratering efficiencies in crushable, porous materials, from hypervelocity experiments in vacuum, lay a factor of a few below the dry sand line (*Schultz et al.,* 2005). The transition from strength regime (lower values of π_2, when $\pi_{\bar{Y}} > \pi_2$) to gravity regime (higher values of π_2) corresponds to the transition from a horizontal line in a π_V–π_2 plot, when the cratering efficiency is independent of π_2, to a power law with slope $-\alpha$. The cratering efficiency in the strength regime increases with increasing impact velocities, as indicated by the curves for impacts into weak rock targets at 0.5, 2.0, and 7.0 km s^{-1}. Note that the cratering efficiency in dry sand is less than for weak rock in the gravity regime because of energy dissipation in the porous sand. Data from impact cratering experiments conducted in vacuum under Earth's gravity into nonporous ice (▲; *Cintala et al.,* 1985; *Lange and Ahrens,* 1987; *Burchell and Johnson,* 2005) and 50% porous ice (•; *Koschny et al.,* 2001; *Burchell et al.,* 2002) fall in the strength regime. Cratering experiments at 1.86 km s^{-1} in porous mixtures of sand and perlite bonded with fly ash and water under 1 atm of pressure and varying gravity are nearly independent of π_2, indicating strength-dominated behavior with a plausible intersection with the gravity-dominated regime (the dry sand line) (section 5.1.3) (*Housen and Holsapple,* 2003).

Cratering events in the Kuiper belt by a nominal 0.5-m radius body at 1 km s^{-1} onto targets of 0.1–1000-km radii correspond to π_2 values in the range from 10^{-10} to 10^{-6}. Therefore, for the range of impact velocities in the Kuiper belt, Plate 3 demonstrates that the presence of any strength is likely to control the final crater size for the majority of impact events. In the upper range of possible values of π_2, gravity may dominate if cratering is less efficient in KBOs than in dry sand.

After formation of the transient crater cavity by the excavation flow, most craters undergo collapse to a final crater shape (*Melosh and Ivanov,* 1999). For simple, strength-dominated craters, the final crater size is similar to the transient cavity with some collapse of the crater walls. For cratering in soils and rocks, the rim radius of the transient crater is approximately $R_r = 1.73V^{1/3}$ (*Holsapple,* 1993). Complex, gravity-dominated craters undergo significant collapse of the transient cavity, and the final crater rim radius scales with the transient crater rim radius and gravity by $R_{complex}(cm) = 0.37R_r(cm)^{1.086}(g/g_{Earth})^{0.086}$ (*Holsapple,* 1993).

The final state of crater formation in porous, icy bodies is not well understood. In section 5, we discuss some of the laboratory experiments that provide our best guesses at the appropriate crater size scaling laws for the Kuiper belt.

4.2.2. Catastrophic disruption theory. Based on the models of the collisional evolution of the Kuiper belt, it is probable that a large fraction of bodies have suffered both cratering events and disruptive collisions. The catastrophic disruption criteria, Q^*_D, is the ratio of the projectile's kinetic energy to the mass of the target required to disrupt and disperse half the mass of the target.

The criteria has two components (*Davis et al.,* 1979)

$$Q^*_D = Q^*_S + Q_b \tag{11}$$

where Q^*_S is the strength of the body to shattering and Q_b is the gravitational binding energy of the target.

The catastrophic disruption criteria for a rocky body is shown by the thick solid line in Fig. 3 (section 5.2.2) (*Benz and Asphaug,* 1999). The critical energy is averaged over all impact angles. A head-on collision is most efficient, requiring about an order of magnitude less energy compared to the angle average (*Benz and Asphaug,* 1999; *Leinhardt et al.,* 2000; *Leinhardt and Richardson,* 2002). In the strength regime, where Q^*_S dominates, the critical energy decreases with increasing target size because tensile strength, the controlling strength measure, is scale dependent (*Housen and Holsapple,* 1999). The larger the body, the larger the number of preexisting natural flaws and the lower the tensile strength. In the gravity regime, pressure from the self-gravity of the object increases the strength, following the shattered rock curve for head-on impacts (*Melosh and Ryan,* 1997). In the gravity regime, the gravitational dispersal criteria dominates over shattering by orders of magnitude. Note that the standard disruption criteria curves assume that the size of the projectile is small compared to the target (see section 4.4).

The manner in which volatile content and porosity affect the disruption criteria is not well understood. Here, we estimate the effects of each using nonporous and porous H_2O ice as an example. There has been little work on catastrophic disruption of large objects in the gravity regime at impact speeds and compositions that are relevant to the the present day Kuiper belt, thus this discussion is meant to provide general guidance, not detailed values.

In the strength regime, the Q^*_D intercepts for nonporous and porous ice (thin solid line and dashed line, respectively) are tied to results from laboratory disruption experiments (see section 5.2.1) (*Arakawa et al.,* 2002). These values are consistent with other experimental results (*Ryan et al.,*

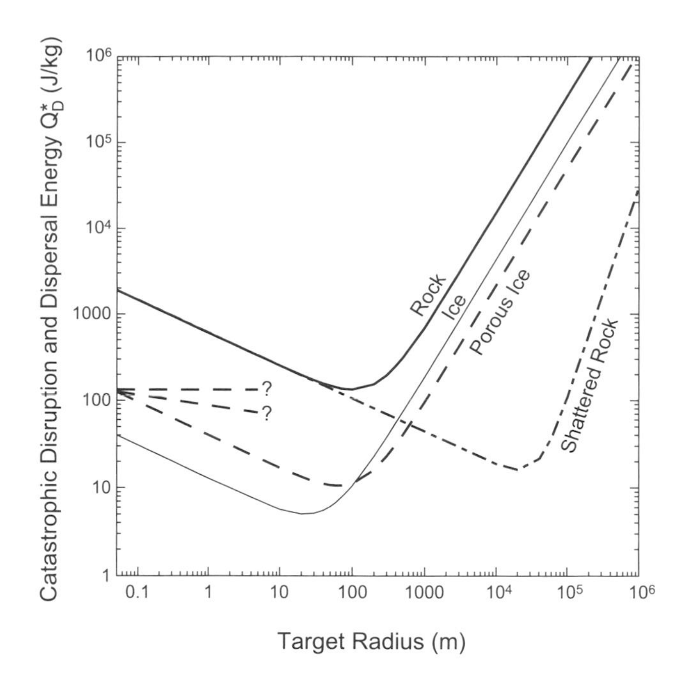

Fig. 3. Catastrophic disruption and dispersal energy (Q^*_D) vs. target radius for rock (thick solid line), nonporous ice (thin solid line), 50% porous ice (dashed line). The negatively sloped portions of the curves are in the strength regime, the positive slopes are in the gravity regime. The criteria for rock is based on angle averaged results from 3 km s^{-1} collisions onto basalt (*Benz and Asphaug,* 1999). The nonporous and porous ice intercepts are based on laboratory experiments (*Arakawa et al.,* 2002) (Q^*_D = 40 J kg^{-1} and Q^*_D = 143 J kg^{-1} for low- and high-porosity targets, respectively, of 5-cm radius). The extrapolation into the gravity regime is highly uncertain for porous materials. These results assume energy coupling by a small projectiles compared to the size of the target.

1999). The slope of Q^*_S for pure ice is assumed to be the same as for rock. For porous ice, on the other hand, the slope in the strength regime is particularly uncertain. The slope for porous materials may be much shallower than for nonporous materials because the size-dependent scaling of flaws may not apply (*Housen and Holsapple,* 1999). This uncertainty is depicted in Fig. 3 by several dashed lines of varying slope. Perhaps counterintuitively, in the strength regime, a porous material is harder to disrupt than a nonporous material due to localization of energy by compaction of pores and/or reflection of the shock wave off of free surfaces (see section 5.2.1).

In the gravity regime, the Q^*_D criteria for nonporous ice lies below the rock criteria by the ratio in mass (for this plot vs. target size). This is consistent with the numerical impact simulations into ice targets by *Benz and Asphaug* (1999). Similarly, a porous target of the same size is easier to disrupt because of its lower total mass. Adjusting the gravity-dominated Q^*_D criteria by the ratio in total mass makes the unreliable assumption that the energy coupling from the collision is similar for each material. Because of the significant dissipative effects of porosity, porosity may have a large affect on energy coupling in the gravity regime (the

Q^*_S term may be more significant). More work is needed to determine exactly how porosity affects the energy coupling for catastrophic disruption events.

As with impact cratering events, it is clear that in order to predict the collision outcome from a disruption event, it is necessary to know something about the material properties of the KBOs. In the case of catastrophic disruption, global properties, rather than surface properties, are more important; e.g., is the target porous, icy, rocky? In section 5.2, we discuss the small amount of work on catastrophic disruption of KBO analog materials.

4.3. Internal Structure and Composition

Kuiper belt objects are likely to possess a wide variety of internal structures, as depicted in Fig. 2. Dynamical excitation and increased collision frequencies from the migration of Neptune removed most of the mass from the original Kuiper belt, leaving a mixture of collision fragments and unmodified material (*Davis and Farinella,* 1997; *Hahn and Malhotra,* 1999; chapter by Morbidelli et al.). Comets may provide clues to the present internal structure of KBOs; however, comets are expected to be diverse themselves (chapter by Barucci et al.; *Weissman et al.,* 2005).

Porosity, either primordial or the result of collision events, is a major complicating factor in predicting the amount of shock deformation. Since KBOs are expected to contain a range of porosities, the outcome of individual collisions could vary widely depending on the properties of the colliding bodies. When the initial porosity is high, the shock impedance (the bulk density times the sound speed) is low, and the peak shock pressures produced for a given impact condition are lower compared to a solid target (equations (1) and (2)). For a given shock pressure, however, the internal energy increase is larger in a porous material because of the greater change in volume during shock compaction (equation (3)). Hence, the temperature rise due to a shock is higher in porous materials, and impacts into porous ices may result in abundant melting or vaporization near the impact site (*Stewart and Ahrens,* 2004, 2005). Porosity is an efficient dissipator of shock energy. As the shock propagates into the target, porosity increases the decay rate of the shock because of the increased energy partitioning into heat (e.g., *Meyers,* 2001). Therefore, the shock-deformed volume in porous materials is smaller compared to a solid.

The length scale of the porosity is also important. Small-scale porosity compared to the shock thickness is described as *microporosity*. The thickness of the shock wave is proportional to the scale of the topography on the surface of the projectile. Large-scale porosity, e.g., a rubble pile of solid pieces, is described as *macroporosity*. In the latter case, the solid (e.g., monolithic) pieces may have high strength, and impact cratering events onto a monolithic piece would reflect the high surface strength. For catastrophic disruption events, however, a rubble pile has low bulk tensile strength. In a rubble pile, the shock wave would reflect upon encountering void space between solid boulders, and as a result, the energy from the shock would be deposited in a smaller volume compared to a shock wave propagating through a monolith of competent rock. On the other hand, a microporous body may have low surface compressive strength, but because of the efficient shock dissipation, a more energetic impact is required to catastrophically disrupt the body (*Asphaug et al.,* 1998). Hence, both macroporous and microporous bodies may have high *disruption strength*.

Compositional variation and surface layers also change the way energy is coupled into the target. The impact energy will be partitioned into more compressible phases and a larger Q^*_D is required to disrupt a more compressible material (*Benz and Asphaug,* 1999). Because some of the energy of the impact is partitioned into heating, each collision event will also result in net devolatilization. Finally, phase changes (melting, vaporization) of volatile materials will result in steeper decay of the shock wave (*Ahrens and O'Keefe,* 1977; *Pierazzo et al.,* 1997; *Pierazzo and Melosh,* 2000) that tends to localize the shock deformation in a manner similar to the effects of porosity.

4.4. Mass Ratio

The mass ratio of the colliding bodies is also an important factor in determining the collision outcome. In simulations of subsonic collisions, *Leinhardt and Richardson* (2002) found that smaller projectiles were not as efficient at disrupting targets as larger projectiles (see Fig. 4) (*Leinhardt and Richardson,* 2002; *Melosh and Ryan,* 1997). The mass ratio affects the volume over which the impact energy and momentum are deposited. When the projectile is much smaller than the target, the impact directly affects a small volume, about the size of the projectile. The rest of the target acts to dampen any material motion. When the projectile mass is similar to the target mass, on the other hand, the projectile comes in direct contact with a significant volume fraction of the target. As a result, the specific energy required to catastrophically disrupt a target decreases by orders of magnitude.

The dependence on mass ratio has not been studied directly for hypervelocity impacts, although the dependence on the coupling of energy and momentum should be similar to the subsonic case. First, the size of the peak pressure region (isobaric core) is proportional to the size of the projectile. Second, the decay of the peak shock pressure with distance depends on the impact velocity. The decay is steeper for higher-impact velocities because more energy is partitioned into phase changes and deformation. Low-impact velocities have a more shallow decay exponent (in the elastic limit). The particle velocities are proportional to the peak shock pressure; thus the shock pressure profile in the target will affect the dispersal of fragments and the catastrophic disruption criteria. Q^*_D should decrease as the projectile size increases for a fixed kinetic energy of the projectile.

Let us now consider a likely impact scenario in the Kuiper belt. For an example 100-km volatile-rich target in

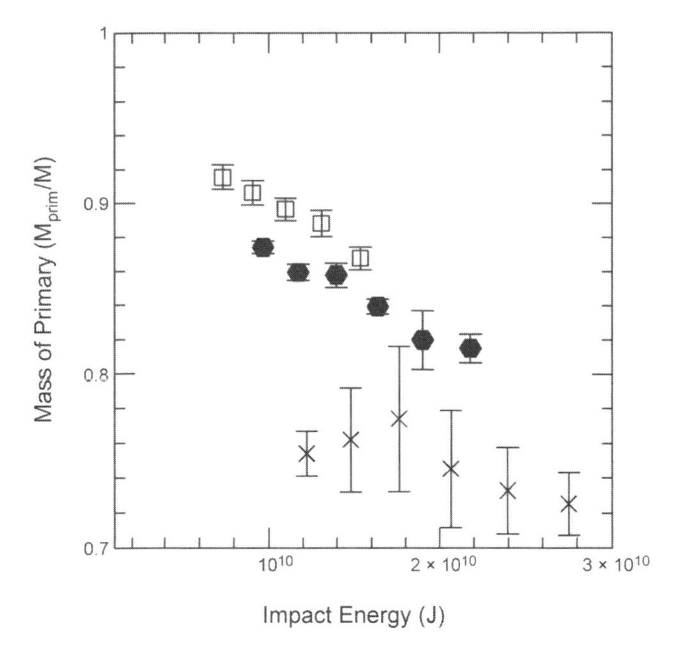

Fig. 4. Mass of the largest reaccumulated body (M_{prim} in units of total system mass m = M_{proj} + M_{targ}) following a large collision event as a function of the kinetic energy of the projectile. Results from N-body impact simulations of km-sized rubble piles in the subsonic regime (*Leinhardt and Richardson,* 2002). The crosses, filled hexagons, and open squares denote projectile to target mass ratios of 1:3, 1:6, and 1:9, respectively. In all cases, targets were identical with a radius of 1 km. All data points are averaged over several simulations at various impact angles (rms error bars). Note that, for the same kinetic energy, the largest projectiles produced the smallest largest post-collision remnant. These results indicate that increasing the projectile size increases the disruption efficiency.

the Kuiper belt, Fig. 3 predicts that about 10^5 J kg^{-1} is necessary for catastrophic disruption. At 1 km s^{-1}, the projectile would have a radius two-thirds that of the target. However, Fig. 3 assumes small (point source) projectiles. As the projectile to target mass ratio approaches unity, the amount of energy per target mass needed to disrupt the target drops (*Holsapple,* 1993; *Melosh and Ryan,* 1997; *Leinhardt and Richardson,* 2002). Therefore, a 100-km target may indeed be catastrophically disrupted by a smaller projectile than predicted in Fig. 3, and the larger objects in the Kuiper belt may have suffered more catastrophic or near-catastrophic impacts than inferred in previous studies. More work is needed to determine how Q_D^* behaves with mass ratio in hypervelocity collisions.

5. STUDIES OF COLLISIONS IN ANALOGS TO KUIPER BELT OBJECTS

We now turn to laboratory and numerical experiments on the major factors that affect the collision outcomes described above. Laboratory experiments and numerical models of collisions between icy and/or porous bodies serve as the best analogs at present for impacts into KBOs. There is a large body of laboratory work on cratering impacts into volatile ices and mixtures as well as porous material (section 5.1). Although a coherent theory will require additional experiments, the laboratory results are a good guide to the possible outcomes of cratering collisions on KBOs. In comparison, the laboratory and numerical experiments on catastrophic disruption of KBO analogs are more limited (section 5.2). This is due in part to the inability to study catastrophic disruption in the gravity regime in the laboratory and the difficulties in modeling collisions between porous, volatile bodies.

5.1. Cratering

We begin with the results of several impact cratering studies into nonporous H_2O ice and ice-silicate mixtures (section 5.1.1). Then, the effects of porosity are introduced (section 5.1.2). However, it is difficult to deconvolve the effects of porosity and low material strength in laboratory studies. Possible outcomes include cratering events that result in compaction rather than the normal crater excavation flow (section 5.1.3). Relatively little work has been conducted on ices more volatile than H_2O, which have been observed on the surfaces of the largest KBOs (see chapters by Barucci et al. and Brown) (section 5.1.4). Finally, experiments into targets with surface layers of different strength materials can also have a significant affect on the crater morphology (section 5.1.5). Because of the influence of an atmosphere on the final crater form, explosion cratering studies (*Holsapple and Housen,* 2004) are not included in this discussion.

5.1.1. Cratering in nonporous H_2O ice and ice-silicate mixtures. As a low-density and volatile material, solid H_2O ice represents a very simple model for the bulk properties of KBOs. Depending on the evolution of KBOs, some surfaces may be dominated by solid ice. Most laboratory impact experiments in ice are conducted in the strength regime. Generally, cratering efficiencies in solid ice are similar to a dry soil or weak rock (Plate 3) (*Chapman and McKinnon,* 1986). Solid ice cratering experiments span impact velocities of 0.1–7.3 km s^{-1} using a wide range of projectile materials (*Burchell and Johnson,* 2005; *Croft et al.,* 1979; *Lange and Ahrens,* 1982, 1987; *Shrine et al.,* 2002; *Grey and Burchell,* 2003; *Kawakami et al.,* 1983; *Kato et al.,* 1995; *Iijima et al.,* 1995; *Cintala et al.,* 1985). For a given impact energy, the crater volume is more than an order of magnitude larger than craters formed in a typical hard silicate rock. In many of these experiments, the measured volumes of the craters include a component of spalled material (near-surface material ejected under tensile failure), forming a terraced crater morphology with a central pit. Hence, the reported volumes are larger than the transient crater volume and comparisons to π-scaling laws must be made with caution. Also, differences in ice target preparation contribute to scatter between experiments.

Impact cratering experiments in solid H_2O ice have investigated the effects of target temperature on the cratering efficiency (*Lange and Ahrens,* 1982, 1987; *Grey and Burchell,* 2003). Low-temperature ice has a cratering efficiency between temperate ice and hard rock. Under fixed impact conditions, the crater depth and volume decreased by factors of 2 and 4, respectively, as the ice temperature decreased from 253 K to 100 K (*Grey and Burchell,* 2003). It is well established that the yield strength of ice increases as the temperature decreases (e.g., *Sammonds et al.,* 1998); however, the magnitude of the effect is not well predicted (*Grey and Burchell,* 2003).

Thermodynamic analyses of shock wave experiments on solid H_2O ice at 100 K indicate that peak shock pressures of 1.6 and 4.1 GPa are required to produce incipient and complete shock-induced melting, respectively (*Stewart and Ahrens,* 2005). For pure ice on ice impacts, these pressures are achieved at impact velocities of about 1 and 2 km s^{-1}. If the bulk shock impedance of solid ice is similar to porous volatile-refractory mixtures, mass melting of solid ice within KBOs is only expected at the upper end of the range of collision velocities within the Kuiper belt (0.5–3 km s^{-1}) (*Dell'Oro et al.,* 2001).

Nonporous mixtures of ice and silicates (e.g., ice-saturated sand) have also been studied in the strength regime (*Croft et al.,* 1979; *Koschny and Grün,* 2001). The cratering efficiency decreases with increasing silicate content.

5.1.2. Cratering in porous material. Several groups have conducted cratering experiments into porous materials (*Schultz et al.,* 2005; *Schultz and Gault,* 1985; *Schmidt,* 1980; *Housen et al.,* 1999; *Housen and Holsapple,* 2003; *Koschny et al.,* 2001). The porous targets include porous ice, sand, Ottawa flint shot, pumice, and vermiculite. Hypervelocity impact experiments, conducted under vacuum, into pumice powder with porosities between 35% and 50% follow a single gravity-controlled crater scaling that is slightly less efficient than dry sand (Plate 3) (*Schultz et al.,* 2005). However, for lower-velocity impacts, in the strength regime or in the transition between strength and gravity-dominated cratering, the effects of porosity can be significant (section 5.1.3).

Results from cratering experiments into ~50% porous (*Koschny et al.,* 2001; *Burchell et al.,* 2002) and solid ice targets indicate that the displaced and ejected mass scales linearly with impact energy. In other words, the crater volume is proportionally larger by the difference in target density. However, it is unlikely that this result can be extrapolated to events with much larger impact energies because of the considerable effects of vaporization on the final crater size (*Holsapple and Housen,* 2007; *Schultz et al.,* 2005).

Because of the large increase in internal energy associated with shock compaction of porous H_2O ice, the critical pressures required for shock-induced melting are lower compared to solid ice. From shock wave experiments, *Stewart and Ahrens* (2004) find that shock pressures of only 0.3 to 0.5 GPa initiate melting in 40–45% porous ice, and complete melting is reached by 2 GPa. These pressures correspond to impact velocities in the range of 1–2.5 km s^{-1} for collisions between porous ice bodies. Although porous ice has lower shock impedance than solid ice, the increase in internal energy from pore compaction results in similar critical impact velocities for shock-induced melting. If KBOs have shock impedances greater than pure porous H_2O ice, as expected if they are ice-refractory mixtures, then mutual collisions under the present dynamical environment will result in abundant melting of H_2O ice.

Therefore, shock-induced melting in porous targets may produce crater cavities lined with quenched melt (rapidly cooled liquid that solidifies as a glass). Quenched melt lined craters have been observed in laboratory impact experiments into 50% porous H_2O ice (nylon projectiles at 0.9–3.8 km s^{-1}) (*Koschny et al.,* 2001) and 5–60% porous soda lime glass (glass projectiles at 4.9–6.1 km s^{-1}) (*Love et al.,* 1993). In some cases, all the impact-generated melt was ejected from the crater. Hence, cratering events onto porous KBOs may produce solid ice ejecta fragments.

The depth of penetration of the projectile plays a significant role in the cratering efficiency in porous materials. In hypervelocity impacts, the impact angle determines the outcome. For example, the cratering efficiency in compressible porous perlite granules ($\rho = 0.2$ g cm^{-3}) increases as the impact angle decreases from 90° to 30° (*Schultz et al.,* 2005). Vertical impacts into porous materials penetrate deeply into the target, resembling a deeply buried explosion. Low-angle impacts, in contrast, reach a more optimal shallow depth of burial to produce a larger crater. In the low-velocity regime, an impedance mismatch between the target and projectile will also influence the depth of penetration. A dense projectile may experience little deformation and penetrate deeply, resulting in less-efficient cratering compared to a projectile with density that matches the target. Interpretation of the cratering record on KBOs will need to include the role of impact angle and the depth of penetration in the final crater size.

In the case where the projectile is more dense than the target (e.g., a solid rock meteoroid impacting a porous KBO), the impact conditions may be supersonic for the target but subsonic for the projectile. In this case, the projectile is not significantly disrupted by the impact event. Laboratory experiments show that the penetration depth increases as the density contrast between the projectile and target increases (*Love et al.,* 1993). Intact or melted nylon ($\rho =$ 1.14 g cm^{-3}) and copper ($\rho = 8.92$ g cm^{-3}) projectiles were recovered after impacts at velocities up to 7 km s^{-1} into 50% porous H_2O ice (*Koschny et al.,* 2001; *Burchell et al.,* 2002). The experimental results suggest that dense meteoroids may embed themselves into the surfaces of KBOs and comets. In a pathological example, a population of compacted, devolatilized projectiles might be found embedded in the surfaces of very porous, volatile KBOs.

Reliable numerical models of crater formation in porous materials have been hindered by the difficulty in modeling the shock compaction of porous materials (e.g., *Herrmann,* 1969; *Johnson,* 1991). Some general results can be drawn from the relatively few simulations to date: (1) a proxy model for porosity using layers of solid ice and void and the Autodyne code (*Burchell and Johnson,* 2005), (2) a P-alpha crush-up model for sand using the CTH code (*Housen and Holsapple,* 2000), and (3) a new ε-alpha compaction model using the iSALE code (*Wünnemann et al.,* 2006).

In summary, the transient crater diameters in porous materials are smaller but the crater is deeper than those in non-porous media. The lower bulk density of the porous target allows the projectile to penetrate more deeply. The shock wave is attenuated more quickly in porous material because energy is partitioned into crushing pores. These numerical experiments show that porous crushable objects are more resilient to large impact than nonporous objects because the damage from the impact is much more localized. With these more advanced models of porosity, future work can address the volume of material that experiences deformation (fragmentation, devolatilization) from impact events in the Kuiper belt.

5.1.3. Compaction cratering. Observations of an unusual main-belt asteroid, 253 Mathilde, have incited several studies on the role of porosity on impact cratering. Imaged during a flyby of the NEAR spacecraft, Mathilde has a low bulk density (~1.3 g cm^{-3}) and exhibits four large impact craters with diameters larger than the mean diameter of the asteroid (*Veverka et al.,* 1997). The large craters have no visible ejecta blankets or raised rims. As a result of their size, the craters are very close to each other and yet seem to show no evidence of interaction. The unique characteristics of Mathilde suggest that the internal structure of this C-type asteroid is different from other classes of main-belt asteroids in a fundamental way.

Housen et al. (1999) and *Housen and Holsapple* (2003) conducted a series of cratering experiments into compressible, porous material in an attempt to explain the craters on asteroid 253 Mathilde. The authors suggest that high microporosity (40–60%) and high compressibility lead to a phenomena they termed compaction cratering.

In their studies, the projectile and impact velocity was held constant, and the target porosity and gravity (using a centrifuge) were varied. In the high-gravity environment, the craters had no raised rims and minimal ejecta outside the crater because most of the ejecta never escaped the crater cavity. A computed tomography scan of the crater showed a region of pore compaction approximately one crater radius below the crater. *Housen et al.* (1999) and *Housen and Holsapple* (2003) also impacted one of the used targets close to the original crater and confirmed that there was little interaction between the craters. For example, the first crater did not collapse as a result of the second impact, nor was the first crater erased as a result of shaking or ejecta filling in the first crater.

The authors conclude that large impacts onto compressible, highly porous targets may not reach the gravity regime in which the gravity scaling laws can be employed to predict crater diameter and depth. In Plate 3, the compaction craters in perlite and mixtures of sand, perlite and fly ash (open symbols) are strength-dominated. As a result, a porous, compressible object may have a very high resistance to disruption even if both the tensile and compressive strengths are low.

The occurrence of compaction cratering in nature is not understood and presently a subject of debate. More work on the compaction cratering phenoma is needed. If compaction cratering is prevalent, the bulk density of a porous compressible object may be significantly increased over the age of the solar system by compaction from impacts.

5.1.4. Cratering in other volatile materials. Kuiper belt objects show wide diversity in volatile content. Large KBOs that are bright enough for detailed spectroscopic study show evidence of significant volatile content (e.g., methane and ethane ices) (chapter by Barucci et al.; *Brown et al.,* 2007a; *Barucci et al.,* 2005). Laboratory experiments have found that the addition of material more volatile than H_2O ice, such as CO_2 and NH_3 (*Burchell et al.,* 1998; *Burchell and Johnson,* 2005; *Schultz,* 1996), can increase the strength of the target and, as a result, decrease the cratering efficiency.

The phase of the volatile is also important. Comet nuclei and their precursors may contain trapped pockets of gas under high internal pressures. If an impact event releases trapped gas, the vapor expansion may aid in the ejection of more mass than would be possible from the kinetic energy of the impact itself (*Durda et al.,* 2003; *Schultz et al.,* 2005; *Holsapple and Housen,* 2007).

5.1.5. Cratering in layered targets. *Belton et al.* (2007) suggest that all three Jupiter-family-comet nuclei (believed to originate in the scattered disk component of the Kuiper belt) that have been closely observed to date (Wild 2, Borrelly, and Tempel 1) show evidence of layering. *Belton et al.* (2007) propose that this layering is primordial and a result of the accretion process. By extrapolation, the precursor objects in the Kuiper belt may also have layered structures. Whether the observed layering is primordial or not is a matter of debate; however, surface layering (a devolatilized "crust") was predicted for comets based on thermal evolution models (*Belton and A'Hearn,* 1999). Layering of different strength materials does explain features seen on other objects in the solar system. For example, concentric crater morphology on the Moon and slightly filled in linear structures on the asteroid Eros can be explained by regolith overlying more competent rock.

Oberbeck and Quaide (1967) and *Ryan et al.* (1991) conducted experiments on regolith-covered targets and determined that the morphology of the resulting crater changed depending on the depth of the regolith. This result has been confirmed with numerical experiments by *Senft and Stewart* (2007). *Ryan et al.* (1991) conducted drop tests to study the collision outcome of aggregate projectiles impacting different depths of regolith (fine particles overlaying a concrete surface). When the depth of the regolith was at least the size of the projectile, the aggregate lost <10% of its mass when dropped from a height that resulted in catastophic disruption when the surface was not covered with a layer of fine particles. The porous regolith was very efficient at dissipating the impact energy.

In addition, impact experiments into granular mixtures of H_2O ice, CO_2 ice, and pyrophylite that have experienced thermal stratification produce craters with very different morphologies (*Arakawa et al.,* 2000). Finally, *Schultz* (2003) looked at the effect of layering on crater scaling. Craters retain their original diameter until the layer becomes less than twice the projectile diameter (for vertical impacts) or less than a projectile diameter (for oblique impacts). Even though the final crater depth is limited by the substrate, the diameter remains unaffected. Imagery of craters on the surfaces of KBOs would provide information about near-surface layering.

5.2. Catastrophic Disruption

As mentioned above, there has been much less work in the catastrophic disruption regime than the cratering regime. We begin this section with a brief summary of the laboratory experiments of catastrophic disruption either using ice targets or investigating a range of porosities. Next, numerical experiments on ice or porous targets are presented.

5.2.1. Catastrophic disruption laboratory experiments. Strength-regime laboratory experiments have investigated the catastrophic disruption of icy and porous targets. Both *Love et al.* (1993) and *Ryan et al.* (1999) conducted catastrophic disruption impact experiments into macroporous targets. *Arakawa et al.* (2002) performed impact experiments into nonporous and porous pure ice and ice silicate mixtures. In most of the experiments, the porous targets were more difficult to disrupt because the kinetic energy of the projectile is partitioned into crushing energy to fill void spaces (*Love et al.,* 1993) and the shock wave reflects off the large number of free surfaces. The result is significant attenuation of the shock wave compared to solid materials.

Ryan et al. (1999) conducted 20 low-speed (100 m s^{-1}) impact experiments into solid and porous ice targets. They found that porous ice targets, although significantly weaker than the solid ice targets under static conditions, had a disruption strength equivalent to the solid targets with similar total mass. The authors attribute this behavior to the efficient dissipation of energy in void spaces.

Love et al. (1993) ran a series of hypervelocity experiments (4.8–6.0 km s^{-1}) into glass targets of varying porosity and strength. They found that the specific energy needed to catastrophically disrupt the target was proportional to (1-porosity)$^{-3.6}$. Their results suggest that the porosity of the target is more important for the collision outcome than the compressive strength of the target. More work is needed to separate the effects of porosity and strength. Impacts into the more porous targets resulted in deeper penetration of the projectile but the excavated volume was about the same as in less porous targets. With higher porosity, the damage from the impact was more localized. These results suggest that porous objects in the solar system would have longer lifetimes against collisional disruption than monoliths of the same material.

Arakawa et al. (2002) performed moderate-speed impact experiments (150–670 m s^{-1}) into ice and ice-silicate mixtures to quantify the effect of porosity on disruption strength. In pure ice targets, the disruption strength increased with increasing porosity. Puzzlingly, in mixed material targets, the disruption strength decreased with increasing porosity. These experiments suggest that the nature of the material bonding (and material strength) can be as important as the bulk porosity.

The work to date demonstrates that porosity plays a significant role in the outcome of catastrophic disruption experiments. However, more work is needed to understand how porosity strengthens a material and how to predict disruption strength as a function of porosity.

5.2.2. Catastrophic disruption numerical simulations. Investigations of catastrophic disruption in the gravity regime rely upon numerical experiments. Studies including KBO analog materials are limited. *Asphaug et al.* (1998) have considered km s^{-1} impacts into macroporous targets. *Benz and Asphaug* (1999) and Leinhardt and Stewart (in preparation) have studied the disruption of solid ice targets. More complex simulations of KBOs including microporosity and mixed silicates with ice have yet to be conducted.

Using a SPH code, *Asphaug et al.* (1998) investigated how different internal configurations affect the collision outcome. They considered 5 km s^{-1} rocky impacts onto a target shaped like asteroid Castalia, which appears to be a contact binary. The possible internal structures considered were (1) a solid rock, (2) a global rubble pile with 50% bulk porosity, and (3) two solid rock pieces separated by a zone of highly damaged rock. In all three cases, the mass of the target was constant (the density of the rock was changed). The model included material strength and self-gravity.

In the rubble-pile case, some of the energy generated by the impact is partitioned into collapsing void space. In addition, the shock wave reflects off the free surfaces of the rubble pieces. As a result, shock effects were focused close to the impact site and the shock pressures were dissipated much more quickly compared to the solid rock target, in agreement with laboratory impact experiments (section 5.2.1). The velocities of the ejected material were higher in the rubble-pile configuration than the solid-rock case, resulting in a small ejecta blanket or none at all. In the two-solid-piece model, the damaged region in the middle of the body reflects the shock wave so that the damage is localized to the piece that was impacted. This study elegantly demonstrates the importance of internal structure in the outcome of collision events.

A significant problem limiting numerical studies of hypervelocity catastrophic disruption events is the vast difference in dynamical times between the shock propagation and gravitational reaccumulation. To make the problem more tractable, Leinhardt and Stewart (in preparation) have begun using a hybridized shock physics–gravity method to study KBO analog objects (Plate 4). The impact and deformation stage is modeled using a shock-physics code, CTH (*McGlaun et al.,* 1990), and the results are handed to a N-body gravity code, pkdgrav (*Stadel,* 2001; *Richardson et al.,* 2000; *Leinhardt et al.,* 2000). This method allows detailed modeling of the shock deformation including heating, phase changes, and mixing of material as well as the final gravitational reaccumulation of fragments.

These simulations record the provenance of the material in the largest remnants and track the degree to which the reaccumulated material is processed by the initial impact event. For example, the peak shock pressure (and hence the amount of melting or vaporization) experienced by each mass element is recorded. In a catastrophic impact, a large fraction of the original surface is lost, and the original surface of the target is only maintained at the antipode of the largest postcollision remnant. The surface materials on the largest remnant reflect heterogeneous shock processing (Plate 4). Both highly and weakly shocked material lines the surface, while the interior material has a more homogeneous history of moderate shock levels. This suggests that the surface materials on KBOs that have suffered a catastrophic impacts could be heterogeneously devolatilized in comparison to the interior. The surface heterogeneity may also lead to color variations.

6. SUMMARY AND FUTURE DIRECTIONS

In this chapter, we have discussed the present state of knowledge about the possible physical effects of collisions in the Kuiper belt. The body of work on impact cratering and disruption events in KBO analog materials demonstrate that composition and internal structure (particularly porosity) have a significant affect on the final outcome of collision events. Understanding the role of collisions in changing the composition and structure of KBOs is important because KBOs are the best representatives of the planetesimals that accreted into the outer solar system planets.

The range of possible outcomes of collisions in the Kuiper belt region is more complicated than in the asteroid belt. In particular, the low relative impact velocities, the low mean density of KBOs, and the likely presence of a variety of internal structures are not fully accounted for in present impact models. Impact cratering scaling laws and catastrophic disruption criteria that have been developed for hypervelocity collisions on solid planetary surfaces and within the asteroid belt may not be widely applicable to KBOs. For example, the unanticipated large amount of mass ejected from Comet Tempel 1 from the Deep Impact mission revealed that important physical processes are missing from the crater scaling laws (*Holsapple and Housen,* 2007).

We close with recommendations for areas of future work to advance our knowledge of the properties of KBOs:

1. What is the role of porosity in the outcome of cratering and catastrophic disruption collisions? And how do we separate the effects of porosity and strength? Predicting the outcome of collisions into porous materials of various strengths requires an improved understanding of (a) energy coupling into the target and (b) shock-induced damage (degradation of strength). As both are difficult to model accurately in codes, clever laboratory experiments that include direct measurements of shock wave decay, damage, and final crater sizes, in targets that vary porosity and strength independently, are necessary. In cratering events, the residual strength of the damaged target ultimately determines the final crater shape and the transition between strength- and gravity-dominated regimes. For disruption events, laboratory results will need to be incorporated into numerical simulations.

2. What is the correct way to scale laboratory experiments on porous materials to larger scales? Two length scales appear to dominate the problem: (a) the length scale of the porosity with respect to the deforming shock wave (microporosity vs. macroporosity), and (b) the depth of energy coupling of the projectile. Laboratory and numerical experiments can directly address the effects of varying each length scale as the problem size increases from laboratory targets to planets. More information about the scale of porosities in KBOs can be obtained from studies of comet nuclei and inferences of rubble pile vs. differentiated internal structures in large KBOs.

3. How should volatility be incorporated into scaling laws? Vapor generation (or release) from a collision event would affect the momentum of the flow of excavated or dispersed material. This difficult problem requires more information on the actual composition of KBOs.

4. How does differentiation or layering affect the catastrophic disruption threshold? The propagation of the impact shock wave through the target is influenced by a layered internal structure. This tractable problem can be addressed through laboratory and numerical experiments of plausible internal configurations in KBOs.

5. How does the mass ratio of the colliding bodies change the catastrophic disruption criteria in the hypervelocity regime? The shock pressure profile through the target depends on the size and velocity of the projectile. Laboratory and numerical experiments can directly address this problem for solid bodies. Solutions to the first question in this list are required for highly porous bodies.

6. How can we validate numerical simulations in the gravity regime? Crater scaling laws have been validated by high-gravity (centrifuge) experiments. In the study of highly porous and weak materials, experiments in vacuum and under low gravity are also needed. Validation of catastrophic disruption simulations in the gravity regime will require new techniques.

7. What is the magnitude of modifications of KBOs from mutual collisions compared to other "weathering" processes? How different are present-day KBOs from the primordial planetesimals in the outer solar system? Cumulative changes in observable properties of KBOs, including densities, colors, composition, and internal structures, can be addressed by updating collisional evolution models of the Kuiper belt with the latest understanding of collisional processes in porous, icy bodies. Given the wide range of possible physical properties of KBOs, studies of individual collisions are warranted to examine common impact scenarios. At present there is no certain answer, and our understanding will be driven by observations to come.

Acknowledgments. The authors thank K. Housen, S. Kenyon, and E. Asphaug for careful review of this manuscript. We would also like to thank A. Morbidelli, L. Senft, M. Holman, J.-L. Margot, M. Brown, E. Schaller, and D. Raggozine for helpful discussions.

REFERENCES

Ahrens T. J. and Johnson M. L. (1995a) Shock wave data for minerals. In *Mineral Physics and Crystallography, A Handbook of Physical Constants, Vol. 2* (T. J. Ahrens, ed.), pp. 143–184. American Geophysical Union, Washington, DC.

Ahrens T. J. and Johnson M. L. (1995b) Shock wave data for rocks.

In *Rock Physics and Phase Relations, A Handbook of Physical Constants, Vol. 3* (T. J. Ahrens, ed.), pp. 35–44. American Geophysical Union, Washington, DC.

Ahrens T. J. and O'Keefe J. D. (1977) Equations of state and impact-induced shock-wave attenuation on the Moon. In *Impact and Explosion Cratering: Planetary and Terrestrial Implications* (D. J. Roddy et al., eds.), pp. 639–656. Pergamon, New York.

Ahrens T. J. and O'Keefe J. D. (1987) Impact on the Earth, ocean and atmosphere. *Intl. J. Impact Eng., 5,* 13–32.

Arakawa M., Higa M., Leliwa-Kopystyński J., and Maeno N. (2000) Impact cratering of granular mixture targets made of H_2O ice-CO_2 ice-pyrophylite. *Planet. Space Sci., 48,* 1437–1446.

Arakawa M., Leliwa-Kopystyński J., and Maeno N. (2002) Impact experiments on porous icy-silicate cylindrical blocks and the implication for disruption and accumulation of small icy bodies. *Icarus, 158,* 516–531.

Asphaug E., Ostro S. J., Hudson R. S., Scheeres D. J., and Benz W. (1998) Disruption of kilometre-sized asteroids by energetic collisions. *Nature, 393,* 437–440.

Asphaug E., Ryan E. V., and Zuber M. T. (2002) Asteroid interiors. In *Asteroids III* (W. F. Bottke Jr. et al., eds.), pp. 463–484. Univ. of Arizona, Tuscon.

Barucci M. A., Cruikshank D. P., Dotto E., Merlin F., Poulet F., Dalle Ore C., Fornasier S., and de Bergh C. (2005) Is Sedna another Triton? *Astron. Astrophys., 439,* L1–L4.

Belton M. J. S. and A'Hearn M. F. (1999) Deep sub-surface exploration of cometary nuclei. *Adv. Space Res., 24,* 1175–1183.

Belton M. J. S., Thomas P., Veverka J., Schultz P., A'Hearn M. F., Feaga L., Farnham T., Groussin O., Li J. Y., Lisse C., McFadden L., Sunshine J., Meech K. J., Delamere W. A., and Kissel J. (2007) The internal structure of Jupiter family cometary nuclei from Deep Impact observations: The "talps" or "layered pile" model. *Icarus, 187,* 332.

Bendjoya P. and Zappalà V. (2002) Asteroid family identification. In *Asteroids III* (W. F. Bottke Jr. et al., eds.), pp. 613–618. Univ. of Arizona, Tuscon.

Benz W. and Asphaug E. (1999) Catastrophic disruptions revisited. *Icarus, 142,* 5–20.

Bernstein G. M., Trilling D. E., Allen R. L., Brown M. E., Holman M., and Malhotra R. (2004) The size distribution of trans-neptunian bodies. *Astron. J., 128,* 1364–1390.

Brown M. E., Barkume K. M., Blake G. A., Schaller E. L., Rabinowitz D. L., Roe H. G., and Trujillo C. A. (2007a) Methane and ethane on the bright Kuiper belt object 2005 FY9. *Astron. J., 133,* 284–289.

Brown M. E., Barkume K. M., Ragozzine D., and Schaller E. L. (2007b) A collisional family of icy objects in the Kuiper belt. *Nature, 446,* 231–346.

Burchell M. J. and Johnson E. (2005) Impact craters on small icy bodies such as icy satellites and comet nuclei. *Mon. Not. R. Astron. Soc., 360,* 769–781.

Burchell M. J., Brooke-Thomas W., Leliwa-Kopystyński J., and Zarnecki J. C. (1998) Hypervelocity impact experiments on solid CO_2 targets. *Icarus, 131,* 210–222.

Burchell M. J., Johnson E., and Grey I. D. S. (2002) Hypervelocity impacts on porous ices. In *Asteroids, Comets, and Meteors: ACM 2002* (B. Warmbein, ed.), pp. 859–862. ESA SP-500, Noordwijk, The Netherlands.

Chapman C. R. and McKinnon W. B. (1986) Cratering of planetary satellites. In *Satellites* (J. A. Burns and M. S. Matthews, eds.), pp. 492–580. Univ. of Arizona, Tuscon.

Chen W. P., Alcock C., Axelrod T., Bianco F. B., Byun Y. I., Chang Y. H., Cook K. H., Dave R., Giammarco J., Kim D. W., King S. K., Lee T., Lehner M., Lin C. C., Lin H. C., Lissauer J. J., Marshall S., Meinshausen N., Mondal S., de Pater I., Porrata R., Rice J., Schwamb M. E., Wang A., Wang S. Y., Wen C. Y., and Zhang Z. W. (2006) Search for small trans-neptunian objects by the TAOS Project. In *Near-Earth Objects, Our Celestial Neighbors: Opportunity and Risk* (A. Milani et al., eds.), pp. 65–68. IAU Symposium 236, Cambridge Univ., Cambridge.

Cintala M. J., Smrekar S., Hörz F., and Cardenas F. (1985) Impact experiments in H_2O ice, I: Cratering (abstract). In *Lunar and Planetary Science XVI,* pp. 131–132. Lunar and Planetary Institute, Houston.

Croft S. K., Kieffer S. W., and Ahrens T. J. (1979) Low-velocity impact craters in ice and ice-saturated sand with implications for martian crater count ages. *J. Geophys. Res., 84,* 8023–8032.

Davis D. R. and Farinella P. (1997) Collisional evolution of Edgeworth-Kuiper belt objects. *Icarus, 125,* 50–60.

Davis D. R., Chapman C. R., Greenberg R., Weidenschilling S. J., and Harris A. W. (1979) Collisional evolution of asteroids — Populations, rotations, and velocities. In *Asteroids* (T. Gehrels, ed.), pp. 528–557. Univ. of Arizona, Tuscon.

Dell'Oro A., Marzari F., Paolicchi P., and Vanzani V. (2001) Updated collisional probabilities of minor body populations. *Astron. Astrophys., 366,* 1053–1060.

dell'Oro A., Paolicchi P., Cellino A., Zappalà V., Tanga P., and Michel P. (2001) The role of families in determining collision probability in the asteroid main belt. *Icarus, 153,* 52–60.

Dohnanyi J.W. (1969) Collisional models of asteroids and their debris. *J. Geophys. Res., 74,* 2531–2554.

Durda D. D., Flynn G. J., and van Veghten T. W. (2003) Impacts into porous foam targets: Possible implications for the disruption of comet nuclei. *Icarus, 163,* 504–507.

Fujiwara A., Kawaguchi J., Yeomans D. K., Abe M., Mukai T., Okada T., Saito J., Yano H., Yoshikawa M., Scheeres D. J., Barnouin-Jha O., Cheng A. F., Demura H., Gaskell R. W., Hirata N., Ikeda H., Kominato T., Miyamoto H., Nakamura A. M., Nakamura R., Sasaki S., and Uesugi K. (2006) The rubble-pile asteroid Itokawa as observed by Hayabusa. *Science, 312(5778),* 1330–1334.

Furnish M. D. and Remo J. L. (1997) Ice issues, porosity, and snow experiments for dynamic NEO and comet medeling. In *Near-Earth Objects: The United Nations International Conference* (J. L. Remo, ed.), pp. 566–582. New York Academy of Science, New York.

Grey I. D. S. and Burchell M. J. (2003) Hypervelocity impact cratering on water ice targets at temperatures ranging from 100 K to 253 K. *J. Geophys. Res., 108,* 6-1.

Gross J., Reichenauer G., and Fricke J. (1988) Mechanical-properties of SiO_2 aerogels. *J. Phys. D, 21(9),* 1447–1451.

Gurnett D. A., Ansher J. A., Kurth W. S., and Granroth L. J. (1997) Micron-sized dust particles detected in the outer solar system by the Voyager 1 and 2 plasma wave instruments. *Geophys. Res. Lett., 24,* 3125–3128.

Hahn J. M. and Malhotra R. (1999) Orbital evolution of planets embedded in a planetesimal disk. *Astron. J., 117,* 3041–3053.

Herrmann W. (1969) Constitutive equation for the dynamic compaction of ductile porous materials. *J. Appl. Phys., 40(6),* 2490–2499.

Holsapple K. A. (1987) The scaling of impact phenomenon. *Intl. J. Impact Eng., 5,* 343–355.

Holsapple K. A. (1993) The scaling of impact processes in planetary sciences. *Annu. Rev. Earth Planet. Sci., 21,* 333–373.

Holsapple K. A. (2007) Spin limits of solar system bodies: From the small fast-rotators to 2003 EL61. *Icarus, 187,* 500–509.

Holsapple K. A. and Housen K. R. (2004) The cratering database: Making code jockeys honest (abstract). In *Lunar and Planetary Science XXXV,* Abstract #1779. Lunar and Planetary Institute, Houston (CD-ROM).

Holsapple K. A. and Housen K. R. (2007) A crater and its ejecta: An interpretation of deep impact. *Icarus, 187(1),* 345.

Holsapple K. A. and Schmidt R. M. (1987) Point source solutions and coupling parameters in cratering mechanics. *J. Geophys. Res., 92,* 6350–6376.

Holsapple K., Giblin I., Housen K., Nakamura A., and Ryan E. (2002) Asteroid impacts: Laboratory experiments and scaling laws. In *Asteroids III* (W. F. Bottke Jr. et al., eds.), pp. 443–462. Univ. of Arizona, Tucson.

Housen K. R. and Holsapple K. A. (1999) Scale effects in strength-dominated collisions of rocky asteroids. *Icarus, 142(1),* 21–33.

Housen K. R. and Holsapple K. A. (2000) Numerical simulations of impact cratering in porous materials (abstract). In *Lunar and Planetary Science XXXI,* Abstract #1498. Lunar and Planetary Institute, Houston.

Housen K. R. and Holsapple K. A. (2003) Impact cratering on porous asteroids. *Icarus, 163,* 102–119.

Housen K. R., Holsapple K. A., and Voss M. E. (1999) Compaction as the origin of the unusual craters on the asteroid Mathilde. *Nature, 402,* 155–157.

Humes D. H. (1980) Results of Pioneer 10 and 11 meteoroid experiments — Interplanetary and near-Saturn. *J. Geophys. Res., 85,* 5841–5852.

Iijima Y. i., Kato M., Arakawa M., Maeno N., Fujimura A., and Mizutani H. (1995) Cratering experiments on ice: Dependence of crater formation on projectile materials and scaling parameter. *Geophys. Res. Lett., 22,* 2005–2008.

Jeanloz R. (1989) Shock wave equation of state and finite strain theory. *J. Geophys. Res., 94(B5),* 5873–5886.

Jewitt D. C. and Luu J. X. (2000) Physical nature of the Kuiper belt. In *Protostars and Planets IV* (V. Mannings et al., eds.), pp. 1201–1230. Univ. of Arizona, Tucson.

Johnson J. B. (1991) Simple model of shock-wave attenuation in snow. *J. Glaciol., 37(127),* 303–312.

Kato M., Iijima Y., Arakawa M., Okimura Y., Fujimura A., Maeno N., and Mizutani H. (1995) Ice-on-ice impact experiments. *Icarus, 113,* 423–441.

Kawakami S., Mizutani H., Takagi Y., Kato M., and Kumazawa M. (1983) Impact experiments on ice. *J. Geophys. Res., 88,* 5806–5814.

Kenyon S. J. and Bromley B. C. (2004) The size distribution of Kuiper belt objects. *Astron. J., 128,* 1916–1926.

Kenyon S. J. and Luu J. X. (1999) Accretion in the early Kuiper belt. II. Fragmentation. *Astron. J., 118,* 1101–1119.

Koschny D. and Grün E. (2001) Impacts into ice-silicate mixtures: Crater morphologies, volumes, depth-to-diameter ratios, and yield. *Icarus, 154,* 391–401.

Koschny D., Kargl G., and Rott M. (2001) Experimental studies of the cratering process in porous ice targets. *Adv. Space Res., 28,* 1533–1537.

Lacerda P. and Jewitt D. (2007) Densities of solar system objects from their rotational lightcurves. *Astron. J.,* in press.

Landgraf M., Liou J. C., Zook H. A., and Grün E. (2002) Origins of solar system dust beyond Jupiter. *Astron. J., 123,* 2857–2861.

Lange M. A. and Ahrens T. J. (1982) Impact cratering in ice and ice-silicate targets: An experimental assessment (abstract). In *Lunar and Planetary Science XIII,* pp. 415–416. Lunar and Planetary Institute, Houston.

Lange M. A. and Ahrens T. J. (1983) The dynamic tensile strength of ice and ice silicate mixtures. *J. Geophys. Res., 88,* 1197–1208.

Lange M. A. and Ahrens T. J. (1987) Impact experiments in low-temperature ice. *Icarus, 69,* 506–518.

Leinhardt Z. M. and Richardson D. C. (2002) N-body simulations of planetesimal evolution: Effect of varying impactor mass ratio. *Icarus, 159,* 306–313.

Leinhardt Z. M., Richardson D. C., and Quinn T. (2000) Direct N-body simulations of rubble pile collisions. *Icarus, 146,* 133–151.

Love S. G. and Ahrens T. J. (1997) Origin of asteroid rotation rates in catastrophic impacts. *Nature, 386,* 154–156.

Love S. G., Hörz F., and Brownlee D. E. (1993) Target porosity effects in impact cratering and collisional disruption. *Icarus, 105,* 216–224.

Margot J. L. (2002) Astronomy: Worlds of mutual motion. *Nature, 416,* 694–695.

McGlaun J. M., Thompson S. L., and Elrick M. G. (1990) CTH: A 3-Dimensional Shock-Wave Physics Code. *Intl. J. Imp. Eng., 10,* 351–360.

Mellor M. (1975) A review of basic snow mechanics. In *Snow Mechanics, Proceedings of the Grindelwald Symposium,* pp. 251–291. IAHS Publ. No. 114, Oxfordshire, UK.

Melosh H. J. (1977) Crater modification by gravity — A mechanical analysis of slumping. In *Impact and Explosion Cratering: Planetary and Terrestrial Implications* (D. J. Roddy et al., eds.), pp. 1245–1260. Pergamon, New York.

Melosh H. J. (1989) *Impact Cratering.* Oxford Univ., New York.

Melosh H. J. and Ivanov B. A. (1999) Impact crater collapse. *Annu. Rev. Earth Planet. Sci., 27,* 385–415.

Melosh H. J. and McKinnon W. B. (1978) The mechanics of ringed basin formation. *Geophys. Res. Lett., 5,* 985–988.

Melosh H. J. and Ryan E. V. (1997) Asteroids: Shattered but not dispersed. *Icarus, 129,* 562–564.

Merk R. and Prialnik D. (2006) Combined modeling of thermal evolution and accretion of trans-neptunian objects — Occurrence of high temperatures and liquid water. *Icarus, 183,* 283–295.

Meyers M. A. (2001) *Dynamic Behavior of Materials.* Wiley, New York.

Müller T. G., Ábrahám P., and Crovisier J. (2005) Comets, asteroids and zodiacal light as seen by ISO. *Space Sci. Rev., 119,* 141–155.

Oberbeck V. R. and Quaide W. L. (1967) Estimated thickness of a fragmental surface layer of Oceanus Procellarum. *J. Geophys. Res., 72,* 4697–4704.

O'Brien D. P. and Greenberg R. (2003) Steady-state size distributions for collisional populations: Analytical solution with size-dependent strength. *Icarus, 164(2),* 334.

O'Brien D. P. and Greenberg R. (2005) The collisional and dynamical evolution of the main-belt and NEA size distributions. *Icarus, 178,* 179–212.

Pan M. and Sari R. (2005) Shaping the Kuiper belt size distribution by shattering large but strengthless bodies. *Icarus, 173,* 342–348.

Paolicchi P., Burns J. A., and Weidenschilling S. J. (2002) Side effects of collisions: Spin rate changes, tumbling rotation states, and binary asteroids. In *Asteroids III* (W. F. Bottke Jr. et al., eds.), pp. 517–526. Univ. of Arizona, Tucson.

Petit J. M., Holman M. J., Gladman B. J., Kavelaars J. J., Scholl H., and Loredo T. J. (2006) The Kuiper belt luminosity function from m_R = 22 to 25. *Mon. Not. R. Astron. Soc., 365,* 429–438.

Petrenko V. F. and Whitworth R. W. (1999) *The Physics of Ice.* Oxford Univ., New York.

Pierazzo E. and Melosh H. J. (2000) Melt production in oblique impacts. *Icarus, 145,* 252–261.

Pierazzo E., Vickery A. M., and Melosh H. J. (1997) A reevaluation of impact melt production. *Icarus, 127,* 408–423.

Poirier J. P. (2000) *Introduction to the Physics of the Earth's Interior.* Cambridge Univ., New York.

Rabinowitz D. L., Barkume K., Brown M. E., Roe H., Schwartz M., Tourtellotte S., and Trujillo C. (2006) Photometric observations constraining the size, shape, and albedo of 2003 EL61, a rapidly rotating, Pluto-sized object in the Kuiper belt. *Astrophys. J., 639,* 1238–1251.

Rice M. H., McQueen R. G., and Walsh J. M. (1958) Compression of solids by strong shock waves. *Solid State Phys., 6,* 1–63.

Richardson D. C., Quinn T., Stadel J., and Lake G. (2000) Direct large-scale N-body simulations of planetesimal dynamics. *Icarus, 143,* 45–59.

Roques F., Doressoundiram A., Dhillon V., Marsh T., Bickerton S., Kavelaars J. J., Moncuquet M., Auvergne M., Belskaya I., Chevreton M., Colas F., Fernandez A., Fitzsimmons A., Lecacheux J., Mousis O., Pau S., Peixinho N., and Tozzi G. P. (2006) Exploration of the Kuiper belt by high-precision photometric stellar occultations: First results. *Astron. J., 132,* 819–822.

Ruoff A. L. (1967) Linear shock-velocity-particle-velocity relationship. *J. Appl. Phys., 38(13),* 4976–4980.

Ryan E. V., Hartmann W. K., and Davis D. R. (1991) Impact experiments. III — Catastrophic fragmentation of aggregate targets and relation to asteroids. *Icarus, 94,* 283–298.

Ryan E. V., Davis D. R., and Giblin I. (1999) A laboratory impact study of simulated Edgeworth-Kuiper belt objects. *Icarus, 142,* 56–62.

Sammonds P. R., Murrell S. A. F., and Rist M. A. (1998) Fracture of multiyear sea ice. *J. Geophys. Res., 103(C10),* 21795–21815.

Schmidt R. M. (1980) Meteor Crater: Energy of formation — Implications of centrifuge scaling (abstract). In *Lunar and Planetary Science XI,* pp. 2099–2128. Lunar and Planetary Institute, Houston.

Schultz P. H. (1996) Effect of impact angle on vaporization. *J. Geophys. Res., 101,* 21117–21136.

Schultz P. H. (2003) Impacts into porous volatile-rich substrates on Mars (abstract). In *Sixth International Conference on Mars,* Abstract #3263. Lunar and Planetary Institute, Houston (CD-ROM).

Schultz P. H. and Gault D. E. (1985) Clustered impacts — Experiments and implications. *J. Geophys. Res., 90,* 3701–3732.

Schultz P. H., Ernst C. M., and Anderson J. L. B. (2005) Expectations for crater size and photometric evolution from the Deep Impact collision. *Space Sci. Rev., 117,* 207–239.

Senft L. E. and Stewart S. T. (2007) Modeling impact cratering in layered surfaces. *J. Geophys. Res., 112,* E11002, DOI: 10.1029/2007JE002894.

Shrine N. R. G., Burchell M. J., and Grey I. D. S. (2002) Velocity scaling of impact craters in water ice over the range 1 to 7.3 km s^{-1}. *Icarus, 155,* 475–485.

Stadel J. G. (2001) Cosmological N-body simulations and their analysis. Ph.D. thesis. Univ. of Washington, Seattle. 141 pp.

Stern S. A. (1996) On the collisional environment, accretion time scales, and architecture of the massive, primordial Kuiper belt. *Astron. J., 112,* 1203–1211.

Stern S. A. (2003) The evolution of comets in the Oort cloud and Kuiper belt. *Nature, 424,* 639–642.

Stern S. A. and Colwell J. E. (1997) Collisional erosion in the primordial Edgeworth-Kuiper belt and the generation of the 30–50 AU Kuiper gap. *Astrophys. J., 490,* 879–882.

Stewart S. T. and Ahrens T. J. (2004) A new H_2O Ice Hugoniot: Implications for planetary impact events. In *Shock Compression of Condensed Matter — 2003* (M. D. Furnish et al., ed.), pp. 1478–1483. American Institute of Physics, New York.

Stewart S. T. and Ahrens T. J. (2005) Shock properties of H_2O ice. *J. Geophys. Res., 110(E9),* 3005.

Strom R. G., Malhotra R., Ito T., Yoshida F., and Kring D. A. (2005) The origin of planetary impactors in the inner solar system. *Science, 309,* 1847–1850.

Trujillo C. A., Jewitt D. C., and Luu J. X. (2001) Properties of the trans-neptunian belt: Statistics from the Canada-France-Hawaii Telescope Survey. *Astron. J., 122,* 457–473.

Veverka J., Thomas P., Harch A., Clark B., Bell J. F. III, Carcich B., Joseph J., Chapman C., Merline W., Robinson M., Malin M., McFadden L. A., Murchie S., Hawkins S. E. III, Farquhar R., Izenberg N., and Cheng A. (1997) NEAR's flyby of 253 Mathilde: Images of a C asteroid. *Science, 278,* 2109–2114.

Weissman P. R., Asphaug E., and Lowry S. C. (2005) Structure and density of cometary nuclei. In *Comets II* (M. C. Festou et al., eds.), pp. 337–357. Univ. of Arizona, Tuscon.

Williams D. R. and Wetherill G. W. (1994) Size distribution of collisionally evolved asteroidal populations: Analytical solution for self-similar collision cascades. *Icarus, 107(1),* 117.

Wünnemann K., Collins G. S., and Melosh H. J. (2006) A strain-based porosity model for use in hydrocode simulations of impacts and implications for transient crater growth in porous targets. *Icarus, 180,* 514–527.

Structure and Evolution of Kuiper Belt Objects and Dwarf Planets

William B. McKinnon
Washington University in St. Louis

Dina Prialnik
Tel Aviv University

S. Alan Stern
NASA Headquarters

Angioletta Coradini
INAF, Istituto di Fisica dello Spazio Interplanetario

Kuiper belt objects (KBOs) accreted from a mélange of volatile ices, carbonaceous matter, and rock of mixed interstellar and solar nebular provenance. The transneptunian region, where this accretion took place, was likely more radially compact than today. This and the influence of gas drag during the solar nebula epoch argue for more rapid KBO accretion than usually considered. Early evolution of KBOs was largely the result of heating due to radioactive decay, the most important potential source being ^{26}Al, whereas long-term evolution of large bodies is controlled by the decay of U, Th, and ^{40}K. Several studies are reviewed dealing with the evolution of KBO models, calculated by means of one-dimensional numerical codes that solve the heat and mass balance equations. It is shown that, depending on parameters (principally rock content and porous conductivity), KBO interiors may have reached relatively high temperatures. The models suggest that KBOs likely lost ices of very volatile species during early evolution, whereas ices of less-volatile species should be retained in cold, less-altered subsurface layers. Initially amorphous ice may have crystallized in KBO interiors, releasing volatiles trapped in the amorphous ice, and some objects may have lost part of these volatiles as well. Generally, the outer layers are far less affected by internal evolution than the inner part, which in the absence of other effects (such as collisions) predicts a stratified composition and altered porosity distribution. Kuiper belt objects are thus unlikely to be "the most pristine objects in the solar system," but they do contain key information as to how the early solar system accreted and dynamically evolved. For large (dwarf planet) KBOs, long-term radiogenic heating alone may lead to differentiated structures — rock cores, ice mantles, volatile-ice-rich "crusts," and even oceans. Persistence of oceans and (potential) volcanism to the present day depends strongly on body size and the melting-point depression afforded by the presence of salts, ammonia, etc. (we review the case for Charon in particular). The surface color and compositional classes of KBOs are usually discussed in terms of "nature vs. nurture," i.e., a generic primordial composition vs. surface processing, but the true nature of KBOs also depends on how they have evolved. The broad range of albedos now found in the Kuiper belt, deep water-ice absorptions on some objects, evidence for differentiation of Pluto and 2003 EL_{61}, and a range of densities incompatible with a single, primordial composition and variable porosity strongly imply significant, intrinsic compositional differences among KBOs. The interplay of formation zone (accretion rate), body size, and dynamical (collisional) history may yield KBO compositional classes (and their spectral correlates) that recall the different classes of asteroids in the inner solar system, but whose members are broadly distributed among the KBO dynamical subpopulations.

1. INTRODUCTION

The Kuiper belt is the solar system's "third zone" (*Stern,* 2003). Beyond the terrestrial planets, beyond the giant planets, it is the roughly coplanar, prograde-orbiting extension of the "classical" solar system. Since 1930, more than 1000 bodies have been discovered there (as of early 2007). Thought to be the main source of the Centaurs and Jupiter-family comets (e.g., *Jewitt,* 2004), it is also home to bodies the mass of Pluto and larger (*Schaller and Brown,* 2007; cf. *Stern,* 1991). The transneptunian population is thus a bridge between small, volatile-rich bodies that impact the Earth and other planets (the comets) and the most important worlds presently known in deep solar space (Pluto and Eris). As such, their structure and evolution are important for understanding cometary properties, observations of Kuiper belt objects (KBOs), and spacecraft mission data to come (such as planned from New Horizons). Understanding the structure and evolution of KBOs also provides important constraints on the dynamical environment in which the outer solar system accreted and evolved. And as a new class of solar system body, sometimes referred to as "ice dwarfs" (*Stern and Levison,* 2002), their structure and evolution are interesting in their own right.

In this chapter we will briefly review the formation environment of KBOs, especially as regards location and timescale. Composition and potential chemistries, which are essential input and clues to evolution, will then be updated. We will also examine the critical issues of porosity and bulk density. With these in hand, theoretical models of KBO internal evolution will be presented, covering a range of sizes (necessarily including that of comets) and timescales. Finally, we will review existing work on the internal structure and evolution of the largest known KBOs, and take stock of the potential relationships between structure and evolution and what is actually observed of KBOs.

2. FORMATION IN THE TRANSNEPTUNIAN REGION

The planets, their satellites, and KBOs formed out of the solar nebula starting approximately 4.57 b.y. ago (*Amelin et al.,* 2002). A straightforward view of accretion in the Kuiper belt naturally focuses on *in situ* formation in the 30–50-AU, or transneptunian, region. Detailed modeling by *Kenyon and Luu* (1998, 1999) predicted that large KBOs (100–1000 km in radius) could have grown in the region between 30 and 42 AU by traditional solid-body binary accretion in ~10 to 100 m.y., but only for nebular disk surface mass densities greatly in excess of what exists today (see also review by *Farinella et al.,* 2000). A primordial, dynamically cold (low orbital eccentricity and inclination) disk containing anywhere from 10 to 30 Earth masses ($M_\oplus$) of solid material is required, which is 2–3 orders of magnitude greater than the present estimated mass of 0.01 to 0.1 $M_\oplus$ (*Bernstein et al.,* 2004; cf. *Luu and Jewitt,* 2002). Such extended timescales are consistent with gas-free accretion, as the solar nebula is not expected to have survived beyond ~1–10 m.y. after initial collapse of the protostellar cloud, based on studies of nebular disk lifetimes around nearby solar-type stars (e.g., *Haisch et al.,* 2001).

It has been long recognized, however, that once the giant planets formed, there must have been strong gravitational interactions between these planets and any massive disk of solid bodies (*Fernandez and Ip,* 1984). Such dynamical interactions cause the now well-appreciated phenomenon of planetary migration, which for Neptune was strongly outward, and which has been implicated in a host of solar system phenomena, including the capture of Pluto and other bodies (the Plutinos) in the 3:2 mean-motion resonance with Neptune (*Malhotra,* 1993, 1995). With increasing levels of numerical sophistication, planetary migration models have been proposed to explain virtually all features of the present Kuiper belt, from the various populations and subpopulations (i.e., the dynamically hot and cold classical populations, resonant groups, and scattered disk), the "edge" at 50 AU, and the size-frequency distribution of KBOs themselves (e.g., *Levison and Morbidelli,* 2003). This development has culminated in the most successful Kuiper belt formation model to date, the "Nice model," and is described in *Levison et al.* (2007b) and the chapter by Morbidelli et al. Besides reproducing the dynamical attributes of the Kuiper belt of today with good fidelity, it also reproduces the orbital architecture of the giant planets.

It is notable that the Nice model initially has Neptune forming inside 20 AU from the Sun. In this case KBO accretion timescales should be shorter than those calculated in *Kenyon and Luu* (1998, 1999). Indeed, in the chapter by Kenyon et al., accretion of 1000-km-radius objects between 20 and 25 AU requires only ~5–10 m.y. (see their Fig. 7). Moreover, *Weidenschilling* (2004) has modeled the accretion of KBOs starting from very small (submeter) sizes and accounting for gas drag during the solar nebula epoch. For the 30–90-AU region he finds that gas-drag-driven orbital decay so enhances planetesimal sweep-up that bodies 50 km in radius may form in under 1 m.y. in those inner nebular regions (closer to 30 AU) that are not depleted by radial migration of solids. *Weidenschilling* (2004) further finds that his modeled Kuiper belt is so depleted of solids in its outer portion that it effectively forms a steep gradient in solid density, or "edge," between 40 and 50 AU. Such an edge is needed in the Nice model — and in particular, one at 30–35 AU — because it is there that Neptune's outward migration halts so that the position of the 2:1 mean-motion resonance with Neptune freezes at 48 AU, the observed limit of the classical Kuiper belt (see chapter by Morbidelli et al.).

The Nice model in fact predicts that the *entire* Kuiper belt is, in effect, scavenged from inside the edge at ~30 AU, with bodies captured, driven outward, released, or otherwise expelled to the transneptunian region as Neptune's orbit expands. Planetesimal accretion may have initiated well outside 30 AU as in Weidenschilling's work, or the edge may have been created in some other manner. For example, if the Sun was born in a stellar cluster, photoevaporation by hot OB stars may have truncated the primordial solar nebula (e.g., *Adams et al.,* 2004). In such a photoevaporation scenario, all KBOs should have accreted *in toto* in the 15–30-AU range.

As for the Sun's birth environment, the evidence for live ^{60}Fe in the earliest solar system seems well established (see *Bizzarro et al.,* 2007, and references therein). Recent data suggest a late (~1 m.y.) injection of this short-lived radioisotope, which appears to require a supernova event (*Bizzarro et al.,* 2007), which would in turn clearly indicate that the Sun *was* born in a rich cluster.

We remark that the chapter by Morbidelli et al. does not rule out some remnant *in situ* bodies contributing to the dynamically cold, classical population. Perhaps this occurred as the result of slow accretion by a remnant population of very small bodies, stranded beyond 30 AU once the solar nebula had dispersed. In this case the KBO population as a whole would have formed over a broad radial range, perhaps as great as 15–50 AU, which may account for some of the surface albedo and color differences seen today (*Grundy et al.,* 2005; *Lykawka and Mukai,* 2005; chapter by Doressoundiram et al.).

Finally, with respect to all the above theoretical models, there is greater room for linkage between them. For example, how do the timescales change in the *Weidenschilling* (2004) model when the more compact 15–30-AU region is modeled? Although binary accretion is faster at smaller heliocentric distances, gas drag drift may be slower (the deviation from Keplerian velocity declines in a relative sense, compared with orbital speed), and once a certain planetesimal size threshold is passed (~1 km to a few kilometers), gas drag may no longer be important. Looking beyond the nebular era, how long do large KBOs of all sizes take to grow in the more compact, primordial transneptunian zone? These issues of accretional timescale are important because they determine the degree to which large KBOs may have been heated by accretional collisions, and the degree to which *all* KBOs may have been heated by the decay of short-lived radionuclides, principally ^{26}Al (half-life = 0.73 m.y.) and ^{60}Fe (half-life = 1.5 m.y.).

3. THE COMPOSITION OF KUIPER BELT OBJECTS

Based on study of comets, asteroids, meteorites, interstellar dust particles (IDPs), interstellar molecular clouds, and star-forming regions, KBOs are thought to be composed of subequal amounts of volatile ices and organics, carbonaceous matter, and refractory "rock" (silicates and related). The adjective subequal, drawn from petrology, expresses the idea that the mass balance between the three main components is not yet precisely known (*Greenberg,* 1998; cf. *McKinnon et al.,* 1997). Nevertheless, recent cometary exploration and observations, plus the return of samples from 81P/Wild 2 by Stardust, are revolutionizing our understanding of the nature of each of these components. In this section we discuss each in turn, as well as summarize spectroscopic evidence pertaining directly to KBO surface compositions.

3.1. Volatile Ices and Organics

A recent, comprehensive summary of observations of cometary volatiles is given in *Bockelée-Morvan et al.* (2004). Figure 1, updated from this work, illustrates the range of abundances (production rates) for detected species, relative to water. We assume, naturally, that cometary volatiles are fundamentally related to KBO volatiles, but we do caution that any such cometary inventory necessarily incorporates possible evolutionary effects on cometary volatile/water ratios. The most important non-water-ice volatile, based on

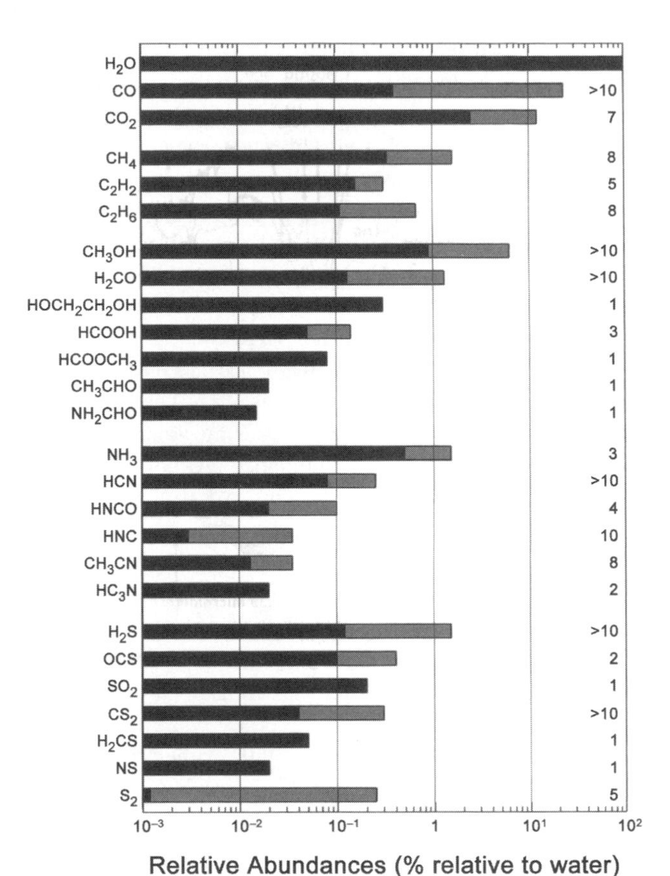

Fig. 1. Abundances relative to water in comets. The range of measured values is shown by the gray portions. The number of comets for which data is available is given on the right. For CO, abundances refer to total CO [both nuclear and distributed (coma) sources]. From *Crovisier* (2006).

the numerous observations summarized in Fig. 1, remains CO, although the range of measured values is quite large, from <1% to >20% [this range may reflect condensation temperature (*Notesco et al.,* 2003) or the aforementioned evolutionary effects]. Nearly as important, at the several to 10% level, is CO_2. These adundances can be compared to the much smaller abundance of reduced carbon, principally CH_4 (methane), at ~1%. The predominance of oxidized carbon in the volatile ice component of comets presumably reflects the compositional importance of CO in stellar winds and star-forming regions [regions similar to that in which the solar system formed (*Ehrenfreund et al.,* 2004)]. Generally, the volatile inventory of comets (and logically, that of KBOs) is consistent with the interstellar inventory, subject to various degrees of processing and modification in the solar nebula (see *Ehrenfreund et al.,* 2004).

The dominant nitrogen-bearing species is NH_3 (ammonia), but it is only present at the 1% level [and only for Comet Halley does the abundance exceed 1%, based on Giotto measurements (*Meier et al.,* 1994)]. Molecular nitrogen has yet to be reliably detected at all, with severe upper limits on its abundance (≤10^{-5} to 10^{-4}) (*Bockelée-Morvan et al.,* 2004). Ammonia ice is, of course, potentially important geologically for KBOs, as it is a hydrogen-bonded ice that forms stochiometric compounds, as well as a low-melting-point "cryovolcanic" melt, with water ice (e.g., *Kargel,* 1992). Based on cometary abundances, though, it appears that the most volumetrically important hydrogen-bonded ice "co-conspirator" is CH_3OH (methanol), at the few percent level (compared with H_2O). This amount of CH_3OH is sufficient to form a mobile methanol-water melt when ice temperatures exceed 171 K (*Kargel,* 1992).

Other volatiles of note in Fig. 1 are H_2CO (formaldehyde) and H_2S, both with abundances $\lesssim$1%. The latter is the most abundant of a number of S-bearing volatile species, none of whom rise above an adundance level of ~0.001, but whose aggregate could be important in KBOs that chemically resemble comets, at the 1% level.

3.2. Carbonaceous Matter

Relatively involatile, macromolecular carboncompounds, hereafter carbonaceous matter, has long been thought important in comets, but the true diversity was first revealed by the Giotto and Vega encounters with Comet 1P/Halley in 1986. Complex compounds of C, H, O, and N (CHON) were found to comprise, wholly or in combination with silicates, ~75% of all the particles measured *in situ* (see, e.g., *Fomenkova,* 1999, and references therein).

Subsequently, early analyses of organic-rich particles returned by Stardust from Wild 2 in 2006 (Wild 2 is a Jupiter-family comet, and thus of direct relevance to the Kuiper belt) have proved enlightening (*Sandford et al.,* 2006). Figure 2 shows that five of the six Wild 2 organic-rich particles measured in their study exhibit O and N abundances that are high relative to chondritic organic matter *and* the average composition of Halley as measured by Giotto. The ensemble of Wild 2 O/C and N/C ratios is not dissimilar to the IDP average. *Sandford et al.* (2006) conclude that Stardust organic samples are, being highly heterogeneous and unequilibrated, more "primitive" than those in meteorites and IDPs.

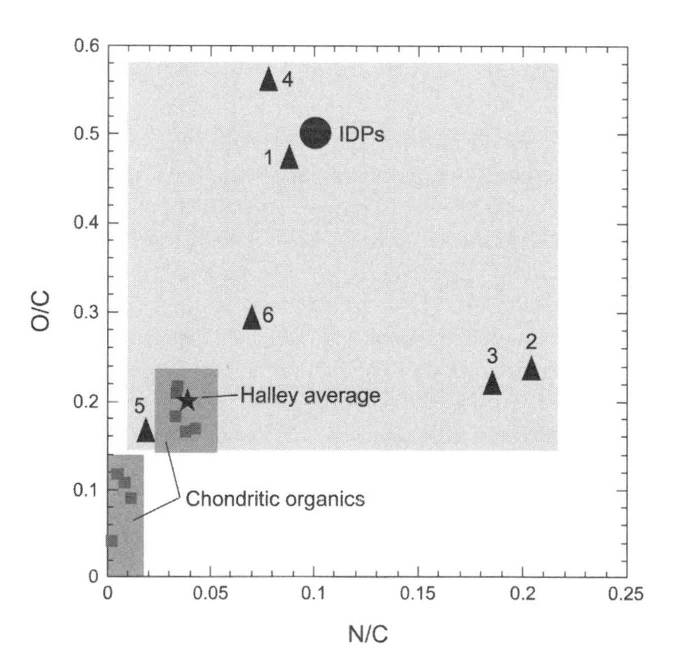

Fig. 2. Atomic O/C and N/C for six Stardust samples derived from X-ray absorption near-edge spectroscopy (XANES) analysis (black triangles), compared with chondritic organic matter (gray squares) and averages for Comet Halley particles (star) and stratospheric IDPs (large circle). Modified from *Sandford et al.* (2006).

Much further work remains with regard to Stardust samples. Figure 2 hints at what may turn out to be an important aspect of KBO chemistry and evolution: If the N/C ratio of carbonaceous matter in KBOs is ~0.1, and carbonaceous matter is about 25% of a given KBO by mass (*Greenberg,* 1998; *McKinnon et al.,* 1997), then carbonaceous matter is the dominant N reservoir in KBOs (when compared with the sum of N-bearing volatile ices, e.g., Fig. 1). A bulk abundance of ~2.5% by mass could translate into ~7.5% with respect to water ice, if the bulk ice/rock ratio in KBOs is ~30/70 (see section 4) *and* thermochemical evolution within the KBO drives release of N from carbonaceous matter, either as N_2 or NH_3 (section 6).

We note that recent detailed analyses of midinfrared emission from cometary dust comae identify an important fraction of cometary dust particles as "amorphous carbon" (*Lisse et al.,* 2006, 2007), but *Lisse et al.* (2007) point out that what they term dark amorphous carbon corresponds to the CHON seen at Halley. In terms of physical modeling, using the properties of amorphous carbon as a proxy for the carbonaceous fraction would not be unjustified, to first order.

3.3 Rock

Silicates and related phases (sulfides, oxides, etc.) make up the rock component of comets and, presumably, KBOs (e.g., *Hanner and Bradley,* 2004). Long considered to be a diverse mixture of crystalline and amorphous phases, the relative contributions of interstellar vs. nebular and altered vs. pristine have been much debated (e.g., *Wooden et al.,* 2007). The recent return of Stardust samples (*Brownlee et al.,* 2006) and sophisticated analyses of Spitzer and the Infrared Space Observatory (ISO) midinfrared emission spectra (*Lisse et al.,* 2006, 2007) have taken this subject to a new level.

A major portion of Stardust particles greater than 1 μm is composed of crystalline olivine and pyroxene. By virtue of their major- and minor-element and isotopic compositions, these particles are not likely to be interstellar in origin, whether annealed or not. These high-temperature minerals, plus fragments of calcium-aluminum-rich inclusions (CAIs) and (possibly) chondrules (*Brownlee et al.,* 2006; D. Brownlee, personal communication, 2007) strongly imply large-scale radial mixing in the solar nebula. At the very least, initial KBO "rock" should have consisted of a nonequilibrium mélange of interstellar silicates *as well as* materials processed and altered, if not wholly recondensed, in the inner solar nebula.

In contrast, the infrared emission spectra from Comets 9P/Tempel 1 and C/Hale-Bopp have been interpreted in terms of a broad array of solid materials, including olivine and pyroxene, but also their amorphous equivalents, as well as sulfides, clays (hydrated silicates), and carbonates (*Lisse et al.,* 2007). Amorphous silicates and sulfides are found in some Stardust samples, but not hydrated silicates or carbonates. Whether this is a discrepancy is unclear. *Brownlee et al.* (2006) offer that impact heating on Tempel 1 may have created the conditions necessary, however transient, for aqueous alteration (and carbonation) of anhydrous precursors. We are inclined to accept these compositional differences as largely real, as it is not unreasonable that the KBO precursors of some of the comets we see today may have undergone thermal processing up to and past the temperature of water-ice melting (*de Bergh et al.,* 2004) (see also section 5).

3.4. Compositional Inferences from Spectroscopy

The observed compositions of KBOs themselves are of course of direct relevance to any chemical or compositional model of KBOs, although it should always be remembered that we are, strictly speaking, discussing the composition of the optical surface. [The first detected KBO species was methane ice, on Pluto (*Cruikshank,* 1976), and it was often assumed, explicitly or implicitly, that Pluto was dominated by solid methane in bulk (e.g., *Lupo and Lewis,* 1980).] Kuiper belt objects are notoriously difficult to observe, because of their remoteness and faintness, but we have probably turned a corner in terms of spectroscopy, owing to the amount of "big glass" being devoted and to the discovery of more intrinsically bright objects. As is usual in infrared remote sensing, the most infrared-active species dominate spectral identifications, at least initially. For the Kuiper belt, three broad classes of surface types have emerged: surfaces dominated by methane-ice absorptions, surfaces dominated by water-ice absorptions, and surfaces that are featureless in the infrared to the level of observational noise; even in the latter case the spectral slopes range from flat (gray) to very red (see *Barucci et al.,* 2004; *Barucci and Peixinho,* 2006; *Cruikshank et al.,* 2007; *Brown et al.,* 2007a; chapter by Barucci et al.).

As emphasized by *Brown et al.* (2007a,b) and references therein, methane ice is seen on the largest KBOs: Eris, Pluto, 2005 FY_9, Sedna, and Triton, if we count Triton as a captured KBO (*McKinnon et al.,* 1995; *Agnor and Hamilton,* 2006). Water-band band depths on other KBOs range from undetectable to deep; the latter being associated (although not exclusively) with the large KBO 2003 EL_{61}, its brightest satellite, and a dynamical cluster (or family) of KBOs plausibly linked to EL_{61} (*Brown et al.,* 2007a).

The large KBOs Pluto and Triton are in the dwarf planet class, and along with Charon, are the best studied. Triton and Pluto each have surfaces dominated spectrally by methane, but dominated physically by hard-to-detect solid N_2 (*Cruikshank,* 2005). CO ice is also detected on both, CO_2 ice only on Triton, and ethane ice recently detected on Triton and possibly on Pluto (*Nakamura et al.,* 2000; *Cruikshank et al.,* 2006). A complete spectral model requires water ice on Triton's surface, and possibly on Pluto (and Sedna), as well (*Cruikshank,* 2005; *Emery et al.,* 2007). On this basis, it is not unreasonable to expect that N_2 ice also exists on the rest of the largest KBOs (Eris, 2005 FY_9, Sedna), along with CH_4 ice already seen, but N_2 ice has so far only been detected on Sedna (*Barucci et al.,* 2005b; *Brown et al.,* 2007b) and circumstantially on Eris (*Licandro et al.,* 2006); ethane ice, though, *is seen* on 2005 FY_9 (*Brown et al.,* 2007b).

The surface of Pluto's moon, Charon, in contrast, is dominated physically (and spectrally) by water ice (similar to 2003 EL_{61}). Remarkably, ammonia-water ice has been detected (*Brown and Calvin,* 2000; *Dumas et al.,* 2001; *Cook et al.,* 2007; *Verbiscer et al.,* 2007); found by several groups, this detection seems secure. Ammonia-water ice may also exist on Quaoar (*Jewitt and Luu,* 2004). The Plutino 2002 VE_{95} displays a spectral signature similar to that of the Centaur Pholus (*Barrucci et al.,* 2006). Pholus' surface reflectance has been modeled with a mixture of water ice and *methanol ice* (the latter covering perhaps 15% of the surface) along with olivine, tholin, and carbon black (*Cruikshank et al.,* 1998), and a similar model works well for 2002 VE_{95} (*Barucci et al.,* 2006). Absorption bands in the visible possibly indicating the presence of phyllosilicates have been reported for three Plutinos as well (*de Bergh et al.,* 2004; see also *Cruikshank et al.,* 2007).

The different colors and compositions of KBOs have generally been attributed to variations in collisional evolution and impact gardening competing with different degrees of surface alteration (insolation, UV radiation, and charged particle bombardment) (e.g., *Luu and Jewitt,* 2002; *Stern,* 2002; *Barucci et al.,* 2004). It is emerging, however, that larger objects have generally icier (higher-albedo) surfaces (*Barucci et al.,* 2006; *Schaller and Brown,* 2007; chapter by Stansberry et al.). Larger and more distant transneptunian bodies can better retain volatile ices such as N_2, CH_4, and CO (*Schaller and Brown,* 2007). Larger bodies are also more likely to have been internally active, either now or in the geologic past (e.g., *McKinnon,* 2002) (sections 5 and 6). However controlled by the physics of volatile retention or escape, for surfaces to be *dominated* by N_2 ice or CH_4 ice, or to contain nontrivial amounts of ammonia-hydrate or methanol-hydrate ice, requires that these volatile components be supplied in appropriate abundance to the surfaces. For surfaces to be intrinsically optically bright and display deep water-ice bands requires a mechanism to separate ice from the cosmogonically more abundant dark carbonaceous matter and silicates accreted. This is the point of view of this chapter, to elucidate the contribution of KBO structure and evolution to the range of surface compositions (and bulk densities, discussed next) now being detected, and to place this contribution in the context of Kuiper belt dynamics and exogenic surface processes.

4. THE CRUCIAL QUESTION OF POROSITY

4.1. Densities of Kuiper Belt Objects

The largest KBOs, Triton, Eris, and the Pluto-Charon binary, are of similarly high density, 2.061 ± 0.007 g cm^{-3} (*Person et al.,* 2006, Table 4), 2.3 ± 0.3 g cm^{-3} (*Brown and Schaller,* 2007), and 1.94 ± 0.09 g cm^{-3} (*Buie et al.,* 2006), respectively. [For cosmogonic purposes we consider the Pluto-Charon system as a whole, and adopt a more generous radius range for Pluto, 1143–1183 km, than in *Buie et al.* (2006).] All these densities agree within the errors, when self-compression is accounted for. When interpreted in terms of rock/water-ice ratio, such densities imply a ≈ 70/30 mix, which has long been thought a signature of accretion in the outer solar system (*McKinnon et al.,* 1997; *Wong et al.,* 2007, and references therein), because abundant and largely uncondensed or equilibrated CO sequestered oxygen that otherwise would have gone into forming water ice [a concept whose genesis goes back to the kinetic inhibition model of *Lewis and Prinn* (1980)].

The masses and densities of the Pluto and Eris are known because they both have satellites (Charon and Dysnomia, respectively); Triton's parameters are well known from the 1989 Voyager 2 flyby. An additional observational technique has yielded densities of other large KBOs, values that have turned out to be surprising. For example, the rotational lightcurve of the large KBO (20000) Varuna (~450 km mean diameter) is highly symmetrical (double-peaked) and of large amplitude (0.42-mag variation) (*Jewitt and Sheppard,* 2002); these authors interpreted the lightcurve as a shape effect, in which Varuna has relaxed to a triaxial Jacobi figure, rotating about its short (c) axis as seen from Earth [or at least an acute projection (large aspect angle) of the c-axis]. The beauty of this interpretation is that the figure distortion (triaxial radius ratios) is solely a function of rotation rate and density if the density is uniform and the body is in hydrostatic equilibrium; it does not depend on size.

This technique was extended to several other KBOs by *Sheppard and Jewitt* (2002) and *Lacerda and Jewitt* (2007). Lightcurves were modeled both in terms of Jacobi ellipsoids and close or contact Roche binaries (the latter are highly deformed by tidal and rotational forces and can also fulfill the projected geometry requirements of the lightcurves). The work of *Lacerda and Jewitt* (2007) in particular considered the surface scattering properties of the KBOs in question when fitting the lightcurves. Figure 3 is taken from their paper and illustrates their bulk density results in comparison with values for Pluto and Charon. Densities range from ~0.6 g cm^{-3} for 2001 QG_{298} to 2.6 g cm^{-3} for 2003 EL_{61}. The important question is how much of this density variation is due to intrinsic compositional differences, and how much might simply be due to variable amounts of porosity?

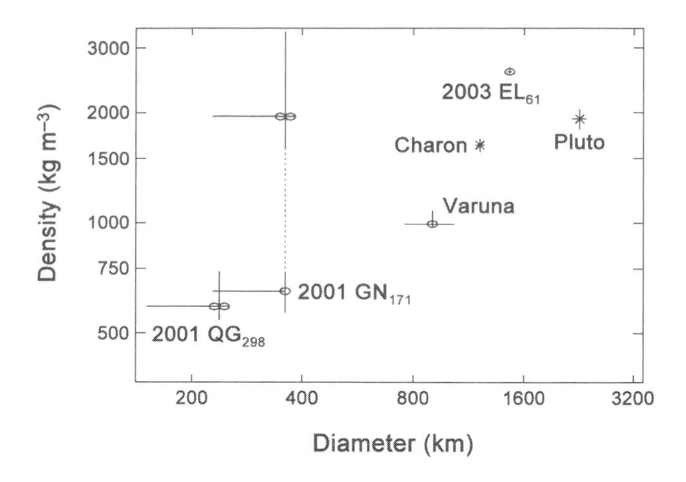

Fig. 3. Log density vs. log equivalent circular diameter for KBOs modeled as Jacobi ellipsoids (single ellipsoid symbols) and/or Roche binaries (double ellipsoid symbols). Pluto and Charon (from *Person et al.,* 2006) are plotted for comparison. KBO 2000 GN_{171} is plotted twice (dotted line). Modified from *Lacerda and Jewitt* (2007).

For the largest known KBOs, in the dwarf planet class, porosity cannot be important. The large inferred density of EL_{61}, based on the lightcurve of *Rabinowitz et al.* (2006), almost certainly implies a greater rock/ice ratio than possessed by Pluto and Triton. A density of 2.6 g cm^{-3} implies a rock/ice ratio of ≈85/15 when modeled in terms of the same anhydrous rock model as in *McKinnon et al.* (1997). The lightcurve of 2003 EL_{61} and the inferred rock-rich nature of the body, contrasted with the strong spectral signature of water ice on its surface, on the surface of its major satellite, and on the surfaces of the dynamically linked EL_{61} family, are compelling evidence for the scenario offered by *Brown et al.* (2007a): A massive collision in the geologic past both spun up a differentiated precursor and shattered and dispersed a good fraction (perhaps approximately half) of its ice mantle (cf. *Morbidelli,* 2007). Quite aside from the rich collisional and gravitational dynamics to be explored, 2003 EL_{61} offers compelling evidence for the global differentiation of a large KBO, one in the dwarf planet class [perhaps >800 km in radius originally (*Brown et al.,* 2007a)]. The state of differentiation of other KBOs will be taken up in section 6.

Both *Jewitt and Sheppard* (2002) and *Lacerda and Jewitt* (2007) find (20000) Varuna's density to be ≈1.0 g cm^{-3}. If interpreted in terms of a solid (or grain) density of 1.93 g cm^{-3} [Triton's uncompressed density (*McKinnnon et al.,* 1997)], this implies a porosity of almost 50%, which intuitively seems implausible for a body of this scale. Even if Varuna is presumed to be rotating at the *minimum* rotation rate (and maximum spin angular momentum) for a Jacobi ellipsoid, its density cannot exceed 1.27 g cm^{-3} (see Fig. 10 in *McKinnon et al.,* 1997). [We note that this implies an aspect angle less than the 75° lower limit assumed by *Lacerda and Jewitt* (2007).] In this limiting case a porosity of ≈35% is required, which is much more geologically reasonable (see below), although it is also required that this porosity be maintained against internal pressures (section 4.3) and thermally driven, creep densification over time.

At the "small" end of Fig. 3 lies 2001 QC_{298}, with a modeled density of just ~0.6 g cm^{-3} (2000 GN_{171} is too uncertain to merit further discussion). This density is similar to that inferred for the progenitor of Shoemaker-Levy 9 (*Asphaug and Benz,* 1996), a body 2 orders of magnitude smaller in scale, and is consistent with the density determined for the nucleus of somewhat larger Comet 9P/Tempel 1 from ballistic modeling of the Deep Impact ejecta plume (*Richardson et al.,* 2007). Essentially arbitrarily large porosities are thought possible for typical comets (~1 to 10 km in diameter), owing to their microgravity environments and (exceedingly modest) internal strengths (*Weissman et al.,* 2004; *Blum et al.,* 2006). For much larger bodies, though, bulk densities substantially below that of pure water ice are usually taken to imply an ice-rich composition (meaning *more* ice rich than Pluto) *and* porosity.

The lesson of 2001 QC_{298} may be that substantial porosity (probably macroporosity) is a reality for some KBOs of up to at least ~125 km in radius. This "lesson" does not depend on a single datum: Two additional KBOs analyzed by *Sheppard and Jewitt* (2002), (33128) 1998 BU_{48} and (40314) 1999 KR_{16}, yielded lightcurve-derived densities of ~0.5 and ~0.3 g cm^{-3}, respectively. These bodies are at least as large as 2001 QC_{298}. Furthermore, observations of KBO binary orbits yield secure estimates of system masses, which when combined with Spitzer-based size estimates (chapter by Stansberry et al.) provide system densities. Two determinations to date, for Plutino (47171) 1999 TC_{36} (*Stansberry et al.,* 2006) and "Twotino" (26308) 1998 SM_{165} (*Spencer et al.,* 2006), both yield ~0.5 g cm^{-3} for bodies whose effective (primary and secondary combined) radii are in the 150–200-km range. In contrast, similar measurements of the Centaur binary (65489) Ceto-Porcys yield a density of ~1.4 g cm^{-3} for an effective radius of ~100 km (*Grundy et al.,* 2007), so the implication is that not *all* KBOs in this size class have densities ≪1 g cm^{-3}.

4.2. The Nature of Porosity, on Earth and in Kuiper Belt Objects

In terrestrial geology, porosity is the natural result of damage and disaggregation of coherent rock masses, accumulation of particles fine or otherwise, or chemical attack and dissolution of susceptible minerals. The porosity of aggregates of loose, noncohesive (dry) rock fragments (e.g., boulders, cobbles, sand), generically termed rubble, depends on particle packing, the shape or irregularity of the particles, and their size distribution (e.g., *Lambe and Whitman,* 1979). Loose, high-porosity aggregates subjected to virtually any source of mechanical agitation will, however, evolve by particle sliding and rotation toward a relatively *maximally dense* (minimum porosity) state (see also *Britt et al.,* 2002). The lunar regolith is maximally dense, with a porosity of ≈40% (*Carrier et al.,* 1991). This porosity is *consistent* with the irregular nature of lunar regolith particles (which promotes greater porosity) and the broad size distribution of these particles (which reduces porosity by allowing small particles to fill in the spaces between larger ones).

Greater porosities (≥50%) generally require unusual circumstances, such as highly irregular particles (snowflakes, volcanic ash), or sufficient strength (cohesion) at interparticle contacts to resist collapse due to overburden pressure (snow, welded ash). Low overburden pressures also favor such underdense configurations, all other things being equal. [In this regard, the anomalously high macroscopic porosities inferred for the M-type asteroids 16 Psyche and 22 Kalliope by *Britt et al.* (2002), ~70%, depend on assuming a pure metal composition and that the IRAS-derived diameters are correct. *Lupishko* (2006) argues that the latter are too high for asteroids with metal-rich surfaces, and that 16 Psyche's macroporosity in particular is actually closer to 30–40%.]

Porosity can be lost by mechanical crushing at particle contact points, or asperities. On Earth, however, porosity in crustal layers is largely lost by other physical and chemical means, by compaction due to loss of aqueous pore fluids, and through the dissolution and precipitation of minerals by these fluids. These processes are important components of sediment *diagenesis*, the lowest temperature and pressure type of metamorphism.

We expect KBOs to have accreted initially as porous bodies (e.g., *Blum et al.,* 2006). We further expect them, however they evolved internally, to acquire substantial porosity when mechanically disrupted and reaccumulated. Accretion and large-scale collisions provide sufficient mechanical agitation for KBOs to evolve to their maximal relative (minimally porous) density, hence the porosities of 30–40% referred to in section 4.1 are judged to be geologically "reasonable." Shattering collisions that do not disrupt a body are expected to generate porosity along fracture surfaces, but of a lesser amount (≪40%) (*Britt et al.,* 2002).

High porosities (≥50%) are perfectly sensible in the ultralow gravity and overburden environments of typical comets and in the very low overburden environments of the near-surfaces of typical KBOs. The compressive strengths needed for comets to resist crushing are exceedingly small, ~10–100 Pa (weaker than dirt), and are probably achieved by the van der Waals attraction between the volatile ices in comets and KBOs alone (cf. *Weissman et al.,* 2004). For larger KBOs (>100-km diameter), internal overburden pressures are much larger (*McKinnon,* 2002), and high porosities (≥50%) are probably unsustainable unless the icy contact points among the constituent particles have welded together. Such internal welding, or sintering, would, however, likely indicate that a major, "diagenetic" loss of porosity had already occurred.

Maximally dense KBOs (up to 30–40% porosity) are permissible at any size, however, as long as the ice particles themselves resist crushing (discussed below) or thermally activated viscous creep (both lead to porosity loss and bulk densification). In the case of maximal relative density, welding or sintering among particle contacts is irrelevant, and interparticle strength is provided by internal friction (*Holsapple,* 2007). Nonequilibrium shapes are also supportable by internal friction (*Holsapple,* 2007), but at least for midsized icy satellites there appears to be a clear, empirical transition from nonequilibrium shapes in the ~100-km-radius class (Hyperion, Phoebe) to equilibrium or near-equilibrium shapes (spheres or triaxial ellipsoids) in the ~200–250-km-radius class (Mimas, Miranda). Whether these latter bodies relaxed to their roughly hydrostatic ("fluid") shapes by transient episodes of low internal friction during major collisions or by viscous creep of internal ice at high homologous temperature is not entirely clear [especially for Mimas, and comparably sized Proteus (around Neptune) remains irregular]. In addition, *Romanishin and Tegler* (2001) argued from KBO lightcurves that those KBOs below ~125 km radius (for 4% albedo) were more likely to be irregular in shape [although presumably not symmetrically so, as in *Lacerda and Jewitt* (2007)], whereas larger KBOs would adopt more spherical or biaxial hydrostatic equilibrium shapes. Thus on an empirical basis, and especially considering the icy satellites for which we have direct observations, assuming that *very large* KBOs such as (20000) Varuna and 2003 EL_{61} are in hydrostatic equilibrium hardly seems as fraught with uncertainty as *Holsapple* (2007) implies.

With respect to KBOs, we do not expect diagenesis by aqueous fluid transport as on Earth, unless the KBO is able to differentiate its rock from ice via ice melting. Such differentiation is clearly plausible for the largest KBOs, of the dwarf planet class, but as we will see in section 5 may also have occurred in small or "typical" KBOs for sufficient early heating by ^{26}Al decay. Perhaps more importantly, KBOs of all sizes contain (or at least accreted) abundant volatile ices, and with sufficient heating these volatile ices can sublimate, or desorb from crystallizing amorphous water ice, move through permeable ice-rock layers, and recondense at lower temperatures elsewhere [or perhaps form clathrates (see *Blake et al.,* 1991)]. These volatile gases are the mobile fluids of KBO internal evolution; their sublimation can enhance porosity in some regions, while their condensation can cement rock-ice particles and lower if not eliminate porosity elsewhere (e.g., *De Sanctis et al.,* 2001).

If KBO water ice itself reaches temperatures typical of cometary activity (~180 K) (*Meech and Svoreň,* 2004), or perhaps several tens of K lower (*Shoemaker et al.,* 1982), internal sublimation and condensation should drive diagenetic metamorphism. Larger grains and pores should grow at the expense of smaller grains and pores, while overburden pressure drives compaction. Through sublimation and condensation and, if warm enough, solid-state creep, the ice fraction within a KBO should sinter and densify, and ultimately eliminate its porosity (and all at temperatures well below that of water ice melting).

4.3. Pressure-induced Densification

The above discussion illustrates some of the complexities involved when considering the possible structure and evolution of KBO porosity. Some fundamental physical limits on the internal porosity of large KBOs can, however, be placed (*McKinnon,* 2002). The radial pressure distribution, P(r), within a self-gravitating uniform sphere of density ρ and radius R is

$$P(r) = \frac{2\pi}{3}\rho^2 G R^2\left(1-\frac{r^2}{R^2}\right) = 2.0\ \text{MPa}\left(\frac{\rho}{1.2\ \text{g/cm}^3}\right)^2\left(\frac{R}{100\ \text{km}}\right)^2\left(1-\frac{r^2}{R^2}\right) \quad (1)$$

where we use a fiducial density of 1.2 g cm^{-3} to represent the combination of a solid rock-ice density of 2 g cm^{-3} and 40% porosity. The pressure within the inner 50% of the sphere's volume exceeds $P_{50} \approx \pi\rho^2 G R^2/4 = 0.75$ MPa $(\rho/1.2\ \text{g cm}^{-3})^2\ (R/100\ \text{km})^2$. For R > 500 km, $P_{50} \geq 20$ MPa. This is a nontrivial pressure for ice and, when multiplied by the inverse of the "Hertz factor," the relative interparticle contact area per grain (or boulder), easily exceeds the unconfined compressive strength of cold (77 K) ice (*Durham et al.,* 1983).

The effect of confining pressure, and the inferred geometrically enhanced pressures at grain contacts, can be experimentally demonstrated. Figure 4 illustrates the results of a series of tests on the crushing of cold (77–120 K), granular ice (*Durham et al.,* 2005). These experiments indicate substantial reduction in porosity in cold, granular water ice (and by implication in cold ice or ice-rock rubble as long as the ice is volumetrically dominant) over a hydrostatic pressure range of ~1–150 MPa. Somewhat surprisingly, substantial residual porosity (~10%) persists at pressures in excess of 100 MPa. Scanning electron microscope analysis of the samples shows that this residual porosity is contained mostly in micropores supported by small shards of fractured ice.

Sample 503 in Fig. 4 started with the widest initial range in ice grain size, and had the lowest initial (lightly tamped) porosity, $\phi = 0.37 \pm 0.03$, as expected for packing of particles of different sizes, and the lowest final porosity. Thus, of this set of experiments, this sample run is probably the most geologically "realistic," and the one most applicable to KBO interiors. The initial crushing and loss of porosity over the first few MPa of external, or macroscopic, pressure is not well resolved, but by P = 10 MPa porosity decreases to 20%. We can conclude that from these experiments that non-negligible porosity can be sustained over solar system history in even large KBOs, but of course only in the *absence* of significant heating and sintering, annealing, and creep-driven pore collapse. Moreover, the maximum sustainable porosity in the ice fraction of a large KBO (e.g., Varuna) is limited, and is likely ≪30–40%. Porosity in the anhydrous rock (or silicate) fraction, if separate, should

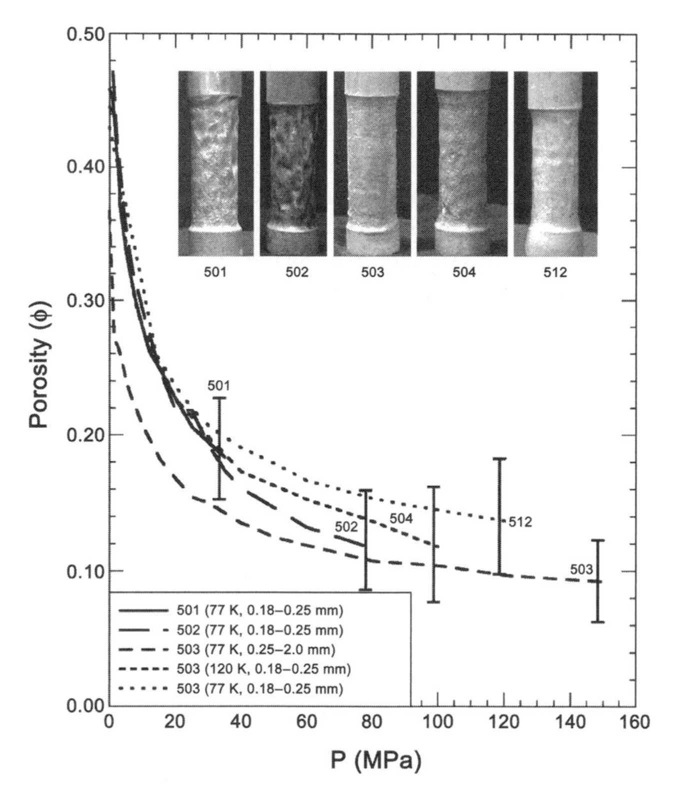

Fig. 4. Compaction of cold, granular ice under hydrostatic pressure. All five samples (inset), compacted to pressures up to 150 MPa at 77–120 K, showed nonlinear porosity reduction with increasing pressure and residual porosities after highest pressurization of at least 0.10. Porosity is corrected for the elastic effects of pressure on volume; error bars shown represent the absolute uncertainty in porosity at the end of each run, and the relative uncertainty of porosity along any given curve is ~±0.005. Inset shows the samples after testing, still inside their 0.5-mm-thick soft indium encapsulation. For scale, the diameter of the end caps is 26.4 mm. Modified from *Durham et al.* (2005).

be essentially unaffected by KBO pressures, although even for a 70/30 dry-rock/water-ice mixture the rock volume fraction is only 35%. Hydrated rock is weaker (more collapsible or crushable), as is presumably the carbonaceous fraction, but the relevant density and porosity behavior of the latter is poorly constrained (section 6.3).

With the above discussion of formation location, timescale, composition, and porosity, we now turn to explicit models of Kuiper belt object evolution.

5. STRUCTURE AND THERMAL EVOLUTION

Kuiper belt objects bear resemblance both to asteroids and to comets, and the numerous studies of asteroid and comet internal evolution published over the years are of direct relevance to understanding KBOs (see, e.g., the chapter by Coradini et al.). Kuiper belt objects also bear resemblance to the midsized icy satellites of the giant planets, and the fruits of research on these bodies are also relevant. The sizes of KBOs span a wide range, but on average they are considered larger than comets (cf. *Lamy et al.,* 2004), although this is essentially an observational selection effect. The size, composition, and location of KBOs all affect their structure and thermal evolution.

5.1. Composition and Energy Sources

Models of the solar nebula, or of protoplanetary disks around solar-type stars, predict rather low temperatures in their outer ("transneptunian") regions (>20 AU), usually ≪100 K (e.g., *D'Alessio et al.,* 1999, 2001; *Lecar et al.,* 2006; *Garaud and Lin,* 2007), but above the ~20–30 K required to directly condense the most-volatile species, such as CO, N_2, or CH_4 (although see *Hersant et al.,* 2004, for an exception). Temperatures are, however, low enough for the water ice to be amorphous at formation (whether condensed in the solar nebula or interstellar molecular clouds), and for moderately volatile species, such as CO_2, HCN, or NH_3, to condense. Amorphous water ice may trap highly volatile species and release them upon crystallization (*Bar-Nun et al.,* 1987, 1988), although the roles of direct condensation vs. trapping in amorphous water ice probably need to be critically reexamined (cf. *Notesco et al.,* 2003).

Indirect support for the originally amorphous nature of the H_2O ice in KBOs is provided by the hypothesis that Jupiter-family comets originate from the Kuiper belt and by observations of highly volatile species in the comae of these (and Oort cloud) comets and their release pattern (e.g., *Meech and Svoreň,* 2004; *Dello Russo et al.,* 2007). Thus, it is reasonable to assume that the initial composition of KBOs includes "dust" (rock + carbonaceous matter) and gas-laden amorphous water ice, possibly mixed with other, more volatile, ices, as described in section 2. As in asteroids and other solid objects of the solar system, the dust will include radioactive species, and their relative abundances should be similar presumably to those found in chondritic meteorites, an inference supported by the solar abundances of refractory elements in Halley dust (*Jessberger et al.,* 1988; *Schulze et al.,* 1997).

This composition leads us directly to the *internal* energy sources and sinks available to KBOs (impact heat being an external source). These internal energy sources are of three kinds: radiogenic energy, in proportion to the rock fraction; heat released upon crystallization, in proportion to the H_2O ice content; and latent heat of phase transition of the various volatiles. Clearly, the first two are energy *sources,* associated with irreversible processes. Latent heat, however, may be either released or absorbed, and if gases can flow through the porous medium, it can be released in one place and absorbed in another, thereby acting as an effective means of heat conduction. In this respect, there is also a significant difference between species that are included in the original composition as ices or as trapped or absorbed gas. Overall, these ices represent a potential heat *sink,* since they require an outside heat source in order to melt or sublimate, even if later they return all or part of the absorbed heat. Trapped gases, on the other hand, once released from amorphous ice, are a potential heat *source,* because they will release latent heat upon eventual condensation, even if they may reabsorb part or all of it, if the volatile ice later sublimates [although to an extent this distinction depends on where one "bookkeeps" the energy of desorption from the amorphous ice (*Kouchi and Sirono,* 2001)]. This difference is particularly important for KBOs, since they reside in a cold environment, and hence have cold outer layers, where gases released in the interior are bound to refreeze.

Some of these complex effects have been studied and illustrated by numerical simulations and will be addressed in section 5.3. However, because they depend crucially on initial abundances assumed, as well as on a number of uncertain physical parameters, and to the extent that no systematic parameter study is as yet possible, it is instructive to investigate KBOs in a more general, albeit less accurate manner, by some basic analytical considerations.

5.2. Analytical Considerations

5.2.1. Heating versus cooling. Assuming that KBOs have been affected by radioactive heat sources, much as larger bodies of the solar system still are, but considering that small bodies cool far more effectively than large ones due to their large surface to volume ratio, only high-power (short-lived) radioactive isotopes are of importance for smaller KBOs (what is small is clarified in section 5.3). The radioactive isotope ^{26}Al (with an exponential decay time $\tau = 1.06 \times 10^6$ yr) was recognized as a potential heat source capable of melting bodies of radii between 100 and 1000 km half a century ago (*Urey,* 1955). Evidence for its existence was supplied by ^{26}Mg-enhanced abundances found in Ca-Al inclusions in meteorites (e.g., *MacPherson et al.,* 1995). Further support, from an independent source, was provided by the detection of interstellar 1.809-MeV γ-rays from the decay of ^{26}Al (*Diehl et al.,* 1997). All this evidence points toward an interstellar isotopic ratio $^{26}Al/^{27}Al \approx 5 \times 10^{-5}$, implying an initial mass fraction $X_0 \approx 7 \times 10^{-7}$ in the rock dust fraction (but in the remainder of this section "dust," for modeling purposes, we will not distinguish between the rocky and carbonaceous dust fractions).

The potential effect of radiogenic heating during the early evolution of icy bodies of different radii R may be roughly estimated based on global energy considerations (cf. *Prialnik,* 1998). The bulk heating rate, or power (energy per unit time), for a body of mass M, is

$$Q_h = \frac{1}{\tau} X_0 M H e^{-t/\tau} \tag{2}$$

where H is the radioactive heat release by ^{26}Al decay per unit mass (1.48×10^{13} J kg^{-1}), and its maximum, obtained at time t = 0 (identified with the crystallization age of CAIs in carbonaceous meteorites, or t_{CAI}), is $\tau^{-1}X_0MH$. A rough estimate for the bulk cooling rate is $Q_c \approx 4\pi kT/R$, where the cooling flux is approximated by kT/R, T representing the internal temperature and k the thermal conductivity. For the internal temperature to rise, the balance Q_h–Q_c must be positive at t = 0, a requirement that is determined by the body's radius and by the thermal conductivity of its material, for given initial conditions. As the heating rate declines with time, it eventually becomes lower than the rate of cooling. For suitably large KBOs the internal temperature rises at the beginning up to a maximum value and then falls off, tending to the local equilibrium temperature for $t \gg \tau$.

Based on this simple, heuristic model, *Prialnik et al.* (2007) calculate T(t) and show that above a certain initial radius an icy body made of amorphous ice will start crystallizing (say, for $T \gtrsim 100$ K), and above a still larger radius, melting of ice eventually occurs (see also *Jewitt et al.,* 2007). [We emphasize that because amorphous water ice is metastable at all temperatures, there is no fixed crystallization temperature. Conversion to crystalline ice is a thermally activated process that depends on both time and temperature (*Schmitt et al.,* 1989).]

5.2.2. Accretion and radioactive heating. The above considered radioactive heating to take place in a body of fixed size, whereas in reality heating also occurs while the body grows. The processes may be separated and treated in turn only if accretion is very rapid as compared with the characteristic decay time of the radioactive species. Thus, for example, a body growing into a comet in the giant-planet region should not be affected by the decay of, say, ^{40}K. For bodies potentially accreting more slowly, like KBOs, and for short-lived radioactive nuclei, like ^{26}Al, the situation is different. Consider an accreting body of mass M(t) and assume that the accreted material contains a mass fraction X(t) of radioactive ^{26}Al, initially X_0 (cf. *Merk and Prialnik,* 2003). The heating rate is now given by

$$Q_h = \frac{1}{\tau} X_0 H e^{-t/\tau}(M(t) - \tau\dot{M}) \tag{3}$$

and its variation with time differs considerably from that obtained at fixed mass. Because the accretion rate is lowest at the beginning (e.g., *Weidenschilling,* 2000), it is possible that an accreting body will first cool (in an average sense) and later, as the mass increases and the surface to volume ratio becomes smaller, heat up (*Merk and Prialnik,* 2006).

We conclude that if the characteristic timescales of accretion and radioactive decay are comparable, a growing body may be affected by radioactivity while it grows: At first, when the body is small, efficient cooling is compensated by the high radioactivity; later, diminished radioactivity is compensated by the less efficient cooling of a larger body. Applications of these concepts to asteroid thermal evolution can be found in *Merk et al.* (2002) and *Ghosh et al.* (2003).

5.2.3. Accretional impact heating. Heating due to the accumulation of gravitational potential energy is not usually considered important for KBOs. For large comets and small KBOs it is simply too small, and even for large bodies, such as Pluto, it has not been judged to be significant when compared with the heat available from the decay of long-lived U, Th, and ^{40}K (*McKinnon et al.,* 1997). In the latter work, however, significance was measured by whether accretional heat would melt Pluto or Charon's ice and so

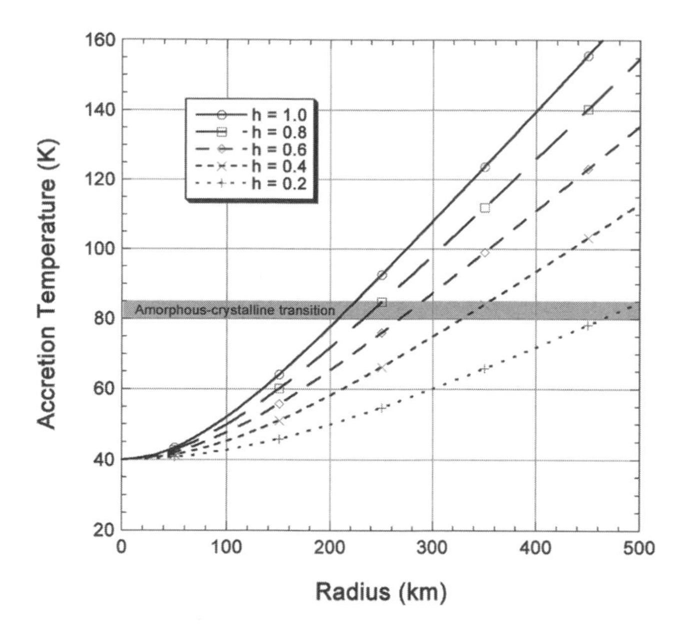

Fig. 5. Accretional temperature profiles as a function of KBO radius, for various impact heat retention fractions, h. Density is taken to be 1 g cm^{-3}, representing a porous mixture of ice, rock, and carbonaceous matter. Growth of temperatures is linear at large radii because of the increase in the heat capacity with T. From *McKinnon* (2002).

drive rock-from-ice differentiation. Kuiper belt objects are of course composed of much more than water ice, so the question of accretional heat should be reconsidered in this light.

Figure 5 shows the result of such a heating calculation, a classic radial temperature profile of an accreting uniform sphere. Contributions from planetesimal kinetic energy prior to gravitational interaction are ignored, so the temperatures illustrated are the minimum a KBO may acquire. The temperature profiles are parameterized in terms of h, the fraction of impact heat retained (cf. Fig. 8 in *McKinnon et al.,* 1997, for a similar figure covering a broader size range). This parameter depends on the balance of impact heat burial vs. radiative (and possibly evaporative) cooling, but conservative values of $h \lesssim 0.5$ are often assumed. The approximate temperature range for amorphous ice to crystallize on a million-year (approximately accretional) timescale, based on *Schmitt et al.* (1989), is also shown.

From Fig. 5, we can see that large ($R \gtrsim 300$ km) KBOs should have begun to crystallize their original amorphous water ice during accretion if h ~ 0.5. The exothermic nature of the amorphous-to-crystalline transition could then in principle initiate an inwardly propagating wave of crystallization (cf. *Haruyama et al.,* 1993), although release of trapped gases from the crystallizing ice may mitigate or eliminate the exothermic heat pulse (*Kouchi and Sirono,* 2001).

With regard to h, the characteristic length scale for thermal conduction during accretion is κ/u_{acc}, where κ is the thermal diffusivity ($\equiv \frac{k}{\rho C_p}$, where C_p is the heat capacity) and u_{acc} is the rate of growth of the body, dr/dt. For a low conductivity representative of porous rock or amorphous ice, k ~ 0.1 W m^{-1} K^{-1} (*Ross and Kargel,* 1998)

$$\frac{\kappa}{u_{acc}} \sim 0.1 \text{ km} \times \left(\frac{100 \text{ km}}{R}\right) \times \left(\frac{\tau_{acc}}{1 \text{ m.y.}}\right) \quad (4)$$

where R is the final radius of the KBO and τ_{acc} is its accretion time. As long as accreting impacts bury heat deeper than this depth, h should be non-negligible and larger KBOs should have experienced substantial initial temperature rises, relative to their initial temperatures. The controlling variables are the accretion rate and, especially, the planetesimal size distribution.

5.2.4. Self-gravity. When we consider larger KBOs, self-gravity is important and is bound to affect the internal structure, as discussed in section 4.3. The question is when should self-gravity taken into account in modeling? Based on equation (1), and the experimental results described in section 4.3, hydrostatic compression should be considered for KBOs with radii above ~100 km. In this case an equation of state (EOS) is required in order to calculate the density distribution throughout the body.

An EOS based on the properties of rock, ice, and carbonaceous matter, along with porosity, could be developed, but here we present a simpler, more generic approach. Kuiper belt objects are, from a condensed matter point-of-view, "cold" bodies of modest scale compared to the major planets, and a simple, temperature-independent equation of state may suffice. Specifically, we ignore all high-pressure phases of water ice, even though these can be structurally important for bodies as small as Charon (*McKinnon et al.,* 1997). Adopting a simple Birch-Murnaghan form for the P(ρ) dependence (see, e.g., *Poirier,* 2000)

$$P(\rho) = K_e\left[\left(\frac{\rho}{\rho_0}\right)^{7/3} - \left(\frac{\rho}{\rho_0}\right)^{5/3}\right] \quad (5)$$

where K_e is an effective bulk modulus, we can solve numerically the hydrostatic equation together with the mass conservation equation, yielding

$$\frac{d}{dr}\left(\frac{r^2}{\rho}\frac{dP(\rho)}{dr}\right) = -4\pi r^2 \rho G \quad (6)$$

for ρ(r). In order to obtain a solution we have to determine two parameters: K_e and ρ_0, and one boundary condition for the above equation (in addition to the trivial condition that the mass at the center vanishes).

If the radius and average density (or mass) of a body are known from observations, there remains one free parameter or condition. If we assume internal pressures are not sufficiently high for compressing the solid material (rock and ice), but can only reduce (but not eliminate) the

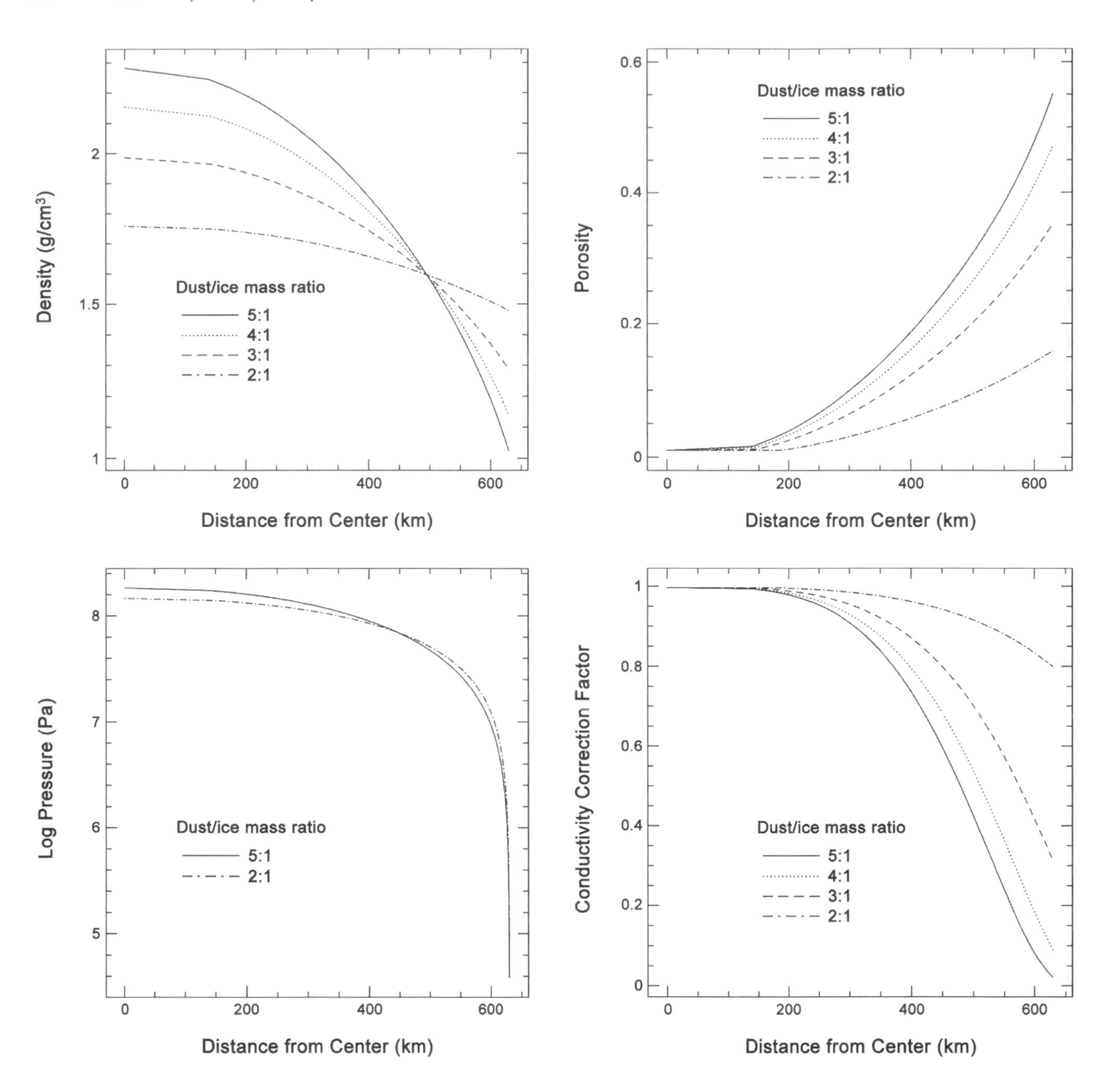

Fig. 6. Structure of a self-gravitating body having the assumed mass and radius of (50000) Quaoar, for different compositions defined by the dust/ice mass ratio, as marked: top left, density; top right, porosity; bottom left, pressure; bottom right, correction factor for the thermal conductivity coefficient resulting from porosity (based on the model of *Shoshany et al.,* 2002). The greater the rock/ice ratio, the greater the solid (nonporous), central density, and thus the more porous on average and compressible the rest of the interior. Modified from *Prialnik et al.* (2007).

porosity, the highest central pressure results by assuming the central porosity (just) vanishes. Adopting this as the additional condition, we present in Fig. 6 an example based on the radius and estimated mass of 50000 Quaoar, 630 km (*Brown and Trujillo,* 2004) and ~1.7×10^{21} kg (1.65 g cm^{-3} assumed), respectively. Results are presented for four different rock/ice mass ratios: 2, 3, 4, and 5 (a unity mass ratio does not allow vanishing porosity at the center). We emphasize that these equations of state are intended to illustrate the potential effects of compressibility; they are not based directly on data such as in Fig. 4.

This caveat aside, it is interesting to note that such a body has a far from uniform structure, which is bound to affect its thermal evolution as well. We note, in particular, the change in the multiplicative correction factor for the thermal conductivity due to porosity (Fig. 6, lower right), from the model of *Shoshany et al.* (2002), ranging from unity at the center to less than 3% (0.03) near the surface. We emphasize that despite the relatively low average porosity, the porosity of the outer layers may be considerable, comparable to that encountered in comet nuclei (cf. section 4).

5.3. Numerical Simulations

The extent to which small outer solar system bodies may have been affected by early radiogenic heating has been debated for four decades (small here meaning comets and sometimes KBOs, but not planets). The pioneering work toward answering this question is that of *Whipple and Stefanik* (1966), who considered the decay of long-lived radionuclides and found that it would lead to loss of the most-volatile species. It was followed by the studies of *Wallis* (1980) and *Irvine et al.* (1980), who showed by simple calculations that liquid water could be obtained in the core of comet nuclei. The subject was further pursued by *Prialnik et al.* (1987), *Prialnik and Bar-Nun* (1990), *Yabushita* (1993), *Haruyama et al.* (1993), and *Prialnik and Podolak* (1995), each study focusing on one of the governing parameters, such as thermal conductivity, ^{26}Al content, or radius. More recently, models including mixtures of volatiles and gas flow through the porous medium have been considered by *De Sanctis et al.* (2001) and by *Choi et al.* (2002), and *McKinnon* (2002) explored the effects of long-term radiogenic heating on larger KBOs. *Merk and Prialnik* (2003, 2006) have shown that radioactive heating must be considered together with accretion of the bodies to their final size, because these processes can proceed on comparable timescales. Their numerical calculations show that the combined effect of accretion and radioactive heating is far more complex.

Full-scale long-term simulations of the evolution of KBOs are extremely demanding due to the interaction between the various thermal processes, which interfere with each other. Hence, to date, such calculations have focused on one aspect of the evolution, neglecting or simplifying other effects. Even so, these calculations shed light and impose constraints on the interior structure of these bodies.

Numerical modeling involves the simultaneous solution of energy and mass conservation for all the components considered, to date assuming a spherical body (see *Prialnik et al.,* 2004, for details). Since solar energy is not the major heat source, and since the orbital period is long and the orbital eccentricity can be relatively low, the assumption of spherical symmetry ("fast rotator") is acceptable for KBOs (far more than it is for comets). Thus a system of nonlinear partial differential equations is solved numerically on a one-dimensional spherical grid. Further details on modeling of the porous medium, and on the method of computing the sublimation rate may be found in *Mekler et al.* (1990); other parameters, as well as numerical procedures are discussed by *Huebner et al.* (1999).

5.3.1. Heating of Kuiper belt objects during accretion. Radioactive heating during the earliest stages of evolution of KBOs, concomitant with accretion, was studied by *Merk and Prialnik* (2003, 2006). Their numerical studies consider a composition of intimately mixed amorphous ice and dust in various proportions and allow for crystallization and melting of the ice. They *do not* consider, however, other volatiles nor allow for flow (gas or liquid) through the porous medium, although thermal properties (such as conductivity) take porosity into account. For a growing, spherically symmetric body of mass M(t) and uniform density ρ, composed of H_2O ice and dust in various proportions, the mass $0 \leq m \leq M(t)$ is chosen as the "spatial" coordinate (*Merk et al.,* 2002), and adaptively gridded during calculation to allow for the mass increase. Time is referenced to t_{CAI} (section 5.2.1).

Using the rates of growth ($\dot{M}$) supplied by an accretion algorithm [the coagulation model in *Merk and Prialnik* (2003)], a reasonably broad parameter-space survey was carried out for the (a, R_{max}) plane, spanned by heliocentric distance and final (maximum) radius of the accreting object in the outer region of the solar system: $20\ AU \leq a \leq 44\ AU$ and $2\ km \leq R_{max} \leq 32\ km$, respectively (i.e, comet to perhaps typical KBO in size). Now, most *observed* KBOs are $\geq$50 km in radius (*Jewitt,* 2004), so these particular calculations pertain less directly to KBOs as observed *in situ,* but more to their cometary descendants and fragments thereof. The calculations are nevertheless indicative of some of the relevant physics and chemistry of KBO internal evolution, especially if one focuses on the largest sizes considered (64 km diameter).

Of the three different compositions considered by *Merk and Prialnik* (2003) for the homogeneously growing body, the ice-poor one (dust/ice by mass ~3.4) is perhaps more relevant to KBOs and results obtained for this value are shown in Plate 5 in the form of contour plots. The top left panel shows the accretion times, which span a wide range, and considerably exceed the lifetime of ^{26}Al at the higher radius end. The main features that emerge from these calculations may be summarized as follows.

5.3.1.1. Crystallization: The initially amorphous ice crystallizes upon heating, and in this study is taken to be fully exothermic. The fractions of crystalline ice obtained when the crystallization process initiated by the decay of ^{26}Al is completed are shown in Plate 5 (bottom left). These fractions, ranging between 0 and 1, correspond to the relative radius (r/R_{max}) of the crystalline/amorphous ice boundary. We note that larger bodies formed in the present Kuiper belt zone retain pristine amorphous ice throughout a considerable fraction of their mass (i.e., if the amorphous/crystalline boundary is located at half the radius, implying that above it the ice is amorphous, it means that only one-eighth of the ice mass has crystallized). Smaller bodies (comets), as well as KBOs formed closer to the Sun, are bound to undergo complete crystallization, *if* their initial dust/ice ratio is as high as illustrated. The results change significantly if the ice content is much higher.

5.3.1.2. Liquid water: Plate 5 (bottom right) shows the maximum extent of *liquid water* (in terms of relative radius r/R_{max}). The occurrence of liquid water is limited in the parameter space considered, where Neptune's present-day orbit (30 AU) seems to describe a borderline. Beyond this heliocentric distance, the fraction of liquid water becomes small regardless of R_{max}, due to longer accretion times. For heliocentric distances smaller than about 30 AU, melting is still not found for the 2-km bodies, even if located as close to the Sun as 20 AU (and hence quickly accreting); these bodies remain permanently frozen. [In addition, internal pressures for the smallest bodies considered in *Merk and Prialnik* (2003, 2006) do not exceed the water-ice-vapor triple point of the pure substance, so melting (as opposed to sublimation) is not possible for these bodies in any case.] In the intermediate distance region, melt fractions between 10% and ~90% are possible. The different progress of these processes — crystallization and melting — is due to the large difference, about 170 K, between the melting temperature and the temperature range of the transition to crystalline ice and to the fact that the two processes are different energetically: Crystallization of (pure) ice is an exothermic transition, whereas melting of ice is endothermic. Comparing the bottom panels of Plate 5, it is very interesting to note the existence of bodies that have liquid water in the deep interior, while retaining pristine amorphous ice in their outer layers. These are bodies with radii in excess of 20 km, formed beyond 40 AU, therefore potentially representative of dynamically "cold," classical KBOs (section 2).

5.3.1.3. Preservation of liquid water: The top-right panel of Plate 5 shows the period of time during which liquid water is preserved, between melting as the temperature rises, and refreezing, as the temperature drops with the decay of ^{26}Al. We note a relatively narrow intermediate region between ~35 and 40 AU, near the present Kuiper belt's inner boundary, where liquid water persists for the longest time (for the particular dust/ice ratio depicted). Closer to the Sun, internal temperatures become too high and the water may boil (pressures are low) and evaporate; farther away accretion is too slow and hence cooling is too fast.

With regard to observations, if KBOs and short-period comets formed within 30 AU, as in the Nice model, we would expect all small bodies (and their collisional descendants) to have been thoroughly crystallized, if not melted and devolatilized. Even larger KBOs (the ones we see), formed well after the accretion times illustrated in Plate 5, would be melted/refrozen if they accreted from an ensemble of somewhat smaller (earlier-formed) bodies. Clearly, this violates observational constraints on the measured volatile production rates of short-period, Jupiter-family comets and Centaurs (*Meech and Svoreň,* 2004; cf. *Dello Russo et al.,* 2007), although the evidence for amorphous ice in comets and Centaurs — exothermally crystallizing and driving cometary outbursts — is largely circumstantial. [Amorphous ice *may* have been detected in Oort cloud comets (*Davies et al.,* 1997; *Kawakita et al.,* 2004).] Nor have Stardust samples been in contact with liquid water (we return to rock alteration in section 5.4). Either comets and KBOs formed farther out from the Sun (>30 AU), the accretion algorithm used to create Plate 5 is inaccurate (too rapid), or the KBO thermophysical model needs adjusting.

Of all the input assumptions, the *rock/ice* ratio may be the most critical. The dust/ice ratio in the model shown (Plate 5), ≈78/22, may in fact be too high. If we consider a carbonaceous mass fraction of ~25%, and use amorphous carbon, with a density of ~1.7 g cm^{-3} (*Wong et al.,* 2007), as a thermophysical proxy, then a mix of 54% dry rock, 21% water ice, and 25% amorphous carbon by mass yields the canonical solid density of 1.93 g cm^{-3}. This latter rock mass fraction is essentially identical to the dust mass fraction used (55%) in the "nominal composition" case in *Merk and Prialnik* (2003, 2006), and because radiogenic heating in the dust in these models is based on chondritic meteorite values, from an energy perspective dust mass = rock mass. Importantly for our discussion, the 55/45 dust/ice models in *Merk and Prialnik* (2003, 2006) indicate a more tempered early thermal evolution, with some KBOs forming within 30 AU with appreciable amorphous ice, and substantial water ice melting only for KBOs formed inside ~25 AU. Moreover, none of the models in *Merk and Prialnik* (2003, 2006) incorporate enhanced conductivity by water vapor flow or sintering when temperatures rise above 180 K, which could further limit the amount of melting predicted.

5.3.2. Survival of volatile ices. We now turn to other volatiles and consider the possible changes that are expected to occur due to internal heating. The main question is whether and to what extent can volatile ices survive in the interior of KBOs. Based on Plate 5, for comets and small KBOs formed within 30 AU, the question is moot; for more distant formation, or lower rock/ice ratios, the situation is more interesting. This question was investigated by *Choi et al.* (2002), with body size and heliocentric distance again being the free parameters of the study. In this work, however, heating during accretion is neglected and a diminished initial abundance of ^{26}Al is assumed instead ($t_{acc} > t_{CAI}$; see section 5.2.2), as another free parameter. Evolutionary calculations are carried out for very long time periods, on the order of 10^9 years, so heat from decay of ^{40}K is considered as well. The composition includes free (not trapped) CO and CO_2 ices, mixed with dust and amorphous water ice, and volatiles are allowed to sublimate and flow through the porous medium. The working assumption is that gases released in the interior can easily escape out of the interior and to the surface (high effective permeability). The detailed numerical solution yields temperature profiles and follows the evolution of temperature-dependent processes and the resulting changes in structure. An example of the results is shown in Fig. 7, for bodies of 10 km and 100 km radius, where the latter case is clearly representative of KBOs as usually considered.

For both radii considered, an outer layer about 1 km deep remains relatively cold at all times and thus preserves the ice in amorphous form. In all cases, however, the CO ice is lost, but this is bound to occur in the Kuiper belt region even in the absence of ^{26}Al (cf. *De Sanctis et al.,* 2001), due to the influence of insolation. The complete loss of free

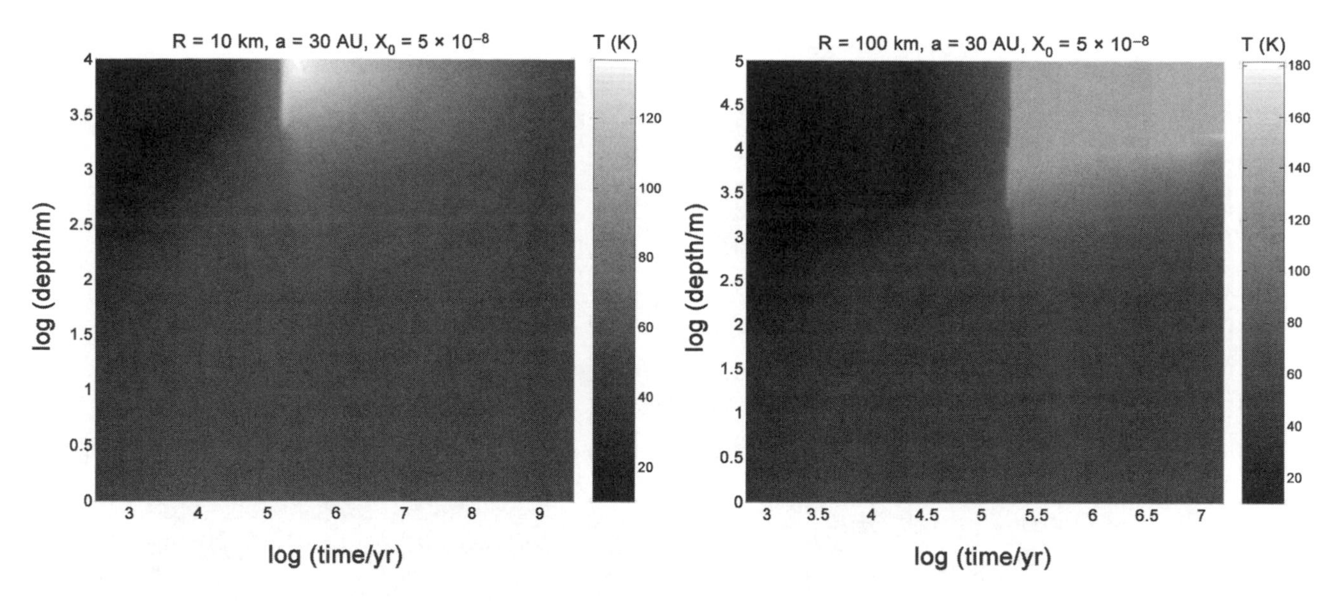

Fig. 7. Grayscale map of the evolving temperature (T) profile for a KBO radius of 10 km (left) and 100 km (right); a and X_0 are given, initial temperature T_0 = 10 K, rock mass fraction is 0.5, and porosity is 53%. The change in tone along a vertical line represents the radial temperature variation at the corresponding time. The change in tone along a horizontal line represents the change of temperature with time at the corresponding (fixed) depth. The time and depth scales are logarithmic, as changes are more pronounced during the early stages of evolution, and sharper, generally, toward the surface of the body. Adapted from *Choi et al.* (2002).

CO ice — and by analogy, all volatiles that sublimate below ~50 K, which were initially included as free ices — is common to all cases considered in this study. Moreover, KBOs may have partially lost less-volatile ices as well. The loss of volatiles is illustrated in Plate 6: The sublimation fronts advance inward from the surface due to solar heating (as the initial temperature, T_0, is set at 10 K for these models), but at larger depths, sublimation is driven by radioactive heating and the ice content gradually declines throughout the interior.

Loss of volatiles from the interior is accompanied by changes in porosity, and presumably, in tensile and compressive strength of the material as well. Internal stresses caused by steep temperature gradients and high gas pressures may contribute to weakening of the porous material. The main conclusion of all the numerical simulations considered so far is that the internal composition of KBOs is most probably not homogeneous, but potentially stratified, with the outer layers being *less* altered by volatile loss. Volatile addition through condensation may affect limited regions as well.

5.3.3. Gases trapped in amorphous ice. The conclusion that volatile ices, even if included in the original composition of KBOs, should be completely lost due to radiogenic heating leads us to consider the possibility that such volatiles may still be present, trapped in the amorphous water ice, which — we have seen — can survive radiogenic heating, at least in a non-negligible outer layer. Otherwise it would be difficult to reconcile observations of CO, for example, in the comae of Jupiter-family comets, with the hypothesis that these comets originate in the Kuiper belt.

As mentioned in section 5.1, trapped gases that are released from amorphous ice, and migrate toward colder regions of the body where they refreeze, constitute an additional local heat source. The processes of crystallization, freezing of the released gases, and their possible subsequent sublimation, are all extremely temperature dependent and they all occur in the same temperature range on comparable timescales, thus interfering with each other. The result is a very complicated composition pattern, where layers enriched in various volatile ices alternate.

A recent numerical study of this complex phenomenon (*Prialnik et al.,* 2007), which considers the evolution of relatively large KBOs and includes self-gravity, shows that in this case the outer zone is the most affected. There, volatiles that have been released in the interior and have migrated toward the surface refreeze, the more-volatile ones closer to the cold surface than the less-volatile. A layered structure is obtained, where different subsurface layers, composed mainly of pristine amorphous ice and rocky and carbonaceous dust, are enriched in different volatile ices. The deep interior may be affected by loss of water ice, even to the extent that a water-depleted core may form. The porosity is significantly affected by these processes, as illustrated in Fig. 8. If the surface layers are heated at a later stage, either by a collision or by a change in orbit that brings them closer to the Sun, sublimation of these ices may lead to enhanced activity. This may occur even if the temperature is not sufficiently high to trigger crystallization of remaining amorphous ice.

5.4. Cryovolcanism and Differentiation

The models described in this section so far, with their emphasis on early heating, amorphous ice crystallization, and sublimation and transport of volatiles, have stressed

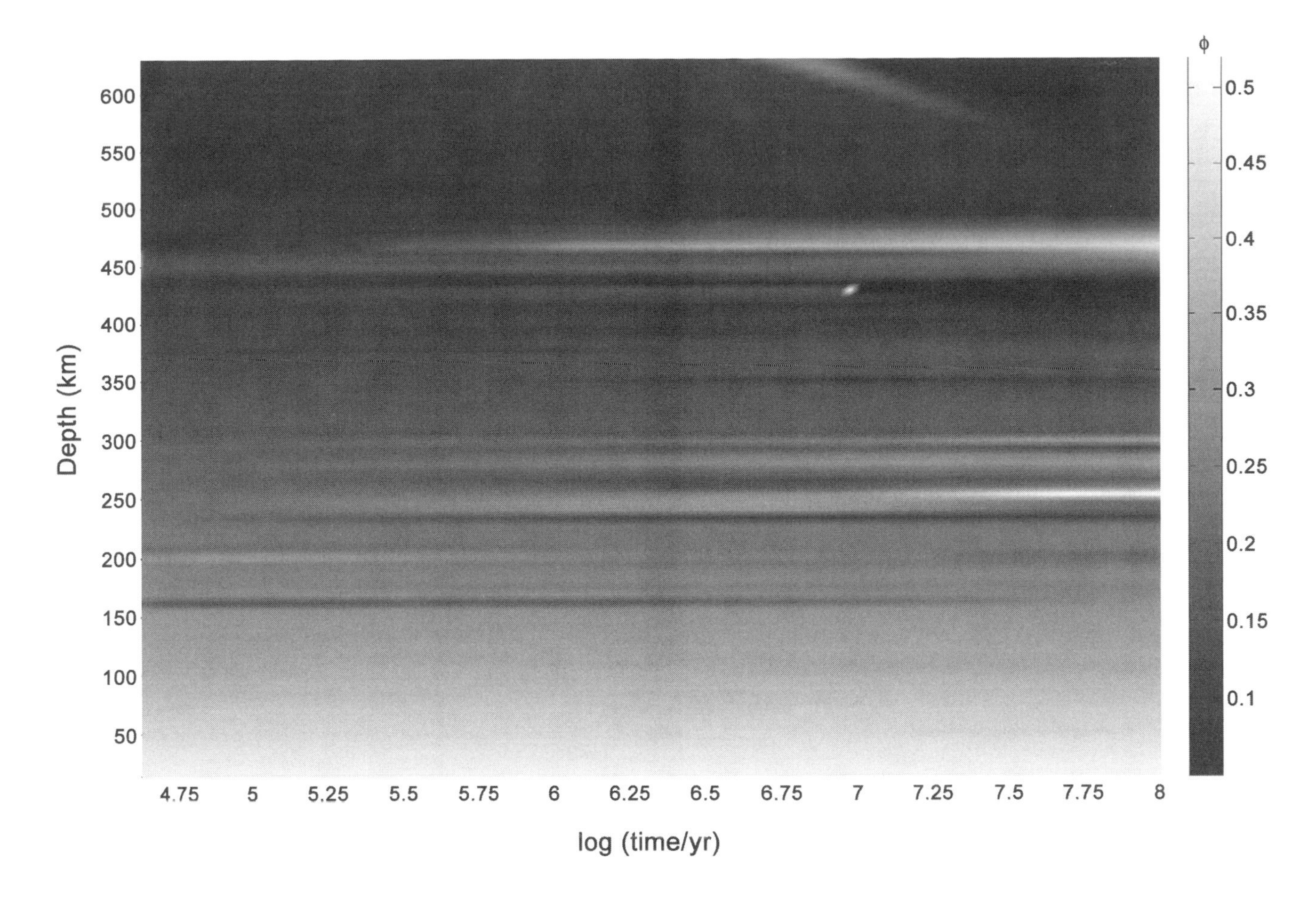

Fig. 8. Grayscale map of the porosity (ϕ) of a self-gravitating body having bulk and orbital properties inspired by 50000 Quaoar (R = 630 km; M = 1.6×10^{21} kg, average porosity = 0.25, dust mass fraction = 0.75, occluded volatiles: $CO/CO_2/HCN/H_2O$ = 0.005/0.005/0.0025/1; $X_0(^{26}Al) = 3 \times 10^{-8}$, equivalent to accreting at $t - t_{CAI}$ = 3.3 m.y., but no long-lived radionuclides; orbital eccentricity = 0.035, a = 42.95 AU) as porosity changes with time throughout the spherical body. Over longer timescales (>100 m.y.) in the presence of long-lived radiogenic heating, the porosities that evolve at great depth due to volatile sublimation and loss should be eliminated by water ice sublimation and viscous creep. Modified from *Prialnik et al.* (2007).

heritage from cometary research. Issues of concern to the icy satellite community, such internal differentiation, volcanic and tectonic activity, and so forth, have received much less attention. Now, however, with real constraints on surface compositions and bulk densities, as well as inferences for differentiation of some bodies (discussed in earlier sections), there is greater interest in the evolution of larger KBOs and dwarf planets as individual worlds.

Figure 9 shows the results of a simple, long-term thermal evolution calculation for a 450-km-radius KBO, Varuna (from *McKinnon,* 2002), and can be compared to both Plate 6 (lower right panel) and Fig. 8. Similar to much of the work already discussed, the model in Fig. 9 utilizes a rock mass fraction of 0.58 and 40% porosity, and the ice is initially amorphous (T_0 = 40 K), but only long-term radiogenic heating is considered (X_0 = 0). Whatever the importance of ^{26}Al (and ^{60}Fe) to early KBO evolution, radiogenic heating by U, Th, and ^{40}Ar over geologic time is inevitable, and is most important for those KBOs whose conductive cooling times R^2/κ exceed ~4.5 G.y. For a thermal diffusivity $\kappa \sim 10^{-6}$ m^2 s^{-1}, appropriate to rock or crystalline ice, this means bodies of ~400 km radius or larger.

One significant departure from the previous models is a lack of exothermic energy release upon amorphous ice crystallization. While obviously a model simplification, it is also experimentally justified (*Kouchi and Sirono,* 2001), when the ice is impure (i.e., contains occluded gases). Ice crystallization increases the thermal conductivity of the ice phase, promotes grain-to-grain thermal contact, and releases gases that by transporting sensible and latent heat increase the conductivity still more (unless the gases are able to vent to space). Stresses caused by the volume change accompanying crystallization (*Petrenko and Whitworth,* 1999, chapter 11) could, in principle, promote pore collapse. For these reasons, and to *minimize* the internal temperatures reached in the model in Fig. 9, the conductivity of crystallized ice + rock was assumed to convert to nonporous values at the time of crystallization. Despite this substantial increase in

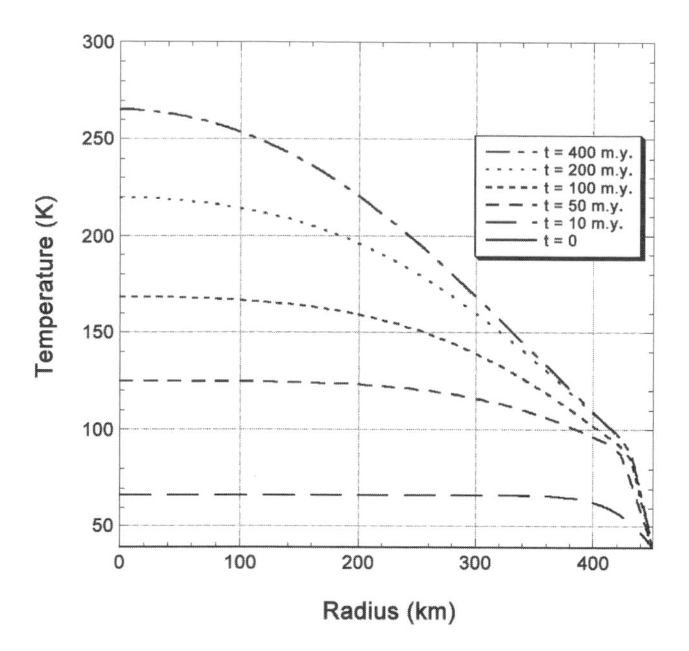

Fig. 9. Temperature profiles for a 450-km-radius KBO (Varuna) as a function of time. The ice fraction (0.42) is initially amorphous, but once it crystallizes at ~90 K the effects of porosity on the overall conductivity are assumed to vanish. Only an outer shell or layer remains "primordial" and uncrystallized. Central temperatures continue to rise after 400 m.y., and in the absence of solid-state convection, would reach the water-ice melting point. Modified from *McKinnon* (2002).

conductivity throughout most of the model volume, the center of the body reaches quite high temperatures after several 100 m.y. The steep temperature gradient indicates, however, that conductive cooling has penetrated to the center.

Similar calculations predict the critical KBO sizes necessary to reach important thermal benchmarks:

1. Large-scale crystallization of KBO interiors occurs between ~75 and 225 km radius, depending on the details of the conductivity model (and in the absence of ^{26}Al). This is an interesting radius range in that it is the same range in which the color properties of classical KBOs change, i.e., the "cold" classical KBOs are smaller and predominantly red (see discussions in *Barucci et al.,* 2004; *Morbidelli and Brown,* 2004; cf. *Peixinho et al.,* 2004), an effect sometimes attributed to collisional resurfacing because of the color-inclination correlation. More to the point, and more importantly, the discovery of crystalline ice on larger KBOs, for spectra that possess sufficient signal-to-noise (*Cruikshank et al.,* 2007), should come as no surprise. All KBOs greater than a certain size (which depends on their initial endowment of ^{26}Al) should be predominantly if not entirely crystalline (factoring in impact erosion and regolith overturn).

2. Mobilization of ammonia-methanol-water eutectic melt (potential cryolava) occurs for KBOs of R ≥ 300 km. As noted in section 3.1, methanol forms a lowest-melting temperature (or eutectic) melt with water ice at 171 K (at 1 bar). The equivalent ammonia-water peritectic is at 175 K, and the ternary eutectic between ammonia dihydrate, methanol hydrate, and water ice (the hydrates being the predicted stable chemical forms within a KBO) is near ~155 K (*Kargel,* 1992). Mobilization means that the centers of KBOs of the critical size will reach the eutectic temperature.

For larger-sized KBOs, greater internal volumes should exceed the eutectic temperature or temperatures, and the resulting melt, being buoyant with respect to the unmelted water ice and rock that remains, should percolate upward; if the melt collects in a sufficiently large layer or lens, it should be able to "hydrofracture" its way to the surface and erupt as a cryovolcanic sheet (*Stevenson,* 1982). It is tempting to treat the relatively large amounts of methanol ice on the surfaces of Pholus and (55638) 2002 VE_{95} (section 3.4) as evidence of such cryovolcanism, but both bodies are about R ~ 200 km, significantly smaller than the minimum predicted size (and thus melting is not easily achieved by tweaking thermophysical model parameters). On the other hand, it is not obvious that ~10–15% CH_3OH on the surface of a body is evidence of a "chemically primitive nature" (*Barucci et al.,* 2006) either. Perhaps a contribution from ^{26}Al is necessary to bring bodies of the Pholus scale to the required internal temperatures (≈155–170 K, depending on ammonia content).

3. Substantial sublimation of water-ice, sintering, and densification by solid-state creep of the ice grains should set in by ~180 K (at least), temperatures first reached by bodies of R ≥ 350 km. Kuiper belt objects above this size should evolve to eliminate any residual porosity in their warm central regions not previously lost to pressure-densification alone (section 4.3). This process has not yet been incorporated in KBO models yet, but an approach along the lines of *Eluszkiewicz et al.* (1998) using modern compression data (section 4.3) and ice rheology (*Durham and Stern,* 2001) would seem warranted.

4. In models similar to that in Fig. 9, the water-ice melting temperature (≈270 K) is reached for KBOs of R ≥ 425 km. The critical question for large KBO evolution is whether melting and differentiation proceeded at the centers of large KBOs, hundreds of millions of years after they accreted, or whether the internal temperatures were moderated by the onset of solid-state convection in the ice-rock mixture (effectively creating a thermal short circuit). This issue was addressed in detail in *McKinnon et al.* (1997) with regard to Pluto's internal evolution, but those authors did not converge on an answer. Given accumulating evidence that the very largest KBOs are indeed differentiated (section 6), this important issue is also worth reexamining, and for all large, dwarf-planet class (R ≥ 500 km) KBOs. The presence of liquid water may also be key to formation of phyllosilicates and carbonates (sections 3.3 and 3.4) from alteration of crystalline or amorphous mafic precursors, although sustained contact with water ice near its melting point (*Reitmeijer,* 1985; B. Fegley, personal communication) or with the water-bearing cryolavas just described may suffice.

5.5. Conclusions from Thermal Evolution Models

The calculations presented here, both as analytical estimates and as simple or complex numerical simulations, should be taken to indicate evolutionary trends and the results should be regarded more as qualitative than quantitative. For example, for large KBOs the predicted thickness of the primordial, amorphous-ice-rich outer crust depends critically on the conductivity model employed and the intensity (and therefore timing) of radiogenic heat release (compare Plate 6, lower right panel, and Fig. 9), but all models agree that the interiors should be overwhelmingly crystalline.

We also note that mobilization of moderately volatile methanol and ammonia and even water vapor or liquid are all predicted for the larger (>300 km radius) KBOs (section 5.4), even in the absence of early short-lived radiogenic heating. The bodies to the right side of Fig. 3, then, should be thought of as fundamentally endogenically evolved bodies, regardless of the specifics of the particular thermal model. The extent of internal evolution for midsized KBOs, at the left side of Fig. 3 (and discussed in section 5.3), is much more dependent on the details of the thermal model employed, and/or whether ^{26}Al was important.

Indeed, many additional factors may affect the structure of KBOs — among them an uneven shape, rotation, orbital migration, and collisions. It may be difficult, perhaps impossible, for all these factors to be accounted for in a systematic manner. Nevertheless, we may state with a reasonable degree of confidence that, if KBOs experienced radioactive heating, their structure and composition were altered, with considerable loss of volatiles and significant departure from internal homogeneity likely.

6. DWARF PLANET STRUCTURE AND GEOLOGICAL ACTIVITY

For the largest KBOs (perhaps $R \gtrsim 500$ km) we may adopt the premise that accretional and radiogenic heating were likely more than sufficient to have driven these bodies to an advanced state of chemical separation and internal differentiation (and no porosity, except in near-surface impact regoliths or "cryoclastic" volcanic layers). In reality this is an open question, and should be studied further, but it is now a defensible point of view. We noted earlier (in section 4.1) the evidence that the precursor to 2003 EL_{61} was differentiated. To this we may add the refined system parameters for Pluto-Charon, which definitively indicate that Charon has a lower density (1.65 ± 0.06 g cm^{-3}) than Pluto or the Pluto system as a whole (*Buie et al.,* 2006; *Sicardy et al.,* 2006). In terms of the paradigm of Charon's origin by large-scale ("giant") impact of two cosmochemically similar bodies (*Canup,* 2005), this implies that one or both bodies had already differentiated, so that there was a greater contribution from one or both icy mantles to the debris that remained in orbit to form Charon. That Charon is as dense as it is might be taken to indicate that the impactor was undifferentiated, or otherwise its rock core might have merged with the body that became Pluto.

Deep water-ice absorptions in the near-IR might also be taken as evidence for KBO differentiation (or the collisional products of such differentiation), although this interpretation has had a checkered history in its application to icy satellites. Given the amount of rock and dark carbonaceous matter likely present in KBOs, and the low albedos of many of the smaller bodies (*Grundy et al.,* 2005; chapter by Stansberry et al.), this interpretation may in fact be correct [and especially if spectral modeling does not require any carbon black or other "dark neutral absorber" to fit the albedo (e.g., *Trujillo et al.,* 2007)].

If differentiation is assumed, questions such as the presence or absence of icy convection, or internal oceans ("aquaspheres"), or even metallic core formation naturally arise. *Hussmann et al.* (2006) have examined one aspect of ocean formation within KBOs, that of thermal stability. Figure 10 is taken from their work, and shows possible internal configurations for ammonia-bearing oceans within the largest KBOs, for the present epoch. In general the preservation of such oceans is facilitated by the freezing-point depression offered by dissolved ammonia, methanol, or salts; in the presence of solid-state convection in the ice layer (which suppresses ice temperatures), such freezing-point depression may be necessary to allow the ocean at all.

The models in *Hussmann et al.* (2006) are highly simplified (details are provided in the caption to Fig. 10), but the key parameter is ammonia content. For all but Sedna, the initial NH_3 abundance is 0.05 with respect to water. From section 3.1 and Fig. 1, this is much higher than the expected NH_3 abundance in KBOs of ~0.01, and thus is not realistic insofar as we understand cometary abundances. One speculative pathway to provide more ammonia would be for carbonaceous, CHON-like material to be hydrothermally processed in the hydrated, rocky cores of these bodies, which in principle could provide a variety of N-bearing organic or inorganic compounds (*Shock and McKinnon,* 1993). For sufficiently reducing conditions in the core, NH_3 would be the most abundant nitrogenous compound produced. From an icy petrology point of view, methanol behaves much like ammonia, and their combined abundances could be close to 0.05 (section 3.1), so we could alternatively view the models in Fig. 10 as proxies for ammonia- and methanol-bearing oceans.

Another aspect of the models in *Hussmann et al.* (2006) is that they assume all the available "antifreeze" is contained in the ocean of a given KBO, and it is implicit then that the ice shell froze from the top down, from an initially completely molten state. This is certainly justifiable for Triton, given its likely strong, capture-related tidal heating (*McKinnon et al.,* 1995), but the situation for other KBOs is not as favorable. As described in section 5.4, for large KBOs that evolve from an undifferentiated, cold start, the first cryovolcanic events are the eutectic/peritectic melting of ammonia and/or methanol hydrate ices. This melting, at relatively low temperatures, will consume all the locally

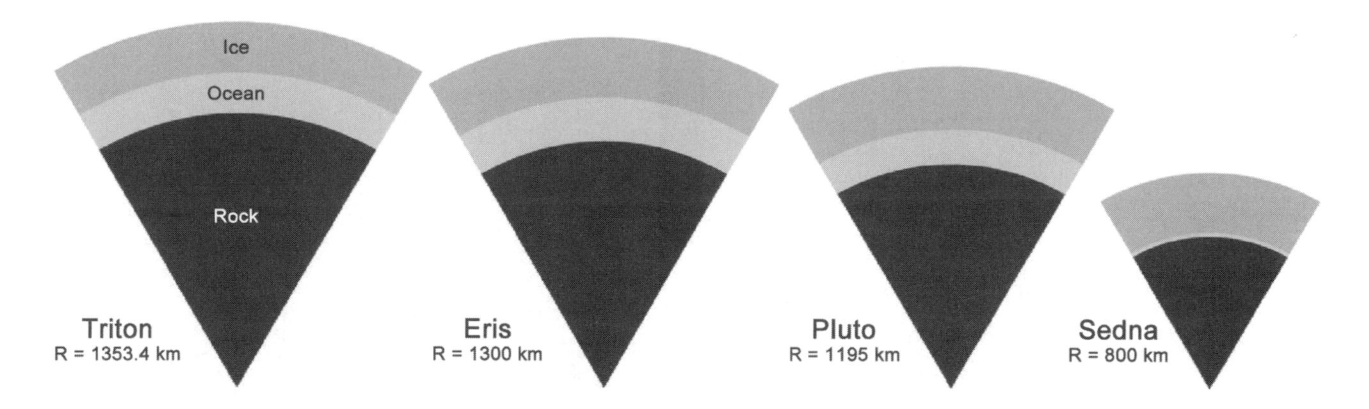

Fig. 10. Interior structure models for dwarf planets and similar worlds. Models are constructed from rock cores (3.5 g cm^{-3}), ice shells, and internal ammonia-water oceans (both 1 g cm^{-3}), with a rock mass fraction of 0.65 assumed for Eris and Sedna, and the ice/ocean boundary determined by thermodynamic equilibrium between the ice and ammonia-bearing ocean for present steady-state heat flows (conductive in the cases shown). The examples shown utilize global NH_3/H_2O ratios of 0.05, except for Sedna, which uses 0.014. Modified from *Hussmann et al.* (2006).

available ammonia and methanol and only a fraction of the water ice (e.g., if only ammonia is present, 5% ammonia implies that ≈10% of the original water ice enters the melt). This melt is very buoyant compared to the rock-ice residuum, and as the body warms the melt should make its way to the colder near-surface regions, eventually erupting to the surface or freezing in subsurface masses (e.g., *Stevenson,* 1982). Rock-from-ice differentiation is not triggered by these early cryovolcanic events, and requires substantially higher temperatures (by ~100 K), but even in this case is likely to have proceeded in a piecemeal fashion as opposed to wholesale (all at once) ice melting. The latter is only possible for large KBOs that are heated much more quickly than conductive or convective losses can compensate, such as by strong, early heating by ^{26}Al (*Busarev et al.,* 2003) or by massive impacts when the progenitor is already close to the water-ice melting point (*McKinnon,* 1989; *Canup,* 2005). Without wholesale ice melting it is not clear how efficiently early ammonia- and/or methanol-rich "crusts" are recycled into deep, oceanic layers.

6.1. Pluto-Class Kuiper Belt Objects

With the above as background, we turn to remarks about individual bodies, and for Pluto and Charon comparisons with models previously published in *McKinnon et al.* (1997). Discovery of Nix and Hydra have allowed a refinement of the position of the Pluto-Charon barycenter, and thus the Pluto/Charon mass ratio (0.1165 ± 0.066) (*Buie et al.,* 2006). The mass of Pluto and Charon are now known to within 0.5% and 5%, respectively, a remarkable improvement over 10 years. The size of Pluto remains uncertain, but its atmosphere was probed down to a radius of ≈1192 km by the stellar occultation of June 12, 2006 (*Elliot et al.,* 2007), and extrapolation of an atmospheric model consistent with vapor-pressure equilibrium with N_2 surface ice gives an upper-limit radius of 1184 km. *Elliot et al.* (2007) note that Pluto's atmosphere could be tens of kilometers deeper, however, and Pluto's radius smaller. For a solid-body radius of, say, 1175 km, Pluto's density is 1.92 g cm^{-3} and its inferred rock mass fraction is ≈0.67 (calculated on an anhydrous basis). Using a smaller radius, such as the 1153 ± 10 km adopted by *Buie et al* (2006), Pluto's density is 2.03 ± 0.06 g cm^{-3} and its inferred dry rock fraction is ≈0.71 ± 0.02.

Bodies of Pluto's size and density mark a transition in KBO structure type; bodies that are smaller (or more rock rich) permit differentiated models of ice I overlying rock cores, whereas larger (or less rock-rich) bodies achieve sufficient pressures at the base of their ice shells (>200 MPa) to require higher-pressure polymorphs of ice at those pressures. This transition pressure also marks the minimum melting point for ice (the liquid-ice I–ice III triple point), and is the natural place for ammonia, methanol, or salty oceans to "perch," between less-dense ice I above and (depending on temperature) denser ice II or III below. If there is enough ammonia or salt, etc., the ocean may be thick enough to extend to the base of the icy layer, and be in direct contact with the rock core as in Fig. 10 [although, technically, *Hussmann et al.* (2006) ignored higher-pressure phases of ice in their models].

Earlier in solar system history, heat flows would have been sufficiently high that the floating ice layers in Fig. 10 would have been much thinner, or they may have been of similar thickness but with convecting lower portions (sublayers), depending on ice viscosity and ocean temperature (cf. *McKinnon,* 2006). Over time radiogenic heating declines, however, and the ocean and sublayer may ultimately cool too far. In this case convection may become subcritical and then shut down altogether (*McKinnon,* 2006). The conductive temperature profile that subsequently establishes implies a thinner ice shell and thicker ocean than before.

Kuiper belt objects the size of Triton are safely above this convective shutdown threshold; due to their relatively

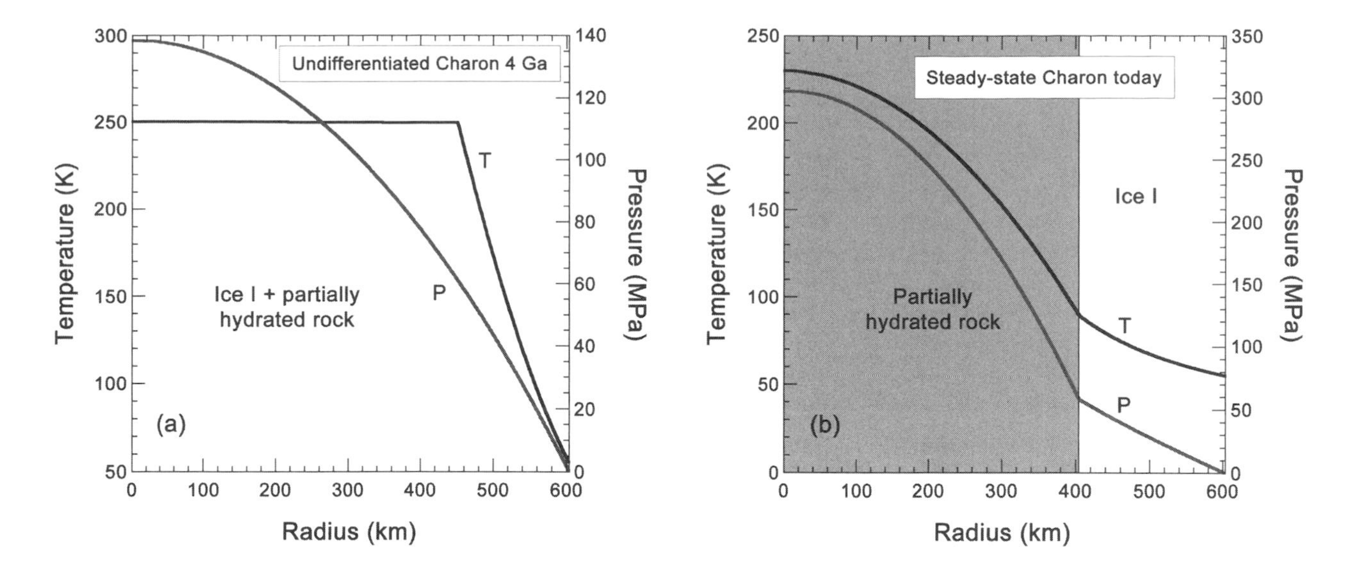

Fig. 11. Temperature and pressure profiles through Charon at two stages of its potential internal evolution [calculated using ICYMOON (*Mueller and McKinnon,* 1988)]. **(a)** Early, undifferentiated Charon approximately 550 m.y. after formation and initial radiogenic heating (cf. Fig. 9). Interior temperatures are moderated by convection (the convective adiabatic temperature shown, ≈250 K, is for illustration), but convection depends on ice viscosity; if the ice is too viscous then convection is inefficient and the ice melts, ultimately resulting in a differentiated Charon. **(b)** A differentiated present-day Charon. A steady-state thermal profile is illustrated, starting at the surface (55 K), and utilizing the thermal conductivities of crystalline ice I in the shell and 3 W m^{-1} K^{-1}, a value typical of igneous and metamorphic rock [such as serpentinite (e.g., *Clauser and Huenges,* 1995)] in the rock core. In these models, Charon today is a cold and geologically inactive body.

elevated heat flows, Pluto-scale KBOs are near the tipping point for convective shutdown. For Pluto itself, the radius limits above imply a rock mass fraction not too far below that of Triton, so it is likely that convection persists (or could persist) today (see *McKinnon,* 2006, for details).

If KBO oceans bear sufficient ammonia or methanol, the physics of convective shutdown may be considerably more complex. In this case, there may be no steady state at shutdown, but an oscillation between a subcritically convecting but thickening shell, and one that is conducting and thinning (*Mitri and Showman,* 2007). The threshold for convective shutdown is in this case a threshold for thermomechanical shell instability, and depends on, e.g., ammonia content, not just body size and the rock/ice ratio.

Ultimately, the ice layers of dwarf planets such as Pluto and Eris, and Triton, could cool sufficiently over solar system history to freeze oceans containing *only* salts (alkali halides, sulfates, and so forth), but cooling to the point of complete freezing of aqueous ammonia or methanol is exceedingly unlikely (if not impossible). If the deep ice shells of these worlds (as opposed to their cold surface layers) contain ammonia or methanol, they almost certainly have oceans.

6.2. Charon

Pluto's moon Charon may not officially be a dwarf planet, but it is larger than Ceres (although not quite as massive), and so we may consider it to be one-half of the first known dwarf planet binary. It is also the first solid body in the solar system where ammonia-ice (probably hydrated) has been spectroscopically identified (*Brown and Calvin,* 2000). Ammonia ice, often taken as the *sine qua non* of icy satellite volcanism, has been sought since the earliest work of J. S. Lewis on the subject (see *Kargel,* 1992), and it is remarkable that it has been found on a KBO, where all other things being equal its abundance should be low (e.g., *Prinn and Fegley,* 1989) (see also section 3.1). An important question is whether the spectroscopic presence of ammonia ice implies recent cryovolcanism on Charon (*Brown and Calvin,* 2000; *Cook et al.,* 2007).

McKinnon et al. (1997) argued that Charon was likely undifferentiated and cold today, but this inference was tempered by the satellite's then uncertain density. Charon's density is now known to within ≈4% (*Buie et al.,* 2006), which implies a dry, solar-composition rock fraction in the 0.53 to 0.57 range, depending on whether Charon is undifferentiated or differentiated, respectively. There is more than enough energy from long-term radiogenic heating alone to bring Charon to the threshold of differentiation (Fig. 11a), but as discussed in *McKinnon et al.* (1997), the issue is whether solid-state convection in the volumetrically ice-rich interior efficiently transports this heat to the near-surface lithosphere (or "stagnant lid," in convective parlance), where it is conducted to the surface and radiated to space. The rock volume percent implied from the mass fraction above (~30%) is substantial, but not quite enough as to demand that the viscosity of primordial Charon ice + rock was too high for efficient convective transport. This depends critically on the ice viscosity (as rock particles remain rigid at the temperatures in question), which in turn depends on ice grain size at the convective stress levels inside icy satellites and KBOs. This is an area of active research for icy satellites (*Barr and McKinnon,* 2007). We can say from Fig. 11a is that Charon's differentiation is a plausible evolutionary outcome. We can also say that even if undifferentiated, Charon became hot enough internally that all the ammonia and methanol within the bulk of the body (≥60% of the volume) should have formed an aqueous melt, which then percolated into the bottom of the stagnant lid, if not erupted to the surface.

Over the long term, however, Charon is a small enough body that it should cool substantially. If it remains undifferentiated it should have become quite cold throughout its interior, notwithstanding its nontrivial rock fraction, undergoing strong global compression as deep ice I converted to more dense ice II. If Charon did differentiate, then the conversion to ice II can be avoided (Fig. 11b), but internal temperatures today should still be modest. In particular, the temperatures in the ice layer should be well below the ammonia-water peritectic temperature of 175 K, even accounting for somewhat elevated heat flows compared with present steady-state values (perhaps by 25%) (*McKinnon et al.,* 1997).

Cook et al. (2007) argue that a very ammonia-rich ice shell ($NH_3/H_2O \approx 0.15$) would be of low enough thermal conductivity to permit an ammonia-rich liquid layer at depth to survive to the present, but this is an extreme compositional choice [essentially the fully equilibrated, solar ammonia abundance, and thus incompatible with the large rock/ice ratios of dwarf planet KBOs (e.g., *Wong et al.,* 2007)]. Rather, we suspect that the lower conductivity of ammonia-water ice (in comparison with water ice) in surface and subsurface cryovolcanic units (*Ross and Kargel,* 1998) only modestly increases internal temperatures, for *likely* ammonia/water abundances (unit thicknesses). As discussed in *McKinnon et al.* (1997), though, the effect of a porous, and thus low-conductivity, regolith may be substantial. Such a low-porosity layer would have to be several tens of kilometers deep (assuming an order of magnitude reduction in conductivity) to sustain an ammonia-rich melt layer (and thus might be termed a megaregolith), but there would be a synergistic effect with deeper water ice — warmer ice is less conductive, which implies comparatively warmer conductive temperature profiles. Unfortunately, internal temperature regulation by icy regoliths, first proposed by *Shoemaker et al.* (1982), has never been adequately quantitatively modeled.

It is probably premature to conclude that the presence of ammonia-water ice on Charon's surface implies recent cryovolcanism. It is, however, reasonable to conclude that there exist ammonia-bearing cryovolcanic units on Charon. These may be continually, spectroscopically exposed by impacts. Implantation of N_2 escaping from Pluto is an alternate explanation for ammonia ice on Charon, but ammonia ice may also exist on Quaoar, for which there is no corresponding external N_2 source (*Jewitt and Luu,* 2004). By implication, this weakens the exogenous N_2 hypothesis for Charon. It would be most valuable to confirm the Quaoar 2.2-μm spectral feature at higher S/N.

It has been argued by *Cook et al.* (2007) that the presence of crystalline, as opposed to amorphous, water ice in Charon's optical surface also implies recent cryovolcanic activity. We will not go into this topic in depth, but we will note, as have others (*Brown and Calvin,* 2000; *Grundy et al.,* 2006; *Jewitt et al.,* 2007; *Trujillo et al.,* 2007), that the crystalline ice feature at 1.65 mm is ubiquitous in the spectra of outer planet satellites and KBOs in which water-ice spectral features can be seen. This includes all five of the midsized satellites of Uranus, none of which have been obviously active in the geologically recent past, and some of which are clearly inactive (heavily cratered), based on Voyager 2 imagery. We reiterate that the issue is not the mere presence of crystalline water ice, as we expect the overwhelming bulk of all larger KBOs to have crystallized over time if they accreted from amorphous ice. Rather it is the expectation that ionizing UV and charged particle bombardment will reamorphotize the optical surface on geological timescales competitive with, say, regolith turnover.

Ultimately, however, nature will answer the question of recent geological activity on Charon, as long as the planned New Horizons encounter with the Pluto system in July 2015 is successful.

6.3. Carbonaceous Caveat

We note in closing that the carbonaceous fraction of KBOs has all but been ignored in this section. *McKinnon et al.* (1997) offered some hypotheses as to how a carbonaceous or graphite-rich layer might affect Pluto's internal structure, but lack of constraints on the chemical or physical makeup and characteristics of such a layer (or its distributed equivalent) have impeded incorporation into either KBO structural models or evolutionary calculations.

7. STILL LARGER BODIES BEYOND NEPTUNE?

The history of KBO discovery has, since the discovery of ~100-km-diameter 1992 QB_1, consistently yielded new, record-setting discoveries of larger and larger bodies. The current (mid-2007) largest KBO is Eris, which has an ~2400 km diameter, some 5–10% larger than Pluto but smaller than 2700-km-diameter Triton, which was formed, before its capture, in heliocentric orbit. It is also simple to calculate, based on brightness alone, that a 10% albedo "Mars" could be detected with current detection capabilities to over 800 AU, and that a 10% albedo "Earth" could be detected to over 1200 AU. The real difficulty with such detections is that such bodies would move very slowly and have very small daily parallaxes at opposition, but such difficulties could be overcome. These various facts beg the question of whether still larger bodies are to be discovered among the KBOs or perhaps beyond, in the Oort cloud. One can approach this topic from any of several directions. As we will see, all lead to generally similar conclusions.

One approach to this question is the "forensic" one taken by *Stern* (1991). In this study, the collisional probabilities of creating the tilts of Uranus and Neptune, of making a Pluto-Charon pair, and of capturing Triton into orbit around Neptune were each found to be quite low, even over the entire age of the solar system. Given that all three of these independent events have probabilities of 10^{-2} to 10^{-4} of having occurred, Stern concluded that there was strong circumstantial evidence for a substantial former population of bodies that once populated the Uranus-Neptune and KBO regions, with sizes ranging from objects just large enough to create Charon by a "giant" impact with Pluto, to Tritons, to Earth-mass bodies capable of tilting Uranus and Neptune. Such bodies, Stern concluded, would have been removed in the clearing of this region after the formation of Uranus and Neptune, and are now to be found in the Kuiper belt or its distant extensions, the Oort cloud, or in interstellar space.

Another approach is based on the results of KBO accretion studies that were initiated some years later after the discovery of KBOs. Indeed, a series of such studies, each with increasing fidelity (e.g., *Stern,* 1996; *Stern and Colwell,* 1997; *Kenyon and Luu,* 1999; *Kenyon,* 2002), have all shown that the large KBOs observed today must have been grown in an environment with considerably (e.g., 100–300 times) higher surface mass density than the current-day Kuiper belt. These studies have also consistently predicted growth times for Pluto-sized bodies of 50–100 m.y., depending on starting assumptions (or less if accretion takes place much closer to the Sun; see section 2 and chapter by Kenyon et al.). Most importantly, however, these studies also demonstrated that such an environment could grow still much larger bodies, at least up to Mars (Earth) size, if left *dynamically undisturbed* for additional amounts of time, i.e., ~200 m.y. (1 G.y.). Given this information, and the existence of Triton (with twice Pluto's mass), one thus easily concludes that unless the accretion process was efficiently aborted owing to some dynamical removal of the growing bodies from the feedstock just *as bodies the scale of Pluto, Eris, and Triton were reached,* then larger bodies were grown and have since been removed to greater distances by some later dynamical clearing process. It has in fact been postulated that just such a large, Mars-sized body, in a more distant orbit, could have been responsible for the "edge" to the Kuiper belt (*Brunini and Melita,* 2002).

There is likely a limit, however, to how large bodies in the Kuiper belt might have grown. *Gomes et al.* (2004) concluded, based on numerical simulations of theirs and in *Morbidelli et al.* (2001), that Mars-mass embryos or larger could not have existed in the planetary disk beyond Neptune unless the disk was truncated near the present position of Neptune, at ~30 AU. And even in this case only a single embryo is permitted, one that was ultimately scattered or lost. The problem with extended disks, and multiple embryos, is that one or more of the embryos should still be in the remnant disk (Kuiper belt) beyond Neptune (and we don't see them), because embryos are massive enough to migrate independently through the disk (*Gomes et al.,* 2004), and if there were more than one, interact with each other so as to dynamically detach themselves from Neptune (*Mordibelli et al.,* 2001). That is, such embryos are not simply low-mass KBOs, subject to resonant "catch and release" as Neptune moves outward.

Yet another approach to the question of large objects is based principally on dynamical arguments. In the "Nice model" of outer planet evolution (e.g., *Tsiganis et al.,* 2005; chapter by Morbidelli et al.), Uranus and Neptune were suddenly evolved, after a long interval of dynamical stability, from their original compact orbits between 10 and 20 AU in a dynamical scattering event that transported them to various farther-removed orbits. This transport, combined with the later clearing of the current Uranus-Neptune zone, would then have presumably emplaced numerous bodies with a size spectrum reaching upward from KBOs, in the deep outer solar system beyond the zone that Uranus and Neptune could clear.

Now, in the Nice model, the upper limit on the size of bodies scattered into the Kuiper belt region (or farther afield) is unclear, but a large (approximately Mars-sized or greater) embryo, following the above discussion, even if scattered to the Oort cloud, might have upset the resonant structure of the Kuiper belt in the process of being scattered (H. Levison, personal communication, 2007). On the other hand, the obliquities of Uranus and Neptune do argue for interaction with Earth-mass-scale embryos (*Stern,* 1991). The Nice model starts with a very compact solar system, in terms of the giant planets. There is not much room in the model between Saturn and the outer planetary disk to fit more Earth-mass or greater planetary embryos and still maintain stability for hundreds of millions of years. *Levison et al.* (2007a) in particular argue that all four giant planets essentially reached their full masses during the lifetime of the solar nebula ($\lesssim$10 m.y.); the nebula provides the needed stability and implies that any Earth-sized embryos or additional giant-planet cores that were scattered out did so during this early epoch. If they are to be found, they probably lie in the Oort cloud or in the extended scattered disk as "detached objects" ("super-Sednas").

As stated at the beginning of this section, all three of these approaches yield consonant results, i.e., one should not be surprised to see the pantheon of record-setting size discoveries in the Kuiper belt and Oort cloud continue for some time. Indeed, it is possible to conclude that objects the diameter of the Moon, Mars, and perhaps even Earth or larger may await our discovery, although the largest-mass bodies are most likely to be found as dynamically detached objects or in the Oort cloud. We eagerly anticipate the discoveries that will come with broader and deeper studies of the population of the deep outer solar system.

8. DISCUSSION AND OUTLOOK

McKinnon (2002) strongly criticized the often unstated assumptions in cometary models incorporating ^{26}Al, in particular highlighting the problematic issue of accretion timescale, and wondering whether such strong heating as ^{26}Al might supply is consistent with the retention of volatiles such as CO in comets or other evidence for the maintenance of low temperatures within cometary nuclei. Indeed, *Prialnik et al.* (1987) had already argued the amount of ^{26}Al originally incorporated in comets must have been limited ($X_0 \leq 4 \times 10^{-9}$) in order to preserve amorphous water ice. Furthermore, if comets and KBOs had heated to the point of water-ice melting and differentiation, where was the evidence of that differentiation? Where was the evidence of water ice mantles or crusts on KBOs and Centaurs? Given collisional processing (see *Farinella et al.,* 2000), where were the more-or-less pure ice comets? *McKinnon* (2002) noted that ^{26}Al was live and important in the inner and main asteroid belt, but asked whether the C, P, and D asteroids and Ceres, which presumably accreted more rapidly than KBOs, were differentiated as well.

Many of these questions are in the process of being answered. Accretion timescales have come down, but as discussed in section 2, there is still a wide range of model values. Whether ^{26}Al was live and important in the Kuiper belt is, from a dynamical point of view, still uncertain. What is clear, though, is that accretion and ^{26}Al decay can coexist in such as way as to provide both thoroughly altered and relatively pristine comets and KBOs (section 2.3) (see also Fig. 5 in *Merk and Prialnik,* 2006). Spectroscopic evidence of water-ice surfaces on several KBOs now exists (section 3.4). Furthermore, the first spectroscopic evidence for possible hydrated silicates on a D-type asteroid has been reported (*Kanno et al.,* 2003), and aqueous alteration is inferred for both Comets Halley and Tempel 1 (section 3.3) and several KBOs (section 3.4). Even Ceres is now thought to be differentiated (from its shape and density), with a rock core, ice mantle, and possible subsurface ocean (*Thomas et al.,* 2005). This does not imply that early heating by ^{26}Al decay was necessarily responsible, but it does mean that models of asteroid accretion (*McSween et al.,* 2002) do not have to avoid early melting and differentiation of Ceres. To this we might add that the D-type asteroids that dominate the outer asteroid belt and Trojan clouds could actually be refugees from the Kuiper belt (*Morbidelli et al.,* 2005).

The question remains, however, whether live ^{26}Al and ^{60}Fe were *required* during accretion of the Kuiper belt, i.e., what observations today demand that early radiogenic heating was important? We will come back to this question at several points below.

The surface colors and compositional variation classes of KBOs, such as derived by *Barucci et al.* (2005a), are usually discussed in terms of "nature vs. nurture," i.e., in terms of intrinsic, primordial compositional variations or different and diverse degrees of surface processing (*Cruikshank et al.,* 2007). Specifically, the redness of the low-inclination classical KBO population may indicate the dynamical uniqueness and isolation of this group, or it may indicate that the group is somehow protected from or less affected by "collisional resurfacing." The latter refers to the hypothesis of *Luu and Jewitt* (1996) and *Jewitt and Luu* (2001), in which continuous radiolytic reddening competes with sporadic impacts bringing grayer material to the surface. Collisional resurfacing explicitly assumes that all KBOs are born equal, and implicitly assumes that (1) KBO interiors and subsurface layers remain primordial; (2) the primordial mixture of rock, water ice, refractory organics, and volatile ices and organics is spectrally neutral in the visible and near-IR; and (3) this primordial mixture only reddens under the influence of solar wind and UV and cosmic-ray bombardment, as new, tholin-like organic compounds are created from simpler precursors and organic compounds devolatilize and evolve to higher carbon fractions.

One problem with this hypothesis, which is otherwise attractive in terms of its simplicity (Occam's principle), is that long-term radiolysis should completely carbonize the organic fraction, as well as preferentially sputter away water ice, rendering old surfaces black (spectrally gray and featureless) (e.g., *Moroz et al.,* 2004). This and the lack of rotational color/spectral variations among individual KBOs or consistent color/spectral correlations with either KBO size or mean collisional speed (see, e.g., *Thébault and Doressoundiram,* 2003; *Peixinho et al.,* 2004; *Delsanti et al.,* 2006; *Cruikshank et al.,* 2007) imply that the collisonal resurfacing hypothesis is, by itself, insufficent.

Delsanti et al. (2004) added outgassing to the mix of processes that might affect KBO surfaces, considering that KBOs might respond as comets to external stimuli. For their modeled KBOs, though, only the sublimation of free CO ice was considered (as long as the bodies remain in the Kuiper belt), and then only when impacts exposed buried, CO-ice-bearing, primordial KBO ice plus rock and carbonaceous matter. Of course, CO ice is "supervolatile" throughout KBO interiors for internal temperatures appropriate to semimajor axes ≤50 AU and low albedos, i.e., $T \geq 40$ K, so it is unclear why removal of meter-scale surface radiation mantles by impact (as in the model of *Delsanti et al.,* 2004) should be a prerequisite for outgassing. Outgassing or, more generally, internal activity (especially past internal activity) should nevertheless be considered as potentially important for KBO surfaces. Even within the "nature" of KBOs, there may be substantial distinctions between KBOs as they are born, and how they evolve or mature.

The broad range of albedos now found for KBOs (*Grundy et al.,* 2005; chapter by Stansberry et al.) do not obviously or easily fit within the primordial vs. radiation mantle paradigm of *Luu and Jewitt* (1996). That all large KBOs [$R \geq 350$ km in the statistical analysis of *Barucci et al.* (2006) (see also the chapter by Barucci et al.)] show the presence of H_2O, CH_4, or N_2 ice spectral features only reinforces this point. While the presence of CH_4 or N_2 might be as much a question of retention as of outgassing (*Schaller and Brown,* 2007), deep water ice absorptions would seem to require some differentiation or separation of water ice from the (presumably low-albedo) primordial composition. When we add to that the body of evidence reviewed in section 4 for a range of intrinsic solid KBOs densities, from rock-dominated 2003 EL_{61} to somewhat smaller, porous, and almost certainly ice-rich KBOs, then the evidence for different KBO compositions, if not compositional classes, becomes almost undeniable.

Precise definition of KBO compositional classes (taxonomy) awaits more and higher-quality spectral and albedo measurements. In this regard, it is probably useful to examine the compositional and taxonomic structure of the asteroid belt, the closest analog to the Kuiper belt in our solar system. The major classes of asteroids, types S, M, C, P, and D, owe their existence to three major factors (e.g., *McSween et al.,* 2002). The first is composition, or more specifically ice content (and to a degree, carbon content), which was controlled by distance from the proto-Sun through the condensation temperature of water ice (and the vaporization temperature of carbonaceous matter). S- and metallic M-type asteroids are thought to be nearly anhydrous, while increasing amounts of water and water ice are thought to have been incorporated into the C-, P-, and D-type asteroids (based on spectra and study of meteoritic analogs). The second major factor is early heating and differentiation. Many asteroids of the inner belt were melted and differentiated, most likely via heating by ^{26}Al and ^{60}Fe decay, based on radiometric dating of igneous meteorites, whereas the heating that drove the aqueous alteration of the C-type asteroids (based on carbonaceous chondrites) was not as intense, and ostensibly did not even lead to ice melting among the more distant D types (*Rivkin et al.,* 2002). The leading theory for this spatial localization is the decrease in accretion rate with distance from the Sun, limiting the effects of early radiogenic heating for later accreting asteroids. The third major factor was collisional evolution, for it is the major collisions (once Jupiter formed and dynamically stirred up the asteroid belt) that broke apart the differentiated rock + metal asteroids, and exposed the iron cores we recognize as metallic M types as well as some subsets of the S-type asteroids (the sources of iron and pallasite meteorites, respectively).

With regard to the Kuiper belt, there does not seem to be any obvious primordial compositional gradient expected, even across as great a range of accretional distances as 20–50 AU. Models of the solar nebula generally predict very cold temperatures at these distances, consistent with complete accretion of rock, amorphous water ice (and absorbed volatile ices), carbonaceous matter, volatile organics, and moderately volatile free ices (such as CO_2). The only compositional gradient we might expect, for a sufficiently cold nebula, would be increasing amounts of independently condensing volatile ices (such as CO and N_2) with greater solar distance. The effects of insolation (once the nebula clears) and any radiogenic heating (early or long-term) all but guarantee that these very volatile gases should be outgassed and lost (section 5), though, except from the largest and most distant KBOs, which can retain them as surface ice deposits.

Evolutionary changes within KBOs are also expected. These depend on body size and, for early radiogenic heating, on formation distance through the accretion rate. The actual effects and patterns predicted depend strongly on accretion rate and on the bulk rock content, which as described in section 5 are simply not that well known or constrained at present. As an example, all KBOs large and small may have crystallized their amorphous water ice (except in thin outer layers), releasing their occluded gases. CO and N_2 may be lost to outgassing, but less-volatile CO_2 (and possibly CH_4) may be retained in the cold outer layers. Or, possibly, only those KBOs originally formed within, e.g., 30 AU (*Merk and Prialnik,* 2006), or KBOs above a certain size (*McKinnon,* 2002), have largely crystallized. If so, uncrystallized KBOs might be the closest to primordial in composition, accounting for reddest spectra (e.g., class RR in *Barucci et al.,* 2005a), while primordial layers have been impact eroded from the rest, exposing undifferentiated but essentially permanently spectrally gray ice-plus-rock-plus-organics. This interpretation implies that the "ultra-red matter" of *Jewitt* (2002) is intrinsic to primordial KBO material (radiolytic alteration may also be necessary), but requires that something important is lost during crystallization and degassing. The state of water ice crystallinity in the very red (RR) and very blue (BB) spectral classes (*Barucci et al.,* 2005a) are important tests of this interpretation. The preferential redness of the largest Plutinos (*Peixinho et al.,* 2004), in contrast, presents a puzzle.

Greater degrees of internal heating should have mobilized other moderately volatile ice formers, such as methanol and ammonia, within KBOs. These would then concentrate in colder near-surface or surface regions. Ice rich in CH_4, methanol, or other moderately volatile organics would presumably strongly redden when bombarded by charged particles or UV radiation (e.g., Pholus and 2002 VE_{95}). Indeed, the gray, spectrally neutral surfaces of many KBOs may require that internal evolution progress to this cryovolcanic stage. That is, crystallization of amorphous ice and release and loss of occluded gases may not be sufficient for the residuum to resist radiation reddening if methanol ice and other moderately volatile organic ices remain. It is difficult if not impossible to raise the internal temperatures of smaller KBOs ($R \leq 200$ km) to the threshold of methanol mobilization (~150–170 K) by long-term radiogenic heating alone (*McKinnon,* 2002), and even in this case the colder, unaffected surface layer or shell may be thick and thus not easily impact eroded or stripped. This may be an instance where interpretation of KBO spectra in terms of internal evolution (heating) does require modest early heating by ^{26}Al.

A major milestone in the thermal evolution of any KBO would be water-ice melting and formation of a rocky core by differentiation. We now have evidence that two of the largest, dwarf-planet-class KBOs, Pluto and 2003 EL_{61}, are differentiated (section 6). The evidence in the case of 2003 EL_{61} seems particularly compelling. For bodies of this scale, either short- or long-term radiogenic heating is probably sufficient for full rock-from-ice differentiation, given the cosmogonic rock-rich nature (or more precisely, ice poor nature) of KBOs.

Interestingly, the Pluto and EL_{61} systems may imply different collisional environments. The formation of Pluto's relatively massive moon, Charon, may require a relatively low-velocity encounter, one characteristic of the accretional era in the Kuiper belt. On the other hand, the extreme impact spin-up of 2003 EL_{61}, plus the shattering and dispersal of much its icy mantle to form the E_{61} family, argues for a very violent, relatively high-velocity event, one consistent with the orbital expansion of Neptune and the dynamical sculpting of the Kuiper belt. These inferences may prove critically important. If it can be conclusively shown (from modeling) that one or both bodies in the Charon-forming impact were differentiated prior to the impact, then heating by decay of ^{26}Al (and ^{60}Fe) would most likely have been the cause, because heating by the decay of U, Th, and ^{40}Ar over the first ~100 m.y. of solar system history is comparatively modest.

The cases of Pluto-Charon and 2003 EL_{61} also illustrate the importance of collisions and dynamical environment, as in the asteroid belt. The shattering and dispersal of previously differentiated KBOs by impact should be the principal mechanism for creating rock-rich and rock-poor KBOs [tidal stripping during close passages to Neptune may also contribute (cf. *Asphaug et al.,* 2006)]. Such catastrophic disruptions are a pathway to create KBOs (and comets) that are essentially pure ice in composition. [And the water ice will already be crystalline; there is no need to invoke shock effects (*Merlin et al.,* 2007).] The fate of primordial and volatile-enriched surface layers is less clear, but presumably some of these survive on their parent KBOs (much as basalt survives on Vesta) or on major fragments thereof. Most material collisionally stripped in the ultimately erosive Kuiper belt environment probably ended up in the collisional cascade of small KBOs, and ultimately pulverized or expelled as comets.

The fate of refractory carbonaceous matter within thermally and structurally evolving KBOs is also unclear. If such material can differentiate from rock upon heating, it may form a separate carbonaceous structural layer within the largest KBOs (see Fig. 4 in *McKinnon et al.,* 1997). Such layers may be revealed by collisional disruption. Alternatively, refractory carbonaceous matter may remain bound up with core-forming rock, and subject to its thermal history. With devolatilization, it may ultimately crystallize into anthracite-like rock, or even graphite. All of this is speculative of course, and outside the bounds of practical constraint, but the presence of abundant, highly thermally conductive graphite in the cores of "giant" KBOs and dwarf planets would be important to their thermal histories.

As for the future, studies of the solar system beyond Neptune are advancing rapidly. Much will become clearer about the formation and evolution of the Kuiper belt and KBOs in the next few years. But it is the constraints that observations and interpretations provide on theoretical models, such as discussed in this chapter, that should prove most important, for nature's imagination exceeds our own.

Acknowledgments. We thank the editors, and especially General Editor R. P. Binzel, for their forbearance, the reviewers for their comments, K. Singer for technical assistance, H. Levison for nth-hour discussions of the Nice model, and V. Bellini for pasta Norma. W.B.M.'s research is supported by grants from NASA's Planetary Geology and Geophysics and Outer Planets Research programs and by the New Horizons mission to Pluto and beyond. D.P. acknowledges support from the Israeli Science Foundation, through grant 942/04.

REFERENCES

Adams F. C., Hollenbach D., Laughlin G., and Gorti U. (2004) Photoevaporation of circumstellar disks due to external far-ultraviolet radiation in stellar aggregates. *Astrophys. J., 611,* 360–379.

Agnor C. B. and Hamilton D. P. (2006) Neptune's capture of its moon Triton in a binary-planet gravitational encounter. *Nature, 192,* 192–194.

Amelin Y., Krot A. N., Hutcheon I. D., and Ulyanov A. (2002) Lead isotopic ages of chondrules and calcium-aluminum-rich inclusions. *Science, 297,* 1678–1683.

Asphaug E. and Benz W. (1996) Size, density, and structure of Comet Shoemaker-Levy 9 inferred from the physics of tidal breakup. *Icarus, 121,* 225–248.

Asphaug E., Agnor C. B., and Williams Q. (2006) Hit-and-run planetary collisions. *Nature, 439,* 155–160.

Bar-Nun A., Dror J., Kochavi E., and Laufer D. (1987) Amorphous water ice and its ability to trap gases. *Phys. Rev. B, 35,* 2427–2435.

Bar-Nun A., Kleinfeld I., and Kochavi E. (1988) Trapping of gas mixtures by amorphous water ice. *Phys. Rev. B, 38,* 7749–7754.

Barr A. C. and McKinnon W. B. (2007) Convection in ice I shells and mantles with self-consistent grain size. *J. Geophys. Res., 112,* E02012, DOI: 10.1029/2006JE002781.

Barucci M. A. and Peixinho N. (2006) Trans-neptunian objects' surface properties. In *Asteroids, Comets, Meteors 2005* (D. Lazzaro et al., eds.), pp. 171–190. IAU Symposium No. 229, Cambridge Univ., Cambridge.

Barucci M. A., Doressoundiram A., and Cruikshank D. P. (2004) Surface characteristics of transneptunian objects and centaurs from photometry and spectroscopy. In *Comets II* (M. C. Festou et al., eds.), pp. 647–658. Univ. of Arizona, Tucson.

Barucci M. A., Belskaya I. N., Fulchignoni M., and Birlan M. (2005a) Taxonomy of Centaurs and trans-neptunian objects. *Astron. J., 130,* 1291–1298.

Barucci M. A., Cruikshank D. P., Dotto E., Merlin F., Poulet F., Dalle Ore C., Fornasier S., and de Bergh C. (2005b) Is Sedna another Triton? *Astron. Astrophys., 439,* L1–L4.

Barucci M. A., Merlin F., Dotto E., Doressoundiram A., and de Bergh C. (2006) Observations of the TNO 55638 (2002 VE_{95}) and analyses of the population's spectral properties. *Astron. Astrophys., 455,* 725–730.

Bernstein G. M., Trilling D. E., Allen R. L., Brown M. E., Holman M., and Malhotra R. (2004) The size distribution of trans-neptunian bodies. *Astron. J., 128,* 1364–1390.

Bizzarro M., Ulfbeck D., Trinquier A., Thrane K., Connelly J. N., and Meyer B. S. (2007) Evidence for a late supernova injection of ^{60}Fe into the protoplanetary disk. *Science, 318,* 1178–1181.

Blake D., Allamandola L., Sandford S., Hudgins D., and Freund F. (1991) Clathrate hydrate formation in amorphous cometary ice analogs in vacuo. *Science, 254,* 548–574.

Blum J., Schräpler R., Davidsson B. J. R., and Trigo-Rodríguez J. M. (2006) The physics of protoplanetesimal dust agglomerates. I. Mechanical properties and relations to primitive bodies in the solar system. *Astrophys. J., 652,* 1768–1781.

Bockelée-Morvan D., Crovisier J., Mumma M. J., and Weaver H. A. (2004) The composition of cometary volatiles. In *Com-*

ets II (M. C. Festou et al., eds.), pp. 391–423. Univ. of Arizona, Tucson.

Britt D. T., Yoemans D., Housen K., and Consolmagno G. (2002) Asteroid density, porosity, and structure. In *Asteroids III* (W. F. Bottke Jr. et al., eds.), pp. 485–500. Univ. of Arizona, Tucson.

Brown M. E. and Calvin W. M. (2000) Evidence for crystalline water and ammonia ices on Pluto's satellite Charon. *Science, 287,* 107–109.

Brown M. E. and Schaller E. L. (2007) The mass of dwarf planet Eris. *Science, 316,* 1585.

Brown M. E. and Trujillo C. A. (2004) Direct measurement of the size of the large Kuiper belt object (50000) Quaoar. *Astron. J., 127,* 2413–2417.

Brown M. E., Barkume K. M., Ragozzine, and Schaller E. L. (2007a) A collisional family of icy objects in the Kuiper belt. *Nature, 446,* 294–296.

Brown M. E., Barkume K. M., Blake G. A., Schaller E. L., Rabinowitz D. L., Roe H. G., and Trujillo C. A. (2007b) Methane and ethane on the bright Kuiper belt object 2005 FY_9. *Astron. J., 133,* 284–289.

Brownlee D. and 183 colleagues (2006) Comet 81P/Wild 2 under a microscope. *Science, 314,* 1711–1719.

Brunini A. and Melita M. D. (2002) The existence of a planet beyond 50 AU and the distribution of the classical Edgeworth-Kuiper belt objects. *Icarus, 160,* 32–43.

Buie M. W., Grundy W. M., Young E. F., Young L. A., and Stern S. A. (2006) Orbits and photometry of Pluto's satellites: Charon, S/2005 P1, and S/2005 P2. *Astron. J., 132,* 290–298.

Busarev V. V., Dorofeeva V. A., and Makalkin A. B. (2003) Hydrated silicates on Edgeworth-Kuiper objects — Probable ways of formation. *Earth Moon Planets, 92,* 345–357.

Canup R. (2005) A giant impact origin of Pluto-Charon. *Science, 307,* 546–550.

Carrier W. D., Olhoeft G. R., and Mendell W. (1991) Physical properties of the lunar surface. In *Lunar Sourcebook, A User's Guide to the Moon* (G. Heiken et al., eds.), pp. 475–594. Cambridge Univ., New York.

Choi Y.-J., Cohen M., Merk R., and Prialnik D. (2002) Long-term evolution of objects in the Kuiper belt zone — Effects of insolation and radiogenic heating. *Icarus, 160,* 300–312.

Clauser C. and Huenges E. (1995) Thermal conductivity of rocks and minerals. In *Rock Physics and Phase Relations, A Handbook of Physical Constants* (T. J. Ahrens, ed.), pp. 105–126. AGU Reference Shelf 3, Washington, DC.

Cook J. C., Desch S. J., Roush T., Trujillo C., and Geballe T. (2007) Near-infrared spectroscopy of Charon: Possible evidence for cryovolcanism on Kuiper belt objects. *Astrophys. J., 663,* 1406–1419.

Crovisier J. (2006) The molecular composition of comets and its interrelation with other small bodies of the solar system. In *Asteroids, Comets, Meteors 2005* (D. Lazzaro et al., eds.), pp. 133–152. IAU Symposium No. 229, Cambridge Univ., Cambridge.

Cruikshank D. P. (2005) Triton, Pluto, Centaurs, and trans-neptunian bodies. *Space Sci. Rev., 116,* 421–439.

Cruikshank D. P., Pilcher C. B., and Morrison D. (1976) Pluto: Evidence for methane frost. *Science, 194,* 835–837.

Cruikshank D. P. and 14 colleagues (1998) The composition of Centaur 5145 Pholus. *Icarus, 135,* 389–407.

Cruikshank D. L., Mason R. E., Dalle Ore C. M., Bernstein M. P., Quirico E., Mastrapa R. M., Emery J. P., and Owen T. C. (2006) Ethane on Pluto and Triton. *Bull. Am. Astron. Soc., 38,* #21.03.

Cruikshank D. P., Barucci M. E., Emery J. P., Fernández Y. R., Grundy W. M., Noll K. A., and Stansberry J. A. (2007) Physical properties of transneptunian objects. In *Protostars and Planets V* (B. Reipurth et al., eds.), pp. 879–893. Univ. of Arizona, Tucson.

D'Alessio P., Calvet N., Hartmann L., Lizano S., and Canto J. (1999) Accretion disks around young objects. II. Tests of well-mixed models with ISM dust. *Astrophys. J., 527,* 893–909.

D'Alessio P., Calvet N., and Hartmann L. (2001) Accretion disks around young objects. III. Grain growth. *Astrophys. J., 553,* 321–334.

Davies J. K., Roush T. L., Cruikshank D. P., Bartholomew M. J., Geballe T. R., Owen T., and de Bergh C. (1997) The detection of water ice in Comet Hale-Bopp. *Icarus, 127,* 238–245.

de Bergh C., Boehnhardt H., Barucci M. A., Lazzarin M., Fornasier S., Romon-Martin J., Tozzi G. P., Doressoundiram A., and Dotto E. (2004) Aqueous altered silicates at the surface of two Plutinos? *Astron. Astrophys., 416,* 791–798.

Dello Russo N., Vervack R. J., Weaver H. A., Biver N., Bockelée-Morvan D., Crovisier J., and Lisse C. M. (2007) Compositional homogeneity in the fragmented Comet 73P/Schwassmann-Wachmann 3. *Nature, 448,* 172–175.

Delsanti A., Hainaut O., Jourdeuil E., Meech K. J., Boehnhardt H., and Barrera L. (2004) Simultaneous visible-near IR photometric study of Kuiper belt objects surfaces with the ESO/Very Large Telescopes. *Astron. Astrophys., 417,* 1145–1158.

Delsanti A., Peixinho N., Boehnhardt H., Barruci A., Merlin F., Doressoundiram A., and Davies J. K. (2006) Near-infrared color properties of Kuiper belt objects and Centaurs: Final results from the ESO Large Program. *Astron. J., 131,* 1851–1863.

De Sanctis M. C., Capria M. T., and Coradini A. (2001) Thermal evolution and differentiation of Edgeworth-Kuiper belt objects. *Astron. J., 121,* 2792–2799.

Diehl R. and 10 colleagues (1997) Models for COMPTEL ^{26}Al data. In *Proceedings of the Fourth Compton Symposium* (C. D. Dermer et al., eds.), pp. 1114–1118. AIP Conf. Proc. 410, New York.

Dumas C., Terrile R. J., Brown R. H., Schneider G., and Smith B. A. (2001) Hubble Space Telescope NICMOS spectroscopy of Charon's leading and trailing hemispheres. *Astron. J., 112,* 1163–1170.

Durham W. B. and Stern L. A. (2001) Rheological properties of water ice — Applications to the satellites of the outer planets. *Annu. Rev. Earth Planet. Sci., 29,* 295–330.

Durham W. B., Heard H. C., and Kirby S. H. (1983) Experimental deformation of polycrystalline H_2O ice at high pressure and low temperature: Preliminary results. *J. Geophys. Res., 88,* B377–B392.

Durham W. B., McKinnon W. B., and Stern L. A. (2005) Cold compaction of water ice. *Geophys. Res. Lett., 32,* L18202, DOI: 10.1029/2005GL023484.

Ehrenfreund P., Charnley S. B., and Wooden D. H. (2004) From interstellar material to cometary particles and molecules. In *Comets II* (M. C. Festou et al., eds.), pp. 115–133. Univ. of Arizona, Tucson.

Elliot J. L. and 19 colleagues (2007) Changes in Pluto's atmosphere: 1988–2006. *Astron. J., 134,* 1–13.

Eluszkiewicz J., Leliwa-Kopystyński J., and Kossacki K. J. (1998) Metamorphism of solar system ices. In *Solar System Ices* (B. Schmitt et al., eds.), pp. 119–138. Kluwer, Dordrecht.

Emery J. P., Dalle Ore C. M., Cruikshank D. P., Fernández Y. R., Trilling D. E., and Stansberry J. A. (2007) Ices on (90377)

Sedna: Confirmation and compositional constraints. *Astron. Astrophys., 466,* 395–398.

Farinella P., Davis D. R., and Stern S. A. (2000) Formation and collisional evolution of the Edgeworth-Kuiper belt. In *Protostars and Planets IV* (V. Mannings et al., eds.), pp. 1255–1282. Univ. of Arizona, Tucson.

Fernandez J. A. and Ip W.-H. (1984) Some dynamical aspects of the accretion of Uranus and Neptune — The exchange of orbital angular momentum with planetesimals. *Icarus, 58,* 109–120.

Fomenkova M. N. (1999) On the organic refractory component of cometary dust. *Space Sci. Rev., 90,* 109–114.

Garaud P. and Lin D. N. C. (2007) The effect of internal dissipation and surface irradiation on the structure of disks and the location of the snow line around Sun-like stars. *Astrophys. J., 654,* 606–624.

Ghosh A., Weidenschilling S. J., and McSween H. Y. Jr. (2003) Importance of the accretion process in asteroid thermal evolution: 6 Hebe as an example. *Meteoritics & Planet. Sci., 38,* 711–724.

Gomes R. S, Morbidelli A., and Levison H. F. (2004) Planetary migration in a planetesimal disk: Why did Neptune stop at 30 AU? *Icarus, 170,* 492–507.

Greenberg J. M. (1998) Making a comet nucleus. *Astron. Astrophys., 330,* 375–380.

Grundy W. M., Noll K. S., and Stephens D. C. (2005) Diverse albedos of small trans-neptunian objects. *Icarus, 176,* 184–191.

Grundy W. M., Young L. A., Spencer J. R., Johnson R. E., Young E. F., and Buie M. W. (2006) Distributions of H_2O and CO_2 ices on Ariel, Umbriel, Titania, and Oberon from IRTF/SpeX observations. *Icarus, 184,* 543–555.

Grundy W. M., Stansberry J. A., Noll K. S., Stephens D. C., Trilling D. E., Kern S. D., Spencer J. R., Cruikshank D. P., and Levison H. F. (2007) The orbit, mass, size, albedo, and density of (65489) Ceto/Phorcys: A tidally-evolved binary Centaur. *Icarus, 191,* 286–297.

Haisch K. E., Lada E. A., and Lada C. J. (2001) Disk frequencies and lifetimes in young clusters. *Astrophys. J. Lett., 553,* L153–L156.

Hanner M. S. and Bradley J. P. (2004) Composition and mineralogy of cometary dust. In *Comets II* (M. C. Festou et al., eds.), pp. 555–564. Univ. of Arizona, Tucson.

Haruyama J., Yamamoto T., Mizutani H., and Greenberg J. M. (1993) Thermal history of comets during residence in the Oort cloud: Effect of radiogenic heating in combination with the very low thermal conductivity of amorphous ice. *Geophys. Res., 98,* 15079–15090.

Hersant F., Gautier D., and Lunine J. I. (2004) Enrichment in volatiles in the giant planets of the solar system. *Planet. Space Sci., 52,* 623–641.

Holsapple K. A. (2007) Spin limits of solar system bodies: From the small fast-rotators to 2003 EL61. *Icarus, 107,* 500–509.

Huebner W. F., Benkhoff J., Capria M. T., Coradini A., de Sanctis M. C., Enzian A., Orosei R., and Prialnik D. (1999) Results from the comet nucleus model team at the International Space Science Institute, Bern, Switzerland. *Adv. Space Res., 23,* 1283–1298.

Hussmannn H., Sohl F., and Spohn T. (2006) Subsurface oceans and deep interiors of medium-sized outer planet satellites and large trans-neptunian objects. *Icarus, 185,* 258–273.

Irvine W. M., Leschine S. B., and Schloerb F. P. (1980) Thermal history, chemical composition and relationship of comets to the origin of life. *Nature, 283,* 748–749.

Jessberger E. K., Christoforidis A., and Kissel J. (1988) Aspects of the major element composition of Halley's dust. *Nature, 332,* 691–695.

Jewitt D. C. (2002) From Kuiper belt object to cometary nucleus: The case of the missing ultrared matter. *Astron. J., 123,* 1039–1049.

Jewitt D. C. (2004) From cradle to grave: The rise and demise of comets. In *Comets II* (M. C. Festou et al., eds.), pp. 659–676. Univ. of Arizona, Tucson.

Jewitt D. C. and Luu J. X. (2001) Colors and spectra of Kuiper belt objects. *Astron. J., 122,* 2099–2014.

Jewitt D. C. and Luu J. (2004) Crystalline water ice on the Kuiper belt object (50000) Quaoar. *Nature, 432,* 731–733.

Jewitt D. C. and Sheppard S. S. (2002) Physical properties of trans-neptunian object (20000) Varuna. *Astron. J., 123,* 2110–2120.

Jewitt D., Chizmadia L., Grimm R., and Prialnik D. (2007) Water in the small bodies of the solar system. In *Protostars and Planets V* (B. Reipurth et al., eds.), pp. 863–878. Univ. of Arizona, Tucson.

Kanno A. and 9 colleagues (2003) The first detection of water absorption on a D type asteroid. *Geophys. Res. Lett., 30,* 17, DOI: 10.1029/2003GL017907.

Kargel J. S. (1992) Ammonia-water volcanism on icy satellites: Phase relations at 1 atmosphere. *Icarus, 100,* 556–574.

Kawakita H., Watanabe J.-I., Ootsubo T., Nakamura R., Fuse T., Takano N., Sasaki S., and Sasaki T. (2004) Evidence of icy grains in Comet C/2002 T7 (LINEAR) at 3.52 AU. *Astrophys. J. Lett., 601,* L191–L194.

Kenyon S. J. (2002) Planet formation in the outer solar system. *Publ. Astron. Soc. Pac., 114,* 265–283.

Kenyon S. J. and Luu J. X. (1998) Accretion in the early Kuiper belt. I. Coagulation and velocity evolution. *Astron. J., 115,* 2136–2160.

Kenyon S. J. and Luu J. X. (1999) Accretion in the early Kuiper belt. II. Fragmentation. *Astron. J., 118,* 1101–1119.

Kouchi A. and Sirono S. (2001) Crystallization heat of impure amorphous H_2O ice. *Geophys. Res. Lett., 28,* 827–830.

Lacerda P. and Jewitt D. C. (2007) Densities of solar system objects from their rotational light curves. *Astron. J., 133,* 1393–1408.

Lambe T. W. and Whitman R. V. (1979) *Soil Mechanics, SI Version.* Wiley, New York. 553 pp.

Lamy P. L., Toth I., Fernández Y. R., and Weaver H. A. (2004) The sizes, shapes, albedos, and colors of cometary nuclei. In *Comets II* (M. C. Festou et al., eds.), pp. 213–264. Univ. of Arizona, Tucson.

Lecar M., Podolak M., Sasselov D., and Chaing E. (2006) On the location of the snowline in a protoplanetary disk. *Astrophys. J., 640,* 1115–1118.

Levison H. F. and Morbidelli A. (2003) The formation of the Kuiper belt by the outward transport of bodies during Neptune's migration. *Nature, 426,* 419–421.

Levison H. F., Morbidelli A., Gomes R., and Backman D. (2007a) Planet migration in planetesimal disks. In *Protostars and Planets V* (B. Reipurth et al., eds.), pp. 669–684. Univ. of Arizona, Tucson.

Levison H. F., Morbidelli A, Gomes R., and Tsiganis K. (2007b) Origin of the structure of the Kuiper belt during a dynamical instability in the orbits of Uranus and Neptune. *Icarus,* in press.

Lewis J. S. and Prinn R. G. (1980) Kinetic inhibition of CO and N_2 reduction in the solar nebula. *Astrophys. J., 238,* 357–364.

Licandro J., Grundy W. M., Pinilla-Alonso N., and Leisy P. (2006) Visible spectroscopy of 2003 UB_{313}: Evidence for N_2 ice on the surface of the largest TNO? *Astron. Astrophys., 458,* L5–L8.

Lisse C. M. and 16 colleagues (2006) Spitzer spectral observations of the Deep Impact ejecta. *Science, 313,* 635–640.

Lisse C. M., Kraemer K. E., Nuth J. A., Li. A., and Joswiak D. (2007) Comparison of the composition of the Tempel 1 ejecta to the dust in Comet C/Hale-Bopp 1995 O1 and YSO HD 100546. *Icarus, 187,* 69–86.

Lupishko D. F. (2006) On the bulk density of the M-type asteroid 16 Psyche. *Solar Sys. Res., 40,* 214–218.

Lupo M. J. and Lewis J. S. (1980) Mass-radius relationships and constraints on the composition of Pluto. *Icarus, 42,* 29–34.

Luu J. X. and Jewitt D. C. (1996) Color diversity among the Centaurs and Kuiper belt objects. *Astron. J., 112,* 2310–2318.

Luu J. X. and Jewitt D. C. (2002) Kuiper Belt objects: Relics from the accretion disk of the Sun. *Annu. Rev. Astron. Astrophys., 40,* 63–101.

Lykawka P. and Mukai T. (2005) Higher albedo and size distribution of large transneptunian objects. *Planet. Space Sci., 53,* 1319–1330.

Malhotra R. (1993) The origin of Pluto's peculiar orbit. *Nature, 365,* 819–821.

Malhotra R. (1995) The origin of Pluto's orbit: Implications for the solar system beyond Neptune. *Astron. J., 110,* 420–429.

MacPherson G. J., Davis A. M., and Zinner E. K. (1995) The distribution of aluminum-26 in the early solar system — A reappraisal. *Meteoritics, 30,* 365–386.

McKinnon W. B. (1989) On the origin of the Pluto-Charon binary. *Astrophys. J. Lett., 344,* L41–L44.

McKinnon W. B. (2002) On the initial thermal evolution of Kuiper belt objects. *Proc. Asteroids, Comets, Meteors (ACM 2002)* (B. Warmbein, ed.), pp. 29–38. ESA SP-500, Noordwijk, The Netherlands.

McKinnon W. B (2006) On convective instability in the ice I shells of outer solar system bodies, with detailed application to Callisto. *Icarus, 183,* 435–450.

McKinnon W. B, Lunine J. I., and Banfield D. (1995) Origin and evolution of Triton. In *Neptune and Triton* (D. P. Cruikshank, ed.), pp. 807–877. Univ. of Arizona, Tucson.

McKinnon W. B., Simonelli D., and Schubert G. (1997) Composition, internal structure, and thermal evolution of Pluto and Charon. In *Pluto and Charon* (S. A. Stern and D. J. Tholen, eds.), pp. 295–343. Univ. of Arizona, Tucson.

McSween H. Y., Ghosh A., Grimm R. E., Wilson L., and Young E. D. (2002) Thermal evolution models of asteroids. In *Asteroids III* (W. F. Bottke Jr. et al., eds.), pp. 559–571. Univ. of Arizona, Tucson.

Meech K. J. and Svoreň J. (2004) Using cometary activity to trace the physical and chemical evolution of cometary nuclei. In *Comets II* (M. C. Festou et al., eds.), pp. 317–335. Univ. of Arizona, Tucson.

Meier R., Eberhardt P., Krankowsky D., and Hodges R. R. (1994) Ammonia in comet P/Halley. *Astron. Astrophys., 287,* 268–278.

Mekler Y., Prialnik D., and Podolak M. (1990) Evaporation from a porous comet nucleus. *Astrophys. J., 356,* 682–686.

Merk R. and Prialnik D. (2003) Early thermal and structural evolution of small bodies in the trans-Neptunian zone. *Earth Moon Planets, 92,* 359–374

Merk R. and Prialnik D. (2006) Combined modeling of thermal evolution and accretion of trans-neptunian objects: Occurrence of high temperatures and liquid water. *Icarus, 183,* 283–295.

Merk R., Breuer D., and Spohn T. (2002) Numerical modeling of ^{26}Al-induced radioactive melting of asteroids considering accretion. *Icarus, 159,* 183–191.

Merlin F., Guilbert A., Dumas C., Barucci M. A., de Bergh C., and Vernazza P. (2007) Properties of the icy surface of the TNO 136108 (2003 EL_{61}). *Astron. Astrophys., 466,* 1185–1188.

Mitri G. and Showman A. P. (2007) Thermal convection in ice-I shells of Titan and Enceladus. *Icarus,* in press. DOI: 10.1016/j.icarus.2007.07.016.

Morbidelli A. (2007) Portrait of a suburban family. *Nature, 446,* 273–274.

Morbidelli A. and Brown M. E. (2004) The Kuiper belt and the primordial evolution of the solar system. In *Comets II* (M. C. Festou et al., eds.), pp. 175–191. Univ. of Arizona, Tucson.

Morbidelli A., Jacob C., and Petit J. M. (2001) Planetary embryos never formed in the Kuiper belt. *Icarus, 157,* 241–248.

Morbidelli A., Levison H., Tsiganis K., and Gomes R. (2005) Chaotic capture of Jupiter's Trojan asteroids in the early solar system. *Nature, 435,* 462–465.

Moroz L., Baratta G., Strazzulla G., Starukhina L., Dotto E., Barucci M. A., Arnold G., and Distefano E. (2004) Optical alteration of complex organics induced by ion irradiation: 1. Laboratory experiments suggest unusual space weathering trend. *Icarus, 170,* 214–228.

Mueller S. and McKinnon W. B. (1988). Three-layered models of Ganymede and Callisto: Compositions, structures, and aspects of evolution. *Icarus, 76,* 437–464.

Nakamura R. and 14 colleagues (2000) Subaru infrared spectroscopy of the Pluto-Charon system. *Publ. Astron. Soc. Japan, 52,* 551–556.

Notesco G., Bar-Nun A., and Owen T. C. (2003) Gas trapping in water ice at very low deposition rates and implications for comets. *Icarus, 162,* 183–189.

Peixinho N., Boehnhardt H., Belsakaya I., Doressoundiram A., Barucci M. A., and Delsanti A. (2004) ESO large program on Centaurs and TNOs: Visible colors — final results. *Icarus, 170,* 153–166.

Person M. J., Elliot J. L., Gulbis A. A. S., Pasachoff J. M., Babcock B. A., Souza S. P., and Gangestad J. (2006) Charon's radius and density from the combined data sets of the 2005 July 11 occultation. *Astron. J., 132,* 1575–1580.

Petrenko V. F. and Whitworth R. W. (1999) *Physics of Ice.* Oxford Univ., Oxford. 373 pp.

Poirier J.-P. (2000) *Introduction to the Physics of the Earth's Interior.* Cambridge Univ., Cambridge. 312 pp.

Prialnik D. (1998) Physical characteristics of distant comets. In *Minor Bodies in the Outer Solar System* (A. Fitzsimmons et al., eds.), pp. 33–50. Springer-Verlag, Berlin.

Prialnik D. and Bar-Nun A. (1990) Heating and melting of small icy satellites by the decay of ^{26}Al. *Astrophys. J., 355,* 281–286.

Prialnik D. and Podolak M. (1995) Radioactive heating of porous comet nuclei. *Icarus, 117,* 420–430.

Prialnik D., Bar-Nun A., and Podolak M. (1987) Radiogenic heating of comets by ^{26}Al and implications for their time of formation. *Astrophys. J., 319,* 993–1002.

Prialnik D., Benkhoff J., and Podolak M. (2004) Modeling the structure and activity of comet nuclei. In *Comets II* (M. C. Festou et al., eds.), pp. 359–387. Univ. of Arizona, Tucson.

Prialnik D., Sarid G., Rosenberg E. D., and Merk R. (2007) Thermal and chemical evolution of comet nuclei and Kuiper belt objects. *Space Sci. Rev.,* in press.

Prinn R. G. and Fegley B. Jr. (1989) Solar nebula chemistry: Origin of planetary, satellite, and cometary volatiles. In *Origin and Evolution of Planetary and Satellite Atmospheres* (S. K. Atreya et al., eds.), pp. 78–136. Univ. of Arizona, Tucson.

Rabinowitz D. L., Barkume K., Brown M. E., Roe H., Schwartz M., Tourtellote S., and Trujillo C. (2006) Photometric observations constraining the size, shape, and albedo of 2003 EL61, a rapidly rotating, Pluto-sized object in the Kuiper belt. *Astrophys. J., 639,* 1238–1251.

Reitmeijer F. J. M. (1985) A model for diagenesis in proto-planetary bodies. *Nature, 313,* 293–294.

Richardson J. E., Melosh H. J., Lisse C. M., and Carcich B. (2007) A ballistics analysis of the Deep Impact ejecta plume: Determining comet Tempel 1's gravity, mass, and density. *Icarus, 191,* 176–209.

Rivkin A. S., Howell E. S., Vilas F., and Lebofsky L. A. (2002) Hydrated minerals on asteroids. In *Asteroids III* (W. F. Bottke Jr. et al., eds.), pp. 33–62. Univ. of Arizona, Tucson.

Romanishin W. and Tegler S. C. (2001) Rotation rates of Kuiper-belt objects from their light curves. *Nature, 398,* 129–132.

Ross R. G. and Kargel J. S. (1998) Thermal conductivity of solar system ices, with special reference to martian polar caps. In *Solar System Ices* (B. Schmitt et al., eds.), pp. 33–62. Kluwer, Dordrecht.

Sandford S. and 54 colleagues (2006) Organics captured from Comet 81P/Wild 2 by the Stardust spacecraft. *Science, 314,* 1720–1724.

Schaller E. L. and Brown M. E. (2007) Volatile loss and retention on Kuiper belt objects. *Astrophys. J. Lett., 659,* L61–L64.

Schmitt B., Espinasse S., Grim R. J. A., Greenberg J. M., and Klinger J. (1989) Laboratory studies of cometary ice analogues. In *Physics and Mechanics of Cometary Materials* (J. Hunt and T. D. Guyenne, eds.), pp. 65–69. ESA SP-302, Noordwijk, The Netherlands.

Schulze H., Kissel J., and Jessberger E. K. (1997) Chemistry and mineralogy of Comet Halley's dust. In *From Stardust to Planetesimals* (Y. J. Pendleton and A. G. G. M. Tielens, eds.), pp. 397–414. ASP Conf. Series 22, San Francisco.

Sheppard S. S. and Jewitt D. C. (2002) Time-resolved photometry of Kuiper belt objects: Rotations, shapes, and phase functions. *Astron J., 124,* 1757–1775.

Shock E. L. and McKinnon W. B. (1993) Hydrothermal processing of cometary volatiles — Applications to Triton. *Icarus, 106,* 464–477.

Shoemaker E. S., Lucchitta B. K., Wilhelms D. E., Plescia J. B., and Squyres S. W. (1982) The geology of Ganymede. In *Satellites of Jupiter* (D. Morrison, ed.), pp. 435–520. Univ. of Arizona, Tucson.

Shoshany Y., Prialnik D., and Podolak M. (2002) Monte Carlo modeling of the thermal conductivity of porous cometary ice. *Icarus, 157,* 219–227.

Sicardy B. and 44 colleagues (2006) Charon's size and an upper limit on its atmosphere from a stellar occultation. *Nature, 439,* 52–54.

Spencer J. R., Stansberry J. A., Grundy W. M., and Noll K. S. (2006) A low density for binary Kuiper belt object (26308) 1998 SM_{165}. *Bull. Am. Astron. Soc., 38,* 546.

Stansberry J. A., Grundy W. M., Margot J. L., Cruikshank D. P., Emery J. P., Riecke G. H., and Trilling D. E. (2006) The albedo, size, and density of binary Kuiper belt object (47171) 1999 TC_{36}. *Astrophys. J., 643,* 556–566.

Stern S. A. (1991) On the number of planets in the outer solar system: Evidence of a substantial population of 1000-km bodies. *Icarus, 90,* 271–281.

Stern S. A. (1996) On the collisional environment, accretion time scales, and architecture of the massive, primordial Kuiper belt. *Astron. J., 112,* 1203–1210.

Stern S. A. (2002) Evidence for a collisional mechanism affecting Kuiper belt object colors. *Astron. J., 124,* 2297–2299.

Stern S. A. (2003) The third zone: Exploring the Kuiper belt. *Sky and Telescope, 106,* 31–36.

Stern S. A. and Colwell J. E. (1997) Accretion in the Edgeworth-Kuiper Belt: Forming 100–1000 km radius bodies at 30 AU and beyond. *Astron. J., 114,* 841–849.

Stern S. A. and Levison H. F. (2002) Regarding the criteria for planethood and proposed planetary classification schemes. In *Highlights of Astronomy, Vol. 12* (H. Rickman, ed.), pp. 205–213. Astronomical Society of the Pacific, San Francisco.

Stevenson D. J. (1982) Volcanism and igneous processes in small icy satellites. *Nature, 298,* 142–144.

Thébault P. and Doressoundiram A. (2003) Colors and collision rates within the Kuiper belt: Problems with the collisional resurfacing scenario. *Icarus, 162,* 27–37.

Thomas P. C., Parker J. W., McFadden L. A., Russell C. T., Stern A. S., Sykes M. V., and Young E. F. (2005) Differentiation of the asteroid Ceres as revealed by its shape. *Nature, 437,* 224–226.

Trujillo C. A., Brown M. E., Barkume K. M., Schaller E. L., and Rabinowitz D. L. (2007) The surface of 2003 EL_{61} in the near-infrared. *Astron. J., 655,* 1172–1178.

Tsiganis K., Gomes R., Morbidelli A., and Levison H. F. (2005) Origin of the orbital architecture of the giant planets of the solar system. *Nature, 435,* 459–461.

Urey H. C. (1955) The cosmic abundances of potassium, uranium, and thorium and the heat balances of the Earth, the Moon, and Mars. *Proc. Natl. Acad. Sci., 41,* 127–144.

Verbiscer A. J., Peterson D. E., Skrutskie M. F., Cushing M., Nelson M. J., Smith J. D., and Wilson J. C. (2007) Solid nitrogen and simple hydrocarbons on Charon. *Workshop on Ices, Oceans, and Fire: Satellites of the Outer Solar System,* Abstract #6070. Lunar and Planetary Institute, Houston.

Wallis M. K. (1980) Radiogenic heating of primordial comet interiors. *Nature, 284,* 431–433.

Weidenschilling S. J. (2000) Formation of planetesimals and accretion of the terrestrial planets. *Space Sci. Rev., 92,* 295–310.

Weidenschilling S. W. (2004) From icy grains to comets. In *Comets II* (M. C. Festou et al., eds.), pp. 97–104. Univ. of Arizona, Tucson.

Weissman P. R., Asphaug E., and Lowry S. C. (2004) Structure and density of cometary nuclei. In *Comets II* (M. C. Festou et al., eds.), pp. 335–357. Univ. of Arizona, Tucson.

Whipple F. L. and Stefanik R. P. (1966) On the physics and splittting of comet nuclei. *Mem. R. Soc. Leige (Ser. 5), 12,* 33–52.

Wong M. H., Lunine J. I., Atreya S. K., Johnson T., Mahaffy P. C., Owen T. C., and Encrenaz T. (2007) Oxygen and other volatiles in the giant planets and their satellites. In *Oxygen in Earliest Solar System Materials and Processes* (G. MacPherson and W. Huebner, eds.), in press. Reviews in Mineralogy and Geochemistry, Mineralogical Society of America.

Wooden D., Desch S., Harker D., Gail H.-P., and Keller L. (2007) Comet grains and implications for heating and radial mixing in the protoplanetary disk. In *Protostars and Planets V* (B. Reipurth et al., eds.), pp. 815–833. Univ. of Arizona, Tucson.

Yabushita S. (1993) Thermal evolution of cometary nuclei by radioactive heating and possible formation of organic chemicals. *Mon. Not. R. Astron. Soc., 260,* 819–825.

The Structure of Kuiper Belt Bodies: Link with Comets

A. Coradini
INAF/Istituto di Fisica dello Spazio Interplanetario

M. T. Capria and M. C. De Sanctis
INAF/Istituto di Astrofisica Spaziale e Fisica Cosmica

W. B. McKinnon
Washington University

The population of small bodies of the outer solar system is made up of objects of different kind and type, such as comets, Kuiper belt objects, and Centaurs, all sharing a common characteristic: They are rich in ices and other volatiles. The knowledge of the composition and properties of these bodies would help in understanding the processes that shaped the solar nebula at large heliocentric distances and determined the formation and evolution of the planets. A large number of observational results are now available on these bodies, due to successful space missions and increasingly powerful telescopes, but all our instruments are unable to probe the interiors. However, we are beginning to see how these seemingly different populations are related to each other by dynamical and genetic relationships. In this chapter we try to see what their thermal evolution could be, how it could bring about their internal differentiation, and how it could be affected by the orbital evolution. This is a way to link the surface properties, as probed by instruments, with the internal properties. We note that the comet activity is well interpreted if we assume that the comets are small, fragile, porous, volatile-rich, and low-density objects. This view, despite the strong differences noted in the few comet nuclei observed *in situ*, has not been disproven. On the other hand, the observations of the Kuiper belt objects indicate that it is possible that they are large, probably collisionally evolved objects (*Farinella and Davis,* 1996), perhaps with larger densities. We are now facing a kind of paradox: We have on the one hand the comets, and on the other a population of objects larger and possibly denser. We know that a dynamical link exists between them, but how can we go from one type of population to another? In this chapter the current status of our knowledge on the subject is reviewed, taking into account the results of thermal modeling and the results of observations.

1. INTRODUCTION

In the recent years a tremendous effort has been put forth to understand the origin and evolution of the Kuiper belt objects (KBOs). There are many unknown factors in attempting to identify the link between origin and evolution; however, there is one factor that is widely accepted as fact, namely that these objects are the source of short-period comets. This is not a minor constraint, since, at present, we have developed a noticeable amount of knowledge of comets, thanks to groundbased observations and to both current and past space missions.

Here we will try to identify the link between comets and bodies that are large enough to undergo a nonnegligible amount of differentiation. We will deal with the problem of the relation between KBOs and comets as seen from the point of view of thermal evolution models. What are the characteristics of comets deriving from their previous life in the outer part of the solar system? Is the gradual inward displacement of short-period comets able to totally obliterate their previous history? We are convinced that this is not the case, and we will try to demonstrate this through the reexamination of our previous results.

After the discovery of the first KBO, a large number of KBOs have been directly detected at different and increasing distances, thereby increasing the area of the solar system in which they are found. The orbits of the so-called classical Kuiper belt objects (see the chapter by Morbidelli et al.) fall into two main categories: objects with semimajor axis $a < 41$ AU and $e > 0.1$ (like Pluto and Charon) that are in mean-motion resonances with Neptune, and objects with $41 < a < 50$ AU and $e < 0.1$ (like 1992 QB_1) that are not found in resonant orbits. Another component of the transneptunian region, the so-called scattered disk, has been added in the last few years: These objects are characterized by highly eccentric orbits extending up to ~130 AU and could be planetesimals that were scattered out of the Uranus-Neptune region into eccentric orbits. The Kuiper belt probably extends much farther than we presently know, and some objects could be found all the way to the Oort cloud.

Given the large extension of this region and the different thermodynamic conditions that can be present, it is impossible to exclude some variability in the structure of these bodies, both original (probably different volatile content) and due to different thermal histories (related in turn to the different composition and different sizes of the objects).

Overlapping with these local differences, the effect of impact evolution could have affected the different bodies, modifying their surfaces and contributing to their thermal evolution. Large impacts could have broken the large bodies apart, resulting in the origin of families of objects that are genetically related, but possibly different in composition, if the original body was already differentiated. Could this be the origin of some of the short-period comets? We are not certain, but our present knowledge seems to suggest this genetic relation. If so, the short-period comets could be an important source of knowledge about the KBOs. It cannot be excluded, however, that short-period comets are simply the tail of the original size distribution. The size distribution of KBOs is not well known, but it should be related to the primordial phases of solar system evolution, even if it has been changed by the resonances and the progressive depletion due to different phenomena. The size distribution of a population of objects can be considered a useful diagnostic tool for understanding the processes leading to the erosion and/or accretion of planetary bodies. From recent observations and theoretical studies, it is emerging that objects in the transneptunian region probably follow a complex size distribution (*Gladman et al.,* 2001). As far as the masses (and densities) are concerned, we have few data for the larger objects. Varuna, for which a density has been estimated (*Jewitt and Sheppard,* 2002) using the analysis of its lightcurve, could be a rotationally distorted rubble-pile object, so it would be porous at an unknown scale and low density (~1000 kg/m^3). Very recently, for the large KBO 2003 EL$_{61}$ a mean density of 2600–3340 kg/m^3 and a visual albedo greater than 0.6 have been estimated (*Rabinowitz et al.,* 2006). Using the new lightcurve data *Trilling and Bernstein* (2006) concluded that the bulk densities of KBOs and Centaurs likely lie in the range 500–1500 kg/m^3. This is roughly consistent with the average bulk density of short-period comets. This agreement, together with the dynamical considerations, may strengthen the proposed genetic link between KBOs and short-period comets.

On the contrary, surprisingly, *Jewitt and Luu* (2004) discovered that the Quaoar spectrum reveals the presence on the surface of crystalline ice. Crystalline ice is formed only at temperatures above 110 K, well above the present temperature of Quaoar, which is about 50 K. This discovery was followed by many other similar discoveries. This observation can be interpreted in different ways: as an indication of internal activity leading to the generation of ice volcanism, similar to that presently observed on Enceladus, or as the exposure of the underlying layers of crystalline ice, the upper layers of amorphous ice having been removed by impact.

Another possibility is that the ice on the surface has been heated above 110 K by meteorite impacts. In the first case, the crystallinity could be an indication of the differentiation that the object undergoes, probably due to the combined effects of radioactive decay, primordial bombardment, and compaction due to the body self-gravity. This will possibly be tested by laboratory measurement and accurate modeling.

Another example of the relevance of collision is the surprisingly high frequency of binaries (see chapter by Noll et al.). The formation of binaries is explained by two competing theories. One entails the physical collision of bodies (*Weidenschilling,* 2002) while the other utilizes dynamical friction or a third body to dissipate excess momentum and energy from the system (*Goldreich et al.,* 2002; *Astakhov et al.,* 2005). In both cases the formation of multiple systems asks for a higher density of the KBO disk that allowed the formation of binary and multiple bodies (*Nazzario and Hyde,* 2005). This implies that the probability of collisions was higher than at present.

It is to be stressed again that KBO observations indicate a contradictory situation: On the one hand, the analysis of Varuna and other KBOs (*Jewitt and Sheppard,* 2002) seems to indicate a porous interior, while the presence of crystalline ice on Quaoar spectrum seems to indicate a differentiation process, leading to porosity reduction and internal evolution. Density (porosity) is an observationally derived property having cosmogonical significance. We can also obtain hints on the internal structure from the modeling of the thermal evolution of ice-rich bodies, and from the new data collected both by planetary missions and by ground-based observations of the different objects belonging to this category.

In this chapter we will try to combine the different sources of information and to see how they can be used to improve our theoretical approach and to reduce and limit the number of free parameters in the modeling of cometary activity. After the previous, very general definitions, we will briefly discuss the objects included in our review from the point of view of what is known about their interiors from observation; after that, we will discuss the hints that we can obtain from formation models and from thermal evolution models; finally, we will try to reach some conclusions.

2. STRUCTURE AND COMPOSITION OF KUIPER BELT OBJECTS AS A RESULT OF THEIR ORIGIN

We will now describe the analogies with other objects and will discuss the link between these objects and the processes in the protosolar nebula that bring about KBO formation (the legacy of planetesimals) (see the chapter by Kenyon et al.). We focus our attention on processes that can bring about the formation of cold and fragile objects. The objects that we have to consider are KBOs, Centaurs, and short-period comets. In fact, even if their genetic relations are not perfectly understood, the work done over the years, mainly from a dynamical point of view, indicates that the objects in the Kuiper belt can be the source of both Centaurs and short-period comets (*Fernandez,* 1980; *Morbidelli,* 2004). In this framework Centaurs, with their instable orbits, represent bodies caught "on the way."

From a physical point of view all these bodies, having originated in the same place, should be closely related and should have the same intrinsic physical nature. There are obviously effects, related to the size of these bodies, that can finally lead to a different evolution, but we should be able to decipher their main evolutionary path and establish common genetic relationships. The small bodies present in the outer solar system are characterized by a high content of volatile elements, which can under certain thermodynamic conditions lead to the development of an intrinsic activity due to the sublimation and loss of water ice and high-volatility carbon compounds. The properties of these bodies can be the result of the physical and chemical conditions prevailing in the solar nebula to the moment of their accretion and of the processes acting on them during the subsequent evolution. Moreover, the high degree of "mobility" (*Morbidelli,* 2004) probably strongly influenced the subsequent history and hence the present structure and composition of many objects. The present structure and appearance of these bodies has been affected by their dynamical history, by the surface aging (reddening of surfaces due to irradiation), by their activity (when present, as in the case of comets), and by their collisional evolution. This last process, in particular, could have heavily shaped them: The comets could even be collisional fragments directly ejected from the Kuiper belt.

Cosmogonical theories usually predict that the first condensates grow through different accumulation processes, which include low-velocity mutual collisions. In the process of adhesion different parameters play important roles, affecting both the velocity and the mass distribution of grains. Among those affecting mainly the velocity distribution of particles, we have to mention gas turbulence and gas-dust drag forces, while the mass distribution depends not only on the relative particle velocity but also on their sticking efficiencies. The relative importance of gravitational instability with respect to collisional coagulation can have consequences on the final structure of the cometesimals and on the porosity of the resulting bodies. Following *Gladman* (2005) one can say that our planetary system is embedded in a disk of asteroids and comets, remnants of the original planetesimal population. The outer solar system is dominated by ices of different kind, H_2O ice being the most important.

The kind of chemistry strongly depends on the reference model of the protosolar nebula. Our understanding of the chemical processes taking place in the primitive solar nebula has increased considerably as more detailed models of the dynamic evolution of such nebulae have become available. Early models (*Grossman,* 1972) assumed that a mixture of hot gases present in the solar nebula cooled slowly, maintaining thermodynamic equilibrium. In the beginning, the more refractory vapors condensed, followed by the lower-melting-point materials. The model suggested that the major textural features and mineralogical composition of the Ca, Al-rich inclusions in the C3 chondrites were produced during condensation in the nebula, characterized by slight departures from chemical equilibrium due to incomplete reaction of high-temperature condensates. Fractionation of such a phase assemblage is sufficient to produce part of the lithophile-element depletion of the ordinary chondrites relative to the cosmic abundances. This result is surprisingly good, given the very strong assumption of thermodynamic equilibrium made by Grossman. *Morfill et al.* (1985), instead, introduced the concept that localized turbulence could be the most probable source of viscosity in accretion disks. Other authors, such as *Fegley and Prinn* (1989), challenged the idea that the nebula was quiescent, demonstrating that even major-gas-phase species such as N_2 and NH_3 could fail to achieve equilibrium due to the low temperatures and the concurrent slow chemical reaction rates in the region of the outer planets. At the low temperatures characteristic of the outer solar system, kinetics may mean that carbon remains as CO, and therefore less oxygen is available to form water ice. The predicted rock/ice mass ratio in this case is 70/30, which gives a density of ~2000 kg/m^3, similar to that observed for both Triton and Pluto. In the hotter nebula, carbon tends to be incorporated in methane and the oxygen is then available to form water ice; the rock/ice mass ratio in this case should be close to 1, giving a density of ~1500 kg/m^3. Detection of CO is also consistent with low temperatures during the formation of bodies such as Triton, Pluto, and other icy bodies. However, *Fegley and Prinn* (1989) point out that several processes can overlap, modifying the original cometary chemistry as a certain mixing of the protosolar nebula material with material formed in circumplanetary nebulae; homogeneous and heterogeneous thermochemical and photochemical reactions; and disequilibration resulting from fluid transport, condensation, and cooling. Therefore, the interplay between chemical, physical, and dynamical processes should be taken into account if one wants to decipher the origin and evolution of the abundant chemically reactive volatiles (H, O, C, N, S) observed in comets.

This type of considerations can be the basis for inferring the composition of KBOs and that of comets, in which we expect therefore to find a large amount of volatiles, with carbon compounds such as CO being the dominant species, but not excluding a small amount of CH_4 of circumplanetary nebulae. N_2 is also more probable than NH_3. The Halley data on CO/CH_4 and N_2/NH_3, which are intermediate between those typical of the interstellar medium and those expected in a hotter nebula, seem to support this hypothesis. The original chemical evolution is only responsible for the initial chemistry of icy bodies; the further evolution could have partially altered it. The process of agglomerate formation by gradual accretion of submillimeter solid grains has been studied both experimentally and numerically (e.g., *Donn and Duva,* 1994; *Blum et al.,* 2000), and this investigation is in agreement with the idea that the primordial solar nebula was a suitable environment for the production of ice-rich grain clusters with a highly porous and fractal structure. These objects are accumulation of fluffy aggregates. If so, we have to expect that the present comets are remnants of this primordial situation. The subsequent growth of these small clusters has been investigated by the use of sophisticated numerical modeling (*Weidenschilling,* 1997), up to the point at which ~10-km-sized planetesimals are formed. If comets originated as icy planetesimals in the outer solar system, their nuclei have low strength, consistent with "rubble-pile" structure and inhomogeneities on scales of tens to hundreds of meters.

Weidenschilling (1997) presented results of numerical simulation of the growth of cometesimals, beginning with a uniform mixture of microscopic grains in the nebular gas. Coagulation and settling yield a thin, dense layer of small aggregates in the central plane of the nebula. The further evolution is dominated by collisions, and the relative collisional velocity is due mainly to the radial drift of the "cometesimals" interacting with the nebular gas. Bodies accreted in this manner should have low mechanical strength and macroscopic voids in addition to small-scale porosity. They will be composed of structural elements having a variety of scales, but with some tendency for preferential sizes in the range ~10–100 m. Weidenschilling states that these properties are in good agreement with inferred properties of comets (*Donn,* 1991; *Weissman,* 1986; *Asphaug and Benz,* 1994), which may preserve a physical record of their accretion. However, the *Weidenschilling* (1997) scheme does not seem to be in agreement with the observation that the KBOs underwent a noticeable degree of collision (*Farinella and Davies,* 1996). Laboratory simulation experiments, performed using micrometer-sized dust particles impacting solid targets at various velocities, seem to indicate the formation of open aggregates (*Blum et al.,* 2000). Slow bombardment of the target generally results in the formation of fluffy dust layers. At higher impact velocities, compact dust-layer growth is observed. Above a certain collision energy, the dust aggregates are disrupted. It has also been shown that heating and evaporation during a collision are rather limited even for collisions between large (about 100 m) cometesimals, even though local thermal and possibly chemical alterations cannot be excluded (as in the primordial rubble-pile model). Furthermore, bodies with sizes below a few tens of kilometers are not affected by gravitational compression. As a result, comets can be seen as low-density objects, formed slowly at low temperature, but possibly characterized by a complex internal structure that can allow their fragmentation under high- to medium-velocity impact conditions. We have verified that the presence of a limited amount of radioactive elements does not change their evolution. Larger bodies, however, if formed early in the evolution of the solar system, can undergo different histories, due to the contribution of short- (as ^{26}Al) and long-life radioactive decay, degassing, and impact compaction.

Detailed modeling of accretion in a massive primordial Kuiper belt was performed by *Stern* (1996), *Stern and Colwell* (1997a,b), and *Kenyon and Luu* (1998, 1999a,b). While each model includes different aspects of the relevant physics of accretion, fragmentation, and velocity evolution, the basic results are in approximate agreement. In general, all models naturally produce a few objects the size of Pluto and approximately the right number of 100-km objects, on a timescale ranging from 10^7 to 10^8 yr. The models suggest that the majority of mass in the disk was found in bodies approximately 10 km and smaller. An upper limit for accretion timescales in the Kuiper belt region seems to be the formation time of Neptune, since it is assumed that the formation of Neptune efficiently terminated the growth in the Kuiper belt region (*Farinella et al.,* 2000), inducing eccentricities and inclinations in the population high enough to move the collisional evolution from the accretional to the erosive regime (*Stern,* 1996). The formation timescale is crucially important in determining the thermal evolution of a body. The strong heliocentric distance dependence of the growth time is consistent with a large radial gradient in their hydration properties across rather modest radial distance differences (*Grimm and McSween,* 1993). Detailed models of KBO accretion show that these objects can form on a timescale of 10–100 m.y., provided very-low-velocity dispersions are maintained (*Kenyon,* 2002). In this case the effect of radioactive elements can be negligible. For this reason, in what follows we have treated the thermal evolution in the presence or the absence of short-lived radioactive elements.

3. THE EFFECT OF RADIOACTIVE ELEMENTS AND POROSITY

We have described the possible composition of KBOs on the basis of their origin. In this section, we consider the effect of two parameters that can strongly condition the evolution of KBOs, beginning with the effects of radioactive element decay. First, we have to take into account the timescale of evolution. In fact, if the formation time is on the order of 10–100 m.y., the KBOs are probably heated by trapped ^{26}Al or other short-lived nuclei only at the beginning of their lives. Short-lived radionuclides are characterized by half-lives that are significantly shorter (i.e., $\lesssim$100 m.y.) than the 4.56-Ga age of the solar system. Based on recent data, there is definitive evidence for the presence of two new short-lived radionuclides (^{10}Be and ^{36}Cl) and a compelling case can be made for revising the estimates of the initial solar system abundances of several others (e.g., ^{26}Al, ^{60}Fe, and ^{182}Hf). The presence of ^{10}Be, which is produced only by spallation reactions, is either the result of irradiation within the solar nebula (a process that possibly also resulted in the production of some of the other short-lived radionuclides) or of trapping of galactic cosmic rays in the protosolar molecular cloud. On the other hand, the most accurate estimates for the initial solar system abundance of ^{60}Fe, which is produced only by stellar nucleosynthesis, indicate that this short-lived radionuclide (and possibly significant proportions of others with mean lives ≤10 m.y.) was injected into the solar nebula from a nearby stellar source. As such, at least two distinct sources (e.g., irradiation and stellar nucleosynthesis) are required to account for the abundances of the short-lived radionuclides estimated to be present in the early solar system.

The levels at which the short-lived radionuclides ^{26}Al, ^{41}Ca, and ^{60}Fe (and probably ^{36}Cl) are maintained in the galaxy are significantly lower than those inferred from meteorites in the primordial solar nebula, and after a delay of ~10^8 yr essentially none of these radionuclides remains in the molecular cloud from which the solar system formed (*Harper,* 1996; *Wasserburg et al.,* 1996; *Meyer and Clayton,* 2000). As such, some nearby processes were creating radionuclides within ~10^6 yr of the birth of the solar system. It is clear that more than one process was involved, since there is no proposed source that can simultaneously produce enough ^{10}Be and ^{60}Fe. The most plausible source of ^{60}Fe in the early solar system is a Type II supernova. When that supernova injected ^{60}Fe into the material that formed the solar system, it is also likely to have injected other short-lived radionuclides such as ^{26}Al, ^{36}Cl, ^{41}Ca, and ^{53}Mn (*Meyer and Clayton,* 2000; *Goswami and Vanhala,* 2000; *Meyer,* 2005). Therefore, while the inferred initial abundance of ^{60}Fe in the early solar system places its formation near a massive star that became a supernova, the timing of this event and the distance from this supernova are uncertain. However, heating by ^{26}Al may well take place during the accretion process itself; moreover, the accretion rate is not linear in time. Thus, the core of an accreting body may be significantly heated before the body reaches its final size long after the decay of ^{26}Al.

The effect of combined growth and internal heating by ^{26}Al decay was investigated by *Merk and Prialnik* (2003) for an amorphous ice and dust composition, without other volatiles. They found that small objects remain almost unaffected by radioactive heating while the effect on larger bodies is not linear with size. There is an intermediate size range (around 25 km), where the melt fraction and duration of liquid water are maximal, and this range depends strongly on formation distance (ambient temperature). This internal evolution should bring about a kind of recompaction of these objects. In fact, if liquid water ice is present, the density of the layer containing it is certain to be higher than the density of ice.

The studies of KBO lightcurve data (*Trilling and Bernstein,* 2006) indicate that the bulk densities of KBOs and Centaurs likely lie in the range 500–1500 kg/m^3. The study by *Consolmagno et al.* (2006) of KBO spins suggests that the mean density of these objects is approximately 450 kg/m^3. These estimates are roughly consistent with the average bulk density for short-period comets. This agreement may strengthen the proposed genetic link between KBOs and short-period comets. The KBOs would likely have bulk compositions similar to comets and thus similar nominal grain densities. As a result, KBO bulk porosities are likely to be in the 60–70% range. However, this analysis seems to be limited to the "medium-sized" KBOs. The largest KBOs are substantially denser. Pluto has a bulk density of 2.0 × 10^3 kg/m^3. The Pluto-sized 2003 EL_{61} is a rapid rotator (with a period of 3.9 h) and may have a bulk density in the range of 2.6–3.3 × 10^3 kg/m^3 (*Rabinowitz et al.,* 2006). These objects, like the largest asteroids, are probably coherent with low macroporosity.

Continuing work on comets has produced some indication of their bulk density. Analysis of ejecta trajectories observed during the Deep Impact encounter with Comet 9P/Tempel 1 indicates a low bulk density (*A'Hearn et al.,* 2005). *Davidsson and Gutierrez* (2004, 2006) estimated densities for Comets 19P/Borrelly and 81P/Wild 2 by analyzing nongravitational orbital changes. Wild 2 density is estimated between 380 and 600 kg/m^3 and 19P/Borrelly between 180 and 300 kg/m^3. Comet rotation period data also support a strengthless rubble-pile model with average low bulk densities. While all these estimates are model-dependent and have large error bars, it appears safe to say that comets have very low bulk densities. To put these numbers in perspective, we need to look at comet composition and the grain density (porosity-free density) of those materials. To first order, comets are mixtures of water ice with a dust composed of hydrated silicates, mafic silicates, and organics. While there are a number of other volatile species, water ice dominates the mass balance of the volatiles. Water ice has a grain density of 930 kg/m^3. Cometary dust compositions are not yet well known, but a reasonable analog may be CI carbonaceous chondrites, which are composed of the same sort of silicate and organic mixture thought to dominate the cometary dust. CI carbonaceous chondrites have a grain density of 2.27 × 10^3 kg/m^3. Dust to ice ratios are thought to be on the order of 2 to 1, which would make the theoretical grain density of a comet approximately 1.8 × 10^3 kg/m^3. It is unlikely that cometary materials will have grain densities much lower than this number. Methane and nitrogen ices have densities in the 0.8–0.9 × 10^3 kg/m^3 range, not much lower than water ice, and their low mass balance would not strongly affect the overall bulk composition of the comet. The dust is unlikely to be much less dense since the hydrated silicates have grain densities in the 2.2–3.0 × 10^3 kg/m^3 range and mafic silicates are much denser. If the "grain density" of a cometary mix of materials is 1.8 × 10^3 kg/m^3 and comet bulk densities range around 0.5 × 10^3 kg/m^3, the implication is that comets have very large porosities. For Tempel 1, a 0.62 × 10^3 kg/m^3 bulk density would translate into a bulk porosity of 60%. For a nominal cometary bulk density of 0.5 × 10^3 kg/m^3 the bulk porosity would be approximately 65%. This level of porosity indicates that cometary structures are, not surprisingly, essentially fluffy balls with more empty space than solid material. Therefore it could be reasonable to assume that small and intermediate KBOs are completely, or at least in large part, porous (*Capria and Coradini,* 2006).

If we assume that short-period comets are the smallest members of the KBO family, in order to use their properties as an example of KBOs, we have also to take into account the effect of the slow migration of short-period comets in the inner solar system. Comets lose their volatiles differentially, and unfortunately the abundance ratio of ejected volatiles in the coma does not represent the nucleus abundances (*Huebner and Benkhoff,* 1999). It should be mentioned that nucleus models succeeded in explaining the release of gases along cometary orbit only when the effect of gas diffusion taking place in a porous medium was simulated (*Huebner et al.,* 2006).

The recent findings of the Deep Impact mission strongly reinforce the arguments cited above. The Deep Impact instruments revealed that, even if the surface of Tempel 1 is remarkably homogeneous in albedo and color, three discrete areas have the spectral signature of water ice. These regions cover a small fraction of the surface, only 0.5% (*A'Hearn et al.,* 2005; *Sunshine et al.,* 2006). Moreover, it is significant that the extent of this ice on the surface of Tempel 1 is not sufficient to produce the observed abundance of water flux observed in the comet's coma, meaning that there are sources of water from beneath the comet's surface. This is an important discovery that confirms the porosity of the surface layers. The comet surface of 9P/Tempel 1 seems to be covered by fine particles because the impact excavated a large volume of very fine (microscopic) grains, probably preexisting either as very fine particles or as weak aggregates of such particles. This fine material layer must be tens of meters deep (*A'Hearn et al.,* 2005). Studying the ejecta at very late stages, the overall strength of the excavated material was determined to be <65 Pa, and the bulk density of the nucleus is estimated at roughly 0.6×10^3 kg/m^3. This low density implies a porous structure of the comet nucleus. The porosity of dust mantle and comet layers has been assumed in some thermal evolution models of comet nuclei (*Capria et al.,* 2001, 2002; *De Sanctis et al.,* 1999, 2003, 2005, 2007; *Prialnik et al.,* 2004; *Huebner,* 2003; *Huebner et al.,* 2006).

4. THERMAL MODELS OF KUIPER BELT OBJECTS

Given the previous considerations, we are now able to describe the kind of modeling that we can develop. We assume therefore that KBOs are volatile-rich, porous objects, as a result of the limited observational data on KBOs and the indications from comets to which KBOs are genetically related. The thermal evolution models of KBOs were treated with two kinds of approaches, corresponding to two different points of view, both legitimate given the great uncertainty that exists about the internal structure of KBOs: models originally developed for comet nuclei and models originally developed for icy satellites (see chapter by McKinnon et al.). One could say, following *McKinnon* (2002), that in one case we are scaling up from the traditional small cometary sizes, while in the other case we are starting from mid-sized icy satellites and moving downward to smaller sizes.

If we also think that objects larger than comets, such as KBOs, can be porous and ice-rich bodies, it is straightforward to apply to them the models initially developed to study the thermal evolution of cometary nuclei. In fact, it is very difficult to draw a clear line between compact and differentiated objects of a certain size and noncompact, porous, almost homogeneous icy bodies. The approach commonly used for comets can be applied to a large variety of objects. This has been done by, for example, *Capria et al.* (2000), *De Sanctis et al.,* (2000, 2001), and *Choi et al.* (2002), who used models derived from comet nuclei models to study Centaurs and KBOs. The underlying idea is that, if the link between comets and KBOs is real, then the observed properties of the comets can be used to constrain KBO models, including low formation temperature, low density (high porosity), and high volatile content; this means, in turn, that it is possible to study both kinds of bodies with the same theoretical models.

In order to study the thermal evolution and differentiation of porous, ice-rich bodies, many models have been developed over the last few years; a complete discussion on the subject and an exhaustive reference list can be found in the book by the ISSI Comet Nucleus Team (*Huebner et al.,* 2006). We will give here only a few details.

In the currently used thermal evolution models, heat diffusion and gas diffusion equations are solved in a porous medium, in which sublimating gas can flow through the pores. A mixture of ices and dust is considered, and the flux from surface and subsurface regions is simulated for different gas and dust compositions and properties. The temperature on the surface is obtained by a balance between the solar energy reaching the surface, the energy reemitted in the infrared, the heat conducted to the interior, and the energy used to sublimate surface ices. When the temperature rises, ices can start to sublimate, beginning from the more volatile ones, and the initially homogeneous nucleus can differentiate, giving rise to a layered structure in which the boundary between different layers is a sublimation front. Due to the larger sizes of KBOs with respect to comet nuclei, and to the consequently higher content of refractories, the heating effect of radiogenic elements, both short and long-lived, is usually taken into account. So these models consider two heating sources of comparable importance, one acting from the surface (solar input) and one present in the entire body; this can give rise to more complex thermal evolution patterns than in the case of comet nuclei.

To give an example, some results from a thermal evolution model that can be applied both to comets and to larger KBO bodies, developed by our group (*Capria et al.,* 2000; *De Sanctis et al.,* 2000, 2001), will be briefly described here.

The nucleus is assumed to be spherical and composed of ices (water, CO_2, and CO) and a refractory component. Water ice can be initially amorphous, and in this case more volatile gases can be trapped in the amorphous matrix and released during the transition to crystalline phase. The refractory material is assumed to be in grains, spherical in shape. The initial grain size distribution can be given as well as the grain physical properties. Energy and mass conservations are expressed by a system of coupled differential equations, solved for the whole nucleus. The heat equation is

$$\rho c \frac{\partial T}{\partial t} = \nabla(K\nabla T) + \sum_i Q_i + Q_{tr} + Q_{rad}$$

where T is the temperature, t the time, K the heat conduction coefficient, ρ the density of the solid matrix, c the specific heat of the material, Q_i the energy exchanged by the solid matrix in the sublimation and recondensation of the ith ice, Q_{tr} the heat released during the transition from amorphous to crystalline form, and Q_{rad} the energy released by the decay of radioisotopes. The gas equation is

$$\frac{\partial \rho}{\partial t} = -\nabla\Phi_i + Q_i'$$

where Φ_i is the gas flux and Q_i' is the source term that is coupled with the heat equation. For radiogenic heating, the effects of ^{40}K, ^{232}Th, ^{235}U, and ^{238}U radioisotopes, and in some cases ^{26}Al. have been considered. The rate of radioactive energy release, Q_{rad}, is given by

$$Q_{rad} = \rho_{dust} \sum \lambda_j X_{0j} H_j e^{-\lambda_j t}$$

where ρ_{dust} is the bulk dust density, λ_j is the decay constant of the jth radioisotope, X_{0j} is its mass fraction within the dust, and H_j is the energy released per unit mass upon decay.

The amount of radioisotopes is unknown and there is no way to measure it; in these models it has been assumed that the abundances of long-lived radioisotopes are in the same proportion as in the C1 chondrites (*Anders and Grevesse,* 1989), while the amount of ^{26}Al is variable in the different cases studied. Here we will describe the results of this model applied to two different kinds of bodies, corresponding to two different hypotheses about the composition and internal structure of KBOs: a body whose composition and density are inherited from the typical ones of comet nuclei, and another one much more dense and rich in refractories.

4.1. Low-Density Ice-rich Kuiper Belt Objects

In this case we consider the thermal evolution of KBOs with the assumption that they are similar to cometary nuclei, so we are using parameters that are considered as standard in cometary models. When we say standard parameters, we refer to a range of values that are considered typical for comets, derived by observations, laboratory experiments, and *in situ* measurements. Because it is impossible to change and test all the parameters, we have analyzed which parameters are critical for these models and have built our cases around them.

We have seen that there is a limited number of key parameters: the amount and type of radioisotopes, the body composition (especially the amount of dust), the size, and the thermal conductivity. These factors affect the evolution in different ways. The amount and kind of radioisotopes provide different heating rates that are also a function of time. The ^{26}Al radioisotope is a very intense heat source, and its abundance strongly affects the evolution of the body. The presence of ^{26}Al in KBOs is debated: Due to the short half-life of this radioisotope (10^5 yr), the formation should have to take place within a few million years. The total amount of radioisotopes is a function of the amount of the refractory materials (dust) in the nucleus. At the same time, the dust affects the overall thermal conductivity: The larger the dust/ice ratio, the larger the thermal conductivity. The combination of these two parameters strongly increases the overall process of heat transfer.

The structure of water ice also influences the thermal evolution of the body. Amorphous ice can be a very inefficient heat conductor. The crystallization process is a strong internal heat source that under particular conditions (very low conductivity) gives a runaway increase of internal temperature. The structure of the body in terms of porosity and pore radius has a strong influence on the thermal conductivity and, consequently, on the internal temperature; porous media are inefficient conductors. Low conductivity results in higher temperature.

The size of the body is important. Earlier work has shown that radiogenic heating is not efficient for small bodies. The values adopted in these models to describe the cometary nuclei composition have been largely discussed in several reviews (*Festou et al.,* 1993; *Rickman,* 1998) and also by the ISSI Comet Nucleus Working Group (*Huebner et al.,* 2006), and they are broadly accepted. For these models (*De Sanctis et al.,* 2001) we have assumed an initial temperature of 30 K throughout the whole body, which is a plausible solar nebula temperature in cometary formation regions (*Yamamoto,* 1985; *Yamamoto and Kozasa,* 1988). A value of 1000 kg m^{-3} for the dust density takes into account the fact that grains are the result of an accumulation process and are therefore highly porous. The emissivity value is 1, the initial porosity is 0.8, and the initial pore radius has a value of 10^{-5} m. In these models we are assuming a fairly low initial amount of CO, but we are considering only the fraction of CO existing as its own ice.

The combined effect of radioisotopes and solar heating, the latter coming from outside and the former uniformly distributed through the whole body, leads to an increase in the overall temperature of the nucleus. The internal temperature gradually increases but never reaches a value high enough to permit the crystalline phase transition: The amorphous ice is preserved. It must be recalled that the central temperature strongly depends on the thermal diffusivity.

The main results of the thermal evolution models applied to "comet-like" classical KBOs is that the internal heating due to radio-decay can be sufficient to mobilize volatiles, giving rise to a compositionally layered structure. In the upper 100 m below the surface, the most volatile ices (such as CO) are completely absent due to the combined effects of solar and internal heating. Due to radiogenic heating, the internal temperature may become high enough to permit the sublimation of CO (or similar hypervolatile ices) from the inner layers. The gas is free to circulate in the body pore system and can recondense in those layers at lower temperature. The nucleus that results from the radiogenic heating has a layered structure made of interlaced layers of frozen CO and layers depleted (with respect to the initial amount) of CO ice. This layered structure with "CO-enriched" and "CO-depleted" zones is due to the fact that the central temperature tends to rise above that of the sublimation temperature of CO.

If the amount of the short-lived radioisotopes, such as ^{26}Al, is low or negligible, the models foresee that the CO ice is confined in the nucleus interior. The depth at which the volatiles are confined depends on the type of radioactive elements and on their quantity. Obviously, small amounts of radioactive elements have little influence on the thermal evolution of KBOs.

From these simulations (*De Sanctis et al.*, 2001) (see Fig. 1) it can be seen that if the body is ice-rich and of low density an undifferentiated core can survive, depending mainly on the type and amount of radiogenic elements contained in the body, but also on the physical parameters assumed, such as thermal conductivity, porosity, radius, etc. The bodies emerging from this scenario retain amorphous ice because the central temperature does not rise too much. It must be recalled that the overall thermal conductivity computed for our models is quite large, and this is a key parameter for the increase of the internal temperature. Much smaller conductivities can give different results with a runaway temperature increase.

In these models we do not consider the presence of trapped volatile molecules. Laboratory experiments on sublimation of mixtures of amorphous water ice and volatiles suggest that a substantial fraction of these volatiles can sublimate when the phase transition occurs. From our results the volatiles, bound within the amorphous ice, are preserved and are present through the whole nucleus until the body is in the Kuiper belt. In these cases, one source of volatile molecules, such as CO, observed in comet comae, could be the trapped volatiles liberated when the phase transition of the amorphous ice occurs.

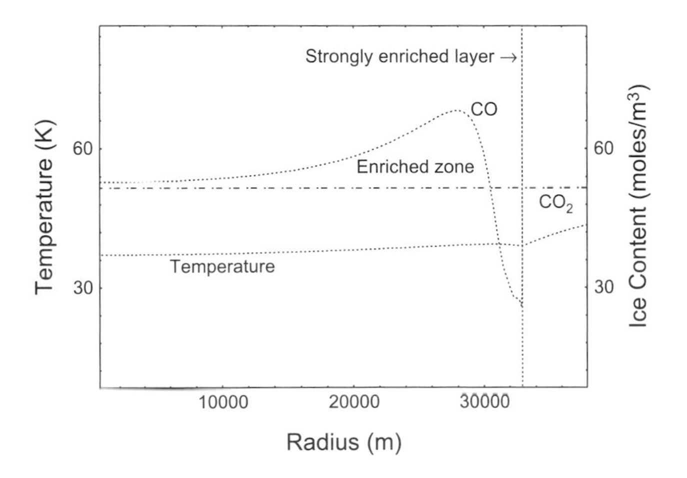

Fig. 1. Thermal profile, CO_2, and CO profiles of a typical KBO model. The scale on the left refers to the temperature; the scale on the right refers to the volatile ice content in the body. It can be seen that the amount of CO_2 is constant and unchanged throughout the entire nucleus, while the CO amount shows enrichment and depletion at different depths (from *De Sanctis et al.*, 2001).

These results are strongly dependent on the amount and kind of radioisotopes assumed. In fact, assuming a larger quantity of ^{26}Al, *Choi et al.* (2002) found that, depending on the initial parameters, the interior may reach quite high temperatures, losing the ices of very volatile species during early evolution. Their models indicate that, in some cases, the amorphous ice crystallizes in the interior, and hence some objects may also lose part of the volatiles trapped in amorphous ice.

According to their models, KBOs may have lost entirely all volatiles that sublimate below ~40–50 K, which were initially included as ices, and may have partially lost less-volatile ices as well. However, in this case the conclusion is valid only on the assumption that the entire surface area is, on average, equally heated. As a result, the internal structure of KBOs is most probably not uniform; rather, density, porosity, H_2O ice phase, and strength all vary with depth.

All the developed models of KBOs indicate that the internal temperature profile may have been substantially affected by both short- and long-lived radionuclides, with accompanying changes in composition and structure (see Figs. 2a,b). Moreover, the models indicate that regions enriched in volatile species, as compared with the initial abundances assumed, arise due to gas migration and refreezing.

These changes in structure and composition should have significant consequences for the short-period comets, which are believed to be descendants of Kuiper belt objects. The evolution of such a body when injected in the inner solar system will be characterized by outbursts of volatiles, when the volatile-enriched layers reach the sublimation temperature. However, the evolution of the temperature profile and the structural modifications are a function of the accretion times of KBOs (the amount of radioactive elements), dust-to-ice mass fraction, density, etc. Nevertheless, we may state that if KBOs did experience radioactive heating, their structure and composition were altered mainly to the extent of considerable loss of volatiles and significant departure from internal homogeneity.

4.2. More-Dense Dust-rich Kuiper Belt Objects

We have applied the model to a Kuiper belt body with larger density with respect to the previous one, in order to show possible differences in the evolution history. In this case, most of the model parameters assumed as reference are the values commonly used for cometary nuclei composition (*Huebner et al.*, 2006). The body has a relatively small radius (100 km), is made up of dust and ices of water and CO_2, and the initial temperature is 20 K through-

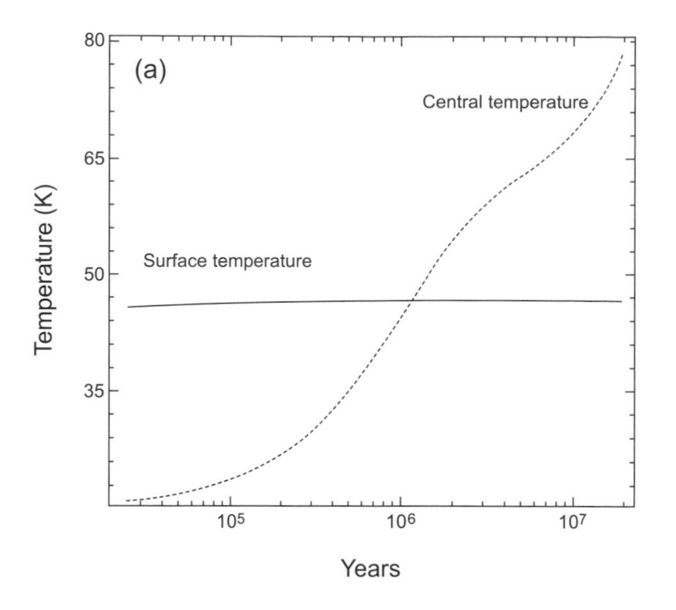

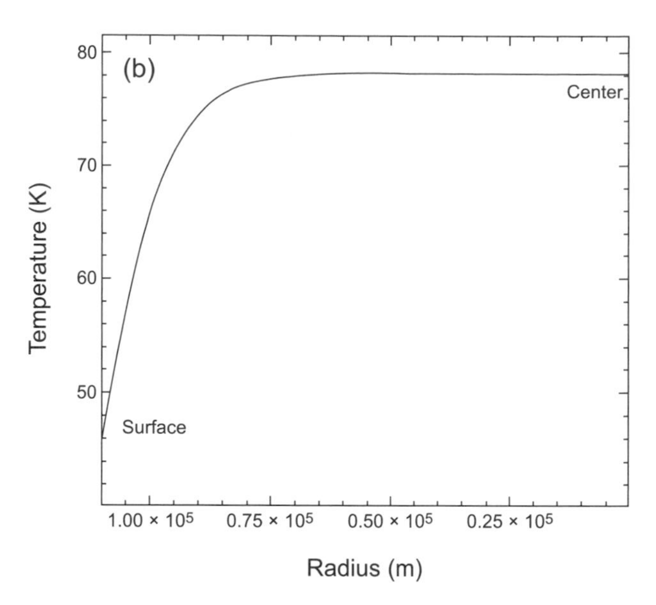

Fig. 2. **(a)** Evolution of the surface and central temperature for an icy-dense KBO. In this case, we have followed the evolution of the body for 10^7 yr, a timescale in which the effect of the short-lived elements becomes apparent. In fact, while the surface temperature remains almost constant, the central temperature increases up to 80 K. **(b)** Thermal profile of the same body shown in **(a)**. In this case the thermal profile is opposite of that shown in Fig. 1. Here the maximum temperature is in the center, and the maximum gradient, after 20 m.y., is close to the surface, which remains almost isothermal due to the limited amount of heat received from the Sun. Volatiles lost from the central part are quenched again when moving outward, where they meet very cold layers.

out the entire nucleus. The ice is initially amorphous. The overall density of the body is 1600 kg/m^3 and has a porosity of 0.3. The orbit has a semimajor axis of 43 AU and an eccentricity value of 0.05. In this model a small amount of the short-lived radioisotope, ^{26}Al, is included in the dust composition. The combined effect of radiogenic and solar heating — the latter coming from outside and the former uniformly distributed throughout the entire nucleus — leads to an increase in the overall temperature of the nucleus. The central temperature increases, reaching the sublimation temperature of the most-volatile ices. The amorphous-crystalline transition can be activated at such a temperature at a very low rate, possibly releasing the trapped gas. However, the internal temperature is not high enough to have liquid water. We can speculate that bodies like this can be deeply altered due to the radiogenic heating losing the hypervolatile ice, as CO (see Fig. 3).

Moreover, in order to verify the long-term behavior of an object poor in high-volatility elements and less porous than typical comets, we have also considered a class of models in which the conditions are closely similar to those of satellites. Table 1 reports the different models that we have developed. These models differ in dust/ice ratio, density, and content of radioactive elements. A model without ^{26}Al was run in order to have a reference case. The results of these models are also reported in Fig. 3, in which the different curves are labeled with the same names as those used in Table 1.

As should be expected, bodies having higher dust content are also characterized by a higher thermal conductivity. The heat generated inside the body is transported toward the surface, and a large part of the body is characterized by a noticeable increase in temperature. In the body in which the dust/ice ratio is higher, the central temperature increases, in about 10^5 yr, from 20 to 230 K because of the decay of ^{26}Al.

We have therefore verified if our results are compatible with solid convection. This has been done by assuming that the convection can take place if the critical Rayleigh number has been overcome. We have used two different definitions

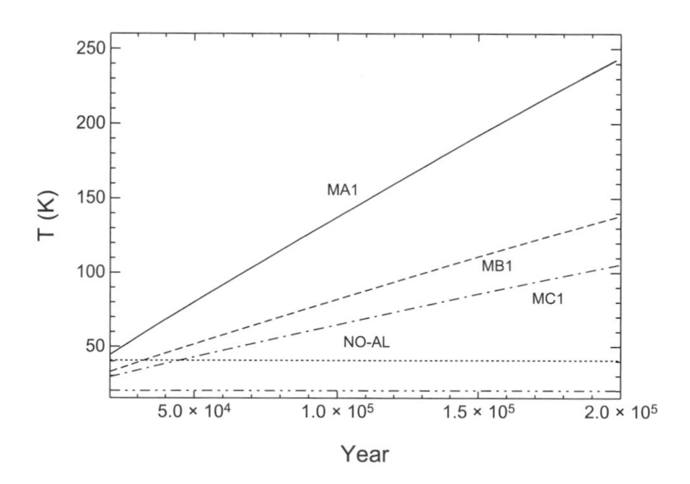

Fig. 3. Central temperature evolution for the low-volatile KBO models described in Table 1. Here we see that when the amount of dust increases, if short-lived elements are present, the central temperature increases up to values that allow the crystallization of water ice.

TABLE 1. Low-volatile KBO models.

Models	MA	MB	MC	NO-AL
Composition	Dust + H_2O	Dust + H_2O	Dust + H_2O	Dust + H_2O
Dust/Ice	5	2	1	5
ρ bulk (g cm^{-3})	1583	871	287	1583
Al_{26}	Yes	Yes	Yes	No
T_0 (K)	20	20	20	20
Radius (km)	200	200	200	200
a	50	50	50	50
e	0.03	0.03	0.03	0.03
Albedo	0.06	0.06	0.06	0.06
Porosity	0.3	0.45	0.7	0.3

of the number: in the presence (R_{a1}) or in the absence (R_{a2}) of radioactive elements. We can assume

$$R_{a1} = \frac{\alpha\rho g\Delta T_{max}\Delta r^3}{k_d\eta}$$

$$R_{a2} = \frac{\alpha\rho gH\Delta r^5}{K\eta k_d}$$

$$\eta = \eta_0 \exp\left[A\left(\frac{T_m}{T} - 1\right)\right]$$

$$g = \frac{4}{3}\pi GR$$

where ρ is the bulk density, α the coefficient of thermal expansion, k_d the thermal diffusivity, K the thermal conductivity, η the solid viscosity, g the gravitational acceleration, H the heat generation due to the presence of radioactive elements, ΔT the thermal gradient in the layer considered as Δr, T_m the melting temperature, A a dimensionless coefficient, and $(Ra)_{Crit}$ the critical Rayleigh number, which following *Schubert et al.* (2001) could be assumed to be between 1000 and 2000.

R_{a1} and R_{a2} are respectively the definition of the Rayleigh number in the absence or in the presence of radioactive elements. We have also assumed that convection can take place if Ra > $(Ra)_{Crit}$, following *Schubert et al.* (2001): This approach is the so-called parametric convection. The most difficult parameter to evaluate is the layer thickness, Δr. In our case we have considered a layer that is compatible with the discretization of the process that we have assumed. Since we have a grid, divided in steps of finite dimensions, we have verified in which step the condition of convective instability is satisfied. Then we have considered two steps around the value where the vertical gradient is maximum, and it corresponds to ~10 km. We have verified that, assuming H = 4.3 × 10^{-3}, $(Ra)_{Crit} \approx 1000$ only when radioactive elements are present and η_0 = 1014 Pa, then R_{a2} ~ 1500–3000 and convection is possible.

We can therefore state that, when a small amount of radioactive elements is present, the combined effect of solar radiation and radiogenic heating leads to KBOs that are strongly volatile depleted, at least in their upper layers. The KBOs are also highly differentiated: A typical result is that interlaced layers that are CO-depleted and CO-enriched are found, particularly when very cold, porous icy bodies are considered. If this result is confirmed, the evolution of KBOs injected in hotter parts of the solar system will be characterized by an outburst of volatiles when the enriched layers reach sublimation temperature. Finally, an undifferentiated core can survive, depending on the size and radiogenic-element content of the body.

When KBOs are characterized by a high content of ^{26}Al, chondritic value, we have a strong heating of the internal layers, surmounted by a layer still remaining at lower temperature; only these layers can be still enriched in volatiles. The two classes of models behave in a completely different way. In this second case, the amount of volatiles is much lower and can be further affected by the dynamical evolution: The layer very close to the surface can be further depleted when the body is injected in the inner part of the solar system, so only a shell of volatile-rich ice could be present. Again, we form interlaced layers of different composition and characteristics, but in the first case high-temperature, volatile-rich ice is dominant, while in the second case the shell of volatile-rich ice is limited. These bodies can be also characterized by the onset of convection.

In both cases, collision in differentiated bodies can lead to fragments of different volatile content and different structural characteristics. Comets can be generated by the fragmentation of a volatile-rich body, or by the destruction of a largely differentiated object, still preserving in its interior a certain amount of volatile-rich ice.

5. THE JOURNEY IN THE INNER SOLAR SYSTEM: THE KUIPER BELT OBJECT–COMET LEGACY

In conclusion, the results obtained up to now are promising, although some uncertainties are still present. The amount of short-lived radioactive elements is not easy to determine, as it is strongly related to the formation timescale. The shorter the time of formation, the higher the amount of radioactive elements can be. Different initial conditions can result in completely different situations; however, thermal processes related to the solar radiation, occurring close to the surface, and heating processes related to the radioactive element decay affecting the deeper layers can bring to complex internal structures. If the amount of short-lived radioactive elements is chondritic, then the global amount of volatiles is depleted, and without them the cometary activity cannot be explained. However, the more-depleted objects still preserve some gas in the intermediate layers, as shown by our models. In a previous paper (*Capria and Coradini,* 2006), we have left this problem open. Here, after a series of efforts to model comets and KBOs, we reach the conclusion that the tremendous variability observed in comets can either be pristine or can be related to the collisional disruption of previously differentiated large objects, which can give rise to bodies with different volatile content. Overlapping these differences are the effects of the collisional and thermal evolution of these bodies, which again can strongly alter the overall structure or volatile content. In any case, in order to explain consistently the present behavior of comets, two main characteristics shall be preserved: porosity and the presence of high-volatility gases. Moreover, great variability in the object can be the result of further dynamical evolution.

We know that there exist bodies located on unstable orbits that are surely linked to KBOs; these are referred to as Centaurs. A growing number of these bodies have been identified with orbits crossing those of Saturn, Uranus, and Neptune. These bodies can be seen as transition bodies between the KBOs and the comets (*Levison and Duncan,* 1994; *Hahn and Bailey,* 1990); the fact that their orbits, on the basis of dynamical calculations, are not stable over the lifetime of the solar system, suggests that the Centaurs formerly resided in the Kuiper belt and only recently have they been delivered into their current orbits. The different behavior of Centaurs has been tentatively attributed not only to different compositions and volatile contents, but also to the presence or absence of a crust (primordial irradiation mantle?) on their surface. Such a crust could inhibit activity and, in the case of organic compounds, redden the spectra.

This common origin with KBOs makes the Centaurs very interesting, because they could provide compositional information on the more distant Kuiper belt objects and could also provide information about their subsequent processing. If Centaurs are bodies that are coming from the Kuiper belt and are waiting to become short-period comets, we can use them to infer the characteristics of KBOs. Thermal evolution models of Centaurs can provide information on the possible structure of these bodies before their injection into the inner solar system. If Centaurs are covered by an organic crust, and if the crust survives, we can ask the following questions: How did the thermal history in the Kuiper belt influence their internal structure? What is the evolution of a differentiated Kuiper belt object when it arrives in the Centaurs' zone? Does the body still exhibit activity due to the residual presence of volatiles? We have computed the evolution of Centaur bodies assuming different parameters and initial conditions, with the aim of answering at least some of these questions. From the modeling of KBOs (*De Sanctis et al.,* 2001), we have verified that amorphous ice could be preserved for very long timescales even in the presence of long-lived radioactive elements. From these simulations, it was seen that Kuiper belt objects can be strongly depleted in hypervolatiles in the outer layers, down to several hundred meters below the surface. The resulting bodies are thus differentiated.

What is the evolution of such a body when it arrives on a Centaur-like orbit? According to *De Sanctis et al.* (2000), the evolution strongly depends on the dynamical path of the object. In the case of objects such as Pholus the dust flux, driven by gas activity, is negligible: Pholus does not develop a dusty coma. If Centaurs are bodies coming from the Kuiper belt, they should be partially differentiated and possibly covered by organic crusts. Pholus could be an example of this evolution. When a crust is present, gas molecules cannot flow freely through the dust layer, diffusing instead from the sublimation front through the crust. If we consider CO only in ice form (no gas trapped in amorphous ice), the CO flux depends on the depth at which CO ice is located. If CO ice is present several kilometers below the surface, gas emission is negligible and this kind of object can be considered inactive (with an activity level below the detection threshold). However, the volatile ices could still be present in the body under the organic crust; the gas flux is not sufficiently strong to remove dust particles from the surface (see Fig. 4).

Until recently, Chiron's activity was considered very unusual and induced by some "exotic" and "episodic" mechanism, such as outbursts due to crystallization (*Prialnik et al.,* 1995). The recent discovery of activity on some Centaurs, such as C/NEAT 2001 T_4, 174P/2000 EC_{98} (60558), P/2004 A_1 (LONEOS), and 2004 PY_{42} (*Epifani et al.*, 2006; *Bauer et al.,* 2003), tells us that active Centaurs are quite common. We do not know if the activity is sustained by the same physical process for all these Centaurs, but in any case, we must begin to consider a possible common sublimation mechanism for all of them.

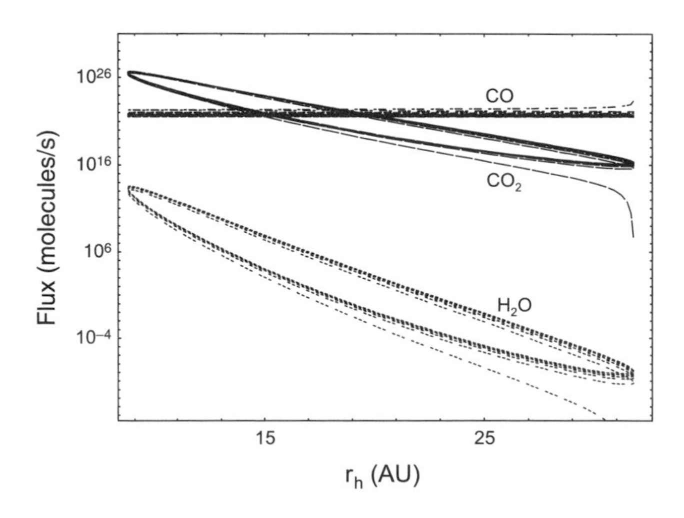

Fig. 4. Gas flux vs. heliocentric distance for Pholus model (from *De Sanctis et al.,* 2000).

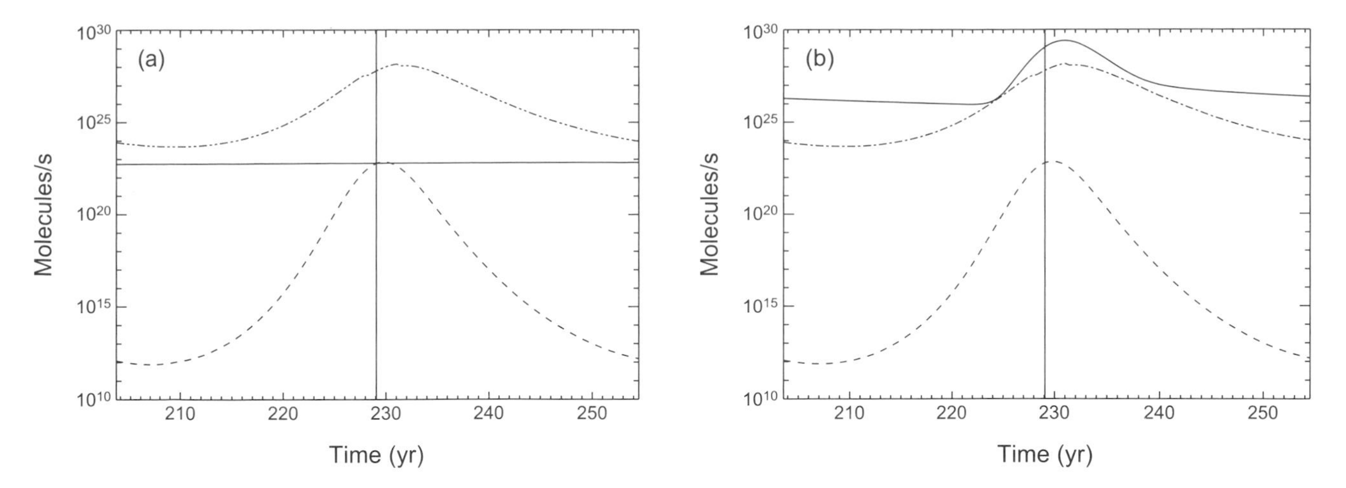

Fig. 5. Chiron models. **(a)** Gas fluxes from an old object when CO is not trapped in amorphous ice (line, CO; dash-dotted line, CO_2; dashed line, H_2O). **(b)** Same as **(a)**, but when the CO is trapped (from *Capria et al.*, 2000). The reason why some Centaurs are active and others are not is one of the main question about these class of objects, the answer to which can provide new and important constraints on the internal structure of KBOs.

Thermal models of active Centaurs indicate that the emission can be driven by volatile gas, as CO (*Capria et al.,* 2000; *Prialnik et al.,* 1995). Moreover, according to *Capria et al.* (2000), the present behavior of Chiron in terms of supervolatile (CO) emission can be explained only if the body is found in its present orbit with CO ice or CO gas trapped not too far from the surface (see Fig. 5). Given its probable origin in the Kuiper belt, Chiron should have arrived on its present orbit with the uppermost layers depleted of very volatile ices for a depth of some kilometers (*De Sanctis et al.,* 2001). If this is true, we could still explain observations of activity by supposing that water ice is mostly amorphous and some CO, trapped as a gas, is released at the transition to the crystalline phase. An alternative possibility could be some rejuvenation event, such as an impact or an orbit change, that ablated the first layers of the body. In any case Chiron should not have been on its present orbit for more than a few thousands of years. Chiron's activity implies that its internal structure preserves volatiles (trapped or as ices) in layers located not very far from the surface and that these layers must be at low temperatures.

The importance of Centaurs in the overall description of the evolution of icy bodies in the solar system is that they provide proof of the existence of a temporary storage of bodies, where they can undergo moderate thermochemical evolution. Further dynamical evolution can bring them either inside (giving origin to short-period comets) or outside the inner solar system.

6. OPEN POINTS

Given the previous discussion, we can state that the new observations available, the better comprehension of the dynamical evolution of the solar system through time (see all the work done in the framework of the "Nice Model"), and the efforts made in developing physically consistent models have allowed us to reasonably infer the overall behavior of icy bodies in the solar system and, in some cases, to predict the way in which activity develops; however, many open points still remain. In particular, we need to improve our knowledge in the following areas:

1. *Formation mechanism:* We have discussed the current theories of the formation of comets and KBOs, from which the overall chemical composition can be inferred; however, the formation mechanism mainly through binary collisions is very slow and the formation timescale is not sufficiently known to constrain the overall amount of short-lived radioactive elements that can be considered typical.

2. *Original composition and structure:* The amount of volatile elements is not known with accuracy, and can only be partially inferred for comets through their activity and the relative ratios of different gases. Cometary behavior became better understood when the concept of the diffusion of gases through a porous medium was developed. However, the interplay between macro- and microporosity in bodies with different sizes and overall strength is difficult to evaluate. It is also risky to assume that the composition of KBOs, Centaurs, and comets is the same. The path from KBO to comet is complex and partially unpredictable because of the role of impacts.

3. *Original mass distribution function:* The original mass distribution function is also related to the formation mechanism; however, this cannot be confirmed or disproved on the sole basis of observations due to the strong observational bias.

4. *Collisional evolution:* Collisional evolution can strongly modify the internal and surface evolution of different objects (*Orosei et al.,* 2001), but the energy distribution in a collision of icy bodies is not well known (*Durham et al.,* 2005).

REFERENCES

A'Hearn M. F., Belton M. J. S., Delamere W. A., Kissel J., Klaasen K. P., et al. (2005) Deep Impact: Excavating Comet Tempel 1. *Science, 310,* 258–264.

Anders E. and Grevesse N. (1989) Abundances of the elements — Meteoritic and solar. *Geochim. Cosmochim. Acta., 53,* 197–214.

Asphaug E. and Benz W. (1994) Density of comet Shoemaker-Levy 9 deduced by break-up of parent 'rubblepile.' *Nature, 370,* 120–124.

Astakhov S. A., Lee E. A., and Farrelly D. (2005) Formation of Kuiper-belt binaries through multiple chaotic scattering encounters with low-mass intruders. *Mon. Not. R. Astron. Soc., 360,* 401–415.

Bauer J. M., Meech K. J., Fernández Y. R., Pittichova J., Hainaut O. R., et al. (2003) Physical survey of 24 Centaurs with visible photometry. *Icarus, 166,* 195–211.

Blum J., Wurm G., Kempf S., Poppe T., Klahr H., et al. (2000) Growth and form of planetary seedlings: Results from a microgravity aggregation experiment. *Phys. Rev. Lett., 85,* 2426–2429.

Capria M. T. and Coradini A. (2006) The interior of outer solar system bodies. In *Asteroids, Comets, Meteors* (D. Lazzaro et al., eds.), pp. 395–411. IAU Symposium 229, Cambridge Univ., Cambridge.

Capria M. T., Coradini A., De Sanctis M. C., and Orosei R. (2000) Chiron activity and thermal evolution. *Astron. J., 119,* 3112–3118.

Capria M. T., Coradini A., De Sanctis M. C., and Blecka M. I. (2001) P/Wirtanen thermal evolution: Effects due to the presence of an organic component in the refractory material. *Planet. Space Sci., 49,* 907–918.

Capria M. T., Coradini A., and De Sanctis M. C. (2002) A model of the activity of comet Wild 2. *Adv. Space Res., 29,* 709–714.

Choi Y.-J., Cohen M., Merk R., and Prialnik D. (2002) Long-term evolution of objects in the Kuiper belt zone — Effects of insolation and radiogenic heating. *Icarus, 160,* 300–312.

Consolmagno G. J., Tegler S. C., Romanishin W., and Britt D. T. (2006) Shape, spin, and the structure of asteroids, Centaurs, and Kuiper belt objects (abstract). In *Lunar and Planetary Science XXXVII,* Abstract #1222. Lunar and Planetary Institute, Houston (CD-ROM).

Davidsson B. J. R. and Gutierrez P. J. (2004) Estimating the nucleus density of Comet 19P/Borrelly. *Icarus, 168,* 392–408.

Davidsson B. J. R. and Gutierrez P. J. (2006) Non-gravitational force modeling of Comet 81P/Wild 2. *Icarus, 180,* 224–242.

De Sanctis M. C., Capaccioni F., Capria M. T., Coradini A., Federico C., et al. (1999) Models of P/Wirtanen nucleus: Active regions versus non-active regions. *Planet. Space Sci., 47,* 855–872.

De Sanctis M. C., Capria M. T., Coradini A., and Orosei R. (2000) Thermal evolution of the Centaur object 5145 Pholus. *Astron. J., 120,* 1571–1578.

De Sanctis M. C., Capria M. T., and Coradini A. (2001) Thermal evolution and differentiation of Edgeworth-Kuiper belt objects. *Astron. J., 121,* 2792–2799.

De Sanctis M. C., Capria M. T., and Coradini A. (2003) Models of P/Borrelly: Activity and dust mantle formation. *Adv. Space Res., 31,* 2519–2525.

De Sanctis M. C., Capria M. T., and Coradini A. (2005) Thermal evolution model of 67P/Churyumov-Gerasimenko, the new Rosetta target. *Astron. Astrophys., 444,* 605–614.

De Sanctis M. C., Capria M. T., and Coradini A. (2007) Thermal evolution model of 9P/Tempel 1. *Astron J., 133,* 1836–1846.

Donn B. (1991) The accumulation and structure of comets. In *Comets in the Post-Halley Era* (R. L. Newburn et al., eds.), pp. 335–339. Kluwer, Dordrecht.

Donn B. and Duva J. (1994) Formation and properties of fluffy planetesimals. *Astrophys. Space Sci., 212,* 43–47.

Durham W. B., McKinnon W. B., and Stern L. (2005) Cold compaction of water ice. *Geophys. Res., 432,* L18230.

Epifani E., Palumbo P., Capria M. T., Cremonese G., Fulle M., and Colangeli L. (2006) The dust coma of the active Centaur P/2004 A1 (LONEOS): A co-driven environment? *Astron. Astrophys., 460,* 935–944.

Farinella P. and Davis D. R. (1996) Short-period comets: Primordial bodies or collisional fragments? *Science, 273,* 938–941.

Farinella P., Davis D. R., and Stern S. A. (2000) Formation and collisional evolution of the Edgeworth-Kuiper belt. In *Protostars and Planets IV* (V. Mannings et al., eds.), p. 1255. Univ. of Arizona, Tucson.

Fegley B. and Prinn R. G. (1989) Solar nebula chemistry — Implications for volatiles in the solar system. In *The Formation and Evolution of Planetary Systems,* pp. 171–205. Cambridge Univ., Cambridge.

Fernandez J. A. (1980) On the existence of a comet belt beyond Neptune. *Mon. Not. R. Astron. Soc., 192,* 481–491.

Festou M. C., Rickman H., and West R. M. (1993) Comets I: Concepts and observations. *Astron. Astrophys. Rev., 4,* 363–447.

Gladman B. (2005) The Kuiper belt and the solar system's comet disk. *Science, 307,* 71–75.

Gladman B., Kavelaars J. J., Petit J. M., Morbidelli A., Holman M. J., and Loredo T. (2001) The structure of the Kuiper belt: Size distribution and radial extent. *Astron. J., 122,* 1051–1066.

Goldreich P., Lithwick Y., and Sari R. (2002) Formation of Kuiper belt binaries by dynamical friction and three-body encounters. *Nature, 420,* 643–646.

Goswami J. N. and Vanhala H. A. T. (2000) Extinct radionuclides and the origin of the solar system. In *Protostars and Planets IV* (V. Mannings et al., eds.), p. 963. Univ. of Arizona, Tucson.

Grimm R. E. and McSween H. Y. (1993) Heliocentric zoning of the asteroid belt by aluminum-26 heating (abstract). In *Lunar and Planetary Science XXIV,* pp. 577–578. Lunar and Planetary Institute, Houston.

Grossman L. (1972) Condensation in the primitive solar nebula. *Geochim. Cosmochim. Acta, 36,* 597–619.

Hahn G. and Bailey M. E. (1990) Rapid dynamical evolution of giant comet Chiron. *Nature, 348,* 132–136.

Harper C. L. (1996) Astrophysical site of the origin of the solar system inferred from extinct radionuclide abundances. *Astrophys. J., 466,* 1026.

Huebner W. F. (2003) A quantitative model for comet nucleus topography. *Adv. Space Res., 31(12),* 2555–2562.

Huebner W. F. and Benkhoff J. (1999) From coma abundances to nucleus composition. *Space Sci. Rev., 90,* 117–130.

Huebner W. F., Benkhoff J., Capria M. T., Coradini A., De Sanctis M. C., et al. (2006) *Heat and Gas Diffusion in Comet Nuclei.* International Space Science Institute SR-004.

Jewitt D. C. and Luu J. (2004) Crystalline water ice on the Kuiper belt object (50000) Quaoar. *Nature, 432,* 731–733.

Jewitt D. C. and Sheppard S. S. (2002) Physical properties of trans-neptunian object (20000) Varuna. *Astron. J., 23,* 2110–2120.

Kenyon S. J. (2002) Planet formation in the outer solar system. *Publ. Astron. Soc. Pacific, 114,* 265–283.

Kenyon S. J. and Luu J. S. (1998) Accretion in the early Kuiper belt. I. Coagulation and velocity evolution. *Astron. J., 115,* 2136–2160.

Kenyon S. J. and Luu J. S. (1999a) Accretion in the early Kuiper belt. II. Fragmentation. *Astron. J., 118,* 1101–1119.

Kenyon S. J. and Luu J. S. (1999b) Accretion in the early outer solar system. *Astrophys. J., 526*, 465–470.

Levison H. F. and Duncan M. J. (1994) The long-term dynamical behavior of short-period comets. *Icarus, 108,* 18–36.

McKinnon W. B. (2002) On the initial thermal evolution of Kuiper belt objects. In *Proceedings of Asteroids, Comets, Meteors — ACM 2002* (B. Warmbein, ed.), pp. 29–38. ESA SP-500, Noordwijk, The Netherlands.

Merk R. and Prialnik D. (2003) Early thermal and structural evolution of small bodies in the trans-neptunian zone. *Earth Moon Planets, 92,* 359–374.

Meyer B. S. (2005) Synthesis of short-lived radioactivities in a massive star. In *Chondrites and the Protoplanetary Disk* (A. N. Krot et al., eds.), p. 515. ASP Conf. Series 341, San Francisco.

Meyer B. S. and Clayton D. D. (2000) Short-lived radioactivities and the birth of the Sun. *Space Sci. Rev., 92,* 133–152.

Morbidelli A. (2004) How Neptune pushed the boundaries of our solar system. *Science, 306,* 1302–1304.

Morfill P., Tscharnuter H., and Volk H. J. (1985) Dynamical and chemical evolution of the protoplanetary nebula. In *Protostars and Planets II* (D. C. Black and M. S. Matthews, eds.), pp. 493–533. Univ. of Arizona, Tucson.

Nazzario R. C. and Hyde T. W. (2005) Numerical investigations of Kuiper belt binaries (abstract). In *Lunar and Planetary Science XXXVI,* Abstract #1254. Lunar and Planetary Institute, Houston (CD-ROM).

Orosei R., Coradini A., De Sanctis M. C., and Federico C. (2001) Collisional-induced thermal evolution of a comet nucleus in the Edgeworth-Kuiper belt. *Adv. Space Res., 28,* 1563–1569.

Prialnik D., Brosch N., and Ianovici D. (1995) Modelling the activity of 2060 Chiron. *Mon. Not. R. Astron. Soc., 276,* 1148–1154.

Prialnik D., Benkhoff J., and Podolak M. (2004) Modeling the structure and activity of comet nuclei. In *Comets II* (M. C. Festou et al., eds.), pp. 359–387. Univ. of Arizona, Tucson.

Rabinowitz D. L., Barkume K., Brown M. E., Roe H., Schwartz M., et al. (2006) Photometric observations constraining the size, shape, and albedo of 2003 EL61, a rapidly rotating, Pluto-sized object in the Kuiper belt. *Astrophys. J., 639,* 1238–1241.

Rickman H. (1998) Composition and physical properties of comets. In *Solar System Ices* (B. Schmitt et al., eds.), p. 395. Kluwer, Dordrecht.

Schubert G., Turcotte D. L., and Olson P. (2001) *Mantle Convection in the Earth and Planets.* Cambridge Univ., Cambridge. 956 pp.

Stern S. A. (1996) On the collisional environment, accretion time scales, and architecture of the massive, primordial Kuiper belt. *Astron. J., 112,* 1203.

Stern S. A. and Colwell J. E. (1997a) Collisional erosion in the primordial Edgeworth-Kuiper belt and the generation of the 30–50 AU Kuiper gap. *Astrophys. J., 490,* 879.

Stern S. A. and Colwell J. E. (1997b) Accretion in the Edgeworth-Kuiper belt: Forming 100–1000 km radius bodies at 30 AU and beyond. *Astron. J., 114,* 841.

Sunshine J. M., A'Hearn M. F., Groussin O., Li J.-Y., and Belton M. J. S. (2006) Exposed water ice deposits on the surface of Comet 9P/Tempel 1. *Science, 311,* 1453–1455.

Trilling D. E. and Bernstein G. M. (2006) Light curves of 20–100 km Kuiper belt objects using the Hubble Space Telescope. *Astron. J., 131,* 1149–1162.

Wasserburg G. J., Busso M., and Gallino R. (1996) Abundances of actinides and short-lived nonactinides in the interstellar medium: Diverse supernova sources for the r-processes. *Astrophys. J. Lett., 466,* L109.

Weidenschilling S. J. (1997) The origin of comets in the solar nebula: A unified model. *Icarus, 127,* 290–306.

Weidenschilling S. J. (2002) On the origin of binary transneptunian objects. *Icarus, 160,* 212–215.

Weissman P. R. (1986) Are cometary nuclei primordial rubble piles? *Nature, 320,* 242–244.

Yamamoto T. (1985) Formation environment of cometary nuclei in the primordial solar nebula. *Astron. Astrophys., 142,* 31–36.

Yamamoto T. and Kozasa T. (1988) The cometary nucleus as an aggregate of planetesimals. *Icarus, 75,* 540–551.

Part V:
Formation and Evolution

The Scattered Disk: Origins, Dynamics, and End States

Rodney S. Gomes
Observatório Nacional

Julio A. Fernández and Tabaré Gallardo
Departamento de Astronomía, Facultad de Ciencias

Adrián Brunini
Universidad de La Plata

From a review of the methods that have been used so far, we estimate the present mass of the scattered disk to be in the range 0.01 to 0.1 $M_{\oplus}$. We review the dynamics of the scattered disk, paying special attention to the mean-motion and Kozai resonances. We discuss the origin of the scattered objects both as coming from the Kuiper belt and as remnants of the scattering process during Neptune's migration. We stress the importance of the mean-motion resonance coupled with the Kozai resonance in raising the perihelia of scattered disk objects, emphasizing that fossil and live high perihelion objects could thus have been produced. We analyze other mechanisms that could have implanted the detached scattered objects onto their current orbits, focusing on a few that demand specific explanations. We explore the different end states of scattered disk objects and highlight the importance of their transfer to the Oort cloud.

1. INTRODUCTION

The transneptunian (TN) population has a very complex dynamical structure whose details are being uncovered as more transneptunian objects (TNOs) are discovered. At the beginning, two dynamical groups appeared as dominant: (1) the *classical* belt, composed of objects in nonresonant orbits with semimajor axes in the range $42 \lesssim a \lesssim 48$ AU in low inclination and low eccentricity orbits; and (2) objects in mean-motion resonances (MMRs) with Neptune, the *Plutinos* in the 2:3 resonance being the most populous group, which present higher inclinations (*Jewitt et al.,* 1998). Then, object 1996 TL_{66} was discovered, belonging to a new category of bodies on highly eccentric orbits, perihelia beyond Neptune ($q > 30$ AU), and semimajor axes beyond the 1:2 resonance with Neptune, here considered for simplicity as $a > 50$ AU (*Luu et al.,* 1997). These bodies were called *scattered disk objects* (SDOs). The sample of discovered SDOs has increased to 96 objects (September 2006). This sample includes several objects with high perihelia (seven objects with $q > 40$ AU). These objects may form one or more subpopulations with respect to their possible origins. These subgroups have been referred to as high perihelion scattered disk, extended scattered disk, detached objects, or inner Oort cloud, the latter nomenclature usually associated with the detached objects with the largest semimajor axes (see chapter by Gladman et al. for a review on nomenclature). Although a couple of these objects deserve a specific explanation, covered elsewhere in this book, we in principle include all of them as SDOs according to the definition above and make a more comprehensive or a more specific analysis wherever suitable throughout this chapter.

From numerical integrations, *Levison and Duncan* (1997) were able to reproduce such a scattered disk from Kuiper belt objects (KBOs) (or TNOs) strongly perturbed by close encounters with Neptune. According to that work, the scattered disk (SD) would thus represent a transit population from the Kuiper belt to other regions of the solar system or beyond. On the other hand, *Duncan and Levison* (1997) suggested that the scattered disk could be a relic population of a primordial population of objects scattered by Neptune since the time of the early solar system. As we will analyze in this chapter, the question is to what extent a remnant population and a transient population coexist.

Although it is not a unanimous concept that the objects with $a > 50$ AU and $q > 30$ AU all share the same dynamical origin, we will however adopt this view since it is associated with a quite coherent global dynamical scenario. Since we are mostly concerned with origins, we also continue the use of the term *scattered* rather than *scattering,* the latter term suggested in the chapter by Gladman et al. Also, resonant orbits and those with $e < 0.24$ will be considered globally as scattered orbits provided the use of the basic definition above. Detached objects are here basically considered according to the definition in the chapter by Gladman et al., although we will give special attention to those with $q > 40$ AU.

This chapter is organized as follows. Section 2 describes the orbital configuration of the observed scattered objects and discusses the disk's mass estimate. Section 3 is devoted to the specific dynamics experienced by SDOs. In section 4, we analyze the processes that could have produced the orbits of SDOs and draw a conclusion as to where most of them should have come from. The particular subpopulation of detached SDOs, particularly their possible origins, is studied in section 5. We give special attention to dynamical processes within the known solar system. We also include the dynamical effect of a putative solar companion or rogue planet in driving the perihelia of SDOs beyond the neighborhood of Neptune. For mechanisms that invoke perturbations from an early dense galactic environment, we refer the reader to the chapter by Duncan et al. In section 6 we describe three different main end states for scattered disk objects, and our review is summarized in section 7.

2. ORBITAL CONFIGURATION AND MASS OF THE SCATTERED DISK

Figure 1 plots the different outer solar system populations in the parametric plane semimajor axis vs. perihelion distance. The SDOs occupy the upper right portion of the figure limited by the rightmost full vertical line and the thick dashed horizontal line.

The basic feature in the orbital distribution of scattered objects is that the perihelia are not much beyond Neptune's orbit. This is most likely related to their very origin as discussed in section 4. Semimajor axes are distributed from just beyond 50 AU to near 500 AU. It is expected that the distribution of semimajor axes shows a concentration (Fig. 1) of objects with relatively small semimajor axes due to the difficulty of observing distant objects. Although the perihelia of SDOs are usually not much above 30 AU, it is also a remarkable feature that a substantial number of objects do not come closer to the Sun than 36 AU, and very likely a nonnegligible fraction of them never comes closer than 40 AU. Above the thinner dashed line in Fig. 1, we see objects belonging to the *extended scattered disk* (*Gladman et al.,* 2002) or, following the nomenclature in the chapter by Gladman et al., detached objects with $q > 40$ AU. Although most of them have semimajor axes not much above 50 AU, two of them (Sedna and 2000 CR_{105}) have semimajor axes above 200 AU. These two objects may have had a specific formation process, as addressed in section 5 (see also the chapter by Duncan et al.). The inclinations of SDOs can be as low as 0.2° or as high as 46.8°. High inclinations can be attained by close encounters with the planets and/or by the Kozai mechanism inside a MMR (section 3).

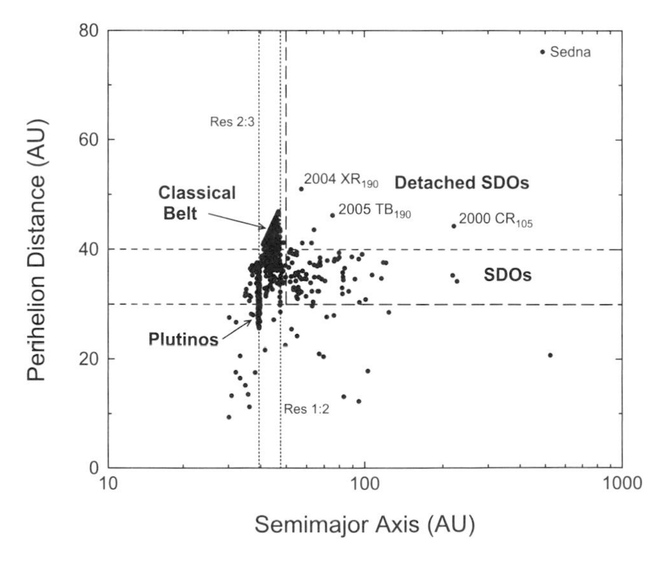

Fig. 1. Distribution of the different populations of outer solar system bodies in the plane a vs. q. The scattered disk objects occupy the upper right portion of the figure limited by the thick dashed line. The zone of the high-perihelion (or detached) SDOs ($q > 40$ AU) is also indicated and the names of the most prominent members are labeled. The objects were taken from the Minor Planet Center's website (*cfa-www.harvard.edu/iau/Ephemerides/Distant/Soft00Distant.txt*).

The population of SDOs with radius $R > 50$ km has been estimated by *Trujillo et al.* (2000) at $(3.1^{+1.9}_{-1.3}) \times 10^4$ bodies (1σ errors) and the total mass at 0.05 $M_\oplus$. Trujillo et al. considered the sample of four discovered SDOs at that time, which all had $q \leq 36$ AU. If we consider instead the SDOs up to $q = 40$ AU, Trujillo et al.'s estimate has to be multiplied by at least a factor of 2. Therefore, in the following, we will adopt a SD population of $\sim 6 \times 10^4$ objects with $R > 50$ km. An independent survey conducted by *Larsen et al.* (2001) led to the discovery of five Centaurs/SDOs and other two recoveries. From this survey they estimated a population of 70 SDOs brighter than apparent red magnitude $m_R = 21.5$. Applying appropriate bias corrections for distance in the detection probability, the estimated total population is in good agreement with that derived above. *Trujillo et al.* (2001) find that the differential size distribution of classical TNOs follows a power-law $dN \propto R^{-s}dR$, where the index $s = 4.0^{+0.6}_{-1.3}$ (1σ errors). If we assume that this size distribution also applies to SDOs and that the same exponent s holds down to a typical comet radius $R = 1$ km, the total population of SDOs is estimated to be

$$N_{SDO}(R > 1 \text{ km}) = 6 \times 10^4 \times 50^{(s-1)} \quad (1)$$

Taking $s = 4.0$ as the most likely value, we obtain $N_{SDO} = 7.5 \times 10^9$, but it may go up to (within 1σ) 7.8×10^{10}, or down to 1.1×10^9 bodies for $s = 4.6$ and 3.5 respectively. Therefore, there is an uncertainty larger than an order of magnitude in the estimated SD population. A recent deep survey with the HST/ACS camera carried out by *Bernstein et al.* (2004) suggests a smaller population of small TNOs than that predicted by an extrapolation of a power-law index $s = 4.0$. The turnover of the size distribution occurs at $D \lesssim 100$ km. The shallower size distribution for smaller bodies would imply a smaller total mass ≈ 0.01 $M_\oplus$ (Earth mass) for the classical belt, and a mass for the high inclination objects perhaps a few times larger (*Bernstein et al.,* 2004). Yet Bernstein et al. surveyed a very small sky area of 0.02 deg² and discovered only three TNOs with diameters between 25 and 44 km (for an assumed albedo 0.04). We have here the problem of small-number statistics, so these results should be considered with caution.

Delsanti and Jewitt (2006) argue that *Trujillo et al.*'s (2001) index near 4.0 ± 0.5 for the differential size distribution will slightly flatten toward *Dohnanyi*'s (1969) value of 3.5 due to collisional shattering. *Delsanti and Jewitt* (2006) estimate for the classical disk a mass of a few percent $M_\oplus$ taking into account that large TNOs have high albedos. They also conclude that the mass of the scattered disk should be larger than the mass of the other components of the TN population.

Given the uncertainties, we conclude that the mass of the scattered disk might be somewhere in between 0.01 and 0.1 $M_\oplus$, i.e., comparable to the mass of the classical belt. This does not include the mass of objects belonging to the inner core of the Oort cloud, of which Sedna appears to be the most promising candidate (see section 5).

3. DYNAMICS OF THE SCATTERED DISK

In the transneptunian region the bodies are so weakly linked to the Sun that the osculating heliocentric orbits show important short-period oscillations due to the gravitational effects of the giant planets on the Sun. The semimajor axis is the most affected orbital element, oscillating around a mean value that coincides with the barycentric a. For this reason the barycentric orbital elements better represent their dynamical states.

A small body orbiting in the transneptunian region can experience basically two types of dynamical evolutions: a stochastic evolution driven by encounters with the planets and a more regular evolution driven by the continuous and regular gravitational effect due to the planets. The first one, associated with a random evolution of the body's semimajor axis, can only be studied by numerical simulations or by statistical methods but for the second one, which conserves a constant mean value for a, several known results of the secular theory can be applied.

The stochastic evolution of a SDO in general occurs when $q < 36$ AU and is mainly due to encounters with Neptune (*Duncan and Levison,* 1997; *Fernández et al.,* 2004). In analogy with encounters of comets with Jupiter, after each encounter the SDO's orbital elements should conserve the Tisserand parameter with respect to Neptune

$$T = \frac{a_N}{a} + 2\sqrt{\frac{a}{a_N}(1 - e^2)}\cos i$$

where a_N and a are respectively Neptune's and the body's semimajor axes, e is the body's eccentricity, and i its inclination with respect to Neptune's orbit. Strictly speaking, the conservation of T is only valid in the circular three-body problem, although it may be a good approximation when the object approaches Neptune to less than a few AU. In this case the evolution of the body will proceed stochastically (unless it is in a MMR), random-walking in the energy space and keeping its perihelion close to Neptune's orbit. However, conservation of T does not apply strictly to the objects with the largest perihelia in the SD, because the gravitational effects of the other giant planets with respect with those of Neptune are no longer negligible. Moreover, if after close approaches to Neptune, the body is transferred inward, falling under the gravitational influence of Uranus, Saturn, or Jupiter, the near constancy of T with respect to Neptune will also break down.

Concerning the more regular evolution, the theory allows us to distinguish here between two kinds of dynamics according to the terms that dominate the Lagrange-Laplace planetary equations (*Murray and Dermott,* 1999). In one case, the evolution is dominated by long-period terms that appear in the planetary equations and yields a slow time evolution of the orbital elements, the so-called secular dynamics. In the other case, for specific values of the semimajor axis, a different kind of evolution appears due to long-period terms involving mean longitudes λ and λ_N. In this case, we say that the dynamics are dominated by a MMR. Once in resonance, the body can experience secular dynamics, which are in general different from that of nonresonant bodies. Below we will examine in more detail these two types of analytically predictable motions.

3.1. Secular Dynamics

For $q \geq 36$ AU, near encounters with the planets are not possible and the object experiences a more regular evolution, very similar to the secular theory predictions. In particular, its barycentric semimajor axis a oscillates around a constant mean value $\bar{a}$ and the longitude of the perihelion ϖ and the longitude of the node Ω have constant rates of precession.

In general, the time evolution of the argument of the perihelion $\omega = \varpi - \Omega$ is a circulation coupled with low-amplitude oscillations of e and i. Besides the invariance of $\bar{a}$, the secular evolution imposes the preservation of $H = \sqrt{1 - e^2}\cos i$ (when planetary eccentricities and inclinations are neglected), known as Kozai dynamics (*Kozai,* 1962), and where the inclination is measured with respect to the invariable plane of the planetary system. If $i \simeq 63°$ (also known as *critical inclination*), then $\dot{\omega} \simeq 0$. In this case, ω oscillates. If e is large enough, then high-amplitude coupled terms appear in the time evolution of (e, i) (Kozai resonance). The conservation of H is analogous to the conservation of the Tisserand parameter but the former is a property of the dynamics induced by all the planets and not just by Neptune.

If we consider only this secular evolution, the secular theory tells us that the circulation frequencies of ϖ and Ω in the SD are very small compared with the fundamental frequencies of the solar system, so that no secular resonances (or at least no first-order secular resonances) are possible in the SD. Secular resonances occur when the proper frequencies of the SDOs are commensurable with the fundamental frequencies of the planetary system. When the invariable plane of the solar system is used as reference plane instead of the ecliptic, the fundamental frequencies are related to the circulation frequencies of the elements ϖ and Ω of the planets. For instance, in the classical belt the region between 40 and 42 AU is occupied by the secular reso-

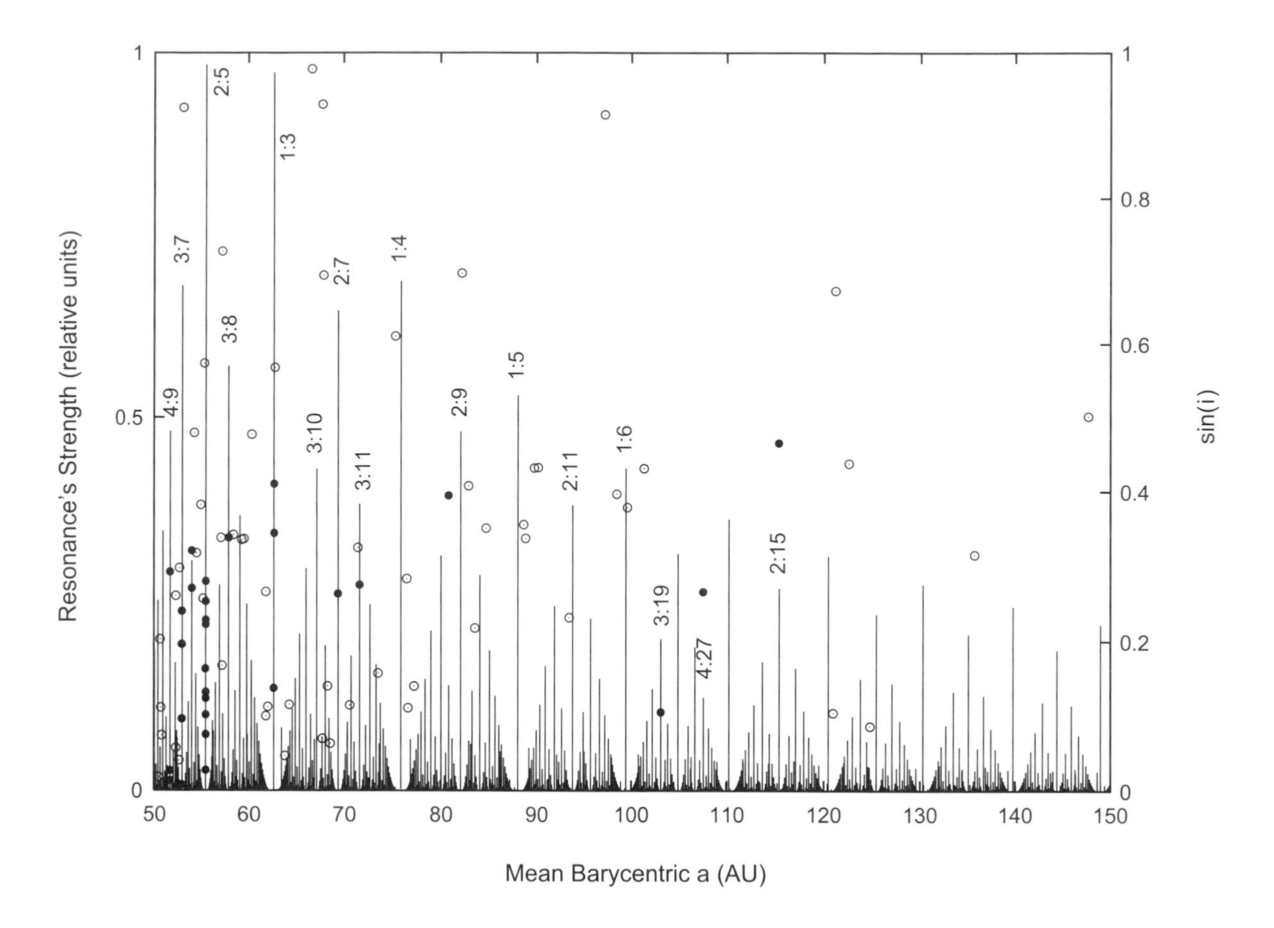

Fig. 2. Localization and strength of the MMRs in the region 50 < a < 150 AU. The strengths of all resonances were calculated assuming orbits with q = 32 AU, i = 20°, and ω = 60°. The known SDOs are plotted in the (a, sin i) parametric plane in order to appreciate their distribution with respect to the resonance's positions. Full circles indicate objects identified in MMRs with librating critical angles (see Table 1) and open circles indicate objects that were not associated with a resonance. In this plot a is the mean barycentric value after a numerical integration of 10^5 yr.

nance ν_8, in which the precession rate of the body's perihelion matches that of Neptune, and ν_{18}, which involves the precession of the nodes (*Kneževič et al.,* 1991). Secular resonances are important to excite larger eccentricities and inclinations.

Due to the absence of secular resonances the only mechanism that can substantially modify the orbital elements (e, i) of a SDO is the Kozai resonance (KR), and this occurs only for high-inclination orbits (say, i ≳ 50°) (*Thomas and Morbidelli,* 1996; *Gallardo,* 2006a). Then, typical SDOs with perihelion distances q > 36 AU and with low-inclination orbits would have a stable time evolution with quasi-constant (a, e, i) values provided that the body is outside a MMR. If the body is in a MMR, the KR can occur for much lower inclinations (see below).

3.2. Mean-Motion Resonances

Numerical simulations have shown that SDOs, having a stochastic evolution in their semimajor axes, experience temporary captures in high-order MMRs with Neptune (*Duncan and Levison,* 1997; *Gladman et al.,* 2002; *Gomes,* 2003; *Fernández et al.,* 2004; *Gomes et al.,* 2005b), suggesting that MMRs can play an important role in the dynamics of the SD. A MMR designed as |p + q| : |p| occurs when the particle's and Neptune's mean motions are commensurable and their locations in barycentric semimajor axis coordinate are given by

$$a_{res} \simeq \frac{a_N}{(1 + m_N)^{1/3}} \left(\frac{p}{p + q} \right)^{2/3}$$

where m_N is Neptune's mass in solar masses. The integer q > 0 is the *order* of the resonance and p is the *degree,* which is negative for exterior resonances. All resonances in the SD are exterior to Neptune, therefore p < 0. It is possible to associate a strength with each resonance, which depends on the elements (e, i, ω) of the resonant orbit (*Gallardo,* 2006a,b). These are shown in Fig. 2 for the region between 50 and 150 AU together with the known SD population in that region. The resonance's strength grows with the eccentricity in such a way that, for very high eccentricity orbits, all resonances are strong enough to be relevant.

TABLE 1. Candidate SDOs in MMR with Neptune.

Resonance	a (AU)	Name
4:9	51.72	(42301) 2001 UR_{163}, 2001 KG_{76}, 2001 QW_{297}
3:7	52.99	(95625) 2002 GX_{32}, 2001 XT_{254}, 1999 CV_{118}, 2004 DK_{71}
5:12	53.99	(79978) 1999 CC_{158}, (119878) 2002 CY_{224}
2:5	55.48	(69988) 1998 WA_{31}, (38084) 1999 HB_{12}, (119068) 2001 KC_{77}, (26375) 1999 DE_9, (60621) 2000 FE_8, 2000 SR_{331}, 2002 GP_{32}, 2002 GG_{32}, 2003 UY_{117}, 2004 EG_{96}, 2001 XQ_{254}
3:8	57.92	(82075) 2000 YW_{134}
1:3	62.65	2003 LG_7, 2000 YY_1, 2005 EO_{297}
2:7	69.43	2001 KV_{76}
3:11	71.62	(126619) 2002 CX_{154}
5:22	80.88	2004 TF_{282}
3:19	103.10	(29981) 1999 TD_{10}
4:27	107.58	2004 PB_{112}
2:15	115.40	1999 CZ_{118}
1:18	206.87	2002 GB_{32}
4:79	220.07	2000 CR_{105}

Then, very high eccentricity orbits will be in general affected by a MMR regardless of the specific semimajor axis.

Using frequency map analysis, *Robutel and Laskar* (2001) identified several high-order exterior MMRs in the region $a < 90$ AU and found the existence of a chaotic region dominated by the superposition of resonances at high eccentricity. They also found that for high inclinations the resonances are wider than for low inclinations. This is in agreement with *Gallardo* (2006a,b), who finds that, in general, with the exception of Neptune's Trojans and first-order MMRs, resonances are stronger for high-inclination orbits.

Gladman et al. (2002) found the empirical result that, for SDOs with $q < 30 + 0.085(a - 30)$ AU and Pluto-like inclination orbits, the dynamics are typically dominated by the mechanism known as *resonance sticking*. This is a dynamical evolution characterized by jumping between different, but near, MMRs, which is possible thanks to the superposition of several high-order MMRs. In this case the semimajor axis evolves chaotically between resonances but in a very long timescale, so that the body remains essentially confined to semimajor axes $a \lesssim 150$ AU over timescales comparable to the age of the solar system (see section 6.2).

A possible mechanism for the capture into a resonant motion is given by the already referred stochastic evolution of the semimajor axis caused by planetary perturbations. Eventually $a \simeq a_{res}$ and the resonant terms of the disturbing function start to dominate the orbital evolution, thus halting the chaotic evolution of a. It must be noted that a MMR offers a natural protection mechanism against close encounters between a pair of bodies. For instance, because Pluto is in the 2:3 MMR resonance with Neptune, the two objects never get closer than about 18 AU to each other, thus preventing Pluto from receiving strong energy kicks during its perihelion passages. If the resonance strength is small, the object will not likely be captured into the resonance since planetary perturbations can overcome the resonance strength and shift a from a_{res}. Resonances of the type 1:n and 2:n with Neptune are relatively strong and isolated from others so they should be the most populated (see Fig. 2). Other resonances could be strong but surrounded by other strong resonances, so the SDO that falls into any of them will evolve by the mechanism of resonance sticking. Table 1 brings a list of the observed SDOs identified in a MMR with Neptune, as coming from a numerical integration based on their nominal orbits. All of them show libration of the critical angle for at least 10^5 yr. The second column is the mean barycentric semimajor axis corresponding to the resonance. Table 1 is in agreement with Table 2 of the chapter by Gladman et al., which brings a more restricted but more accurate sample of resonant SDOs obtained by a more rigorous process.

Once the SDO is evolving inside an isolated MMR the resonant theory predicts that the body's orbital elements will show very small amplitude oscillations (librations) in (e, i) and somewhat more evident oscillations in a. All these are librating with the same frequency of the critical angle $\sigma = (p + q)\lambda_N - p\lambda - q\varpi$. This is an important difference with respect to librations of asteroids in MMR with Jupiter where the proximity and mass of the planet make the amplitude of the librations clearly greater than in the case of resonances with Neptune. But contrary to what we can expect from the theory of the resonant motion, it has been found (*Duncan and Levison,* 1997) that most commonly the eccentricity and the inclination show notable and slow variations superposed to the almost negligible and relatively quick oscillations due to the resonant motion. This is not due purely to the MMR but also to a secular evolution of the angular element ω that imposes notable excursions of (e, i) (*Gallardo,* 2006a). These excursions are due to the Kozai resonance inside the MMR and will be discussed in section 5.1.

Other secular resonances inside a high-order MMR with Neptune could occur but this point has not yet been investigated. Three-body resonances involving mean motions of Uranus, Neptune, and a SDO are also present in the SD (*Morbidelli,* 2002) and they can possibly contribute to the chaotic evolution of some SDOs but this is not a well-investigated issue so far.

4. THE SCATTERED DISK: PRIMORDIAL RECORD OR TRANSIENT POPULATION?

We shall discuss now the origin of the SD. One possibility is that it is a byproduct of the formation of the jovian planets and of the planet migration that was caused by a massive scattering of planetesimals. On the other hand, SDOs may come from the classical belt through processes including collisions (*Davis and Farinella,* 1997) and a slow diffusion in which MMRs and secular resonances play a fundamental role.

4.1. Migration Through a Planetesimal Disk

According to the classical conjecture (*Kuiper,* 1951), there should be a disk of objects beyond Neptune that, due to its low surface density, could not accrete to form planet-sized bodies. This naturally yields the idea that, at some primordial time, there should have been a disk of planetesimals that extended from the giant planets region up to the putative Kuiper belt outer edge. This orbital configuration of planets and planetesimal disk allows the exchange of energy and angular momentum between particles and planets at close encounters. This process induces a planetary migration during which Neptune, Uranus, and Saturn migrate outward while Jupiter migrates inward (*Fernández and Ip,* 1984; *Hahn and Malhotra,* 1999). So if we assume that at some time in the past the giant planets coexisted with a disk of planetesimals extending fairly beyond the outermost planet, then we have to conclude that the giant planets were originally on orbits with much smaller mutual separations. The original planetary separations are discussed in several works (*Hahn and Malhotra,* 1999; *Gomes,* 2003) and it is possible that Neptune was originally below 20 AU, probably around 15 AU (*Gomes,* 2003; *Tsiganis et al.,* 2005).

The migration mechanism is fueled by a huge number of close encounters between the planetesimals and the planets. When Neptune reaches the edge of the disk (*Gomes et al.,* 2004) its radial drift is halted (not abruptly but asymptotically). At this point, a great number of planetesimals will be on orbits that already suffered scattering by Neptune, although their cumulative mass is now small enough not to produce any more significant migration of the planet. It is thus a reasonable conclusion that a population of objects scattered by Neptune existed just as the migration calmed down. It is also intuitive to conclude that the present scattered population is composed of those objects that managed to survive up to the present time even though experiencing at some level close encounters with Neptune. In other words, the present scattered objects are the remnants of a much larger scattered population of objects that had perihelia just beyond Neptune's orbit when planetary migration calmed down (*Duncan and Levison,* 1997).

Estimates of the original planetesimal disk mass range from 50 $M_\oplus$ (*Hahn and Malhotra,* 1999) to 35 $M_\oplus$ (*Tsiganis et al.,* 2005; *Gomes et al.,* 2005a; *Morbidelli et al.,* 2005). This last number is based on the migration dynamics in a truncated disk that yields good final positions for the giant planets (*Tsiganis et al.,* 2005). Numerical simulations of the migration process extended to the solar system age yield 0.2–0.4% (*Gomes,* 2003; *Gomes et al.,* 2005b) of the original planetesimal disk mass as left over in the present SD. These simulations also produce a roughly equal number of (hot) classical Kuiper belt objects. A model simulation originally designed to explain the origin of the late heavy bombardment (LHB) on the terrestrial planets (*Gomes et al.,* 2005a), extended for the solar system age, yielded as much as 0.14 $M_\oplus$ in the total transneptunian population, with 0.08 $M_\oplus$ in the scattered disk. These numbers agree fairly well with observational estimates (see section 2) confirming that the current scattered disk could be produced by the interaction of the primordial planetesimal disk with a migrating Neptune.

Besides the total mass of the scattered disk, it is natural to ask whether the orbital distribution of the SDOs is in agreement with the final outcome of the migration process. As the scattering process proceeds, the average semimajor axis of a body steadily departs from the original value in the protoplanetary disk. However, the semimajor axis is not a good parameter to use for comparisons of simulations with observations, since there must be a great bias in the observational data that favors the discovery of the bodies on the lowest semimajor axes orbits (*Morbidelli et al.,* 2004). Although not quite bias-free, it may be interesting to compare the distribution of perihelion distances and inclinations of the real scattered objects with those coming from a numerical integration extended to the solar system age. This comparison is shown in Fig. 3 where the black dots refer to the real objects and the gray ones to a numerical integration extended to solar system age, following the LHB model (*Gomes et al.,* 2005a). It is interesting to note that both black and gray dots occupy about the same region in the q vs. i parametric space. Sedna is surely a remarkable exception, pointing to another explanation for its origin in addition to the perturbations from the known planets (see section 5). Although 2000 CR_{105} may have an origin similar to Sedna's (*Morbidelli and Levison,* 2004), this is not clearly concluded from Fig. 3. Another difference between the real and simulated distributions concerns the greater number of low-inclination objects in the real population as compared with the simulated one. However, this is probably due to observational bias that favors the discovery of low-inclination objects.

4.2. From the Classical Kuiper Belt

In the picture outlined before, the present scattered disk is the remnant of a much more numerous primordial population, stored on orbits with perihelion distances near Neptune, early in the history of the solar system.

In this section we will discuss also a possible contribution from the Kuiper belt to the scattered disk, and, more specifically, from the *present* Kuiper belt to the scattered disk.

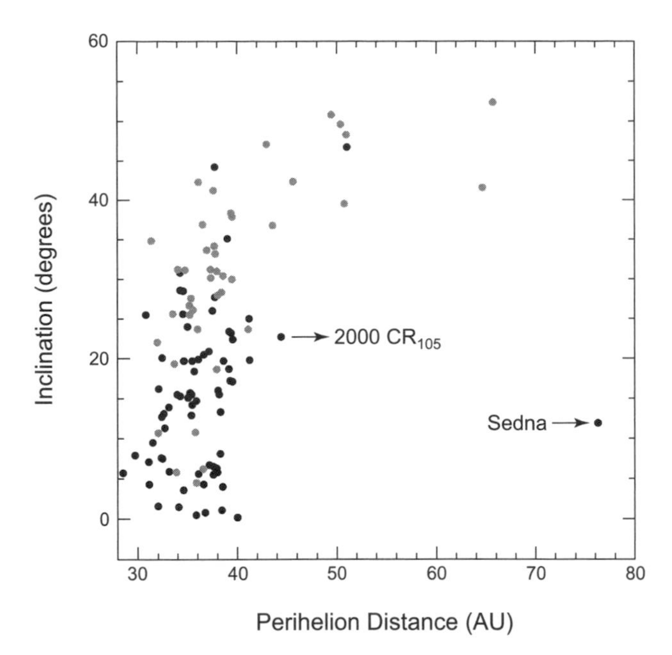

Fig. 3. Distribution of perihelion distances with inclinations for real scattered objects (black) and objects obtained in a numerical simulation extended to solar system age (gray). The numerical simulation followed the LHB model (*Tsiganis et al.,* 2005; *Gomes et al.,* 2005a).

Duncan et al. (1995) have analyzed the stability of test particles in the TN region. They showed that weak dynamical instabilities are capable of producing an influx of objects from the KB to the SD, and also that the chaotic diffusion in the Kuiper belt is complex and associated with the superposition of MMRs and secular resonances. In the region occupied by the Plutinos, namely the 2:3 MMR with Neptune, the instability timescale ranges from less than a million years near the borders of the resonance, to longer than the age of the solar system, deep inside the resonance (*Morbidelli,* 1997). Performing long-term dynamical simulations of test particles on initially low-eccentricity orbits, under the gravitational perturbation of the giant planets, *Nesvorný and Roig* (2001) also found escape routes from the Kuiper belt to the scattered disk via the chaotic borders of the strong 2:3 and 1:2 MMRs with Neptune, which are known to be occupied by KBOs. They also showed that weaker resonances, such as the 5:7, 8:11, 7:10, and 9:13, are possible routes of chaotic diffusion to the scattered disk, and that large-eccentricity excitation may also occur for KBOs located at some other high-order resonances with Neptune and three-body resonances with Uranus and Neptune. The region just beyond the 2:3 resonance and up to 43 AU presents a drop in the number of known KBOs. *Nesvorný and Roig* (2001) computed the maximum Lyapunov Characteristic Exponent of thousands of particles in this region. They have shown that the dynamics are very complex there, and that the characteristic diffusion times are nearly 2 orders of magnitude shorter than beyond 43 AU. This result was recently confirmed by long-term numerical integrations (*Jones et al.,* 2005). The instability in the 40–42 AU region is due to secular resonances (*Duncan et al.,* 1995).

Lykawka and Mukai (2006) found that inside the 4:7 resonance with Neptune (semimajor axis a = 43.6 AU) there are regions of large mobility in phase space, and that bodies inside these regions typically leave the resonance and are subsequently scattered by Neptune. This result is particularly interesting because recent direct numerical integration of trajectories of real KBOs have shown that the 4:7 and 2:5 resonances are inhabited (*Chiang et al.,* 2004).

The chaotic diffusion associated with the resonances has produced a gradual erosion of the primordial Kuiper belt that continues at present. In addition, objects inside regions dynamically stable for timescales comparable or even longer than the age of the solar system can be moved to unstable places in phase space by two mechanisms:

1. Encounters with 1000-km KBOs: *Yu and Tremaine* (1999) have shown that Pluto plays some role in shaping the orbital distribution of Plutinos, in the sense that encounters with Pluto can drive some of them out of the 2:3 MMR. They argued that this mechanism may be a source of Jupiter-family comets. The existence of an object of comparable size within other resonances would have similar significant effects. Nevertheless, none of the 1000-km-sized objects known so far in the Kuiper belt are resonant objects. If any, they await to be discovered.

2. Collisional activity: The other possibility to populate in a significant way the scattered disk from the present Kuiper belt is via the production of fragments by collisional activity. *Davis and Farinella* (1997) and *Stern and Colwell* (1997) have performed the first time-dependent collisional evolution simulations in the Kuiper belt region. They conclude that the Kuiper belt is an active collisionally evolved population. Although *Chiang et al.* (2004) found that no rigorously convincing collisional family can be identified among the nonresonant KBOs they tested, it does not mean that the collisional activity in the Kuiper belt is not intense, because most collisional families are probably dispersed by the slow chaotic diffusion of the numerous narrow resonances present in the Kuiper belt (*Nesvorný and Roig,* 2001). On the other hand, *Brown et al.* (2007) have recently presented dynamical and spectroscopic evidence that the large KBO 2003 EL_{61}, and five other much smaller KBOs, belong to a family that resulted from a near catastrophic collision of the proto-2003 EL_{61}. The typical impact velocity in the Kuiper belt is 1–2 km s^{-1}, a value that largely exceeds the surface velocity for a typical KBO of size 100–200 km. As a result, *Davis and Farinella* (1997) estimated that about 10 fragments per year of 1–10 km in size are currently produced in the Kuiper belt region. This number has an uncertainty factor of ~4 depending on the assumed collisional response parameter (*Davis and Farinella,* 1997). The relative velocities of these fragments are enough to change their semimajor axes by an amount of 0.1–1.0 AU relative to those of their parent bodies. This is sufficient to inject at least a fraction of these fragments into the unstable paths associated with MMRs or secular resonances, and drive them out of the Kuiper belt. Furthermore, *Pan and Sari* (2005) argued that KBOs are virtually strengthless bodies held together mainly by gravity (i.e., rubble piles), enhancing the capability of producing fragments from disruptive

collisions, and the flux of them to the scattered disk. Yet, we still know very little about the internal strength of KBOs, so we should be very cautious about predictions regarding the outcomes of collisions.

We can conclude that, in addition to the scattered disk coming from a primordial scattering process by Neptune, there is a certain flux to the SD of objects coming continuously from the Kuiper belt, either by dynamical mobility within or near the numerous MMRs or by the injection of small fragments produced in the intense collisional activity that the Kuiper belt presents.

In a steady-state scenario, the mass of the scattered disk objects coming from the Kuiper belt must be a small fraction of that of the KB population. However, observational evidence indicates that the SD and the KB populations have roughly the same mass (cf. section 2), so we can conclude that most large scattered objects should be remnants of a primordial, much larger population scattered by Neptune during its primordial migration. Moreover, migration simulations (section 3.1) roughly reproduce the mass in the SD estimated from observations (*Gomes,* 2003; *Gomes et al.,* 2005b) and its orbital distribution.

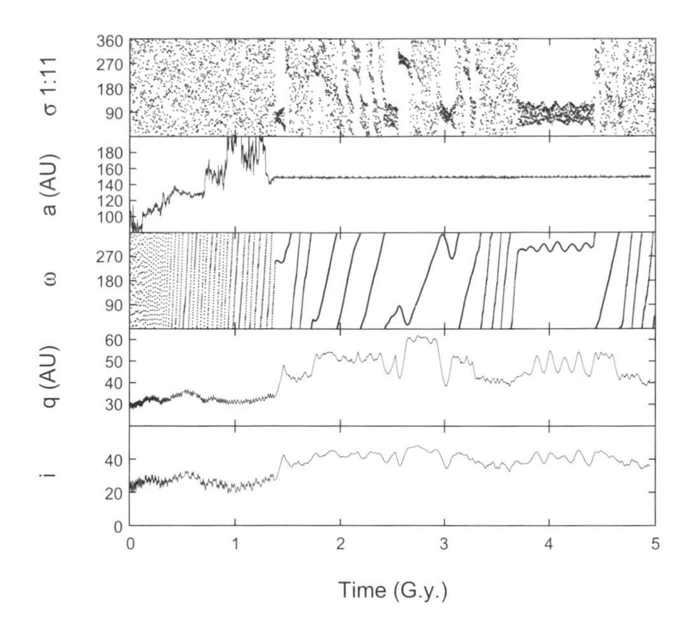

Fig. 4. Orbital evolution of a test particle captured into the 1:11 resonance and then entering into the Kozai resonance. The test particle is a clone of SDO 1999 RZ_{215} (*Gomes et al.,* 2005b).

5. THE FORMATION OF A HIGH-PERIHELION SCATTERED DISK

There are at present (September 2006) seven objects with a > 50 AU and q > 40 AU. These are 2003 VB_{12} (Sedna), 2004 XR_{190}, 2004 PD_{112}, 2000 CR_{105}, 2000 YW_{134}, 2005 EO_{297}, and 2005 TB_{190} (Fig. 1). These objects represent a conspicuous population among the detached objects and will thus be given special attention in this section. Although it is not unanimous that all detached objects come from the scattered disk, we will adopt this view here to be coherent with the scope of this chapter. In the following three subsections we discuss some mechanisms that can raise the perihelia of SDOs and give a tentative classification of each of these seven objects according to its formation mechanism.

5.1. Transfer of Scattered Disk Objects from Small q to High q via Secular Dynamics Inside Mean-Motion Resonances

Inside a MMR the most important secular effect is the Kozai resonance, which seems to be the rule for resonant orbits with inclinations comparable to that of Pluto or larger (*Gallardo,* 2006a). We have stated before that outside a MMR, the KR can only occur at very high inclinations, but inside a MMR the KR can act at lower inclinations. For resonant orbits with Pluto-like or greater inclination, the time evolution of ω slows down, generating nonnegligible terms in the resonant equations. These terms cause notable changes of the orbit's (e, i). Figure 4, reproduced from *Gomes et al.* (2005b), shows a particle that is captured in resonance 1:11 for which the KR starts to act almost immediately. However, for very low inclination orbits, the Kozai resonance does not appear, so in those cases capture into a MMR is possible, but not associated with strong variations in (e, i).

The coupled MMR + KR mechanism is able to create scattered objects with a quite high perihelion up to ~60 AU, as suggested by Fig. 4. Four observed objects with q > 40 AU can be associated with this mechanism. These are 2004 XR_{190}, 2000 YW_{134}, 2005 EO_{297}, and 2005 TB_{190}. The first one is near the 3:8 MMR with Neptune, with, however, a semimajor axis a little smaller than the value corresponding to the resonance center. It is possible that 2004 XR_{190} has escaped the resonance while Neptune was still migrating (see section 6.1). On the other hand, 2000 YW_{134} is well inside the same 3:8 resonance and is presently experiencing the MMR + KR mechanism (see chapter by Gladman et al.). Numerical integrations of the object 2005 EO_{297} indicate that it is in the 1:3 resonance and could also be experiencing the KR. This can also be the case for 2005 TB_{190}, which is near the 1:5 resonance. As seen in Fig. 5, the eccentricity and inclination of a body under the combined effect of a MMR + KR are coupled through the condition $H = \sqrt{1 - e^2} \cos i$, where H is constant. This allows us to see the minimum perihelion distance allowed for the body and check if it could come from the SD by the MMR + KR mechanism. The answer is negative for 2004 PD_{112} and also for 1995 TL_8, whose q is 39.986 AU (and therefore close to our arbitrary boundary of the detached population, placed at q = 40 AU). The origin of these objects is presently obscure but they seem to share the same origin as the cold population in the classical Kuiper belt (see chapter by Morbidelli et al.).

5.2. Perturbations from External Agents

The dynamical mechanism that can raise the perihelia of scattered objects discussed in section 5.1 can account for many high-perihelion orbits of SDOs. Exceptions are, at

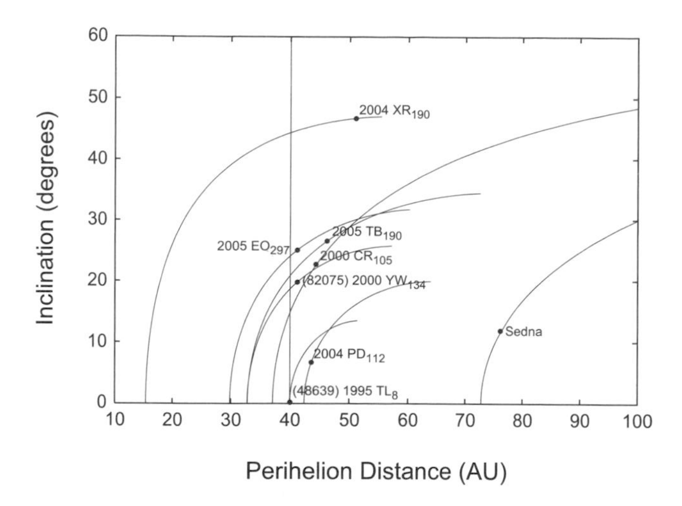

Fig. 5. Location of seven observed detached objects plus 1995 TL_8 in the parametric plane (q, i) with the corresponding curves H = *constant*. These curves show the possible evolution of e and i due to the KR.

small a, the low-inclination objects (see above comments); at the other extreme, there are two detached objects with large semimajor axes whose orbital origins also cannot be explained by the KR perihelion raising mechanism. One of these is 2000 CR_{105}. It is located at an average distance from the Sun of 221 AU and has a perihelion distance at 44.4 AU. Although numerical integrations may show, at solar system age, objects with semimajor axis and perihelion near 2000 CR_{105} values (*Gomes et al.,* 2005b), the probability of producing objects on orbits with CR_{105}-like perihelion distance but smaller semimajor axis is much larger. This implies that a greater number of CR_{105}-type objects should have been discovered on orbits with smaller a. Furthermore, our numerical experiments also show that the semimajor axes of SDOs with q > 36 AU hardly exceeds a ~ 150 AU (see section 6.2 below). Thus one can conclude with some confidence that the mechanism that raised the perihelion of 2000 CR_{105} must be different from the one described in section 5.1. The second object with large perihelion and semimajor axis is 2003 VB_{12} (Sedna). Its mean barycentric semimajor axis is 505 AU and the object gets no closer to the Sun than 76 AU. Undoubtedly Sedna demands another explanation for its orbit since the mechanism in section 5.1 cannot by any means explain such a high perihelion. *Brown et al.* (2004) estimated a total mass of 5 $M_\oplus$ for the population of objects on Sedna-like orbits. This estimate is still highly uncertain because it rests on a single discovery, but it is very likely that Sedna hints at the existence of a substantial population, at least an order of magnitude more massive than that of the Kuiper belt (see chapter by Brown).

Several theories have been proposed to account for Sedna's high-perihelion orbit. A theory based on a probable primordial scenario that yields generally good results can be referred to as the "Sun in a star cluster model" (see chapter by Duncan et al.). It is based on the assumption that the Sun was formed in a star cluster embedded in a large molecular cloud (*Lada and Lada,* 2003). This theory considers that objects scattered by the giant planets at the primordial time could have the perihelia lifted through the effect of passing stars and tides from the molecular gas (*Fernández and Brunini,* 2000; *Brasser et al.,* 2006). It thus generalizes the effect of a single star passage considered in *Morbidelli and Levison* (2004). The best result is obtained for a cluster with an average density of 10^5 $M_\odot$ pc^{-3} (*Brasser et al.,* 2006). In this case both Sedna and 2000 CR_{105} are well located inside the cluster-produced high-perihelion scattered population.

5.2.1. Perturbations from a solar companion. This section is devoted to a particular mechanism (not included in the chapter by Duncan et al.) that can create very high perihelion orbits from the scattered disk. Perihelion lifting can be experienced by scattered objects through the perturbation of a solar companion (*Gomes et al.,* 2006). This effect is produced by secular and/or Kozai resonances induced by the companion. The precessions of perihelia and nodes of distant objects are very slow and can often match the precession of the companion's node and perihelion, thus raising secular perturbations. Populations of detached objects, which naturally include Sedna, can be produced by a distant planet orbiting the Sun with semimajor axes ranging roughly from 10^3 to 10^4 AU and masses from a fraction of an Earth mass to several Jupiter masses. There is a relationship among semimajor axis, eccentricity, and mass of the companion (more concisely semiminor axis and mass) that induces similar orbital distributions for the detached population. This strength parameter is given by $\rho_c \equiv M_c/b_c^3$ where $b_c \equiv a_c\sqrt{1-e_c^2}$ is the companion's semiminor axis and M_c its mass. The companion's inclination also affects the distribution of inclinations of the detached scattered population. In general, companions with inclination near 90° create a less-inclined extended population. However, numerical simulations show that companions near the solar system invariable plane also induce a low-inclination high-perihelion scattered population. In this case, close encounters with the companion are responsible for lifting the perihelion of several SDOs from Neptune's orbit. Figure 6 shows the distribution of semimajor axes and perihelia, after 2 b.y. of integration, of scattered objects, some of which had their perihelia raised by the secular perturbation of a solar companion with $M_c = 10^{-4}$ $M_\odot$, $a_c = 1500$ AU, $q_c = 900$ AU, and $i_c = 90°$. Numerical simulations of a scenario like that of the LHB model (*Gomes et al.,* 2005a) but also including a companion with the same parameters as the one above (except for a 40° orbital inclination) yield a total mass of the detached SDOs of roughly 1 $M_\oplus$ at solar system age. This is about of the same order of magnitude of the Sedna-like population estimated by *Brown et al.* (2004).

A basic difference between the star cluster model and the solar companion model is that the latter creates mostly a "live" population as opposed to the "fossil" population produced by the star cluster model. The secular effects imposed by the companion continue as long as the companion exists. In this way, objects can move from the scattered disk to the detached population, then back to the scattered

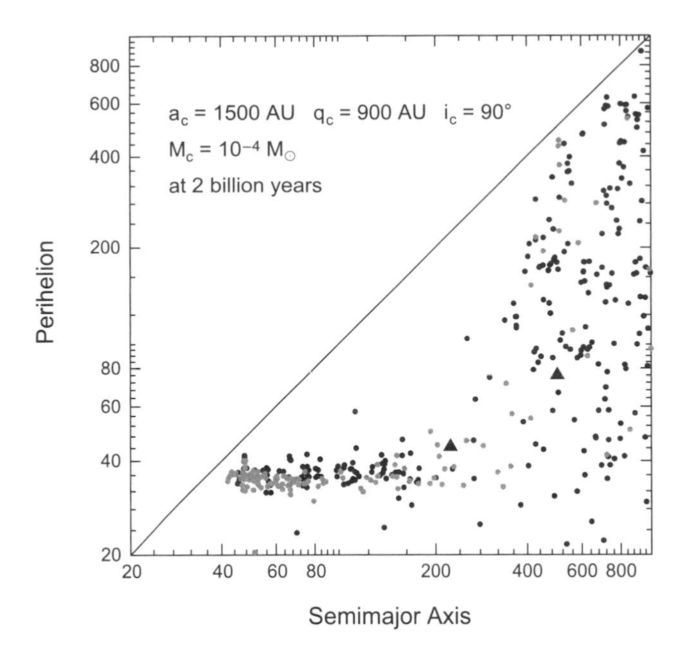

Fig. 6. Distributions of semimajor axes and perihelia of particles started near Neptune. Direct numerical integration were undertaken with all major planets and a companion with parameters $M_c = 10^{-4}$, $M_\odot$, a_c = 1500 AU, e_c = 0.4, and i_c = 90°. Gray dots ↔ i < 15°, black dots ↔ i > 15°, triangles ↔ Sedna and 2000 CR_{105}.

disk or they can get a lower perihelion and become Centaurs, defined as those objects that cross Neptune's orbit, i.e., that get q < 30 AU. Consequently, the influx of comets in the solar companion scenario must be significantly different from those in the no-companion scenario, and these dynamics deserve specific investigation.

A Jupiter- or higher-mass companion beyond 5000 AU could have been formed as a small distant binary-star like companion. Relatively smaller companions (Earth to Neptune size at average distances from 10^3 to 2×10^3 AU) could have been scattered by Jupiter or Saturn at very early times of solar system formation and have their perihelia raised by the action of passing starts in a putative dense galactic environment around the primordial Sun, as in the scenario discussed in section 5.2.

5.2.2. Other external perturbations. *Gladman and Chan* (2006) present another mechanism that cannot be strictly classified within the external perturber scenario but shares some common features with it. These authors consider that one or more Earth-sized bodies were scattered by Neptune in the early solar system, as was previously suggested by *Ip* (1989) and *Petit et al.* (1999). From numerical simulations, *Gladman and Chan* (2006) show that a "rogue" planet interacting with the SDO population, while the planet was still bound to the solar system, would have been able to raise the perihelia of some SDOs to values above 40 AU. This theory is intended to account for the detached objects as a whole, but it does not produce detached objects with a large semimajor axis like Sedna as effectively as those with a smaller semimajor axis.

It is also noteworthy to mention a mechanism that suggests an extrasolar origin for the large semimajor axis detached objects. As noted above, during the early evolution of the solar system, the Sun was likely in a dense primordial star cluster. This cluster might also contain substellar objects like brown dwarfs (BDs). If such a BD had an extended planetesimal disk surrounding it, part of this disk could be captured by the Sun during a putative close encounter between the stars, thus producing a population of detached objects including Sedna-like orbits (*Morbidelli and Levison,* 2004; *Kenyon and Bromley,* 2004).The problem with this model is that we do not know if BDs have such extended disks of planetesimals as big as Sedna.

6. END STATES OF SCATTERED OBJECTS

6.1. A Primordial Reservoir of Detached Objects Formed by a Migrating Neptune

Scattered object orbits are intrinsically unstable by virtue of their very formation process. That is why the current SD population accounts for less than 1% of the original population (section 4.1). Two natural fates for a scattered object can be easily predicted. In fact, Neptune can either scatter out the object by increasing its semimajor axis or it can scatter it into the region with $a < a_N$ if there is a close encounter. The outcomes of these processes are either feeding the Oort cloud or becoming a Centaur and possibly a Jupiter-family comet (JFC). These are the subjects of the following two sections. A third possible fate for a scattered object is obtained if its perihelion is lifted in an irreversible way, so that it enters the detached population and no longer experiences close encounters with Neptune. A fourth less likely fate is the collision with one of the planets (or the Sun).

The mechanisms that induce a perihelion increase by resonant perturbations from Neptune (section 5.1) has a reversible character so that a high-perihelion scattered object can eventually again experience close encounters with Neptune. The timescale for temporary decoupling from Neptune can be as high as 100 m.y. for objects with large semimajor axis (around 200 AU). The irreversibility appears when the conservative character of the planet-particle dynamics is broken by the migration of Neptune or, in other words, by its close interactions with other planetesimals. So, the reservoir of detached objects may be considered as an end state of SDO dynamics given either the irreversibility of the process that increased their perihelion distance or the very long dynamical lifetimes as compared to the solar system age.

Figure 7 shows an example of that dynamical behavior. A scattered object is trapped into the 2:5 resonance with a migrating Neptune. Also experiencing the Kozai resonance inside the MMR, the pair (e, i) starts to follow a variation typical of KR dynamics. At some point when its eccentricity is relatively low and the resonance strength diminishes, the resonance relationship is broken and the particle gets

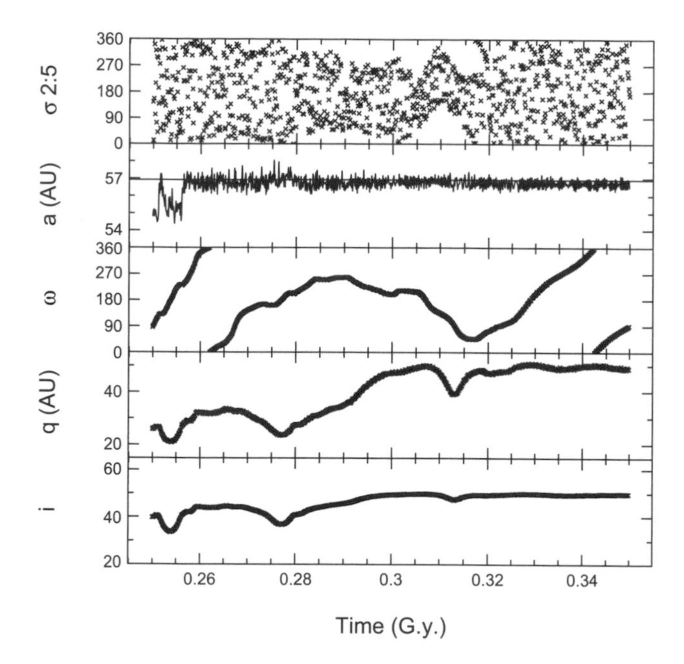

Fig. 7. Orbital evolution of a SDO captured into the 2:5 MMR with Neptune, also experiencing the Kozai resonance for some time. The particle leaves the 2:5 resonance during Neptune's migration when the eccentricity is low, being fossilized near but outside the MMR.

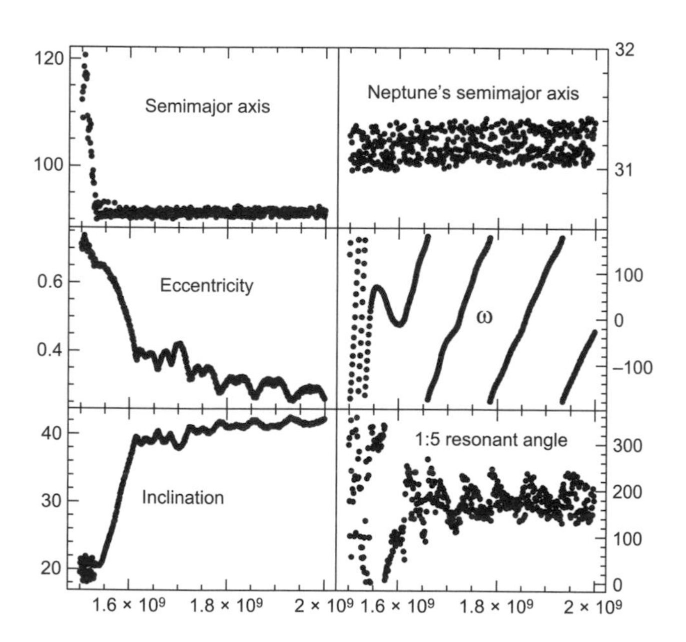

Fig. 8. Orbital evolution of a SDO captured into the 1:5 MMR with Neptune, also experiencing the Kozai resonance for some time. The particle leaves the Kozai resonance but not the MMR, being fossilized inside the MMR.

fossilized outside, but in the vicinity of the 2:5 resonance. Another interesting example in shown in Fig. 8. In this case the particle is trapped into the 1:5 resonance with Neptune and starts to experience the Kozai resonance. However, the Kozai resonance is broken when the eccentricity is low, but the MMR remains active until the end of the integration at 4.5 b.y. In this case, the capture into the MMR and the capture/escape process from the Kozai resonance take place when Neptune migration is very slow. At this point, escape from the Kozai resonance is possible, but escape from the MMR seems unlikely. For semimajor axis above 100 AU, numerical simulations show trapping into MMR + KR, but no escape from any of these resonances seems likely.

During the primordial scattering process, when Neptune still experiences a fairly fast migration, particles are first trapped into the strongest resonances, characterized by small semimajor axes. In this case, escape from the MMR is possible. There must have been a greater fraction of scattered particles with lower semimajor axes as compared with larger semimajor axes in the first 100 m.y. while Neptune still experienced some migration. This must be responsible for fossilized detached objects outside MMRs, which therefore must be found preferentially at relatively small semimajor axes (e.g., a < 60–70 AU). The object 2004 XR_{190} may well be a detached object that escaped from the 3:8 MMR with Neptune. A fossil object obtained in a numerical simulation, as shown in Fig. 9, ended at a semimajor axis slightly smaller than that of the 2:5 resonance. The semimajor axis of 2004 XR_{190} is by a similar amount smaller than that of the 3:8 resonance with Neptune. Inclinations for both 2004 XR_{190} and the simulation object are also of the same order. This suggests that 2004 XR_{190} is a fossil detached object that escaped the 3:8 resonance early in the solar system evolution when Neptune still experienced migration.

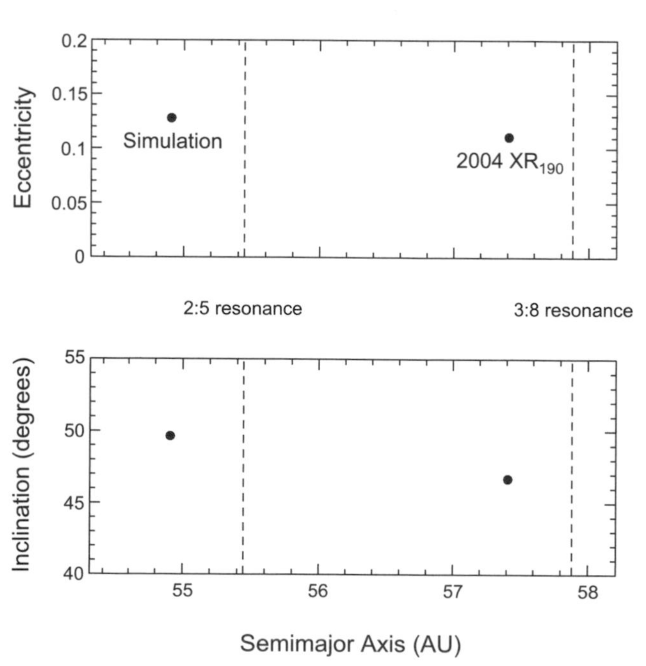

Fig. 9. Semimajor axis, eccentricity, and inclination for an object coming from a numerical simulation (*Gomes et al,* 2005b) and for 2004 XR_{190}. This figure suggests that, like the simulated object that escaped from the 2:5 resonance, 2005 XR_{190} is a fossil detached object escaped from the 3:8 resonance.

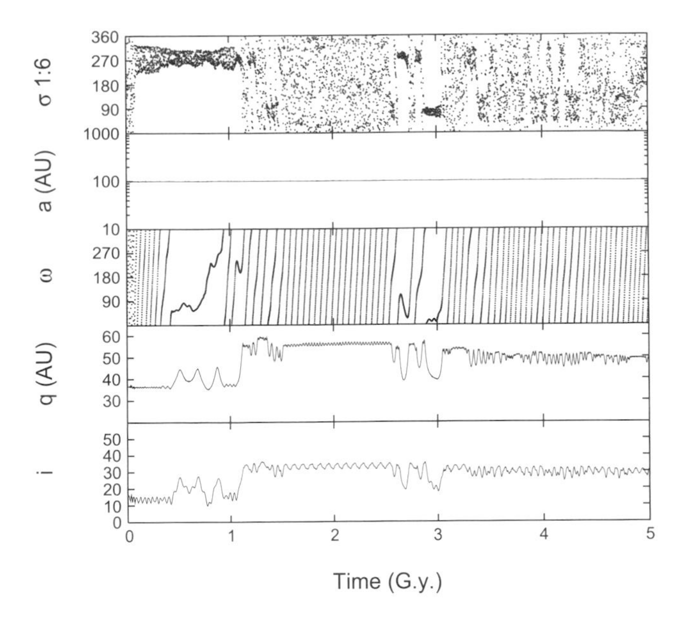

Fig. 10. An example of particle injected in a high q orbit via MMR + KR and conserved there after the breakup of the resonant motion despite the fact that no migration of Neptune occurs in this simulation (*Gomes et al.,* 2005b).

Fossil objects inside a resonance are more likely to be found among middle-valued semimajor axes (e.g., 70 < a < 100 AU). For larger semimajor axes (e.g., a > 100 AU), numerical simulations show interesting trappings into the mean motion/Kozai resonances with low-eccentricity excursions. However, in these cases, the particle always gets back to its primordial low-perihelion SDO state. Nevertheless, due to the very slow secular evolutions at these remote regions, a particle can show a high-perihelion orbit for quite a long time, on the order of several hundred million years (*Gomes et al.,* 2005b).

It is important to note that although we have blamed Neptune's migration for the breakup of a particle's resonance, when its perihelion is high enough, numerical integrations with massless particles (and no induced migration of the planets) also show instances of resonance breakup. When e is low enough, the resonance's strength drops and the resonant relationship can be broken, or at least the libration of the critical angle can be transformed into circulation. At this point the KR stops and only small oscillations in (e, i) coupled with circulation of ω are left. Figure 10 shows an example where a particle is injected into the high q region due to MMR + KR. Once with high q, the resonance's strength, calculated following *Gallardo* (2006a,b), drops to one-sixth of its original value and the librations are broken. The object can be stored for billion of years in high-q orbits by this mechanism. It must be noted that this is not in a strict sense a case of a dynamical end state since the particle can always return to its original low-q orbit, even though this may take a time longer than the solar system age. A real end state always needs the action of migration to break the reversibility.

6.2. Feeding the Oort Cloud

Scattered disk objects will slowly diffuse outward under the action of planetary perturbations. We can see in Fig. 11 the dynamical evolution of one of such body, 1999 DP_8, which ends up in the Oort cloud after 3.35 G.y. We note that the Kozai resonance plays a role in the evolution of these bodies, when ω slows down or starts librating around 180°. As seen in Fig. 11, the perihelion distance of the body increases for a while to q ~ 50 AU, due to the KR, so that the body avoids close encounters with Neptune. In fact, diffusion of the SDOs in the energy space cannot be described as a random-walk process, since bodies very often fall in different resonances, a process known as resonance sticking (see section 3.2), which helps to enhance the dynamical longevity of SDOs. The dynamical half-life of SDOs can be expressed as (*Fernández et al.,* 2004)

$$t_{dyn} \simeq 10^{\frac{(q-33.5)}{4.7}} \text{ G.y.} \qquad (2)$$

where q is expressed in AU. From equation (2) we get an average lifetime $\bar{t}_{dyn} \sim 1.8 \times 10^9$ yr.

To endure strong interactions with Neptune, SDOs must first decrease their q to values close to Neptune's orbital radius. At the beginning, a remains more or less constant as q decreases (thus increasing its eccentricity) (*Holman and Wisdom,* 1993). When SDOs get close to Neptune's orbit they suffer strong perturbations from this planet, so they can be scattered onto orbits with larger semimajor axes.

Neptune acts as a dynamical barrier that prevents scattering outward as compared to a slow decrease of the object's perihelion distances near or below Neptune's orbital radius. *Fernández et al.* (2004) found that about 60% of the

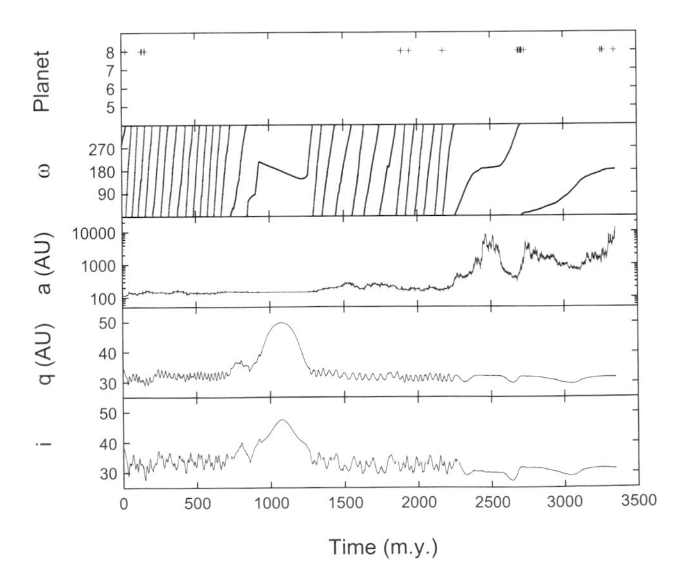

Fig. 11. Dynamical evolution of the SDO 1999 DP_8, which ends up in the Oort cloud. Close encounters with any of the jovian planets are indicated in the upper panel, where the numbers 5 . . . 8 stand for Jupiter . . . Neptune (*Fernández et al.,* 2004).

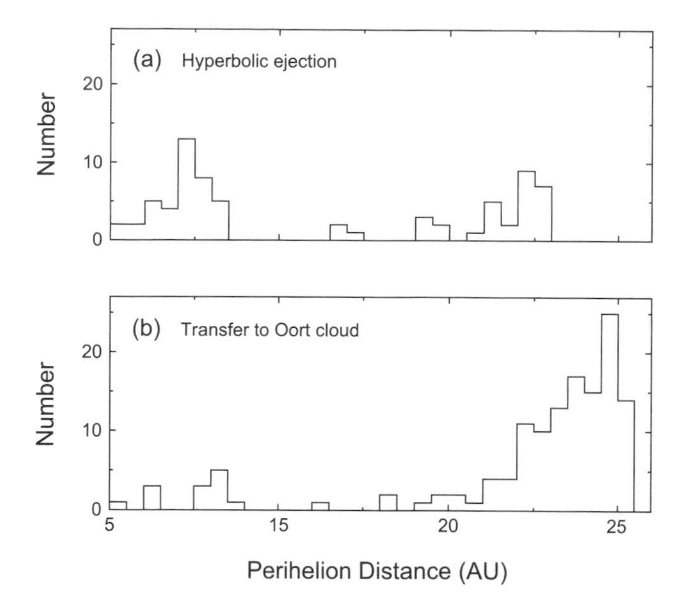

Fig. 12. The distribution of perihelion distances of a sample of 399 objects, consisting of the real SDOs + clones, at the moment they reach their final states: **(a)** hyperbolic ejection, or **(b)** insertion in the Oort cloud (*Fernández et al.,* 2004).

bodies inserted in the Oort cloud have perihelia in the range $31 < q < 36$ AU (Fig. 12).

For bodies reaching Neptune-crossing or closely approaching orbits, close interactions with this planet will favor transfer to the inner planetary region or hyperbolic ejection, instead of insertion into the Oort cloud.

Figure 13 shows the dynamical evolution of fictitious bodies in the parametric plane (a, q). We can see that the transfer to the Oort cloud takes place for bodies with $q < 36$ AU. The sharp upper limit at $q \simeq 36$ AU for bodies diffusing outward is quite remarkable. We can advance the hypothesis that when SDOs with $q \gtrsim 36$ AU fall in MMRs, planetary perturbations are too weak to dislodge the bodies from such resonances, thus preventing their further evolution in the energy space. Within the MMRs the KR may also act to raise the perihelia of the bodies. In such a dynamical state the bodies can be stored for very long timescales.

From the estimated SD population (cf. section 2) and the dynamical lifetime of SDOs, we can compute the current injection rate of SDOs with radii R > 1 km into the Oort cloud (*Fernández et al.,* 2004)

$$\nu \simeq \frac{N_{SDO}}{\bar{t}_{dyn}} \simeq 4 \text{ yr}^{-1} \quad (3)$$

The average rate $\bar{\nu}$ over the age of the solar system should be greater bearing in mind that the primordial SD population could have been up to $10^2\times$ greater, so a value $\bar{\nu} \sim 10$ should give at least the correct order of magnitude. Adopting this value, we determine that the total number of SDOs incorporated into the Oort cloud over the solar system age is $N_{oort} \sim 4.6 \times 10^9 \times 10 = 4.6 \times 10^{10}$. This population has been subject to external perturbers (passing starts, galactic tidal forces) that caused the reinjection of a fraction of it into the planetary region. Most of the objects injected into the planetary region were then ejected to interstellar space.

The previous result shows that the transneptunian belt (via the scattered disk) could have been a major supplier of bodies to the cloud, even rivaling other sources within the planetary region, e.g., the Uranus-Neptune zone. And even at present SDOs may still supply the Oort cloud with a significant population.

6.3. Back to the Inner Solar System: Centaurs and Jupiter-Family Comets

In addition to the outwards dynamical evolution, SDOs are also able to evolve to the planetary region, becoming Centaurs, and possibly JFCs. Although there is not a unique definition of Centaurs, it is generally accepted that they are objects that enter the planetary region from beyond Neptune (*Fernández,* 1980; *Duncan et al.,* 1988; *Levison and Duncan,* 1997). The observed Centaur population, strongly biased to low-perihelion distances (70% of the known Centaurs have $q < 17$ AU), has a mean lifetime of 9 m.y. (*Tiscareno and Malhotra,* 2003) with a large dispersion, ranging from 1 m.y. up to lifetimes larger than 100 m.y. *Levison and Duncan* (1997), through numerical simulations, estimated the number of JFCs with $H_T < 9$ ($R \gtrsim 1$ km) as 1.2×10^7. *Sheppard et al.* (2000) conducted a wide-field CCD survey for Centaurs. They concluded that if the differential size distribution is a power law with $s \sim 4$, and assuming an albedo of 0.04, the number of Centaurs should be on the order of 10^7. Therefore, assuming that the population of Centaurs is in steady state, the rate of injection of Centaurs from the scattered disk would be ~1 object larger than 1 km per year (assuming that the SD is the source of all Centaurs).

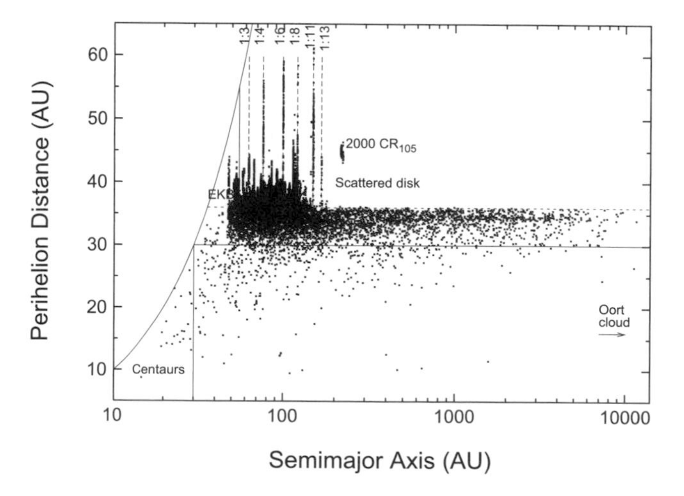

Fig. 13. Perihelion distance vs. semimajor axis of all the objects (real SDOs + clones) plotted every 50 m.y. (*Fernández et al.,* 2004).

7. CONCLUSIONS

The best estimate for the scattered disk mass is in the range 0.01–0.1 $M_\oplus$, thus comparable to the Kuiper belt mass. Due to our poor knowledge of the albedo and size distribution of the SDOs, the SD mass cannot be more precisely determined.

Most SDOs must be relics from a much more numerous population of objects, scattered from a primordial disk by Neptune during its migration. However, some SDOs may have escaped from the Kuiper belt by some chaotic dynamical process. Since SDOs very frequently fall in MMRs, due to resonance sticking, their diffusion in the energy space cannot be properly described as a random-walk. We note that in the region $a < 200$ AU no objects with $q > 36$ AU are found to diffuse to the Oort cloud in timescales of G.y. In a MMR, a SDO may also experience the Kozai resonance, inducing an important increase of its perihelion distance, temporarily detaching the objects from Neptune's close perturbation. If this mechanism takes place while Neptune is still migrating, fossil detached objects can be created by the rupture of the reversibility of the resonant dynamics. Sedna and possibly 2000 CR_{105} could not have their perihelia increased solely by the effect of the known planets. An external agent is needed to detach these objects from Neptune's close influence. The formation of the Sun in a dense star cluster could have raised the perihelia of distant detached objects like Sedna. A solar companion at roughly 10^3 AU to 10^4 AU from the Sun with a mass from 1 $M_\oplus$ to several Jupiter masses can also raise the perihelia of scattered objects, thus producing a "live population" of detached objects that naturally includes Sedna and 2000 CR_{105}.

Neptune constitutes a dynamical barrier that prevents most SDOs from diffusing inward. This prevents the outward scattering of SDOs that eventually feed the Oort cloud. At present, there may still be a significant contribution of SDOs to the Oort cloud. However, some SDOs may nevertheless escape to the solar system inside Neptune's orbit, becoming Centaurs or possibly Jupiter-family comets.

Acknowledgments. R.G. acknowledges financial support by Conselho Nacional de Desenvolvimento Científico e Tecnológico, J.A.F. and T.G. from Comisión Sectorial de Investigación Científica, and A.B. from Agencia Nacional de Promoción Científica y Tecnológica. We thank V. Emel'yanenko for reviewing this chapter. We are indebted to A. Morbidelli for his many comments and suggestions for improving this chapter.

REFERENCES

Bernstein G. M., Trilling D. E., Allen R. L., Brown M. E., Holman M., and Malhotra R. (2004) The size distribution of trans-Neptunian bodies. *Astron. J., 128,* 1364–1390.

Brasser R., Duncan M. J., and Levison H. F. (2006) Embedded star clusters and the formation of the Oort cloud. *Icarus, 184,* 59–82.

Brown M. E., Trujillo C., and Rabinowitz D. (2004) Discovery of a candidate inner Oort cloud planetoid. *Astrophys. J., 617,* 645–649.

Brown M. E., Barkume K. M., Ragozzine D., and Schaller E. L. (2007) A collisional family of icy objects in the Kuiper belt. *Nature, 446,* 294–296.

Chiang E. I., Lovering J. R., Millis R. L., Buie M. W., Wasserman L. H., and Meech K. J. (2004) Resonant and secular families of the Kuiper belt. *Earth Moon Planets, 92,* 49–62.

Davis D. R. and Farinella P. (1997) Collisional evolution of Edgeworth-Kuiper belt objects. *Icarus, 125,* 50–60.

Delsanti A. and Jewitt D. (2006) The solar system beyond the planets. In *Solar System Update* (Ph. Blondel and J. Mason, eds.), pp. 267–294. Springer-Praxis, Germany.

Dohnanyi J. W. (1969) Collisional models of asteroids and their debris. *J. Geophys. Res., 74,* 2531–2554.

Duncan M. J. and Levison H. F. (1997) A disk of scattered icy objects and the origin of Jupiter-family comets. *Science, 276,* 1670–1672.

Duncan M., Quinn T., and Tremaine S. (1988) The origin of short-period comets. *Astrophys. J., Lett., 328,* L69–L73.

Duncan M. J., Levison H. F., and Budd S. M. (1995) The dynamical structure of the Kuiper belt. *Astron. J., 110,* 3073–3081.

Fernández J. A. (1980) On the existence of a comet belt beyond Neptune. *Mon. Not. R. Astron. Soc., 192,* 481–491.

Fernández J. A. and Brunini A. (2000) The buildup of a tightly bound comet cloud around an early Sun immersed in a dense galactic environment: Numerical experiments. *Icarus, 145,* 580–590.

Fernández J. A. and Ip W.-H. (1984) Some dynamical aspects of the accretion of Uranus and Neptune: The exchange of orbital angular momentum with planetesimals. *Icarus, 58,* 109–120.

Fernández J. A., Gallardo T., and Brunini A. (2004) The scattered disk population as a source of Oort cloud comets: Evaluation of its current and past role in populating the Oort cloud. *Icarus, 172,* 372–381.

Gallardo T. (2006a) The occurrence of high order mean motion resonances and Kozai mechanism in the scattered disk. *Icarus, 181,* 205–217.

Gallardo T. (2006b) Atlas of mean motion resonances in the solar system. *Icarus, 184,* 29–38.

Gladman B. and Chan C. (2006) Production of the extended scattered disk by rogue planets. *Astrophys. J. Lett., 643,* L135–L138.

Gladman B., Holman M., Grav T., Kavelaars J., Nicholson P., Aksnes K., and Petit J.-M. (2002) Evidence for an extended scattered disk. *Icarus, 157,* 269–279.

Gomes R. S. (2003) The origin of the Kuiper belt high-inclination population. *Icarus, 161,* 404–418.

Gomes R. S., Morbidelli A., and Levison H. F. (2004) Planetary migration in a planetesimal disk: Why did Neptune stop at 30 AU? *Icarus, 170,* 492–507.

Gomes R. S., Levison H. F., Tsiganis K., and Morbidelli A. (2005a) Origin of the cataclysmic late heavy bombardment period of the terrestrial planets. *Nature, 435,* 466–469.

Gomes R. S., Gallardo T., Fernández J. A., and Brunini A. (2005b) On the origin of the high-perihelion scattered disk: The role of the Kozai mechanism and mean motion resonances *Cel. Mech. Dyn. Astron., 91,* 109–129.

Gomes R. S., Matese J., and Lissauer J. (2006) A distant planetary-mass solar companion may have produced distant detached objects. *Icarus, 184,* 589–601.

Hahn J. M. and Malhotra R. (1999) Orbital evolution of planets embedded in a planetesimal disk. *Astron. J., 117,* 3041–3053.

Holman M. J. and Wisdom J. (1993) Dynamical stability in the outer solar system and the delivery of short period comets. *Astron. J., 105,* 1987–1999.

Ip W.-H. (1989) Dynamical processes of macro-accretion of Uranus and Neptune: A first look. *Icarus, 80,* 167–178.

Jewitt D. G., Luu J., and Trujillo C. (1998) Large Kuiper belt objects: The Mauna Kea 8K CCD survey. *Astron. J., 115,* 2125–2135.

Jones D. C., Williams I. P., and Melita M. D. (2005) The dynamics of objects in the inner Edgeworth-Kuiper belt. *Earth Moon Planets, 97,* 435–458.

Kenyon S. J. and Bromley B. C. (2004) Stellar encounters as the origin of distant solar system objects in highly eccentric orbits. *Nature, 432,* 598–602.

Knežević Z., Milani A., Farinella P., Froeschle Ch., and Froeschle Cl. (1991) Secular resonances from 2 to 50 AU. *Icarus, 93,* 316–330.

Kozai Y. (1962) Secular perturbations of asteroids with high inclination and eccentricity. *Astron. J., 67,* 591–598.

Kuiper G. (1951) On the origin of the solar system. In *Astrophysics: A Topical Symposium* (J. A. Hynek, ed.), pp. 457–414. McGraw-Hill, New York.

Lada C. J. and Lada E. A. (2003) Embedded clusters in molecular clouds. *Annu. Rev. Astron. Astrophys., 41,* 57–115.

Larsen J. A., Gleason A. E., Danzi N. M., Descour A. S., McMillan R. S., Gehrels T., Jedicke R., Montani J. L., and Scotti J. V. (2001) The Spacewatch wide-area survey for bright Centaurs and trans-Neptunian objects. *Astron. J., 121,* 562–579.

Levison H. F. and Duncan M. J. (1997) From the Kuiper belt to Jupiter-family comets: The spatial distribution of ecliptic comets. *Icarus, 127,* 13–32.

Luu J., Marsden B. G., Jewitt D., Trujillo C. A., Hergenrother C. W., Chen J., and Offutt W. B. (1997) A new dynamical class of object in the outer solar system. *Nature, 387,* 573–575.

Lykawka P. S. and Mukai T. (2006) Exploring the 7:4 mean motion resonance — II: Scattering evolutionary paths and resonance sticking. *Planet. Space Sci., 54,* 87–100.

Morbidelli A. (1997) Chaotic diffusion and the origin of comets from the 2/3 resonance in the Kuiper belt. *Icarus, 127,* 1–12.

Morbidelli A. (2002) *Modern Celestial Mechanics: Dynamics in the Solar System.* Taylor & Francis, New York.

Morbidelli A. and Levison H. F. (2004) Scenarios for the origin of the orbits of the trans-Neptunian objects 2000 CR105 and 2003 VB12 (Sedna). *Astron. J., 128,* 2564–2576.

Morbidelli A., Emel'yanenko V. V., and Levison H. F. (2004) Origin and orbital distribution of the trans-Neptunian scattered disc. *Mon. Not. R. Astron. Soc., 355,* 935–940.

Murray C. D. and Dermott S. F. (1999) *Solar System Dynamics.* Cambridge Univ., Cambridge.

Nesvorný D. and Roig F. (2001) Mean motion resonances in the trans Neptunian region. Part II: 1:2, 3:4 and weaker resonances. *Icarus, 150,* 104–123.

Pan M. and Sari R. (2005) Shaping the Kuiper belt size spectrum by shattering large but strengthless bodies. *Icarus, 173,* 342–348.

Petit J.-M., Morbidelli A., and Valsecchi G. B. (1999) Large scattered planetesimals and the excitation of the small body belts. *Icarus, 141,* 367–387.

Robutel P. and Laskar J. (2001) Frequency map and global dynamics in the solar system I. *Icarus, 152,* 4–28.

Sheppard S. S., Jewitt D. C., Trujillo C. A., Brown M. J. I., and Ashley M. C. B. (2000) A wide-field CCD survey for Centaurs and Kuiper belt objects. *Astron. J., 120,* 2687–2694.

Stern S. A. and Colwell J. E. (1997) Accretion in the Edgeworth-Kuiper belt: Forming 100–1000 km radius bodies at 30 AU and beyond. *Astron. J., 114,* 841–849.

Thomas F. and Morbidelli A. (1996) The Kozai resonance in the outer solar system and the dynamics of long-period comets. *Cel. Mech. Dyn. Astron., 64,* 209–229.

Tiscareno M. S. and Malhotra R. (2003) The dynamics of known Centaurs. *Astron. J., 126,* 3122–3131.

Trujillo C. A., Jewitt D. C., and Luu J. X. (2000) Population of the scattered Kuiper belt. *Astrophys. J. Lett., 529,* L103–L106.

Trujillo C. A., Jewitt D. C., and Luu J. X. (2001) Properties of the trans-Neptunian belt: Statistics from the Canada-France-Hawaii telescope survey. *Astron. J., 122,* 457–473.

Tsiganis K., Gomes R. S., Morbidelli A., and Levison H. F. (2005) Origin of the orbital architecture of the giant planets of the solar system. *Nature, 435,* 459–461.

Yu Q. and Tremaine S. (1999) The dynamics of plutinos. *Astron. J., 118,* 1873–1881.

The Dynamical Structure of the Kuiper Belt and Its Primordial Origin

Alessandro Morbidelli
Observatoire de la Côte d'Azur

Harold F. Levison
Southwest Research Institute

Rodney Gomes
Observatório Nacional/MCT-Brazil

This chapter discusses the dynamical properties of the Kuiper belt population. Then, it focuses on the characteristics of the Kuiper belt that cannot be explained by its evolution in the framework of the current solar system. We review models of primordial solar system evolution that have been proposed to reproduce the Kuiper belt features, outlining advantages and problems of each of them.

1. INTRODUCTION

Since its discovery in 1992, the Kuiper belt has slowly revealed a stunningly complex dynamical structure. This structure has been a gold mine for those of us interested in planet formation because it provides vital clues about this process. This chapter is a review of the current state of knowledge about these issues. It is divided into two parts. The first part (section 2) is devoted to the description of the current dynamics in the Kuiper belt. This will be used in section 3 to highlight the properties of the Kuiper belt population that cannot be explained by the current dynamical processes, but need to be understood in the framework of a scenario of primordial evolution of the outer solar system. In the second part, we will review the models that have been proposed so far to explain the various puzzling properties of the Kuiper belt. More precisely, section 4 will focus on the origin of the outer edge of the belt; section 5 will describe the effects of the migration of Neptune on the orbital structure of the Kuiper belt objects (KBOs), and section 6 will discuss the origin of the mass deficit of the transneptunian population. In section 7 we will present the consequences on the Kuiper belt of a model of outer solar system evolution that has been recently proposed to explain the orbital architecture of the planets, the Trojan populations of both Jupiter and Neptune, and the origin of the late heavy bombardment of the terrestrial planets. A general discussion on the current state-of-the-art in Kuiper belt modeling will conclude the chapter in section 8.

2. CURRENT DYNAMICS IN THE KUIPER BELT

Plate 7 shows a map of the dynamical lifetime of transneptunian bodies as a function of their initial semimajor axis and eccentricity, for an inclination of 1° and with their orbital ellipses oriented randomly (*Duncan et al.,* 1995). Additional maps, referring to different choices of the initial inclination or different projections on orbital element space, can be found in *Duncan et al.* (1995) and *Kuchner et al.* (2002). These maps have been computed numerically, by simulating the evolution of massless particles from their initial conditions, under the gravitational perturbations of the giant planets. The planets were assumed to be on their current orbits throughout the integrations. Each particle was followed until it suffered a close encounter with Neptune. Objects encountering Neptune would then evolve in the scattered disk (see chapter by Gomes et al.).

In Plate 7, the colored strips indicate the timespan required for a particle to encounter Neptune, as a function of its initial semimajor axis and eccentricity. Yellow strips represent objects that survive for the length of the simulation, 4×10^9 yr (the approximate age of the solar system) without encountering the planet. Plate 7 also reports the orbital elements of the known Kuiper belt objects. Green dots refer to bodies with inclination $i < 4°$, consistent with the low inclination at which the stability map has been computed. Magenta dots refer to objects with larger inclination and are plotted only for completeness.

As shown in Plate 7, the Kuiper belt has a complex dynamical structure, although some general trends can be easily identified. If we denote the perihelion distance of an orbit by q, and we note that $q = a(1 - e)$, where a is the semimajor axis and e is the eccentricity, Plate 7 shows that most objects with $q \lesssim 35$ AU (in the region $a < 40$ AU) or $q \lesssim 37$–38 AU (in the region with $42 < a < 50$ AU) are unstable. This is due to the fact that they pass sufficiently close to Neptune to be destabilized. It may be surprising that Neptune can destabilize objects passing at a distance of 5–8 AU, which corresponds to ~10× the radius of Neptune's gravitational sphere of influence, or Hill radius. The instability, in fact, is not due to close encounters with the planet, but to the overlapping of its outer mean-motion resonances. It is well known that mean-motion resonances become wider at larger eccentricity (see, e.g., *Dermott and Murray,* 1983; *Morbidelli,* 2002) and that resonance overlapping produces large-scale chaos (*Chirikov,* 1960). The overlapping of resonances produces a chaotic band whose extent in perihelion distance away from the planet is proportional to the planet mass at the 2/7 power (*Wisdom,* 1980). To date, the most extended analytic calculation of the width of the mean-motion resonances with Neptune up to on the order of 50 has been done by D. Nesvorný, and the result — in good agreement with the stability boundary observed in Plate 7 — is published online at *www.boulder.swri.edu/~davidn/kbmmr/kbmmr.html.*

The semimajor axis of the objects that are above the resonance-overlapping limit evolves by "jumps," passing from the vicinity of one resonance to another, mostly during a conjunction with the planet. Given that the eccentricity of Neptune's orbit is small, the Tisserand parameter

$$T = \frac{a_N}{a} + 2\sqrt{\frac{a}{a_N}(1 - e^2)} \cos i$$

(a_N denoting the semimajor axis of Neptune; a, e, i the semimajor axis, eccentricity, and inclination of the object) is approximately conserved. Thus, the eccentricity of the object's orbit has "jumps" correlated with those of the semimajor axis, and the perihelion distance remains roughly constant. Consequently, the object wanders over the (a, e) plane and is effectively a member of the scattered disk. In conclusion, the boundary between the black and the colored region in Plate 7 marks the boundary of the scattered disk and has a complicated, fractal structure, which justifies the use of numerical simulations in order to classify the objects (see chapters by Gladman et al. and Gomes et al.).

Not all bodies with $q < 35$ AU are unstable, though. The exception is those objects deep inside low-order mean-motion resonances with Neptune. These objects, despite approaching (or even intersecting) the orbit of Neptune at perihelion, never get close to the planet. The reason for this can be understood with a little algebra. If a body is in a k_N:k resonance with Neptune, the ratio of its orbital period P to Neptune's P_N is, by definition, equal to k/k_N. In this case, and assuming that the planet is on a quasicircular orbit and the motions of the particle and of the planet are coplanar, the angle

$$\sigma = k_N\lambda_N - k\lambda + (k - k_N)\varpi \qquad (1)$$

(where λ_N and λ denote the mean longitudes of Neptune and of the object and ϖ is the object's longitude of perihelion) has a time derivative that is zero on average, so that it librates around an equilibrium value, say σ_{stab} (see Fig. 1). The radial distance from the planet's orbit is minimized when the object passes close to perihelion. Perihelion passage happens when $\lambda = \varpi$. When this occurs, from equation (1) we see that the angular separation between the planet and the object, $\lambda_N - \lambda$, is equal to σ/k_N. For small-amplitude librations, $\sigma \sim \sigma_{stab}$; because σ_{stab} is typically far from 0 (see Fig. 1), we conclude that close encounters cannot occur (*Malhotra,* 1996). Conversely, if the body is not in resonance, σ circulates (Fig. 1). So, eventually it has to pass through 0, which causes the object to be in conjuction with Neptune during its closest approach to the planet's orbit. Thus, close encounters are possible, if the object's perihelion distance is small enough.

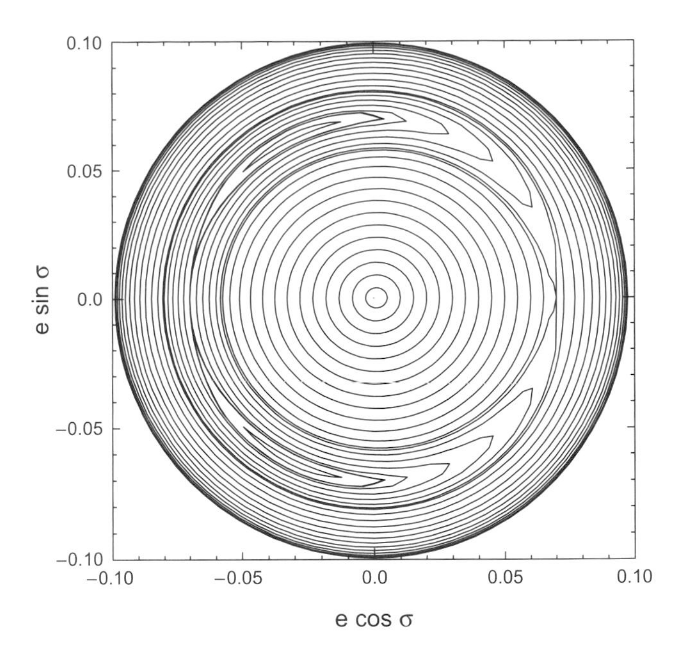

Fig. 1. The dynamics in the 1:2 resonance with Neptune, in e cos σ, e sin σ coordinates. The motions follow the closed curves plotted in the figure. Librations occur on those curves that do not enclose the origin of the figure. Notice two unstable equilibrium points on the e sin $\sigma = 0$ line. Each unstable equilibrium is the origin of a critical curve called *separatrix,* plotted in bold in the figure. The one with origin at the unstable point at $\sigma = 0$ separates resonant from nonresonant orbits, for which σ respectively librates or circulates. The separatrix with origin at the unstable point in $\sigma = 180°$ delimits two islands of libration around each of the asymmetric stable equilibria.

For most mean-motion resonances, $\sigma_{stab} = 180°$. However, this is not true for the resonances of type 1:k. In these resonances, if the eccentricity of the body is not very small, there are two stable equilibria at $\sigma_{stab} = 180 \pm \delta$, with $\delta \sim 60°$, while $\sigma = 180°$ is an unstable equilibrium (*Message,* 1958; *Beaugé,* 1994) (see Fig. 1). Thus, bodies with small amplitudes of libration necessarily librate asymmetrically in σ relative to the (0, 2π) interval. Symmetric librations are possible only for large-amplitude librators.

A detailed exploration of the stability region inside the two main mean-motion resonances of the Kuiper belt, the 2:3 and 1:2 resonances with Neptune, has been done in *Nes-*

vorný and Roig (2000, 2001). In general, they found that orbits with large-amplitude librations and moderate to large eccentricities are chaotic, and eventually escape from the resonance, joining the scattered disk population. Conversely, orbits with small eccentricity or small libration amplitude are stable over the age of the solar system. At large eccentricity, only asymmetric librations are stable in the 1:2 resonance.

Mean-motion resonances are not the only important agent structuring the dynamics in the Kuiper belt. In Plate 7, one can see that there is a dark region that extends down to e = 0 when 40 < a < 42 AU. The instability in this case is due to the presence of the secular resonance that occurs when $\dot{\varpi} \sim \dot{\varpi}_N$, where ϖ_N is the perihelion longitude of Neptune. More generally, secular resonances occur when the precession rate of the perihelion or of the longitude of the node of an object is equal to the mean precession rate of the perihelion or the node of one of the planets. The secular resonances involving the perihelion precession rates excite the eccentricities, while those involving the node precession rates excite the inclinations (*Williams and Faulkner,* 1981; *Morbidelli and Henrard,* 1991).

The location of secular resonances in the Kuiper belt has been computed in *Knežević et al.* (1991). This work showed that both the secular resonance with the perihelion and that with the node of Neptune are present in the 40 < a < 42 AU region, for i < 15°. Consequently, a low-inclination object in this region undergoes large variations in orbital eccentricity so that — even if the initial eccentricity is zero — the perihelion distance eventually decreases below 35 AU, and the object enters the scattered disk (*Holman and Wisdom,* 1993; *Morbidelli et al.,* 1995). Conversely, a large inclination object in the same semimajor axis region is stable. Indeed, Plate 7 shows that many objects with i > 4° (magenta dots) are present between 40 and 42 AU. Only large dots, representing low-inclination objects, are absent.

Another important characteristic revealed by Plate 7 is the presence of narrow regions, represented by brown bands, where orbits become Neptune-crossing only after billions of years of evolution. The nature of these weakly unstable orbits remained mysterious for several years. Eventually, it was found (*Nesvorný and Roig,* 2001) that they are, in general, associated either with high-order mean-motion resonances with Neptune (i.e., resonances for which the equivalence $k\dot{\lambda} = k_N\dot{\lambda}_N$ holds only for large values of the integer coefficients k, k_N) or three-body resonances with Uranus and Neptune (which occur when $k\dot{\lambda} + k_N\dot{\lambda}_N + k_U\dot{\lambda}_U = 0$ occurs for some integers k, k_N, and k_U).

The dynamics of objects in these resonances is chaotic due to the nonzero eccentricity of the planetary orbits. The semimajor axes of the objects remain locked at the corresponding resonant value, while the eccentricity of their orbits slowly evolves. In an (a, e) diagram like Fig. 2, each object's evolution leaves a vertical trace. This phenomenon is called *chaotic diffusion.* Eventually the growth of the eccentricity can bring the diffusing object to decrease the

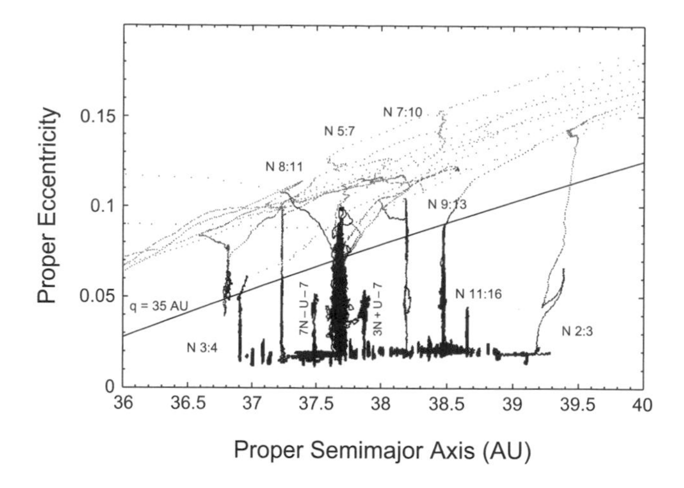

Fig. 2. The evolution of objects initially at e = 0.015 and semimajor axes distributed in the 36.5–39.5-AU range. The dots represent the proper semimajor axis and the eccentricity of the objects — computed by averaging their a and e over 10-m.y. time intervals — over the age of the solar system. They are plotted in gray after the perihelion has decreased below 32 AU for the first time. Labels Nk_N:k denote the k_N:k two-body resonances with Neptune. Labels $k_NN + k_UU + k$ denote the three-body resonances with Uranus and Neptune, corresponding to the equality $k_N\dot{\lambda}_N + k_U\dot{\lambda}_U + k\dot{\lambda} = 0$. From *Nesvorný and Roig* (2001).

perihelion distance below 35 AU. These resonances are too weak to offer an effective protection against close encounters with Neptune (σ_{stab}/k_N is a small quantity because k_N is large), unlike the low-order resonances considered above. Thus, once the perihelion distance becomes too low, the encounters with Neptune start to change the semimajor axis of the objects, which leave their original resonance and evolve — from that moment on — in the scattered disk.

Notice from Fig. 2 that some resonances are so weak that, despite forcing the resonant objects to diffuse chaotically, they cannot reach the q = 35 AU curve within the age of the solar system. Therefore, although these objects are not stable from a dynamics point of view, they can be consider that way from an astronomical perspective.

Notice also that chaotic diffusion is effective only for selected resonances. The vast majority of the simulated objects are not affected by any macroscopic diffusion. They preserve their initial small eccentricity for the entire age of the solar system. Thus, the current moderate/large eccentricities and inclinations of most of the Kuiper belt objects cannot be obtained from primordial circular and coplanar orbits by dynamical evolution in the framework of the current orbital configuration of the planetary system. Likewise, the region beyond the 1:2 mean-motion resonance with Neptune is totally stable up to an eccentricity of ~0.3 (Plate 7). As a result, the absence of bodies beyond 48 AU cannot be explained by current dynamical instabilities. Therefore, these (and other) intriguing properties of the Kuiper belt's structure must, instead, be explained within the framework of the formation and primordial evolution of the solar system. These topics will be treated in the following sections.

3. KUIPER BELT PROPERTIES ACQUIRED DURING A PRIMORDIAL AGE

From the current dynamical structure of the Kuiper belt, we conclude that the properties that require an explanation in the framework of the primordial solar system evolution are:

1. The existence of conspicuous populations of objects in the main mean-motion resonances with Neptune (2:3, 3:5, 4:7, 1:2, 2:5, etc.). The dynamical analysis presented above shows that these resonances are stable, but does not explain how and why objects populated these resonances on orbits with eccentricities as large as allowed by stability considerations.

2. The excitation of the eccentricities in the classical belt, which we define here as the collection of nonresonant objects with $42 < a < 48$ AU and $q > 37$ AU. The median eccentricity of the classical belt is ~0.07. It should be noted, however, that the upper eccentricity boundary of this population is set by the long-term orbital stability of the Kuiper belt (see Plate 7), and thus this semimajor axis region could have contained at some time in the past objects with much larger eccentricities. In any case, even if the current median eccentricity is small, it is nevertheless much larger (an order of magnitude or more) than the one that must have existed when the KBOs formed. The current dynamics are stable, so that, without additional stirring mechanisms, the primordial small eccentricities should have been preserved to modern times.

3. The peculiar (a, e) distribution of the objects in the classical belt (see Plate 7). In particular, the population of objects on nearly circular orbits ($e \lesssim 0.05$) effectively ends at about 44 AU, and beyond this location the eccentricity tends to increase with semimajor axis. If this were simply the consequence of an observational bias that favors the discovery of objects on orbits with smaller perihelion distances, we would expect that the lower bound of the a–e distribution in the 44–48 AU region follows a curve of constant q. This is not the case. Indeed, the eccentricity of this boundary grows more steeply with semimajor axis than this explanation would predict. Thus, the apparent relative underdensity of objects at low eccentricity in the region $44 < a < 48$ AU is likely to be a real feature of the Kuiper belt distribution. This underdensity cannot be explained by a lack of stability in this region.

4. The outer edge of the classical belt (Plate 7). This edge appears to be precisely at the location of the 1:2 mean-motion resonance with Neptune. Only large eccentricity objects, typical of the scattered disk or of the detached population (see chapter by Gladman et al. for a definition of these populations) seem to exist beyond this boundary (Plate 7). Again, the underdensity (or absence) of low-eccentricity objects beyond the 1:2 mean-motion resonance cannot be explained by observational biases (*Trujillo and Brown,* 2001; *Allen et al.,* 2001, 2002; see also chapter by Kavelaars et al.). As Plate 7 shows, the region beyond the 1:2 mean-motion resonance looks stable, even at moderate eccentricity. So, a primordial distant population should have remained there.

5. The inclination distribution in the classical belt. The observations (see Fig. 3) show a clump of objects with $i \lesssim 4°$. However, there are also several objects with much larger inclinations, up to $i \sim 30°$, despite the fact that an object's inclination does not change much in the current solar system. Observational biases definitely enhance the low-inclination clump relative to the large inclination population [the probability of discovery of an object in an ecliptic survey is roughly proportional to $1/\sin(i)$]. However, the clump persists even when the biases are taken into account. *Brown* (2001) argued that the debiased inclination distribution is bimodal and can be fitted with two Gaussian functions, one with a standard deviation $\sigma \sim 2°$ for the low-inclination core, and the other with $\sigma \sim 12°$ for the high-inclination population (see also chapter by Kavelaars et al.). Since the work of *Brown* (2001), the classical population with $i < 4°$ is called the "cold population," and the remaining one is called the "hot population" (see chapter by Gladman et al. for nomenclature issues).

6. The correlations between physical properties and orbital distribution. The cluster of low-inclination objects visible in the (a, i) distribution disappears if one selects only objects with absolute magnitude $H < 6$ (*Levison and Stern,* 2001). [The absolute magnitude is the brightness that the object would have if it were viewed at 1 AU, with the observer at the Sun. It relates to size by the formula $\mathrm{Log}D^2 = 6.244 - 0.4H - \mathrm{Log}(p)$, where D is the diameter in kilometers and p is the albedo.] This implies that intrinsically bright objects are underrepresented in the cold population. *Grundy et al.* (2005) have shown that the objects of the cold population have a larger albedo, on average, than those of the hot population. Thus, the correlation found by *Levison and Stern* (2001) implies that the hot population contains bigger objects. *Bernstein et al.* (2004) showed that the hot population has a shallower H distribution than the cold population,

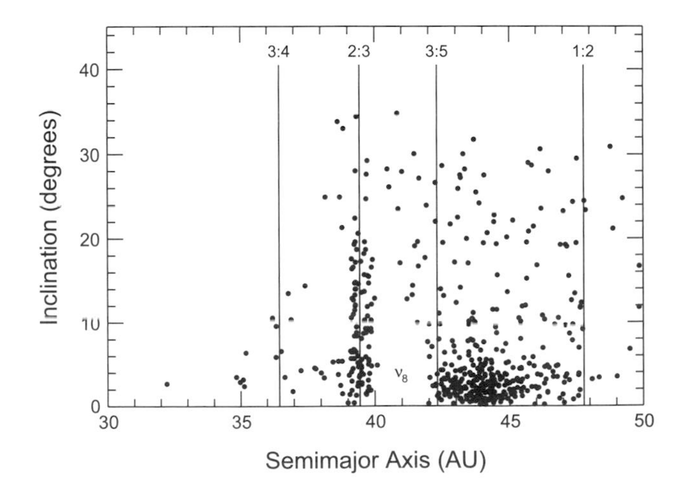

Fig. 3. The semimajor axis–inclination distribution of all well-observed Kuiper belt objects. The important mean-motion resonances are also shown.

which is consistent with the absence of the largest objects in the cold belt. In addition, there is a well-known correlation between color and inclination (see chapter by Doressoundiram et al.). The hot-population objects show a wide range of colors, from red to gray. Conversely, the cold-population objects are mostly red. In other words, the cold population shows a significant deficit of gray bodies relative to the hot population. The differences in physical properties argue that the cold and hot populations have different origins.

7. The mass deficit of the Kuiper belt. The current mass of the Kuiper belt is very small. Estimates range from 0.01 $M_\oplus$ (*Bernstein et al.,* 2004) to 0.1 $M_\oplus$ (*Gladman et al.,* 2001). The uncertainty is due mainly to the conversion from absolute magnitudes to sizes, assumptions about bulk density, and ambiguities in the size distribution (see chapter by Petit et al.). Whatever the exact real total mass, there appears to be a significant mass deficit (of 2–3 orders of magnitude) with respect to what models say is needed in order for the KBOs to accrete *in situ.* In particular, in order to grow the objects that we see within a reasonable time (10^7–10^8 m.y.), the Kuiper belt must have consisted of about 10–30 $M_\oplus$ of solid material in a dynamically cold disk (*Stern,* 1996; *Stern and Colwell,* 1997a,b; *Kenyon and Luu,* 1998, 1999a,b; *Kenyon and Bromley,* 2004a). If most of the Kuiper belt is currently stable, and therefore objects do not escape, what depleted ≥99.9% of the Kuiper belt primordial mass?

All these issues provide us with a large number of clues to understand what happened in the outer solar system during the primordial era. Potentially, the Kuiper belt might teach us more about the formation of the giant planets than the planets themselves. This is what makes the Kuiper belt so important and fascinating for planetary science.

4. ORIGIN OF THE OUTER EDGE OF THE KUIPER BELT

The existence of an outer edge of the Kuiper belt is very intriguing. Several mechanisms for its origin have been proposed, none of which have resulted yet in a general consensus among the experts in the field. These mechanisms can be grouped into three classes.

4.1. Class I: Destroying the Distant Planetesimal Disk

It has been argued in *Brunini and Melita* (2002) that a martian-mass body residing for 1 G.y. on an orbit with a ~ 60 AU and e ~ 0.15–0.2 could have scattered most of the Kuiper belt bodies originally in the 50–70 AU range into Neptune-crossing orbits, leaving this region strongly depleted and dynamically excited. It might be possible (see chapter by Kavelaars et al.) that the apparent edge at 50 AU is simply the inner edge of such a gap in the distribution of Kuiper belt bodies. A main problem with this scenario is that there are no evident dynamical mechanisms that would ensure the later removal of the massive body from the system. In other words, the massive body should still be present, somewhere in the ~50–70 AU region. A Mars-sized body with 4% albedo at 70 AU would have apparent magnitude brighter than 20. In addition, its inclination should be small, both in the scenario where it was originally a scattered disk object whose eccentricity (and inclination) were damped by dynamical friction (as envisioned by *Brunini and Melita,* 2002), and in the one where the body reached its required heliocentric distance by migrating through the primordially massive Kuiper belt (see *Gomes et al.,* 2004). Thus, in view of its brightness and small inclination, it is unlikely that the putative Mars-sized body could have escaped detection in the numerous wide-field ecliptic surveys that have been performed up to now, and in particular in that described in *Trujillo and Brown* (2003).

A second possibility for destroying the Kuiper belt beyond the observed edge is that the planetesimal disk was truncated by a close stellar encounter. The eccentricities and inclinations of the planetesimals resulting from a stellar encounter depend critically on a/D, where a is the semimajor axis of the planetesimal and D is the closest heliocentric distance of the stellar encounter (*Ida et al.,* 2000; *Kobayashi and Ida,* 2001). An encounter with a solar-mass star at ~200 AU would make most of the bodies beyond 50 AU so eccentric that they intersect the orbit of Neptune, which would eventually produce the observed edge (*Melita et al.,* 2002). An interesting constraint on the time at which such an encounter occurred is set by the existence of the Oort cloud. It was shown in *Levison et al.* (2004) that the encounter had to occur much earlier than ~10 m.y. after the formation of Uranus and Neptune, otherwise most of the existing Oort cloud would have been ejected to interstellar space. Moreover, many of the planetesimals in the scattered disk at that time would have had their perihelion distance lifted beyond Neptune, decoupling them from the planet. As a consequence, the detached population, with $50 \leq a \leq 100$ AU and $40 < q < 50$ AU, would have had a mass comparable to or larger than that of the resulting Oort cloud, hardly compatible with the few detections of detached objects achieved up to now. Finally, this mechanism predicts a correlation between inclination and semimajor axis that is not seen.

A way around the above problems could be achieved if the encounter occurred during the first million years of solar system history (*Levison et al.,* 2004). At this time, the Sun was still in its birth cluster, making such an encounter likely. However, the Kuiper belt objects were presumably not yet fully formed (*Stern,* 1996; *Kenyon and Luu,* 1998), and thus an edge to the belt would form at the location of the disk where eccentricities are ~0.05. Interior to this location, collisional damping is efficient and accretion can recover from the encounter; beyond this location the objects rapidly grind down to dust (*Kenyon and Bromley,* 2002). If this scenario is true, the stellar passage cannot be responsible for exciting the Kuiper belt because the objects that we observe there did not form until much later.

According to the analysis done in *Levison et al.* (2004), an edge-forming stellar encounter should not be responsible for the origin of the peculiar orbit of Sedna (a = 484 AU and q = 76 AU), unlike what was proposed in *Kenyon and Bromley* (2004b). In fact, such a close encounter would also produce a relative overabundance of bodies with perihelion distance similar to that of Sedna but with semimajor axes in the 50–200 AU range (*Morbidelli and Levison,* 2004). These bodies have never been discovered, despite the fact that they should be easier to find than Sedna because of their shorter orbital period (see chapter by Gomes et al.).

4.2. Class II: Forming a Bound Planetesimal Disk from an Extended Gas-Dust Disk

In *Weidenschilling* (2003), it was shown that the outer edge of the Kuiper belt might be the result of two factors: (1) accretion takes longer with increasing heliocentric distance and (2) small planetesimals drift inward due to gas drag. According to Weidenschilling's models, this leads to a steepening of the radial surface density gradient of solids. The edge effect is augmented because, at whatever distance large bodies can form, they capture the approximately meter-sized bodies spiraling inward from farther out. The net result of the process, as shown by numerical modeling by Weidenschilling (see Fig. 4), is the production of an effective edge, where both the surface density of solid matter and the mean size of planetesimals decrease sharply with increasing distance.

A somewhat similar scenario has been proposed in *Youdin and Shu* (2002). In their model, planetesimals formed by gravitational instability, but only in regions of the solar nebula where the local solid/gas ratio was ~3× that of the Sun (*Sekiya,* 1983). According to the authors, this large ratio occurs because of a radial variation of orbital drift speeds of millimeter-sized particles induced by gas drag. This drift also acts to steepen the surface density distribution of the disk of solids. This means that at some point in the nebula, the solid/gas ratio falls below the critical value to form planetesimals, so that the resulting planetesimal disk would have had a natural outer edge.

A third possibility is that planetesimals formed only within a limited heliocentric distance because of the effect of turbulence. If turbulence in protoplanetary disks is driven by magneto-rotational instability (MRI), one can expect that it was particularly strong in the vicinity of the Sun and at large distances (where solar and stellar radiation could more easily ionize the gas), while it was weaker in the central, optically thick region of the nebula, known as the "dead zone" (*Stone et al.,* 1998). The accretion of planetesimals should have been inhibited by strong turbulence, because the latter enhanced the relative velocities of the grains. Consequently, the planetesimals could have formed only in the dead zone, with well-defined outer (and inner) edge(s).

4.3. Class III: Truncating the Original Gas Disk

The detailed observational investigation of star formation regions has revealed the existence of many *proplyds* (anomalously small protoplanetary disks). It is believed that

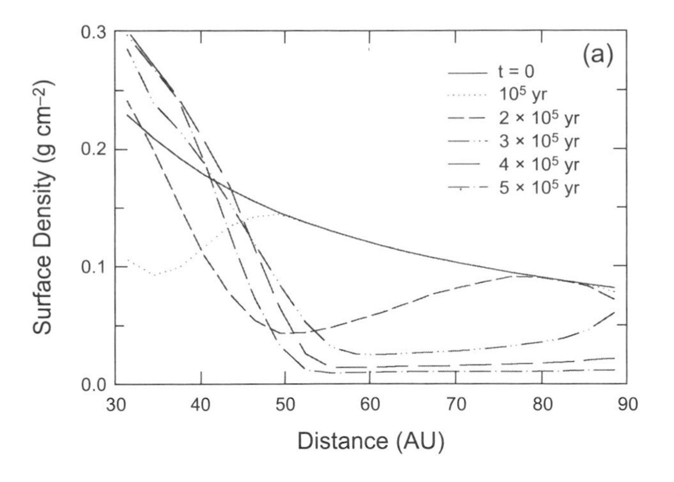

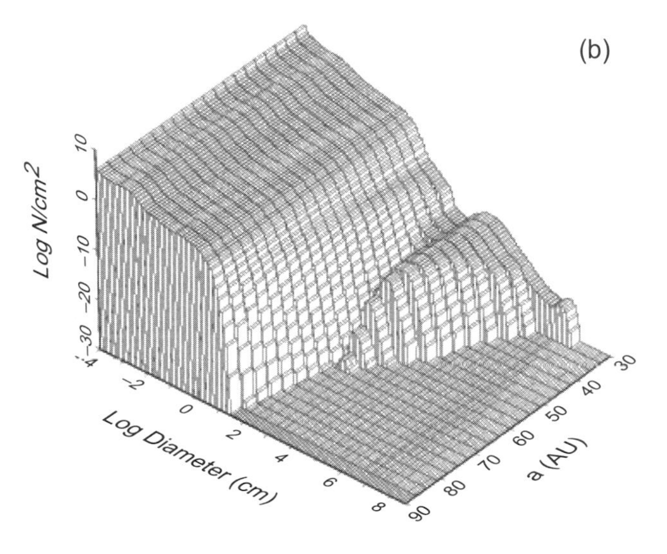

Fig. 4. **(a)** The time evolution of the surface density of solids. **(b)** The size distribution as a function of heliocentric distance. From *Weidenschilling* (2003).

these disks were originally much larger, but in their distant regions the gas was photoevaporated by highly energetic radiation emitted by the massive stars of the cluster (*Adams et al.,* 2004). Thus, it has been proposed that the outer edge of the Kuiper belt reflects the size of the original solar system proplyd (*Hollenbach and Adams,* 2004).

There is also the theoretical possibility that the disk was born small and did not spread out substantially under its own viscous evolution (*Ruden and Pollack,* 1991). In this case, no truncation mechanism is needed. Observations, however, do not show many small disks, other than in clusters with massive stars.

In all the scenarios discussed above, the location of the edge can be adjusted by tuning the relevant parameters of the corresponding model. In all cases, however, Neptune played no direct role in the edge formation. In this context, it is particularly important to remark (as seen in Plate 7) that the edge of the Kuiper belt appears to coincide precisely with the location of the 1:2 mean-motion resonance with

Neptune. This suggests that, whatever mechanism formed the edge, the planet was able to adjust the final location of the outer boundary through gravitational interactions. We will return to this in section 6.2. Notice that a planetesimal disk truncated at ~34 AU has been recently postulated in order to explain the dust distribution in the AU Mic system (*Augereau and Beust,* 2006).

5. THE ROLE OF NEPTUNE'S MIGRATION

It was shown in *Fernandez and Ip* (1984) that, in the absence of a massive gas disk, while scattering away the primordial planetesimals from their neighboring regions, the giant planets had to migrate in semimajor axis as a consequence of angular momentum conservation. Given the configuration of the giant planets in our solar system, this migration should have had a general trend (for a review, see *Levison et al.,* 2006). Uranus and Neptune have difficulty ejecting planetesimals onto hyperbolic orbits. Apart from the few percent of planetesimals that can be permanently stored in the Oort cloud or in the scattered disk, the remaining planetesimals (the large majority) are eventually scattered inward, toward Saturn and Jupiter. Thus, the ice giants, by reaction, have to move outward. Jupiter, on the other hand, eventually ejects from the solar system almost all of the planetesimals that it encounters, and therefore has to move inward. The fate of Saturn is more difficult to predict, *a priori*. However, numerical simulations show that this planet also moves outward, although only by a few tenths of an AU for reasonable disk masses (*Hahn and Malhotra,* 1999; *Gomes et al.,* 2004).

5.1. The Resonance-Sweeping Scenario

In *Malhotra* (1993, 1995) it was realized that, following Neptune's migration, the mean-motion resonances with Neptune also migrated outward, sweeping through the primordial Kuiper belt until they reached their present positions. From adiabatic theory (see, e.g., *Henrard,* 1982), some of the Kuiper belt objects over which a mean-motion resonance swept were captured into resonance; they subsequently followed the resonance through its migration, with ever-increasing eccentricities. In fact, it can be shown (*Malhotra,* 1995) that, for a k_N:k resonance, the eccentricity of an object grows as

$$\Delta e^2 = \frac{(k - k_N)}{k} \log \frac{a}{a_i}$$

where a_i is the semimajor axis that the object had when it was captured in resonance and a is its current semimajor axis. This relationship neglects secular effects inside the mean-motion resonance that can be important if Neptune's eccentricity is not zero and its precession frequencies are comparable to those of the resonant particles (*Levison and Morbidelli,* 2003). This model can account for the existence of the large number of Kuiper belt objects in the 2:3 mean-motion resonance with Neptune (and also in other resonances), and can explain their large eccentricities (see Fig. 5). Assuming that all objects were captured when their eccentricities were close to zero, the above formula indicates that Neptune had to have migrated ~7 AU in order to quantitatively reproduce the observed range of eccentricities (up to ~0.3) of the resonant bodies.

In *Malhotra* (1995), it was also shown that the bodies captured in the 2:3 resonance can acquire large inclinations, comparable to those of Pluto and other objects. The mechanisms that excite the inclination during the capture process have been investigated in detail in *Gomes* (2000), who concluded that, although large inclinations can be achieved, the resulting proportion of high-inclination vs. low-inclination bodies, as well as their distribution in the e–i plane, does not reproduce the observations well. We will return to this issue in section 5.2.

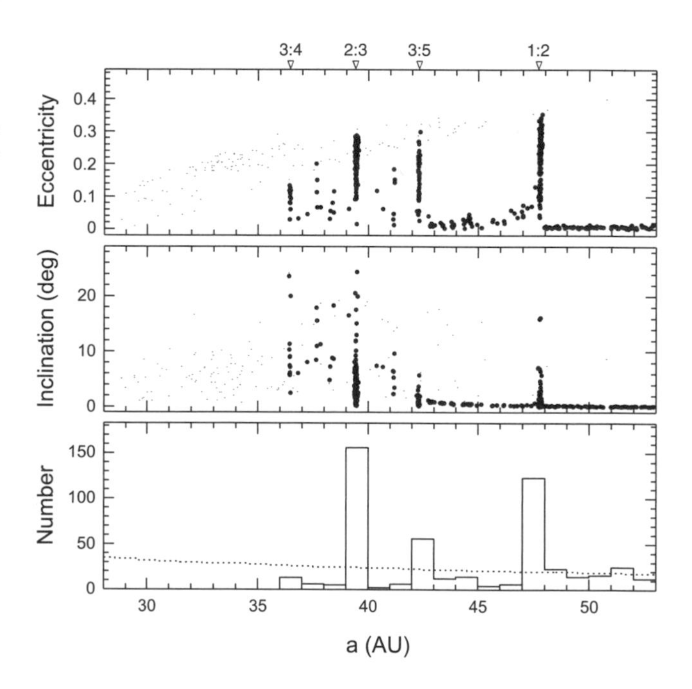

Fig. 5. The final distribution of Kuiper belt bodies according to the sweeping resonances scenario (courtesy of R. Malhotra). This simulation was done by numerically integrating, over a 200-m.y. time span, the evolution of 800 test particles on initially quasi-circular and coplanar orbits. The planets are forced to migrate by a quantity Δa (equal to –0.2 AU for Jupiter, 0.8 AU for Saturn, 3 AU for Uranus, and 7 AU for Neptune) and approach their current orbits exponentially as $a(t) = a_\infty - \Delta a \exp(-t/4 \text{ m.y.})$, where a_∞ is the current semimajor axis. Large solid dots represent "surviving" particles (i.e., those that have not suffered any planetary close encounters during the integration time); small dots represent the "removed" particles at the time of their close encounter with a planet (e.g., bodies that entered the scattered disk and whose evolution was not followed further). In the lowest panel, the solid line is the histogram of semimajor axes of the "surviving" particles; the dotted line is the initial distribution. The locations of the main mean-motion resonances are indicated above the top panel.

The mechanism of adiabatic capture into resonance requires that Neptune's migration happened very smoothly. If Neptune had encountered a significant number of large bodies, its jerky migration would have jeopardized the capture into resonances. For instance, direct simulations of Neptune's migration in *Hahn and Malhotra* (1999) — which modeled the disk with lunar- to martian-mass planetesimals — did not result in any permanent captures. Adiabatic captures into resonance can be seen in numerical simulations only if the disk is modeled using many more planetesimals with smaller masses (*Gomes,* 2003; *Gomes et al.,* 2004). The constraint set by the capture process on the maximum size of the planetesimals that made up the bulk of the mass in the disk has been recently estimated in *Murray-Clay and Chiang* (2006). They found that resonance capture due to Neptune's migration is efficient if the bulk of the disk particles was smaller than ~100 km and the fraction of disk mass in objects with sizes ≥1000 km was less than a few percent. This result appears too severe, because the results in *Gomes* (2003) and *Gomes et al.* (2004) show that resonant captures occur in disks entirely constructed of Pluto-mass objects, although probably with a smaller efficiency than required in *Murray-Clay and Chiang* (2006).

If migration really happened smoothly, Murray-Clay and Chiang worked out a constraint on the migration rate. Remember that there are two islands of libration in the 1:2 resonance with Neptune (see Fig. 1). If Neptune's migration occurs in less than 10 m.y., they showed that objects captured in the 1:2 resonance should preferentially be in the trailing island (that where σ librates around a value $\sigma_{stab} > \pi$). Converesely, most of the observed objects are in the leading island. This, at first sight, points to a slow Neptune migration. As we will see below, however, resonant objects can also be captured from the scattered disk, and Neptune's migration might have been very different from what was originally envisioned. Thus, it is unclear which kind of constraint is provided by the internal distribution of the 1:2 resonant objects.

As shown in Fig. 5, if the resonance-sweeping scenario can explain the existence of the resonant populations, it cannot explain the orbital distribution in the classical belt, between 40 and 48 AU, nor the mass depletion of the Kuiper belt. The eccentricity excitation and, in particular, the inclination excitation obtained in the simulation in that region are far too small compared to those inferred from the observations. Thus, *Hahn and Malhotra* (2005) suggested that resonance sweeping occurred after some perturbation excited, and perhaps depleted, the planetesimal disk. In this case, the eccentricity and inclination distribution in the classical belt would not have been sculpted by the sweeping process, but would be the relic of such previous excitation mechanism(s). A similar conclusion was reached recently by *Lykawka and Mukai* (2007a), who found that the populations of objects in distant mean-motion resonances with Neptune (i.e., with a > 50 AU) and their eccentricity-inclination-libration amplitude distributions can be explained by resonance sweeping only if the disk in the 40–48 AU region was already preexcited in both e and i. Possible mechanisms of excitation and their problems will be briefly discussed in section 6.

5.2. The Origin of the Hot Population

The observation that the largest objects in the hot population are bigger than those in the cold population led *Levison and Stern,* (2001) to suggest that the hot population formed closer to the Sun and was transported outward during the final stages of planet formation. *Gomes* (2003) showed that the simple migration of Neptune described in the previous section could accomplish this process.

In particular, *Gomes* (2003) studied the migration of the giant planets through a disk represented by 10,000 particles, a much larger number than had previously been attempted. In Gomes' simulations, during its migration Neptune scattered the planetesimals and formed a massive scattered disk. Some of the scattered bodies decoupled from the planet, decreasing their eccentricities through interactions with some secular or mean-motion resonance (see chapter by Gomes et al. for a detailed discussion of how resonances can decrease the eccentricities). If Neptune were not migrating, the decoupled phases would have been transient. In fact, the dynamics are reversible, so that the eccentricity would have eventually increased back to Neptune-crossing values. However, Neptune's migration broke the reversibility, and some of the decoupled bodies managed to escape from the resonances and remained permanently trapped in the Kuiper belt. As shown in Fig. 6, the current Kuiper belt would therefore be the result of the superposition in (a, e)-space of these bodies with the local population, originally formed beyond 30 AU.

The local population stayed dynamically cold, in particular in inclination, because its objects were only moderately excited by the resonance-sweeping mechanism, as in Fig. 5. Conversely, the population captured from the scattered disk had a much more extended inclination distribution, for two reasons: (1) the inclinations got excited during the scattered disk phase before capture and (2) there was a dynamical bias in favor of the capture of high-inclination bodies, because at large i the ability of mean-motion resonances to decrease the eccentricity is enhanced (see chapter by Gomes et al.). Thus, in Gomes' model the current cold and hot populations should be identified respectively with the local population and with the population trapped from the scattered disk.

This scenario is appealing because, assuming that the bodies' color varied in the primordial disk with heliocentric distance, it qualitatively explains why the scattered objects and hot classical belt objects — which mostly come from regions inside ~30 AU — appear to have similar color distributions, while the cold classical objects — the only ones that actually formed *in situ* — have a different distribution. Similarly, assuming that at the time of Neptune's migration

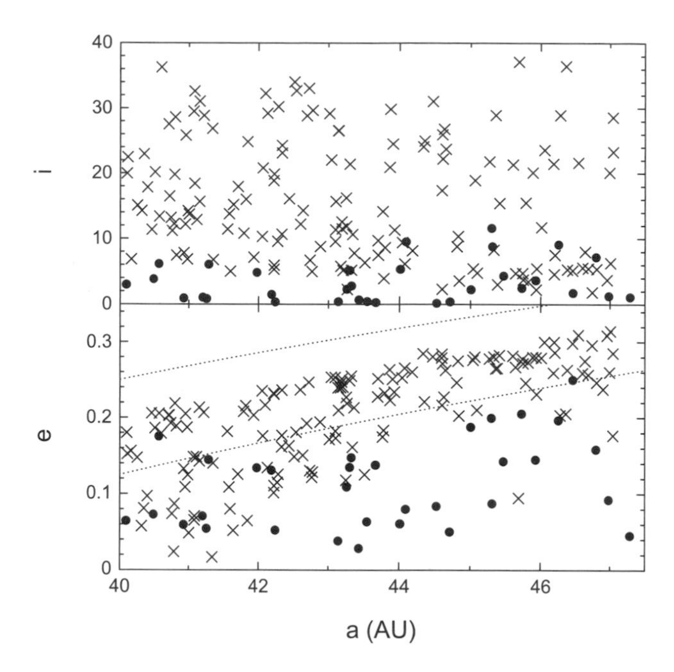

Fig. 6. The orbital distribution in the classical belt according to the simulations in *Gomes* (2003). The dots denote the population that formed locally, which is only moderately dynamically excited. The crosses denote the bodies that were originally inside 30 AU. Therefore, the resulting Kuiper belt population is the superposition of a dynamically cold population and a dynamically hot population, which gives a bimodal inclination distribution. The dotted curves in the eccentricity vs. semimajor axis plot correspond to q = 30 AU and q = 35 AU.

the maximum size of the objects was a decreasing function of their initial heliocentric distance, the scenario also explains why the biggest Kuiper belt objects are all in the hot population.

As Fig. 6 shows, there may be some quantitative problems in the reproduction of the orbital distribution of the hot Kuiper belt in Gomes' scenario (e.g., the perihelion distances appear to be somewhat too low). However, an issue of principle concerns the relative weight between the hot and cold populations. In Gomes' simulations, only a fraction of a percent of the original scattered disk remained trapped in the hot population. On the other hand, the cold population was not depleted by the resonance sweeping, so that it retained most of the original objects. Thus, if the local population was similar in size distribution and number density to the planetesimal disk from which the scattered disk was extracted, it should outnumber the hot population by a huge factor (~1000). So, in order to obtain a final inclination distribution that quantitatively reproduces the debiased inclination distribution of *Brown* (2001), *Gomes* (2003) had to scale down the number of objects in the cold population by an appropriate factor, assuming that some mechanism, not included in the simulation, caused a decimation, and hence a mass depletion, of the local population. These mechanisms are reviewed in the next section.

Before concluding this section, we note that the work by Gomes also has important implications for the origin of the detached population. This issue is addressed in detail in the chapter by Gomes et al. and therefore we do not discuss it here.

6. THE MASS DEFICIT PROBLEM

As we described in section 3, the Kuiper belt only contains roughly 0.1% of the mass that is required to grow the objects that we see. So, the natural question is, what happened to all that mass? We refer to this issue as the "mass deficit problem." We now review ideas that are currently in the literature.

6.1. Mass Removal

Two general scenarios have been proposed for the mass depletion: (1) a strong dynamical excitation of the Kuiper belt, which caused the ejection of most of the bodies from the Kuiper belt to the Neptune-crossing region; and (2) the collisional comminution of most of the mass of the Kuiper belt into dust. We start our discussion with scenario (1).

Because dynamics are size-independent, a dynamical-depletion scenario requires that the primordial population in the Kuiper belt had a size distribution similar to the one that (currently) exists, but with a number of objects at each size multiplied by the ratio between the primordial mass and the current mass. Remember that, in the current solar system configuration, most of the Kuiper belt is stable, so dynamical erosion cannot signficantly reduce the total number of objects. The idea is, therefore, that some perturbation, which acted in the past and is no longer at work, strongly excited the orbital distribution of the Kuiper belt population. Most of the original objects acquired Neptune-crossing eccentricities, so that they were subsequently eliminated by the scattering action of the planets. Only a small fraction of the original population, corresponding to the surviving mass fraction, remained in the Kuiper belt on excited orbits like those of the observed objects. Thus, this scenario aims to simultaneously explain both the mass depletion of the Kuiper belt and its orbital excitation.

A first dynamical depletion mechanism was proposed in *Morbidelli and Valsecchi* (1997) and later revisited in *Petit et al.* (2001). This mechanism invokes the existence of a planetary embryo, with mass comparable to that of Mars or Earth, in the scattered disk for ~10^8 yr. Another mechanism was proposed by *Nagasawa and Ida* (2000) and invokes the sweeping of secular resonances through the Kuiper belt during the dispersion of the primordial gas disk.

The problem with the dynamical-depletion scenario, which was not immediately recognized, is that the ejection of a massive population of objects from the Kuiper belt to the Neptune-crossing region would cause Neptune to migrate into the Kuiper belt. After all, this scenario invokes a ~15 $M_\oplus$ object to remove ≥15 $M_\oplus$ of disk material and angular momentum must be conserved. For instance, revisiting the *Petit et al.* (2001) work with simulations that account for the effect of the planetesimals on the dynamics of the massive bodies, *Gomes et al.* (2004) showed that even a disk containing ~4 $M_\oplus$ of material between 40 and 50 AU drives Neptune beyond 30 AU. This is much less than the mass required (10–30 $M_\oplus$) by models of the accretion of Kuiper belt bodies (*Stern and Colwell,* 1997a; *Kenyon and Luu,* 1999b).

The sole possibility for a viable dynamical model of Kuiper belt depletion is if the objects were kicked directly to hyperbolic or Jupiter-crossing orbits and were eliminated without interacting with Neptune. Only the passage of a star through the Kuiper belt seems to be capable of such an extreme excitation (*Kobayashi et al.,* 2005). However, the cold Kuiper belt would not survive in this case.

We note in passing that, even if we ignore the problem of Neptune's migration, massive embryos or secular resonance sweeping are probably not able to reproduce the inclination distribution observed in the Kuiper belt. Thus, these mechanisms are unlikely to be an alternative to the scenario proposed in *Gomes* (2003) for producing the hot classical belt. Consequently, the idea of *Hahn and Malhotra* (2005) and *Lykawka and Mukai* (2007a) that the classical belt acquired its current excitation before Neptune's migration is not supported, so far, by an appropriate excitation mechanism.

The collisional grinding scenario was proposed in *Stern and Colwell* (1997b) and *Davis and Farinella* (1997, 1998), and then pursued in *Kenyon and Luu* (1999a) and *Kenyon and Bromley* (2002, 2004a). It is reviewed in detail in the chapter by Kenyon et al. In essence, a massive Kuiper belt with large eccentricities and inclinations would experience very intense collisional grinding. Consequently, most of the mass originally in bodies smaller than several tens of kilometers could be comminuted into dust, and then evacuated by radiation pressure and Poynting-Robertson drag. This would lead to a substantial depletion in mass.

To work, the collisional erosion scenario requires that two essential conditions be fulfilled. First, it requires a peculiar primordial size distribution, such that all the missing mass was contained in small, easy-to-break objects, while the number of large objects was essentially identical to that in the current population. Some models support the existence of such a size distribution at the end of the accretion phase (*Kenyon and Luu,* 1998, 1999b). However, there are several arguments in favor of a completely different size distribution in the planetesimal disk. The collisional formation of the Pluto-Charon binary (*Canup,* 2005) and of the 2003 EL_{61} family (*Brown et al.,* 2007), the capture of Triton onto a satellite orbit around Neptune (*Agnor and Hamilton,* 2006), and the fact that the Eris, the largest known Kuiper belt object, is in the detached population (*Brown et al.,* 2005), all suggest that the number of big bodies was much larger in the past, with as many as 1000 Pluto-sized objects (*Stern,* 1991). Moreover, we have seen above that the mechanism of *Gomes* (2003) for the origin of the hot population also requires a disk's size distribution with ~1000× more large objects than currently present in the Kuiper belt. Finally, *Charnoz and Morbidelli* (2007) showed that, if the size distribution required for collisional grinding in the Kuiper belt is assumed for the entire planetesimal disk (5–50 AU), the Oort cloud and the scattered disk would not contain enough comet-sized objects to supply the observed fluxes of long-period and Jupiter-family comets: The cometesimals would have been destroyed before being stored in the comet reservoirs (also see *Stern and Weissman,* 2001). So, to fulfill all these constraints and still have an effective collisional grinding in the Kuiper belt, one has to assume that the size distributions were totally different in the region of the protoplanetary disk swept by Neptune and in the region of the disk that became the Kuiper belt. This, *a priori,* seems unlikely, given the proximity between the two regions; however, we will come back to this in section 8.

The second essential condition for substantial collisional grinding is that the energy of collisions is larger than the energy required for disruption of the targets. Thus, either the KBOs are extremely weak [the successful simulations in *Kenyon and Bromley* (2004a) had to assume a specific energy for disruption that is at least an order of magnitude lower than predicted by the smooth particle hydrodynamical (SPH) simulations of fragmentation by *Benz and Asphaug* (1999)], or the massive primordial Kuiper belt had a large dynamical excitation, with $e \sim 0.25$ and/or $i \sim 7°$ (as assumed in *Stern and Colwell,* 1997b). However, if, as we argued above, the hot population was implanted in the Kuiper belt via the low-efficiency process of *Gomes* (2003), then it was never very massive and would not have had much effect on the collisional evolution of the cold population. Thus, the cold population must have ground itself down. This is unlikely because the excitation of the cold-population is significantly smaller than the required values reported above. There is the possibility that the collisional erosion of the cold belt was due to the high-velocity bombardment by projectiles in the scattered disk. The scattered disk was initially massive, but its dynamical decay was probably too fast (~100 m.y.) (see *Duncan and Levison,* 1997). The collisional action of the scattered disk onto the cold belt was included in *Charnoz and Morbidelli* (2007) but turned out to be a minor contribution.

6.2. Pushing Out the Kuiper Belt

Given the problems explained just above, an alternative way of solving the mass deficit problem was proposed in *Levison and Morbidelli* (2003). In this scenario, the primordial edge of the massive protoplanetary disk was somewhere around 30–35 AU and the *entire* Kuiper belt population — not only the hot component as in *Gomes* (2003) — formed within this limit and was transported to its current location during Neptune's migration. The transport process for the cold population had to be different from the one found in *Gomes* (2003) for the hot population (but still work in parallel with it), because the inclinations of the hot population were excited, while those of the cold population were not.

In the framework of the classical migration scenario (*Malhotra,* 1995; *Gomes et al.,* 2004), the mechanism proposed in *Levison and Morbidelli* (2003) was the following:

The cold population bodies were initially trapped in the 1:2 resonance with Neptune; then, as they were transported outward by the resonance, they were progressively released due to the nonsmoothness of the planetary migration. In the standard adiabatic migration scenario (*Malhotra*, 1995), there would be a resulting correlation between the eccentricity and the semimajor axis of the released bodies. However, this correlation was broken by a secular resonance embedded in the 1:2 mean-motion resonance. This secular resonance was generated by the objects in the resonance themselves. In particular, unlike previous studies of migration, *Levison and Morbidelli* (2003) included the mass of the objects in the resonance, which modified the precession rate of Neptune's orbit.

Simulations of this process matched the observed (a, e) distribution of the cold population fairly well, while the initially small inclinations were only very moderately perturbed. In this scenario, the small mass of the current cold population is simply due to the fact that only a small fraction of the massive disk population was initially trapped in the 1:2 resonance and then released on stable nonresonant orbits. The final position of Neptune would simply reflect the primitive truncation of the protoplanetary disk (see *Gomes et al.,* 2004, for a more detailed discussion). Most importantly, this model explains why the current edge of the Kuiper belt is at the 1:2 mean-motion resonance with Neptune, despite the fact that none of the mechanisms proposed for the truncation of the planetesimal disk involves Neptune in a direct way (see section 4). The location of the edge was modified by the migration of Neptune, via the migration of the 1:2 resonance.

On the flip side, the model in *Levison and Morbidelli* (2003) reopened the problem of the origin of the different physical properties of the cold and hot populations, because both would have originated within 35 AU, although in somewhat different parts of the disk. Moreover, *Lykawka and Mukai* (2007a) showed that this model cannot reproduce the low-to-moderate inclination objects in the distant (i.e., beyond 50 AU) high-order mean-motion resonances with Neptune.

7. EFFECTS OF A DYNAMICAL INSTABILITY IN THE ORBITS OF URANUS AND NEPTUNE

The models reviewed in the previous sections assume that Neptune migrated outward on a nearly circular orbit. However, substantially different models of the evolution of the giant planets have been recently proposed.

7.1. The Nice Model

This model — whose name comes from the city in France where it has been developed — reproduces, for the first time, the orbital architecture of the giant planet system (orbital separations, eccentricities, inclinations) (*Tsiganis et al.,* 2005) and the capture of the Trojan populations of Jupiter (*Morbidelli et al.,* 2005) and Neptune (*Tsiganis et al.,* 2005; *Sheppard and Trujillo,* 2006). It also naturally supplies a trigger for the late heavy bombardment (LHB) of the terrestrial planets (*Gomes et al.,* 2005), and quantitatively reproduces most of the LHB's characteristics.

In the Nice model, the giant planets are assumed to be initially on nearly circular and coplanar orbits, with orbital separations significantly smaller than the ones currently observed. More precisely, the giant planet system is assumed to lie in the region from ~5.5 AU to ~14 AU, and Saturn is assumed to be closer to Jupiter than their mutual 1:2 mean-motion resonance. A planetesimal disk is assumed to exist beyond the orbits of the giant planets, on orbits whose dynamical lifetime is at least 3 m.y. (the supposed lifetime of the gas disk). The outer edge of the planetesimal disk is assumed to lie at ~34 AU and the total mass is ~35 $M_\oplus$ (see Fig. 7a).

With the above configuration, the planetesimals at the inner edge of the disk evolve onto Neptune-scattering orbits on a timescale of a few million years. Consequently, the migration of the giant planets proceeds at very slow rate, governed by the slow planetesimal escape rate from the disk. Because the planetary system would be stable in absence of interactions with the planetesimals, this slow mi-

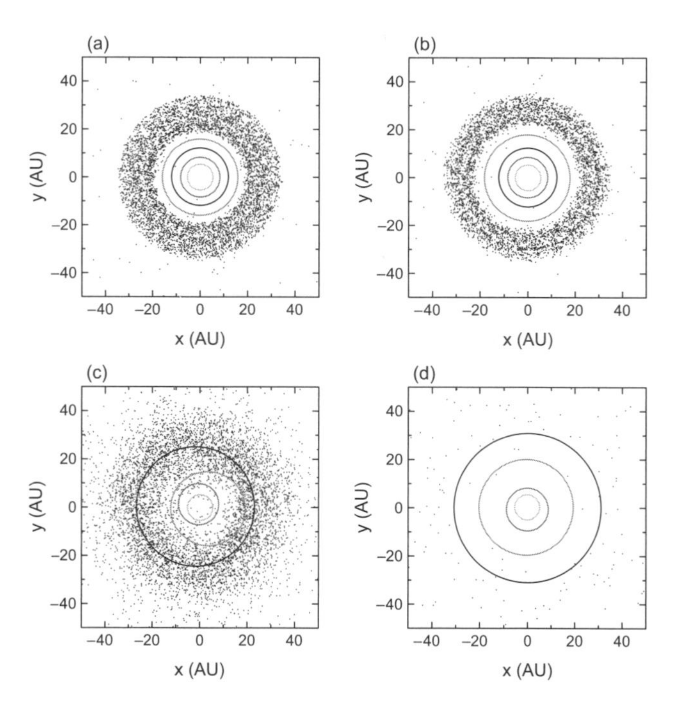

Fig. 7. Solar system evolution in the Nice model. **(a)** At a time close to the beginning of the evolution. The orbits of the giant planets (concentric circles) are very close to each other and are quasicircular. They are surrounded by a disk of planetesimals, whose inner edge is due to the perturbations from the planets and the outer edge is assumed to be at 34 AU. **(b)** Immediately before the great instability. Saturn is approximately crossing the 1:2 resonance with Jupiter. **(c)** At the time of the instability. Notice that the orbits of the planets have become eccentric and now penetrate the planetesimal disk. **(d)** After the LHB. The planets are parked on orbits very similar (in terms of separation, eccentricity, and inclination) to their current ones. The massive planetesimal disk has been destroyed. Only a small fraction of the planetesimals remain in the system on orbits typical of the scattered disk, Kuiper belt, and other small body reservoirs. From *Gomes et al.* (2005).

gration continues for a long time, slightly damping out as the unstable disk particles are removed from the system (Fig. 7). After a long time, ranging from 350 m.y. to 1.1 G.y. in the simulations of *Gomes et al.* (2005) — which is consistent with the timing of the LHB, approximately 650 m.y. after planet formation — Jupiter and Saturn eventually cross their mutual 1:2 mean-motion resonance (Fig. 7b). This resonance crossing excites their eccentricities to values slightly larger than those currently observed. The small jump in Jupiter's and Saturn's eccentricities drives up the eccentricities of Uranus and Neptune, however. The ice giant's orbits become chaotic and start to approach each other. Thus, a short phase of encounters follows the resonance-crossing event. Consequently, both ice giants are scattered outward, onto large eccentricity orbits (e ~ 0.3–0.4) that penetrate deeply into the disk (Fig. 7c). This destabilizes the full planetesimal disk and disk particles are scattered all over the solar system. The eccentricities of Uranus and Neptune and — to a lesser extent — of Jupiter and Saturn, are damped on a timescale of a few million years due to the dynamical friction exerted by the planetesimals. Thus, the planets decouple from each other, and the phase of mutual encounters rapidly ends. During and after the eccentricity damping phase, the giant planets continue their radial migration, and eventually reach final orbits when most of the disk has been eliminated (Fig. 7d).

The temporary large eccentricity phase of Neptune opens a new degree of freedom for explaining the orbital structure of the Kuiper belt. The new key feature to the dynamics is that, when Neptune's orbit is eccentric, the full (a, e) region up to the location of the 1:2 resonance with the planet is chaotic, even for small eccentricities. This allows us to envision the following scenario. We assume, in agreement with several of the simulations of the Nice model, that the large eccentricity phase of Neptune is achieved when the planet has a semimajor axis of ~28 AU, after its last encounter with Uranus. In this case, a large portion of the current Kuiper belt is already interior to the location of the 1:2 resonance with Neptune. Thus, it is unstable, and can be invaded by objects coming from within the outer boundary of the disk (i.e., ≲34 AU). When the eccentricity of Neptune damps out, the mechanism for the onset of chaos disappears. The Kuiper belt becomes stable, and the objects that happen to be there at that time remain trapped for the eternity. Given that the invasion of the particles is fast and the damping of Neptune's eccentricity is also rapid, there is probably not enough time to excite significantly the particles' orbital inclinations if Neptune's inclination is also small. Therefore, we expect that this mechanism may be able to explain the observed cold population. The hot population is then captured later, when Neptune is migrating up to its final orbit on a low-eccentricity orbit, as in *Gomes* (2003).

The numerical simulations of this process are presented in *Levison et al.* (2007). Figure 8 compares with the observations the semimajor axis vs. eccentricity distribution resulting from one of the simulations, 1 G.y. after the giant planet instability. The population of quasicircular objects at low inclination extends to ~45 AU, in nice agreement with the observations. The deficit of low-eccentricity objects between 45 and 48 AU is reproduced, and the outer edge of the classical belt is at the final location of the 1:2 mean-motion resonance with Neptune as observed. Moreover, the real Kuiper belt shows a population of objects with q ~ 40 AU beyond the 1:2 resonance with Neptune, which are known to be stable (*Emel'yanenko et al.,* 2003). This population has been known by several names in the literature, which include the fossilized scattered disk, the extended scattered disk, and the detached population (see chapter by Gladman et al.). This model reproduces this population quite well.

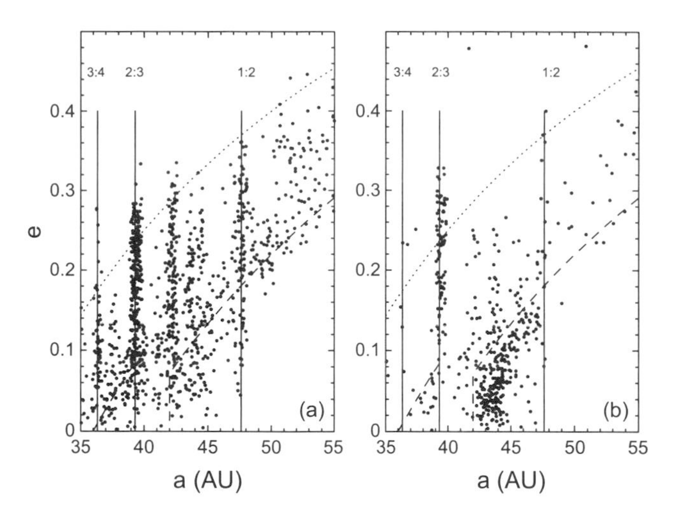

Fig. 8. The distribution of semimajor axes and eccentricities in the Kuiper belt. **(a)** Result of a simulation based on the Nice model. **(b)** The observed distribution (three oppositions objects only). The vertical solid lines mark the main resonance with Neptune. The dotted curve denotes perihelion distance equal to 30 AU and the dashed curve delimits the region above which only high-inclination objects or resonant objects can be stable over the age of the solar system.

Three main differences are also noticeable, though: (1) All the mean motion resonances are overpopulated relative to the classical belt. This is probably the consequence of the fact that in the simulations the migration of Neptune's orbit and its eccentricity damping were forced smoothly, through fake analytic terms of the equations of motion. As we said above, a migration with some stochastic component (due to the encounters with massive objects in the disk) would have produced fewer surviving bodies in the resonances. (2) The region above the long-dashed curve is overpopulated in the simulation. The curve represents approximately the boundary between the stable (yellow) and the unstable (black) regions in Plate 7. Thus, if the final orbits of the giant planets were exactly the same as the real ones and the simulations were extended for the age of the solar system, most of the population above the curve would be depleted as a consequence of chaotic dynamics. (3) The cold Kuiper belt has eccentricities that are slightly too large. The median eccentricity of the real objects with $42 < a < 48$ AU and $q > 37$ AU is 0.07, while the model produces a value of 0.10.

Figure 9 shows the cumulative inclination distribution of objects trapped in the classical belt at the end of a simulation and compares it with the observed distribution. For the comparison to be meaningful, the simulated distribution was run through a survey bias-calculator, following the approach of *Brown* (2001). The two curves are very similar. Indeed, they are almost indistinguishable for inclinations less than 6°. This means that the cold population, and the inclination distribution within it, has been correctly reproduced, as well as the distribution in the lower part of the hot population (i.e., that with $4° < i < 10°$). We note, however, a dearth of large-inclination objects. This deficit is intriguing and unexplained, in particular given that the raw simulations of the Nice model (namely those in which the planets are not forced to migrate, but are let free to respond to their interactions with massive planetesimals) produce objects captured in the classical belt or in the detached population with inclinations up to 50° (see Fig. 3 in the chapter by Gomes et al.). In general, we would expect an inclination distribution in the hot population that is equivalent to that of *Gomes* (2003), or even more excited. In fact, as pointed out in *Lykawka and Mukai* (2007b), the inclinations in the scattered disk, from which the hot population is derived, are restricted to be less than ~40° by the conservation of the Tisserand parameter with respect to Neptune, which holds if the planet is on a quasicircular orbit as in the simulations of *Gomes* (2003). In the Nice model, the eccentricity of Neptune breaks the conservation of the Tisserand parameter, and hence, in principle, inclinations can be larger.

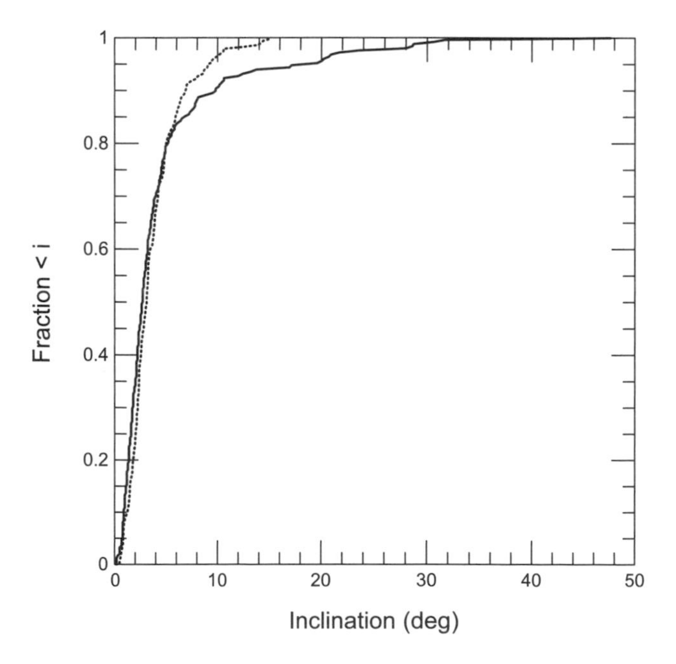

Fig. 9. The cumulative inclination distribution of the observed classical belt objects (solid curve) and that expected from the result of our simulation, once the observational biases are taken into account (dotted curve).

The results of the simulations based on the Nice model also provide a qualitative explanation for the observed correlations between inclination and physical properties. The particles that are trapped in the cold classical belt come, almost exclusively, from the outermost parts of the planetesimal disk — in particular beyond 29 AU. Conversely, a significant fraction of those trapped in the hot population come from the inner disk. Thus, if one assumes that the largest objects could form only in the inner part of the disk, then these objects can only (or predominantly) be found in the hot population. Similarly, if one assumes that (for some unknown reason) the objects from the outer part of the disk are red and those from the inner part are gray, the cold population would be composed almost exclusively of red objects, whereas the hot population would contain a mixture of red and gray bodies.

The simulations in *Levison et al.* (2007) show that 50 to 130 particles out of 60,000 are trapped in the classical belt (cold and hot populations together in roughly equal proportion). According to the Nice model, the original planetesimal disk contained 35 $M_\oplus$, thus this model predicts that the classical Kuiper belt should currently contain between ~0.02 and ~0.08 $M_\oplus$, in good agreement with observational estimates. Of course, to be viable, the model needs to explain not only the total mass of the belt, but also the total number of bright, detectable bodies. It does this quite nicely if one assumes that the original disk size distribution is similar to the one currently observed. As we explained in section 6.1, this is consistent with other constraints like the formation of the Pluto-Charon binary. Thus, the Nice model explains, for the first time, the mass deficit of the Kuiper belt and the ratio between the hot and the cold population, in the framework of an initial planetesimal size distribution that fulfills all the constraints enumerated in section 6.

Finally, the Nice model reproduces in a satisfactory way the orbital distributions of the populations in the main mean-motion resonances with Neptune. Figure 10 compares the (e, i) distribution of the Plutinos obtained in one of *Levison et al.*'s (2007) simulations, against the observed distribution. The overall agreement is quite good. In particular, this is the first model that does not produce an overabundance of resonant objects with low inclinations.

Moreover, the left panel of Fig. 10 uses different symbols to indicate the particles captured from the inner ($a < 29$ AU) and the outer ($a > 29$ AU) parts of the disk. As one sees, the particles are very well mixed, which is in agreement with the absence of correlations between colors and inclinations among the Plutinos. Conversely, a very strong correlation was expected in the original *Gomes* (2003) scenario because a large number of low-inclination bodies were captured from the cold disk. The reason that the Nice model is so

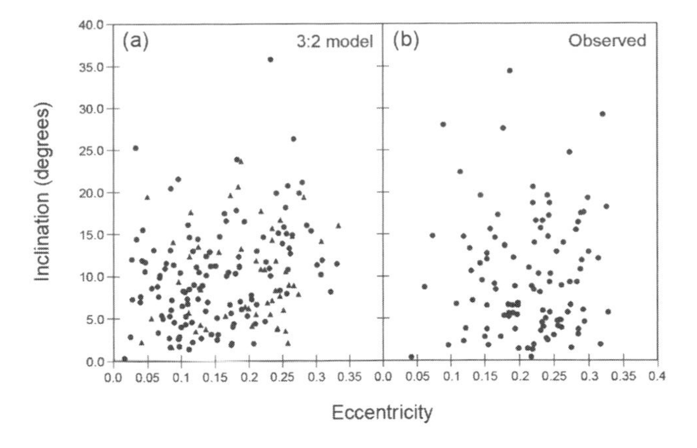

Fig. 10. The eccentricity–inclination distribution of the Plutinos. **(a)** The simulated distribution; black dots refer to particles from the outer disk and gray triangles to particles from the inner disk. **(b)** The observed distribution. The relative deficit of observed Plutinos at low eccentricity with respect to the model is probably due to observational biases and to the criterion used to select "resonant objects" from the simulation (see *Levison et al.,* 2007).

much more successful than previous models is that the 2:3 resonance cannot capture any objects via the mechanism of *Malhotra* (1995). This is due to the fact that the resonance is already beyond the disk's outer edge at the beginning of the simulation (i.e., after the last encounter of Neptune with Uranus). We believe that this success strongly supports the idea of a planetesimal disk truncated at 30–35 AU and of a "jump" of Neptune toward the outer edge of the disk.

As for the higher-order resonances beyond 50 AU, the simulations of *Levison et al.* (2007) produce populations with moderate libration amplitudes and inclinations consistent with observations, thus satisfying the constraint posed by *Lykawka and Mukai* (2007a).

In summary, the strength of the Nice model is that it is able to explain *all* the intriguing properties of the Kuiper belt, at least at a semiquantitative level, in the framework of a single, unique event. That the same scenario also explains the orbital architecture of the giant planets, the Trojans of both Jupiter and Neptune, and the LHB is, of course, a nonnegligible additional bonus that should give credence to the model.

Of course, the Nice model is not perfect. As we have seen above, the simulations thus far performed have not been able to simultaneously produce (1) the very highest inclinations that we see and (2) a cold belt that is cold enough. The numerical simulations contain some simplifications that might affect the results. In particular, mutual collisions and collective gravitational effects among the planetesimals are neglected. Moreover, as discussed in *Levison et al* (2007), only a subset of the evolutions of the giant planets observed in the simulations of the *Tsiganis et al.* (2005) will produce a cold classical belt. Some experiments overly excite inclinations so that a cold belt is not formed, despite producing good final planetary orbits. Nonetheless, we feel that the Nice model's strengths outweigh its weaknesses, particularly given that other models of Kuiper belt formation have had much more limited success at reproducing the observations.

7.2. Other Planetary Instability Models

The Nice model is not the first model to make use of a temporary dynamical instability of the giant planet system (and will probably not be the last!).

Thommes et al. (1999) proposed that Uranus and Neptune formed in between the orbits of Jupiter and Saturn. They were subsequently destabilized and scattered onto orbits with larger semimajor axis and eccentricities. The dynamical excitation was eventually damped by the dynamical friction exerted by a massive planetesimal disk, and the planets achieved stable orbits. Their simulations showed an interesting sculpting of the (a, e) distribution in the region corresponding to the classical Kuiper belt. However, the planetesimal disk was extended to 60 AU. Thus, no outer edge was produced at the 1:2 resonance with Neptune and there was not enough mass depletion in the Kuiper belt. Moreover, since Uranus and Neptune started between Jupiter and Saturn they suffered much stronger encounters with the gas giants than occurred in the Nice model. As a result, the planetesimal disk needed to be much more massive — so massive that if the simulations had been run to completion, Neptune would have migrated well beyond 30 AU.

Chiang et al. (2006) have recently speculated on a scenario based on recent work by *Goldreich et al.* (2004a,b), who, from analytic considerations, predicted the formation of five planets between 20 and 40 AU. These planets remained stable during their formation because their orbits were continuously damped by the dynamical friction exerted by a disk of planetesimals that contained more mass than the planetary system. These planetesimals were very small (submeter in size) and thus remain dynamically cold due to collisional damping. When the planets reached Neptune mass, the mass of the planets and the mass of the disk became comparable, so that the planets became unstable. Goldreich et al. conjectured that three of the five planets were ejected and the two remaining ones stabilized on orbits comparable to those of Uranus and Neptune.

Chiang et al. (2006) suggested that at the time of the instability, the disk contained two populations: one made up of ~100-km objects and one consisting of submeter-sized objects. The current Kuiper belt structure would be the result of the orbital excitation suffered during the multiplanet instability. The hot population would be made up of the larger objects, which were permanently excited during the instability. In contrast, the smallest planetesimals would suffer a significant amount of collisional damping, which would have led to the eventual accretion of the cold population. Numerical simulations made by *Levison and Morbi-*

delli (2007) with a new code that accounts for a planetesimal disk with strong internal collisional damping invalidate the *Goldreich et al.* (2004a,b) proposal. It is found that a system of five unstable Neptune-mass planets systematically leads to a system with more than two planets, spread in semimajor axis well beyond 30 AU. Thus, the architecture of the solar planetary system seems to be inconsistent with Goldreich et al.'s idea.

8. CONCLUSIONS AND DISCUSSION

In this chapter we have tried to understand which kind of solar system evolution could have produced the most important properties of the Kuiper belt: its mass deficit, its outer edge, the coexistence of a cold and a hot classical population with different physical properties, and the presence of resonant populations. We have proceeded by basic steps, trying to narrow the number of possibilities by considering one Kuiper belt feature after the other, and starting from the most accepted dynamical process (planet migration) and eventually ending with a more extravagant one (a temporary instability of the giant planets).

We have converged on a basic scenario with three ingredients: The planetesimal disk was truncated close to 30 AU and the Kuiper belt was initially empty; the size distribution in the planetesimal disk was similar to the current one in the Kuiper belt, but the number of objects at each size was larger by a factor of ~1000; and the Kuiper belt objects are just a very small fraction of the original planetesimal disk population and were implanted onto their current orbit from the disk during the evolution of the planets. A temporary high-eccentricity phase of Neptune, when the planet was already at ~28 AU — as in the Nice model — seems to be the best way to implant the cold population.

The mass deficit problem is the main issue that drove us to this conclusion. In particular, we started from the consideration that, if a massive disk had extended into the Kuiper belt, neither collisional grinding nor dynamical ejection could have depleted its mass to current levels. Dynamical ejection seems to be excluded by the constraint that Neptune did not migrate past 30 AU. Collisional grinding seems to be excluded by the arguments that the size distribution was about the same everywhere in the disk and that ~1000 Pluto-sized bodies had to exist in the planetary region.

Is there a flaw in this reasoning? Are we really sure that the cold population did not form *in situ*? The argument that the size distribution of the *in situ* population should be similar to that in the region spanned by Neptune's migration neglects possible effects due to the presence of an edge. After all, an edge is a big discontinuity in the size and mass distribution, so that it may not be unreasonable that the region adjacent to the edge had very different properties from the region further away from the edge.

Our view of the Kuiper belt evolution could radically change if a model of accretion were developed that produces a disk of planetesimals with a size distribution that changes drastically with distance, such that (1) beyond 45 AU all objects are too small to be detected by telescope surveys, (2) in the 35–45 AU region the distribution of the largest objects is similar to that observed in the current cold population while most of the mass is contained in small bodies, and (3) within 35 AU most of the mass is contained in large bodies and the size distribution culminates with ~1000 Pluto-sized objects. If this were the case, the disk beyond 35 AU could lose most of its mass by collisional grinding before the beginning of Neptune's migration, particularly if the latter was triggered late as in our LHB scenario (see section 3.4 of the chapter by Kenyon et al.). Within 35 AU, because of the different size distribution, collisional grinding would have been ineffective (*Charnoz and Morbidelli,* 2007). Therefore, at the time of the LHB, the system would have been similar to the one required by the Nice model, in that Neptune would have seen an effective edge in the planetesimal mass distribution that would have kept it from migrating beyond 30 AU.

Whether the spatial variation we described above of the size distribution in the planetesimal disk is reasonable or not is beyond our current understanding. The chapter by Kenyon et al. nicely shows that the coagulation/erosion process is always on the edge of an instability. In fact, the dispersion velocity of the small bodies is on the order of the escape velocity from the largest bodies. Depending on the details of the collisional cascade (see section 3.5 in that chapter) the dispersion velocity can be slightly smaller than the escape velocities (favoring the accretion of a large number of massive bodies and producing a top-heavy size distribution that does not allow an effective collisional grinding), or can be slightly larger (stalling runaway accretion and leaving most of the mass in small, easy-to-break bodies). Perhaps the inner part of the disk was in the first regime and the outer part was in the second one, with a relatively sharp transition zone between the two parts. More work is needed to clarify the situation, with a close collaboration between experts of accretion and of dynamical evolution.

From a purely dynamical point of view, the Nice model is not inconsistent with the existence of a local, low-mass Kuiper belt population, extended up to 44–45 AU. In fact, Fig. 11 compares the (a, e) distribution observed in the classical belt with q > 38 AU (Fig. 11c) with the one that the Nice model predicts assuming that the disk was truncated at 34 AU (Fig. 11a, which is an enlargement of Fig. 8a) or assuming that the disk was truncated at 44 AU (Fig. 11b). The two model distributions are statistically equivalent, and are both very similar to the observed distribution. In the case where the outer edge was placed at 44–45 AU about 7% of the particles initially in the Kuiper belt (a > 40 AU, q > 38 AU) remain there, although their orbits have been modified. The others escape to larger eccentricities during the phase when the Kuiper belt is globally unstable (see section 7). If 90% of the local mass escapes, the local belt had to have been significantly depleted before the time of the LHB, probably accounting for only a few tens of an Earth

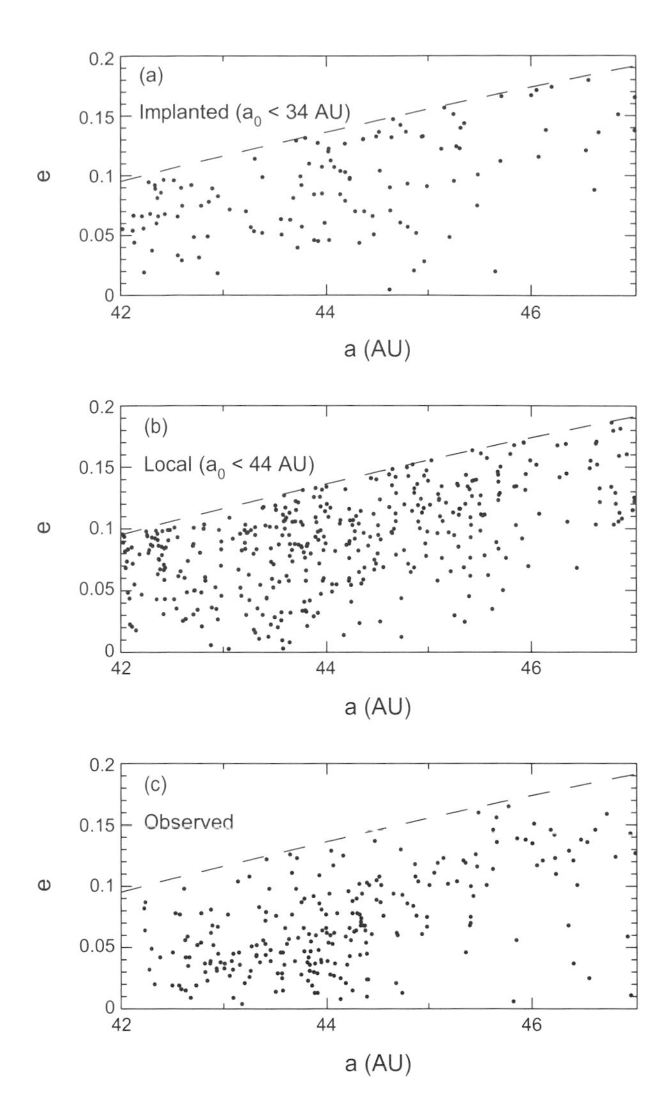

Fig. 11. The semimajor axis vs. eccentricity distribution in the classical belt (q > 38 AU). **(a)** The distribution of the objects that originated within 34 AU, implanted in the Kuiper belt at the time of the LHB. **(b)** The distribution of the bodies assuming the disk originally extended to 44 AU. **(c)** The observed distribution.

mass, otherwise, presumably (see section 6.1), Neptune would have been driven past 30 AU.

We end this chapter by encouraging observers to attempt to probe the regions beyond 50 AU. In particular, even if the absence of a population of objects similar to that in the 40–50 AU region is now secure, nothing is known about the possible existence of small objects. Recent claims on a large number of stellar occultation events by distant ~100-m objects (*Roques et al.*, 2006; *Chiang et al.*, 2006; *Georgevits*, 2006) may suggest, if confirmed [see *Jones et al.* (2006) for a rebuttal of the *Chiang et al.* (2006) results], the existence of an extended disk of small planetesimals that did not grow to directly detectable sizes. If true, this would give us extremely valuable information about the origin of the Kuiper belt edge. The model by *Weidenschilling* (2003) predicts the existence of an extended disk of exclusively small bodies, although of sizes up to ~1 m rather than ~100 m (see Fig. 4). Conversely, other models of edge formation, such as disk stripping by passing stars, photoevaporation, or turbulent stirring, would predict the total absence of objects of any size. Therefore we encourage the pursuit of stellar occultation programs until the real situation is clarified.

So far, the Kuiper belt has taught us a lot about the evolution of the planets. Planet migration, for instance, had been totally overlooked, despite pioneering work by Fernandez and Ip, until the Kuiper belt was discovered. Moreover, as we mentioned above, we can find in the Kuiper belt further evidence that the giant planets passed through a temporary phase of violent instability. But, the Kuiper belt can do more. It can potentially teach us about planetesimal formation and the growth of larger objects because the objects that inhabit it most likely probe different regions of the protoplanetary disk where accretion proceeded in very different ways. Thus, it is a dreamed-of laboratory to test and calibrate the accretion models. This is probably the main venue for the future.

Acknowledgments. H.F.L. thanks NASA's Origins and Planetary Geology and Geophysics programs for supporting his involvement in the research related to this chapter. A.M. is also grateful to the French National Council of Scientific Research (CNRS) and National Programme of Planetology (PNP) for support. R.S.G. thanks the Brasilian National Council for Science and Technology (CNPq) for support.

REFERENCES

Adams F. C., Hollenbach D., Laughlin G., and Gorti U. (2004) Photoevaporation of circumstellar disks due to external far-ultraviolet radiation in stellar aggregates. *Astrophys. J., 611,* 360–379.

Agnor C. B. and Hamilton D. P. (2006) Neptune's capture of its moon Triton in a binary-planet gravitational encounter. *Nature, 441,* 192–194.

Allen R. L., Bernstein G. M., and Malhotra R. (2001) The edge of the solar system. *Astrophys. J. Lett., 549,* L241–L244.

Allen R. L., Bernstein G. M., and Malhotra R. (2002) Observational limits on a distant cold Kuiper belt. *Astron. J., 124,* 2949–2954.

Augereau J.-C. and Beust H. (2006) On the AU Microscopii debris disk. Density profiles, grain properties, and dust dynamics. *Astron. Astrophys., 455,* 987–999.

Beauge C. (1994) Asymmetric liberations in exterior resonances. *Cel. Mech. Dyn. Astron., 60,* 225–248.

Benz W. and Asphaug E. (1999) Catastrophic disruptions revisited. *Icarus, 142,* 5–20.

Bernstein G. M., Trilling D. E., Allen R. L., Brown M. E., Holman M., and Malhotra R. (2004) The size distribution of trans-neptunian bodies. *Astron. J., 128,* 1364–1390.

Brown M. (2001) The inclination distribution of the Kuiper belt. *Astron. J., 121,* 2804–2814.

Brown M. E., Trujillo C. A., and Rabinowitz D. L. (2005) Discovery of a planetary-sized object in the scattered Kuiper belt. *Astrophys. J. Lett., 635,* L97–L100.

Brown M. E., Barkume K. M., Ragozzine D., and Schaller E. L. (2007) A collisional family of icy objects in the Kuiper belt. *Nature, 446,* 294–296.

Brunini A. and Melita M. (2002) The existence of a planet beyond 50 AU and the orbital distribution of the classical Edgeworth Kuiper belt objects. *Icarus, 160,* 32–43.

Canup R. M. (2005) A giant impact origin of Pluto-Charon. *Science, 307,* 546–550.

Charnoz S. and Morbidelli A. (2007) Coupling dynamical and collisional evolution of small bodies. II: Forming the Kuiper Belt, the scattered disk and the Oort cloud. *Icarus,* in press.

Chiang E., Lithwick Y., Murray-Clay R., Buie M., Grundy W., and Holman M. (2006) A brief history of transneptunian space. In *Protostars and Planets V* (B. Reipurth et al., eds.), pp. 895–911. Univ. of Arizona, Tucson.

Chirikov B. V. (1960) Resonance processes inmagnetic traps. *J. Plasma Phys., 1,* 253–260.

Davis D. R. and Farinella P. (1997) Collisional evolution of Edgeworth-Kuiper belt objects. *Icarus, 125,* 50–60.

Davis D. R. and Farinella P. (1998) Collisional erosion of a massive Edgeworth-Kuiper belt: Constraints on the initial population (abstract). In *Lunar and Planetary Science XXIX,* pp. 1437–1438. Lunar and Planetary Institute, Houston.

Dermott S. F. and Murray C. D. (1983) Nature of the Kirkwood gaps in the asteroid belt. *Nature, 301,* 201–205.

Duncan M. J. and Levison H. F. (1997) A scattered comet disk and the origin of Jupiter family comets. *Science, 276,* 1670–1672.

Duncan M. J., Levison H. F., and Budd S. M. (1995) The long-term stability of orbits in the Kuiper belt. *Astron. J., 110,* 3073–3083.

Emel'yanenko V. V., Asher D. J., and Bailey M. E. (2003) A new class of trans-Neptunian objects in high-eccentricity orbits. *Mon. Not. R. Astron. Soc., 338,* 443–451.

Fernandez J. A. and Ip W.-H. (1984) Some dynamical aspects of the accretion of Uranus and Neptune — The exchange of orbital angular momentum with planetesimals. *Icarus, 58,* 109–120.

Georgevits G. (2006) Detection of small Kuiper belt objects by stellar occultation. *AAS/Division for Planetary Sciences Meeting Abstracts 38,* #37.07.

Gladman B., Kavelaars J. J., Petit J. M., Morbidelli A., Holman M. J., and Loredo Y. (2001) The structure of the Kuiper belt: Size distribution and radial extent. *Astron. J., 122,* 1051–1066.

Goldreich P., Lithwick Y., and Sari R. (2004a) Final stages of planet formation. *Astrophys. J., 614,* 497–507.

Goldreich P., Lithwick Y., and Sari R. (2004b) Planet formation by coagulation: A focus on Uranus and Neptune. *Annu. Rev. Astron. Astrophys., 42,* 549–601.

Gomes R. S. (2000) Planetary migration and Plutino orbital inclinations. *Astron. J., 120,* 2695–2707.

Gomes R. S. (2003) The origin of the Kuiper Belt high-inclination population. *Icarus, 161,* 404–418.

Gomes R. S., Morbidelli A., and Levison H. F. (2004) Planetary migration in a planetesimal disk: Why did Neptune stop at 30 AU? *Icarus, 170,* 492–507.

Gomes R., Levison H. F., Tsiganis K., and Morbidelli A. (2005) Origin of the cataclysmic late heavy bombardment period of the terrestrial planets. *Nature, 435,* 466–469.

Grundy W. M., Noll K. S., and Stephens D. C. (2005) Diverse albedos of small trans-neptunian objects. *Icarus, 176,* 184–191.

Hahn J. M. and Malhotra R. (1999) Orbital evolution of planets embedded in a planetesimal disk. *Astron. J., 117,* 3041–3053.

Hahn J. M. and Malhotra R. (2005) Neptune's migration into a stirred-up Kuiper belt: A detailed comparison of simulations to observations. *Astron. J., 130,* 2392–2414.

Henrard J. (1982) Capture into resonance — An extension of the use of adiabatic invariants. *Cel. Mech., 27,* 3–22.

Hollenbach D. and Adams F. C. (2004) Dispersal of disks around young stars: Constraints on Kuiper belt formation. In *Debris Disks and the Formation of Planets* (L. Caroff et al., eds.), p. 168. ASP Conf. Series 324, San Francisco.

Holman M. J. and Wisdom J. (1993) Dynamical stability in the outer solar system and the delivery of short period comets. *Astron. J., 105,* 1987–1999.

Ida S., Larwood J., and Burkert A. (2000) Evidence for early stellar encounters in the orbital distribution of Edgeworth-Kuiper belt objects. *Astrophys. J., 528,* 351–356.

Jones T. A., Levine A. M., Morgan E. H., and Rappaport S. (2006) Millisecond dips in Sco X-1 are likely the result of high-energy particle events. *The Astronomer's Telegram, 949,* 1.

Kenyon S. J. and Bromley B. C. (2002) Collisional cascades in planetesimal disks. I. Stellar flybys. *Astron. J., 123,* 1757–1775.

Kenyon S. J. and Bromley B. C. (2004a) The size distribution of Kuiper belt objects. *Astron. J., 128,* 1916–1926.

Kenyon S. J. and Bromley B. C. (2004b) Stellar encounters as the origin of distant solar system objects in highly eccentric orbits. *Nature, 432,* 598–602.

Kenyon S. J. and Luu J. X. (1998) Accretion in the early Kuiper belt: I. Coagulation and velocity evolution. *Astron. J., 115,* 2136–2160.

Kenyon S. J. and Luu J. X. (1999a) Accretion in the early Kuiper belt: II. Fragmentation. *Astron. J., 118,* 1101–1119.

Kenyon S. J. and Luu J. X. (1999b) Accretion in the early outer solar system. *Astrophys. J., 526,* 465–470.

Knežević Z., Milani A., Farinella P., Froeschlé Ch., and Froeschlé C. (1991) Secular resonances from 2 to 50 AU. *Icarus, 93,* 316–330.

Kobayashi H. and Ida S. (2001) The effects of a stellar encounter on a planetesimal disk. *Icarus, 153,* 416–429.

Kobayashi H., Ida S., and Tanaka H. (2005) The evidence of an early stellar encounter in Edgeworth Kuiper belt. *Icarus, 177,* 246–255.

Kuchner M. J., Brown M. E., and Holman M. (2002) Long-term dynamics and the orbital inclinations of the classical Kuiper belt objects. *Astron. J., 124,* 1221–1230.

Levison H. F. and Morbidelli A. (2003) The formation of the Kuiper belt by the outward transport of bodies during Neptune's migration. *Nature, 426,* 419–421.

Levison H. F. and Morbidelli A. (2007) Models of the collisional damping scenario for ice-giant planets and Kuiper belt formation. *Icarus, 189,* 196–212.

Levison H. F. and Stern S. A. (2001) On the size dependence of the inclination distribution of the main Kuiper belt. *Astron. J., 121,* 1730–1735.

Levison H. F., Morbidelli A., and Dones L. (2004) Sculpting the Kuiper belt by a stellar encounter: Constraints from the Oort cloud and scattered disk. *Astron. J., 128,* 2553–2563.

Levison H. F., Morbidelli A., Gomes R., and Backman D. (2006) Planet migration in planetesimal disks. In *Protostars and Planets V* (B. Reipurth et al., eds.), pp. 669–684. Univ. of Arizona, Tucson.

Levison H. F., Morbidelli A., Gomes R., and Tsiganis K. (2007) Origin of the structure of the Kuiper belt during a dynamical instability in the orbits of Uranus and Neptune. *Icarus,* in press.

Lykawka P. S. and Mukai T. (2007a) Evidence for an excited Kuiper belt of 50 AU radius in the first Myr of solar system history. *Icarus,* in press.

Lykawka P. S. and Mukai T. (2007b) Dynamical classification of trans-neptunian objects: Probing their origin, evolution and interrelation. *Icarus,* in press.

Malhotra R. (1993) The origin of Pluto's peculiar orbit. *Nature, 365,* 819.

Malhotra R. (1995) The origin of Pluto's orbit: Implications for the solar system beyond Neptune. *Astron. J., 110,* 420.

Malhotra R. (1996) The phase space structure near Neptune resonances in the Kuiper belt. *Astron. J., 111,* 504.

Melita M., Larwood J., Collander-Brown S, Fitzsimmons A., Williams I. P., and Brunini A. (2002) The edge of the Edgeworth-Kuiper belt: Stellar encounter, trans-Plutonian planet or outer limit of the primordial solar nebula? In *Asteroids, Comet, Meteors,* pp. 305–308. ESA Spec. Publ. 500, Noordwijk, The Netherlands.

Message P. J. (1958) Proceedings of the Celestial Mechanics Conference: The search for asymmetric periodic orbits in the restricted problem of three bodies. *Astron. J., 63,* 443.

Morbidelli A. (2002) *Modern Celestial Mechanics: Aspects of Solar System Dynamics.* Advances in Astronomy and Astrophysics Series, Taylor and Francis, London.

Morbidelli A. and Henrard J. (1991) The main secular resonances nu6, nu5 and nu16 in the asteroid belt. *Cel. Mech. Dyn. Astron., 51,* 169–197.

Morbidelli A. and Levison H. F. (2004) Scenarios for the origin of the orbits of the trans-neptunian objects 2000 CR_{105} and 2003 VB_{12} (Sedna). *Astron. J., 128,* 2564–2576.

Morbidelli A., Thomas F., and Moons M. (1995) The resonant structure of the Kuiper belt and the dynamics of the first five trans-neptunian objects. *Icarus, 118,* 322.

Morbidelli A. and Valsecchi G. B. (1997) Neptune scattered planetesimals could have sculpted the primordial Edgeworth-Kuiper belt, *Icarus, 128,* 464–468.

Morbidelli A., Levison H. F., Tsiganis K., and Gomes R. (2005) Chaotic capture of Jupiter's Trojan asteroids in the early solar system. *Nature, 435,* 462–465.

Murray-Clay R. A. and Chiang E. I. (2006) Brownian motion in planetary migration. *Astrophys. J., 651,* 1194–1208.

Nagasawa M. and Ida S. (2000) Sweeping secular resonances in the Kuiper belt caused by depletion of the solar nebula. *Astron. J., 120,* 3311–3322.

Nesvorný D. and Roig F. (2000) Mean motion resonances in the trans-neptunian region: Part I: The 2:3 resonance with Neptune. *Icarus, 148,* 282–300.

Nesvorný D. and Roig F. (2001) Mean motion resonances in the trans-neptunian region: Part II: The 1:2, 3:4 and weaker resonances. *Icarus, 150,* 104–123.

Petit J.-M., Morbidelli A., and Chambers J. (2001) The primordial excitation and clearing of the asteroid belt. *Icarus, 153,* 338–347.

Roques F., and 17 colleagues (2006) Exploration of the Kuiper belt by high-precision photometric stellar occultations: First results. *Astron. J., 132,* 819–822.

Ruden S. P. and Pollack J. B. (1991) The dynamical evolution of the protosolar nebula. *Astrophys. J., 375,* 740–760.

Sekiya M. (1983) Gravitational instabilities in a dust-gas layer and formation of planetesimals in the solar nebula. *Progr. Theor. Phys., 69,* 1116–1130.

Sheppard S. S. and Trujillo C. A. (2006) A thick cloud of Neptune Trojans and their colors. *Science, 313,* 511–514.

Stern S. A. (1991) On the number of planets in the outer solar system — Evidence of a substantial population of 1000-km bodies. *Icarus, 90,* 271–281.

Stern S. A. (1996) On the collisional environment, accretion time scales, and architecture of the massive, primordial Kuiper belt. *Astron. J., 112,* 1203–1210.

Stern S. A. and Colwell J. E. (1997a) Accretion in the Edgeworth-Kuiper belt: Forming 100–1000 km radius bodies at 30 AU and beyond. *Astron. J., 114,* 841–849.

Stern S. A. and Colwell J. E. (1997b) Collisional erosion in the primordial Edgeworth-Kuiper belt and the generation of the 30–50 AU Kuiper gap. *Astrophys. J., 490,* 879–885.

Stern S. A. and Weissman P. R. (2001) Rapid collisional evolution of comets during the formation of the Oort cloud. *Nature, 409,* 589–591.

Stone J. M., Gammie C. F., Balbus S. A., and Hawley J. F. (1998) In *Protostars and Planets IV* (V. Mannings et al., eds.), p. 589. Univ. of Arizona, Tucson.

Thommes E. W., Duncan M. J., and Levison H. F. (1999) The formation of Uranus and Neptune in the Jupiter-Saturn region of the solar system. *Nature, 402,* 635–638.

Trujillo C. A. and Brown M. E. (2001) The radial distribution of the Kuiper belt. *Astrophys. J., 554,* 95–98.

Trujillo C. A. and Brown M. E. (2003) The Caltech Wide Area Sky Survey. *Earth Moon Planets, 92,* 99–112.

Tsiganis K., Gomes R., Morbidelli A., and Levison H. F. (2005) Origin of the orbital architecture of the giant planets of the solar system. *Nature, 435,* 459–461.

Weidenschilling S. (2003) Formation of planetesimals/cometesimals in the solar nebula. In *Comets II* (M. C. Festou et al., eds.), pp. 97–104. Univ. of Arizona, Tucson.

Williams J. G. and Faulkner J. (1981) The positions of secular resonance surfaces. *Icarus, 46,* 390–399.

Wisdom J. (1980) The resonance overlap criterion and the onset of stochastic behavior in the restricted three-body problem. *Astron. J., 85,* 1122–1133.

Youdin A. N. and Shu F. H. (2002) Planetesimal formation by gravitational instability. *Astrophys. J., 580,* 494–505.

Formation and Collisional Evolution of Kuiper Belt Objects

Scott J. Kenyon
Smithsonian Astrophysical Observatory

Benjamin C. Bromley
University of Utah

David P. O'Brien and Donald R. Davis
Planetary Science Institute

This chapter summarizes analytic theory and numerical calculations for the formation and collisional evolution of Kuiper belt objects (KBOs) at 20–150 AU. We describe the main predictions of a baseline self-stirring model and show how dynamical perturbations from a stellar flyby or stirring by a giant planet modify the evolution. Although robust comparisons between observations and theory require better KBO statistics and more comprehensive calculations, the data are broadly consistent with KBO formation in a massive disk followed by substantial collisional grinding and dynamical ejection. However, there are important problems reconciling the results of coagulation and dynamical calculations. Contrasting our current understanding of the evolution of KBOs and asteroids suggests that additional observational constraints, such as the identification of more dynamical families of KBOs (like the 2003 EL_{61} family), would provide additional information on the relative roles of collisional grinding and dynamical ejection in the Kuiper belt. The uncertainties also motivate calculations that combine collisional and dynamical evolution, a "unified" calculation that should give us a better picture of KBO formation and evolution.

1. INTRODUCTION

Every year in the galaxy, a star is born. Most stars form in dense clusters of thousands of stars, as in the Orion nebula cluster (*Lada and Lada,* 2003; *Slesnick et al.,* 2004). Other stars form in small groups of 5–10 stars in loose associations of hundreds of stars, as in the Taurus-Auriga clouds (*Gomez et al.,* 1993; *Luhman,* 2006). Within these associations and clusters, most newly formed massive stars are binaries; lower-mass stars are usually single (*Lada,* 2006).

Large, optically thick circumstellar disks surround nearly all newly formed stars (*Beckwith and Sargent,* 1996). The disks have sizes of ~100–200 AU, masses of ~0.01–0.1 $M_\odot$, and luminosities of ~0.2–1 $L_\star$, where $L_\star$ is the luminosity of the central star. The masses and geometries of these disks are remarkably similar to the properties of the minimum mass solar nebula (MMSN), the disk required for the planets in the solar system (*Weidenschilling,* 1977b; *Hayashi,* 1981; *Scholz et al.,* 2006).

As stars age, they lose their disks. For solar-type stars, radiation from the opaque disk disappears in 1–10 m.y. (*Haisch et al.,* 2001). Many older stars have optically thin *debris disks* comparable in size to the opaque disks of younger stars but with much smaller masses, $\lesssim 1\ M_\oplus$, and luminosities, $\lesssim 10^3\ L_\star$ (chapter by Moro-Martín et al.). The lifetime of this phase is uncertain. Some 100-m.y.-old stars have no obvious debris disk; a few 1–10-G.y.-old stars have massive debris disks (*Greaves,* 2005).

In the current picture, planets form during the transition from an optically thick protostellar disk to an optically thin debris disk. From the statistics of young stars in molecular clouds, the timescale for this transition, ~10^5 yr, is comparable to the timescales derived for the formation of planetesimals from dust grains (*Weidenschilling,* 1977a; *Youdin and Shu,* 2002; *Dullemond and Dominik,* 2005) and for the formation of lunar-mass or larger planets from planetesimals (*Wetherill and Stewart,* 1993; *Weidenschilling et al.,* 1997; *Kokubo and Ida,* 2000; *Nagasawa et al.,* 2005; *Kenyon and Bromley,* 2006). Because the grains in debris disks have short collision lifetimes, $\lesssim 1$ m.y., compared to the ages of their parent stars, $\gtrsim 10$ m.y., high-velocity collisions between larger objects must maintain the small grain population (*Aumann et al.,* 1984; *Backman and Paresce,* 1993). The inferred dust production rates for debris disks around 0.1–10-G.y.-old stars, ~10^{20} g yr^{-1}, require an initial mass in 1-km objects, M_i ~ 10–100 $M_\oplus$, comparable to the amount of solids in the MMSN. Because significant long-term debris production also demands gravitational stirring by an ensemble of planets with radii of 500–1000 km or larger (*Kenyon and Bromley,* 2004a; *Wyatt et al.,* 2005), debris disks probably are newly formed planetary systems (*Aumann et al.,* 1984; *Backman and Paresce,* 1993; *Artymowicz,* 1997; *Kenyon and Bromley,* 2002b, 2004a,b).

Kuiper belt objects (KBOs) provide a crucial test of this picture. With objects ranging in size from 10–20 km to ~1000 km, the size distribution of KBOs yields a key comparison with theoretical calculations of planet formation (*Davis and Farinella,* 1997; *Kenyon and Luu,* 1998, 1999a,b). Once KBOs have sizes of 100–1000 km, collisional grinding, dynamical perturbations by large planets and passing stars, and self-stirring by small embedded planets, produce features in the distributions of sizes and dynamical elements that observations can probe in detail. Although complete calculations of KBO formation and dynamical evolution are not available, these calculations will eventually yield a better understanding of planet formation at 20–100 AU.

The Kuiper belt also enables a vital link between the solar system and other planetary systems. With an outer radius of ≳1000 AU (Sedna's aphelion) and a current mass of ~0.1 $M_\oplus$ (*Luu and Jewitt,* 2002; *Bernstein et al.,* 2004, chapter by Petit et al.), the Kuiper belt has properties similar to those derived for the oldest debris disks (*Greaves et al.,* 2004; *Wyatt et al.,* 2005). Understanding planet formation in the Kuiper belt thus informs our interpretation of evolutionary processes in other planetary systems.

This paper reviews applications of coagulation theory for planet formation in the Kuiper belt. After a brief introduction to the theoretical background in section 2, we describe results from numerical simulations in section 3, compare relevant KBO observations with the results of numerical simulations in section 4, and contrast the properties of KBOs and asteroids in section 5. We conclude with a short summary in section 6.

2. COAGULATION THEORY

Planet formation begins with dust grains suspended in a gaseous circumstellar disk. Grains evolve into planets in three steps. Collisions between grains produce larger aggregates that decouple from the gas and settle into a dense layer in the disk midplane. Continued growth of the loosely bound aggregates leads to planetesimals, gravitationally bound objects whose motions are relatively independent of the gas. Collisions and mergers among the ensemble of planetesimals form planets. Here, we briefly describe the physics of these stages and summarize analytic results as a prelude to summaries of numerical simulations.

We begin with a prescription for the mass surface density Σ of gas and dust in the disk. We use subscripts "d" for the dust and "g" for the gas and adopt

$$\Sigma_{d,g} = \Sigma_{0d,0g}\left(\frac{a}{40\ \mathrm{AU}}\right)^{-n} \quad (1)$$

where a is the semimajor axis. In the MMSN, n = 3/2, $\Sigma_{0d} \approx$ 0.1 g cm^{-2}, and $\Sigma_{0d} \approx$ 5–10 g cm^{-2} (*Weidenschilling,* 1977b; *Hayashi,* 1981). For a disk with an outer radius of 100 AU, the MMSN has a mass of ~0.03 $M_\odot$, which is comparable to the disk masses of young stars in nearby star-forming regions (*Natta et al.,* 2000; *Scholz et al.,* 2006).

The dusty midplane forms quickly (*Weidenschilling,* 1977a, 1980; *Dullemond and Dominik,* 2005). For interstellar grains with radii, r ~ 0.01–0.1 μm, turbulent mixing approximately balances settling due to the vertical component of the star's gravity. As grains collide and grow to r ~ 0.1–1 mm, they decouple from the turbulence and settle into a thin layer in the disk midplane. The timescale for this process is ~10^3 yr at 1 AU and ~10^5 yr at 40 AU.

The evolution of grains in the midplane is uncertain. Because the gas has some pressure support, it orbits the star slightly more slowly than the Keplerian velocity. Thus, orbiting dust grains feel a headwind that drags them toward the central star (*Adachi et al.,* 1976; *Weidenschilling,* 1984; *Tanaka and Ida,* 1999). For meter-sized objects, the drag timescale at 40 AU, ~10^5 yr, is comparable to the growth time. Thus, it is not clear whether grains can grow by direct accretion to kilometer sizes before the gas drags them into the inner part of the disk.

Dynamical processes provide attractive alternatives to random agglomeration of grains. In ensembles of porous grains, gas flow during disruptive collisions leads to planetesimal formation by direct accretion (*Wurm et al.,* 2004). Analytic estimates and numerical simulations indicate that grains with r ~ 1 cm are also easily trapped within vortices in the disk (e.g., *de la Fuente Marcos and Barge,* 2001; *Inaba and Barge,* 2006). Large enhancements in the solid-to-gas ratio within vortices allows accretion to overcome gas drag, enabling formation of kilometer-sized planetesimals in 10^4–10^5 yr.

If the dusty midplane is calm, it becomes thinner and thinner until groups of particles overcome the local Jeans criterion — where their self-gravity overcomes local orbital shear — and "collapse" into larger objects on the local dynamical timescale, ~10^3 yr at 40 AU (*Goldreich and Ward,* 1973; *Youdin and Shu,* 2002; *Tanga et al.,* 2004). This process is a promising way to form planetesimals; however, turbulence may prevent the instability (*Weidenschilling,* 1995, 2003, 2006). Although the expected size of a collapsed object is the Jeans wavelength, the range of planetesimal sizes the instability produces is also uncertain.

Once planetesimals with r ~ 1 km form, gravity dominates gas dynamics. Long-range gravitational interactions exchange kinetic energy (dynamical friction) and angular momentum (viscous stirring), redistributing orbital energy and angular momentum among planetesimals. For 1-km objects at 40 AU, the initial random velocities are comparable to their escape velocities, ~1 m s^{-1} (*Weidenschilling,* 1980; *Goldreich et al.,* 2004). The gravitational binding energy (for brevity, we use energy as a shorthand for specific energy), E_g ~ 10^4 erg g^{-1}, is then comparable to the typical collision energy, E_c ~ 10^4 erg g^{-1}. Both energies are smaller than the disruption energy — the collision energy needed to remove half the mass from the colliding pair of objects — which is $Q_D^* \sim 10^5$–10^7 erg g^{-1} for icy material (*Davis et al.,* 1985; *Benz and Asphaug,* 1999; *Ryan et al.,* 1999; *Michel et al.,* 2001; *Leinhardt and Richardson,* 2002; *Giblin et al.,* 2004). Thus, collisions produce mergers instead of debris.

Initially, small planetesimals grow slowly. For a large ensemble of planetesimals, the collision rate is $n\sigma v$, where n is the number of planetesimals, σ is the cross-section, and v is the relative velocity. The collision cross-section is the geometric cross-section, πr^2, scaled by the gravitational focusing factor f_g

$$\sigma_c \sim \pi r^2 f_g \sim \pi r^2 (1 + \beta (v_{esc}/ev_K)^2) \quad (2)$$

where e is the orbital eccentricity, v_K is the orbital velocity, v_{esc} is the escape velocity of the merged pair of planetesimals, and $\beta \approx 2.7$ is a coefficient that accounts for three-dimensional orbits in a rotating disk (*Greenzweig and Lissauer,* 1990; *Spaute et al.,* 1991; *Wetherill and Stewart,* 1993). Because $ev_K \approx v_{esc}$, gravitational focusing factors are small, and growth is slow and orderly (*Safronov,* 1969). The timescale for slow, orderly growth is

$$t_s \approx 30\left(\frac{r}{1000\ \text{km}}\right)\left(\frac{P}{250\ \text{yr}}\right)\left(\frac{0.1\ \text{g cm}^{-2}}{\Sigma_{0d}}\right)\ \text{G.y.} \quad (3)$$

where P is the orbital period (*Safronov,* 1969; *Lissauer,* 1987; *Goldreich et al.,* 2004).

As larger objects form, several processes damp particle random velocities and accelerate growth. For objects with $r \sim 1$–100 m, physical collisions reduce particle random velocities (*Ohtsuki,* 1992; *Kenyon and Luu,* 1998). For larger objects with $r \gtrsim 0.1$ km, the smaller objects damp the orbital eccentricity of larger particles through dynamical friction (*Wetherill and Stewart,* 1989; *Kokubo and Ida,* 1995; *Kenyon and Luu,* 1998). Viscous stirring by the large objects excites the orbits of the small objects. For planetesimals with $r \sim 1$ m to $r \sim 1$ km, these processes occur on short timescales, $\lesssim 10^6$ yr at 40 AU, and roughly balance when these objects have orbital eccentricity $e \sim 10^{-5}$. In the case where gas drag is negligible, *Goldreich et al.* (2004) derive a simple relation for the ratio of the eccentricities of the large ("l") and the small ("s") objects in terms of their surface densities $\Sigma_{l,s}$ (see also *Kokubo and Ida,* 2002; *Rafikov,* 2003a,b,c,d)

$$\frac{e_l}{e_s} \sim \left(\frac{\Sigma_l}{\Sigma_s}\right)^{\gamma} \quad (4)$$

with $\gamma = 1/4$ to $1/2$. Initially, most of the mass is in small objects. Thus $\Sigma_l/\Sigma_s \ll 1$. For $\Sigma_l/\Sigma_s \sim 10^{-3}$ to 10^{-4}, $e_l/e_s \approx$ 0.1–0.25. Because $e_s v_K \ll v_{l,esc}$ gravitational focusing factors for large objects accreting small objects are large. Runaway growth begins.

Runaway growth relies on positive feedback between accretion and dynamical friction. Dynamical friction produces the largest f_g for the largest objects, which grow faster and faster relative to the smaller objects and contain an ever-growing fraction of the total mass. As they grow, these protoplanets stir the planetesimals. The orbital velocity dispersions of small objects gradually approach the escape velocities of the protoplanets. With $e_s v_K \sim v_{l,esc}$, collision rates decline as runaway growth continues (equations (2) and (4)). The protoplanets and leftover planetesimals then enter the oligarchic phase, where the largest objects — oligarchs — grow more slowly than they did as runaway objects but still faster than the leftover planetesimals. The timescale to reach oligarchic growth is (*Lissauer,* 1987; *Goldreich et al.,* 2004)

$$t_o \approx 30\left(\frac{P}{250\ \text{yr}}\right)\left(\frac{0.1\ \text{g cm}^{-2}}{\Sigma_{0d}}\right)\ \text{m.y.} \quad (5)$$

For the MMSN, $t_o \propto a^{-3}$. Thus, collisional damping, dynamical friction and gravitational focusing enhance the growth rate by 3 orders of magnitude compared to orderly growth.

Among the oligarchs, smaller oligarchs grow the fastest. Each oligarch tries to accrete material in an annular "feeding zone" set by balancing the gravity of neighboring oligarchs. If an oligarch accretes all the mass in its feeding zone, it reaches the "isolation mass" (*Lissauer,* 1987; *Kokubo and Ida,* 1998, 2002; *Rafikov,* 2003a; *Goldreich et al.,* 2004)

$$m_{iso} \approx 28\left(\frac{a}{40\ \text{AU}}\right)^3\left(\frac{\Sigma_{0d}}{0.1\ \text{g cm}^{-2}}\right)M_{\oplus} \quad (6)$$

Each oligarch stirs up leftover planetesimals along its orbit. Smaller oligarchs orbit in regions with smaller Σ_l/Σ_s. Thus, smaller oligarchs have larger gravitational focusing factors (equations (2) and (4)) and grow faster than larger oligarchs (*Kokubo and Ida,* 1998; *Goldreich et al.,* 2004).

As oligarchs approach m_{iso}, they stir up the velocities of the planetesimals to the disruption velocity. Instead of mergers, collisions then yield smaller planetesimals and debris. Continued disruptive collisions lead to a collisional cascade, where leftover planetesimals are slowly ground to dust (*Dohnanyi,* 1969; *Williams and Wetherill,* 1994). Radiation pressure from the central star ejects dust grains with $r \lesssim 1$–10 μm; Poynting-Robertson drag pulls larger grains into the central star (*Burns et al.,* 1979; *Artymowicz,* 1988; *Takeuchi and Artymowicz,* 2001). Eventually, planetesimals are accreted by the oligarchs or ground to dust.

To evaluate the oligarch mass required for a disruptive collision, we consider two planetesimals with equal mass m_p. The center-of-mass collision energy is

$$Q_i = v_i^2/8 \quad (7)$$

where the impact velocity $v_i^2 = v^2 + v_{esc}^2$ (*Wetherill and Stewart,* 1993). The energy needed to remove half the combined mass of two colliding planetesimals is

$$Q_D^* = Q_b\left(\frac{r}{1\ \mathrm{cm}}\right)^{\beta_b} + \rho Q_g\left(\frac{r}{1\ \mathrm{cm}}\right)^{\beta_g} \quad (8)$$

where $Q_b r^{\beta_b}$ is the bulk (tensile) component of the binding energy and $\rho Q_g r^{\beta_g}$ is the gravity component of the binding energy (*Davis et al.,* 1985; *Housen and Holsapple,* 1990, 1999; *Holsapple,* 1994; *Benz and Asphaug,* 1999). We adopt $v \approx v_{esc,o}$, where $v_{esc,o} = (Gm_o/r_o)^{1/2}$ is the escape velocity of an oligarch with mass m_o and radius r_o. We define the disruption mass m_d by deriving the oligarch mass where $Q_i \approx Q_D^*$. For icy objects at 30 AU

$$m_d \sim 3 \times 10^{-6}\left(\frac{Q_D^*}{10^7\ \mathrm{erg\ g^{-1}}}\right)^{3/2} M_\oplus \quad (9)$$

Figure 1 illustrates the variation of Q_D^* with radius for several variants of equation (8). For icy objects, detailed numerical collision simulations yield $Q_b \lesssim 10^7$ erg g^{-1}, $-0.5 \lesssim \beta_b \lesssim 0$, $\rho \approx 1$–2 g cm^{-3}, $Q_g \approx 1$–2 erg cm^{-3}, and $\beta_g \approx 1$–2 (solid line in Fig. 1) (*Benz and Asphaug,* 1999; see also chapter by Leinhardt et al.). Models for the breakup of Comet Shoemaker-Levy 9 suggest a smaller component of the bulk strength, $Q_b \sim 10^3$ erg g^{-1} (e.g., *Asphaug and Benz,* 1996), which yields smaller disruption energies for smaller objects (Fig. 1, dashed and dot-dashed curves). Because nearly all models for collisional disruption yield similar results for objects with $r \gtrsim 1$ km (e.g., *Kenyon and Bromley,* 2004d), the disruption mass is fairly independent of theoretical uncertainties once planetesimals become large. For typical $Q_D^* \sim 10^7$–10^8 erg g^{-1} for 1–10-km objects (Fig. 1), leftover planetesimals start to disrupt when oligarchs have radii $r_o > 200$–500 km.

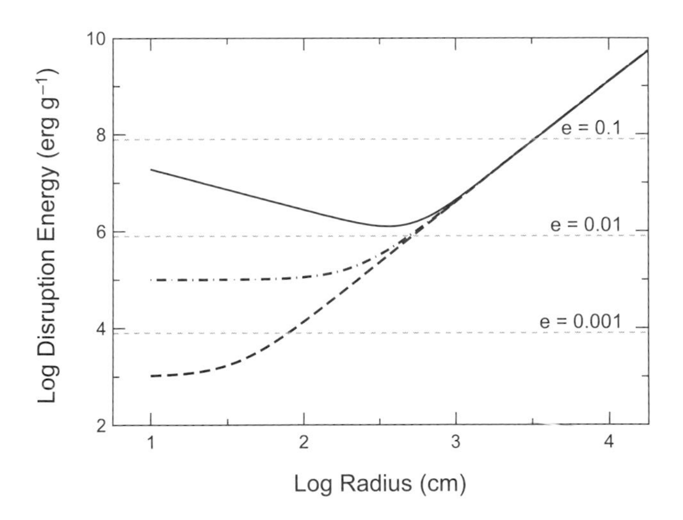

Fig. 1. Disruption energy, Q_D^*, for icy objects. The solid curve plots a typical result derived from numerical simulations of collisions that include a detailed equation of state for crystalline ice ($Q_b = 1.6 \times 10^7$ erg g^{-1}, $\beta_b = -0.42$, $\rho = 1.5$ g cm^{-3}, $Q_g = 1.5$ erg cm^{-3}, and $\beta_b = 1.25$) (*Benz and Asphaug,* 1999). The other curves plot results using Q_b consistent with model fits to comet breakups ($\beta_b \approx 0$; $Q_b \sim 10^3$ erg g^{-1}, dashed curve; $Q_b \sim 10^5$ erg g^{-1}, dot-dashed curve) (*Asphaug and Benz,* 1996). The dashed horizontal lines indicate the center of mass collision energy (equation (7)) for equal-mass objects with e = 0.001, 0.01, and 0.1. Collisions between objects with $Q_i \ll Q_D^*$ yield merged remnants; collisions between objects with $Q_i \gg Q_D^*$ produce debris.

Once disruption commences, the final mass of an oligarch depends on the timescale for the collisional cascade (*Kenyon and Bromley,* 2004a,b,d; *Leinhardt and Richardson,* 2005). If disruptive collisions produce dust grains much faster than oligarchs accrete leftover planetesimals, oligarchs with mass m_o cannot grow much larger than the disruption radius (maximum oligarch mass $m_{o,max} \approx m_d$). However, if oligarchs accrete grains and leftover planetesimals effectively, oligarchs reach the isolation mass before collisions and radiation pressure remove material from the disk (equation (6)) (*Goldreich et al.,* 2004). The relative rates of accretion and disruption depend on the balance between collisional damping and gas drag — which slow the collisional cascade — and viscous stirring and dynamical friction — which speed up the collisional cascade. Because deriving accurate rates for these processes requires numerical simulations of planetesimal accretion, we now consider simulations of planet formation in the Kuiper belt.

3. COAGULATION SIMULATIONS

3.1. Background

Safronov (1969) invented the current approach to planetesimal accretion calculations. In his particle-in-a-box method, planetesimals are a statistical ensemble of masses with distributions of orbital eccentricities and inclinations (*Greenberg et al.,* 1978; *Wetherill and Stewart,* 1989, 1993; *Spaute et al.,* 1991). This statistical approximation is essential: N-body codes cannot follow the $n \sim 10^9$–10^{12} 1-km planetesimals required to build Pluto-mass or Earth-mass planets. For large numbers of objects on fairly circular orbits (e.g., $n \gtrsim 10^4$, $r \lesssim 1000$ km, and $e \lesssim 0.1$), the method is also accurate. With a suitable prescription for collision outcomes, solutions to the coagulation equation in the kinetic theory yield the evolution of n(m) with arbitrarily small errors (e.g., *Wetherill,* 1990; *Lee,* 2000; *Malyshkin and Goodman,* 2001).

In addition to modeling planet growth, the statistical approach provides a method for deriving the evolution of orbital elements for large ensembles of planetesimals. If we (1) assume the distributions of e and i for planetesimals follow a Rayleigh distribution and (2) treat their motions as perturbations of a circular orbit, we can use the Fokker-Planck equation to solve for small changes in the orbits due to gas drag, gravitational interactions, and physical collisions (*Hornung et al.,* 1985; *Wetherill and Stewart,* 1993; *Ohtsuki et al.,* 2002). Although the Fokker-Planck equation cannot derive accurate orbital parameters for planetesimals and oligarchs near massive planets, it yields accurate solutions for the ensemble average e and i when orbital resonances and other dynamical interactions are not important (e.g., *Wetherill and Stewart,* 1993; *Weidenschilling et al.,* 1997; *Ohtsuki et al.,* 2002).

Several groups have implemented *Safronov*'s (1969) method for calculations relevant to the outer solar system (*Greenberg et al.,* 1984; *Stern,* 1995, 2005; *Stern and Colwell,* 1997a,b; *Davis and Farinella,* 1997; *Kenyon and Luu,* 1998, 1999a,b; *Davis et al.,* 1999; *Kenyon and Bromley,* 2004a,d, 2005). These calculations adopt a disk geometry and divide the disk into N concentric annuli with radial width Δa_i at distances a_i from the central star. Each annulus is seeded with a set of planetesimals with masses m_{ij}, eccentricities e_{ij}, and inclinations i_{ij}, where the index i refers to one of N annuli and the index j refers to one of M mass batches within an annulus. The mass batches have mass spacing $\delta \equiv m_{j+1}/m_j$. In most calculations, $\delta \approx 2$; $\delta \leq 1.4$ is optimal (*Ohtsuki et al.,* 1990; *Wetherill and Stewart,* 1993; *Kenyon and Luu,* 1998).

Once the geometry is set, the calculations solve a set of coupled difference equations to derive the number of objects n_{ij} and the orbital parameters e_{ij} and i_{ij} as functions of time. Most studies allow fragmentation and velocity evolution through gas drag, collisional damping, dynamical friction, and viscous stirring. Because Q_D^* sets the maximum size $m_{c,max}$ of objects that participate in the collisional cascade, the size distribution for objects with $m < m_{c,max}$ depends on the fragmentation parameters (equation (8)) (*Davis and Farinella,* 1997; *Kenyon and Bromley,* 2004d; *Pan and Sari,* 2005). The size and velocity distributions of the merger population with $m > m_{c,max}$ are established during runaway growth and the early stages of oligarchic growth. Accurate treatment of velocity evolution is important for following runaway growth and thus deriving good estimates for the growth times and the size and velocity distributions of oligarchs.

When a few oligarchs contain most of the mass, collision rates depend on the orbital dynamics of individual objects instead of ensemble averages. *Safronov*'s (1969) statistical approach then fails (e.g., *Wetherill and Stewart,* 1993; *Weidenschilling et al.,* 1997). Although N-body methods can treat the evolution of the oligarchs, they cannot follow the evolution of leftover planetesimals, where the statistical approach remains valid (e.g., *Weidenschilling et al.,* 1997). *Spaute et al.* (1991) solve this problem by adding a Monte Carlo treatment of binary interactions between large objects to their multiannulus coagulation code. *Bromley and Kenyon* (2006) describe a hybrid code, which merges a direct N-body code with a multiannulus coagulation code. Both codes have been applied to terrestrial planet formation, but not to the Kuiper belt.

Current calculations cannot follow collisional growth accurately in an entire planetary system. Although the 6 order of magnitude change in formation timescales from ~0.4 AU to 40 AU is a factor in this statement, most modern supercomputers cannot finish calculations involving the entire disk with the required spatial resolution on a reasonable timescale. For the Kuiper belt, it is possible to perform a suite of calculations in a disk extending from 30–150 AU following 1-m and larger planetesimals. These calculations yield robust results for the mass distribution as a function of space and time and provide interesting comparisons with observations. Although current calculations do not include complete dynamical interactions with the giant planets or passing stars (see, e.g., *Charnoz and Morbidelli,* 2007), sample calculations clearly show the importance of external perturbations in treating the collisional cascade. We begin with a discussion of self-stirring calculations without interactions with the giant planets or passing stars and then describe results with external perturbers.

3.2. Self-Stirring

To illustrate *in situ* KBO formation at 40–150 AU, we consider a multiannulus calculation with an initial ensemble of 1-m to 1-km planetesimals in a disk with $\Sigma_{0d} = 0.12$ g cm^{-2}. The planetesimals have initial radii of 1 m to 1 km (with equal mass per logarithmic bin), $e_0 = 10^{-4}$, $i_0 = e_0/2$, mass density $\rho = 1.5$ g cm^{-3}, and fragmentation parameters $Q_b = 10^3$ erg g^{-1}, $Q_g = 1.5$ erg cm^{-3}, $\beta_b = 0$, and $\beta_g = 1.25$ (dashed curve in Fig. 1) (*Kenyon and Bromley,* 2004d, 2005). [Our choice of mass density is a compromise between pure ice ($\rho = 1$ g cm^{-3}) and the measured density of Pluto ($\rho \approx 2$ g cm^{-3}) (*Null et al.,* 1993). The calculations are insensitive to factor of 2 variations in the mass density of planetesimals.] The gas density also follows a MMSN, with $\Sigma_g/\Sigma_d = 100$ and a vertical scale height $h = 0.1\ r^{9/8}$ (*Kenyon and Hartmann,* 1987). The gas density is $\Sigma_g \propto e^{-t}/t_g$, with $t_g = 10$ m.y.

This calculation uses an updated version of the *Bromley and Kenyon* (2006) code that includes a Richardson extrapolation procedure in the coagulation algorithm. As in the Eulerian (*Kenyon and Luu,* 1998) and fourth-order Runge-Kutta (*Kenyon and Bromley,* 2002a,b) methods employed previously, this code provides robust numerical solutions to kernels with analytic solutions (e.g., *Ohtsuki et al.,* 1990; *Wetherill,* 1990) without producing the wavy size distributions described in other simulations with a low-mass cutoff (e.g., *Campo Bagatin et al.,* 1994). Once the evolution of large ($r > 1$ m) objects is complete, a separate code tracks the evolution of lower-mass objects and derives the dust emission as a function of time.

Figure 2 shows the evolution of the mass and eccentricity distributions at 40–47 AU for this calculation. During the first few million years, the largest objects grow slowly. Dynamical friction damps the orbits of the largest objects; collisional damping and gas drag circularize the orbits of the smallest objects. This evolution erases many of the initial conditions and enhances gravitational focusing by factors of 10–1000. Runaway growth begins. A few (and some-

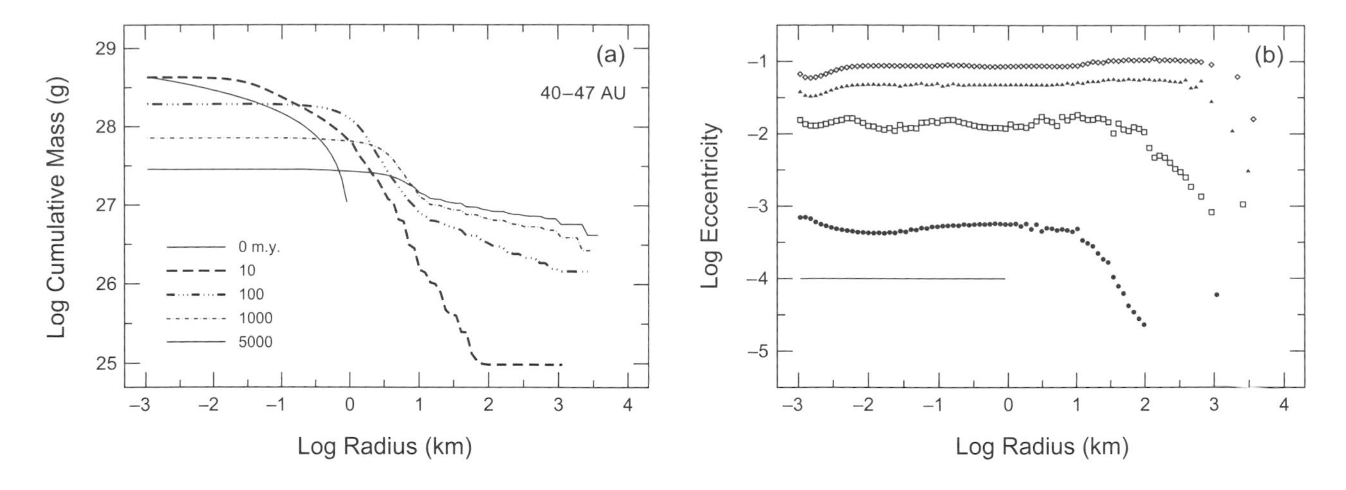

Fig. 2. Evolution of a multiannulus coagulation model with $\Sigma = 0.12(a_i/40\ AU)^{-3/2}$ g cm^{-2}. **(a)** Cumulative mass distribution at times indicated in the legend. **(b)** Eccentricity distributions at t = 0 (light solid line), t = 10 m.y. (filled circles), t = 100 m.y. (open boxes), t = 1 G.y. (filled triangles), and t = 5 G.y. (open diamonds). As large objects grow in the disk, they stir up the leftover planetesimals to e ~ 0.1. Disruptive collisions then deplete the population of 0.1–10-km planetesimals, which limits the growth of the largest objects.

times only one) oligarchs then grow from r ~ 10 km to r ~ 1000 km in ~30 m.y. at 40 AU and in ~1 G.y. at 150 AU (see equation (5)). Throughout runaway growth, dynamical friction and viscous stirring raise the random velocities of the leftover planetesimals to e ≈ 0.01–0.1 and i ≈ 2°–4° (v ~ 50–500 m s^{-1} at 40–47 AU; Fig. 2b). Stirring reduces gravitational focusing factors and ends runaway growth. The large oligarchs then grow slowly through accretion of leftover planetesimals.

As oligarchs grow, collisions among planetesimals initiate the collisional cascade. Disruptive collisions dramatically reduce the population of 1–10-km objects, which slows the growth of oligarchs and produces a significant debris tail in the size distribution. In these calculations, disruptive collisions remove material from the disk faster than oligarchs can accrete the debris. Thus, growth stalls and produces ~10–100 objects with maximum sizes r_{max} ~ 1000–3000 km at 40–50 AU (*Stern and Colwell,* 1997a,b; *Kenyon and Bromley,* 2004d, 2005; *Stern,* 2005).

Stochastic events lead to large dispersions in the growth time for oligarchs, t_o (equation (5)). In ensembles of 25–50 simulations with identical starting conditions, an occasional oligarch will grow up to a factor of 2 faster than its neighbors. This result occurs in simulations with δ = 1.4, 1.7, and 2.0, and thus seems independent of mass resolution. These events occur in ~25% of the simulations and lead to factor of ~2 variations in t_0 (equation (5)).

In addition to modest-sized icy planets, oligarchic growth generates copious amounts of dust (Fig. 3). When runaway growth begins, collisions produce small amounts of dust from "cratering" (see, e.g., *Greenberg et al.,* 1978; *Wetherill and Stewart,* 1993; *Stern and Colwell,* 1997a,b; *Kenyon and Luu,* 1999a). Stirring by growing oligarchs leads to "catastrophic" collisions, where colliding planetesimals lose more than 50% of their initial mass. These disruptive collisions produce a spike in the dust production rate that coincides with the formation of oligarchs with r ≳ 200–300 km (equation (9)). As the wave of runaway growth propagates outward, stirring produces disruptive collisions at ever-larger heliocentric distances. The dust mass grows in time and peaks at ~1 G.y., when oligarchs reach their maximum mass at 150 AU. As the mass in leftover planetesimals de-

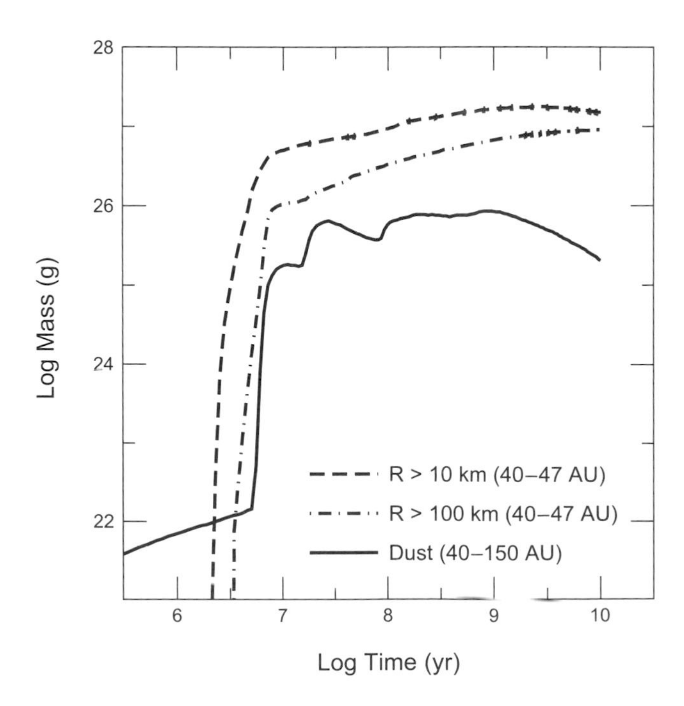

Fig. 3. Time evolution of the mass in KBOs and dust grains. Solid line: dust mass (r ≲ 1 mm) at 40–150 AU. Dashed (dot-dashed) lines: total mass in small (large) KBOs at 40–47 AU.

clines, Poynting-Robertson drag removes dust faster than disruptive collisions produce it. The dust mass then declines with time.

3.3. External Perturbation

Despite the efficiency of self-stirring models in removing leftover planetesimals from the disk, other mechanisms must reduce the derived mass in KBOs to current observational limits. In self-stirring calculations at 40–50 AU, the typical mass in KBOs with r ~ 30–1000 km at 4–5 G.y. is a factor of 5–10 larger than currently observed (*Luu and Jewitt,* 2002; chapter by Petit et al.). Unless Earth-mass or larger objects form in the Kuiper belt (*Chiang et al.,* 2007; *Levison and Morbidelli,* 2007), external perturbations must excite KBO orbits and enhance the collisional cascade.

Two plausible sources of external perturbation can reduce the predicted KBO mass to the desired limits. Once Neptune achieves its current mass and orbit, it stirs up the orbits of KBOs at 35–50 AU (*Levison and Duncan,* 1990; *Holman and Wisdom,* 1993; *Duncan et al.,* 1995; *Kuchner et al.,* 2002; *Morbidelli et al.,* 2004). In ~100 m.y. or less, Neptune removes nearly all KBOs with a $\lesssim$ 37–38 AU. Beyond a ~ 38 AU, some KBOs are trapped in orbital resonance with Neptune (*Malhotra,* 1995, 1996); others are ejected into the scattered disk (*Duncan and Levison,* 1997). In addition to these processes, Neptune stirring increases the effectiveness of the collisional cascade (*Kenyon and Bromley,* 2004d), which removes additional mass from the population of 0.1–10-km KBOs and prevents growth of larger KBOs.

Passing stars can also excite KBO orbits and enhance the collisional cascade. Although Neptune dynamically ejects scattered disk objects with perihelia q $\lesssim$ 36–37 AU (*Morbidelli et al.,* 2004), objects with q $\gtrsim$ 45–50 AU, such as Sedna and Eris, require another scattering source. Without evidence for massive planets at a $\gtrsim$ 50 AU (*Morbidelli et al.,* 2002), a passing star is the most likely source of the large q for these KBOs (*Morbidelli and Levison,* 2004; *Kenyon and Bromley,* 2004c).

Adams and Laughlin (2001) (see also chapter by Duncan et al.) examined the probability of encounters between the young Sun and other stars. Most stars form in dense clusters with estimated lifetimes of ~100 m.y. To account for the abundance anomalies of radionuclides in solar system meteorites (produced by supernovae in the cluster) and for the stability of Neptune's orbit at 30 AU, the most likely solar birth cluster has ~2000–4000 members, a crossing time of ~1 m.y., and a relaxation time of ~10 m.y. The probability of a close encounter with a distance of closest approach a_{close} is then ~60% $(a_{close}/160\ AU)^2$ (*Kenyon and Bromley,* 2004c).

Because the dynamical interactions between KBOs in a coagulation calculation and large objects like Neptune or a passing star are complex, here we consider simple calculations of each process. To illustrate the evolution of KBOs after a stellar flyby, we consider a very close pass with a_{close} = 160 AU (*Kenyon and Bromley,* 2004c). This corotating flyby produces objects with orbital parameters similar to those of Sedna and Eris. For objects in the coagulation calculation, the flyby produces an e distribution

$$e_{KBO} = \begin{cases} 0.025(a/30\ AU)^4 & a < a_0 \\ 0.5 & a > a_0 \end{cases} \qquad (10)$$

with $a_0 \approx 65$ AU (see *Ida et al.,* 2000; *Kenyon and Bromley,* 2004c; *Kobayashi et al.,* 2005). This e distribution produces a dramatic increase in the debris production rate throughout the disk, which freezes the mass distribution of the largest objects. (The i distribution following a flyby depends on the relative orientations of two planes, the orbital plane of KBOs and the plane of the trajectory of the passing star. Here, we assume the flyby produces no change in i, which simplifies the discussion without changing any of the results significantly.) Thus, to produce an ensemble of KBOs with r $\gtrsim$ 300 km at 40–50 AU, the flyby must occur when the Sun has an age $t_\odot \gtrsim$ 10–20 m.y. (Fig. 2). For $t_\odot \gtrsim$ 100 m.y., the flyby is very unlikely. As a compromise between these two estimates, we consider a flyby at $t_\odot$ ~ 50 m.y.

Figure 4 shows the evolution of the KBO surface density in three annuli as a function of time. At early times (t $\lesssim$ 50 m.y.), KBOs grow in the standard way. After the flyby, the disk suffers a dramatic loss of material. At 40–47 AU, the disk loses ~90% (93%) of its initial mass in ~1 G.y. (4.5 G.y.). At ~50–80 AU, the collisional cascade removes ~90% (97%) of the initial mass in ~500 m.y. (4.5 G.y.). Beyond ~80 AU, KBOs contain less than 1% of the initial mass. Compared to self-stirring models, flybys that produce Sedna-like orbits are a factor of 2–3 more efficient at removing KBOs from the solar system.

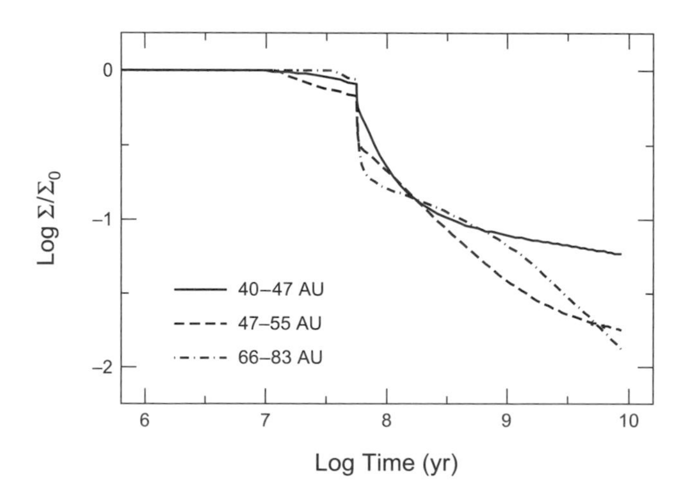

Fig. 4. Evolution of Σ after a stellar flyby. After 50 m.y. of growth, the close pass excites KBOs to large e (equation (10)) and enhances the collisional cascade.

To investigate the impact of Neptune on the collisional cascade, we parameterize the growth of Neptune at 30 AU as a simple function of time (*Kenyon and Bromley,* 2004d)

$$M_{Nep} \approx \begin{cases} 6 \times 10^{27}\, e^{(t-t_N)/t_1}\ \mathrm{g} & t < t_N \\ 6 \times 10^{27}\ \mathrm{g} + C(t - t_1) & t_N < t < t_2 \\ 1.0335 \times 10^{29}\ \mathrm{g} & t > t_2 \end{cases} \quad (11)$$

where $C_{Nep} = 1.927 \times 10^{21}$ g yr^{-1}, $t_N = 50$ m.y., $t_1 = 3$ m.y., and $t_2 = 100$ m.y. These choices enable a model Neptune to reach a mass of 1 $M_\oplus$ in 50 m.y., when the largest KBOs form at 40–50 AU, and reach its current mass in 100 m.y. (This prescription is not intended as an accurate portrayal of Neptune formation, but it provides a simple way to investigate how Neptune might stir the Kuiper belt once massive KBOs form.) As Neptune approaches its final mass, its gravity stirs up KBOs at 40–60 AU and increases their orbital eccentricities to e ~ 0.1–0.2 on short timescales. In the coagulation model, distant planets produce negligible changes in i, so self-stirring sets i in these calculations (*Weidenschilling,* 1989). This evolution enhances debris production by a factor of 3–4, which effectively freezes the mass distribution of 100–1000-km objects at 40–50 AU. By spreading the leftover planetesimals and the debris over a larger volume, Neptune stirring limits the growth of the oligarchs and thus reduces the total mass in KBOs.

Figure 5 shows the evolution of the surface density in small and large KBOs in two annuli as a function of time. At 40–55 AU, Neptune rapidly stirs up KBOs to e ~ 0.1 when it reaches its current mass at ~100 m.y. Large collision velocities produce more debris, which is rapidly ground to dust and removed from the system by radiation pressure at early times and by Poynting-Robertson drag at later times. Compared to self-stirring models, the change in Σ is dramatic, with only ~3% of the initial disk mass remaining at ~4.5 G.y.

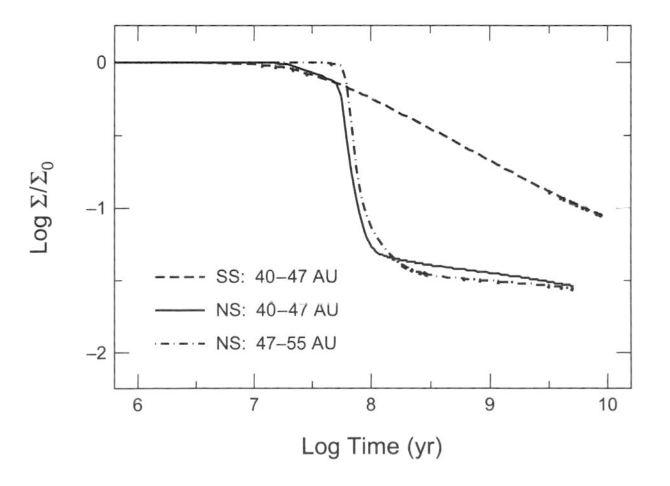

Fig. 5. Evolution of Σ (KBO) in models with Neptune stirring. Compared to self-stirring models (SS; dashed curve), stirring by Neptune rapidly removes KBOs at 40–47 AU (NS; solid curve) and at 47–55 AU (NS; dot-dashed curve).

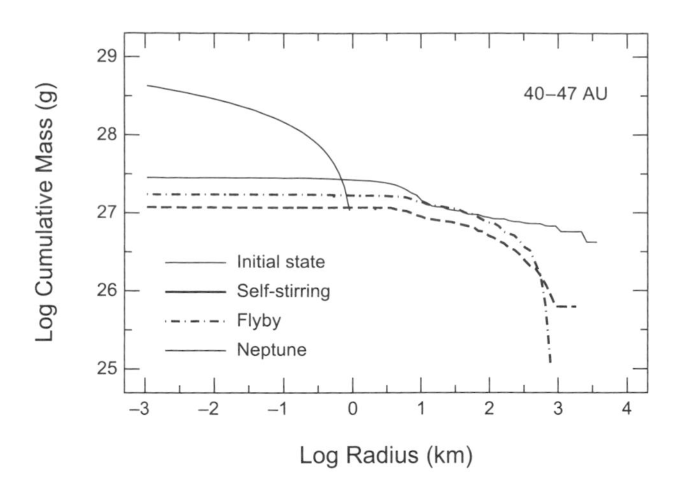

Fig. 6. Mass distributions for evolution with self-stirring (heavy solid line), stirring from a passing star (dot-dashed line), and stirring from Neptune at 30 AU (dashed line). After 4.5 G.y., the mass in KBOs with r ≥ 50 km is 5% (self-stirring), ~3.5% (flyby), and ~2% (Neptune stirring) of the initial mass. The number of objects with r ≥ 1000 km is ~100 (self-stirring), 1 (flyby), and 10 (Neptune stirring). The largest object has r_{max} ~ 3000 km (self-stirring), r_{max} ~ 500–1000 km (flyby), and r_{max} ~ 1000–2000 km (Neptune stirring).

From these initial calculations, it is clear that external perturbations dramatically reduce the mass of KBOs in the disk (see also *Charnoz and Morbidelli,* 2007). Figure 6 compares the mass distributions at 40–47 AU and at 4.5 G.y. for the self-stirring model in Fig. 2 (solid line) with results for the flyby (dot-dashed line) and Neptune stirring (dashed line). Compared to the self-stirring model, the close flyby reduces the mass in KBOs by ~50%. Neptune stirring reduces the KBO mass by almost a factor of 3 relative to the self-stirring model. For KBOs with r ≳ 30–50 km, the predicted mass in KBOs with Neptune stirring is within a factor of 2–3 of the current mass in KBOs.

These simple calculations for the stellar flyby and Neptune stirring do not include dynamical depletion. In the stellar flyby picture, the encounter removes nearly all KBOs beyond a truncation radius, a_T ~ 48 (a_{close}/160 AU) AU (*Kenyon and Bromley,* 2004c). Thus, a close pass with a_{close} ~ 160 AU can produce the observed outer edge of the Kuiper belt at 48 AU. Although many objects with initial a > a_T are ejected from the solar system, some are placed on very elliptical, Sedna-like orbits. [*Levison et al.* (2004) consider the impact of the flyby on the scattered disk and Oort cloud. After analyzing a suite of numerical simulations, they conclude that the flyby must occur before Neptune reaches its current orbit and begins the dynamical processes that populate the Oort cloud and the scattered disk. If Neptune forms *in situ* in 1–10 m.y., then the flyby cannot occur after massive KBOs form. If Neptune migrates to 30 AU after massive KBOs form, then a flyby can truncate the Kuiper belt without much impact on the Oort cloud or the scattered disk.] In the Neptune-stirring model, dynamical interactions will eject some KBOs at 40–47 AU. If the dynamical interactions that produce the scattered disk reduce the mass in KBOs by a factor of 2 at 40–47 AU (e.g., *Duncan et al.,* 1995; *Kuchner et al.,* 2002), the Neptune-stirring model yields a KBO mass in reasonably good agreement with observed limits (for a different opinion, see *Charnoz and Morbidelli,* 2007).

3.4. Nice Model

Although *in situ* KBO models can explain the current amount of mass in large KBOs, these calculations do not address the orbits of the dynamical classes of KBOs. To explain the orbital architecture of the giant planets, the "Nice group" centered at Nice Observatory developed an inspired, sophisticated picture of the dynamical evolution of the giant planets and a remnant planetesimal disk (*Tsiganis et al.,* 2005; *Morbidelli et al.,* 2005; *Gomes et al.,* 2005, and references therein). The system begins in an approximate equilibrium, with the giant planets in a compact configuration (Jupiter at 5.45 AU, Saturn at ~8.2 AU, Neptune at ~11.5 AU, and Uranus at ~14.2 AU) and a massive planetesimal disk at 15–30 AU. Dynamical interactions between the giant planets and the planetesimals lead to an instability when Saturn crosses the 2:1 orbital resonance with Jupiter, which results in a dramatic orbital migration of the gas giants and the dynamical ejection of planetesimals into the Kuiper belt, scattered disk, and the Oort cloud. Comparisons between the end state of this evolution and the orbits of KBOs in the "hot population" and the scattered disk are encouraging (chapter by Morbidelli et al.).

Current theory cannot completely address the likelihood of the initial state in the Nice model. *Thommes et al.* (1999, 2002) demonstrate that n-body simulations can produce a compact configuration of gas giants, but did not consider how fragmentation or interactions with low-mass planetesimals affect the end state. *O'Brien et al.* (2005) show that a disk of planetesimals has negligible collisional grinding over 600 m.y. if most of the mass is in large planetesimals with $r \gtrsim 100$ km. However, they did not address whether this state is realizable starting from an ensemble of 1-km and smaller planetesimals. In terrestrial planet simulations starting with 1–10-km planetesimals, the collisional cascade removes ~25% of the initial rocky material in the disk (*Wetherill and Stewart,* 1993; *Kenyon and Bromley,* 2004b). Interactions between oligarchs and remnant planetesimals are also important for setting the final mass and dynamical state of the terrestrial planets (*Bromley and Kenyon,* 2006; *Kenyon and Bromley,* 2006). Because complete hybrid calculations of the giant planet region are currently computationally prohibitive, it is not possible to make a reliable assessment of these issues for the formation of gas giant planets.

Here, we consider the evolution of the planetesimal disk outside the compact configuration of giant planets, where standard coagulation calculations can follow the evolution of many initial states for 1–5 G.y. in a reasonable amount of computer time. Figure 7 shows the time evolution for the surface density of planetesimals in three annuli from one

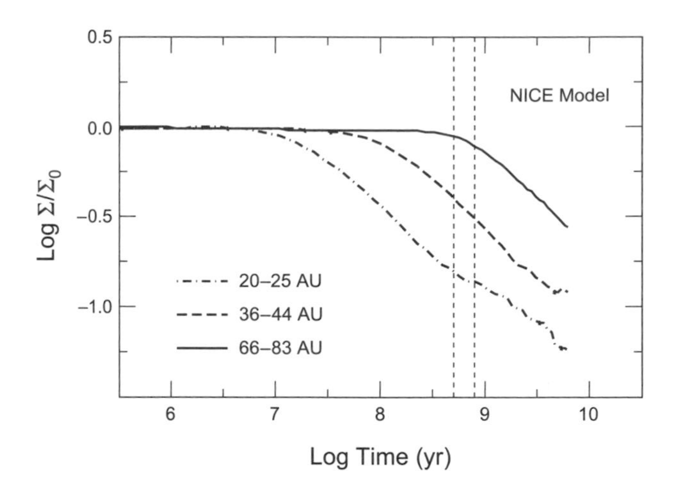

Fig. 7. Evolution of Σ in a self-stirring model at 20–100 AU. At 20–25 AU, it takes 5–10 m.y. to form 1000-km objects. After ~0.5–1 G.y., there are ~100 objects with r ~ 1000–2000 km and ~10^5 objects with r ~ 100–200 km at 20–30 AU. As these objects grow, the collisional cascade removes 90% of the mass in remnant planetesimals. The twin vertical dashed lines bracket the time of the late heavy bombardment at 300–600 m.y.

typical calculation at 20–25 AU (dot-dashed curve; $M_i = 6\ M_\oplus$), 36–44 AU (dashed curve; $M_i = 9\ M_\oplus$), and 66–83 AU (solid curve; $M_i = 12\ M_\oplus$). Starting from the standard surface density profile (equation (1)), planetesimals at 20–25 AU grow to 1000-km sizes in a few million years. Once the collisional cascade begins, the surface density slowly declines to ~10% to 20% of its initial value at the time of the late heavy bombardment, when the Nice model predicts that Saturn crosses the 2:1 orbital resonance with Jupiter.

These results provide a strong motivation to couple coagulation calculations with the dynamical simulations of the Nice group (see also *Charnoz and Morbidelli,* 2007). In the Nice model, dynamical interactions with a massive planetesimal disk are the "fuel" for the dramatic migration of the giant planets and the dynamical ejection of material into the Kuiper belt and the scattered disk. If the mass in the planetesimal disk declines by ~80% as the orbits of the giant planets evolve, the giant planets cannot migrate as dramatically as in the *Gomes et al.* (2005) calculations. Increasing the initial mass in the disk by a factor of 3–10 may allow coagulation and the collisional cascade to produce a debris disk capable of triggering the scattering events of the Nice model.

3.5. A Caveat on the Collisional Cascade

Although many of the basic outcomes of oligarchic growth and the collisional cascade are insensitive to the initial conditions and fragmentation parameters for the planetesimals, several uncertainties in the collisional cascade can modify the final mass in oligarchs and the distributions of r and e. Because current computers do not allow coagulation calculations that include the full range of sizes (1 μm to 10^4 km), published calculations have two pieces, a solution for large objects (e.g., *Kenyon and Bromley,* 2004a,b) and a separate solution for smaller objects (e.g., *Krivov et al.,* 2006). Joining these solutions assumes that (1) collision fragments continue to collide and fragment until particles are removed by radiative processes and (2) mutual (destructive) collisions among the fragments are more likely than mergers with much larger oligarchs. These assumptions are reasonable but untested by numerical calculations (*Kenyon and Bromley,* 2002a). Thus, it may be possible to halt or to slow the collisional cascade before radiation pressure rapidly removes small grains with $r \approx 1$–100 μm.

In current coagulation calculations, forming massive oligarchs at 5–15 AU in a massive disk requires an inefficient collisional cascade. When the cascade is efficient, the most massive oligarchs have $m \lesssim M_\oplus$. Slowing the cascade allows oligarchs to accrete planetesimals more efficiently, which results in larger oligarchs that contain a larger fraction of the initial mass. If collisional damping is efficient, halting the cascade completely at sizes of ~1 mm leads to rapid *in situ* formation of Uranus and Neptune (*Goldreich et al.,* 2004) and early stirring of KBOs at 40 AU.

There are two simple ways to slow the collisional cascade. In simulations where the cascade continues to small sizes, $r \sim 1$–10 μm, the radial optical depth in small grains is $\tau_s \sim 0.1$–1 at 30–50 AU (*Kenyon and Bromley,* 2004a). Lines-of-sight to the central star are not purely radial, so this optical depth reduces radiation pressure and Poynting-Robertson drag by small factors, $\sim e^{-0.2\tau_s} \sim 10$–$30\%$, and has little impact on the evolution of the cascade. With $\tau_s \propto a^{-s}$ and $s \sim 1$–2, however, the optical depth may reduce radiation forces significantly at smaller a. Slowing the collisional cascade by factors of 2–3 could allow oligarchs to accrete leftover planetesimals and smaller objects before the cascade removes them.

Collisional damping and gas drag on small particles may also slow the collisional cascade. For particles with large ratios of surface area to volume, $r \lesssim 0.1$–10 cm, collisions and the gas effectively damp e and i (*Adachi et al.,* 1976; *Goldreich et al.,* 2004) and roughly balance dynamical friction and viscous stirring. Other interactions between small particles and the gas — such as photophoresis (*Wurm and Krauss,* 2006) — also damp particles random velocities and thus might help to slow the cascade. Both collisions and interactions between the gas and the solids are more effective at large volume density, so these processes should be more important inside 30 AU than outside 30 AU. The relatively short lifetime of the gas, ~3–10 m.y., also favors more rapid growth inside 30 AU. If damping maintains an equilibrium $e \sim 10^{-3}$ at $a \sim 20$–30 AU, oligarchs can grow to the sizes required in the Nice model ($r \gtrsim 2000$ km). Rapid growth at $a \sim 5$–15 AU might produce oligarchs with the isolation mass ($r \sim 10$–$30\ R_\oplus$; equation (6)) and lead to the rapid formation of gas giants.

Testing these mechanisms for slowing the collisional cascade requires coagulation calculations with accurate treatments of collisional damping, gas drag, and optical depth for particle radii $r \sim 1$–10 μm to $r \sim 10{,}000$ km. Although these calculations require factors of 4–6 more computer time than published calculations, they are possible with multiannulus coagulation codes on modern parallel computers.

3.6. Model Predictions

The main predictions derived from coagulation models are n(r), n(e), and n(i) as functions of a. The cumulative number distribution consists of three power laws (*Kenyon and Bromley,* 2004d; *Pan and Sari,* 2005)

$$n(r) = \begin{cases} n_d r^{-\alpha_d} & r \le r_1 \\ n_1 & r_1 \le r \le r_0 \\ n_m r^{-\alpha_m} & r \ge r_0 \end{cases} \qquad (12)$$

The debris population at small sizes, $r \le r_1$, always has $\alpha_d \approx 3.5$. The merger population at large sizes, $r \ge r_0$, has $\alpha_m \approx 2.7$–4. Because the collisional cascade robs oligarchs of material, calculations with more stirring have steeper size distributions. Thus, self-stirring calculations with $Q_b \gtrsim 10^5$ erg g^{-1} ($Q_b \lesssim 10^3$ erg g^{-1}) typically yield $\alpha_m \approx 2.7$–3.3 (3.5–4). Models with a stellar flyby or stirring by a nearby gas giant also favor large α_m.

The transition radii for the power laws depend on the fragmentation parameters (see Fig. 1) (see also *Pan and Sari,* 2005). For a typical $e \sim 0.01$–0.1 in self-stirring models, $r_0 \approx r_1 \approx 1$ km when $Q_b \gtrsim 10^5$ erg g^{-1}. When $Q_b \lesssim 10^3$ erg g^{-1}, $r_1 \approx 0.1$ km and $r_0 \approx 10$–20 km. Thus the calculations predict a robust correlation between the transition radii and the power law exponents: large r_0 and α_m or small r_0 and α_m.

Because gravitational stirring rates are larger than accretion rates, the predicted distributions of e and i at 4–5 G.y. depend solely on the total mass in oligarchs (see also *Goldreich et al.,* 2004). Small objects with $r \lesssim r_0$ contain a very small fraction of the mass and cannot stir themselves. Thus e and i are independent of r (Fig. 2). The e and i for larger objects depends on the total mass in the largest objects. In self-stirring models, dynamical friction and viscous stirring between oligarchs and planetesimals (during runaway growth) and among the ensemble of oligarchs (during oligarchic growth) set the distribution of e for large objects with $r \gtrsim r_0$. In self-stirring models, viscous stirring among oligarchs dominates dynamical friction between oligarchs and leftover planetesimals, which leads to a shallow relation between e and r, $e \propto r^{-\gamma}$ with $\gamma \approx 3/4$. In the flyby and Neptune-stirring models, stirring by the external perturber dominates stirring among oligarchs. This stirring yields a very shallow relation between e and r with $\gamma \approx 0$–0.25.

Other results depend little on the initial conditions and the fragmentation parameters. In calculations with different initial mass distributions, an order of magnitude range in e_0, and $Q_b = 10^0$–10^7 erg g^{-1}, $\beta_b = 0.5$–0, $Q_g = 0.5$–5 erg cm^{-3}, $\beta_b \ge 1.25$, r_{max} and the amount of mass removed by the collisional cascade vary by ≤10% relative to the evolution of the models shown in Figs. 2–7. Because collisional damping among 1-m to 1-km objects erases the initial orbital distribution, the results do not depend on e_0 and i_0. Damping and dynamical friction also quickly erase the initial mass distribution, which yields growth rates that are insensitive to the initial mass distribution.

The insensitivity of r_{max} and mass removal to the fragmentation parameters depends on the rate of collisional disruption relative to the growth rate of oligarchs. Because the collisional cascade starts when $m_o \sim m_d$ (equation (9)), calculations with small Q_b ($Q_b \lesssim 10^3$ erg g^{-1}) produce large amounts of debris before calculations with large Q_b ($Q_b \gtrsim 10^3$ erg g^{-1}). Thus, an effective collisional cascade should yield lower mass oligarchs and more mass removal when Q_b is small. However, oligarchs with $m_o > m_d$ still have fairly large gravitational focusing factors and accrete leftover planetesimals more rapidly than the cascade removes them. As oligarchic growth continues, gravitational focusing factors fall and collision disruptions increase. All calculations then reach a point where the collisional cascade removes leftover planetesimals more rapidly than oligarchs can accrete them. As long as most planetesimals have r ~ 1–10 km, the timing of this epoch is more sensitive to gravitational focusing and the growth of oligarchs than the collisional cascade and the fragmentation parameters. Thus, r_{max} and the amount of mass processed through the collisional cascade are relatively insensitive to the fragmentation parameters.

4. CONFRONTING KUIPER BELT OBJECT COLLISION MODELS WITH KUIPER BELT OBJECT DATA

Current data for KBOs provide two broad tests of coagulation calculations. In each dynamical class, four measured parameters test the general results of coagulation models and provide ways to discriminate among the outcomes of self-stirring and perturbed models. These parameters are r_{max}, the size of the largest object; α_m, the slope of the size distribution for large KBOs with $r \gtrsim 10$ km; r_0, the break radius, which measures the radius where the size distribution makes the transition from a merger population ($r \gtrsim r_0$) to a collisional population ($r \lesssim r_0$) as summarized in equation (12); and M_t, the total mass in large KBOs.

For all KBOs, measurements of the dust mass allow tests of the collisional cascade and link the Kuiper belt to observations of nearby debris disks. We begin with the discussion of large KBOs and then compare the Kuiper belt with other debris disks.

Table 1 summarizes the mass and size distribution parameters derived from recent surveys. To construct this table, we used online data from the Minor Planet Center (*cfa-www.harvard.edu/iau/lists/MPLists.html*) for r_{max} (see also *Levison and Stern,* 2001) and the results of several detailed analyses for α_m, r_{max}, and r_0 (e.g., *Bernstein et al.,* 2004; *Elliot et al.,* 2005; *Petit et al.,* 2006; chapter by Petit et al.).

TABLE 1. Data for KBO size distribution.

KBO Class	M_l ($M_\oplus$)	r_{max} (km)	r_0 (km)	q_m
Cold cl	0.01–0.05	400	20–40 km	≥4
Hot cl	0.01–0.05	1000	20–40 km	3–3.5
Detached	n/a	1500	n/a	n/a
Resonant	0.01–0.05	1000	20–40 km	3
Scattered	0.1–0.3	700	n/a	n/a

Because comprehensive KBO surveys are challenging, the entries in the table are incomplete and sometimes uncertain. Nevertheless, these results provide some constraints on the calculations.

Current data provide clear evidence for physical differences among the dynamical classes. For classical KBOs with a = 42–48 AU and q > 37 AU, the cold population ($i \lesssim 4°$) has a steep size distribution with $\alpha_m \approx 3.5$–4 and $r_{max} \sim$ 300–500 km. In contrast, the hot population ($i \gtrsim 10°$) has a shallow size distribution with $\alpha_m \approx 3$ and $r_{max} \sim 1000$ km (*Levison and Stern,* 2001). Both populations have relatively few objects with optical brightness $m_R \approx 27$–27.5, which implies $r_0 \sim 20$–40 km for reasonable albedo ~0.04–0.07. The detached, resonant, and scattered disk populations contain large objects with $r_{max} \sim 1000$ km. Although there are too few detached or scattered disk objects to constrain α_m or r_0, data for the resonant population are consistent with constraints derived for the hot classical population, $\alpha_m \approx 3$ and $r_0 \approx 20$–40 km.

The total mass in KBOs is a small fraction of the ~10–30 $M_\oplus$ of solid material in a MMSN from ~35–50 AU (*Gladman et al.,* 2001; *Bernstein et al.,* 2004; *Petit et al.,* 2006; see also chapter by Petit et al.). The classical and resonant populations have $M_l \approx 0.01$–0.1 $M_\oplus$ in KBOs with $r \gtrsim 10$–20 km. The scattered disk may contain more material, $M_l \sim 0.3$ $M_\oplus$, but the constraints are not as robust as for the classical and resonant KBOs.

These data are broadly inconsistent with the predictions of self-stirring calculations with no external perturbers. Although self-stirring models yield inclinations, $i \approx 2°$–4°, close to those observed in the cold, classical population, the small r_{max} and the large α_m of this group suggest that an external dynamical perturbation — such as a stellar flyby or stirring by Neptune — modified the evolution once r_{max} reached ~300–500 km. The observed break radius, $r_0 \sim 20$–40 km, also agrees better with the $r_0 \sim 10$ km expected from Neptune-stirring calculations than the 1 km achieved in self-stirring models (*Kenyon and Bromley,* 2004d; *Pan and Sari,* 2005). Although a large r_{max} and a small α_m for the resonant and hot, classical populations agree reasonably well with self-stirring models, the observed $r_{max} \sim 1000$ km is much smaller than the $r_{max} \sim 2000$–3000 km typically achieved in self-stirring calculations (Fig. 1). Both of these populations appear to have large r_0, which is also more consistent with Neptune-stirring models than with self-stirring models.

The small M_l for all populations provides additional evidence against self-stirring models. In the most optimistic scenario, where KBOs are easily broken, self-stirring models leave a factor of 5–10 more mass in large KBOs than currently observed at 40–48 AU. Although models with Neptune stirring leave a factor of 2–3 more mass in KBOs at 40–48 AU than is currently observed, Neptune ejects half the KBOs at 40–48 AU into the scattered disk (e.g., *Duncan et al.,* 1995; *Kuchner et al.,* 2002). With an estimated mass of 2–3× the mass in classical and resonant KBOs, the scattered disk contains enough material to bridge the difference between the KBO mass derived from Neptune-stirring models and the observed KBO mass.

The mass in KBO dust grains provides a final piece of evidence against self-stirring models. From an analysis of data from Pioneer 10 and 11, *Landgraf et al.* (2002) estimate a dust production rate of 10^{15} g yr^{-1} in 0.01–2-mm particles at 40–50 AU. The timescale for Poynting-Robertson drag to remove these grains from the Kuiper belt is ~10–100 m.y. (*Burns et al.,* 1979), which yields a mass of ~10^{22}–10^{24} g. Figure 8 compares this dust mass with masses derived from mid-IR and submillimeter observations of several nearby solar-type stars (*Greaves et al.,* 1998, 2004; *Williams et al.,* 2004; *Wyatt et al.,* 2005) and with predictions from the self-stirring, flyby, and Neptune-stirring models. The dust masses for nearby solar-type stars roughly follow the predictions of self-stirring models and flyby models with $Q_b \sim 10^3$ erg g^{-1}. The mass of dust in the Kuiper belt is 1–3 orders of magnitude smaller than predicted in self-stirring models and is closer to the predictions of the Neptune-stirring models.

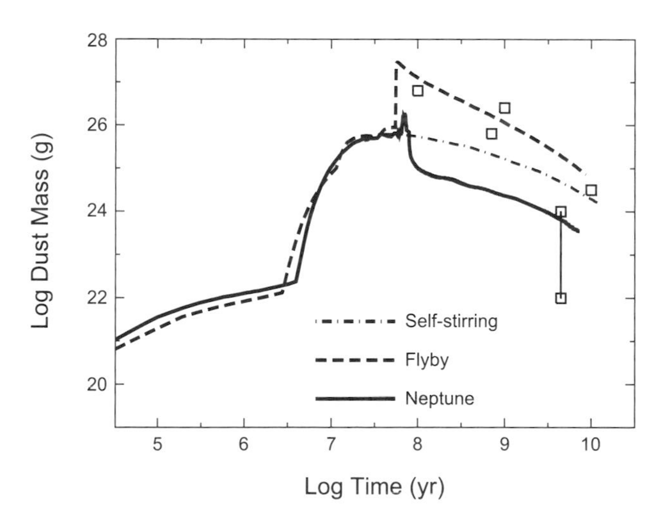

Fig. 8. Evolution of mass in small dust grains (0.001–1 mm) for models with self-stirring (dot-dashed line), stirring from a passing star (dashed line), and stirring from Neptune at 30 AU (solid line) for $Q_b = 10^3$ erg g^{-1}. Calculations with smaller (large) Q_b produce more (less) dust at t ≲ 50 m.y. and somewhat more (less) dust at t ≳ 100 m.y. At 1–5 G.y., models with Neptune stirring have less dust than self-stirring or flyby models. The boxes show dust mass estimated for four nearby solar-type stars (from left to right in age: HD 107146, ε Eri, η Crv, and τ Cet) (*Greaves et al.,* 1998, 2004; *Williams et al.,* 2004; *Wyatt et al.,* 2005) and two estimates for the Kuiper belt (boxes connected by solid line) (*Landgraf et al.,* 2002).

To combine the dynamical properties of KBOs with these constraints, we rely on results from N-body simulations that do not include collisional processing of small objects (see chapter by Morbidelli et al.). For simplicity, we consider coagulation in the context of the Nice model, which provides a solid framework for interpreting the dynamics of the gas giants and the dynamical classes of KBOs. In the Nice model, Saturn's crossing of the 2:1 resonance with Jupiter initiates the dynamical instability that populates the Kuiper belt. As Neptune approaches a ≈ 30 AU, it captures resonant KBOs, ejects KBOs into the scattered disk and the Oort cloud, and excites the hot classical KBOs. Although Neptune might reduce the number of cold, classical KBOs formed roughly *in situ* beyond 30 AU, the properties of these KBOs probably reflect conditions in the Kuiper belt when the instability began.

The Nice model requires several results from coagulation calculations. Once giant planets form at 5–15 AU, collisional growth must produce thousands of Pluto-mass objects at 20–30 AU. Unless the planetesimal disk was massive, growth of oligarchs must dominate collisional grinding in this region of the disk. To produce the cold classical population at ~45 AU, collisions must produce 1–10 Pluto-mass objects and then efficiently remove leftover planetesimals. To match the data in Table 1, KBOs formed at 20–30 AU should have a shallower size distribution and a larger r_{max} than those at 40–50 AU.

Some coagulation results are consistent with the trends required in the Nice model. In current calculations, collisional growth naturally yields smaller r_{max} and a steeper size distribution at larger a. At 40–50 AU, Neptune-stirring models produce a few Pluto-mass objects and many smaller KBOs with e ~ 0.1 and i ≈ 2°–4°. Although collisional growth produces more Plutos at 15–30 AU than at 40–50 AU, collisional erosion removes material faster from the inner disk than from the outer disk (Fig. 7). Thus, collisions do not produce the thousands of Pluto-mass objects at 15–30 AU required in the Nice model.

Reconciling this aspect of the Nice model with the coagulation calculations requires a better understanding of the physical processes that can slow or halt the collisional cascade. Producing gas giants at 5–15 AU, thousands of Plutos at 20–30 AU, and a few or no Plutos at 40–50 AU implies that the outcome of coagulation changes markedly from 5 AU to 50 AU. If the collisional cascade can be halted as outlined in section 3.5, forming 5–10 $M_\oplus$ cores at 5–15 AU is straightforward. Slowing the collisional cascade at 20–30 AU might yield a large population of Pluto mass objects at 20–30 AU. Because α_m and r_{max} are well-correlated, better constraints on the KBO size distributions coupled with more robust coagulation calculations can test these aspects of the Nice model in more detail.

To conclude this section, we consider constraints on the Kuiper belt in the more traditional migration scenario of *Malhotra* (1995), where Neptune forms at ~20–25 AU and slowly migrates to 30 AU. To investigate the relative importance of collisional and dynamical depletion at 40–50 AU,

Charnoz and Morbidelli (2007) couple a collision code with a dynamical code and derive the expected distributions for size and orbital elements in the Kuiper belt, the scattered disk, and the Oort cloud. Although collisional depletion models can match the observations of KBOs, these models are challenged to provide enough small objects into the scattered disk and Oort cloud. Thus, the results suggest that dynamical mechanisms dominate collisions in removing material from the Kuiper belt.

Although *Charnoz and Morbidelli* (2007) argue against a dramatic change in collisional evolution from 15 AU to 40 AU, the current architecture of the solar system provides good evidence for this possibility. In the MMSN, the ratio of timescales to produce gas giant cores at 10 AU and at 25 AU is $\xi = (25/10)^3 \sim 15$. In the context of the Nice model, formation of Saturn and Neptune at 8–11 AU in 5–10 m.y. thus implies formation of other gas giant cores at 20–25 AU in 50–150 m.y. If these cores *had* formed, they would have consumed most of the icy planetesimals at 20–30 AU, leaving little material behind to populate the outer solar system when the giant planets migrate. The apparent lack of gas giant core formation at 20–30 AU indicates that the collisional cascade changed dramatically from 5–15 AU (where gas giant planets formed) to 20–30 AU (where gas giant planets did not form). As outlined in section 3.5, understanding the interaction of small particles with the gas and the radiation field may provide important insights into the evolution of oligarchic growth and thus into the formation and structure of the solar system.

5. KUIPER BELT OBJECTS AND ASTEROIDS

In many ways, the Kuiper belt is similar to the asteroid belt. Both are populations of small bodies containing relatively little mass compared to the rest of the solar system; the structure and dynamics of both populations have been influenced significantly by the giant planets; and both have been and continue to be significantly influenced by collisions. Due to its relative proximity to Earth, however, there are substantially more observational data available for the asteroid belt than the Kuiper belt. While the collisional and dynamical evolution of the asteroid belt is certainly not a solved problem, the abundance of constraints has allowed for the development of reasonably consistent models. Here we briefly describe what is currently understood about the evolution of the asteroid belt, what insights that may give us with regards to the evolution of the Kuiper belt, and what differences might exist in the evolution of the two populations.

It has long been recognized that the primordial asteroid belt must have contained hundreds or thousands of times more mass than the current asteroid belt (e.g., *Lecar and Franklin,* 1973; *Safronov,* 1979; *Weidenschilling,* 1977c; *Wetherill,* 1989). Reconstructing the initial mass distribution of the solar system from the current masses of the planets and asteroids, for example, yields a pronounced mass deficiency in the asteroid belt region relative to an otherwise smooth distribution for the rest of the solar system (*Weidenschilling,* 1977c). To accrete the asteroids on the timescales inferred from meteoritic evidence would require hundreds of times more mass than currently exists in the main belt (*Wetherill,* 1989).

In addition to its pronounced mass depletion, the asteroid belt is also strongly dynamically excited. The mean proper eccentricity and inclination of asteroids larger than ~50 km in diameter are 0.135 and 10.9° (from the catalog of *Knežević and Milani,* 2003), significantly larger than can be explained by gravitational perturbations among the asteroids or by simple gravitational perturbations from the planets (*Duncan,* 1994). The fact that the different taxonomic types of asteroids (S-type, C-type, etc.) are radially mixed somewhat throughout the main belt, rather than confined to delineated zones, indicates that there has been significant scattering in semimajor axis as well (*Gradie and Tedesco,* 1982).

Originally, a collisional origin was suggested for the mass depletion in the asteroid belt (*Chapman and Davis,* 1975). The difficulty of collisionally disrupting the largest asteroids, coupled with the survival of the basaltic crust of the ~500-km-diameter asteroid Vesta, however, suggest that collisional grinding was not the cause of the mass depletion (*Davis et al.,* 1979, 1985, 1989, 1994; *Wetherill,* 1989; *Durda and Dermott,* 1997; *Durda et al.,* 1998; *Bottke et al.,* 2005a,b; *O'Brien and Greenberg,* 2005). In addition, collisional processes alone could not fully explain both the dynamical excitation and the radial mixing observed in the asteroid belt, although *Charnoz et al.* (2001) suggest that collisional diffusion may have contributed to its radial mixing.

Several dynamical mechanisms have been proposed to explain the mass depletion, dynamical excitation, and radial mixing of the asteroid belt. As the solar nebula dissipated, the changing gravitational potential acting on Jupiter, Saturn, and the asteroids would lead to changes in their precession rates and hence changes in the positions of secular resonances, which could "sweep" through the asteroid belt, exciting e and i, and coupled with gas drag, could lead to semimajor axis mobility and the removal of material from the belt (e.g., *Heppenheimer,* 1980; *Ward,* 1981; *Lemaitre and Dubru,* 1991; *Lecar and Franklin,* 1997; *Nagasawa et al.,* 2000, 2001, 2002). It has also been suggested that sweeping secular resonances could lead to orbital excitation in the Kuiper belt (*Nagasawa and Ida,* 2000). However, as reviewed by *Petit et al.* (2002) and *O'Brien et al.* (2007), secular resonance sweeping is generally unable to simultaneously match the observed e and i excitation in the asteroid belt, as well as its radial mixing and mass depletion, for reasonable parameter choices (especially in the context of the Nice model).

Another possibility is that planetary embryos were able to accrete in the asteroid belt (e.g., *Wetherill,* 1992). The fact that Jupiter's ~10-$M_\oplus$ core was able to accrete in our solar system beyond the asteroid belt suggests that embryos were almost certainly able to accrete in the asteroid belt, even accounting for the roughly 3–4× decrease in the mass density of solid material inside the snow line. The scattering of asteroids by those embryos, coupled with the jovian and saturnian resonances in the asteroid belt, has been shown to be able to reasonably reproduce the observed e and i excitation in the belt as well as its radial mixing and mass depletion (*Petit et al.,* 2001, 2002; *O'Brien et al.,* 2007). In the majority of simulations of this scenario by both groups, the embryos are completely cleared from the asteroid belt.

Thus, the observational evidence and theoretical models for the evolution of the asteroid belt strongly suggest that dynamics, rather than collisions, dominated its mass depletion. Collisions, however, have still played a key role in sculpting the asteroid belt. Many dynamical families, clusterings in orbital element space, have been discovered, giving evidence for ~20 breakups of 100-km or larger parent bodies over the history of the solar system (*Bottke et al.,* 2005a,b). The large 500-km-diameter asteroid Vesta has a preserved basaltic crust with a single large impact basin (*McCord et al.,* 1970; *Thomas et al.,* 1997). This basin was formed by the impact of a roughly 40-km projectile (*Marzari et al.,* 1996; *Asphaug,* 1997).

The size distribution of main-belt asteroids is known or reasonably constrained through observational surveys down to ~1 km in diameter (e.g., *Durda and Dermott,* 1997; *Jedicke and Metcalfe,* 1998; *Ivezić et al.,* 2001; *Yoshida et al.,* 2003; *Gladman et al.* 2007). Not surprisingly, the largest uncertainties are at the smallest sizes, where good orbits are often not available for the observed asteroids, which makes the conversion to absolute magnitude and diameter difficult (e.g., *Ivezić et al.,* 2001; *Yoshida et al.,* 2003). Recent results from the Sub-Kilometer Asteroid Diameter Survey (SKADS) (*Gladman et al.,* 2007), the first survey since the Palomar-Leiden Survey designed to determine orbits as well as magnitudes of main-belt asteroids, suggest that the asteroid magnitude-frequency distribution may be well represented by a single power law in the range from H = 14.0 to 18.8, which corresponds to diameters of 0.7 to 7 km for an albedo of 0.11. These observational constraints are shown in Fig. 9 alongside the determination of the TNO size distribution from *Bernstein et al.* (2004).

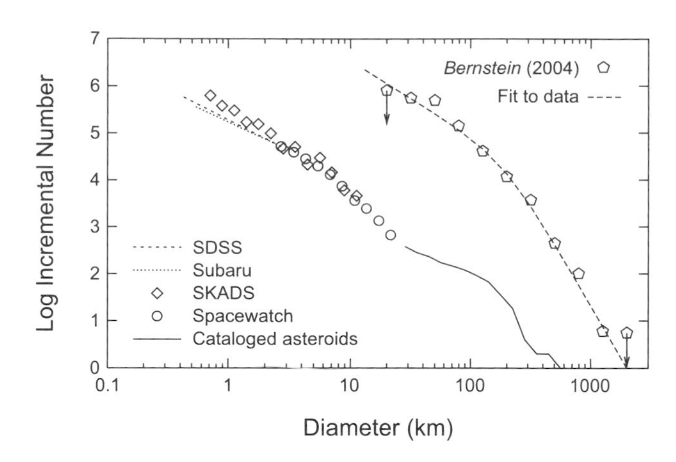

Fig. 9. Observational estimates of the main belt and TNO size distributions. The pentagons (with dashed best-fit curve) show the total TNO population as determined from the *Bernstein et al.* (2004) HST survey, converted to approximate diameters assuming an albedo of 0.04. Points with arrows are upper limits given by nondetections. The solid line is the population of observed asteroids, and open circles are from debiased Spacewatch main-belt observations (*Jedicke and Metcalfe,* 1998). These data, converted to diameters, were provided by D. Durda. The two dashed lines are extrapolations based on the Sloan Digital Sky Survey (*Ivezić et al.,* 2001) and the Subaru Subkilometer Main Belt Asteroid Survey (*Yoshida et al.,* 2003), and diamonds show the debiased population estimate from the SKADS survey (*Gladman et al.,* 2007). Error bars are left out of this plot for clarity. Note that the TNO population is substantially more populous and massive, by roughly a factor of 1000, than the asteroid population.

While over some size ranges, the asteroid size distribution can be fit by a single power law, over the full range of observed asteroid diameters from ~1–1000 km there are multiple bumps or kinks in the size distribution (namely around 10 and 100 km in diameter). The change in slope of the size distribution around 100 km is due primarily to the fact that asteroids larger than this are very difficult to disrupt, and hence the size distribution of bodies larger than 100 km is likely primordial. The change in slope around 10 km has a different origin — such a structure is produced as a result of a change in the strength properties of asteroids, namely the transition from when a body's resistance to disruption is dominated by material strength to when it is dominated by self-gravity. This transition in strength properties occurs at a size much smaller than 10 km, but results in a structure that propagates to larger sizes (see, e.g., *Durda et al.,* 1998; *O'Brien and Greenberg,* 2003). The presence of this structure in the asteroid size distribution is consistent with the asteroids being a collisionally relaxed population, i.e., a population in which the size distribution has reached an approximate steady state where collisional production and collisional destruction of bodies in each size range are in balance.

The collisional evolution of the asteroid belt has been studied by many authors (e.g., *Davis et al.,* 1985, 1994; *Durda,* 1993; *Durda and Dermott,* 1997; *Durda et al.,* 1998; *Campo Bagatin et al.,* 1993, 1994, 2001; *Marzari et al.,* 1999). The most recent models of collisional evolution of the asteroid belt incorporate aspects of dynamical evolution as well, such as the removal of bodies by resonances and the Yarkovsky effect, and the enhancement in collisional activity during its massive primordial phase (*O'Brien and Greenberg,* 2005; *Bottke et al.,* 2005a,b). In particular, *Bottke et al.* (2005a) explicitly incorporate the results of dynamical simulations of the excitation and clearing of the main belt by embedded planetary embryos performed by *Petit et al.* (2001). Such collisional/dynamical models can be constrained by a wide range of observational evidence such as the main-belt size distribution, the number of observed asteroid families, the existence of Vesta's basaltic crust, and the cosmic-ray exposure ages of ordinary chondrite meteorites, which suggest that the lifetimes of meter-scale stony bodies in the asteroid belt are on the order of 10–20 m.y. (*Marti and Graf,* 1992).

One of the most significant implications of having an early massive main belt, which was noted in early collisional models (e.g., *Chapman and Davis,* 1975) and recently emphasized in the case of collisional evolution plus dynamical depletion (e.g., *Bottke et al.,* 2005b), is that the majority of the collisional evolution of the asteroid belt occurred during its early, massive phase, and there has been relatively little change in the main-belt size distribution since then. The current, wavy main-belt size distribution, then, is a "fossil" from its first few hundred million years of collisional and dynamical evolution.

So how does the Kuiper belt compare to the asteroid belt in terms of its collisional and dynamical evolution? Evidence and modeling for the asteroid belt suggest that dynamical depletion, rather than collisional erosion, was primarily responsible for reducing the mass of the primordial asteroid belt to its current level. In the case of the Kuiper belt, this is less clear. As shown in section 3, collisional erosion, especially when aided by stellar perturbations or the formation of Neptune, can be very effective in removing mass. At the same time, dynamical models such as the Nice model result in the depletion of a large amount of mass through purely dynamical means and are able to match many observational constraints. Recent modeling that couples both collisional fragmentation and dynamical effects suggests that collisional erosion cannot play too large a role in removing mass from the Kuiper belt, otherwise the scattered disk and Oort cloud would be too depleted to explain the observed numbers of short- and long-period comets (*Charnoz and Morbidelli,* 2007). That model currently does not include coagulation. Further modeling work, which self-consistently integrates coagulation, collisional fragmentation, and dynamical effects, is necessary to fully constrain the relative contributions of collisional and dynamical depletion in the Kuiper belt.

We have noted that the asteroid belt has a collisionally relaxed size distribution that is not well-represented by a single power law over all size ranges. Should we expect the same for the Kuiper belt size distribution, and is there evidence to support this? The collision rate in the Kuiper belt should be roughly comparable to that in the asteroid belt, with the larger number of KBOs offsetting their lower intrinsic collision probability (*Davis and Farinella,* 1997), and as noted earlier in this chapter, the primordial Kuiper belt, like the asteroid belt, would have been substantially more massive than the current population. This suggests that the Kuiper belt should have experienced a degree of collisional evolution roughly comparable to the asteroid belt, and thus is likely to be collisionally relaxed like the asteroid belt. Observational evidence thus far is not detailed enough to say for sure if this is the case, although recent work (*Kenyon and Bromley,* 2004d; *Pan and Sari,* 2005) suggests that the observational estimate of the TNO size distribution by *Bernstein et al.* (2004), shown in Fig. 9, is consistent with a collisionally relaxed population.

While the Kuiper belt is likely to be collisionally relaxed, it is unlikely to mirror the exact shape of the asteroid belt size distribution. The shape of the size distribution is determined, in part, by the strength law Q_D^*, which is likely to differ somewhat between asteroids and KBOs. This is due to the difference in composition between asteroids, which are primarily rock, and KBOs, which contain a substantial amount of ice, as well as the difference in collision velocity between the two populations. With a mean velocity of ~5 km/s (*Bottke et al.,* 1994), collisions between asteroids are well into the supersonic regime (relative to the sound speed in rock). For the Kuiper belt, collision velocities are about a factor of 5 or more smaller (*Davis and Farinella,* 1997), such that collisions between KBOs are close to the subsonic/supersonic transition. For impacts occurring in these different velocity regimes, and into different materials, Q_D^* may differ significantly (e.g., *Benz and Asphaug,* 1999).

The difference in collision velocity can influence the size distribution in another way as well. With a mean collision velocity of ~5 km/s, a body of a given size in the asteroid belt can collisionally disrupt a significantly larger body. Thus, transitions in the strength properties of asteroids can lead to the formation of waves that propagate to larger sizes and manifest themselves as changes in the slope of the size distribution, as seen in Fig. 9. For the Kuiper belt, with collision velocities that are about a factor of 5 or more smaller than in the asteroid belt, the difference in size between a given body and the largest body it is capable of disrupting is much smaller than in the asteroid belt, and waves should therefore be much less pronounced or nonexistent in the KBO size distribution (e.g., *O'Brien and Greenberg,* 2003). There is still likely to be a change in slope at the largest sizes where the population transitions from being primordial to being collisionally relaxed, and such a change appears in the debiased observational data of *Bernstein et al.* (2004) (shown here in Fig. 9), although recent observations suggest that the change in slope may actually occur at smaller magnitudes than found in that survey (*Petit et al.,* 2006).

Is the size distribution of the Kuiper belt likely to be a "fossil" like the asteroid belt? The primordial Kuiper belt would have been substantially more massive than the current population. Thus, regardless of whether the depletion of its mass was primarily collisional or dynamical, collisional evolution would have been more intense early on and the majority of the collisional evolution would have occurred early in its history. In either case, its current size distribution could then be considered a fossil from that early phase, although defining exactly when that early phase ends and the size distribution becomes "fossilized" is not equally clear in both cases. In the case where the mass depletion of the Kuiper belt occurs entirely through collisions, there would not necessarily be a well-defined point at which one could say that the size distribution became fossilized, as the collision rate would decay continuously with time. In the case of dynamical depletion, where the mass would be removed fairly rapidly as in the case of the Nice model described in section 3.4, the collision rate would experience a correspondingly rapid drop, and the size distribution could be considered essentially fossilized after the dynamical depletion event.

As noted earlier in this section, an important observable manifestation of collisions in the asteroid belt is the formation of families, i.e., groupings of asteroids with similar orbits. Asteroid families are thought to be the fragments of collisionally disrupted parent bodies. These were first recognized by *Hirayama* (1918), who found 3 families among the 790 asteroids known at that time. The number increased to 7 families by 1926 when there were 1025 known asteroids (*Hirayama,* 1927). Today, there are over 350,000 known asteroids while the number of asteroid families has grown to about 30.

Given that the Kuiper belt is likely a collisionally evolved population, are there collisional families to be found among these bodies? Families are expected to be more difficult to recognize in the Kuiper belt than in the asteroid belt. Families are identified by finding statistically significant clusters of asteroid orbit elements — mainly the semimajor axis, eccentricity, and inclination. The collisional disruption of a parent body launches fragments with speeds of perhaps a few hundred meters per second relative to the original target body. This ejection speed is small compared with the orbital speeds of asteroids, hence the orbits of fragments differ by only small amounts from that of the original target body and, more importantly, from each other. Thus, the resulting clusters of fragments are easy to identify.

However, in the Kuiper belt, where ejection velocities are likely to be about the same but orbital speeds are much lower, collisional disruption produces a much greater dispersion in the orbital elements of fragments. This reduces the density of the clustering of orbital elements and makes the task of distinguishing family members from the background population much more difficult (*Davis and Farinella,* 1997). To date, there are over 1000 known KBOs, many of which have poorly determined orbits or are in resonances that would make the identification of a family difficult or impossible. *Chiang et al.* (2003) applied lowest-order secular theory to 227 nonresonant KBOs with well-determined orbits and found no convincing evidence for a dynamical family. Recently, however, *Brown et al.* (2007) found evidence for a single family with at least five members associated with KBO 2003 EL_{61}. This family was identified based on the unique spectroscopic signature of its members, and confirmed by their clustered orbit elements.

Given the small numbers involved, it cannot be said whether or not finding a single KBO family at this stage is statistically that different from the original identification of 3 asteroid families when there were only 790 known asteroids (*Hirayama,* 1918). However, the fact that the KBO family associated with 2003 EL_{61} was first discovered spectroscopically, and its clustering in orbital elements was later confirmed, while nearly all asteroid families were discovered based on clusterings in orbital elements alone, suggests that even if comparable numbers of KBO families and asteroid families do exist, the greater dispersion of KBO families in orbital element space may make them more difficult to identify unless there are spectroscopic signatures connecting them as well.

Perhaps when the number of nonresonant KBOs with good orbits approaches 1000, more populous Kuiper belt families will be identified, and as can be done now with the asteroid belt, these KBO families can be used as constraints on the interior structures of their original parent bodies as well as on the collisional and dynamical history of the Kuiper belt as a whole.

6. CONCLUDING REMARKS

Starting with a swarm of 1-m to 1-km planetesimals at 20–150 AU, the growth of icy planets follows a standard pattern (*Stern and Colwell,* 1997a,b; *Kenyon and Luu,* 1998, 1999a,b; *Kenyon and Bromley,* 2004a,d, 2005). Collisional damping and dynamical friction lead to a short period of runaway growth that produces 10–100 objects with r ~ 300–1000 km. As these objects grow, they stir the orbits of leftover planetesimals up to the disruption velocity. Once disruptions begin, the collisional cascade grinds leftover planetesimals into smaller objects faster than the oligarchs can accrete them. Thus, the oligarchs always contain a small fraction of the initial mass in solid material. For self-stirring models, oligarchs contain ~10% of the initial mass. Stellar flybys and stirring by a nearby gas giant augment the collisional cascade and leave less mass in oligarchs. The two examples in section 3.3 suggest that a very close flyby and stirring by Neptune leave ~2% to 5% of the initial mass in oligarchs with r ~ 100–1000 km.

This evolution differs markedly from planetary growth in the inner solar system. In ~0.1–1 m.y. at a few AU, runaway growth produces massive oligarchs, $m \geq 0.01\ M_\oplus$, that contain most of the initial solid mass in the disk. Aside from a few giant impacts like those that might produce the Moon (*Hartmann and Davis,* 1975; *Cameron and Ward,* 1976), collisions remove little mass from these objects. Although the collisional cascade removes many leftover planetesimals before oligarchs can accrete them, the lost material is much less than half the original solid mass (*Wetherill and Stewart,* 1993; *Kenyon and Bromley,* 2004b). For $a \gtrsim 40$ AU, runaway growth leaves most of the mass in 0.1–10-km objects that are easily disrupted at modest collision velocities. In 4.5 G.y., the collisional cascade removes most of the initial disk mass inside 70–80 AU.

Together with numerical calculations of orbital dynamics (see chapter by Morbidelli et al.), theory now gives us a foundation for understanding the origin and evolution of the Kuiper belt. Within a disk of planetesimals at 20–100 AU, collisional growth naturally produces objects with r ~ 10–2000 km and a size distribution reasonably close to that observed among KBOs. As KBOs form, migration of the giant planets scatters KBOs into several dynamical classes (see chapter by Morbidelli et al.). Once the giant planets achieve their current orbits, the collisional cascade reduces the total mass in KBOs to current levels and produces the break in the size distribution at r ~ 20–40 km. Continued dynamical scattering by the giant planets sculpts the inner Kuiper belt and maintains the scattered disk.

New observations will allow us to test and to refine this theoretical picture. Aside from better measures of α_m, r_{max}, and r_0 among the dynamical classes, better limits on the total mass and the size distribution of large KBOs with a ~ 50–100 AU should yield a clear discriminant among theoretical models. In the Nice model, the Kuiper belt was initially nearly empty outside ~50 AU. Thus, any KBOs found with a ~ 50–100 AU should have the collisional and dynamical signatures of the scattered disk or detached population. If some KBOs formed *in situ* at a $\gtrsim$ 50 AU, their size distribution depends on collisional growth modified by self-stirring and stirring by ~30 $M_\oplus$ of large KBOs formed at 20–30 AU and scattered through the Kuiper belt by the giant planets. From the calculations of Neptune stirring (section 3.3), stirring by scattered disk objects should yield a size distribution markedly different from the size distribution of detached or scattered disk objects formed at 20–30 AU. Wide-angle surveys on 2–3-m-class telescopes (e.g., Pan-Starrs) and deep probes with 8–10-m telescopes can provide this test.

Information on smaller size scales — α_d and r_1 — place additional constraints on the bulk properties (fragmentation parameters) of KBOs and on the collisional cascade. In any of the stirring models, there is a strong correlation between r_0, r_1, and the fragmentation parameters. Thus, direct measures of r_0 and r_1 provide a clear test of KBO formation calculations. At smaller sizes ($r \lesssim 0.1$ km), the slope of the size distribution α_d clearly tests the fragmentation algorithm and the ability of the collisional cascade to remove KBOs with r ~ 1–10 km. Although the recent detection of KBOs with $r \ll 1$ km (*Chang et al.*, 2006) may be an instrumental artifact (*Jones et al.*, 2006; *Chang et al.*, 2007), optical and X-ray occultations (e.g., TAOS) will eventually yield these tests.

Finally, there is a clear need to combine coagulation and dynamical calculations to produce a "unified" picture of planet formation at a $\gtrsim$ 20 AU. *Charnoz and Morbidelli* (2007) provide a good start in this direction. Because the collisional outcome is sensitive to internal *and* external dynamics, understanding the formation of the observed n(r), n(e), and n(i) distributions in each KBO population requires treating collisional evolution and dynamics together. A combined approach should yield the sensitivity of α_m, r_{max}, and r_0 to the local evolution and the timing of the formation of giant planets, Neptune migration, and stellar flybys. These calculations will also test how the dynamical events depend on the evolution during oligarchic growth and the collisional cascade. Coupled with new observations of KBOs and of planets and debris disks in other planetary systems, these calculations should give us a better understanding of the origin and evolution of KBOs and other objects in the outer solar system.

Acknowledgments. We thank S. Charnoz, S. Kortenkamp, A. Morbidelli, and an anonymous reviewer for comments that considerably improved the text. We acknowledge support from the NASA Astrophysics Theory Program (grant NAG5-13278. B.C.B. and S.J.K.), the NASA Planetary Geology and Geophysics Program (grant NNX06AC50G, D.P.O.), and the JPL Institutional Computing and Information Services and the NASA Directorates of Aeronautics Research, Science, Exploration Systems, and Space Operations (B.C.B. and S.J.K.). This paper is PSI Contribution 417.

REFERENCES

Adachi I., Hayashi C., and Nakazawa K. (1976) The gas drag effect on the elliptical motion of a solid body in the primordial solar nebula. *Progr. Theor. Phys., 56,* 1756–1771.

Adams F. C. and Laughlin G. (2001) Constraints on the birth aggregate of the solar system. *Icarus, 150,* 151–162.

Artymowicz P. (1988) Radiation pressure forces on particles in the Beta Pictoris system. *Astrophys. J. Lett., 335,* L79–L82.

Artymowicz P. (1997) Beta Pictoris: An early solar system? *Annu. Rev. Earth Planet. Sci., 25,* 175–219.

Asphaug E. (1997) Impact origin of the Vesta family. *Meteoritics & Planet. Sci., 32,* 965–980.

Asphaug E. and Benz W. (1996) Size, density, and structure of Comet Shoemaker-Levy 9 inferred from the physics of tidal breakup. *Icarus, 121,* 225–248.

Aumann H. H., Beichman C. A., Gillett F. C., de Jong T., Houck J. R., Low F. J., Neugebauer G., Walker R. G., and Wesselius P. R. (1984) Discovery of a shell around Alpha Lyrae. *Astrophys. J. Lett., 278,* L23–L27.

Backman D. E. and Paresce F. (1993) Main-sequence stars with circumstellar solid material — The VEGA phenomenon. In *Protostars and Planets III* (E. H. Levy and J. I. Lunine, eds.), pp. 1253–1304. Univ. of Arizona, Tucson.

Beckwith S. V. W. and Sargent A. I. (1996) Circumstellar disks and the search for neighbouring planetary systems. *Nature, 383,* 139–144.

Benz W. and Asphaug E. (1999) Catastrophic disruptions revisited. *Icarus, 142,* 5–20.

Bernstein G. M., Trilling D. E., Allen R. L., Brown M. E., Holman M., and Malhotra R. (2004) The size distribution of trans-neptunian bodies. *Astron. J., 128,* 1364–1390.

Bottke W. F., Nolan M. C., Greenberg R., and Kolvoord R. A. (1994) Velocity distributions among colliding asteroids. *Icarus, 107,* 255–268.

Bottke W. F., Durda D. D., Nesvorný D., Jedicke R., Morbidelli A., Vokrouhlický D., and Levison H. F. (2005a) Linking the collisional history of the main asteroid belt to its dynamical excitation and depletion. *Icarus, 179,* 63–94.

Bottke W. F., Durda D. D., Nesvorný D., Jedicke R., Morbidelli A., Vokrouhlický D., and Levison H. (2005b) The fossilized size distribution of the main asteroid belt. *Icarus, 175,* 111–140.

Bromley B. C. and Kenyon S. J. (2006) A hybrid N-body-coagulation code for planet formation. *Astron. J., 131,* 2737–2748.

Brown M. E., Barkume K. M., Ragozzine D., and Schaller E. L. (2007) A collisional family of icy objects in the Kuiper belt. *Nature, 446,* 294–297.

Burns J. A., Lamy P. L., and Soter S. (1979) Radiation forces on small particles in the solar system. *Icarus, 40,* 1–48.

Cameron A. G. W. and Ward W. R. (1976) The origin of the Moon (abstract). In *Lunar Science VII,* pp. 120–121. The Lunar Science Institute, Houston.

Campo Bagatin A., Farinella P., and Paolicchi P. (1993) Collisional evolution of the asteroid size distribution: A numerical simulation. *Cel. Mech. Dyn. Astron., 57,* 403–404.

Campo Bagatin A., Cellino A., Davis D. R., Farinella P., and Paolicchi P. (1994) Wavy size distributions for collisional systems with a small-size cutoff. *Planet. Space Sci., 42,* 1079–1092.

Campo Bagatin A., Petit J.-M., and Farinella P. (2001) How many rubble piles are in the asteroid belt? *Icarus, 149,* 198–209.

Chang H.-K., King S.-K., Liang J.-S., Wu P.-S., Lin L. C.-C., and Chiu J.-L. (2006) Occultation of X-rays from Scorpius X-1 by small trans-neptunian objects. *Nature, 442,* 660–663.

Chang H.-K., Liang J.-S., Liu C.-Y., and King S.-K. (2007) Millisecond dips in the RXTE/PCA light curve of Sco X-1 and TNO occultation. *Mon. Not. R. Astron. Soc., 378,* 1287.

Chapman C. R. and Davis D. R. (1975) Asteroid collisional evolution — Evidence for a much larger early population. *Science, 190,* 553–556.

Charnoz S. and Morbidelli A. (2007) Coupling dynamical and collisional evolution of small bodies II: Forming the Kuiper belt, the scattered disk and the Oort cloud. *Icarus, 188,* 468–480.

Charnoz S., Thébault P., and Brahic A. (2001) Short-term collisional evolution of a disc perturbed by a giant-planet embryo. *Astron. Astrophys., 373,* 683–701.

Chiang E. I., Lovering J. R., Millis R. L., Buie M. W., Wasserman L. H., and Meech K. J. (2003) Resonant and secular families of the Kuiper belt. *Earth Moon Planets, 92,* 49–62.

Chiang E., Lithwick Y., Murray-Clay R., Buie M., Grundy W., and Holman M. (2007) A brief history of transneptunian space. In *Protostars and Planets V* (B. Reipurth et al., eds.), pp. 895–911. Univ. of Arizona, Tucson.

Davis D. R. and Farinella P. (1997) Collisional evolution of Edgeworth-Kuiper belt objects. *Icarus, 125,* 50–60.

Davis D. R., Chapman C. R., Greenberg R., Weidenschilling S. J., and Harris A. W. (1979) Collisional evolution of asteroids — Populations, rotations, and velocities. In *Asteroids* (T. Gehrels, ed.), pp. 528–557. Univ. of Arizon, Tucson.

Davis D. R., Chapman C. R., Weidenschilling S. J., and Greenberg R. (1985) Collisional history of asteroids: Evidence from Vesta and the Hirayama families. *Icarus, 63,* 30–53.

Davis D. R., Weidenschilling S. J., Farinella P., Paolicchi P., and Binzel R. P. (1989) Asteroid collisional history — Effects on sizes and spins. In *Asteroids II* (R. P. Binzel et al., eds.), pp. 805–826. Univ. of Arizona, Tucson.

Davis D. R., Ryan E. V., and Farinella P. (1994) Asteroid collisional evolution: Results from current scaling algorithms. *Planet. Space Sci., 42,* 599–610.

Davis D. R., Farinella P., and Weidenschilling S. J. (1999) Accretion of a massive Edgeworth-Kuiper belt (abstract). In *Lunar and Planetary Science XXX,* pp. 1883–1884. Lunar and Planetary Institute, Houston.

de la Fuente Marcos C. and Barge P. (2001) The effect of long-lived vortical circulation on the dynamics of dust particles in the mid-plane of a protoplanetary disc. *Mon. Not. R. Astron. Soc., 323,* 601–614.

Dohnanyi J.W. (1969) Collisional models of asteroids and their debris. *J. Geophys. Res., 74,* 2531–2554.

Dullemond C. P. and Dominik C. (2005) Dust coagulation in protoplanetary disks: A rapid depletion of small grains. *Astron. Astrophys., 434,* 971–986.

Duncan M. (1994) Orbital stability and the structure of the solar system. In *Circumstellar Dust Disks and Planet Formation* (R. Ferlet and A. Vidal-Madjar, eds.), pp. 245–255. Editions Frontieres, Gif-sûr-Yvette.

Duncan M. J. and Levison H. F. (1997) A scattered comet disk and the origin of Jupiter family comets. *Science, 276,* 1670–1672.

Duncan M. J., Levison H. F., and Budd S. M. (1995) The dynamical structure of the Kuiper belt. *Astron. J., 110,* 3073–3081.

Durda D. D. (1993) The collisional evolution of the asteroid belt and its contribution to the zodiacal cloud. Ph.D. Thesis, Univ. of Florida.

Durda D. D. and Dermott S. F. (1997) The collisional evolution of the asteroid belt and its contribution to the zodiacal cloud. *Icarus, 130,* 140–164.

Durda D. D., Greenberg R., and Jedicke R. (1998) Collisional models and scaling laws: A new interpretation of the shape of the main-belt asteroid size distribution. *Icarus, 135,* 431–440.

Elliot J. L., Kern S. D., Clancy K. B., Gulbis A. A. S., Millis R. L., Buie M. W., Wasserman L. H., Chiang E. I., Jordan A. B., Trilling D. E., and Meech K. J. (2005) The Deep Ecliptic Survey: A search for Kuiper belt objects and Centaurs. II. Dynamical classification, the Kuiper belt plane, and the core population. *Astron. J., 129,* 1117–1162.

Giblin I., Davis D. R., and Ryan E. V. (2004) On the collisional disruption of porous icy targets simulating Kuiper belt objects. *Icarus, 171,* 487–505.

Gladman B., Kavelaars J. J., Petit J., Morbidelli A., Holman M. J., and Loredo T. (2001) The structure of the Kuiper belt: Size distribution and radial extent. *Astron. J., 122,* 1051–1066.

Gladman B. J., Davis D. R., Neese N., Williams G., Jedicke R., Kavelaars J. J., Petit J.-M., Scholl H., Holman M., Warrington B., Esquerdo G., and Tricarico P. (2007) SKADS: A Sub-Kilometer Asteroid Diameter Survey. *Icarus,* in press.

Goldreich P. and Ward W. R. (1973) The formation of planetesimals. *Astrophys. J., 183,* 1051–1062.

Goldreich P., Lithwick Y., and Sari R. (2004) Planet formation by coagulation: A focus on Uranus and Neptune. *Annu. Rev. Astron. Astrophys., 42,* 549–601.

Gomes R., Levison H. F., Tsiganis K., and Morbidelli A. (2005) Origin of the cataclysmic late heavy bombardment period of the terrestrial planets. *Nature, 435,* 466–469.

Gomez M., Hartmann L., Kenyon S. J., and Hewett R. (1993) On the spatial distribution of pre-main-sequence stars in Taurus. *Astron. J., 105,* 1927–1937.

Gradie J. and Tedesco E. (1982) Compositional structure of the asteroid belt. *Science, 216,* 1405–1407.

Greaves J. S. (2005) Disks around stars and the growth of planetary systems. *Science, 307,* 68–71.

Greaves J. S., Holland W. S., Moriarty-Schieven G., Jenness T., Dent W. R. F., Zuckerman B., McCarthy C., Webb R. A., Butner H. M., Gear W. K., and Walker H. J. (1998) A dust ring around epsilon Eridani: Analog to the young solar system. *Astrophys. J. Lett., 506,* L133–L137.

Greaves J. S., Wyatt M. C., Holland W. S., and Dent W. R. F. (2004) The debris disc around τ Ceti: A massive analogue to the Kuiper belt. *Mon. Not. R. Astron. Soc., 351,* L54–L58.

Greenberg R., Hartmann W. K., Chapman C. R., and Wacker J. F. (1978) Planetesimals to planets — Numerical simulation of collisional evolution. *Icarus, 35,* 1–26.

Greenberg R., Weidenschilling S. J., Chapman C. R., and Davis D. R. (1984) From icy planetesimals to outer planets and comets. *Icarus, 59,* 87–113.

Greenzweig Y. and Lissauer J. J. (1990) Accretion rates of protoplanets. *Icarus, 87,* 40–77.

Haisch K. E. Jr., Lada E. A., and Lada C. J. (2001) Disk frequencies and lifetimes in young clusters. *Astrophys. J. Lett., 553,* L153–L156.

Hartmann W. K. and Davis D. R. (1975) Satellite-sized planetesimals and lunar origin. *Icarus, 24,* 504–514.

Hayashi C. (1981) Structure of the solar nebula, growth and decay of magnetic fields and effects of magnetic and turbulent viscosities on the nebula. *Progr. Theor. Phys. Suppl., 70,* 35–53.

Heppenheimer T. A. (1980) Secular resonances and the origin of eccentricities of Mars and the asteroids. *Icarus, 41,* 76–88.

Hirayama K. (1918) Groups of asteroids probably of common origin. *Astron. J., 31,* 185–188.

Hirayama K. (1927) Families of asteroids: Second paper. *Ann. Observ. Astron. Tokyo, 19,* 1–26.

Holman M. J. and Wisdom J. (1993) Dynamical stability in the outer solar system and the delivery of short period comets. *Astron. J., 105,* 1987–1999.

Holsapple K. A. (1994) Catastrophic disruptions and cratering of solar system bodies: A review and new results. *Planet. Space Sci., 42,* 1067–1078.

Hornung P., Pellat R., and Barge P. (1985) Thermal velocity equilibrium in the protoplanetary cloud. *Icarus, 64,* 295–307.

Housen K. R. and Holsapple K. A. (1990) On the fragmentation of asteroids and planetary satellites. *Icarus, 84,* 226–253.

Housen K. R. and Holsapple K. A. (1999) Scale effects in strength-dominated collisions of rocky asteroids. *Icarus, 142,* 21–33.

Ida S., Larwood J., and Burkert A. (2000) Evidence for early stellar encounters in the orbital distribution of Edgeworth-Kuiper belt objects. *Astrophys. J., 528,* 351–356.

Inaba S. and Barge P. (2006) Dusty vortices in protoplanetary disks. *Astrophys. J., 649,* 415–427.

Ivezić Ž., Tabachnik S., Rafikov R., Lupton R. H., Quinn T., Hammergren M., Eyer L., Chu J., Armstrong J. C., Fan X., Finlator K., Geballe T. R., Gunn J. E., Hennessy G. S., Knapp G. R., Leggett S. K., Munn J. A., Pier J. R., Rockosi C. M., Schneider D. P., Strauss M. A., Yanny B., Brinkmann J., Csabai I., Hindsley R. B., Kent S., Lamb D. Q., Margon B., McKay T. A., Smith J. A., Waddel P., York D. G., and the SDSS Collaboration (2001) Solar system objects observed in the Sloan Digital Sky Survey commissioning data. *Astron. J., 122,* 2749–2784.

Jedicke R. and Metcalfe T. S. (1998) The orbital and absolute magnitude distributions of main belt asteroids. *Icarus, 131,* 245–260.

Jones T. A., Levine A. M., Morgan E. H., and Rappaport S. (2006) Millisecond dips in Sco X-1 are likely the result of high-energy particle events. *The Astronomer's Telegram, 949.*

Kenyon S. J. and Bromley B. C. (2002a) Collisional cascades in planetesimal disks. I. Stellar flybys. *Astron. J., 123,* 1757–1775.

Kenyon S. J. and Bromley B. C. (2002b) Dusty rings: Signposts of recent planet formation. *Astrophys. J. Lett., 577,* L35–L38.

Kenyon S. J. and Bromley B. C. (2004a) Collisional cascades in planetesimal disks. II. Embedded planets. *Astron. J., 127,* 513–530.

Kenyon S. J. and Bromley B. C. (2004b) Detecting the dusty debris of terrestrial planet formation. *Astrophys. J. Lett., 602,* L133–L136.

Kenyon S. J. and Bromley B. C. (2004c) Stellar encounters as the origin of distant solar system objects in highly eccentric orbits. *Nature, 432,* 598–602.

Kenyon S. J. and Bromley B. C. (2004d) The size distribution of Kuiper belt objects. *Astron. J., 128,* 1916–1926.

Kenyon S. J. and Bromley B. C. (2005) Prospects for detection of catastrophic collisions in debris disks. *Astron. J., 130,* 269–279.

Kenyon S. J. and Bromley B. C. (2006) Terrestrial planet formation. I. The transition from oligarchic growth to chaotic growth. *Astron. J., 131,* 1837–1850.

Kenyon S. J. and Hartmann L. (1987) Spectral energy distributions of T Tauri stars — Disk flaring and limits on accretion. *Astrophys. J., 323,* 714–733.

Kenyon S. J. and Luu J. X. (1998) Accretion in the early Kuiper belt. I. Coagulation and velocity evolution. *Astron. J., 115,* 2136–2160.

Kenyon S. J. and Luu J. X. (1999a) Accretion in the early Kuiper belt. II. Fragmentation. *Astron. J., 118,* 1101–1119.

Kenyon S. J. and Luu J. X. (1999b) Accretion in the early outer solar system. *Astrophys. J., 526,* 465–470.

Knežević Z. and Milani A. (2003) Proper element catalogs and asteroid families. *Astron. Astrophys., 403,* 1165–1173.

Kobayashi H., Ida S., and Tanaka H. (2005) The evidence of an early stellar encounter in Edgeworth Kuiper belt. *Icarus, 177,* 246–255.

Kokubo E. and Ida S. (1995) Orbital evolution of protoplanets embedded in a swarm of planetesimals. *Icarus, 114,* 247–257.

Kokubo E. and Ida S. (1998) Oligarchic growth of protoplanets. *Icarus, 131,* 171–178.

Kokubo E. and Ida S. (2000) Formation of protoplanets from planetesimals in the solar nebula. *Icarus, 143,* 15–27.

Kokubo E. and Ida S. (2002) Formation of protoplanet systems and diversity of planetary systems. *Astrophys. J., 581,* 666–680.

Krivov A. V., Löhne T., and Sremčević M. (2006) Dust distributions in debris disks: Effects of gravity, radiation pressure and collisions. *Astron. Astrophys., 455,* 509–519.

Kuchner M. J., Brown M. E., and Holman M. (2002) Long-term dynamics and the orbital inclinations of the classical Kuiper belt objects. *Astron. J., 124,* 1221–1230.

Lada C. J. (2006) Stellar multiplicity and the initial mass function: Most stars are single. *Astrophys. J. Lett., 640,* L63–L66.

Lada C. J. and Lada E. A. (2003) Embedded clusters in molecular clouds. *Annu. Rev. Astron. Astrophys., 41,* 57–115.

Landgraf M., Liou J.-C., Zook H. A., and Grün E. (2002) Origins of solar system dust beyond Jupiter. *Astron. J., 123,* 2857–2861.

Lecar M. and Franklin F. A. (1973) On the original distribution of the asteroids I. *Icarus, 20,* 422–436.

Lecar M. and Franklin F. A. (1997) The solar nebula, secular resonances, gas drag, and the asteroid belt. *Icarus, 129,* 134–146.

Lee M. H. (2000) On the validity of the coagulation equation and the nature of runaway growth. *Icarus, 143,* 74–86.

Leinhardt Z. M. and Richardson D. C. (2002) N-body simulations of planetesimal evolution: Effect of varying impactor mass ratio. *Icarus, 159,* 306–313.

Leinhardt Z. M. and Richardson D. C. (2005) Planetesimals to protoplanets. I. Effect of fragmentation on terrestrial planet formation. *Astrophys. J., 625,* 427–440.

Lemaitre A. and Dubru P. (1991) Secular resonances in the primitive solar nebula. *Cel. Mech. Dyn. Astron., 52,* 57–78.

Levison H. F. and Duncan M. J. (1990) A search for proto-comets in the outer regions of the solar system. *Astron. J., 100,* 1669–1675.

Levison H. F. and Morbidelli A. (2007) Models of the collisional damping scenario for ice giant planets and Kuiper belt formation. *Icarus, 189,* 196–212.

Levison H. F. and Stern S. A. (2001) On the size dependence of the inclination distribution of the main Kuiper belt. *Astron. J., 121,* 1730–1735.

Levison H. F., Morbidelli A., and Dones L. (2004) Sculpting the Kuiper belt by a stellar encounter: Constraints from the Oort cloud and scattered disk. *Astron. J., 128,* 2553–2563.

Lissauer J. J. (1987) Timescales for planetary accretion and the structure of the protoplanetary disk. *Icarus, 69,* 249–265.

Luhman K. L. (2006) The spatial distribution of brown dwarfs in Taurus. *Astrophys. J., 645,* 676–687.

Luu J. X. and Jewitt D. C. (2002) Kuiper belt objects: Relics from the accretion disk of the Sun. *Annu. Rev. Astron. Astrophys., 40,* 63–101.

Malhotra R. (1995) The origin of Pluto's orbit: Implications for the solar system beyond Neptune. *Astron. J., 110,* 420–429.

Malhotra R. (1996) The phase space structure near Neptune resonances in the Kuiper belt. *Astron. J., 111,* 504–516.

Malyshkin L. and Goodman J. (2001) The timescale of runaway stochastic coagulation. *Icarus, 150,* 314–322.

Marti K. and Graf T. (1992) Cosmic-ray exposure history of ordinary chondrites. *Annu. Rev. Earth Planet. Sci., 20,* 221–243.

Marzari F., Cellino A., Davis D. R., Farinella P., Zappala V., and Vanzani V. (1996) Origin and evolution of the Vesta asteroid family. *Astron. Astrophys., 316,* 248–262.

Marzari F., Farinella P., and Davis D. R. (1999) Origin, aging, and death of asteroid families. *Icarus, 142,* 63–77.

McCord T. B., Adams J. B., and Johnson T. V. (1970) Asteroid Vesta: Spectral reflectivity and compositional implications. *Science, 178,* 745–747.

Michel P., Benz W., Tanga P., and Richardson D. C. (2001) Collisions and gravitational reaccumulation: Forming asteroid families and satellites. *Science, 294,* 1696–1700.

Morbidelli A. and Levison H. F. (2004) Scenarios for the origin of the orbits of the trans-neptunian objects 2000 CR_{105} and 2003 VB_{12} (Sedna). *Astron. J., 128,* 2564–2576.

Morbidelli A., Jacob C., and Petit J.-M. (2002) Planetary embryos never formed in the Kuiper belt. *Icarus, 157,* 241–248.

Morbidelli A., Emel'yanenko V. V., and Levison H. F. (2004) Origin and orbital distribution of the trans-neptunian scattered disc. *Mon. Not. R. Astron. Soc., 355,* 935–940.

Morbidelli A., Levison H. F., Tsiganis K., and Gomes R. (2005) Chaotic capture of Jupiter's Trojan asteroids in the early solar system. *Nature, 435,* 462–465.

Nagasawa M. and Ida S. (2000) Sweeping secular resonances in the Kuiper belt caused by depletion of the solar nebula. *Astron. J., 120,* 3311–3322.

Nagasawa M., Tanaka H., and Ida S. (2000) Orbital evolution of asteroids during depletion of the solar nebula. *Astron. J., 119,* 1480–1497.

Nagasawa M., Ida S., and Tanaka H. (2001) Origin of high orbital eccentricity and inclination of asteroids. *Earth Planets Space, 53,* 1085–1091.

Nagasawa M., Ida S., and Tanaka H. (2002) Excitation of orbital inclinations of asteroids during depletion of a protoplanetary disk: Dependence on the disk configuration. *Icarus, 159,* 322–327.

Nagasawa M., Lin D. N. C., and Thommes E. (2005) Dynamical shake-up of planetary systems. I. Embryo trapping and induced collisions by the sweeping secular resonance and embryo-disk tidal interaction. *Astrophys. J., 635,* 578–598.

Natta A., Grinin V., and Mannings V. (2000) Properties and evolution of disks around pre-main-sequence stars of intermediate mass. In *Protostars and Planets IV* (V. Mannings et al., eds.), pp. 559–588. Univ. of Arizona, Tucson.

Null G. W., Owen W. M., and Synnott S. P. (1993) Masses and densities of Pluto and Charon. *Astron. J., 105,* 2319–2335.

O'Brien D. P. and Greenberg R. (2003) Steady-state size distributions for collisional populations: Analytical solution with size-dependent strength. *Icarus, 164,* 334–345.

O'Brien D. P. and Greenberg R. (2005) The collisional and dynamical evolution of the main-belt and NEA size distributions. *Icarus, 178,* 179–212.

O'Brien D. P., Morbidelli A., and Bottke W. F. (2005) Collisional evolution of the primordial trans-neptunian disk: Implications for planetary migration and the current size distribution of TNOs. *Bull. Am. Astron. Soc., 37,* 676.

O'Brien D. P., Morbidelli A., and Bottke W. F. (2007) Re-evaluating the primordial excitation and clearing of the asteroid belt. *Icarus, 191,* in press.

Ohtsuki K. (1992) Evolution of random velocities of planetesimals in the course of accretion. *Icarus, 98,* 20–27.

Ohtsuki K., Nakagawa Y., and Nakazawa K. (1990) Artificial acceleration in accumulation due to coarse mass-coordinate divisions in numerical simulation. *Icarus, 83,* 205–215.

Ohtsuki K., Stewart G. R., and Ida S. (2002) Evolution of planetesimal velocities based on three-body orbital integrations and growth of protoplanets. *Icarus, 155,* 436–453.

Pan M. and Sari R. (2005) Shaping the Kuiper belt size distribution by shattering large but strengthless bodies. *Icarus, 173,* 342–348.

Petit J., Morbidelli A., and Chambers J. (2001) The primordial excitation and clearing of the asteroid belt. *Icarus, 153,* 338–347.

Petit J., Chambers J., Franklin F., and Nagasawa M. (2002) Primordial excitation and depletion of the main belt. In *Asteroids III* (W. F. Bottke Jr. et al., eds.), pp. 711–738. Univ. of Arizona, Tucson.

Petit J.-M., Holman M. J., Gladman B. J., Kavelaars J. J., Scholl H., and Loredo T. J. (2006) The Kuiper Belt luminosity function from m_R = 22 to 25. *Mon. Not. R. Astron. Soc., 365,* 429–438.

Rafikov R. R. (2003a) Dynamical evolution of planetesimals in protoplanetary disks. *Astron. J., 126,* 2529–2548.

Rafikov R. R. (2003b) Planetesimal disk evolution driven by embryo-planetesimal gravitational scattering. *Astron. J., 125,* 922–941.

Rafikov R. R. (2003c) Planetesimal disk evolution driven by planetesimal-planetesimal gravitational scattering. *Astron. J., 125,* 906–921.

Rafikov R. R. (2003d) The growth of planetary embryos: Orderly, runaway, or oligarchic? *Astron. J., 125,* 942–961.

Ryan E. V., Davis D. R., and Giblin I. (1999) A laboratory impact study of simulated Edgeworth-Kuiper belt objects. *Icarus, 142,* 56–62.

Safronov V. S. (1969) *Evoliutsiia Doplanetnogo Oblaka (Evolution of the Protoplanetary Cloud and Formation of the Earth and Planets),* Nauka, Moscow. (Translated in 1972, NASA TT F-677.)

Safronov V. S. (1979) On the origin of asteroids. In *Asteroids* (T. Gehrels, ed.), pp. 975–991. Univ. of Arizona, Tucson.

Scholz A., Jayawardhana R., and Wood K. (2006) Exploring brown dwarf disks: A 1.3 mm survey in Taurus. *Astrophys. J., 645,* 1498–1508.

Slesnick C. L., Hillenbrand L. A., and Carpenter J. M. (2004) The spectroscopically determined substellar mass function of the Orion nebula cluster. *Astrophys. J., 610,* 1045–1063.

Spaute D., Weidenschilling S. J., Davis D. R., and Marzari F. (1991) Accretional evolution of a planetesimal swarm. I — A new simulation. *Icarus, 92,* 147–164.

Stern S. A. (1995) Collisional time scales in the Kuiper disk and their implications. *Astron. J., 110,* 856–868.

Stern S. A. (2005) Regarding the accretion of 2003 VB_{12} (Sedna) and like bodies in distant heliocentric orbits. *Astron. J., 129,* 526–529.

Stern S. A. and Colwell J. E. (1997a) Accretion in the Edgeworth-Kuiper belt: Forming 100–1000 km radius bodies at 30 AU and beyond. *Astron. J., 114,* 841–849.

Stern S. A. and Colwell J. E. (1997b) Collisional erosion in the primordial Edgeworth-Kuiper belt and the generation of the 30–50 AU Kuiper gap. *Astrophys. J., 490,* 879–882.

Takeuchi T. and Artymowicz P. (2001) Dust migration and morphology in optically thin circumstellar gas disks. *Astrophys. J., 557,* 990–1006.

Tanaka H. and Ida S. (1999) Growth of a migrating protoplanet. *Icarus, 139,* 350–366.

Tanga P., Weidenschilling S. J., Michel P., and Richardson D. C. (2004) Gravitational instability and clustering in a disk of planetesimals. *Astron. Astrophys., 427,* 1105–1115.

Thomas P. C., Binzel R. P., Gaffey M. J., Storrs A. D., Wells E. N., and Zellner B. H. (1997) Impact excavation on asteroid 4 Vesta: Hubble Space Telescope results. *Science, 277,* 1492–1495.

Thommes E. W., Duncan M. J., and Levison H. F. (1999) The formation of Uranus and Neptune in the Jupiter-Saturn region of the solar system. *Nature, 402,* 635–638.

Thommes E. W., Duncan M. J., and Levison H. F. (2002) The formation of Uranus and Neptune among Jupiter and Saturn. *Astron. J., 123,* 2862–2883.

Tsiganis K., Gomes R., Morbidelli A., and Levison H. F. (2005) Origin of the orbital architecture of the giant planets of the solar system. *Nature, 435,* 459–461.

Ward W. R. (1981) Solar nebula dispersal and the stability of the planetary system. I — Scanning secular resonance theory. *Icarus, 47,* 234–264.

Weidenschilling S. J. (1977a) Aerodynamics of solid bodies in the solar nebula. *Mon. Not. R. Astron. Soc., 180,* 57–70.

Weidenschilling S. J. (1977b) The distribution of mass in the planetary system and solar nebula. *Astrophys. Space Sci., 51,* 153–158.

Weidenschilling S. J. (1980) Dust to planetesimals — Settling and coagulation in the solar nebula. *Icarus, 44,* 172–189.

Weidenschilling S. J. (1984) Evolution of grains in a turbulent solar nebula. *Icarus, 60,* 553–567.

Weidenschilling S. J. (1989) Stirring of a planetesimal swarm — The role of distant encounters. *Icarus, 80,* 179–188.

Weidenschilling S. J. (1995) Can gravitation instability form planetesimals? *Icarus, 116,* 433–435.

Weidenschilling S. J. (2003) Radial drift of particles in the solar nebula: Implications for planetesimal formation. *Icarus, 165,* 438–442.

Weidenschilling S. J. (2006) Models of particle layers in the midplane of the solar nebula. *Icarus, 181,* 572–586.

Weidenschilling S. J., Spaute D., Davis D. R., Marzari F., and Ohtsuki K. (1997) Accretional evolution of a planetesimal swarm. *Icarus, 128,* 429–455.

Wetherill G. W. (1989) Origin of the asteroid belt. In *Asteroids II* (R. P. Binzel et al., eds.), pp. 661–680. Univ. of Arizona, Tucson.

Wetherill G. W. (1990) Comparison of analytical and physical modeling of planetesimal accumulation. *Icarus, 88,* 336–354.

Wetherill G. W. (1992) An alternative model for the formation of the asteroids. *Icarus, 100,* 307–325.

Wetherill G. W. and Stewart G. R. (1989) Accumulation of a swarm of small planetesimals. *Icarus, 77,* 330–357.

Wetherill G. W. and Stewart G. R. (1993) Formation of planetary embryos — Effects of fragmentation, low relative velocity, and independent variation of eccentricity and inclination. *Icarus, 106,* 190–209.

Williams D. R. and Wetherill G. W. (1994) Size distribution of collisionally evolved asteroidal populations — Analytical solution for self-similar collision cascades. *Icarus, 107,* 117–128.

Williams J. P., Najita J., Liu M. C., Bottinelli S., Carpenter J. M., Hillenbrand L. A., Meyer M. R., and Soderblom D. R. (2004) Detection of cool dust around the G2 V star HD 107146. *Astrophys. J., 604,* 414–419.

Wurm G. and Krauss O. (2006) Concentration and sorting of chondrules and CAIs in the late solar nebula. *Icarus, 180,* 487–495.

Wurm G., Paraskov G., and Krauss O. (2004) On the importance of gas flow through porous bodies for the formation of planetesimals. *Astrophys. J., 606,* 983–987.

Wyatt M. C., Greaves J. S., Dent W. R. F., and Coulson I. M. (2005) Submillimeter images of a dusty Kuiper belt around η Corvi. *Astrophys. J., 620,* 492–500.

Yoshida F., Nakamura T., Watanabe J., Kinoshita D., Yamamoto N., and Fuse T. (2003) Size and spatial distributions of sub-km main-belt asteroids. *Publ. Astron. Soc. Japan, 55,* 701–715.

Youdin A. N. and Shu F. H. (2002) Planetesimal formation by gravitational instability. *Astrophys. J., 580,* 494–505.

The Role of the Galaxy in the Dynamical Evolution of Transneptunian Objects

Martin J. Duncan
Queen's University at Kingston, Canada

Ramon Brasser
Queen's University at Kingston and
Canadian Institute for Theoretical Astrophysics

Luke Dones and Harold F. Levison
Southwest Research Institute

Our understanding of the past and present dynamical imprint of the galaxy on the outer reaches of our solar system has evolved considerably since the pioneering work of *Oort* (1950). In particular, the recent discoveries of objects with perihelia well beyond Neptune (*Gladman et al.,* 2002; *Brown et al.,* 2004) may be our first *in situ* glimpse of members of the hitherto hypothetical inner Oort cloud. This chapter reviews our current understanding of the formation and evolution of the Oort cloud and summarizes aspects of the problem needing to be explored in the near future.

1. INTRODUCTION

Much of the current interest in the transneptunian region arises from the possibility that the cometary reservoirs therein have preserved some record of the formational processes that led to the origin of our solar system. In this chapter we begin by describing what are thought to be the key ingredients in understanding the interplay between galactic and solar system perturbations on minor bodies in the outer solar system that may have led to the formation and dynamical evolution of the Oort cloud (henceforth referred to as the OC).We then focus on the dramatic observational discoveries that have prompted a reexamination of our models and describe recent efforts to include some of the processes that were important during the solar system's formative years.

The chapter is organized as follows. In section 2 we briefly review the history of key concepts associated with the OC. Section 3 discusses the major perturbations on cometary orbits and summarizes the key timescales associated with each. Section 4 discusses how one can estimate the mass of the outer cloud and section 5 discusses clues about the structure interior to the outer OC provided by observations of Halley-type comets. In section 6, we present the results of an idealized model of OC formation called the "reference model," in which the outer planets are assumed to have their current masses and orbits, the protoplanetary gas nebula has dissipated and the Sun's galactic environment is assumed to be its current one. Section 7 describes the results of recent simulations that model OC formation in cases in which the Sun formed in an embedded star cluster, and in some cases including the effects of the primordial solar nebula. Finally, Section 8 discusses future avenues of exploration, both observational and theoretical, that may help us further understand the formation and evolution of the the cometary reservoirs in the transneptunian region.

2. CONCEPTS OF THE OORT CLOUD

Before proceeding to a detailed review of current models of OC formation and evolution, we will briefly review key concepts concerning its structure. Details of the orbital characteristics of long-period comets (those with periods longer than 200 years, hereafter called LPCs) and of the early history of this field are summarized in several extensive reviews (e.g., *Bailey et al.,* 1990; *Fernández and Ip,* 1991; *Weissman,* 1990, 1996, 1999; *Festou et al.,* 1993a,b; *Wiegert and Tremaine,* 1999; *Dones et al.,* 2004 — hereafter *DWLD* — and other papers in the *Comets II* book). We do not describe studies of the formation of individual transneptunian objects (TNOs); this topic is discussed by, e.g., *Greenberg et al.* (1984) and *Weidenschilling* (1997, 2004), and reviewed in the chapter by Kenyon et al.

In his historic paper, *Oort* (1950) proposed that the Sun was surrounded by a spherical cloud of comets with semimajor axes of tens of thousands of AU. In Oort's model, planetary perturbations (primarily by Jupiter) acting on small bodies placed the comets onto large, highly eccentric orbits, after which perturbations by passing stars raised the comets' perihelia from the planetary region. Oort showed that comets in the cloud are so far from the Sun that perturbations from random passing stars can change their orbital angular momenta significantly and occasionally send some comets back into the planetary system as potential

LPCs. [Although *Öpik* (1932) first showed that stellar perturbations could raise the perihelia of minor bodies being scattered by the giant planets, he specifically rejected the idea that comets in the cloud could ever be observed, even indirectly, because he did not recognize that stellar perturbations would also cause some orbits to diffuse back into the planetary region.] The relative importance of the four giant planets in populating the OC was later studied by many authors (e.g., *Safronov,* 1972; *Duncan et al.,* 1987, henceforth *DQT87*).

Oort already recognized that the energy distribution of observed LPCs did not agree with his model. Specifically, there were too few "returning" comets with semimajor axes $a \lesssim 10^4$ AU relative to the number of "new" comets with $a \gtrsim 10^4$ AU. Oort therefore had to invoke an empirical "fading" law to hide many of the comets that had recently passed through the planetary region. Cometary fading has been modeled by *Whipple* (1962), *Weissman* (1980), *Bailey* (1984), *Wiegert and Tremaine* (1999), and *Levison et al.* (2001, 2002), while *Jewitt* (2004) reviewed physical loss mechanisms for comets. No one has found a dynamical mechanism that resolves the fading problem (*Wiegert and Tremaine,* 1999). The only known way to make models agree with the observed orbital distribution is to allow the comets to become extinct, i.e., they must stop their production of gas and dust. A comet can become extinct if a mantle forms on its surface, thereby shutting off outgassing (*Brin and Mendis,* 1979), or if the nucleus breaks into many small pieces (*Levison et al.,* 2002). Catastrophic disruption was observed, for example, for Comet C/1999 S4 (LINEAR) (*Weaver et al.,* 2001, and other papers in the May 18, 2001, issue of *Science*). However, *Neslusan* (2007) has recently argued that an underestimation of cometary fading in previous studies has led to an overestimation of the flux of new comets by as much as a factor of 10. This, in turn, may help to alleviate the problem of what has previously been thought to be the prediction of many more old "dead" comets than are observed. Therefore, the relative importance of extinction vs. fragmentation and disruption remains an issue that is under investigation.

Hills (1981) showed that the apparent inner edge of the OC at a semimajor axis $a = a_I \approx (1–2) \times 10^4$ AU could be a selection effect due to the rarity of close stellar passages capable of perturbing comets with $a < a_I$ onto orbits with perihelion distances small enough to make them observable. Most comets (and perhaps the great majority of comets) might reside in the unseen inner OC at semimajor axes of a few thousand AU. However, during rare close stellar passages, "comet showers" could result (*Hills,* 1981; *Heisler et al.,* 1987; *Fernández,* 1992; *Dybczyński,* 2002a,b). Comet showers caused by various astronomical mechanisms, such as passages of the Sun's hypothetical companion star, Nemesis, through the inner OC, were invoked to explain claimed periodicities in crater formation and mass extinction events on Earth (*Shoemaker and Wolfe,* 1986, and other papers in *The Galaxy and the Solar System*; *Hut et al.,* 1987; *Bailey et al.,* 1987; *Heisler et al.,* 1987). These periodicities have not held up under scrutiny (*Jetsu and Pelt,* 2000; but cf. *Muller,* 2002). However, despite the lack of direct evidence for the inner OC, its existence is still generally accepted. Some authors (e.g., *Bailey et al.,* 1990) have stated that the inner OC must exist in order to replenish the outer OC, which might be completely stripped by passages of molecular clouds. [But cf. *Hut and Tremaine* (1985); we return to this claim in the next section.]

Indeed, two members of the inner OC's population may recently have been found: the unusual body (90377) Sedna (a = 501 AU, q = 76 AU) (*Brown et al.,* 2004) may be a member of the inner OC as may be the object 2000 CR_{105} (a = 224 AU, q = 44 AU) (*Gladman et al.,* 2002). Whether the unusually large perihelia of these objects was produced by a passing star (*Morbidelli and Levison,* 2004; *Kenyon and Bromley,* 2004) is a matter of some debate since other models for their origin exist (*Matese et al.,* 2005; *Gomes et al.,* 2006; *Gladman and Chan,* 2006). However, if Sedna in particular is representative of the inner regions of the OC, then the inner OC may be rather massive [containing as much as 5 $M_\oplus$ in the rough estimation of *Brown et al.* (2004)]. In section 7 we shall explore models of OC formation that produce a massive inner cloud as a byproduct of the Sun's presumed early history in the denser environment of an embedded star cluster. But first we review the key features of the dynamical evolution of comets in the current solar system environment.

3. PERTURBATIONS ON COMETARY ORBITS

The process by which small bodies evolve from the planetary region involves several stages during which the body's path evolves under the gravitational effects of the Sun and the following potential perturbers.

3.1. Planets

Assuming that comets formed in the region of the giant planets, their orbits initially evolve due to gravitational scattering by the planets. At first, planetary perturbations produce comparable changes in the comets' semimajor axes (a) and perihelion distances (q) when cometary eccentricities (e) are small. Eventually, most comets are placed onto highly eccentric orbits with perihelia still in the planetary region. The planets continue to change the comets' orbital energies (i.e., a) via a random walk (*Yabushita,* 1980), while leaving their angular momenta (or, equivalently, q for highly eccentric orbits) nearly unchanged.

3.2. Stars

Stars with masses above the hydrogen-burning limit of 0.07 $M_\odot$ pass within 1 pc (~2×10^5 AU) of the Sun about once per 10^5 yr (*Garcia-Sánchez et al.,* 1999, 2001). The average mass of these stars is ~0.5 $M_\odot$ (*Chabrier,* 2001).

Brown dwarfs (stellar objects with masses between 0.01 and 0.07 $M_\odot$) are about as numerous as stars in the solar neighborhood, but their average mass is only ~0.05 $M_\odot$ (*Chabrier,* 2002, 2003). Thus brown dwarfs probably do not perturb the OC much. In the impulse approximation, the change in velocity $\Delta\mathbf{v}$ of an OC comet with respect to the Sun due to a stellar passage is given by

$$\Delta\mathbf{v} = \frac{2GM_*}{V_*}(\hat{\mathbf{d}}_C/d_C - \hat{\mathbf{d}}_S/d_S)$$

where G is the gravitational constant, M_* is the mass of the star, V_* is its speed relative to the Sun, and $\mathbf{d}_C$, $\mathbf{d}_S$ are the impact parameters for the encounter with respect to the comet and the Sun, respectively (*Oort,* 1950). The strongest encounters are those that have $d_S \ll d_C$ or $d_C \ll d_S$, i.e., those in which the star perturbs either the Sun or the comet much more than it does the other object. In fractional terms, stars change q much more than they change a. This is a consequence of the long lever arm and slow speed of comets on highly eccentric orbits near aphelion [see Appendix B of *Eggers* (1999) for an elaboration of this argument]. However, as we discuss next, tides from the galactic disk are slightly more effective than stars in producing systematic changes in q. Nonetheless, passing stars do produce a random walk, or "diffusive" change, in cometary semimajor axes (*Weinberg et al.,* 1987).

3.3. The Galactic Tidal Field

The importance of galactic tides, i.e., the differential gravitational acceleration of OC comets relative to the Sun due to the disk and bulge of the Milky Way, was pointed out by *Byl* (1983, 1986, 1990), *Smoluchowski and Torbett* (1984), *Heisler and Tremaine* (1986), *Delsemme* (1987), and *Matese et al.* (1995). Slightly less than half of the local galactic mass density is in stars. Thus at most times (i.e., at times other than during a strong comet shower), the rate at which comets are fed into the planetary region from the OC due to the galactic tide is probably slightly larger than the influx due to stellar passages (*Heisler and Tremaine,* 1986). Indeed, *Heisler* (1990) concluded that when stellar impulses are added to the galactic tidal interaction, the long-timescale average increase in the steady state flux due to the tide alone is only ~20%. Tides change comets' q at nearly constant a. The disk ("z") component of the galactic tide causes q to oscillate in and out of the planetary region with a period T_z that is on the order of 1 G.y. for comets with a ~ 10,000 AU and initial q ~ 25 AU (equation (18) in *Heisler and Tremaine,* 1986; *DQT87*). The value of T_z scales as $a^{-3/2}$, i.e., inversely as the comet's orbital period. In addition, there is a "radial" component of the galactic tide due to the mass interior to the Sun's orbit around the galaxy. The amplitude of the radial tide is a factor of ~8 smaller than the disk tide, but the radial tide is still important because it breaks conservation of J_z, the component of a comet's angular momentum perpendicular to the galactic plane. Thus the radial tide modulates cometary perihelion distances and must be considered in studies of the influx of LPCs and Halley-type comets (*Matese and Whitmire,* 1996; *Levison et al.,* 2006).

3.4. Molecular Clouds

While stars and tides have been the only perturbers included in most models of the OC, it has been suggested that molecular clouds (MCs) (see *Williams et al.,* 2000, for a review) might be very destructive of the outer OC (*Biermann,* 1978; *Napier and Clube,* 1979; *Bailey,* 1983, 1986; *Hut and Tremaine,* 1985; *Weinberg et al.,* 1987; *Bailey et al.,* 1990). *Napier and Staniucha* (1982) claimed that giant molecular clouds (GMCs) have removed ~99.9% of the comets from the outer OC. *Hut and Tremaine* (1985) found that MCs were about as effective as stars in stripping the OC; for either type of perturber, acting alone, the half-life of a comet at a = 25,000 AU (0.12 pc) is about 3 G.y., implying a net half-life of 1.5 G.y. The half-life is shorter at larger semimajor axes. *Weinberg et al.* (1987) found slightly shorter timescales. At face value, these results imply that most of the outer OC is lost, so that perhaps only 10% of the comets survive for 4 G.y. However, the local mass density in molecular clouds and other parameters for clouds remain uncertain. *Hut and Tremaine* (1985) argue that the existence of wide binary stars with separations on the order of 20,000 AU implies that molecular clouds cannot be vastly more destructive to the OC than stars. Thus we will neglect molecular clouds in what follows.

3.5. Other Galactic Perturbers

Giant molecular clouds were actually postulated by *Spitzer and Schwarzschild* (1953) before molecular clouds were discovered. Spitzer and Schwarzschild were trying to explain "disk heating," i.e., the observation that older stars within the galactic disk typically have larger random velocities than younger stars. In their model, gravitational scattering of stars by GMCs progressively excites stellar velocities over time. However, it appears that additional sources of disk heating are required, since GMCs appear incapable of producing the velocity dispersion of the dynamically hottest disk stars, such as white dwarfs (*Binney and Tremaine,* 1987). Thus additional perturbers of stars, and potentially the OC, such as spiral arms (*Jenkins and Binney,* 1990) and possibly massive black holes in the galactic halo (*Hänninen and Flynn,* 2002), have been invoked. We ignore spiral arms and galactic perturbers other than the smooth component of the tide and passing stars in this paper.

3.6. Other Perturbers Within the Solar System

There have been suggestions that the OC may contain a red dwarf (*Davis et al.,* 1984; *Whitmire and Jackson,* 1984), a brown dwarf or giant planet (*Matese et al.,* 1999; *Murray,* 1999; *Horner and Evans,* 2002), or an Earth-mass planet

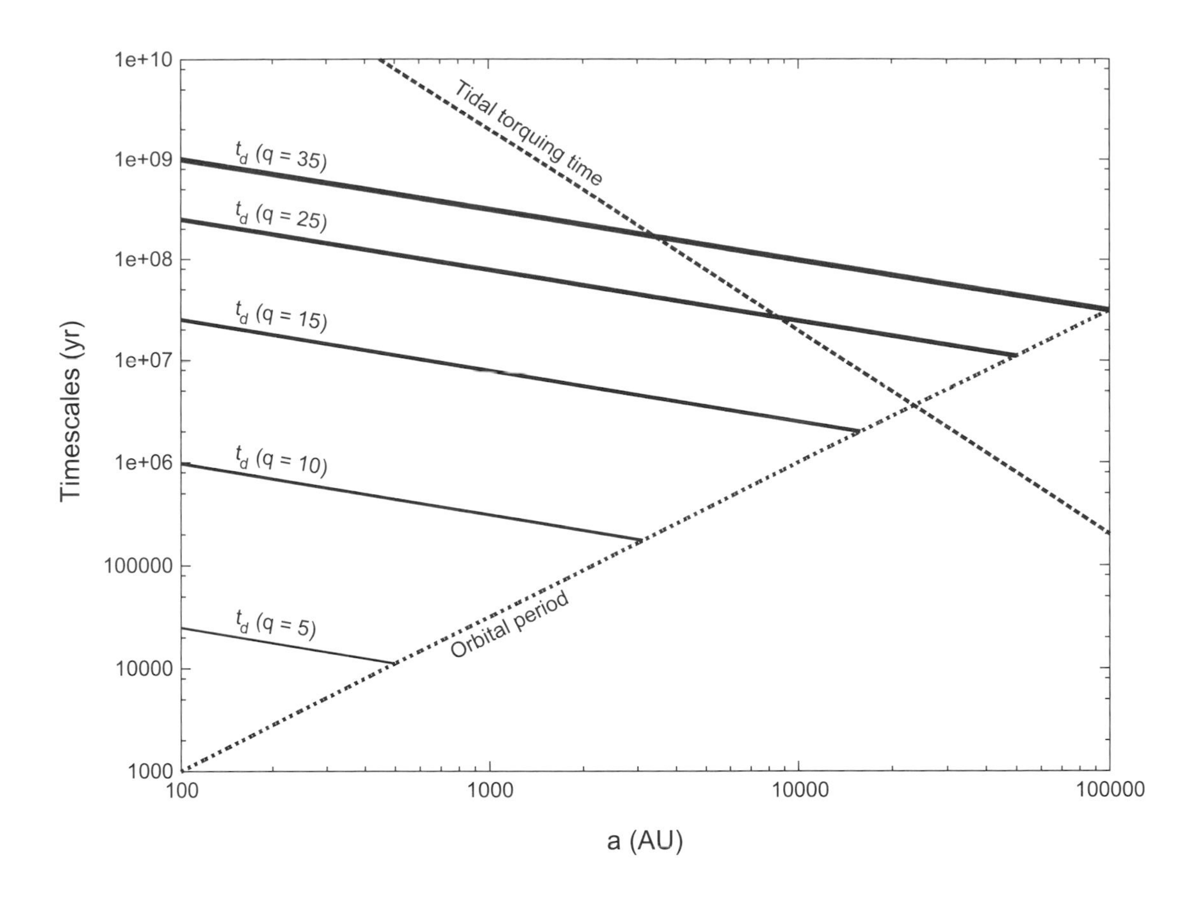

Fig. 1. Key timescales relevant to the dynamical evolution of a comet in the current solar system are plotted against semimajor axis. These include (1) orbital period (dotted line), (2) the energy diffusion time t_d (timescale for cumulative planetary perturbations near pericenter to change the comet's semimajor axis by roughly a factor of 2) for several pericentric distances q (solid lines), and (3) the tidal torquing time t_q (timescale for the galactic tidal field to change q by roughly 10 AU) for the current galactic disk density (dashed line).

(*Goldreich et al.,* 2004). *Wiegert and Tremaine* (1999) included solar companions and massive circumsolar disks in some of their simulations. While the existence of extra perturbers cannot be disproved, we will not include them in this review, in the absence of compelling evidence for their reality (see the chapter by Gomes et al.).

3.7. Key Timescales

As discussed above, to a good first approximation the main perturbers in the current solar system are the four giant planets and the z-component of the galactic tide. The timescales shown in Fig. 1 (adapted from *DQT87*) can then be used to follow the expected evolutionary path of cometary orbits once the comets are scattered onto eccentric orbits with $a \gg q$. Such comets tend to diffuse outward in semimajor axis along lines of constant q, so that they evolve approximately along the lines of t_d, which are slowly decreasing with increasing a. Two things can happen to the comet, depending on which line it crosses first: If it crosses the period line (increasing with a) before it crosses the line of t_q, the comet has a high probability of being ejected on the next passage through the planetary region; if the other case occurs, i.e., the line of t_q is crossed before the period line, the comet's perihelion is often lifted out of the planetary region by the tide and stars quickly enough to be saved from being ejected. The intersection of the lines of t_d and t_q indicates the value of a at which the lifting is likely to begin and is a measure of the inner edge of the OC.

Thus, Fig. 1 shows that comets with pericenters in the Jupiter-Saturn zone (q ~ 5–15 AU) will diffuse on timescales of 10^4–10^7 yr to the point where the energy diffusion time t_d is comparable to the orbital period P (a ~ several hundred to 10,000 AU, depending on q), at which point the diffusion approximation breaks down since the energy kick in one orbit is comparable to the binding energy of the orbit. Most comets in this range of q will subsequently be ejected, although due to the wide range in possible planetary "kicks" in energy, a small fraction (typically a few percent) can be launched to the region where the time for the pericenter to be lifted away from the Jupiter-Saturn region is less than an orbital period ($a \gtrsim 20,000$ AU, as can be seen in Fig. 1). On the other hand, comets with pericenters in the Uranus-Neptune zone (q ~ 25 AU) will diffuse on 10^8-yr timescales until they reach the region where the tidal torquing time is comparable to the period (a ≲ 10,000 AU from Fig. 1). In the next orbit, depending on the argument of perihelion with respect to the galactic plane, roughly half the comets will be torqued outward to the relative safety of the OC and roughly half will be drawn into the Jupiter-Saturn zone, where they are likely to be ejected. Similarly, objects with q ~ 35 AU — i.e., those in the scattered disk — will evolve on billion-year timescales to a ~ 3000 AU where their perihelia will be torqued in or out.

Figure 1 can also be used to understand the key observational bias concerning the OC noted by *Hills* (1981). Observed LPCs come directly from the "outer" OC (a > 20,000 AU), where the galactic tidal field can drive the perihelion distance from q ~ 15 AU to q ≲ 3 AU (where they produce a visible coma) in one cometary orbital period. Comets with perihelia driven inward from the "inner" OC (a < 20,000 AU) in the current galactic environment will do so over several orbits and thus at some time will pass through perihelion when q is near Jupiter and/or Saturn and the planetary perturbations almost inevitably drive such comets back into interstellar space (or occasionally into the outer OC) or into much more tightly bound orbits where tidal torquing is negligible. Thus, Jupiter (and to a lesser extent Saturn) acts like a barrier preventing most objects from the inner OC from directly being observed. This effect is called the "Jupiter barrier." Since inner OC comets are rarely directly injected into visible orbits, the inner cloud could be quite massive, as we discussed in section 2.

Note that since the tidal torquing time is inversely proportional to the local galactic density, the "lifting" of cometary perihelia and hence the inner edge of the OC will occur at smaller radii for larger densities. We shall return to this issue is section 7.

4. THE MASS OF THE OORT CLOUD

The present-day mass of the OC may provide interesting constraints on models of planet formation, under the assumption that comets represent planetesimals from the region of the giant planets. If the OC's mass can be estimated from observations and the efficiency of OC formation can be determined by dynamical models, we can hope to infer the total mass in the planetesimal disk during the epoch when the OC formed. The formation timescale and migration histories of the giant planets, in turn, depend critically on the mass of the planetesimal disk (*Hahn and Malhotra,* 1999; *Gomes et al.,* 2004).

From the observed flux of new comets, *Heisler* (1990) used her model-determined cometary influx rate, and the assumption that the current flux equals the long-term average, to infer that the present-day outer OC contains N_O = 5 × 10^{11} comets with a > 20,000 AU. This estimate refers to comets with "absolute magnitude" H_{10} < 11. *Weissman* (1996) reviews estimates of the number of comets in the OC and attempts to relate H_{10}, which is a measure of a comet's total brightness (generally dominated by coma) to cometary masses. In Weissman's calibration, which is based upon 1P/Halley's measured size and total brightness and an assumed density of 0.6 g cm^{-3} [cf. the determination of a density of 0.6 (+0.5, –0.3) g cm^{-3} for Comet 9P/Tempel 1 by *A'Hearn et al.* (2005)], the diameter of a comet with H_{10} = 11 is d_{11} = 2.3 km, and the mass of a comet with H_{10} = 11 is m_{11} = 4 × 10^{15} g. *Weissman* (1996) assumes, based on *Everhart* (1967), that LPCs have a shallow size distribution (with index 2 for the cumulative distribution) up to a size d_{crit} ~ 20 km, and follow a steep distribution at larger sizes. For these assumptions, most of the mass in the OC is in bodies with diameters between d_{11} and d_{crit}, and the average mass of a comet $\bar{M}$ is on the order of $d_{crit}/d_{11} \times m_{11}$, or $\bar{m}$ ~ 4 × 10^{16} g. Weissman assumes N_O = 1 × 10^{12}, giving a mass for the outer OC of $N_O\bar{m}$ = 4 × 10^{28} g, i.e., 7 $M_\oplus$. This estimate is extremely uncertain; since the size of a dynamically new LPC has never been directly measured, cometary densities may be even smaller than 0.6 g cm^{-3} (*Davidsson and Gutiérrez,* 2004) and uncertainties in the extent of cometary fading may have lead to an overestimation of the flux of new comets (*Neslusan,* 2007).

Recently, *Francis* (2005) has used the data from the LINEAR survey to provide revised estimates of the flux of LPCs, the dependence on perihelion distance and the absolute magnitude distribution. He estimates that the outer OC contains ~5 × 10^{11} comets down to absolute magnitude H_{10} = 17 and ~2 × 10^{11} comets down to absolute magnitude H_{10} = 11. The latter estimate is 5× lower than that used by *Weissman* (1996), which would predict an outer OC mass of 1.4 $M_\oplus$ if all of the other parameters of Weissman's fit remained unchanged. Like Weissman, *Francis* (2005) assumes a two-component magnitude distribution, with a break at magnitude H_b = 6 or 6.5 (cf. *Everhart,* 1967; *Hughes,* 2001). The magnitude distribution for "faint" comets (i.e., comets with H_{10} > H_b) that *Francis* (2005) finds is shallower (more top-heavy) than *Weissman* (1996) assumes. However, the mass of the outer cloud is likely dominated by "bright" comets (H_{10} ≲ 6–6.5), which are poorly constrained by the LINEAR data.

We find that the mass of the outer OC could be between ~1 and 60 ME, depending upon the assumed magnitude distribution for bright comets (*Everhart,* 1967; *Hughes,* 2001) and the relationship between absolute magnitude and cometary mass (*Weissman,* 1996; *Bailey and Stagg,* 1988). We obtain a low value for the outer cloud's mass if we assume that it contains very few comets with H_{10} < 3, the magnitude of the brightest comet seen by LINEAR thus far (*Francis,* 2005). This assumption yields an outer cloud mass of 3 ME using *Weissman*'s (1996) scaling or 0.6 ME using *Bailey and Stagg*'s (1988). However, an abrupt cutoff for any value of H_{10} ≲ 0 seems unlikely, since Hale-Bopp, which had H_{10} = –0.8 (Andreas Kammerer, *kometen.fg-vds.de/koj_1997/c1995o1/95o1eaus.htm*) probably originated in the OC. Given the very poorly calibrated relationship between H_{10} and cometary mass, we cannot place a lower (or upper) limit on the mass of the outer cloud from the observed rate of cometary passages through the inner solar system. Low outer cloud masses, on the order of 1 $M_{\oplus}$, are more consistent with current models of the dynamical evolution of the outer solar system (see section 6).

5. CLUES FROM THE CAPTURE OF HALLEY-TYPE COMETS

Due to the "Jupiter barrier" (or more accurately the "Jupiter-Saturn barrier") discussed in section 3.7, bodies from the inner OC for which the galactic tide is driving their perihelion distances into the region of the giant planets will suffer perturbations that will inevitably drive such comets back into the outer OC (or interstellar space) or into much more tightly bound orbits where tidal torquing is negligible. A small fraction of the latter comets will eventually diffuse down into orbits with sufficiently small semimajor axes that their perihelia will begin to be substantially affected and they can then come sufficiently close to the Sun to produce comas. Numerical simulations (*Duncan et al.,* 1988; *Quinn et al.,* 1990; *Emel'yanenko and Bailey,* 1998) have suggested that these objects may be an important source of Halley-type comets (short-period comets — periods less than 200 yr — with Tisserand parameters with respect to Jupiter, T, less than 2) (*Carusi et al.,* 1987; *Levison,* 1996). Thus, it has been suggested that Halley-type comets (hereafter HTCs) may represent our only currently observable link to the inner OC.

Levison et al. (2001, henceforth *LDD01*) attempted to constrain the structure of the inner OC by modeling the "capture" process by which OC comets evolve onto orbits like those of HTCs. While some HTCs, such as Halley and Swift-Tuttle, follow retrograde orbits, most (19 of 26 currently known with q < 1.3 AU — see *www.physics.ucf.edu/~yfernandez/cometlist.html*) revolve on prograde orbits. *LDD01* found that cometary inclinations were roughly conserved during the capture process, so that the source region had to be somewhat flattened. [An alternative, proposed by *Fernández and Gallardo* (1994), is that the observed preponderance of prograde orbits is due to cometary fading: When the number of perihelion passages is limited by physical causes, fewer retrograde comets are found in evolved states since they evolve dynamically more slowly than their prograde counterparts. However, *LDD01* modeled this possibility in their detailed numerical integrations and found that for the range of fading times proposed by Fernández and Gallardo, the simulated HTCs generally had larger semimajor axes than are observed.] Since the outer OC is roughly spherical, *LDD01* concluded that most HTCs must derive from a flattened inner core which they associated with the inner OC. *LDD01* found that models in which the median inclination of inner OC comets, i', is between 10° and 50° can fit the HTC orbital distribution.

However, *Levison et al.* (2006, henceforth *LDDG06*) subsequently realized that there was a problem in identifying the flattened outer source with a "fossilized" inner OC. Recall that the other, more numerous class of short-period comets are the Jupiter-family comets (hereafter JFCs; short-period comets with Tisserand parameters with respect to Jupiter, T, greater than 2) (*Carusi et al.,* 1987; *Levison,* 1996). Most of the JFCs are believed to leak in from the inner edge of the scattered disk (*Duncan and Levison,* 1997; *Duncan et al.,* 2004), initially under the dynamical control of Neptune. As might be expected, the integrations of *LDD01* showed that low-inclination objects with initial semimajor axes in the inner OC also evolve into JFCs if their perihelia originated near Neptune. However, those low-inclination comets from the inner OC with initial q ≲ 25 AU generally evolved into HTCs. Thus, in order for a flattened inner OC to be the source of the low-inclination HTCs, inner OC comets must be able to evolve onto orbits with q ≲ 25 AU and remain on low-inclination orbits in the inner OC for the age of the solar system.

LDDG06 showed that (1) only if a ≲ 3000 AU are galactic tides weak enough that orbital inclinations do not change significantly over the age of the solar system (i.e., the OC remains disk-like), and (2) only if a ≳ 12,000 AU are galactic tides strong enough that an object can evolve from an orbit with q > 30 AU to one with q ≲ 25 AU before perturbations from Neptune remove it from the OC. This obvious inconsistency is resolved if the flattened source is itself being dynamically replenished, rather than being a fossilized remnant of the early history of the solar system. Thus, *LDDG06* studied the dynamical evolution of objects that evolve off the outer edge of the scattered disk and have their perihelion distances driven inward by galactic tides. They found that roughly 0.01% of these objects evolve onto HTC-like orbits. The orbital element distribution of the resulting HTCs is consistent with observations, including the requisite number of retrograde orbits, that are produced by the tidally driven precession of the line of nodes in galactic coordinates.

In order for the scattered disk to supply enough HTCs, the model predicts that it needs to contain 3 billion comets with diameters larger than 10 km. This value is larger than estimates inferred from observations of the scattered disk and it may be larger than that needed for the JFCs. However, the structure of the scattered disk at large heliocentric distances, where the HTCs would come from, is not constrained by either the JFC models or observations. In addition, *LDDG06*'s HTC model lacked passing stars and molecular clouds, which could affect the delivery rates.

It should be noted that, in the simulations of *LDDG06*, comets coming off the outer edge of the scattered disk not only become HTCs, but contribute to the LPC population, especially to those LPCs known as dynamically new comets (DNCs) [a ≳ 10,000 AU; see *Levison* (1996), although *Dybczyński* (2006) argues that a ≳ 25,000 AU is a better criterion]. In particular, the models produce a population of DNCs with a median inclination of 40° and in which only 30% are retrograde. The observed inclination distribution of the DNCs is isotropic. However, one interesting aspect of the model DNCs is that they have large semimajor axes:

75% of DNCs in *LDDG06* have a > 30,000 AU. This is relevant because *Fernández* (2002) performed an analysis of the inclination distribution of LPCs in different ranges of semimajor axes and concluded that although the DNCs are isotropic, 68% of the DNCs with a > 32,000 AU are prograde. This is consistent with *LDDG06*'s model if most of these objects are from the scattered disk. To test the latter requirement, future investigations combining the simulations just discussed together with a complete evolutionary model of the OC will be needed.

6. SIMULATIONS OF OORT CLOUD FORMATION: THE REFERENCE MODEL

In his 1950 paper, Oort did not consider the formation of the comet cloud in detail, but speculated that the comets were scattered from the asteroid belt due to planetary perturbations and lifted by stellar perturbations into a large cloud surrounding the solar system. Oort proposed the asteroid belt as the source region for the LPCs on the grounds that (1) asteroids and cometary nuclei are fundamentally similar in nature and (2) the asteroid belt was the only stable reservoir of small bodies in the planetary region known at that time. *Kuiper* (1951) was the first to propose that the icy nature of comets required that they be from a more distant part of the solar system, among the orbits of the giant planets. Thus, ever since Oort and Kuiper's work, the roles of the four giant planets in populating the comet cloud have been debated. *Kuiper* (1951) proposed that Pluto, which was then thought to have a mass similar to that of Mars or the Earth, scattered comets that formed between 38 and 50 AU (i.e., in the Kuiper belt!) onto Neptune-crossing orbits, after which Neptune, and to a lesser extent the other giant planets, placed comets in the OC.

Later work (*Whipple,* 1962; *Safronov,* 1972) indicated that Jupiter and Saturn tended to eject comets from the solar system, rather than placing them in the OC. The gentler perturbations by Neptune and Uranus (if these planets were assumed to be fully formed) thus appeared to be more effective in populating the cloud. However, their role was unclear because the ice giants took a very long time to form in Safronov's orderly accretion scenario. *Fernández* (1978) used a Monte Carlo, Öpik-type code and suggested that "Neptune, and perhaps Uranus, could have supplied an important fraction of the total mass of the cometary cloud." *Fernández* (1980) extended this work by following the subsequent evolution of comets on plausible near-parabolic orbits for bodies that had formed in the Uranus-Neptune region. He concluded that about 10% of the bodies scattered by Uranus and Neptune would occupy the OC at present, and that the implied amount of mass scattered by the ice giants was cosmogonically reasonable, i.e., not vastly greater than the masses of Uranus and Neptune themselves.

Shoemaker and Wolfe (1984) performed an Öpik-type simulation to follow the ejection of Uranus-Neptune planetesimals to the OC. They found that ~9% of the original population survived over the history of the solar system, with ~90% of those comets in orbits with semimajor axes between 500 and 20,000 AU.

The first study using direct numerical integrations to model the formation of the OC was that of *DQT87*. To save computing time, *DQT87* began their simulations with comets on low-inclination, but highly eccentric, orbits with perihelia in the region of the giant planets. Gravitational perturbations due to the giant planets and the disk (z) component of the galactic tide were included. A Monte Carlo scheme from *Heisler et al.* (1987) was used to simulate the effects of stellar encounters.

The main prediction of *DQT87* was that the OC is centrally condensed, with roughly 4–5× as many comets in the inner OC (a ≲ 20,000 AU) as in the classical outer OC. In their model, comets with $q_0 \gtrsim 15$ AU are much more likely to reach the OC and survive for billions of years than are comets with smaller initial perihelia. For example, only 2% of the comets with $q_0 = 5$ AU should occupy the OC at present, while 24% of the comets with $q_0 = 15$ AU and 41% with $q_0 = 35$ AU should do so. This result appeared to confirm that Neptune and Uranus, which have semimajor axes of 30 and 19 AU, respectively, are primarily responsible for placing comets in the OC. However, this finding is questionable, since the highly eccentric starting orbits had the consequence of pinning the perihelion distances of the comets at early stages. This, in turn, allowed Neptune and Uranus to populate the OC efficiently because they could not lose objects to the control of Jupiter and Saturn.

DWLD present results of a similar study to that of *DQT87*, but starting with "comets" with semimajor axes between 4 and 40 AU and initially small eccentricities and inclinations. These initial conditions are more realistic than the highly eccentric initial orbits assumed by *DQT87*. The study integrated the orbits of 3000 comets for times up to 4 b.y. under the gravitational influence of the Sun, the four giant planets, the galaxy, and random passing stars. Their model of the galaxy included both the "disk" and "radial" components of the galactic tide. The disk tide is proportional to the local density of matter in the solar neighborhood and exerts a force perpendicular to the galactic plane, while the radial tide exerts a force within the galactic plane (see section 3.3). These simulations did not include other perturbers such as molecular clouds, a possible dense early environment if the Sun formed in a cluster (*Gaidos,* 1995; *Fernández,* 1997), or the effects of gas drag (*de la Fuente Marcos and de la Fuente Marcos,* 2002; *Higuchi et al.,* 2002). Studies involving the latter two effects are discussed below.

DWLD describe two sets of runs with dynamically "cold" and "warm" initial conditions. The results were very similar, so they focus on the "cold" runs, which included 2000 particles with root-mean-square initial eccentricity, e_0, and inclination to the invariable plane, i_0, equal to 0.02 and 0.01 radians, respectively.

DWLD take the results of these calculations at 4 G.y. to refer to the present time. For a comet to be considered a member of the OC, they required that its perihelion distance exceeded 45 AU at some point in the calculation. For the "cold" runs, the percentage of objects that were integrated that currently occupy the classical "outer" OC (20,000 AU $\leq$ a < 200,000 AU) is only 2.5%, about a factor of 3 smaller than found by *DQT87*. The percentage of objects in the inner OC (2000 AU $\leq$ a < 20,000 AU) is 2.7%, almost an order of magnitude smaller than calculated by *DQT87*. *DQT87* found a density profile $n(r) \propto r^{-\gamma}$ with $\gamma \sim 3.5$ for 3000 AU < r < 50,000 AU, so that in their model most comets reside in the (normally unobservable) inner OC. Fitting the entire OC at 4 G.y. in the models reported in *DWLD* to a single power law yields $\gamma \sim 3$, shallower than the value found by *DQT87*. A value of $\gamma \sim 3$ implies that the inner and outer OCs contain comparable numbers of comets at present in this model. This result holds because most comets that begin in the Uranus-Neptune zone evolve inward and are ejected from the solar system by Jupiter or Saturn. A small fraction of these are placed in the OC, most often by Saturn. However, all four of the giant planets place comets in the OC, albeit with different efficiencies.

The OC is built in two distinct stages in the *DWLD* model. In the first few tens of million years, primarily the outer OC is built by Jupiter and Saturn; subsequently, most of the inner OC is built, mainly by Neptune and Uranus, with the population peaking about 800 m.y. after the beginning of the simulation (Fig. 2). Objects that enter the OC during this second phase typically first spend time in the "scattered disk" [45 AU $\leq$ a < 2000 AU, with perihelion distance <45 AU at all times (*Duncan and Levison,* 1997; see the chapter by Gomes et al.)] and then end up in the inner OC. Figure 2 shows the time evolution of the populations of the OC and scattered disk in the simulation. The scattered disk is initially populated by comets scattered by Jupiter and Saturn, and peaks in number at 10 m.y. (off-scale on the plot). After this time the population of the scattered disk declines with time t approximately as a power law, $N(t) \propto t^{-\alpha}$, with $\alpha \sim 0.7$. The predicted population of the scattered disk in this model at the present time is roughly 10% the population of the outer OC.

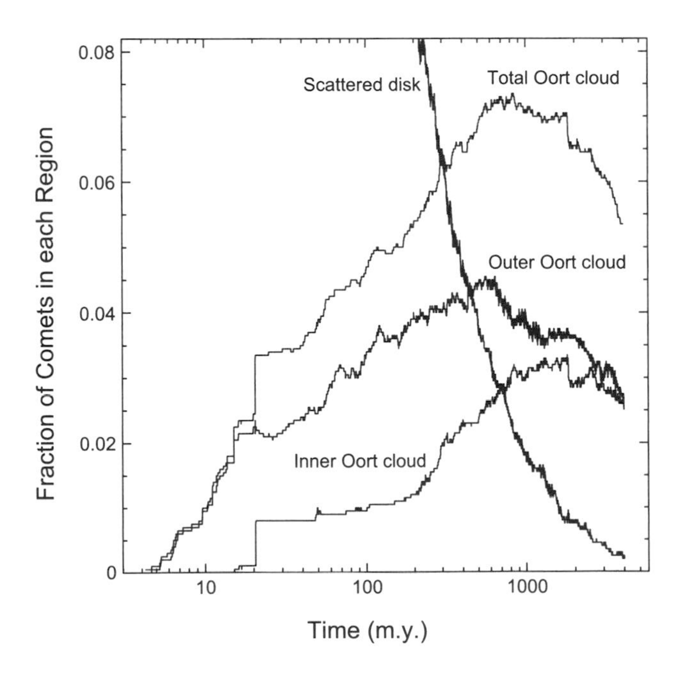

Fig. 2. In the simulations reported in *DWLD* the outer OC, which is originally populated by comets injected by Jupiter and Saturn, forms more rapidly than the inner OC. These simulations predict that the present populations of the inner and outer OCs should be comparable and that of the scattered disk should contain roughly 10% as many comets as the outer OC.

Figure 2 also shows the populations of the inner and outer OCs individually. The population of the outer OC peaks around 600 m.y. while the inner OC peaks around 1.8 G.y. Because of the faster decline of the outer OC, the ratio of numbers of inner to outer OC comets increases with time, to 1.1 at present. As noted above, only 2.5% of the comets that were initially in the simulation occupy the outer OC at 4 G.y.

At face value, the low efficiency of OC formation in the simulations reported in *DWLD* implies a massive primordial protoplanetary disk. Assuming an outer OC mass of 2–40 $M_\oplus$ (*Francis,* 2005) (see section 4), the simulation efficiency implies that the original mass in planetesimals between 4 and 40 AU was ~80–1600 $M_\oplus$, some 2–40× the mass in solids in a "minimum mass" solar nebula. The amounts of mass at anything but the low end of this estimate would likely have produced excessive migration of the giant planets and/or formation of additional giant planets (*Hahn and Malhotra,* 1999; *Thommes et al.,* 2002; *Gomes et al.,* 2004). However, the uncertainties discussed at the end of section 4 suggest that it is possible that the outer OC may currently contain ~1 $M_\oplus$, in which case the disk mass inferred in *DWLD* does not present a problem.

The results may be inconsistent with observations in another way. The population of the scattered disk that *DWLD* predicted, on the order of 10% the population of the OC, may be much larger than the actual population of the scattered disk inferred from observations of large bodies (*Trujillo et al.,* 2000). Again, however, the large uncertainties in the observational estimates of both populations mean that there may be no problem.

However, we must certainly come to terms with the fact that none of the models described above will produce the orbits of the inner OC objects described at the end of section 2. For that, we will likely need the types of models discussed next.

7. SIMULATIONS OF OORT CLOUD FORMATION IN A STAR CLUSTER

Tremaine (1993), *Gaidos* (1995), *Fernández* (1997), *Eggers et al.* (1997, 1998), *Eggers* (1999), and *Fernández and Brunini* (2000) have discussed star formation in different galactic environments. These authors point out that the Sun may have formed in a denser environment than it now occupies (i.e., in a molecular cloud or star cluster), and found that a more tightly bound OC would form.

7.1. Simulations Without a Primordial Solar Nebula

A first attempt to simulate the formation of the OC when the Sun is still in a star cluster was done by *Eggers* (1999). In that work, the formation of the OC was simulated for 20 m.y. using a Monte Carlo method with two star clusters, in which the stellar encounters occurred at constant time intervals and were computed analytically and the cluster tidal field was neglected. The first cluster had an effective density of 625 stars pc^{-3} and the other had an effective density of 6.25 stars pc^{-3}, as compared with the current density of ~0.1 stars/pc^3. Both clusters had a velocity dispersion of 1 km s^{-1}. Eggers started with a population of comets on nearly circular, low-eccentricity orbits in the region of the giant planets, and defined a comet to be in the OC if it evolved under the combined planetary and stellar perturbations onto an orbit with q > 33 AU and a > 110 AU. With these definitions, he obtained efficiencies of 1.7% and 4.8% for the loose and dense clusters respectively. Most objects for the low-density cluster had a = 6–7 × 10^3 AU and the resulting OC was fairly isotropic. For the high-density cluster, most objects were in the range a = 2–4 × 10^3 AU and again had a fairly isotropic inclination distribution.

It is worth noting that *Eggers* (1999) also used an N-body simulation to investigate the interesting proposal of *Zheng et al.* (1990) that some fraction of OC comets might have been captured from the intracluster medium during star–star encounters. Eggers found that at most a few percent of outer OC comets are likely to be of extrasolar origin. Subsequently, motivated by the discovery of *Sedna et al.* (2004) and *Morbidelli and Levison* (2004), he independently investigated the capture into the Sun's inner OC of comets from the protoplanetary disks of other stars during very close stellar encounters.

Fernández and Brunini (2000), henceforth *FB2K*, performed simulations of the evolution of comets starting on eccentric orbits (e ~ 0.9) with semimajor axes 100–300 AU and included an approximate model of the tidal field of the gas and passing stars from the cluster in their model. The clusters had densities ranging from 10–100 stars pc^{-3}, and the density of the core of the molecular cloud in their models ranged from 500 to 5000 $M_\odot$ pc^{-3}. Their simulations formed a dense inner OC with semimajor axes of a few hundred to a few thousand AU. The outer edge of this cloud was dependent on the density of gas and stars in the cluster. *FB2K* reported they were able to successfully save material scattered by Jupiter and particularly Saturn, which were the main contributors to forming the inner OC, since Uranus and Neptune took too long to scatter material out to large-enough distances (see *DQT87*). However, as they and others (*Gaidos,* 1995; *Adams and Laughlin,* 2001) pointed out, if the Sun remained in this dense environment for long, the passing stars could strip the comets away and portions of the inner OC might not be stable.

A comprehensive set of simulations have recently been performed by *Brasser et al.* (2006, hereafter *BDL06*). Their model assumes that the Sun formed in an embedded star cluster, which are clusters that are very young and heavily obscured by dust since the molecular gas is still present (*Lada and Lada,* 2003). Most stars in the galaxy probably form in embedded clusters with between 100 and 1000 members (*Adams et al.,* 2006); indeed, typical populations of embedded clusters within 2 kpc of the Sun today are 50–1500 stars (*Lada and Lada,* 2003). The lifetime of the gas in these clusters is typically 1–5 m.y.: Only 10% of embedded clusters last for 10 m.y. (*Lada and Lada,* 2003). Since the formation of unbound stellar clusters is the rule and not the exception (*Lada et al.,* 1984), it is probable that the Sun formed in such a cluster and escaped from it within ≲5 m.y.

Recently, *Gutermuth et al.* (2005) have observed three embedded clusters using near-IR data. The clusters in their sample were chosen because they are rich and relatively young. One of these clusters exhibits clumping of stars into three different regions. The peak volume densities in these clusters in gas and stars are on the order of 10^4–10^5 $M_\odot$ pc^{-3}, with mean volume densities ranging from 10^2–10^3 $M_\odot$ pc^{-3}. The highest peak volume density quoted is 3 × 10^5 $M_\odot$ pc^{-3}. In two of the clusters, their observations show that 72% and 91% of the stars are in locations with stellar densities of 10^4 $M_\odot$ pc^{-3} or larger, respectively. For the third cluster this fraction is 24%. These constraints on the stellar density were used by *BDL06* to select a range in central densities for the clusters in their simulations.

BDL06 adopted a model for the cluster that is often used (e.g., *Kroupa et al.,* 2001) and assume the ratio of stars to gas is constant throughout the cluster. In this so-called Plummer model, the potential Φ(r) and density ρ(r) are given as a function of distance by (e.g., *Binney and Tremaine,* 1987)

$$\Phi(r) = \frac{GM}{\sqrt{(r^2 + c^2)}}; \ \rho(r) = \frac{\rho_0}{(r^2 + c^2)^{5/2}} \tag{1}$$

where M is the total mass of gas and stars combined in the cluster, r is the distance from the centre of the cluster, ρ_0 is the central density of the gas and stars combined, and c is the Plummer radius (which defines the length scale of the model; half the cluster mass is contained within a radius of 1.3c). *BDL06* studied Plummer models with 200–400 stars, with a wide range of initial central densities. Since the simulations spanned only the first few million years of the Sun's lifetime, only the planets Jupiter and Saturn were included and the planets were taken to have their current masses on orbits appropriate to the period before the late heavy bombardment (LHB) (*Tsiganis et al.,* 2005). In each simulation, the Sun's orbit was integrated in the cluster potential together with Jupiter and Saturn and 2200 test particles. The first 2000 test particles were given semimajor axes uniformly spaced from 4 to 12 AU. The remaining 200 had a ∈ (20, 50) AU and resembles a primordial Kuiper belt. The stirring of this belt by stellar passages was a measure of the damage that close stellar passages did to the system

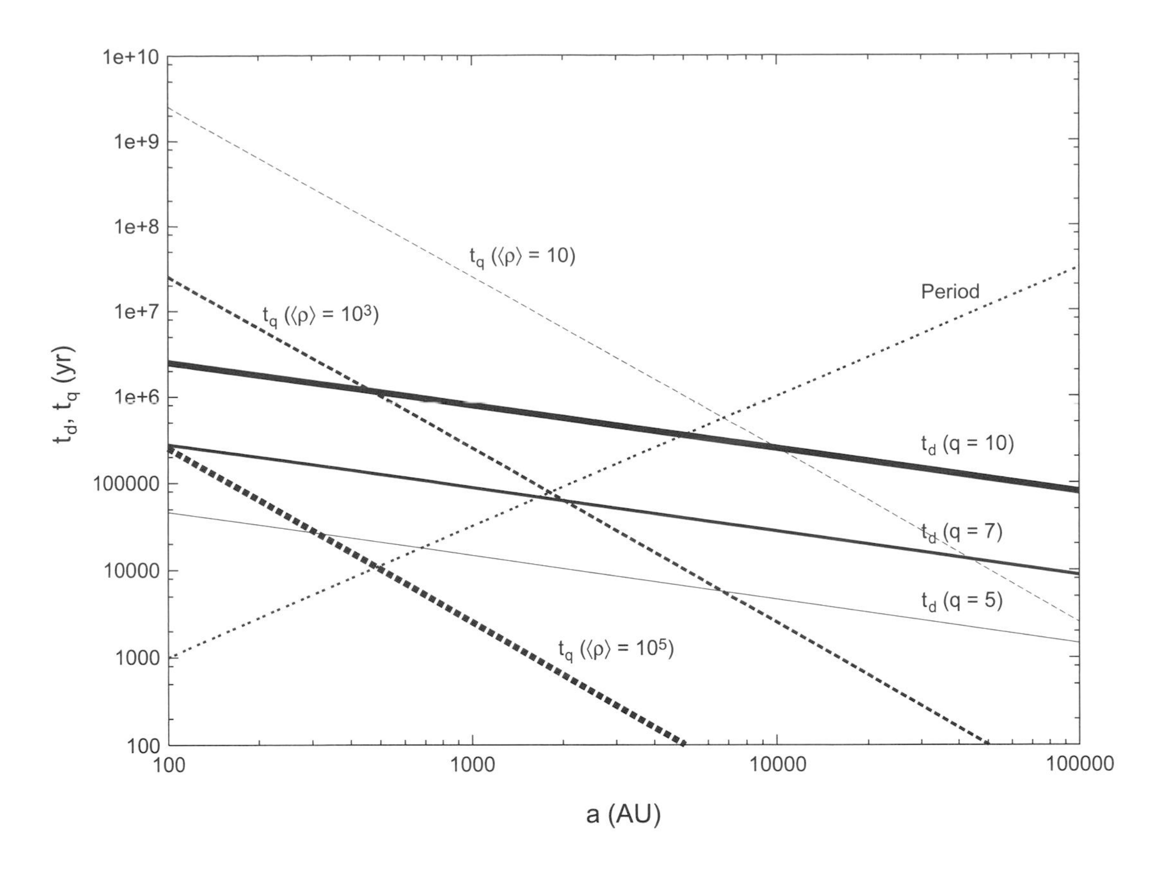

Fig. 3. Plot of the relevant timescales for a comet's evolution as a function of semimajor axis for the models of *BDL06*. Symbols are the same as in Fig. 1. Three values of t_d are plotted (near-horizontal solid lines): q = 5 (thin), 7 (medium), and 10 AU (thick). Three values of t_q are plotted (downward-sloping broken lines): $\langle\rho\rangle$ = 10 (thin), 10^3 (medium), and 10^5 $M_\odot$ pc^{-3} (thick). The orbital period is also plotted (upward-sloping, dotted line).

and is discussed below. The rms values of eccentricity and inclination were 0.02 and 0.01 radians respectively. Stellar encounters were incorporated by directly integrating the effects of stars passing within a sphere centered on the Sun of radius equal to the Plummer radius for low-density clusters and half a Plummer radius for high-density clusters. The gravitational influence of the cluster gas was modeled using the tidal force of the cluster potential. For a given solar orbit, the mean density, $\langle\rho\rangle$, was computed by orbit averaging the density of material encountered. This parameter proved to be a good measure for predicting the properties of the resulting OC. Many of the results can be understood by referring to Fig. 3, which is modeled after Fig. 1 and in which the symbols have the same meaning. In Fig. 3, the tidal torquing time from the Plummer potential is computed in *BDL06* (and is confirmed by numerical integrations) and the energy diffusion timescales for the pre-LHB planetary configuration are numerically obtained as a function of q.

Recall (see, e.g., section 3.7) that once a comet is scattered to semimajor axes a ≫ q, it tends to diffuse outward at fixed q along the lines of t_d, until it hits one of two lines. If it crosses the period line first, it tends to be ejected. If it crosses the tidal torquing time line first, the comet is usually lifted by the tide and stars and can thus be saved from being ejected. The intersection of the lines of t_d and t_q indicates the value of a at which the lifting is likely to begin and is a measure of the inner edge of the OC.

As it turns out, a mean density of at least $\langle\rho\rangle \sim 2000$ $M_\odot$ pc^{-3} is needed to save the comets from Saturn. As can be seen from Fig. 3, even when $\langle\rho\rangle$ = 10^5 $M_\odot$ pc^{-3}, the comets with pericenters close to Jupiter can barely be saved, because the mean kick in energy from this planet is too strong. However, for all the densities shown, a fair number of comets from around Saturn can be saved, with the subsequent OCs being formed ranging in size from a few hundred to several thousand AU.

Figure 4 shows snapshots at the end of five different runs in a–q space. The panels show one run selected from each of the different central densities respectively, with the lowest density in the bottom panel and the highest in the top left panel. On average 2–18% of the initial sample of comets end up in the OC after 1–3 m.y. A comet is defined to be part of the OC if it is bound and has q > 35 AU. The models show that the median distance of an object in the OC scales approximately as $\langle\rho\rangle^{-1/2}$ when $\langle\rho\rangle \gtrsim 10$ $M_\odot$ pc^{-3}.

The models of *BDL06* easily produce objects on orbits like that of (90377) Sedna (*Brown et al.,* 2004) within ~1 m.y. in cases where the mean density is 10^3 $M_\odot$ pc^{-3} or

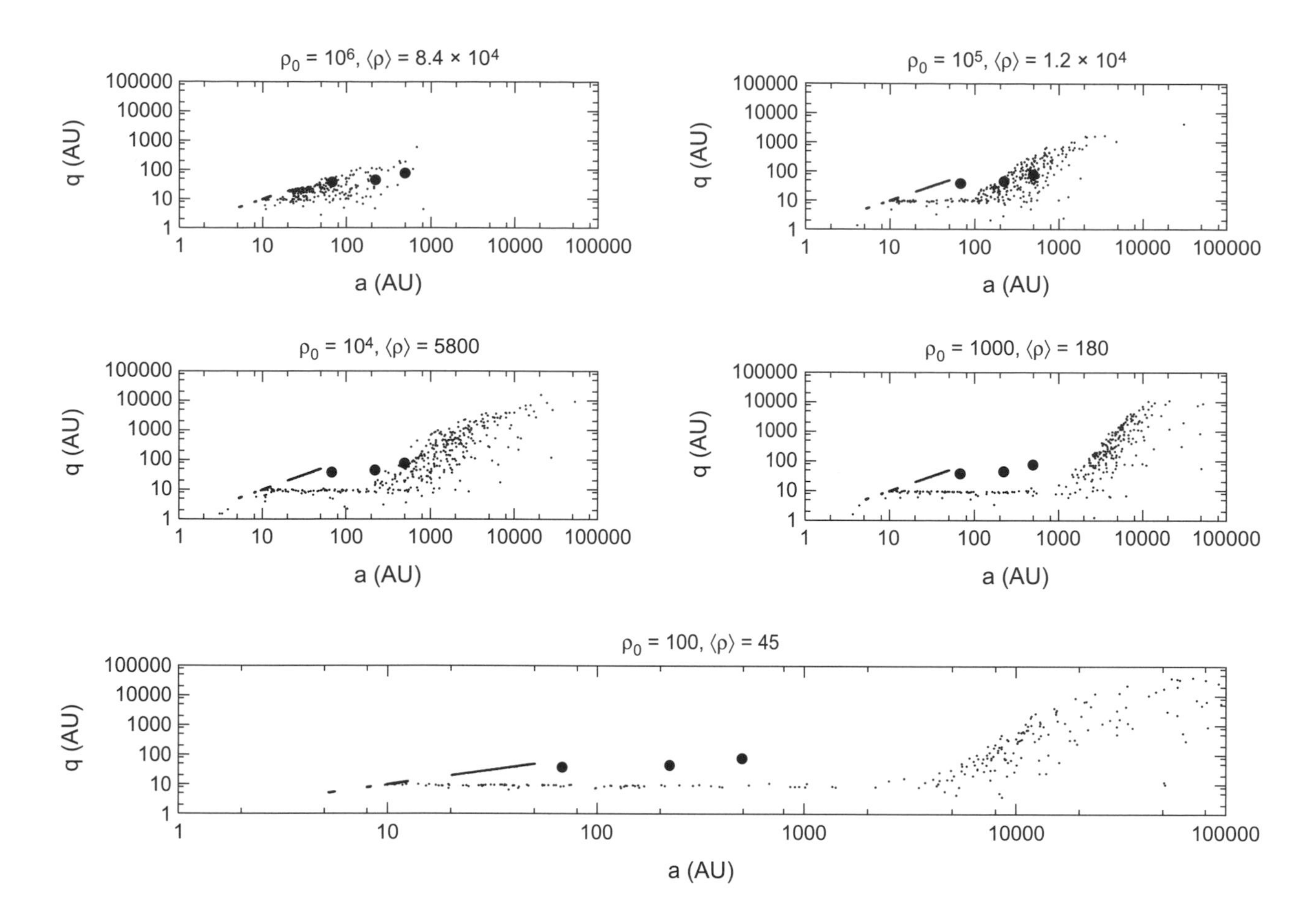

Fig. 4. Snapshots in a–q space are shown at the end of five different runs from *BDL06*, one from each set of runs with a different central density. The mean densities the Sun encountered are shown above each panel. The lowest density is at the bottom. Note that the extent and median values of a of the members of the OC increase with decreasing density, as is expected. The positions of (90377) Sedna, 2000 CR_{105}, and (136199) Eris in order of descending semimajor axis are marked with bullets. Thus it can be seen that objects with orbits like (90377) Sedna and 2000 CR_{105} only form through this mechanism when the density is high.

higher; one needs mean densities on the order of 10^4 $M_\odot$ pc^{-3} to create objects like 2000 CR_{105} by this mechanism, which are reasonable given the observations of *Guthermuth et al.* (2005). Thus the latter object may also be part of the OC.

Close stellar passages can stir the primordial Kuiper belt to sufficiently high eccentricities ($e \gtrsim 0.05$) (*Kenyon and Bromley,* 2002) that collisions become destructive. From the simulations performed it is determined that there is a 50% or better chance to stir the primordial Kuiper belt to eccentricities $e \geq 0.05$ at 50 AU when $\langle\rho\rangle \gtrsim 10^5$ $M_\odot$ pc^{-3}. Note also that in the case of a close stellar encounter that actually truncates the belt at 50 AU, there would be a population of objects on orbits with perihelia like Sedna's, but much smaller semimajor axes. This seems inconsistent with the lack of detections of bodies on these orbits.

The orbit of the new object (136199) Eris (*Brown et al.,* 2005) is only reproduced for mean cluster densities on the order of 10^5 $M_\odot$ pc^{-3}, but in the simulations of *BDL06* it could not come to be on its current orbit by this mechanism without close stellar passages causing disruptive collisions among bodies in the primordial Kuiper belt down to 20 AU. *BDL06* concluded that it is therefore improbable that the latter object is created by this mechanism. It is possible, although by no means proven, that the Kozai mechanism (coupling between the argument of perihelion, eccentricity, and inclination) associated with mean-motion resonances with Neptune may be responsible for raising both the perihelion distances and the inclinations of relatively close-in objects such as Eris (*Gomes et al.,* 2005; see the chapter by Gomes et al.). If so, then the combined simulations of *BDL06* and Gomes et al. may provide a rough delineation in (q–a) space between the inner OC and the region relatively unaffected by stellar perturbations.

7.2. The Effect of the Primordial Solar Nebula

Drag due to gas in the solar nebula may have been very important in the formation of the OC (*de la Fuente Marcos and de la Fuente Marcos,* 2002; *Higuchi et al.,* 2002). *Bras-*

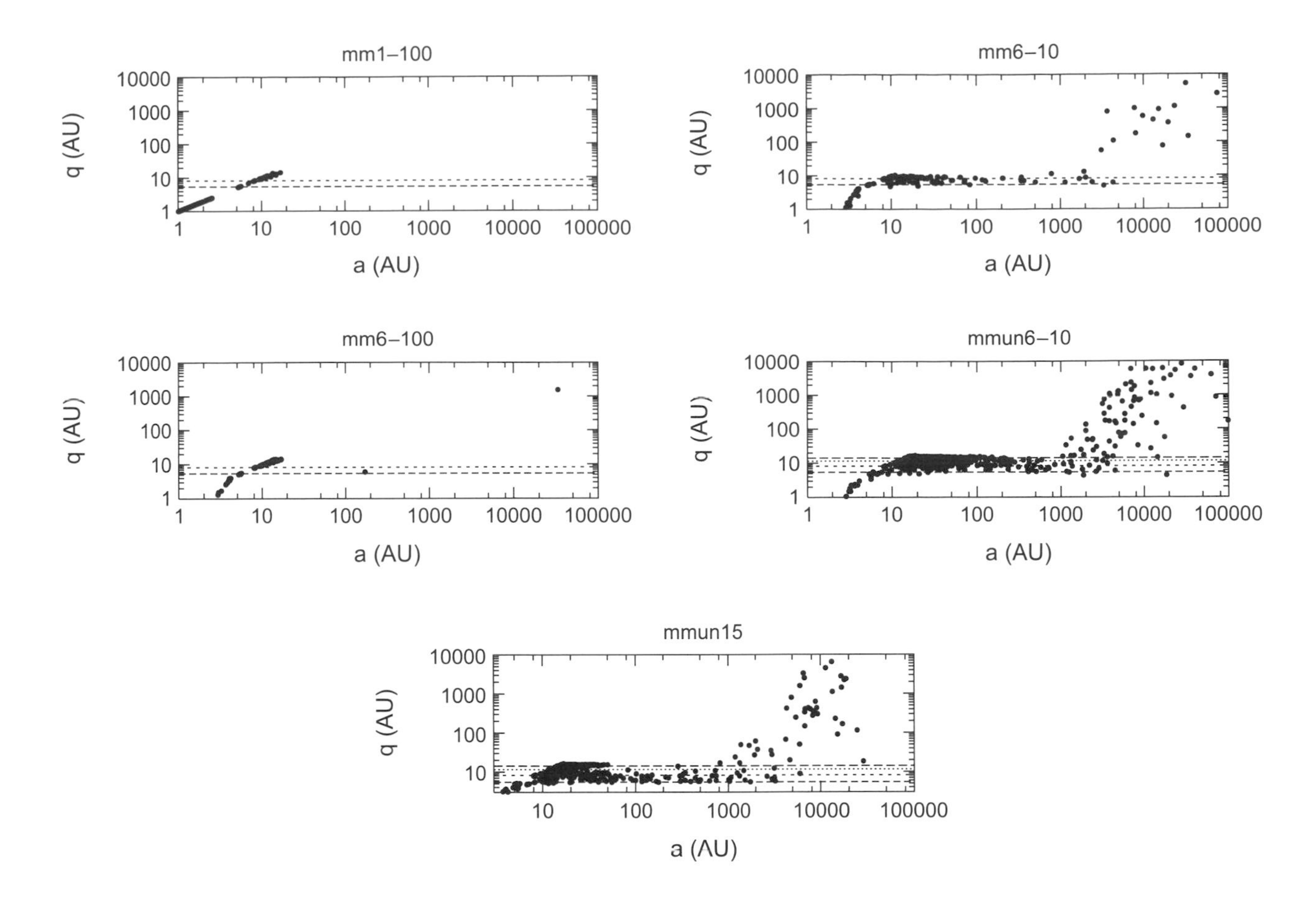

Fig. 5. Snapshots of the endstates in a–q space are shown for five runs from *BDL07*. All runs have a cluster central density of 10^3 $M_\odot$ pc^{-3}. The headers above each caption are indicative of the parameters used. The horizonal lines correspond to the orbital radius of the planets used. See text for details.

ser et al. (2007, hereafter *BDL07*) built upon the work of *BDL06* by incorporating the aerodynamic drag (adapting the results of *Adachi et al.,* 1976) and gravitational potential of the primordial solar nebula. The solar nebula was approximated by the minimum-mass Hayashi model (*Hayashi et al.,* 1985) with scale height $0.047s^{5/4}$ AU, where s is the cylindrical distance (in AU) from the Sun, and in which the inner and outer radii were truncated at various distances from the Sun. In all the simulations, the density of the primordial solar nebula decayed exponentially with an e-folding time of 2 m.y. Since the deceleration due to gas drag experienced by a comet is inversely proportional to its size, a typical comet radius of 1.7 km was adopted for most of the simulations [consistent with the LINEAR observations of LPCs of *Francis* (2005) and using the relation between mass and absolute magnitude of *Weissman* (1996)]. The numerical simulations of *BDL07* followed the evolution of comets subject to the gravitational influence of the Sun, Jupiter, Saturn, star cluster, and primordial solar nebula; some of the simulations included the gravitational influence of Uranus and Neptune as well.

Figure 5 shows a vs. q at the end of the simulation for runs 1–5 of *BDL07,* i.e., after 5 m.y. of evolution. The top-left panel refers to run 1 (disk truncated at 1 and 100 AU, using a minimum-mass model: mm1–100), the top-right panel is for run 2 (disk truncated at 6.2 and 10 AU, minimum-mass model: mm6–10), the middle-left panel refers to run 3 (disk truncated at 6.2 and 100 AU, minimum-mass model: mm6–100), the middle-right panel to run 4 (disk truncated at 6 and 10 AU with Uranus and Neptune present, minimum-mass model: mmun6–10), and run 5 is the bottom panel (disk truncated at 6.2 and 15 AU with Uranus and Neptune present, minimum-mass model: mmun6–15). The horizontal lines show the pericenter distances corresponding to the orbital distances of Jupiter and Saturn, as well as those of Uranus and Neptune where these are present.

The panel for run 1 shows that the comets end up either on circular orbits inside of Jupiter, as co-orbitals of Jupiter and Saturn or on nearly circular orbits outside of Saturn. No comets end up in the OC: Close to 60% of the comets end up on circular orbits inside of Jupiter and about 20% end up in slightly eccentric orbits at exterior mean-motion resonances with Saturn for reasons that are described in *Weidenschilling and Davis* (1985). The material that ends up inside of Jupiter, however, is not in resonance with this planet. Instead, the density of material peaks at a ~ 1.5 AU and quickly drops to zero at a = 1.8 AU. In other words, the material is inside the current asteroid belt and in the region where the terrestrial planets are now. Material ends up in this region due to a scattering by Jupiter that places the comets on an orbit with q ~ 2 AU. For this value of q the time to decrease the apocenter away from Jupiter's orbit is much shorter than the time to receive a strong scattering by Jupiter, so that the orbit quickly circularizes and subsequently slowly spirals inward.

For the other runs, in which the inner edge of the disk is truncated at 6.2 AU, almost no material ends up inside of Jupiter. For run 3, in which the outer disk extends to100 AU, virtually all the material ends up on low-eccentricity orbits outside Saturn. For runs in which the outer disk is truncated at much smaller radii (e.g., by photoevaporation) (*Hollenbach et al.,* 2000), the absence of drag near aphelion produces comets on eccentric orbits, often trapped in exterior mean-motion resonances. For example, in run 3 (upper right panel of Fig. 5), having the outer edge truncated at 10 AU means that the material does not end up on (nearly) circular orbits outside of Saturn due to gas drag. Instead, the majority of the comets have q interior to 10 AU (78%) with almost all these on eccentric orbits with q ~ 9 AU in exterior mean-motion resonances with Saturn.

In general, the simulations of *BDL07* show that when the primordial solar nebula extends much beyond Saturn or Neptune, virtually no kilometer-sized comets will end up in the inner OC during this phase. Instead, the majority of the material will be on circular orbits inside of Jupiter if the inner edge of the disk is well inside Jupiter's orbit. If the disk's inner edge is beyond Jupiter's orbit, most comets end up on orbits in exterior mean-motion resonances with Saturn when Uranus and Neptune are not present. In those cases where the outer edge of the disk is close to Saturn or Neptune, the fraction of material that ends up in the subsequently formed inner OC is much less than that found in *BDL06* for the same cluster densities.

All this implies that the presence of the primordial solar nebula greatly reduces the population of kilometer-sized comets to be expected in the inner OC. In order to determine the effect of the size of the comets on inner OC formation efficiency, a set of runs with the same initial conditions but different cometary radii have been performed by *BDL07*. It was determined that the threshold comet size to begin producing significant inner OCs is roughly 20 km if the minimum-mass Hayashi model is present for ~2 m.y. This implies that the presence of the primordial solar nebula in the models studied acts as a size-sorting mechanism, with large bodies (such as Sedna) unaffected by the gas drag and ending up in the inner OC while kilometer-sized comets remain in the planetary region (or in some models in the Saturn- or Uranus-Neptune scattered disks). It should be noted, however, that comets left on dynamically cold orbits just beyond the outermost planet are likely to be eventually scattered (typically to large semimajor axes in a manner reminiscent of the models described in section 6) in the relatively gas-free environment in which Uranus and Neptune are thought subsequently to migrate (see, e.g., *Tsiganis et al.,* 2005).

8. AVENUES TO BE EXPLORED

The models of OC formation described above are still highly idealized. The "reference model" of section 6 assumes that the OC formed by the scattering of residual planetesimals by the giant planets in their current configuration with the Sun in its current galactic environment. However, as noted in that section, there appear to be several difficulties with this simple model, not the least of which is that it does not explain the origins of orbits such as that of (90377) Sedna. Thus, the formation of the OC needs to be studied in the context of the Sun's early galactic environment and with a realistic model for planet formation. That is, models must incorporate the fact that the planets were still forming during at least the early stages of the formation of the OC. Planetary migration in the early solar system (*Fernández and Ip,* 1984) appears to have been important in shaping the Kuiper belt (*Malhotra,* 1995; *Gomes,* 2003; *Levison and Morbidelli,* 2003; *Gomes et al.,* 2004; *Tsiganis et al.,* 2005), and the same is likely true for the OC. Uranus and Neptune may even have formed in the Jupiter-Saturn region (*Thommes et al.,* 1999, 2002), likely changing the fraction of comets that ended up in the OC (see section 2). Also, as noted in section 7, gas drag may have played a major role in inner OC formation and collisions may have been important in determining which regions of the protoplanetary disk could populate the OC (*Stern and Weissman,* 2001; *Stern,* 2003; but cf. *Charnoz and Morbidelli,* 2003, 2007).

As noted in section 3.4, it is also quite likely that the OC that was formed while the Sun was in its stellar nursery has been subjected to pruning and possibly more dramatic shaking by passage through GMCs and stochastic spiral arms during its multiple circuits around the galactic center over the past 4.5 G.y. Future investigations of these effects will doubtless cast further light on this issue.

On the observational side, several groundbased telescopes [e.g., SkyMapper, PanSTARRS and LSST (see *Francis,* 2005, for a review)], each capable of studying large areas to deeper than 22nd magnitude, may yield several hundred new LPC detections over the next few years. Many of these will have perihelia out to and beyond 10 AU and will provide critically needed information about the inner OC. In addition, occultation observations (e.g., *Chen et al.,* 2007; *Roques et al.,* 2006; *Cheng et al.,* 2006) will provide constraints on the TNO population in general while proposals for occultation measurements from space (*Lehner,* 2006) offer the tantalizing possibility of detecting comet-sized bodies within the OC itself! Given the dramatic developments of the past few years and the theoretical and observational prospects over the next few years, it is clear that exciting times are ahead in this field.

Acknowledgments. This work was partially supported by the NASA Planetary Geology and Geophysics Program. M.D. and

R.B. are grateful to Canada's NSERC grant program for continued support and R.B. is grateful for a CITA National Postdoctoral Fellowship.

REFERENCES

Adachi I., Hayashi C., and Nakazawa K. (1976) The gas drag effect on the elliptical motion of a solid body in the primordial solar nebula. *Progr. Theor. Phys., 56,* 1756–1771.

Adams F. C. and Laughlin G. (2001) Constraints on the birth aggregate of the solar system. *Icarus, 150,* 151–162.

Adams F. C., Proszkow E. M., Fatuzzo M., and Myers P. C. (2006) Early evolution of stellar groups and clusters: Environmental effects on forming planetary systems. *Astrophys. J., 641,* 504–525.

A'Hearn M. and 32 colleagues (2005) Deep Impact: Excavating Comet Tempel 1. *Science, 310,* 255–264.

Bailey M. E. (1983) The structure and evolution of the solar system comet cloud. *Mon. Not. R. Astron. Soc., 204,* 603–633.

Bailey M. E. (1984) The steady-state 1/a distribution and the problem of cometary fading. *Mon. Not. R. Astron. Soc., 211,* 347–368.

Bailey M. E. (1986) The mean energy transfer rate to comets in the Oort cloud and implications for cometary origins. *Mon. Not. R. Astron. Soc., 218,* 1–30.

Bailey M. E. and Stagg C. R. (1988) Cratering constraints on the inner Oort cloud — Steady-state models. *Mon. Not. R. Astron. Soc., 235,* 1–32.

Bailey M. E., Wilkinson D. A., and Wolfendale A. W. (1987) Can episodic comet showers explain the 30 m.y. cyclicity in the terrestrial record? *Mon. Not. R. Astron. Soc., 227,* 863–885.

Bailey M. E., Clube S. V. M., and Napier W. M. (1990) *The Origin of Comets.* Pergamon, Oxford, England. 599 pp.

Biermann L. (1978) Dense interstellar clouds and comets. In *Astronomical Papers Dedicated to Bengt Stromgren* (A. Reiz and T. Anderson, eds.), pp. 327–335. Copenhagen Observatory, Copenhagen.

Binney J. and Tremaine S. (1987) *Galactic Dynamics.* Princeton Univ., Princeton, New Jersey. 747 pp.

Brasser R., Duncan M. J., and Levison H. F. [BDL06] (2006) Embedded star clusters and the formation of the Oort Cloud. *Icarus, 184,* 59–82.

Brasser R., Duncan M. J., and Levison H. F. [BDL07] (2007) Embedded star clusters and the formation of the Oort cloud: II. The effect of the primordial solar nebula. *Icarus,* in press.

Brin G. D. and Mendis D. A. (1979) Dust release and mantle development in comets. *Astrophys. J., 229,* 402–408.

Brown M. E., Trujillo C., and Rabinowitz D. (2004) Discovery of a candidate inner Oort cloud planetoid. *Astrophys. J. Lett., 617,* 645–649.

Brown M. E., Trujillo C., and Rabinowitz D. (2005) Discovery of a planet-sized body in the scattered Kuiper Belt. *Astrophys. J. Lett., 635,* L97–L100.

Byl J. (1983) Galactic perturbations on nearly parabolic cometary orbits. *Moon and Planets, 29,* 121 137.

Byl J. (1986) The effect of the galaxy on cometary orbits. *Earth Moon Planets, 36,* 263–273.

Byl J. (1990) Galactic removal rates for long-period comets. *Astron. J., 99,* 1632–1635.

Carusi A., Kresak L., Perozzi E., and Valsecchi G. B. (1987) High-order librations of Halley-type comets. *Astron. Astrophys., 187,* 899–890.

Chabrier G. (2001) The galactic disk mass budget. I. Stellar mass function and density. *Astrophys. J., 554,* 1274–1281.

Chabrier G. (2002) The galactic disk mass budget. II. Brown dwarf mass function and density. *Astrophys. J., 567,* 304–313.

Chabrier G. (2003) Galactic stellar and substellar initial mass function. *Publ. Astron. Soc. Pac., 115,* 763–795.

Chang H.-K., King S.-K., Liang J.-S., Wu P.-S., Lin L. C.-C., and Chiu J.-L. (2006) Occultation of X-rays from Scorpius X-1 by small trans-neptunian objects. *Nature, 442,* 660–663.

Charnoz S. and Morbidelli A. (2003) Coupling dynamical and collisional evolution of small bodies: An application to the early ejection of planetesimals from the Jupiter-Saturn region. *Icarus, 166,* 141–156.

Charnoz S. and Morbidelli A. (2007) Coupling dynamical and collisional evolution of small bodies II: Forming the Kuiper belt, the scattered disk and the Oort cloud. *Icarus, 188,* 468–480.

Chen W. P. and 26 colleagues (2007) Search for small trans-neptunian objects by the TAOS Project. In *Near Earth Objects, Our Celestial Neighbors: Opportunity and Risk* (G. B. Valsecchi and D. Vokrouhlický, eds.), pp. 65–68. IAU Symposium No. 236, Cambridge Univ., Cambridge.

Davidsson B. J. R. and Gutiérrez P. J. (2004) Estimating the nucleus density of Comet 19P/Borrelly. *Icarus, 168,* 392–408.

Davis M., Hut P., and Muller R. A. (1984) Extinction of species by periodic comet showers. *Nature, 308,* 715–717.

de la Fuente Marcos C. and de la Fuente Marcos R. (2002) On the origin of comet C/1999 S4 LINEAR. *Astron. Astrophys., 395,* 697–704.

Delsemme A. H. (1987) Galactic tides affect the Oort cloud: An observational confirmation. *Astron. Astrophys., 187,* 913–918.

Dones L., Weissman P. R., Levison H. F., and Duncan M. J. [DWLD] (2004) Oort cloud formation and dynamics. In *Comets II* (M. C. Festou et al., eds.), pp. 153–174. Univ. of Arizona, Tucson.

Duncan M. J. and Levison H. F. (1997) A scattered comet disk and the origin of Jupiter family comets. *Science, 276,* 1670–1672.

Duncan M., Quinn T., and Tremaine S. [DQT87] (1987) The formation and extent of the solar system comet cloud. *Astron. J., 94,* 1330–1338.

Duncan M., Quinn T., and Tremaine S. (1988) The origin of short period comets. *Astrophys. J. Lett., 328,* L69–L73.

Duncan M. J., Levison H. F., and Dones L. (2004) Dynamical evolution of ecliptic comets. In *Comets II* (M. C. Festou et al., eds.), pp. 193–204. Univ. of Arizona, Tucson.

Dybczyński P. A. (2002a) On the asymmetry of the distribution of observable comets induced by a star passage through the Oort cloud. *Astron. Astrophys., 383,* 1049–1053.

Dybczyński P. A. (2002b) Simulating observable comets. I. The effects of a single stellar passage through or near the Oort cometary cloud. *Astron. Astrophys., 396,* 283–292.

Dybczyński P. A. (2006) Simulating observable comets. III. Real stellar perturbers of the Oort cloud and their output. *Astron. Astrophys., 449,* 1233–1242.

Edgeworth K. E. (1949) The origin and evolution of the solar system. *Mon. Not. R. Astron. Soc., 109,* 600–609.

Eggers S. (1999) Cometary dynamics during the formation of the solar system. Ph.D. thesis, Max-Planck-Institut für Aeronomie.

Eggers S., Keller H. U., Kroupa P., and Markiewicz W. J. (1997) Origin and dynamics of comets and star formation. *Planet. Space Sci., 45,* 1099–1104.

Eggers S., Keller H. U., Markiewicz W. J., and Kroupa P. (1998) Cometary dynamics in a star cluster (abstract). *Astronomische*

Gesellschaft Meeting Abstracts, 14, 5.

Emel'yanenko V. V. and Bailey M. E. (1998) Capture of Halley-type comets from the near-parabolic flux. *Mon. Not. R. Astron. Soc., 298,* 212–222.

Everhart E. (1967) Intrinsic distributions of cometary perihelia and magnitudes. *Astron. J., 72,* 1002–1011.

Fernández J. A. (1978) Mass removed by the outer planets in the early solar system. *Icarus, 34,* 173–181.

Fernández J. A. (1980) Evolution of comet orbits under the perturbing influence of the giant planets and nearby stars. *Icarus, 406,* 406–421.

Fernández J. (1992) Comet showers. In *Chaos, Resonance and Collective Dynamical Phenomena in the Solar System* (S. Ferraz-Mello, ed.), pp. 239–254. IAU Symposium No. 152, Kluwer, Dordrecht.

Fernández J. A. (1997) The formation of the Oort cloud and the primitive galactic environment. *Icarus, 129,* 106–119.

Fernández J. A. (2002) Changes in the inclination-distribution of long-period comets with the orbital energy. In *Asteroids, Comets, and Meteors (ACM 2002)* (B. Warmbein, ed.), pp. 303–304. ESA SP-500, Noordwijk, The Netherlands.

Fernández J. A. and Brunini A. (2000) The buildup of a tightly bound comet cloud around an early Sun immersed in a dense galactic environment: Numerical experiments. *Icarus, 145,* 580–590.

Fernández J. A. and Gallardo T. (1994) The transfer of comets from parabolic orbits to short-period orbits: Numerical studies. *Astron. Astrophys., 281,* 911–922.

Fernández J. A. and Ip W.-H. (1984) Some dynamical aspects of the accretion of Uranus and Neptune — The exchange of orbital angular momentum with planetesimals. *Icarus, 58,* 109–120.

Fernández J. A. and Ip W.-H. (1991) Statistical and evolutionary aspects of cometary orbits. In *Comets in the Post-Halley Era* (R. L. Newburn et al., eds.), pp. 487–535. Astrophysics and Space Science Library Vol. 167, Kluwer, Dordrecht.

Festou M. C., Rickman H., and West R. M. (1993a) Comets. I — Concepts and observations. *Astron. Astrophys. Rev., 4,* 363–447.

Festou M. C. Rickman H., and West R. M. (1993b) Comets. II — Models, evolution, origin and outlook. *Astron. Astrophys. Rev., 5,* 37–163.

Francis P. J. (2005) The demographic of long-period comets. *Astrophys. J., 635,* 1348–1361.

Gaidos E. J. (1995) Paleodynamics: Solar system formation and the early environment of the sun. *Icarus, 114,* 258–268.

García-Sánchez J., Preston R. A, Jones D. L., Weissman P. R., Lestrade J. F., Latham D. W., and Stefanik R. P. (1999) Stellar encounters with the Oort cloud based on Hipparcos data. *Astron. J., 117,* 1042–1055. (Erratum in *Astron. J., 118,* 600.)

García-Sánchez J., Weissman P. R., Preston R. A., Jones D. L., Lestrade J.-F., Latham D. W., Stefanik R. P., and Paredes J. M. (2001) Stellar encounters with the solar system. *Astron. Astrophys., 379,* 634–659.

Gladman B. and Chan C. (2006) Production of the extended scattered disk by rogue planets. *Astrophys. J. Lett., 643,* L135–L138.

Gladman B., Holman M., Grav T., Kavelaars J., Nicholson P., Aksnes K., and Petit J.-M. (2002) Evidence for an extended scattered disk. *Icarus, 157,* 269–279.

Goldreich P., Lithwick Y., and Sari R. (2004) Final stages of planet formation. *Astrophys. J., 614,* 497–507.

Gomes R. S. (2003) The origin of the Kuiper belt high-inclination population. *Icarus, 161,* 404–418.

Gomes R. S., Morbidelli A., and Levison H. F. (2004) Planetary migration in a planetesimal disk: Why did Neptune stop at 30 AU? *Icarus, 170,* 492–507.

Gomes R. S., Gallardo T., Fernandez J. A., and Brunini A. (2005) On the origin of the high-perihelion scattered disk: The role of the Kozai mechanism and mean motion resonances. *Cel. Mech. Dyn. Astron., 91,* 109–129.

Gomes R. S., Matese J. J., and Lissauer J. J. (2006) A distant planetary-mass solar companion may have produced distant detached objects. *Icarus, 184,* 589–601.

Greenberg R., Weidenschilling S. J., Chapman C. R., and Davis D. R. (1984) From icy planetesimals to outer planets and comets. *Icarus, 59,* 87–113.

Gutermuth R. A., Megeath S. T., Pipher J. L., Williams J. P., Allen L. E., Myers P. C. and Raines S. N. (2005) The initial configuration of young stellar clusters: A K-band number counts analysis of the surface density of stars. *Astrophys. J., 632,* 397–420.

Hahn J. M. and Malhotra R. (1999) Orbital evolution of planets embedded in a planetesimal disk. *Astron. J., 117,* 3041–3053.

Hänninen J. and Flynn C. (2002) Simulations of the heating of the galactic stellar disc. *Mon. Not. R. Astron. Soc., 337,* 731–742.

Hayashi C., Nakazawa K., and Nakagawa Y. (1985) Formation of the solar system. In *Protostars and Planets II* (D. C. Black and M. S. Matthews, eds.), pp. 1100–1153. Univ. of Arizona, Tucson.

Heisler J. (1990) Monte Carlo simulations of the Oort comet cloud. *Icarus, 88,* 104–121.

Heisler J. and Tremaine S. (1986) The influence of the galactic tidal field on the Oort comet cloud. *Icarus, 65,* 13–26.

Heisler J., Tremaine S., and Alcock C. (1987) The frequency and intensity of comet showers from the Oort cloud. *Icarus, 70,* 269–288.

Higuchi A., Kokubo E., and Mukai T. (2002) Cometary dynamics: Migration due to gas drag and scattering by protoplanets. *Asteroids, Comets, and Meteors (ACM 2002)* (B. Warmbein, ed.), pp. 453–456. ESA SP-500, Noordwijk, The Netherlands.

Hills J. G. (1981) Comet showers and the steady-state infall of comets from the Oort cloud. *Astron. J., 86,* 1730–1740.

Hollenbach D. J., Yorke H. W., and Johnstone D. (2000) Disk dispersal around young stars. In *Protostars and Planets IV* (V. Mannings et al., eds.), pp. 401–429. Univ. of Arizona, Tucson.

Horner J. and Evans N. W. (2002) Biases in cometary catalogues and Planet X. *Mon. Not. R. Astron. Soc., 335,* 641–654.

Hughes D. W. (2001) The magnitude distribution, perihelion distribution and flux of long-period comets. *Mon. Not. R. Astron. Soc., 326,* 515–523.

Hut P. and Tremaine S. (1985) Have interstellar clouds disrupted the Oort comet cloud? *Astron. J., 90,* 1548–1557.

Hut P., Alvarez W., Elder W. P., Kauffman E. G., Hansen T., Keller G., Shoemaker E. M., and Weissman P. R. (1987) Comet showers as a cause of mass extinction. *Nature, 329,* 118–126.

Jenkins A. and Binney J. (1990) Spiral heating of galactic discs. *Mon. Not. R. Astron. Soc., 245,* 305–317.

Jetsu L. and Pelt J. (2000) Spurious periods in the terrestrial impact crater record. *Astron. Astrophys., 353,* 409–418.

Jewitt D. (2004) From cradle to grave: The rise and demise of the comets. In *Comets II* (M. C. Festou et al., eds.), pp. 659–676. Univ. of Arizona, Tucson.

Kenyon S. J. and Bromley B. C. (2002) Collisional cascades in planetesimal disks. I. Stellar flybys. *Astron. J., 123,* 1757–1775.

Kenyon S. J. and Bromley B. C. (2004) Stellar encounters as the origin of distant solar system objects in highly eccentric orbits. *Nature, 432,* 598–602.

Kroupa P., Aarseth S., and Hurley J. (2001) The formation of a bound star cluster: From the Orion nebula cluster to the Pleiades. *Mon. Not. R. Astron. Soc., 321,* 699–712.

Kuiper G. P. (1951) On the origin of the solar system. In *Proceedings of a Topical Symposium, Commemorating the 50th Anniversary of the Yerkes Observatory and Half a Century of Progress in Astrophysics* (J. A. Hynek, ed.) pp. 357–414. McGraw-Hill, New York.

Lada C. J. and Lada E. A. (2003) Embedded clusters in molecular clouds. *Annu. Rev. Astron. Astrophys. 41,* 57–115.

Lada C. J., Margulis M., and Dearborn D. (1984) The formation and early dynamical evolution of bound stellar systems. *Astrophys. J., 285,* 141–152.

Lehner M. J. (2006) The Whipple mission. Talk presented at 2006 TNO Conference in Catania, Italy.

Levison H. F. (1996) Comet taxonomy. In *Completing the Inventory of the Solar System* (T. W. Rettig and J. M. Hahn, eds.), pp. 173–191. ASP Conf. Series 107, San Francisco.

Levison H. F. and Morbidelli A. (2003) Forming the Kuiper belt by the outward transport of bodies during Neptune's migration. *Nature, 426,* 419–421.

Levison H. F., Dones L., and Duncan M. J. [LDD01] (2001) The origin of Halley-type comets: Probing the inner Oort cloud. *Astron. J., 121,* 2253–2267.

Levison H. F., Morbidelli A., Dones L., Jedicke R., Wiegert P. A., and Bottke W. F. (2002) The mass disruption of Oort cloud comets. *Science, 296,* 2212–2215.

Levison H. F., Duncan M. J., Dones L., and Gladman B. J. [LDDG06] (2006) The scattered disk as a source of Halley-type comets. *Icarus, 184,* 619–633.

Malhotra R. (1995) The origin of Pluto's orbit: Implications for the solar system beyond Neptune. *Astron. J., 110,* 420–429.

Matese J. and Whitmire D. (1996) Tidal imprint of distant galactic matter on the Oort comet cloud. *Astrophys. J. Lett., 472,* L41–L43.

Matese J. J., Whitman P. G., Innanen K. A., and Valtonen M. J. (1995) Periodic modulation of the Oort cloud comet flux by the adiabatically changing galactic tide. *Icarus, 116,* 255–268.

Matese J. J., Whitman P. G., and Whitmire D. P. (1999) Cometary evidence of a massive body in the outer Oort cloud. *Icarus, 141,* 354–366.

Matese J. J., Whitmire D. P. and Lissauer J. L. (2005) A wide binary solar companion as a possible origin of Sedna-like objects. *Earth Moon Planets, 97,* 459–470.

Morbidelli A. and Levison H. (2004) Scenarios for the origin of the orbits of the trans-Neptunian objects 2000 CR_{105} and 2003 VB_{12}. *Astron. J., 128,* 2564–2576.

Muller R. A. (2002) Measurement of the lunar impact record for the past 3.5 b.y. and implications for the Nemesis theory. In *Catastrophic Events and Mass Extinctions: Impacts and Beyond* (C. Koeberl and K. G. MacLeod, eds.), pp. 659–665. GSA Special Paper 356, Boulder, Colorado.

Murray J. B. (1999) Arguments for the presence of a distant large undiscovered solar system planet. *Mon. Not. R. Astron. Soc., 309,* 31–34.

Napier W. M. and Clube S. V. M. (1979) A theory of terrestrial catastrophism. *Nature, 282,* 455–459.

Napier W. M. and Staniucha M. (1982) Interstellar planetesimals. I — Dissipation of a primordial cloud of comets by tidal encounters with massive nebulae. *Mon. Not. R. Astron. Soc., 198,* 723–735.

Neslusan L. (2007) The fading problem and the population of the Oort cloud. *Astron. Astrophys., 461,* 741–750.

Oort J. H. (1950) The structure of the cloud of comets surrounding the solar system and a hypothesis concerning its origin. *Bull. Astron. Inst. Neth., 11,* 91–110.

Öpik E. J. (1932) Note on stellar perturbations of nearly parabolic orbits. *Proc. Am. Acad. Arts Sci., 67,* 169–183.

Quinn T., Tremaine S., and Duncan M. (1990) Planetary perturbations and the origins of short-period comets. *Astrophys. J., 355,* 667–679.

Roques F. and 17 colleagues (2006) Exploration of the Kuiper belt by high-precision photometric stellar occultations: First results. *Astrophys. J., 132,* 819–822.

Safronov V. S. (1972) Ejection of bodies from the solar system in the course of the accumulation of the giant planets and the formation of the cometary cloud. In *The Motion, Evolution of Orbits, and Origin of Comets* (G. A. Chebotarev et al., eds.), pp. 329–334. IAU Symposium No. 45, Kluwer, Dordrecht.

Shoemaker E. M. and Wolfe R. F. (1984) Evolution of the Uranus-Neptune planetesimal swarm (abstract). In *Lunar and Planetary Science XV,* pp. 780–781. Lunar and Planetary Institute, Houston.

Shoemaker E. M. and Wolfe R. F. (1986) Mass extinctions, crater ages, and comet showers. In *The Galaxy and the Solar System* (R. Smoluchowski et al., eds.), pp. 338–386. Univ. of Arizona, Tucson.

Smoluchowski R. and Torbett M. (1984) The boundary of the solar system. *Nature, 311,* 38–39.

Spitzer L. J. and Schwarzschild M. (1953) The possible influence of interstellar clouds on stellar velocities. II. *Astrophys. J., 118,* 106–112.

Stern S. A. (2003) The evolution of comets in the Oort cloud and Kuiper belt. *Nature, 424,* 639–642.

Stern S. A. and Weissman P. R. (2001) Rapid collisional evolution of comets during the formation of the Oort cloud. *Nature, 409,* 589–591.

Thommes E. W., Duncan M. J., and Levison H. F. (1999) The formation of Uranus and Neptune in the Jupiter-Saturn region of the solar system. *Nature, 402,* 635–638.

Thommes E. W., Duncan M. J., and Levison H. F. (2002) The formation of Uranus and Neptune among Jupiter and Saturn. *Astron. J., 123,* 2862–2883.

Tremaine S. (1993) The distribution of comets around stars. In *Planets Around Pulsars* (J. A. Phillips et al., eds.), pp. 335–344. ASP Conf. Series 36, San Francisco.

Trujillo C. A., Jewitt D. C., and Luu J. X. (2000) Population of the scattered Kuiper belt. *Astrophys. J. Lett., 529,* L103–L106.

Tsiganis K., Gomes R., Morbidelli A., and Levison H. F. (2005) Origin of the orbital architecture of the giant planets in the solar system. *Nature, 402,* 635–638.

Weaver H. A. and 20 colleagues (2001) HST and VLT investigations of the fragments of Comet C/1999 S4 (LINEAR). *Science, 292,* 1329–1334.

Weidenschilling S. J. (1997) The origin of comets in the solar nebula: A unified model. *Icarus, 127,* 290–306.

Weidenschilling S. J. (2004) From icy grains to comets. In *Comets II* (M. C. Festou et al., eds.), pp. 97–104. Univ. of Arizona, Tucson.

Weidenschilling S. J. and Davis D. (1985) Orbital resonances in the solar nebula: Implications for planetary accretion. *Icarus, 62,* 16–29.

Weinberg M. D., Shapiro S. L., and Wasserman I. (1987) The dynamical fate of wide binaries in the solar neighborhood. *Astrophys. J., 312,* 367–389.

Weissman P. R. (1980) Physical loss of long-period comets. *Astron. Astrophys., 85,* 191–196.

Weissman P. R. (1990) The Oort cloud. *Nature, 344,* 825–830.

Weissman P. R. (1996) The Oort cloud. In *Completing the Inventory of the Solar System* (T. W. Rettig and J. M. Hahn, eds.), pp. 265–288. ASP Conf. Series 107, San Francisco.

Weissman P. R. (1999) Diversity of comets: Formation zones and dynamical paths. *Space Sci. Rev., 90,* 301–311.

Whipple F. L. (1962) On the distribution of semi-major axes among comet orbits. *Astron. J., 67,* 1–9.

Whitmire D. P. and Jackson A. A. (1984) Are periodic mass extinctions driven by a distant solar companion? *Nature, 308,* 713–715.

Wiegert P. and Tremaine S. (1999) The evolution of long-period comets. *Icarus, 137,* 84–121.

Williams J. P., Blitz L., and McKee C. F. (2000) The structure and evolution of molecular clouds: From clumps to cores to the IMF. In *Protostars and Planets IV* (V. Mannings et al., eds.), pp. 97–120. Univ. of Arizona, Tucson.

Yabushita S. (1980) On exact solutions of diffusion equations in cometary dynamics. *Astron. Astrophys., 85,* 77–79.

Zheng J. Q., Valtonen M. J., and Valtaoja L. (1990) Capture of comets during the evolution of a star cluster and the origin of the Oort cloud. *Cel. Mech. Dyn. Astron., 49,* 265–272.

Part VI:

Individualities and Peculiarities

The Largest Kuiper Belt Objects

Michael E. Brown
California Institute of Technology

While for the first decade of the study of the Kuiper belt, a gap existed between the sizes of the relatively small and faint Kuiper belt objects (KBOs) that were being studied and the largest known KBO, Pluto, recent years have seen that gap filled and the maximum size even expanded. These large KBOs occupy all dynamical classes of the Kuiper belt with the exception of the cold classical population, and one large object, Sedna, is the first member of a new more distant population beyond the Kuiper belt. Like Pluto, most of the large KBOs are sufficiently bright for detailed physical study, and, like Pluto, most of the large KBOs have unique dynamical and physical histories that can be gleaned from these observations. The four largest known KBOs contain surfaces dominated in methane, but the details of the surface characteristics differ on each body. One large KBO is the parent body of a giant impact that has strewn multiple fragments throughout the Kuiper belt. The large KBOs have a significantly larger satellite fraction than the remainder of the Kuiper belt, including the only known multiple satellite systems and the relatively smallest satellites known. Based on the completeness of the current surveys, it appears that approximately three more KBOs of the same size range likely still await discovery, but that tens to hundreds more exist in the more distant region where Sedna currently resides.

1. INTRODUCTION

While once Pluto appeared as a unique object in the far reaches of the solar system, the discovery of the Kuiper belt caused the immediate realization that Pluto is a member of a much larger population. But while Pluto's orbit makes it a typical member of the Kuiper belt population dynamically, Pluto itself has still remained special as one of the few transneptunian objects bright enough for detailed studies. Much of what we understand of the composition, density, and history of objects in the Kuiper belt ultimately derives from detailed studies of Pluto.

Recently, however, surveys of the Kuiper belt began to discover Kuiper belt objects (KBOs) of comparable and now even larger size than Pluto. The largest survey to date has used the 48-inch Palomar Schmidt telescope to cover almost 20,000 deg^2 of sky to a limiting magnitude of R ~ 20.5 (Fig. 1). This survey has uncovered most of the known large KBOs (i.e., *Trujillo and Brown,* 2003; *Brown et al.,* 2004, 2005b). A total of 71 objects beyond 30 AU have been detected, of which 21 were previously known (or have been independently discovered subsequently). Recovery of objects is still underway to define the dynamics of the large objects; to date 54 of the 71 objects have secure orbits.

This survey for the largest KBOs serves both as a search for individual objects bright enough for detailed study and also as the first modern widefield survey of the outer solar system to more fully define the dynamical properties of the entire region. In this chapter we will first survey the dynamical properties of the largest KBOs and compare them to the population as a whole, then we will examine the largest individual KBOs, and finally we will review the bulk properties of these largest objects, summarized in Table 1.

2. POPULATION PROPERTIES OF THE LARGEST KUIPER BELT OBJECTS

2.1. Dynamical Distribution

As first noticed by *Levison and Stern* (2001), KBOs brighter than an absolute magnitude of about 6.0 are distributed with a much broader inclination distribution than those fainter. Physically this trend is better stated that the low-inclination population is missing the largest objects (or at least the brightest objects) that are found in the high-inclination population. This effect is easily visible in a simple plot of inclination vs. absolute magnitude, but such a direct comparison of simple discovery statistics is thoroughly biased by the fact that most surveys for fainter KBOs have been restricted much more closely to the ecliptic and thus preferentially find low-inclination objects.

One method for examining a population relatively unbiased by differences in latitudinal coverage of surveys is to consider only objects detected at a restricted range of latitudes. In such debiased examinations, no statistically significant difference can be discerned between the size distribution of the high- and low-inclination populations. From the discovery statistics alone no definitive indication exists that the populations differ. The question must be addressed with actual measurements of sky densities of KBOs of different brightnesses rather than simple discovery statistics. Fortunately, the Palomar survey for large KBOs is complete for low inclinations (with the exception of the galactic plane), so we now know that there are no objects brighter than absolute magnitude 4.5 in the low-inclination population [or, more pertinently, in the cold classical region of the Kuiper belt, defined by *Morbidelli and Brown* (2005) as the dy-

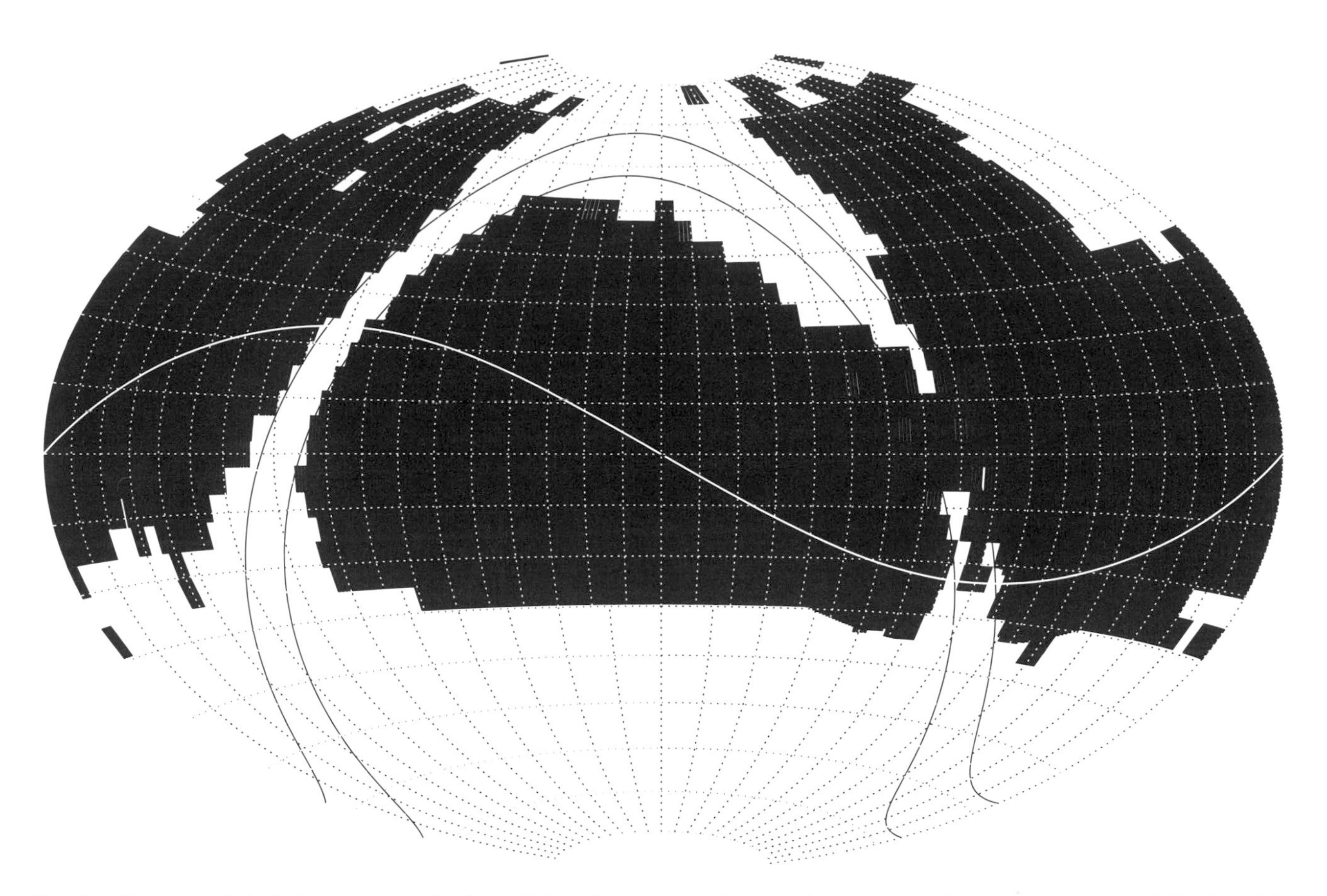

Fig. 1. Coverage of the Palomar survey for large Kuiper belt objects. The map is centered at RA and declination of 0°. The white line shows the ecliptic. Approximately 20,000 deg^2 north of –30° declination, mostly avoiding the galactic plane, have been covered to a limiting magnitude of R ~ 20.5. Seventy-one large KBOs have been found in the survey, including most of the large KBOs discussed here.

TABLE 1. Properties of the largest Kuiper belt objects.

	Eris	Pluto	2005 FY_9	2003 EL_{61}	Sedna	Quaoar	Orcus	Ixion
Diameter (km)	2400 ± 100	2290	1500 ± 300	~2000 × 1500 × 1000	1300–1800	1260 ± 190	950 ± 70	590 ± 190
a (AU)	67.8	39.6	45.7	43.2	488	43.1	39.4	39.3
e	0.44	0.25	0.15	0.19	0.84	0.04	0.22	0.25
i (deg)	44.0	17.1	29.0	28.2	11.9	8.00	20.5	19.7
r (AU)	96.8	31.2	52.0	51.1	88.5	43.3	47.8	42.1
H	–1.2	–1.0	–0.3	0.3	1.6	2.7	2.3	3.4
Surface composition	CH_4 + ?	$CH_4 + CO + N_2$	$CH_4 + C_2H_6$	H_2O	$CH_4 + N_2$	H_2O + ?	H_2O	?
Albedo (%)	86 ± 7	50–65	80^{+10}_{-20}	~73	15–30	9 ± 3	20 ± 3	15^{+15}_{-6}
Mass (10^{20} kg)	166 ± 2	130.5 ± 0.6	—	42 ± 1	—	—	9 ± 1	—
Density (g cm^{-3})	2.3 ± 0.3	2.03 ± 0.06	—	~2.6	—	—	1.9 ± 0.4	—
Satellite frac. brightness (%)	0.4	18, 0.018, 0.015	—	5.9, 1.5	—	0.6	8	—
Satellite period (days)	15.8	6.4, 38.2, 24.8	—	49.1, ?		?	9.8	—
Additional sat. limit (%)	0.04	0.001	0.01	0.5	0.2	0.2	0.1	0.5

References for all data can be found throughout the text.

namically and physically distinct subpopulation of classical KBOs with uniquely uniform red colors and inclinations lower than about 4°], while in the excited population (defined as the resonant, scattered, and hot classical population) 29 objects brighter than that absolute magnitude are currently known to exist, with the current brightest (Eris) known having an absolute magnitude of –1.2. The difference in maximum brightness and presumably maximum size between the cold classical and excited populations is vast.

This difference in maximum size places a powerful constraint on the dynamical rearrangement of the outer solar system. No dynamical process can preferentially damp the

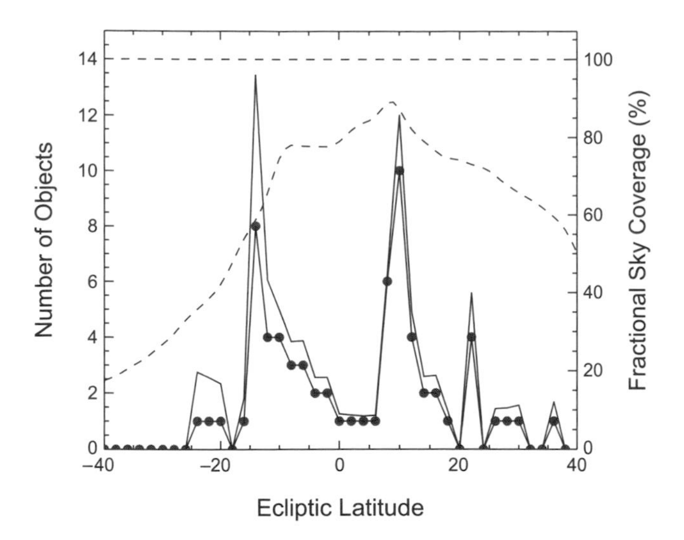

Fig. 2. The latitudinal distribution of objects found in the Palomar survey for large KBOs. The lower line with dots shows the number of KBO detections in 2° bins. The dashed line shows the fractional sky coverage as a function of ecliptic latitude. Sky coverage is incomplete because of galactic plane avoidance (substantial) gaps between CCDs in the mosaic camera, and occasional lack of sky coverage. The solid line above the dots shows the expected number of large KBOs per latitude bin corrected for sky coverage. The prominent peaks in sky density at –10 and +10 ecliptic latitude are likely a general property of the high-inclination Kuiper belt rather than a property of only the large KBOs.

inclinations of only the small KBOs nor preferentially excite the inclinations of only large KBOs, so the high- and low-inclination populations must have either formed at different times or in different places. A current working hypothesis for the larger sizes of the high-inclination population was suggested by *Levison and Stern* (2001) and examined in detail by *Gomes* (2003). They noted that the difference in size distribution can be explained if the largest objects formed in the solar nebula closer to the Sun where nebular densities were higher and growth times were faster and that the objects closer to the Sun suffered more extreme scattering by Neptune and thus acquired higher inclinations. Other forces may be at play, however, and a fully convincing explanation remains elusive.

A survey of the largest KBOs, then, is only a survey of the excited populations of the Kuiper belt. With this caveat, we can now examine the spatial distribution of the largest KBOs. Figure 2 shows the latitudinal distribution, corrected for coverage completeness, of the KBOs from our survey. The prominent peaks around 10° north and south ecliptic latitude cannot be modeled with any simple inclination distribution of objects in circular orbits. Even if all objects in the sky had inclinations of 10° or higher, more objects would appear at lower latitudes than are seen in the survey. While such a latitudinal distribution is impossible for objects with circular or even randomly oriented orbits, many of the objects are consistent with being resonant objects and thus can have preferential orientations in the sky. Pluto, for example, as well as many other KBOs in 3:2 resonance with Neptune, comes to perihelion near its maximum excursion above the ecliptic. This effect will cause a magnitude-limited survey to preferentially detect resonant objects at large distances above the ecliptic. A full examination of this effect awaits full dynamical characterization of the survey population, but from the preliminary data it appears that resonances are likely able to explain these high-latitude concentrations. If true, the high-latitude concentrations are not likely a characteristic of the largest KBOs, but a general property of the high-inclination Kuiper belt, which has not been adequately surveyed until now. The resonant population may be significantly more populated than low-latitude surveys have indicated.

2.2. Beyond the Kuiper Belt

Among the large objects detected, one appears dynamically distinct from the entire Kuiper belt population. Sedna has a perihelion well beyond the main concentration of KBOs and an extreme eccentric orbit with a aphelion around 900 AU (*Brown et al.,* 2004). Although the discovery of Sedna presages a large population in this distant region beyond the Kuiper belt, no surveys for fainter objects have yet succeeded in detecting such distant objects. While some bias against the slow motions of these objects presumably exists in the main KBO surveys, it is also possible that Sedna has an albedo higher than the more numerous smaller members of the population. Sedna could thus be, like Pluto, an atypically bright member of its population, which allows us to detect it much more easily than would have been otherwise possible.

Sedna exists in a dynamical region of the solar system that was not expected to be occupied. It has been proposed to be part of a fossilized inner Oort cloud (*Brown et al.,* 2004; *Brasser et al.,* 2006), a product of a single anomalous stellar encounter (*Morbidelli and Levison,* 2004), an object captured from a passing star (*Kenyon and Bromley,* 2004), a consequence of scattering by now-ejected Kuiper belt planets (*Gladman and Chan,* 2006), a signature of perturbation by a distant massive planet (*Gomes et al.,* 2006), and others. Each of these processes creates a dynamically unique population in this region beyond the Kuiper belt. Finding even a handful more of these distant objects should give powerful insights into some of the earliest processes operating at the beginning of the solar system.

This distant population could be significantly more massive than that of the Kuiper belt. Sedna is currently near perihelion of its 11,000-yr orbit. It would have been detected in the Palomar survey only during a ~150-yr period surrounding perihelion, suggesting that the total number of Sedna-sized or larger objects in the distant population is between about 40 and 120. The total number of Sedna-sized or larger objects in the Kuiper belt is ~5–8. If the distant population has the same size distribution as the Kuiper belt — which seems likely given that the Kuiper belt is the most likely source region for this population — this number of Sedna-sized objects suggests a total mass at least an order of magnitude higher than that in the Kuiper belt.

2.3. Size Distribution

A finite discrete population that generally follows a power-law size distribution cannot maintain this distribution at the largest sizes. Early surveys of the Kuiper belt expected that for the brightest objects the number of detections would fall significantly below the power-law prediction. Figure 3 shows the opposite. For objects brighter than R ~ 19.8 the power law found by *Bernstein et al.* (2004) for the excited population falls well short of the actual numbers of detections. This increase in the numbers of bright objects over that expected is a consequence of the general increase in albedo with size occurring for these objects (see chapter by Stansberry et al.). A plot of number of objects vs. absolute magnitude shows the same trend (with a bias toward higher absolute magnitude because of the flux-limited nature of the survey), and the location of the deviation from the power law is a useful indicator of the approximate location where albedo changes begin to be important. Eight objects brighter than H ~ 3 deviate most strongly from the power law and are a convenient dividing line between the largest individual KBOs and the remaining population. Each of these largest KBOs has interesting unique properties that we describe below.

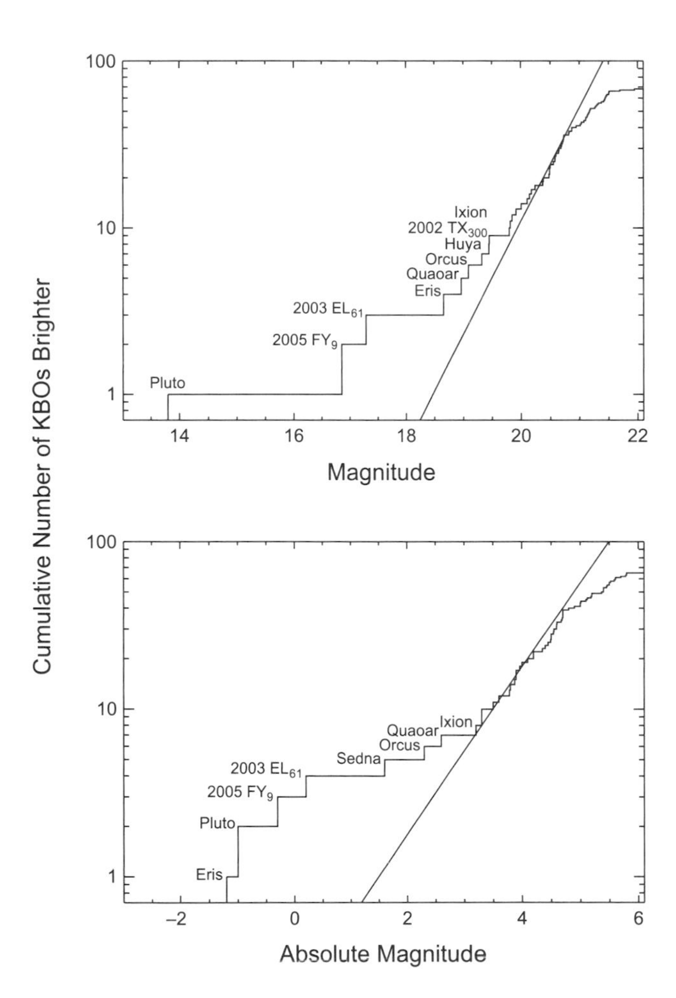

Fig. 3. The cumulative magnitude distribution of the large KBOs found in the Palomar survey. The upper plot shows the total number of KBOs detected brighter than a limiting R magnitude, while the straight line shows the slope of the *Berstein et al.* (2004) power law fit to the distribution of the excited population. The deviation from the power law at magnitudes fainter than ~20.5 is an indication of where the survey begins to become incomplete. The deviation from the power law for the brightest objects, also seen in the distribution of absolute magnitude in the lower plot, is an indication of the increase in albedo of the largest objects.

3. INDIVIDUAL PROPERTIES OF THE LARGEST KUIPER BELT OBJECTS

3.1. Eris

Eris is currently the largest known object in the Kuiper belt. Direct measurement of the size with the Hubble Space Telescope (HST) gives a diameter of 2400 ± 100 km (*Brown et al.,* 2006a), while radiometric measurement with IRAM gives 3000 ± 400 (*Bertoldi et al.,* 2006). While the two measurements appear discrepant, they only differ by 1.5σ owing to the large uncertainty in the radiometric measurement. We will take the measurement with the smaller uncertainty for the remainder of the discussion but comment on the larger diameter at the end. This size measurement implies a remarkably high V-band albedo of 0.86 ± 0.07.

The infrared spectrum of Eris is dominated by absorption from methane similar to the spectrum of Pluto (*Brown et al.,* 2005b) (Fig. 4). Unlike Pluto, however, the infrared spectrum of Eris shows no evidence for the small shifts in the wavelength of the methane absorption associated with methane being dissolved in a nitrogen matrix. The weakest methane absorptions in the visible, however, do possibly show a small shift (*Licandro et al.,* 2006a), perhaps suggesting that methane and nitrogen are layered, with mostly pure methane on the surface (where it is probed by the strong absorption features, which show no shifts) and dissolved methane below (where it is probed by the weak absorption features, which require long path lengths to appear). The weak 2.15-μm absorption feature of nitrogen ice has not been definitively identified, but at the low temperature expected on Eris nitrogen should be in its α, rather than β, form as it is on Pluto. The α form has an absorption even weaker than that of the β form (*Grundy et al.,* 1993; *Tryka et al.,* 1995), so detection may be extremely difficult even if nitrogen is indeed abundant. The visible properties of Eris also differ from those of Pluto. Eris is less red than Pluto, and, while Pluto has one of the highest-contrast surfaces in the solar system and varies in brightness by 36% over a single rotation (see *Brown,* 2002), no variation has been seen on Eris to an upper limit of 0.05 mag. [*Carraro et al.* (2006) report a photometric variation on one of five nights

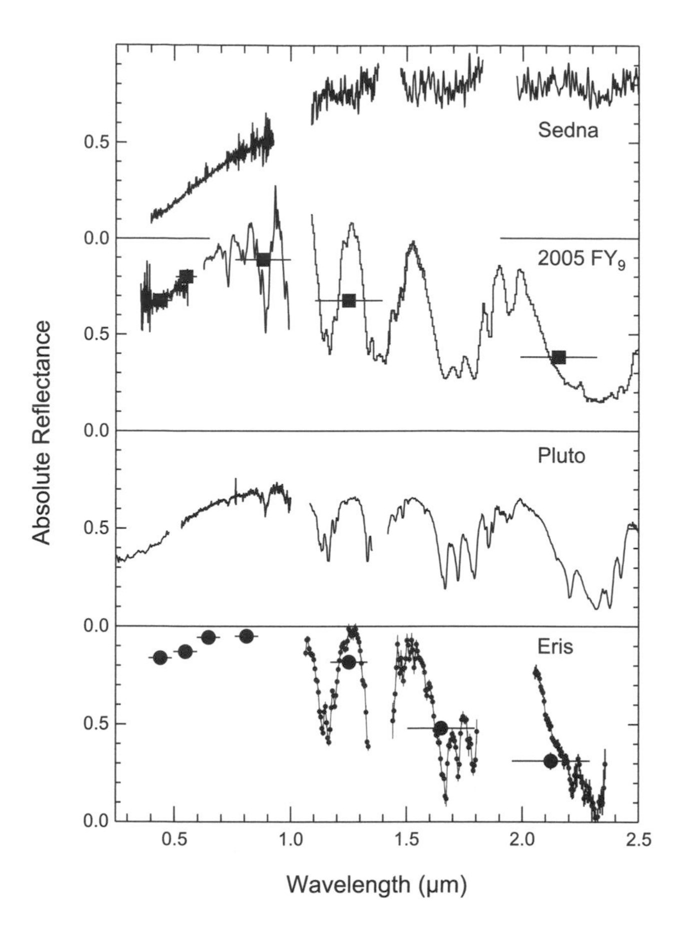

Fig. 4. Visible-to-infrared spectra of the four methane-covered objects (*Barucci et al.,* 2005; *Brown et al.,* 2005b, 2007a; *Brown,* 2002). While each of the objects is dominated by the signature of methane (with the exception of Sedna where the signal is weak but convincing), major differences appear in the objects' surface compositions. Methane on Eris and 2005 FY_9 appears to be dominantly in pure form, while on Pluto much of the methane is dissolved in N_2, whose spectral signature can be seen at 2.15 μm. On 2005 FY_9, large path lengths through pure methane give rise to broad saturated bands, and absorption due to ethane can be seen at what should be the at bottom of the 2.3-μm methane absorption. The low signal-to-noise of the Sedna spectra prevents detailed analysis, but the weakness of the methane and the possible presence of a broad N_2 line show a different surface character.

of ~0.02 mag, but no additional observations have confirmed this potential long-term variability].

The high albedo, lower red coloring, and lack of rotational variation on Eris are all consistent with a surface dominated by seasonal atmospheric cycling. With Eris currently near aphelion at 97 AU the radiative equilibrium temperature is ~20 K and nitrogen and methane have essentially zero vapor pressures, compared to vapor pressures of 17 μbar and 2 nbar at the ~36-K equilibrium temperature at the 38-AU perihelion. At the current aphelion position, the perihelion atmosphere should be collapsed onto the surface as 0.6 μm of methane and 2 mm of nitrogen. The darker and redder regions such as those on Pluto, which give Pluto its strong contrast, red color, and lower average albedo, should be covered, giving Eris a more-uniform, brighter, and less-red surface. Indeed, the high albedo of Eris appears similar to individual regions on Pluto where no dark material appears to be present (*Young et al.,* 2001). As Eris proceeds from aphelion and the surface warms we should expect that darker regions will become uncovered and the surface will appear darker, redder, and more Pluto-like. While this story for the seasonal evolution of surface of Eris consistently explains many aspects of the observations, the pure methane ice on the surface remains unexplained. Methane will freeze out before nitrogen, so the surface might be expected to be layered with methane below nitrogen, with perhaps a mixed Pluto-like layer from perihelion below, but better constrained observations and more detailed modeling will be required to understand the surface state and evolution.

Eris is orbited by an apparently single satellite, Dysnomia, approximately 500× fainter than Eris (*Brown et al.,* 2006b; *Brown and Schaller,* 2007). More distant satellites up to 10× fainter than Dysnomia can be ruled out from deep HST observations. Models for satellite capture that appear successful at describing many of the large satellites detected around many KBOs (*Goldreich et al.,* 2002) cannot account for the presence of such a small satellite. The most likely creation mechanism appears to be impacts such as modeled by *Canup* (2005) who, while attempting to find models describing the Charon-forming impact, found many cases in which the impact generated a disk that could coalesce to form a much smaller satellite. The near-circular orbit of Dysnomia ($e < 0.013$) is also consistent with the idea of formation from a disk and outward tidal evolution. From the orbit of Dysnomia, the mass of Eris is found to be $(1.67 \pm 0.02) \times 10^{22}$ kg or 27 ± 2% greater than that of Pluto (*Brown and Schaller,* 2007). Using the HST size measurement the density is then 2.3 ± 0.3 g cm^{-3}, with almost all the uncertainty due to the uncertainty in the size measurement. The density is consistent on the lower end with the 2.03 ± 0.06 g cm^{-3} density of Pluto and on the high end with the ~2.6 g cm^{-3} density of 2003 EL_{61} (see below). Note that the larger size measurement from IRAM would give a density of 1.1 ± 0.6 g cm^{-3}, which, when compared to other large KBOs and icy satellites, appears unreasonably low for an object of this size (see below).

We might expect that the disk-forming impact that generated Dysnomia would have removed some ice from Eris, leading to a higher density than that of Pluto. A more accurate measurement of the density, which would require a more accurate measurement of the size, is clearly warranted. It appears that only an occultation is likely to give an improved size estimate for Eris, and, with Eris far from the galactic plane, opportunities will be limited.

3.2. Pluto

Pluto, discussed in detail in the chapter by Stern and Trafton, is the largest object in the highly populated 3:2 mean-motion resonance with Neptune. Its high albedo and current position near perihelion make it the brightest object in the Kuiper belt and thus the first discovered and most heavily observed. Physically, it appears to be a slightly smaller twin of Eris. The main visible differences appear to be the redder color, the presence of dark areas on the surface, and the different state of methane on the surface. As discussed above, most of these differences can be explained as an expected consequence of the closer heliocentric distance of Pluto. Pluto is surrounded by a system of one large (Charon) (*Christy and Harrington,* 1978) and two small satellites (Nix and Hydra) (*Weaver et al.,* 2006). Modeling by *Canup* (2005) suggests that the large satellite Charon can be explained as a consequence of a grazing collision between the proto-Pluto and Charon in which little exchange or heating takes place. While no detailed modeling of the formation of the smaller satellites has been performed, their similar orbital plane to Charon and near-circular orbits (*Buie et al.,* 2006) suggest that they were formed in the same collision.

3.3. Sedna

While the size of Sedna remains uncertain, an upper limit can be placed from Spitzer observations (see chapter by Stansberry et al.), and a more tenuous lower limit can be placed by assuming that the geometric albedo at all wavelengths is lower than 100% (which need not necessarily be true). These limits constrain the V albedo of Sedna to be between 0.16 and 0.30 and the diameter to be between 1200 and 1600 km. A deep HST search for satellites has revealed no candidates to a limit of about 500× fainter than the primary (Brown and Suer, in preparation).

Sedna is one of the reddest KBOs known, and in moderate signal-to-noise data, the infrared spectrum appears to contain methane and perhaps nitrogen (*Barucci et al.,* 2005) (Fig. 4). The visible-to-infrared spectrum and moderate albedo is consistent with an object covered in dark red organic tholins but with some covering of methane and nitrogen frosts. Sedna is currently at 90 AU and 70 years away from its 76-AU perihelion in its 11,000-yr orbit, which takes it to 900 AU. It is currently warming and developing whatever limited atmosphere it will have. A 76-AU equilibrium temperature atmosphere of ~160 nbar of nitrogen will correspond to a ~40-μm solid layer of nitrogen ice at aphelion and a ~36-μm layer at its current position of 90 AU.

The darker and redder surface of Sedna is consistent in albedo and color with the darker regions on Pluto. The long orbital period and high eccentricity mean that Sedna spends very little time near perihelion, so much more time is available for solid-state processing of the material than there is for surface regeneration. The extremely low temperature of Sedna prevents much of an atmosphere even near perihelion and thus no extensive frost surface should ever develop.

3.4. 2005 FY_9

2005 FY_9 is the brightest KBO after Pluto, and radiometric measurements from the Spitzer Space Telescope (see chapter by Stansberry et al.) suggest a diameter of 1500 ± 300 and an albedo of 80^{+10}_{-20}%. Like Eris, Pluto, and Sedna, 2005 FY_9 has a surface spectrum dominated by methane (*Barkume et al.,* 2005; *Licandro et al.,* 2006b; *Brown et al.,* 2007a), but the methane absorption features on 2005 FY_9 are significantly deeper and broader than those on the other objects (Fig. 4). The depth and breadth of solid-state absorption features is a function of optical path length through the absorbing material, so the features on 2005 FY_9 can be interpreted as being due to extremely large (~1 cm) methane grains on 2005 FY_9, or, likely more appropriately, as due to a slab of methane ice with scattering impurities separated by ~1 cm. Methane grain sizes on the other bodies are closer to 100 μm in contrast.

In addition to the large methane path lengths, 2005 FY_9 differs from Pluto in that even moderately high signal-to-noise spectra show no evidence for the presence of the 2.15-μm nitrogen ice absorption feature (*Brown et al.,* 2007a). Nitrogen appears depleted on 2005 FY_9 relative to methane by at least an order of magnitude compared to Pluto. Visible spectroscopy shows evidence, however, for slight shifts in the wavelengths of the methane absorptions features that could be indicative of a small amount of surface coverage of methane dissolved inside nitrogen (*Tegler et al.,* 2007).

Finally, 2005 FY_9 has a clear signature of the presence of small grains of ethane, in addition to the methane (*Brown et al.,* 2007a). Ethane is one of expected dissociation products of both gaseous and solid-state methane.

All these unique characteristics of 2005 FY_9 can be interpreted as being due to a large depletion of nitrogen on the object. The depletion of nitrogen would make methane the dominant volatile on the surface and allow grains of relatively pure methane to grow large as the grains of nitrogen do on Pluto. In addition, the presence of methane in pure rather than diluted form would allow the solid-state degradation of methane to ethane that would not be possible with methane diluted in small concentrations in nitrogen. 2005 FY_9 may be a transition between the larger surface-volatile-rich objects and the smaller surface-volatile-depleted objects.

2005 FY_9 is the largest KBO to have no known satellite. Deep observations from HST place an upper limit for the brightness of faint distant satellites of one part in 10,000 (Brown and Suer, in preparation).

3.5. 2003 EL_{61}

2003 EL_{61} was first found to be unusual due to its rapid rotation and large light curve variation. *Rabinowitz et al.* (2006) inferred that 2003 EL_{61} was a rapidly rotating ellipsoid with a 4-h rotation period. Assuming that the primary spins in the same plane as the first satellite discovered (*Brown et al.,* 2005a), the lightcurve and period suggest a body with a density of 2.6 g cm^{-3}, a size (based on the density and mass determined from the satellite orbit) of 2000 × 1500 × 1000 km, and a visual albedo (based on the derived size and on the brightness) of 0.73 (the formal uncertainties on these parameters are small, but probably do not reflect the true uncertainties in our understanding of the interior state of large icy bodies and the degree to which uniform density hydrostatic equilibrium holds), consistent with infrared spectra showing deep water-ice absorption (*Trujillo et al.,* 2007).

Infrared spectroscopy of the satellite revealed the deepest water-ice absorption features of any body detected in the outer solar system (*Barkume et al.,* 2006), which effectively ruled out a capture origin, as capture of a spectrally unique body appears implausible. The rapid rotation, high density, unusual satellite spectrum, and discovery of a second inner satellite (*Brown et al.,* 2006b) all strongly point to a collisional origin for this system.

A large infrared survey showed that a small number of KBOs have deep water ice absorptions similar to that of 2003 EL_{61} and almost as deep as its satellite (see Fig. 4 in chapter by Barucci et al.). Remarkably, these KBOs are all dynamically clustered near the dynamical position of 2003 EL_{61} itself. Determination of the proper orbital elements of these objects shows that they represent a tight dynamical family separated by only 140 m s^{-1} (*Brown et al.,* 2007b). Such a tight dynamical clustering is itself unusual enough; coupled with the spectral similarity and the additional evidence for a giant impact, it becomes clear that the objects in this family are the collisional fragments of a giant impact on the proto-2003 EL_{61}. While the fragments themselves are tightly clustered, 2003 EL_{61} itself has a velocity difference of approximately 500 m s^{-1} from the fragments. This difference is easily explained by the residence of 2003 EL_{61} with the 12:7 mean-motion resonance with Neptune, which causes long-term eccentricity and inclination evolution that can take an object from near the center of the cluster to the position of 2003 EL_{61} on a timescale of ~1 G.y.

While a giant impact on the proto-2003 EL_{61} appears capable of explaining each of the individual observations, some mysteries remain. In modeling to date, impacts are seen to either disperse fragments or create a disk out of which satellites can form. 2003 EL_{61} appears to have done both. In addition, the very small velocity dispersion of the family implies that the fragments left the surface of 2003 EL_{61} with velocities a small fraction above the 1 km s^{-1} escape velocity. Detailed modeling will be required for a further understanding of the 2003 EL_{61} system.

3.6. Other Large Objects

The three other objects in our collection of large KBOs each also have unique properties. Quaoar and Orcus each have water-ice absorption among the deepest of non-2003 EL_{61} fragment KBOs (*Jewitt and Luu,* 2004; *de Bergh et al.,* 2005; *Trujillo et al.,* 2005). Ixion is the largest known object with a nearly featureless infrared spectrum (*Brown et al.,* 2007b).

The infrared spectrum of Quaoar has an absorption feature at 2.2 μm that has been interpreted as being due to ammonia (*Jewitt and Luu,* 2004) in analogy to an absorption feature on Charon (*Brown and Calvin,* 2000), although the two spectra appear different. The absorption feature is also, however, consistent with the position of one of the strongest absorptions for methane. More detailed observations to constrain the composition of the surface of Quaoar are clearly warranted. Quaoar is, in addition, the smallest object known to have a faint satellite (fractional brightness of 0.6%) like those of Eris, Pluto, and 2003 EL_{61} (Brown and Suer, in preparation).

Orcus is a Plutino with an orbit that is nearly a mirror image of that of Pluto. It is the largest KBO with an (apparently) single large (fractional brightness of 8%) satellite; deep HST images show that any more distant satellites must be fainter than Orcus by at least a factor of 1000 (Brown and Suer, in preparation). Outer satellites of the relative faintness of those of Pluto would remain undetected. The satellite of Orcus is on a near-circular orbit with a 9.5-d period, consistent with outward evolution from an initially tighter orbit. Ixion is the brightest object in absolute magnitude with a nearly featureless infrared spectrum, although it is not clear that it is the largest such object. Spitzer observations (see chapter by Stansberry et al.) only moderately constrain the size to 590 ± 190 km and the albedo to between ~9 and 30%. A handful of other KBOs have Spitzer measurements of a similar or greater size, including Varuna, Huya, 2002 AW_{197}, 2002 UX_{25}, 2004 GV_9, 2002 MS_4, and 2003 AZ_{84}, and their derived albedos range from 6% to 30%. Some of these objects (2002 UX_{25} and 2003 AZ_{84}) are known to have moderately large close satellites, and one — Varuna — is known to be a rapid rotator with similarities to 2003 EL_{61} (*Jewitt and Sheppard,* 2002), but, in general, these objects appear to share few of the properties of the unique larger KBOs.

4. ENSEMBLE PROPERTIES

4.1. Surface Composition

The most striking visible difference between the largest KBOs and the remainder of the population is the presence of volatiles such as methane, nitrogen, and CO in the spectra of the large objects compared to relatively featureless spectra of the remaining objects. The transition from small objects with volatile-free to large objects with volatile-rich surfaces appears to be explainable with a simple model of atmospheric escape shown in Fig. 5 (*Schaller and Brown,* 2007).

Most KBOs are too small and too hot to be able to retain volatiles against atmospheric escape over the life of the solar system, a few objects are so large or so cold that they easily retain volatiles, and a small number are in the potential transition region between volatile-free and volatile-rich surfaces. 2003 EL_{61} is sufficiently large that it could retain volatiles, but it seems likely that the giant impact that removed much of its water ice would have removed much of the volatile mass as well, either through direct ejection or heating. 2005 FY_9 and Quaoar are both sufficiently hot that the low-vapor-pressure nitrogen should all have escaped, but

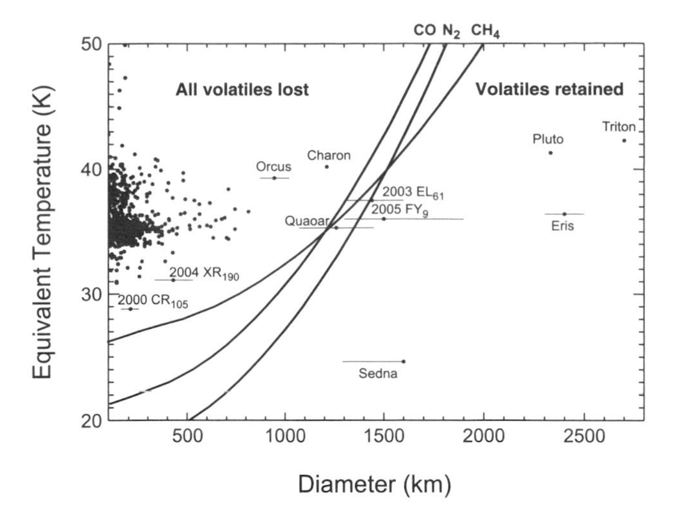

Fig. 5. A model of surface volatile loss on objects in the Kuiper belt (*Schaller and Brown,* 2007). Most objects in the Kuiper belt are sufficiently small or sufficiently hot that atmospheric loss will remove all accessible surface volatiles over the lifetime of the solar system. No volatiles have been detected on any of these objects. A small number of objects are large enough or cold enough to easily retain surface volatiles, and each of these has indeed had surface volatiles detected. Three objects are in the transition region between certain volatile loss and possible volatile retention. 2003 EL_{61} has no volatiles detected on the surface, but the mantle-shattering impact that it likely experienced would likely have removed many of the volatiles along with much of the water ice. 2005 FY_9 indeed appears to be a transition object as the model predicts, with methane clearly present, but a large depletion of nitrogen relative to methane. Quaoar has a dominantly water ice spectra, but an absorption feature at 2.2 µm could be interpreted as being due to the strongest band of methane being weakly present, implying that Quaoar, too, is currently undergoing the transition from a volatile-rich to volatile-poor surface.

the lower-vapor-pressure methane could still be retained. This depletion of nitrogen relative to methane is precisely what is observed on 2005 FY_9. On Quaoar, if the 2.2-µm absorption is interpreted as being due to methane instead of ammonia, it would appear that Quaoar is in the last stages of volatile loss.

The model shown in Fig. 5 provides the first basic framework for understanding the surface compositions of the objects in the Kuiper belt. The vast majority of the known objects are too small and/or too hot to have the possibility of retaining any surface volatiles. Surfaces dominated by relatively featureless involatile heavier organics or exposures of water ice (see chapter by Barucci et al.) are therefore expected on such objects. Volatile-rich surfaces are only possible on these largest of the bodies in the Kuiper belt. In the region beyond the Kuiper belt inhabited by bodies such as Sedna, we should expect that most of the bodies — even relatively small ones — will have the capability of retaining surface volatiles.

The largest nonmethane objects have the deepest water-ice-absorption features (ignoring the presumably special case of 2003 EL_{61} and its fragments), even taking into account the lower signal-to-noise of the spectra of the fainter objects (see Fig. 4 of the chapter by Barucci et al.). Unlike for the presence or absence of surface volatiles, no clear explanation of this trend is apparent, although a partial explanation could include the initially higher temperatures of the larger objects leading to greater internal volatile loss and perhaps differentiation. Fewer organic volatiles could then lead to less creation of dark organic tholins. Such a process would lead to higher albedos for these larger objects, which is indeed observed, but also bluer colors, which is not observed. An alternative explanation could invoke the satellite-forming impacts that these objects experience in an attempt to explain their surface compositions. Our understanding of the processes affecting the colors and compositions of all the objects in the Kuiper belt is still primitive.

4.2. Satellites

The largest KBOs appear to have a different style of satellite formation than the other objects. These objects have a greater frequency of satellites, the only two known multiple satellite systems, and the possibility of much smaller satellites. *Brown et al.* (2006b) found in an adaptive optics survey of the four largest KBOs that the probability that three out of four of these would have detectable satellites suggests that they are drawn from a different population than the remainder of the Kuiper belt at the 98.2% confidence level. Updating this calculation for our currently defined population, we find that the probability that five or more out of eight in our sample of large objects are drawn from the same population as the remainder of the Kuiper belt is less than 1%.

The presence of relatively small satellites around Eris, Pluto, 2003 EL_{61}, and Quaoar suggests formation by impact, rather than dynamical friction-aided capture. The moderate size and tight circular orbit of the satellite of Orcus could also indicate a collisional rather than capture origin. After the early discovery of near-equal brightness well-separated eccentrically orbiting KBO binaries (see chapter by Noll et al.), much emphasis was placed on trying to explain the genesis of these unusual systems through some sort of capture mechanism. Collisions, however, appear a dominant satellite-creating process among the largest KBOs and perhaps also for the now numerous known closely spaced binaries.

4.3. Densities

The abundance of satellites and the ability to make accurate size measurements (see chapter by Stansberry et al.) allows determination of the density for many of the largest KBOs. While the handful of smaller KBOs with known densities appear to have unexpectedly low densities of ~1 g cm^{-3} and even lower (*Stansberry et al.,* 2006), the largest KBOs have densities between ~1.9 and 2.5 g cm^{-3} as expected from cosmochemical abundances in the outer solar system (*McKinnon and Mueller,* 1989). Figure 6 shows the measured densities of the large KBOs, including Triton and Charon.

Within the largest KBOs, no statistically significant trend exists in KBO density vs. radius. Similarly, no significant trend is seen in the densities of the icy satellites of the outer three planets through this size range. A rank correlation test shows that the KBOs are more dense than the icy satellites, however, at the 95% confidence level. These higher densities could be the result of the different formation environment between the protosolar and protogiant planet nebulae, although a bias in KBO densities caused by impacts cannot be ruled out, as density measurements of KBOs (with the exception of Triton) require the presence of a satellite.

5. CONCLUSIONS

Each of the largest KBOs has a unique dynamical and physical history that can be gleaned from detailed observations such as those described here. As a whole, the largest KBOs appear distinct in surface composition, satellite frequency and style, and density. Impacts appear to have played a more discernible role among the largest KBOs than among the population at large.

Based on the latitudinal completeness of the Palomar survey, it appears that two or three more KBOs of the size range of those described here likely await discovery, although many more large objects must exist in the distant regions beyond the Kuiper belt. The most likely location to find large undiscovered KBOs is in the band at 10° south ecliptic latitude where the sky densities are highest and the completeness is lowest, although with the low numbers remaining to be found, they could be almost anywhere within the Kuiper belt.

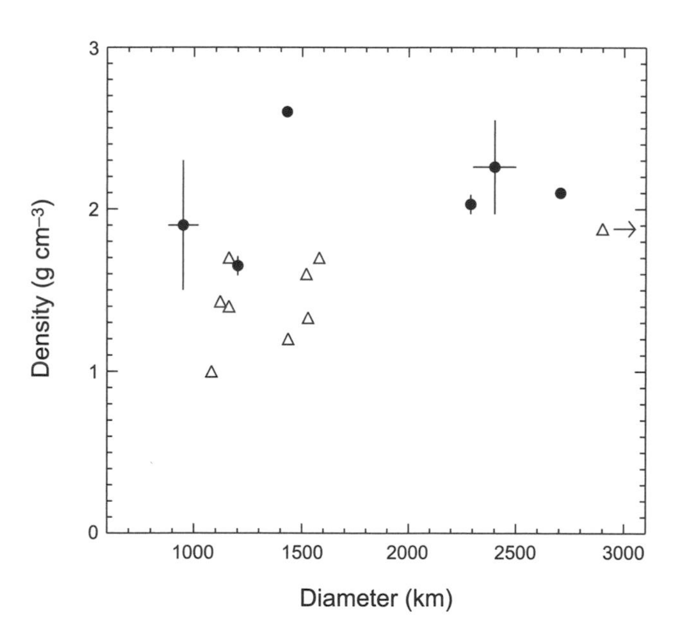

Fig. 6. Densities of the largest KBOs, shown with solid circles. No clear trend exists in density with size, although several KBOs with smaller sizes are known to have significantly lower densities. No statistically significant trend is seen among the densities of icy satellite of the outer three planets over this same size range (open triangles; the triangle with an arrow represents Titan, with a diameter of 5150 km). The KBOs are more dense than the satellites at the 95% confidence level.

Several outstanding questions remain about the largest KBOs:

1. Why are there no large KBOs among the cold classical population?
2. What does Sedna's dynamical location tell us about the history of the solar system?
3. What causes the density enhancements at ±10° ecliptic latitude and what implication does this have for the formation of the Kuiper belt?
4. How does atmospheric cycling affect the presence and layering of species on volatile-rich large KBOs?
5. Why are the water-absorption features of 2003 EL_{61} and its satellites and fragments distinctly deeper than those of other water-rich KBOs?
6. Are multiple satellite systems common among the large KBOs?
7. Is Quaoar at the transition from having a volatile-rich to a volatile-poor surface?
8. Are any active sources of methane, such as serpentinization of ultramafic rock, necessary to explain the volatiles on the largest KBOs?
9. Is the impact frequency required to explain all the presumably impact-related features of the large KBOs higher than expected?
10. Do impacts such as those experienced by 2003 EL_{61} raise the densities on other KBOs?
11. What explains the difference between the water-ice-rich surfaces of some moderate-sized KBOs and the spectrally featureless surfaces of others?

The recent discoveries of these largest KBOs ensures an accessible population for addressing these questions and promises a slew of new questions as more details of these objects are discerned.

REFERENCES

Barkume K. M., Brown M. E., and Schaller E. L. (2005) Near infrared spectroscopy of icy planetoids. *AAS/Division for Planetary Sciences Meeting Abstracts 37,* 52.11.

Barkume K. M., Brown M. E., and Schaller E. L. (2006) Water ice on the satellite of Kuiper belt object 2003 EL61. *Astrophys. J. Lett., 640,* L87–L89.

Barucci M. A., Cruikshank D. P., Dotto E., Merlin F., Poulet F., Dalle Ore C., Fornasier S., and de Bergh C. (2005) Is Sedna another Triton? *Astron. Astrophys., 439,* L1–L4.

Bernstein G. M., Trilling D. E., Allen R. L., Brown M. E., Holman M., and Malhotra R. (2004) The size distribution of trans-neptunian bodies. *Astron. J., 128,* 1364–1390.

Bertoldi F., Altenho W., Weiss A., Menten K. M., and Thum C. (2006) The transneptunian object UB313 is larger than Pluto. *Nature, 439,* 563–564.

Brasser R., Duncan M. J., and Levison H. F. (2006) Embedded star clusters and the formation of the Oort Cloud. *Icarus, 184,* 59–82.

Brown M. E. (2002) Pluto and Charon: Formation, seasons, composition. *Annu. Rev. Earth Planet. Sci., 30,* 307–345.

Brown M. E. and Calvin W. M. (2000) Evidence for crystalline water and ammonia ices on Pluto's satellite Charon. *Science, 287,* 107–109.

Brown M. E. and Schaller E. L. (2007) The mass of dwarf planet Eris. *Science, 316,* 1585.

Brown M. E., Trujillo C., and Rabinowitz D. (2004) Discovery of a candidate inner Oort cloud planetoid. *Astrophys. J., 617,* 645–649.

Brown M. E., Bouchez A. H., Rabinowitz D., Sari R., Trujillo C. A., van Dam M., Campbell R., Chin J., Hartman S., Johansson E., Lafon R., Le Mignant D., Stomski P., Summers D., and Wizinowich P. (2005a) Keck Observatory laser guide star adaptive optics discovery and characterization of a satellite to the large Kuiper belt object 2003 EL61. *Astrophys. J. Lett., 632,* L45–L48.

Brown M. E., Trujillo C. A., and Rabinowitz D. L. (2005b) Discovery of a planetary-sized object in the scattered Kuiper belt. *Astrophys. J. Lett., 635,* L97–L100.

Brown M. E., Schaller E. L., Roe H. G., Rabinowitz D. L., and Trujillo C. A. (2006a) Direct measurement of the size of 2003 UB313 from the Hubble Space Telescope. *Astrophys. J. Lett., 643,* L61–L63.

Brown M. E., van Dam M. A., Bouchez A. H., Le Mignant D., Campbell R. D., Chin J. C. Y., Conrad A., Hartman S. K., Johansson E. M., Lafon R. E., Rabinowitz D. L., Stomski P. J. Jr., Summers D. M., Trujillo C. A., and Wizinowich P. L. (2006b) Satellites of the largest Kuiper belt objects. *Astrophys. J. Lett., 639,* L43–L46.

Brown M. E., Barkume K. M., Blake G. A., Schaller E. L., Rabinowitz D. L., Roe H. G., and Trujillo C. A. (2007a) Methane and ethane on the bright Kuiper belt object 2005 FY9. *Astron. J., 133,* 284–289.

Brown M. E., Barkume K. M., Ragozzine D., and Schaller E. L. (2007b) Discovery of an icy collisional family in the Kuiper belt. *Nature, 446,* 294–296.

Buie M. W., Grundy W. M., Young E. F., Young L. A., and Stern S. A. (2006) Orbits and photometry of Pluto's satellites: Charon, S/2005 P1, and S/2005 P2. *Astron. J., 132,* 290–298.

Canup R. M. (2005) A giant impact origin of Pluto-Charon. *Science, 307,* 546–550.

Carraro G., Maris M., Bertin D., and Parisi M. G. (2006) Time series photometry of the dwarf planet ERIS (2003 UB313). *Astron. Astrophys., 460,* L39–L42.

Christy J. W. and Harrington R. S. (1978) The satellite of Pluto. *Astron. J., 83,* 1005, 1007–1008.

de Bergh C., Delsanti A., Tozzi G. P., Dotto E., Doressoundiram A., and Barucci M. A. (2005) The surface of the transneptunian object 90482 Orcus. *Astron. Astrophys., 437,* 1115–1120.

Gladman B. and Chan C. (2006) Production of the extended scattered disk by rogue planets. *Astrophys. J. Lett., 643,* L135–L138.

Goldreich P., Lithwick Y., and Sari R. (2002) Formation of Kuiper-belt binaries by dynamical friction and three-body encounters. *Nature, 420,* 643–646.

Gomes R. (2003) The common origin of the high inclination TNO's. *Earth Moon Planets, 92,* 29–42.

Gomes R. S., Matese J. J., and Lissauer J. J. (2006) A distant planetary-mass solar companion may have produced distant detached objects. *Icarus, 184,* 589–601.

Grundy W. M., Schmitt B., and Quirico E. (1993) The temperature-dependent spectra of α and β nitrogen ice with application to Triton. *Icarus, 105,* 254–258.

Jewitt D. C. and Luu J. (2004) Crystalline water ice on the Kuiper belt object (50000) Quaoar. *Nature, 432,* 731–733.

Jewitt D. C. and Sheppard S. S. (2002) Physical properties of trans-neptunian object (20000) Varuna. *Astron. J., 123,* 2110–2120.

Kenyon S. J. and Bromley B. C. (2004) Stellar encounters as the origin of distant solar system objects in highly eccentric orbits. *Nature, 432,* 598–602.

Levison H. F. and Stern S. A. (2001) On the size dependence of the inclination distribution of the main Kuiper belt. *Astron. J., 121,* 1730–1735.

Licandro J., Grundy W. M., Pinilla-Alonso N., and Leisy P. (2006a) Visible spectroscopy of 2003 UB313: Evidence for N_2 ice on the surface of the largest TNO? *Astron. Astrophys., 458,* L5–L8.

Licandro J., Pinilla-Alonso N., Pedani M., Oliva E., Tozzi G. P., and Grundy W. M. (2006b) The methane ice rich surface of large TNO 2005 FY9: A Pluto-twin in the transneptunian belt? *Astron. Astrophys., 445,* L35–L38.

McKinnon W. B. and Mueller S. (1989) The density of Triton — A prediction. *Geophys. Res. Lett., 16,* 591–594.

Morbidelli A. and Brown M. E. (2005) The Kuiper belt and the primordial evolution of the solar system. In *Comets II* (M. C. Festou et al., eds.), pp. 175–191. Univ. of Arizona, Tucson.

Morbidelli A. and Levison H. F. (2004) Scenarios for the origin of the orbits of the trans-neptunian objects 2000 CR105 and 2003 VB12 (Sedna). *Astron. J., 128,* 2564–2576.

Rabinowitz D. L., Barkume K., Brown M. E., Roe H., Schwartz M., Tourtellotte S., and Trujillo C. (2006) Photometric observations constraining the size, shape, and albedo of 2003 EL61, a rapidly rotating, Pluto-sized object in the Kuiper belt. *Astrophys. J., 639,* 1238–1251.

Schaller E. L. and Brown M. E. (2007) Volatile loss and retention on Kuiper belt objects. *Astrophys. J. Lett., 659,* L61–L64.

Stansberry J. A., Grundy W. M., Margot J. L., Cruikshank D. P., Emery J. P., Rieke G. H., and Trilling D. E. (2006) The albedo, size, and density of binary Kuiper belt object (47171) 1999 TC36. *Astrophys. J., 643,* 556–566.

Tegler S. C., Grundy W. M., Romanishin W., Consolmagno G. J., Mogren K., and Vilas F. (2007) Optical spectroscopy of the large Kuiper belt objects 136472 (2005 FY9) and 136108 (2003 EL61). *Astron. J., 133,* 526–530.

Trujillo C. A. and Brown M. E. (2003) The Caltech wide area sky survey. *Earth Moon Planets, 92,* 99–112.

Trujillo C. A., Brown M. E., Rabinowitz D. L., and Geballe T. R. (2005) Near-Infrared surface properties of the two intrinsically brightest minor planets: (90377) Sedna and (90482) Orcus. *Astrophys. J., 627,* 1057–1065.

Trujillo C. A., Brown M. E., Barkume K. M., Schaller E. L., and Rabinowitz D. L. (2007) The surface of 2003 EL61 in the near-infrared. *Astrophys. J., 655,* 1172–1178.

Tryka K. A., Brown R. H., and Anicich V. (1995) Near-infrared absorption coefficients of solid nitrogen as a function of temperature. *Icarus, 116,* 409–414.

Weaver H. A., Stern S. A., Mutchler M. J., Steffl A. J., Buie M. W., Merline W. J., Spencer J. R., Young E. F., and Young L. A. (2006) Discovery of two new satellites of Pluto. *Nature, 439,* 943–945.

Young E. F., Binzel R. P., and Crane K. (2001) A two-color map of Pluto's sub-Charon hemisphere. *Astron. J., 121,* 552–561.

Binaries in the Kuiper Belt

Keith S. Noll
Space Telescope Science Institute

William M. Grundy
Lowell Observatory

Eugene I. Chiang
University of California, Berkeley

Jean-Luc Margot
Cornell University

Susan D. Kern
Space Telescope Science Institute

Binaries have played a crucial role many times in the history of modern astronomy and are doing so again in the rapidly evolving exploration of the Kuiper belt. The large fraction of transneptunian objects that are binary or multiple, 48 such systems now known, has been an unanticipated windfall. Separations and relative magnitudes measured in discovery images give important information on the statistical properties of the binary population that can be related to competing models of binary formation. Orbits, derived for 13 systems, provide a determination of the system mass. Masses can be used to derive densities and albedos when an independent size measurement is available. Angular momenta and relative sizes of the majority of binaries are consistent with formation by dynamical capture. The small satellites of the largest transneptunian objects, in contrast, are more likely formed from collisions. Correlations of the fraction of binaries with different dynamical populations or with other physical variables have the potential to constrain models of the origin and evolution of the transneptunian population as a whole. Other means of studying binaries have only begun to be exploited, including lightcurve, color, and spectral data. Because of the several channels for obtaining unique physical information, it is already clear that binaries will emerge as one of the most useful tools for unraveling the many complexities of transneptunian space.

1. HISTORY AND DISCOVERY

Ever since Herschel noticed, 200 years ago, that gravitationally bound stellar binaries exist, the search for binaries has followed close on the heels of the discovery of each new class of astronomical object. The reasons for such searches are, of course, eminently practical. Binary orbits provide determinations of system mass, a fundamental physical quantity that is otherwise difficult or impossible to obtain. The utilization of binaries in stellar astronomy has enabled countless applications, including Eddington's landmark mass-luminosity relation. Likewise, in planetary science, bound systems have been extensively exploited; they have been used, for example, to determine the masses of planets and to make Roemer's first determination of the speed of light. In addition to providing mass, the statistics of binaries in astronomical populations can be related to formation and subsequent evolutionary and environmental conditions.

Searches for bound systems among the small-body populations in the solar system has a long and mostly fruitless history that has been summarized in several recent reviews (*Merline et al.,* 2002; *Noll,* 2003, 2006; *Richardson and Walsh,* 2006). But the discovery of the second transneptunian binary (TNB), 1998 WW_{31} (*Veillet et al.,* 2001), marked the start of a landslide of discovery that shows no signs of abating.

1.1. Discovery and Characterization of Charon

The first example of what we would now call a TNB was discovered during a very different technological epoch than the present, prior to the widespread astronomical use of CCD arrays. The discovery of Charon (*Christy and Harrington,* 1978) on photographic plates taken for Pluto astrometry heralded a spectacular flourishing of Pluto science, and offered a glimpse of the tremendous potential of TNBs to contribute to Kuiper belt science in general.

Charon's orbit (*Christy and Harrington,* 1978, 1980) revealed the system mass, which up until then had been estimated via other methods, with wildly divergent results.

About the same time, spectroscopy revealed the presence of methane on Pluto (*Cruikshank et al.,* 1976) indicating a high albedo, small size, and the possible existence of an atmosphere. Observations of occultations of stars by Pluto confirmed its small size (*Millis et al.,* 1993) and enabled direct detection of the atmosphere (*Hubbard et al.,* 1988).

The orbit plane of Charon, as viewed from Earth, was oriented edge-on within a few years of Charon's discovery. This geometry happens only twice during Pluto's 248-year orbit, so its occurrence just after Charon's discovery was fortuitous. Mutual events, when Charon (or its shadow) passed across the face of Pluto, or Pluto (or its shadow) masked the view of Charon, were observable from 1985 through 1990 (*Binzel and Hubbard,* 1997). From the timing of these events and from the changes in observable flux during them, much tighter constraints on the sizes and albedos of Pluto and Charon were derived (e.g., *Young and Binzel,* 1994). Mutual events also made it possible to distinguish the surface compositions of Pluto and Charon (e.g., *Buie et al.,* 1987; *Fink and DiSanti,* 1988) by comparing reflectance spectra of the two objects blended together with spectra of Pluto alone, when Charon was completely hidden from view. From subtle variations in flux as Charon blocked different regions of Pluto's surface (and vice versa), maps of albedo patterns on the faces of the two objects were constructed (e.g., *Buie et al.,* 1997; *Young et al.,* 1999). The mutual events are now over and will not be repeated during our lifetimes, but telescope and detector technology continue their advance. For relatively well-separated, bright TNBs like Pluto and Charon, it is now feasible to study them as separate worlds even without the aid of mutual events (e.g., *Brown and Calvin,* 2000; *Buie and Grundy,* 2000). Just as the discovery of Charon propelled Pluto science forward, the recent study of two additional moons of Pluto (*Weaver et al.,* 2006) can be expected to give another boost to Pluto science, by enabling detailed studies of the dynamics of the system (*Buie et al.,* 2006; *Lee and Peale,* 2006) and providing new constraints on formation scenarios (e.g., *Canup,* 2005; *Ward and Canup,* 2006).

1.2. Discovery of Binaries

The serendipitous discovery of the second TNB, 1998 WW_{31} (*Veillet et al.,* 2001), marked a breakthrough for binaries in the Kuiper belt. It immediately provided a context for Pluto/Charon as a member of a group of similar systems rather than as a unique oddity. The relatively large separation and size of the secondary dispelled the notion that Kuiper belt satellites would all be small, faint, and difficult-to-resolve collision fragments. Some, at least, were detectable from the ground with moderately good observing conditions, as had been foreseen by *Toth* (1999). The next two discovered TNBs were just such systems (*Elliot et al.,* 2001; *Kavelaars et al.,* 2001).

The first conscious search for satellites of transneptunian objects (TNOs) was carried out by M. Brown and C. Trujillo

Fig. 1. Images of 2001 FL_{185} (top left), (42355) Typhon/Echidna (top right), 1999 OJ_4 (bottom left), and 2002 GZ_{31} (bottom right). The images shown are each combinations of four separate 300-s exposures taken with the High Resolution Camera on HST. The dithered exposures have been combined using multidrizzle. The images are shown with a linear grayscale normalized to the peak pixel. The pixels in the drizzled images are 25 milliarcsec on a side in a nondistorted coordinate frame. Each of the four panels is 1 arcsec square. Images are oriented in detector coordinates.

using the Space Telescope Imaging Spectrograph (STIS) on the Hubble Space Telescope (HST) starting in August 2000 (*Trujillo and Brown,* 2002; *Brown and Trujillo,* 2002). A series of large surveys with HST followed, producing the discovery of most of the known TNBs (e.g., *Noll et al.,* 2002a,b,c, 2003, 2006a,b,c,d,e,f; *Stephens et al.,* 2004; *Stephens and Noll,* 2006) (Fig. 1).

Large groundbased surveys — Keck (*Schaller and Brown,* 2003), Deep Ecliptic Survey (DES) followup at Magellan (*Millis et al.,* 2002; *Elliot et al.,* 2005), and the Canada-France Ecliptic Plane Survey (CFEPS) (*Jones et al.,* 2006) — have produced a few detections. Although significantly less productive because of the limited angular resolution possible from the ground compared to HST, the sheer number of objects observed by these surveys makes them a valuable statistical resource (*Kern and Elliot,* 2006). Both space- and groundbased discoveries are described in detail in section 2.2.

Binaries may also be "discovered" theoretically. *Agnor and Hamilton* (2006) have shown that the most likely explanation for the origin of Neptune's retrograde satellite Triton is the capture of one component of a binary that encountered the giant planet. The viability of this model is enabled by the paradigm-shifting realization that binaries in the transneptunian region are common.

2. INVENTORY

Much can be learned about binaries and the environment in which they formed from simple accounting. The fraction of binaries in the transneptunian population is far higher than anyone guessed a decade ago when none were yet recognized (except for Pluto/Charon), and considerably higher than was thought even four years ago as the first spate of discoveries was being made. As the number of binaries has climbed, it has become apparent that stating the fraction of binaries is not a simple task. The fraction in a given sample is strongly dependent on a number of observational factors, chiefly angular resolution and sensitivity. To add to the complexity, TNOs can be divided into dynamical groups with possibly differing binary fractions. Thus, the fraction of binaries in a particular sample also depends on the mix of dynamical classes in the sample. Perversely, perhaps, the brightest TNOs, and thus the first to be sampled, belong to dynamical classes with a lower overall fraction of sizable binary companions. Other dependencies may also help determine the fraction and nature of binaries and multiples; for example, very small, possibly collision-produced companions appear to be most likely around the largest of the TNOs. The current inventory of TNBs, while impressive compared to just a few years ago, remains inadequate to address all the questions one would like to ask.

2.1. Current Inventory of Transneptunian Binaries

As of February 2007, more than 40 TNO and Centaur binaries had been announced through the International Astronomical Union Circulars (IAUC) and/or other publications. Additional binaries are not yet documented in publications. All the binaries of which we are aware are compiled in Table 1. References listed are generally the discovery announcement, when available. Values for separation and relative magnitude were recalculated for many of the objects and supersede earlier published values. The osculating heliocentric orbital parameters $a_{\odot}$, $e_{\odot}$, and $i_{\odot}$ are listed as well as the dynamical class. For the latter we have followed the DES convention (*Elliot et al.,* 2005), with the resonant objects identified by their specific resonance as n:m, where n refers to the mean motion of Neptune. Classical objects on orbits of low inclination and eccentricity are designated C. classical objects with an integrated average inclination $i > 5°$ relative to the invariable plane are denoted by H. Objects in the scattered disk are labeled S or X depending on whether their Tisserand parameter, T, is less than or greater than 3. The Tisserand parameter relative to Neptune is defined as $T_N = a_N/a + 2[(1 - e^2)a/a_N]^{1/2} \cos(i)$ where a, e, and i are the heliocentric semimajor axis, eccentricity, and inclination of the TNO and a_N is the semimajor axis of Neptune. Objects in the extended scattered disk, X, have the additional requirement of a time-averaged eccentricity greater than 0.2. Centaurs and Centaur-like objects, labeled ¢, are on unstable, nonresonant, planet-crossing orbits and are, therefore, dynamically young. (Centaur-like objects are in unstable, nonresonant, giant-planet-crossing orbits just like the Centaurs, but have a semimajor axis greater than 30.1 AU. There is currently disagreement on what this class of objects should be called. Because of their similarly unstable orbits, we follow the DES convention and group them with Centaurs.) In Table 1 we list the objects in three broad dynamical groupings: classical, scattered, and resonant. The classical grouping includes all classical objects regardless of inclination, i.e., both hot and cold classicals. The resonant group includes all objects verified to be in mean-motion resonances by numerical integration. The scattered group includes both near and extended scattered objects and the Centaurs. Within each group we have ordered the objects by absolute magnitude, H_V.

In addition to the dynamical class and osculating heliocentric orbital elements of the binaries, we list in Table 1 three additional measurements available for all the known binaries: (1) The reported separation at discovery (in arcsec) is shown with the error in the final significant digit in parentheses. Separations reported without an error estimate are shown in italics. As we discuss in more detail below in section 3.3, the separation at discovery is not an intrinsic property of binary orbits, but can be useful for estimating the distribution of binary semimajor axes. (2) The magnitude difference, Δ_{mag}, can be used to derive the size ratio of the components (with the customary assumption of equal albedos). Once again, errors in the final digit are shown in parentheses, and estimated quantities are in italics. (3) The absolute magnitude, H_V, is taken from the Minor Planet Center (MPC) and applies to the combined light of the unresolved binary. Better measurements of H_V are available for some objects (*Romanishin and Tegler,* 2005), but for the sake of uniformity we use the MPC values for all objects. H_V can provide a determination of the size if the albedo is known or can be estimated. However, the range of albedo in the transneptunian population is large (*Grundy et al.,* 2005), as is the phase behavior (*Rabinowitz et al.,* 2007), making any such estimate risky (see also chapters by Stansberry et al. and Belskaya et al.).

2.2. Large Surveys, Observational Limits, and Bias

Several large surveys have produced the discovery of large fractions of the known TNBs. Observational limits are, to first order, a function of the telescope and instrument used for the observations. This is more easily characterized for spacebased instruments than for groundbased surveys, but approximate limits for the latter can be estimated.

The largest semiuniform groundbased survey of TNOs that has been systematically searched for binaries is the Deep Ecliptic Survey (*Millis et al.,* 2002; *Elliot et al.,* 2005). *Kern and Elliot* (2006) searched 634 unique objects from the DES survey and identified 1. These observations were made with 4-m telescopes at CTIO and KPNO utilizing widefield mosaic cameras (*Muller et al.,* 1998) with 0.5 arcsec pix-

TABLE 1. Transneptunian binaries.

Object	Dynamical Class	Heliocentric $a_{\odot}$ (AU)	Orbit $e_{\odot}$	Elements $i_{\odot}$ (°)	Separation s_0 (arcsec)	Δ_{mag}	H_V	Reference
Classical								
(50000) Quaoar	H	43.609	0.037	8.0	0.35(1)	5.6(2)	2.6	[1]
(79360) 1997 CS_{29}	C	43.876	0.013	2.2	0.07(1)	0.09(9)	5.1	[2]
(148780) 2001 UQ_{18}	C	44.545	0.057	5.2	0.177(7)	0.7(2)	5.1	[†]
2003 QA_{91}	C	44.157	0.067	2.4	0.056(4)	0.1(6)	5.3	[†]
2001 QY_{297}	C	43.671	0.081	1.5	0.091(2)	0.42(7)	5.4	[†]
(88611) 2001 QT_{297}	C	44.028	0.028	2.6	0.61(2)	0.7(2)	5.5	[3]
2001 XR_{254}	C	43.316	0.023	1.2	0.107(2)	0.09(6)	5.6	[†]
2003 WU_{188}	C	44.347	0.039	3.8	0.042(4)	0.7(3)	5.8	[†]
(66652) 1999 RZ_{253}	C	42.779	0.090	0.6	0.21(2)	0.33(6)	5.9	[4]
(134860) 2000 OJ_{67}	C	42.840	0.023	1.1	0.08(1)	0.8	6.0	[2]
2001 RZ_{143}	C	44.282	0.068	2.1	0.046(3)	0.1(3)	6.0	[†]
1998 WW_{31}	C	44.485	0.089	6.8	*1.2*	*0.4*	6.1	[5]
2005 EO_{304}	C	45.966	0.080	3.4	2.67(6)	1.2(1)	6.2	[6]
2003 QR_{91}	H	46.361	0.183	3.5	0.062(2)	0.2(3)	6.2	[†]
(80806) 2000 CM_{105}	C	42.236	0.064	6.7	0.059(3)	0.6(1)	6.3	[2]
2003 QY_{90}	C	42.745	0.052	3.8	0.34(2)	0.1(2)*	6.3	[7]
(123509) 2000 WK_{183}	C	44.256	0.044	2.0	0.080(4)	0.4(7)	6.4	[8]
(58534) Logos/Zoe	C	45.356	0.119	2.9	0.20(3)	0.4(1)*	6.6	[9]
2000 CQ_{114}	C	46.230	0.110	2.7	0.178(5)	0.4(2)	6.6	[10]
2000 CF_{105}	C	43.881	0.037	0.5	0.78(3)	0.7(2)	6.9	[11]
1999 OJ_4	C	38.067	0.023	4.0	0.097(4)	0.16(9)	7.0	[2]
2001 FL_{185}	C	44.178	0.077	3.6	0.065(14)	0.8(6)	7.0	[†]
2003 UN_{284}	C	42.453	0.010	3.1	2.0(1)	0.6(2)	7.4	[12]
2001 QW_{322}	C	44.067	0.027	4.8	*4*	0.0(1)	7.8	[13]
1999 RT_{214}	C	42.711	0.052	2.6	0.107(4)	0.81(9)	7.8	[14]
2003 TJ_{58}	C	44.575	0.089	1.0	0.119(2)	0.50(7)	7.8	[†]
Scattered								
(136199) Eris	S	67.695	0.441	44.2	0.53(1)	4.43(5)	–1.2	[15]
(136108) 2003 EL_{61}	S	43.316	0.190	28.2	0.63(2)	3.1(1)	0.2	[16]
					0.52(3)	4.6(4)		[17]
(55637) 2002 UX_{25}	S	42.551	0.141	19.5	0.164(3)	2.5(2)	3.6	[1]
(120347) 2004 SB_{60}	S	42.032	0.104	24.0	0.107(3)	2.2(1)	4.4	[18]
(48639) 1995 TL_8	X	52.267	0.235	0.2	0.01(1)	*1.7*	5.4	[2]
2001 QC_{298}	S	46.222	0.123	30.6	0.130(7)	0.58(3)	6.1	[19]
2004 PB_{108}	S	44.791	0.096	20.3	0.172(3)	1.2(1)	6.3	[20]
(65489) Ceto/Phorcys	¢	102.876	0.821	22.3	0.085(2)	0.6(1)	6.3	[21]
2002 GZ_{31}	X	50.227	0.237	1.1	0.070(9)	1.0(2)	6.5	[22]
(60458) 2000 CM_{114}	S	59.838	0.407	19.7	0.074(6)	0.57(7)	7.0	[23]
(42355) Typhon/Echidna	¢	38.112	0.540	2.4	0.109(2)	1.47(4)	7.2	[24]

els. Median seeing for the entire dataset was 1.65 arcsec (*Kern,* 2006) and effectively sets the detection limit on angular separation. Magnitude limits in the broad V–R filter for a well-separated secondary vary, depending on seeing, from V–R = 23 to V–R = 24.

Follow-up observations of 212 DES objects made with the Magellan telescopes to improve astrometry were also searched for undetected binaries (*Kern and Elliot,* 2006). Most observations were made with the MagIC camera (*Osip et al.,* 2004), which has a pixel scale of 0.069 arcsec. *Kern* (2006) reports the median seeing for these observations was 0.7 arcsec with a magnitude limit similar to the DES survey. Of the 212 objects observed with Magellan, 3 were found to be binaries (*Osip et al.,* 2003, *Kern and Elliot,* 2005, 2006).

The Keck telescope survey reported by *Schaller and Brown* (2003) observed 150 objects and found no new binaries. The observational limits for this survey have not been published; they are probably similar to the DES as the primary limiting factor is seeing.

The target lists for the three large groundbased surveys have not been published and it is unclear how much overlap there may be. However, even duplicate observations of an individual target can be useful since some TNBs are known to have significantly eccentric or edge-on orbits and are, therefore, variable in their detectability. The one firm conclusion, however, that can be reached from these data is that binaries separated sufficiently for detection with uncorrected groundbased observations are uncommon, occurring around 1–2% of TNOs at most.

The most productive tool for finding TNBs is the HST, which has found 41 of the 52 companions listed in Table 1. The combination of high angular resolution, high sensitivity, and stable point-spread-function (PSF) make it ideally matched to the requirements for finding and studying TNBs.

TABLE 1. (continued).

Object	Dynamical Class	Heliocentric $a_{\odot}$ (AU)	Orbit $e_{\odot}$	Elements $i_{\odot}$ (°)	Separation s_0 (arcsec)	Δ_{mag}	H_V	Reference
Resonant								
(134340) Pluto/Charon	3:2	39.482	0.248	17.1	*0.9*	*2–3*	–1.0	[25]
Nix					1.85(4)	9.26(2)		[26]
Hydra					2.09(4)	8.65(2)		[26]
(90482) Orcus	3:2	39.386	0.220	20.6	0.256(2)	*2.5*	2.3	[1]
2003 AZ_{84}	3:2	39.414	0.181	13.6	0.22(1)	5.0(3)	3.9	[1]
(47171) 1999 TC_{36}	3:2	39.270	0.222	8.4	0.367(4)	2.21(1)	4.9	[27]
(82075) 2000 YW_{134}	8:3	57.779	0.287	19.8	0.06(1)	*1.3*	5.0	[2]
(119979) 2002 WC_{19}	2:1	47.625	0.260	9.2	0.090(8)	2.5(4)	5.1	[28]
(26308) 1998 SM_{165}	2:1	47.501	0.370	13.5	0.205(1)	2.6(3)	5.8	[29]
2003 QW_{111}	7:4	43.659	0.111	2.7	0.321(3)	1.47(8)	6.2	[30]
2000 QL_{251}	2:1	47.650	0.216	3.7	0.25(6)	0.05(5)	6.3	[31]
(60621) 2000 FE_8	5:2	55.633	0.404	5.9	0.044(3)	0.6(3)	6.7	[20]
(139775) 2002 QG_{298}	3:2	39.298	0.192	6.5	*contact?*	N/A	7.0	[32]

*The brightness of Logos varies significantly as described in section 3.6.

Objects are sorted into three dynamical groups: classical (both hot, H, and cold C), scattered (includes scattered-near, S; scattered-extended, X; and Centaurs, ¢), and resonant. Objects in each grouping are sorted by absolute magnitude, H_V. Uncertainties in the last digit(s) of measured quantities appear in parentheses. Table entries in italics indicate quantities that have been published without error estimates or that have been computed by the authors from estimated quantities. The H_V column lists the combined absolute magnitude of the system as tabulated by the Minor Planet Center (MPC).

References: [1] *Brown and Suer* (2007); [2] *Stephens and Noll* (2006); [3] *Elliot et al.* (2001); [4] *Noll et al.* (2003); [5] *Veillet et al.* (2001); [6] *Kern and Elliot* (2005); [7] *Elliot et al.* (2003); [8] *Noll et al.* (2007a); [9] *Noll et al.* (2002a); [10] *Stephens et al.* (2004); [11] *Noll et al.* (2002b); [12] *Millis and Clancy* (2003); [13] *Kavelaars et al.* (2001); [14] *Noll et al.* (2006f); [15] *Brown* (2005a); [16] *Brown et al.* (2005b); [17] *Brown* (2005b); [18] *Noll et al.* (2006e); [19] *Noll et al.* (2002c); [20] *Noll et al.* (2007d); [21] *Noll et al.* (2006g), *Grundy et al.* (2006); [22] *Noll et al.* (2007c); [23] *Noll et al.* (2006b); [24] *Noll et al.* (2006a); [25] *Smith et al.* (1978), *Christy and Harrington* (1978); [26] *Weaver et al.* (2005); [27] *Trujillo and Brown* (2002); [28] *Noll et al.* (2007b); [29] *Brown and Trujillo* (2002); [30] *Noll et al.* (2006c); [31] *Noll et al.* (2006d); [32] *Sheppard and Jewitt* (2004); [†] unpublished as of February 28, 2007.

The first conscious search for TNBs using HST was carried out in August 2000–January 2001 in a program that looked at just two TNOs and found no companions. Two other small programs executed between October 1997 and September 1998 observed eight TNOs with the potential to identify a binary, had there been one among the objects observed.

The first moderate-scale program to search for binaries used STIS in imaging mode to search for binaries around 25 TNOs from August 2001 to August 2002. This program found two binaries, both relatively faint companions to the resonant TNOs (47171) 1999 TC_{36} and (26308) 1998 SM_{165} (*Trujillo and Brown,* 2002; *Brown and Trujillo,* 2002). The STIS imaging mode has a pixel scale of 50 milliarcsec, making it possible to directly resolve objects separated by approximately 100 milliarcsec or more. In principle, PSF analysis can detect binaries at significantly smaller separations in HST data because of the stability of the PSF (*Stephens and Noll,* 2006). The STIS data were obtained without moving the telescope to track the target's motion. The TNOs observed this way drift measurably during exposures, complicating the PSF analysis for these data.

From July 2001 to June 2002, 75 separate TNOs were observed with WFPC2 in a program designed to obtain V-, R-, and I-band colors (*Stephens et al.,* 2003); 3 of these were found to be binary. To achieve better sensitivity and because of the relatively large uncertainties in TNO orbits at the time, the targets were observed with the WF camera with 100 milliarcsec pixels. The sensitivity to faint companions for these data has not been fully quantified and exposure times were varied depending on the anticipated brightness of the TNO so that, in any case, the limits vary. Typical photometric uncertainties ranged from 3% to 8% for V magnitudes that ranged from 21.9 to 25.1 with a median magnitude of 23.6.

NICMOS was used to observe 82 separate TNOs from August 2002 through June 2003. Observations were made with the NIC2 camera at a 75 milliarcsec pixel scale. Two broad filters, the F110W and F160W, approximating the J- and H-band filters, were used in the observations. A total of nine new binary systems were identified from this data set, three of which were resolved and visible in the unprocessed data. The other six binaries were identified from a process of PSF fitting that enabled a significant increase in detectivity beyond the usual Nyquist limit with a separation/relative brightness/total brightness function that is complex and modeled with simulated binaries. Several of the binaries identified in this way have been subsequently resolved with the higher-resolution HRC, verifying the accuracy of the analysis of the NICMOS data (*Stephens and Noll,* 2006).

From July 2005 through January 2007, HST's HRC was used to look at more than 100 TNOs. This program identified a significant number of new binaries (*Noll et al.,*

2006a,b,c,d,e,f). The pixels of the HRC, while operational, were a distorted 25 × 28 milliarcsec quadrilateral. Observations with the clear filter were able to reach a faint limiting magnitude of at least V = 27, significantly deeper than most other HST observations of TNOs (*Noll et al.,* 2006h).

The wide variety of instruments and inherent observing limitations for each makes analysis from a concatenation of these data extremely problematic. Indeed, even observations taken with the same instrument vary in their limits based on exposure time, focus, seeing, and other systematics. This variability is reduced for spacebased observations, but not entirely eliminated. Additionally, the detection of close pairs depends on the ability to model and subtract the PSF of the primary with a resulting spatially dependent detection limit.

2.3. Binary Frequency

Does the fraction of TNBs vary with any of the observable properties of TNOs? In the search for correlations, the most promising quantities are those that can be associated with formation or survival.

Dynamical class is one such quantity because objects in different dynamical classes may have had different origins and dynamical histories. On one extreme, objects on unstable, nonresonant, planet-crossing orbits (the Centaurs and similar objects) have many close encounters with giant planets during their lifetimes. These close encounters can potentially disrupt weakly bound binaries (*Noll et al.,* 2006g; *Petit and Mousis,* 2004) (see also section 3.3). At the other extreme, objects in the classical disk may have persisted largely undisturbed from the protoplanetary disk and thus be a more congenial environment for survival of binaries. *Stephens and Noll* (2006) have shown that, indeed, the binary frequency in the cold classical disk is significantly higher than for other dynamical classes at the observational limits attainable by the NIC2 camera. They found that $22 \pm ^{10}_{5}\%$ of classical TNOs with inclinations less than 5° had detectable binaries at magnitude-dependent separations typically >60 milliarcsec. The rate of binaries for all other dynamical classes combined was $5.5 \pm ^{4}_{2}\%$ for the same limits. The number of objects and binaries in this dataset precluded any meaningful conclusions on the binary frequency in the various dynamically "hot" populations. Including more recent HRC observations strengthens these conclusions, as shown in Fig. 2.

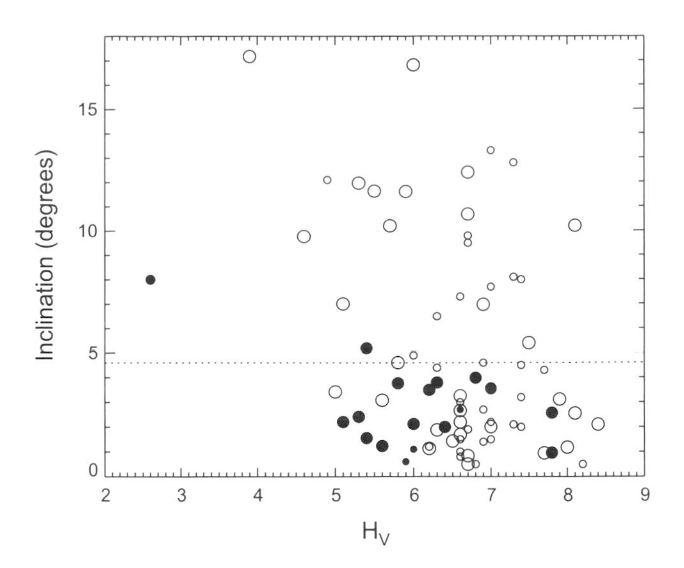

Fig. 2. Classical TNOs observed by HST are plotted with their absolute magnitude, H_V, on the horizontal axis and their inclination to the ecliptic, i, plotted vertically. Objects observed with the HRC (30 milliarcsec resolution) are shown as large circles; less-sensitive observations made with NIC2 (75 milliarcsec resolution) are shown as smaller circles. Binaries are shown as filled circles. The dotted line at 4.6° is the boundary between hot and cold classical populations proposed by *Elliot et al.* (2005). The extremely strong preference for low-inclination binaries in this sample is evident.

The size of the primary is another parameter that could, potentially, be correlated with binary status. To the extent that size correlates with dynamical class, the two sorting criteria can be confused for one another. There are, in fact, differences in size distributions as a function of dynamical class with the resonant objects, scattered disk, and high-inclination classical objects all having significantly larger upper size limits than the cold classical population (*Levison and Stern,* 2001; *Bernstein et al.,* 2004). The published binary searches to date cover an insufficient range in size (as measured by H_V) to be able to reach any strong conclusions with regard to binary frequency as a function of size. *Brown et al.* (2006a) have noted an apparently higher fraction of bound systems among the largest TNOs based on the satellites known for Pluto, Eris, and 2003 EL_{61}. With the size bins chosen in this work (the four largest TNOs vs. smaller TNOs) the difference in binary frequency is mathematically significant. The subsequent detection of small satellites around four more large TNOs (*Brown and Suer,* 2007) strengthens this apparent trend. In considering whether small satellites exist around smaller TNOs, it is important to consider that companions as faint as those of the largest TNOs would only be detectable for widely separated companions in a subset of the most recent deep observations (Fig. 3) (*Noll et al.,* 2006g) and would not have been detectable in most earlier surveys. At the same fractional Hill radius, small satellites of 100-km-class TNOs would be extremely difficult to detect in any existing observations. Caveats aside, however, it seems reasonable to hypothesize that the small satellites of the largest TNOs may be collisional in origin (*Brown et al.,* 2006a; *Stern et al.,* 2006), while the nearly equal sized binaries of smaller TNOs may form from dynamical capture, and thus there may be real differences in the frequencies of these two types of bound systems. We discuss these two modes of formation in more detail in section 4.

An obvious, but sometimes neglected, qualifier that must accompany any description of binary frequency is the limit in magnitude and separation imposed by the observational method used to obtain the data. This is especially true because, as discussed in more detail below and as shown in

Fig. 4, the number of binaries detectable in a given sample appears to be a strong function of separation (*Kern and Elliot,* 2006). The variation of binary frequency with dynamical class and/or size means that statements about binary frequency must be further qualified by a description of the sample. Thus, it is impossible to state a unique "frequency of transneptunian binaries." The need for a more nuanced description of binary frequency among the transneptunian populations is both a challenge and an opportunity waiting to be exploited.

3. PHYSICAL PARAMETERS

3.1. Relative Sizes of Binary Components

The relative sizes of the primary and secondary components is an important physical parameter that is usually available from a single set of observations (with the important caveat of possible nonspherical shapes and lightcurve variations as we discuss in section 3.6). The observed magnitude difference between the two components of a binary, Δ_{mag}, can be used to constrain the ratios of their radii R_1/R_2 and surface areas A_1/A_2 according to

$$\frac{R_1}{R_2} = \sqrt{\frac{A_1}{A_2}} = \sqrt{\frac{p_2}{p_1}}10^{0.2\Delta_{mag}} \quad (1)$$

where p_1 and p_2 are the albedos of the TNB components. It is usual, but not necessary, to assume that both components have the same albedo. As we note in section 3.6, the similarity of the colors of TNB components suggests common surface materials with similar albedos may be the norm for TNBs. However, we also note that in the one instance where separate albedos have been measured, the Pluto/Charon system, they are not identical. Given the large range of albedos of TNOs of all sizes (e.g., *Grundy et al.,* 2005; chapter by Stansberry et al.), it is reasonable to keep this customary simplification in mind.

The relative sizes of TNBs found so far is heavily skewed to nearly equal-sized systems as can be seen in Table 1 and Fig. 3. The prevalence of nearly equal-sized systems is a unique feature of TNBs compared to binaries in the main belt or near-Earth populations (e.g., *Richardson and Walsh,* 2006; *Noll,* 2006). To some extent, this conclusion is limited by observational bias, but deep surveys with the HST's HRC show an apparent lack of asymmetric binaries (*Noll et al.,* 2006h) (Fig. 3). A preference for similar-sized components is a natural outcome of dynamical capture models for the formation of binaries (*Astakhov et al.,* 2005) and may explain the observed distribution of relative sizes after

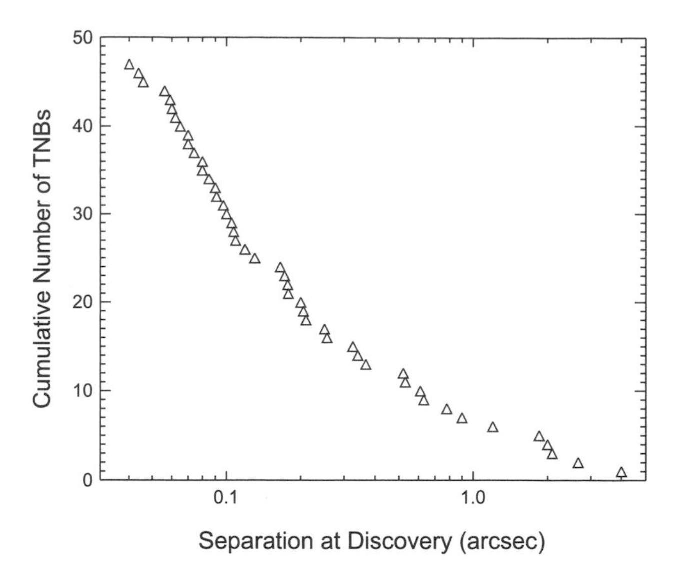

Fig. 4. The number of objects with a separation greater than or equal to a given separation is shown. The number of objects at small separation increases faster than an exponential as can be seen in this logarithmic plot. Observational bias favors detection of widely separated binaries, suggesting that the underlying distribution may rise even more rapidly with decreased separation.

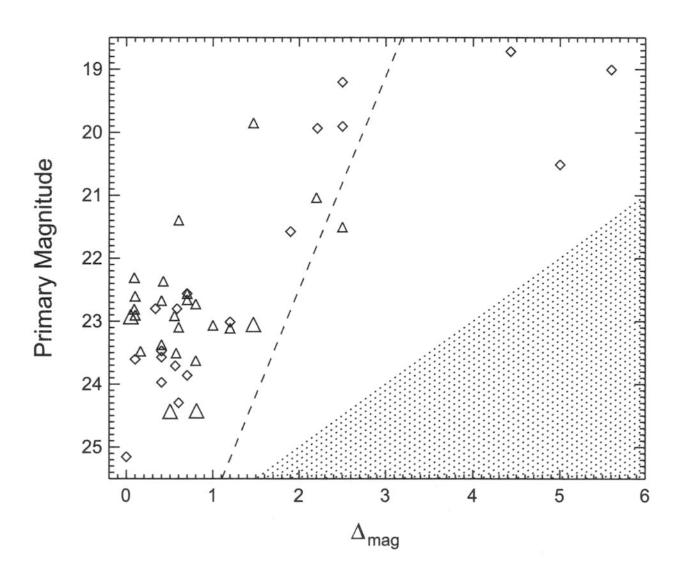

Fig. 3. In this figure the y-axis is the observed magnitude of the primary component of observed TNBs. The x-axis shows the range of observed and detectable magnitude differences, Δ_{mag}, for secondaries under several different observing circumstances. Binaries detected with HST's HRC are shown as triangles; binaries found with other instruments are shown as diamonds. The stippled area shows the 7σ detection limit for companions that are sufficiently separated that their local background is dominated by sky noise and dark current. Objects that are observed by the HRC and are background dominated are shown as large triangles. The background and detection limits for the objects shown by the small triangles are limited to a varying degree by the PSF of the primary. The dashed line defines an approximate empirical detection limit for objects separated by 3 pixels from their primary, i.e., 75 milliarcsec. The background at this separation is dominated by the PSF of the primary to a degree that varies as a function of the primary. Both detection limits are dependent on the details of the observations and therefore do not apply to the observations from other instruments (diamonds) that are less sensitive. The clustering of binaries at $\Delta_{mag} < 1$ appears to be an intrinsic property of TNBs and not an observational bias. From *Noll et al.* (2006h).

accounting for the small number of large object satellites that appear to have been formed from collisions (e.g., *Stern et al.,* 2006).

3.2. Orbit Determination

Determining the mutual orbit of a spatially resolved pair of gravitationally bound objects is a classic problem of celestial mechanics. Solutions are reviewed in numerous textbooks (e.g., *Aitken,* 1964; *Heintz,* 1978; *Smart,* 1980; *Danby,* 1992).

The Keplerian orbit of a pair of bound point masses can be described by seven independent quantities. There is considerable flexibility in the choice of the seven. One possible set is the system mass plus the three-dimensional relative position and velocity vectors at a specific time. This form is generally preferred for specifying inputs to orbital integration routines. A second set of elements is more frequently used to specify binary orbits in the scientific literature: period T, semimajor axis a, eccentricity e, inclination i, mean longitude at a specific time ε, longitude of the ascending node Ω, and longitude of periapsis ϖ. Other sets of elements are possible and are occasionally encountered in the literature.

Each observation of a resolved binary pair provides two constraints, the separation and position angle, or equivalently, the relative positions in right ascension and declination. To constrain seven orbital elements requires at least four observations, in order to have more constraints than unknowns. In practice, four observations are often not enough to uniquely determine the seven unknowns, because the constraints from the separate observations are not necessarily independent of one another. For example, four observations that happen to sample the same orbital longitude do little to constrain the shape and orientation of the orbit. Many more than four observations may be needed, if they are poorly timed.

Instruments capable of doing relative astrometry on the extremely close-spaced and faint components of TNBs are a scarce and valuable resource. To make the most efficient use of these facilities requires strategic timing of follow-up observations to maximize the additional constraints provided by each successive observation. Several groups have applied Monte Carlo techniques to this task (e.g., *Margot et al.,* 2004, 2005; *Hestroffer et al.,* 2005; *Grundy et al.,* 2007). The general approach is to generate random orbits consistent with the existing set of observations, which do not yet uniquely determine the binary orbit. The collection of orbits produced by this exercise is used to map out regions of orbital element space consistent with the constraints already available, and to identify times when follow-up observations would be most effective for collapsing the cloud of possible orbits.

An additional complication involves near-equal brightness binaries, which are not uncommon among the TNBs. When the "primary" and "secondary" have similar brightness, or large-amplitude lightcurve variations, they become difficult to distinguish from one another, leading to uncertainties that require additional observations to resolve. The binary system 2003 QY_{90} provides a recent example of this situation (*Kern and Elliot,* 2006).

Various nonlinear least-squares techniques such as the Levenburg-Marquardt and AMOEBA algorithms (*Press et al.,* 1992) can be used to fit a set of orbital elements to the observational data by iteratively minimizing the residuals (as measured by the χ^2 statistic) between the astrometric data and the positions computed from the orbital elements. In situations where observational data are particularly abundant, linear least-squares fitting techniques may be applicable. Knowledge of the uncertainties in the fitted orbital elements is just as important as knowledge of the elements themselves. The Monte Carlo techniques mentioned earlier can be used to directly investigate uncertainties in orbital elements (e.g., *Virtanen et al.,* 2001, 2003; *Hestroffer et al.,* 2005). Another approach involves varying each parameter around its best fit value and allowing the remaining parameters to readjust themselves to compensate, resulting in a new fit with different χ^2. This approach can be used to map out a seven-dimensional χ^2 space, within which contours of 1σ or 3σ likelihood can be traced (*Lampton et al.,* 1976; *Press et al.,* 1992).

For TNBs, relative motion between Earth and the orbiting pair leads to changing viewing geometry over time. The motion has two components: the comparatively rapid motion of Earth around the Sun, and the much slower motion of the TNB along its heliocentric orbit. The resulting changes in viewing geometry complicate the orbit fitting process compared with procedures developed for binary stars, which assume fixed geometry between the observer and the binary system. However, there are advantages to observing a binary from different angles. Relative astrometry is otherwise not able to distinguish between the actual orbit and its mirror reflection through the sky plane. For TNBs, the ambiguity between these two "mirror" orbits can be broken by observations spanning as little as a few years (e.g., *Hestroffer and Vachier,* 2006), but this has only actually been done for a small minority of orbits to date.

For larger and closer binaries, departures from spherical symmetry of the primary's mass distribution (parameterized by the gravitational harmonic coefficients, primarily J_2) can exert torques on the secondary leading to secular precession of the line of nodes and/or of the line of apsides (e.g., *Brouwer and Clemence,* 1961). Observations of this secular evolution can be used to measure J_2 as has been done for a binary asteroid (*Marchis et al.,* 2005), but this has not yet been done for any TNBs except Pluto [and that measurement, by *Descamps* (2005), neglected possible perturbations on the orbit of Charon by Pluto's smaller satellites]. Comparable effects can be anticipated for the orbits of the more-distant satellites of close pairs, such as the orbits of Hydra and Nix around the Pluto/Charon binary (*Buie et al.,* 2006; *Lee and Peale,* 2006).

3.3. Semimajor Axis Distribution

The separation of components in binaries is a fundamental datum that can be diagnostic in comparing systems and in understanding general processes affecting populations. Semimajor axes are known for only the subset of TNBs with measured orbits (Table 2). Unfortunately, from a statistical perspective, the available data are too few and are heavily biased against small semimajor axes by limitations of both discovery and follow-up observations.

As a proxy for the semimajor axis, we can use the separation at discovery, s_0, listed in Table 1. In any individual case this is only an approximate substitute for a measurement of the semimajor axis from an orbit because of the unknown orientation of the orbit plane with respect to the line of sight and the unknown eccentricity of the binary orbit. Observational biases tend to favor detection of objects close to their widest separation from the primary independent of the orientation of the orbit plane. For the 13 systems in Table 2, the ratio of the semimajor axis to the distribution of separation at discovery, a/s_0, ranges from 0.5 to 2. The median value of a/s_0 is 0.96. Thus, for a moderate-sized ensemble, the distribution of separations at discovery provides an excellent statistical approximation for the semimajor axis distribution. The separations at discovery from Table 1 are plotted in Fig. 4. This log plot shows a stronger than exponential increase at smaller separations. The trend of increasing binary frequency at small separation is robust because observational bias will decrease the number of detectable companions near the limits of instrument resolution. Interestingly, this trend is in qualitative agreement with the binary formation model discussed in more detail in section 4.5.

It is natural to wonder if the trend of binary frequency continues at even smaller separations or if there is some eventual cutoff. The separation of TNB components in terms of "typical" values for the distance to Earth in AU, $\Delta_\oplus$, and an angular separation in arcsec, θ, can be expressed as

$$s \simeq 2900\left(\frac{\theta}{0.1}\right)\left(\frac{\Delta_\oplus}{40}\right) \text{km} \qquad (2)$$

or nearly 60 primary radii for a 100-km-diameter TNO. This leaves a significant amount of separation phase space where a sizable population of stable binaries could exist undiscovered.

At the extreme of small separation are bilobed objects and contact binaries. *Sheppard and Jewitt* (2004) proposed that the TNO 2001 QG_{298} may be a very close, possibly contact, binary based on its large-amplitude lightcurve (1.14 mag) and its relatively slow rotation (13.77 h). The rotation is too slow and the amplitude too large to explain the lightcurve as a fluid deformation into a Jacobi ellipsoid,

TABLE 2. Physical properties of transneptunian binaries.

Object	a (km)	e	T (d)	M (10^{18} kg)	p_λ	ρ (g cm^{-3})	J/J'	References
(136199) Eris	*36,000*	*0*	*14*	*16,400*	0.86(7) (V)	2.26(25)	*0.16*	[1]
(134340) Pluto/Charon				14,570(9)			*0.40*	[2]
Pluto				13,050(620)	0.51–0.71 (V)	2.03(6)		[2]
Charon	19,571(4)	0.00000(7)	6.387230(1)	1,521(65)	0.38 (V)	1.65(6)	—	[4]
Nix	48,670(120)	0.002(2)	24.8562(13)	*0.1–2.7*	*0.01–0.35*	*2.0*	—	[4]
Hydra	64,780(90)	0.005(1)	38.2065(14)	*0.2–4.9*	*0.01–0.35*	*2.0*	—	[4]
(136108) 2003 EL_{61}	49,500(400)	0.050(3)	49.12(3)	4,210(100)	0.7(1) (v)	2.9(4)	*0.53*	[3]
(47171) 1999 TC_{36}	7,720(460)	0.22(2)	50.4(5)	14.4(2.5)	0.08(3) (v)	0.5(3/2)	*0.31*	[5]
2001 QC_{298}	3,690(70)	0.34(1)	19.2(2)	10.8(7)	0.08(V606)	*1.0*	*1.16*	[6]
(26308) 1998 SM_{165}	11,310(110)	0.47(1)	130(1)	6.78(24)	0.08(3/2) (V)	0.7(3/2)	*0.56*	[7]
(65489) Ceto/Phorcys	1840(50)	<0.014	9.557(8)	5.4(4)	0.08(2) (V)	1.4(6/3)	*0.89*	[8]
(66652) 1999 RZ_{253}	4660(170)	0.46(1)	46.263(6/74)	3.7(4)	0.17 (R)	*1.0*	*1.56*	[9]
1998 WW_{31}	22,300(800)	0.82(5)	574(10)	2.7(4)	0.054 (R)	*1.0*	2.22	[10]
(88611) 2001 QT_{297}	27,880(150)	0.241(2)	825(1)	2.51(5)	*0.13 (r')*	*1.0*	3.41	[11]
(42355) Typhon/Echidna	1830(30)	0.53(2)	18.971(1)	0.96(5)	0.05 (V)	0.47(18/10)	*2.13*	[12]
(58534) Logos/Zoe	8,010(80)	0.45(3)	312(3)	0.42(2)	0.37(4) (R)	*1.0*	*2.65*	[13]
2003 QY_{90}	*7,000–13,000*	*0.44–0.93*	*306–321*	*0.3–1.7*	*0.19–0.35* (r')	*1.0*	*2.14*	[14]

Objects are listed in order of decreasing system mass. Uncertainties in the last digit(s) of measured quantities appear in parentheses. Table entries in italics indicate quantities that have been published without error estimates or that have been computed by the authors from estimated quantities. The components of the Pluto system are listed separately because they have had independent determinations of radius, mass, albedo, and density that are not yet available for other binary or multiple systems. The wavebands used to determine geometric albedos are shown in parentheses (v is for references citing "visual" albedo, V_{606} is the HRC's F606W filter). Densities of 1.0 g cm^{-3} are assumed for objects without an independent size measurement. The corresponding geometric albedos are for this assumed unit density.

References: [1] *Brown et al.* (2006a,b); [2] *Buie et al.* (1997), *Buie et al.* (2006), *Rabinowitz et al.* (2006); [3] *Brown et al.* (2005a); [4] *Buie and Grundy* (2000), *Buie et al.* (2006), *Lee and Peale* (2006); [5] *Stansberry et al.* (2006); [6] *Margot et al.* (2004); [7] *Margot et al.* (2004), *Spencer et al.* (2006); [8] *Grundy et al.* (2007); [9] *Noll et al.* (2004a); [10] *Veillet et al.* (2002); [11] *Osip et al.* (2003), *Kern* (2006); [12] *Grundy et al.* (2007); [13] *Noll et al.* (2004b), *Margot et al.* (2004); [14] Kern et al., in preparation.

as is postulated for some other TNOs (*Sheppard and Jewitt,* 2002; *Jewitt and Sheppard,* 2002). That leaves only albedo variation, strength-dominated shape, or a close binary to explain the lightcurve of 2001 QG_{298}. *Sheppard and Jewitt* (2004) conclude that a close binary is the most likely explanation for the observed lightcurve. They further extrapolate from this single object to suggest that as many as 10–15% of TNOs could be close binaries. While unproven, a 10–15% fraction of TNOs as contact binaries is not incompatible with the frequency of wider binaries and the trend of frequency as a function of separation.

The natural dimension for scaling the separations of TNBs is the Hill radius, R_H, given by the equation

$$R_H \cong a_\odot(M_1/3M_\odot)^{1/3} \quad (3)$$

where M_1 is the mass of the largest component of the binary, $M_\odot$ is the mass of the Sun, and $a_\odot$ is the semimajor axis of the primary's heliocentric orbit. Satellite orbits well inside the primary's Hill sphere are generally stable with respect to external perturbations. However, the instantaneous Hill radius, calculated by replacing $a_\odot$ above with the instantaneous heliocentric distance, can deviate significantly from the average Hill radius for TNBs with large heliocentric eccentricities such as (65489) Ceto/Phorcys. This result must be taken into account when considering, for example, the impact of gravitational perturbations from major planets. The Hill radius scales with the radius of the primary, and thus the large size range of the primary objects in Table 2 implies a similarly large range in Hill radii measured in absolute units. In dimensionless units, a typical Hill radius for a TNB is on the order of 7000× the radius of the primary. Objects with measured orbits have semimajor axes that range from ~0.1% to ~8% of the Hill radius (*Noll et al.,* 2004b; *Kern,* 2006), well within the stable portion of the Hill sphere. Whether this represents an intrinsic outcome of formation (*Astakhov et al.,* 2005), or is a signature of a surviving remnant population where more weakly bound systems have been disrupted while more tightly bound systems have become even tighter in the wake of encounters with third bodies (*Petit and Mousis,* 2004), or some combination of the two, remains to be determined.

3.4. Mass, Albebo, and Density

A particularly valuable piece of information that can be derived from the mutual orbit of a binary system is the total mass, M_{sys}, of the system, according to the equation

$$M_{sys} = \frac{4\pi^2 a^3}{GT^2} \quad (4)$$

where a is the semimajor axis, G is the gravitational constant, and T is the orbital period. Knowledge of the semimajor axis, a, tends to be limited by the spatial resolution of the telescope, while knowledge of T is limited by the timespan over which observations are carried out (modulo the binary orbit period). It is possible to extend the timespan of observations, whereas the spatial resolution cannot generally be improved. Thus, typically, T is determined to much higher fractional precision than a, and the uncertainty in M_{sys} is dominated by the uncertainty in a. M_{sys} can often be calculated before all seven elements of the binary orbit are fully determined, since the elements T, a, and e are relatively insensitive to the ambiguity between orbits mirrored through the instantaneous sky plane (see section 3.1).

For a system with a known mass, it is possible to derive other parameters that offer potentially valuable compositional constraints. For instance, if one assumes a bulk density ρ, the bulk volume of the system V_{sys} can be computed as $V_{sys} = M_{sys}/\rho$. How the volume and mass is partitioned between the two components remains unknown. Assuming the components share the same albedo, the individual radii of the primary and secondary can be obtained from

$$R_1 = \left(\frac{3V_{sys}}{4\pi(1 - 10^{-0.6\Delta_{mag}})}\right)^{1/3} \quad (5)$$

and equation (1) simplifies to give $R_2 = R_1 10^{-0.2\Delta_{mag}}$. An effective radius R_{eff}, equal to the radius of a sphere having the same total surface area as the binary system, can be computed as

$$R_{eff} = \sqrt{R_1^2 + R_2^2} \quad (6)$$

Combining R_{eff} and the absolute magnitude of the system H_λ, one obtains the geometric albedo p_λ

$$p_\lambda = \left(\frac{C_\lambda}{R_{eff}}\right)^2 10^{-0.4H_\lambda} \quad (7)$$

where C_λ is a wavelength-dependent constant (*Bowell et al.,* 1989; *Harris,* 1998). For observations in the V band, C_V = 664.5 km. This approach has been used to estimate albedos for a number of TNBs by assuming their bulk densities must lie within a plausible range, typically taken to be 0.5–2 g cm^{-3} (e.g., *Noll et al.,* 2004a,b; *Grundy et al.,* 2005; *Margot et al.,* 2005). These efforts demonstrate that the TNBs have very diverse albedos, but those albedos are not clearly correlated with size, color, or dynamical class. The calculation could also be turned around such that an assumed range of albedos leads to a range of densities.

When the sizes of the components of a binary system can be obtained from an independent observation, that information can be combined with the system mass to obtain the bulk density, providing a fundamental constraint on bulk composition and interior structure. Sizes of TNOs are extremely difficult to obtain, although a variety of methods can be used, ranging from direct observation (e.g., *Brown et al.,* 2006b) to mutual events and stellar occultations (e.g., *Gulbis et al.,* 2006). For rotationally deformed bodies it is possible to constrain the density directly from the observed lightcurve assuming the object is able to respond as a "fluid" rubble pile (*Jewitt and Sheppard,* 2002; *Takahashi and Ip,* 2004; chapter by Sheppard et al.). Spitzer Space Telescope observations of thermal emission have recently led to a number of TNO size estimates (e.g., *Cruikshank et al.,* 2007; chapter by Stansberry et al.). Unfortunately, many of the known binaries are too small and distant to be detected at thermal infrared wavelengths by Spitzer or directly resolved by HST.

Systems with density estimates include three large TNBs: Pluto and Charon, with $\rho = 2.0 \pm 0.06$ and $\rho = 1.65 \pm 0.06$ g cm^{-3}, respectively (*Buie et al.,* 2006); 2003 EL_{61}, with $\rho = 3.0 \pm 0.4$ g cm^{-3} (*Rabinowitz et al.,* 2006); and Eris, with $\rho = 2.26 \pm 0.25$ g cm^{-3} (*Brown,* 2006). The relatively high densities of the large TNOs are indicative of substantial amounts of rocky and/or carbonaceous material in the interiors of these objects, quite unlike their ice-dominated surface compositions.

Four smaller TNBs have recently had their densities determined from Spitzer radiometric sizes. These are (26308) 1998 SM_{165}, with $\rho = 0.70 \pm ^{0.32}_{0.21}$ g cm^{-3} (*Spencer et al.,* 2006), (47171) 1999 TC_{36}, with $\rho = 0.5 \pm ^{0.3}_{0.2}$ g cm^{-3} (*Stansberry et al.,* 2006); (65489) Ceto/Phorcys, with $\rho = 1.38 \pm ^{0.65}_{0.32}$ g cm^{-3} (*Grundy et al.,* 2007); and (42355) Typhon/Echidna, with $\rho = 0.47 \pm ^{0.18}_{0.10}$ g cm^{-3} (Grundy et al., in preparation). *Takahashi and Ip* (2004) estimate a density of $\rho < 0.7$ g cm^{-3} for 2001 QG_{298}. The very low densities of four of these five require little or no rock in their interiors, and even for pure H_2O ice compositions, call for considerable void space. The higher density of Ceto/Phorcys is consistent with a mixture of ice and rock. It is clear from these results that considerable diversity exists among densities of TNOs, but it is not yet known whether densities correlate with externally observable characteristics such as color, lightcurve amplitude, or dynamical class.

3.5. Eccentricity Distribution and Tidal Evolution

The eccentricities of binary orbits known to date span the range from values near zero to a high of 0.8 (Table 2), with perhaps a clustering in the range 0.3–0.5. With the usual caveats about small number statistics (N ~ 10), no obvious correlation between eccentricity and semimajor axis is present in the data obtained to date.

Keplerian two-body motion assumes point masses orbiting one another. The finite size of real binary components allows differential gravitational acceleration between nearer and more distant parts of each body to stretch them along the line connecting them. If their mutual orbit is eccentric, this tidal stretching varies between apoapsis and periapsis, leading to periodic flexing. The response of a body to such flexing is characterized by the parameter Q, which is a complex function of interior structure and composition (*Goldreich and Soter,* 1966; *Farinella et al.,* 1979). In a body with low Q, tidal flexing creates more frictional heating, which dissipates energy. A body with higher Q can flex with less energy dissipation. For a body having a rotation state different from its orbital rotation Q is related to the angular lag, δ, of the tidal bulge behind the line of centers: $Q^{-1} = \tan(2\delta)$. As with orbital eccentricity, this situation produces time-variable flexing, and thus frictional heating and dissipation of energy. Typical values of Q for rocky planets and icy satellites are in the 10–500 range (*Goldreich and Soter,* 1966; *Farinella et al.,* 1979; *Dobrovolskis et al.,* 1997).

The energy dissipated by tidal flexing comes from orbital and/or rotational energy, leading to changes in orbital parameters and rotational states over time. Tides raised on the primary tend to excite eccentricity, while tides raised on the secondary result in damping. The general trend is toward circular orbits, with both objects' spin vectors aligned and spinning at the same rate as their orbital motion. The timescale for circularization of the orbit is given by

$$\tau_{circ} = \frac{4Q_2M_2}{63M_1}\sqrt{\frac{a^3}{G(M_1+M_2)}}\left(\frac{a}{R_2}\right)^5 \qquad (8)$$

where a is the orbital semimajor axis, M_1 is the mass of the primary, and M_2, R_2, and Q_2 are the mass, radius, and Q of the secondary (e.g., *Goldreich and Soter,* 1966). It is important to recognize that this formula (1) assumes the secondary to have zero rigidity, and (2) assumes that the eccentricity evolution due to tides raised on the primary is ignorable. Neither of these assumptions may be justified, especially in the case of nearly equal-sized binaries (for a thorough discussion, see *Goldreich and Soter,* 1966). The timescale in equation (8) is sensitive to the ratio of the semimajor axis to the size of the secondary. Larger and/or closer secondaries are likely to have their orbits circularized much faster than more widely separated systems. For example, the close binary (65489) Ceto/Phorcys has a = 1840 km, $M_1 = 3.7 \times 10^{18}$ kg, $M_2 = 1.7 \times 10^{18}$ kg, and $R_2 = 67$ km (*Grundy et al.,* 2007). For Q = 100 (a generic value for solid bodies), its orbit should circularize on a relatively short timescale on the order of ~10^5 yr. A more widely separated example, (26308) 1998 SM_{165}, has a = 11,300 km, $M_1 = 6.5 \times 10^{18}$ kg, $M_2 = 2.4 \times 10^{17}$ kg, and $R_2 = 48$ km (*Spencer et al.,* 2006), leading to a much longer circularization timescale on the order of ~10^{10} yr, consistent with the observation that it still has a moderate orbital eccentricity of 0.47 (*Margot,* 2004).

Tidal effects can also synchronize the spin rate of the secondary to its orbital period (as in the case of the Earth/Moon system) and, on a longer timescale, synchronize the primary's spin rate as well (as for Pluto/Charon). The timescale for spin-locking the primary (slowing its spin to match the mutual orbital period) is given by

$$\tau_{despin,1} = \frac{Q_1R_1^3\omega_1}{GM_1}\left(\frac{M_1}{M_2}\right)^2\left(\frac{a}{R_1}\right)^6 \qquad (9)$$

where ω_1 is the primary's initial angular rotation rate and R_1 is its radius (*Goldreich and Soter,* 1966). The initial angular rotation rate is not known, but an upper limit is the breakup rotation rate, which is within 50% of 3.3 h for the 0.5–2 g cm^{-3} range of densities discussed. For (65489) Ceto/Phorcys and (26308) 1998 SM_{165} with primary radii (R_1) of 86 km (*Grundy et al.,* 2007) and 147 km (*Spencer et al.,* 2006), we find the spin-locking timescale to be ~10^4 and ~10^5 yr respectively, slightly faster than the circularization timescale. The timescale for despinning the secondary, $\tau_{despin,2}$, is given by swapping subscripts 1 and 2 in equation (9).

The general case for both tidal circularization and tidal despinning in binaries is complex (e.g., *Murray and Dermott,* 1999). Binaries in the transneptunian population include many where the secondary is of comparable size to the primary. For these systems, it is not safe to make the common assumption that the tide raised by the secondary on the primary is ignorable. Furthermore, there are many systems where the binary orbit has moderate to high eccentricity, invalidating the simplifying assumptions possible for nearly circular orbits. A full treatment of the tidal dynamics for the kinds of systems we find in the transneptunian population would make an interesting addition to the literature. In the meantime, observation is likely to lead the way in understanding the dynamics of these systems.

3.6. Colors and Lightcurves

The spectral properties of TNOs and their temporal variation are fundamental probes of the surfaces of these objects (see chapters by Doressoundiram et al., Tegler et al., Belskaya et al., Sheppard et al., and Barucci et al.). The colors of TNOs have long been known to be highly variable (*Jewitt and Luu,* 1998) and some correlations of color with other physical or dynamical properties have been claimed (e.g., *Peixinho et al.,* 2004). A natural question is whether this variability can be used to constrain either the origin of TNBs, the origin of color diversity, or both. For example, one can ask whether TNB components are similarly or differently colored. Because TNBs are thought to be primordial, differences in color could be either due to mixing of different composition populations in the protoplanetary disk before the bound systems were formed or different collisional and evolutionary histories of components after they are bound.

A handful of TNBs have reported single-epoch resolved color measurements. Some of these objects are solar-colored [2000 CF_{105}, (58534) Logos/Zoe, (47171) 1999 TC_{36}, (66652) 1999 RZ_{253}, (88611) 2001 QT_{297}], while others are more gray [2001 QC_{298}, (65489) Ceto/Phorcys]. However, so far, the components have colors that are consistent with each other within the uncertainties of the measurements, 0.1–0.3 mag (*Margot et al.,* 2005; *Noll et al.,* 2004a,b; *Osip et al,* 2003; *Grundy et al.,* 2007). This similarity implies that the components are composed of similar material, at least on the surface. It also suggests that the assumption of equal albedo and density usually made for binaries may have some basis in fact.

Spectra are even better composition diagnostics than color measurements. Separate spectra of binary components are currently available only for the Pluto/Charon pair (*Buie et al.,* 1987; *Fink and DiSanti,* 1988; *Buie and Grundy,* 2000) and for 2003 EL_{61} (*Barkume et al.,* 2006). Pluto and Charon have well-known spectral differences that may be primarily related to the size threshold for retaining the very volatile CH_4 and N_2 ices found on Pluto but not on Charon. 2003 EL_{61} and its larger satellite, by contrast, both have spectra that are dominated by water ice.

Lightcurves are diagnostic of both compositional variation on surfaces and of nonspherical shapes. They also give rotation rates. In binary systems, the rotation state of the components is subject to tidal evolution (see section 3.5). Both unresolved and resolved lightcurves can be useful for addressing these issues.

Unresolved lightcurves of short duration for a number of TNBs have been obtained, sometimes with incomplete or inconsistent results. Lightcurves of (47171) 1999 TC_{36} and (42355) Typhon/Echidna showed variations on the order of 0.10–0.15 mag, but no period was determinable from the data (*Ortiz et al.,* 2003). Similarly, observations of (66652) 1999 RZ_{253} and 2001 QC_{298} showed small-amplitude, but nonsystematic, variation over a 4–6-h duration (*Kern,* 2006). *Romanishin et al.* (2001) reported a lightcuve for the unresolved binary (26308) 1998 SM_{165}, obtained from the 1.8-m Vatican Advanced Technology Telescope in 1999 and 2000, with a moderate amplitude of 0.56 mag. The period was determined to be either 3.983 h (single-peaked, caused by an albedo spot) or 7.966 h (double-peaked, caused by nonspherical shape). The single-peaked period is near the breakup period (3.3 h) for a solid ice body. Because the unresolved lightcurve of (26308) 1998 SM_{165} did not show any color variation with time, Romanishin et al. argued for the longer, double-peaked period as the most likely. Lightcurve measurements of the same binary made at Lowell Observatory in 2006 found a slightly longer period of 8.40 ± 0.05 h (*Spencer et al.,* 2006).

Resolved groundbased lightcurves of binaries are challenging and can only be obtained under excellent conditions at a few facilities, and only for objects with the widest separations. Discovery observations of the binary (88611) 2001 QT_{297} at Las Campanas Observatory with Magellan revealed brightness changes in the secondary component of 0.3 mag in 30 mins (*Osip et al.,* 2003). Follow-up observations revealed the secondary to have a single-peaked period of 5–7 h, while the magnitude of the primary remained constant. Additional resolved color lightcurve measurements found the two surfaces to share similar colors throughout the rotation, indicating homogeneous, similar surfaces (*Kern,* 2006). Similar observations showed both components of 2003 QY_{90} to be variable. The primary and secondary components were observed to change by 0.34 ± 0.12 and 0.9 ± 0.36 mag, respectively, over 6 h of observation (*Kern and Elliot,* 2006). The large amplitude of the secondary component sometimes results in the secondary being brighter than the primary. Both components of the wide binary 2005

EO_{304} are variable with variations on the order of 0.3 mag over a period of 4 h (*Kern,* 2006).

Space-based observations from HST resolve the components of binaries and can constrain the variability of components in these systems. The best-studied system, by far, is the Pluto/Charon binary where detailed lightcurve measurements have been made (*Buie et al.,* 1997). Transneptunian binaries that have had their orbits measured by HST have multiple-epoch photometric measurements, although frequently the temporal sampling is poor. (58534) Logos/Zoe shows variability in the primary of at least ~0.8 mag, making it challenging at times to distinguish the primary from the secondary (*Noll et al.,* 2004b). However, with only a few widely spaced samples, this remains, for the moment, only an intriguing suggestion of a lightcurve. Three objects, (47171) 1999 TC_{36}, 2001 QC_{298}, and (65489) Ceto/Phorcys have virtually no variation in flux, implying they may be relatively spherical, homogeneous, and/or pole-on [although we note the contradictory groundbased observations of (47171) 1999 TC_{36}]. Once again, the sampling density is far to small to allow anything more than informed speculation.

3.7. Orbit Plane and Mutual Events

The tremendous scientific benefit that can derive from mutual occultations or eclipses between a poorly resolved object and its satellite was abundantly illustrated by the series of mutual events between Pluto and Charon during the 1980s (*Binzel and Hubbard,* 1997). As discussed before, these events enabled measurement of the sizes and albedos of both objects, of their distinct surface compositions, and even of albedo patterns on their surfaces.

For observable mutual events to happen either the observer or the Sun (or both) must be temporarily aligned with a TNB's orbit plane. An "occultation-type" event occurs when one component of the TNB passes in front of, and fully or partially occults, the other component from the observer's point of view. An "eclipse-type" event takes place when the TNB components are aligned with the Sun and the shadow of one falls on the other. Because the Sun and Earth have nearly equal lines of sight to TNBs, most mutual events observable from Earth are combinations of occultation-type and eclipse-type events.

The larger the objects are compared with their separation, the farther the orbit plane can deviate from either of these two types of alignments and still produce an observable mutual event. The criteria for both types of events can be expressed as $R_1 + R_2 > s \sin(\phi)$, where R_1 and R_2 are previously defined, s is their separation during a conjunction (equal to the semimajor axis, for the case of a circular orbit), and ϕ is the angle between the observer or the Sun and the plane of the binary orbit. During any conjunction when either criterion is satisfied, a mutual event can be observed. The period during which the orbit plane is aligned closely enough to the Sun's or to Earth's line of sight to satisfy the criteria for events can be thought of as a mutual event season.

Each orbit of a transneptunian secondary brings a superior conjunction (when the primary is closer to the observer) and an inferior conjunction (when the secondary is closer), so shorter orbital periods lead to more frequent conjunctions and associated opportunities for mutual events. The most recent mutual event season of Pluto and Charon lasted from 1985 through 1990, and, since their mutual orbit period is only 6.4 d, there were hundreds of observable events during that season. For more widely separated, smaller pairs, with longer orbital periods, the mutual event seasons may be shorter and conjunctions may be less frequent, leading to far fewer observable events, or even none at all. For example, (26308) 1998 SM_{165} has a reasonably well-determined orbit with a period of 130 d, a semimajor axis of 11,170 km, and an eccentricity of 0.47 (*Margot,* 2004). From Spitzer thermal observations, the diameters of the primary and secondary bodies are estimated to be 294 and 96 km, respectively (*Spencer et al.,* 2006). Ignoring uncertainties in the current orbital elements, the mutual event season will last from 2020 through 2026, with 12 mutual events being observable at solar elongations of 90° or more. Of these 12 events, 2 are purely occultation-type events and one is purely an eclipse-type event. The rest involve combinations of both occultation and eclipse.

4. BINARY FORMATION

When the Pluto/Charon binary was the only example of a true binary in the solar system (true binary ≡ two objects orbiting a barycenter located outside either body), it could be discounted as just another of the peculiarities associated with this yet-to-be-dwarf planet. The discovery of numerous similar systems among TNOs, however, has changed this calculus. Any successful model for producing TNBs cannot rely on low-probability events, but must instead employ processes that were commonplace in the portion of the preplanetary nebula where these objects were formed. Formation models must also account for the observed properties of TNBs, including the prevalence of similar-sized binaries, the range of orbital eccentricities, and the steeply increasing fraction of binaries at small angular separations. Survival of binaries, once they are formed, is another important factor that must be considered when, for instance, comparing the fraction of binaries found in different dynamical populations or their distribution as a fraction of Hill radius.

Several possible modes for the formation of solar system binaries have been discussed in the literature, including fission, dynamical capture, and collision (cf. reviews by *Richardson and Walsh,* 2006; *Dobrovolskis et al.,* 1997). For TNOs, capture and/or collision models have been the most thoroughly investigated. Fission is unlikely to be important for objects as large as the currently known TNBs. Other possible mechanisms for producing binaries, e.g., volatile-driven splitting, as is observed in comets, have not been explored. Interestingly, both capture and collisional formation models share the requirement that the number of objects in the primordial Kuiper belt (at least the small ones) be at least a couple of orders of magnitude higher than currently found in transneptunian space. It follows that all the TNBs observed today are primordial.

4.1. Capture

Capture models rely, in one form or another, on three-body interactions to remove angular momentum and produce a bound pair. As we show in detail below in section 4.5, they are also very sensitive to the assumed velocity distribution of planetesimals. *Goldreich et al.* (2002, hereafter *G02*) described two variations of the three-body model, L^3, involving three discrete bodies, and L^2s, where the third body is replaced by a dynamical drag coefficient corresponding to a "sea" of weakly interacting smaller bodies. In *G02*'s analysis, the L^2s channel was more efficient at forming binaries by roughly an order of magnitude.

Astakhov et al. (2005) extended the capture model by exploring how a weakly and temporarily bound ($a \sim R_H$) pair of big bodies hardens when a third small body ("intruder") passes within the Hill radius (equation (3)) of the pair. Their calculations assume the existence of transitory binaries that can complete up to ~10 mutual orbits before the third body approaches. They find that a binary hardens most effectively when the intruder mass is a few percent of that of a big body [this result likely depends on their assumed approach velocities of up to 5 v_H where $v_H = \Omega_K R_H$ is the Hill velocity, with $\Omega_K \simeq 2\pi/(200\ \text{yr})$ denoting the local Kepler frequency].

The two capture formation channels L^3 and L^2s require that binaries form — and formation times increase with decreasing separation a — before $v > v_H$. Given the preponderance of classical binaries (*Stephens and Noll,* 2006) (Fig. 2), we can speculate that the primordial classical belt may have enjoyed dynamically cold (sub-Hill or marginally Hill) conditions for a longer duration than the primordial scattered population.

Dynamical capture is the only viable formation scenario for many TNBs with high angular momentum (section 4.4) and is a possible formation scenario for most, if not all, known TNBs. Given the apparent importance of capture models, we explore them in detail in section 4.5 with a specific focus on the case of binaries with similar mass components.

4.2. Collision

Collisional models were proposed for the Pluto/Charon binary early on based on the angular momentum of the system (*McKinnon,* 1984, 1989), and one such model has recently been shown to be numerically feasible (*Canup,* 2005). However, as we note below, angular momentum arguments alone are not sufficient to prove an impact origin. Nix and Hydra, the outermost satellites of Pluto, plausibly resulted from the same impact that generated Charon (*Stern et al.,* 2006). *Ward and Canup* (2006) propose that collisional debris within the exterior 4:1 and 6:1 resonances of Charon — resonances stabilized by Charon's initially large eccentricity — accumulated to form Nix and Hydra. According to their scenario, as Charon's orbit tidally expanded, the small satellites would have been forced outward to their current locations in resonant lockstep with Charon. Tidal circularization of Charon's orbit would have eventually released Nix and Hydra from resonance. In this scenario, the nearly circular orbits of Nix and Hydra result from their coalescence from a dissipative, nearly circular disk; their eccentricities are not altered by resonant migration because the resonances involved are of the co-rotation type.

The small satellites of Pluto-sized TNOs 2003 EL_{61} and (136199) Eris, characterized by satellite-to-primary mass ratios of ~1%, might also have formed by impacts (*Stern,* 2002). Some collisional simulations (*Durda et al.,* 2004; *Canup,* 2005) reproduce such low mass ratios. Tidal expansion of satellite orbits can explain, to within factors of a few, the current semimajor axes of the companions of 2003 EL_{61} (*Brown et al.,* 2005a). An unresolved issue is the origin of the small, but significant, orbital eccentricity, 0.05 ± 0.003, for the outermost satellite of 2003 EL_{61}. Tides should have reduced the eccentricity to values much smaller. Also, the mutual orbital inclination of the satellites of 2003 EL_{61}, which might be as large as 39° (*Brown et al.,* 2006a), has yet to be explained. Formation by collisionless capture along the lines of *G02*, although not ruled out, remains poorly explored for unequal mass components (*Brown et al.,* 2005a).

To occur with reasonable frequency, collisions between Pluto-sized (R ~ 1000 km) TNOs must be gravitationally focused. Transneptunian space may have been populated by a few dozen such objects (*Kenyon and Luu,* 1999). If their relative velocities were less than the Hill velocity, v_H, then the collision timescale would be ~6 m.y. (equation (18)). Otherwise the collision time exceeds ~500 m.y. Like collisionless capture (see section 4.1), binary formation by giant impacts must have taken place while the disk was dynamically cold.

4.3. Hybrids

Hybrid collision/capture models are possible as well; two variants on this theme have been proposed.

Weidenschilling (2002) considered a model in which a third big body collides with one member of the scattering pair. Since physical collisions have smaller cross-sections than gravitational interactions, this mechanism requires $\sim 10^2$ more big (R ~ 100 km) bodies than are currently observed to operate at the same rate as L^3. It also predicts an unobserved prevalence of widely separated binaries.

Funato et al. (2004) proposed that observed binaries form by the exchange reaction $Ls + L \rightarrow L^2 + s$ wherein a small body of mass M_{sm}, originally orbiting a big body of mass M_{big}, is ejected by a second big body. In the majority of ejections, the small body's energy increases by its orbital binding energy $\sim M_{sm} v_{esc}^2/2$, leaving the big bodies bound with separation $a \sim (M_{big} = M_{sm})R$. As formulated, this model predicts a prevalence of very-high-eccentricity binaries that is not observed. The rate-limiting step for the exchange reaction model is the formation of the preexisting

(Ls) binary, which requires two big bodies to collide and fragment (equation (18)).

4.4. Angular Momentum

It is possible, in some cases, to distinguish between capture and collision based on the angular momentum of the binary. Early theoretical arguments in favor of a collisional origin for Pluto/Charon were based on the fact that the total angular momentum of the system exceeds breakup for a single, reconstituted object (e.g., *McKinnon,* 1989). However, it can also be shown that to have formed from a fragmentary collision, binary components cannot have too much angular momentum. It is conventional to express angular momentum as J/J', where the combined orbital and spin angular momentum of the binary J is normalized by

$$J' = \sqrt{GM_{tot}^3 R_{eff}}$$

where G is the gravitational constant, M_{tot} is the total system mass, and R_{eff} is the radius of an equivalent spherical object containing the total system mass. *Canup* (2005) found that binary systems produced by single collisions have $J/J' < 0.8$ [for an order-of-magnitude derivation of this result, see *Chiang et al.* (2007)]. For instance, the Earth/Moon system has $J/J' \simeq 0.1$ and the Pluto/Charon system has $J/J' \simeq 0.4$.

In Table 2, we list J/J' for TNBs. We use actual measurements where they exist, otherwise the calculation assumes that both binary components have spin periods of 8 h and densities of 1 g cm^{-3}. For about half the TNBs documented in Table 2, J/J' exceeds unity, so much angular momentum that formation via two-body collisions can be ruled out.

4.5. Detailed Capture Models

Capture models are of particular importance for TNBs as the only class of models that can explain the existence of high-angular-momentum systems. We review and expand, in detail, on capture models in this section.

Transneptunian objects can become bound ("fuse") by purely gravitational means while they are still dynamically cold. Following *G02*, we consider how "big" TNOs, having sizes $R_{big} \sim 100$ km, fuse when immersed in a primordial sea of "small" bodies, each of size R_{sm}. We assume that the small bodies contain the bulk of the disk mass: the surface density of small bodies $\sigma \sim \sigma_{MMSN}$, where $\sigma_{MMSN} \sim 0.2$ g cm^{-2} is the surface density of solids in the minimum-mass solar nebula at a heliocentric distance of 30 AU. We assume that the surface density of big bodies was the same then as it is now: $\Sigma \sim 0.01\ \sigma_{MMSN}$. This last condition agrees with the output of numerical simulations of coagulation by *Kenyon and Luu* (1999). The velocity dispersion of small bodies is $u > v_H$. For convenience we define $\alpha \equiv R_{big}/R_H \simeq 1.5 \times 10^{-4}$ and note that the surface escape velocity from a big body $v_{esc} \simeq v_H \alpha^{-1/2}$. The velocity dispersion of big bodies is $v < u$.

Small bodies have their random velocities u amplified by gravitational stirring by big bodies and damped by inelastic collisions with other small bodies. Balancing stirring with damping sets u (*G02*)

$$\frac{u}{v_H} \sim \left(\frac{R_{sm}}{R_{big}}\frac{\Sigma}{\sigma}\right)^{1/4} \alpha^{-1/2} \sim 3\left(\frac{R_{sm}}{20\ m}\right)^{1/4} \qquad (10)$$

Big bodies have their random velocities v amplified by gravitational stirring by other big bodies and damped by dynamical friction with small bodies. When $v > v_H$, this balance yields (*Goldreich et al.,* 2004, hereafter *G04*)

$$\frac{v}{u} \sim \left(\frac{\Sigma}{\sigma}\right)^{1/4} \sim \frac{1}{3} \qquad (11)$$

Combining equation (10) with equation (11), we have

$$\frac{v}{v_H} \sim \left(\frac{\Sigma}{\sigma}\right)^{1/2}\left(\frac{R_{sm}}{R_{big}}\right)^{1/4} \alpha^{-1/2} \sim 1\left(\frac{R_{sm}}{20\ m}\right)^{1/4} \qquad (12)$$

valid for $R_{sm} > 20$ m. If $R_{sm} < 20$ m, then $v < v_H$, neither equation (11) nor equation (12) holds up, but equation (10) still does.

By allowing for the possibility that $v > v_H$, we depart from *G02.* When $v > v_H$, inclinations and eccentricities of the big bodies' heliocentric orbits can be comparable (*G04*). If, prior to fusing, big bodies have an isotropic velocity dispersion, then the resultant mutual binary orbits will be randomly inclined, in agreement with observation. Otherwise, if $v < v_H$, big bodies collapse into a vertically thin disk (*G04*) and mutual orbit normals, unless subsequently torqued, will be parallel, contrary to observation. Invoking $v > v_H$ comes at a cost: Efficiencies for fusing decrease with increasing v. To quantify this cost, we define a normalized velocity parameter F as

$$F = \begin{cases} 1 & \text{if } v < v_H \\ v/v_H & \text{if } v > v_H \end{cases} \qquad (13)$$

and derive how the rates of fusing depend on F. How F increased to its current value of $\sim 10^3$ — i.e., how the Kuiper belt was dynamically heated — remains contested (*Chiang et al.,* 2007; *Levison et al.,* 2007).

Both the L^3 and L^2s scenarios described by *G02* begin when one big body (L) enters a second big body's (L) sphere of influence. This sphere has radius $R_I \sim R_H/F^2$. Per big body, the entry rate is

$$\dot{N}_I \sim \frac{\Sigma\Omega_K}{\rho R_{big}} \alpha^{-2} F^{-4} \qquad (14)$$

If no other body participates in the interaction, the two big bodies pass through their spheres of influence in a time $t_{enc} \sim R_I/v \sim \Omega_K^{-1}F^{-3}$ (assuming they do not collide). The two bodies fuse if they transfer enough energy to other participants during the encounter. In L^3, transfer is to a third big body: $L + L + L \rightarrow L^2 + L$. To just bind the original pair, the third body must come within R_I of the pair. The probability for this to happen in time t_{enc} is $P_{L^3} \sim \dot{N}_I t_{enc}$. If the third body succeeds in approaching this close, the probability that two bodies fuse is estimated to be on the order of unity (this probability has yet to be precisely computed). Therefore the timescale for a given big body to fuse to another by L^3 is

$$t_{fuse,L^3} \sim \frac{1}{\dot{N}_1 P_{L^3}} \sim \left(\frac{\rho R_{big}}{\Sigma}\right)^2 \frac{\alpha^4}{\Omega_K} F^{11} \sim 2F^{11} \text{ m.y.} \quad (15)$$

The extreme sensitivity to F in equation (15) implies that no binaries form by L^3 once v exceeds v_H. This estimate is pessimistic because it neglects big bodies on the low-velocity tail of the velocity distribution (R. Sari, personal communication).

In L^2s, energy transfer is to small bodies by dynamical friction: $L + L + s^\infty \rightarrow L^2 + s^\infty$. In time t_{enc}, the pair of big bodies undergoing an encounter lose a fraction $(\sigma\Omega_K/\rho R_{big})(v_{esc}/u)^4 t_{enc}$ of their energy, under the assumption $v_{esc} > u > v_H$ (*G04*). This fraction is on the order of the probability P_{L^2s} that they fuse, whence

$$t_{fuse,L^2s} \sim \frac{1}{\dot{N}_1 P_{L^2s}} \sim \left(\frac{\rho R_{big}}{\sigma}\right)^2 \frac{R_{sm}}{R_{big}} \frac{\alpha^2}{\Omega_K} \sim 1\left(\frac{R_{sm}}{20 \text{ m}}\right) \text{m.y.} \quad (16)$$

where we have used equation (10). Equation (16) is only valid for $v < v_H$. For $v > v_H$, L^2s cannot operate at all, since dynamical friction cannot reduce the energy of the pair of big bodies by as much as $\sim v^2 > v_H^2$ during the brief encounter (R. Sari, personal communication). In sum, explaining random binary inclinations by appealing to $v \gtrsim v_H$ spawns a fine-tuning problem: Why should $v \approx v_H$ during binary formation?

Given the difficulty in forming binaries when $v > v_H$, we restrict this next discussion to $v < v_H$. Having formed with semimajor axis $a \sim R_I \sim 7000\ R_{big}$, the mutual orbit shrinks by further energy transfer. If L^3 is the more efficient formation process, passing big bodies predominantly harden the binary; if L^2s is more efficient, dynamical friction dominates hardening. The probability P per orbit that a shrinks from $\sim R_I$ to $\sim R_I/2$ is on the order of either PL^3 or P_{L^2s}. We equate the formation rate of binaries, N_{all}/t_{fuse}, with the shrinkage rate, $\Omega_K P N_{bin}|_{x \sim R_I}$, to conclude that the steady-state fraction of TNOs that are binaries with separation R_I is

$$f_{bin}(a \sim R_I) \equiv \frac{N_{bin}|_{a \sim R_I}}{N_{all}} \sim \frac{\Sigma}{\rho R_{big}} \alpha^{-2} \sim 0.4\% \quad (17)$$

As a decreases below R_I, shrinkage slows. Therefore f_{bin} increases with decreasing a. Scaling relations can be derived by arguments similar to those above. If L^2s dominates, $f_{bin} \propto a^0$ for $a < R_H(v_H/u)^2$ and $f_{bin} \propto a^{-1}$ for $a > R_H(v_H/u)^2$ (*G02*). If L^3 dominates, $f_{bin} \propto a^{-1/2}$. For reference, resolved TNBs typically have $a \sim 100\ R_{big}$. The candidate close binary reported by *Sheppard and Jewitt* (2004), and any similar objects, may be the hardened end-products of L^3 and L^2s (although a collisional origin cannot be ruled out since for these binaries J/J' might be less than unity).

These values for $f_{bin}(a)$ characterize the primordial disk. Physical collisions with small bodies over the age of the solar system, even in today's rarefied environment, can disrupt a binary. For this reason, *Petit and Mousis* (2004) find that the widest and least-massive binaries, having $a \gtrsim 400\ R_{big}$, may originally have been 10× more numerous than they are today.

Using the formalism we have developed for capture, it is also possible to develop expressions for the collision timescale. As with capture, this rate of collision is sensitive to the relative velocity of potential impactors, F. This results from the dependence of the collisional rate on gravitational focusing. For the two different velocity regimes, the timescales are given by

$$t_{fuse,exchange} \sim \begin{cases} \frac{\rho R}{\Sigma\Omega_K}\alpha^{3/2} \sim 0.6 \text{ m.y.} & \text{if } v < v_H \\ \frac{\rho R}{\Sigma\Omega_K}\alpha F^2 \sim 50F^2 \text{ m.y.} & \text{if } v > v_H \end{cases} \quad (18)$$

5. THE FUTURE OF BINARIES

Binaries offer unique advantages for the study of the Kuiper belt and are likely to be among the most intensively studied TNOs. Statistical studies will continue to refine the range of properties of binaries, their separations, their relative sizes, and their frequency as a function of dynamical class, size, and other physical variables. The frequency of multiple systems will be constrained. As binary orbits are measured, orbital parameters such as semimajor axis, eccentricity, and orbit plane orientation will also become the subject of statistical investigations. System masses derived from orbits will fuel vigorous studies of physical properties of objects, both singly and in ensemble. Albedo and density are constrained by the measurement of system mass alone, and can be separated with the addition of thermal infrared measurements. The internal structure of TNOs can be inferred from densities. The extremely low densities of a few objects measured so far requires structural models with a high fraction of internal void space. The study of lightcurves of binaries is at a very early stage and can be expected to shed light on the shapes, pole orientations, and tidal evolution of binaries. This, in turn, may yield additional information on the internal structure of TNOs by constraining the possible values of Q.

There are probably more than 100 binary systems that could be discovered in the currently known transneptunian population of 600+ objects with well-established heliocentric orbits. As the TNO population expands, the number of binaries can be expected to expand with it. Most of these discoveries will be made with HST or other instruments with equivalent capabilities. There are also yet-to-be explored areas of interest where theory and modeling can be expected to make significant progress. As progress is made in understanding how binaries formed in the Sun's protoplanetary disk, these principles can be extended to other circumstellar disks that are now found in abundance. If binary protoplanets are common, as seems to be the case for the solar system, we may even expect to find binary planets as we explore extrasolar planetary systems.

Acknowledgments. This work was supported in part by grants GO 9746, 10508, 10514, and 10800 from the Space Telescope Science Institute, which is operated by AURA under contract from NASA. Additional support was provide through a NASA Planetary Astronomy grant, NNG04GN31G.

REFERENCES

Agnor C. B. and Hamilton D. P. (2006) Neptune's capture of its moon Triton in a binary-planet gravitational encounter. *Nature, 441,* 192–194.

Aitken R. G. (1964) *The Binary Stars.* Dover, New York.

Astakhov S. A., Lee E. A., and Farrelly D. (2005) Formation of Kuiper-belt binaries through multiple chaotic scattering encounters with low-mass intruders. *Mon. Not. R. Astron. Soc., 360,* 401–415.

Barkume K. M., Brown M. E., and Schaller E. L. (2006) Water ice on the satellite of Kuiper Belt object 2003 EL_{61}. *Astrophys. J. Lett., 640,* L87–L89.

Bernstein G. M., Trilling D. E., Allen R. L., Brown M. E., Holman M., and Malhotra R. (2004) *Astrophys. J., 128,* 1364–1390.

Binzel R. P. and Hubbard W. B. (1997) Mutual events and stellar occultations. In *Pluto and Charon* (S. A. Stern and D. J. Tholen, eds.), pp. 85–102. Univ. of Arizona, Tucson.

Bowell E., Hapke B., Domingue D., Lumme K., Peltoniemi J., and Harris A. (1989) Application of photometric models to asteroids. In *Asteroids II* (R. P. Binzel et al., eds.), pp. 524–556. Univ. of Arizona, Tucson.

Brouwer D. and Clemence G. M. (1961) *Methods of Celestial Mechanics.* Academic, New York.

Brown M. E. (2005a) S/2005 (2003 UB_{313}) 1. *IAU Circular 8610,* 1.

Brown M. E. (2005b) S/2005 (2003 EL_{61}) 2. *IAU Circular 8636,* 1.

Brown M. E. (2006) The largest Kuiper belt objects. *Bull. Am. Astron. Soc., 38,* #37.01.

Brown M. E. and Calvin W. M. (2000) Evidence for crystalline water and ammonia ices on Pluto's satellite Charon. *Science, 287,* 107–109.

Brown M. E. and Suer T.-A. (2007) Satellites of 2003 AZ_{84}, (50000), (55637), and (90482). *IAU Circular 8812,* 1.

Brown M. E. and Trujillo C. A. (2002) (26308) 1998 SM_{165}. *IAU Circular 7807,* 1.

Brown M. E., Bouchez A. H., Rabinowitz D., Sari R., Trujillo C. A., et al. (2005a) Keck observatory laser guide star adaptive optics discovery and characterization of a satellite to the large Kuiper Belt object 2003 EL_{61}. *Astrophys. J. Lett., 632,* L45–L48.

Brown M. E., Trujillo C. A., and Rabinowitz D. (2005b) 2003 EL_{61}, 2003 UB_{313}, and 2005 FY_9. *IAU Circular 8577,* 1.

Brown M. E., van Dam M. A., Bouchez A. H., Le Mignant D., Campbell R. D., et al. (2006a) Satellites of the largest Kuiper Belt objects. *Astrophys. J., 639,* L43–L46.

Brown M. E., Schaller E. L., Roe H. G., Rabinowitz D. L., and Trujillo C. A. (2006b) Direct measurement of the size of 2003 UB_{313} from the Hubble Space Telescope. *Astrophys. J. Lett., 643,* L61–L64.

Buie M. W. and Grundy W. M. (2000) The distribution and physical state of H_2O on Charon. *Icarus, 148,* 324–339.

Buie M. W., Cruikshank D. P., Lebofsky L. A., and Tedesco E. F. (1987) Water frost on Charon. *Nature, 329,* 522–523.

Buie M. W., Tholen D. J., and Wasserman L. H. (1997) Separate lightcurves of Pluto and Charon. *Icarus, 125,* 233–244.

Buie M. W., Grundy W. M., Young E. F., Young L. A., and Stern S. A. (2006) Orbits and photometry of Pluto's satellites: Charon, S/2005 P1, and S/2005 P2. *Astron. J., 132,* 290–298.

Canup R. M. (2005) A giant impact origin of Pluto-Charon. *Science, 307,* 546–550.

Chiang E., Lithwick Y., Murray-Clay R., Buie M., Grundy W., and Holman M. (2007) A brief history of transneptunian space. In *Protostars and Planets V* (B. Reipurth et al., eds.), pp. 895–911. Univ. of Arizona, Tucson.

Christy J. W. and Harrington R. S. (1978) The satellite of Pluto. *Astron. J., 83,* 1005–1008.

Christy J. W. and Harrington R. S. (1980) The discovery and orbit of Charon. *Icarus, 44,* 38–40.

Cruikshank D. P., Pilcher C. B., and Morrison D. (1976) Pluto: Evidence for methane frost. *Science, 194,* 835–837.

Cruikshank D. P., Barucci M. A., Emery J. P., Fernndez Y. R., Grundy W. M., Noll K. S., and Stansberry J. A. (2007) Physical properties of transneptunian objects. In *Protostars and Planets V* (B. Reipurth et al., eds.), pp. 879–893. Univ. of Arizona, Tucson.

Danby J. M. A. (1992) *Fundamentals of Celestial Mechanics, 2nd edition.* Willmann-Bell, Inc., Richmond, Virginia.

Descamps P. (2005) Orbit of an astrometric binary system. *Cel. Mech. Dyn. Astron., 92,* 381–402.

Dobrovolskis A. R., Peale S. J., and Harris A. W. (1997) Dynamics of the Pluto-Charon binary. In *Pluto and Charon* (S. A. Stern and D. Tholen, eds.), pp. 159–190. Univ. of Arizona, Tucson.

Durda D. D., Bottke W. F., Enke B. L., Merline W. J., Asphaug E., Richardson D., and Leinhardt Z. M. (2004) The formation of asteroid satellites in large impacts: Results from numerical simulations. *Icarus, 167,* 382–396.

Elliot J. L., Kern S. D., Osip D. J., and Burles S. M. (2001) 2001 QT_{297}. *IAU Circular 7733,* 2.

Elliot J. L., Kern S. D., and Clancy K. B. (2003) 2003 QY_{90}. *IAU Circular 8235,* 2.

Elliot J. L., Kern S. D., Clancy K. B., Gulbis A. A. S., Millis R. L., Buie M. W., Wasserman L. H., Chiang E. I., Jordan A. B., Trilling D. E., and Meech K. J. (2005) The Deep Ecliptic Survey: A search for Kuiper belt objects and Centaurs. II. Dynamical classification, the Kuiper belt plane, and the core population. *Astrophys. J., 129,* 1117–1162.

Farinella P., Milani A., Nobili A. M., and Valsecchi G. B. (1979) Tidal evolution and the Pluto-Charon system. *Moon and Planets, 20,* 415–421.

Fink U. and DiSanti M. A. (1988) The separate spectra of Pluto and its satellite Charon. *Astron. J., 95,* 229–236.

Funato Y., Makino J., Hut P., Kokubo E., and Kinoshita D. (2004) The formation of Kuiper-belt binaries through exchange reactions. *Nature, 427,* 518–520.

Goldreich P. and Soter S. (1966) Q in the solar system. *Icarus, 5,* 375–389.

Goldreich P., Lithwick Y., and Sari R. [G02] (2002) Formation of Kuiper belt binaries by dynamical friction and three-body encounters. *Nature, 420,* 643–646.

Goldreich P., Lithwick Y., and Sari R. [G04] (2004) Planet formation by coagulation: A focus on Uranus and Neptune. *Annu. Rev. Astron. Astrophys., 42,* 549–601.

Grundy W. M., Noll K. S., and Stephens D. C. (2005) Diverse albedos of small trans-neptunian objects. *Icarus, 176,* 184–191.

Grundy W. M., Stansberry J. A., Noll K. S, Stephens D. C., Trilling D. E., Kern S. D., Spencer J. R., Cruikshank D. P., and Levison H. F. (2007) The orbit, mass, size, albedo, and density of (65489) Ceto-Phorcys: A tidally-evolved binary Centaur. *Icarus, 191,* 286–297.

Gulbis A. A. S., Elliot J. L., Person M. J., Adams E. R., Babcock B. A., et al. (2006) Charon's radius and atmospheric constraints from observations of a stellar occultation. *Nature, 439,* 48–51.

Harris A. W. (1998) A thermal model for near-earth asteroids. *Icarus, 131,* 291–301.

Heintz W. D. (1978) *Double Stars.* Reidel, Dordrecht.

Hestroffer D. and Vachier F. (2006) Orbit determination of binary TNOs. Paper presented at Trans-Neptunian Objects: Dynamical and Physical Properties, Catania, Italy, 2006 July 3–7.

Hestroffer D., Vachier F., and Balat B. (2005). Orbit determination of binary asteroids. *Earth Moon Planets, 97,* 245–260.

Hubbard W. B., Hunten D. M., Dieters S. W., Hill K. M., and Watson R. D. (1988) Occultation evidence for an atmosphere on Pluto. *Nature, 336,* 452–454.

Jewitt D. and Luu J. (1998) Optical-infrared spectral diversity in the Kuiper belt. *Astrophys. J., 115,* 1667–1670.

Jewitt D. C. and Sheppard S. S. (2002) Physical properties of trans-neptunian object (20000) Varuna. *Astron. J., 123,* 2110–2120.

Jones R. L., Gladman B., Petit J-M., Rousselot P., Mousis O., Kavelaars J. J., Parker J., Nicholson P., Holman M., Grav T., Doressoundiram A., Veillet C., Scholl H., and Mars G. (2006) The CFEPS Kuiper Belt Survey: Strategy and pre-survey results. *Icarus, 185,* 508–522.

Kavelaars J. J., Petit J.-M., Gladman B., and Holman M. (2001) 2001 QW_{322}. *IAU Circular 7749,* 1.

Kenyon S. J. and Luu J. X. (1999) Accretion in the early Kuiper belt. II. Fragmentation. *Astron. J., 118,* 1101–1119.

Kern S. D. (2006) A study of binary Kuiper belt objects. Ph.D. thesis, Massachussetts Institute of Technology, Cambridge.

Kern S. D. and Elliot J. L. (2005) 2005 EO_{304}. *IAU Circular 8526,* 2.

Kern S. D. and Elliot J. L. (2006) The frequency of binary Kuiper Belt objects. *Astrophys. J. Lett., 643,* L57–L60.

Lampton M., Margon B., and Bowyer S. (1976) Parameter estimation in X-ray astronomy. *Astron. J., 208,* 177–190.

Lee M. H. and Peale S. J. (2006) On the orbits and masses of the satellites of the Pluto-Charon system. *Icarus, 184,* 573–583.

Levison H. F. and Stern S. A. (2001) On the size dependence of the inclination distribution of the main Kuiper belt. *Icarus, 121,* 1730–1735.

Levison H. F., Morbidelli A., Gomes R., and Backman D. (2007) Planet migration in planetesimal disks. In *Protostars and Planets V* (B. Reipurth et al., eds.), pp. 669–684. Univ. of Arizona, Tucson.

Marchis F., Hestroffer D., Descamps P., Berthier J., Laver C., and de Pater I. (2005) Mass and density of asteroid 121 Hermione from an analysis of its companion orbit. *Icarus, 178,* 450–464.

Margot J. L. (2004) Binary minor planets. Urey Prize Lecture presented at DPS/AAS meeting 2004 November 9, Louisville, Kentucky.

Margot J. L., Brown M. E., Trujillo C. A., and Sari R. (2004) HST observations of Kuiper belt binaries. *Bull. Am. Astron. Soc., 36,* 1081.

Margot J. L., Brown M. E., Trujillo C. A., Sari R., and Stansberry J. A. (2005) Kuiper belt binaries: Masses, colors, and a density. *Bull. Am. Astron. Soc., 37,* 737.

McKinnon W. B. (1984) On the origin of Triton and Pluto. *Nature, 311,* 355–358.

McKinnon W. B. (1989) On the origin of the Pluto-Charon binary. *Astrophys. J. Lett., 344,* L41–L44.

Merline W. J., Weidenschilling S. J., Durda D. D., Margot J. L., Pravec P., and Storrs A. D. (2002) Asteroids *do* have satellites. In *Asteroids III* (W. F. Bottke Jr. et al., eds.), pp. 289–312. Univ. of Arizona, Tucson.

Millis R. L. and Clancy K. B. (2003) 2003 UN_{284}. *IAU Circular 8251,* 2.

Millis R. L., Wasserman L. H., Franz O. G., Nye R. A., Elliot J. L., et al. (1993) Pluto's radius and atmosphere: Results from the entire 9 June 1988 occultation data set. *Icarus, 105,* 282–297.

Millis R. L., Buie M. W., Wasserman L. H., Elliot J. L., Kern S. D., and Wagner R. M. (2002) The Deep Ecliptic Survey: A search for Kuiper belt objects and Centaurs. I. Description of methods and initial results. *Astron. J., 123,* 2083–2109.

Muller G. P., Reed R., Armandroff, T., Boroson T. A., and Jacoby G. H. (1998) What is better than an 8192 × 8192 CCD mosaic imager: Two wide field imagers, one for KPNO and one for CTIO. In Optical Astronomical Instrumentation (S. D'Odorico, ed.), pp. 577–585. SPIE Proceedings Vol. 3355, Bellingham, Washington.

Murray C. D. and Dermott S. F. (1999) *Solar System Dynamics.* Cambridge Univ., Cambridge.

Noll K. S. (2003) Trans-neptunian binaries. *Earth Moon Planets, 92,* 395–407.

Noll K. S. (2006) Solar system binaries. In *Asteroids, Comets, Meteors 2005* (D. Lazzaro et al., eds.), pp. 301–318. IAU Symposium No. 229, Cambridge Univ., Cambridge.

Noll K., Stephens D., Grundy W., Spencer J., Millis R., Buie M., Cruikshank D., Tegler S., and Romanishin W. (2002a) 1997 CQ_{29}. *IAU Circular 7824,* 2.

Noll K., Stephens D., Grundy W., Spencer J., Millis R., Buie M., Cruikshank D., Tegler S., and Romanishin W. (2002b) 2000 CF_{105}. *IAU Circular 7857,* 1.

Noll K., Stephens D., Grundy W., Cruikshank D., Tegler S., and Romanishin W. (2002c) 2001 QC_{298}. *IAU Circular 8034,* 1.

Noll K. S., Stephens D. C., Cruikshank D., Grundy W., Romanishin W., and Tegler S. (2003) 1999 RZ_{253}. *IAU Circular 8143,* 1.

Noll K. S., Stephens D. C., Grundy W. M., and Griffin I. (2004a) The orbit, mass, and albedo of (66652) 1999 RZ_{253}. *Icarus, 172,* 402–407.

Noll K. S., Stephens D. C., Grundy W. M., Osip D. J., and Griffin I. (2004b) The orbit and albedo of trans-neptunian binary (58534) 1997 CQ_{29}. *Astron. J., 128,* 2547–2552.

Noll K. S., Grundy W. M., Stephens D. C., and Levison H. F. (2006a) (42355) 2002 CR_{46}. *IAU Circular 8689,* 1.

Noll K. S., Grundy W. M., Levison H. F., and Stephens D. C. (2006b) (60458) 2000 CM_{114}. *IAU Circular 8689,* 2.

Noll K. S., Grundy W. M., Stephens D. C., and Levison H. F. (2006c) 2003 QW_{111}. *IAU Circular 8745,* 1.

Noll K. S., Stephens D. C., Grundy W. M., and Levison H. F. (2006d) 2000 QL_{251}. *IAU Circular 8746,* 1.

Noll K. S., Levison H. F., Stephens D. C., and Grundy W. M. (2006e) (120347) 2004 SB_{60}. *IAU Circular 8751,* 1.

Noll K. S., Grundy W. M., Levison H. F., and Stephens D. C. (2006f) 1999 RT_{214}. *IAU Circular 8756,* 2.

Noll K. S., Levison H. F., Grundy W. M., and Stephens D. C. (2006g) Discovery of a binary Centaur. *Icarus, 184,* 611–618.

Noll K. S., Grundy W. M., Levison H. F., and Stephens D. C. (2006h) The relative sizes of Kuiper belt binaries. *Bull. Am. Astron. Soc., 38,* #34.03.

Noll K. S., Stephens D. C., Grundy W. M., Levison H. F., and Kern S. D. (2007a) (123509) 2000 WK_{183}. *IAU Circular 8811,* 1.

Noll K. S., Kern S. D., Grundy W. M., Levison H. F., and Stephens D. C. (2007b) (119979) 2002 WC_{19}. *IAU Circular 8814,* 1.

Noll K. S., Kern S. D., Grundy W. M., Levison H. F., and Stephens D. C. (2007c) 2002 GZ_{31}. *IAU Circular 8815,* 1.

Noll K. S., Kern S. D., Grundy W. M., Levison H. F., and Stephens D. C. (2007d) 2004 PB_{108} and (60621) 2000 FE_8. *IAU Circular 8816,* 1.

Ortiz J. L., Gutiérrez P. J., Casanova V., and Sota A. (2003) A study of short term rotational variability in TNOs and Centaurs from Sierra Nevada Observatory. *Astron. Astrophys., 407,* 114–1155.

Osip D. J., Kern S. D., and Elliot J. L. (2003)Physical characterization of the binary Edgeworth-Kuiper Belt object 2001 QT_{297}. *Earth Moon Planets, 92,* 409–421.

Osip D. J., Phillips D. M., Bernstein R., Burley G., Dressler A. , Elliot J. L., Persson E., Shectman S. A., and Thompson I. (2004) First-generation instruments for the Magellan telescopes: Characteristics, operation, and performance. In *Ground-based Instrumentation for Astronomy* (A. F. M. Moorwood and I. Masanori, eds.), pp. 49–59. SPIE Proceedings Vol. 5492, Bellingham, Washington.

Peixinho N., Boehnhardt H., Belskaya I., Doressoundiram A., Barucci M. A., and Delsanti A. (2004) ESO large program on Centaurs and TNOs. *Icarus, 170,* 153–166.

Petit J.-M. and Mousis O. (2004) KBO binaries: How numerous were they? *Icarus, 168,* 409–419.

Press W. H., Teukolsky S. A., Vetterling W. T., and Flannery B. P. (1992) *Numerical Recipes in C.* Cambridge Univ., New York.

Rabinowitz D. L., Barkume K., Brown M. E., Roe H., Schwartz M., Tourtellotte S., and Trujillo C. (2006) Photometric observations constraining the size, shape, and albedo of 2003 EL_{61}, a rapidly rotating, Pluto-sized object in the Kuiper belt. *Astrophys. J., 639,* 1238–1251.

Rabinowitz D. L., Schaefer B. E., and Tourtellotte S. (2007) The diverse solar phase curves of distant icy bodies. I. Photometric observations of 18 trans-neptunian objects, 7 Centaurs, and Nereid. *Astron. J., 133,* 26–43.

Richardson D. C. and Walsh K. J. (2006) Binary minor planets. *Annu. Rev. Earth Planet. Sci., 34,* 47–81.

Romanishin W. and Tegler S. C. (2005) Accurate absolute magnitudes for Kuiper belt objects and Centaurs. *Icarus, 179,* 523–526.

Romanishin W., Tegler S. C., Rettig T. W., Consolmagno G., and Botthof B. (2001) 1998 SM_{165}: A large Kuiper belt object with an irregular shape. *Publ. Natl. Acad. Sci., 98,* 11863–11866.

Schaller E. L. and Brown M. E. (2003) A deep Keck search for binary Kuiper belt objects. *Bull. Am. Astron. Soc., 35,* 993.

Sheppard S. S. and Jewitt D. C. (2002) Time-resolved photometry of Kuiper belt objects: Rotations, shapes, and phase functions. *Astron. J., 124,* 1757–1775.

Sheppard S. S. and Jewitt D. (2004) Extreme Kuiper belt object 2001 QG_{298} and the fraction of contact binaries. *Astron. J., 127,* 3023–3033.

Smart W. M. (1980) *Textbook on Spherical Astronomy.* Cambridge Univ., New York.

Smith J. C., Christy J. W., and Graham J. A. (1978) 1978 P_1. *IAU Circular 3241,* 1.

Spencer J. R., Stansberry J. A., Grundy W. M., and Noll K. S. (2006) A low density for binary Kuiper belt object (26308) 1998 SM_{165}. *Bull. Am. Astron. Soc., 38,* 564.

Stansberry J. A., Grundy W. M., Margot J. L., Cruikshank D. P., Emery J. P., Rieke G. H., and Trilling D. E. (2006) The albedo, size, and density of binary Kuiper belt object (47171) 1999 TC_{36}. *Astrophys. J., 643,* 556–566.

Stephens D. C. and Noll K. S. (2006) Detection of six trans-neptunian binaries with NICMOS: A high fraction of binaries in the cold classical disk. *Astron. J., 131,* 1142–1148.

Stephens D. C., Noll K. S., Grundy W. M., Millis R. L., Spencer J. R., Buie M. W., Tegler S. C., Romanishin W., Cruikshank D. P. (2003) HST photometry of trans-neptunian objects. *Earth Moon Planets, 92,* 251–260.

Stephens D. C., Noll K. S., and Grundy W. (2004) 2000 CQ_{114}. *IAU Circular 8289,* 1.

Stern S. A. (2002) Implications regarding the energetics of the collisional formation of Kuiper belt satellites. *Astron. J., 124,* 2300–2304.

Stern S. A., Weaver H. A., Steffl A. J., Mutchler M. J., Merline W. J., Buie M. W., Young E. F., Young L. A., and Spencer J. R. (2006) A giant impact origin for Pluto's small moons and satellite multiplicity in the Kuiper belt. *Nature, 439,* 946–948.

Takahashi S. and Ip W.-H. (2004) A shape-and-density model of the putative binary EKBO 2001 QG_{298}. *Publ. Astron. Soc. Japan, 56,* 1099–1103.

Toth I. (1999) NOTE: On the detectability of satellites of small bodies orbiting the Sun in the inner region of the Edgeworth-Kuiper belt. *Icarus, 141,* 420–425.

Trujillo C. A. and Brown M. E. (2002) 1999 TC_{36}. *IAU Circular 7787,* 1.

Veillet C., Doressoundiram A., Shapiro J., Kavelaars J. J., and Morbidelli A. (2001) S/2000 (1998 WW_{31}) 1. *IAU Circular 7610,* 1.

Veillet C., Parker J. W., Griffin I, Marsden B., Doressoundiram A., Buie M., Tholen D. J., Connelley M., and Holman M. J. (2002) The binary Kuiper-belt object 1998 WW_{31}. *Nature, 416,* 711–713.

Virtanen J., Muinonen K., and Bowell E. (2001) Statistical ranging of asteroid orbits. *Icarus, 154,* 412–431.

Virtanen J., Tancredi G., Muinonen K., and Bowell E. (2003) Orbit computation for trans-neptunian objects. *Icarus, 161,* 419–430.

Ward W. R. and Canup R. M. (2006) Forced resonant migration of Pluto's outer satellites by Charon. *Science, 313,* 1107–1109.

Weaver H. A., Stern S. A., Mutchler M. J., Steffl A. J., Buie M. J., et al. (2005) S/2005 P 1 and S/2005 P 2. *IAU Circular 8625,* 1.

Weaver H. A., Stern S. A., Mutchler M. J., Steffl A. J., and Buie M. (2006) Discovery of two new satellites of Pluto. *Nature, 439,* 943–945.

Weidenschilling S. J. (2002) On the origin of binary trans-neptunian objects. *Icarus, 160,* 212–215.

Young E. F., and Binzel R. P. (1994) A new determination of radii and limb parameters for Pluto and Charon from mutual event lightcurves. *Icarus, 108,* 219–224.

Young E. F., Galdamez K., Buie M. W., Binzel R. P., and Tholen D. J. (1999) Mapping the variegated surface of Pluto. *Astron. J., 117,* 1063–1076.

On the Atmospheres of Objects in the Kuiper Belt

S. Alan Stern
Southwest Research Institute
(now at NASA Headquarters)

Laurence M. Trafton
University of Texas

Atmospheres around solar system bodies reveal key insights into the origins, chemistry, thermal evolution, and surface/interior interaction of their parent bodies. Atmospheres are also themselves of intrinsic interest for understanding the physics and chemistry of gaseous envelopes. Furthermore, atmospheres also reveal information about primordial nebular materials trapped in accreting bodies. For these reasons and others, the detection and study of atmospheres on objects in the Kuiper belt (KB) is of interest. Here we review what is known about the atmosphere of both Kuiper belt objects (KBOs) and Pluto; we then go on to more generally examine the source and loss processes relevant to KBO atmospheres, the likely kinds of vertical and horizontal structure of such atmospheres, and then briefly reflect on KBO atmospheric detection techniques.

1. MOTIVATION

In our solar system, atmospheres range from the tenuous, surface boundary exosphere of the Moon, Mercury, icy satellites, and asteroids (e.g., *Stern,* 1999), to the freely escaping, collisionally thick atmospheres of comets, to the classical atmospheres of Mars, Venus, Earth, Titan, and the giant planets.

We set the stage for this review chapter by considering the requirements for atmospheric generation on Kuiper belt (KB) bodies, which are simple: All that is required is the presence of gases or sublimating/evaporating materials on, or sufficiently near, the surface, and a source of energy that can maintain substantial vapor pressures generated from the available reservoir of volatiles.

For atmospheres to persist over geologic time there is an additional requirement that the loss of volatiles be less than complete by the present epoch. So in the case of low-gravity bodies like Kuiper belt objects (KBOs), which we will show to have prodigious atmospheric escape rates; this in turn implies some resupply mechanism to the surface, such as internal activity, or the import or up-dredging of volatiles from impactors.

Atmospheres can be very generally defined to include both gravitationally unbound gaseous envelopes such as exospheres, or unbound cometary comae, as well as gravitationally bound collisional ensembles of gas. An exosphere is the portion of an atmosphere that is sufficiently rarefied that it escapes directly to space by one or more processes; i.e., it is the portion of an atmosphere where the mean free path exceeds the density scale height, so that molecules with sufficient energy can be expected to escape directly to space. In atmospheres that are too rarefied to have anything but an exosphere, their gases escape at the speed of sound, and the atmospheric density falls off as $1/r^2$.

In what follows we discuss the likelihood and characteristics of what one expects for KBO atmospheres. We begin in section 2 with a concise overview of evidence for atmospheres around KBOs, and Centaurs (inward-scattered KBOs). We then proceed in section 3 to give a very brief overview of what is known about Pluto's atmosphere — the only presently established atmosphere around a bona fide KBO. In section 4 we discuss production mechanisms for atmospheres in the KB region. In section 5 we then describe atmospheric loss processes. We follow that with a discussion of expected KBO atmospheric vertical and horizontal structure in section 6. We conclude in section 7 with a brief look at future KBO/Centaur atmospheric detection and study prospects.

2. KUIPER BELT OBJECT ATMOSPHERIC DETECTIONS AND RELATED EVIDENCE

In the KB region, only one body is presently known to have an extant atmosphere: Pluto (although Neptune's moon Triton has an atmosphere and almost certainly had some former relation to the KB).

Perhaps foreshadowing the eventual discovery of the KBO atmospheres we will discuss below, the existence of Pluto's atmosphere was speculated about long before it was observationally established. Early arguments for an atmosphere around Pluto were based entirely on theoretical considerations (e.g., *Hart,* 1974), but these were supplanted by strong circumstantial evidence for a vapor pressure equilibrium atmosphere after the discovery of the volatile CH_4 ice on Pluto's surface (*Cruikshank et al.,* 1976; *Stern et al.,*

1988). The actual detection of Pluto's atmosphere only occurred later, when rare but formally diagnostic stellar occultations revealed the unmistakable signature of atmospheric refraction at roughly microbar pressure levels (*Hubbard et al.,* 1988; *Elliot et al.,* 1989; *Brosch,* 1995).

At Pluto, where the surface temperature varies from ~35 K in volatile-rich regions cooled by latent heat sinks, to ~55 K in volatile-free regions warmed to their pure radiative equilibrium temperature (e.g., *Stern et al.,* 1993; *Jewitt,* 1994), candidate volatiles that may comprise parent species in the atmosphere include Ne, Ar, O_2, CO, N_2, and CH_4 (*Stern,* 1981; *Stern and Trafton,* 1984). H_2 and He easily escape Pluto, and are therefore not expected to remain in sufficient quantity to be important at the present day, approximately 4+ G.y. after Pluto's formation. Heavy noble gases, like Kr and Xe, are volatile at the relevant temperatures for Pluto, but are so cosmogonically rare that they are not expected to be abundant in significant quantities (*Stern,* 1981). Of the species just reviewed, N_2, CO, and CH_4 have since been detected on Pluto's surface and must play significant roles in its atmosphere. The same volatile species are likely to be important for other large worlds in the KB.

In the case of Pluto's large satellite, none of the volatiles mentioned just above have been discovered in its surface spectrum (*Grundy and Buie,* 2000), so one might expect the prospects for an atmosphere at Charon to be dim. Consistent with this hypothesis, stellar occultations by Charon (e.g., *Sicardy et al.,* 2006) have failed to detect any atmospheric signature, with derived upper limits on the atmospheric pressure of 110 nbar for N_2 and 15 nbar for CH_4.

In contrast to the known atmosphere around Pluto, only upper limits have been placed on possible atmospheres around other KB bodies. To date, no published reports of stellar occultations of KBOs report atmospheric detections. Similarly, only negative results have been obtained in searches for light-scattering particulate comae.

Despite this, the discoveries of N_2, CH_4, and C_2H_6 (ethane) ices on the KBO 2005 FY_9 (*Licandro et al.,* 2006) and the tentative detection of N_2 and CH_4 on KBO Sedna and CH_4 on Eris (*Barucci et al.,* 2005; *Licandro et al.,* 2006) indicate that KBO atmospheres may be discovered in the future. And the report by *Hainaut et al.* (2000) that found a change in the lightcurve of KBO 1996 TO_{66} from a double peak to a single peak over the course of a year, which they suggested may have resulted from an episode of cometary activity, is of related interest.

Additional circumstantial evidence for atmospheres comes from the detections (see chapter by Brown) of albedos of 60–90% for the large KBOs Eris, 2005 FY_9, and 2003 EL_{61}. Following the same logic that connected the detection of surface volatiles and a high albedo on Pluto to a strong case for an associated atmosphere, these clues strongly suggest that these KBOs have, or recently had, atmospheres. However, we stress that definitive evidence for atmospheres must come from occultations or direct spectroscopic detections of gas phase constituents — both of which are difficult. Model inferences alone, while suggestive, cannot be considered definitive.

We now turn to Centaurs, which are objects that have recently escaped from the KB to orbits among the giant planets, and therefore enjoy significantly increased insolation to drive ice sublimation. A variety of these objects have shown activity at large heliocentric distances (e.g., 6–25 AU). In most cases, this activity manifests itself as an extended coma.

The first and most well-known case for atmospheric phenomena at a Centaur revolves around the Centaur 2060 Chiron. Chiron has long been known to exhibit both photometric variability (*Hartmann et al.,* 1990) and a sporadic cometary coma (*Meech and Belton,* 1989). Analysis of archival images reveal Chiron's activity as far back to the 1940s, and have shown that this activity occurs at all distances of its orbit, including near aphelion, with no clear correlation between the level of cometary activity and heliocentric distance (*Bus et al.,* 1991a). Indeed, Chiron was more active at its 1970 aphelion than it was near its 1989 perihelion.

What generates Chiron's atmospheric activity? *Bus et al.* (1991b) detected CN in Chiron's coma, and *Womack and Stern* (1999) detected CO at low signal-to-noise ratio (SNR). Large variability in the magnitude and timing of Chiron's outbursts and the observations of discrete jet-like features observed during a stellar occultation (*Elliot et al.,* 1995) provide further clues. Together these various facts imply that Chiron's surface may contain an uneven distribution of surface or near-surface volatile ices, likely including CO, with both distributed and discrete surface sources of atmospheric gas and particulates. The variation in Chiron's activity with time may be related to a complex interaction between the sites of volatile frosts near the surface, obliquity effects, and heliocentric distance. Alternatively, it could be related to residual trapped heat at localized impact sites that can remain bottled up by low thermal conductivity for decades to centuries (*Capria et al.,* 2000). Similar phenomenology probably powers other active Centaurs with atmospheres.

Finally, it is worth noting that the captured dwarf planet Triton, now a satellite of Neptune, is known to possess an N_2-dominated atmosphere with some CH_4 and a base pressure not unlike Pluto's. Since Triton is also roughly Pluto's size and density, and is widely thought to have originated in the same region of the solar system as Pluto, it serves as a possible guide to future discoveries that may be made around dwarf planets in the KB and farther out as well.

3. PLUTO'S ATMOSPHERE

In what follows, space limitations only allow us to provide an overview of what is known about Pluto's atmosphere. Our goal is to give an illustrative example of what might be anticipated regarding the nature and phenomenology expected in atmospheres that may later be discovered elsewhere in the KB. [Much more complete information about Pluto's atmosphere is contained in the suite of review articles published by *Yelle and Elliot* (1997), *Spencer et al.* (1997), and *Trafton et al.* (1997).]

As described in section 2, despite earlier circumstantial arguments and the 1976 discovery of CH_4 frost on Pluto's surface, it was not until June 1988 that Pluto's atmosphere was definitively detected. This discovery was made by stellar occultation observations at eight separate sites in, around, and over Australia and New Zealand; see Fig. 1. These various sites probed different chords across Pluto, showing that the atmosphere is global and exhibits a characteristic scale height H of 55.7 ± 4.5 km and a pressure of 2.3 μbar at a distance of 1250 km from Pluto's center.

A CH_4-dominated atmosphere, corresponding to $\mu = 16$ and an ~60 K upper atmospheric temperature, was ruled out when both N_2 and CO were detected in high-quality IR spectra of Pluto on Pluto's surface in greater quantities than CH_4 (*Owen et al.,* 1993). For the assumption of $\mu = 28$ (i.e., an N_2- or CO-dominated atmosphere), an upper atmospheric temperature near 102 K can be derived from these data, knowing that the scale height $H = kT/(m_p \mu g)$, where m_p is the mass of a proton, μ is the average molecular weight of the atmosphere, and T is its temperature and g is the local gravity.

The upper atmosphere probed by the 1988 occultation was modeled by an isothermal (102 ± 9 K) temperature structure above the nominal half-light radius of 1215 km from Pluto's center and a composition of pure CO or N_2 (e.g., *Yelle and Elliot,* 1997). A near-surface thermal gradient of 10–30 K/km was predicted to link the surface temperature to the upper atmosphere. *Yelle and Lunine* (1989) realized that CH_4 can act as a thermostat in Pluto's upper atmosphere by absorbing energy in the ν_3 band at 3.3 μm, cooling via the ν_4 band at 7.6 μm, and conducting heat to the surface, thus potentially explaining the high temperatures at the microbar pressure level in light of Pluto's colder (35–55 K) surface temperatures. *Strobel et al.* (1996) later showed that most of this heating (80%) actually occurs in the CH_4 band at 2.3 μm. For this "CH_4-thermostat" model to work, however, there must be sufficient quantities (~0.1–1%) of atmospheric CH_4. [It is worth noting that this model does not work for Triton, at least without some modifications as discussed by *Elliot et al.* (2000). CO plays an important role in cooling these atmospheres by rotational line emission as was shown for Pluto (*Strobel et al.,* 1996) and Triton (*Elliot et al.,* 2000). This will be discussed further in section 5.]

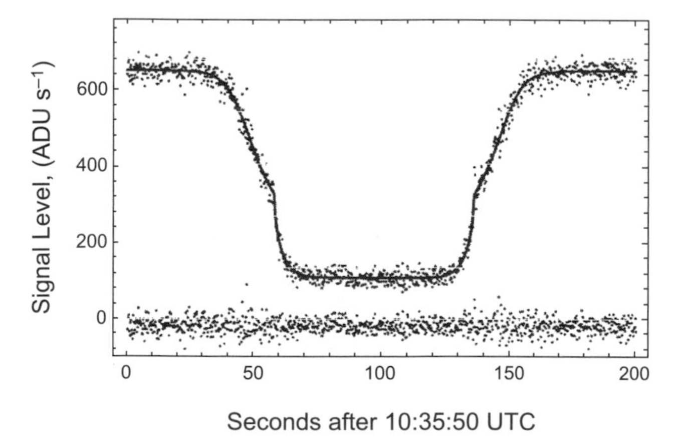

Fig. 1. Data (points) (*Elliot et al.,* 1989) and model (line) (*Elliot and Young,* 1992) resulting from a KAO light curve from Pluto's 1988 stellar occultation. The points along the bottom are the residuals that result when the model and the data are differenced.

The direct spectroscopic detection of gaseous CH_4 in Pluto's atmosphere was finally obtained from high-resolution IR spectroscopy in 1994 (*Young et al.,* 1997). These workers found that the partial pressure of CH_4 was too high for methane in solid solution with nitrogen, but consistent with methane in vapor pressure equilibrium with pure CH_4 frost at 41 K to 45 K, as shown in Fig. 2. This work also confirmed the T/μ finding from the 1988 occultation datasets that either CO or N_2 must dominate Pluto's atmospheric composition and that gaseous CH_4 is only a minor constituent compared to the total, with a mole fraction of perhaps 1–9%. Despite being only a minor constituent, Pluto's CH_4 abundance is clearly sufficient to make the CH_4-thermostat work.

Importantly, the 1988 stellar occultation lightcurves exhibit a sharp change of slope, or "knee," during immersion and emersion (*Hubbard et al.,* 1988; *Elliot et al.,* 1989) (see also Fig. 1). The change in slope of these lightcurves below the half-light level indicates either nonisothermal temperature structure in the atmosphere (which changes the refractive index) (e.g., *Eshelman,* 1989; *Hubbard et al.,* 1990; *Stansberry et al.,* 1994), or an extinguishing photochemical haze layer (*Elliot et al.,* 1989), or both.

Extrapolations of the 1988 occultation lightcurves to the surface depend on the assumed radius of Pluto. Pluto's radius is uncertain to approximately ±3%, so the surface pressure and atmospheric column are also both uncertain. Pluto's radius values are constrained by both an uncertainty in interpreting the stellar occultation data, which may not probe to the surface, and an uncertainty in the surface ice temperature, which is constrained but not fixed by the vapor pressure of N_2. The various constraints suggest bracketing radii and surface pressures, respectively, of 1200 km and 3 μbar and 1145 km and 28 μbar (e.g., *Spencer et al.,* 1997). The corresponding Pluto atmospheric column densities are 39 and 285 cm-am, respectively. Further, both hazes and refractive effects (e.g., mirages) (*Stansberry et al.,* 1994) introduce considerable (10–30 km) uncertainties in radius determinations. These and other effects in turn create significant uncertainties in resulting estimates of the surface pressure, with plausible values ranging from ~2 to perhaps ~60 μbar. As an aside, assuming complete atmospheric condensation near aphelion, this range of surface pressure corresponds to a seasonally deposited pure N_2 frost

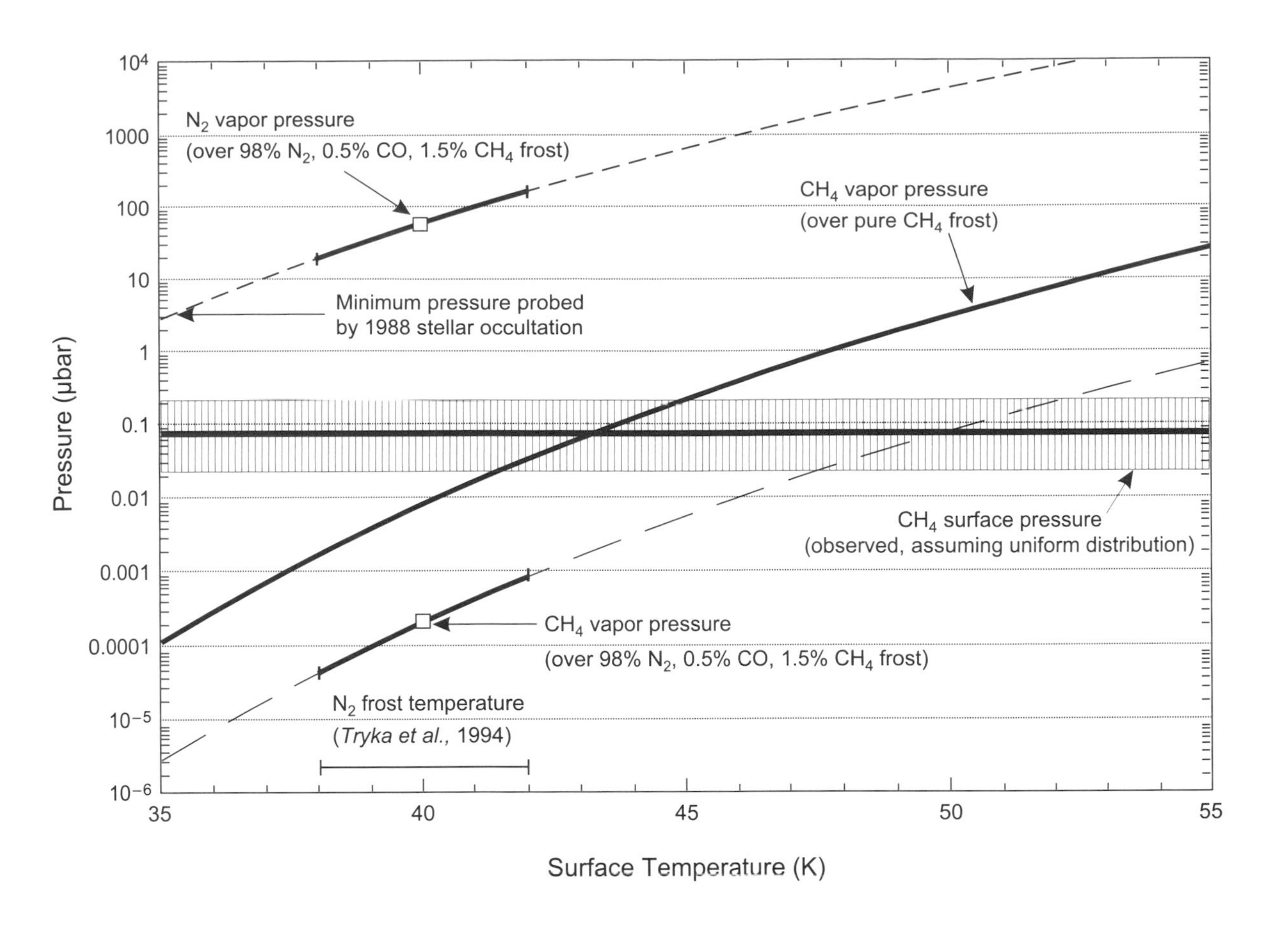

Fig. 2. Surface and vapor pressures on Pluto (*Young et al.,* 1997). The upper curve for N_2 should be close to the total atmospheric surface pressure, and the lower curve for CH_4 is for the same mixture as for the N_2, where as the middle curve is for pure CH_4. The solid parts of the curves show the vapor pressures for a frost temperature of 40 ± 2 K.

layer that would have a depth on the surface of 0.5 to perhaps 10 mm, depending on the actual base pressure and the frost porosity.

Stellar occultations by Pluto are rare, and no subsequent event was observed until August 2002. From that well-observed event it was found that Pluto's atmospheric pressure had significantly increased from 1988 (*Elliot et al.,* 2003; *Sicardy et al.,* 2003). This pressure increase — a full factor of 2 — was initially surprising to some, since Pluto had been receding from the Sun as it moved away from its 1989 perihelion. However, the observed factor of 2 pressure doubling only requires an increase in N_2 surface frost temperature of 1.3 K (*Elliot et al.,* 2003), which could be the result of a thermal phase lag following Pluto's 1989 perihelion maximum in insolation (e.g., *Stern et al.,* 1988; *Trafton,* 1990; see also review by *Spencer et al.,* 1997). It can also be explained by the exposure of new, volatile-rich terrains to sunlight as the southern (IAU convention) polar cap moved into summer, or a combination of both effects.

In addition to the increase in atmospheric pressure, a notable increase in the number of strong refractive spikes was also observed in the occultation lightcurves of 2002 compared with 1988 (e.g., *Pasachoff et al.,* 2005); this has been associated with an increase of turbulence or waves in Pluto's atmosphere. And yet another important change seen from 1988 to 2002 was the dramatic muting of the kink in the occultation light curves seen near the half-light level in 1988. This indicated that a change in the atmosphere's vertical structure had also taken place. Together these various findings make clear that Pluto's atmosphere is time variable, likely due to a combination of seasonal and heliocentric distance effects.

One final result from the 2002 event is worth mentioning here. *Elliot et al.* (2003) reported that observations at a variety of wavelengths from 0.75 μm to 2.2 μm showed that the minimum flux of the lightcurve varies with wavelength, indicating that there is extinction in Pluto's lower atmosphere. While this does not rule out a thermal gradient also being in effect, it is clear evidence for a haze layer.

An even more recent stellar occultation of Pluto was observed at several sites on June 12, 2006. Data analysis is currently in progress, but initial results reveal that the 2006 atmospheric pressure and temperature structure in 2006 is far more similar to what was observed in the 2002 occultation (see Fig. 3) than to that in 1988. Given the much shorter time base (4 years vs. 14), this is not particularly surprising,

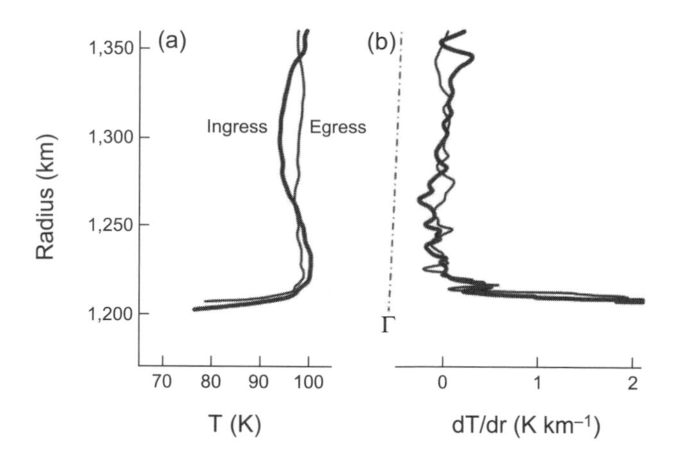

Fig. 3. Pluto's vertical thermal structure and thermal gradient as retrieved from the 2002 stellar occultation (from *Sicardy et al.,* 2003).

but it is nonetheless noteworthy that the pace of change was not further accelerating.

Regarding orbital and seasonal change on Pluto, dramatic atmospheric change on Pluto was predicted as early as the beginning of the 1980s (*Stern,* 1981; *Trafton and Stern,* 1983) because Pluto's surface sees large changes in insolation and insolation distribution during its orbit around the Sun. These are due to both its large variation in heliocentric distance (29.5 to 49.6 AU) and its high (~120°) axial obliquity. Such insolation changes result in surface temperature variations that probably drive strong vapor pressure variations, in turn resulting in the transport of ices across the surface on seasonal and orbital timescales.

Unfortunately, quantitative models of Pluto's so-called "seasonal change," although adequately sophisticated, are not well constrained because numerous model input parameters (e.g., the compositional distribution, the albedo distribution, the surface temperature distribution, the thermal inertia) are not adequately known. Various models predict differing atmospheric pressure and volatile transport histories, and suggest that Pluto's surface/atmosphere interactions are more dramatic than any planet or satellite that orbits closer to the Sun, except perhaps Triton. And predictions of the timing and degree of postperihelion atmospheric collapse (i.e., a factor of >100 decline in mass) as a condensation frost onto the surface vary widely (e.g., *Stern et al.,* 1988; *Yelle and Lunine,* 1989; *Trafton,* 1990; *Hansen and Paige,* 1996; *Lykawka and Mukai,* 2005; see also *Spencer et al.,* 1997, for a review). Why? The process of atmospheric collapse is complicated by both the patchy nature of Pluto's volatile distribution and thermal emission, as well as the complex, nonlinear feedbacks that volatile transport drives on atmospheric and surface properties, and ultimately, the volatile distribution itself (e.g., *Spencer et al.,* 1997; *Trafton et al.,* 1997). It is likely that good atmospheric collapse (and subsequent regeneration as Pluto next approaches perihelion) predictions will require the surface albedo, temperature, and composition maps that NASA's New Horizons Pluto flyby mission will deliver in 2015. Whether the atmosphere is extant at that time, however, is not clear.

We now conclude this section by briefly considering the escape of gases from Pluto's atmosphere as a precursor to a more general discussion relating to KBO atmospheric escape in section 5.

Pluto's combination of low surface gravity (~55 cm/s^2) and comparatively high stratospheric temperature (~100 K) conspire to create a circumstance where a much greater fraction of the initially Maxwellian distribution of molecular energies is capable of escape than in a typical planetary atmosphere. Indeed, unlike the terrestrial and giant planets and Triton, where Jeans escape and photochemical/ion pickup processes dominate, on Pluto, the hydrodynamic (i.e., streaming, bulk) escape of atoms and molecules likely dominates (e.g., *Trafton et al.,* 1997, and references therein).

Hunten and Watson (1982) (hereinafter *HW82*) pointed out that hydrodynamic escape of Pluto is energy-throttled by the adiabatic cooling of the expanding atmosphere, which is regulated by the downward conduction of solar FUV/EUV heat absorbed in the thermosphere. The *HW82* formulation provides an upper limit to the escape flux because it approximates the subthermosphere temperature minimum to be 0 K.

Other investigators have used related methods to estimate escape rates for Pluto at perihelion, finding rates between 3×10^{26} s^{-1} and 2×10^{28} s^{-1} (*McNutt,* 1989; *Hubbard et al.,* 1990; *Trafton et al.,* 1989, 1997). All are upper limits except for Trafton, who solved the hydrostatic escape equations for a CH_4 atmosphere, estimating 3.3×10^{27} s^{-1}. *McNutt* (1989) first considered a gas other than CH_4 (namely, CO) in hydrodynamic escape, using a self-consistent analytic approach. *Yelle* (1993) presented numerical solutions of the Navier-Stokes equations for Pluto's hydrodynamically escaping atmosphere for N_2 and CO that included solar EUV heating, energy transport by thermal conduction, and viscous dissipation of mechanical energy. *Krasnopolsky* (1999) extended McNutt's analytic approach to include a neglected term in the hydrodynamic flow equations; he also included the hitherto neglected solar UV heating of Pluto's upper atmosphere, which he showed to be six times stronger than the solar EUV heating. He then applied these results to the hydrodynamic escape of Pluto's N_2, with CH_4 diffusing upward through it, deriving a perihelion N_2 escape rate of 2.3×10^{27} s^{-1} at mean solar activity. Recently, *Tian and Toon* (2005) were the first to solve the time-dependent hydrodynamic escape equations for a planetary atmosphere and to apply them to the hydrodynamic escape of N_2 from Pluto, treating the spatial distribution of UV-EUV energy deposition realistically over a range of depths in the atmosphere. They derived a corresponding perihelion escape rate around 1×10^{28} s^{-1}, about an order of magnitude higher than the Krasnopolsky value. They argue that this discrepancy arises from Krasnopolsky's single-altitude heating approximation. This has merit in a case where the effective atmospheric depth of UV absorption is significantly lower than that of EUV absorption.

As we noted above, characteristic escape fluxes at perihelion of 3×10^{26} to 2×10^{28} N_2 s^{-1} have been predicted by various modelers. Over the 4.5-G.y. age of the solar system, this corresponds to the potential loss of 1 km to perhaps 10 km of surface ice, depending again in part on the surface ice porosity. This in turn implies either (1) an essentially 100% pure, volatile crust devoid of involatile constituents that would create, even in tiny amounts, a lag deposit that chokes off sublimation and therefore prevents atmospheric regeneration over time; (2) a very recent source of surface volatiles and therefore atmosphere; or (3) some kind of endogenic (e.g., geologic) or exogenic (cratering) activity that replenishes the source of volatiles available to the surface. Such considerations also apply to many KBOs, which may someday be discovered to have past or extant atmospheres that suffer similarly high escape rates.

4. KUIPER BELT OBJECT ATMOSPHERIC PRODUCTION

We now turn to the subject of atmospheric generation on KBOs. In section 5 and section 6, respectively, we will discuss KBO atmospheric loss mechanisms and then structure. Three important atmospheric generation processes will be discussed here: ice sublimation, internal outgassing, and impacts. We take each in turn.

4.1. Ice Sublimation

Sublimation is the change of phase of a substance from the solid directly into the gaseous state. Sublimation is responsible for generating the atmospheres of Pluto, Triton, and comets. Sublimation is an endothermic process: Energy input is required to supply the latent heat of sublimation. In general, the kinetic energy of the molecules in the solid will have a distribution due to thermal lattice motions. As a result, some fraction of the molecules near the surface will always have enough kinetic energy to overcome the binding potential energy of the lattice and escape into the gas phase, even though the temperature is below freezing. If the system is allowed to come to thermal equilibrium, then the rate at which gas molecules stick to the surface will equal the rate at which they "evaporate" from the surface, and the net latent heat transfer will be zero.

The equilibrium vapor pressure of any given frost is exponentially sensitive to the ratio of binding energy of the molecular matrix L, called the latent heat of sublimation, to its thermal energy kT; this requires the condition that the gas density be high enough for the sticking rate to match the evaporation rate. Each of these rates is proportional to $e^{-L/kT}$, the equilibrium vapor pressure. In the case of more volatile frosts like N_2, CO, and CH_4, which are seen on Pluto, Triton, and some KBOs, L/k is such that an increase in temperature of only 1 K is enough to double the vapor pressure.

The saturation vapor pressures for three of these ices are shown in Fig. 4 over a temperature range relevant to much of the KB. Over this range the vapor pressures change dramatically, i.e., by 5–7 orders of magnitude. The diurnally averaged insolation and resulting surface temperature vary with latitude. Consequently, atmospheric vapor pressure and density can vary dramatically over the surface for an atmosphere too tenuous to be in hydrostatic equilibrium (*Stern and Trafton,* 1984; *Trafton,* 1990).

Sublimation of an ice layer into a vacuum is a rapid, nonequilibrium process. The net evaporation flux (i.e., the difference between the sublimation and condensation rates of the gas) will cause the ice to cool (or heat, if the evaporation flux is negative). In the absence of other heat sources, an upper limit on the rate of sublimation is set by the absorbed insolation.

When there are volatile ices of different species on the surface, the vapor pressures of their gas phase depend on how intimately they are mixed (see *Trafton et al.,* 1997). If they exist separately or as a mixture of their pure grains, then each ice's vapor pressure is specified by its temperature according to the saturation vapor pressure relation for that ice. When two or more volatile ices are mixed together intimately as in a solid solution, the less-volatile ice will become enriched at the surface as the more-volatile ice preferentially sublimes. From a surface evolution standpoint, this can eventually reduce the sublimation rate of the surface to that of the less-volatile ice. Differential sublimation will be controlled by Raoult's Law, and if any refractory impurity exists in an ice, then an involatile lag deposit will form from impurities that cannot sublime (e.g., *Stern,* 1989). Even ice mixtures with impurities of <1% can be sufficient to create a lag deposit that will eventually choke off sublimation by burying the ice beneath an involatile overburden.

In the 30–50-AU KB zone, a rapidly rotating body warmed only by the Sun, with spin axis normal to the direction to the Sun and unit emissivity ε, will have an effective temperature between 39 K and 51 K, or between 31 K and 40 K, for bolometric albedos of 0 and 0.6, respectively. For either a slow rotator, or a fast or slow rotator with spin axis pointed toward the Sun, the mean dayside effective temperature would be some 8–10 K, higher. If $\varepsilon < 1$, then the physical temperature would be higher than the above values by a factor of $1/\varepsilon^{1/4}$. In the extreme, surface temperatures warmed only by insolation might reach ~70 K on KBOs. Tidal heating, radioactive decay, and other energy sources can further raise these temperatures, but volatile ices such as O_2, N_2, CO, and CH_4 already have significant vapor pressures in the 30–70-K temperature range. As a result, gaseous envelopes can accompany KBOs having volatile reservoirs in contact with the insolation field. The bulk atmospheric composition in such cases can be expected to reflect that of the illuminated reservoir ice, as it does for comets.

Additionally, for KBOs having eccentric orbits, an orbital effect will result from the change in planetary-averaged insolation over the orbit, which in general will not be in phase with obliquity-driven seasons, although both will

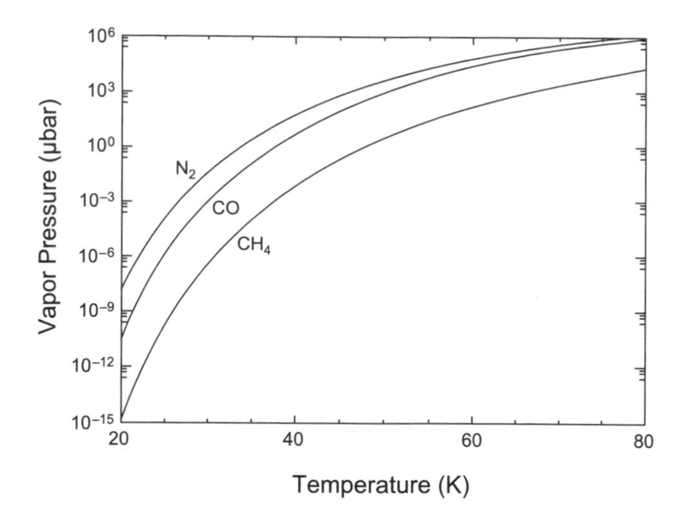

Fig. 4. Equilibrium vapor pressure curves for CH_4, CO, and N_2 as a function of temperature.

have the annual period. As a result, many KBOs with atmospheres are likely to have highly time-variable atmospheric bulk, composition, and structure.

4.2. Internal Outgassing

Outgassing from the interior is another important process, not only owing to the high volatile content of KBOs, but also because they may be unusually porous (see the chapter by McKinnon et al.). Such porosity increases the conductivity of volatiles to the surface and therefore the effective size of the reservoir that supports an escaping atmosphere or coma. The importance of internal release, whether from the near surface (as for Triton's geysers) or deeper, cannot be underestimated since without a resupply of volatiles, an impure surface will eventually choke itself off as an ever-increasing thickness of involatile lag deposit accumulates.

To better understand interior evolution, *McKinnon* (2002) and *Choi et al.* (2002) conducted studies of the thermal evolution of KBOs of various sizes up to 500 km in diameter. *Choi et al.* (2002) found that the long-term evolution of the temperature profile and the structural modifications are a strong function of the KBO's accretion time, size, and dust/ice mass fraction rather than its heliocentric distance. Based on their one-dimensional, 10–500-km models with an initial composition of a porous mixture of H_2O, CO, and CO_2 ices and dust, and at distances of 30–120 AU, they expect CO, as well as N_2, and possibly CH_4, to be lost entirely. In contrast, the less-volatile ices CO_2, H_2CO, and NH_3 should be partially retained. Other important thermal benchmarks include the temperature 155 K, where a eutectic CH_3OH-H_2O melt can form, and 190 K, which is sufficient to melt ice mixtures such as H_2O-NH_3, which would be buoyant and make its way upward.

Choi et al.'s calculations also show that the central temperature of a 100-km-radius KBO may reach temperatures up to 180 K, and that the internal heating can result in a compositionally layered (differentiated) structure, with interlaced layers depleted and enriched in volatiles. For 150-km KBOs, *McKinnon* (2002) found the central temperature limit to be only 105 K, but for a 450-km-radius KBO, central temperatures can exceed 270 K, which opens the possibility of liquid H_2O in large KBOs and Pluto.

In this regard, cryogenic volcanism has been suggested to resurface Quaoar in response to the reported detection of crystalline water ice and possibly ammonia hydrate as well, both of which should have been destroyed by energetic particle irradiation on a timescale of 10 m.y. unless recently resupplied (*Jewitt and Luu,* 2004). The reported crystallinity also indicates that the surface ice on Quaoar has been heated to at least 110 K.

4.3. Collisions

Objects in the KB have characteristic mutual collisional speeds of 1–1.5 km s^{-1}. As such, collisions represent a source of energy to drive sublimation. The role of collisions was investigated by *Orosei et al.* (2001) who found that in some cases, depths over 1 km were altered by KBO impacts while for some other collisions, very small effects were produced. *Durda and Stern* (2000) found that over the 3.5-G.y. age of the classical KB, impacts of 1-km-radius comets onto individual 100-km-radius KBOs occur some ~8–54 times, producing craters ~6 km in diameter.

By excavating deeply buried materials that are beyond the influence of solar heating, these collisions also have the potential to replenish surface ice previously depleted through escape or buried under a refractory lag deposit (*Stern,* 1989). For example, a nitrogen ice deposit excavated from a 6-km-diameter crater 1 km deep (presumably with an involatile material just above it) would correspond to a column abundance of 23 g cm^{-2}, or 180 m-am of gas on a 100-km-radius KBO.

Additionally, impacts import energy and can either promptly or over time (depending on the impact speed and the depth of burial of the impactor) power the sublimation of volatiles at the impact site. At the 1–1.5-km s^{-1} impact speeds characteristic of KBO collisions, H_2O is not promptly vaporized in large quantities. As for more-volatile ices like N_2, CH_4, and CO, it is unfortunate that neither experiments nor adequate thermophysical modeling has been undertaken (E. Pierazzo, personal communication), so it is not presently possible to quantify the efficacy of this process.

5. KUIPER BELT OBJECT ATMOSPHERIC ESCAPE

Escape, as opposed to chemical destruction, is expected to be the dominant long-term loss process for atmospheres on KBOs. Unlike the cyclic effects of seasonal/orbital collapse of an atmosphere, escape results in the permanent loss of volatiles.

The escape rate from a KBO depends on both upper atmospheric density and temperature. The former controls the altitude of solar UV heating and the escape level; the latter controls the escape energetics. Because of the low escape speed of KBOs, typically 0.05 km/s to 1.2 km/s (corresponding to KBO radii of 50 km and 1270 km, respectively, for a reference density of 1.6 g cm^{-3}), their gas envelopes are likely to be predominantly escaping hydrodynamically. This means that a significant fraction of the Maxwellian is sufficiently energetic to escape, rather than just the high-energy tail as in more classical, Jeans escape. This is illustrated by the fact that the mean thermal speed of N_2 at 20 K is 0.12 km s^{-1}, which is comparable to the escape speed from a KBO of radius 125 km having a density of 1.6 g cm^{-3}.

To illustrate this point further, a good general measure of the degree of boundedness of an atmosphere is the ratio of the gravitational potential of a molecule near the surface to kT. This dimensionless gravitational parameter is

$$\lambda = GMm_p\mu/RkT \quad (1)$$

Here G is the universal gravitational constant, M is the KBO's mass, m_p is the mass of a proton, μ is the mean molecular weight of the atmosphere, R is the KBO's effective radius, and T the exospheric temperature.

An atmosphere with $\lambda = 2$ is hardly bound since it blows off (escapes hydrodynamically) at sonic speed. Over the range $3 < \lambda < 140$ that is expected on larger KBOs, atmospheric escape is likely to span the short atmospheric lifetime hydrodynamic streaming regime to the long atmospheric lifetime thermal escape regime.

Evaluating the constants in the above equation, we obtain the dimensionless relation

$$\lambda = 3.361 \times 10^{-5}\mu\rho R^2/T \quad (2)$$

where density ρ replaces M assuming a homogeneous sphere of radius R in kilometers. In Table 1 we provide the surface value of λ as a function of KBO size, assuming N_2-, CO-, and CH_4-dominated atmospheres, an exospheric temperature range of 25–100 K, and $\rho = 1.6$ g cm^{-3}.

We now discuss in turn the various modes of gravitational escape that KBO atmospheres can be expected to experience, based on their individual λ and atmospheric structure.

5.1. Jeans Thermal Escape

The lowest escape rates occur for atmospheres that are in the Jeans escape regime, where escape occurs by molecular evaporation of the Maxwellian high-velocity tail. This thermal escape takes place from an exobase, the altitude where the mean free path equals the local scale height. Jeans escape is limited by the energy input available to heat the exosphere.

Jeans escape must take place near the planet because as the atmospheric temperature rises, or as λ otherwise becomes smaller, the exobase altitude rises toward $\lambda = 2$, while the escape transitions to the hydrodynamic regime (see below) as the gravitational binding energy of the gas molecules declines to approach the mean thermal kinetic energy.

TABLE 1. λ estimates for KBOs having CH_4-, CO-, or CH_4-dominated atmospheres.

Radius (km)	μ	λ range (over T = 25–100 K)
250	16	0.6–2.2
250	28	1.0–3.8
500	16	2.2–8.6
500	28	3.6–15
1000	16	16–32
1000	28	16–60
1500	16	20–78
1500	28	34–137

Here μ = 16 represents a CH_4-dominated atmosphere and μ = 28 represents an N_2 or CO-dominated atmosphere.

It is important to note that Jeans escape, in effect, cools the exosphere by removing its highest-velocity molecules and atoms. As a result, the velocity distribution of the exosphere — the region above the exobase — will deviate from a strict Maxwellian and the escape rate will drop below the classical Jeans value if the escape rate is too high for atmospheric diffusion to readily replace the higher-energy escaping molecules. One must therefore be careful in calculating Jeans escape rates to properly account for this and other effects that modify the bulk gas temperature, and hence kT.

The Jeans escape flux at the exobase is given by

$$F_e = n_eU(1 + \lambda_e)\exp((-\lambda_e)/(2\sqrt{\pi})) \quad (3)$$

Here $U = (2kT_e/m)^{1/2}$, m is the mass of the escaping molecular species of interest, and T_e, n_e, λ_e, are the temperature, number density, and gravitational parameter, respectively, at the exobase. One thus sees that the Jeans escape rate is highly sensitive to the value of λ at the exobase, varying exponentially with it.

The Jeans escape regime typically obtains for atmospheres around massive planets, where it results in very low escape rates. In fact, the Jeans escape timescale of such planetary atmospheres can be longer than the age of the solar system.

Atmospheres in the Jeans escape regime are unlikely to occur on most KBOs, owing to the low mass of most KBOs. However, depending on the composition (hence μ), Jeans escape could dominate on Pluto/Triton-sized or larger KBOs.

As an example of one type of KBO atmosphere that may be encountered, Triton's atmosphere is entirely in the Jeans regime, including the escape of H, H_2, and N generated by CH_4 photolysis and ion chemistry driven by precipitating electrons trapped in Neptune's magnetosphere (*Summers and Strobel,* 1991; *Krasnopolsky,* 1993; *Strobel et al.,* 1996). For a Triton-like exobase height of 900 km and exobase temperature of 100 K, the escape-level $\lambda = 21.5$ and N_2 number density is 1.1×10^7 cm^{-3}, according to the preferred Triton model of *Krasnopolsky* (1993, his Table 9). In his basic models, the number density of N_2 is insensitive to the flux of magnetospheric electrons; the CO mixing ratio is

10^{-3} in his preferred model. The N_2 Jeans escape flux is only 2140 cm^{-2} s^{-1} or 5×10^{20} s^{-1}, far below the hydrodynamic regime.

In closing, Jeans escape can occur for any value of λ at the exobase greater than ~2. This escape regime is most likely for cold, high-molecular-weight atmospheres around dense, massive bodies. The loss of volatiles from small KBOs having lower λ values is likely to occur through the direct sublimation of volatile ice to space. The sublimation flux into a vacuum is then governed by the speed of sound for the escaping molecules and a molecular density that is constrained by the saturation vapor pressure.

5.2. Hydrodynamic Escape

As we stated above, hydrodynamic escape involves the wholesale escape of a large fraction of the bulk Maxwellian, and it occurs when λ is very low. While Jeans escape is an evaporative process from an essentially static atmosphere, hydrodynamic escape is a collisional process that maintains a non-zero bulk outflow speed throughout the atmosphere.

This process may occur on KBOs having λ only a bit greater than 2 (e.g., 3–6), but it fails when escape is limited by a process other than energy deposition into a thermosphere. One example of this is an escaping KBO atmosphere that is optically thin to EUV radiation. It also fails when the absorbed insolation occurs over a wide range of altitudes, as may be the case when multiple gases are present with very different FUV/EUV absorption cross sections. Hydrodynamic escape may still occur in these situations, but its treatment requires a less-approximate solution of the escape equations.

When the hydrodynamic escape regime is achieved, it can be subclassified according to whether the exobase of a secondary gas that is diffusing through a primary (i.e., hydrodynamically) escaping constituent lies below or above the height at which the primary species becomes supersonic. Following *Krasnopolsky* (1999), one can distinguish between two hydrodynamic escape regimes, "slow" and "fast" hydrodynamic escape.

In slow hydrodynamic escape, the upwardly diffusing secondary gas escapes thermally from an exobase that lies below the sonic level of the primary, hydrodynamically escaping constituent. The hydrodynamic speed at the exobase adds to the radial component of the speeds of the thermally escaping species. This occurs in the case of the minor constituent CH_4 in Pluto's atmosphere, which buoyantly diffuses upward through the primary gas, N_2. Since its exobase lies below the sonic level, it arrives at the exobase still having a quasithermal velocity distribution that is not wholly relegated to streamline flow. It thus escapes from this level quasithermally, with escape favoring the fastest molecules. This is opposite to classical hydrodynamic escape where the exobase lies above the sonic level, and the upward diffusing species fully participates in the bulk hydrodynamic escape.

In contrast, in the fast hydrostatic escape regime the thermal energy of all escaping species has been effectively converted through cooling collisions to streamlined, radial, transonic velocities with escape rate close to the UV-EUV solar energy input limit (*Hunten and Watson,* 1982; *Trafton et al.,* 1997).

5.3. Application to Kuiper Belt Objects

Kuiper belt object atmospheres having large scale heights that are a significant fraction of the KBO radius have small λ values (as low as ~2), and so should exhibit hydrodynamic escape since they are gravitationally weakly bound. In contrast, KBO atmospheres with small scale heights that lie deep in the body's gravitational potential should be escaping thermally from an exobase. With an exobase λ of 21, Triton is such an example. The crossover escape flux between the two regimes depends on the exobase value of λ and the mean thermal speed as well as the surface λ and amount of EUV heating. With H = 56 km, λ = 22 at the occultation level (1250 km), and a CH_4 exobase lying below the sonic altitude, Pluto is an example of a KBO in the transition region between Jeans and hydrodynamic escape. [Notice this is the occultation level, not the exobase level, which is where λ was calculated for Triton above. Pluto's N_2 does not have an exobase. The regimes are different because Triton's atmosphere is evaporating slowly at the exobase, while Pluto's N_2 is escaping hydrodynamically, with a much higher flux. Hydrodynamic escape depends on the flux of solar UV-EUV heating of the thermosphere, as well as λ; hence, equality of λ does not imply equality of escape regime.] However, it is also possible for KBOs to be in the transition region between Jeans and hydrodynamic escape.

An important factor in whether a given KBO atmosphere will be in the Jeans, transitional, or full hydrodynamic escape regime is the atmospheric CO/CH_4 ratio. CO is a net coolant in thin, cold, vapor-pressure-supported atmospheres that do not have large optical depths in the CO lines. In contrast, when the optical depth of sunlight to CH_4 is high, CH_4 is a net heating source. The relative amounts of these gases therefore determine the radiative equilibrium temperature structure in such KBO atmospheric layers, and thus the density and scale height variation. For example, the surface CH_4 mixing ratio is 2 orders of magnitude higher for Pluto than for Triton, so CH_4 is primarily responsible for heating Pluto's low stratosphere to 100 K. Triton's lower CH_4 mixing ratio accounts for Triton's significantly colder atmosphere at the microbar level (*Krasnopolsky,* 1993; *Strobel et al.,* 1996). Pluto's higher CH_4/CO ratio results in Pluto's higher temperature and larger scale height, and therefore its greater, hydrodynamic, escape rate.

6. KUIPER BELT OBJECT ATMOSPHERIC STRUCTURE

In this section we will discuss both the vertical and lateral structure of atmospheres one may find around KBOs.

The column abundance and base pressure of an ice-supported KBO atmosphere will depend sensitively on the ice

TABLE 2. KBO atmospheric characteristics.

R (km)	T (K)	λ	H (km)	Pressure (μbar)	N (cm-am)
CH_4 (μ = 16)					
400	32	4.3	93	5.0×10^{-6}	5.7×10^{-4}
400	50	2.8	145	2.9	402
1400	32	53	27	5.0×10^{-6}	1.2×10^{-4}
1400	60	28	50	153	3700
N_2 (μ = 28)					
400	25	9.6	42	1.1×10^{-4}	5.8×10^{-3}
400	50	4.8	83	3960	$2.5 \times 10^{+5}$
1400	25	12	12	1.1×10^{-4}	1.4×10^{-3}
1400	60	49	29	$6.3 \times 10^{+4}$	$8.5 \times 10^{+5}$

Quantities refer to values at or near the surface. A KBO density of 1.6 g cm^{-3} is assumed.

temperature (through the saturation vapor pressure relations) and on the scale height. In the approximation of an isothermal atmosphere, the column abundance N of gas in cm-am units (1 cm-am is the number of molecules in 1 cm^3 of gas at STP) is the product of the equilibrium vapor pressure of the ice $P_s(T)$, converted to density, and the atmospheric scale height corrected to first order for sphericity. Thus, the column is, in cm-am, is

$$N = 10^5 P_s(273/T_{ice})(R/\lambda)(1 + 2/\lambda) \quad (4)$$

Here T_{ice} is the ice temperature, $P_s(T_{ice})$ is in bars, R is in km, and the dimensionless gravitational parameter λ is evaluated at the KBO's surface.

Table 2 provides estimates for the gas column N, and a range of other atmospheric parameters, for two KBO radii and various relevant surface temperatures; it assumes a KBO density of 1.6 g cm^{-3}. The other parameters are the surface λ, scale height, and pressure. For the temperature and radius ranges listed, λ varies between 2.8 and 53, and H varies from 12 to 145 km. The surface pressure and column abundance estimates range over 10 orders of magnitude. Higher values are expected on KBOs with polar illumination because their equilibrium temperatures will be higher, generating higher vapor pressures. The N values are least certain at the lowest λ value, where the scale height is an appreciable fraction of the KBO radius.

With this information in hand, we now turn to vertical structure considerations for KBO atmospheres.

6.1. Vertical Structure

The vertical structure of KBO atmospheres is most strongly affected by four primary kinds of physics: hydrostatic equilibrium, vapor pressure, atmospheric escape, and radiative transfer.

The pressure scale height H = R/λ (same as the density scale height in an isothermal atmosphere) characterizes the atmospheric vertical pressure structure, which is determined by hydrostatic equilibrium or escape. For KBOs smaller than ~100 km in radius, the scale height will usually exceed the KBOs radius, and even for KBOs as large as 500 km in radius, scale heights will be a significant fraction of the KBO radius. As a result, atmospheres around all but the very largest KBOs are expected to have significant sphericity effects, making the plane parallel approximation inadequate.

Figure 5 shows pressure vs. altitude curves for simple isothermal KBO atmospheres for two bracketing sizes, temperatures, and compositions. The barometric approximation with hydrostatic equilibrium was assumed, so the curves do not account for the effects of atmospheric escape. A KBO density of 1.6 g cm^{-3} and surface pressure of 15 μbar was also assumed; curves for other surface pressures can be estimated by sliding the curves or pressure scale horizontally. The effect of molecular weight, temperature, and radius on the scale height is evident. Any KBO atmosphere can be expected to be quite extended, especially for CH_4 on a smaller KBO.

The thermal structure of a KBO atmosphere will depend on the distance from the Sun, on the atmospheric composition, on the internal radiative transfer, on adiabatic expansion and cooling due to escape, and on the balance between the atmospheric absorption of heat from the Sun plus surface, and the heat lost from net sublimation and radiation to space.

With atmospheric surface pressures expected in the microbar regime and lower, the atmospheric thermal opacity is too low to drive lower atmosphere convection. Sublimation winds that transport volatiles to lower-temperature regions of the globe will be most effective in the lowest scale height of the atmosphere. Just above this regime, the thermal structure will depend on whether the atmosphere experiences net heating at these levels.

Both CH_4 and CO have absorption bands capable of absorbing near-infrared sunlight and radiating it at longer wavelengths, which affects the temperature and density structure. In the case of CH_4 alone, this results in a net heating of the upper atmosphere. For a Pluto-like atmospheric pressure (microbars or tens thereof), a CH_4 mixing ratio of 1–2% is required to elevate a KBO thermospheric tempera-

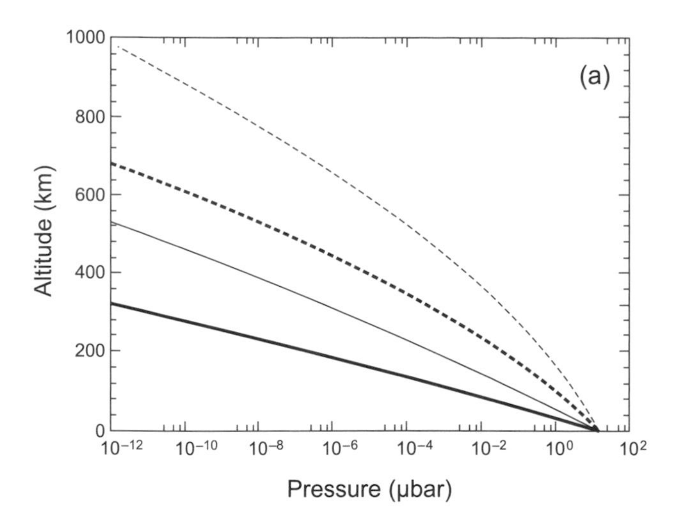

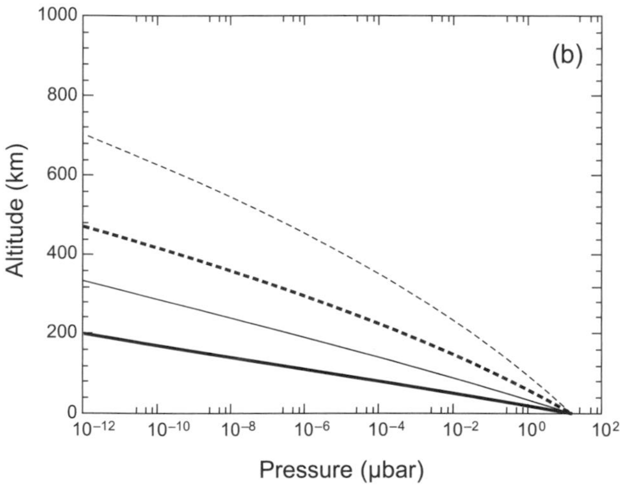

Fig. 5. Pressure vs. altitude curves for simple, isothermal atmospheres for KBOs with various sizes and surface temperatures with pure N_2 and CH_4 atmospheres. **(a)** T = 50 K; **(b)** T = 30 K. The solid curves refer to a radius of R = 1400 km, and the dashed curves to R = 400 km. The heavier curves refer to N_2 and the lighter ones to CH_4. A surface pressure of 15 μbar and a KBO density of 1.6 g cm^{-3} are assumed.

ture to 100 K in ~10 km (*Yelle and Elliot,* 1997). To raise the temperature to 100 K but with a smaller temperature gradient, a mixing ratio of 0.1% is sufficient. N_2 gas does not have this ability. On the other hand, CO is a net coolant of KBO thermospheres via millimeter and submillimeter radiation (*Strobel et al.,* 1996). Such cooling is more important than near-IR heating and could be sufficient to inhibit hydrodynamic escape if the CO/CH_4 mixing ratio is large enough, as is the case for Triton (see also section 5).

For KBO atmospheres escaping in the hydrodynamic regime, the atmosphere cools adiabatically, which limits the escape rate. Escape will then be regulated by EUV + UV solar heating of the upper thermosphere, if one exists. Above these absorption levels, the exospheric temperature will peak, then drop off while the escape speed becomes transonic. The heat conducted downward from this level offsets the adiabatic cooling below to regulate the escape rate, much like a throttle. This causes a temperature minimum in the lower atmosphere, which affects the density structure (*Hunten and Watson,* 1982; *Trafton et al.,* 1997; *Tian and Toon,* 2005).

6.2. Lateral Structure and Conditions for Atmospheric Globality

An important point about the low-pressure atmospheres expected around KBOs is that they may not in all cases be global. Instead, in some cases, local atmospheric "domes" may occur (e.g., over a subsolar region). Thus, globality is an important KBO atmospheric structural attribute.

Pluto has a thermal lightcurve (*Lellouch et al.,* 2000), even though its volatile ice is expected to be globally isothermal, because the sublimating ices do not cover the surface uniformly. Thus, its thermal lightcurve arises because the nonvolatile surface does not sublimate and therefore does not cool (e.g., *Stern et al.,* 1993; *Spencer et al.,* 1997). For a steady-state KBO atmosphere to be global, it must have sufficiently small lateral gradients. This in turn implies that the reservoir of sublimating surface ice must be globally isothermal (neglecting topographic elevation differences), even if the ice itself is not distributed globally.

Isothermality is achieved and maintained by "sublimation winds" created by the net sublimation of ice from regions of higher temperature or insolation to icy regions of lower temperature or insolation. This transfer of volatiles through the atmosphere is accompanied by a latent heat flux that effectively redistributes the ice-absorbed insolation heat evenly to all the exposed volatile ices, which then radiates this heat away (*Trafton and Stern,* 1983). The resulting radiative balance, averaged over a diurnal cycle, determines the global ice temperature. For ice isothermality to occur, the mass of the atmosphere has to be high enough for the sublimation winds to transport sufficient latent heat flux to effect this redistribution at wind speeds low enough for the atmosphere to approximate hydrostatic equilibrium. This in turn sets a lower limit to the vapor pressure and temperature of the ices.

In quantitatively evaluating this criterion, we have followed the formalism given in the Appendix of *Trafton* (1984) that was applied to Triton to estimate the conditions under which a KBO atmosphere is supported hydrostatically on a global basis such that the volatile surface ice is isothermal. That work considered the net sublimation of ice covering Triton's surface for two bracketing geometries: solar latitude 0° and 90°, and allowed for radiative losses and the polar insolation geometry. The insolation geometry is relevant because insolation over the equator generates two sublimation winds, each moving poleward in opposite hemispheres while polar illumination generates a single wind moving toward the opposite pole. Atmospheric escape and its associated cooling are neglected in this model. Sublimation winds originating at low subsolar latitudes were assumed to travel to polar regions and freeze out. Sublimation

TABLE 3. KBO atmospheres for equatorial illumination.

R (km)	A	r (AU)	T_{crit} (K)	λ	H (km)	Pressure (μbar)	N_{min} (cm-am)
CH_4 (μ = 16)							
400	0.0	56.3	37.2	3.71	108	8.17×10^{-4}	0.099
400	0.6	35.6	37.2	3.71	108	8.26×10^{-4}	0.099
1400	0.0	45.3	41.4	40.7	34.4	2.14×10^{-2}	0.505
1400	0.6	28.6	41.2	40.7	34.4	2.17×10^{-2}	0.505
N_2 (μ = 28)							
400	0.0	109	26.7	9.04	44.2	1.29×10^{-3}	0.052
400	0.6	69.1	26.7	9.04	44.2	1.00×10^{-3}	0.052
1400	0.0	90.4	29.3	100.7	13.9	1.95×10^{-2}	0.256
1400	0.6	57.2	29.3	100.7	13.9	1.95×10^{-2}	0.256

Ice temperature, column density, and globality pressures given all refer to the minimum required to satisfy the globality constraint discussed in the text; r is the calculated distance beyond which the specified KBO atmospheric case cannot remain global. N_{min} is the resulting corresponding minimum column for globality (for insolation-limited sublimation).

winds originating in a polar region at high latitudinal insolation were assumed to travel to the opposite pole after reaching minimum speed and maximum density at the equator — the inverse solution. A necessary condition for globality was found to be that the equatorial sublimation wind speed v_o must be small compared to the speed of sound $v_s = (\gamma kT/\mu m_p)^{1/2}$; here γ is the ratio of the specific heats of the gas.

This condition, that the speed of the sublimation wind must be adequately subsonic (see also *Trafton and Stern,* 1983; *Ingersoll,* 1990; *Trafton et al.,* 1997), ensures that the accompanying latent heat flux is sufficient to redistribute the absorbed insolation in order to eliminate significant horizontal temperature and pressure gradients. In this circumstance, atmospheric structure has a high degree of symmetry and regularity, except for the sublimation wind itself, which is affected significantly by the subsolar latitude.

This regulation occurs for Pluto, Triton, and Mars, but does not occur for ices supporting an exosphere or a very thin collisional atmosphere, such as Io's, that is not close to overall hydrostatic equilibrium. Such thin atmospheres tend to be "patchy"; i.e., clustered around isolated volatile ice sublimation sources on the surface.

So what obtains on KBOs? Following *Trafton and Stern* (1983), we adopt the criterion for a significant deviation from hydrostatic equilibrium of a KBO atmosphere to be a 10% drop in pressure going from the equator to co-latitude 10°. Equation (A6) of *Trafton and Stern* (1983) then gives $v_o = 0.072v_s$. This wind speed leads one to the minimal column abundance for global hydrostatic equilibrium for the case of insolation-limited sublimation

$$N_{min} = \xi(30/r)^2(1 - A)R/L(T_{ice})/(\gamma\mu T)^{1/2} \quad (5)$$

Here, N_{min} is in cm-am, r is the solar distance of the KBO in AU, A the surface albedo, R the radius in km, $L(T_{ice})$ the latent heat of sublimation of the ice in erg g^{-1} [given along with the vapor pressures, e.g., in *Brown and Ziegler* (1980)], and T the lower-atmospheric temperature, approximated by the ice temperature. [Note that we found a minus sign missing for the α-N_2 heat of sublimation coefficient A_3 in *Brown and Ziegler*'s (1980) Table V, and a spurious discontinuity in their algorithmic fit to the CH_4 heat of sublimation vs. T. We used their source CH_4 data instead (*Ziegler et al.,* 1962) to construct Figs. 6 and 7. The source data for N_2 are available in *Ziegler and Mullins* (1963).] The constant $\xi = 1.497 \times 10^8$ here for a subsolar latitude of 0°; and $\xi = 1.459 \times 10^9$ for a subsolar latitude of 90°. Diurnal ice temperature variations are assumed negligible due to latent heat transfer.

Requiring equality of the two expressions for column abundance, i.e., $N = N_{min}$, yields the limiting ice temperature T_{ice} for global hydrostatic equilibrium and surface ice isothermality. Sample values of N_{min} for various solar distances are given in Tables 3 and 4 for some representative KBO parameters. Both the equatorial and polar insolation cases assume that the KBO is covered with volatile ice radiating heat uniformly to space.

An important caveat to these calculations is that, in the case of sufficiently thin atmospheres or large solar distances, one must take into account that the net sublimation rate of KBO ices is likely to be limited by the one-way sublimation flux, such as would occur into a vacuum. This one-way flux is the maximum at which absorbed insolation can endothermically drive sublimation through supplying the required latent heat. It is proportional to the product of the equilibrium saturation density and mean thermal speed of the gas. Depending on the sticking coefficient and KBO albedo, we find this regime does not occur until 60–80 AU for CH_4 and beyond 100–130 AU for N_2. In this case, it would not be correct to equate the two column abundances above. These distances are large enough that the values in Tables 3 and 4 remain valid.

The structure of a given KBO global atmosphere depends on the ice and atmospheric temperature, KBO radius, and spin orientation. In about half the cases shown in Tables 3 and 4, λ is high enough for the atmosphere to be "tightly

TABLE 4. KBO atmospheres for polar illumination.

R (km)	A	r (AU)	T_{crit} (K)	λ	H (km)	Pressure (μbar)	N_{min} (cm-am)
CH_4 (μ = 16)							
400	0.0	47.6	40.4	3.41	117	1.03×10^{-2}	1.29
400	0.6	30.1	40.4	3.41	117	1.04×10^{-2}	1.29
1400	0.0	37.4	45.6	37.0	37.8	2.91×10^{-1}	6.88
1400	0.6	23.7	45.6	37.0	37.8	2.95×10^{-1}	6.88
N_2 (μ = 28)							
400	0.0	93.3	28.9	8.35	47.9	1.20×10^{-2}	0.67
400	0.6	59.0	28.9	8.35	47.9	1.21×10^{-2}	0.67
1400	0.0	75.6	32.1	92.1	15.2	2.62×10^{-1}	3.42
1400	0.6	47.8	32.1	92.1	15.2	2.62×10^{-1}	3.43

Ice temperature, column density, and globality pressures given all refer to the minimum required to satisfy the globality constraint discussed in the text; r is the calculated distance beyond which the specified KBO atmospheric case cannot remain global. N_{min} is the resulting corresponding minimum column for globality (for insolation-limited sublimation).

bound" (λ > 25) although still escaping hydrodynamically. Smaller KBOs with only CH_4 ice can have at best loosely bound atmospheres. The solar distance over which the atmosphere is global then depends on the KBO albedo.

In order to determine the boundary of the regime where globality obtains, we compared the equatorial sublimation wind speed against 0.072 v_s based on the surface density obtained from the saturation vapor pressure $P_s(T_{ice})/kT_{ice}$. Following *Trafton* (1984), the equatorial wind speed is

$$V = 1519\xi[\lambda k T_{ice}(1 - A)(30/r)^2]/[(m_p \mu L(T_{ice}))P_s(T_{ice})] \quad (6)$$

where ξ = 0.2812 for polar insolation and ξ = 0.0289 for equatorial insolation. T_{ice} is assumed to vary for an ice-covered KBO as $r^{-1/2}$ according to $T_{ice} = [1.367 \times 10^6(1 - A)/(4\sigma r^2)]^{1/4}$. This expresses the radiation balance for a volatile-ice-covered KBO of arbitrary orientation for which sublimation winds redistribute the solar flux absorbed by the disk evenly over the globe through latent heat transfer, so that the ice isothermally and isotropically radiates this heat to space. As for Tables 3 and 4, T_{ice} in that model is not affected by the orientation of the spin axis relative to the insolation direction. The main difference in the wind speeds for the different insolation geometries arises from the different net sublimation rates. For KBOs with a patchy ice distribution, different values of N could obtain. We neglect ice patchiness here.

Hence, our approach is to begin by assuming such conditions for a KBO atmosphere and then investigating where globality breaks down, e.g., by moving the KBO farther from the Sun. We conservatively neglect ice emissivities less than unity since they would result in higher temperatures and thicker atmospheres.

These equatorial sublimation wind speeds vs. heliocentric distance are compared for various KBO attributes in Figs. 6 and 7 for the equatorial and polar geometries, respectively, for the two plausible ices, CH_4 and N_2, which differ significantly in volatility. A uniform KBO density of ρ = 1.6 g cm^{-3} was assumed. KBO atmospheres will be in global hydrostatic equilibrium, with surface pressure varying by less than 10% and volatile ice being nearly isothermal, when they are thick enough that the sublimation wind speed is below the indicated scaled sonic speed. So the hydrostatic regime of the KB is the solar distance domain where the curves lie below the appropriate nearly horizontal curve. Lower than limiting speeds in this regime will give denser global atmospheres; higher ones will not be global.

Figures 6 and 7 show that KBOs with lower albedo and radius can retain global atmospheres and ice isothermality farther away from the Sun than those KBOs with higher albedo or radius, well into the scattered KB. This is because the total mass sublimated per unit time is proportional to the disk area of the KBO, while the corresponding sublimation wind mass crossing the equator (or a latitude circle) per unit time is proportional to the product of R, volatile gas column density, and wind speed. Equating these, one sees that the wind speed is directly proportional to R and inversely proportional to the column; hence, small R and large vapor pressure favor subsonic winds and so favor globality.

Also, KBOs with lower axial obliquity can have globally distributed atmospheres deeper into the KB. This is because the net diurnally averaged sublimation rate is greater in the high-obliquity configuration of continuous daylight, resulting in faster sublimation winds. Figures 6 and 7 show that KBOs with CH_4 ice would have global atmospheres for solar distances less than 30–50 AU, depending on albedo, size, and obliquity, but KBOs with N_2 ice would have global atmospheres for solar distances less than 45–110 AU, again depending on size and obliquity.

We emphasize that much larger columns are possible on KBOs than these minimum values for globality. For example, according to Table 3, an R = 400-km KBO at r = 109 AU with effectively zero albedo and equatorial illumination having a limiting global N_2 atmosphere will display

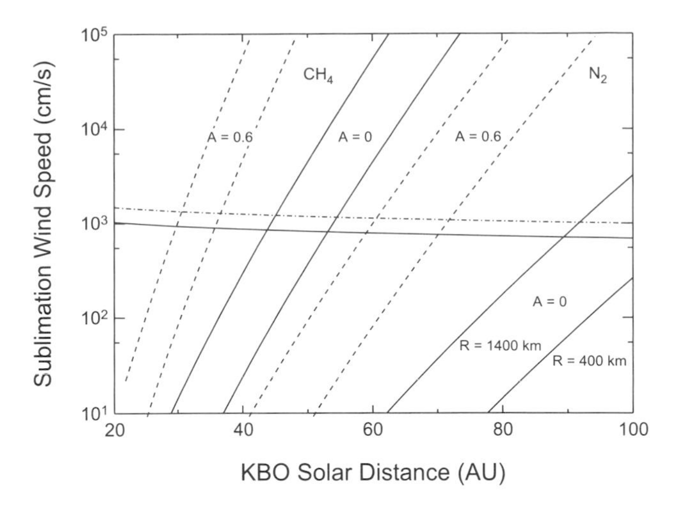

Fig. 6. Hemispherical sublimation wind speeds in KBO atmospheres for subsolar latitude 0° vs. solar distance. Curves are shown for KBO albedo A and KBO radius R of two gases of very different volatility. The four curves on the left are for CH_4 and the four on the right are for N_2. The heavy curves are plotted for albedo A = 0 and the dashed ones for A = 0.6. The upper (or leftmost) curve for each isoalbedo pair is for R = 1400 km and the lower (or rightmost) is for R = 400 km. The nearly horizontal curves indicate 7.2% of the speed of sound for atmospheric N_2 (solid) and CH_4 (dot-dash). KBO atmospheres will be in global hydrostatic equilibrium, with surface pressure varying by less than 10% and volatile ice being nearly isothermal, when they are thick enough (see Tables 3 and 4) that the sublimation wind speed is below the indicated scaled sonic speed. Rapidly rotating, volatile-ice-covered, spherical KBOs are assumed at the radiative balance temperature; atmospheric escape is neglected.

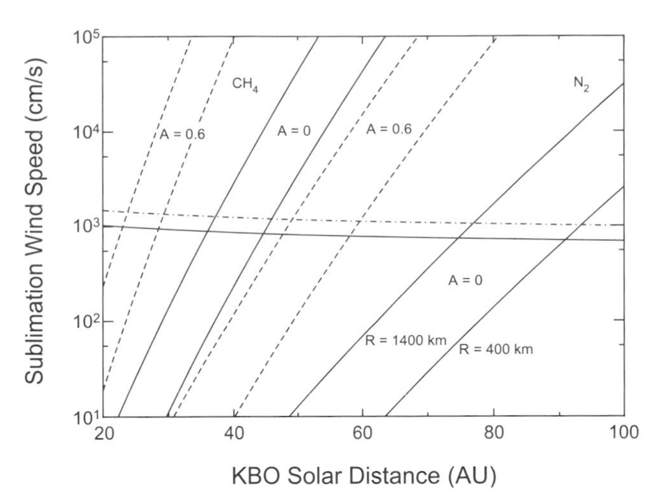

Fig. 7. Same as Fig. 6 for subsolar latitude 90° without regard to rotation rate. KBOs with high obliquity are less likely to have globally distributed atmospheres deep into the Kuiper belt than low-obliquity KBOs.

an ice temperature near 26.7 K, $\lambda = 9.0$, $P = 1.3 \times 10^{-3}$ μbar, and a critical column N = 0.052 cm-am. This same KBO at a closer distance to the Sun will exhibit a higher column abundance and its atmosphere will still be global. The speed of the sublimation wind will be less than at r = 109 AU. This condition is illustrated by the rightmost solid curve on Fig. 6. This curve, as shown, is in the globality regime. Moving down this curve to a solar distance of r = 78 AU results in an atmosphere with 2 orders of magnitude lower sublimation wind speed and a correspondingly greater column abundance. Specifically, the solar equilibrium ice temperature will be 31.6 K, $\lambda = 7.6$, P = 0.17 μbar, and the vapor-supported column abundance will be N = 9.6 cm-am. This is much higher than the column of 0.052 cm-am listed in Table 3 for the limiting wind case.

In summary, KBOs are more likely to exhibit global vs. local atmospheres when the generated sublimation winds are less than 10 m s^{-1}, which corresponds to a volatile gas column abundance at least on the order of 0.05 cm-am to 7 cm-am. Accordingly, equation (6) implies that higher density and volatility favor globality. And indeed, the limiting atmospheric column abundance for CH_4 globality varies from 0.1 to 7 cm-am for the solar distances listed in Tables 3 and 4, and 0.05 to 3.5 cm-am for N_2, neglecting ice patchiness. Lower albedo, obliquity, and radius also favor globality, although a low radius also results in a loosely bound atmosphere. Globality is therefore normally restricted to solar distances less than about 55 AU for KBOs with CH_4 ice and less than 120 AU for N_2 ice, depending on the KBO size, albedo, and obliquity. Without a nonsolar heat source, such as radioactivity or a transient event, like an impact, a global KBO atmosphere is highly improbable beyond 120 AU.

7. DETECTION PROSPECTS

As we stated early in this paper, the detection of surface volatiles and high albedos on some KBOs indicates to us that the existence of at least transient (e.g., seasonal) atmospheres on KBOs other than Pluto is highly likely.

Short of sending spacecraft to such KBOs, how might such atmospheres be detected? Stellar occultations observed from Earth are probably the most powerful tools available for the detection of KBOs atmospheres. Although such occultations are capable of detecting atmospheres below microbar pressure levels, stellar occultations of KBOs are rare owing to their tiny angular sizes and slow angular speeds. Moreover, given the fact that astrometric uncertainties for KBO occultations can be substantial, combined with the fact that KBO shadow paths are narrow, makes it necessary to deploy dense grids of observers normal to the occultation path. To capitalize on the rare but valuable KBO occultation events, we urge occultation groups to make event predictions of the kind discussed by *Elliot and Kern* (2003). We further recommend particular emphasis be placed on occultation predictions for large KBOs transiting across dense star fields, such as the galactic center, where occultation event rates are higher than elsewhere on the sky.

A second approach to KBO atmosphere detection is to obtain high-resolution IR spectra to directly search for atmospheric absorptions, as *Young et al.* (1997) did in detecting CH_4 in Pluto's atmosphere. Yet another detection possibility would be to search for faint, extended coma around smaller KBOs that would have atmospheres freely escaping to space. This is the technique that revealed coma around the Centaur Chiron.

In urging observations such as these, we look forward to the day when KBO atmospheres move from the realm of expectation to realization.

Acknowledgments. This work was supported by the New Horizons mission and NASA's Planetary Astronomy program (grant NNG04G131G). We thank M. Bullock, J. Eliot, C. Olkin, and J. Lovering for helpful comments. We thank E. Lellouch and D. Strobel for their insightful reviews of this manuscript.

REFERENCES

Barucci M. A., Cruikshank D. P., Dotto E. Merlin F., Poulet F., Dalle Ore C., Fornasier S., and de Bergh C. (2005) Is Sedna another Triton? *Astron. Astrophys., 439,* L1–L4.

Brosch N. (1995) The 1985 stellar occultation by Pluto. *Mon. Not. R. Astron. Soc., 276,* 571–575.

Brown G. N. and Ziegler W. T. (1980) Vapor pressure and heats of sublimation of liquids and solids of interest in cryogenics below 1-atm pressure. In *Advances in Cryogenic Engineering, Vol. 25* (K. Timmerhaus and H. A. Snyder, eds.), pp. 662–670. Plenum, New York.

Bus S. J., A'Hearn M. F., Bowell E., and Stern S. A. (1991a) Chiron: Evidence for historic activity (abstract). In *Lunar and Planetary Science XXIII,* p. 34. Lunar and Planetary Institute, Houston.

Bus S. J., A'Hearn M. F., Schleicher D. G., and Bowell E. (1991b) Detection of CN emission from (2060) Chiron. *Science, 251,* 774–777.

Capria M. T., Coradini A., De Sanctis M. C., and Orosei R. (2000) Chiron activity and thermal evolution. *Astron. J., 119,* 3112–3118.

Choi Y., Cohen M., Merk R., and Prialnik D. (2002) Long-term evolution of objects in the Kuiper belt zone — Effects of insolation and radiogenic heating. *Icarus, 160,* 300–312.

Cruikshank D. P., Pilcher C. B., and Morrison D. (1976) Pluto — Evidence for methane frost. *Science, 194,* 835–837.

Doute S., Schmitt B., Quirico E., Owen T., Cruikshank D., deBergh C., Geballe T., and Roush T. (1999) Evidence for methane segregation at the surface of Pluto. *Icarus, 142,* 421–444.

Durda D. D. and Stern S. A. (2000) Collision rates in the present-day Kuiper belt and Centaur regions: Applications to surface activation and modification on comets, Kuiper belt objects, Centaurs and Pluto-Charon. *Icarus, 145,* 220–229.

Elliot J. L. and Kern S. D. (2003) Pluto's atmosphere and a targeted occultation search for other bound KBO atmospheres. *Earth Moon Planets, 92,* 375–393.

Elliot J. L., Dunham E. W., Bosh A. S., Slivan S. M., Young L. A., Wasserman L. H., and Millis R. L. (1989) Pluto's atmosphere. *Icarus, 77,* 148–170.

Elliot J. L., Olkin C. B., Dunham E. W., Ford C. H., Gilmore D. K., Kurtz D., Lazzaro D., Rank D. M., Temi P., Bandyopadhyay R. M., Barroso J., Barrucci A., Bosh A. S., Buie M. W., Bus S. J., Dahn C. C., Foryta D. W., Hubbard W. B., Lopes D. F., and Marcialis R. L. (1995) Jet-like features near the nucleus of Chiron. *Nature, 373,* 46–46.

Elliot J. L., Strobel D. F., Zhu X., Stansberry J. A., Wasserman L. H., and Franz O. G. (2000) The thermal structure of Triton's middle atmosphere. *Icarus, 143,* 425–428.

Elliot J. L., Ates A., Babcock B. A., Bosh A. S., Buie M. W., Clancy K. B., Dunham E. W., Eikenberry S. S., Hall D. T., Kern S. D., Leggett S. K., Levine S. E., Moon D.-S., Olkin C. B., Osip D. J., Pasachoff J. M., Penprase B. E., Person M. J., Qu S., Rayner J. T., Roberts L. C., Salyk C. V., Souza S. P., Stone R. C., Taylor B. W., Tholen D. J., Thomas-Osip J. E., Ticehurst D. R., and Wasserman L. H. (2003) The recent expansion of Pluto's atmosphere. *Nature, 424,* 165–168.

Eshelman V. R. (1989) Pluto's atmosphere: Models based on refraction, inversion, and vapor-pressure equilibrium. *Icarus, 80,* 439–443.

Grundy W. M. and Buie M. W. (2000) The distribution and physical state of H_2O on Charon. *Icarus, 148,* 324–339.

Hart M. H. (1974) A possible atmosphere for Pluto. *Icarus, 21,* 242–247.

Hainaut O. R., Delahodde C. E., Boehnhardt H., Dotto E., Barucci M. A., Meech K. J., Bauer J. M., West R. M., and Doressoundiram A. (2000) Physical properties of TNO 1996 TO66. Light curves and possible cometary activity. *Astron. Astrophys., 356,* 1076–1088.

Hansen C. J. and Paige D. A. (1996) Seasonal nitrogen cycles on Pluto. *Icarus, 120,* 247–265.

Hartmann W. K., Tholen D. J., Meech K. J., and Cruikshank D. P. (1990) 2060 Chiron — Colorimetry and cometary behavior. *Icarus, 83,* 1–15.

Hubbard W. B., Hunten D. M., Dieters S. W., Hill K. M., and Watson R. D. (1988) Occultation evidence for an atmosphere on Pluto. *Nature, 336,* 452–454.

Hubbard W. B., Yelle R. V., and Lunine J. I. (1990) Non-isothermal Pluto atmosphere models. *Icarus, 84,* 1–11.

Hunten D. M. and Watson A. J. (1982) Stability of Pluto's atmosphere. *Icarus, 51,* 665–667.

Ingersoll A. P. (1990) Dynamics of Triton's atmosphere. *Nature, 344,* 315–317.

Jewitt D. C. (1994) Heat from Pluto. *Astron. J., 107,* 372–378.

Jewitt D. C. and Luu J. (2004) Crystalline water ice on the Kuiper belt object (50000) Quaoar. *Nature, 432,* 731–733.

Krasnopolsky V. A. (1993) On the haze model for Triton. *J. Geophys. Res., 98,* 17123–17124.

Krasnopolsky V. A. (1999) Hydrodynamic flow of N_2 from Pluto. *J. Geophys. Res., 104,* 5955–5962.

Lellouch E., Laureijs R., Schmitt B., Quirico E., de Bergh C., Crovisier J., and Coustenis A. (2000) Pluto's non-isothermal surface. *Icarus, 147,* 220–250.

Licandro J., Pinilla-Alonso N., Pedani M., Oliva E., Tozzi G. P., and Grundy W. M. (2006) The methane ice rich surface of large TNO 2005 FY9: A Pluto-twin in the trans-neptunian belt? *Astron. Astrophys., 445,* L35–L38.

Lykawka P. S. and Mukai T. (2005) Higher albedos and size distribution of large transneptuninan objects. *Planet. Space Sci., 53,* 1319–1330.

McKinnon W. B. (2002) On the initial thermal evolution of Kuiper belt objects. In *Proceedings of Asteroids, Comets, Meteors — ACM 2002* (B. Warmbein, ed.), pp. 29–38. ESA SP-500, Noordwijk, Netherlands.

McNutt R. L. (1989) Models of Pluto's upper atmosphere. *Geophys. Res. Lett., 16,* 1225–1228.

Meech K. J. and Belton M. J. S. (1989) (2060) Chiron. *IAU Circular 4770,* 1.

Orosei R., Coradini A., de Sanctis M. C., and Federico C. (2001) Collision-induced thermal evolution of a comet nucleus in the Edgeworth-Kuiper Belt. *Adv. Space Res., 28,* 1563–1569.

Owen T. C., Roush T. L., Cruikshank D. P., Elliot J. L., Young L. A., de Bergh C., Schmitt B., Geballe T. R., Brown R. H., and Bartholomew M. J. (1993) Surface ices and the atmospheric composition of Pluto. *Science, 261,* 745–748.

Pasachoff J. M., Souza S. P., Babcock B. A., Ticehurst D. R., Elliot J. L., Person M. J., Clancy K. B., Roberts L. C., Hall D. T., and Tholen D. J. (2005) The structure of Pluto's atmosphere from the 2002 August 21 stellar occultation. *Astron. J., 129,* 1718–1723.

Sicardy B., Widemann T., Lellouch E., Veillet C., Cuillandre J.-C., et al. (2003) Large changes in Pluto's atmosphere as revealed by recent stellar occultations. *Nature, 424,* 168–170.

Sicardy B., Bellucci A., Gendron E., Lacombe F., Lacour S., et al. (2006) Charon's size and an upper limit on its atmosphere from a stellar occultation. *Nature, 439,* 52–54.

Spencer J. R., Stansberry J. A., Trafton L. M., Young E. F., Binzel R. P., and Croft S. K. (1997) Volatile transport, seasonal cycles, and atmospheric dynamics on Pluto. In *Pluto and Charon* (S. A. Stern and D. J. Tholen, eds.), pp. 435–473. Univ. of Arizona, Tucson.

Stansberry J. A., Lunine J. I., Hubbard W. B., Yelle R. V., and Hunten D. M. (1994) Mirages and the nature of Pluto's atmosphere. *Icarus, 111,* 503–513.

Stern S. A. (1981) Theoretical investigations of the atmospheric environment of Pluto. Masters' thesis, University of Texas.

Stern S. A. (1989) Pluto — Comments on crustal composition, evidence for global differentiation. *Icarus, 81,* 14–23.

Stern S. A. (1999) The lunar atmosphere. *Space Sci. Rev., 37,* 453–520.

Stern S. A. and Trafton L. (1984) Constraints on bulk composition, seasonal variation, and global dynamics of Pluto's atmosphere. *Icarus, 57,* 231–240.

Stern S. A, Trafton L., and Gladstone G. (1988) Why is Pluto bright? Implications of the albedo and lightcurve behavior of Pluto. *Icarus, 75,* 485–498.

Stern S. A., Weintraub D. A., and Festou M. C. (1993) Evidence for a low surface temperature on Pluto from millimeter-wave thermal emission measurements. *Science, 261,* 1713.

Strobel D. F., Zhu X., Summers M. E., and Stevens M. H. (1996) On the vertical thermal structure of Pluto's atmosphere. *Icarus, 120,* 266–289.

Summers M. E. and Strobel D. F. (1991) Triton's atmosphere — A source of N and H for Neptune's magnetosphere. *Geophys. Res. Lett., 18,* 2309–2312.

Tian F. and Toon O. B. (2005) Hydrodynamic escape of nitrogen from Pluto. *Geophys. Res. Lett., 32,* L18201.

Trafton L. (1984) Large seasonal variations in Triton's atmosphere. *Icarus, 58,* 312–324.

Trafton L. (1990) A two-component volatile atmosphere for Pluto. I — The bulk hydrodynamic escape regime. *Astrophys. J., 329,* 512–523.

Trafton L. and Stern S. A. (1983) On the global distribution of Pluto's atmosphere. *Astrophys. J., 267,* 872–881.

Trafton L. M., Whipple A. L., and Stern S. A. (1989) Hydrodynamic calculations of Pluto's atmosphere. *Ninth Planet News, 7,* 1–2.

Trafton L. M., Hunten D. M., Zahnle K. J., and McNutt R. L. Jr. (1997) Escape processes at Pluto and Charon. In *Pluto and Charon* (S. A. Stern and D. J. Tholen, eds.), pp. 475–522. Univ. of Arizona, Tucson.

Womack M. and Stern S. A. (1999) Detection of carbon monoxide in (2060) Chiron. *Solar Sys. Res. (Astron. Vest.), 33,* 187–192.

Yelle R. V. (1993) Hydrodynamic escape of Pluto's atmosphere. In *Abstracts Presented at the Meeting on Pluto and Charon* (July 6–9, 1993, Flagstaff, Arizona), p. 74.

Yelle R. V. and Elliot J. L. (1997) Atmospheric structure and composition: Pluto and Charon. In *Pluto and Charon* (S. A. Stern and D. J. Tholen, eds.), pp. 347–390. Univ. of Arizona, Tucson.

Yelle R. V. and Lunine J. I. (1989) Evidence for a molecule heavier than methane in the atmosphere of Pluto. *Nature, 339,* 288–290.

Young L. A., Elliot J. L., Tokunaga A., de Bergh C., and Owen T. (1997) Detection of gaseous methane on Pluto. *Icarus, 127,* 258–270.

Ziegler W. T. and Mullins J. C. (1963) *Calculation of the Vapor Pressure and Heats of Vaporization and Sublimation of Liquids and Solids, Especially Below One Atmosphere, IV. Nitrogen and Fluorine.* Tech. Report No. 1, Project No. A-663, Engineering Experiment Station, Georgia Institute of Technology, April 15, 1963 (Contract No. CST-7404, National Bureau of Standards, Boulder, Colorado). (See also *Corrections to Nitrogen and Fluorine Report,* minor corrections to this report issued July 2, 1963.)

Ziegler W. T., Mullins J. C., and Kirk B. S. (1962) *Calculation of the Vapor Pressure and Heats of Vaporization and Sublimation of Liquids and Solids, Especially Below One Atmosphere, III. Methane.* Tech. Report No. 3, Project No. A-460, Engineering Experiment Station, Georgia Institute of Technology, August 31, 1962 (Contract No. CST-7238, National Bureau of Standards, Boulder, Colorado).

Part VII:

Links with Other Solar System Populations

De Troianis: The Trojans in the Planetary System

Elisabetta Dotto
INAF, Osservatorio Astronomico di Roma

Joshua P. Emery
NASA Ames Research Center/SETI Institute

Maria A. Barucci
LESIA, Observatoire de Paris

Alessandro Morbidelli
Observatory of Nice

Dale P. Cruikshank
NASA Ames Research Center

Trojan objects are minor bodies having stable orbits in the L_4 and L_5 Lagrangian points of a planet. Mars, Jupiter, and Neptune are known to support Trojans, but Saturn and Uranus are also believed to share their orbits with similar populations of small bodies. Recent dynamical modeling suggests a genetic relationship among transneptunian objects (TNOs) and Jupiter and Neptune Trojans: All these bodies are believed to have formed at large heliocentric distances in a region rich in frozen volatiles. In this context, the analysis and the comparison of the physical properties of Trojans, Centaurs, and TNOs can help us to constrain the link among them and the scenario of the planetary formation in the outer solar system. This chapter presents an overview of current knowledge of the physical properties of Trojans. Since the Jupiter Trojans are the most well studied of the Trojan populations, discussion is centered on the analysis of the properties of this group and comparison with asteroids, comets, Centaurs, and TNOs. The physical characteristics of Jupiter Trojans share some similarities with those of the other populations of small bodies of the outer solar system, but also some notable differences. Some analogies with neutral/less-red Centaurs suggest that Jupiter Trojans are more similar to the active and post-active comets than to the non-active icy bodies. This may support a genetical link among these objects, but the complete puzzle is still far from being understood.

1. INTRODUCTION

Why a chapter devoted to the Trojans in a book on transneptunian objects (TNOs)? The answer to this question comes from the observational evidence of some similarities in the physical characteristics of Jupiter Trojans, TNOs, short-period comets, and Centaurs, and from recent dynamical modeling that suggests that Jupiter Trojans originated in the primordial transneptunian disk.

The objects located in the L_4 and L_5 Lagrangian points of a planet's orbit are called Trojans. To date, Lagrangian bodies have been discovered in the orbits of Mars, Jupiter, and Neptune. The identification of Mars Trojans is still a matter of debate: Only four objects have been confirmed to be in the Mars Lagrangian points (*Scholl et al.,* 2005), and several other bodies have been identified as potential Mars Trojans. Although the population of Neptune Trojans is expected to be 20 times larger than that of Jupiter Trojans (*Sheppard and Trujillo,* 2006a), only six objects are known so far. At present, the most numerous group of known Trojans is in the orbit of Jupiter. It includes more than 2250 objects, about 1230 in the L_4 cloud and about 1050 in the L_5 cloud.

The absolute magnitude distribution of Jupiter Trojans has been the object of a study by *Jewitt et al.* (2000), with observations targeted at the discovery of faint objects. They found that the absolute magnitude (H) distribution of objects with $11 < H < 16$ is exponential, with an exponent $\alpha = 0.4 \pm 0.3$. Assuming that albedo is independent of size, this implies that the cumulative size distribution is a power law with an exponent $q = -2.0$. On the bright end of the distribution, the observations cataloged at the time allowed the authors to infer that the absolute magnitude distribution is much steeper, with an exponent $\alpha = 5.5 \pm 0.9$. In order to estimate a total population, the measurements have to be corrected for the incompleteness of the survey, which is a difficult process. Jewitt et al. only surveyed a small fraction of the L_4 swarm. They estimated the surface density distribution of Trojans (as a function of the angle from the L_4 point) as a Gaussian fit to their data, and used this fit to correct for the incompleteness of their survey. They then multiplied by a factor of 2 to account for the L_5 swarm as well. Jewitt et al. concluded that the rollover from the steep distribution at the bright end to the shallow distribution at the faint end occurs around H ~ 10, and that the total number of Trojans brighter than H = 16 is ~10^5. The situation has evolved since 2000, thanks to the enhanced discovery rate due to the new asteroid surveys such as LINEAR. An updated catalog (see, e.g., *hamilton.dm.unipi.it/cgi-bin/astdys/astibo*) shows more objects than considered by Jewitt et al. for H < 9 and fewer objects at fainter magnitudes. *Yoshida and Nakamura* (2005) performed a survey of L_4 for faint Trojans similar to that of *Jewitt et al.* (2000). They find a cumulative H-distribution slope of 1.89 ± 0.1, which agrees very well with the value found by Jewitt et al. Furthermore, they note an apparent change in slope at H ~ 16 (D ~ 5 km for p_v = 0.04), which is similar to a change in slope discovered for small main-belt asteroids (*Ivezić et al.,* 2001; *Yoshida et al.,* 2003). The *Yoshida and Nakamura* (2005) search area was small, and they used the same surface density correction for the incompleteness of their survey as that used by *Jewitt et al.* (2000). Their final distribution results in about three times more Trojans larger than 1 km than the Jewitt et al. estimate, although it is not clear what distribution they used for the bright (H < 14) objects. According to recent results from the Sloan Digital Sky Survey (SDSS) (*Szabó et al.,* 2007), the L_4 swarm seems to contain significantly more asteroids than the L_5 one. The Trojan catalog is complete up to absolute magnitude H = 13.8 (corresponding to a diameter of about 10 km). Beyond this threshold, SDSS detections confirm the slope of the H distribution found by *Jewitt et al.* (2000). This implies that the number of Jupiter Trojans is about the same as that of the main-belt asteroids down to the same size limit.

About 3% of Jupiter Trojans are on unstable orbits having eccentricities larger than 0.10 and inclinations greater than 55°. Nevertheless, the orbits of the majority of Jupiter Trojans are stable over the age of the solar system (*Levison et al.,* 1997; *Giorgilli and Skokos,* 1997). The population as a whole is widely believed to be as collisionally evolved as the asteroid main belt, and the recent discovery of dynamical families (*Shoemaker et al.,* 1989; *Milani,* 1993; *Milani and Knežević,* 1994; *Beaugé and Roig,* 2001) confirms this hypothesis.

While the dynamical characteristics are quite well determined, the physical properties of the Mars, Jupiter, and Neptune Trojan populations are not as well known. *Rivkin et al.* (2003) carried out visible and near-infrared spectroscopy of three out of the four confirmed Mars Trojans, finding large spectral differences: 5261 Eureka and 101429 1998 VF_{31} have been classified as Sr (or A) type and Sr (or Sa) type, respectively, while 121514 1999 UJ_7 belongs to the X (or T) class. These results seem to suggest that these objects did not all form in their current locations, or alternatively they suffered a strong variation in their sizes.

Color measurements of Neptune Trojans have shown that they are statistically indistinguishable from one another with slightly red colors, similar to the Jupiter Trojans and neutral/less-red Centaurs. On the basis of this result, *Sheppard and Trujillo* (2006b) argued that Neptune Trojans had a common origin with Jupiter Trojans, irregular satellites, and the dynamically excited gray Kuiper belt population, and are distinct from the classical Kuiper belt objects. For Jupiter Trojans, we have visible color indexes of about 300 objects, visible spectra of less than 150 bodies, near-infrared spectra of a sample of about 50 objects (see section 3.4), and thermal-IR spectra of only 3 bodies (see section 3.5). Albedo values are known for a few tens of objects, mainly published by *Fernández et al.* (2003), while only two measurements of the density are available in the literature so far (*Lacerda and Jewitt,* 2006; *Marchis et al.,* 2006a,b). On the basis of this still incomplete sample of information, the population of Jupiter Trojans shows some similarities, together with some differences, with the other populations of minor bodies of the outer solar system. Comparison among the physical and dynamical properties of the Jupiter Trojans, and those of Centaurs, TNOs, and outer dwarf planets, although challenging, is necessary to constrain the scenario of the formation and early evolution of the outer part of the solar system, and give an answer to the still open questions of where these bodies formed and how they evolved.

2. ORIGIN AND POSSIBLE DYNAMICAL LINK BETWEEN JUPITER TROJANS AND TRANSNEPTUNIAN OBJECTS

There are two models for the origin of Jupiter Trojans. Each model has distinct implications for the composition of these objects, and therefore distinct implications for similarities and differences between Trojans and TNOs.

The first model, which we will call "classical" as it remained unchallenged until 2005, considers that the Trojans originally were planetesimals formed in the vicinity of Jupiter's orbit. They were captured on tadpole orbits (namely on orbits that librate around the Lagrange equilateral equilibrium points L_4 and L_5) when Jupiter's gravity abruptly increased due to the accretion of a massive atmosphere (pull-down mechanism) (*Marzari and Scholl,* 1998a,b; *Fleming and Hamilton,* 2000). Assuming a time evolution of Jupiter's mass as in *Pollack et al.* (1996), *Marzari and Scholl* (1998a,b) showed with numerical simulations that this capture mechanism is very efficient: Between 40% and 50% of the planetesimals populating a ring extending 0.4 AU around Jupiter's orbit can be captured as Trojans. After capture, the angular amplitude of libration shrinks by a factor $(M_J = M_{J,c})^{-1/4}$, as Jupiter's mass continues to grow, M_J and $M_{J,c}$ denoting the mass of Jupiter at the current time and at the time of capture, respectively (*Fleming and Hamilton,* 2000). Gas drag could also help in the capture of Trojans from the local planetesimal population, but is effective only for the small objects (*Peale,* 1993).

The problem with the classical model is that the resulting orbital distribution of the Trojans is not, at first sight, very similar to the observed one. The captured Trojans typically have large libration amplitudes, despite the partial damping process mentioned above. Conversely, the observed objects have a fairly uniform libration amplitude distribution. Using a Monte Carlo method, *Marzari and Scholl* (1998b) showed that collisions can significantly alter the distribution of libration amplitudes by injecting initially large librators into more-stable, small-libration-amplitude orbits. In addition, the Trojans with the largest libration amplitudes would tend to escape by chaotic diffusion over the age of the solar system (*Levison et al.,* 1997). Thus, the libration amplitude distribution resulting from the pull-down mechanism might be reconciled with the observed distribution, invoking the subsequent collisional and dynamical evolution.

A more serious disagreement concerns the inclination distributions. The pull-down mechanism does not significantly affect the eccentricities and the inclinations of the planetesimals. Thus the eccentricity and inclination distributions of the captured Trojans should be reminiscent of those of the planetesimal disk. Because the disk was dynamically stirred by the presence of Jupiter's core, the eccentricities and inclinations of the local planetesimals are not expected to be very small. However, they are not expected to be large either, because the bodies kicked to large eccentricity/inclination orbits by encounters with the proto-Jupiter were most likely displaced (in semimajor axis) from Jupiter's orbit, so that they could not be captured by the pull-down mechanism. The observed eccentricity distribution of Trojans ranges from 0 to ~0.15 [the latter being a sort of dynamical stability limit (*Rabe,* 1965; *Levison et al.,* 1997; *Robutel and Gabern,* 2006)], so it might not be a problem. However, the observed inclination distribution ranges up to about 40°, well beyond expectations from the local capture model. *Marzari and Scholl* (2000) showed that the inclination can be excited up to 20°–30° by the secular resonance ν_{16}, which occurs when the longitude of the node of a Trojan precesses at the same rate of those of Jupiter and Saturn (which are equal to each other). The problem is that this resonance operates only on Trojans with a libration amplitude of about 30° (here the libration amplitude is defined as the half-difference between the minimal and the maximal value of $\lambda_T - \lambda_J$, where λ denotes the mean longitude of a body and the subscript T and J refer to the Trojan and Jupiter, respectively). Another possibility, in analogy with the asteroid belt excitation/depletion model of *Wetherill* (1992) and *Petit et al.* (2001), is that the primordial Trojan population contained massive planetary embryos, which excited the inclinations of the smaller objects by repeated encounters, up to the time when they eventually escaped from the Trojan region due to their mutual interactions. The problem with this model is that, because the inclination distribution of the Trojans around the L_4 and L_5 points are similar, almost equal populations of embryos should have orbited in the two Trojan regions, and for about the same time, which seems unlikely from a probabilistic point of view.

A full simulation of the Trojan capture process by the pull-down mechanism, including the effects of collisional damping, secular resonance excitation, and/or the presence of massive planetary embryos, has never been done. Thus, it has never been shown that the local capture model can satisfactorily reproduce the orbital distribution of the observed Trojans.

An alternative model for the origin of the Trojans has been recently proposed (*Morbidelli et al.,* 2005). It also invokes the capture of Trojans, but from a more distant disk. The latter should be identified with the primordial transneptunian disk, which is also the ancestor of the Kuiper belt. The model assumes that initially Saturn was closer to Jupiter than their mutual 1:2 mean-motion resonance, and invokes the well-known migration of the giant planets due to their interaction with the disk of planetesimals (*Fernández and Ip,* 1984; *Malhotra,* 1993, 1995; *Hahn and Malhotra,* 1999, 2005; *Gomes et al.,* 2004). During their migration in divergent directions, Jupiter and Saturn eventually had to cross the 1:2 resonance. It is known from *Gomes* (1997) and *Michtchenko et al.* (2001) that, if and when this happened, the jovian Trojan region had to become fully unstable. Consequently, any preexisting jovian Trojans would have left the coorbital region.

However, the dynamical evolution of a gravitating system of objects is time-reversible. Thus, if the original objects can escape the Trojan region when the latter becomes unstable, other bodies can enter the same region and be temporarily trapped. Consequently, a transient Trojan population can be created if there is an external source of objects. In the *Morbidelli et al.* (2005) scenario, the source consists of the very bodies that are forcing the planets to migrate, which must be a very large population given how much the planets had to move. When Jupiter and Saturn get far enough from the 1:2 resonance, so that the coorbital region becomes stable again, the population that happens to be there at that time remains trapped. It becomes the population of permanent jovian Trojans still observable today.

This possibility has been tested with numerical simulations in *Morbidelli et al.* (2005). Among the particles that were Jupiter or Saturn crossers during the critical period of Trojan instability, between 2.4×10^{-6} and 1.8×10^{-5} remained permanently trapped as jovian Trojans. Given the mass of the planetesimals that is required in order to move Jupiter and Saturn over the semimajor axis range corresponding to Trojan's instability, this corresponds to a captured Trojan population of total mass between $\sim 4 \times 10^{-6}$ and $\sim 3 \times 10^{-5}$ $M_\oplus$. Previous estimates (*Jewitt et al.,* 2000) from detection statistics concluded that the current mass of the Trojan population is $\sim 10^{-4}$ $M_\oplus$. However, taking into account modern, more-refined knowledge of the Trojans' absolute magnitude distribution (discussed in *Morbidelli et al.,* 2005), mean albedo (*Fernández et al.,* 2003), and density (see section 3.3), the estimate of the current mass of Trojan population is reduced to 7×10^{-6} $M_\oplus$, consistent with the mass achieved in the capture simulations.

More importantly, at the end of the simulations, the distribution of the trapped Trojans in the space of the three fundamental quantities for Trojan dynamics — the *proper* eccentricity, inclination, and libration amplitude (*Milani,*

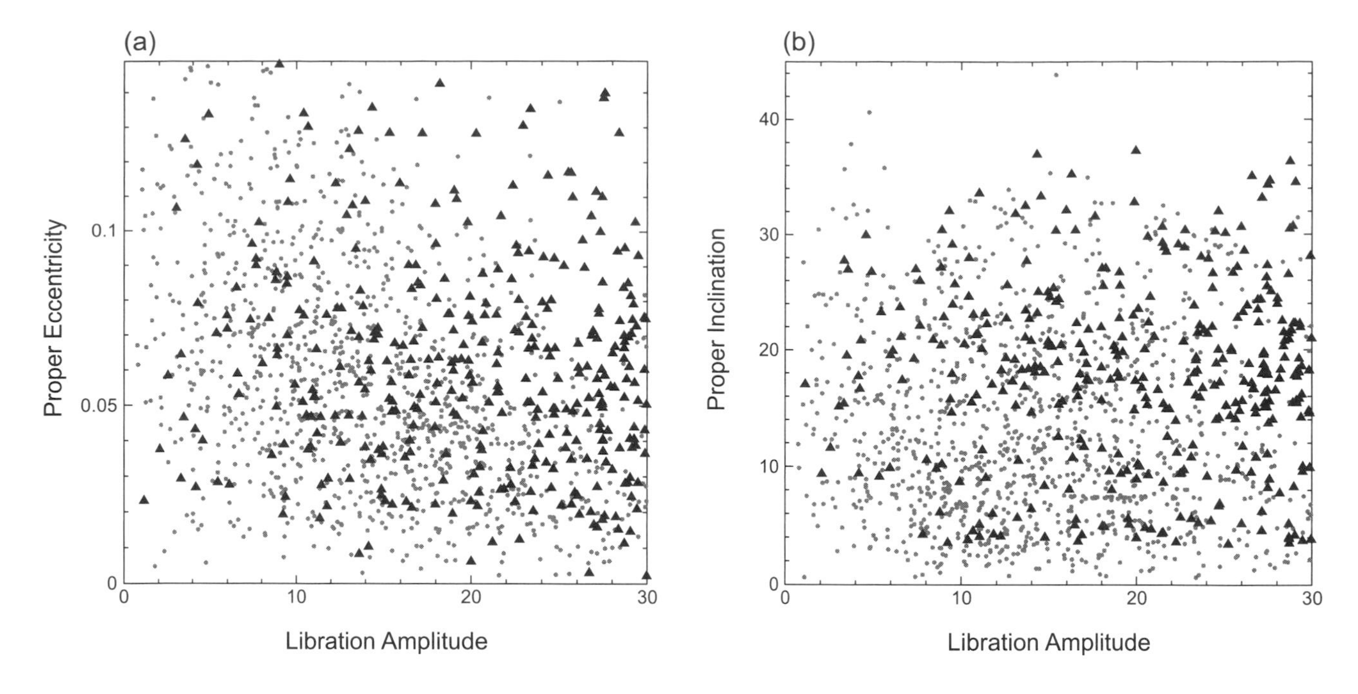

Fig. 1. Comparison of the orbital distribution of Trojans between the simulations in *Morbidelli et al.* (2005) and observations. The simulation results are shown as black triangles and the observations as gray dots in the planes of **(a)** proper eccentricity vs. libration amplitude and **(b)** proper inclination vs. libration amplitude. The distribution of the simulated Trojans is somewhat skewed toward large libration amplitudes, relative to the observed population. However, this is not a serious problem because a fraction of the planetesimals with the largest amplitudes would leave the Trojan region during the subsequent 4 G.y. of evolution (*Levison et al.,* 1997), leading to a better match. The similarity between the two inclination distributions provides strong support for this model of the origin of Trojans.

1993) — was remarkably similar to the current distribution of the observed Trojans, as illustrated in Fig. 1. In particular, this is the only model proposed thus far that explains the inclination distribution of jovian Trojans. This model also predicts that, before being captured in the Trojan region, the objects typically evolved through a large eccentricity phase that brought them relatively close to the Sun. In fact, in the simulations all particles reached temporarily perihelion distance q less than ~3 AU before capture. Of them, 72% spent more than 10,000 yr on orbits with $q < 3$ AU, and 68% even reached $q < 2$ AU. Since it may take roughly 10,000 yr for an active Jupiter-family comet to become dormant (*Levison and Duncan,* 1997), it is possible that the surfaces of the Trojans could have been devolatilized during their high-eccentricity phase. We will return to this issue in section 4.

The positive results of the *Morbidelli et al.* (2005) simulations provide by themselves a strong argument in favor of the passage of Jupiter and Saturn through their mutual 1:2 mean-motion resonance. Additional support comes from the fact that this transition through the resonance explains the orbital excitation of the giant planets' orbits, starting from perfectly circular ones (*Tsiganis et al.,* 2005). Moreover, *Gomes et al.* (2005) showed that, with reasonable assumptions, the passage through the resonance could have occurred after hundreds of millions of years of slow planetary migration, which provides a mechanism for the origin of the otherwise mysterious late heavy bombardment of the terrestrial planets (see *Hartmann et al.,* 2000, for a review). Finally, the orbital architecture of the Kuiper belt also seems to be consistent with the orbital evolution of the planets subsequent to the 1:2 mean-motion resonance crossing (see chapter by Morbidelli et al.). Therefore, the strength of *Morbidelli et al.* (2005) scenario is that it is cast in a more general framework, which is consistent with a large body of constraints given by the solar system's structure.

3. THE PHYSICAL PROPERTIES OF JUPITER TROJANS

As mentioned above, the analysis of the physical similarities and differences between Jupiter Trojans, Centaurs, and TNOs is of fundamental importance to investigations of the possible link among these populations of the outer solar system.

3.1. Rotational Properties

Planetary rotation is the result of the angular momentum added by mutual collisions to the initial angular momentum determined by formation processes. For this reason measurements of rotational properties can provide important clues about the history and evolution of the small-body population. Even though radar and adaptative optics have recently emerged as powerful sources of information, lightcurve observations still represent the basic tool for determining the rotational properties of small bodies, allowing determination of the rotation rate, axis direction, and an approximation of the body shape.

Starting in 1969, Dunlap and Gehrels observed 624 Hektor and revealed a body with a very elongated shape. *Hartmann et al.* (1988) published lightcurves of 18 Trojans, which, on average, had higher amplitudes than main-belt asteroids. They suggested that elongated shapes are characteristic of Trojans, possibly reflecting a difference in composition and collisional evolution with respect to the main-belt asteroid population. *Binzel and Sauter* (1992) also reported the presence of high lightcurve amplitudes from a sample of 31 objects. In particular, they found that the amplitudes were significantly larger than in the main belt but only for objects larger than about 90 km. *Barucci et al.* (2002a) reported the results of a large survey obtained by a team of observers on an unbiased sample of 72 Trojans down to an absolute magnitude H ~ 10.2. Combined with existing data, these increase the number of known periods and amplitudes to 75 Trojans, most of which are in the diameter range 70–150 km. The mean rotation frequency of this sample (f = 2.14 ± 0.12 rev/day) is statistically indistinguishable from the main belt (f = 2.26 ± 0.14 rev/day), but the Trojan distribution is well fit by a Maxwellian, unlike the main belt.

The few known spin axes of Trojans seem to be randomly distributed, and include both prograde- and retrograde-sense rotations (*Barucci et al.,* 2002a). All these results suggest that the Trojan population has undergone a higher degree of collisional evolution than the main belt.

The lightcurve amplitude also gives some indication of the elongation of the body. Assuming a triaxial ellipsoid shape with semiaxes a > b > c and no albedo variation, the estimation of the lower limit of the semiaxis ratio can be derived

$$a/b = 10^{0.4\Delta m}$$

The amplitude, however, varies considerably depending on the unknown aspect angle under which the observations are made, with the amplitude being largest for an equatorial aspect, and smallest with a polar aspect. The simple inversion tends to underestimate the maximum amplitudes, and therefore the a/b ratio, for objects that are observed only once. This ambiguity can be removed by obtaining lightcurves at multiple epochs. Nevertheless, even reducing the amplitudes to the aspect of 60° in the case of multiple observations to eliminate bias effects, the Trojans appear (at the 99% confidence level) to have a larger mean amplitude than the main-belt objects. This implies more elongated shapes for the Trojan population.

3.2. Albedo and Diameters

Albedo and diameters of Jupiter Trojans are still relatively poorly known. The widest data sample in this field was published by *Fernández et al.* (2003). On the basis of midinfrared and visible observations, they radiometrically derived V-band geometric albedo and radii of 32 Jupiter Trojans. No statistically significant correlation between albedo and radius has been found. The midinfrared colors seem to support a thermal behavior of "slow rotators" with a thermal inertia no greater than about half that of the Moon and similar to the limits found for some of the Centaurs. Figure 2 shows the albedo distribution of Jupiter Trojans, together with those of active comets and dead comet candidates. The mean value, as well as the standard deviation, depends on the value of an empirical term, η, called the beaming parameter, which enters into the equations for the temperature of the asteroid's surface. The beaming parameter acts as a proxy of both surface roughness and thermal inertia (a measure of the resistance of the surface to changes in temperature), and can vary over factors of several (see additional discussion in chapter by Stansberry et al.). Lack of knowledge of the albedo and η leads to uncertainty in the surface temperature, and dominates the uncertainties in size as estimated from a thermal flux measurement. In the case of Jupiter Trojans, with η = 0.94, the computed mean albedo value is 0.041 ± 0.002 (with a standard deviation of 0.007), while with the standard value of 0.756 the mean albedo increases to 0.056 ± 0.003. The albedo distribution, found by *Fernández et al.* (2003), is narrower than the one derived from IRAS measurements. With a beaming parameter close to 1 it becomes consistent with that of comets, but it does not match the albedo values of Centaurs and TNOs published since 2003 (reported in the chapter by Stansberry et al.). According to *Fernández et al.* (2003) this could imply that the Jupiter Trojan surfaces are probably more like those of active and post-active comets, than like those of the pre-active ones (e.g., Centaurs). Only one object, 4079 Ennomos, has been found to have a high albedo, about 14σ away from the mean value. This could be due to the presence on the surface of this body of pristine ices, excavated from a subsurface layer by a recent collision. Alternatively, Ennomos could have a more "standard" albedo but a very unusual thermal inertia.

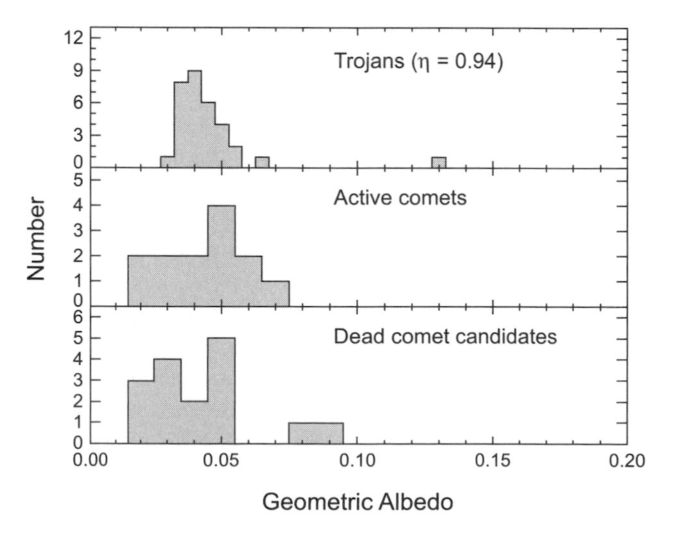

Fig. 2. Comparison of Trojan albedos with those of active comets and dead comet candidates (by *Fernández et al.,* 2003).

3.3. Density

Whereas the aspects treated in the previous sections concern only the surface properties, the densities can give information on the similarities (or differences) between the objects' bulk physical properties. However, one should keep in mind the caveat that bodies with related origin can have different densities. For instance, in the Kuiper belt, the density of Pluto is larger than that of Varuna, because the former probably underwent some sort of differentiation and lost some volatile material.

We know the bulk density of two Trojan objects: 617 Patroclus and 624 Hektor. Patroclus is the first discovered binary Trojan (*Merline et al.,* 2001). A preliminary orbital determination allowed a first estimate of a bulk density of ~1.3 g/cm^3 (*Merline et al.,* 2002; *Noll,* 2006), similar to that of C-type asteroids (see *Britt et al.,* 2002, for a review). *Marchis et al.* (2006a) have taken several observations of Patroclus using the laser guide star adaptive optics at Keck. These observations provided better measurements of the period and the orbital distance of the two components, which in turn allows the determination of the masses. Using the thermal measurements of *Fernández et al.* (2003) to estimate the sizes of the components, *Marchis et al.* (2006a) concluded that the bulk density of this object is ~0.8 g/cm^3. This is a very low density, compared to any other asteroid known so far. However, it is close to the bulk densities of 1–2 g/cm^3 inferred for Kuiper belt objects (*Jewitt and Sheppard,* 2002; *Lacerda and Luu,* 2006).

Recent observations of 624 Hektor with adaptive optics techniques have revealed a moonlet orbiting a primary that appears to be a close-in or contact binary (*Marchis et al.,* 2006b). Masses determined from the orbit of the moonlet and from modeling the stability of the binary primary agree, and both result in a density of ~2.4 g/cm^3 for Hektor.

The densities of Trojans appear to span a broad range that includes estimated densities of main-belt asteroids [0.6–3.8 g/cm^3 (*Noll,* 2006)], comet nuclei [0.1–1.5 g/cm^3 (chapter by Lowry et al.)], and TNOs [0.6–2.5 g/cm^3 (chapter by McKinnon et al.)]. This range of densities may indicate that the Trojans are a mixture of objects from different source populations, or could reflect the collisional environment in the Trojan swarms (i.e., some low-density rubble piles and some coherent impact fragments or undisrupted primitive bodies).

3.4. V + NIR Photometry and Spectroscopy

The physical properties and the surface composition of Jupiter Trojans are not at present well known. Visible photometry is available from SDSS data for about 300 objects (*Szabó et al.,* 2007). Visible spectra are available for less than 150 objects (*Jewitt and Luu,* 1990; *Vilas et al.,* 1993; *Fitzsimmons et al.,* 1994; *Lazzarin et al.,* 1995; *Bendjoya et al.,* 2004; *Fornasier et al.,* 2004; *Lazzaro et al.,* 2004; *Fornasier et al.,* 2007), while near-infrared spectra have been published for about 50 bodies (*Jones et al.,* 1990; *Luu et al.,* 1994; *Dumas et al.,* 1998; *Cruikshank et al.,* 2001; *Emery and Brown,* 2003; *Dotto et al.,* 2006; *Yang and Jewitt,* 2006). All the observed Trojans appear spectrally featureless: The large majority of them can be classified in the asteroid taxonomy (*Tholen and Barucci,* 1989) as belonging to the D class, but P and C types are also present among them. In particular, no indication of hydration bands, as seen in some asteroids (*Vilas and Gaffey,* 1989; *Vilas et al.,* 1994) and in several small bodies of the solar system (see review by *de Bergh et al.,* 2004), is in the Trojan spectra. The visible spectral slopes range from –1% to 25%/10^3 Å, while for TNOs the visible slopes span between –1% and 55%/10^3 Å (chapter by Barucci et al.).

Although Jupiter Trojans are believed to be formed in a region rich in frozen volatiles, water ice is still undetected in their spectra. *Emery and Brown* (2003, 2004) published 0.3–4.0-μm spectra of 17 bodies and also presented models of the surface composition (see Fig. 3). They did not detect water ice and hydrated silicate features in their V + NIR spectra and they estimated upper limits of a few percent and up to 30% respectively for these materials at the surface. More recently, *Yang and Jewitt* (2006) published near-infrared spectra and models of the surface composition of five Jupiter Trojans, assessing at less than 10% the total amount of water ice present on their surface.

Several mechanisms can be invoked to explain this lack of water ice on the surface of the observed objects (assuming they contained water ice to begin with). Laboratory experiments have shown that space-weathering processes on the icy surfaces of atmosphereless bodies can produce an irradiation mantle spectrally red and with low albedo (*Moore et al.,* 1983; *Thompson et al.,* 1987; *Strazzulla,* 1998; *Hudson and Moore,* 1999). In the scenario suggested by *Morbidelli et al.* (2005), Jupiter Trojans could have been devolatized during their high-eccentricity phase, when cometary activity should have been intense. Alternatively, they could have formed a dust mantle as suggested by *Tancredi et al.* (2006) for kilometer-sized comet nuclei. Therefore, water ice, originally present on the surface of Jupiter Trojans, would be now completely covered and ice signatures would be now detectable only if inner fresh material would be exposed by recent collisions.

Unfortunately, the observations of Jupiter Trojans belonging to dynamical families have shown no spectral features related to the presence of ices on the surface of the observed bodies. *Dotto et al.* (2006) published visible and near-infrared (0.5–2.5 μm) spectra of 24 Jupiter Trojans belonging to dynamical families, also presenting models of the surface composition (as an example, Fig. 4 shows the spectra obtained for the Makhaon family). The most important characteristic they found is the uniformity of the Trojan population. All the investigated dynamical families appear quite similar in surface composition, without any peculiar difference. No relation exists between spectral properties and dimensions of the bodies, and some small differences in the spectral behaviors can be explained by different degrees of space-weathering alteration. All the investigated

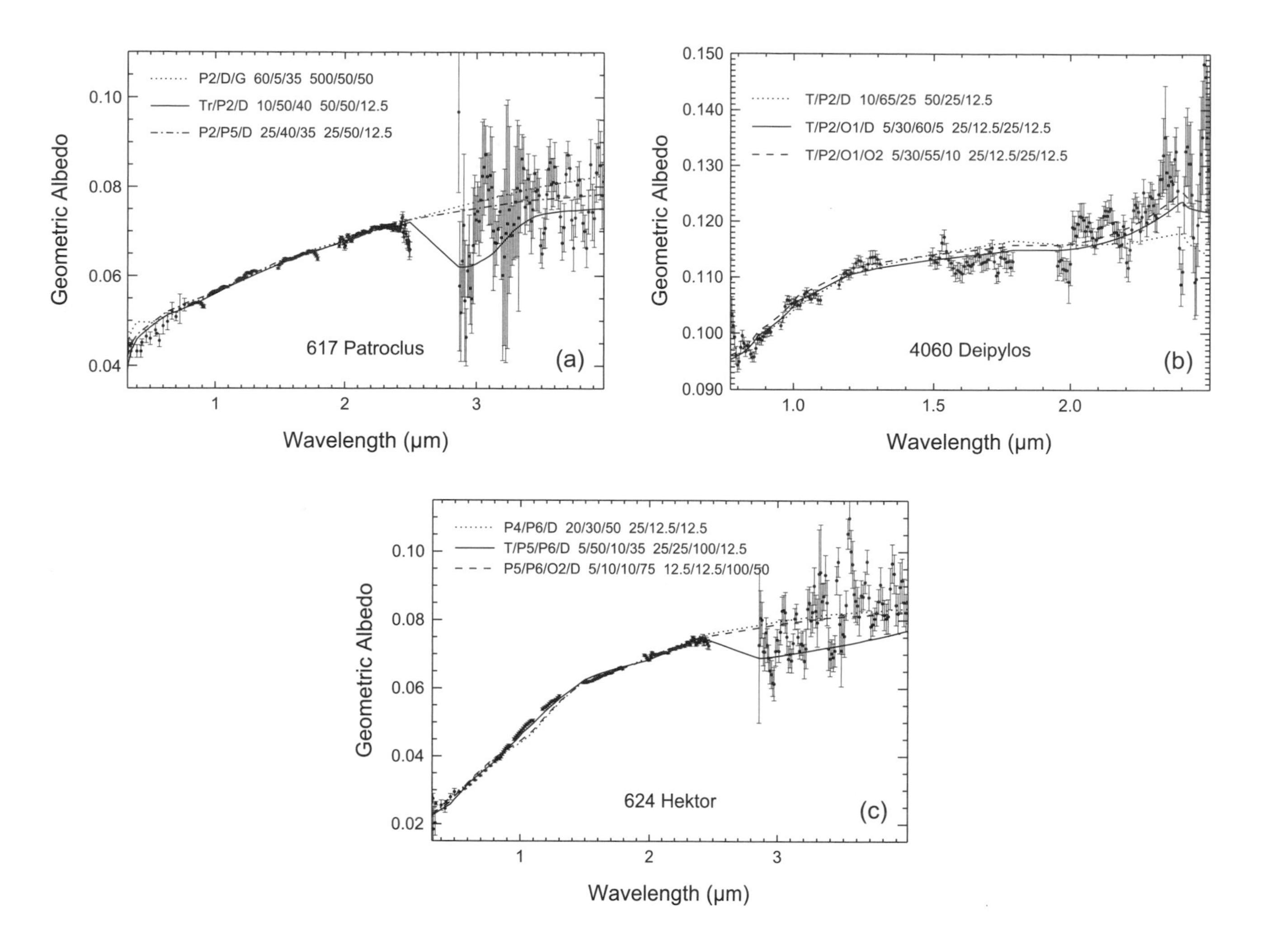

Fig. 3. V + NIR spectra and models of the surface composition of three Jupiter Trojans. The one- or two-digit codes represent the following materials: P2 to P8 — glassy silicates with pyroxene stoichiometry, O1 and O2 — glassy silicate with olivine stoichiometry, D — amorphous carbon, G — graphite, T — Titan tholin. The first set of numbers following the codes are mixing ratios, and the second set are grain diameters in micrometers (see *Emery and Brown,* 2004, for more details).

Trojans have featureless spectra. No diagnostic features that would enable distinguishing the family members from the background objects of the Trojan population have been detected. Importantly, no signatures of water ice have been observed in the spectra of these bodies.

Fornasier et al. (2007), analyzing the spectral slopes of Jupiter Trojans as a function of the orbital elements, found a color-inclination trend with bluer objects at lower i. In their sample this trend is completely dominated by the L_4 Eurybates family, a compact core inside the Menelaus family (*Beaugé and Roig,* 2001) (see also the P.E.Tr.A. Project at *www.daf.on.br/froig/petra/*). Eurybates constitutes a peculiar case among the families analyzed so far, since the spectral behavior of its members is quite homogeneous: The spectral slopes are strongly clustered around S = 2%/10^3 Å, with the highest S values corresponding to the smaller objects (D < 25 km). The visible spectra of the Eurybates family members (see Fig. 5) are very similar to those of C-type main-belt asteroids, of the less-red Centaurs, and of cometary nuclei. This family could be produced by the fragmentation of a very peculiar parent body, whose origin must be still assessed or, alternatively, could be an old family, where space-weathering processes have flattened all the spectra, covering any original differences in composition among the different members. In this last case we would have the first observational evidence of objects whose spectra have been flattened by space-weathering processes and, according to the scenario suggested by *Moroz et al.* (2004), the composition of the parent body of such a family would have been rich in complex hydrocarbons. Unfortunately, we do not know the age of this dynamical family and we do not have infrared spectra of Eurybates members. Further observations in an enlarged wavelength range and numerical simulations are absolutely needed to investigate and definitively assess the nature and the origin of this very peculiar family. Dynamical families belonging to the L_4 swarm seem to have a more heterogeneous composition than those of the L_5 swarm, since a higher presence of C and P types is

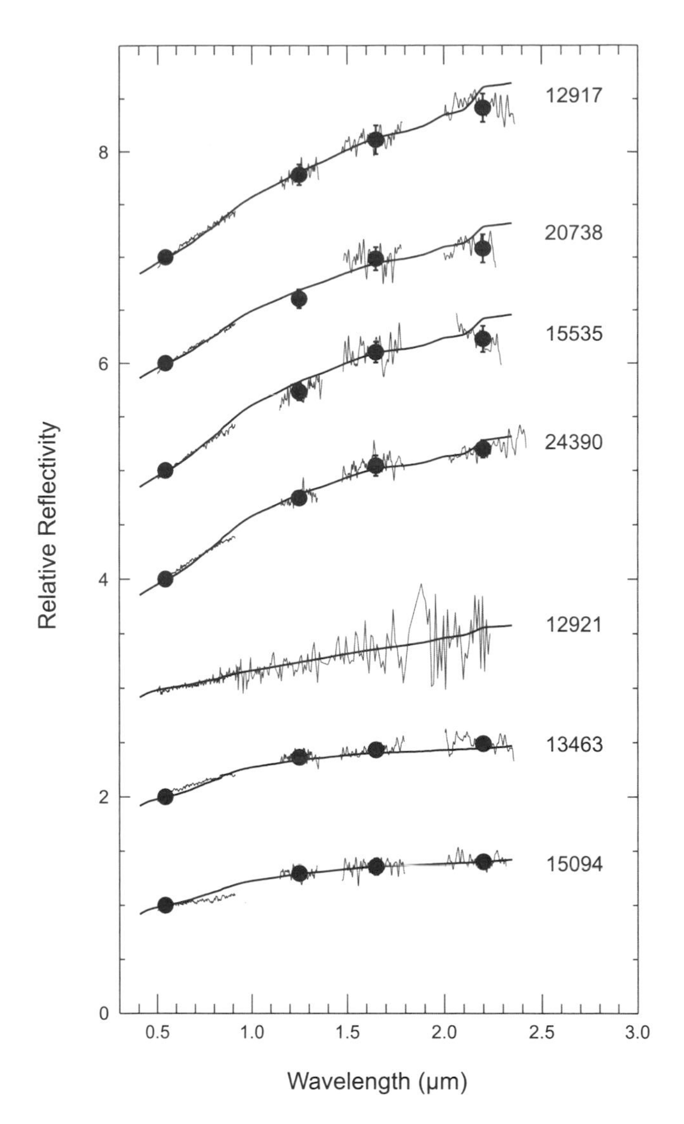

Fig. 4. V + NIR spectra and models of the surface composition of the members of the Makhaon family (by *Dotto et al.,* 2006). All the spectra are normalized at 0.55 μm and shifted for clarity.

observed among the L_4 objects (*Fornasier et al.,* 2007). Moreover, the dynamical families belonging to the L_4 cloud are more robust than those of the L_5, surviving as densely populated clustering at low relative velocity cutoff. This could suggest that the L_4 cloud is more collisionally evolved than the L_5, but it is still too early to give an interpretation of this in terms of the composition of the two cloud populations, since we cannot exclude that still unobserved C- and P-type families are present also in the L_5 cloud.

Szabó et al. (2007), on the basis of the SDSS observations, found also that the color of Trojans is correlated with the orbital inclination (with redder objects and larger inclination) and did not detect any difference between the L_4 and L_5 swarms.

Fornasier et al. (2007) also performed a comparison of the whole sample of Jupiter Trojans' spectral slopes and (B, V, R, I, J, H, and K) colors available in the literature, with those of small bodies of the outer solar system: comets, scattered TNOs, classical disk objects, and Plutinos. They found that the Trojan mean colors are compatible with those of the short-period comets. Nevertheless, the widths of their color distributions are incompatible, as well as the shapes of the distributions. The compatibility in color is possibly caused by the small size of the short-period comet sample rather than by a physical similarity. Trojans do not have any of the ultrared slopes seen on many Centaurs and TNOs. Their average colors are fairly similar to those of the neutral/less-red Centaurs, but the overall distributions are not compatible.

3.5. Thermal Emission Observations

The majority of measurements of thermal emission from Trojan asteroids are broadband photometric observations for the purpose of determining sizes and albedos, as described in section 3.2. Spectroscopic observations of Trojans in the thermal-IR have not been possible from the ground due to strong telluric absorptions, bright and rapidly varying sky background, and the inherent faintness of Trojans due to their distance from the Sun. The sensitivity of the ISO satellite was also insufficient for thermal-IR spectroscopy of Trojans. The Infrared Spectrograph (IRS) on the Spitzer Space Telescope is more sensitive, however, and *Emery et al.* (2006a) have recently reported 5.2–37-μm thermal emission spectroscopy of three Trojan asteroids: 624 Hektor, 911 Agamemnon, and 1172 Aneas.

The flux density at each wavelength measured by IRS (also called the spectral energy distribution, or SED) de-

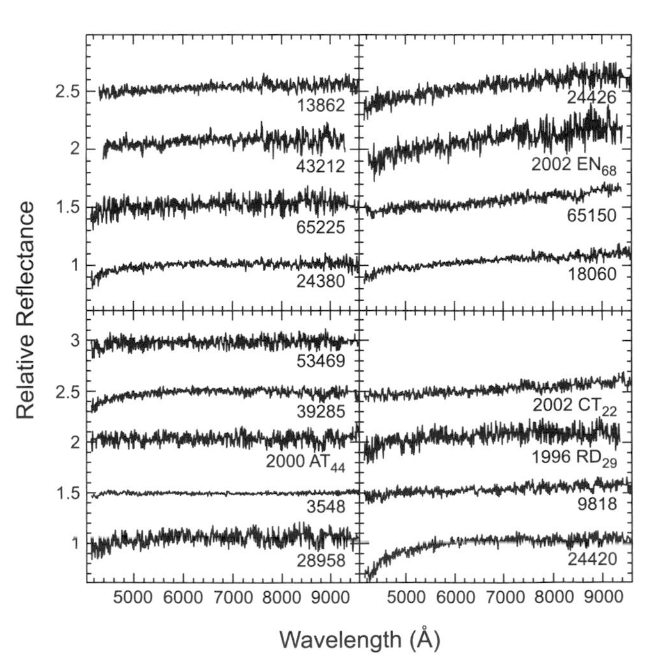

Fig. 5. Visible spectra of the members of the Eurybates family (by *Fornasier et al.,* 2007). All the spectra are normalized at 0.55 μm and shifted for clarity.

pends on the Trojan's size, composition, and surface temperature distribution. This last term is in turn dependent on several factors, including distance from the Sun, albedo, surface roughness, and thermal inertia. Spectral features in the SED are superposed on the thermal continuum, which must be removed with the use of a physical model. *Emery et al.* (2006a) employed a modified version of the standard thermal model (STM) as well as a more advanced thermophysical model, which includes the effects of thermal inertia. Fits of these models to the data result in estimates of physical parameters such as size, albedo, and thermal inertia (see chapter by Stansberry et al.). The sizes and albedos derived from the IRS spectra are in agreement with previous estimates. The Trojan data are consistent with zero thermal inertia, although the thermophysical models allow thermal inertias of these Trojan asteroids of up to about 5 J m^{-2} $s^{-1/2}$ K^{-1} for reasonable values of surface roughness (in these units, thermal inertia is ~50 for the Moon, ~15 for large main-belt asteroids, and ~2500 for bare rock).

Emissivity spectra are derived by dividing the measured SED by the modeled thermal continuum. The emissivity spectra of Hektor, Agamemnon, and Aneas are shown in Fig. 6. Compositional features evident in these spectra include an emission plateau at about 9.1–11.5 μm and a broader emission high from about 18–28 μm. More subtle features include possible peaks near 19 μm and near 24 μm and another emissivity rise near 34 μm. The Trojan spectra broadly resemble the emissivity spectra of some carbonaceous meteorites and fine-grained silicates (Fig. 7).

Coarse-grained silicates exhibit emissivity lows instead of highs near 10 and 20 μm, and therefore cannot explain the data. Upon closer comparison with fine-grained materials, however, several differences are also apparent: The 10-μm plateau is narrower for the Trojans, their spectra do not rise as sharply near 15 μm. Emissivity spectra of two low-albedo main-belt asteroids (10 Hygiea and 308 Polyxo) from ISO exhibit similarly narrow 10-μm emission plateaus (*Barucci et al.,* 2002b; *Dotto et al.,* 2004). No minerals in available spectral libraries resolve these differences, nor do linear mixtures of up to five components.

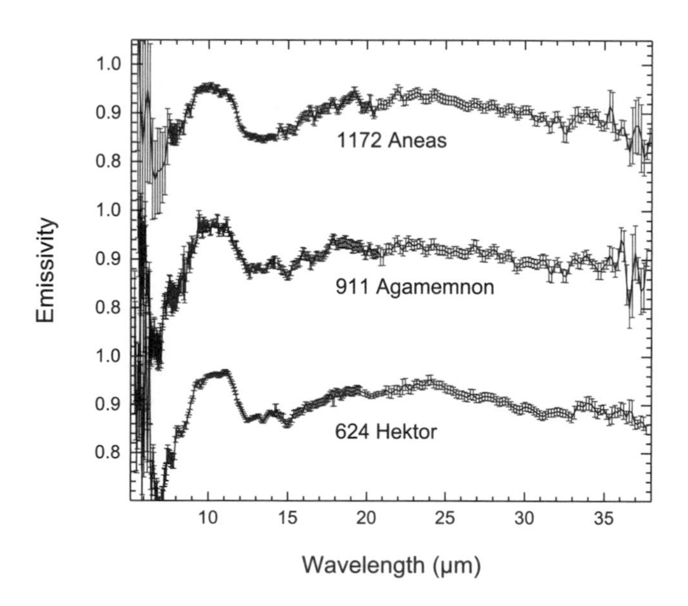

Fig. 6. Emissivity spectra of Aneas, Agamemnon, and Hektor from *Emery et al.* (2006a). The shortest wavelength portion (λ < 7.5 μm) of the Aneas spectrum has been binned by a factor of 5 to improve the S/N.

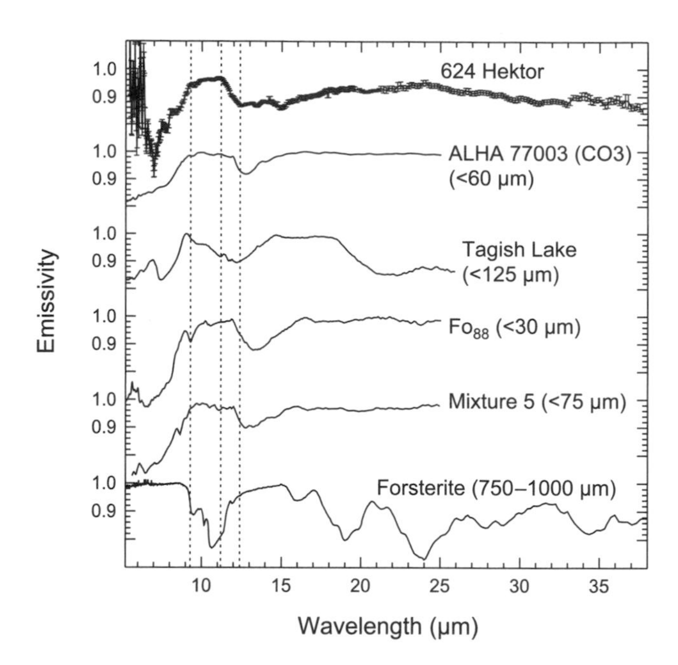

Fig. 7. Emissivity spectra of meteorites, minerals, and mixtures. Grain sizes are listed in parentheses. Mixture 5 is 63% Fo_{92} + 28% enstatite + 4% muscovite + 2% pyrophyllite + 3% calcite.

Emery et al. (2006a) suggest three hypotheses for the differences between the Trojan asteroid data and expected regolith emissivity spectra. The first is that the Trojans support comet-like comae. This hypothesis is rejected because no comae are apparent in deep optical and thermal-IR images of these objects. The second possibility is that a fine-grained, low-density regolith with a fairy castle structure emits in a manner similar to an extended coma, perhaps from extreme porosity. The third hypothesis is that fine-grained silicates are imbedded in a matrix of material that is relatively transparent in the midinfrared. Both of these latter two hypotheses imply a significant fraction of fine-grained silicates on the surfaces, and the last requires an additional matrix material. The presence and spectral dominance of silicates on the surfaces of Trojan asteroids is consistent with some modeling of V + NIR reflectance spectra of Trojans (*Cruikshank et al.,* 2001; *Emery and Brown,* 2004) that rely on silicates rather than organics to provide the red spectral slope, but *Emery et al.* (2006b) show that extension of the V + NIR models to the thermal-IR does not match the measured Trojan emissivity spectra. Additional analysis exploring the effects of surface structure (porosity and particles embedded in a transparent matrix) while simultaneously conforming to the constraints imposed by both V + NIR reflectance spectra and the thermal-IR emissivity spectra will likely provide further insight into the composition of Trojan surfaces.

4. DISCUSSION AND CONCLUSION

On the basis of the presently available data, it seems that Jupiter Trojans constitute a very homogeneous population whose members have featureless and neutral to moderately red V + NIR spectra. No differences have been found among members of dynamical families and background objects and no relations have been found among spectral properties and dimensions. A very peculiar case is given by the Eurybates family, which shows a peculiar abundance of spectrally flat objects, similar to C-type main-belt asteroids or to the neutral/less-red Centaurs. A correlation seems to exist between colors and inclinations, with redder objects at higher inclinations. The presence of more robust dynamical families in the L_4 cloud seems to suggest that it is more collisionally evolved than the L_5.

According to the most recent models on the origin and early evolution of Jupiter Trojans, Centaurs, and TNOs, some similarities among the different populations of minor bodies of the outer solar system would be expected. Nevertheless, the rotational properties, albedo distributions, and spectral characteristics of Jupiter Trojans are only partially in agreement with the analogous characteristics of the other populations. Jupiter Trojans are characterized by larger-amplitude lightcurves, implying elongated shapes and a higher degree of collisional evolution compared to the population of larger TNOs (see chapter by Sheppard et al.), but this is probably just a size effect. Albedo, color, and visible spectral slope distributions of Jupiter Trojans are very different from those of TNOs, comets, and Centaurs: Jupiter Trojans are among the less-red objects within these populations and have the narrowest color distribution. While their mean colors are compatible with those of the short-period comets, the width of their color distributions is not, nor is the shape. Likewise, the Trojan average colors are similar to those of the neutral/less-red Centaurs, but the overall color distributions are not statistically compatible. Unfortunately, V + NIR reflectance spectra of Trojan asteroids do not exhibit absorption features that would provide direct clues to their surface composition. These featureless spectra rule out significant amounts of water ice, hydrated silicates, crystalline anhydrous silicates such as those apparent on main-belt asteroids, certain organic materials (those with strong K- and L-band absorptions, typically from many aliphatic bonds), and other simple ices (e.g., CH_3OH, H_2O, CH_4, SO_2). But for determination of specific composition, we are left to ask what materials produce featureless spectra with neutral to moderately red spectral slopes. *Gradie and Veverka* (1980) suggested macromolecular organic materials, and subsequent spectral modeling has proven that such materials do an excellent job of matching Trojan asteroid spectra shortward of 2.5 μm (e.g., *Dotto et al.,* 2006). However, *Emery and Brown* (2004) reported that they could not simultaneously match the red spectral slope at $\lambda < 2.5$ μm and the absence of absorptions in the L band (2.8–4.0 μm) with these organics (tholins), and conclude that organics cannot be responsible for the red spectral slopes unless some organic exists that has a red slope, but no L-band absorptions. They used amorphous silicates to model the red slopes of Trojans, but *Emery et al.* (2006b) noted that these models do not successfully reproduce midinfrared emissivity spectra of Trojans.

Fine-grained silicates have recently been detected in thermal emission spectra of three Trojan asteroids (*Emery et al.,* 2006a). These represent the first discrete mineralogical signatures detected for Trojans. Models of emissivity spectra of regoliths are not yet sophisticated enough to determine specific silicate mineralogy, but the Trojan emissivity spectra between 5.2 and 37 μm are qualitatively similar to those of comets (e.g., *Crovisier et al.,* 1997; *Stansberry et al.,* 2004; *Lisse et al.,* 2006) and some Centaurs (chapter by Barucci et al.), but distinct from many main-belt asteroids (*Emery et al.,* 2005). The grain size for the Trojan silicates (less than a few micrometers) is smaller than can be modeled with the techniques generally used to model reflectance spectra (*Hapke,* 1981, 1993; *Shkuratov et al.,* 1999) (both based on geometric optics). This indicates that new techniques are necessary for proper modeling of reflectance data.

As already noted by *Fernández et al.* (2003), it seems that Jupiter Trojans are more similar to the active and post-active comets than to the non-active icy bodies (Centaurs and TNOs). This is compatible with the scenario suggested by *Morbidelli et al.* (2005), where Jupiter Trojans, before being captured in the region where they currently reside, temporarily had large eccentricities that brought them relatively close to the Sun, where cometary activity should have been intense. The major problem with this scenario is due to the information we have on Jupiter Trojans belonging to dynamical families. These objects were separated from the parent body when it was already in one of the Lagrangian clouds, where they are still observable. As a consequence, if a family is not very old, we must be able to see on the surface of the fragments the internal composition of the progenitor. In this context is hard to explain why we do not see any ice signature on the spectra of family members, if the progenitor originally contained ices in the interior, as expected. Objects formed at large heliocentric distances must contain ices on their interior and it is still unknown what mechanism could completely hide the ice content on the surface of the small fragments.

Knowledge of the ages of the Trojan dynamical families would be helpful on this topic and it is of fundamental importance in the interpretation of the data on the C-type spectrally neutral objects belonging to the Eurybates family. We still cannot assess if (1) this is a very old family, where space-weathering processes flattened the spectra covering the primordial ices; (2) it is a young family produced by an object spectrally similar to C-type asteroids or neutral/less-red Centaurs; or (3) it is a young family where irradiation mantles formed in a timescale shorter than the family age.

The comparison among the spectral properties of Jupiter Trojans, TNOs, cometary nuclei, and Centaurs is difficult to interpret. These objects do not all look the same. These differences could mean that there are no relationships among these populations or that their surfaces have been modified in different ways. From dynamical modeling we are quite confident that there is a link among three of these populations: TNOs are the source, Centaurs are the transient population, and comets are the end members in the transfer chain. Unfortunately, the observational constraints are at present too weak to determine the origin of Jupiter Trojans, and these objects remain among the most intriguing bodies of the solar system.

As anticipated in section 2, the two models on the origin of Jupiter Trojans, local capture vs. capture from a distant disk, allow us to interpret similarities and differences between Trojans and TNOs in very different ways. If the Trojans have been captured from the local planetesimal population, then they represent relatively unaltered samples of the middle part of the solar nebula. We have no other direct samples from this region that not only fed a growing Jupiter, but also probably contained the "snow-line" marking the onset of H_2O condensation and may have supported formation of nebular organics via Fischer-Tropsch-type catalytic reactions. In this scenario, we might expect Trojan compositions consistent with the *Gradie and Tedesco* (1982) paradigm of the trend of asteroid composition with heliocentric distance (i.e., low-temperature silicates, organics, some water ice). It is somewhat unclear what primordial TNO compositions are implied by this scenario, but perhaps some TNOs could also have originated in this region and/or at slightly larger distances and would therefore have similar compositions to Trojans, while more distant objects were more compositionally distinct. Or, conversely, the various types of space weathering could have affected the evolution of surfaces in this region differently than other regions. These scenarios are qualitatively consistent with observational results that Trojans are similar to some taxonomic classes of Centaurs and TNOs as well as some classes of outer main-belt asteroids. It also predicts compositional differences between Trojans and TNOs formed at larger distances. Conversely, if one accepts the *Morbidelli et al.* (2005) scenario, both Trojans and TNOs come from the primordial transneptunian disk. They are therefore genetically related, although they might have formed in slightly different parts of the aforementioned disk. Therefore, the physical similarities between Trojans and TNOs appear normal, whereas the differences need to be explained on the basis of the subsequent physical evolutions of bodies stored at different places.

Of course, a detailed comparison between the physical properties of Trojans and TNOs can also help to distinguish between the two formation models, although it is probably too early to reach a conclusion at this stage. Further observations are absolutely needed to constrain the composition of Jupiter Trojans and to look for Eurybates-like families in other regions of the orbital parameter phase space. A larger sample of V + NIR spectra would be useful to investigate the nature of more objects, especially looking for spectral features related to the presence of water ice on their surfaces. Polarimetric observations could help to investigate the surface structure (grain size and porosity). Numerical simulations would be useful to assess the age of the known families, in order to investigate the effects of space-weathering processes on the surface of these atmosphereless bodies formed at large heliocentric distances and, as a consequence, to constrain their primordial composition.

REFERENCES

Barucci M. A., Cruikshank D. P., Mottola S., and Lazzarin M. (2002a) Physical properties of Trojan and Centaur asteroids. In *Asteroids III* (W. F. Bottke Jr. et al., eds.), pp. 273–287. Univ. of Arizona, Tucson.

Barucci M. A. and 12 colleagues (2002b) 10 Hygiea: ISO infrared observations. *Icarus, 156,* 202–210.

Beaugé C. and Roig F. (2001) A semianalytical model for the motion of the Trojan asteroids: Proper elements and families. *Icarus, 153,* 391–415.

Bendjoya P., Cellino A., Di Martino M., and Saba L. (2004) Spectroscopic observations of Jupiter Trojans. *Icarus, 168,* 374–384.

Binzel R. P. and Sauter L. M. (1992) Trojan, Hilda, and Cybele asteroids: New lightcurve observations and analysis. *Icarus, 95,* 222–238.

Britt D. T., Yeomans D., Housen K., and Consolmagno G. (2002) Asteroid density, porosity, and structure. In *Asteroids III* (W. F. Bottke Jr. et al., eds.), pp. 485–500. Univ. of Arizona, Tucson.

Crovisier J., Leech K., Bockelée-Morvan D., Brook T. Y., Hanner M. S., Altieri B., Keller H. U., and Lellouch E. (1997) The spectrum of Comet Hale-Bopp (C/1995 O1) observed with the Infrared Space Observatory at 2.9 astronomical units from the sun. *Science, 275,* 1904–1907.

Cruikshank D. P., Dalle Ore C. M., Roush T. L., Geballe T. R., Owen T. C., de Bergh C., Cash M. D., and Hartmann W. K. (2001) Constraints on the composition of Trojan asteroid 624 Hektor. *Icarus, 153,* 348–360.

de Bergh C., Boehnhardt H., Barucci M. A., Lazzarin M., Fornasier S., Romon-Martin J., Tozzi G. P., Doressoundiram A., and Dotto E. (2004) Aqueous altered silicates at the surface of two Plutinos? *Astron. Astrophys., 416,* 791–798.

Dotto E., Barucci M. A., Brucato J. R., Mueller T. G., and Carvano J. (2004) Polyxo: ISO-SWS spectrum up to 26 micron. *Astron. Astrophys., 427,* 1081–1084.

Dotto E., Fornasier S., Barucci M. A., Licandro J., Boehnhardt H., Hainaut O., Marzari F., de Bergh C., and De Luise F. (2006) The surface composition of Jupiter Trojans: Visible and near-infrared survey of dynamical families. *Icarus, 183,* 420–434.

Dumas C., Owen T., and Barucci M. A. (1998) Near-infrared spectroscopy of low-albedo surfaces of the solar system: Search for the spectral signature of dark material. *Icarus, 133,* 221–232.

Dunlap J. L. and Gehrels T. (1969) Minor planets. III. Lightcurves of a Trojan asteroid. *Astron. J., 74,* 797–803.

Emery J. P. and Brown R. H. (2003) Constraints on the surface composition of Trojan asteroids from near-infrared (0.8–4.0 μm) spectroscopy. *Icarus, 164,* 104–121.

Emery J. P. and Brown R. H. (2004) The surface composition of Trojan asteroids: Constraints set by scattering theory. *Icarus, 170,* 131–152.

Emery J. P., Cruikshank D. P., and Van Cleve J. (2005) Thermal emission spectroscopy of asteroids with the Spitzer Space Telescope. *Bull. Am. Astron. Soc., 37,* 15.07.

Emery J. P., Cruikshank D. P., and Van Cleve J. (2006a) Thermal emission spectroscopy (5.2–38 μm) of three Trojan asteroids with the Spitzer Space Telescope: Detection of fine-grained silicates. *Icarus, 182,* 496–512.

Emery J. P., Cruikshank D. P., and Van Cleve J. (2006b) Structure and composition of the surfaces of Trojan asteroids from reflection and emission spectroscopy (abstract). In *Lunar and Planetary Science XXXVII,* Abstract #2075. Lunar and Planetary Institute, Houston (CD-ROM).

Fernández J. A. and Ip W.-H. (1984) Some dynamical aspects of the accretion of Uranus and Neptune — The exchange of orbital angular momentum with planetesimals. *Icarus, 58,* 109–120.

Fernández Y. R., Sheppard S. S., and Jewitt D. J. (2003) The albedo distribution of jovian Trojan asteroids. *Astron. J., 126,* 1563–1574.

Fitzsimmons A., Dahlgren M., Lagerkvist C.-I., Magnusson P., and Williams I. P. (1994) A spectroscopic survey of D-type asteroids. *Astron. Astrophys., 282,* 634–642.

Fleming H. J. and Hamilton D. P. (2000) On the origin of the Trojan asteroids: Effects of Jupiter's mass accretion and radial migration. *Icarus, 148,* 479–493.

Fornasier S., Dotto E., Marzari F., Barucci M. A., Boehnhardt H., Hainaut O., and de Bergh C. (2004) Visible spectroscopic and photometric survey of L_5 Trojans: Investigation of dynamical families. *Icarus, 172,* 221–232.

Fornasier S., Dotto E., Hainaut O., Marzari F., Boehnhardt H., De Luise F., and Barucci M. A. (2007) Visible spectroscopic and photometric survey of Jupiter Trojans: Final results on dynamical families. *Icarus, 190,* 622–642.

Giorgilli A. and Skokos C. (1997) On the stability of Trojan asteroids. *Astron. Astrophys., 317,* 254–261.

Gomes R. S. (1997) Dynamical Effects of planetary migration on the primordial asteroid belt. *Astron. J., 114,* 396–401.

Gomes R. S., Morbidelli A., and Levison H. F. (2004) Planetary migration in a planetesimal disk: Why did Neptune stop at 30 AU? *Icarus, 170,* 492–507.

Gomes R., Levison H. F., Tsiganis K., and Morbidelli A. (2005) Origin of the cataclysmic late heavy bombardment period of the terrestrial planets. *Nature, 435,* 466–469.

Gradie J. C. and Tedesco E. F. (1982) Compositional structure of the asteroid belt. *Science, 216,* 1405–1407.

Gradie J. C. and Veverka J. (1980) The composition of the Trojan asteroids. *Nature, 283,* 840–842.

Hahn J. M. and Malhotra R. (1999) Orbital evolution of planets embedded in a planetesimal disk. *Astron. J., 117,* 3041–3053.

Hahn J. M. and Malhotra R. (2005) Neptune's migration into a stirred-up Kuiper belt: A detailed comparison of simulations to observations. *Astron. J., 130,* 2392–2414.

Hapke B. (1981) Bidirectional reflectance spectroscopy. I — Theory. *J. Geophys. Res., 86,* 3039–3054.

Hapke B. (1993) *Theory of Reflectance and Emittance Spectroscopy.* Cambridge Univ., New York. 455 pp.

Hartmann W. K., Tholen D. J., Goguen J., Binzel R. P., and Cruikshank D. P. (1988) Trojan and Hilda asteroid lightcurves. I. Anomalously elongated shapes among Trojans (and Hildas?). *Icarus, 73,* 487–498.

Hartmann W. K., Ryder G., Dones L., and Grinspoon D. (2000) The time-dependent intense bombardment of the primordial Earth/Moon system. In *Origin of the Earth and Moon* (R. M. Canup and K. Righter, eds.), pp. 493–512. Univ. of Arizona, Tucson.

Hudson R. L. and Moore M. H. (1999) Laboratory studies of the formation of methanol and other organic molecules by water + carbon monoxide radiolysis: Relevance to comets, icy satellites, and interstellar ices. *Icarus, 140,* 451–461.

Ivezić Ž. and 32 colleagues (2001) Solar system objects observed in the Sloan Digital Sky Survey commissioning data. *Astron. J., 122,* 2749–2784.

Jewitt D. C. and Luu J. X. (1990) CCD spectra of asteroids. II — The Trojans as spectral analogs of cometary nuclei. *Astron. J., 100,* 933–944.

Jewitt D. C. and Sheppard S. S. (2002) Physical properties of trans-neptunian object (20000) Varuna. *Astron. J., 123,* 2110–2120.

Jewitt D. C., Trujillo C. A., and Luu J. X. (2000) Population and size distribution of small jovian Trojan asteroids. *Astron. J., 120,* 1140–1147.

Jones T. D., Lebofsky L. A., Lewis J. S., and Marley M. S. (1990) The composition and origin of the C, P, and D asteroids — Water as a tracer of thermal evolution in the outer belt. *Icarus, 88,* 172–192.

Lacerda P. and Jewitt D. (2006) Densities from lightcurves. *Bull. Am. Astron. Soc., 38,* 34.02.

Lacerda P. and Luu L. (2006) Analysis of the rotational properties of Kuiper belt objects. *Astron. J., 131,* 2314–2326.

Lazzarin M., Barbieri C., and Barucci M. A. (1995) Visible spectroscopy of dark, primitive asteroids. *Astron. J., 110,* 3058–3072.

Lazzaro D., Angeli C. A., Carvano J. M., Mothé-Diniz T., Duffard R., and Florczak M. (2004) S3OS2: The visible spectroscopic survey of 820 asteroids. *Icarus, 172,* 179–220.

Levison H. F. and Duncan M. J. (1997) From the Kuiper belt to Jupiter-family comets: The spatial distribution of ecliptic comets. *Icarus, 127,* 13–32.

Levison H., Shoemaker E. M., and Shoemaker C. S. (1997) The dispersal of the Trojan asteroid swarm. *Nature, 385,* 42–44.

Lisse C. M. and 16 colleagues (2006) Spitzer spectral observations of the Deep Impact ejecta. *Science, 313,* 635–640.

Luu J. X., Jewitt D., and Cloutis E. (1994) Near-infrared spectroscopy of primitive solar system objects. *Icarus, 109,* 133–144.

Malhotra R. (1993) The origin of Pluto's peculiar orbit. *Nature, 365,* 819–821.

Malhotra R. (1995) The origin of Pluto's Orbit: Implications for the solar system beyond Neptune. *Astron. J., 110,* 420–429.

Marchis F. and 17 colleagues (2006a) A low density of 0.8 g cm^{-3} for the Trojan binary asteroid 617 Patroclus. *Nature, 439,* 565–567.

Marchis F., Wong M. H., Berthier J., Descamps P., Hestroffer D., Vachier F., Le Mignant D., and De Pater I. (2006b) S/2006 (624) 1. *IAU Circular 8732.*

Marzari F. and Scholl H. (1998a) The growth of Jupiter and Saturn and the capture of Trojans. *Astron. Astrophys., 339,* 278–285.

Marzari F. and Scholl H. (1998b) Capture of Trojans by a growing proto-Jupiter. *Icarus, 131,* 41–51.

Marzari F. and Scholl H. (2000) The role of secular resonances in the history of Trojans. *Icarus, 146,* 232–239.

Merline W. J., Close L. M., Menard F., Dumas C., Chapman C. R., and Slater D. C. (2001) Search for asteroid satellites. *Bull. Am. Astron. Soc., 33,* 1133.

Merline W. J., Weidenschilling S. J., Durda D. D., Margot J. L., Pravec P., and Storrs A. D. (2002) Asteroids do have satellites.

In *Asteroids III* (W. F. Bottke Jr. et al., eds.), pp. 289–312. Univ. of Arizona, Tucson.

Milani A. (1993) The Trojan asteroid belt: Proper elements, stability, chaos and families. *Cel. Mech. Dyn. Astron., 57,* 59–94.

Milani A. and Knežević Ž. (1994) Asteroid proper elements and the dynamical structure of the asteroid main belt. *Icarus, 107,* 219–254.

Michtchenko T. A., Beaugé C., and Roig F. (2001) Planetary migration and the effects of mean motion resonances on Jupiter's Trojan asteroids. *Astron. J., 122,* 3485–3491.

Moore M. H., Donn B., Khanna R., and A'Hearn M. F. (1983) Studies of proton-irradiated cometary-type ice mixtures. *Icarus, 54,* 388–405.

Morbidelli A., Levison H. F., Tsiganis K., and Gomes R. (2005) Chaotic capture of Jupiter's Trojan asteroids in the early solar system. *Nature, 435,* 462–465.

Moroz L., Baratta G., Strazzulla G., Starukhina L., Dotto E., Barucci M. A., Arnold G., and Distefano E. (2004) Optical alteration of complex organics induced by ion irradiation: 1. Laboratory experiments suggest unusual space weathering trend. *Icarus, 170,* 214–228.

Noll K. S. (2006) Solar system binaries. In *Asteroids Comets Meteors 2005* (D. Lazzaro et al., eds.), pp. 301–318. IAU Symposium 229, Cambridge Univ., Cambridge.

Peale S. J. (1993) The effect of the nebula on the Trojan precursors. *Icarus, 106,* 308–322.

Petit J.-M., Morbidelli A., and Chambers J. (2001) The primordial excitation and clearing of the asteroid belt. *Icarus, 153,* 338–347.

Pollack J. B., Hubickyj O., Bodenheimer P., Lissauer J. J., Podolak M., and Greenzweig Y. (1996) Formation of the giant planets by concurrent accretion of solids and gas. *Icarus, 124,* 62–85.

Rabe E. (1965) Limiting eccentricities for stable Trojan librations. *Astron. J., 70,* 687–688.

Rivkin A. S., Binzel R. P., Howell E. S., Bus S. J., and Grier J. A. (2003) Spectroscopy and photometry of Mars Trojans. *Icarus, 165,* 349–354.

Robutel P. and Gabern F. (2006) The resonant structure of Jupiter's Trojan asteroids — I. Long-term stability and diffusion. *Mon. Not. R. Astron. Soc., 372,* 1463–1482.

Scholl H., Marzari F., and Tricarico P. (2005) Dynamics of Mars Trojans. *Icarus, 175,* 397–408.

Sheppard S. S. and Trujillo C. (2006a) A survey for Trojan asteroids of Saturn, Uranus and Neptune. *Bull. Am. Astron. Soc., 38,* 44.03.

Sheppard S. S. and Trujillo C. (2006b) A thick cloud of Neptune Trojans and their colors. *Science, 313,* 511–514

Shoemaker E. M., Shoemaker C. S., and Wolfe R. F. (1989) Trojan asteroids: Populations, dynamical structure and origin of the L_4 and L_5 swarms. In *Asteroids II* (R. P. Binzel et al., eds.), pp. 487–523. Univ. of Arizona, Tucson.

Shkuratov Y., Starukhina L., Hoffmann H., and Arnold G. (1999) A model of spectral albedo of particulate surfaces: Implications for optical properties of the Moon. *Icarus, 137,* 235–246.

Stansberry J. A. and 17 colleagues (2004) Spitzer observations of the dust coma and nucleus of 29P/Schwassmann-Wachmann 1. *Astrophys. J. Suppl. Ser., 154,* 463–468.

Strazzulla G. (1998) Chemistry of ice induced by bombardment with energetic charged particles. In *Solar System Ices* (B. Schmitt et al., eds.), pp. 281–301. Kluwer, Dordrecht.

Szabó Gy. M., Ivezić Ž., Jurić M., and Lupton R. (2007) The properties of jovian Trojan asteroids listed in SDSS Moving Object Catalog 3. *Mon. Not. R. Astron. Soc., 377,* 1393–1406.

Tancredi G., Fernández J. A., Rickman H., and Licandro J. (2006) Nuclear magnitudes and size distribution of Jupiter family comets. *Icarus, 182,* 527–549.

Tholen D. J. and Barucci M. A. (1989) Asteroid taxonomy. In *Asteroids II* (R. P. Binzel et al., eds.), pp. 298–315. Univ. of Arizona, Tucson.

Thompson W. R., Murray B. G. J. P. T., Khare B. N., and Sagan C. (1987) Coloration and darkening of methane clathrate and other ices by charged particle irradiation — Applications to the outer solar system. *J. Geophys. Res., 92,* 14933–14947.

Tsiganis K., Gomes R., Morbidelli A., and Levison H. F. (2005) Origin of the orbital architecture of the giant planets of the solar system. *Nature, 435,* 459–461.

Vilas F. and Gaffey M. J. (1989) Phyllosilicate absorption features in main-belt and outer-belt asteroid reflectance spectra. *Science, 246,* 790–792.

Vilas F., Larson S. M., Hatch E. C., and Jarvis K. S. (1993) CCD reflectance spectra of selected asteroids. II. Low-albedo asteroid spectra and data extraction techniques. *Icarus, 105,* 67–78.

Vilas F., Jarvis K. S., and Gaffey M. J. (1994) Iron alteration minerals in the visible and near-infrared spectra of low-albedo asteroids. *Icarus, 109,* 274–283.

Wetherill G. W. (1992) An alternative model for the formation of the asteroids. *Icarus, 100,* 307–325.

Yang B. and Jewitt D. (2006) Spectroscopic search for water ice on jovian Trojan asteroids. *Bull. Am. Astron. Soc., 38,* 50.03.

Yoshida F. and Nakamura T. (2005) Size distribution of faint jovian L_4 Trojan asteroids. *Astron. J., 130,* 2900–2911.

Yoshida F., Nakamura T., Watanabe J., Kinoshita D., Yamamoto N., and Fuse T. (2003) Size and spatial distributions of sub-km main-belt asteroids. *Publ. Astron. Soc. Japan, 55,* 701–715.

Kuiper Belt Objects in the Planetary Region: The Jupiter-Family Comets

Stephen Lowry and Alan Fitzsimmons
Queen's University Belfast

Philippe Lamy
Laboratoire d'Astrophysique de Marseille

Paul Weissman
NASA Jet Propulsion Laboratory

Jupiter-family comets (JFCs) are a dynamically distinct group with low orbital inclinations and orbital periods ≤20 yr. Their origin has been shown computationally to be the Kuiper belt region beyond Neptune. Therefore studying the nuclei of these comets, as well as their coma species, can provide valuable insights into the nature of the kilometer-sized Kuiper belt objects (KBOs). These include their size distribution, internal structure, and composition, as well as some hints at their likely surface features. Although JFCs are much closer to the Sun than KBOs, they are still very difficult to observe due to their intrinsic faintness and outgassing comae. However, observational studies are advancing rapidly and we are now starting to place valuable constraints on the bulk physical properties of these nuclei. In this chapter, we review some of the more important findings in this field and their relevance to KBO studies.

1. KUIPER BELT OBJECTS: PROGENITORS OF THE JUPITER-FAMILY COMETS

Considerable progress has been made in understanding the dynamical histories of the low-inclination ecliptic comets (ECs) (*Duncan et al.,* 2004). It is generally accepted that most if not all of the ECs, consisting of the Jupiter-family comets (JFCs), Encke-type comets, and Centaurs, must have originated from the Kuiper belt. Indeed, it was dynamical studies of JFCs (*Fernández,* 1980; *Duncan et al.,* 1988) that suggested that the most efficient source for them was a low-inclination reservoir beyond the giant planets, in order to match the low-inclination distribution of the JFC orbits. These results then stimulated the first confirmed Kuiper belt object discovery (*Jewitt and Luu,* 1993; see also chapter by Davies et al.) after Pluto and Charon. Kuiper belt objects (KBOs) can be perturbed into Neptune-crossing orbits or out of stable resonances by gravitational interactions with the giant planets, collisions, or perhaps by nongravitational forces due to surface outgassing. Once this happens the KBOs can be handed down through the giant planets region toward the terrestrial planets zone (*Horner et al.,* 2004) and end up as JFCs.

Jupiter-family comets have orbital periods ≤20 yr, and low-inclination, direct orbits. Their orbital behavior is chaotic due to strong gravitational interactions with Jupiter, hence their name. Their aphelia are generally around 5–6 AU from the Sun, although some can eventually evolve to orbits entirely within the orbit of Jupiter, such as Comet 2P/Encke. Encke-type comets are simply old JFCs, and are treated as JFCs for the purposes of this review. Jupiter-family comets are defined by their dynamical Tisserand parameter T_J (with respect to Jupiter), which is conserved in the circular restricted three-body problem, and can provide a measure of the relative velocity of approach to Jupiter. T_J is defined as

$$T_J = \frac{a_J}{a} + 2\cos(i)\sqrt{(1 - e^2)a/a_J} \qquad (1)$$

where a, e, and i are the comet's orbital semimajor axis, eccentricity, and inclination, respectively, and a_J is the orbital semimajor axis of Jupiter. Jupiter-family comets are defined as those comets having $2 < T_J < 3$. Dynamical studies by *Levison and Duncan* (1994) found that T_J does not vary substantially for JFCs, i.e., only ~8% of comets moved in or out of this dynamical class throughout the computer simulations. Thus, the JFCs are dynamically distinct.

Recent dynamical studies have shown that the most likely source for most of the JFCs is the scattered disk objects (SDOs) (*Duncan and Levison,* 1997; see chapter by Gomes et al.). Scattered disk objects are in orbits with perihelia close to Neptune, and are actively interacting dynamically with that planet. This leads to a much higher probability that the SDOs will be thrown into the planetary region, as compared with classical KBOs (CKBOs), which are in more distant and stable orbits. The source of the SDOs is likely objects from the inner Kuiper belt, close to Neptune, and remnant icy planetesimals from the Uranus-Neptune zone.

The importance of collisions in moving small KBOs into the various dynamical resonances has been discussed by *Stern* (1995) and *Farinella and Davis* (1996). Collisions play an important role particularly at diameters <20 km. In fact, as noted in these papers, most of the observed JFCs are likely collisional fragments from the Kuiper belt and thus may not represent the primordial state of the smallest, kilometer-sized end of the size distribution beyond Neptune. Whether primordial or not, studying the small KBOs and their collisional products is valuable for understanding formation and evolution in this size regime, and their thermal histories. As a daughter population of the KBOs, the JFCs provide a unique data source for understanding the physical properties of their progenitors.

As noted above, the JFCs are one of several dynamically distinct groups of comets in the solar system. The other major groups are the long-period comets (LPCs) and the Halley-type comets (HTCs). Long-period comets have orbital periods >200 yr (and up to $\sim 10^7$ yr) and have random orbital inclinations as well as very high eccentricities. Dynamical simulations place their likely formation zone in the giant planets region, from which they were scattered out of the planetary region to form the distant Oort cloud. The Oort cloud consists of $\sim 10^{12}$ comets in gravitationally bound but very distant orbits, with semimajor axes between ~3000 and 100,000 AU. Occasionally, Oort cloud comets are perturbed back toward the planetary region where they appear as LPCs. Approximately one-third of observed LPCs are on their first return to the planetary region. Halley-type comets have orbital periods $20 < P < 200$ yr. Their orbits are more inclined than JFCs but not totally randomized like the LPCs, and have eccentricities also between that of the JFCs and LPCs. Their source region is not determined but they are likely a mix of Oort cloud and Kuiper belt comets, the latter again showing a preference for SDOs.

Recently, several objects with comet-like tails have been found in stable, low-inclination orbits in the outer asteroid belt, referred to as main-belt comets (MBCs) (*Hsieh and Jewitt,* 2006). All three members of this group are associated with the Themis collisional family at ~3.16 AU, and therefore are likely to be volatile-rich asteroids, where volatiles buried beneath the surface have been exposed by recent impacts. The relatively stable orbits of these objects suggest that they formed at their current location in the outer asteroid belt, and thus they have little or no connection with the JFCs or the Kuiper belt.

Jupiter-family comets are the most observationally accessible of the comet groups, with perihelion distances in the realm of the terrestrial planets, and relatively short periods that result in frequent and predictable returns. They are typically much "older" than LPCs (which make an average of only five returns) (*Weissman,* 1979), having likely been in their current orbits for many hundreds of returns or more. This has the advantage that their surfaces are less active, allowing the nucleus to be observed directly in many cases. However, it also means that the observed surfaces are now substantially evolved from their presumably primitive state in the Kuiper belt.

Given our extensive knowledge of JFCs, which comprise more than 200 known objects, it is impossible to provide a complete, detailed description of the properties of this population in a single chapter. For indepth discussions on numerous aspects of the JFCs, we refer the reader to the comprehensive reviews by *Lamy et al.* (2004), *Weissman et al.* (2004), *Samarasinha et al.* (2004), and *Bockelée-Morvan et al.* (2004), among many others, in the recent *Comets II* book. Instead, we focus herein on the broad ensemble properties of both JFCs and KBOs. We make comparisons that offer insights into the nature of the parent KBOs, such as their likely size distribution at kilometer sizes, discussed in section 2. The JFC population is currently unobservable from Earth, so provide a valuable proxy for understanding the nature of the KBO size distribution at diameters <20 km. If the KBOs are the parent bodies of JFCs then it is reasonable to suggest that their internal structures are very similar (with the exception of larger KBOs where gravity dominates their internal structure). Information on internal structure can be inferred from the rotational properties of JFC nuclei (section 3), using methods similar to those developed to study the asteroid population.

Surface imaging of JFC nuclei obtained by recent spacecraft flybys is potentially very powerful in providing representative, closeup views of the surfaces of KBOs. However, we must also recognize that comet nucleus surfaces have likely been modified from their initial state in the Kuiper belt by processes such as sublimation and space weathering (section 4). Some KBOs are active, displaying visual comae, and JFCs may provide clues to understanding this activity, although the volatiles involved are likely very different, given the substantially different thermal regimes in which they occur. In this regard, other cometary populations, in particular the LPCs, may provide more valuable insights as the LPCs often display activity at relatively large solar distances. Jupiter-family comets are particularly valuable for understanding the potential future evolution of KBO surfaces when they are perturbed out of the Kuiper belt toward the terrestrial planets region (section 4). The proximity of JFCs allows for much easier study of outflowing comae, and thus their molecular composition. This is covered in section 5.

The study of cometary nuclei is rapidly advancing and data on their physical properties continues to grow, in some cases well beyond the scope of the *Comets II* reviews mentioned previously. As well as utilizing ever-larger ground-based telescope facilities, the Hubble Space Telescope (HST) has proved most fruitful in probing cometary nuclei, given the high spatial resolution that allows for better coma-removal during nucleus imaging. Additionally, the NASA Spitzer Space Telescope is opening up new areas of investigation through its ability to study cometary nuclei and comae in the infrared.

2. PROBING THE KUIPER BELT SIZE DISTRIBUTION AT THE KILOMETER SIZE RANGE

Jupiter-family comets have typical nucleus radii of 1–5 km. Thus, they are smaller than any detected KBOs. Since it is largely agreed that the JFCs derive primarily from the KBO region, including the high-inclination scattered disk population, the JFCs provide a ready means of sampling both the size distribution and physical nature of small KBOs.

Size estimates for cometary nuclei have always been difficult to obtain. When the comets are in the terrestrial planets region and close to Earth, they are active, and their bright comae obscure the signal from their relatively small, dark nuclei. When the comets are far from the Sun and presumably inactive, they are faint objects, with typical apparent magnitudes $m_R \geq 22$, and require large-aperture telescopes to observe them.

A variety of techniques are used to estimate the sizes of cometary nuclei. These include (1) direct imaging by spacecraft; (2) simultaneous optical and IR photometry of a distant or low-activity nucleus that permits a solution for both the size and albedo; (3) IR photometry alone of a distant or low-activity nucleus, where it can be assumed that almost all sunlight incident on the low-albedo object is re-radiated as thermal energy; (4) HST imaging of comets close to Earth with modeling and subtraction of the coma signal; (5) CCD photometry of distant nuclei, far from the Sun where they are likely to be inactive, and using an assumed albedo of typically 4%; and (6) radar imaging. Of these techniques, (5) is the most widely used, followed closely by (4). Although both techniques rely on an assumed albedo, the repeatability of observed nucleus absolute magnitudes by numerous observers, as well as the confirmation of size and shape estimates from flyby spacecraft, show that they are indeed reliable. Flyby spacecraft have only imaged four cometary nuclei to date: 1P/Halley in 1986 (Giotto, Vega), 19P/Borrelly in 2001 (Deep Space 1), 81P/Wild 2 in 2004 (Stardust), and 9P/Tempel 1 in 2005 (Deep Impact).

There are generally two types of observations: snapshot and lightcurve. Snapshot observations are comprised of several exposures of a nucleus taken in quick succession. They capture the brightness of the nucleus at an instant in time, but there is no knowledge of where the images are in the rotation lightcurve of the presumably irregularly shaped nucleus. More complete coverage is provided by lightcurve observations that image the nucleus over many hours, and even on several successive nights, or orbits in the case of the HST and the Spitzer Space Telescope. This much more complete temporal coverage allows one to obtain the rotation period of the nucleus, and a lower limit on its axial ratio.

An important question in observing distant nuclei is the possible presence of coma. The observer's goal is to image the bare nucleus with no coma contamination of the signal. To do this, cometary targets are usually chosen when they are far from the Sun, beyond 3 AU (preferably >4 AU), and on the inbound leg of their orbits. Many JFCs display more activity postperihelion on the outbound legs of their orbits, even beyond 3 AU where water ice sublimation in theory becomes negligible. A technique to check for coma contamination is to compare the image profile of the nucleus to that of nearby stars in the field. The coma will make itself known as a widening of the comet's radial brightness profile as compared to the on-chip background stars. If the nucleus appears stellar then the likelihood of significant coma contamination is fairly minimal.

The success of these techniques is demonstrated by the repeatability of nucleus size estimates by multiple observers, and by comparison with direct imaging in the four cases where comets have been encountered by spacecraft. For example, *Weissman et al.* (1999) and *Lamy et al.* (2001) found dimensions for the nucleus of 9P/Tempel 1 of 3.8 × 2.9 km and 3.9 × 2.8 km, respectively, each assuming an albedo of 0.04. The dimensions derived from the Deep Impact flyby in 2005 are 3.8 × 2.5 km with a measured albedo of 0.04 (*A'Hearn et al.,* 2005), in excellent agreement.

An important quantity to estimate is the slope of the cumulative size distribution, which can be expressed as a power law of the form

$$N(>r) \propto r^{-\alpha} \qquad (2)$$

where r is the radius, N is the number of nuclei with radius >r, and α is the slope parameter. The cumulative brightness distribution can similarly be expressed by an equation of the form

$$N(<H) \propto 10^{\beta H} \qquad (2)$$

where H is the absolute magnitude, N is the number of nuclei with absolute magnitude <H, and β is the slope parameter. For populations with the same albedo, the two equations are related by $\alpha = 5\beta$.

Several groups have assembled size estimates of cometary nuclei and derived the size distribution. *Weissman and Lowry* (2003) compiled a catalog of CCD, IR, HST, and spacecraft measurements of the dimensions of cometary nuclei. The catalog presently contains 120 measurements of 57 JFCs and 4 HTCs. The data were normalized to an assumed albedo of 0.04 except in cases where the albedo was directly measured. Weissman and Lowry found that the cumulative number of JFCs at or larger than a given radius can be described by a power law with a slope parameter of $\alpha = 1.73 \pm 0.06$ (Fig. 1) for nuclei with radii between 1.4 and 6 km. This corresponds to $\beta = 0.35 \pm 0.01$.

As seen in Fig. 1, the cumulative size distribution has two parts: a steeply ascending series of points that is fitted to give the slope parameter, followed by a roll-off at smaller sizes that is indicative of observational incompleteness. The

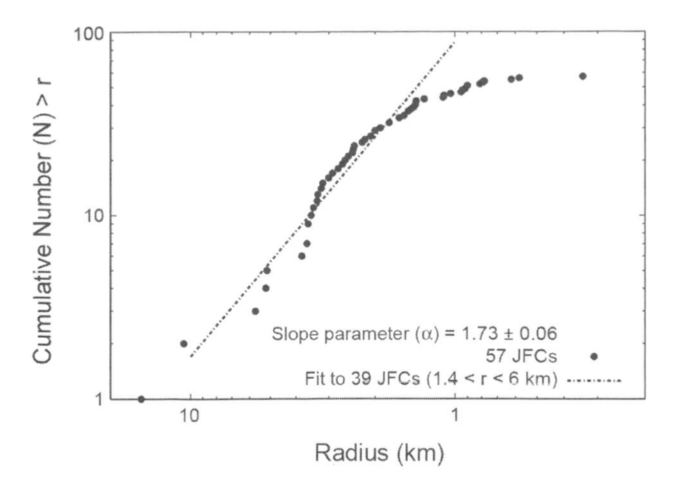

Fig. 1. One of several size distribution estimates for JFCs (*Weissman and Lowry,* 2003). Current estimates of the slope of this distribution are much shallower than for large KBOs, which we believe is reasonable evidence for a broken power-law size distribution within the Kuiper belt.

choice of the lower limit for the fitting of the size distribution has a strong effect on the value of the fitted slope. Since about one-fourth of all JFC nuclei have had size estimates to date, it is difficult to estimate at what size the determination of the distribution is complete. This is illustrated by *Meech et al.* (2004), who found a slope value of $\alpha = 1.91 \pm 0.06$ for JFC nuclei with radii between 2 and 5 km, but a shallower slope of $\alpha = 1.45 \pm 0.05$ for nuclei with radii between 1 and 10 km. *Lamy et al.* (2004) found a slope parameter $\alpha = 1.9 \pm 0.3$ for nuclei larger than 1.6 km in radius. *Lowry et al.* (2003) found a slope of $\alpha = 1.6 \pm 0.1$. These values correspond to $\beta = 0.29$–0.38.

In contrast to the three groups cited above, *Fernández et al.* (1999) and *Tancredi et al.* (2006) found much steeper values of $\alpha = 2.65 \pm 0.25$ and $\alpha = 2.7 \pm 0.3$ ($\beta = 0.53$ and 0.54), respectively, for nuclei brighter than H = 16.7, corresponding to a radius of ~1.7 km. The inclusion of uncalibrated magnitudes reported in the Minor Planet Circulars may seriously compromise their sample and thus contribute to the differences in slope parameter. However, even if these higher values of α are correct, it is clear that the slope of the JFC size distribution is shallower than similar estimates for the larger KBOs.

Weissman and Lowry (2003) pointed out that the size distribution estimate for JFCs was probably not the primordial size distribution when these nuclei first evolved inward from the Kuiper belt. Sublimation mass loss and fragmentation events have likely decreased the sizes of the nuclei over time. Because sublimation loss is a surface process and thought to be independent of nucleus radius, Weissman and Lowry estimated that smaller nuclei would proportionately lose a greater fraction of their initial radius over time than larger nuclei. They estimated that due to this effect the primordial slope α of the cumulative size distribution would be ~0.1 greater than the currently estimated values. Thus, based on the work of Weissman and Lowry, *Meech et al.* (2004), and *Lamy et al.* (2004), the primordial slope parameter likely has a value between 1.83 and 2.01.

Early theoretical estimates of the slope parameter were largely based on *Dohnanyi* (1969), who showed that for constant material strength vs. size, the cumulative size distribution of a collisionally evolved population should have a slope of $\alpha = 2.5$. However, we now know that strength is a function of size (e.g., *Asphaug et al.,* 2002). *O'Brien and Greenberg* (2003) applied current strength models to show that the expected cumulative slope parameter α in the gravity-dominated regime for a collisionally evolved population is 2.04. This is similar to the shallower values cited above, and also to the value found for near-Earth objects of $\alpha = 1.96$ (*Stuart,* 2001) ($\beta = 0.39$).

We can compare the brightness distribution of JFCs with that of KBOs. In the case of the KBOs, the estimated quantity is the slope of the cumulative luminosity function (CLF), which is defined in terms of apparent magnitudes but can be related to the KBO size distribution, and compared with other populations, if several assumptions are applied (see chapter by Petit et al.). Typical KBO β values range from 0.63 to 0.69 (*Trujillo et al.,* 2001; *Gladman et al.,* 2001; *Bernstein et al.,* 2004). Note that a more accurate comparison could be made if the KBO distribution employed absolute magnitudes, as is done for the JFCs. The shallower slope parameter of the JFCs, which are considerably smaller than the observed KBOs, is likely due to a change in the slope of the KBO size distribution at smaller sizes (*Weissman and Levison,* 1997; *Bernstein et al.,* 2004). Barring any unusual processes that would sharply change the size distribution of small KBOs as they evolve inward to JFC orbits, the size distribution of the JFCs can be considered as reasonable proof that the KBO size distribution must be much shallower at smaller sizes. *Sheppard et al.* (2000) found a CLF slope for Centaurs of 0.6 ± 0.1, consistent with the CLF for the KBOs. Thus, the break in the KBO size distribution likely occurs at radii lower than the faintest detected Centaurs, which is about 10–20 km.

3. ROTATION PROPERTIES, BULK DENSITY, AND INTERNAL STRUCTURE

Rotational properties of cometary nuclei are difficult to obtain. While being much closer than KBOs, the smaller size of the JFCs means that they are inevitably faint. When close to the Sun the nucleus is effectively shielded due to the presence of a masking coma, which acts to reduce the brightness amplitude of the rotational lightcurve should the modulation be detected at all.

As mentioned above, time-series photometric measurements can be made from which periodicities can be evaluated. The lightcurve amplitude is related to the elongation of the nucleus, and by combining this with the measured period, one can set limits on the nucleus density, i.e., the minimum density required in order to withstand centrifugal disruption under the assumption of negligible cohesive strength (*Luu and Jewitt,* 1992; *Weissman et al.,* 2004). The density estimate is a lower limit because we use the projected axial ratio, a/b, which is a lower limit to the true axial ratio since the orientation of the rotation axis is unknown. Also, the nucleus does not necessarily need to be spinning at its rotational disruption limit. Lightcurves for ~22 JFC nuclei have been obtained to date (Fig. 2). The derived rotation periods range from ~5.2 to 40.8 h, while the projected axial ratios range from 1.02 to 2.60. The inferred, lower-limit nucleus bulk densities are given by the position of each comet in the figure and are limited to ≤0.56 g cm^{-3}, consistent with density values determined through other methods (*Weissman et al.,* 2004).

Lowry et al. (2003) also noted that a 5.2-h cutoff in rotation period exists for cometary nuclei, corresponding roughly to a density cutoff of 0.6 g cm^{-3}. As more comet data is acquired that result seems to be holding up quite well. A similar spin-period cutoff is unambiguously seen for the asteroid population, albeit for a much larger sample, at the faster period of 2.2 h and thus higher density of ~2.5 g cm^{-3} (see *Pravec et al.,* 2002). Pravec et al. interpret this result as evidence for small asteroids being loosely bound, gravity-dominated aggregates with negligible tensile strength. The same description should then apply to JFCs. There exists a distinct population of small (<150 m) asteroids that can spin well above this limit, believed to be monolithic-rock fragments, perhaps individual fragments of larger rubble-pile asteroids.

Figure 2 compares all cometary lightcurve data with those available for KBOs and Centaurs. Cometary nuclei are shown as open circles, Centaurs as filled triangles, and KBOs as filled circles. One can see that the rotation period vs. shape distributions overlap quite nicely, with exceptional objects labeled. It is thus reasonable, even at this early stage,

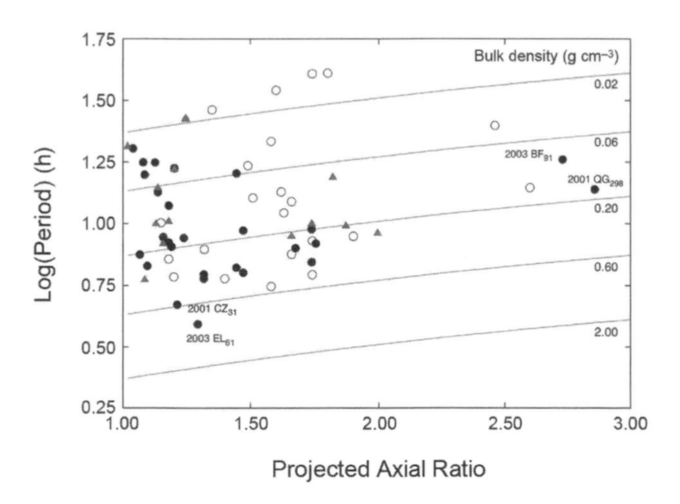

Fig. 2. Available cometary nucleus lightcurve data (*Snodgrass,* 2006). Comet data are shown as open circles, KBOs as filled circles, and Centaurs as filled triangles. KBO and Centaur data are from *Lacerda and Luu* (2006), *Trilling and Bernstein* (2006), *Ortiz et al.* (2006), and *Rabinowitz et al.* (2006). The distributions for JFCs and KBOs are similar on this plot. Like comets, KBOs all lie above the 0.6 g cm^{-3} threshold, with the exception of the large object 2003 EL_{61}. Curves are lines of constant density, for a simple centrifugal breakup model.

to apply a similar interpretation as done for the small asteroid population with regard to their rubble-pile internal structure. Further evidence for the rubble-pile structure of comets comes from their observed disruption, such as in the case of D/1999 S4 (LINEAR) and D/1993 F2 (Shoemaker-Levy 9). Further discussion on the likely internal structure of KBOs can be seen in the chapter by Coradini et al.

A cutoff density of ~0.6 g cm^{-3} implies that comets are remarkably porous (~70%) (*Snodgrass,* 2006) for nuclei with similar proportions of dust and ice to that found for 9P/Tempel 1 by Deep Impact. Snodgrass also points out that there is a trend for the fastest-rotating nuclei to have lower values of a/b, which may reflect the inability of the rubble-pile nuclei to maintain extended shapes near the rotational disruption limit, again similar to small asteroids and JFCs.

There is only one object that appears to be rotating faster that the proposed spin period cutoff at 5.2 h. That is the large object 2003 EL_{61}, which rotates at 3.9 h (*Rabinowitz et al.,* 2006). The presence of such a body should not necessarily rule out the rubble-pile interpretation of the available lightcurve data, but its existence requires highlighting. The higher density of this large object (~1500–2500 km in length) can likely be explained by gravitational compression. A major goal in the study of cometary nuclei is to populate the graph in Fig. 2, to build up a more robust picture of the rotation period distribution.

Alternatively, *Holsapple* (2003) has suggested that small cohesive strengths, on the order of 10^4 dynes cm^{-2}, could allow even fast-rotating, elongated asteroids to survive as rubble piles. This value is similar to the tensile strengths inferred for cometary nuclei, although these calculations do not explain the sharp edge seen for the vast majority of small asteroids in the rotation period vs. axial ratio plot of *Pravec et al.* (2002). The existence of that edge strongly implies that many of these bodies are indeed acting as if they are strengthless.

Analyzing the actual distribution of spin periods for JFCs may not reveal much information about the spin rate distribution of small KBOs as cometary spin rates can be modified. Outgassing can alter the period over timescales less than the orbital period. Other mechanisms are planet encounters or even the Yarkovsky-O'Keefe-Radzievskii-Paddack (YORP) effect (*Rubincam,* 2000). The YORP effect is a torque due to both incident solar radiation pressure and the recoil effect from the anisotropic emission of thermal photons on small bodies in the solar system. The effect was directly detected for the first time on near-Earth asteroid (54509) 2000 PH5 (*Lowry et al.,* 2007; *Taylor et al.,* 2007). However, YORP timescales are long compared with other period-changing mechanisms. The available lightcurve data for JFCs and KBOs show the rotation-period distributions to be flat (*Snodgrass,* 2006), expected for comets but not necessarily for KBOs, which should be similar to the observed collisionally relaxed Maxwellian distribution seen for asteroids. This implies that the spin rates of KBOs and JFCs are indeed being altered over time, although more data are required to confirm these preliminary results based on limited data.

4. SURFACE CHARACTERISTICS AND EVOLUTION OF JUPITER-FAMILY COMETS

4.1. The Search for Compositional Links to Kuiper Belt Objects from Colors and Albedos

Colors and albedos are important properties of solar system small bodies as they can constrain composition and surface processes. Common patterns or trends in these properties among KBOs and cometary nuclei could reveal information about their putative relationships and subsequent evolution. Broadband colors are not as diagnostic in terms of surface composition as spectroscopy, but systematic correlations between different color indices, and between color indices and dynamical parameters could suggest evolutionary trends or compositional groupings. Such groupings are seen in the asteroid population, which radically advanced our understanding of the nature of the asteroid belt (e.g., *Tholen,* 1984).

Several forms of color groupings in the Centaur and KBO populations have been reported. A color bimodality was reported within both populations by *Tegler and Romanishin* (1998) and *Tegler and Romanishin* (2003). However, after independent analysis of the same sample, *Peixinho et al.* (2003) put forward that if one separates the complete sample into Centaurs and KBOs, then the color bimodality exists only in the Centaur population, while the KBOs exhibit a continuous spread. More recent data from a large-scale homogeneous survey at the European Southern Observatory (ESO) imply the existence of a KBO compositional taxonomy (*Barucci et al.,* 2005; *Fulchignoni et al.,* 2006; see also chapter by Fulchignoni et al.). With this in mind, it is a major goal of cometary nucleus observers to search for potential compositional groupings or trends in JFC nuclei, to develop compositional links with KBOs, as only dynamical links have been firmly established so far.

Broadband color data is normally acquired over spectroscopic data simply because cometary nuclei are small and faint, as noted above, and sometimes can only be observed using CCD imaging techniques with broadband filters that effectively integrate the observed flux over a large wavelength range, which improves S/N substantially. Imaging over several bandpasses can result in a broadband spectrum. Cometary nucleus spectroscopy ideally requires 5–8-m-class telescopes to attain the required S/N to reveal surface compositional spectroscopic signatures. To date, cometary nuclei have been primarily observed using groundbased 3–4-m telescopes and the HST.

The following discussion of the comparison of colors between KBOs and cometary nuclei is based on two new extended treatments of this topic by *Lamy and Toth* (2005) and *Snodgrass* (2006) (hereafter *LT05* and *S06*, respectively). The former is based on a compilation of color indices for 282 KBOs and 35 nuclei of ECs, where the EC observations were carried out with the HST. The *LT05* EC sample includes both JFCs and Encke-types ($T_J > 3$, $a < a_J$). *S06* is based on their groundbased survey of cometary nuclei at large heliocentric distances. Although nucleus color data are still limited at present, these datasets are the largest homogeneous datasets of their kind. Therefore, conclusions drawn will naturally be on firmer footing than from just a collation of observations presented throughout the literature. In the following discussion we describe what is seen in the available data, and what this actually means for studying surface properties of the small KBOs and potential future evolutionary states.

4.1.1. The distributions of color. An important measurement is the color distribution of JFCs and how they compare to either the ensemble KBO color distribution, or that of various KBO dynamical subgroups. Figure 3 displays the histograms of the four color indices (B–V), (V–R), (R–I), and (B–R). The top row includes all KBOs, while the bottom row includes all ECs (excluding Centaurs). Both populations exhibit a large range of colors, in fact much larger for cometary nuclei than first anticipated by *Luu* (1993). However, there is a clear trend for KBOs to be globally redder than comets. A few objects bluer than the Sun exist in both populations, but we point out that, in the case of the nuclei, large uncertainties affect their indices so that they could well be less blue than implied by Fig. 3. This global perception hides a more diverse situation when considering the different dynamical classes of KBOs. To a large extent, the extreme red color comes from the classical KBOs in both low- and high-inclination orbits (CKBO-LI, CKBO-HI). The bimodality of the distribution of colors for the Centaurs [best seen on the (B–R) index] is suspected for Plutinos but is totally absent for comets, although the number of comet nucleus color measurements is small. The closest associations for the cometary nuclei based on the presently available distributions of colors (Fig. 3) would be the SDOs and the Centaurs, which interestingly are their most likely parent bodies based on dynamical studies (section 1).

The situation of the color-color correlations is highly contrasted among KBOs. There exists different partial correlations, i.e., involving only two color indices (e.g., Plutinos and Centaurs) with implications for the multiplicity of coloring agents and/or processes (*Peixinho et al.,* 2004; *Doressoundiram et al.,* 2005). The situation is radically different for cometary nuclei as a nonparametric statistical test using the Spearman rank correlation has indicated that there are no statistically significant correlations in the global set of colors (*LT05*). *S06* combined the available (V–R) and (R–I) data from the published literature with many of their own new measurements and found no color groupings and no evidence for the ultrared matter on JFC surfaces, as seen on KBOs (Fig. 4), consistent with *Jewitt* (2002) and *Delsante et al.* (2004). The mean (V–R) color index was found to be ~0.45 for the JFCs (from 31 measurements), as compared to ~0.59 for KBOs (based on 62 data points).

Also, *LT05* conducted systematic Komolgorov-Smirnov tests between the cumulative distributions (CDs) of color indices of the different populations. They found that the original CDs do not reveal compositional relationships between KBOs and ECs, as also noted by *Hainaut and Delsanti* (2002) and *Doressoundiram et al.* (2005). The highest probabilities come from the (B–V) index and favor first

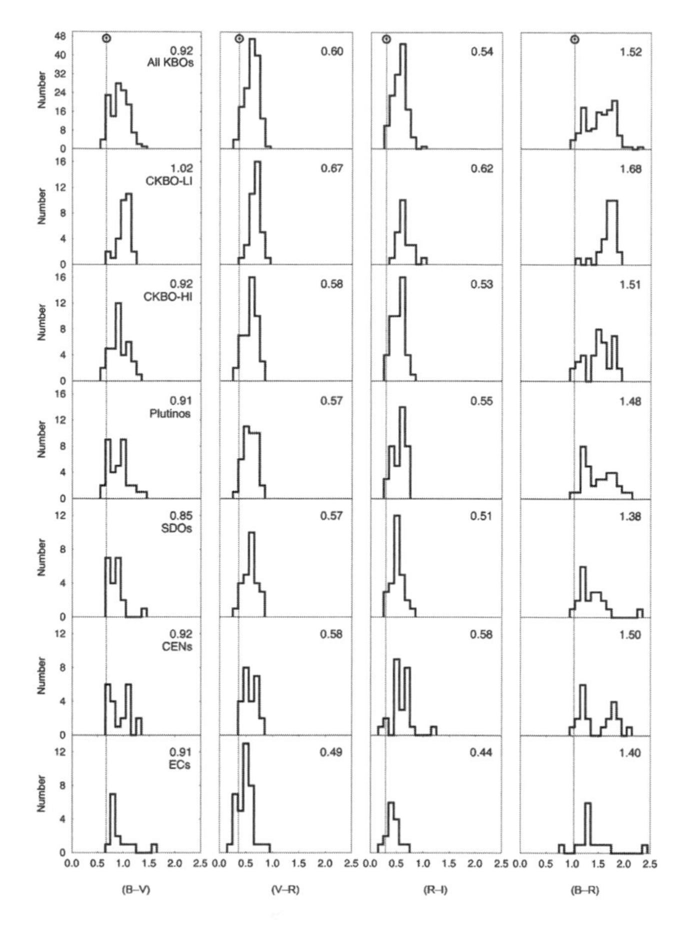

Fig. 3. Distributions of the color indices for different families of primitive bodies of the solar system: (B–V) (first column), (V–R) (second column), (R–I) (third column), and (B–R) (fourth column). From the top row to the bottom are all KBOs, the classical KBOs in low-inclination orbits, the classical KBOs in high-inclination orbits, the Plutinos, the scattered disk objects, the Centaurs, and finally the nuclei of ecliptic comets at the bottom. The means of the distributions are indicated. Solar colors are marked by vertical lines. Data are from *Lamy and Toth* (2005).

the Centaur–EC connection, followed by the SDO–EC and the Plutino–EC relationships, but these probabilities never exceed 42%. Color differences between the different families of primitive objects are most likely due to their formation and evolution at widely different places in the solar system. Considering the observed color distribution of Centaurs and ECs (and extinct cometary candidates within the asteroid population), ultrared matter has to be progressively removed or buried as these objects evolve toward the inner part of the solar system. *LT05* favor thermal alteration of different organic compounds because it can differentially affect the colors as observed.

4.1.2. Correlations between color, orbit, and size. It is also worthwhile investigating potential correlations between color and distance, as well as color vs. size. The case of the color-distance relationship is highly contrasted among the individual families of KBOs. On the basis of the analysis of *Peixinho et al.* (2004), the CKBOs do not show a color-semimajor axis relation. The larger objects do exhibit a strong correlation, with the colors increasing with semimajor axis and perihelion distance. In agreement with *Doressoundiram et al.* (2005), *Peixinho et al.* (2004) did not find any color-distance trend, neither for the SDOs nor for the Centaurs, contrary to *Bauer et al.* (2003), who found strong correlations of the (V–R) and (R–I) indices of Centaurs with semimajor axis. Regarding cometary nuclei, *LT05* have determined that (1) for the red nuclei, the (V–R) index appears to vary quasilinearly with perihelion distance while the other two indices do not, and (2) for blue nuclei, their colors are independent of perihelion distance. As for size, there now seems to be a consensus that the larger Plutinos are redder than the small ones (*Hainaut and Delsanti,* 2002; *Peixinho et al.,* 2004). The case of CKBOs remains confused because of possible multiple correlations between colors and perihelion distances, inclinations, and sizes. Based on the present data, colors are uncorrelated with size for SDOs, Centaurs, and cometary nuclei. This was also the case for the *S06* cometary dataset.

4.1.3. Comparing the albedos of Kuiper belt objects and Jupiter-family comets. Kuiper belt objects show a remarkable diversity of albedos ranging from ~1% to larger than 70%, and this property applies among both small and large, and gray and red objects. In fact, there is currently no evidence of correlations between albedo and either object size, color, or dynamical properties (with the possible exception of orbital inclination) (*Grundy et al.,* 2005). Cometary nuclei exhibit a very narrow range of albedos, ranging from ~3% to ~5% (*Lamy et al.,* 2004) so that a canonical value of 4% can be safely applied to derive sizes from magnitudes, contrary to KBOs. Asteroids in comet-like orbits, many of which are thought to be extinct comet candidates, are also very dark with albedos as low as 2% (*Fernández et al.,* 2001). Figure 5 displays the distribution of albedos of

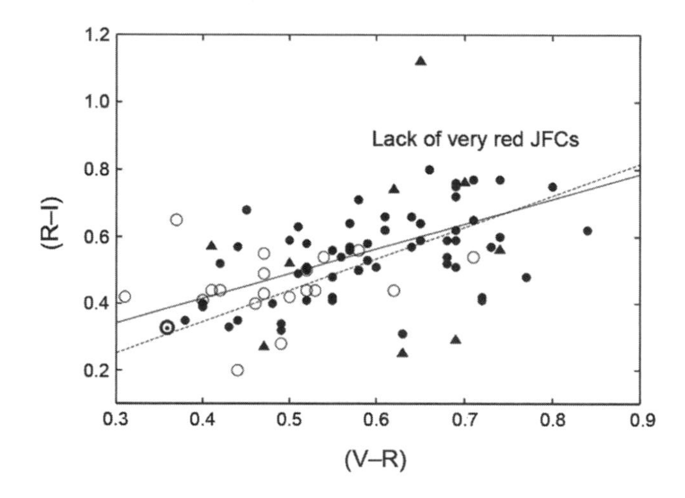

Fig. 4. (R–I) vs. (V–R) for all JFCs, KBOs, and Centaurs with known colors. Open circles = JFCs; filled circles = KBOs; filled triangles = Centaurs. The position of the Sun on these axes is marked. The solid line shows the best fit to the comet data; the dashed line is the fit to the KBOs. KBO and Centaur data from *Jewitt and Luu* (2001) and *Peixinho et al.* (2004). From *Snodgrass* (2006).

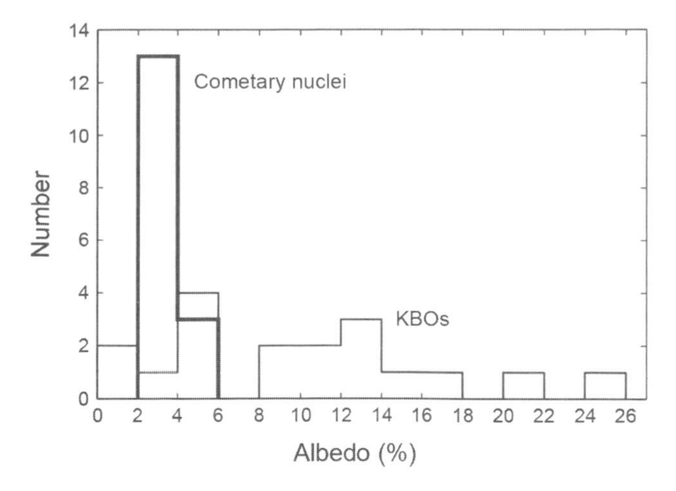

Fig. 5. Distributions of albedo (V- and R-band values combined) for the KBOs and cometary nuclei. Data from *Lamy et al.* (2004) and chapter by Stansberry et al.

KBOs (see chapter by Stansberry et al.) restricted to ≤30%, and of cometary nuclei as compiled by *Lamy et al.* (2004) augmented by the recent result of *Fernández et al.* (2006) for Comet 162P/Siding-Spring. For the present purpose of illustrating the widely different behaviors, we have merged V- and R-band determinations.

The colors and albedos of primitive bodies of the solar system result from their intrinsic initial composition (which may vary with their place of formation in the protoplanetary nebula) and many competing processes that subsequently alter their surfaces like collisional erosion and cratering, thermal processes such as volatile transport or differentiation, radiolysis and photolysis aging, impact by micrometeroids, and mantling from either impact ejecta or cometary activity. The dependence of collisional erosion rates upon size and orbital parameters (inclination and eccentricity) is often invoked to explain the diversity of albedos among KBOs since more pristine subsurface materials are expected to be brighter than space-weathered surfaces. Such a process is absent for cometary nuclei, and that may explain their narrow range of albedos. Alternatively, KBOs may possess surface frosts, which simply do not survive the journey to the planetary region.

4.2. Surface Morphology from Spacecraft Flybys

The four cometary nuclei observed to date by flyby spacecraft show vastly different shape and surface morphologies, although this may be due in part to the different spatial resolutions of the imagery for each nucleus (Fig. 6). Comet 1P/Halley's nucleus most clearly appears to be a rubble-pile structure, with large topographic features and, at least, a binary shape. About 30% of the illuminated surface is active, with large, apparently collimated jets (*Keller et al.*, 1986). The remainder of the surface is inactive and likely covered by a lag deposit crust of large particles that serve to insulate the icy-conglomerate material at depth.

The nucleus of 19P/Borrelly also has a binary shape but has a smoother surface with less topography and some evidence of erosional processes (*Soderblom et al.*, 2002). In addition to chaotic terrain, Borrelly displays mesa-like structures on its surface with smooth, flat tops and steep walls. It has been suggested that the walls of the mesas are where sublimation is currently taking place. In contrast to Halley, only a few percent of the nucleus surface appears active. Comet 81P/Wild 2 has a fairly ellipsoidal shape but a very unusual surface morphology, covered by numerous shallow and deep depressions that may be either eroded impact craters or sublimation pits, or some combination of the two (*Brownlee et al.*, 2004). Large blocks protruding from the surface also suggest an underlying rubble-pile structure. The orbital history of 81P/Wild 2 suggests that it may be a relatively young JFC, having been thrown into the terrestrial planets region after a close encounter with Jupiter in 1971, and thus the surface may preserve features that are truly

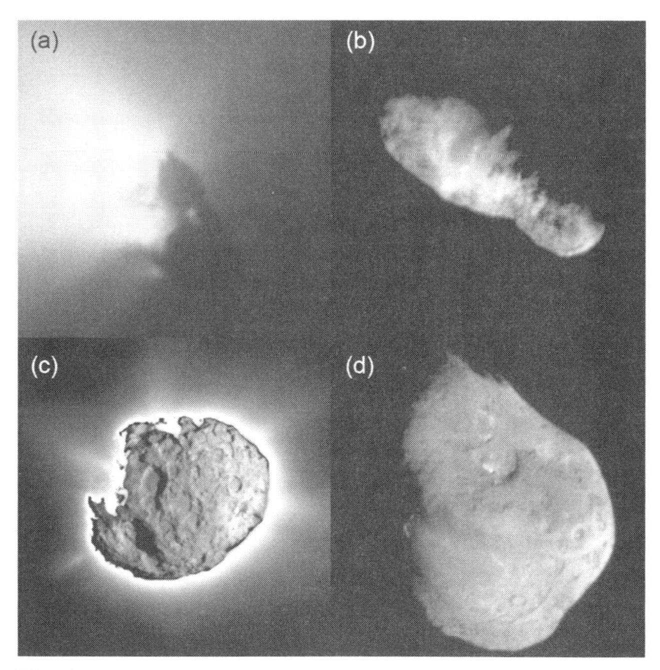

Fig. 6. There have been four cometary nuclei imaged through spacecraft flybys to date, with each flyby revealing very different surface features. Although many of these features could have resulted from evolutionary processes occuring after their departure from the Kuiper belt, clues to their internal structure and surface composition could mimic KBOs (see text). **(a)** Comet 1P/Halley from the Giotto encounter in 1986 (*Keller et al.*, 1986); **(b)** Comet 19P/Borrelly from Deep Space 1 taken in 2001 (*Soderblom et al.*, 2002); **(c)** Comet 81P/Wild 2 imaged in 2004 by the Stardust sample return mission (*Brownlee et al.*, 2004); **(d)** the most detailed images to date have come from the Deep Impact mission to Comet 9P/Tempel 1 in 2005 (*A'Hearn et al.*, 2005). Image credits: ESA/Max-Planck Institute for Aeronomy (1P), NASA/JPL (19P), NASA/JPL/Univ. of Washington (81P), and NASA/JPL/Univ. of Maryland (9P).

primitive. The coma images of Wild 2 also show numerous jets but they have not yet been identified clearly with surface features.

The highest-resolution images to date are of the nucleus of Comet 9P/Tempel 1. These images reveal a complex surface morphology with strong evidence for erosional and geologic processes (*A'Hearn et al.,* 2005). There also appear to be two relatively well-defined and large impact craters on the surface, somewhat surprising since it was assumed that impacts were rare on such a small body, and that sublimation would quickly erode such features. Apparent layering in the surface images may be primitive, but more likely is further evidence of erosional processes acting on the nucleus. Also, there are features that suggest material flowing across the nucleus surface, in particular flowing "downhill" (*Veverka et al.,* 2006). Some surface features on Tempel 1 resemble those on Borrelly and this may be consistent with both nuclei being older and more evolved, having had a long residence time in the terrestrial planet zone.

5. COMPOSITIONAL IMPLICATIONS FOR KUIPER BELT OBJECTS FROM COMETARY OBSERVATIONS

As discussed in the chapter by Barucci et al., optical and near-infrared spectroscopy of KBOs has provided important information on their composition. Decades of studies of JFCs have produced a vast body of knowledge on their chemical makeup, giving valuable insights into (at least) kilometer-sized KBOs. Sublimation of volatiles on the surface of a JFC nucleus leads to an unbound gas and dust coma where the gas pressure gradient lifts once-embedded dust grains from the surface. The gas and dust decouple beyond distances of ~10^3 km and hence the coma appears collisionless and freely expanding to the majority of ground-based observations.

Historically the gas-phase coma has been studied via optical and near-UV spectroscopy, where molecules such as OH, C_2, C_3, and CN have bright emission bands due to resonance fluoresence with solar radiation. However, these are all at least secondary daughter species created by photodissociation and other processes from their parent molecules (i.e., $H_2O \rightarrow OH + H$), and in turn are destroyed by the same processes (i.e., $OH \rightarrow O + H$), so knowledge of the various creation and destruction pathways are necessary to accurately deduce the original sublimation rate at the nucleus. Parent molecules released directly from the nucleus such as H_2O, CO, CH_3OH, C_2H_6 (ethane), CH_4 (methane), and HCN (hydrogen cyanide) can now be observed in bright comets with high-resolution radio or near-infrared spectroscopy from the ground. Finally, the opening up of the submillimeter spectrum has allowed a large number of carbon- and sulfur-based molecules to be detected. The list of detected species, the transitions involved, and their relative abundances are given in the comprehensive review of *Bockelée-Morvan et al.* (2004). All comets studied have been shown to possess H_2O as their dominant volatile constituent; CO and CO_2 have relative abundances of ~5–20% and all other species have abundances of $X/H_2O \leq 1\%$, with the possible exception of CH_3OH (methanol).

5.1. Molecular Abundances

Although many comets appear to share similiar compositions, the existence of comets with anomalous abundance ratios of trace species has long been known, e.g., 21P/Giacobinni-Zinner (*Cochran and Barker,* 1987). The largest population study performed to date has been by *A'Hearn et al.* (1995), who used narrowband optical photometry to measure sublimation rates for several gas-phase species in 85 comets. One of their most surprising findings was that 30% of comets they observed were highly depleted (by a factor ~5) in the carbon-chain molecules C_2 and C_3. Importantly, nearly all these depleted comets were JFCs and in fact accounted for ~50% of the observed JFCs. *Fink and Hicks* (1996) obtained gaseous productions rates for 21 comets via optical spectrophotometry, and upper limits for another 18. In their sample of seven JFCs they found two to be depleted in CN and C_2. Importantly, both studies agreed on whether individual comets were "typical" or "depleted." Additionally, the *A'Hearn et al.* (1995) analysis also hinted at other possible chemically distinct groupings, including comets with enhanced C_2 and C_3 and another group with high NH/OH abundances.

A'Hearn et al. argued for this being a primordial signature, from which one concludes that there must have been at least two chemically distinct regions in the Kuiper belt during formation. They speculated that this may mark a radial distance in the protoplanetary disk, where beyond a certain distance either the creation of the parents of carbon-chain molecules was inhibited due to some process, or perhaps they were destroyed due to an increasing flux of cosmic rays. However, any interpretation must now also take into account current thinking that many JFCs may have derived from the SDO population and thus originated in the Uranus-Neptune zone. Therefore, there may be considerable overlap between the formation zones of LPCs and JFCs, which potentially confounds the explanation as to why there is such large compositional variation in just the JFCs. Furthermore, *Fink and Hicks* (1996) found carbon-chain depletions for the LPC C/1988 Y1 Yanaka, while *A'Hearn et al.* (1995) also measured depletions in the LPC C/1986 V1 Sorrels and the HTC C/1984 U2 Shoemaker. An additional point is that these studies both looked at the photodissociative carbon daughter products in the coma, and it is not clear how to accurately map these depletions onto parent molecules. However, the first comparative study of parent species in 24 comets was reported by *Biver et al.* (2003) using millimeter and submillimeter observations. They found that H_2O/HCN varied by a factor of 3 within their sample, while ratios of other carbon carrying species varied by factors of 5 to at least 9. While this is on the same order as that obtained from the optical studies above, Biver et al. did not find any clear correlation with orbital classification. Hence it now appears that a similar chemical diversity is present in JFCs, HTCs, and LPCs.

An important point that should not be overlooked is that the JFCs found to be "carbon-normal" shared similar abundances with the majority of HTCs and LPCs, implying that LPC abundances may be applicable to JFCs to first order. This is important as the majority of detailed groundbased studies are of bright HTCs or LPCs (1P/Halley, C/1996 B2 Hyakutake, C/1995 O1 Hale-Bopp). Current dynamical theories imply that LPCs formed closer to the Sun than KBOs in the region of Jupiter-Neptune over a prolonged period (*Dones et al.,* 2004). The higher temperatures and densities in this region may have produced bodies with lower volatile abundances compared to the Kuiper belt, and differences between LPCs and JFCs could be expected. The studies performed to date imply that any such differences are not striking and are less than the variations observed from comet to comet.

Studies of JFCs can also address the question of the internal heterogeneity of comets. As pointed out by *A'Hearn et al.* (1995), the similar molecular abundance ratios implies comets share the same global compositions on the largest scales (0.1–10 km). Jupiter-family comets have a known tendency to fragment and break apart, thereby allowing ices in the deep interior to sublimate (*Boehnhardt,* 2004). Molecular abundances in the combined coma of these split comets have shown no significant differences from measurements before the disruption event, also implying large-scale homogeneity within the nucleus. Recently, *Villanueva et al.* (2006) and *Dello Russo et al.* (2006) performed a detailed study using high-resolution near-IR spectoscopy of components B and C of 73P/Schwassmann-Wachmann 3 and found both components to share a C-depleted composition. This agreed with optical observations of C_2 obtained by *Schleicher* (2006).

While comets may be regarded as globally homogeneous, there exists some evidence for small-scale chemical variations within cometary nuclei. *A'Hearn et al.* (1985) observed that the outgassing rate of water on 2P/Encke varied in a different manner to other observed species, and suggested this was due to different volatile compositions at two primary sublimation sites. Intriguingly, *Lowry and Weissman* (2007) reported what appeared to be a significant variation with rotation in the optical color of Comet Encke's nucleus, which could be related to chemically distinct units on the surface. In their population study, *A'Hearn et al.* (1995) found that several other comets exhibited a change in their C-X/OH abundance ratio as a function of orbital position, implying that different vents have different relative amounts of carbon-based species. Unfortunately, their data was not dense enough to allow a search for rotationally driven variations. *Mumma et al.* (1993) reported that the detections of H_2CO in 1P/Halley appeared to be periodic, implying that production was concentrated at a single location on the rotating nucleus, although it was not possible to confirm this. Finally, *in situ* spectra of the near-nucleus coma of 9P/Tempel 1 from the Deep Impact spacecraft appeared to show a significant difference in the CO_2/H_2O gas abundance ratio on alternate sides of the nucleus (*Feaga et al.,* 2006), although further modeling of optical depth effects is required.

To summarize, it is probable that the Kuiper belt possesses two chemically distinct populations, one having depleted abundances of carbon-chain molecules with respect to HTCs and LPCs. Other chemically similar groupings also probably exist. Within these groups, there is significant evidence of small-scale chemical inhomogeneities within comets. One should therefore expect that KBOs will also exhibit localized variations in their compositions, as well as differences from object to object.

5.2. Noble Gases, Ortho-Para Ratios, and Formation Temperatures

Detection of noble gases (He, Ne, Ar) in JFCs would provide a direct link to their formation sites, as these species are chemically inert and highly volatile. Assuming a solar composition throughout the protosolar nebula, significant depletions will occur above a threshold temperature. With a model nebula, measured abundances of these species could suggest the heliocentric distance at which they formed. Unfortunately, there is yet to be a single detection of Ar or Ne in a JFC; indeed a study of three LPCs by *Weaver et al.* (2002) only produced upper limits, implying formation temperatures of ≥40–60 K. The single reported detection of Ar/O was in Hale-Bopp by *Stern et al.* (2000), where a marginal detection gave an enrichment over solar abundances of a factor of 1.8, implying that the core of Hale-Bopp has never risen above 40 K, although the measurement uncertainties were also consistent with a solar abundance. Given the unusual brightness and activity of this comet, it is likely that JFC noble gas abundances will require *in situ* measurements as planned with the Rosetta mission (*Slater et al.,* 2001).

Molecules containing two H atoms can have their nuclear spins parallel or orthogonal. The relative populations are set by the statistical weights of the levels and the excitation temperature at formation. Observations of ortho-para ratios of H_2O in Comet 103P/Hartley 2 with the Infrared Space Observatory by *Crovisier et al.* (1999) corresponded to a spin temperature of 36 ± 3 K. Values for LPCs from observations of H_2O and NH_2 range from >50 K for C/1986 P1 Wilson (*Mumma et al.,* 1993) to 28 K for C/1995 O1 Hale-Bopp (*Crovisier et al.,* 1997). *Kawakita et al.* (2001) showed that it was possible to use the relatively bright optical lines of NH_2 to derive the ortho-para ratio of the parent NH_3 to high precision. As summarized by *Bockelée-Morvan et al.* (2004), these cold temperatures imply that no re-equilibrium occurs in the nucleus, as they are consistent with heliocentric distances >50 AU. If they are primordial, these temperatures are an indication that ice formation took place in the solid phase on the surfaces of interstellar dust grains in the interstellar medium rather than in the gas phase. This paradigm supports the viewpoint that JFCs and KBOs manage to retain signatures of their formation environment even through their subsequent dynamical and physical evolution (*Stern,* 2003), although some uncertainties remain over whether ortho-para ratios remain unchanged over 4.5 b.y.

5.3. Dust-Phase Abundances

Mid-infrared spectroscopy of cometary dust particles over the years have shown that many comets exhibit distinct emission features due to silicate grains, so one should expect the presence of silicates on the surfaces of KBOs as well. Detailed spectra of Comet C/1995 O1 (Hale-Bopp) obtained by *Crovisier et al.* (1997) with the European Space Agency (ESA) Infrared Space Observatory revealed the presence of crystalline Mg-rich olivine and pyroxene, together with amorphous pyroxene. Spectra of the ejecta from the Deep Impact encounter with 9P/Tempel 1 were obtained by *Lisse et al.* (2006) using the Spitzer Space Telescope. Apart from containing similar minerals to Hale-Bopp, spectral fitting also revealed the presence of crystalline Fe-rich olivines and pyroxenes, together with phyllosillicates, carbonates, polyaromatic hydrocarbons (PAHs), and amorphous carbon. Lisse et al. conclude that carbonaceous material makes up ~15% of the observed dust, less than that found for other comets but consistent with the classification of 9P/Tempel 1 as a carbon-depleted comet.

In situ sampling of cometary dust particles by the ESA Giotto spacecraft and the two Soviet Vega spacecraft at 1P/Halley revealed a large population of organic dust particles composed primarily of H, C, N, and O (*Fomenkova et al.,* 1992). At Halley these "CHON" grains formed ~25% of the dust particles detected, while 50% were a mixture of CHON and silicate grains. The presence of dust grains containing large amounts of volatiles is also implied by the observation of distributed sources of CN, CO, and OCS in the inner comae of several comets. Hence cometary dust grains have been shown to contain material from high-temperature and low-temperature environments, implying that KBOs formed in a well-mixed environment. Furthermore, at the time of writing, results of the analysis of dust particles captured by the NASA Stardust sample-return mission (*Brownlee et al.,* 2004) became available. One of the many highlights was that Comet 81P/Wild 2 contains material formed at high temperature in the inner solar system. Therefore, if this comet accreted in the colder Kuiper belt region then this material must have been transported there, consistent with several protosolar disk models (see *McKeegan et al.,* 2006).

5.4. Nucleus Spectroscopy

Ground-based spectroscopy of cometary nuclei have not yet revealed any spectral signatures (*Lamy et al.,* 2004). In the past this was possibly due to their extreme faintness in the near-IR where ice absorption bands lie. However, the discovery by current near-Earth object surveys of JFCs with weak or intermittent outgassing at small heliocentric distances now allows more detailed studies. High signal-tonoise near-infrared spectra of 162P/Siding-Spring obtained by *Campins et al.* (2006) showed no sign of any absorption bands with depths >2% of the continuum. A similar null result was obtained for C/2001 OG108 by *Abell et al.* (2005). This is possibly understandable as these are weakly active comets, which implies that their surfaces are heavily mantled by dust particles, but it is still disappointing.

Spectroscopy of resolved nuclei by spacecraft have been more successful. Spectroscopy with Deep Space 1 observed an absorption band at 2.39 μm on the surface of 19P/Borrelly, whose origin is currently unknown. Three areas exhibiting water-ice absorption bands at 1.5 μm and 2.0 μm were identified on the nucleus of 9P/Tempel 1 with the Deep Impact spacecraft (*Sunshine et al.,* 2006). These covered an extremely small fraction of the visible surface area, and remote disk-integrated spectra would not have revealed them. If other JFCs share similar surface characteristics, it is not surprising that ground-based spectroscopy efforts have been unsuccessful. Of course, JFCs will have undergone substantial erosion and alteration of their surfaces through sublimation processes (*Meech and Svoren,* 2004), so it is probably unwise to attempt comparisons with the extant spectra of the much larger and less eroded KBOs.

6. DISCUSSION

Throughout this chapter we have shown that studying certain aspects of JFC nuclei and their comae can reveal information on the bulk physical properties of small KBOs, and their composition, and just as importantly what cannot be learned. The size distribution of JFCs is particularly valuable for ascertaining the KBO size distribution at the small km-size end. Although the various size distribution estimates have not yet converged on a single solution, we can say for sure that the JFC size distribution is much shallower than for the KBOs measured at the much larger (>100 km) size regime. As pointed out by *Lowry et al.* (2003), this could imply that there is some process acting on cometary nuclei that is significantly altering the size distribution. For example, the rate of complete disintegration of small nuclei could be more frequent than once thought (i.e. smaller objects are being removed from the system, which could reduce the slope's steepness). Alternatively, there is a broken power law to the size distribution within the belt. A record is preserved of this within the JFCs, as surface sublimation does not significantly affect the size distribution as the small KBOs evolve inwards through the planetary region. This broken power law will need to be accounted for in any models of the formation and evolution within the Kuiper belt.

Studying the rotational properties of JFCs and KBOs is also very revealing and indicates the presence of a spin-period or density cut-off which naturally has implications for the internal structure of KBOs. It is likely that internal properties are preserved within the nuclei of JFCs as they evolve dynamically. Therefore the spin period cut-off should be common to both groups, which we believe it is although statistics are small. If such a spin-period cut-off is unambiguously confirmed then this would act as a *physical link* to KBOs, supporting the established dynamical links. Although we can infer KBO compositional information from cometary observations, as discussed in section 5, establishing compositional links are still in their early stages. Of course, an obvious caveat to this and the arguments above is that many JFCs may have originated from within the giant planets region. Awaiting investigation are the isotopic ratios in JFCs, for which no remote observations yet exist. The first preliminary measurements have recently been reported for dust particles collected *in situ* by the Stardust mission (*McKeegan et al.,* 2006).

It is now widely accepted from both the early and latest observations that surface colors and albedos of comets do not reflect the surface properties of KBOs, perhaps due to surface sublimation and mantle formation processes acting on JFCs, and one must proceed with caution when making comparisons. Of course, we learn a great deal about how JFC surfaces are evolving as they migrate to the planetary region. The search for possible color-color groupings within the JFC population will go on. As for albedo, systematic surveys of JFCs (as well as Centaurs and KBOs) using the recently launched Spitzer Infrared Space Telescope will allow their size and albedo distributions to be constrained with much superior accuracy, hopefully allowing comet-nucleus observers to reach a consensus on the JFC size-distribution slope.

Close-up spacecraft images reveal details on the likely internal structure, such as the presence of large boulders protruding through the surface of Comet 81P/Wild 2, and the binary appearance of Comet 19P/Borrelly, both supporting the rubble-pile model for the internal structure of JFCs (*Weissman,* 1986), and thus the KBOs. The very detailed images from Deep Impact show evidence of material "flowing" on the surface of comet 9P/Tempel 1.

The next spacecraft mission to a JFC nucleus is the Rosetta mission by the ESA (*Schwehm,* 2003). Not only will the nucleus of 67P/Churyumov-Gerasimenko be imaged in fine detail with the OSIRIS optical imaging camera (*Keller et al.,* 2007), but the evolution of surface features will be monitored for ~1 yr. Of the many questions this mission will answer, some that are obviously relevant to this discussion will be the first highly accurate density measurement for a cometary nucleus. Also, if similar material flows are seen, one could look for changes in their structure to assess if they are being produced today, thus increasing their likelihood of occurring on KBO surfaces. We look forward to seeing if these details are reproduced on the surfaces of KBOs when the New Horizons probe goes on to explore the Kuiper belt after its main mission at the Pluto and Charon system (see chapter by Weaver and Stern), and seeing how a "fresh" JFC should look.

Acknowledgments. We thank J. Crovisier and an anonymous referee for their very helpful reviews. We gratefully acknowledge support from the Leverhulme Trust (S.C.L.) and the UK Particle Physics and Astronomy Research Council (S.C.L., A.F.). This work was performed in part at the Jet Propulsion Laboratory under a contract with NASA, and partly funded by the NASA Rosetta and Planetary Astronomy Programs (P.R.W.).

REFERENCES

Abell P. A., Fernández Y. R., Pravec P., French L. M., Farnham T. L., et al. (2005) Physical characteristics of comet nucleus C/2001 OG108 (LONEOS). *Icarus, 179,* 174–194.

A'Hearn M. F., Birch P. V., Feldman P. D., and Millis R. L. (1985) Comet Encke: Gas production and lightcurve. *Icarus, 64,* 1–10.

A'Hearn M. F., Millis R. L., Schleicher D. G., Osip D. J., and Birch P. V. (1995) The ensemble properties of comets: Results from narrowband photometry of 85 comets, 1976–1992. *Icarus, 118,* 223–270.

A'Hearn M. F., Belton M. J. S., Delamere W. A., Kissel J., Klaasen K. P., et al. (2005) Deep Impact: Excavating Comet Tempel 1. *Science, 310,* 258–264.

Asphaug E., Ryan E. V., and Zuber M. T. (2002) Asteroid interiors. In *Asteroids III* (W. F. Bottke Jr. et al., eds.), pp. 463–484. Univ. of Arizona, Tucson.

Barucci M. A., Belskaya I. N., Fulchignoni M., and Birlan M. (2005) Taxonomy of Centaurs and trans-neptunian objects. *Astron. J., 130,* 1291–1298.

Bauer J. M., Meech K. J., Fernández Y. R., Pittichova J., Hainaut O. R., Boehnhardt H., and Delsanti A. C. (2003) Physical survey of 24 Centaurs with visible photometry. *Icarus, 166,* 195–211.

Bernstein G. M., Trilling D. E., Allen R. L., Brown M. E., Holman M., and Malhotra R. (2004) The size distribution of trans-Neptunian bodies. *Astron. J., 128,* 1364–1390.

Biver N., Bockelée-Morvan D., Crovisier J., Colom P., Henry F., Moreno R., Paubert G., Despois D., and Lis D. C. (2003) Chemical composition diversity among 24 comets observed at radio wavelengths. *Earth Moon Planets, 90,* 323–333.

Bockelée-Morvan D., Crovisier J., Mumma M. J., and Weaver H. A. (2004) The composition of cometary volatiles. In *Comets II* (M. C. Festou et al., eds.), pp. 391–423. Univ. of Arizona, Tucson.

Boehnhardt H. (2004) Split comets. In *Comets II* (M. C. Festou et al., eds.), pp. 301–316. Univ. of Arizona, Tucson.

Brownlee D. E., Horz F., Newburn R. L., Zolensky M., Duxbury T. C., et al. (2004) Surface of young Jupiter family Comet 81P/Wild 2: View from the Stardust spacecraft. *Science, 304,* 1764–1769.

Campins H., Ziffer J., Licandro J., Pinilla-Alonso N., Fernandez Y., de León J., Mothé-Diniz T., and Binzel R. P. (2006) Nuclear spectra of Comet 162P/Siding Spring (2004 TU12). *Astron J., 132,* 1346–1353.

Crovisier J., Leech K., Bockelée-Morvan D., Brooke T. Y., Hanner M. S., Altieri B., Keller H. U., and Lellouch E. (1997) The spectrum of Comet Hale-Bopp (C/1995 01) observed with the Infrared Space Observatory at 2.9 AU from the Sun. *Science, 275,* 1904–1907.

Crovisier J., Encrenaz T., Lellouch E., Bockelée-Morvan D., Altieri B., et al. (1999) ISO spectroscopic observations of short-period comets. In *The Universe as Seen by ISO* (P. Cox and M. F. Kessler, eds.), p. 161. ESA SP-427, Noordwijk, The Netherlands.

Dello Russo N., Vervack R. J. Jr., Weaver H. A., Biver N., Bockelée-Morvan D., Crovisier J., and Lisse C. M. (2006) High-resolution infrared spectroscopy of Comet 73P/Schwassmann-Wachmann 3: A test for chemical heterogeneity within a cometary nucleus (abstract). *Bull. Am. Astron. Soc., 38,* #03.05.

Delsante A., Hainaut O., Jourdeuil E., Meech K. J., Boehnhardt H., and Barrera L. (2004) Simultaneous visible-near IR photometric study of Kuiper belt object surfaces with the ESO/Very Large Telescopes. *Astron. Astrophys., 417,* 1145–1158.

Dohnanyi J. W. (1969) Collisional models of asteroids and their debris. *J. Geophys. Res., 74,* 2531–2554.

Dones L., Weismann P. R., Levison H. F., and Duncan M. J. (2004) Oort cloud formation and dynamics. In *Comets II* (M. C. Festou et al., eds.), pp. 153–174. Univ. of Arizona, Tucson.

Doressoundiram A., Peixinho N., Doucet C., Mousis O., Barucci A. M., Petit J. M., and Veillet C. (2005) The Meudon Multicolour Survey (2MS) of Centaurs and trans-neptunian objects: Extended dataset and status on the correlations reported. *Icarus, 174,* 90–104.

Duncan M. and Levison H. F. (1997) A scattered comet disk and the origin of Jupiter-family comets. *Science, 276,* 1670–1672.

Duncan M., Levison H., and Dones L. (2004) Dynamical evolution of ecliptic comets. In *Comets II* (M. C. Festou et al., eds.), pp. 193–204. Univ. of Arizona, Tucson.

Farinella P. and Davis D. R. (1996) Short-period comets: Primordial bodies or collisional fragments? *Science, 273,* 938–941.

Feaga L. M., A'Hearn M. F., Sunshine J. M., Groussin O., and the Deep Impact Science Team (2006) Asymmetry of gaseous CO_2 and H_2O in the inner coma of Comet Tempel 1 (abstract). In *Lunar and Planetary Science XXXVII,* Abstract #2149. Lunar and Planetary Institute, Houston (CD-ROM).

Fernández J. A., Tancredi G., Rickman H., and Licandro J. (1999) The population, magnitudes, and sizes of Jupiter family comets. *Astron. Astrophys., 352,* 327–340.

Fernández Y. R., Jewitt D. C., and Sheppard S. S. (2001) Low albedos among extinct comet candidates. *Astrophys. J. Lett., 553,* L197–L200.

Fernández Y. R., Campins H., Kassis M., Hergenrother C. W., Binzel R. P., Licandro J., Hora J. L., and Adams J. D. (2006) Comet 162P/Siding Spring: A surprisingly large nucleus. *Astrophys. J., 132,* 1354–1360.

Fink U. and Hicks M. D. (1996) A survey of 39 comets using CCD spectroscopy. *Astrophys. J., 459,* 729–743.

Fomenkova M., Kerridge J. F., Marti K., and McFadden L. (1992) Classification of carbonaceous components in Comet Halley CHON particles (abstract). In *Lunar and Planetary Science XXIII,* p. 379. Lunar and Planetary Institute, Houston.

Fulchignoni M., Barucci M. A., Belskaya I., and Doressoundiram A. (2006) TNOs' taxonomy confirmed (abstract). *Bull. Am. Astron. Soc., 38,* #40.05.

Gladman B., Kavelaars J. J., Petit J.-M., Morbidelli A., Holman M. J., and Loredo T. (2001) The structure of the Kuiper belt: Size distribution and radial extent. *Astron. J., 122,* 1051–1066.

Grundy W. M., Noll K. S., and Stephens D. C. (2005) Diverse albedos of small trans-neptunian objects. *Icarus, 176,* 184–191.

Hainaut O. and Delsanti A. C. (2002) Colors of minor bodies in the outer solar system. A statistical analysis. *Astron. Astrophys., 389,* 641–664.

Holsapple K. A. (2003) Could fast rotator asteroids be rubble piles? (abstract). In *Lunar and Planetary Science XXXIV,* Abstract #1792. Lunar and Planetary Institute, Houston (CD-ROM).

Horner J., Evans N. W., and Bailey M. E. (2004) Simulations of the population of Centaurs — I. The bulk statistics. *Mon. Not. R. Astron. Soc., 354,* 798–810.

Hsieh H. H. and Jewitt D. (2006) A population of comets in the main asteroid belt. *Science, 312,* 561–563.

Jewitt D. C. (2002) From Kuiper Belt object to cometary nucleus: The missing ultrared matter. *Astron. J., 123,* 1039–1049.

Jewitt D. and Luu J. (1993) Discovery of the candidate Kuiper belt object 1992 QB1. *Nature, 362,* 730–732.

Jewitt D. C. and Luu J. X. (2001) Colors and spectra of Kuiper belt objects. *Astron. J., 122,* 2099–2114.

Kawakita H., Watanabe J.-I., Ando H., Aoki W., Fuse T., et al. (2001) The spin temperature of NH_3 in comet C/1994 S4 (LINEAR). *Science, 294,* 1089–1091.

Keller H. U., Arpigny C., Barbieri C., Bonnet R. M., Cazes S., et al. (1986) First Halley multicolour camera imaging results from Giotto. *Nature, 321,* 320–326.

Keller H. U., Barbieri C., Lamy P., Rickman H., Rodrigo R., et al. (2007) OSIRIS — The Optical, Spectroscopic, and Infrared Remote Imaging System. *Space Sci. Rev., 128,* 433–506.

Lacerda P. and Luu J. X. (2006) Analysis of the rotational properties of Kuiper belt objects. *Astron. J., 131,* 2314–2326.

Lamy P. L. and Toth I. (2005) The colors of cometary nuclei and other primitive bodies (abstract). In *IAU Symposium No. 229: Asteroids, Comets, Meteors,* August 7–12, 2005, Rio de Janeiro, Brazil.

Lamy P. L., Toth I., A'Hearn M. F., Weaver H. A., and Weissman P. R. (2001) Hubble Space Telescope observations of the nucleus of Comet 9P/Tempel 1. *Icarus, 154,* 337–344.

Lamy P. L., Toth I., Fernández Y. R., and Weaver H. A. (2004) The sizes, shapes, albedos, and colours of cometary nuclei. In *Comets II* (M. C. Festou et al., eds.), pp. 223–264. Univ. of Arizona, Tucson.

Levison H. and Duncan M. (1994) The long-term dynamical behavior of short-period comets. *Icarus, 108,* 18–36.

Lisse C. M., Van Cleave J., Adams A. C., A'Hearn M. F., Fernández Y. R., et al. (2006) Spitzer spectral observations of the Deep Impact ejecta. *Science, 313,* 635–640.

Lowry S. C. and Weissman P. R. (2007) Rotation and color properties of the nucleus of Comet 2P/Encke. *Icarus, 188,* 212–223.

Lowry S. C., Fitzsimmons A., and Collander-Brown S. (2003) CCD photometry of distant comets III: Ensemble properties of Jupiter family comets. *Astron. Astrophys., 397,* 329–343.

Lowry S. C., Fitzsimmons A., Pravec P., Vokrouhlický D., Boehnhardt H., Taylor P. A., Margot J-L., Galád A., Irwin M., Irwin J., and Kušnirák P. (2007) Direct detection of the asteroidal YORP effect (abstract). In *Lunar and Planetary Science XXXVIII,* Abstract #2438. Lunar and Planetary Institute, Houston (CD-ROM).

Luu J. X. (1993) Spectral diversity among the nuclei of comets. *Icarus, 104,* 138–148.

Luu J. X. and Jewitt D. C. (1992) Near-aphelion CCD photometry of Comet P/S Schwassmann-Wachmann 2. *Astron. J., 104,* 2243–2249.

McKeegan D., Aléon J., Bradley J., Brownlee D., Busemann H., et al. (2006) Isotopic compositions of cometary matter returned by Stardust. *Science, 314,* 1724–1728.

Meech K. J. and Svoren J. (2004) Using cometary activity to trace the physical and chemical evolution of cometary nuclei. In *Comets II* (M. C. Festou et al., eds.), pp. 223–264. Univ. of Arizona, Tucson.

Meech K. J., Hainaut O. R., and Marsden B. G. (2004) Comet nucleus size distribution from HST and Keck telescopes. *Icarus, 170,* 463–491.

Mumma M. J., Weissman P. R., and Stern S. A. (1993) Comets

and the origin of the solar system: Reading the Rosetta Stone. In *Protostars and Planets III* (E. H. Levy and J. I. Lunine, eds.), pp. 1177–1252. Univ. of Arizona, Tucson.

O'Brien D. P. and Greenberg R. (2003) Steady-state size distributions for collisional populations: Analytical solution with size dependent strength. *Icarus, 164,* 334–345.

Ortiz J. L., Gutiérrez P. J., Santos-Sanz P., Casanova V., and Sota A. (2006) Short-term rotational variability of eight KBOs from Sierra Nevada Observatory. *Astron. Astrophys., 447,* 1131–1144.

Peixinho N., Doressoundiram A., Delsanti A., Boehnhardt H., Barucci M. A., and Belskaya I. (2003) Reopening the TNOs color controversy: Centaurs bimodality and TNOs unimodality. *Astron. Astrophys., 410,* L29–L32.

Peixinho N., Boehnhardt H., Belskaya I., Doressoundiram A., Barucci M. A., and Delsanti A. C. (2004) ESO Large Program on Centaurs and KBOs: Visible colours — Final results. *Icarus, 170,* 153–166.

Pravec P., Harris A. W., and Michalowski T. (2002) Asteroid rotations. In *Asteroids III* (W. F. Bottke Jr. et al., eds.), pp. 113–122. Univ. of Arizona, Tucson.

Rabinowitz D. L., Barkume K., Brown M. E., Roe H., Schwartz M., Tourtellotte S., and Trujillo C. (2006) Photometric observations constraining the size, shape, and albedo of 2003 EL61, a rapidly rotating, Pluto-sized object in the Kuiper belt. *Astrophys. J., 639,* 1238–1251.

Rubincam D. P. (2000) Radiative spin-up and spin-down on small asteroids. *Icarus, 148,* 2–11.

Samarasinha N.H., Mueller B. E. A., Belton M. J. S., and Jorda L. (2004) Rotation of cometary nuclei. In *Comets II* (M. C. Festou et al., eds.), pp. 281–299. Univ. of Arizona, Tucson.

Schleicher D. (2006) Comet 73P/Schwassmann-Wachmann. *IAU Circular 8681.*

Schwehm G. (2003) The Rosetta mission — The new mission scenario (abstract). *Bull. Am. Astron. Soc., 35,* #41.06.

Sheppard S. S., Jewitt D. C., Trujillo C. A., Brown M. J. I., and Ashley M. C. B. (2000) A wide-field CCD survey for Centaurs and Kuiper belt objects. *Astron. J., 120,* 2687–2694.

Slater D. C., Stern S. A., Booker T., Scherrer J., A'Hearn M. F., Bertaux J.-L., Feldman P. D., Festou M. C., and Siegmund O. H. (2001) Radiometric and calibration performance of the Rosetta UV imaging spectrometer ALICE. In *UV/EUV and Visible Space Instrumentation for Astronomy and Solar Physics* (O. H. W. Siegmund et al., eds.), pp. 239–247. Proc. SPIE Vol. 4498.

Snodgrass C. (2006) Forms and rotational states of the nuclei of ecliptic comets. Ph.D. thesis, Queen's Univ. Belfast.

Soderblom L. A., Becker T. L., Bennett G., Boice D. C., Britt D. T., et al. (2002) Observations of Comet 19P/Borrelly by the Miniature Integrated Camera and Spectrometer aboard Deep Space 1. *Science, 296,* 1087–1091.

Stern S. A. (1995) Collisional time scales in the Kuiper disk and their implications. *Astron. J., 110,* 856–868.

Stern S. A. (2003) The evolution of comets in the Oort cloud and Kuiper belt. *Nature, 426,* 639–642.

Stern S. A., Slater D. C., Festou M. C., Parker J. Wm., Gladstone G. R., A'Hearn M. F., and Wilkinson E. (2000) The discovery of argon in Comet C/1995 O1 (Hale-Bopp). *Astrophys. J. Lett., 544,* L169–L172.

Stuart J. S. (2001) A near-Earth asteroid population estimate from the LINEAR survey. *Science, 294,* 1691–1693.

Sunshine J. M., A'Hearn M. F., Groussin O., Li J.-Y., Belton M. J. S., et al. (2006) Exposed water ice deposits on the surface of Comet 9P/Tempel 1. *Science, 311,* 1453–1455.

Tancredi G., Fernández J. A., Rickman H., and Licandro J. (2006) Nuclear magnitudes and the size distribution of Jupiter family comets. *Icarus, 182,* 527–549.

Taylor P. A., Margot J.-L., Vokrouhlický D., Scheeres D. J., Pravec P., et al. (2007) The increasing spin rate of asteroid (54509) 2000 PH5: A result of the YORP effect (abstract). In *Lunar and Planetary Science XXXVIII,* Abstract #2229. Lunar and Planetary Institute, Houston (CD-ROM).

Tegler S. C. and Romanishin W. (1998) Two distinct populations of Kuiper-belt objects. *Nature, 392,* 49–51.

Tegler S. C. and Romanishin W. (2003) Resolution of the Kuiper belt object color controversy: Two distinct color populations. *Icarus, 161,* 181–191.

Tholen D. J. (1984) Asteroid taxonomy from cluster analysis of photometry. Ph.D. thesis, Univ. of Arizona, Tucson.

Trilling D. E. and Bernstein G. M. (2006) Light curves of 20–100 km Kuiper belt objects using the Hubble Space Telescope. *Astron. J., 131,* 1149–1162.

Trujillo C. A., Jewitt D. C., and Luu J. X. (2001) Properties of the trans-Neptunian belt: Statistics from the Canada-France-Hawaii telescope survey. *Astron. J., 122,* 457–473.

Veverka J., Thomas P., and Hidy A. (2006) Tempel 1: Surface processes and the origin of smooth terrains (abstract). In *Lunar and Planetary Science XXXVII,* Abstract #1364. Lunar and Planetary Institute, Houston (CD-ROM).

Villaneuva G. L., Bobnev B. P., Mumma M. J., Magee-Sauer K., DiSanti M. A., Salyk M. A., and Blake G. A. (2006) The volatile composition of the split ecliptic Comet 73P/Schwassmann-Wachmann 3: A comparison of fragments B and C. *Astrophys. J. Lett., 650,* L87–L90.

Weaver H. A., Feldman P. D., Combi M. R., Krasnopolsky V., Lisse C. M., and Shemansky D. E. (2002) A search for argon in three comets using the Far Ultraviolet Spectroscopic Explorer. *Astrophys J. Lett., 576,* L95–L98.

Weissman P. R. (1979) Physical and dynamical evolution of long-period comets. In *Dynamics of the Solar System* (R. L. Duncabe, ed.), pp. 277–282. IAU Symposium No. 81, Reidel, Dordrecht.

Weissman P. R. (1986) Are cometary nuclei primordial rubble piles? *Nature, 320,* 242–244.

Weissman P. R. and Levison H. F. (1997) The population of the trans-Neptunian region. In *Pluto and Charon* (S. A. Stern and D. Tholen, eds.), pp. 559–604. Univ. of Arizona, Tucson.

Weissman P. R. and Lowry S. C. (2003) The size distribution of Jupiter-family cometary nuclei (abstract). In *Lunar and Planetary Science XXXIV,* Abstract #2003. Lunar and Planetary Institute, Houston (CD-ROM).

Weissman P., Doressoundiram A., Hicks M., Chamberlin A., Sykes M., Larson S., and Hergenrother C. (1999) CCD photometry of comet and asteroid targets of spacecraft missions (abstract). *Bull. Am. Astron. Soc., 31,* #31.03.

Weissman P. R., Asphaug E., and Lowry S. C. (2004) Structure and density of cometary nuclei. In *Comets II* (M. C. Festou et al., eds.), pp. 337–358. Univ. of Arizona, Tucson.

Irregular Satellites of the Giant Planets

Philip D. Nicholson
Cornell University

Matija Ćuk
University of British Columbia

Scott S. Sheppard
Carnegie Institution of Washington

David Nesvorný
Southwest Research Institute

Torrence V. Johnson
Jet Propulsion Laboratory

The irregular satellites of the outer planets, whose population now numbers over 100, are likely to have been captured from heliocentric orbit during the early period of solar system history. They may thus constitute an intact sample of the planetesimals that accreted to form the cores of the jovian planets. Ranging in diameter from ~2 km to over 300 km, these bodies overlap the lower end of the presently known population of transneptunian objects (TNOs). Their size distributions, however, appear to be significantly shallower than that of TNOs of comparable size, suggesting either collisional evolution or a size-dependent capture probability. Several tight orbital groupings at Jupiter, supported by similarities in color, attest to a common origin followed by collisional disruption, akin to that of asteroid families. But with the limited data available to date, this does not appear to be the case at Uranus or Neptune, while the situation at Saturn is unclear. Very limited spectral evidence suggests an origin of the jovian irregulars in the outer asteroid belt, but Saturn's Phoebe and Neptune's Nereid have surfaces dominated by water ice, suggesting an outer solar system origin. The short-term dynamics of many of the irregular satellites are dominated by large-amplitude coupled oscillations in eccentricity and inclination and offer several novel features, including secular resonances. Overall, the orbital distributions of the irregulars seem to be controlled by their long-term stability against solar and planetary perturbations. The details of the process(es) whereby the irregular satellites were captured remain enigmatic, despite significant progress in recent years. Earlier ideas of accidental disruptive collisions within Jupiter's Hill sphere or aerodynamic capture within a circumplanetary nebula have been found wanting and have largely given way to more exotic theories involving planetary migration and/or close encounters between the outer planets. With the Cassini flyby of Phoebe in June 2004, which revealed a complex, volatile-rich surface and a bulk density similar to that of Pluto, we may have had our first closeup look at an average-sized Kuiper belt object.

1. INTRODUCTION

Planetary satellites are conventionally divided into two major classes, based on their orbital characteristics and presumed origins (*Burns,* 1986a; *Stevenson et al.,* 1986; *Peale,* 1999). The inner, regular satellites move on short-period, near-circular orbits in or very close to their parent planets' equatorial planes, and are generally believed to have formed *in situ* via accretion from a protoplanetary nebula [see, e.g., *Canup and Ward* (2002) and *Mosqueira and Estrada* (2003) for recent models]. By contrast, the outer irregular satellites move on orbits with periods on the order of 1–10 yr, which are frequently highly eccentric and inclined. The latter bodies are generally believed to have been captured from heliocentric orbit during the final phases of planetary accretion (*Pollack et al.,* 1979; *Heppenheimer and Porco,* 1977) or via a collisional mechanism at some later time (*Colombo and Franklin,* 1971), but there is no consensus on a single model. Recent years have seen an explosion in the number of known outer satellites of the jovian planets, due to deep imaging searches carried out with wide-field CCD cameras on large-aperture telescopes (*Gladman et al.,* 1998a, 2000, 2001a; *Sheppard and Jewitt,* 2003; *Holman et al.,* 2004; *Kavelaars et al.,* 2004; *Sheppard et al.,*

2005, 2006). At the time of writing, a total of 106 irregular satellites have been catalogued, ~70 of which have been assigned permanent numbers and names by the International Astronomical Union (IAU).

2. DISCOVERY

2.1. Photographic Surveys

The advent of photographic plates in the late nineteenth century led to a burst of satellite discoveries, in a field that had lain dormant since the visual discoveries of Hyperion, Ariel, Umbriel, and Triton by W. Lassell and W. Bond in 1846–1851 and Phobos and Deimos by A. Hall in 1877. Almost all the faint, newly found objects proved to be what came to be known as irregular satellites: first Saturn's Phoebe (m_V = 16.5), discovered by W. Pickering at Harvard's southern station in 1898, followed over the next 16 years by Jupiter's Himalia (m_V = 14.6), Elara (16.3), Pasiphae (17.0), and Sinope (18.1) discovered by C. Perrine, P. Melotte, and S. Nicholson. Working at Mount Wilson, S. Nicholson discovered three additional 18th-mag jovian irregulars between 1938 and 1951: Lysithea, Carme, and Ananke. In the 1940s and 1950s, G. Kuiper undertook photographic surveys for new satellites at Uranus and Neptune using the McDonald 82-inch telescope, resulting in the discovery of Nereid in 1949. All these surveys depended on manual comparison of pairs of photographic plates, generally using blink comparators to superimpose the two images. The final discovery of the photographic era was Leda (J XIII), discovered by *Kowal et al.* (1975) on plates taken with the 48-inch Schmidt telescope at Palomar. At mean opposition magnitudes of 19.5 and 19.7, Leda and Nereid represented the practical limit for the detection of moving objects on long-exposure plates.

Much of this early survey work is summarized by *Kuiper* (1961), including his own extensive investigations, while some later but unsuccessful searches are reviewed by *Burns* (1986a). In a postscript to the era of photographic discoveries, we note the report by Kowal et al. of a possible ninth irregular jovian satellite, later designated S/1975 J1 (see IAU Circular 2855). This object remained unconfirmed for a quarter century until it was unexpectedly and independently rediscovered by Holman et al. and Sheppard et al. in CCD images in 2000; it was subsequently named J XVIII Themisto (see IAU Circular 7525).

2.2. CCD Surveys

Photographic plates are very inefficient for detecting moving objects. They do not reach faint magnitudes on short timescales, are very time intensive to handle, and are hard to compare in order to find moving objects. Additional irregular satellite discoveries would require more sensitive detectors that covered large fields of view (Fig. 1). In the mid 1990s digital CCDs finally were able to cover significant areas of the sky per exposure (10–30 arcmin on a side compared to only a few arcminutes previously). Since 1997, with the discovery of Caliban and Sycorax at Uranus by *Gladman et al.* (1998a), 95 irregular satellites have been discovered around the giant planets (*Gladman et al.*, 2000, 2001a; *Sheppard and Jewitt,* 2003; *Holman et al.,* 2004; *Kavelaars et al.,* 2004; *Sheppard et al.,* 2005, 2006). With these new discoveries Jupiter's known irregular satellite population has grown from 8 to 55, Saturn's from 1 to 35, Uranus' from 0 to 9, and Neptune's from 2 (if we include Triton) to 7.

Table 1 provides some basic information about the current state of the surveys for irregular satellites around the giant planets. The known irregular satellites (106 as of October 2006) now greatly outnumber the known regular satellites (Fig. 1). Because of its proximity to the Earth, Jupiter currently has the largest irregular satellite population but it is expected that each giant planet's environs may contain similar numbers of such objects (*Jewitt and Sheppard,* 2005).

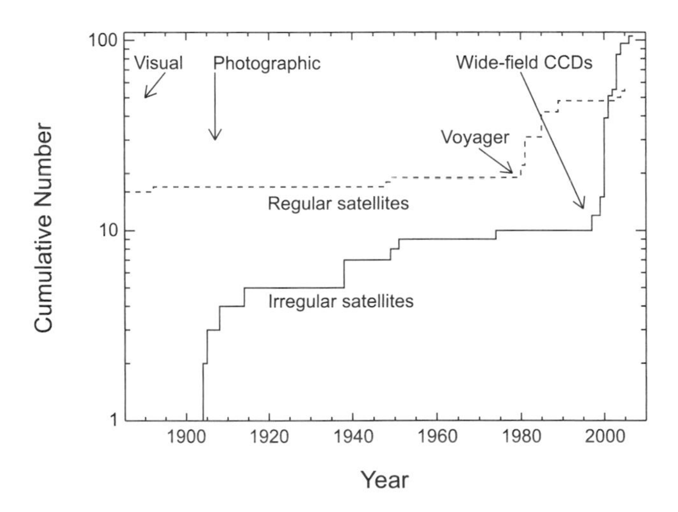

Fig. 1. The number of irregular and regular satellites discovered since the late 1800s. Several key technological advances that resulted in a jump in the pace of discoveries are identified.

TABLE 1. Irregular satellites of the planets.

Planet	#*	m_p† (10^{25}kg)	r_{min}‡ (km)	R_H§ (deg)	R_H (10^7 km)
Jupiter	55	190	1	4.7	5.1
Saturn	35	57	2	3.0	6.9
Uranus	9	8.7	6	1.5	7.3
Neptune	7¶	10.2	16	1.5	11.6

*The number of known irregular satellites, as of December 2006.
†The mass of the planet.
‡Minimum radius of a satellite that current surveys would have detected to date, for an assumed albedo of ~0.05.
§Radius of the Hill sphere as seen from Earth at opposition.
¶Including Triton.

CCDs are not only more sensitive than photographic plates but they allow for the use of computers, which can process the vast amount of data generated with relative ease. Two main techniques have been used with the images from CCDs to discover irregular satellites. Both techniques require observations of the planets to be made when they are near opposition. This ensures that the dominant apparent motion is parallactic, allowing the distance to the object to be calculated. At opposition the movement of an object is simply related to its heliocentric distance, r, by (*Luu and Jewitt,* 1988)

$$\frac{d\theta}{dt} \approx 148\left(\frac{1 - r^{-0.5}}{r - 1}\right) \quad (1)$$

where r is expressed in AU and $d\theta/dt$ is the rate of change of apparent position of the object relative to the stellar background in arcsec hr^{-1}. This allows foreground main-belt asteroids and background Kuiper belt objects (KBOs) to be distinguished easily from possible outer planetary satellites, which have motions typically within a few arcseconds per day of their host planet.

The first technique requires the observer to take three images of a region of sky near the planet spread out over a few hours. Each set of three images per field is then searched using computer programs that detect objects moving relative to the background stars and galaxies near the rate determined from equation (1). Once flagged, these objects are followed up during the next few weeks and months to confirm their satellite status and to obtain preliminary orbits.

The second technique, first used for deep transneptunian object (TNO) searches and described by *Gladman et al.* (1998b) and *Petit et al.* (2004), is more computer-intensive and can only be used efficiently to search for satellites of Uranus and Neptune since these satellites have apparent motions almost exactly the same as those of their host planets (*Holman et al.,* 2004; *Kavelaars et al.,* 2004). In this technique several tens of images are taken of the same field over a period of a few hours. These images are then shifted at the planet's rate and combined using a median filter. Satellites will show up as point sources while foreground and background objects will disappear or be smeared out by the shifting and combining. This technique prevents the CCDs from saturating on the sky while increasing the signal-to-noise of any satellite detection. Fainter satellites can be found with this method, but as it is very telescope-time-intensive, only limited areas of sky can be searched on each night.

Unsuccessful searches for irregular satellites of Mars have been carried out by *Showalter et al.* (2001) and *Sheppard et al.* (2004), while *Stern et al.* (1991, 1994) and *Nicholson and Gladman* (2006) have reported similar negative searches for irregular satellites at Pluto. [Two newly discovered satellites of Pluto, II Nix and III Hydra (*Weaver et al.,* 2006) both orbit close to the planet's equatorial plane and are best classified as regular satellites.]

2.3. Orbital Groupings

Even a cursory inspection of the mean orbital elements of the irregular satellites of Jupiter and Saturn reveals several obvious groupings, illustrated in Plate 8. Prior to 1999, Jupiter had four known prograde satellites, dominated by 160-km-diameter Himalia and tightly clustered in a, e, and i, plus a much looser grouping of four retrograde satellites centered on Pasiphae (D = 60 km) and Carme (45 km). These were generally considered to represent fragments from the disruption — perhaps incidental to their capture — of two parent objects (*Burns,* 1986b; *Colombo and Franklin,* 1971).

With the current census of 55 irregular jovian satellites, at least 3 and perhaps as many as 4 distinct groups can be identified in a–e–i space (*Nesvorný et al.,* 2003; *Sheppard and Jewitt,* 2003): one prograde and two to three retrograde. The original prograde group with $a_6 \simeq 11{:}5$ and $i \simeq 28°$ remains, consisting of Himalia, Elara, Lysithea and Leda, possibly augmented by S/2000 J11, although this object has not been recovered and is now considered "lost." (We use the shorthand notation $a_6 = a/10^6$ km.) In addition, there are two isolated high-inclination objects: Themisto ($a_6 = 7.5$, $i = 43°$) and Carpo ($a_6 = 17.0$, $i = 51°$). All remaining jovians are retrograde, and were divided by *Sheppard and Jewitt* (2003) into three groups centered on Ananke ($a_6 = 21$, $i = 149°$), Pasiphae ($a_6 = 24$, $i = 151°$), and Carme ($a_6 = 23$, $i = 165°$). With 24 additional satellites discovered in 2002 and 2003, the Ananke and Carme groups have become more sharply delineated, but the Pasiphae group remains rather diffuse. The Ananke group has at least 8 smaller named members (Euanthe, Harpalyke, Hermippe, Iocaste, Mneme, Praxidike, Thelxinoe, and Thyone), while Carme has at least another 11 (Aitne, Arche, Chaldene, Erinome, Eukelade, Isonoe, Kale, Kallichore, Kalyke, Pasithee, and Taygete). Euporie and Orthosie, grouped with Ananke by *Sheppard and Jewitt* (2003), now appear more likely to be outliers (*Nesvorný et al.,* 2003). The Pasiphae group may contain as many as 10 other named objects, including 40-km-diameter Sinope, Autonoe, Cyllene, Eurydome, Megaclite, and Sponde (*Sheppard et al.,* 2003), but its exact membership is unclear. Questionable members are Helike (whose latest orbit puts it nearer the Ananke group), Hegemone, Aode, and Callirrhoe.

The saturnian irregulars comprise two relatively tight prograde groups in a–e–i space: Albiorix, Erriapo, Tarvos, and S/2004 S11 at $a_6 \simeq 17$, $i \simeq 34°$ and the Kiviuq/Ijiraq pair at $a_6 = 11.4$, $i = 46°$. Paaliaq and Siarnaq share the latter group's inclination, but are at much larger semimajor axes of $a_6 = 15$ and 18, respectively. Excepting Skathi (S/2000 S8), all the retrograde saturnian satellites were lumped together by *Gladman et al.* (2001a) into a rather loose Phoebe inclination group with $i \simeq 170°$. Unlike the situation at Jupiter, the saturnian retrograde satellites do not form well-defined groups in a–e–i space. Instead, we may group them roughly by inclination alone. Including a dozen new satellites reported by Jewitt et al. in IAU Circular 8523, a smaller, tighter Phoebe group has emerged at i ≃ 175° (with Suttungr, Thrymr, Ymir, and S/2004 S8), while Mundilfari at i = 168° is joined by S/2004 S7, S10, S12, S13, S14, S16, and S17. Skathi at i = 153° has acquired four comrades (Narvi = S/2003 S1, S/2004 S9, S15 and S18). These retrograde saturnian inclination groups encompass a wide range of semimajor axes and eccentricities, raising considerable doubt as to whether they represent actual "families" in the genetic sense (see section 5.3). Nine new retrograde saturnian satellites reported by Sheppard et al. (see IAU Circular 8727) all appear to fall within the above three inclination groups.

At Uranus the situation is rather different. First, all but one (Margaret) of the nine known irregulars are retrograde. Second, both *Kavelaars et al.* (2004) and *Sheppard et al.* (2005) concluded that while the inclination distribution at Uranus is essentially random, with the possible exception of the pair Caliban and Stephano at $a_6 \simeq 7.5$, i ≃ 142°, there is a statistically-significant separation into two groups in a–e space. The inner, more circular group at $a_6 \simeq 7$, e ≃ 0.2 consists of Caliban, Stephano, Francisco, and Trinculo, while the outer, more eccentric group at $a_6 \simeq 16$, e ≃ 0.5 consists of Sycorax, Setebos, Prospero, and Ferdinand.

With only seven irregular satellites currently known (five if Triton and Nereid are excluded as atypically large and possibly sharing a unique dynamical history, as discussed in section 6.2), the classification of Neptune's outer satellites remains problematic (*Holman et al.,* 2004; *Sheppard et al.,* 2006). The three known prograde satellites, Nereid, Laomedeia = S/2002 N3, and Sao = S/2002 N2, have very different inclinations of i = 7°, 35°, and 48° respectively. Three of the four retrograde satellites (Halimede = S/2002 N1, Neso = S/2002 N4, and Psamathe = S/2003 N1) form a fairly tight grouping in inclination at i ~ 135°, while the latter pair also have quite similar values of a and e.

As discussed further in section 3.1, the rapid decrease in brightness of satellites with increasing heliocentric distance means that current surveys at Uranus and Neptune are almost certainly much less complete than those at Jupiter and Saturn, raising the distinct possibility that we are seeing only the largest member or two of each orbital group. It is worth recalling that what appeared a decade ago to be a single loose cluster of four retrograde jovian irregulars is now seen as three distinct and much tighter groups.

2.4. Nomenclature

Historically, different conventions have been followed in naming the irregular satellites of the various planets. The outer satellites of Jupiter have names derived from classical Roman or Greek mythology and are lovers or descendants of Jupiter or Zeus. The jovian prograde satellites with inclinations near 28° and presumably related to Himalia have names ending in "a"; more highly inclined prograde objects have names ending in "o"; and retrograde objects have names ending in "e." At Saturn, where the inner satellites are named after Titans, the outer irregulars are named after giants from Gallic (i ≃ 35°), Inuit (i ≃ 45°), and Norse (retrograde) mythology, Phoebe excepted. At Uranus, all names are derived from English literary (Shakespearean) sources, in keeping with the regular satellites, with no orbital subdivisions. At Neptune the small outer satellites are named after the 50 Nereids but generally follow the jovian scheme of "a" and "e" endings for prograde and retrograde satellites, with an "o" ending reserved for those with unusually high inclinations.

3. PHYSICAL PROPERTIES

The irregular satellites inhabit a unique niche in the solar system. They lie in an otherwise mostly empty region for stable small solar system bodies. They may be some of the only small bodies remaining that are still relatively near their formation locations within the giant planet region rather than being incorporated into the planets or ejected from the area. Their compositions may be intermediate between the rocky main-belt asteroids and volatile-rich TNOs.

3.1. Albedos and Size Distribution

Jupiter's largest irregular satellites have very low albedos of about 0.04 to 0.05, which, along with their colors, are consistent with dark C-, P-, and D-type carbon-rich asteroids in the outer main belt (*Cruikshank,* 1977) and very similar to the jovian Trojans (*Fernandez et al.,* 2003). The Cassini spacecraft obtained resolved images of Himalia that showed it to be an elongated object with axes of 150 × 120 km and an albedo of ~0.05 (*Porco et al.,* 2003). Himalia's relatively irregular shape makes it typical of small objects and satellites with internal pressures so low that they do not assume a hydrostatic shape [e.g., satellites with volumes $\lesssim 10^7$ km^3 exhibit axial ratios significantly greater than 1, according to *Thomas et al.* (1986)].

Saturn's Phoebe has an average albedo of about 0.08 (*Simonelli et al.,* 1999). Cassini obtained high-resolution images of Phoebe that showed it to be intensively cratered with many high albedo patches — possibly fresh ice exposures — located on crater walls (*Porco et al.,* 2005). Further discussion of these results may be found in section 6.1. From Voyager data, Neptune's Nereid was found to have an albedo of 0.16 (*Thomas et al.,* 1991). These albedos for Phoebe and Nereid are more similar to the higher albedos found for large KBOs (*Grundy et al.,* 2005; *Cruikshank et al.,* 2005).

When measured to a given size, the populations and size distributions of the irregular satellites of each of the giant planets appear to be very similar (Fig. 2) (*Sheppard et al.,* 2006). In order to model the size distributions we use a differential power law of the form $n(r)dr = \Gamma r^{-q}dr$, where Γ and q are constants, r is the radius of the satellite, and n(r)dr is the number of satellites with radii in the range r to r + dr.

In the size range of 10 < r < 100 km the irregular satellites of all four giant planets appear to have shallow power-law size distributions with q ≃ 2. Because of their proximity to Earth, only Jupiter and Saturn have known irregular satellites with r < 5 km. These smaller satellites appear to follow a steeper power law (q > 3.5), which may be a sign of collisional evolution. In the Kuiper belt the size distribution is not well known at sizes similar to the irregular satellites, but KBOs with r > 50 km appear to have a much steeper power law (q ~ 4) than the irregular satellites. This distribution appears to become shallower for r < 30 km, but more data are needed to determine a reliable Kuiper belt size distribution (*Trujillo et al.,* 2001; *Gladman et al.,* 2001b; *Bernstein et al.,* 2004; *Elliot et al.,* 2005; see also the chapter by Petit et al.).

If we do not include Triton, the largest irregular satellite at each planet is ~100 km in radius, with about 100 irregular satellites expected around each planet with radii larger than 1 km. This similarity between all four giant planet irregular satellite systems is unexpected considering the different formation scenarios envisioned for the gas giants vs. the ice giants (see section 5). Either the capture mechanism(s) for irregular satellites are independent of the planet formation process (*Jewitt and Sheppard,* 2005), or their size distributions are dominated by subsequent collisional evolution (see section 5.3).

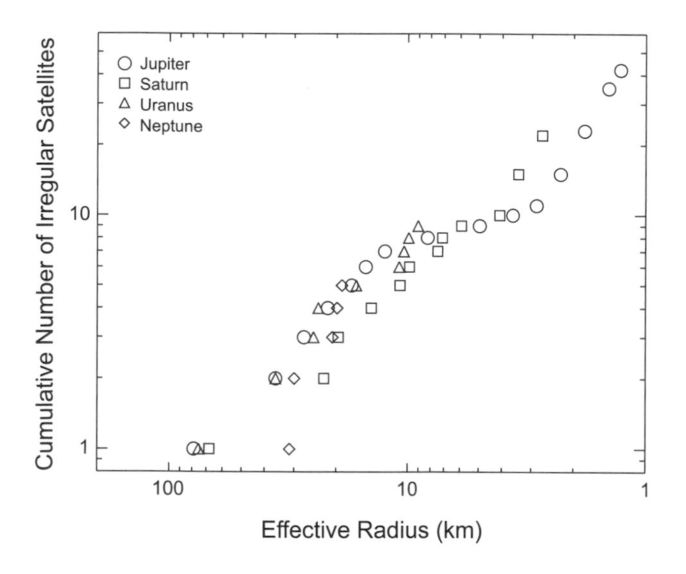

Fig. 2. The cumulative radius function for the irregular satellites with r < 100 km of Jupiter, Saturn, Uranus, and Neptune. This figure directly compares the sizes of the satellites of all the giant planets assuming all satellite populations have albedos of about 0.04. Jupiter, Saturn, and Uranus all have shallow irregular satellite size distributions of q ~ 2 for satellites with 100 > r > 10 km. Neptune's limited number of known small outer irregular satellites with 100 > r > 10 km show a steeper size distribution of q ~ 4, but if Nereid and/or Triton are included we find a much shallower size distribution of q ~ 1.5. Both Jupiter and Saturn appear to show a steeper size distribution for irregular satellites with r < 5 km, which may be a sign of collisional processing. To date neither Uranus' nor Neptune's Hill spheres have been surveyed to these smaller sizes. Further discoveries of irregular satellites around Neptune are needed to obtain a reliable size distribution. After *Sheppard et al.* (2006).

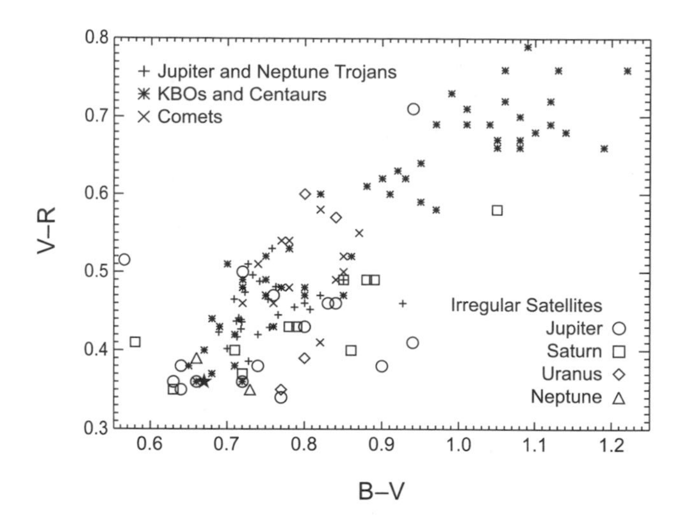

Fig. 3. The colors of the irregular satellites of Jupiter, Saturn, Uranus, and Neptune compared to the Jupiter and Neptune Trojans, Kuiper belt objects, Centaurs, and comet nuclei. The jovian irregular satellites are fairly neutral in color and very similar to the nearby Jupiter Trojans and possibly the comets. Saturn's irregulars are significantly redder than Jupiter's but do not reach the extreme red colors seen in the KBOs. Uranus' irregular satellites are very diverse in color, with some being the bluest known while others are the reddest known irregular satellites. Only two of Neptune's irregulars have measured colors and not much can yet be said except that they don't show the very red colors seen in the Kuiper belt. Irregular satellite colors are from *Rettig et al.* (2001), *Grav et al.* (2003, 2004), and *Grav and Bauer* (2007). Jupiter Trojan colors are from *Fornasier et al.* (2004) while the Neptune Trojan colors are from *Sheppard and Trujillo* (2006). Comet nuclei and dead comet colors are from *Jewitt* (2002, 2005) and references therein. Centaur and KBO colors are from *Barucci et al.* (2005), *Peixinho et al.* (2001), *Jewitt and Luu* (2001), and references therein.

3.2. Broadband Colors

Colors of the irregular satellites are neutral to moderately red (*Tholen and Zellner,* 1984; *Luu,* 1991; *Rettig et al.,* 2001; *Maris et al.,* 2001; *Grav et al.,* 2003, 2004; *Grav and Bauer,* 2007). Most do not show the very red material found in the distant Kuiper belt (Fig. 3). The jovian irregular satellites' colors are very similar to those of the carbonaceous C-, P-, and D-type outer main-belt asteroids (*Degewij et al.,* 1980), as well as to the Trojans and dead comets. Colors of the Jupiter irregular satellite dynamical groupings are consistent with, but do not prove, the notion that each group originated from a single undifferentiated parent body. Optical colors of the eight bright outer satellites of Jupiter show that the prograde (or Himalia) group appears redder and more tightly clustered in color space than the retrograde irregulars (*Rettig et al.,* 2001; *Grav et al.,* 2003). Near-infrared colors recently obtained for the brighter satellites agree with this scenario and suggest that the jovian irregulars' colors are consistent with D- and C-type asteroids (*Sykes et al.,* 2000; *Grav and Holman,* 2004).

The saturnian irregular satellites are on average redder than Jupiter's but still do not show the very red material observed in the Kuiper belt. *Buratti et al.,* (2005) show that the color of the dark reddish side of Iapetus is consistent with dust from the small outer satellites of Saturn but not from Phoebe. Interestingly, most of Saturn's smaller irregular satellites do not appear to have similar spectrophotometry to the neutral-colored Phoebe (*Grav et al.,* 2003; *Buratti et al.,* 2005), with the exception of Mundilfari (*Grav and Bauer,* 2007). Although the Saturn irregulars do not show such obvious dynamical groupings in semimajor axis and inclination phase space as those at Jupiter, the optical colors and phase curves of several prograde saturnians do seem to be correlated for objects of similar inclination (*Bauer et al.,* 2006; *Grav and Bauer,* 2007). The retrograde satellites, on the other hand, exhibit a wide range of optical colors with few strong orbital correlations (*Grav and Bauer,* 2007), consistent with several different parent bodies.

The irregular satellites of Uranus show a wide range of colors, especially in V–R (*Maris et al.,* 2001; *Romon et al.,* 2001; *Grav et al.,* 2004), but little else can be said at present. There are only very limited observational data for Neptune's irregulars, but to date they also do not appear to show the extreme red colors seen in the Kuiper belt.

Figure 3 shows that in general the irregular satellites, the Jupiter and Neptune Trojans, and cometary nuclei have similar colors that are quite different from the overall color distribution of Centaurs and KBOs. It has also been well established that the Kuiper belt appears to have two color populations, with the dynamically excited objects being less red on average (*Tegler and Romanishin,* 2000; *Trujillo and Brown,* 2002). *Sheppard and Trujillo* (2006) suggest that all the "dispersed populations," i.e., those objects that are currently on stable orbits but were likely transported and trapped there, were derived from a similar location in the solar nebula because of their similar moderately red colors. These dispersed objects include the irregular satellites of the four giant planets as well as the Jupiter and Neptune Trojans and the dynamically "hot" — or excited — population of KBOs. If this scenario is true, it would suggest that all these populations were transported and trapped in their current locations at a similar time in the early solar system.

3.3. Spectra and Surface Compositions

Because the vast majority of the irregular satellites are very faint, little has been done in terms of spectroscopy except for the largest few objects. Optical and near-infrared spectra of the brighter jovian satellites are mostly featureless with moderately red slopes (*Luu,* 1991; *Brown,* 2000; *Jarvis et al.,* 2000; *Chamberlain and Brown,* 2004; *Geballe et al.,* 2002). *Jarvis et al.* (2000) find a possible 0.7-μm absorption feature in Jupiter's Himalia and attribute this to oxidized iron in phyllosilicate minerals, which are typically produced by aqueous alteration. Cassini obtained a largely featureless near-infrared spectrum of Jupiter's Himalia (*Chamberlain and Brown,* 2004). The neutral to moderately red linear spectra of Jupiter's irregular satellites, like their albedos and colors, are consistent with those of dark, carbonaceous outer main-belt asteroids.

The one irregular satellite at Saturn and one at Neptune with measured spectra appear to be remarkably different from Jupiter's irregular satellites. Saturn's Phoebe and Neptune's Nereid both show volatile-rich surfaces with strong water-ice signatures (*Brown et al.,* 1998; *Owen et al.,* 1999). In addition to almost ubiquitous water ice, the near-infrared spectra of Phoebe obtained by Cassini show ferrous-iron-bearing minerals (perhaps phyllosilicates), bound water, trapped CO_2, organics, and possibly nitriles and cyanide compounds (*Clark et al.,* 2005). Phoebe's diverse, volatile-rich surface suggests strongly that this object was formed beyond the main asteroid belt and that it may be more similar in composition to cometary nuclei. Further discussion is deferred to section 6.1.

4. DYNAMICS

4.1. Short-Term Dynamics

The largest perturbations on irregular satellite orbits invariably come from the Sun. Solar tides place the upper limit on the distance between the planet and the satellite. In the simplest approximation, the planet's gravity will be stronger than solar tides within its so-called "Hill sphere," the radius of which is given by

$$R_H = \left(\frac{m}{3M}\right)^{1/3} r \tag{2}$$

where m and M are the planet's and the Sun's masses, and r is the mean distance between them. However, the region of long-term stability is somewhat smaller than the Hill sphere, and will be discussed in section 4.2.

On the other hand, most irregulars are sufficiently far from their parent planets not to be appreciably affected by the planet's oblateness. The critical distance for the transition between oblateness-dominated and Sun-dominated dynamics is

$$a_c = \left(2J_2 \frac{m}{M} R^2 r^3\right)^{1/5} \tag{3}$$

where J_2 and R are the planet's second gravitational moment and radius. When we calculate this distance for Jupiter, Saturn, Uranus, and Neptune, we get 2.3, 2.5, 1.4, and 1.8 million km, respectively, although these distances are increased by ~30% when allowance is made for the contribution to each planet's effective J_2 by its equatorial, regular satellites. With the notable exception of Triton, all the satellites that were likely captured from heliocentric orbit have $a > a_c$.

The most important short-period solar perturbation of the orbit of a distant satellite is evection. It is associated with the argument $2\lambda' - 2\varpi$, where λ' is the solar true longitude and ϖ is the satellite's longitude of pericenter. The period of this perturbation is somewhat longer (or shorter, for retrograde satellites) than half the planet's year. Evection induces variability in all orbital elements, but its most notable effect is on the eccentricity. The eccentricity is largest when the Sun is along the line of the moon's apsides, and lowest when the Sun is perpendicular to that line (this is again reversed for retrograde orbits). Lunar evection is well described by *Brouwer and Clemence* (1961), and *Ćuk and Burns* (2004b) have applied evection to high-e and high-i orbits.

The Lidov-Kozai (LK) (*Lidov,* 1962; *Kozai,* 1962) mechanism is an extremely important consequence of solar perturbations, but was not discovered until the Space Age. This perturbation affects only objects with significantly inclined orbits. In the simplest treatment of the LK mechanism (e.g., *Innanen et al.,* 1997), the interactions are purely secular, and the planet's eccentricity has no direct role in the dynamics. The only relevant angle is the argument of pericenter ω, which describes the orientation of the satellite's line of apsides relative to its line of nodes on the planet's orbit. During the precession cycle of ω, eccentricity and inclination are coupled in such a way that the normal component of the angular momentum, $H = \sqrt{1 - e^2} \cos i$, is conserved. Eccentricity is at maximum (and inclination at minimum) when the line of apsides is perpendicular to the line of nodes ($\omega = 90°$ or $\omega = 270°$), while the minimum e and maximum i coincide with $\omega = 0°$ or $\omega = 180°$. Satellites with inclinations above 39.2° and below 140.8° can exhibit librations of ω about 90° or 270°. The first real satellites found to be in such a resonance were Saturn's Ijiraq and Kiviuq (*Vashkov'yak,* 2001; *Ćuk et al.,* 2002; *Nesvorný et al.,* 2003), but LK librators are now known around all four giant planets, including Euporie and Carpo at Jupiter, Margaret (S/2003 U3) at Uranus, and possibly Sao (S/2002 N2) and Neso (S/2002 N4) at Neptune (*Carruba et al.,* 2004).

Finally, the traditional secular interactions involving the longitudes of pericenter (ϖ) of both the perturber and perturbee also can have a significant effect on the irregulars. Most distant irregulars (especially the retrograde jovians) show a perturbation in eccentricity that is governed by the angle between the satellite's and the planet's (or apparent solar) line of apsides, $\Psi = \varpi - \varpi'$, where ϖ' is the planet's longitude of perihelion. While such perturbations are generally weaker than the evection and LK mechanisms, some irregular satellites are known to be in secular resonance, i.e., the argument Ψ librates for some or all of the time. Jupiter's Pasiphae (*Whipple and Shelus,* 1993) and Sinope (*Saha and Tremaine,* 1993) have Ψ librating around 180°, while Saturn's Siarnaq and Narvi exhibit complex circulations of Ψ (*Ćuk et al.,* 2002; *Nesvorný et al.,* 2003; *Ćuk and Burns,* 2004b). While Pasiphae's librations can be very long-lived, none of these resonances appears to be primordial; instead, the critical argument undergoes chaotic variation over many millions of years (*Nesvorný et al.,* 2003).

4.2. Long-Term Stability

Plate 8 clearly shows two trends among the irregular satellite orbits: They avoid high inclinations ($60° < i < 130°$), and the prograde satellites are never found outside about a third of the Hill sphere, while the retrogrades extend out to about $R_H/2$.

The paucity of high-inclination satellites at Jupiter was explored in detail by *Carruba et al.* (2002). They show analytically that the LK mechanism will cause an initially high-inclination, circular orbit to become very eccentric at certain times during its secular cycle. This can lead to the orbit intersecting those of the Galilean satellites for a part of the irregular's precession period, leading to an almost certain collision over the age of the solar system. *Carruba et al.* (2002) also performed numerical simulations for a grid of high-i orbits and showed that the range of inclinations unstable to the above effect is in reality even wider than predicted by analytical theory, due to the perturbations from the other three giant planets. *Nesvorný et al.* (2003) performed analogous integrations for the other three giant planets, with similar results.

Carruba et al. (2002) also find that the center of the "high-inclination hole" is slightly shifted away from 90° toward the retrograde region. This stems from the apsidal precession not being symmetric for prograde and retrograde bodies, despite LK theory predicting identical behavior. The asymmetric term is principally caused by the incomplete averaging of the evection inequality (section 4.1). This extra term was first found to have major effects on the precession of the Moon by Clairaut in the eighteenth century [it almost doubles the predicted precession rate based strictly on secular perturbations (*Baum and Sheehan,* 1997)]. The "secular evection" term always causes additional prograde precession of the apsides, which accelerates and decelerates the precession of prograde and retrograde orbits, respectively. *Ćuk and Burns* (2004b) derived this term for orbits of any eccentricity and inclination using classical celestial mechanics. *Beaugé et al.* (2006) further improved the secular theory for irregulars using a more rigorous approach based on *Hori* (1966).

Averaged evection also leads to the reduced stability of prograde orbits. The feedback between the solar perturbation and apsidal precession of prograde orbits leads to stronger coupling and large swings in eccentricity. At sufficiently large a, the apsidal precession rate will become comparable to the Sun's mean motion, turning the evection inequality into a destabilizing resonance (*Nesvorný et al.,* 2003). This does not happen for retrograde orbits because their basic secular apsidal precession rate is retrograde. However, the competition between the "normal" secular and "secular evection" apsidal precession terms can lead to very slow precession of distant retrograde orbits, which enables their capture into secular resonances (see section 4.1). Solar octupole perturbations that cause the secular resonance lock for Pasiphae are relatively weak (*Yokoyama et al.,* 2003), and the resonance is possible only because the more important apsidal precession terms nearly cancel each other (*Ćuk and Burns,* 2004b).

5. ORIGINS

5.1. Capture via Nebular Gas Drag

The first detailed hypothesis on the capture of irregular satellites was formulated by *Colombo and Franklin* (1971). They proposed that the two then-known groups of jovian irregulars are remnants of two asteroids that collided while passing through Jupiter's Hill sphere. Such an outcome is inconsistent with modern understanding of collisional disruption, and this hypothesis has been mostly abandoned. *Heppenheimer and Porco* (1977) argued that capture could have occurred while Jupiter's primordial gas envelope was collapsing, greatly increasing its mass in a short period of time and leading to permanent capture of passing small bodies. However, this scenario requires the future satellite to share the Hill sphere of the planet with many Earth masses of accreting gas. It can be shown (*Nesvorný et al.,* 2007) that the effects of aerodynamic drag on a small body will generally be more important than those of the planet's changing mass, making the above scenario unrealistic.

Pollack et al. (1979) proposed that it was this aerodynamic drag in the collapsing envelope that produced the capture. However, since the orbital in-spiraling due to drag does not stop with capture, this method also requires a relatively rapid collapse of the circumplanetary nebula in order to remove the gas. Since gas drag affects smaller objects more strongly, surviving objects must have just the right size in order to be captured by drag but evolve slowly enough to outlive the collapse of the gaseous envelope. Additionally, any evolution by gas drag must have happened early, before the formation of dynamical families, as there is no indication of any relationship between size and either semimajor axis or eccentricity for the irregulars (e.g., smaller objects might be expected to be affected more strongly by gas drag), as illustrated in Fig. 4.

Ćuk and Burns (2004a) tested the *Pollack et al.* (1979) scenario numerically and found that it could work for certain satellites, with the largest jovian irregular, Himalia, being the most likely candidate for such a capture. If the small body first goes through a temporary capture lasting a few tens of orbits, its permanent capture can be achieved in a much less dense gas environment than would be possible if the capture has to happen during a single passage. Nevertheless, the nebula still needs to disappear on timescales of 10^4–10^5 yr in order for a Himalia-like body to survive, as there is no other way to stop its orbital evolution.

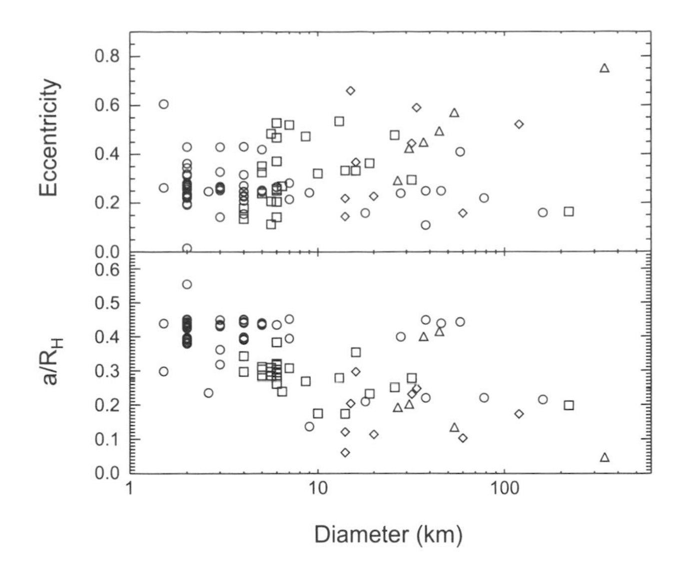

Fig. 4. The distribution of eccentricity and semimajor axis (scaled by the Hill sphere radius, R_H) as a function of estimated size for all irregular satellites. Diameters are calculated assuming geometric albedoes of 0.04–0.07, depending on author and planet. We see that there is no positive correlation between satellite size and either eccentricity or semimajor axis, such as might be expected if gaseous drag had significantly modified the current satellites' orbits. Symbols are as in Plate 1 and Figs. 2–3.

In principle, resonances can either capture a decaying orbit or lift its pericenter (by giving the body in question a negative eccentricity kick) and therefore save the irregulars from collapse. Unfortunately, the only resonance that is currently strong enough to induce large changes in the satellite's eccentricity is the "great inequality," in which the pericenters of prograde saturnians are caught in the 1:2 commensurability with the 900-yr near-resonant perturbation of Jupiter and Saturn. This perturbation arises from the proximity of these planets' orbits to their mutual 2:5 mean-motion resonance. *Ćuk and Burns* (2004b) find that this resonance can induce large secular variations in a satellite's eccentricity that eventually lead to instability, and note that resonances of this type could have have had important effects on irregular satellites during planetary migration (e.g., *Hahn and Malhotra,* 1999; cf. *Carruba et al.,* 2004).

Tsiganis et al. (2005; see also chapter by Morbidelli et al.) proposed that Jupiter and Saturn crossed their mutual 1:2 resonance at some point in the past. Around the time of this resonance crossing, their irregular satellites would be subject to rapidly changing perturbations similar to the great inequality, but likely stronger. *Ćuk and Gladman* (2006) postulated that continuous gas-drag capture and decay of bodies caught in the inner disk (where the regular satellites are now) would provide a steady-state population of bodies on continuously decaying eccentric orbits that can then

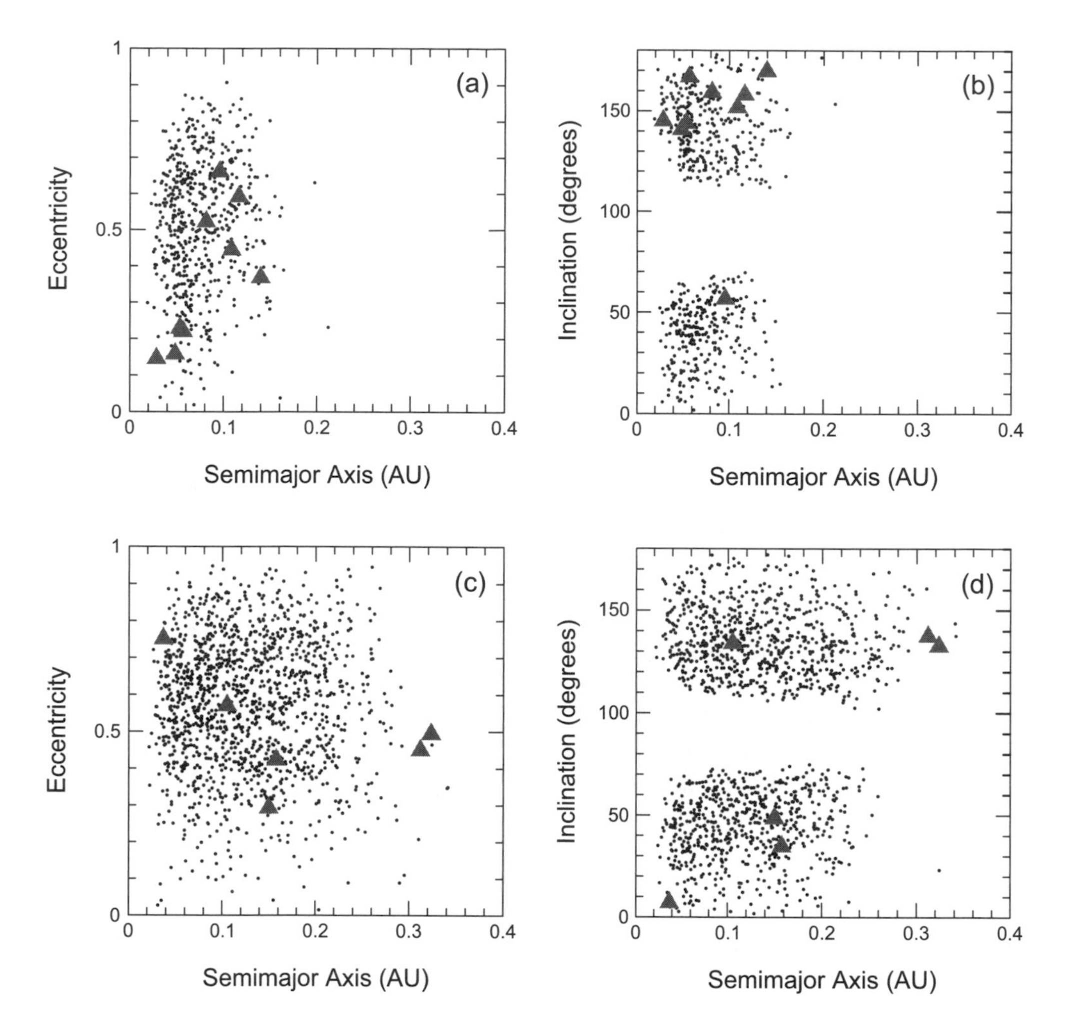

Fig. 5. A comparison between the orbits of objects captured in a numerical calculation (dots) with those of the known irregular satellites (solid triangles). **(a,b)** Satellites of Uranus; **(c,d)** satellites of Neptune. In total, 568 and 1368 stable satellites were captured at Uranus and Neptune, respectively. From *Nesvorný et al.* (2007).

have their pericenters raised by the resonance passage. They numerically tested this hypothesis and found that a significant fraction of very high-e satellites of Saturn would have their eccentricities lowered by such a resonance passage. Their results show that the a and i distributions of bodies so stabilized strongly resemble the known irregular satellites, while the theoretical eccentricities are a little too high compared to the observed ones. A similar mechanism could also help capture permanent irregulars around Uranus and Neptune if they have ever crossed a resonance with Saturn (the orbits of the outer irregulars of Uranus also match the theoretical predictions for this mechanism). This mechanism does not work well for Jupiter, and it seems likely that other mechanisms helped capture its irregulars (see next section).

While *Ćuk and Gladman* (2006) use the basic framework of *Tsiganis et al.* (2005), namely the Jupiter-Saturn resonance crossing, their model is not consistent with the full Nice model (see chapter by Morbidelli et al. and section 5.2 below). The scattering phase following the resonance crossing that *Ćuk and Gladman* (2006) use to capture irregulars would also destroy any prior distant satellites, including those captured during the resonance crossing itself.

5.2. Purely Dynamical Capture Mechanisms

While a permanent capture is impossible in the gravitational three-body problem (Sun, planet, satellite), the presence of a fourth body can change this. *Nesvorný et al.* (2007) find that in the context of *Tsiganis et al.* (2005; also called the "Nice model," see chapter by Morbidelli et al.), there are two distinct classes of encounters that could in principle result in irregular satellites. One involves encounters between a giant planet and a binary planetesimal, while the other involves encounters between two giant planets during the scattering phase of the Nice model, with the latter being more promising. One of the ice giants (Uranus or Neptune) typically has a few encounters with Saturn, while the two can have hundreds of mutual close fly-bys. While the Hill spheres of the two planets overlap, planetesimals passing through that region can get enough of a velocity kick relative to one of the planets to become its permanent satellites.

Nesvorný et al. (2007) tested this idea numerically, using the results of *Gomes et al.* (2005) to generate their initial conditions. They found this process to be efficient, producing a large number of stable irregulars at Saturn and an even larger number at Uranus and Neptune (see Fig. 5). The orbital distribution of these irregulars is mostly random within the stable region, except that very distant retrograde satellites rarely form. This is consistent with the present uranian irregular system, but only partially consistent with the saturnian one, where there are distinct "inclination groups" (*Gladman et al.,* 2001a) (see section 2.3). While the overall mass of captured irregulars in their model is more than adequate to account for the present irregulars, the observed size distribution is much shallower (has fewer small bodies) than that of present-day TNOs. This might be a consequence of a different past size distribution of TNOs, observational bias, or other unmodeled processes.

Since Jupiter does not experience any close encounters with the other planets in the Nice model, none of its irregular satellites could have been captured by this mechanism. Since the mechanism of *Ćuk and Gladman* (2006) cannot explain most jovians either, there is a discrepancy between theoretical expectations (which predict that jovians should be different) and observations of similar size distributions for irregular satellites around all four giant planets (cf. section 3.1). Either this similarity is just a coincidence or is due to significant collisional evolution, as discussed below, or the models of both *Ćuk and Gladman* (2006) and *Nesvorný et al.* (2007) need modfication, with the overall "Nice model" possibly requiring some fine-tuning.

5.3. Collisional Evolution and Families

The osculating orbits of irregular satellites are not constant on century timescales due to gravitational perturbations from the Sun and the other planets (see section 4.1). To determine which irregular satellites have similar orbits and may thus share a common origin, more constant orbital elements must first be defined. Several numerical and analytical methods have been developed for this purpose. The simplest method is to integrate numerically the satellites' orbits over a suitably long timescale and determine the mean of their semimajor axes, eccentricities, and inclinations (hereafter denoted ⟨a⟩, ⟨e⟩, and ⟨i⟩). This is the method adopted by *Nesvorný et al.* (2003), as well as by R. A. Jacobson in generating the widely used mean orbital elements listed on the JPL HORIZONS website (*http://ssd.jpl.nasa.gov/?horizons*). Precise analytic definition of constant orbit elements is more difficult to achieve but is more flexible and can be easily repeated at low CPU cost when the orbit determinations improve (*Beaugé and Nesvorný,* 2007).

Figure 6 illustrates the distribution of numerically defined ⟨a⟩, ⟨e⟩, and ⟨i⟩ for the jovian retrograde satellites. As first noted by *Sheppard and Jewitt* (2003) and *Nesvorný et al.* (2003), two groups of tightly clustered orbits are apparent around Ananke and Carme. A third group around Pasiphae has also been proposed, although in this case the statistical significance of the concentration is low due to a small number of known orbits in the cluster. These satellite groups are reminiscent of the distribution of orbits in the main asteroid belt, where disruptive collisions between asteroids produced groups of fragments sharing similar orbits [the so-called asteroid families (*Hirayama,* 1918; *Zappalá et al.,* 1994)].

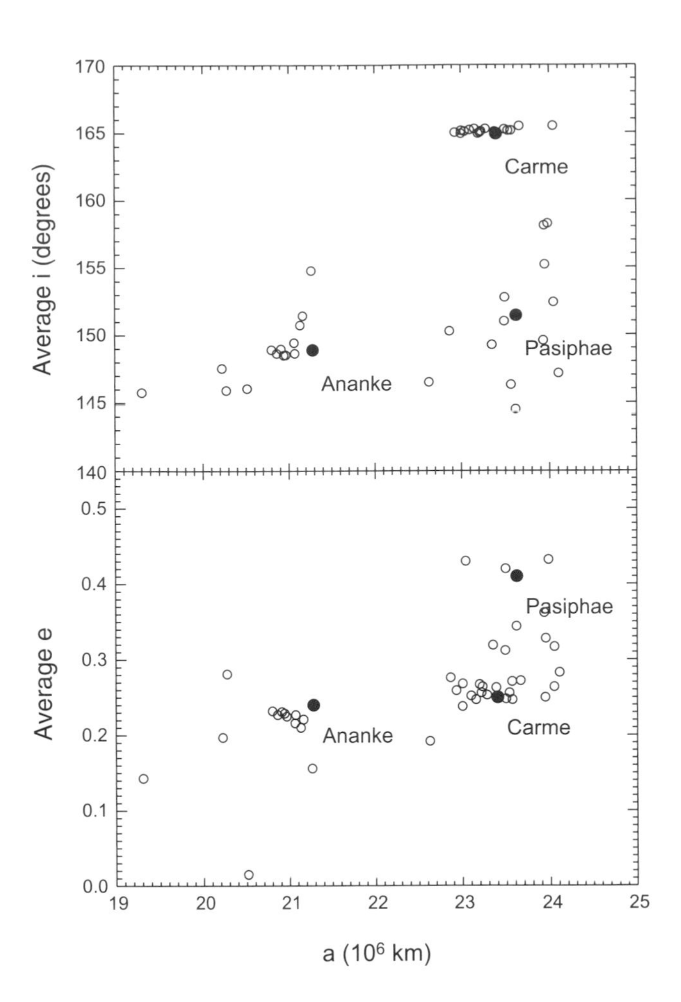

Fig. 6. Averaged orbital elements for the retrograde irregular satellites of Jupiter, from unpublished work by R. A. Jacobson (see *http://ssd.jpl.nasa.gov*). The orbits of many moons are tightly clustered around the orbits of Ananke and Carme, suggesting the possibility of common origins. Many of the remaining objects are more loosely grouped about the orbits of Pasiphae and Sinope (a_6 = 23.9; e = 0.250; i = 158°), but the statistical significance of this concentration is unclear.

The dispersion of orbits in the Ananke and Carme groups (hereafter families) corresponds to ejection speeds of 50 m/s, which nicely corresponds to values expected for the collisional breakup of putative 50-km-diameter parent satellites. Conversely, the prograde Himalia group at Jupiter and the so-called inclination groups at Saturn (e.g., the Phoebe group; see section 2.3) would indicate much larger ejection speeds that may be difficult to reconcile with what we know about large-scale disruptive collisions (*Nesvorný et al.,* 2003; *Grav and Bauer,* 2007). Therefore, either the orbits in these groups evolved significantly after their formation (*Christou et al.,* 2005) or they were captured separately; Christou et al. argue for the collisional origin of the Himalia group. These results are in broad agreement with photometric observations, described in section 3.2, which show that objects in the Himalia, Ananke, and Carme families have similar colors, indicating similar physical properties (as expected for breakup of mineralogically homogeneous objects), while those in the Phoebe group at Saturn show substantial color diversity (*Grav et al.,* 2003; *Grav and Bauer,* 2007). As discussed in section 2.3 above, no unambiguous orbital clusters have been found at Uranus and Neptune, perhaps due to the smaller numbers of currently known irregular moons at these planets.

Nesvorný et al. (2003), following *Kessler* (1981), calculated the rates of disruptive collisions between irregular moons. They found that (1) the large irregular moons must have collisionally eliminated many small irregular moons, thus shaping their population to the currently observed structures; (2) Phoebe's surface must have been heavily cratered by impacts from an extinct population of saturnian irregular moons, much larger than the present one; and (3) disruptive collisions between jovian irregular moons cannot explain their orbital groupings. The first of these findings may account for the similarity in the size distributions noted in section 3.1.

The current impact rate on these moons from kilometer-sized comets and escaped Trojan asteroids is negligible (*Nakamura,* 1993; *Zahnle et al.,* 2003). It has therefore been proposed that the origin of the Carme and Ananke families (and the Pasiphae family, if confirmed) dates back to early epochs of the solar system when impactors were more numerous (*Sheppard and Jewitt,* 2003). *Nesvorný et al.* (2004) analyzed the scenario whereby the satellite families form early by collisions between large parent moons and planetesimals. They found that the Ananke and Carme families at Jupiter could have been produced by these collisions unless the residual disk of planetesimals in heliocentric orbit was already severely depleted when the irregular satellites formed. Conversely, they found that formation of the Himalia group of prograde jovian satellites by the same mechanism was unlikely unless a massive residual planetesimal disk was still present when the progenitor moon of the Himalia group was captured. These results help to place constraints on the mass of the residual disk when satellites were captured, and when the Ananke and Carme families formed. Unfortunately, these constraints also depend sensitively on the assumed size-frequency distribution of planetesimals in the disk at 5–30 AU.

6. INDIVIDUAL OBJECTS

6.1. Phoebe

To date, Phoebe is the only irregular satellite, besides Triton, that has been studied in detail by spacecraft. Low-resolution Voyager images showed it to be an irregularly shaped body, roughly 100 km in radius, with some brightness variations on its surface. The Cassini flyby of December 12, 2004, provided the first close look at an object of this type. In addition to providing information on this satellite's volatile-rich surface composition, the Cassini data on Phoebe's shape and mass determined from tracking data provide a value for its mean density, a key constraint on its bulk composition.

Phoebe's mean radius is 106.6 ± 1 km and its mean density is 1630 ± 45 kg m^{-3} (*Porco et al.,* 2005). This density suggests a bulk composition consisting of a mixture of water ice and silicate, with the proportions depending on the amount of porosity present in this small body. Porco et al. note that even for zero porosity, Phoebe's density is higher than that of the regular satellites (Titan's uncompressed density is ~1500 kg m^{-3}, while the average density of the smaller icy satellites is only ~1300 kg m^{-3}), consistent with its assumed origin from outside the forming Saturn system. Its actual material density is probably even higher since it is plausible that Phoebe has significant bulk porosity, due to the low pressures in its interior (<4 MPa). Jupiter's moon, Amalthea, for instance, is about the same size as Phoebe, but with a density of 857 ± 99 kg m^{-3} (*Anderson et al.,* 2005).

Johnson and Lunine (2005) calculate that for a porosity of only 15%, Phoebe's material density would be similar to the uncompressed densities of Pluto and Triton (~1900 kg m^{-3}). This would be consistent with an origin in the outer protoplanetary nebula with about 70% of the carbon in the form of CO, given current revised solar abundances of C and O. They point out, however, that pure solar composition equilibrium condensation does not easily explain the low densities of the regular satellites. Another uncertainty is the amount of carbon in solid form during the condensation process. Given these uncertainties, perhaps the best general conclusion about Phoebe's composition is that it is probably a body with at least modest bulk porosity and a material density indicating a silicate-rich composition compared with Saturn's regular satellites. Combined with the observed water ice and volatiles on its surface (*Clark et al.,* 2005), this suggests that it formed originally in the outer parts of the solar nebula from a reservoir of material similar to that which formed Pluto and Triton.

6.2. Triton and Nereid

Triton and Nereid satisfy some but not all criteria for being irregular. Triton's orbit is retrograde, but close to Neptune and circular, while that of Nereid is large and eccentric, although Nereid might not be a captured body.

McCord (1966) and *McKinnon* (1984) proposed that Triton is a captured satellite, whose originally eccentric orbit was circularized due to tidal dissipation within Triton. *Goldreich et al.* (1989) proposed that Triton was captured from heliocentric orbit by a collision with a preexisting satellite, and its initial high-eccentricity orbit then evolved due to tidal dissipation. *McKinnon and Leith* (1995) proposed that Triton was captured and evolved by gas drag, but this hypothesis has been less widely accepted than collisional capture. *Agnor and Hamilton* (2006) recently proposed a three-body capture scenario for Triton. They suggest precapture Triton may have been a member of a binary whose disruption during a Neptune encounter led to Triton's capture and its companion's escape.

Neptune almost certainly had preexisting regular satellites, likely similar to those of Uranus (with a total mass of about 40% that of Triton). *Ćuk and Gladman* (2005) show that the largest (hypothetical) satellites of Neptune would collide with each other after only ~10^3 yr if perturbed by an eccentric and inclined early Triton, and then be ground into a disk. This disk would largely be accreted by Triton, causing rapid orbital decay, with final circularization due to tidal dissipation. In this picture, Triton is not simply a captured TNO, but an amalgamation of a large captured object and a significant amount of regular satellite material.

Goldreich et al. (1989) suggested that Nereid was a regular moon of Neptune, scattered onto an irregular orbit by the newly captured Triton. This agrees with Nereid being prograde, as well as with its icy spectrum (*Brown et al.,* 1998), and opens the possibility that there could exist other, smaller bodies in similar orbits. Even after their orbits decoupled, Triton would have perturbed Nereid (and other relatively close-in irregulars) more strongly than the Sun. *Ćuk and Gladman* (2005) find that test particles perturbed by a massive interior body with large e and i can oscillate between high-i prograde and retrograde orbits. Therefore, high-i retrograde objects could also be fragments of regular satellites.

Acknowledgments. The work described in this paper was carried out with support from NASA's Planetary Astronomy program (P.D.N.) and the National Science Foundation (D.N.). S.S.S. was supported by NASA through Hubble Fellowship grant #HF-01178.01-A awarded by the Space Telescope Science Institute, which is operated by the Association of Universities for Research in Astronomy, Inc., for NASA, under contract NAS 5-26555. M.Ć. acknowledges support from the Canadian Institute for Theoretical Astrophysics (CITA) and the Natural Sciences and Engineering Research Council (NSERC) of Canada. A portion of this work was done at the Jet Propulsion Laboratory, California Institute of Technology, under a contract from NASA.

REFERENCES

Agnor C. B. and Hamilton D. P. (2006) Neptune's capture of its moon Triton in a binary-planet gravitational encounter. *Nature, 441,* 192–194.

Anderson J. D., Johnson T. V., Schubert G., Asmar S., Jacobson R. A., Johnston D., Lau E. L., Lewis G., Moore W. B., Taylor A., Thomas P. C., and Weinwurm G. (2005) Amalthea's density is less than that of water. *Science, 308,* 1291–1293.

Barucci A., Belskaya I., Fulchignoni M., and Birlan M. (2005) Taxonomy of Centaurs and trans-neptunian objects. *Astron. J., 130,* 1291–1298.

Bauer J., Grav T., Buratti B., and Hicks M. (2006) The phase curve survey of the irregular saturnian satellites: A possible method of physical classification. *Icarus, 184,* 181–197.

Baum R. and Sheehan W. (1997) *In Search of Planet Vulcan.* Plenum, New York.

Beaugé C. and Nesvorný D. (2007) Proper elements and secular resonances of irregular satellites. *Astron. J., 133,* 2537–2558.

Beaugé C., Nesvorný D., and Dones L. (2006) A high-order analytical model for the secular dynamics of irregular satellites. *Astron. J., 131,* 2299–2313.

Bernstein G., Trilling D., Allen R., Brown M., and Malhotra R. (2004) The size distribution of trans-neptunian bodies. *Astron. J., 128,* 1364–1390.

Brouwer D. and Clemence G. M. (1961) *Methods of Celestal Mechanics.* Academic, New York.

Brown M. E. (2000) Near-infrared spectroscopy of Centaurs and irregular satellites. *Astron. J., 119,* 977–983.

Brown M. E., Koresko C. D., and Blake G. A. (1998) Detection of water ice on Nereid. *Astrophys. J. Lett., 508,* L175–L176.

Buratti B., Hicks M., and Davies A. (2005) Spectrophotometry of the small satellites of Saturn and their relationship to Iapetus, Phoebe, and Hyperion. *Icarus, 175,* 490–495.

Burns J. A. (1986a) Some background about satellites. In *Satellites* (J. A. Burns and M. S. Matthews, eds.), pp. 1–38. Univ. of Arizona, Tucson.

Burns J. A. (1986b) The evolution of satellite orbits. In *Satellites* (J. A. Burns and M. S. Matthews, eds.), pp. 117–158. Univ. of Arizona, Tucson.

Canup R. M. and Ward W. R. (2002) Formation of the Galilean satellites: Conditions of accretion. *Astron. J., 124,* 3404–3423.

Carruba V., Burns J. A., Nicholson P. D., and Gladman B. J. (2002) On the inclination distribution of the jovian irregular satellites. *Icarus, 158,* 434–449.

Carruba V., Nesvorný D., Burns J. A., Ćuk M., and Tsiganis K. (2004) Chaos and the effects of planetary migration on the orbit of S/2000 S5 Kiviuq. *Astron. J., 128,* 1899–1915.

Chamberlain M. and Brown R. (2004) Near-infrared spectroscopy of Himalia. *Icarus, 172,* 163–169.

Christou A. A. (2005) Gravitational scattering within the Himalia group of jovian prograde irregular satellites. *Icarus, 174,* 215–229.

Clark R., Brown R., Jaumann R., Cruikshank D., et al. (2005) Compositional maps of Saturn's moon Phoebe from imaging spectroscopy. *Nature, 435,* 66–69.

Colombo G. and Franklin F. A. (1971) On the formation of the outer satellite groups of Jupiter. *Icarus, 15,* 186–189.

Cruikshank D. (1977) Radii and albedos of four Trojan asteroids and jovian satellites 6 and 7. *Icarus, 30,* 224–230.

Cruikshank D., Stansberry J., Emery J., Fernandez Y., Werner M., Trilling D., and Rieke G. (2005) The high-albedo of Kuiper belt object (55565) 2002 AW_{197}. *Astrophys. J. Lett., 624,* L53–L56.

Ćuk M. and Burns J. A. (2004a) Gas-drag-assisted capture of Himalia's family. *Icarus, 167,* 369–381.

Ćuk M. and Burns J. A. (2004b) On the secular behavior of the irregular satellites. *Astron. J., 128,* 2518–2541.

Ćuk M. and Gladman B. J. (2005) Constraints on the orbital evolution of Triton. *Astrophys. J. Lett., 626,* L113–L116.

Ćuk M. and Gladman B. J. (2006) Irregular satellite capture during planetary resonance passage. *Icarus, 183,* 362–372.

Ćuk M., Burns J. A., Carruba V., Nicholson P. D., and Jacobson R. A. (2002) New secular resonances involving the irregular satellites of Saturn. *Bull. Am. Astron. Soc., 34,* 943.

Degewij J., Zellner B., and Andersson L. E. (1980) Photometric properties of outer planetary satellites. *Icarus, 44,* 520–540.

Elliot J. L. and 10 colleagues (2005) The deep ecliptic survey: A search for Kuiper belt objects and Centaurs. II. Dynamical classification, the Kuiper belt plane, and the core population. *Astron. J., 129,* 1117–1162.

Fernandez Y., Sheppard S., and Jewitt D. (2003) The albedo distribution of jovian Trojan asteroids. *Astron. J., 126,* 1563–1574.

Fornasier S., Dotto E., Marzari F., Barucci M., Boehnhardt H., Hainaut O., and de Bergh C. (2004) Visible spectroscopic and photometric survey of L5 Trojans: Investigation of dynamical families. *Icarus, 172,* 221–232.

Geballe T., Dalle Ore C., Cruikshank D., and Owen T. (2002) The 1.95–2.50 micron spectrum of J6 Himalia. *Icarus, 159,* 542–544.

Gladman B. J., Nicholson P. D., Burns J. A., Kavelaars J. J., Marsden B. G., Williams G. V., and Offutt W. B. (1998a) Discovery of two distant irregular moons of Uranus. *Nature, 392,* 897–899.

Gladman B., Kavelaars J. J., Nicholson P. D., Loredo T. J., and Burns J. A. (1998b) Pencil-beam surveys for faint trans-neptunian objects. *Astron. J., 116,* 2042–2054.

Gladman B., Kavelaars J., Holman M., Petit J.-M., Scholl H., Nicholson P., and Burns J. A. (2000) NOTE: The discovery of Uranus XIX, XX, and XXI. *Icarus, 147,* 320–324.

Gladman B. and 10 colleagues (2001a) Discovery of 12 satellites of Saturn exhibiting orbital clustering. *Nature, 412,* 163–166.

Gladman B., Kavelaars J., Petit J-M., Morbidelli A., Holman M., and Loredo T. (2001b) The structure of the Kuiper belt: Size distribution and radial extent. *Astron. J., 122,* 1051–1066.

Goldreich P., Murray N., Longaretti P. Y., and Banfield D. (1989) Neptune's story. *Science, 245,* 500–504.

Gomes R., Levison H. F., Tsiganis K., and Morbidelli A. (2005) Origin of the cataclysmic late heavy bombardment period of the terrestrial planets. *Nature, 435,* 466–469.

Grav T. and Bauer J. (2007) A deeper look at the colors of the saturnian irregular satellites. *Icarus, 191,* 267–285.

Grav T. and Holman M. (2004) Near-infrared photometry of the irregular satellites of Jupiter and Saturn. *Astrophys. J. Lett., 605,* L141–L144.

Grav T., Holman M., Gladman B., and Aksnes K. (2003) Photometric survey of the irregular satellites. *Icarus, 166,* 33–45.

Grav T., Holman M., and Fraser W. (2004) Photometry of irregular satellites of Uranus and Neptune. *Astrophys. J. Lett., 613,* L77–L80.

Grundy W., Noll K., and Stephens D. (2005) Diverse albedos of small trans-neptunian objects. *Icarus, 176,* 184–191.

Hahn J. M. and Malhotra R. (1999) Orbital evolution of planets embedded in a planetesimal disk. *Astron. J., 117,* 3041–3053.

Heppenheimer T. A. and Porco C. (1977) New contributions to the problem of capture. *Icarus, 30,* 385–401.

Hirayama K. (1918) Groups of asteroids probably of common origin. *Astron. J., 31,* 185–188.

Holman M. J. and 13 colleagues (2004) Discovery of five irregular moons of Neptune. *Nature, 430,* 865–867.

Hori G. (1966) Theory of general perturbation with unspecified canonical variable. *Publ. Astron. Soc. Japan, 18,* 287–296.

Innanen K. A., Zheng J. Q., Mikkola S., and Valtonen M. J. (1997) The Kozai mechanism and the stability of planetary orbits in binary star systems. *Astron. J., 113,* 1915–1919.

Jarvis K., Vilas F., Larson S., and Gaffey M. (2000) JVI Himalia: New compositional evidence and interpretations for the origin of Jupiter's small satellites. *Icarus, 145,* 445–453.

Jewitt D. (2002) From Kuiper belt object to cometary nucleus: The missing ultrared matter. *Astron. J., 123,* 1039–1049.

Jewitt D. (2005) A first look at the damocloids. *Astron. J., 129,* 530–538.

Jewitt D. and Luu J. (2001) Colors and spectra of Kuiper belt objects. *Astron. J., 122,* 2099–2114.

Jewitt D. and Sheppard S. (2005) Irregular satellites in the context of planet formation. *Space Sci. Rev., 116,* 441–455.

Johnson T. V. and Lunine J. I. (2005) Saturn's moon Phoebe as a captured body from the outer solar system. *Nature, 435,* 69–71.

Kavelaars J. J., Holman M. J., Grav T., Milisavljevic D., Fraser W., Gladman B. J., Petit J.-M., Rousselot P., Mousis O., and Nicholson P. D. (2004) The discovery of faint irregular satellites of Uranus. *Icarus, 169,* 474–481.

Kessler D. J. (1981) Derivation of the collision probability between orbiting objects: The lifetimes of Jupiter's outer moons. *Icarus, 48,* 39–48.

Kowal C., Aksnes K., Marsden B., and Roemer E. (1975) Thirteenth satellite of Jupiter. *Icarus, 80,* 460–464.

Kozai Y. (1962) Secular perturbations of asteroids with high inclination and eccentricity. *Astron J., 67,* 591–598.

Kuiper G. P. (1961) Limits of completeness. In *The Solar System, Vol. III: Planets and Satellites* (G. P. Kuiper and B. M. Middlehurst, eds.), pp. 575–591. Univ. of Chicago, Chicago.

Lidov M. L. (1962) The evolution of orbits of artificial satellites of planets under the action of gravitational perturbations of external bodies. *Planet. Space Sci., 9,* 719–759.

Luu J. (1991) CCD photometry and spectroscopy of the outer jovian satellites. *Astron. J., 102,* 1213–1225.

Luu J. X. and Jewitt D. (1988) A two-part search for slow-moving objects. *Astron. J., 95,* 1256–1262.

Maris M., Carraro G., Cremonese G., and Fulle M. (2001) Multicolor photometry of the Uranus irregular satellites Sycorax and Caliban. *Astron. J., 121,* 2800–2803.

McCord T. B. (1966) Dynamical evolution of the neptunian system. *Astron J., 71,* 585–590.

McKinnon W. B. (1984) On the origin of Triton and Pluto. *Nature, 311,* 355–358.

McKinnon W. B. and Leith A. C. (1995) Gas drag and the orbital evolution of a captured Triton. *Icarus, 118,* 392–413.

Morbidelli A., Levison H. F., Tsiganis K., and Gomes R. (2005) Chaotic capture of Jupiter's Trojan asteroids in the early solar system. *Nature, 435,* 462–465.

Mosqueira I. and Estrada P. R. (2003) Formation of the regular satellites of giant planets in an extended gaseous nebula I: Subnebula model and accretion of satellites. *Icarus, 163,* 198–231.

Nakamura A. M. (1993) *Laboratory Studies on the Velocity of Fragments from Impact Disruptions.* ISAS Report 651, Institute of Space and Astronautical Science, Tokyo.

Nesvorný D., Alvarellos J. L. A., Dones L., and Levison H. F. (2003) Orbital and collisional evolution of the irregular satellites. *Astron. J., 126,* 398–429.

Nesvorný D., Beaugé C., and Dones L. (2004) collisional origin of families of irregular satellites. *Astron. J., 127,* 1768–1783.

Nesvorný D., Vokrouhlický D., and Morbidelli A. (2007) Capture of irregular satellites during planetary encounters. *Astron. J., 133,* 1962–1976.

Nicholson P. D. and Gladman B. J. (2006) Satellite searches at Pluto and Mars. *Icarus, 181,* 218–222.

Owen T., Cruikshank D., Dalle Ore C., Geballe T., Roush T., and de Bergh C. (1999) Detection of water ice on Saturn's satellite Phoebe. *Icarus, 139,* 379–382.

Peale S. J. (1999) Origin and evolution of the natural satellites. *Annu. Rev. Astron. Astrophys., 37,* 533–602.

Peixinho N., Lacerda P., Ortiz J., Doressoundiram A., Roos-Serote M., and Gutierrez P. (2001) Photometric study of Centaurs 10199 Chariklo (1997 CU_{26}) and 1999 UG_5. *Astron. Astrophys., 371,* 753–759.

Petit J.-M., Holman M., Scholl H., Kavelaars J., and Gladman B. (2004) A highly automated moving object detection package. *Mon. Not. R. Astron. Soc., 347,* 471–480.

Pollack J. B., Burns J. A., and Tauber M. E. (1979) Gas drag in primordial circumplanetary envelopes — A mechanism for satellite capture. *Icarus, 37,* 587–611.

Porco C. C. and 23 colleagues (2003) Cassini imaging of Jupiter's atmosphere, satellites, and rings. *Science, 299,* 1541–1547.

Porco C. C. and 34 colleagues (2005) Cassini imaging science: Initial results on Phoebe and Iapetus. *Science, 307,* 1237–1242.

Rettig T., Walsh K., and Consolmagno G. (2001) Implied evolutionary differences of the jovian irregular satellites from a BVR color survey. *Icarus, 154,* 313–320.

Romon J., de Bergh C., Barucci M., Doressoundiram A., Cuby J., Le Bras A., Doute S., and Schmitt B. (2001) Photometric and spectroscopic observations of Sycorax, satellite of Uranus. *Astron. Astrophys., 376,* 310–315.

Saha P. and Tremaine S. (1993) The orbits of the retrograde jovian satellites. *Icarus, 106,* 549–562.

Sheppard S. S. and Jewitt D. C. (2003) An abundant population of small irregular satellites around Jupiter. *Nature, 423,* 261–263.

Sheppard S. and Trujillo C. (2006) A thick cloud of Neptune Trojans and their colors. *Science, 313,* 511–514.

Sheppard S. S., Jewitt D., and Kleyna J. (2004) A survey for outer satellites of Mars: Limits to completeness. *Astron. J., 128,* 2542–2546.

Sheppard S. S., Jewitt D., and Kleyna J. (2005) An ultradeep survey for irregular satellites of Uranus: Limits to completeness. *Astron. J., 129,* 518–525.

Sheppard S., Jewitt D., and Kleyna J. (2006) A survey for "normal" irregular satellites around Neptune: Limits to completeness. *Astron. J., 132,* 171–176.

Showalter M. R., Hamilton D. P., and Nicholson P. D. (2001) A search for martian dust rings. *Bull. Am. Astron. Soc., 33,* 1095.

Simonelli D., Kay J., Adinolfi D., Veverka J., Thomas P., and Helfenstein P. (1999) Phoebe: Albedo map and photometric properties. *Icarus, 138,* 249–258.

Stern S. A., Parker J. W., Fesen R. A., Barker E. S., and Trafton L. M. (1991) A search for distant satellites of Pluto. *Icarus, 94,* 246–249.

Stern S. A., Parker J. W., Duncan M. J., Snowdall J. C. J., and Levison H. F. (1994) Dynamical and observational constraints on satellites in the inner Pluto-Charon system. *Icarus, 108,* 234–242.

Stevenson D. J., Harris A. W., and Lunine J. I. (1986) Origins of satellites. In *Satellites* (J. A. Burns and M. S. Matthews, eds.), pp. 38–88. Univ. of Arizona, Tucson.

Sykes M., Nelson B., Cutri R., Kirkpatrick D., Hurt R., and Skrutskie M. (2000) Near-infrared observations of the outer jovian satellites. *Icarus, 143,* 371–375.

Tegler S. and Romanishin W. (2000) Extremely red Kuiper-belt objects in near-circular orbits beyond 40 AU. *Nature, 407,* 979–981.

Tholen D. and Zellner B. (1984) Multicolor photometry of outer jovian satellites. *Icarus, 58,* 246–253.

Thomas P., Veverka J., and Dermott S. (1986) Small satellites. In *Satellites* (J. A. Burns and M. S. Matthews, eds.), pp. 802–835. Univ. of Arizona, Tucson.

Thomas P., Veverka J., and Helfenstein P. (1991) Voyager observations of Nereid. *J. Geophys. Res., 96,* 19253.

Trujillo C. and Brown M. (2002) A correlation between inclination and color in the classical Kuiper belt. *Astrophys. J. Lett., 566,* L125–L128.

Trujillo C., Jewitt D., and Luu J. (2001) Properties of the trans-Neptunian belt: Statistics from the Canada-France-Hawaii telescope survey. *Astron. J., 122,* 457–473.

Tsiganis K., Gomes R., Morbidelli A., and Levison H. F. (2005) Origin of the orbital architecture of the giant planets of the solar system. *Nature, 435,* 459–461.

Vashkov'yak M. A. (2001) Orbital evolution of Saturn's new outer satellites and their classification. *Astron. Lett., 27,* 455–463.

Whipple A. L. and Shelus P. J. (1993) A secular resonance between Jupiter and its eighth satellite? *Icarus, 101,* 265–271.

Yokoyama T., Santos M. T., Cardin G., and Winter O. C. (2003) On the orbits of the outer satellites of Jupiter. *Astron. Astrophys., 401,* 763–772.

Weaver H. A., Stern S. A., Mutchler M. J., Steffl A. J., Buie M. W., Merline W. J., Spencer J. R., Young E. F., and Young L. A. (2006) Discovery of two new satellites of Pluto. *Nature, 439,* 943–945.

Zahnle K., Schenk P., Levison H. F., and Dones L. (2003) Cratering rates in the outer solar system. *Icarus, 163,* 263–289.

Zappalá V., Cellino A., Farinella P., and Milani A. (1994) Asteroid families. 2: Extension to unnumbered multiopposition asteroids. *Astron. J., 107,* 772–801.

Structure of the Kuiper Belt Dust Disk

J.-C. Liou
Engineering and Sciences Contract Group (ESCG)/ERC, Inc.

David E. Kaufmann
Southwest Research Institute

An overview of the Kuiper belt dust disk is provided in this chapter. Mutual collisions among Kuiper belt objects should produce a dust disk in the outer solar system similar to the observed circumstellar dust disks. As the Kuiper belt dust particles migrate toward the Sun due to Poynting-Robertson drag, they are perturbed by the giant planets. Mean-motion resonances with Neptune and gravitational scattering by Saturn and Jupiter alter their orbital evolution dramatically. As a result, large-scale structures are created in the disk. Descriptions of the dynamics involved, and the numerical simulations required to unveil the disk features, are included. Implications for extrasolar planet detection from circumstellar dust disk modeling are also discussed.

1. KUIPER BELT DUST PARTICLES

Interplanetary dust particles (IDPs) are a natural component of the solar system. They originate primarily from collisions among asteroids and disintegration of comets near the Sun. There is a continuous size distribution from parent objects that are tens or hundreds of kilometers in diameter down to fragments that are submicrometer in size. Particles smaller than ~1 µm, however, the so-called "blow-out" size for the solar system, are removed from the solar system on timescales less than or equal to a few hundred years by solar radiation pressure. The term "dust" in general refers to particles in the micrometer-to-millimeter size regime. The zodiacal light that is visible to the naked eye right after sunset or right before sunrise, when the sky is clear and dark, is simply the reflection of sunlight by IDPs near 1 AU. An estimated 15,000 to 40,000 tons of IDPs fall onto Earth every year (*Grün et al.,* 1985; *Love and Brownlee,* 1993). Although the particle environment in the outer solar system is not well understood, impact data collected by *in situ* sensors, such as those onboard Pioneer 10 and 11 and Cassini, also indicate the existence of IDPs beyond the orbit of Jupiter.

Shortly after the discovery of 1992 QB_1 (*Jewitt and Luu,* 1993), *Stern* (1995, 1996a,b) modeled the collision activities among the Kuiper belt (KB) objects, and arrived at the following conclusions for KB dust particles: (1) the quasi-steady-state collision production rate of objects a few kilometers to a few micrometers in size is between 3×10^{16} and 10^{19} g yr^{-1}; (2) there should be a relatively smooth, quasi-steady-state, longitudinally isotropic, far-infrared emission near the ecliptic in the solar system's invariable plane; and (3) recent impacts should produce short-lived (i.e., with ~10-yr windows of detectability), discrete clouds with significantly enhanced, localized IR emission signatures superimposed on the smooth, invariable plane emission. Similar short-lived dust enhancements are expected in the zodiacal cloud. *Backman et al.* (1995) also modeled the KB dust population, and the resulting far-infrared emission independently at the same time. In addition to mutual collisions among KB objects, they included solar radiation pressure, Poynting-Robertson (PR) drag, and sublimation in the model. The innermost edge of the dust structure was assumed to end at the orbit of Neptune. The modeled KB dust population has a wedge geometry centered at the ecliptic, and is symmetric in longitude.

A second possible mechanism to produce KB dust particles was proposed by *Yamamoto and Mukai* (1998). Based on the interstellar dust (ISD) flux measured by the impact detector on the Ulysses spacecraft (*Grün et al.,* 1993), they analyzed the outcome of ISD impacting the large KB objects. The results showed that the secondary ejecta production rate could be between 1.2×10^{13} g yr^{-1} and 9.8×10^{14} g yr^{-1}, although most of the particles are in the 10-µm or smaller size regime.

Even before 1992 QB_1 was discovered, the survey from the Infrared Astronomical Satellite (IRAS) and the subsequent observations by other telescopes in infrared and submillimeter wavelengths revealed many circumstellar dust disks around nearby stars (e.g., *Aumann et al.,* 1984; *Backman and Paresce,* 1993; *Greaves,* 2005). As the number of discovered KB objects continues to increase, it is not unreasonable to argue that the outer solar system is an analog to those observed systems, but at a more mature evolutionary stage. The existence of a collision-produced dust disk in the KB region appears to be further strengthened.

Once dust particles are created in the KB region, they will spiral slowly toward the Sun due to PR drag (e.g., *Wyatt and Whipple,* 1950; *Burns et al.,* 1979). This unique Sunward motion makes KB dust particles stand out from their much larger parent KB objects. Rather than remaining in the KB region, their dynamical path takes them on a long and very different journey toward the inner solar system. *Liou et al.* (1995a, 1996) performed a detailed numerical simulation of the orbital evolution of KB dust particles. The results indicated that gravitational perturbations from the four giant planets have a profound impact on their orbital evolution. Mean-motion resonances (MMRs), close encounters, and gravitation scatterings often interrupt their Sunward motion. As a result, the distribution of the KB dust particles is anything but uniform. Large-scale structures are part of the KB dust disk, as seen from afar (*Liou and Zook,* 1999).

Small asteroidal IDPs in the zodiacal cloud can be trapped into exterior MMRs with the Earth, and form a ring-like structure with certain distinct patterns (*Jackson and Zook,* 1989). Perturbations by an unseen planet could be used to explain features embedded in the dust disk surrounding β Pictoris, as proposed by *Roques et al.* (1994) and *Lazzaro et al.* (1994) after the infrared and optical observations of the system (*Smith and Terrile,* 1984). Similar structures have also been observed for other systems, such as Vega and ε Eridani (e.g., *Aumann et al.,* 1984; *Greaves et al.,* 1998). Since the solar system is the only planetary system where the exact planet configuration (number, orbits, masses) is known, understanding features of the zodiacal cloud, and correlating features of the KB dust disk with the giant planets, may provide insights into modeling extrasolar planets based on the observed features on circumstellar dust disks (see chapter by Moro-Martín et al. for additional details).

2. ORBITAL EVOLUTION OF KUIPER BELT DUST PARTICLES

The orbital motion of KB dust particles is dominated by the gravitational attraction of the Sun. The resultant elliptical motion for particles on bound orbits is well known. Particles smaller than ~1 μm usually do not remain on bound orbits upon release from their parent bodies due to solar radiation pressure (*Zook,* 1975). In addition to the Sun's gravity, the orbits of KB dust particles respond to the gravitational perturbations of the planets as well as to forces due to solar radiation and solar wind, including solar radiation pressure, PR drag, and solar wind drag (e.g., *Burns et al.,* 1979). Although these perturbing forces are generally much smaller in magnitude than the gravitational attraction of the Sun, they produce, acting consistently and over long time periods, significant evolutionary changes in the orbits of KB dust particles.

To model the structure of the KB dust disk properly, an analytical approach and numerical simulations are both needed. The former provides a key to understanding the physics involved, and can be used to verify results from the numerical simulations. The latter provides realistic and quantitative orbital elements needed to build a model KB dust disk. They are described in some detail below.

2.1. Radiation Pressure, Poynting-Robertson Drag, and Planetary Perturbations

Solar radiation incident on a stationary surface area, A, perpendicular to the solar direction produces the force

$$\mathbf{F}_r = \frac{S_0 A Q_{pr}}{r_s^2 c}\hat{\mathbf{r}}_s \tag{1}$$

where $\hat{\mathbf{r}}_S$ is the unit position vector of the surface element with respect to the Sun, r_s is the distance from the Sun to the surface element, S_0 is the solar constant or radiation flux density at unit distance, c is the speed of light, and Q_{pr} is the efficiency factor for radiation pressure weighted by the solar spectrum. Radiation pressure varies with heliocentric distance as the flux density of sunlight, which has an inverse-square dependence (*Burns et al.,* 1979).

Because the gravitational attraction of the Sun also has an inverse-square dependence on the heliocentric distance, it is customary to eliminate r_s by introducing the dimensionless quantity

$$\beta = \frac{\text{radiation pressure force}}{|\text{solar gravitational force}|} = \frac{S_0 A Q_{pr}}{G M_\odot m c} \tag{2}$$

where G is the Newtonian gravitational constant, $M_\odot$ is the mass of the Sun, and m is the mass of the particle. The Sun radiates nearly all its energy in a narrow waveband around 0.6 μm so that the transition from geometric optics to Rayleigh scattering occurs in the micrometer size range. The efficiency Q_{pr} is usually calculated from Mie theory for homogeneous spheres (*van de Hulst,* 1957).

Given the particle mass and β, then, the resulting force field acting on a stationary particle is

$$\mathbf{F}_{\odot+r} = -\frac{G(1-\beta)M_\odot m}{r_s^2}\hat{\mathbf{r}}_s \tag{3}$$

Thus, the effect of radiation pressure reduces by a factor of (1 – β) the gravitational mass of the Sun as "seen" by the particle. A nonradial force component is also usually present on an arbitrarily oriented surface, but averages out on a sphere or any other body with rotational symmetry about $\hat{\mathbf{r}}_S$.

Kuiper belt dust particles are not stationary, however. Their orbital motion within the electromagnetic and particle radiation fields of the Sun induces drag forces (*Burns et al.,* 1979), namely PR drag and solar wind drag, respectively. Although these drag forces are weak compared to the velocity-independent radial pressure force component, they dissipate energy and momentum and thereby cause the particles to eventually spiral into the Sun. The PR drag can be thought of as arising from an aberration of the sunlight as seen from the particle and a Doppler-shift-induced change in momentum. To the first order in v/c, where v is the magnitude of **v**, the inertial velocity of the particle, the PR drag force acting on a spherical particle is

$$\mathbf{F}_{P\text{-}R} = \frac{S_0AQ_{pr}}{r_s^2c}\left(-\frac{\mathbf{v}\cdot\hat{\mathbf{r}}_s}{c}\hat{\mathbf{r}}_s - \frac{\mathbf{v}}{c}\right) \tag{4}$$

Except for the efficiency factor Q_{pr}, equation (4) does not depend on the wave nature of light and has also been derived in a corpuscular formulation (*Klačka,* 1992). Thus, forces due to collisions with solar wind particles are analogous to forces due to radiation. The ratio of solar wind pressure to radiation pressure is very small, hence solar wind pressure can generally be ignored. However, the ratio of solar wind to PR drag is much larger because of the greater aberration angle and Doppler shift for solar wind particles than for photons. The contribution of the solar wind to the drag forces on dust particles is typically parameterized as a constant fraction sw of the PR drag. The numerical value of sw is commonly taken to be 0.35 (*Gustafson,* 1994). The total drag force on the dust particle can thus be written as

$$\mathbf{F}_{drag} = \frac{S_0AQ_{pr}}{r_s^2c}\left(-(1+sw)\frac{\mathbf{v}\cdot\hat{\mathbf{r}}_s}{c}\hat{\mathbf{r}}_s - (1+sw)\frac{\mathbf{v}}{c}\right) \tag{5}$$

The last perturbing forces to be considered are the gravitational forces due to the planets. This contribution to the total force on a given dust particle, assuming n planets, can be written simply as

$$\mathbf{F}_{pl} = -\sum_{i=1}^{n}\frac{GM_im}{r_i^2}\hat{\mathbf{r}}_i \tag{6}$$

where M_i, r_i, and $\hat{\mathbf{r}}_i$ are, respectively, the mass of the ith planet, the distance from the ith planet to the particle, and the unit position vector of the particle with respect to that planet. Regarding the orbital evolution of KB dust particles, one primary effect of planetary perturbations is to trap the particles into MMRs, particularly the exterior MMRs with Neptune. An MMR occurs when the orbital period of the dust particle is a simple ratio of integers to that of the perturbing planet. For example, a 2:1 exterior MMR with Neptune occurs when the dust particle evolves to an orbit such that its orbital period is twice that of Neptune. While in an MMR, the orbital energy gained by the dust particle from the perturbing planet counterbalances its orbital energy loss due to the drag forces (e.g., *Weidenschilling and Jackson,* 1993). Its sunward motion is temporarily halted. Resonant trapping does not last indefinitely, however. Eventually the trapped particle escapes the MMR and continues its sunward spiral. Not all KB dust particles become trapped in MMRs with the planets. Rather, resonance trapping is a probabilistic phenomenon (e.g., see section 2.3). If a dust particle closely encounters a planet during its sunward trek, it may suffer another primary effect of planetary perturbations. It may be gravitationally ejected from the solar system.

Combining the gravitational forces on the KB dust particle from the Sun and planets, the radiation pressure force, and the PR and solar wind drag forces, the complete equation of motion for the particle can be written as

$$m\dot{\mathbf{v}} = -\left[\frac{GM_\odot m}{r_s^2}\right]\hat{\mathbf{r}}_s - \sum_{i=1}^{n}\frac{GM_im}{r_i^2}\hat{\mathbf{r}}_i + \frac{S_0AQ_{pr}}{r_s^2c}\left[\left(1-(1+sw)\frac{\mathbf{v}\cdot\hat{\mathbf{r}}_s}{c}\right)\hat{\mathbf{r}}_s - (1+sw)\frac{\mathbf{v}}{c}\right] \tag{7}$$

For a given set of initial conditions, the solution of this equation yields the orbital evolution of the corresponding KB dust particle. The next section describes what can be learned about this evolution using an analytical approach.

2.2. Analytical Approach

When a KB dust particle is released from its parent body (with zero relative velocity), it immediately "sees," due to radiation pressure, a less massive Sun. This reduction in centripetal force (or, equivalently, gain in orbital energy) causes its new semimajor axis, a_n, and eccentricity, e_n, to be

$$a_n = a\left[\frac{1-\beta}{1-2\beta(a/r)}\right] \tag{8}$$

$$e_n = \left[1 - \frac{(1-2\beta a/r)(1-e^2)}{(1-\beta)^2}\right]^{0.5} \tag{9}$$

where a and e are the semimajor axis and orbital eccentricity, respectively, of the parent body, and r is the radial distance from the Sun when the dust particle is released. Depending on the point at which the dust particle is released (the factor a/r in the above equation), a_n can become negative. If that happens, the dust particle simply leaves the solar system on a hyperbolic orbit (*Zook,* 1975).

For a dust particle that remains on a bound elliptical orbit around the Sun, its orbit immediately begins to evolve under the influence of PR and solar wind drag. Since the drag forces are in the opposite direction to the particle's velocity, energy and angular momentum are extracted from the particle's orbit, decreasing both the orbital semimajor axis and the eccentricity. Note from equations (1) and (5) that there is no force component of the combined radiation pressure force and drag forces that is normal to the orbit plane. Hence there is no change in the orbital inclination for a particle due to these effects. The time-averaged rates of change for a and e due to the drag forces are given by *Burns et al.* (1979) as

$$\left\langle \frac{da}{dt} \right\rangle = -\left(\frac{\eta}{a}\right) Q_{pr} \frac{(2 + 3e^2)}{(1 - e^2)^{3/2}} \quad (10)$$

$$\left\langle \frac{de}{dt} \right\rangle = -\frac{5}{2}\left(\frac{\eta}{a^2}\right) Q_{pr} \frac{e}{(1 - e^2)^{1/2}} \quad (11)$$

where $\eta = (1 + sw)S_0 r_0^2 A/mc^2$ and r_0 is the heliocentric distance at which the solar constant S_0 is measured. The characteristic orbital decay time for a circular orbit, in the absence of planetary perturbations, is easily found by setting $e = 0$ in equation (10) and integrating.

If we include the planets, the situation becomes more complex. As mentioned previously, the sunward spiral of bound KB dust particles may be temporarily halted by capture into MMRs with the planets, or the particles may be scattered out of the solar system. *Liou and Zook* (1997) studied analytically the evolution of dust particles trapped in MMRs with the planets. Their findings are summarized below.

The radiation pressure force changes the location of an MMR with a planet. The new shifted location is

$$a_R = a_P(1 - \beta)^{1/3}\left(\frac{p}{q}\right)^{2/3} \equiv a_P(1 - \beta)^{1/3}K^{2/3} \quad (12)$$

where a_P is the semimajor axis of the planet, and p and q are two integers that specify a particular, p:q, resonance. The parameter K, defined as the ratio p/q, is larger than one for an exterior MMR, less than one for an interior MMR, and equal to one for a 1:1 MMR. For a given exterior MMR, the larger β is, the closer its location is shifted toward the planet. Conversely, for a given interior or 1:1 MMR, the larger β is, the farther its location is shifted away from the planet. In the circular restricted three-body problem (where the planet is in circular orbit), the resonant angle of the particle, φ_r, is defined as (e.g., *Allan,* 1969)

$$\varphi_r = p\lambda_d - q\lambda_p + l_1\varpi_d + l_2\Omega_d \quad (13)$$

where λ_p is the mean longitude of the planet, and λ_d, ϖ_d, and Ω_d are the mean longitude, longitude of pericenter, and longitude of ascending node of the dust particle, respectively, and the D'Alembert relation requiring that $l_1 + l_2 = q - p$. The resonant angle is related to the dominant terms in the disturbing function that control the motion of the particle in an MMR. The "order" of an MMR is given by the quantity $|q - p|$. In general there are three types of MMR (e.g., *Allan,* 1969): eccentricity type (e type, where $l_2 = 0$), inclination type (I type, where $l_1 = 0$), and mixed type (where l_1 and l_2 are nonzero integers). The lowest-order terms in the corresponding disturbing functions of these three different types of MMR are of the first, second, and third orders in eccentricity and inclination, respectively. The first-order MMRs are all e types, whereas the second-order MMRs can be both e types and I types. The mixed-type resonances occur only at the third- or higher-order resonances. Since the order of an MMR largely determines the strength of the resonance, e-type and I-type MMRs are more common than mixed-type MMRs.

In *Liou et al.* (1995b), the time variation of the Jacobi "constant" due to drag forces was expressed in spherical coordinates and used to study the effects of radiation pressure, PR drag, and solar wind drag, in a restricted three-body system. *Liou and Zook* (1997) further analyzed this method to determine the explicit time evolution of the eccentricity and inclination of a dust particle trapped in an arbitrary MMR with a planet. The three coordinate systems used in these studies are the inertial frame (ξ, η, ζ), the rotating frame (x, y, z), and the spherical rotating frame (r, θ, ϕ). The rotating frame (x, y, z) is rotating with constant angular velocity around the collocated ζ and z axes, with the x axis lying along the line from the center of mass (CM) of the system to the planet. The transformations from (ξ, η, ζ) to (x, y, z) and from (x, y, z) to (r, θ, ϕ) are

$$\begin{cases} \xi = x \cos t - y \sin t \\ \eta = x \sin t + y \cos t \end{cases} \quad (14)$$

and

$$\begin{cases} x = r \cos\theta \cos\phi \\ y = r \cos\theta \sin\phi \\ z = r \sin\theta \end{cases} \quad (15)$$

respectively. Figure 1 shows the geometry and relationships

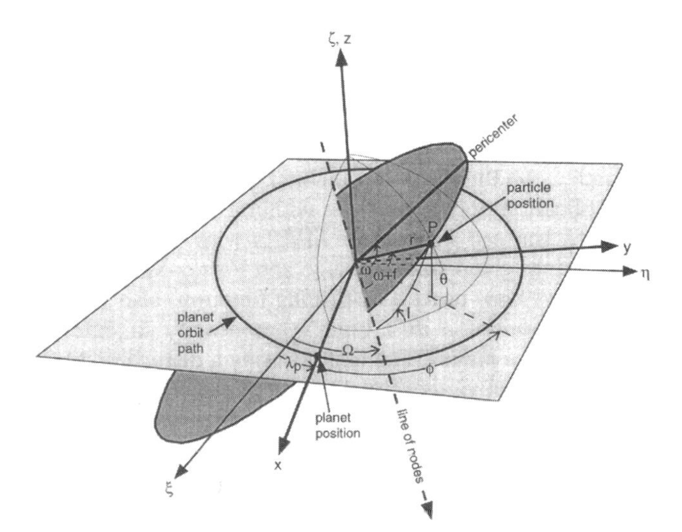

Fig. 1. Geometry and coordinate systems used in the circular restricted Sun-planet-dust three-body system. The coordinate system (x, y, z) is co-rotating with the planet, with the x axis pointing from the origin to the planet. The elements I, Ω, ω, and f are the orbital inclination, longitude of the ascending node, argument of pericenter, and true anomaly of the dust grain, respectively. The angles φ and θ are the longitude and latitude angles of the dust grain in the (x, y, z) coordinate system as defined by equation (15).

between these three coordinate systems. In a circular restricted three-body system, the Jacobi "constant," C, can be approximately described in terms of the orbital elements, when the dust particle is far from the Sun and the perturbing planet, by

$$C \approx \frac{(1-\beta)}{a} + 2\sqrt{(1-\beta)a(1-e^2)}\cos I \equiv C_T \qquad (16)$$

where the identity is traditionally called the Tisserand criterion (e.g., *Moulton,* 1914). Following the derivation in *Liou et al.* (1995b), the time derivative of the Jacobi "constant," C, in a system with radiation pressure, PR drag, and solar wind drag is given by

$$\dot{C} = 2(1+sw)\frac{\beta\mu_1}{c}\left[2\left(\frac{\dot{r}}{r}\right)^2 + \dot{\theta}^2 + (\dot{\phi}\cos\theta)^2 + \dot{\phi}\dot{\lambda}_p\cos^2\theta\right] \qquad (17)$$

where μ_1 is the mass of the Sun and c is the speed of light. The unit of mass is the sum of the Sun (μ_1) and the planet (μ_2). The unit of length is the distance between the Sun and the planet. The unit of time is the orbital period of the planet divided by 2π while $\dot{\lambda}_p = 1$ in these units. As K equals the ratio of the orbital period of a dust particle in resonance with a perturbing planet to the orbital period of that planet, then, in such normalized units, the orbital period of the dust particle, T, is $2\pi K$.

Equation (17) can be used to investigate the stability of a dust particle in different MMRs with a planet. When a dust particle in a prograde orbit is trapped in an interior MMR with a planet, its orbital angular velocity is usually larger than that of the planet. Thus, in the rotating reference frame, the particle's $\dot{\phi}$ is usually positive. This means that all terms on the righthand side of equation (17) are positive and C increases with time. The dust particle is eventually forced, by the expansion of the forbidden region, to leave the resonance. When I = 0 (the planar case) and C increases, a dust particle trapped in an interior MMR with constant semimajor axis must have its eccentricity decreased, as indicated by equation (16). When the eccentricity reaches zero, the ever-increasing C will force the semimajor axis to be decreased. The dust particle is forced out of the resonance and continues its sunward spiral. As the eccentricity decreases, the dust particle's aphelion distance moves farther away from the planet. This means that gravitational scattering is not the mechanism that ejects the dust particle out of the resonance. Rather, it is released from resonance because no net positive orbital energy is given by a planet's perturbation on a dust particle in circular orbit and the drag forces then predominate.

When a dust grain is trapped in a 1:1 MMR with a planet, C also increases with time. This implies that either its orbital eccentricity or its orbital inclination, or both, must decrease, as indicated by equation (16). In actual practice, both occur simultaneously (*Liou et al.,* 1995b). Eventually, the increasing C, and the resultant expanding forbidden regions, force the dust particle to change from a tadpole orbit to a horseshoe orbit and, finally, to escape the resonance via close encounters with the planet.

When a dust particle is trapped in an exterior MMR with a planet, its orbital angular velocity is smaller than that of the planet. The last term on the righthand side of equation (17) is negative on average. It is possible to have a solution such that $\dot{C} = 0$, which might suggest that the particle would remain in a quasistable trap; however, in an inclined orbit, the eccentricity value required to maintain $\dot{C} = 0$ is a function of the inclination. As the inclination of a dust particle in resonance varies in time, it is especially doubtful that this condition can be maintained.

Equation (17) can also be expressed in terms of the orbital elements of the dust particle. This is accomplished by averaging all quantities on the righthand side of equation (17) over one orbital period of the dust particle, yielding

$$\dot{C} = 2(1+sw)\frac{\beta\mu_1}{c}\left[\frac{2\pi^2(3e^2+2)}{T^2(1-e^2)^{3/2}} - \frac{2\pi\cos I}{T}\right] \qquad (18)$$

The condition that $\dot{C} = 0$ when in an exterior MMR requires the value in brackets on the righthand side of equation (18) to be zero, which leads to

$$\frac{T}{2\pi}(=K) = \frac{3e^2+2}{2(1-e^2)^{3/2}}\frac{1}{\cos I} \qquad (19)$$

In the special case where I is zero, equation (19) gives the eccentricity that the orbit of a dust particle will approach while in a given exterior MMR.

When a dust particle is trapped in an MMR with a planet, its semimajor axis oscillates around a fixed value. Thus, $\dot{a} = 0$ on average. The time derivative of equation (16) becomes

$$\frac{dC}{dt} = \frac{\partial C}{\partial e}\left(\frac{de}{dt}\right) + \frac{\partial C}{\partial I}\left(\frac{dI}{dt}\right) = \qquad (20)$$

$$-2(1-\beta)^{1/2}a^{1/2}\left[e(1-e^2)^{-1/2}\cos I\frac{de}{dt} + (1-e^2)^{1/2}\sin I\frac{dI}{dt}\right]$$

Combining equations (18) and (20) and $T = 2\pi a^{3/2}/\sqrt{1-\beta}$, the equation becomes

$$e(1-e^2)^{-1/2}\cos I\frac{de}{dt} + (1-e^2)^{1/2}\sin I\frac{dI}{dt} = \frac{(1+sw)\beta\mu_1}{a^2c}\left[\cos I - \frac{(3e^2+2)(1-\beta)^{1/2}}{2a^{3/2}(1-e^2)^{3/2}}\right] \qquad (21)$$

This expression relates the time rates of change of the eccentricity and inclination of a dust particle in an MMR with a planet. In order to examine the behavior at small eccentricity and inclination, we can expand equation (21) to the second order in the combined powers of eccentricity and inclination. For dust particles in prograde orbits, the second-order expansion is

$$\frac{d}{dt}[e^2 + I^2] = A\left[\left(1 - \frac{1}{K}\right) - \frac{3}{K}e^2 - \frac{1}{2}I^2\right] \quad (22)$$

where

$$A = \frac{2(1 + sw)\beta\mu_1}{a^2c} \quad (23)$$

For a 1:1 MMR (K = 1), and assuming that, as in the classical (no drag involved) secular perturbation theory, the variations in eccentricity and inclination are decoupled from each other, this yields

$$\begin{cases} e = e_0 \exp\left(-\frac{3A}{2}t\right) \\ I = I_0 \exp\left(-\frac{A}{4}t\right) \end{cases} \quad (24)$$

where e_0 and I_0 are the initial eccentricity and inclination when the particle becomes trapped. Therefore, in a 1:1 MMR, both eccentricity and inclination of the dust particle's orbit decrease exponentially with time.

Equation (22) can also be used to investigate the behavior of particle orbits in interior and exterior e-type or I-type MMRs (MMRs of mixed-type require higher-order terms). In an exterior MMR, K is larger than 1. In an interior MMR, K is less than 1. Introducing a constant Q, and again assuming decoupling for small e and I, the general solutions can be written as

$$e^2 = e_0^2 \exp\left(-\frac{3A}{K}t\right) + \frac{K}{3A}\left[\frac{A(K-1)}{K} - Q\right]\left[1 - \exp\left(-\frac{3A}{K}t\right)\right] \quad (25)$$

$$I^2 = I_0^2 \exp\left(-\frac{A}{2}t\right) + \frac{2Q}{A}\left[1 - \exp\left(-\frac{A}{2}t\right)\right]$$

There are three possibilities for the value of Q: (1) Q = 0; (2) Q = A(K – 1)/K; and (3) Q equals some nonzero value other than A(K – 1)/K. For possibility (1), the eccentricity of the dust particle's orbit increases with time and approaches the fixed value $\sqrt{(K-1)/3}$ at exterior MMRs (where K > 1) and decreases with time at interior MMRs (where K < 1). For possibility (2), the eccentricity decreases exponentially with time while the inclination increases with time and approaches the fixed value $\sqrt{2(K-1)/K}$ at exterior MMRs and decreases with time at interior MMRs. For possibility (3), both e and I can increase, depending on the value of Q. The most common resonances for IDPs in prograde orbits in the solar system appear to be e-type resonances (e.g., *Jackson and Zook,* 1992; *Marzari and Vanzani,* 1994). For such resonances, equation (25) simplifies to

$$e^2 = \left[e_0^2 - \frac{K-1}{3}\right]\exp\left(-\frac{3A}{K}t\right) + \frac{K-1}{3}$$
$$I = I_0 \exp\left(-\frac{A}{4}t\right) \quad (26)$$

Equation (18) is valid also for dust particles in retrograde orbits. In such orbits, cos I is negative, and the righthand side of equation (18) is positive. Thus, the value of C for all retrograde orbits always increases with time. In the special case where I = 180° (the planar case), the eccentricity of the particle's orbit increases without limit, regardless of what resonance it is in. No permanent trapping is possible. The eccentricity of a dust particle in retrograde orbit increases much more dramatically with time than that of a dust particle in prograde orbit when they are both trapped in the same MMR. This implies that in the same MMR and with the same starting eccentricity, a dust particle on a prograde orbit tends to be trapped longer than one on a retrograde orbit.

In the three-dimensional case, if the inclination of the particle's orbit is close to 180°, one can define an angle $\sigma = \pi - I$, and expand equation (21) to the second order in e and σ in a way similar to that done above for prograde orbits. Again, assuming decoupling for small e and σ, the solutions can be written as

$$e^2 = e_0^2 \exp\left(\frac{3A}{K}t\right) - \frac{K}{3A}\left[\frac{A(K+1)}{K} - Q\right]\left[1 - \exp\left(\frac{3A}{K}t\right)\right]$$
$$\sigma^2 = \sigma_0^2 \exp\left(-\frac{A}{2}t\right) + \frac{2Q}{A}\left[1 - \exp\left(-\frac{A}{2}t\right)\right] \quad (27)$$

For Q = 0, the eccentricity increases exponentially while σ decreases (I increases toward 180°) exponentially with time. For Q = A(K + 1)/K, the eccentricity also increases exponentially while σ increases (I decreases away from 180°) and approaches a fixed value $\sigma_{max} = \sqrt{2(K+1)/K}$. This places the inclination of the dust particle's orbit in the range less than 90°. This means that the dust particle can evolve from a retrograde orbit into a prograde orbit in such a resonance. At inclination near 90°, equation (21) becomes

$$\frac{dI}{dt} = -\frac{\pi A}{2T}\frac{(3e^2 + 2)}{(1 - e^2)^2} < 0 \quad (28)$$

where A is given by equation (23). This implies that it is impossible for a dust particle on a prograde orbit to evolve into a retrograde orbit while trapped in an MMR.

The predictions made by the analytical theory described above can be tested by means of numerical integrations. *Liou and Zook* (1997) performed such calculations. Two of their examples are shown here. Figure 2 illustrates the orbital evolution of a dust particle with $\beta = 0.26$ trapped in a 1:1 MMR with a Jupiter-like planet. The exact resonance location is at a semimajor axis equal to 4.71 AU. The initial eccentricity and inclination of the dust particle's orbit are 0.1° and 10°, respectively. While trapped in the MMR, the dust particle's orbit evolved from a tadpole-like orbit to a horseshoe-like orbit until close encounters with the planet eventually ejected it out of resonance after about 213,000 years. While trapped, its eccentricity and inclination both decreased in good agreement with the analytical predictions from equation (26).

Figure 3 shows the orbital evolution of a dust particle in a 7:4 exterior MMR (at 42.37 AU) with a Neptune-like planet with semimajor axis of 30.22 AU. It was trapped for more than 21 m.y. While trapped, its eccentricity increased and its inclination decreased gradually with short-term oscillations. Its eccentricity nearly reached the maximum value of 0.43 predicted by equation (19). Since I remains small, the prediction of equation (26) describes well the secular decrease of the inclination; however, as the eccentricity increases, the second-order prediction for e deviates somewhat from the calculated value when e is larger than about 0.3. Since the inclination of the dust particle's orbit remains small and decreases with time, the planar-case prediction (equation (26) of *Liou and Zook,* 1997), which is valid for any eccentricity, gives a better description of the actual variation of the eccentricity.

While the analytic approach described above gives valuable insights into the orbital evolution of dust particles trapped in MMRs with planets, direct numerical simulations are needed to reveal the full nonlinear behavior as well as to assess the importance of phenomena not addressed by

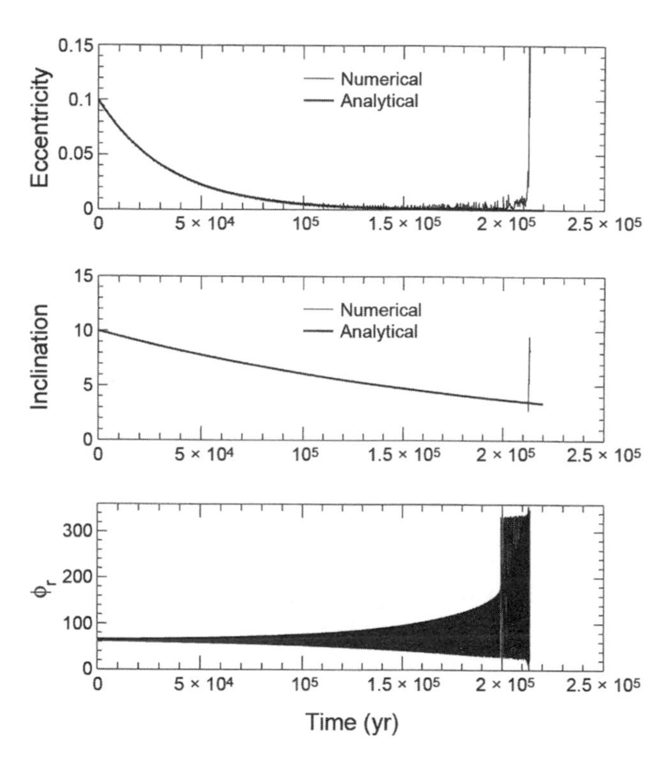

Fig. 2. Evolution of a dust particle ($\beta = 0.26$) in a circular Sun-Jupiter-dust numerical simulation. It was trapped in a 1:1 MMR with Jupiter for about 213,000 yr. While trapped, its eccentricity decreased from 0.1 to about 0.01 and its inclination decreased from 10° to 3.5° when it escaped the resonance. The bold curves in the top and middle panels are calculated using the analytical predictions of equation (26). The analytical predictions agree with those from the numerical simulation (the analytical and numerical results overlap each other from the beginning to about 200,000 yr in eccentricity and to about 213,000 yr in inclination). The bottom panel plots the resonant angle with time. The orbit of the dust particle changed from a tadpole to a horseshoe after 200,000 yr.

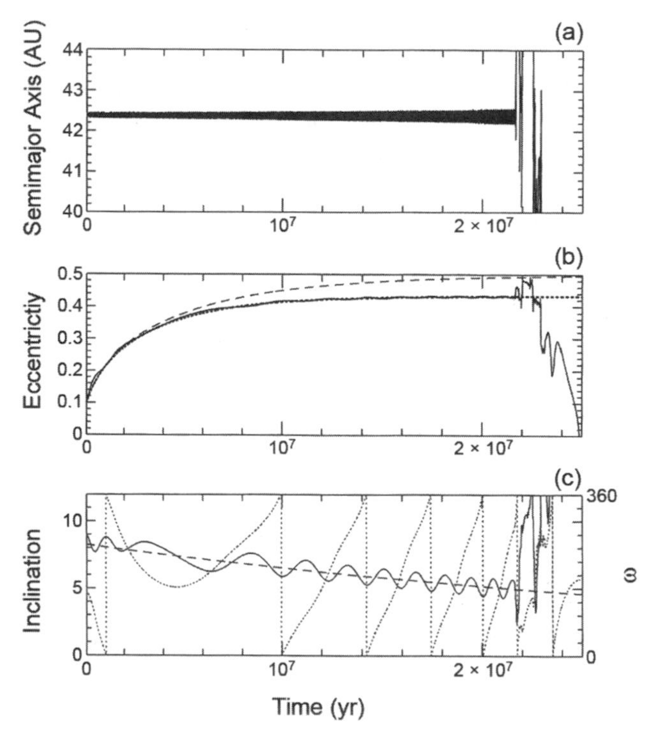

Fig. 3. Evolution of a dust particle ($\beta = 0.1$) in a circular Sun-Neptune-dust numerical simulation. The particle was placed at a 7:4 exterior MMR (at 42.37 AU) initially. It was trapped for more than 21 m.y. **(a)**. Its resonant angle oscillated around 180° with an amplitude of 35° and increased gradually to about 70° before it escaped the resonance. While trapped, its eccentricity increased from 0.1 to about 0.43 and its inclination decreased slowly with large short-term oscillations [solid curves in **(b)** and **(c)**]. The second-order expansion predictions of equation (26) describe well the secular decrease in inclination but not quite so well the eccentricity when e is larger than about 0.3 [dashed curves in the **(b)** and **(c)**]. Since the inclination remains small and decreases with time, actual variation in eccentricity agrees well with the planar prediction dotted curve in the middle panel). The short-term variation in I is due to the precession of the argument of pericenter ω (dotted curve in the bottom panel). The oscillation period of I is half that of the precession period of ω.

the analytic theory, phenomena such as resonance capture and gravitational scattering.

2.3. Numerical Simulations

The principle behind numerical simulations of the KB dust particles is rather simple. A numerical integrator is needed to integrate equation (7). Once the initial conditions of the dust particles and planets are defined, the process is carried out by computers. Depending on the available resources, the simulations can cover hundreds to thousands or more particles. In order to understand how the KB dust particle distribution is affected by planets, *Liou et al.* (1996) and *Liou and Zook* (1999) undertook a series of numerical simulations to follow the dynamical evolution of KB dust particles from their origin in the KB to their ultimate fates, either spiraling into the Sun or being removed by a planet through ejection from the solar system. *Moro-Martín and Malhotra* (2002, 2003) and *Holmes et al.* (2003) have performed similar calculations. Since their results are essentially the same as those of *Liou and Zook* (1999), the summary below is solely based on the results of *Liou et al.* (1996) and *Liou and Zook* (1999).

The RADAU integrator (*Everhart,* 1985) was used to numerically integrate the orbits of KB dust particles characterized by four different values of β: 0.4, 0.2, 0.1, and 0.05. If one assumes a 1 g cm^{-3} density, the particle diameters corresponding to these values of β are approximately 3, 6, 11, and 23 μm, respectively. Included in the simulations were the gravitational perturbations of the seven most massive planets (Venus through Neptune), solar radiation pressure, and PR and solar wind drag. All of the planets gravitationally interacted with each other and acted on the dust particles, while the dust particles were treated as massless test particles and exerted no gravitational forces of their own. Solar wind drag was assumed to be 35% that of PR drag (*Gustafson,* 1994), hence sw = 0.35 in equation (7). Dust particles were assumed to be released from their parent KB objects located at 45 AU with eccentricities of 0.1 and inclinations of 10°. (An additional simulation was also carried out using 50 23-μm dust particles released from parent bodies at 50 AU and including only the four giant outer planets.) Initial longitudes of ascending node, longitudes of pericenter, and mean longitudes of the parent bodies were chosen randomly from 0 to 2π. One hundred particles were integrated for each value of β, except for $\beta = 0.05$, for which 50 particles were used.

The dynamical lifetimes of the KB dust particles from this set of simulations are shown in Fig. 4. The lifetime of a dust particle depends on the PR and solar wind drag rate, its initial semimajor axis displacement from the parent body due to solar radiation pressure, MMR traps with planets, the eccentricity when it escapes an MMR, and ejection by the giant planets. The actual lifetime of a KB dust particle is usually several times longer than that predicted by pure PR and solar wind drag alone.

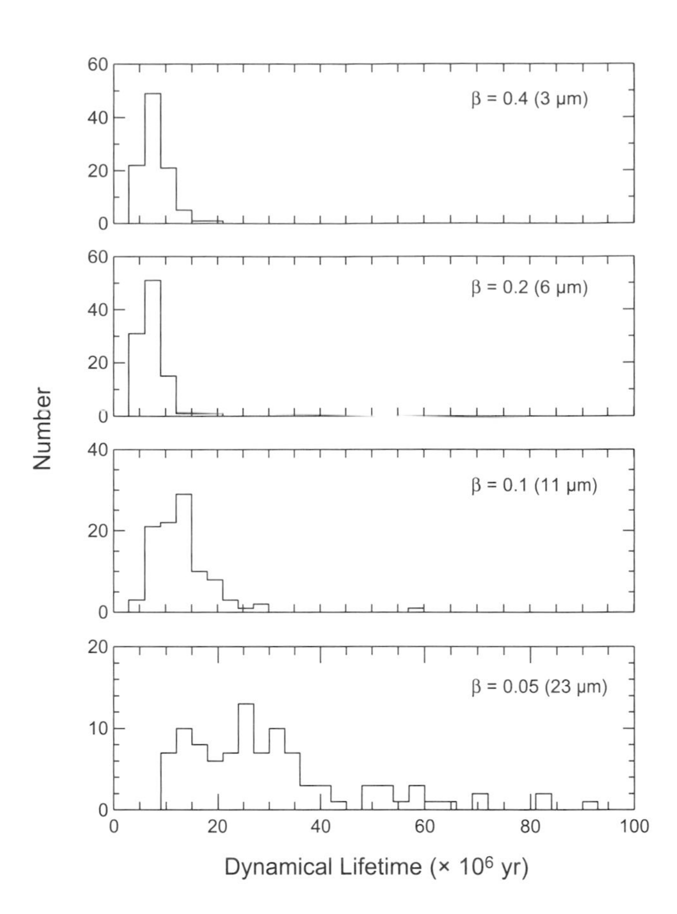

Fig. 4. Simulated lifetimes of IDPs from the numerical simulations. The PR and solar wind drag lifetime for a 23-μm IDP released from a parent body at 45 (50) AU is only about 13.5 (16.5) m.y. However, due to complex planetary perturbations, the actual dynamical lifetimes of IDPs are quite different from those predicted based on PR and solar wind drag alone. One 23-μm IDP that is trapped in an unusually long 7:4 MMR with Neptune has a lifetime of 159 m.y. is not included in the bottom panel.

The numerical simulations show that trapping into exterior MMRs with Neptune dominates the orbital evolution of the KB dust particles. On average, the 3-, 6-, 11-, and 23-μm particles spent 19%, 24%, 37%, and 47%, respectively, of their lifetimes in exterior MMRs with Neptune. The smaller the drag force is on a dust particle, the easier it is for the particle to become trapped in an MMR, and the longer it remains trapped. Trapping into MMRs with interior giant planets is rare because gravitational perturbations from exterior giant planets usually make the resonance traps highly unstable, unless the exterior resonances are very close to the perturbing planet or the resonance is a 1:1 MMR. Eventually all particles are able to escape the MMR traps and continue their journey toward the Sun. However, the majority of them are ejected from the solar system by Jupiter and Saturn. Only about 20% of the particles survive the process and make their way to the inner solar system.

Numerical simulations unveil a far more complicated nature of the orbital evolution of KB dust particles. The simulated results allow us to describe the orbital characteristics of the KB dust particles in a quantitative manner. It leads to a more realistic picture of the overall structure of the KB dust disk described in the following sections.

3. KUIPER BELT DUST DISK

3.1. Expected Distribution of Kuiper Belt Dust Particles in the Solar System

Under PR and solar wind drag perturbations, the predicted steady-state distribution of dust particles is symmetric in longitude, as suggested by the models of *Stern* (1995) and *Backman et al.* (1995). It also has a smooth and continuous radial distribution all the way to the Sun. The radial dependence of the dust particles can be expressed as

$$n(r) \propto r^{-\tau} \tag{29}$$

where n(r) is the spatial density at a heliocentric distance r from the Sun. The index τ varies from 1 for particles with near circular orbits to 2.5 for particles with eccentricities close to 1 (e.g., *Bandermann,* 1968; *Leinert and Grün,* 1990).

The orbital evolution of KB dust particles becomes very complicated when planetary perturbations are considered, as described in section 2. Figure 5 shows the dramatic difference in semimajor axis distribution from numerical simulations with (solid histograms) and without (dotted lines) planets. Seven planets are included (Venus through Neptune) in the "with planets" scenario. It is obvious that to have a reliable description of the structure of the KB dust disk, planetary perturbations must be included. Hence detailed numerical simulations of the orbital evolution of KB dust particles are needed. However, due to the long orbital lifetimes of KB dust particles, such simulations are limited by the available computer resources. The first attempt by *Liou et al.* (1995a, 1996) included only 80 particles in the simulation. It was later increased to 400 particles by *Liou and Zook* (1999). All their simulations were carried out using the implicit Runge-Kutta integrator, RADAU, with an adjusting step size control (*Everhart,* 1985). The particles in the simulations had diameters ranging from 3 to 23 μm.

Moro-Martín and Malhotra (2002, 2003) followed the same approach, using a more efficient multiple time step symplectic integrator based on the algorithm developed by *Duncan et al.* (1998) and with the radiation forces added. Hundreds of dust particles were included in their simulations using updated orbital element distributions of the KB objects as initial conditions. They extended the particle size range to 100 μm in diameter. In addition, they validated the underlying ergodic assumption utilized by various groups to create dust disk models based on a limited number of particles. *Holmes et al.* (2003) used RADAU in their numerical simulations with up to 1017 dust particles in some of their test scenarios. Their simulated KB dust particles ranged from 4 to 100 μm in diameter. They also included the Lorentz force in some of their simulations, although for particles larger than about 10 μm this effect is negligible, and for smaller particles the effect is still marginal unless they are charged to an unreasonably high potential.

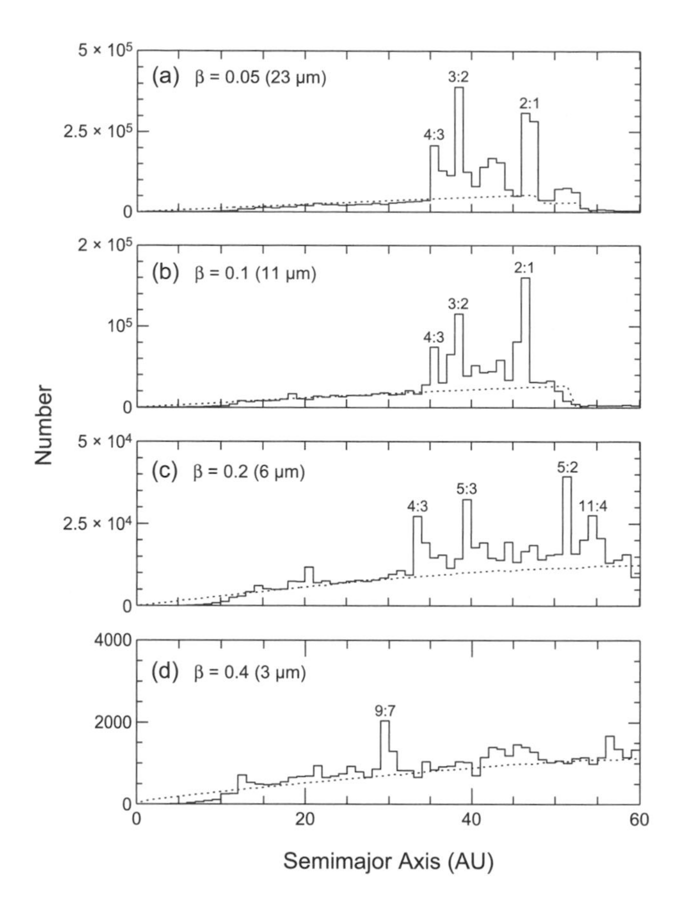

Fig. 5. Semimajor axis distributions from numerical simulations with (solid histograms) and without (dotted lines) planets. The distributions are different for particles of different sizes. One hundred dust particles and seven planets are included in each simulation.

Overall, results from numerical simulations performed by different groups are consistent with one another. The major features unveiled regarding the distribution of KB dust particles in the solar system can be summarized as follows: (1) Trapping into exterior MMRs with Neptune dominates the orbital evolution of dust particles 5 μm and larger. The main resonances are the 2:1 and 3:2 exterior MMRs. (2) Trapping into MMRs with Uranus is rare because gravitational perturbations from Neptune usually make the resonance trap highly unstable. (3) Once particles escape MMRs with Neptune, they continue to spiral toward the Sun. (4) About 80% of the particles are eventually ejected from the solar system by Jupiter and Saturn. (5) The remaining particles evolve all the way to the Sun. None of them are ejected from the solar system by the terrestrial planets. Their orbital elements near 1 AU resemble those of asteroidal IDPs, rather than cometary IDPs. Note all planets are treated as point masses in the simulations. In reality, there is another possible sink mechanism for particles — accretion onto the planets. This effect may reduce the percentage of KB dust particles entering the inner solar system slightly.

3.2. Structure of the Kuiper Belt Dust Disk Beyond Jupiter

A simple way to construct a model KB dust disk from numerical simulations based on a limited number of particles was described in *Liou and Zook* (1999). Positions of dust particles from each small output interval over a long simulation were accumulated. To represent a steady-state distribution, the simulation must cover the lifetimes of the particles, i.e., from the beginning (when they were created) to the end (when they were ejected from the solar system, or when they reached the Sun). The result is a three-dimensional spatial density distribution of KB dust particles in the solar system. It can be processed further to create column density or brightness distribution maps with appropriate thermal property assumptions. Viewing geometry can also be easily implemented from different locations, and with different viewing directions. This approach, examined in detail by *Moro-Martín and Malhotra* (2002) and shown to be both practical and reliable, was adopted by *Moro-Martín and Malhotra* (2002, 2003) and by *Holmes et al.* (2003) for their KB dust disk models, and by others for nearby circumstellar dust systems (e.g., *Quillen and Thorndike,* 2002; *Deller and Maddison,* 2005).

Plates 9a and 9b show the steady-state column density maps of the simulated 23-μm KB dust particles, as viewed from above the north ecliptic pole. Under PR and solar wind drag alone (without planetary perturbations), the column density remains constant as illustrated by Plate 9a. When planetary perturbations are included, however, the resulting column density map is very different (Plate 9b). The orbits and positions of the four giant planets are also added to the map for reference (epoch of June 1, 1998). The dramatic concentrations and gaps correspond directly to the features outlined in section 3.1. The ring-like structure along, and outside, Neptune's orbit consists of particles trapped in MMRs with Neptune. Additional variations along the ring are caused by the overlapping geometric patterns of the two dominant resonances: 2:1 and 3:2 exterior MMRs with Neptune. As a result, two arcs are observed, one leading and the other trailing Neptune along the orbit. Another significant feature of the ring is the lack of particles near where Neptune is located. Since dust particles in stable MMRs tend not to pass near that perturbing planet (in this case, Neptune), a column density minimum (a dark "hole") is formed around Neptune.

Numerical simulations show that trapping into MMRs exterior to a planet is much easier, and the traps last longer, than does trapping into interior MMRs (e.g., *Jackson and Zook,* 1992; *Marzari and Vanzani,* 1994). The smaller the drag force is on a dust particle, the easier it is for the dust particle to become trapped into an MMR, and it will remain trapped longer. Therefore similar but perhaps more prominent structures, such as rings and arcs, around the orbit of Neptune are expected for larger KB dust particles. No obvious ring- or arc-like patterns are seen around the orbit of Uranus. It is simply difficult for dust particles to be trapped in stable and long-term exterior MMRs with Uranus while being perturbed by Neptune. The exception is the 1:1 MMR with Uranus. Only a few particles were observed to be trapped in that resonance for long periods of time from the numerical simulations. The resulting ring-like feature is not as prominent as those caused by exterior MMR traps with Neptune.

A brightness distribution map can also be developed with reasonable assumptions of the thermal properties of the dust particles. A simulated face-on image is shown in Plate 10. The signatures of giant planets are: (1) the deviation of radial brightness profile from the PR and solar wind drag one that monotonically increases toward the Sun, (2) a ring-like structure along the orbit of Neptune, (3) a brightness variation along the ring with an opening (a dark spot) located where Neptune is, (4) a seasonal variation of the dark spot that moves along with Neptune's orbital motion, and (5) a relative lack of particles inside about 10 AU. The radial brightness variation inside 10 AU is controlled by two competing factors. The first factor is that Saturn ejects about 40% of the KB dust particles that approach it from outside its orbit and Jupiter ejects about 67% of the particles that approach its orbit from outside, leading to a combined depletion of about 80%. The second factor is the r^{-2} brightness weighting factor of the particles. The combined effects cause the brightness to temporarily decrease at 10 AU and increase again inside about 5 AU.

If an extraterrestrial intelligence were observing our solar system and had the image of Plate 10, it would know (if its knowledge was similar to or better than ours) that at least one giant planet at about 30 AU exists in our solar system, from signatures (1), (2), and (3). With continuous observations that determine the motion of the dark spot along the ring [signature (4)], it could easily obtain the orbital location of Neptune using Kepler's Laws. With quantitative numerical modeling of signatures (1) to (3), it could place strong constraints on the mass of Neptune. Since Uranus does not trap particles effectively due to its orbital location, nor is it massive enough to eject particles from the solar system, it will not be recognizable from the KB dust disk. From signature (5), it could identify the existence of at least one other giant planet that is preventing dust particles from entering the region inside about 10 AU. It may be difficult to know, however, since Jupiter and Saturn are very close to each other, whether there is only one planet or two planets around that location. The masses of Jupiter and Saturn, or the effective mass of a single planet between 5 and

10 AU, can be estimated based on the radial brightness profile near that region. The terrestrial planets — Mars, Earth, and Venus — are not recognizable from the KB dust disk observation.

3.3. *In Situ* Measurements: Pioneer and Voyager Data

The best evidence to support the existence of the KB dust disk comes from *in situ* data. The dust detector onboard Pioneer 10 measured IDP impacts all the way to 18 AU where the detector ceased to function (*Humes,* 1980). The interpretation is that the spatial density of 10^{-9} g IDPs is essentially constant to 18 AU (*Humes,* 1980). In addition, the plasma wave instruments onboard Voyager 1 and 2 also provided evidence of IDP impacts on the spacecraft out to almost 100 AU (*Gurnett et al.,* 1997, 2005). The two most likely sources of dust particles in the outer solar system are Oort cloud comets and KB objects. Dust particles coming from the Oort cloud comets cannot produce a constant spatial density distribution unless the parent comets release a dust cloud of particles having a perihelion distribution that increases with q^2, where q is the perihelion distance (*Liou et al.,* 1999). The modeled spatial density of the 23-μm KB dust particles (with a mass of about 10^{-9} g) within 10° from the ecliptic, between 5 and 20 AU, is approximately constant (*Liou and Zook,* 1999). The difference is less than a factor of 3, which is consistent with the Pioneer 10 data, and within the detection uncertainty of the instruments (*Humes,* 1980).

Quantitatively, the Voyager data are difficult to model since the instruments were not designed to detect dust. This is not the case for the Pioneer impact detectors. *Landgraf et al.* (2002) modeled the distributions of KB dust particles and particles from two other cometary sources, and showed that the former is needed to account for the actual measurements from the Pioneer 10 and 11 detectors, especially outside 10 AU. The fits to the two sets of data are shown in Fig. 6. The data also provide a constraint on the production rate of the KB dust particles. According to *Landgraf et al.* (2002), it is about 5×10^7 g s^{-1} for particles with sizes between 0.01 and 6 mm. This estimate is consistent with the production rates calculated by *Stern* (1995) and by *Yamamoto and Mukai* (1998), when modeling uncertainties are considered.

3.4. Characteristics of Kuiper Belt Dust Particles Inside 5 AU

The majority of the KB dust particles are ejected from the solar system by Jupiter and Saturn. As a result, the column density inside 5 AU is very low. This pattern is not very sensitive to particle size since the gravitational scattering timescale is much shorter than the drag timescale. When converting to a brightness map, however, the region inside 5 AU is actually brighter than that outside of 5 AU (Plate 10). None of the KB particles inside 5 AU are ejected from the solar system by any terrestrial planets. Based on the simulated distributions in eccentricity, between 0.05 and 0.55, and in inclination, between 3° and 28°, when KB dust particles cross the orbit of Earth, *Liou et al.* (1996) concluded that KB dust particles behaved more like asteroidal IDPs than cometary IDPs.

Moro-Martín and Malhotra (2003) reexamined the distribution of KB dust particles in the inner solar system with more particles included in the simulations. Their results placed the Earth-crossing KB dust particles between 0.05 and 0.75 in eccentricity, and between 0° and 30° in inclination. Although the distributions are similar to those obtained by *Liou et al.* (1996), they stated that KB dust particles were actually more like cometary IDPs than asteroidal IDPs. The cause of this discrepancy is in the cometary IDP benchmark samples *Moro-Martín and Malhotra* (2003) adopted. They used the "cometary" IDP distributions from the study by *Kortenkamp and Dermott* (1998), which followed the orbital evolution of IDPs released from Jupiter-family short-period comets. However, cometary IDPs should also include those from Halley-type short-period comets as well as those from long-period comets. In other words, the cometary IDP distributions at 1 AU should be extended to almost 180° in inclination, and close to unity in eccentricity. Good examples to illustrate the eccentricity and inclination distributions include the annual meteor showers (at Earth) such as Leonids (associated with 55P/Tempel-Tuttle, inclination ~162°, eccentricity ~0.92), Quadrantids (associated with 96P/Machholz, inclination ~72.5°, eccentricity ~0.68), and η Aquarids (associated with 1P/Halley, inclination ~163.5°, eccentricity ~0.96). The inclination distribution is especially important since it can increase dust particles' Earth encounter speeds up to 72 km/s. Qualitative descriptions such as asteroidal- or cometary-like are always somewhat subjective. However, when the total population of comets (rather than just a subgroup) is considered, it is probably reasonable to argue that the eccentricity and inclination distributions obtained from the numerical simulations of *Liou et al.* (1996) and by *Moro-Martín and Malhotra* (2003) are closer to those of asteroidal IDPs rather than cometary IDPs.

The prediction that some KB dust particles can approach Earth with very low eccentricities and inclinations has a major implication. It may be possible to capture these particles intact using aerogel collectors in the near-Earth environment, such as the proposed Large Area Debris Collector (*Liou et al.,* 2007). Although the captured KB dust particles may have orbital characteristics similar to those of asteroidal IDPs, there are ways to identify them from the collected samples. The solar flare track density measurement is one possibility (e.g., *Bradley and Brownlee,* 1984; *Sandford,* 1986). Kuiper belt dust particles should have a much higher track density than asteroidal dust particles simply because they have a much longer orbital lifetime coming from the KB region.

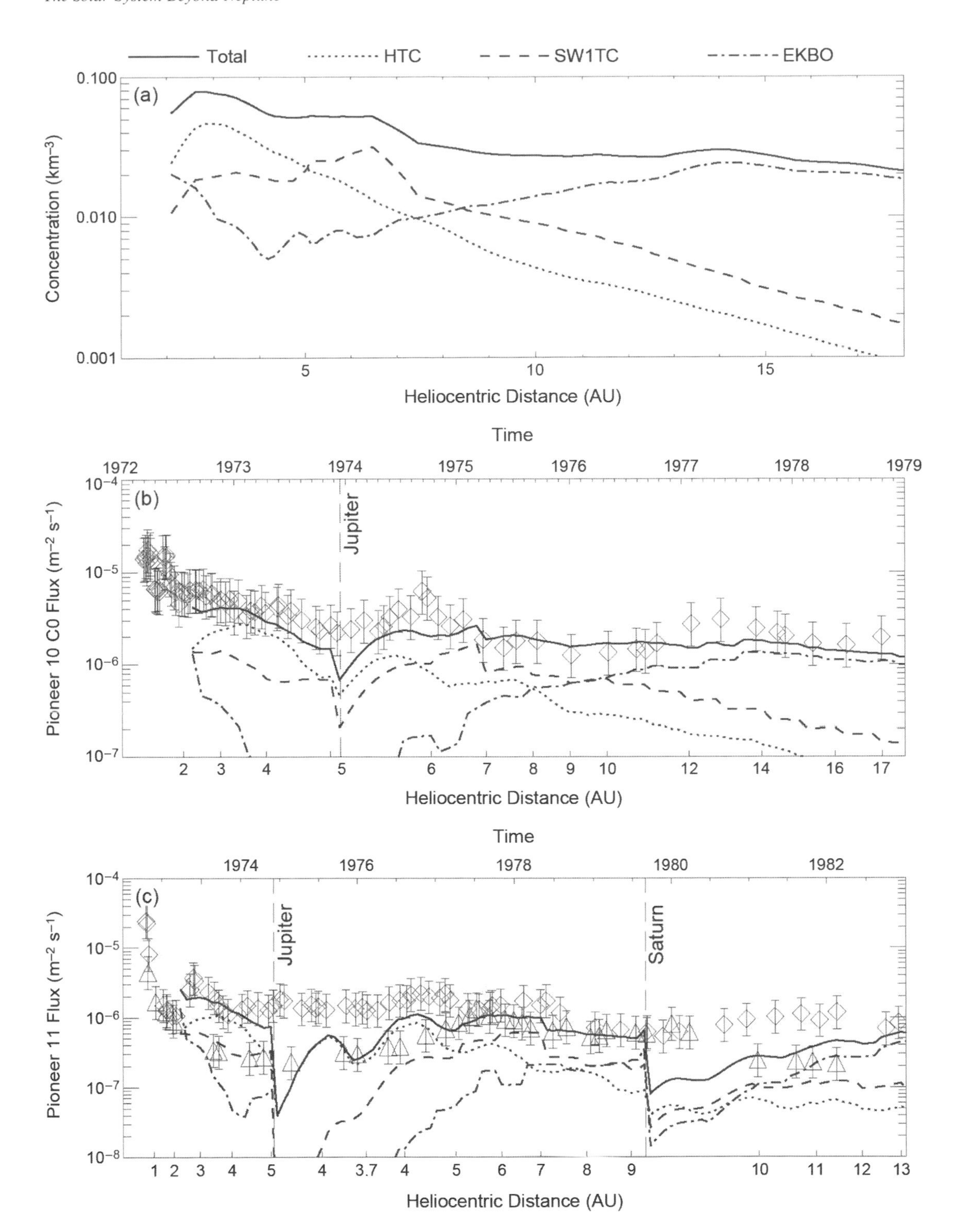

Fig. 6. Dust impact data collected by the Pioneer 10 and 11 detectors. A reasonable fit to the data is accomplished by combining dust particles from three different sources. KB dust particles are needed to maintain the constant flux beyond about 10 AU. This figure is adopted from *Landgraf et al.* (2002).

Do KB dust particles contribute significantly to the interplanetary dust complex at 1 AU? The issue was first raised by *Flynn* (1994). A simple estimate can be made by assuming the Pioneer 10 data fit by *Landgraf et al.* (2002) indeed represents the distribution of KB dust particles in the solar system. By comparing the spatial density of KB dust particles with the IDP spatial density at 1 AU (*Grün et al.*, 1985), it would appear that KB dust particles represent about 5% of the IDPs at 1 AU. Of course, there are uncertainties associated with this estimate. Impact data on the space-facing side of the Long Duration Exposure Facility (LDEF) suggested the flux of 10^{-9} g IDPs at 1 AU was about 50% higher than the *Grün et al.* (1985) model. There are uncertainties in the Pioneer data, and in the model fit as well. This estimate, however, does appear to be consistent with the solar flare track measurements of the collected IDPs (*Bradley and Brownlee,* 1984). Most of the collected IDPs have space exposure ages of approximately 10^4 yr, too short for KB dust particles coming from the outer solar system.

4. IMPLICATIONS FOR EXTRASOLAR PLANET DETECTION FROM CIRCUMSTELLAR DUST DISKS

The structure of the KB dust disk and the known configuration of the giant planets serve as a good model for other circumstellar dust disks. That perturbations of an unseen planet could explain some of the observed features of the β Pictoris dust disk was proposed and analyzed in a series of three papers between 1994 and 1996 (*Roques et al.,* 1994; *Lazzaro et al.,* 1994; *des Etang et al.,* 1996). Their main results include: (1) A planet can trap dust particles into long-term MMRs and create arc-like structures in the disk; (2) a depleted region around the perturbing planet is expected; (3) a central clearing zone inside the orbit of the perturbing planet can be formed; and (4) depending on the spatial density of particles, mutual collisions among dust particles may diminish significantly the signatures of the unseen planet.

ε Eridani is a young star at a distance of only 3.22 pc from the Sun. The dust disk around ε Eridani was revealed and imaged using the Submillimeter Common User Bolometer Array at the James Clerk Maxwell Telescope on Mauna Kea, Hawaii (*Greaves et al.,* 1998). At 850 μm wavelength, the observations clearly show a dust disk around the star with an azimuthal brightness variation along the disk. The disk has a 25° inclination with respect to the plane of sky. The inner and outer edges of the disk are approximately 30 and 90 AU, respectively, from the star with a peak brightness occurring around 60 AU. The brightness variation along the ring is about a factor of 2.5. The brightest spot/arc in the ring is a real feature. The presence of other features (secondary bright spots and a dark gap) along the ring is less certain since they were identified in only half of the observational maps. What makes the ε Eridani disk differ from the β Pictoris disk is its almost face-on orientation. If one compares the pattern of the variation with the modeled structures of the KB dust disk, one finds global similarities between the two. Therefore, a simple explanation for the observed structures in ε Eridani is that they are caused by perturbations from an unseen planet, or planets, orbiting the star.

Liou and Zook (2000) simulated the dust disk with a single planet on a near-circular orbit. They tested the planet at 30 and 40 AU from the star, and tested planet-to-star mass ratios ranging from 1 $M_\oplus$ to 5 M_{Jup}. Hundreds of dust particles, ranging from 5 to 100 μm in diameter, were included in the simulations. They reached the following conclusions: (1) To have a ring-like structure with a strongly depleted inner region, a giant planet (planet-to-star mass ratio ≥ Jupiter-to-Sun mass ratio) is needed to eject dust particles from the system; (2) to have an azimuthal variation along the ring, a planet (Earth-like to Jupiter-like) is needed to trap particles in MMRs; and (3) to have a single bright spot/arc in the ring, the 2:1 MMR with a giant planet has to be the dominant resonance. *Quillen and Thorndike* (2002) modeled the ε Eridani dust disk with a single planet that had a semimajor axis of 40 AU, and with various eccentricities. Three hundred dust particles were included in each numerical simulation. They concluded that the morphology of the ε Eridani dust disk was best reproduced by dust particles trapped in the 5:3 and 3:2 MMRs with a moderate eccentricity (~0.3) near periastron.

Kuchner and Holman (2001, 2003) explored the general geometric patterns of particles trapped in various MMRs with a planet, with different masses and/or eccentricities up to 0.6. The characteristics of the dust disk can be classified into four basic types: (1) a ring with a gap at the location of the perturbing planet; (2) a smooth ring; (3) a blobby, eccentric ring; and (4) an offset ring plus a pair of clumps. *Wilner et al.* (2002) observed Vega in 1.3 mm and 3 mm wavelengths and discovered two emission peaks on the dust disk surrounding Vega, consistent with the pair of clumps described above. Their numerical simulations suggested the features could be the dynamical signatures of an unseen planet with several Jupiter masses, a semimajor axis of 40 AU, and a high eccentricity of 0.6.

Deller and Maddison (2005) explored the dust disk morphology further by creating a catalog of about 300 simulated dust disks. They varied five parameters in the simulations: the mass and eccentricity of the single planet, and the sizes, initial semimajor axes, and eccentricities of the dust particles. The resulting catalog is a collection of disk features that can be used as the first reference for any new circumstellar dust disk observations that show potential signs of planets.

Trying to infer the existence of unseen planets from circumstellar dust disks is much more difficult than trying to model the structure of the KB dust disk (see chapter by Moro-Martín et al. for additional details). The fundamental difference between the two is that while the configuration of the planets is known for the solar system, the information is completely missing for other circumstellar dust disks. The numerically produced dust disk patterns depend on several factors: (1) the number of planets; (2) the mass, semimajor axis, and orbital eccentricity of each planet; (3) the size distribution of the dust particles; and (4) the source region and the orbits of the parent objects of the dust particles. Even with unlimited computer resources to explore every possible region of parameter space, there is a high probability that the final answer may still be ambiguous. Nevertheless, features on circumstellar dust disks will continue to serve as valuable tools to quantify the likelihood of planet(s) in systems where astronomers lack other means for extrasolar planet detection.

5. CONCLUDING REMARKS AND FUTURE WORK

It is ironic that while more and more circumstellar dust disks are being discovered, we have yet to see our own KB dust disk directly. Due to the low optical depth of the KB dust disk and the thermal emission from the local zodiacal cloud particles, direct infrared observations of the KB dust disk will continue to be a major challenge, and may never happen. The Student Dust Counter onboard the New Horizons spacecraft will provide a nice impact dataset (for ~1.6 μm and larger particles) beyond what Pioneer 10 achieved. The data will further enhance our knowledge of the dust environment in the outer solar system, and provide additional constraints on the dust production rate and the collision history of the KB objects. From the modeling perspective, collisions among dust particles will need to be included for future numerical simulations. This is very important for extrasolar circumstellar dust disks since many of them are optically thick.

Acknowledgments. J.-C.L. thanks the generous support of the NASA Orbital Debris Program Office during the preparation of the manuscript. We also thank two anonymous reviewers for their very constructive comments and suggestions.

REFERENCES

Allan R. R. (1969) Evolution of Mimas-Tethys commensurability. *Astron. J., 74,* 497–506.

Aumann H. H., Gillett F. C., Beichman C. A., De Jong T., Houck J. R., et al. (1984) Discovery of a shell around Alpha Lyrae. *Astrophys. J. Lett., 278,* L23–L27.

Backman D. E. and Paresce F. (1993) Main sequence starts with circumstellar solid material: The Vega phenomenon. In *Protostars and Planets III* (E. H. Levy et al., eds.), pp. 1253–1304. Univ. of Arizona, Tucson.

Backman D. E., Dasgupta A., and Stencel R. E. (1995) Model of a Kuiper belt small grain population and resulting far-infrared emission. *Astrophys. J. Lett., 450,* L35–L38.

Bandermann L. W. (1968) Physical properties and dynamics of interplanetary dust. Ph.D. thesis, Univ. of Maryland.

Bradley J. P. and Brownlee D. E. (1984) Discovery of nuclear tracks in interplanetary dust. *Science, 226,* 1432–1434.

Burns J. A., Lamy P. L., and Soter S. (1979) Radiation forces on small particles in the solar system. *Icarus, 40,* 1–48.

de Etangs A. L., Scholl H., Roques F., Sicardy B., and Vidal-Madjar A. (1996) Is there a planet around β Pictoris? Perturbations of a planet on a circumstellar dust disk, 3. Time scale of collisional destruction versus resonance time scale. *Icarus, 123,* 168–179.

Deller A. T. and Maddison S. T. (2005) Numerical modeling of dusty debris disks, *Astrophys. J., 625,* 398–413.

Duncan M. J., Levison H. F., and Lee M. H. (1998) A multiple time step symplectic algorithm for integrating close encounters. *Astron. J., 116,* 2067–2077.

Everhart E. (1985) An efficient integrator that uses Gauss-Radau spacings. In *Dynamics of Comets: Their Origin and Evolution* (A. Carusi and G. B. Valsecchi, eds.), pp. 185–202. Reidel, Boston.

Flynn G. J. (1994) Does the Kuiper belt contribute significantly to the zodiacal cloud and the stratospheric interplanetary dust? (abstract). In *Lunar and Planetary Science XXV,* pp. 379–380. Lunar and Planetary Institute, Houston.

Greaves J. S. (2005) Disks around stars and the growth of planetary systems. *Science, 307,* 68–71.

Greaves J. S., Holland W. S., Moriarty-Schieven G., Jenness T., Dent W. R. F., et al. (1998) A dust ring around ε Eridani: Analog to the young solar system. *Astrophys. J. Lett., 506,* L133–L137.

Grün E., Zook H. A., Fechtig H., and Giese R. H. (1985) Collisional balance of the meteoritic complex. *Icarus, 62,* 244–272.

Grün E., Zook H. A., Baguhl M., Balogh A., Bame S. J., et al. (1993) Discovery of jovian dust streams and interstellar grains by the Ulysses spacecraft. *Nature, 362,* 428–430.

Gurnett D. A., Ansher J. A., Kurth W. S., and Granroth L. J. (1997) Micron-sized dust particles detected in the outer solar system by the Voyager 1 and 2 plasma wave instruments. *Geophys. Res. Lett., 24,* 3125–3128.

Gurnett D. A., Wang Z. Z., Persoon A. M., and Kurth W. S. (2005) Dust particles detected in the outer solar system by Voyager 1 and 2. In *Dust in Planetary Systems,* pp. 63–64. LPI Contribution No. 1280, Lunar and Planetary Institute, Houston.

Gustafson B. Å. S. (1994) Physics of zodiacal dust. *Annu. Rev. Earth Planet. Sci., 22,* 553–595.

Holmes E. K., Dermott S. F., Gustafson B. Å. S., and Grogan K. (2003) Resonant structure in the Kuiper disk: An asymmetric plutino disk. *Astrophys. J., 597,* 1211–1236.

Humes D. H. (1980) Results of Pioneer 10 and 11 meteoroid experiments: Interplanetary and near-Saturn. *J. Geophys. Res., 85,* 5841–5852.

Jackson A. A. and Zook H. A. (1989) A solar system dust ring with the Earth as its shepherd. *Nature, 337,* 629–631.

Jackson A. A. and Zook H. A. (1992) Orbital evolution of dust particles from comets and asteroids. *Icarus, 97,* 70–84.

Jewitt D. and Luu J. (1993) Discovery of the candidate Kuiper belt object 1992 QB_1. *Nature, 362,* 730–732.

Klačka J. (1992) Poynting-Robertson effect. I — Equation of motion. *Earth Moon Planets, 59,* 41–59.

Kortenkamp S. J. and Dermott S. F. (1998) Accretion of interplanetary dust particles by the Earth. *Icarus, 135,* 469–495.

Kuchner M. J. and Holman M. J. (2001) Planets on eccentric orbits, phase-locked dust blobs, and the ring around Epsilon Eridani. *Bull. Am. Astron. Soc., 33,* 1151.

Kuchner M. J. and Holman M. J. (2003) The geometry of resonant signatures in debris disks with planets. *Astrophys. J., 588,* 1110–1120.

Landgraf M., Liou J.-C., Zook H. A., and Grün E. (2002) Origins of solar system dust beyond Jupiter. *Astron. J., 123,* 2857–2861.

Lazzaro D., Sicardy B., Roques F., and Greenberg R. (1994) Is there a planet around β Pictoris? Perturbations of a planet on a circumstellar dust disk, 2. The analytical model. *Icarus, 108,* 59–80.

Leinert C. and Grün E. (1990) Interplanetary dust. In *Physics of the Inner Heliosphere, 1. Large-Scale Phenomena* (R. Schwenn and E. Marsch, eds.), pp. 207–275. Springer-Verlag, Berlin.

Liou J.-C. and Zook H. A. (1997) Evolution of interplanetary dust particles in mean motion resonances with planets. *Icarus, 128,* 354–367.

Liou J.-C., and Zook H. A. (1999) Signatures of the giant planets imprinted on the Edgeworth-Kuiper belt dust disk. *Astron. J., 118,* 580–590.

Liou J.-C., and Zook H. A. (2000) Structure of the Edgeworth-Kuiper belt dust disk and implications for extrasolar planet(s) in ε Eridani. In *Dust in the Solar System and other Planetary Systems* (S. F. Green et al., eds.), pp. 225–228. Pergamon, Amsterdam.

Liou J.-C., Zook H. A., and Dermott S. F. (1995a) Orbital evolution of micron-sized dust grains coming from the Kuiper belt (abstract). In *Lunar and Planetary Science XXVI,* pp. 853–854. Lunar and Planetary Institute, Houston.

Liou J.-C., Zook H. A., and Jackson A. A. (1995b) Radiation pressure, Poynting-Robertson drag, and solar wind drag in the restricted three-body problem. *Icarus, 116,* 186–201.

Liou J.-C., Zook H. A., and Dermott S. F. (1996) Kuiper belt dust grains as a source of interplanetary dust particles. *Icarus, 124,* 429–440.

Liou J.-C., Zook H. A., and Jackson A. A. (1999) Orbital evolution of retrograde interplanetary dust particles and their distribution in the solar system. *Icarus, 141,* 13–28.

Liou J.-C., Giovane F. J., Corsaro R. D., and Stansbery E. G. (2007) LAD-C: A large area cosmic dust and orbital debris collector on the International Space Station. In *Proceedings of Dust in Planetary Systems* (A. Wilson, ed.), pp. 227–230. ESA SP-643, Noordwijk, The Netherlands.

Love S. G. and Brownlee D. E. (1993) A direct measurement of the terrestrial mass accretion rate of cosmic dust. *Science, 262,* 550–553.

Marzari F. and Vanzani V. (1994) Dynamical evolution of interplanetary dust particles. *Astron. Astrophys., 283,* 275–286.

Moro-Martín A. and Malhotra R. (2002) A study of the dynamics of dust from the Kuiper belt: Spatial distribution and spectral energy distribution. *Astron. J., 124,* 2305–2321.

Moro-Martín A. and Malhotra R. (2003) Dynamical models of Kuiper belt dust in the inner and outer solar system. *Astron. J., 125,* 2255–2265.

Moulton F. R. (1914) *An Introduction to Celestial Mechanics,* 2nd edition. Dover, New York.

Quillen A. C. and Thorndike S. (2002) Structure in the Epsilon Eridani dusty disk caused by mean motion resonances with a 0.3 eccentricity planet at periastron. *Astrophys. J. Lett., 578,* L149–L152.

Roques F., Scholl H., Sicardy B., and Smith B. A. (1994) Is there a planet around β Pictoris? Perturbations of a planet on a circumstellar dust disk, 1. The numerical model. *Icarus, 108,* 37–58.

Sanford S. A. (1986) Solar flare tract densities in interplanetary dust particles: The determination of asteroidal versus cometary source of the zodiacal dust cloud. *Icarus, 68,* 377–394.

Smith B. A. and Terrile R. J. (1984) A circumstellar disk around β Pictoris. *Science, 226,* 1421–1424.

Stern S. A. (1995) Collisional time scale in the Kuiper disk and their implications. *Astron. J., 110,* 856–868.

Stern S. A. (1996a) Signatures of collisions in the Kuiper disk. *Astron. Astrophys., 310,* 999–1010.

Stern S. A. (1996b) On the collisional environment, accretion time scales, and architecture of the massive, primordial Kuiper belt. *Astron. J., 112,* 1203–1211.

Van de Hulst H. C. (1957) *Light Scattering by Small Particles.* Wiley, New York.

Weidenschilling S. J. and Jackson A. A. (1993) Orbital resonances and Poynting-Robertson drag. *Icarus, 104,* 244–254.

Wilner D., Holman M. J., Kuchner M. J., and Ho P. T. P. (2002) Structure in the dusty debris around Vega. *Astrophys. J. Lett., 569,* L115–L119.

Wyatt S. P. and Whipple F. L. (1950) The Poynting-Robertson effect on meteor orbits. *Astrophys. J., 111,* 134–141.

Yamamoto S. and Mukai T. (1998) Dust production by impacts of interstellar dust on Edgeworth-Kuiper belt objects. *Astron. Astrophys., 329,* 785–791.

Zook H. A. (1975) Hypervelocity cosmic dust: Its origin and its astrophysical significance. *Planet. Space Sci., 23,* 1391–1397.

Part VIII:

Boundaries and Connections to Other Stellar Systems

The Limits of Our Solar System

John D. Richardson
Massachusetts Institute of Technology

Nathan A. Schwadron
Boston University

The heliosphere is the bubble the Sun carves out of the interstellar medium. The solar wind moves outward at supersonic speeds and carries with it the Sun's magnetic field. The interstellar medium is also moving at supersonic speeds and carries with it the interstellar magnetic field. The boundary between the solar wind and the interstellar medium is called the heliopause. Before these plasmas interact at the heliopause, they both go through shocks that cause the flows to become subsonic and change direction. Voyager 1, which recently crossed the termination shock of the solar wind, is the first spacecraft to observe the region of shocked solar wind called the heliosheath. The interstellar neutrals are not affected by the magnetic fields and penetrate deep into the heliosphere, where they can be observed both directly and indirectly. This chapter discusses these outer boundaries of the solar system and the interactions between the solar wind and the interstellar medium, and describes the current state of knowledge and future missions that may answer the many questions that remain.

1. INTRODUCTION

The limits of the solar system can be defined in many ways. The farthest object orbiting the Sun could be one. The farthest planet could be another. We define this limit as the boundary between the Sun's solar wind and the interstellar medium, the material between the stars in our galaxy. The interstellar medium is not uniform, but is composed of many clouds with different densities, temperatures, magnetic field strengths and directions, and flow speeds and directions. The Sun is now embedded in a region of the interstellar medium called the local interstellar cloud. The solar wind flows outward from the Sun and creates a bubble in the interstellar medium.

The solar wind pushes into the local interstellar cloud until it too weak to push any farther, then turns and moves downstream in the direction of the local interstellar cloud flow. The plasmas in these two winds cannot mix, since they are confined to magnetic field lines and magnetic fields cannot move through one another. Neutrals, however, are not bound by magnetic fields and can cross between regions. So local interstellar cloud neutrals can and do penetrate deep into the solar system. The boundary between the solar and interstellar winds is called the heliopause.

Both the solar wind and the interstellar wind are supersonic; thus a shock forms upstream of the heliopause in each flow. At these shocks, the plasmas in each wind are compressed, heated, become subsonic, and change flow direction so that the plasma can move around the heliopause (for interstellar medium) or down the heliotail (for the solar wind). The shock in the solar wind is called the termination shock. The shock in the local interstellar cloud is called the bow shock. The region between the termination shock and the heliopause is the inner heliosheath and the region between the bow shock and the heliopause is the outer heliosheath.

Plate 11 shows the equatorial plane from a model simulation of the heliospheric system (*Müller et al.,* 2006). The color coding in the top panel gives the plasma temperature and in the bottom panel shows the neutral H density. The plasma flow lines are shown in the top panel. The trajectories of Voyager 1 (V1) and Voyager 2 (V2) are also shown; V1 is headed nearer the heliospheric nose than V2. We note that it is only by serendipity that the Voyager spacecraft are headed toward the nose of the heliosphere; the Voyager trajectories were driven solely by the locations of the planets that were part of Voyager's grand tour of the solar system. The main heliospheric boundaries, termination shock, heliopause, and bow shock are labeled in the upper panel. The local interstellar cloud is only barely supersonic, so the heating that occurs at the bow shock is small; at this boundary the flow of local interstellar cloud plasma begins to divert around the heliosphere. In the solar wind the flow is radially outward until the termination shock is reached. At the termination shock, the plasma is strongly heated and the flow lines bend toward the tail. At the heliopause, the local interstellar cloud and solar wind flow are parallel; plasma generally cannot cross this boundary. The neutral density increases at the nose of the bow shock, forming an enhanced density region known as the hydrogen wall, and decreases at the heliopause.

We are just beginning to learn about these outer regions of our heliosphere and how they interact with the local interstellar cloud. Until V1 crossed the termination shock in 2004 at 94 astronomical units (AU) (1 AU is the distance from the Sun to Earth), we could only guess the boundary locations. Now we have observed one termination shock crossing and have several years of heliosheath data, but we still have much to learn about the other heliosphere boundaries and the global aspects of these boundaries. This chapter discusses what we know about the interaction of the heliosphere with the local interstellar cloud and these boundary regions, the unsolved problems, and the future observations we hope will help answer these questions.

2. THE HELIOSPHERE

The heliosphere is the region dominated by plasma and magnetic field from the Sun. No spacecraft has yet crossed outside this region; estimates of the size of the heliosphere at the nose (in the upwind direction) range from 105 to 150 AU (*Opher et al.,* 2006). In the tail, the heliosphere extends many hundreds and probably thousands of AU. The location of the nose of the heliopause is determined by pressure balance between the outward dynamic pressure of the solar wind and the pressure of the local interstellar cloud, which has contributions from the dynamic pressure, thermal pressure, and magnetic pressure (see *Belcher et al.,* 1993). The dynamic pressures are given by $1/2\rho V^2$ where ρ is the mass density and V is the speed. For the local interstellar cloud the magnetic pressure $B^2/8\pi$ and the thermal pressures nkT could be important but the magnitudes are not well constrained. We discuss first the unperturbed flows from the Sun and in the local interstellar cloud, then discuss the interactions of these two flows.

2.1. The Solar Wind

The Sun is a source of both photons and plasma; the fluxes of each fall off as $1/R^2$. The effects of solar radiation on the heliosphere/local interstellar cloud interaction are small compared to those of the solar wind. The first detections of the solar wind in the early 1960s (*Neugebauer and Snyder,* 1962; *Gringauz,* 1961) confirmed *Parker*'s (1958) theoretical predictions of the basic solar wind structure. The solar wind has been thoroughly measured and monitored by spacecraft since that time. In addition to the near-equatorial spacecraft in the inner heliosphere, Pioneers 10 and 11 and V1 and V2 have observed the solar wind in the outer heliosphere, in the case of V1 past 100 AU. Ulysses has provided measurements in the high-latitude heliosphere. These observations give a good picture of the solar wind parameters throughout the heliosphere.

The solar wind is a supersonic flow of mostly protons, with on average 3–4% He^{++} (this percentage varies over a solar cycle) (*Aellig et al.,* 2001) and a small amount of highly charged heavy ions. The Sun has a magnetic field that reverses polarity every 11-year solar cycle. To first order, plasma cannot cross magnetic field lines, so the solar wind outflow drags the solar magnetic field lines outward. This outward motion combined with the Sun's 27-day rotation causes the magnetic field lines to form spirals; at Earth the spiral angle is about 45°. In the outer heliosphere these spirals become tightly wound and the magnetic field angle is nearly 90° (the field is almost purely tangential). Since the magnetic field lines from the northern and the southern hemispheres of the Sun have opposite polarity, a heliospheric current sheet (HCS) forms near the equator, across which the magnetic field direction reverses. The HCS passes through the termination shock and has been observed in the heliosheath past 100 AU. The average solar wind speed at Earth is about 440 km/s with a density of about 7.5 cm^{-3}, an average temperature of 96,000 K, and an average magnetic field strength of about 5 nT; these values are based on a solar cycle of data from the Wind spacecraft from 1994 to 2005. The average dynamic pressure is about 2.25 nP; this pressure determines how big a cavity the solar wind carves into the local interstellar cloud.

2.2. The Local Interstellar Cloud

The local interstellar cloud surrounds the heliosphere. It contains neutral and ionized low-energy particles, H, He, O, H^+, He^+, and O^+, as well as very-high-energy galactic cosmic rays (GCRs). These GCRs are observed at Earth but their intensities are modulated (reduced) by the solar wind magnetic field as they move through the heliosphere. Plate 11 shows how the low-energy plasma deflects around the heliosphere. Helium is the only major species that makes it to Earth in pristine form and thus provides the best determination of the properties of the interstellar medium. Observations of the H and He emission combined with models of the neutral flow give estimates of the neutral speed, density, and temperature as well as the flow direction.

Hydrogen is the largest component of the interstellar medium, but it is difficult to determine the local interstellar cloud properties from H observations because a large fraction of the H charge exchanges with the local interstellar cloud H^+ before it enters the heliosphere. Charge exchange occurs when an ion and a neutral collide and an electron goes from the neutral to the ion. The former ion is a now neutral and is not bound by magnetic fields, so it continues to move at the speed it had when it was an ion. The former neutral, now an ion, is bound to the magnetic field and accelerated to the speed of the bulk plasma flow. Outside the bow shock the plasma and neutrals have the same speed and temperature so charge exchange has no effect. But inside the bow shock, the plasma parameters are different from those in the interstellar medium since the plasma is heated, compressed, and deflected at the shock. The plasma slows further as it piles up in front of the heliosphere. So when charge exchange occurs, the new neutrals have speeds and temperatures from the shocked plasma, not the pristine interstellar medium. The H observed in the heliosphere has an inflow speed of only 20 km/s, 6 km/s less than the interstellar medium speed (*Lallement et al.,* 1993), confirming that the H has charge exchanged with local interstellar cloud H^+, which has slowed down near the heliosphere boundary. The plasma flow in this region has begun to divert around the heliosphere, so the flow direction of the H is also altered.

Observations with the SOHO/SWAN instrument show that the arrival direction of the interstellar H differs from that of the interstellar He by about 4° (*Lallement et al.,* 2005). This offset results from charge exchange of the H with the deflected H^+ in front of the heliosphere. The direction of the offset tells us the direction of the plasma flow in the local interstellar cloud just outside the heliopause and thus provides information on the magnetic field direction (*Lallement et al.,* 2005). The magnetic field strength in the local interstellar cloud is not known; generally it is thought to be on the order of 1 μG with an upper limit of 3–4 μG.

The interstellar neutrals enter the heliosphere and absorb and emit sunlight at specific wavelengths. Backscattered H Lyman-α (*Bertaux and Blamont,* 1971; *Thomas and Krassa,* 1971) and He 58.4 nM radiation have both been observed (*Weller and Meier,* 1979). As the neutrals move toward the Sun, they are accelerated inward by gravity but pushed outward by the radiation pressure (see review by *Fahr,* 1974). For H, the radiation pressure and gravity are comparable, so the H is not accelerated inward. Helium is heavier and is accelerated inward. Hydrogen also has a much shorter ionization time in the solar wind than He. The result is that most H becomes ionized before it reaches 5–10 AU. The peak in backscattered Lyman-α intensity is roughly in the direction of the local interstellar cloud flow with a small deflection occurring from the charge exchange in the hydrogen wall.

The ionization time of He is long; most He is not ionized until it is inside 1 AU. The He with trajectories that remain outside 1 AU survive and flow downstream out of the heliosphere. The Sun's gravity bends the trajectories so that the He is focused downstream of the Sun, where the maximum emission from He is observed (*Axford,* 1972; *Fahr and Lay,* 1973). The location of the maximum He emission is thus opposite to the direction of the incoming He flow. Observation of the direction with peak He intensity is one method used to determine the local interstellar cloud flow direction.

The interstellar neutrals are ionized in the solar wind via charge exchange with protons (the dominate process for H) or photoionization. These neutrals can be observed directly as H^+ and He^+ pickup ions. He^+ is rarely observed in the solar wind and thus is a good marker of interstellar neutrals. Both H^+ and He^+ pickup ions have distinctive signatures in the solar wind; when the neutral H and He become ionized they are accelerated to the solar wind speed. These ions, called pickup ions because they are picked up (accelerated) by the solar wind, have a gyromotion roughly equal to the solar wind speed, so these ions show a sharp energy cutoff at four times the solar wind energy.

The energy to accelerate the pickup ions comes from the bulk flow of the solar wind, so an indirect way to measure the pickup ion density (discussed further below) is to look at how much the solar wind slows down as it moves outward. These direct and indirect observations can be combined with models of the neutral inflow and loss to estimate speeds, temperatures, and densities of H and He in the interstellar medium.

The gas detector on Ulysses measures the neutral He velocity and temperature directly (*Witte,* 2004; *Witte et al.,* 1992). Even with these direct measurements, determination of the density of the local interstellar cloud still requires modeling since some of the He is lost on its way through the heliosphere. All these measurements taken together provide fairly tight constraints on the local interstellar cloud He neutral speed and temperature (*Möbius et al.,* 2004).

The He moves toward the Sun from an ecliptic longitude of 75° and an ecliptic latitude of –5° with a speed of 26.4 ± 0.3 km/s. The temperature is 6300 ± 340 K and the density of He is 0.015 ± 0.002 cm^{-3}. The collision timescales in the local interstellar cloud are less than the neutral and plasma lifetimes, so the plasma and neutral species should all have the same speed and temperature far from the heliosphere. Since the H is altered by charge exchange with the plasma near the heliosphere, the H density is more difficult to determine. Best estimates are on the order of 0.20 ± 0.03 cm^{-3} in the local interstellar cloud. Since the ionization fraction depends on the electron temperature, which is not well known, the plasma density is also uncertain, with best estimates of 0.06–0.10 cm^{-3} based on the speed of the H measured at Earth (*Lallement et al.,* 1993).

3. THE TERMINATION SHOCK LOCATION

Although some properties of the local interstellar cloud are well constrained, others are not, which makes determining the heliosphere size and shape difficult. We have observed one boundary location; the termination shock at the position of V1 was at 94 AU on December 16, 2004 (*Stone et al.,* 2005; *Decker et al.,* 2005; *Burlaga et al.,* 2005). This crossing provides one point with which to calibrate our models. Since the average solar wind dynamic pressure is 2.25 nP at 1 AU, one can roughly estimate that the total local interstellar cloud pressure must be 2.5×10^{-4} nP. But there are problems with this simple approach, which we describe below.

3.1. Solar Wind Time Dependence

The average solar wind values given above are only a piece of the story. The solar wind pressure is variable on scales from hours to 11-year solar cycles and likely has longer-term variations. Small-scale variations of the pressure do not affect the boundary locations, but larger-scale variations change the balance between the solar wind and local interstellar cloud pressures and cause the termination shock and heliopause to move in and out. The solar wind dynamic pressure changes by a factor of about 2 over a solar cycle, with minimum pressure near solar maximum (when sunspot numbers are highest), then a rise in pressure with a peak a few years after solar maximum, followed by a decrease to the next solar maximum (*Lazarus and McNutt,* 1990). Models of the heliosphere show that these pressure changes drive an inward and outward "breathing" of the heliosphere; the termination shock location changes by about

10 AU over a solar cycle and the heliopause by 3–4 AU (*Karmesin et al.,* 1995; *Wang and Belcher,* 1998). The solar wind structure also changes over the solar cycle. At solar minimum, the solar wind near the equator is slow (400 km/s) and dense (8 cm^{-3}). The solar wind at higher latitudes comes from polar coronal holes and is fast, 700–800 km/s with a density of 3–4 cm^{-3}. At solar maximum, the polar coronal holes are not present and most of the flow is low-speed solar wind. Despite these latitudinal changes in the solar wind, the dynamic pressure changes over the solar cycle are the same at all heliolatitudes (*Richardson and Wang,* 1999), so the heliosphere is not expected to change shape over a solar cycle.

Large events on the Sun can also affect the termination shock position. Active regions on the Sun generate coronal mass ejections (CMEs), explosive discharges of material into the solar wind where they are called interplanetary CMEs (ICMEs). ICMEs propagate through the heliosphere. When a series of ICMEs occurs, the later ones catch up to and merge with previous ICMEs, causing a buildup of plasma and magnetic field known as a merged interaction region (MIR) (*Burlaga et al.,* 1984; *Burlaga,* 1995). These regions are usually preceded by a fast forward shock at which the speed, density, temperature, and magnetic field strength all increase. Shocks are rare in the outer heliosphere; most can be related to a large CME or a series of CMEs on the Sun (see *Richardson et al.,* 2005a). The dynamic pressure in the MIRs can be a factor of 10 higher than that in the ambient solar wind and these structures last 1–3 months (*Richardson et al.,* 2003). The MIRs drive the termination shock outward several AU for 2–3 months, after which the termination shock rebounds inward (*Zank and Müller,* 2003).

3.2. Local Interstellar Cloud Time Dependence

Changes in the pressure or magnetic field of the local interstellar cloud could also cause the boundary locations to move. The scale length for changes in the local interstellar cloud is not known. The Sun is very near the local interstellar cloud boundary and will cross into another interstellar medium environment in a few thousand years, and scale lengths for variations may be smaller near this boundary (*Müller et al.,* 2006). Observations of He by Ulysses are consistent with no change in the interstellar neutral density since the launch of Ulysses in 1990 (*Witte,* 2004).

3.3. Local Interstellar Cloud Effects on the Solar Wind

If we assume that the local interstellar cloud pressure were constant, we can use the observed solar wind pressures to model the heliosphere boundary changes. Models must include the effects of the local interstellar cloud neutrals on the solar wind. These neutrals are the largest (by mass) component of the interstellar space outside about

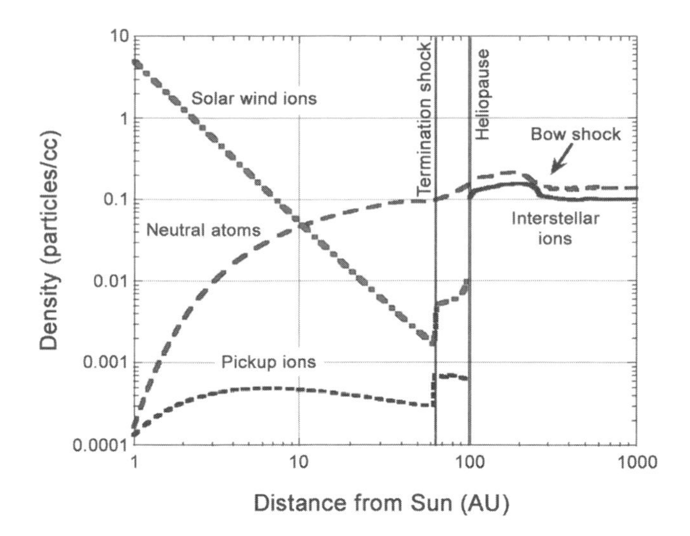

Fig. 1. The density of the plasma and neutral components from 1 to 1000 AU. The solar wind ions come from the Sun. The pickup ions are interstellar neutrals that have been ionized in the solar wind. The densities of the solar wind and of the pickup ions jump at the termination shock. Outside the heliopause, the ions are part of the local interstellar cloud. Both the ion and neutral density increase in front of the heliopause (in the so-called hydrogen wall). The interstellar neutrals dominate the mass density outside 10 AU. Figure courtesy of R. Mewaldt.

10 AU. Figure 1 shows the density of plasma and neutrals vs. distance (we know now the termination shock and heliopause in this figure are located too close to the Sun). The solar wind plasma falls off as R^{-2}, then increases abruptly at the termination shock and heliopause. The H density is about 0.1 at the termination shock and 0.2 in the local interstellar cloud, then decreases rapidly inside 10 AU, the ionization cavity inside which most H is ionized. Both the H and H^+ increase between the bow shock and heliopause and form the hydrogen wall. The pickup ion density/thermal ion density ratio increases to the termination shock, where about 20% of the ions are pickup ions. This ratio is consistent with the observed slowdown of the solar wind. Outside about 30 AU the pickup ions (of local interstellar cloud origin) dominate the plasma thermal pressure. So even though the Sun has carved out its own space in the local interstellar cloud, the local interstellar cloud influence is very strong.

3.3.1. Solar wind slowdown. The neutral interstellar H is continuously being lost, mostly by charge exchange with solar wind protons. The neutrals formed from the solar wind protons escape the solar system at the solar wind speed. The new protons formed from ionized local interstellar cloud neutrals are accelerated from their initial speed of 20 km/s inward to the solar wind speed, 400–700 km/s outward. As mentioned earlier, plasma does not flow across field lines; ions at speeds different from the bulk solar wind speed are accelerated by an electric field E = –VXB. The initial thermal energy of the pickup ions is equal to the solar wind energy, about 1 keV. The energy for the acceleration and heating comes from the bulk flow of the solar wind.

Two consequences of these pickup ions are that the solar wind is heated and slows down. The amount of the slowdown depends on the density of the local interstellar cloud H, so measurements of the slowdown give an estimate of the local interstellar cloud density. This slowdown can be difficult to measure because the solar wind speed varies with time and solar latitude, as well as with distance. In the inner heliosphere, the slowdown is small and difficult to detect. In the outer heliosphere, determining the slowdown from the inner heliosphere to V2 requires (1) another spacecraft in the inner heliosphere at a similar heliolatitude to provide a speed baseline and (2) a model to propagate the solar wind outward to determine how much of the speed change comes from radial evolution of the solar wind and how much is due to pickup ions. Figure 2 shows a compilation of various determinations of the slowdown. The plot shows the solar wind speed normalized to 1 AU. The horizontal lines show the average amount of slowdown; the length of these lines shows the radial range covered in each study. The vertical spread shows the uncertainties in the slowdown determination. The curves shows the predicted solar wind slowdown for densities of interstellar H at the termination shock of 0.07, 0.09, and 0.11 cm^{-3}. The data are best fit with a density of about 0.09 cm^{-3} at the termination shock with an uncertainty of 0.01 cm^{-3}. We note that the H density at the termination shock is not the density in the local interstellar cloud; much filtration, or H loss, occurs between the bow shock and the termination shock.

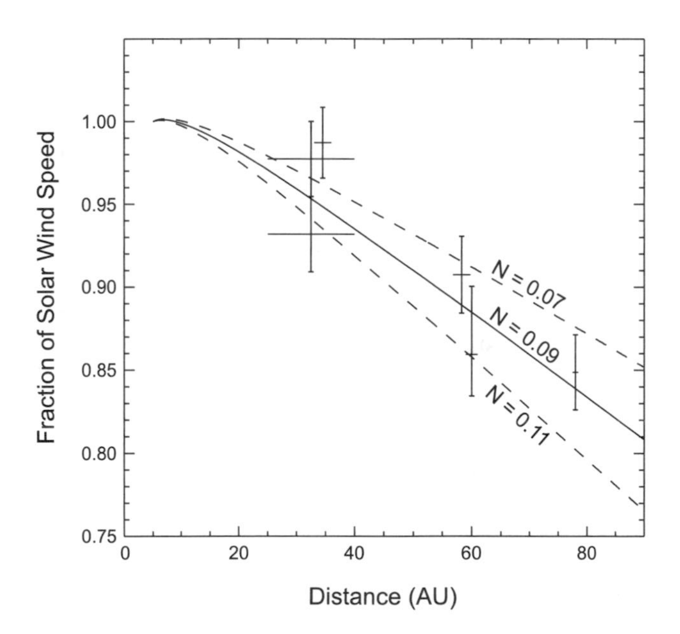

Fig. 2. The plot shows the slowdown of the solar wind caused by the interstellar neutrals. The fraction of the initial solar wind speed is plotted vs. distance. The horizontal lines with error bars show observed speed decreases; the lengths of these lines show the radial width over which the slowdown was observed. The vertical lines are error bars. The curves shows model predictions of the solar wind speed as a function of distance for interstellar neutral H densities of 0.09 cm^{-3} (solid line), 0.07 cm^{-3} (upper dashed line), and 0.11 cm^{-3} (lower dashed line).

By 95 AU, the solar wind speed has decreased by about 20% [and since the mass flux, the speed (v) times the mass density (ρ), is a constant, the density must increase by 20%]. So the dynamic pressure, ρv^2, is only 77% of what it was at 1 AU. The effect of the interstellar H is to lower the solar wind dynamic pressure. This reduced pressure causes the heliosphere boundaries to be closer to the Sun. The solar wind slowdown is related to the pickup ion density by $n_{pu}/n_{sw} = (3\gamma - 1)/(2\gamma - 1)(\delta v/v_0)$, where n_{sw} is the solar wind density, n_{pu} is the pickup ion density, δv is the solar wind slowdown, v_0 is the solar wind speed at 1 AU, and is the ratio of specific heats. For $\gamma = 5/3$, $n_{pu}/n_{sw} = 7\delta v/6v_0$ (*Richardson et al.,* 1995).

3.3.2. Solar wind heating. As the solar wind moves out, it expands and thus might be expected to cool. The solar wind temperature does decrease out to about 30 AU, but not as quickly as adiabatic cooling would predict, then increases with distance with some solar cycle variations superposed. Internally driven processes, such as shocks and dissipation of magnetic fluctuations, provide some heating, but not enough to produce the observed temperature profile (*Richardson and Smith,* 2003; *Smith et al.,* 2001, 2006). The source of this heating is the pickup ions. These ions dominate the thermal energy of the plasma outside about 30 AU, and a small fraction of their energy is transferred to the thermal plasma. The mechanism for this energy transfer is through waves; the pickup ions initially have ring distributions, which are unstable. Low-frequency waves are generated that isotropize the pickup ions, but about 4% of the energy in these waves is transferred to the thermal ions, heating the solar wind (*Smith et al.,* 2006; *Isenberg et al.,* 2005). Since the pickup ion energy and thus the amount of energy transfered to the thermal ions are proportional to the solar wind speed, the solar wind speed and temperature are strongly correlated.

3.4. Termination Shock Model Results

Many models developed to determine the termination shock motion, the asymmetry of the boundaries, and the distance between boundaries (*Webber,* 2005; *Linde et al.,* 1998, *Pogorelov et al.,* 2004; *Izmodenov et al.,* 2005; *Opher et al.,* 2006). We have developed a two-dimensional model that includes the effect of the local interstellar cloud neutrals on the solar wind and use it to track the termination shock location. Figure 3 shows predictions of the termination shock location based on a two-dimensional model that includes the effect of the local interstellar cloud neutrals on the solar wind. The model input is the solar wind pressure observed by V2; the profile is normalized to the V1 termination shock crossing (*Richardson et al.,* 2006a). The model shows that V1 and the shock were both moving radially outward start-

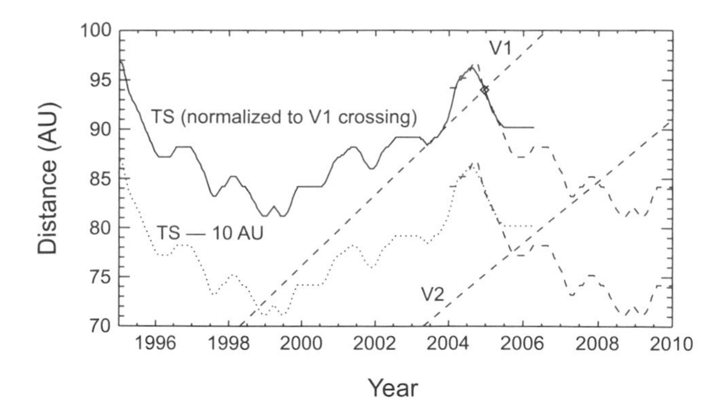

Fig. 3. The predicted position of the termination shock using V2 observations of the solar wind as input to a two-dimensional time-dependent hydrodynamic model. The termination shock locations are extended into the future by repeating the pressure profile from the previous solar cycle. The V1 and V2 trajectories are shown by the dashed diagonal lines. The lower curve shows the termination shock position shifted inward 10 AU to simulate a possible termination shock asymmetry. The termination shock moved inward from 1995 to 1999, then moved outward ahead of V1 until 2004, when it moved inward and crossed V1. The model suggests V2 could cross the termination shock before the end of 2007.

ing in 1999, that V1 made some close encounters to the shock, but that the shock crossing did not occur until the termination shock moved inward in 2004.

3.5. Heliospheric Asymmetries

At Earth, the 45° average angle of the solar magnetic field to the solar wind flow direction results in asymmetries in the magnetosheath plasma (*Paularena et al.,* 2001) and the magnetopause locations (*Dmitriev et al.,* 2004). The draping of magnetic field lines around the magnetosphere results in different field strengths on the dawn and dusk sides, causing these asymmetries. The interstellar magnetic field could result in a similar distortions of the termination shock and heliopause and significantly affect the width of the heliosheath (see, e.g., *Linde et al.,* 1998; *Ratkiewicz et al.,* 1998; *Pogorelov et al.,* 2004). Voyager 2 entered the termination shock foreshock region much closer to the Sun (75 AU) than V1 (85 AU) (*Stone et al.,* 2005). This large a difference probably does not arise entirely from solar wind dynamic pressure changes but also from asymmetries in the boundary locations, with the boundaries closer in the direction of V2.

The nature of the heliosphere distortion depends on the orientation and strength of the interstellar magnetic field. Several suggestions for the magnetic field direction in the local interstellar cloud near the heliosphere have been published. The location of the sources of the heliospheric radio emissions (described below) lies along a line roughly parallel to the galactic plane. *Gurnett et al.* (2006) suggest that the radio emission should occur on magnetic field lines tangential to the shock that excites these emissions. In this case, the magnetic field would be roughly perpendicular to the galactic plane. A field orientation 60° from the galactic plane is derived from differences in the flow velocities of interstellar H and He (*Lallement et al.,* 2005; *Izmodenov et al.,* 2005).

Magnetohydrodynamic (MHD) models of the heliosphere that include the effects of an interstellar magnetic field reveal that the distortion of the heliosphere is sensitive to the angle between the field and the velocity of the interstellar wind. *Opher et al.* (2007) show that a field tilted 60°–90° from the galactic plane is needed to match the flow data and the radio emission data. This tilt creates a large asymmetry in the model heliosphere in both the north/south and east/west directions. The models show that magnetic field lines that intersect the termination shock extend 2 AU inside the termination shock at the position of V1 and 4–7 AU inside the termination shock in the direction of V2 (*Opher et al.,* 2006). These distances represent the average thickness of the foreshock region in front of the termination shock. Small-scale magnetic field variations will affect these distances locally. This model also predicts that the distance of the termination shock in the direction of V2 is 8–11 AU closer than in the direction of V1 and the heliopause distance is 13–24 AU closer in the direction of V2. This model did not include neutrals, which may reduce the asymmetry (*Pogorolev et al.,* 2006). Other models also give asymmetries with ratios of the termination shock distances at V1 and V2 varying from 1.05 to 1.33. The predicted heliopause locations vary from 145–157 AU at V1 and 109–139 AU at V2, with heliosheath thicknesses of 55–67 AU toward V1 and 33–50 AU toward V2 (see *Opher et al.,* 2006).

None of these models include all the important physics; that task is currently too expensive computationally. In particular, the effects of the HCS may be important, since the location of the HCS at the termination shock varies over a solar rotation, but this small-scale structure is hard to include in a global model. Although the models differ quantitatively, the results suggest that the average shock distance at V2 could be ~5–10 AU closer than at V1. In addition, the solar cycle variation of the solar wind dynamic pressure is such that in 2007 the shock should be several AU closer than average, suggesting that V2 might cross the shock in 2007–2008. The models also indicate that the heliosheath is likely narrower at V2, so V2 may cross the heliopause at about the same time as V1. Voyager 1 and V2 observations of the asymmetry of the termination shock and heliosheath should make possible improved models and estimates of the distances to the heliopause.

4. COSMIC RAYS

Cosmic rays are high-energy charged particles that bombard Earth from above the atmosphere. These particles have important effects at Earth. Cloud cover is directly correlated with cosmic ray flux, so these particles may affect Earth's weather (*Svensmark et al.,* 2007). Several thousand cosmic rays pass through a person's body every minute. These can cause biologic damage but also cause mutations that accelerate evolution. A new source of cosmic rays discussed below is generated from the grains in the Kuiper belt (*Schwadron,* 2002). These cosmic rays were likely more intense in the past and the termination shock was at times much closer to the Sun than it is today. Both of these effects may have caused periods when Earth was ensconced in much more severe radiation environments than at present. Several types of cosmic rays come from the outer heliosphere, as discussed below.

4.1. Galactic Cosmic Rays

Galactic cosmic rays (GCRs) possess energies above 100 MeV and are thought to be accelerated in the shock waves associated with supernova in our galaxy. The GCRs are predominately fully stripped nuclei but lesser fluxes of GCR electrons are also observed. Galactic cosmic rays representing most elements in the periodic table have been detected. Since cosmic rays are charged, they are affected by magnetic fields in interplanetary space, the heliosphere, and near Earth. Although GCRs do reach Earth, many are deflected by the heliospheric magnetic field before they reach the inner solar system. Fewer GCRs are observed at Earth at solar maximum, when the solar wind magnetic field is stronger and more turbulent. Large merged interaction regions (MIRs) temporarily reduce the flux of GCRs as these structures contain large and turbulent magnetic fields. An empirical relation between GCRs and the magnitude of B shows that the GCR flux increases when B is above normal and increases when B is below normal; this relationship holds out to the termination shock (*Burlaga et al.,* 2003a, 2005). Spacecraft in the outer solar system see increased GCR fluxes. These particles move so fast and are scattered so thoroughly by the interstellar magnetic field that their flux is likely to be fairly uniform outside the heliosphere.

4.2. Anomalous Cosmic Rays

Early cosmic-ray observers discovered an unusual subset of cosmic rays that consisted of singly ionized ions (instead of fully stripped nuclei) with energies of 1–50 MeV/nuc (see *Mewaldt et al.,* 1994). Most of the anomalous cosmic rays (ACRs) are species that have high ionization thresholds, such as He, N, O, Ne, and Ar. Until recently, ACRs were thought to arise only from neutral atoms in the interstellar medium (*Fisk et al.,* 1974) that drift freely into the heliosphere through a process that has four essential steps: first, the neutral particles from the local interstellar cloud enter the heliosphere; second, the neutrals are converted into pickup ions; third, the pickup ions are preaccelerated by shocks and waves in the solar wind [see also section 5 and *Schwadron et al.* (1998)]; and finally, they are accelerated to their final energies at or beyond the termination shock (*Pesses et al.,* 1981). Easily ionized elements such as C, Si, and Fe are expected to be strongly depleted in ACRs, since these elements are not neutral in the interstellar medium and therefore cannot drift into the heliosphere.

Instruments such as the Solar Wind Ion Composition Spectrometer (SWICS) on Ulysses are able to detect pickup ions directly. Ongoing measurements by cosmic-ray detectors on Voyager and Wind have detected additional ACR components (*Reames,* 1999; *Mazur et al.,* 2000; *Cummings et al.,* 2002). We now understand that, in addition to the traditional interstellar source, grains produce pickup ions throughout the heliosphere: Grains near the Sun produce an "inner source" of pickup ions, and grains from the Kuiper belt provide an "outer" source of pickup ions and anomalous cosmic rays.

Recent observations call into question not only the sources of ACRs, but also the means by which they are accelerated. The prevailing theory, until V1 crossed the termination shock, was that pickup ions were energized at the termination shock to the 10–100-MeV energies observed (*Pesses et al.,* 1981). When V1 crossed the termination shock, it did not see a peak in the ACR intensity as this theory predicts (*Stone et al.,* 2005). Instead, the ACR intensities continued to increase in the heliosheath. One suggestion is that the ACRs are accelerated in the flanks of the heliosheath and the source region is not observed by V1 near the nose (*McComas and Schwadron,* 2006). Another suggestion is that the acceleration region was affected by a series of MIRs before the V1 crossing that diminished the acceleration process (*McDonald et al.,* 2006; *Florinski and Zank,* 2006), in which case V2 may see a different ACR profile when it crosses the termination shock.

4.3. The Interstellar and Inner Source Pickup Ions

The telltale signature of interstellar pickup ions, as seen for H^+ in Fig. 4, is a cutoff in the distribution at ion speeds twice that of the solar wind in the spacecraft reference frame. This cutoff is produced since interstellar ions are initially nearly stationary in the spacecraft reference frame and subsequently change direction, but not energy, in this frame due to wave-particle interactions and gyration, causing them to be distributed over a range of speeds between zero and twice the solar wind speed in the spacecraft frame.

In comparison to the interstellar pickup ion distributions, the C^+ and O^+ distributions in Fig. 4 are puzzling. The distributions do not cut off at twice the solar wind speed and have a peak, contrary to the other ACR ions, which have flat distribution. These ions are singly charged, which precludes them from having been emitted directly by the Sun; if this were the case, they would be much more highly charged, e.g., C^{5+} and O^{6+} are each common solar wind heavy ions. The C^+ and O^+ distributions in Fig. 4 are consistent with a pickup ion source close to the Sun. As the ions travel out in the solar wind, they cool adiabatically in the solar wind reference frame, and instead of having a distribution that cuts off at twice the solar wind speed in the spacecraft reference frame, they have a peak near the solar wind speed. The solid curve that goes through the C^+ and

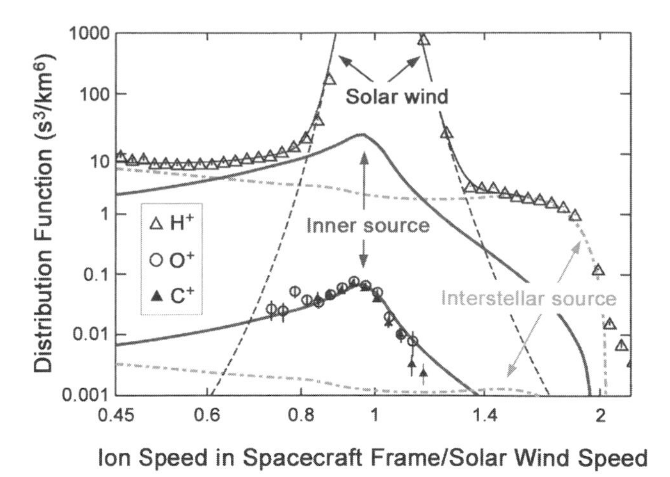

Fig. 4. Observed distributions from Ulysses/SWICS of C^+ (solid triangles), O^+ (open circles), and H^+ (open triangles) vs. ion speed in the spacecraft frame normalized by the solar wind speed (*Schwadron et al.,* 2000). The observations are compared with simulated distributions of solar wind H^+ (dashed black curve), interstellar H^+ (upper gray dash-dot curve), inner source H^+ (upper thick black line), inner source C^+, O^+ (lower thick black line), and interstellar O^+ (lower gray dash-dot curve).

O^+ data points is based on a transport model (*Schwadron,* 1998; *Schwadron et al.,* 2000) for ions picked up close to the Sun, thereby proving the source is in the inner solar system. Data like those presented in Fig. 4 suggested that another pickup ion source was present in addition to interstellar neutrals (*Geiss et al.,* 1995; *Gloeckler and Geiss,* 1998; 2000a; *Schwadron et al.,* 2000). Since the source is peaked near the Sun, as is the distribution of interplanetary grains, it was hypothesized that grains were associated with the source.

The initial, naive expectation was that the inner source composition would resemble that of the grains, i.e., enhanced in C and O with strong depletions of noble elements such as Ne. However, the composition strongly resembled that of the solar wind (*Gloeckler et al.,* 2000a). This conundrum was resolved by assuming a production mechanism whereby solar wind ions become embedded within grains and subsequently reemitted as neutrals. Remarkably, the existence of the inner source was hypothesized many years prior to its discovery (*Banks,* 1971).

The concept that the inner source is generated due to neutralization of solar wind requires that sputtered atoms do not strongly contribute to the source. However, for grains larger than a micrometer, sputtering yields of the grains are much larger than the yields of neutralized solar wind (*Wimmer-Schweingruber and Bochsler,* 2003). This suggests that the grains that give rise to the inner source are extremely small (hundreds of angstroms). A small grain population generated through catastrophic collisions of larger interplanetary grains would also yield a very large filling factor (the filling factor is the net area of the sky filled in by grains divided by the net area of the sky over which they are distributed; so, for example, a large solid object would have a filling factor of 1, whereas a finite set of points would have a filling factor of 0). A large filling factor is consistent with observations of the inner source that require a net geometric cross-section typically 100 times larger than that inferred from zodiacal light observations (*Schwadron et al.,* 2000). The very small grains act effectively as ultrathin foils for neutralizing solar wind (*Funsten et al.,* 1993), but are not effective for scattering light owing to their very small size. The implications of this suggestion are striking: The inner source may come from a large population of very small grains near the Sun generated through catastrophic collisions of larger interplanetary grains (*Wimmer-Schweingruber and Bochsler,* 2003).

4.4. Pickup Ions from Cometary Tails

Comets are also a source of pickup ions in the heliosphere. Until recently, it was thought that the only way to observe cometary pickup ions was to send spacecraft to sample cometary matter directly. However, Ulysses has now had three unplanned crossings of distant cometary tails, Comets McNaught-Hartley (C/1991 T1), Hyakutake (C/1996 B2), and McNaught (C/2006 P1) (*Gloeckler et al.,* 2000b; *Gloeckler,* 2004; *Neugebauer et al.,* 2007). These three unplanned crossings of cometary tails by Ulysses suggest an entirely new way to observe comets through detection of distant cometary tails.

4.5. Outer Source Pickup Ions and Anomalous Cosmic Rays

Recent observations from the Voyager and Wind spacecraft resolved ACR components comprised of easily ionized elements [such as Si, C, Mg, S, and Fe (*Reames,* 1999; *Mazur et al.,* 2000; *Cummings et al.,* 2002)]. An interstellar source for these "additional" ACRs, other than a possible interstellar contribution to C, is not possible (*Cummings et al.,* 2002). Thus, the source for these ACRs must reside within the heliosphere. A number of potential ACR sources are present within the heliosphere (*Schwadron et al.,* 2002). The solar wind particles are easily ruled out as a source since they are highly ionized, whereas ACRs are predominantly singly charged. Discrete sources such as planets are also easily ruled out since their source rate is not sufficient to generate the needed amounts of pickup ions. Another potential source is comets, but the net cometary source rate is sufficiently large only inside 1.5 AU, a location so close to the Sun that the generated pickup ions would be strongly cooled by the time they reach the termination shock, making acceleration to ACR energies extremely difficult. Moreover, a cometary source would naturally be rich in C, which is not consistent with the compositional observations of easily ionized ACRs.

The inner source (discussed in the preceding section) is another possibility, but again, adiabatic cooling poses a sig-

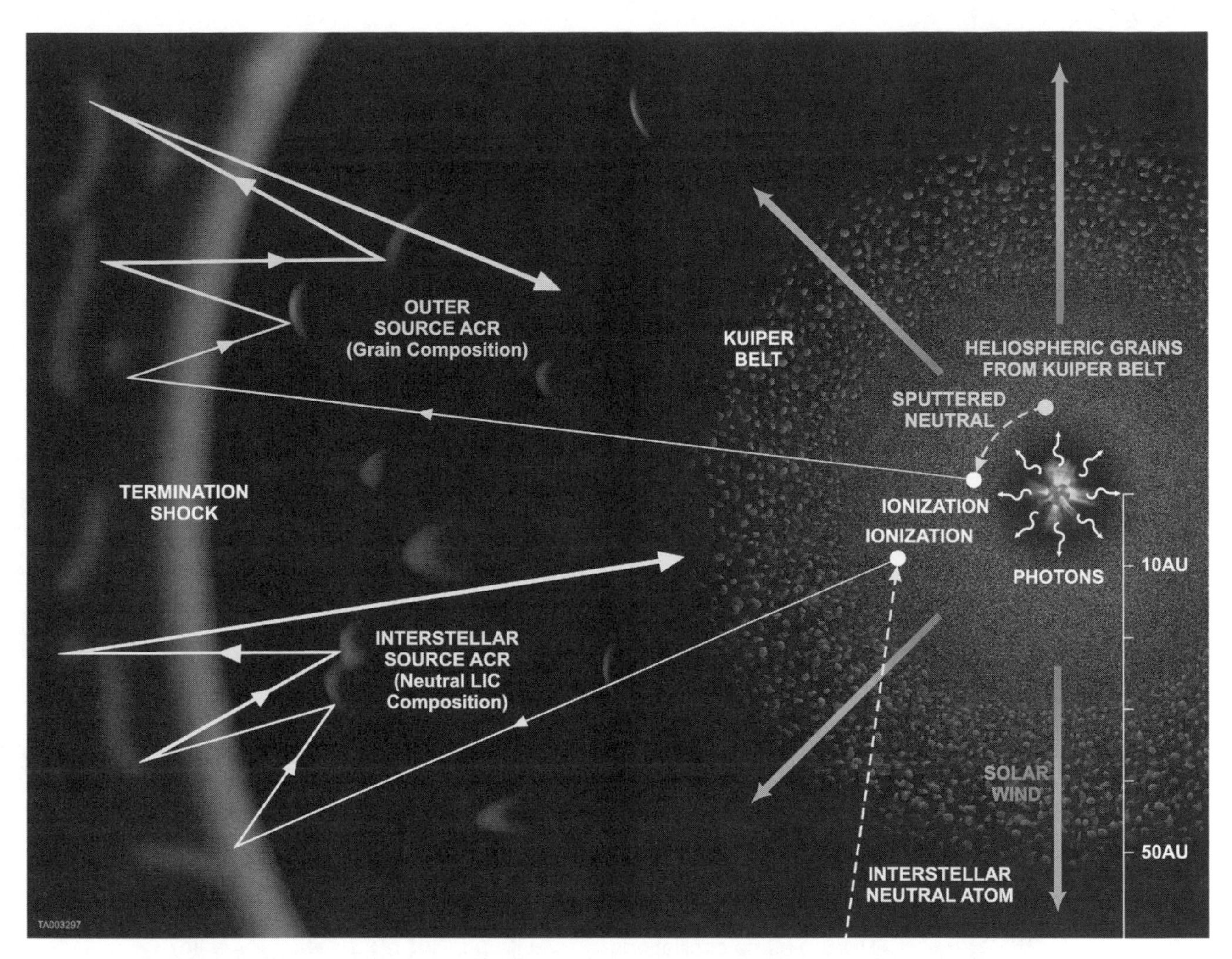

Fig. 5. An illustration of ACR production (*Schwadron et al.,* 2002). Lower curves apply to the known interstellar source ACRs (adapted from *Jokipii and MacDonald,* 1995), while upper curves apply to the outer source, described later in the paper.

nificant problem since these particles are picked up so close to the Sun. It has been suggested that the inner source may be substantially preaccelerated in the inner heliosphere, thereby overcoming the effects of adiabatic cooling. Contrary to this suggestion, inner source observations in slow solar wind show pronounced effects of adiabatic cooling (*Schwadron et al.,* 1999) and charge-state observations of energetic particles near 1 AU indicate no evidence for acceleration of the inner source (*Mazur et al.,* 2002). The additional population of ACRs requires a large source of pickup ions inside the heliosphere and outside 1 AU.

Schwadron et al. (2002) suggest that the source is material extracted from the Kuiper belt through a series of processes shown schematically by the lines in Fig. 5: First, micrometer-sized grains are produced due to collisions of objects within the Kuiper belt; grains spiral in toward the Sun due to the Poynting-Robertson effect; neutral atoms are produced by sputtering and are converted into pickup ions when they become ionized; the pickup ions are transported by the solar wind to the termination shock and, as they are convected, are preaccelerated due to interaction with shocks and due to wave-particle interactions; and finally, they are injected into an acceleration process at the termination shock to achieve ACR energies. The predicted abundances are all within a factor of 2 of observed values, providing strong validation of this scenario. See *Schwadron et al.* (2002) for a detailed presentation of these points.

5. THE TERMINATION SHOCK FORESHOCK

When V1 reached 84 AU in 2002 it observed increases in energetic particles with energies from 40 keV to >50 MeV (*Stone et al.,* 2005; *Decker et al.,* 2005). Figure 6 shows that these ion fluxes increased in mid-2002 by a factor of 100, disappeared in early 2003, came back at high energies in mid-2003 and at low energies in early 2004, and persisted until the termination shock crossing in late 2004. These intensity increases were not associated with solar events and were not observed by V2, which trails V1 by 18.5 AU or about six years. Thus it seemed likely that these particles were associated with the termination shock. But these particles were flowing along the magnetic field lines predomi-

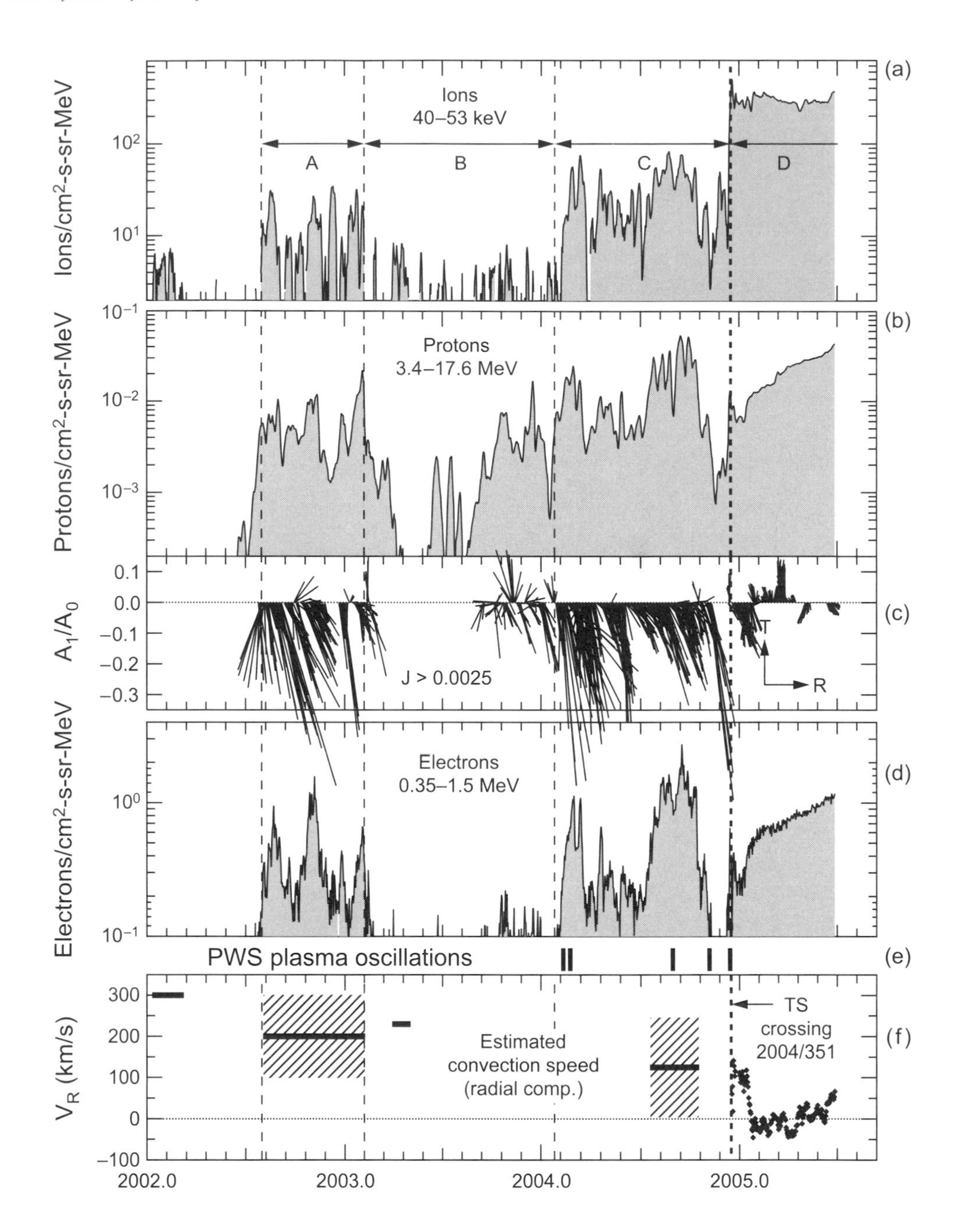

Fig. 6. V1 energetic particle data during 2002.0–2005.7 (83.4–97.0 AU). **(a)**, **(b)**, **(d)** Background-corrected, scan-averaged five-day smoothed daily-averaged intensities of 40–53-keV ions, 3.4–17.6-MeV protons, and 0.35–1.5-MeV electrons. **(c)** First-order anisotropy vector A1/A0 of proton channel in **(b)** when the intensity >0.0025 flux units (inset shows orientation of R and T). Whiskers show direction that particles are traveling. **(e)** Black bars show times when the V1 plasma wave instrument detected electron plasma oscillations (*Gurnett and Kurth,* 2005). **(f)** Estimated plasma radial flow speed V_R based on 40–220-keV ion angular data and a velocity extraction algorithm (*Krimigis et al.,* 2005). Bounds on V_R during period A and the second half of period C (2004.56–2004.78) are indicated by the cross-hatched rectangles and means are shown by the horizontal bars. Horizontal bars during 2002.041–2002.172 and 2003.257–2003.322 are from *Krimigis et al.* (2003). Figure courtesy of R. Decker.

nately in one direction and that direction was away from the Sun (the actual field direction is tangential due to the winding of the Parker spiral with distance, so the particles were moving tangentially but from the direction that is connected to the Sun). If the particle source were the termination shock and V1 was inside this shock, then the particles seemed to be flowing the wrong way. This backward flow direction led to suggestions that V1 had already crossed the termination shock (*Krimigis et al.,* 2003) and that the particles were coming outward from the termination shock. Subsequent work has shown that if the termination shock were blunt, the magnetic field lines first intersect the termination shock closer to the Sun at the distance of V1, so that particles from the termination shock would flow to V1 in the nominally outward direction (*Jokipii et al.,* 2004). This hypothesis predicts that flows observed by V2, on the opposite side of the nose of the termination shock, would be in the opposite direction (nominally toward the Sun). In late 2004, V2 observed these particles and they were in the opposite direction from those at V1, as predicted (*Decker et al.,* 2006). Model predictions (*Opher et al.,* 2006) show that field lines from the termination shock should intersect V1 2–3 AU upstream of the termination shock and V2 4–7 AU upstream of the shock. The amount of time each Voyager spends in the foreshock region depends on the motion of the termination shock. As V1 approached the termination shock the solar wind dynamic pressure was increasing, pushing the termination shock outward, so that V1 was surfing behind the shock and remained in or near the foreshock for almost 9 AU. The solar wind pressure then decreased and the shock moved quickly inward past V1.

Figure 6 shows that the first termination shock particle event ended at 2003.1. The end of this event coincided with the passage of a merged interaction region (MIR) by the spacecraft that pushed the termination shock and thus the foreshock region outward. The first termination shock particle event at V2 also ended when a shock followed by a MIR passed V2, driving the shock outward. The foreshock particle fluxes observed have a lot of structure and many peaks. Some of this structure is connected with the arrival of MIRs at the location of V1 (*Richardson et al.,* 2005b); since the V1 plasma instrument does not work, the MIR arrival times at V1 were estimated by propagating V2 data to the distance of V1. These MIRs can affect the particle fluxes in several different ways; as in the above example, they can push the termination shock outward and sever or weaken the spacecraft connection to the shock. Since the MIRs push through the ambient solar wind, the magnetic field direction may change at the MIR or at the leading shock, thus changing the magnetic connection between the termination shock and the spacecraft, which changes the observed particle flux. Since MIRs are regions of enhanced magnetic field strength and turbulence, energetic particles diffuse through these regions slowly and pile up ahead of the enhanced magnetic field region, forming intensity peaks as the particles are "snowplowed" ahead of the MIRs (*McDonald et al.,* 2000). Only one MIR has occurred since V2 entered the sheath; it was associated with an increase of lower-energy particles (28–540 keV) at the leading shock, probably due to acceleration at the shock. The more-energetic 540–3500 keV particles had intensity peaks after the shock; these could result from snowplowing of particles ahead of the MIR. The particle peak could also be behind the shock if the shock has weakened so that higher-energy particles are no longer accelerated; the particles already energized before the shock source turned off would convect with the solar wind while the shock would propagate ahead through the solar wind at the fast mode speed (*Decker et al.,* 2001).

Both V1 and V2 observe periodic enhancements in the particle fluxes with periods of roughly a solar rotation. In the first V1 foreshock encounter and in the V2 foreshock data through mid-2007, these enhancements are not associated with changes in the plasma or magnetic field. However, when V1 was near the termination shock, in 2004, many of these enhancements were associated with crossings of the heliospheric current sheet (HCS), but only from negative to positive magnetic polarities (*Richardson et al.,* 2006b). One explanation for this correlation is that particles from the termination shock current sheet drift (*Burger and Potgieter,* 1989) along the HCS to V1, and at southern to northern magnetic polarity HCS crossings the connection to the termination shock is much closer. This mechanism would give peaks once per solar rotation. This hypothesis predicts that when V2, at southerly heliolatitudes, is close to the termination shock, particle intensity peaks would be observed only at positive to negative polarity crossings of the HCS. Particle peaks are observed by both V1 and V2 away from the termination shock with periodicities of roughly a solar rotation; the origin of these peaks is not understood.

6. THE TERMINATION SHOCK CROSSING

The termination shock is where the solar wind plasma makes its transition from supersonic to subsonic flow. The solar wind plasma is compressed and heated and the magnetic field strength increases. *Zank* (1999) reviews pre-encounter predictions of the nature and location of the termination shock. Since the average spiral magnetic field upstream of the termination shock is along the T axis (where R is outward from the Sun and T is in the solar equatorial plane and positive in the direction of the Sun's rotation), the termination shock is expected to be a quasi-perpendicular shock. However, energetic particles accelerated or reaccelerated at the termination shock and in the turbulent heliosheath can propagate into the upstream region. When the Voyager spacecraft are near the termination shock and connected to it by the magnetic field, measured particle intensities and anisotropies may be comparable to those at the termination shock if pitch angle scattering is infrequent over the connection length and particles can traverse this length before being convected shockward. Otherwise, intensities and anisotropies observed at Voyager will be reduced relative to those at the termination shock.

The termination shock was crossed in December 2004 on DOY 351, unfortunately during a tracking gap. The Voyager spacecraft are tracked by the DSN 10–12 hours/day due to competition with other spacecraft. The day before the shock crossing, the V1 Low Energy Charged Particle (LECP) instrument saw an intense beam of field-aligned ions and electrons at the same time as the Plasma Wave Spectrometer (PWS) experiment saw electron plasma oscillations; theory predicts that electron beams will drive these waves (*Decker et al.,* 2005; *Gurnett and Kurth,* 2005). After the shock crossing the lower-energy (tens of keV) ions jump in intensity by about an order of magnitude above those in the foreshock and the intensities become steady as shown in Fig. 6. The termination shock particles shown in Fig. 7 are well fit by power laws, consistent with diffusive shock ac-

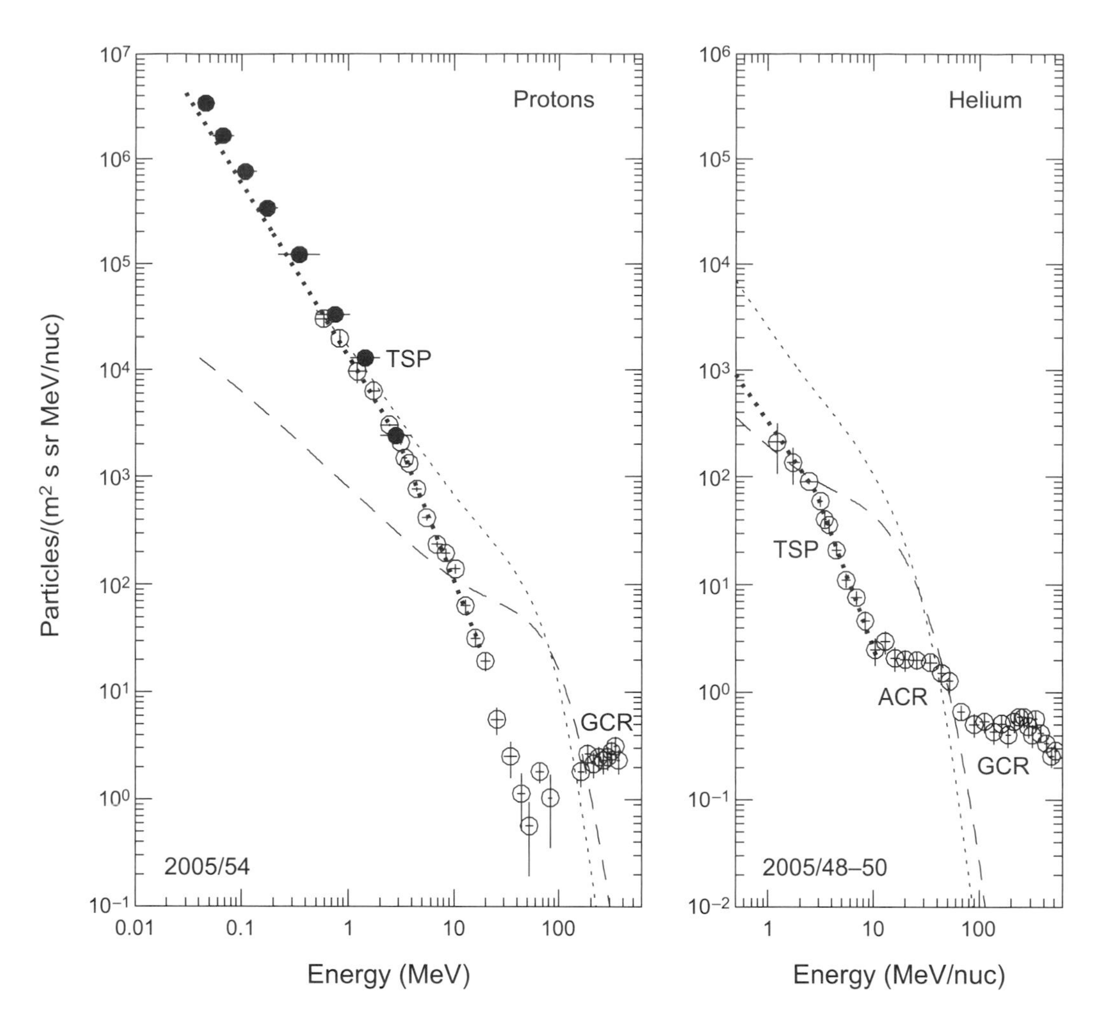

Fig. 7. Typical spectra observed by Voyager in the heliosheath. The solid circles in the left panel are background-corrected ion ($Z \geq 1$) intensities from the LECP instrument and the open circles are from the Cosmic Ray Subsystem (CRS). The spectrum of the termination shock particles is a broken power law with break energy of ~3.5 MeV for protons and a helium. ACR helium dominates the mid-energies in the right panel, with GCRs at higher energies in both panels. The predicted ACR spectra at the shock are shown for a strong (r = 4) and a weak (r = 2.4) shock (*Cummings et al.,* 2002).

celeration (*Decker,* 1988) theory for ions energized at a shock. The spectra of termination shock particles both inside and outside the termination shock are best fit by two power laws with a break energy at 3.5 MeV. The constancy of the break energy suggests that the termination shock particles in the foreshock and heliosheath have the same source. If the termination shock particles were accelerated by diffusive shock acceleration, then the shock strength can be determined from the slope of the termination shock particle power law and is 2.6 (2.4–3.0 with error bars) (*Stone et al.,* 2005). The shock strength is the compression ratio at the shock ($N_{downstream}/N_{upstream}$) and is 4 for a strong shock. An alternative to diffusive shock acceleration has been suggested for the <20-MeV particles; a simple model propagating the measured pickup ions from Ulysses outward and heating them at the termination shock can fit the observed spectra at these energies (*Gloeckler et al.,* 2005). In this model, up to 80% of the solar wind dynamic pressure goes into heating the pickup ions. A relatively weak shock is consistent with the magnetic field observations, which showed the magnetic field strength increased by about a factor of 3 at the shock and stayed high, convincing evidence that V1 was in the heliosheath (*Burlaga et al.,* 2005). The standard deviation of the magnetic field components increased across the termination shock, consistent with predictions for the heliosheath region, and the distribution of the field magnitudes changed from log normal in the solar wind to Gaussian in the heliosheath. After V1 crossed the termination shock into the heliosheath, the intensities of ions of all energies first increased, then fluctuated for 20–30 days, and then became relatively steady, remaining essentially flat at lower energies and increasing slowly at higher energies. At the termination shock crossing, the intensity of 40–53-keV ions increased tenfold, reached a peak, then dropped and thereafter was relatively smooth, varying typically by no more than ≈20% about a mean of ≈300 for the next six months. The anisotropies went from beamlike in the foreshock to nearly isotropic in the heliosheath.

The intensity of the more-energetic ACR particles increases at the shock but remains lower than the peak values in the foreshock. As V1 moved deeper into the heliosheath the ACR intensities increased. These ions are isotropic, not flowing in field-aligned beams as in the foreshock. The major surprise at the shock crossing was that the ACR intensities did not peak and the ACR spectra did not unroll into a power law distribution. For almost 30 years the paradigm had been that ACRs were accelerated at the termination shock through diffusive shock acceleration. This theory predicted a peak in ACRs at the shock and that the acceleration process would result in a power law spectra at the termination shock. Instead, the ACRs intensities at the termination shock did not reach even the level in the foreshock and the ACR energy spectra barely changed. Figure 7 compares the observed ACR spectra to those predicted for strong and weak shocks. For energies less than 30 MeV the observed intensities are far below those predicted. ACRs were not being accelerated where V1 crossed the termination shock. So where are they accelerated? That is still an open question. The intensities did increase further into the heliosheath. Some investigators have suggested that the heliosheath itself was the acceleration region.

Another attractive hypothesis is that the ACRs are accelerated at the flanks of the heliosphere where the connection time of a field line to the termination shock is longer (*McComas and Schwadron,* 2006). Field lines first make contact with a blunt termination shock near its nose. As these field lines are dragged out into the heliosheath by the solar wind, the connection point between the magnetic field and termination shock moves toward the flanks. Anomalous cosmic rays likely require substantial time to be accelerated; insufficient connection time near the nose could prevent acceleration to ACR energies. The much longer connection times toward the distant flanks of the shock may provide the necessary acceleration time to achieve the very high 100–300-MeV ACR energies (*McComas and Schwadron,* 2006; *Schwadron and McComas,* 2007). If the termination shock has a strong nose-to-tail asymmetry, then the flanks would be connected by field lines that are well beyond V1 near the nose, as illustrated in Fig. 8.

Figure 9 documents the strong correlation between modulated ACR and GCR helium. The data points show V1 observations (*McDonald et al.,* 2006) from January 2002 to February 2006; the upper panel, energy range 150–380 MeV/nuc, shows GCRs, and the lower panels, energy ranges 10–

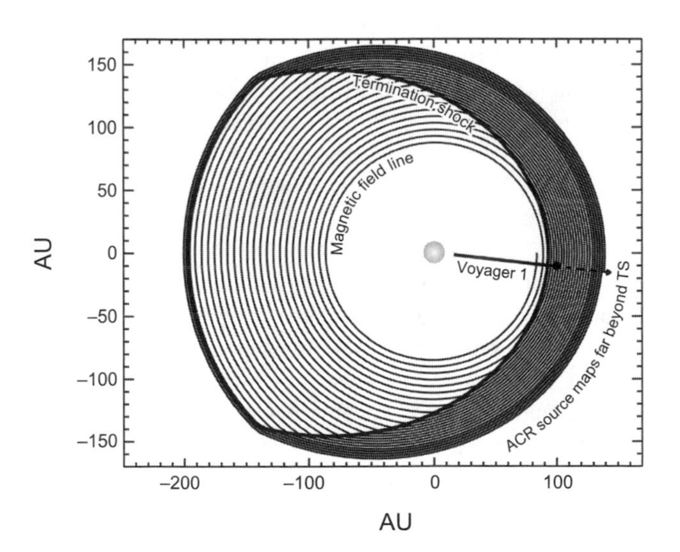

Fig. 8. The configuration of the termination shock and magnetic field in the heliosheath from *Schwadron and McComas* (2007). The bluntness of the termination shock leads to ACR acceleration on the distant flanks (*McComas and Schwadron,* 2006; *Schwadron and McComas,* 2007). The red region shows the outer modulation boundary of ACRs accelerated on the distant flanks of the termination shock.

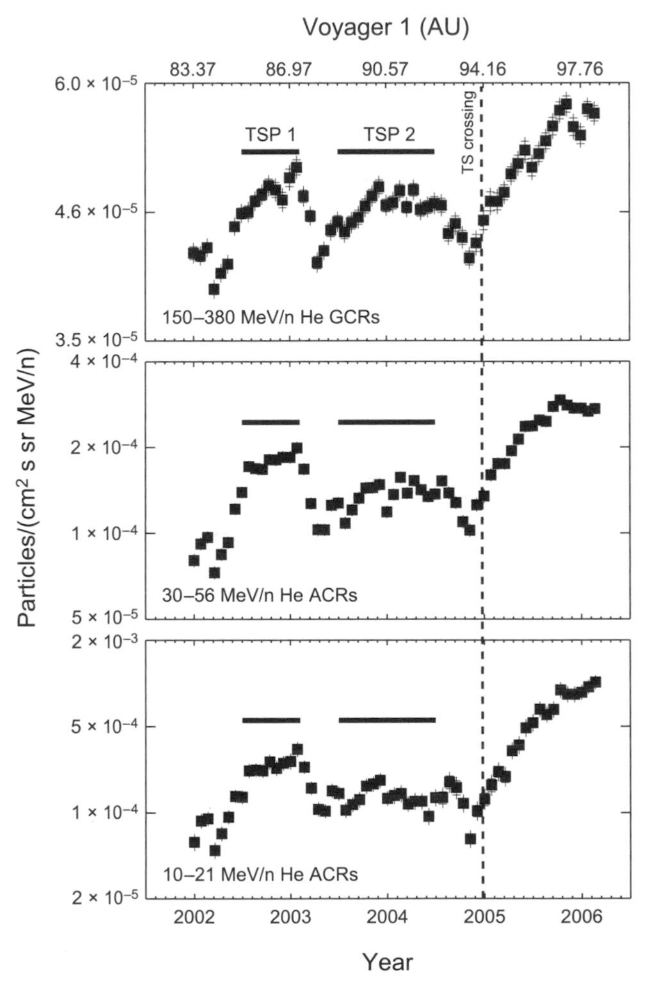

Fig. 9. The correlation between modulation of ACRs and GCRs. The panels show the relative fluxes of ACRs and GCRs observed by Voyager 1. The data (plus signs) were reduced by F. McDonald and B. Heikkila. All fluxes have been normalized by their maximum values and the fluctuations over time (averaged over 26 days) are plotted relative to one another. The respective energy ranges and species are indicated on the axes. The solid curves show theoretical predictions based on a simple Parker-type diffusion-convection modulation model (see text for details).

21 MeV/n and 30–56 MeV/n, measure ACRs. The data are normalized to the maximum value of the differential energy flux in the respective energy ranges. *Schwadron and McComas* (2007) show how the strongly correlated modulation of ACRs and GCRs requires that the ACR accelerator be on field lines that connect far beyond V1 on either the distant reaches of termination shock flanks or far out in the inner heliosheath near the heliopause.

Another possibility for explaining ACR acceleration in the heliosheath is stochastic acceleration (*Fisk,* 2006). However, stochastic mechanisms encounter several problems with explaining ACR acceleration at high energies: (1) the rates of stochastic acceleration are typically lower than the rate of diffusive acceleration at the termination shock, which makes it difficult to generate highly energetic ACRs in the one- to a few-year time frame before they become multiply ionized (*Jokipii,* 1987; *Mewaldt et al.,* 1996); and (2) the mechanism must lead to ACRs generated only far out in the heliosheath, or even toward the tail, whereas a distributed process seems more likely to generate a very distributed source. Although stochastic processes may not act quickly enough to accelerate ACRs to high energies, they are critical for generating the suprathermal seed population, which is naturally favored for diffusive acceleration at the termination shock (*Fisk,* 2006).

McDonald et al. (2006) suggest that the lack of an ACR peak was a time-dependence effect. A MIR passed V1 and encountered the termination shock just before the termination shock crossed V1; perhaps this reduced the acceleration efficiency and the ACR spectra recovered with time, so the increase in intensity observed after the V1 termination shock crossing is a time, not radial distance, effect. This scenario accounts for the larger ACR fluxes in the foreshock than at the termination shock; the MIR reduces the ACR fluxes at the termination shock from the previous, larger intensities that were observed flowing into the foreshock from the termination shock. Over six months later, the ACR fluxes were still recovering from this large temporal change. This hypothesis will be tested when V2 crosses the termination shock.

Another surprise was that after the termination shock crossing V1 stayed in the same magnetic sector for 150 days (*Burlaga et al.,* 2005). Since the Sun rotates every 25.4 days and the HCS is tilted, on average two HCS crossings should occur every 25 days from southern magnetic polarities to northern and back. These data were even more puzzling since V1 was in the southern polarity sector, where V1 should spend less time since it is at northerly latitudes. Sector crossings did resume (*Burlaga et al.,* 2006); the explanation for the initial lack of crossings is that the termination shock and heliosheath were moving inward at a speed comparable to the outward heliosheath flow speed, so V1 remained in the same heliosheath plasma as it moved outward (*Jokipii,* 2005). The solar wind pressure was decreasing, which would make the termination shock and heliosheath move inward. The radial speeds deduced from the LECP data range from –30 to 100 km/s but average 30–50 km/s in this time period, consistent with the termination shock moving inward.

An interesting feature in the magnetic field data are large fluctuations in the field magnitude, a factor of 3, that are purely compressional (the field direction does not change) (*Burlaga et al.,* 2006). These features look qualitatively like the mirror mode waves observed in the magnetosheaths of Jupiter and Saturn (*Joy et al.,* 2006; *Bavassano-Catteneo et al.,* 1998); sometimes they are dips below the background field value as are observed in low β regions of planetary magnetosheaths, and sometimes they are increase above the background field as observed in high β regions. The periods of these waves are roughly an hour. If they are mirror mode waves, they result from an instability that occurs in high β plasmas when $T_{perp} > T_{par}$, where T_{perp} and T_{par} are the temperatures perpendicular and parallel to the magnetic field direction. These conditions often occur behind perpendicular shocks such as the termination shock. As Voyager gets further from the shock, one expects that the plasma will, as a result of these waves, become more stable and the waves less frequent.

7. THE HELIOSHEATH

The spectra of the termination shock ions in the heliosheath are much less variable than in the foreshock. The steadiness of the spectrum suggests that the spectral shape is little affected by modulation in the heliosheath, as expected (*Jokipii and Giacalone,* 2003). The energy spectrum in the heliosheath hardens (has a larger percentage of high-energy ions) with time; the intensities of the higher-energy ions increase while those of the lower-energy ions are roughly constant. The angular distributions of ions from 40 keV to at least 30 MeV were highly anisotropic during the 2.5-year approach of V1 to the termination shock. In the heliosheath the ion anisotropies are much smaller and more variable in direction (*Decker et al.,* 2005; *Krimigis et al.,* 2005).

Figure 6 shows that 5–6 days after the termination shock crossing, V1 began measuring a positive radial flow component fluctuating around $V_R \approx +100$ km s^{-1}, which continued for 27–28 days. Near 2005.045, V_R decreased rapidly (within about 5 days) and flow became inward from 2005.063 to 2005.300. From about 2005.30 to 2005.4, V_R was 0 to +30 km s^{-1}. V_R increased again around 2005.45 and remained near +50 to +100 km s^{-1} until at least 2005.54. These excursions of V_R over a period ≈90–140 days are consistent with V1 sampling the variable flow downstream of an inwardly moving termination shock. Convective flow at V1 that fluctuates about zero implies that V1 would essentially be riding with and sampling the same parcel of plasma for an extended period. Tangential flows are also observed in the direction away from the heliospheric nose direction, as expected for flow around the heliosphere (*Decker et al.,* 2007).

The relationship between ACRs and termination shock particles is unknown. Both are deficient in C ions, indicating that they are accelerated pickup ions (*Krimigis et al.,* 2003).

However, the H/He ratio is ~10 for termination shock particles (*Krimigis et al.,* 2003) and ~5 for ACRs (*Cummings et al.,* 2002), suggesting that He is more easily accelerated to ACR energies than H (*Zank et al.,* 2001).

The flow streamlines in the heliosheath were calculated by *Parker* (1963), assuming subsonic, incompressible, irrotational flow. Using a similar flow model and a kinematic approximation, *Nerney et al.* (1991) and *Washimi and Tanaka* (1996) calculated the variation of the magnetic field B in the heliosheath. Close to the termination shock, the velocity remains nearly radial and B remains nearly azimuthal. The speed decreases as the plasma moves toward the heliopause; consequently, the magnetic field strength B increases. *Cranfill* (1971) and *Axford* (1972) showed that if the heliosheath is thick enough, B might become strong enough to influence the flow. *Nerney et al.* (1993) estimated that B might influence the flow in a restricted region. The flow is deflected away from the radial direction as it approaches the nose of the heliopause, and B develops a radial component. At the heliopause, B must be parallel to the surface (unless significant reconnection occurs at the boundary). The observations of B in the heliosheath show features not described by the simple kinematic models. While B does have a strong azimuthal component, a persistent meridional component (northward) is observed that was not predicted (*Burlaga et al.,* 2005).

The magnetic field direction in the solar wind is often alternately toward and away from the Sun for several days. The existence of these "sectors" is related to extensions of fields from the polar regions of the Sun to the latitude of the observing spacecraft. Sectors have been observed by V1 out to 94 AU (*Burlaga et al.,* 2003b, 2005).

8. THE HELIOPAUSE

As the Voyagers continue their trek through the uncharted heliosheath, the next milepost in their journey will be the heliopause, which we think they will encounter between 2013 and 2021. The crossing of the heliopause will be a truly historic occasion; the first manmade vehicle to leave the solar system. The heliopause is thought to be a tangential discontinuity with a large jump in the plasma density and a rotation and change in the magnitude of the magnetic field. Given the lack of an operational plasma instrument on V1, the first crossing of the heliopause will probably be identified by a durable change in magnitude and/or direction of the magnetic field. If the model asymmetries discussed above are real, then the V1 and V2 crossings could be close to each other in time.

The flow velocity vectors are expected to become tangent to the nominal heliopause surface. However, variations of the heliopause surface and fluctuations in the flow velocities could make it difficult to use these data to identify a heliopause crossing. In analogy with planetary magnetopauses, the heliopause is probably a complex surface that varies locally in thickness and orientation and is the site of patchy reconnection and a variety of plasma instabilities. The appearance of large variations in flow velocities associated with disturbances from such relatively small-scale dynamical processes may indicate that V1 is near the heliopause. Reconnection at the heliopause-LISM interface could accelerate low-energy ions (and electrons) and create large departures from the normal heliosheath intensities and anisotropies. Intermittent reconnection along a slightly corrugated heliopause surface might appear in the low-energy ion (and possibly, electron) intensity-time profiles as gradual intensity increases with superposed, anisotropic intensity spikes as the spacecraft approached the heliopause, not unlike structures encountered by V1 as it approached the termination shock. These comments are speculative, drawing largely upon analogies to planetary magnetopauses; however, as with the termination shock, energetic particle observations may enable us to remotely sense the heliopause well before the Voyagers cross it.

Instabilities on the heliopause have been the subject of theoretical conjecture (cf. *Fahr,* 1986; *Florinski et al.,* 2005). In principle, the Rayleigh-Taylor instability could be important near the nose of the heliosphere where the difference between the flow densities are large. This instability requires an effective "gravity" or destabilizing force. *Liewer et al.* (1996) proposed that coupling between ions and neutrals via charge exchange would produce such a destabilizing force. Near the flanks, away from the nose, the Kelvin-Helmholtz instability is more likely to be important where the shear velocity across the heliopause is large. Either type of instability would produce motion of the heliopause superposed on the "breathing" expected to occur due to varying solar wind pressure. Estimates of the magnitude of the resulting excursions are tens of AU and have periods on the order of 100 years or more (*Liewer et al.,* 1996; *Wang and Belcher,* 1998). For Voyager, the long periods make it unlikely that such wave motion can be detected, but the additional radial motion from such instabilities, should they exist, increases the uncertainty in the distance to the heliopause.

9. THE HYDROGEN WALL

Plate 11 shows that a stagnation point forms in the flow at the nose of the heliosphere. The plasma flow speed in this region slows, so the plasma is compressed and heated. Since the H^+ and the H are coupled via charge exchange, the H slows as well. Since the flux of H must be conserved, slower H speeds translate to larger H densities (see *Baranov and Malama,* 1993, 1995; *Zank et al.,* 1996). Plate 11 shows the region of denser H known as the hydrogen wall in front of the heliosphere. This wall has been observed. Observations of H Lyman-α emission from nearby stars in the nose direction show that the H absorption is red-shifted relative to the other lines by 2.2 km/s; its width is too broad to be consistent with the $T = 5400 \pm 500$ K temperature suggested by the width of the deuterium line (*Linsky and Wood,* 1996). The profiles could only be fit assuming additional absorption by hot H, such as the compressed H in the hydrogen wall. Two H components were needed to fit the data, one from the heliospheric hydrogen wall (which, since the H is moving sunward slower than the local interstellar cloud H, produces absorption on the red side of the H line) and one from the hydrogen wall of the nearby stars (which, since the H is moving sunward faster than the local interstellar cloud H, produces absorption on the blue side of the H line) (*Linsky and Wood,* 1996; *Gayley et al.,* 1997; *Izmodenov et al.,* 1999, 2002; *Wood et al.,* 1996, 2000). Many hydrogen walls have been detected in front of other astrospheres, allowing comparison to the heliosphere and providing validation to our general view of the heliosphere structure.

10. THE HELIOSPHERIC RADIO EMISSIONS

The Voyager spacecraft periodically detect radio emission from the boundaries of our solar system. This radio source is the most powerful in the solar system with an emitted power of over 10^{13} W (*Gurnett and Kurth,* 1996). Plate 12 shows a spectrogram of these emissions, plotting intensity as a function of time and frequency. These waves have two frequency components, a constant frequency 2-kHz component and a component where the frequency rises from 2.4 to 3.5 kHz over a period of 180 days. The two largest events occurred in 1983–1984 and 1992–1994, a few years after solar maxima. Large interplanetary shocks or MIRs were hypothesized to trigger these emissions when they entered the higher-density hydrogen wall beyond the heliopause. But despite the occurrence of several large events after the 2000 solar maximum, only very weak emission was observed in 2002–2003, even though Voyager is presumably much closer to the radio source. The source region and trigger for these emissions is not well understood.

11. THE HELIOSHEATH: THE LOCAL INTERSTELLAR MEDIUM

The local interstellar medium surrounding the heliosphere is not a constant. As the Sun and interstellar clouds move through the galaxy, the Sun has been and will be in very different environments. The interstellar medium has hot tenuous regions and cool dense regions that coexist in pressure balance. The Sun is now in a hot tenuous region of the interstellar medium known as the local bubble. This region is about 200 pc across and has $T = 10^6$ K and an average neutral H density of 0.07 cm^{-3}. This density is about 10 times smaller than the average interstellar H density. The local bubble was created by several supernova explosions that occurred on the order of a million years ago, blowing a hole in the interstellar medium. The local bubble contains several clouds, including the local interstellar cloud in which the Sun now resides, with an H density of about 0.2 cm^{-3}. The Sun is very near the edge of the local interstellar cloud and will enter a different interstellar environment in less than 4000 years. It entered the current local interstellar cloud within the past 40,000 years. Possible values of the densities encountered by the heliosphere in the past and future are 0.005–15 cm^{-3} with Sun-cloud relative velocities of up to 100 km/s (*Müller et al.,* 2006). Given possible values of the interstellar pressure, in the past and future the heliopause could be located anywhere between 12 and 400 AU and the termination shock from between 11 and 250 AU. For a very small heliosphere, cosmic-ray fluxes would dramatically increase, affecting the climate and increasing rates of genetic mutation. When the heliosphere is in the local bubble, it will be greatly expanded. The GCR fluxes at Earth will be similar to present-day fluxes, but the fluxes of ACRS and neutral H will be greatly reduced.

12. FUTURE OBSERVATIONS

12.1. Voyager

The Voyager spacecraft continue their journeys, V1 moving out about 3.5 AU/year and V2 about 3 AU/year. Voyager 2 is in the termination shock foreshock region; models suggest that this region only extends 4–7 AU from the termination shock. In June 2007, V2 had already been in the foreshock region for almost 8 AU. The location of the termination shock and foreshock change with the solar wind pressure and thus the time spent in the foreshock can be longer than its width would suggest, as when the termination shock was moving out just ahead of V1, but V2 will probably cross the termination shock before the end of 2008. This crossing will provide information on the asymmetry of the heliosphere. It will also provide the first plasma data from the heliosheath, providing direct information on the plasma flow, shock compression, and heating at the termination shock. Figure 10 shows the V1 and V2 trajectories in distance and heliolatitude. Model estimates of the locations of the termination shock and heliopause are shown shown by the lines.

The V1 heliopause crossing is likely to occur between 2013 and 2021 and the V2 heliopause crossing should occur in the same time frame. The spacecraft will have sufficient

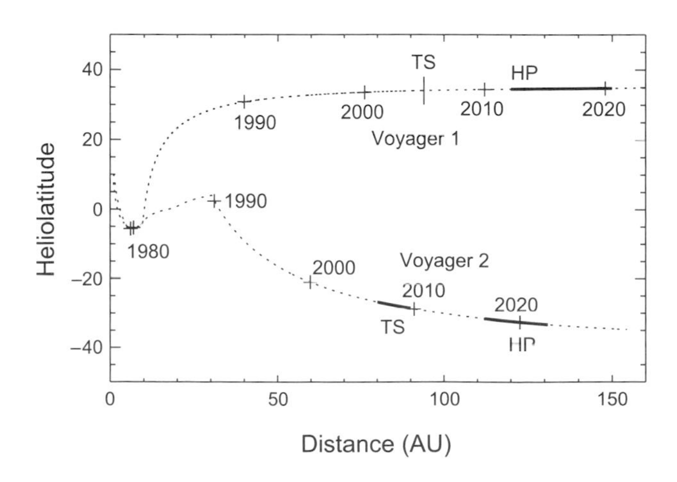

Fig. 10. Trajectories of the Voyager spacecraft in distance and heliolatitude. The range of predicted heliopause (for V1 and V2) and termination shock (for V2) distances are shown by the solid parts of the trajectory.

power to operate all instruments until 2016; after this time, power-sharing will extend the useful life of the spacecraft beyond 2020. Thus the Voyagers are likely to provide the first *in situ* measurements of the local interstellar cloud.

12.2. New Horizons

New Horizons is a NASA mission that will fly by the Pluto-Charon system and then proceed to other Kuiper belt objects. Again, due purely to serendipity, Pluto is now located toward the nose of the heliosphere. Although this mission will not return data from the interstellar medium, it will measure the pickup ions directly, providing new information on the densities and composition of the interstellar neutrals entering the solar system.

12.3. Interstellar Boundary Explorer

The Interstellar Boundary Explorer (IBEX) is an exciting new NASA mission that will observe the large-scale structure of the heliosphere directly. It will be launched into a highly eccentric Earth orbit in June 2008 [see the general description of the mission in *McComas et al.* (2004)]. All-sky energetic neutral atom (ENA) images of the global structure of the termination shock and heliosheath will be accumulated every 6 months. These ENAs are generated by charge-exchange collisions between the inflowing interstellar neutral gas and the protons that are heated/accelerated at the termination shock and that subsequently populate the entire heliosheath. Two IBEX single-pixel telescopes will image these hydrogen ENAs over semiannually precessing great circles in the sky at low energies (10 eV to 2 keV) and in an overlapping band of higher energies (300 keV to 6 keV). The intensities of the ENAs depend strongly on both the shape of the proton spectrum and the plasma convection flow pattern in the heliosheath. Where the anti-Sunward component of the heliosheath convection velocity is high, the ENA intensity directed toward Earth is low, and vice versa. The ENA foreground from the heliosphere between Earth and the termination shock is negligible because the solar wind and pickup ion populations are being convected away from the Sun at the solar wind velocity (which is greater than the Earth-directed velocities of the thermal or pickup ions). Beyond the termination shock, the shocked decelerated flow in the heliosheath allows heated/accelerated protons to reach Earth and be imaged by IBEX.

The ENA emission seen from Earth is accumulated along the entire line of sight outward through the inner heliosheath, across the heliopause, and into the outer heliosheath. The measured ENA intensity integrates the effect of the termination shock heating/acceleration and heliosheath plasma flow over several tens of AU, thus yielding diagnostics of this immense boundary region (*Gruntman et al.,* 2001). Even though V1 and V2 do not directly measure protons in the energy range corresponding to the IBEX ENA hydrogen, these complementary measurements will prove of inestimable value in constraining the theoretical models of the global interaction of the solar wind with the interstellar medium. For example, *Gloeckler et al.* (2005) extrapolated the V1 LECP proton spectra (40–4000 keV) upstream and downstream of the termination shock to lower energies and demonstrated that 80% of the upstream ram pressure goes into heating the pickup protons. Thus V1 provides the input proton spectrum for the global heliosheath (albeit only at the V1 location) that will be imaged in IBEX hydrogen ENAs. And, returning the scientific favor, the very first IBEX half-year images will, at a glance, reveal asymmetries in the large-scale heliosheath configuration that bear on the puzzling directions of the magnetic field and energetic particle streaming observed by V1 and V2. The IBEX all-sky images, in concert with local measurements from Voyager, will provide a quantum leap forward in our understating of the outer heliosphere and its interstellar interaction.

12.4. The Interstellar Probe

Finally, in the era after Voyager and IBEX, we hope to send a new, optimally instrumented mission called the Interstellar Probe out beyond the termination shock, heliopause, and bow shock and into the pristine interstellar medium. Such a mission will require propulsion capable of accelerating a spacecraft to immense speeds if results are to be gained in a reasonable mission lifetime. Recent studies have examined both solar sails and nuclear-powered electric propulsion. While the results of such a mission are still decades off, we need to begin soon if we are to finally send a probe out into the material of the stars.

Acknowledgments. J.D.R. was supported under NASA contract 959203 from the Jet Propulsion Laboratory to the Massachusetts Institute of Technology. N.A.S. was supported by the NASA Interstellar Boundary Explorer mission, which is part of the GSFC Explorer Program, and by NASA's Earth-Moon-Mars Radiation Environment Model (EMMREM) Project.

REFERENCES

Aellig M. R., Lazarus A. J., and Steinberg J. T. (2001) The solar wind helium abundance: Variation with wind speed and the solar cycle. *Geophys. Res. Lett., 28,* 2767–2770.

Axford W. I. (1972) The interaction of the-solar wind with the interstellar medium. In *The Solar Wind* (C. P. Sonnett et al., eds.), pp. 609–631. NASA SP-308, Washington, DC.

Banks P. M. (1971) Interplanetary hydrogen and helium from cosmic dust and the solar wind. *J. Geophys. Res., 76,* 4341–4354.

Baranov V. B. and Malama Y. G. (1993) Model of the solar wind interaction with the local interstellar medium — numerical solution of self-consistent problem. *J. Geophys. Res., 98,* 15157–15163.

Baranov V. B. and Malama Y. G. (1995) Effect of local interstellar medium hydrogen fractional ionization on the distant solar wind and interface region. *J. Geophys. Res., 100,* 14755–14762.

Bavassano-Cattaneo M. B., Basile C., Moreno G., and Richardson J. D. (1998) Evolution of mirror structures in the magnetosheath of Saturn from the bow shock to the magnetopause. *J. Geophys. Res., 103,* 11961–11972.

Belcher J. W., Lazarus A. J., McNutt R. L. Jr., and Gordon G. S. Jr.

(1993) Solar wind conditions in the outer heliosphere and the distance to the termination shock. *J. Geophys. Res., 98,* 15166–15183.

Bertaux J. L. and Blamont J. E. (1971) Evidence for a source of an extraterrestrial hydrogen Lyman-alpha emission. *Astron. Astrophys., 11,* 200–217.

Burger R. A. and Potgieter M. S. (1989) The calculation of neutral sheet drift in two-dimensional cosmic-ray modulation models. *Astrophys. J., 339,* 501–511.

Burlaga L. F. (1995) *Interplanetary Magnetohydrodynamics.* Oxford Univ., Oxford.

Burlaga L. F., McDonald F. B., Ness N. F., Schwenn R., Lazarus A. J., and Mariani F. (1984) Interplanetary flow systems associated with cosmic ray modulation in 1977–1980. *J. Geophys. Res., 89,* 6579–6588.

Burlaga L. F., Ness N. F., Stone E. C., McDonald F. B., Acuna M. H., Lepping R. P., and Connerney J. E. P. (2003a) Search for the heliosheath with Voyager 1 magnetic field measurements. *Geophys. Res. Lett., 30,* 2072–2075, DOI: 10.1029/2003GL018291.

Burlaga L. F., Ness N. F., and Richardson J. D. (2003b) Sectors in the distant heliosphere: Voyager 1 and 2 observations from 1999 through 2002 between 57 and 83 AU. *J. Geophys. Res., 108,* 8028, DOI: 10.1029/2003JA009870.

Burlaga L. F., Ness N. F., Acuna M. H., Lepping R. P., Connerney J. E. P., Stone E. C., and McDonald F. B. (2005) Crossing the termination shock into the heliosheath. Magnetic Fields. *Science, 309,* 2027–2029.

Burlaga L. F., Ness N. F., and Acuna M. H. (2006) Multiscale structure of magnetic fields in the heliosheath. *J. Geophys. Res., 111,* A09112, DOI: 10.1029/2006JA011850.

Cranfill C. W. (1971) Flow problems in astrophysical systems, Ph.D. thesis, Univ. of California, San Diego.

Cummings A. C., Stone E. C., and Steenberg C. D. (2002) Composition of anomalous cosmic rays and other heliospheric ions. *Astrophys. J., 578,* 194–210.

Decker R. B. (1988) Computer modeling of test particle acceleration at oblique shocks. *Space Sci. Rev., 48,* 195–262.

Decker R. B., Paranicas C., Krimigis S. M., Paularena K. I., and Richardson J. D. (2001) Recurrent ion events and plasma disturbances at Voyager 2: 5 to 50 AU. In *The Outer Heliosphere: The Next Frontiers* (K. Scherer et al., eds.), pp. 321–324. Pergamon, Oxford.

Decker R. B., Krimigis S. M., Roelof E. C., Hill M. E., Armstrong T. P., Gloeckler G., Hamilton D. C., and Lanzerotti L. J. (2005) Voyager 1 in the foreshock, termination shock, and heliosheath. *Science, 309,* 2020–2024, DOI: 10.1126/science.1117569.

Decker R. B., Krimigis S. M., Roelof E. C., and Hill M. E. (2006) Low-energy ions near the termination shock. In *Physics of the Inner Heliosheath: Voyager Observations, Theory, and Future Prospects; 5th Annual IGPP International Astrophysics Conference* (J. Heerikhuisen et al., eds.), pp. 73–78. AIP Conf. Proc. 858, New York.

Decker R. B., Krimigis S. M., and Roelof E. C. (2007) Energetic particles at Voyager 1 in the heliosheath and Voyager 2 in the termination foreshock. *Eos Trans. AGU, 88(23)*, Joint Assembly Suppl., Abstract SH43A–02.

Dmitriev A. V., Suvorova A. V., Chao J. K., and Yang Y.-H. (2004) Dawn–dusk asymmetry of geosynchronous magnetopause crossings. *J. Geophys. Res., 109,* A05203, DOI: 10.1029/2003JA010171.

Fahr H. J. (1974) The extraterrestrial UV-background and the nearby interstellar medium. *Space Sci. Rev., 15,* 483–540.

Fahr H. J. (1986) Is the heliospheric interface submagnetosonic? Consequences for the LISM presence in the heliosphere. *Adv. Space Res., 6,* 13–25.

Fahr H. J. (1996) The interstellar gas flow through the interface region. *Space Sci. Rev., 78,* 199–212.

Fahr H. J.and Lay G. (1973) Interplanetary He I 584 A background radiation. *Space Res., XIII(2),* 843–847.

Fisk L. A. (2006) The common spectrum for accelerated ions in the quiet-time solar wind. *Astrophys. J. Lett., 640,* L29–L32.

Fisk L. A., Kozlovski B. and Ramaty R. (1974) An interpretation of the observed oxygen and nitrogen enhancements in low energy cosmic rays. *Astrophys. J. Lett., 190,* L35–L37.

Florinski V. and Zank G. P. (2006) Galactic cosmic ray intensities in response to interstellar environments. In *Solar Journey: The Significance of Our Galactic Environment for the Heliosphere and Earth* (P. C. Frisch, ed.), Chapter 10. Astrophysics and Space Science Library Series Vol. 338, Kluwer, Dordrecht.

Florinski V., Zank G. P., and Pogorelov N. V. (2005) Heliopause stability revisited: Dispersion analysis and numerical simulations. *Adv. Space Res., 35,* 2061–2066.

Funsten H. O., Barraclough B. L., and McComas D. J. (1993) Shell effects observed in exit charge state distributions of 1–30 keV atomic projectiles transiting ultrathin carbon foils, nuclear instruments and methods. *Phys. Res. B, 80/81,* 49–52.

Gayley K. G., Zank G. P., Pauls H. L., Frisch P. C., and Welty D. E (1997) One- versus two-shock heliosphere: Constraining models with GHRS Lyman-α spectra toward α Centauri. *Astrophys. J., 487,* 259–270.

Geiss J., Gloeckler G., Fisk L. A., and von Steiger R. (1995) C^+ pickup ions in the heliosphere and their origin. *J. Geophys. Res., 100,* 23373–23378.

Gloeckler G. and Geiss J. (1998) Measurement of the abundance of helium-3 in the Sun and in the Local Interstellar Cloud with SWICS on Ulysses. *Space Sci. Rev., 84,* 275–284.

Gloeckler G., Fisk L. A., Geiss J., Schwadron N. A., and Zurbuchen T. H. (2000a) Elemental composition of the inner source pickup ions. *J. Geophys. Res., 105,* 7459–7464.

Gloeckler G., Geiss J., Schwadron N. A., Fisk L. A., Zurbuchen T. H., Ipavich F. M., von Steiger R., Balsiger H., and Wilken B. (2000b) Interception of comet Hyakutake's ion tail at a distance of 500 million kilometres. *Nature, 404,* 576–578.

Gloeckler G., Allegrini F., Elliott H. A., McComas D. J., Schwadron N. A., Geiss J., von Steiger R., and Jones G. H. (2004) Cometary ions trapped in a coronal mass ejection. *Astrophys. J. Lett., 604,* L121–L124.

Gloeckler G., Fisk L. A., and Lanzerotti L. J. (2005) Acceleration of solar wind and pickup ions by shocks. In *Solar Wind 11/SOHO 16 Programme and Abstract Book,* p. 52. ESA, Noordwijk, The Netherlands.

Gringauz K. I. (1961) Some results of experiments in interplanetary space by means of charged particle traps on Soviet space probes. *Space Res., 2,* 539–553.

Gruntman M., Roelof E. C., Mitchell D. G. Fahr H. J., Funsten H. O., and McComas D. J. (2001) Energetic neutral atom imaging of the heliospheric boundary region. *J. Geophys. Res., 106,* 15767–15781.

Gurnett D. A. and Kurth W. S. (1996) Radio emissions from the outer heliosphere. *Space Sci. Rev., 78,* 53–66.

Gurnett D. A. and Kurth W. S. (2005) Electron plasma oscillations upstream of the solar wind. Termination shock. *Science, 309,* 2025–2027.

Gurnett D. A., Kurth W. S., Cairns I. H., and Mitchell J. (2006) The local interstellar magnetic field direction from direction-finding measurements of heliospheric 2–3 kHz radio emissions. In *The Physics of the Inner Heliosheath: Voyager Observations, Theory, and Future Prospects; 5th Annual IGPP International Astrophysics Conference* (J. Heerikhuisen et al., eds.), pp. 129–134. AIP Conf. Proc. 858, New York.

Isenberg P. A, Smith C. W., Matthaeus W. H., and Richardson J. D. (2005) Turbulent heating of the distant solar wind by interstellar pickup protons with a variable solar wind speed. In *Proceedings of Solar Wind 11: Connecting Sun and Heliosphere* (B. Fleck and T. H. Zurbuchen, eds.), pp. 347–350. ESA SP-592, Noordwijk, The Netherlands.

Izmodenov V. V., Lallement R., and Malama Y. G. (1999) Heliospheric and astrospheric hydrogen absorption towards Sirius: No need for interstellar hot gas. *Astron. Astrophys., 342,* L13–L16.

Izmodenov V. V., Wood B. E., and Lallement R. (2002) Hydrogen wall and heliosheath Lyman-α absorption toward nearby stars: Possible constraints on the heliospheric interface plasma flow. *J. Geophys. Res., 107,* 1308–1322.

Izmodenov V. V., Malama Y. G., and Ruderman M. (2005) Solar cycle influence on the interaction of the solar wind with local interstellar cloud. *Astron. Astrophys., 429,* 1069–1080, DOI: 10.1051/004-6361:20041348.

Jokipii J. R. (1987) Rate of energy gain and maximum energy in diffusive shock acceleration. *Astrophys. J., 313,* 842–846.

Jokipii J. R. (2005) The magnetic field structure in the heliosheath. *Astrophys. J. Lett., 631,* L163–L165.

Jokipii J. R. and Giacalone J. (2003) Anomalous cosmic rays at a termination shock crossing. *28th Intl. Cosmic Ray Conf., 7,* 3753–3756.

Jokipii J. R. and MacDonald F. B. (1995) Quest for the limits of the heliosphere. *Sci. Am., 273,* 58.

Jokipii J. R., Giacalone J., and Kota J. (2004) Transverse streaming anisotropies of charged particles accelerated at the solar wind termination shock. *Astrophys. J., 611,* L141–L144.

Joy S. P., Kivelson M. G., Walker R. J., Khurana K. K., Russell C. T., and Paterson W. R. (2006) Mirror mode structures in the Jovian magnetosheath. *J. Geophys. Res., 111,* A12212, DOI: 10.1029/2006JA011985.

Karmesin S. R., Liewer P. C., and Brackbill J. U. (1995) Motion of the termination shock in response to an 11 year variation in the solar wind. *Geophys. Res. Lett., 22-25,* 1153–1156.

Krimigis S. M., Decker R. B., Hill M. E., Armstrong T. P., Gloeckler G., Hamilton D. C., Lanzerotti L. J., and Roelof E. C. (2003) Voyager 1 exited the solar wind at a distance of 85 AU from the Sun. *Nature, 426,* 45–48, DOI: 10.1038/nature02068.

Krimigis S. M., Decker R. B., Roelof E. C., and Hill M. D. (2005) Voyager's discovery of region of energetic particles associated with the termination shock. In *Proceedings of Solar Wind 11/ SOHO 16 Conference: Connecting Sun and Heliosphere* (B. Fleck et al., eds.), pp. 23–28. ESA SP-592, Noordwijk, The Netherlands.

Lallement R., Bertaux J. L., and Clarke J. T. (1993) Deceleration of interstellar hydrogen at the heliospheric interface. *Science, 260,* 1095–1098. DOI: 10.1126/science.260.5111.1095.

Lallement R., Quemerais E., Bertaux J. L., Ferron S., Koutroumpa D., and Pellinen R. (2005) Deflection of the interstellar neutral hydrogen flow across the heliospheric interface. *Science, 307,* 1447–1449, DOI: 10.1126/science.1107953.

Lazarus A. J. and McNutt R. L. Jr. (1990) Plasma observations in the distant heliosphere: A view from Voyager. In *Physics of the Outer Heliosphere* (S. Grzedzielski and D. E. Page, eds.), pp. 229–234. Pergamon, New York.

Liewer P. C., Karmesin S. R., and Brackbill J. U. (1996) Hydrodynamic instability of the heliopause driven by plasma-neutral charge exchange interactions. *J. Geophys. Res., 101,* 17119–17128.

Linde T. J., Gombosi T. I., Roe P. L., Powell K. G., and DeZeeuw D. L. (1998) Heliosphere in the magnetized local interstellar medium: Results of a three-dimensional MHD simulation. *J. Geophys. Res., 103,* 1889–1904.

Linsky J. L. and Wood B. E. (1996) The Alpha Centauri line of sight: D/H ratio, physical properties of local interstellar gas, and measurement of heated hydrogen (the 'hydrogen wall') near the heliopause. *Astrophys. J., 463,* 254–270.

Mazur J. E., Mason G. M., Blake J. B., Klecker B., Leske R. A., Looper M. D., and Mewaldt R. A. (2000) Anomalous cosmic ray argon and other rare elements at 1–4 MeV/nucleon trapped within the Earth's magnetosphere. *J. Geophys. Res., 105,* 21015–21023.

Mazur J. E., Mason G. M., and Mewaldt R. A. (2002) Charge states of energetic particles from co-rotating interaction regions as constraints on their source. *Astrophys. J., 566,* 555–561.

McComas D. J. and Schwadron N. A. (2006) An explanation of the Voyager paradox: Particle acceleration at a blunt termination shock. *Geophys. Res. Lett., 33,* L04102, DOI: 10.1029/2005GL025437.

McComas D. J., et al. (2004) The interstellar boundary explorer (IBEX) mission. In *Physics of the Outer Heliosphere* (V. Florinski et al., eds.), pp. 162–181. AIP Conf. Proc. 719, New York.

McDonald F. B., Burlaga L. F., McGuire R. E., and Ness N. F. (2000) The onset of long-term cosmic ray modulation in cycle 23 coupled with a transient increase of anomalous cosmic rays in the distant heliosphere. *J. Geophys. Res., 105,* 20997–21004.

McDonald F. B., Webber W. R., Stone E. C., Cummings A. C., Heikkila B. C., and Lal N. (2006) Voyager observations of galactic and anomalous cosmic rays in the helioshealth. In *The Physics of the Inner Heliosheath: Voyager Observations, Theory, and Future Prospects; 5th Annual IGPP International Astrophysics Conference* (J. Heerikhuisen et al., eds.), pp. 79–85. AIP Conf. Proc. 858, New York.

Mewaldt R. A., Cummings A. C., and Stone E. C. (1994) Anomalous cosmic rays: Interstellar interlopers in the heliosphere and magnetosphere. *Eos Trans. AGU, 75,* 16.

Mewaldt R. A., Selesnick R. S., Cummings J. R., Stone E. C., and von Rosenvinge T. T. (1996) Evidence for multiply charged anomalous cosmic rays. *Astrophys. J. Lett., 466,* L43–L46.

Möbius E., Bzowski M., Chalov S., Fahr H. J., Gloeckler G., Izmodenov V., Kallenbach R., Lallement R., McMullin D., Noda H., Oka M., Pauluhn A., Raymond J., Rucinski D., Skoug R., Terasawa T., Thompson Vallerga W. J., von Steiger R., and Witte M. (2004) Synopsis of the interstellar He parameters from combined neutral gas, pickup ion and UV scattering observations and related consequences. *Astron. Astrophys., 426,* 897–907, DOI: 10.1051/0004-6361:20035834.

Müller H.-R., Frisch P. C., Florinski V., and Zank G. P. (2006) Heliospheric response to different possible interstellar environments. *Astrophys. J., 647,* 1491–1505.

Nerney S., Suess S. T., and Schmahl E. J. (1991) Flow downstream of the heliospheric terminal shock — Magnetic field kinematics. *Astron. Astrophys., 250,* 556–564.

Nerney S. F., Suess S. T., and Schmahl E. J. (1993) Flow downstream of the heliospheric terminal shock: The magnetic field on the heliopause. *J. Geophys. Res., 98,* 15169–15176.

Neugebauer M. and Snyder C. W. (1962) Mariner 2 observations of the solar wind, 1, average properties. *J. Geophys. Res., 71,* 4469–4484.

Neugebauer M., Gloeckler G., Gosling J. T., Rees A., Skoug R., et al. (2007) Encounter of the Ulysses spacecraft with the ion tail of Comet McNaught. *Astrophys. J., 667,* 1262–1266.

Opher M., Stone E. C., and Liewer P. C. (2006) The effects of a local interstellar magnetic field on Voyager 1 and 2 observations. *Astrophys. J. Lett., 640,* L71–L74.

Opher M., Stone E. C., and Gombosi T. I. (2007) The orientation of the local interstellar magnetic field. *Science, 316,* 875–878, DOI: 10.1126/science.1139480.

Parker E. N. (1958) Dynamics of the interplanetary gas and magnetic fields. *Astrophys. J., 128,* 664–676.

Parker E. N. (1963) *Interplanetary Dynamical Processes.* Interscience, New York.

Paularena K. I., Richardson J. D., Kolpak M. A., Jackson C. R., and Siscoe G. L. (2001) A dawn-dusk density asymmetry in Earth's magnetosheath. *J. Geophys. Res., 106,* 25377–25394.

Pesses M. E., Jokipii J. R., and Eichler D. (1981) Cosmic ray drift, shock acceleration, and the anomalous component of cosmic rays. *Astrophys. J. Lett., 246,* L85–L88.

Pogorelov N. V. and Zank G. P. (2006) The direction of the neutral hydrogen velocity in the inner heliosphere as a possible interstellar magnetic field compass field compass. *Astrophys. J. Lett., 636,* L161–L164.

Pogorelov N. V., Zank G. P., and Ogino T. (2004) Three dimensional features of the outer heliosphere due to coupling between the interstellar and interplanetary magnetic fields. I. Magnetohydrodynamic model: Interstellar perspective. *Astrophys. J., 614,* 1007–1021.

Ratkiewicz R., Barnes A., Molvik G. A., Spreiter J. R., Stahara S. S., Vinokur M., and Venkateswaran S. (1998) Effect of varying strength and orientation of local interstellar magnetic field on configuration of exterior heliosphere: 3D MHD simulations. *Astron. Astrophys., 335,* 363–369.

Reames D. V. (1999) Quiet-time spectra and abundances of energetic particles during the 1996 solar minimum. *Astrophys. J., 518,* 473–479.

Richardson J. D. and Smith C. W. (2003) The radial temperature profile of the solar wind. *Geophys. Res. Lett., 30,* 1206–1209, DOI: 10.1029/2002GL016551.

Richardson J. D. and Wang C. (1999) The global nature of solar cycle variations of the solar wind dynamic pressure. *Geophys. Res. Lett., 26,* 561–564.

Richardson J. D., Paularena K. I., Lazarus A. J., and Belcher J. W. (1995) Evidence for a solar wind slowdown in the outer heliosphere? *Geophys. Res. Lett., 22,* 1469–1472.

Richardson J. D., Wang C., and Burlaga L. F. (2003) Correlated solar wind speed, density, and magnetic field changes at Voyager 2. *Geophys. Res. Lett., 30,* 2207–2210, DOI: 10.1029/2003GL018253.

Richardson J. D., Wang C., and Kasper J. C. (2005a) Propagation of the October/November 2003 CMEs through the heliosphere. *Geophys. Res. Lett., 32,* L03S03, DOI: 10.1029/2004GL020679.

Richardson J. D., McDonald F. B., Stone E. C., Wang C., and Ashmall J. (2005b) Relation between the solar wind dynamic pressure at Voyager 2 and the energetic particle events a Voyager 1. *J. Geophys. Res., 110,* A09106, DOI: 10.1029/2005JA011156.

Richardson J. D., Stone E. C., Cummings A. C., Kasper J. C., Zhang M., Burlaga L. F., Ness N. F., and Liu Y. (2006a) Correlation between energetic ion enhancements and heliospheric current sheet crossings in the outer heliosphere. *Geophys. Res. Lett., 33,* L21112, DOI: 10.1029/2006GL027578.

Richardson J. D., Wang C., and Zhang M. (2006b) Plasma in the outer heliosphere and the heliosheath. In *The Physics of the Inner Heliosheath: Voyager Observations, Theory, and Future Prospects; 5th Annual IGPP International Astrophysics Conference* (J. Heerikhuisen et al., eds.), pp. 110–115. AIP Conf. Proc. 858, New York.

Schwadron N. A. (1998) A model for pickup ion transport in the heliosphere in the limit of uniform hemispheric distributions. *J. Geophys. Res., 103,* 20643–20650.

Schwadron N. A. and McComas D. J. (2007) Modulation of galactic and anomalous cosmic rays beyond the termination shock. *Geophys. Res. Lett., 34,* L14105, DOI: 10.1029/2007GL029847.

Schwadron N. A., Fisk L. A., and Gloeckler G. (1998) Statistical acceleration of interstellar pick-up ions in co-rotating interaction regions. *Geophys. Res. Lett., 23,* 2871–2874.

Schwadron N. A., Gloeckler G., Fisk L. A., Geiss J., and Zurbuchen T. H. (1999) The inner source for pickup ions. In *Solar Wind Nine: Proceedings of the Ninth International Solar Wind Conference* (S. R. Habbal et al., eds.), pp. 487–490. AIP Conf. Proc. 471, New York.

Schwadron N. A., Geiss J., Fisk L. A., Gloeckler G., Zurbuchen T. H., and von Steiger R. (2000) Inner source distributions: Theoretical interpretation, implications, and evidence for inner source protons. *J. Geophys. Res., 105,* 7465–7472.

Schwadron N. A., Combi M., Huebner W., and McComas D. J. (2002) The outer source of pickup ions and anomalous cosmic rays. *Geophys. Res. Lett., 29,* 54–57, DOI: 10.1029/2002GL015829.

Smith C. W., Matthaeus W. H., Zank G. P., Ness N. F., Oughton S., and Richardson J. D. (2001) Heating of the low-latitude solar wind by dissipation of turbulent magnetic fluctuations. *J. Geophys. Res., 106,* 8253–8272.

Smith C. W., Isenberg P. A., Matthaeus W. H., and Richardson J. D. (2006) Turbulent heating of the solar wind by newborn interstellar pickup protons. *Astrophys J., 638,* 508–517.

Stone E. C., Cummings A. C., McDonald F. B., Heikkila B., Lal N., and Webber W. R. (2005) Voyager 1 explores the termination shock region and the heliosheath beyond. *Science, 309,* 2017–2020.

Svensmark H., Pedersen J. O. P., Marsh N. D., Enghoff M. B. and Uggerhoj U. I. (2007) Experimental evidence for the role of ions in particle nucleation under atmospheric conditions. *Proc. R. Soc. A: Math., Phys. Eng. Sci., 463(2078),* 385–396, DOI: 10.1098/rspa.2006.1773.

Thomas G. E. and Krassa R. F. (1971) OGO-5 measurements of the Lyman-alpha sky background in 1970 and 1971. *Astron. Astrophys., 30,* 223–232.

Wang C. and Belcher J. W. (1998) The heliospheric boundary response to large scale solar wind fluctuations: A gas dynamic model with pickup ions. *J. Geophys. Res., 104,* 549–556.

Washimi H. and Tanaka T. (1996) 3-D magnetic field and current system in the heliosphere. *Space Sci. Rev., 78,* 85–94.

Webber W. R. (2005) An empirical estimate of the heliospheric termination shock location with time with application to the intensity increases of MeV protons seen at Voyager 1 in 2002–2005. *J. Geophys. Res., 110,* A10103, DOI: 10.1029/2005JA011209.

Weller C. S. and Meier R. R. (1979) Analysis of the helium component of the local interstellar medium. *Astrophys. J., 227,* 816–823.

Wimmer-Schweingruber R. F. and Bochsler P. (2003) On the origin of inner-source pickup ions. *Geophys. Rev. Lett., 30,* 49–52.

Witte M. (2004) Kinetic parameters of interstellar neutral helium — Review of results obtained during one solar cycle with the Ulysses/GAS-instrument. *Astron. Astrophys., 426(3),* 835–844, DOI: 10.1051/0004-6361:20035956.

Witte M., Rosenbauer H., Keppler E., Fahr H., Hemmerich P., Lauche H., Loidl A., and Zwick R. (1992) The interstellar neutral-gas experiment on ULYSSES. *Astron. Astrophys. Suppl. Ser., 92,* 333–348.

Wood B. E., Alexander W. R., and Linsky J. L. (1996) The properties of the local interstellar medium and the interaction of the stellar winds of Epsilon Indi and Lambda Andromedae with the interstellar environment. *Astrophys. J., 470,* 1157–1171.

Wood B. E., Müller H.-R., and Zank G. P. (2000) Hydrogen Lyman-α absorption predictions by Boltzmann models of the heliosphere. *Astrophys. J., 542,* 493–503.

Zank G. P. (1999) Interaction of the solar wind with the local interstellar medium: A theoretical perspective. *Space Sci. Rev., 89,* 413–687.

Zank G. P. and Müller H.-R. (2003) The dynamical heliosphere. *J. Geophys. Res., 108,* 1240–1254, DOI: 10.1029/2002JA009689.

Zank G. P., Pauls H. L., Williams L. L., and Hall D. T. (1996) Interaction of the solar wind with the local interstellar medium: A multi-fluid approach. *J. Geophys. Res., 101,* 21639–21656.

Zank G. P., Rice W. K. M., le Roux J. A., and Matthaeus W. H. (2001) The injection problem for anomalous cosmic rays. *Astrophys. J., 556,* 494–500.

Extrasolar Kuiper Belt Dust Disks

Amaya Moro-Martín
Princeton University

Mark C. Wyatt
University of Cambridge

Renu Malhotra and David E. Trilling
University of Arizona

The dust disks observed around mature stars are evidence that plantesimals are present in these systems on spatial scales that are similar to that of the asteroids and the Kuiper belt objects (KBOs) in the solar system. These dust disks (a.k.a. "debris disks") present a wide range of sizes, morphologies, and properties. It is inferred that their dust mass declines with time as the dust-producing planetesimals get depleted, and that this decline can be punctuated by large spikes that are produced as a result of individual collisional events. The lack of solid-state features indicate that, generally, the dust in these disks have sizes ≥10 μm, but exceptionally, strong silicate features in some disks suggest the presence of large quantities of small grains, thought to be the result of recent collisions. Spatially resolved observations of debris disks show a diversity of structural features, such as inner cavities, warps, offsets, brightness asymmetries, spirals, rings, and clumps. There is growing evidence that, in some cases, these structures are the result of the dynamical perturbations of a massive planet. Our solar system also harbors a debris disk and some of its properties resemble those of extrasolar debris disks. From the cratering record, we can infer that its dust mass has decayed with time, and that there was at least one major "spike" in the past during the late heavy bombardment. This offers a unique opportunity to use extrasolar debris disks to shed some light in how the solar system might have looked in the past. Similarly, our knowledge of the solar system is influencing our understanding of the types of processes that might be at play in the extrasolar debris disks.

1. INTRODUCTION

During the last two decades, space-based infrared observations, first with the Infrared Astronomical Satellite (IRAS) and then with the Infrared Space Observatory (ISO) and the Spitzer Space Telescope, have shown that main-sequence stars are commonly surrounded by dust disks (a.k.a. debris disks), some of which extend to hundreds of AU from the central star. With the recent Spitzer observations, the number of debris disks known to date is approaching 100, of which 11 are spatially resolved.

Dust particles are affected by radiation pressure, Poynting-Robertson and stellar wind drag, mutual collisions, and collisions with interstellar grains. All these processes contribute to make the lifetime of the dust particles significantly shorter than the age of the star. Therefore, it was realized early on that this dust could not be primordial, i.e., part of the original molecular cloud where the star once formed, but it had to be a second generation of dust, likely replenished by a reservoir of (undetected) dust-producing planetesimals like the asteroids, comets, and Kuiper belt objects (KBOs) in our solar system (*Backman and Paresce,* 1993). This represented a major leap in the search for other planetary systems: By 1983, a decade before extrasolar planets were discovered, IRAS observations proved that there is planetary material surrounding nearby stars (*Aumann et al.,* 1984).

How do the extrasolar debris disks compare with our own solar system? The existence of an inner planetary dust complex has long been known from observations of zodiacal light by Cassini in 1683. In the inner solar system, dust is produced by debris from Jupiter-family short-period comets and asteroids (*Liou et al.,* 1995; *Dermott et al.,* 1994). The scattering of sunlight by these grains gives rise to the zodiacal light and its thermal emission dominates the night sky between 5 μm and 500 μm. This thermal emission dust was observed by the IRAS and COBE space telescopes, and the interplanetary dust particles (IDPs) were detected *in situ* by dust detectors on the Pioneer 10 and 11, Voyager, Galileo, and Ulysses spacecrafts. Its fractional luminosity is estimated to be $L_{dust}/L_* \sim 10^{-8}$–$10^{-7}$ (*Dermott et al.,* 2002). In the outer solar system, significant dust production is expected from the mutual collisions of KBOs and collisions with interstellar grains (*Backman and Paresce,* 1993; *Stern,* 1996; *Yamamoto and Mukai,* 1998). The thermal emission of the outer solar system dust is overwhelmed by the much stronger signal from the inner zodiacal cloud (so Kuiper belt dust is not seen in the IRAS and COBE infrared maps). However, evidence of its existence comes from the Pioneer 10 and 11 dust collision events measured beyond the orbit of Saturn (*Landgraf et al.,* 2002). Extrapolating from the size distribution of KBOs, its fractional luminosity is estimated to be $L_{dust}/L_* \sim 10^{-7}$–$10^{-6}$ (*Stern,* 1996).

In this chapter we describe the debris disk phenomenon: how debris disks originate (section 2); how they evolve in time (section 3); what they are made of (section 4); whether or not they are related to the presence of close-in planets (section 5); and how planets can affect their structure (section 6). We then discuss how debris disks compare to the solar system's dust disk in the present and in the past (section 7), and finish with a discussion of the prospects for the future of debris disk studies (section 8). In summary, the goal of the chapter is to review how debris disks can help us place our solar system into context within extrasolar planetary systems.

2. FROM PRIMORDIAL TO DEBRIS DISKS

Stars form from the collapse of dense regions of molecular clouds, and a natural byproduct of this process is the formation of a circumstellar disk (*Shu et al.,* 1987; *Hartmann,* 2000). Observations show that young stars with masses below ~4 $M_\odot$ down to brown dwarfs and planetary-mass objects have disks, while disks around more massive stars are more elusive, due to fast disk dissipation and observational difficulties as they tend to be highly embedded and typically very distant objects. Disk masses are estimated to be in the range 0.003 $M_\odot$–0.3 $M_\odot$, showing a large spread even for stars with similar properties (*Natta,* 2004, and references therein). For 1-$M_\odot$ stars, disk masses are 0.01 $M_\odot$–0.10 $M_\odot$ (*Hartmann,* 2000, and references therein). With regard to the disk sizes, there is evidence for gas on scales from 10 AU to 800 AU (*Simon et al.,* 2000). Both the disk masses and scales are comparable to the minimum mass solar nebula, ~0.015 $M_\odot$. This is the total mass of solar composition material needed to produced the observed condensed material in the solar system planets (~50 $M_\oplus$) (*Hayashi,* 1981; *Weidenschilling,* 1977).

Eventually, infall to the disk stops and the disk becomes depleted in mass: Most of the disk mass is accreted onto the central star; some material may be blown away by stellar wind ablation or by photoevaporation by high-energy stellar photons, or stripped away by interactions with passing stars; the material that is left behind might coagulate or accrete to form planets (only ~10% of the solar nebula gas is accreted into the giant planet's atmospheres). After ~10^7 yr, most of the primordial gas and dust have disappeared (see, e.g., *Hollenbach et al.,* 2005; *Pascucci et al.,* 2006), setting an important time constraint for giant planet formation models.

However, many main-sequence stars older than ~10^7 yr still show evidence of dust. The timescale of dust grain removal due to radiation pressure is on the order of an orbital period, while the Poynting-Robertson (P-R) drag lifetime of a dust grain located at a distance R is given by

$$t_{PR} = 710\left(\frac{b}{\mu m}\right)\left(\frac{\rho}{g/cm^3}\right)\left(\frac{R}{AU}\right)^2\left(\frac{L_\odot}{L_*}\right)\frac{1}{1 + albedo}\ yr$$

where b and ρ are the grain radius and density, respectively (*Burns et al.,* 1979; *Backman and Paresce,* 1993). Grains can also be destroyed by mutual grain collisions, with a collisional lifetime of

$$t_{col} = 1.26 \times 10^4\left(\frac{R}{AU}\right)^{3/2}\left(\frac{M_\odot}{M_*}\right)^{1/2}\frac{10^{-5}}{L_{dust}/L_*}\ yr$$

(*Backman and Paresce,* 1993). Because all the above timescales are generally much shorter than the age of the disk, it is inferred that the observed dust is not primordial but is likely produced by a reservoir of undetected kilometer-sized planetesimals producing dust by mutual collisions or by evaporation of comets scattered close to the star (*Backman and Paresce,* 1993).

At any particular age, observations show a great diversity of debris disks surrounding similar type stars [see section 3.1 and *Andrews and Williams* (2005)]. This may be due to the following factors that can influence the disks at different stages during their evolution: (1) different initial masses and sizes, caused by variations in the angular momentum of the collapsing protostellar cloud; (2) different external environments, causing variations in the dispersal timescales of the outer primordial disks, and therefore strongly affecting the formation of planets and planetesimals in the outer regions; and (3) different planetary configurations, affecting the populations and velocity dispersions of the dust-producing planetesimals.

For example, the formation environment can have an important effect on the disk size and its survival. If the star is born in a sparsely populated Taurus-like association, the possibility of having a close encounter with another star that could truncate the outer protoplanetary disk is very small. In this environment, the probability of having a nearby massive star is also small, so photoevaporation does not play an important role in shaping the disk, and neither does the effect of explosions of nearby supernovae (*Hollenbach and Adams,* 2004). However, if the star is born in a densely populated OB association, the high density of stars results in a high probability of close encounters that could truncate the outer protoplanetary disk. In addition, nearby massive stars and supernovae explosions are likely to be present, affecting the size of the disk by photoevaporation, a process in which the heated gas from the outer disk evaporates into interstellar space, dragging along dust particles smaller than a critical size of 0.1–1 cm before they have time to coagulate into larger bodies. Dust coagulation to this critical size takes ~10^5–10^6 yr at 30–100 AU (*Hollenbach and Adams,* 2004), and therefore occurs rapidly enough for Kuiper belt formation to take place inside 100 AU, even around low-mass stars in OB associations like the Trapezium in Orion. However, in Trapezium-like conditions (*Hillenbrand and Hartman,* 1998), where stars form within groups/clusters containing >100 members, at larger distances from the star photoevaporation takes place on a faster timescale than coagulation, and the dust is carried away by the evaporating gas causing a sharp cutoff in the formation of planetesimals beyond ~100 ($M_\star$/1 $M_\odot$) AU, and therefore suppressing the production of debris dust (*Hollenbach and Adams,* 2004, and references therein). Debris disks can present a wide range of sizes because the distance at which photoevaporation takes place on a faster timescale than coagulation depends not only on the mass of the central star, but also on the initial disk mass and the mass and proximity of the most massive star in the group/cluster.

It is thought that the Sun formed in an OB association: Meteorites show clear evidence that isotopes with short lifetimes (<10^5 yr) were present in the solar nebula, which indicates that a nearby supernova introduced them immediately before the dust coagulated into larger solids (*Cameron and Truran,* 1977; *Tachibana et al.,* 2006); in addition, it has been suggested that the edge of the Kuiper belt may be due to the dynamical interaction with a passing star (*Kobayashi et al.,* 2005), indicating that the Sun may have been born in a high-density stellar environment. In contrast, kinematic studies show that the majority of the nearby spatially resolved debris disks formed in loosely populated Taurus-like associations (see, e.g., *Song et al.,* 2003).

Debris disks found around field stars may be intrinsically different than those found around stars that once belonged to densely populated clusters, and one needs be cautious of the conclusions drawn from comparing these systems directly, as well as the conclusions drawn from stellar samples that include debris disks indiscriminately forming in these two very different environments.

3. DEBRIS DISK EVOLUTION AND FREQUENCY

The study of debris disk evolution, i.e., the dependency on stellar age of the amount of dust around a main-sequence star, is of critical importance in the understanding of the timescales for the formation and evolution of planetary systems, as the dust production rate is thought to be higher during the late stages of planet formation, when planetesimals are colliding frequently, than later on, when mature planetary systems are in place, planet formation is complete, and the planets are not undergoing migration. Because it is obviously not possible to observe in real time the evolution of a particular system during millions to billions of years, the study of debris disk evolution is based on the observations of a large number of stars with different ages, with the goal of determining how the amount of excess emission (related to the dust mass) and the probability of finding an excess depend on stellar age. The assumption is that all the disks will evolve in a similar way (but see caveats in section 2.1).

The age-dependency of the dust emission (a.k.a "excess" with respect to the photospheric values) has been elusive until recently. The limited sensitivity of IRAS allowed only the detection of the brightest and nearest disks, mostly around A stars. In addition, with its limited spatial resolution it was not possible to determine whether the infrared excess emission was coming from the star (i.e., from a debris disk) or from extended galactic cirrus or background galaxies. The ISO, with its improvement of a factor of 2 in spatial resolution and a factor of 10 in sensitivity over IRAS, made a big step forward in the study of debris disk evolution. However, the ISO samples were too small to establish any age-dependency on a sound statistical basis. More recently, the Spitzer/MIPS instrument, with its unprecedented sensitivity at far-IR wavelengths (a factor of ~100–1000 better than IRAS, and at least a factor of 10 in spatial resolution), has extended the search of disks around main-sequence stars to more tenuous disks and to greater distances, providing more homogeneous samples. This is still ongoing research but is leading to new perspective on debris disk evolution. The following subsections summarize the main results so far.

3.1. Observations

3.1.1. A stars. Using Spitzer/MIPS at 24 μm, *Rieke et al.* (2005) carried out a survey of 76 A stars (2.5 $M_\odot$) with ages of 5–580 m.y., with all the stars detected to 7σ relative to their photospheric emission. These observations were complemented with archival data from ISO and IRAS, resulting in a total of 266 A stars in the final sample studied. The results show an overall decline in the average amount of 24-μm excess emission. Large excesses (more than a factor of 2 relative to the photosphere) decline from ~25% in the youngest age bins to only one star (~1%) for ages >190 m.y.; a functional fit to this data suggests a t_0/t decline, with t_0 = 100–200 m.y. Intermediate excesses (factors of 1.25–2) decrease much more slowly and are present in ~7% of stars older than several hundred million years. The persistence of excesses beyond 200 m.y. rules out a fast $1/t^2$ decay. Using a subsample of 160 A stars (including the ones in *Rieke et al.,* 2005), *Su et al.* (2006) confirmed that the 24-μm excess emission is consistent with a t_0/t decay, where t_0 ~ 150 m.y., while the 70-μm excess (tracing dust in the Kuiper belt region) is consistent with t_0/t, where $t_0 \geq$ 400 m.y. Even though there is a clear decay of the excess emission with time, *Rieke et al.* (2005) and *Su et al.* (2006) showed that at a given stellar age there are at least 2 order of magnitude variations in the amount of dust: As many as 50–60% of the younger stars (<30 m.y.) do not show dust emission at 24 μm, while ~25% of disks are still detected at 150 m.y.

3.1.2. FGK stars. For FGK stars, the excess rates at 24 μm decrease from ~30% to 40% for ages <50 m.y., to ~9% for 100–200 m.y., and ~1.2% for ages >1 G.y. (see Fig. 1) (*Siegler et al.,* 2006; *Gorlova et al.,* 2006; *Stauffer et al.,* 2005; *Beichman et al.,* 2005a; *Kim et al.,* 2005; *Bryden et al.,* 2006). At 70 μm, the excess rate is 10–20% and is fairly constant for a wide range of ages (*Bryden et al.,* 2006; Hillenbrand et al., in preparation). At first sight, it appears that for the older stars warm asteroid-belt-like disks

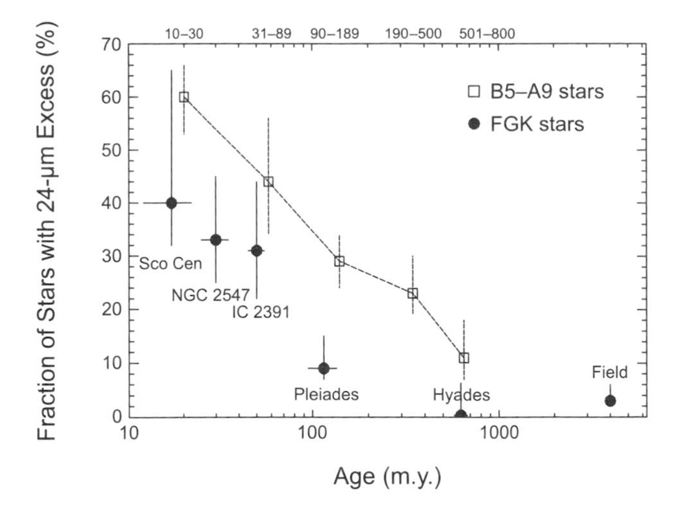

Fig. 1. Fraction of early-type stars (open squares) and FGK stars (circles) with excess emission at 24 μm as a function of stellar age. Figure from *Siegler et al.* (2006) using data from *Chen et al.* (2005), Gorlova et al. (in preparation), *Stauffer et al.* (2005), *Gorlova et al.* (2006), *Cieza et al.* (2005), *Bryden et al.* (2006), *Rieke et al.* (2005), and *Su et al.* (2006). The age bins used in the early-type star survey are shown across the top horizontal axis. Vertical error bars are 1σ binomial distribution uncertainties.

are rare (few percent), while cold Kuiper belt-like disks are common (10–20%). However, one needs to keep in mind that the sensitivity thresholds at 24 μm and 70 μm are different: Spitzer/MIPS is currently able to constrain dust masses at Kuiper belt-like distances (10–100 AU) that are 5–100× the level of dust in our solar system, and at AB-like distances (1–10 AU) that are 1000× our zodiacal emission (*Bryden et al.,* 2006). However, spectroscopy observations with Spitzer/IRS are better suited to search for hot dust. Preliminary results by *Beichman et al.* (2006a) indicated that indeed warm excesses (<25 μm) with luminosities 50–1000× the zodiacal emission are rare for stars >1 G.y., and found that only ~1 out of 40 stars are in agreement with theoretical calculations of disk dispersal by *Dominik and Decin* (2003) that indicate that the fractional luminosity of the warm dust will generally drop below the IRS detectability level after 1 G.y. of evolution. In contrast, colder disks with excesses at 30–34 μm are found for ~5 out of 41 stars, 12 ± 5%, in agreement with *Bryden et al.* (2006).

Even though the Spitzer/MIPS detection rate of excess emission for FGK stars is lower than for A stars (see Fig. 1), this is also a result of a sensitivity threshold: Similar levels of excess emission are more easily detected around hotter stars than around colder stars. Accounting for this, the actual frequency of debris disks does not seem to be a strong function of stellar type (*Siegler et al.,* 2006), but it drops to zero for stars later than K1 (*Beichman et al.,* 2006b).

As for A stars, FGK stars also show large variations in the amount of excess emission at a given stellar age at 24 μm (see Fig. 2) and 70 μm. In addition, *Siegler et al.* (2006) found that the upper envelope of the ratio of the excess emission over the stellar photosphere at 24 μm also decays as t_0/t, with t_0 = 100 m.y. and ages >20 m.y. At younger ages, <25 m.y., the decay is significantly faster and could trace the fast transition of the disk between primordial and debris stages (*Siegler et al.,* 2006). For colder dust (at 70 μm), even though there is a general trend to find less dust at older ages, the decay time is longer than for warmer dust (at 24 μm).

3.2. Theoretical Predictions

3.2.1. Inverse-time decay. If all the dust is derived from the grinding down of planetesimals, and assuming the planetesimals are destroyed after one collision, and that the number of collisions is proportional to the square of the number of planetesimals (N), then $dN/dt \propto -N^{-2}$ and $N \propto 1/t$. Therefore, the dust production rate, $R_{prod} \propto dN/dt \propto N^2 \propto 1/t^2$. To solve for the amount of dust in the disk in steady state, one needs equate the dust production rate to the dust loss rate, R_{loss}, and this gives two different solutions depending on the number density of the dust in the disk (*Dominik and Decin,* 2003): (1) In the collisionally dominated disks ($M_{dust} \gtrsim 10^{-3}\ M_\oplus$), the dust number density is high and the main dust removal process is grain-grain collisions, so that $R_{loss} \propto n^2$, where n is the number of dust grains. From $R_{prod} = R_{loss}$, we get $n \propto 1/t$. (2) In the radiatively dominated disks ($M_{dust} \lesssim 10^{-3}\ M_\oplus$), the dust loss rate is dominated by Poynting-Robertson drag, and therefore is proportional to the number of particles, $R_{loss} \propto n$, and from $R_{prod} = R_{loss}$, we get $n \propto 1/t^2$.

The Kuiper belt disk has little mass and is radiatively dominated. However, all the debris disks observed so far are significantly more massive than the Kuiper belt because the surveys are sensitivity limited. *Wyatt* (2005a) estimates that the observed disks are generally collisionally dominated, so

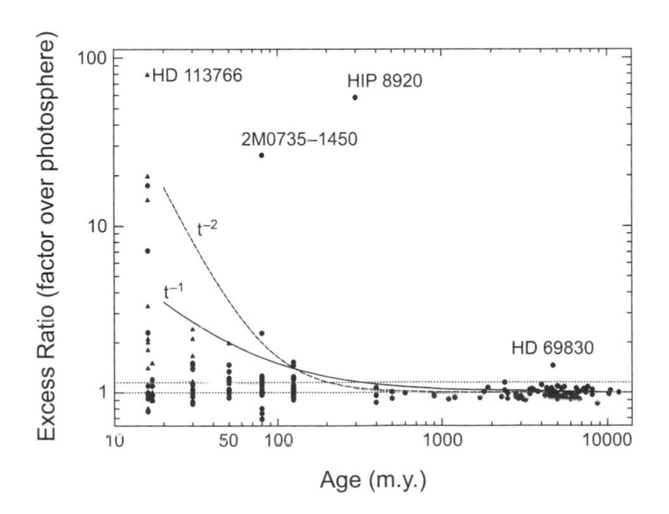

Fig. 2. Ratio of the 24-μm excess emission to the predicted photospheric value for FGK stars as a function of stellar age. Triangles represent F0–F4 stars and circles represent F5–K7 stars (similar to the Sun). Stars aligned vertically belong to clusters or associations. Figure from *Siegler et al.* (2006) using the same data as in Fig. 1 and from *Gorlova et al.* (2004), *Hines et al.* (2006), and *Song et al.* (2005).

one would expect that the dust emission will evolve as 1/t, in agreement with the Spitzer/MIPS observations of debris disks around A and FGK stars.

3.2.2. Episodic stochastic collisions. Numerical simulations of the evolution of dust generated from the collision of planetesimals around solar-type stars by *Kenyon and Bromley* (2005) predict that after 1 m.y. there is a steady decline of the 24-μm excess emission, as the dust-producing planetesimals get depleted, a decay that is punctuated by large spikes produced by individual collisional events (see Fig. 3). Therefore, the high degree of debris disk variability observed by Spitzer/MIPS — seen as spikes in Fig. 2 — may be the result of recent collisional events. It is thought that these events initiate a collisional cascade leading to short-term increases in the density of small grains, which increases the brightness density of the disk by an order of magnitude. Because the clearing time of dust in the 24-μm-emitting zone (10–60 AU) is ~1–10 m.y. (*Dominik and Decin,* 2003; *Kenyon and Bromley,* 2004), these individual events could dominate the properties of debris disks over million-year timescales (*Rieke et al.,* 2005). However, there is a discrepancy between these numerical simulations and the observations because the models do not predict excess ratios larger than two for stars older than 50 m.y., in disagreement with the existence of two of the outliers in Fig. 2 (HIP 8920 and 2M0735-1450).

In addition to the large differences in excess emission found among stars within the same age range (for both A stars and FGK stars), the presence of large amounts of small grains in systems like HIP 8920 and HD 69830 (two of the outliers in Fig. 2), Vega, and in a clump in β Pic (*Telesco et al.,* 2005) indicate that recent collisional events have taken place in these systems (see discussion in section 4). The argument goes as follows: Because small grains are removed quickly by radiation pressure, the dust production rate needed to account for the observations is very high, implying a mass loss that could not be sustained during the full age of the system. For example, the Spitzer/MIPS observations of Vega (350-m.y.-old A star) show that the disk at 24 μm and 70 μm extends to distances of 330 AU and 540 AU from the star, respectively (*Su et al.,* 2006), far outside the ~80-AU ring of dust seen in the submillimeter (*Wilner et al.,* 2002) that probably traces the location of the dust-producing planetesimals. *Su et al.* (2006) suggested that the dust observed in the mid-IR comes from small grains that were generated in a recent collisional event that took place in the planetesimal belt, and are being expelled from the system under radiation pressure. This scenario would explain the large extent of the disk and the unusually high dust production rate (10^{15} g/s), unsustainable for the entire lifetime of Vega.

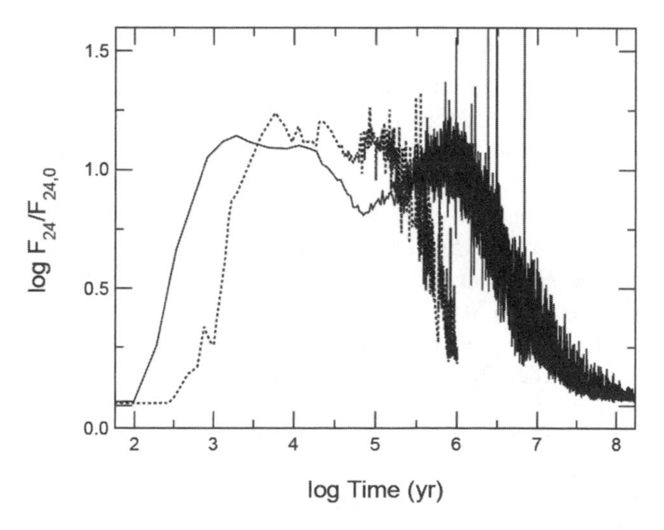

Fig. 3. Evolution of the 24-μm excess as a function of time for two planetesimal disks extending from 0.68 to 1.32 AU (dashed line) and 0.4 to 2 AU (solid line). The central star is solar type. Excess emission decreases as planetesimals grow into Mars-sized or larger objects and collisions become increasingly rare. Figure from *Kenyon and Bromley* (2005).

4. DEBRIS DISK GRAIN SIZE AND COMPOSITION

Most debris disk spectroscopy observations show few or no solid-state features, indicating that at those stages the dust grains have sizes ≥10 μm (*Jura et al.,* 2004; *Stapelfeldt et al.,* 2004), much larger than the submicrometer amorphous silicate grains that dominate the dust emission in young protoplanetary disks. While for A-stars, the lack of features is explained by the ejection of dust grains <10 μm by radiation pressure, the reason why this is also the case in debris disks around solar-type stars is still under debate. However, there are a few debris disks where spectral features have been observed, allowing us to set constraints on the particle size and composition. We briefly describe three of these systems: β Pictoris, for which small quantities of silicates have been observed, and HIP 8920 and HD 69830, showing very strong silicate features.

β Pictoris is one of the youngest and closest (19 pc) stars to Earth harboring a disk. It is an A5V star (2 $M_\odot$) with an estimated age of 12 m.y. probably in the process of clearing out its protoplanetary disk, as the solar system did 4 b.y. ago. The disk is likely in the transition between the primordial and debris stages. Its dust disk, seen edge on, extends to 1000 AU (i.e., ~10× that of the solar system) and contains a few lunar masses in grains that are relatively large (>1 μm), with a large fractional luminosity, $L_{dust}/L_\star \sim 3 \times 10^{-3}$. The break in the surface brightness profile of the disk indicates that the outer edge of the dust-producing planetesimal belt is at ~120 AU (*Heap et al.,* 2000). Small particles produced by collisions in the belt are diffused out by radiation pressure, explaining the power-law index of the brightness profile. On a smaller scale, spatially resolved spectroscopy observations indicate that the disk emission is dominated by grains emitting in the continuum, with moderate silicate emission features (amorphous and crystalline) seen only within 25 AU of the star. This indicates that the ratio of small to large silicate grains decreases with distance (*Weinberger et al.,* 2003). Additional spatially resolved spectroscopy observations by *Okamoto et al.* (2004) showed that the submicrometer amorphous silicate grains have three peaks in their distribution around 6 AU, 16 AU, and 30 AU, and their locations possibly trace three belts of dust-producing planetesimals. Finally, in the innermost system, the gas absorption lines detected toward the star indicate that there is a stable gas component that is located at about 1 AU and can be explained by the replenishment of gas by evaporating comets near the star, which would also give rise to the transient red-shifted absorption events observed in the spectra. The frequency of star-grazing comets needed to explain the observations is several orders of magnitude higher than that found in the solar system (see review in *Lagrange et al.,* 2000).

HIP 8920 (one of the outliers in Fig. 2) is a 300-m.y.-old star with a disk that has a high surface density of small ($\lesssim$2.5 μm) dust grains at 1 AU from the star. Mid-infrared spectroscopy observations of the dust emission at 8–13 μm show a very strong silicate feature with broad peaks at 10 and 11 μm that can be modeled with a mixture of amorphous and crystalline silicate grains (pyroxenes and olivines), with sizes of 0.1–2.5 μm. Because HIP 8920 is too old for the dust to be primordial, it has been suggested that the anomalous large quantities of small grains could be the result of a recent collision (Weinberger et al., personal communication).

HD 69830 is a 2-G.y.-old K0V star (0.8 $M_\odot$, 0.45 $L_\odot$) with an excess emission at 8–35 μm (60% over the photosphere at 35 μm, and with fractional luminosity $L_{dust}/L_\star \sim 2 \times 10^{-4}$) that shows strong silicate features remarkably similar to the ones in Comet C/1995 O1 (a.k.a. Hale-Bopp; see Fig. 4 of *Beichman et al.,* 2005b). The spectral features are identified as arising from mostly crystalline olivine (including fosterite) and a small component of crystalline pyroxene (including enstatite), both of which are also found in interplanetary dust particles and meteorite inclusions (*Yoneda et al.,* 1993; *Bradley,* 2003). Observations show that there is no 70-μm emission, and this indicates that the dust is warm, originating from dust grains with a low long-wavelength emissivity, i.e., with sizes $\lesssim$70 μm/2π ~ 10 μm, located within a few AU of the star, with the strong solid-state features arising from a component of small, possibly submicrometer grains (*Beichman et al.,* 2005b). Upper (3σ) limits to the 70-μm emission ($L_{dust}/L_\star < 5 \times 10^{-6}$) suggest a potential Kuiper belt less than 5× as massive as the solar system's. The emission between the crystalline silicate features at 9–11 μm, 19 μm, and 23.8 μm indicates that there is a source of continuum opacity, possibly a small component of larger grains (*Beichman et al.,* 2005b). The emitting surface area of the dust is large (2.7×10^{23} cm^2, >1000× the zodiacal emission), and the collisional and P-R drag time for submicrometer (0.25 μm) grains is <1000 yr. This indicates that the dust is either produced by the grinding down of a dense asteroid belt (22–64× more massive than the solar system's) located closer to the star, or originates in a transient event. *Wyatt et al.* (2006) ruled out the massive asteroid belt scenario and suggested that it is a transient event, likely the result of recent collisions produced when planetesimals located in the outer regions were scattered toward the star in a late heavy bombardment-type event.

The disk around β Pic seems to be "normal" in terms of its mass content with respect to the stellar age, and does not contain large amounts of small silicate grains; on the other hand, the disks around HIP 8920 and HD 69830 are unusually dusty and show strong silicate emission features, indicating that silicate features may be related to recent collisional events (Weinberger et al., personal communication).

The composition of the disk can also be studied from the colors of the scattered light images. In general, debris disks are found to be red or neutral. Their redness has commonly been explained by the presence of 0.4-μm silicate grains, but except for the two exceptions mentioned above (HIP 8920 and HD 69830), spatially resolved spectra have shown that debris disks do not generally contain large amounts of small silicate grains; a possible explanation for the colors could be that grains are intrinsically red, perhaps due to an important contribution from organic materials (Weinberger et al., personal communication; see also *Meyer et al.,* 2007). For comparison, KBOs present a wide range of surface colors, varying from neutral to very red (see chapter by Doressoundiram et al.).

5. DEBRIS DISKS AND CLOSE-IN PLANETS: RELATED PHENOMENA?

The observation of debris disks indicates that planetesimal formation has taken place around other stars. In these systems, did planetesimal formation proceed to the formation of one or move massive planets, as was the case of the Sun? In the following cases, the answer is yes: HD 33636, HD 50554, HD 52265, HD 82943, HD 117176, and HD 128311 are stars known from radial velocity observations to have at least one planet, and they all show 70-μm excess (with an excess SNR of 15.4, 14.9, 4.3, 17.0, 10.2, and 7.1, respectively) arising from cool material (T < 100 K) located mainly beyond 10 AU, implying the presence of an outer belt of dust-producing plantesimals. Their fractional luminosities, $L_{dust}/L_\star$, in the range $(0.1–1.2) \times 10^{-4}$, are ~100× that inferred for the Kuiper belt (*Beichman et al.,* 2005a). Similarly, HD 38529 is a two-planet system that also shows 70-μm excess emission (with an excess SNR of 4.7) (*Moro-Martín et al.,* 2007b). HD 69830 is a three-planet system with a strong 24-μm excess (see section 4) (*Beichman et al.,* 2005b). And finally, ε Eridani has at least one close-in planet (*Hatzes et al.,* 2000) and a spatially resolved debris disk (*Greaves et al.,* 2005).

The nine systems above confirm that debris disks and planets coexist. But are debris disks and the presence of massive planets related phenomena? *Moro-Martín et al.* (2007a) found that from the observations of the Spitzer Legacy Program FEPS and the GTO results in *Bryden et al.* (2006), there is no sign of correlation between the presence of IR excess and the presence of radial velocity planets (see also *Greaves et al.,* 2004a). This, together with the observation that high stellar metallicities are correlated with the presence of giant planets (*Fischer and Valenti,* 2005) but not correlated with the presence of debris disks (*Greaves et al.,* 2006), may indicate that planetary systems with

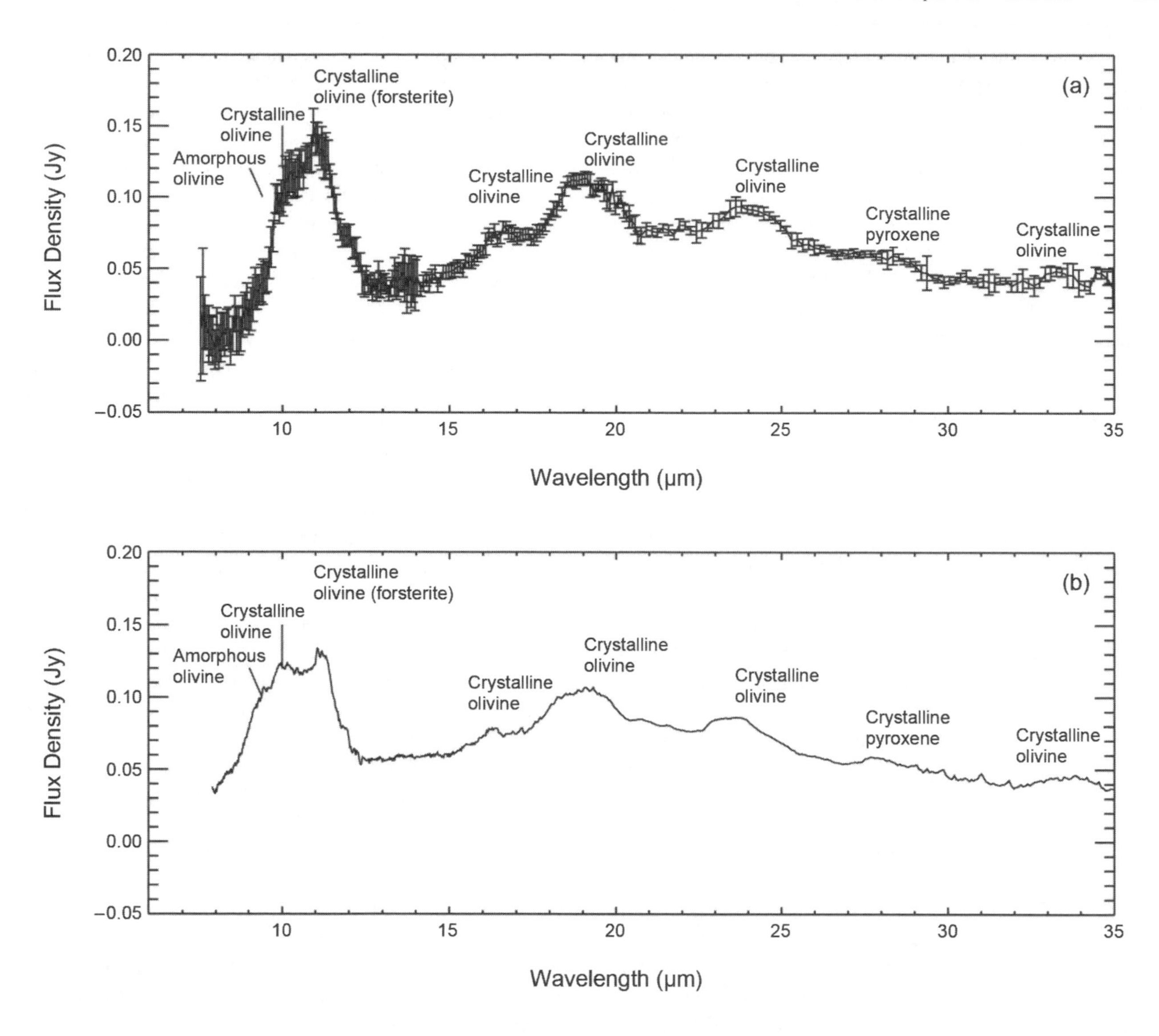

Fig. 4. **(a)** Spectrum of the excess of HD 69830. **(b)** Spectrum of the Comet Hale-Bopp from *Crovisier et al.* (1996) normalized to a blackbody temperature of 400 K to ease the comparison of the two spectra (the observed blackbody temperature is 20 K). Figure from *Beichman et al.* (2005b).

KBOs producing debris dust by mutual collisions may be more common than planetary systems harboring gas giant planets (*Greaves et al.,* 2006; *Moro-Martín et al.,* 2007a).

Most of the debris disks detected with Spitzer emit only at 70 µm, i.e., the dust is mainly located at distances >10 AU, while the giant planets detected by radial velocity studies are located within a few AU of the star, so the dust and the giant planet(s) could be dynamically unconnected (but see *Moro-Martín et al.,* 2007b). What about more distant giant planets? Do debris disk observations contain evidence for long-period planets? We discuss this issue in the next section.

6. DEBRIS DISK STRUCTURE

The gravitational perturbations produced by a massive planet on both the dust-producing planetesimals and the dust particles themselves can create structure in the debris disk that gives rise to observable features (see, e.g., *Roques et al.,* 1994; *Mouillet et al.,* 1997; *Wyatt et al.,* 1999; *Wyatt,* 2005b, 2006; *Liou and Zook,* 1999; *Moro-Martín and Malhotra,* 2002, 2003, 2005; *Moro-Martín et al.,* 2005; *Kuchner and Holman,* 2003).

If the disk is radiatively dominated, $M_{dust} \lesssim 10^{-3}\,M_\oplus$, as in the case of the Kuiper belt dust disk, and if the system contains an outer belt of planetsimals and one or more inner planets, the disk structure is created because the dust grains migrate inward due to the effect of P-R drag, eventually coming in resonance with the planet and/or crossing its orbit. This has important consequences on their dynamical evolution and therefore on the debris disk structure.

If the disk is collisionally dominated, $M_{dust} \gtrsim 10^{-3}\,M_\oplus$, before the dust grains migrate far from their parent bodies, they will suffer frequent collisions that could grind them down into smaller grains that are blown away by radiation pressure. In this case, the dust grains may not survive long enough to come into resonance with an inner planet. However, the structure of the KBOs gives strong evidence that Neptune migrated outward. This process may have also taken place in other planetary systems, where the outward migration of a planet could have scattered planetesimals out of the system or trapped them into Plutino-like orbits. Because the larger dust particles trace the location of the parent bodies, this outward migration can strongly affect the debris disk structure.

In this section we summarize the processes by which planets can affect the debris disk structure and the observational evidence that indicates that planets may be responsible for some of the observed feature.

6.1. Theoretical Predictions

6.1.1. Gravitational scattering. Massive planets can eject planetesimals and dust particles out of the planetary system via gravitational scattering. In the radiatively dominated disks, if the sources of dust are outside the orbit of the planet, this results in an inner cavity, a lower density of dust within the planet's orbit, as the particles drifting inward due to P-R drag are likely to be scattered out of the system when crossing the orbit of the planet (*Roques et al.,* 1994). Similarly, planetesimals can get scattered out by a planet migrating outward, resulting in a depletion of planetesimals and dust inside the orbit of the planet. Planets with masses of 3–10 M_{Jup} located between 1 AU and 30 AU in a circular orbit around a solar-type star eject >90% of the dust grains that go past their orbits by P-R drag; a 1-M_{Jup} planet at 30 AU ejects >80% of the grains, and about 50–90% if located at 1 AU, while a 0.3-M_{Jup} planet is not able to open a gap, ejecting <10% of the grains (*Moro-Martín and Malhotra,* 2005). These results are valid for dust grain sizes in the range 0.7–135 μm, but are probably also applicable to planetesimals (in the case of an outward-migrating planet), because gravitational scattering is a process independent of mass as long as the particle under consideration can be considered a "test particle," i.e., its mass is negligible with respect to that of the planet.

6.1.2. Resonant perturbations. Resonant orbits are locations where the orbital period of the planet is (p + q)/p× that of the particle (which can be either a dust grain or a planetesimals), where p and q are integers, $p > 0$ and $p + q \geq 1$. Each resonance has a libration width that depends on the particle eccentricity and the planet mass, in which resonant orbits are stable. The region close to the planet is chaotic because neighboring resonances overlap (*Wisdom,* 1980). Because of the finite width of the resonant region, resonant perturbations only affect a small region of the parameter space, but this region can be overpopulated compared to the size of that parameter space by the inward migration of dust particles under the effect of P-R drag or by the outward migration of the resonance as the planet migrates (*Malhotra,* 1993, 1995; *Liou and Zook,* 1995; *Wyatt,* 2003). When the particle crosses a mean-motion resonance ($q > 0$), it receives energy from the perturbing planet that can balance the energy loss due to P-R drag, halting the inward motion of the particle and giving rise to planetary resonant rings. Due to the geometry of the resonance, the spatial distribution of material in resonance is asymmetric with respect to the planet, being concentrated in clumps. There are four basic high-contrast resonant structures that a planet with eccentricity ≤0.6 can create in a disk of dust released on low-eccentricity orbits: a ring with a gap at the location of the planet; a smooth ring, a clumpy eccentric ring, and an offset ring plus a pair of clumps, with the appearance/dominance of one of these structures depending on the mass and eccentricity of the planet (*Kuchner and Holman,* 2003).

6.1.3. Secular perturbations. When a planet is embedded in a debris disk, its gravitational field perturbs the orbits of the particles (dust grains or planetesimals). Secular perturbations are the long-term average of the perturbing forces, and act on timescales >0.1 m.y. (see overview in *Wyatt et al.,* 1999). As a result of secular perturbations, the planet tries to align the particles with its orbit. The first particles to be affected are the ones closer to the planet, while the particles further away are perturbed at a later time, therefore if the planet's orbital plane is different from that of the planetesimal disk, secular perturbations will result in the formation of a warp. A warp will also be created if there are two planets on non-coplanar orbits. If the planet is in an eccentric orbit, the secular perturbations will force an eccentricity on the dust particles, and this will create an offset in the disk center with respect to the star and a brightness asymmetry in the reemitted light, as the dust particles near periastron are closer to the star and therefore hotter than the dust particles at the other side of the disk.

Because secular perturbations act faster on the particles closer to the planet, and the forced eccentricities and pericenters are the same for particles located at equal distances from the planet, at any one time the secular perturbations of a planet embedded in a planetesimal disk can result in the formation of two spiral structures, one inside and one outside the planet's orbit (*Wyatt,* 2005b).

6.2. Observations

Some of the structural features described above have indeed been observed in the spatially resolved images of debris disks (see Fig. 5).

6.2.1. Inner cavities. Inner cavities have long been known to exist. They were first inferred from the IRAS spectral energy distributions (SEDs) of debris disks around A stars, and more recently from the Spitzer SEDs of debris disks around AFGK stars. From the modeling of the disk SED, we can constrain the location of the emitting dust by fixing the grain properties. Ideally, the latter can be constrained through the modeling of solid-state features; however, most debris disk spectroscopy observations show little or no features, in which cases it is generally assumed that the grains have sizes ≥10 μm and are composed of "astronomical silicates" [i.e., silicates with optical constants from *Weingartner and Draine* (2001)]. In most cases, the SEDs show a depletion (or complete lack) of mid-infrared thermal emission that is normally associated with warm dust located close to the star, and this lack of emission implies the presence of an inner cavity [or more accurately, a depletion of grains that could be traced observationally (see, e.g., *Meyer*

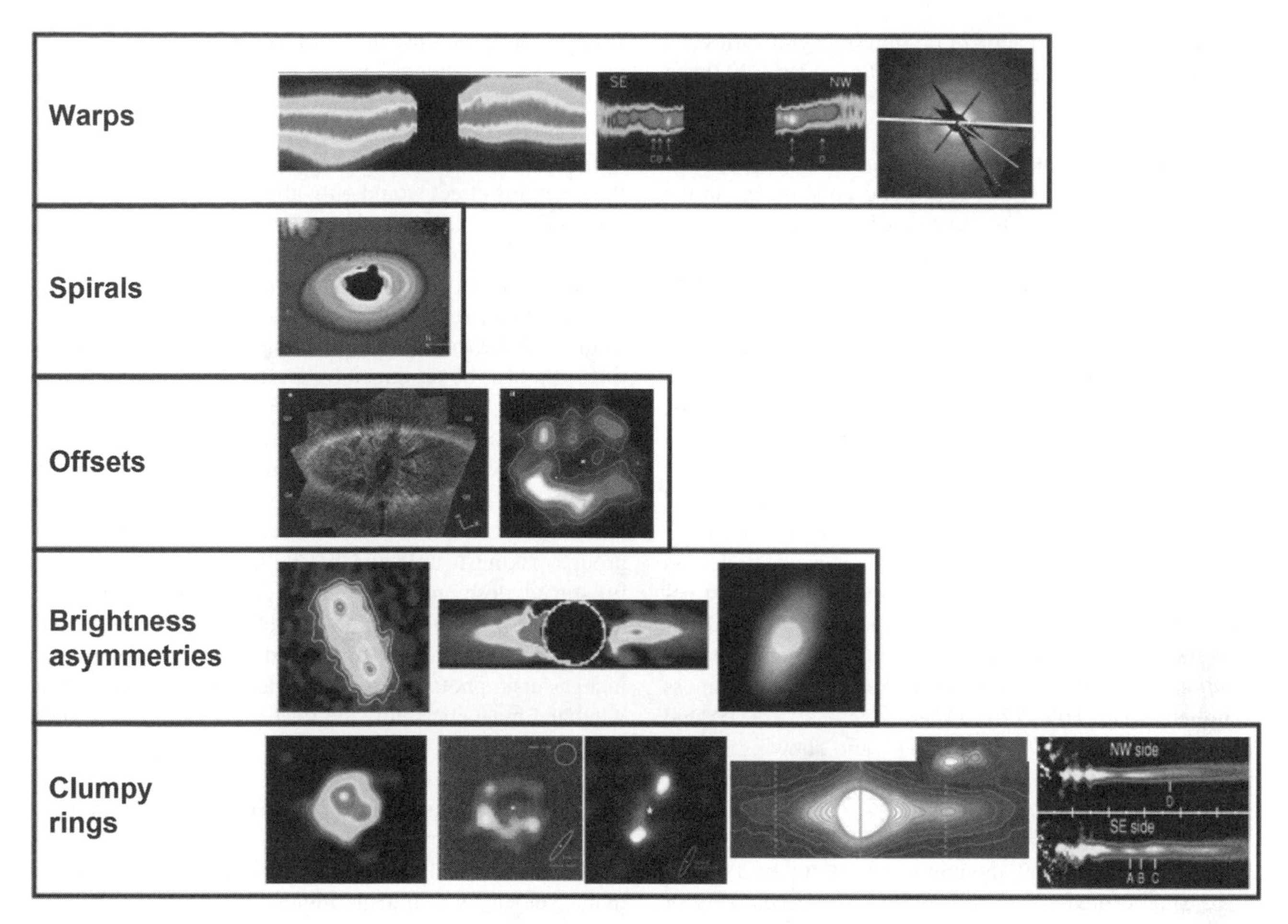

Fig. 5. Spatially resolved images of debris disks showing a wide diversity of debris disk structure. From left to right the images correspond to: *first row* — β Pic (STIS CCD coronography at 0.2–1 μm) (*Heap et al.,* 2000), AU Mic (Keck AO at 1.63 μm) (*Liu,* 2004), and TW Hydra (STIS CCD coronography at 0.2–1 μm) (*Roberge et al.,* 2005); *second row* — HD 141569 (HST/ACS at 0.46–0.72 μm) (*Clampin et al.,* 2003); *third row* — Fomalhaut (HST/ACS at 0.69–0.97 μm) (*Kalas et al.,* 2005) and ε Eri (JCMT/SCUBA at 850 μm) (*Greaves et al.,* 2005); *fourth row* — HR 4796 (Keck/OSCIR at 18.2 μm) (*Wyatt et al.,* 1999), HD 32297 (HST/NICMOS coronography at 1.1 μm) (*Schneider et al.,* 2005), and Fomalhaut (Spitzer/MIPS at 24 and 70 μm) (*Stapelfeldt et al.,* 2004); *fifth row* — Vega (JCMT/SCUBA at 850 μm) (*Holland et al.,* 1998), ε Eri (JCMT/SCUBA at 850 μm) (*Greaves et al.,* 1998), Fomalhaut (JCMT/SCUBA at 450 μm) (*Holland et al.,* 2003), β Pic (Gemini/T-ReCS at 12.3 μm) (*Telesco et al.,* 2005), and Au Mic (HST/ACS at 0.46–0.72 μm) (*Krist et al.,* 2005). All images show emission from tens to hundreds of AU.

et al., 2004; *Beichman et al.,* 2005a; *Bryden et al.,* 2006; *Kim et al.,* 2005; *Moro-Martín et al.,* 2005, 2007b; Hillenbrand et al., in preparation)].

Spatially resolved observations of nearby debris disks have confirmed the presence of central cavities. From observations in scattered light, *Kalas et al.* (2006) concluded that debris disks show two basic architectures, either narrow belts about 20–30 AU wide and with well-defined outer boundaries (HR 4796A, Fomalhaut, and HD 139664), or wide belts with sensitivity limited edges implying widths >50 AU (HD 32297, β Pic, AU Mic, HD 107146, and HD 53143). Millimeter and submillimeter observations show that inner cavities are also present in ε Eri (50 AU) (*Greaves et al.,* 1998), Vega (80 AU) (*Wilner et al.,* 2002), and η Corvi (100 AU) (*Wyatt et al.,* 2005).

Are all these cavities created by the gravitational ejection of dust by massive planets? *Wyatt* (2005a) pointed out that because of the limited sensitivity of the instruments, most of the debris disks observed so far have large number densities of dust particles and therefore are collisionally dominated. In this regime, mutual collision naturally create inner cavities without the need of invoking the presence of a planet to scatter out the dust particles. But this scenario assumes that the parent bodies are depleted from the inner cavity, and the presence of an inner edge to the planetesimal distribution may still require the presence of a planet.

Planet formation theories predict the formation of cavities because the planets form faster closer to the star, depleting planetesimals from the inner disk regions. But planet formation and circumstellar disk evolution are still under debate, so even though cavities may be credible evidence for the presence of planets, the connection is not well understood.

6.2.2. Rings and clumps. Face-on debris disks showing structure that could be associated with resonant trapping are Vega (*Wilner et al.,* 2002), ε Eridani (*Ozernoy et al.,* 2000; *Quillen and Thorndike,* 2002), and Fomalhaut (*Wyatt and Dent,* 2002), while in edge-on debris disks resonant trapping may lead to the creation of brightness asymmetries like those observed in β Pic (*Thommes and Lissauer,* 2003) and AU Mic.

6.2.3. Warps, offsets, spirals, and brightness asymmetries. The debris disk around β Pic has two warps, one in the outer disk (*Heap et al.,* 2000) and another one in the inner disk [with a wavy structure consisting of four clumps with counterparts at the other side of the disk and none of them aligned with each other (*Wahhaj,* 2005)]. New Hubble/ACS observations in scattered light show that the inner "warp" in β Pic is really a secondary disk inclined by 5° with respect to the primary disk. This secondary disk extends to ~80 AU and is probably sustained by a planet that has perturbed planetesimals from the outer primary disk into coplanar orbits. Another debris disk showing a warp is AU Mic, where the outer part of the disk (>80 AU) is tilted by 3°, while the rest of the disk is seen mostly edge-on.

The debris disks around HR 4796 show a 5% brightness asymmetry that could be the result of a small forced eccentricity imposed by the binary companion HR 4796B, or by an unseen planet located near the inner edge of the disk (*Wyatt et al.,* 1999). Other debris disks showing brightness asymmetries are HD 32297 (*Schneider et al.,* 2005) and Fomalhaut (*Stapelfeldt et al.,* 2004), and showing offsets are Fomalhaut (15 AU) (*Kalas,* 2005) and ε Eridani (6.6–16.6 AU) (*Greaves et al.,* 2005).

A spiral structure has been seen at 325 AU in the debris disk around HD 141569, thought to be created by a 0.2–2-M_{Jup} planet located at 235–250 AU with an eccentricity of 0.05–0.2 (*Wyatt,* 2005b).

In summary, dynamical simulations show that gravitational perturbations by a massive planet can result in the formation of the inner cavities, warps, offsets, brightness asymmetries, spirals, rings and clumps, and these features have indeed been observed in several debris disks.

6.3. Other Possible Causes of Debris Disk Structure

Clumps could trace the location of a planetesimal suffering a recent massive collision, instead of the location of dust-producing planetesimals or dust particles trapped in mean-motion resonances with a planet. This alternative interpretation has been proposed to explain the brightness asymmetries seen in the mid-IR observations of the inner β Pic disk (*Telesco et al.,* 2005). The brightness asymmetry could arise from the presence of a bright clump composed of dust particles with sizes smaller than those in the main disk, that could be the result of the collisional grinding of resonantly trapped planetesimals (making the clump long-lived, and likely to be observed), or the recent cataclysmic breakup of a planetesimal with a size >100 km [in which case there is no need to have a massive planet in the system, with the disadvantage that the clump is short-lived and we are observing it at a very particular time, maybe within ~50 yr of its breakup (*Telesco et al.,* 2005)]. However, the clumps seen in the submillimeter in systems like Fomalhaut are not easily explained by catastrophic planetesimal collisions because the dust masses involved are too large, implying the unlikely collision of two ~1400-km-sized planetesimals (*Wyatt and Dent,* 2002). Brightness asymmetries could also be due to "sandblasting" of a debris disk by interstellar dust particles, as the star moves with respect to the ISM, but this effect would only affect (if anything) the outskirts of the disk, ≥400 AU from the central star (*Artymowicz and Clampin,* 1997). Asymmetries and spiral structure can also be produced by binary companions, but, e.g., cannot explain all structure seen in the HD 141569 disk. And spiral structure and subsequent collapse into nested eccentric rings can also be produced by a close stellar flyby (*Kalas et al.,* 2001). This could in principle explain the clumps seen in the northeast of the β Pic disk, however, it would require a flyby on the scale of <1000 AU and these encounters are expected to be very rare. In addition, now the same type of structure is seen in AU Mic, another star of the same stellar group, making it unlikely that both stars suffered such a fine-tuned close encounter. Other effects that could be responsible for some of the disk features include instrumental artifacts, background/foreground objects, dust migration in a gas disk, photoevaporation, interaction with the stellar wind and magnetic field, and dust avalanches (*Grigorieva et al.,* 2006).

6.4. Debris Disks as a Planet-Detection Technique

The two well-established planet-detection techniques are radial velocity and transit studies, and both are sensitive only to close-in planets. Direct detection of massive planets has proven to be very difficult even in their younger (i.e., brighter) stages. This means that old long-period planets are likely to remain elusive in the foreseeable future.

However, we have seen that debris disk structure is sensitive to the presence of massive planets with a wide range of semimajor axis (out to hundreds of AU), complementing the parameter space covered by the other techniques. In this regard, the study of debris disk structure has the potential to characterize the diversity of planetary systems and to set constraints on the outward migration of extrasolar "Neptunes." However, before claiming that a planet is present in a debris disk system, the models should be able to explain observations at different wavelengths and account for dust particles of different sizes. Different wavelengths trace different particles sizes, and different particle sizes have different dynamical evolutions that result in different features. Large particles dominate the emission at longer wavelengths, and their location might resemble that of the dust-producing planetesimals. The small grains dominate at short wavelengths; they interact with the stellar radiation field more strongly so that their lifetime in the disk is shorter, and therefore their presence may signal a recent dust-producing event (like a planetesimal collision). And even shorter wavelengths are needed to study the warm dust produced by asteroid-like bodies in the terrestrial planet region. In addition, some of the dynamical models are able to make testable predictions, as, e.g., the position of resonant structures in multi-epoch imaging, as it is expected that they will orbit the planet with periods short enough to result in detectable changes within a decade. This rotation may have already been detected in ε Eri to a 2σ level (*Greaves et al.,* 2006). Dynamical models can also predict the location of the planets, but detecting the planet directly is not feasible with current technology.

7. THE SOLAR SYSTEM DEBRIS DISK

Our solar system harbors a debris disk, and the inner region is known as the zodiacal cloud. The sources of dust are very heterogeneous: asteroids and comets in the inner region, and KBOs and interstellar dust in the outer region. The relative contributions of each of these sources to the dust cloud is likely to have changed with time, and even the present relative contributions are controversial: From the He content of the interplanetary dust particles collected at Earth, it is possible to distinguish between low- and high-velocity grains, associated with an asteroidal and cometary origin, respectively. The ratio between the two populations is not well known, but is thought to differ by less than a factor of 10. The contribution of the asteroids to the zodiacal cloud is confirmed by the observation of dust bands (associated with the formation of individual asteroidal families), and must amount to at least a few 10%. The contribution from the comets is also confirmed by the presence of dust trails and tails. In the outer solar system, on spatial scales that are more relevant for comparison with other debris disks, significant dust production is expected from the mutual collisions of KBOs and collisions with interstellar grains (*Backman and Paresce,* 1993; *Stern,* 1996; *Yamamoto and Mukai,* 1998). There is evidence for the presence of Kuiper belt dust from the Pioneer 10 and 11 dust collision events that took place beyond the orbit of Saturn (*Landgraf et al.,* 2002), but the dust production rates are still uncertain.

In parallel to the debris disks properties described in the previous sections, we will now review some of the properties of the solar system debris disk. Comparison of these with the extrasolar systems can shed some light into the question of whether or not our solar system is unique.

7.1. Evolution

Debris disks evolve with time. Therefore, the imaging of debris disks at different evolutionary stages could be equivalent to a solar system "time machine." However, one needs to be cautious when comparing different systems because (1) the initial conditions and forming environment of the disks may be significantly different (see section 2.1); (2) the solar system debris disk is radiatively dominated, while the extrasolar debris disks observed so far, being significantly more massive, are collisionally dominated, so they are in different physical regimes; and (3) the physical processes affecting the later evolution of the disks depend strongly on the planetary configuration, e.g., by exciting and/or ejecting planetesimals, and radial velocity observations indicate that planetary configurations are very diverse. With those caveats in mind, we can draw some broad similarities between the time evolution of debris disks and the dust in our solar system.

As we saw in section 3, debris disk evolution consists of a slow decay of dust mass, punctuated by spikes of high activity, possibly associated with stochastic collisional events. Similarly, numerical simulations by *Grogan et al.* (2001) indicated that over the lifetime of the solar system, the asteroidal dust surface area slowly declined by a factor of 10, and that superimposed on this slow decay, asteroidal collisions produced sudden increases of up to an order of magnitude, with a decay time of several million years. Overall, for the 4-G.y.-old Sun, the dust surface area of the zodiacal cloud is about twice its quiescent level for 10% of the time. Examples of stochastic events in the recent solar system history are the fragmentation of the asteroid giving rise to the Hirayama asteroid families, the creation 8.3 m.y. ago of the Veritas asteroid families, which gave rise to a collisional cascade still accounting for ~25% of the zodiacal thermal emission (*Dermott et al.,* 2002), as well as collisional events resulting in the formation of the dust bands observed by IRAS (*Sykes and Greenberg,* 1986). In addition to these small "spikes" in the dust production rate at late times, there has been one major event in the early solar system evolution that produced much larger quantities of dust. Between 4.5 Ga and 3.85 Ga there was a heavy cratering phase that resurfaced the Moon and the terrestrial planets, creating the lunar basins and leaving numerous impact craters in the Moon, Mercury, and Mars (all with little surface erosion). This "heavy bombardment" ended ~3.85 G.y. ago, 600 m.y. after the formation of the Sun. Thereafter, the impact rate decreased exponentially with a time constant ranging from 10 to 100 m.y. (*Chyba,* 1990). *Strom et al.* (2005) argue that the impact crater record of the terrestrial planets show that the late heavy bombardment was an event lasting 20–200 m.y., that the source of the impactors was the main asteroid belt, and that the mechanism for this event was the orbital migration of the giant planets, which caused a resonance sweeping of the asteroid belt and a large scale ejection of asteroids into planet-crossing orbits. This event would have been accompanied by a high rate of asteroid collisions; the corresponding high rate of dust production would have caused a large spike in the warm dust luminosity of the solar system. Although this phenomenon has not been modeled in any detail, it is likely to be similar to the spikes inferred for extrasolar debris disks.

A massive clearing of planetesimals is also thought to have occurred in the Kuiper belt. This is inferred from the observation that the total mass in the Kuiper belt region (30–55 AU) is ~0.1 $M_\oplus$, insufficient to have been able to form the KBOs within the age of the solar system (*Stern,* 1996). It is estimated that the primordial Kuiper belt had a mass of 30–50 $M_\oplus$ between 30 and 55 AU, and was heavily depleted after Neptune formed and started to migrate outward (*Malhotra et al.,* 2000; *Levison et al.,* 2007). This resulted in the clearing of KBOs with perihelion distances near or inside the orbit of Neptune, and in the excitation of the KBOs' orbits, which increased their relative velocities from tens of meters per second to >1 km/s, making their collisions violent enough to result in a significant mass of the KBOs ground down to dust and blown away by radiation pressure.

As we have seen in section 3.2.2 and section 4, detailed studies of nearby debris disks show that unusually high dust production rates are needed to explain the properties of several stars, including Vega, ζ Lep, HIP 8920, HD 69830, and η Corvi. Even though one needs to be cautious about claiming that we are observing all these stars at a very special time during their evolution (possibly equivalent to the late heavy bombardment), this remains to date the most straightforward explanation of their "unusual" properties.

Observations therefore indicate that the solar and extrasolar debris disks may have evolved in broadly similar ways, in the sense that their dust production decays with time but is punctuated by short periods of increased dust production. However, the details of this evolution and the comparison of the absolute quantities of dust produced are difficult to assess. Preliminary results from the Spitzer FGK survey (*Bryden et al.,* 2006) indicated that even though the disks observed have a luminosity of ~100× that of the Kuiper belt dust disk, using the observed cumulative distribution and assuming the distribution of disk luminosities follows a Gaussian distribution, the observations are consistent with the solar system having an order of magnitude greater or less dust than the typical level of dust found around similar nearby stars, with the results being inconsistent with most stars having disks much brighter than the solar system's. However, from the Spitzer FEPS Legacy, *Meyer et al.* (2007) arrives at a different preliminary conclusion, suggesting that at times before the late heavy bombardment (10–300 m.y.), the dust production rate in the solar system was much higher than that found around stars of similar ages, while at times after the late heavy bombardment (1–3 G.y.), the dust production rate was much lower than average. For example, τ Ceti is a G8V (solar-type) star with an estimated age of 10 G.y., surrounded by a debris disk that is 20× dustier than the solar system's Kuiper belt (*Greaves et al.,* 2004b). Which star is "normal," τ Ceti or the Sun? If the present dust production rate in τ Ceti has been going on for the last 10 G.y., shouldn't all these dust-producing planetesimals have been ground down to dust? Have potential planets around τ Ceti undergone a heavy bombardment for the last 10 G.y., or is the dust the result of a recent massive collision?

7.2. Grain Size and Composition

As discussed in section 4, most debris disk spectra show little or no solid-state features, indicating that dust particles have grown to sizes ≥10 μm. The lack of silicate features, resulting from a lack of small dust grains, is also confirmed by the spatially resolved spectroscopy observations of a few nearby debris disks. In this regard, our zodiacal cloud is similar to most debris disks, presenting a predominantly featureless spectrum, thought to arise from dust grains 10–100 μm in size, with a small component of small silicate grains yielding a weak (10% over the continuum) 10-μm emission feature (*Reach et al.,* 2003). The analysis of the impact craters on the Long Duration Exposure Facility indicated that the mass distribution of the zodiacal dust peaks at ~200 μm (*Love and Brownlee,* 1993). The reason why large dust grains are dominant is a direct result from P-R drag because smaller grains evolve more quickly and therefore are removed on shorter timescales than larger grains. However, for the solar system, we only have information from the zodiacal cloud, i.e., the warmer component of the solar system's debris disks, because the emission from the colder Kuiper belt dust component is hidden by the inner cloud foreground.

In section 4, we also mentioned that there seems to be a correlation between the presence of silicate features and large quantities of dust (due possibly to a recent dust-producing event). The solar system, in its quiescent state, seems to be similar (in their lack of small silicate grains) to other debris disks that contain "normal" amounts of dust for their ages. But the solar system went through periods of high activity, like the late heavy bombardment, where dust production was orders of magnitude higher. Even though we do not know how the solar system looked like during those spikes in dust production, the remarkable similarity between the spectra of the dusty disk around HD 69830 (a 2-G.y. solar-type star) and Comet C/1995 O1 (Hale-Bopp) (*Beichman et al.,* 2005b) may indicate that during those stages, the solar system's dust disk could have also been similar to other debris disks experiencing similar spikes in their dust production.

7.3. Structure

The solar system, being filled with interplanetary dust and harboring planets, is an ideal case for investigating the effect of the planets on the dynamics of the dust particles, and consequently on the structure of the debris disks. Dynamical models predict that the Kuiper belt dust disk has a density enhancement in a ring-like structure between 35 and 50 AU, with some azimuthal variation due to the trapping into mean-motion resonances with Neptune and the tendency of the trapped particles to avoid the resonance planet, creating a minimun density at Neptune's position (*Liou and Zook,* 1999; *Moro-Martín and Malhotra,* 2002; *Holmes et al.,* 2003; see chapter by Liou and Kaufmann). The models also predict a depletion of dust inside 10 AU, due to gravitational scattering of dust particles by Jupiter and Saturn. However, the presence of this structure has not yet been observed [but there is clear evidence of the trapping of KBOs in resonance with Neptune (*Malhotra,* 1995; *Jewitt,* 1999; *Elliot et al.,* 2005)].

As we mentioned above, the thermal emission from the colder Kuiper belt dust is hidden by the much brighter inner zodiacal cloud foreground, which has been studied in detail by the IRAS, COBE, and ISO space telescopes (which could also map the spatial structure of the cloud, as their observing geometry changed throughout the year). These observations, together with numerical simulations, revealed that Earth is embedded in a resonant circumsolar ring of asteroidal dust, with a 10% number density enhancement located in Earth's wake, giving rise to the asymmetry observed in the zodiacal emission (*Jackson and Zook,* 1989; *Dermott et al.,* 1994; *Reach et al.,* 1995). In addition, it was found that the zodiacal cloud has a warp, as the plane of symmetry of the cloud depends on heliocentric distance (*Wyatt et al.,* 1999). This ring, the brightness asymmetry, and the warp indicate that even though the solar system debris disk is radiatively dominated, while the extrasolar debris disks observed so far are collisionally dominated, there are some structural features that are common to both.

In terms of disk size, the comparison of the solar system's dust disk with the handful of nearby spatially resolved debris disks observed to date indicates that the solar system is small. This would be consistent with the Sun being born in an OB association, while kinematic studies show that most of the nearby spatially resolved debris disks formed in loosely populated Taurus-like associations (see discussion in section 2.1). However, it may also be the result of an observational bias because so far we have only been able to study large disks. We have to wait until the next generation of interferometers come on line to be able to tell whether or not our solar system debris disk is normal in its size.

8. FUTURE PROSPECTS

Debris disks are evidence that many stars are surrounded by dust-producing planetesimals, like the asteroids and KBOs in our solar system. In some cases, they also provide evidence of the presence of larger bodies: first, because the production of dust requires the stirring of planetesimals, and the minimum mass for an object needed to start a collisional cascade is the mass of Pluto (see chapter by Kenyon et al.); and second, because some debris disks show structural features that may be the result of gravitational perturbations by a Neptune- to Jupiter-mass planet. Due to limits in sensitivity, we are not yet able to detect debris disks with masses similar to that of our solar system, but only those that are >100× more massive. Observations are beginning to indicate that the solar and extrasolar debris disks may have evolved in broadly similar ways, in the sense that their dust production decays with time but is punctuated by short periods of increased dust production, possibly equivalent to the late heavy bombardment. This offers a unique opportunity to use extrasolar debris disks to shed some light in how the solar system might have looked in the past. Similarly, our knowledge of the solar system is influencing our understanding of the types of processes that might be at play in the extrasolar debris disks. In the future, telescopes like ALMA, LBT, JWST, TPF, and SAFIR will be able to image the dust in planetary systems analogous to our own. This will allow the carrying out of large unbiased surveys sensitive down to the level of dust found in our own solar system that will answer the question of whether or not our solar system debris disk is common or rare. But very little information is known directly about the Kuiper belt dust disk, in terms of its mass, its spatial structure, and its composition, mainly because its thermal emission is overwhelmed by the much stronger signal from the inner zodiacal cloud. Any advance in understanding the structure and evolution of the Kuiper belt is directly relevant to our understanding of extrasolar planetary systems. And to that end, there is the need to carry out dust experiments on spacecraft traveling to the outer solar system, like the one onboard New Horizons, and to perform careful modeling of the dynamical evolution of Kuiper belt dust particles and their contribution to the solar system debris disk that takes into account our increased knowledge of the KBOs.

Acknowledgments. A.M.M. is under contract with the Jet Propulsion Laboratory (JPL), funded by NASA through the Michelson Fellowship Program. JPL is managed for NASA by the California Institute of Technology. A.M.M. is also supported by the Lyman Spitzer Fellowship at Princeton University. R.M. acknowledges support from the NASA Origins of Solar Systems and Outer Planets Research Programs.

REFERENCES

Andrews S. M. and Williams J. P. (2005) Circumstellar dust disks in Taurus-Auriga: The submillimeter perspective. *Astrophys. J., 631,* 1134–1160.

Artymowicz P. and Clampin M. (1997) Dust around main-sequence stars: Nature or nurture by the interstellar medium? *Astrophys. J., 490,* 863–878.

Aumann H. H., Beichman C. A., Gillett F. C., de Jong T., Houck J. R., et al. (1984) Discovery of a shell around Alpha Lyrae. *Astrophys. J. Lett., 278,* L23–L27.

Backman D. E. and Paresce F. (1993) Main-sequence stars with circumstellar solid material — The VEGA phenomenon. In *Protostars and Planets III* (E. H. Levy and J. I. Lunine, eds.), pp. 1253–1304. Univ. of Arizona, Tucson.

Beichman C. A., Bryden G., Rieke G. H., Stansberry J. A., Trilling D. E., et al. (2005a) Planets and infrared excesses: Preliminary results from a Spitzer MIPS survey of solar-type stars. *Astrophys. J., 622,* 1160–1170.

Beichman C. A., Bryden G., Gautier T. N., Stapelfeldt K. R., Werner M. W., et al. (2005b) An excess due to small grains around the nearby K0 V star HD 69830: Asteroid or cometary debris? *Astrophys. J., 626,* 1061–1069.

Beichman C. A., Tanner, A., Bryden G., Stapelfeldt K. R., and Gautier T. N. (2006a) IRS spectra of solar-type stars: A search for asteroid belt analogs. *Astrophys. J., 639,* 1166–1176.

Beichman C. A., Bryden G., Stapelfeldt K. R., Gautier T. N., Grogan K., et al. (2006b) New debris disks around nearby main-sequence stars: Impact on the direct detection of planets. *Astrophys. J., 652,* 1674–1693.

Bradley J. (2003) The astromineralogy of interplanetary dust particles. In *Astromineralogy* (T. K. Henning, ed.), pp. 217–235. Lecture Notes in Physics, Vol. 609.

Bryden G., Beichman C. A., Trilling D. E., Rieke G. H., Holmes E. K., et al. (2006) Frequency of debris disks around solar-type stars: First results from a Spitzer MIPS survey. *Astrophys. J., 636,* 1098–1113.

Burns J. A., Lamy P. L., and Soter S. (1979) Radiation forces on small particles in the solar system. *Icarus, 40,* 1–48.

Cameron A. G. W. and Truran J. W (1977) The supernova trigger for formation of the solar system. *Icarus, 30,* 447–461.

Chen C. H., Jura M., Gordon K. D., and Blaylock M. (2005) A Spitzer study of dusty disks in the Scorpius-Centaurus OB association. *Astrophys. J., 623,* 493–501.

Chyba C. F. (1990) Impact delivery and erosion of planetary oceans in the early inner solar system. *Nature, 343,* 129–133.

Cieza L. A., Cochran W. D., and Paulson D. B. (2005) Spitzer observations of the Hyades: Circumstellar debris disks at 625 m.y.s age (abstract). In *Protostars and Planets V,* Abstract #8421. LPI Contribution No. 1286, Lunar and Planetary Institute, Houston (CD-ROM).

Clampin M., Krist J. E., Ardila D. R., Golimowski D. A., Hartig G. F., et al. (2003) Hubble Space Telescope ACS coronagraphic imaging of the circumstellar disk around HD 141569A. *Astron. J., 126,* 385–392.

Crovisier J., Brooke T. Y., Hanner M. S., Keller H. U., Lamy P. L., et al. (1996) Spitzer observations of the Hyades: Circumstellar debris disks at 625 m.y.s age. *Astron. Astrophys., 315,* L385–L388.

Dermott S. F., Jayaraman S., Xu Y. L., Gustafson B. A. S., and Liou J. C. (1994) A circumsolar ring of asteroidal dust in resonant lock with the Earth. *Nature, 369,* 719–723.

Dermott S. F., Kehoe T. J. J., Durda D. D., Grogan K., and Nesvorny D. (2002) Recent rubble-pile origin of asteroidal solar system dust bands and asteroidal interplanetary dust particles. In *Asteroids, Comets, Meteors — ACM 2002* (B. Warmbein, ed.) pp. 319–322. ESA, Noordwijk, The Netherlands.

Dominik C. and Decin G. (2003) Age dependence of the Vega phenomenon: Theory. *Astrophys. J., 598,* 626–635.

Elliot J. L., Kern S. D., Clancy K. B., Gulbis A. A. S., Millis R. L., et al. (2005) The Deep Ecliptic Survey: A search for Kuiper belt objects and Centaurs. II. Dynamical classification, the Kuiper belt plane, and the core population. *Astron. J., 129,* 1117–1162.

Fischer D. A. and Valenti J. (2005) The planet-metallicity correlation. *Astrophys. J., 622,* 1102–1117.

Gorlova N., Padgett D. L., Rieke G. H., Muzerolle J., Stauffer J. R., et al. (2004) New debris-disk candidates: 24 micron stellar excesses at 100 million years. *Astrophys. J. Suppl., 154,* 448–452.

Gorlova N., Rieke G. H., Muzerolle J., Stauffer J. R., Siegler N., et al. (2006) Spitzer 24 micron survey of debris disks in the Pleiades. *Astrophys. J., 649,* 1028–1042.

Greaves J. S., Holland W. S., Moriarty-Schieven G., Jenness T., Dent W. R. F., et al. (1998) A dust ring around epsilon Eridani: Analog to the young solar system. *Astrophys. J. Lett., 506,* L133–L137.

Greaves J. S., Holland W. S., Jayawardhana R., Wyatt M. C., and Dent W. R. F. (2004a) A search for debris discs around stars with giant planets. *Mon. Not. R. Astron. Soc., 348,* 1097–1104.

Greaves J. S., Wyatt M. C., Holland W. S., and Dent W. R. F. (2004b) The debris disc around tau Ceti: A massive analogue to the Kuiper belt. *Mon. Not. R. Astron. Soc., 351,* L54–L58.

Greaves J. S., Holland W. S., Wyatt M. C., Dent W. R. F., and Robson E. I. (2005) Structure in the ε Eridani debris disk. *Astrophys. J. Lett., 619,* L187–L190.

Greaves J. S., Fischer D. A., and Wyatt M. C. (2006) Metallicity, debris discs and planets. *Mon. Not. R. Astron. Soc., 366,* 283–286.

Grigorieva A., Artymowicz P., and Thebault P. (2006) Collisional dust avalanches in debris discs. *Astron. Astrophys., 461,* 537–549.

Grogan K., Dermott S. F., and Durda D. D. (2001) The size-frequency distribution of the zodiacal cloud: Evidence from the solar system dust bands. *Icarus, 152,* 251–267.

Hartmann L. (2000) *Accretion Processes in Star Formation.* Cambridge Univ., Cambridge.

Hatzes A. P., Cochran W. D., McArthur B., Baliunas S. L., Walker G. A. H., et al. (2000) Evidence for a long-period planet orbiting ε Eridani. *Astrophys. J. Lett., 544,* L145–L148.

Hayashi C. (1981) Structure of the solar nebula, growth and decay of magnetic fields and effects of magnetic and turbulent viscosities on the nebula. *Prog. Theor. Phys. Suppl., 70,* 35–53.

Heap S. R., Lindler D. J., Lanz T. M., Cornett R. H., Hubeny I., and Maran et al. (2000) Space Telescope Imaging Spectrograph Coronagraphic observations of beta Pictoris. *Astrophys. J., 539,* 435–444.

Hillenbrand L. A. and Hartmann L. W. (1998) A preliminary study of the Orion nebula cluster structure and dynamics. *Astrophys. J., 492,* 540–553.

Hines D. C., Backman D. E., Bouwman J., Hillenbrand L. A., Carpenter J. M., et al. (2006) The formation and evolution of planetary systems (FEPS): Discovery of an unusual debris system associated with HD 12039. *Astrophys. J., 638,* 1070–1079.

Holland W. S., Greaves J. S., Zuckerman B., Webb R. A., McCarthy C., et al. (1998) Submillimetre images of dusty debris around nearby stars. *Nature, 392,* 788–790.

Holland W. S., Greaves J. S., Dent W. R. F., Wyatt M. C., Zuckerman B., et al. (2003) Submillimeter observations of an asymmetric dust disk around Fomalhaut. *Astrophys. J., 582,* 1141–1146.

Hollenbach D. and Adams F. C. (2004) Dispersal of disks around young stars: Constraints on Kuiper belt formation. In *Debris Disks and the Formation of Planets: A Symposium in Memory of Fred Gillett* (L. Caroff et al., eds.), p. 168. ASP Conf. Series 324, San Francisco.

Hollenbach D., Gorti U., Meyer M., Kim J. S., Morris P., et al. (2005) Formation and evolution of planetary systems: Upper limits to the gas mass in HD 105. *Astrophys. J., 631,* 1180–1190.

Holmes E. K., Dermott S. F., Gustafson B. A. S., and Grogan K. (2003) Resonant structure in the Kuiper disk: An asymmetric Plutino disk. *Astrophys. J., 597,* 1211–1236.

Jackson A. A. and Zook H. A. (1989) A solar system dust ring with Earth as its shepherd. *Nature, 337,* 629–631.

Jewitt D. (1999) Kuiper belt objects. *Annu. Rev. Earth Planet. Sci., 27,* 287–312.

Jura M., Chen C. H., Furlan E., Green J., Sargent B., et al. (2004) Mid-infrared spectra of dust debris around main-sequence stars. *Astrophys. J. Suppl., 154,* 453–457.

Kalas P., Deltorn J.-M., and Larwood J. (2001) Stellar encounters with the beta Pictoris planetesimal system. *Astrophys. J., 553,* 410–420.

Kalas P., Graham J. R., and Clampin M. (2005) A planetary system as the origin of structure in Fomalhaut's dust belt. *Nature, 435,* 1067–1070.

Kalas P., Graham J. R., Clampin M. C., and Fitzgerald M. P. (2006) First scattered light images of debris disks around HD 53143 and HD 139664. *Astrophys. J., 637,* L57–L60.

Kenyon S. J. and Bromley B. C. (2004) Collisional cascades in planetesimal disks. II. Embedded planets. *Astron. J., 127,* 513–530.

Kenyon S. J. and Bromley B. C. (2005) Prospects for detection of catastrophic collisions in debris disks. *Astron. J., 130,* 269–279.

Kim J. S., Hines D. C., Backman D. E., Hillenbrand L. A., Meyer M. R., et al. (2005) Formation and evolution of planetary systems: Cold outer disks associated with Sun-like stars. *Astrophys. J., 632,* 659–669.

Kobayashi H., Ida S., and Tanaka H. (2005) The evidence of an early stellar encounter in Edgeworth Kuiper belt. *Icarus, 177,* 246–255.

Krist J. E., Ardila D. R., Golimowski D. A., Clampin M., and Ford H. C. (2005) Hubble Space Telescope Advanced Camera for Surveys coronagraphic imaging of the AU Microscopii debris disk. *Astron. J., 129,* 1008–1017.

Kuchner M. J. and Holman M. J. (2003) The geometry of resonant signatures in debris disks with planets. *Astrophys. J., 588,* 1110–1120.

Lagrange A.-M., Backman D. E., and Artymowicz P. (2000)Main-sequence stars with circumstellar solid material — The VEGA phenomenon. In *Protostars and Planets IV* (V. Mannings et al., eds.), p. 639. Univ. of Arizona, Tucson.

Landgraf M., Liou J.-C., Zook H. A., and Grün E (2002) Origins of solar system dust beyond Jupiter. *Astron. J., 123,* 2857–2861.

Levison H. F., Morbidelli A., Gomes R., and Backman D. (2007) Planet migration in planetesimal disks. In *Protostars and Planets V* (B. Reipurth et al., eds.), pp. 669–684. Univ. of Arizona, Tucson.

Liou J.-C. and Zook H. A. (1995) An asteroidal dust ring of micron-sized particles trapped in 1:1 mean motion with Jupiter. *Icarus, 113,* 403–414.

Liou J.-C. and Zook H. A. (1999) Signatures of the giant planets imprinted on the Edgeworth-Kuiper belt dust disk. *Astron. J., 118,* 580–590.

Liou J.-C., Dermott S. F., and Xu Y. L. (1995) The contribution of cometary dust to the zodiacal cloud. *Planet. Space Sci., 43,* 717–722.

Liu M. C. (2004) Substructure in the circumstellar disk around the young star AU Microscopii. *Science, 305,* 1442–1444.

Love S. G. and Brownlee D. E. (1993) A direct measurement of the terrestrial mass accretion rate of cosmic dust. *Science, 262,* 550–553.

Malhotra R. (1993) The origin of Pluto's peculiar orbit. *Nature, 365,* 819–821.

Malhotra R. (1995) The origin of Pluto's orbit: Implications for the solar system beyond Neptune. *Astron. J., 110,* 420–430.

Malhotra R., Duncan M. J., and Levison H. F. (2000) Dynamics of the Kuiper belt. In *Protostars and Planets IV* (V. Mannings et al., eds.), p. 1231. Univ. of Arizona, Tucson.

Meyer M. R., Hillenbrand L. A., Backman D. E., Beckwith S. V. W., Bouwman J., et al. (2004) The formation and evolution of planetary systems: First results from a Spitzer Legacy Science Program. *Astrophys. J. Suppl., 154,* 422–427.

Meyer M. R., Backman D. E., Weinberger A. J., and Wyatt M. C. (2007) Evolution of circumstellar disks around normal stars: Placing our solar system in context. In *Protostars and Planets V* (B. Reipurth et al., eds.), pp. 573–588. Univ. of Arizona, Tucson.

Moro-Martín A. and Malhotra R. (2002) A study of the dynamics of dust from the Kuiper belt: Spatial distribution and spectral energy distribution. *Astron. J., 124,* 2305–2321.

Moro-Martín A. and Malhotra R. (2003) Dynamical models of Kuiper belt dust in the inner and outer solar system. *Astron. J., 125,* 2255–2265.

Moro-Martín A. and Malhotra R. (2005) Dust outflows and inner gaps generated by massive planets in debris disks. *Astrophys. J., 633,* 1150–1167.

Moro-Martín A., Wolf S., and Malhotra R. (2005) Signatures of planets in spatially unresolved debris disks. *Astrophys. J., 621,* 1079–1097.

Moro-Martín A., Carpenter J. M., Meyer M. R., Hillenbrand L. A., Malhotra R., et al. (2007a) Are debris disks and massive planets correlated? *Astrophys. J., 658,* 1312–1321.

Moro-Martín A., Malhotra R., Carpenter J. M., Hillenbrand L. A., Wolf S., et al. (2007b) The dust, planetesimals and planets of HD 38529. *Astrophys. J., 668,* in press.

Mouillet D., Larwood J. D., Papaloizou J. C. B., and Lagrange A. M. (1997) A planet on an inclined orbit as an explanation of the warp in the Beta Pictoris disc. *Mon. Not. R. Astron. Soc., 292,* 896–904.

Natta A. (2004) Circumstellar disks in pre-main sequence stars. In *Debris Disks and the Formation of Planets: A Symposium in Memory of Fred Gillett* (L. Caroff et al., eds.), p. 20. ASP Conf. Series 324, San Francisco.

Okamoto Y. K., Kataza H., Honda M., Yamashita T., et al. (2004) An early extrasolar planetary system revealed by planetesimal belts in beta Pictoris. *Nature, 431,* 660–663.

Ozernoy L. M., Gorkavyi N. N., Mather J. C., and Taidakova T. A. (2000) Signatures of exosolar planets in dust debris disks. *Astrophys. J. Lett., 537,* L147–L151.

Pascucci I., Gorti U., Hollenbach D., Najita J., and Meyer M. R. (2006) Formation and evolution of planetary systems: Upper limits to the gas mass in disks around solar-like stars. *Astrophys. J., 651,* 1177–1193.

Quillen A. C. and Thorndike S. (2002) Structure in the ε Eridani Dusty disk caused by mean motion resonances with a 0.3 eccentricity planet at periastron. *Astrophys. J. Lett., 578,* L149–L152.

Reach W. T., Franz B. A., Weiland J. L., Hauser M. G., Kelsall T. N., et al. (1995) Observational confirmation of a circumsolar dust ring by the COBE satellite. *Nature, 374,* 521–523.

Reach W. T., Morris P., Boulanger F., and Okumura K. (2003) The mid-infrared spectrum of the zodiacal and exozodiacal light. *Icarus, 164,* 384–403.

Rieke G. H., Su K. Y. L., Stansberry J. A., Trilling D., Bryden G., et al. (2005) Decay of planetary debris disks. *Astrophys. J., 620,* 1010–1026.

Roberge A., Weinberger A. J., and Malumuth E. M. (2005) Spatially resolved spectroscopy and coronagraphic imaging of the TW Hydrae circumstellar disk. *Astrophys. J., 622,* 1171–1181.

Roques F., Scholl H., Sicardy B., and Smith B. A. (1994) Is there a planet around beta Pictoris? Perturbations of a planet on a circumstellar dust disk. 1: The numerical model. *Icarus, 108,* 37–58.

Schneider G., Silverstone M. D., and Hines D. C. (2005) Discovery of a nearly edge-on disk around HD 32297. *Astrophys. J. Lett., 629,* L117–L120.

Shu F. H., Adams F. C., and Lizano S. (1987) Star formation in molecular clouds — Observation and theory. *Annu. Rev. Astron. Astrophys., 25,* 23–81.

Siegler N., Muzerolle J., Young E. T., Rieke G. H., Mamajek E., et al. (2006) Spitzer 24 micron observations of open cluster IC 2391 and debris disk evolution of FGK stars. *Astrophys. J., 654,* 580–594.

Simon M., Dutrey A., and Guilloteau S. (2000) Dynamical masses of T Tauri stars and calibration of pre-main-sequence evolution. *Astrophys. J., 545,* 1034–1043.

Song I., Zuckerman B., and Bessell M. S. (2003) New members of the TW Hydrae association, beta Pictoris moving group, and Tucana/Horologium association. *Astrophys. J., 599,* 342–350.

Song I., Zuckerman B., Weinberger A. J., and Becklin E. E. (2005) Extreme collisions between planetesimals as the origin of warm dust around a Sun-like star. *Nature, 436,* 363–365.

Stapelfeldt K. R., Holmes E. K., Chen C., Rieke G. H., and Su K. Y. L., et al. (2004) First look at the Fomalhaut debris disk with the Spitzer Space Telescope. *Astrophys. J. Suppl., 154,* 458–462.

Stauffer J. R., Rebull L. M., Carpenter J., Hillenbrand L., Backman D., et al. (2005) Spitzer Space Telescope observations of G dwarfs in the Pleiades: Circumstellar debris Disks at 100 m.y. age. *Astron. J., 130,* 1834–1844.

Stern S. A. (1996) On the collisional environment, accretion time scales, and architecture of the massive, primordial Kuiper belt. *Astron. J., 112,* 1203–1214.

Strom R. G., Malhotra R., Ito T., Yoshida F., and Kring D. A. (2005) The origin of planetary impactors in the inner solar system. *Science, 309,* 1847–1850.

Su K. Y. L., Rieke G. H., Stansberry J. A., Bryden G., Stapelfeldt K. R., et al. (2006) Debris disk evolution around A stars. *Astrophys. J., 653,* 675–689.

Sykes M. V. and Greenberg R. (1986) The formation and origin of the IRAS zodiacal dust bands as a consequence of single collisions between asteroids. *Icarus, 65,* 51–69.

Tachibana S., Huss G. R., Kita N. T., Shimoda G., and Morishita Y. (2006) ^{60}Fe in chondrites: Debris from a nearby supernova in the early solar system? *Astrophys. J. Lett., 639,* L87–L90.

Telesco C. M., Fisher R. S., Wyatt M. C., Dermott S. F., Kehoe T. J. J., et al. (2005) Mid-infrared images of beta Pictoris and the possible role of planetesimal collisions in the central disk. *Nature, 433,* 133–136.

Thommes E. W. and Lissauer J. J. (2003) Resonant inclination excitation of migrating giant planets. *Astrophys. J., 597,* 566–580.

Wahhaj Z. (2005) Planetary signatures in circumstellar debris disks. Ph.D dissertation, Univ. of Pennsylvania, Philadelphia.

Weidenschilling S. J. (1977) The distribution of mass in the planetary system and solar nebula. *Astrophys. Space Sci., 51,* 153–158.

Weinberger A. J., Becklin E. E., and Zuckerman B. (2003) The first spatially resolved mid-infrared spectroscopy of beta Pictoris. *Astrophys. J. Lett., 584,* L33–L37.

Weingartner J. C. and Draine B. T. (2001) Dust grain-size distributions and extinction in the Milky Way, Large Magellanic Cloud, and Small Magellanic Cloud. *Astrophys. J., 548,* 296–309.

Wilner D. J., Holman M. J., Kuchner M. J., and Ho P. T. P (2002) Structure in the dusty debris around Vega. *Astrophys. J. Lett., 569,* L115–L119.

Wisdom J. (1980) The resonance overlap criterion and the onset of stochastic behavior in the restricted three-body problem. *Astron. J., 85,* 1122–1133.

Wyatt M. C. (2003) Resonant trapping of planetesimals by planet migration: Debris disk clumps and Vega's similarity to the solar system. *Astrophys. J., 598,* 1321–1340.

Wyatt M. C. (2005a) The insignificance of P-R drag in detectable extrasolar planetesimal belts. *Astron. Astrophys., 433,* 1007–1012.

Wyatt M. C. (2005b) Spiral structure when setting up pericentre glow: Possible giant planets at hundreds of AU in the HD 141569 disk. *Astron. Astrophys., 440,* 937–948.

Wyatt M. C. (2006) Dust in resonant extrasolar Kuiper belts: Grain size and wavelength dependence of disk structure. *Astrophys. J., 639,* 1153–1165.

Wyatt M. C. and Dent (2002) Collisional processes in extrasolar planetesimal discs — Dust clumps in Fomalhaut's debris disc. *Mon. Not. R. Astron. Soc., 334,* 589–607.

Wyatt M. C., Dermott S. F., Telesco C. M., Fisher R. S., Grogan K., et al. (1999) How observations of circumstellar disk asymmetries can reveal hidden planets: Pericenter glow and its application to the HR 4796 disk. *Astrophys. J., 527,* 918–944.

Wyatt M. C., Greaves J. S., Dent W. R. F., and Coulson I. M. (2005) Submillimeter images of a dusty Kuiper belt around eta Corvi. *Astrophys. J., 620,* 492–500.

Wyatt M. C., Smith R., Greaves J. S., Beichman C. A., Bryden G., and Lisse C. M. (2006) Transience of hot dust around sun-like stars. *Astrophys. J., 658,* 569–583.

Yamamoto S. and Mukai T. (1998) Dust production by impacts of interstellar dust on Edgeworth-Kuiper belt objects. *Astron. Astrophys., 329,* 785–791.

Yoneda S., Simon S. B., Sylvester P. J., Hsu A., and Grossman L. (1993) Large siderophile-element fractionations in Murchison sulfides. *Meteoritics, 28,* 465–516.

Part IX: Laboratory

Laboratory Data on Ices, Refractory Carbonaceous Materials, and Minerals Relevant to Transneptunian Objects and Centaurs

C. de Bergh
LESIA, Observatoire de Paris

B. Schmitt
Laboratoire de Planétologie de Grenoble, CNRS–Université Joseph Fourier

L. V. Moroz
Institut für Planetologie, Universitat Münster, and Institut für Planetenforschung, DLR, Berlin

E. Quirico
Laboratoire de Planétologie de Grenoble, CNRS–Université Joseph Fourier

D. P. Cruikshank
NASA Ames Research Center

Information on the surface compositions of transneptunian objects (TNOs) and Centaurs is derived primarily from spectrophotometric or spectroscopic observations. The identification of the surface constituents and detailed studies of the surface composition depend on the availability of appropriate laboratory data. In particular, optical constants (complex refractive indices) are necessary to calculate synthetic spectra for different grain sizes or deal with components mixed in various ways. Observations of TNOs and Centaurs are made primarily in the wavelength region of reflected sunlight ($0.3 < \lambda < 5$ μm), while some of these objects can be studied in the region of their thermal IR emission ($\lambda > 10$ μm). We review the general spectroscopic characteristics and available optical constants for the ices, refractory carbonaceous materials, and minerals that have been detected or are expected to be present at the surfaces of these objects.

1. INTRODUCTION

Here we review what we know from the laboratory about the surface constituents detected so far at the surfaces of transneptunian objects (TNOs) (Pluto and Charon included) and Centaurs, or that are expected either because they are present in related objects [Triton, irregular satellites of the outer planets, some asteroids, meteorites, comets, and interplanetary dust particles (IDPs)] or because they are easily formed in the laboratory by irradiation or impact of already detected species.

The compounds detected so far with certainty (from well-identified absorption features) at the surfaces of TNOs (including Pluto-Charon) and Centaurs are water ice, methane ice, nitrogen ice, carbon monoxide ice, and ethane ice. In addition, methanol ice (or ice of a photolytic product of methanol) may be present at the surfaces of two objects (absorption feature detected around 2.27 μm). Furthermore, emission features due to fine-grained silicates are detected in thermal-IR spectra of Centaur Asbolus, a signature at 2.2 μm present in spectra of Charon has been tentatively assigned to a mixture of ammonia and ammonia-hydrate, and weak features in visible and near-IR spectra of a few TNOs have been tentatively attributed to hydrated minerals (see chapters by Barucci et al. and Brown).

Additional materials have been introduced in models to account for the general shapes of the spectra, the slopes (or colors) in the visible and near-IR ranges (which are most often different), the possible convexity or concavity of the spectra at the short-wavelength end of the visible spectra, and the (relatively few) measured albedos. These materials include ices, minerals, and complex organic solids (see chapter by Barucci et al.). Since the surfaces are subjected to ultraviolet, solar wind, and cosmic ray irradiation (see chapter by Hudson et al.), irradiated materials have been considered as well as "fresh" materials. The presence of "fresh" materials on TNOs, in spite of continuous exposure to the space environment, could be due to cryovolcanism, seasonal recondensation of the atmosphere, cometary-type activity, or big impacts. Furthermore, although the dust environment in the region of TNOs is unknown, these objects are thought to be subjected to micrometeoroid impacts that affect the uppermost (and optically visible) regolith by slowly exposing less irradiated material from below the more heavily irradiated crust. The suite of plausible materials with which the models can be calculated is severely limited by the paucity of optical constants available. Prominent among the non-ice materials used in modeling the colors and spectra of TNOs and Centaurs are the so-called Titan or Triton tholins, which are produced in the laboratory by irradiation of gaseous methane-nitrogen mixtures. Also widely used are ice tholins, produced by irradiation of simple hydrocarbon-containing H_2O ices, and kerogen-type material (as an analog of the fine-grained matrix material of carbonaceous chondrites).

In this chapter we review the available optical constants or the spectra covering the visible (0.3 μm < λ < 1.0 μm), near-IR (1 < λ < ~3 μm), mid-IR (~3 μm < λ < 5 μm), and in some cases far-IR (beyond 5 μm), spectral regions for various kinds of detected or plausible TNOs surface materials. The optical constants of a material are the real and imaginary components of the complex refractive index. The refractive index, ñ, of a substance describes the interaction of electromagnetic (EM) radiation with that material. The refractive index is a complex number, consisting of a real and an imaginary component, $\tilde{n} = n + ik$. In this representation, n is the "real" index of refraction, and k is the extinction coefficient, which describes the damping of an EM wave in the material. Both n and k are wavelength dependent. As EM radiation passes through the material, some fraction is absorbed per unit distance it travels, according to the Beer-Lambert law, which allows the calculation of the absorption coefficient α. If λ is the (vacuum) wavelength of the EM radiation of interest, then the imaginary index $k = \lambda\alpha/4\pi$.

We also list or give references to the bands (or type of transitions) that are detected, and note the dependence of the spectra with phase and temperature, grain size, composition (mixtures of ices, etc.) and mode of formation. For each class of materials we examine what laboratory work is needed to make further progress in the study of the surfaces of TNOs and Centaurs. For the irradiation experiments required, we refer to the chapter by Hudson et al.

2. REQUIRED LABORATORY DATA

The identification of the materials present at the surface of TNOs requires the knowledge of the spectral properties of these materials over the UV, visible, and IR ranges. Indeed, the only current means of observation of these surfaces is by optical and IR remote sensing (photometry, spectroscopy, and imagery, but also imaging spectroscopy for the New Horizons mission) either from groundbased telescopes or from space.

The first goal of remote sensing is to detect and identify the materials present at the surface, derive some information on the ways they are mixed (at molecular, granular, or geographic and stratigraphic levels), and estimate their relative abundances. Various physical and thermodynamic properties such as temperature, thermal history, the state of the ices (crystalline, amorphous) and the texture of the surface (grain size, surface compactness, roughness) may also be inferred by using some specific spectral or photometric tracers. In order to retrieve all this information from surface spectra, either in reflectance or in emission, laboratory data of relevant materials (ices, solid organics, carbonaceous materials, minerals) in conditions appropriate for TNOs are absolutely required. In addition, optical constants are calculated from laboratory spectra to estimate reflectance and emittance of particulate surfaces by radiative transfer models. The quality of the laboratory data and models should be commensurate with the quality of the observational data in order not to limit their interpretation.

The photon flux coming from a surface has two components: reflected solar radiation and thermal emission by the surface, with relative intensities depending on the wavelength considered. Because of the very low temperatures of TNO surfaces (~20 K < T < 60 K), the transition between the solar reflection and thermal emission regimes occurs at long wavelength, typically at λ > 20–30 μm. Thus reflectance spectra in the region normally accessible for TNOs (wavelengths less than about 5 μm) are not perturbed by thermal emission and therefore give access to part (2.5–5 μm) of the range of the fundamental vibrations of most molecules. Although these very strong bands are frequently saturated in reflectance spectra, they add interesting constraints on the surface composition when analyzed together with the better defined combination and overtone bands controlling the near-IR spectrum. The visible range also contains some spectral information (mostly the wing of strong UV bands with possible structures, some weak CH_4 bands, etc.) but, as discussed later, the constraints they add on specific phases (organics, carbonaceous materials, minerals) in terms of surface composition are not always strong. At the other end of the spectrum the thermal emission dominates the far-IR range (>20–30 μm), but extracting information from emission spectra is more complicated due to a lack of accurate emission radiative transfer models within surfaces and of appropriate laboratory emission spectra of relevant materials.

Direct comparison of a TNO spectrum with laboratory transmission or reflectance spectra may be enough to identify the main components of a surface and to have some first idea of its physical state. Laboratory reflectance spectra can be linearly combined to simulate, as a first approach, a geographic (or spatial) mixture of materials, but this approach is very limited in terms of surface description. Most of the spectral data on minerals are available as reflectance spectra. While for ices only few laboratory reflectance spectra are available, it is possible to calculate these reflectance

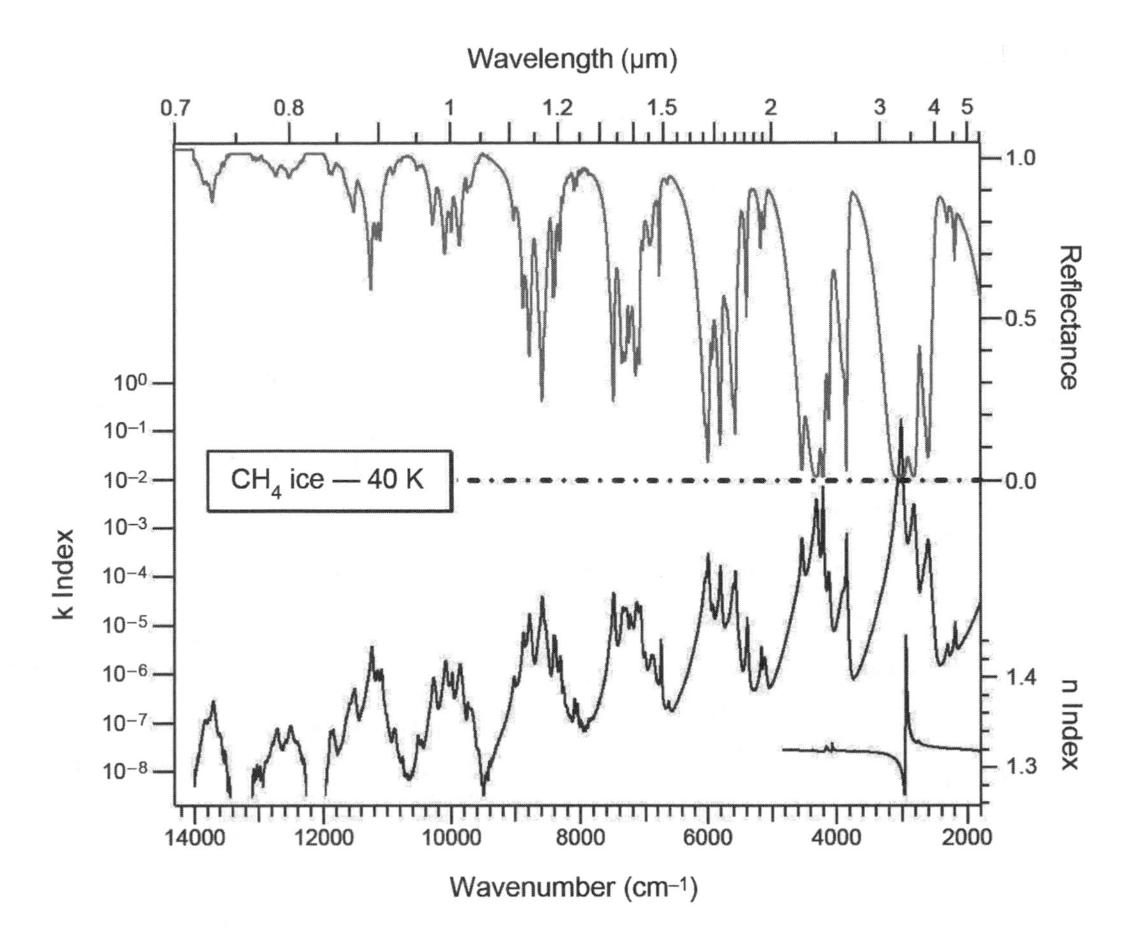

Fig. 1. *Lower curves*: Optical constant (k, n) spectra of CH_4 ice at 40 K in the visible, near-IR, and mid-IR (*Grundy et al.,* 2002; *Schmitt et al.,* 1998). The n index below 2 μm (>5000 cm^{-1}), not shown on the figure, decreases linearly from 1.36 at 15,800 cm^{-1} (0.63 μm) to 1.32 at 5000 cm^{-1} (2 μm). The data have been compiled using six experiments with thicknesses varying from a few micrometers (thin films deposited from the gas phase) to 1 cm (crystal grown from liquid in a closed cell). *Upper curve*: Reflectance spectrum of a CH_4 ice surface with 500-μm grain size, calculated from the optical constants.

spectra in cases where the complex refractive indices have been determined (which requires transmission spectra). Reflectance spectra have also been obtained for a limited number of relevant mixtures of ices with organics and minerals. We note, however, that the physical state of the laboratory samples is often not relevant for TNOs in terms of temperature, phase, grain size, etc. In particular, the samples measured in reflectance are typically at temperatures of 77 K or above.

The determination of the state of mixing, composition, and texture of the surface materials, as well as the extraction and identification of faint signatures, require detailed radiative transfer modeling. Such modeling requires the optical constants of the materials to simulate the reflectance spectrum of the surface, but also some frequently unknown parameters such as the single-scattering phase function of the medium (see chapter by Barucci et al.). However, measuring the optical constants is not a simple matter. In particular, for the major ices of interest here, the variation of the absorption coefficient in the entire visible, near-IR, and mid-IR ranges covers several orders of magnitude (e.g., Fig. 1). A large suite of experiments with various sample thicknesses and measurement techniques is therefore needed to obtain the spectral transmission measurements required to compile an optical constant spectrum over a wide spectral range, for example from 0.8 to 5 μm. Approximate optical constants can, however, be extracted from reflectance measurements (*Hapke and Wells,* 1981; *Clark and Roush,* 1984; *Shkuratov et al.,* 1999; *Roush,* 2003). For materials only available as powders (most minerals and organics) this technique, which requires several assumptions, currently provides the only available option to enable the inclusion of these materials in radiative transfer models. However, these data need to be used with care, especially when considering their absolute values (*Douté et al.,* 2007).

In our review, we will not focus on detailed description of the slopes and shapes of reflectance spectra. Direct comparisons between the spectra of TNOs/Centaurs and their analogue materials in terms of slopes and shapes may be misleading. For many relevant phases, visible and near-IR

(VNIR) spectral slopes strongly depend on grain size and composition within a given class of material (e.g., tholins, kerogens). In addition, a dark material (e.g., amorphous carbon) may be spectrally neutral as a pure particulate phase but may induce spectral reddening if the dark material is fine-grained and intimately mixed with a bright phase (ice or mineral) or forms a thin film or dust layer on a bright phase. Similar effects can alter the spectral shapes. For example, VNIR spectra of pure coals and bitumens are concave due to a broad saturated UV absorption. However, the spectral curves may become convex for intimate mixtures with bright phases, especially if those bright phases have convex visible spectra (e.g., Fe-bearing silicates). The application of an appropriate spectral mixing model may help to establish the presence or absence of a certain phase on the surface of a TNO or Centaur.

In the following, we will review the optical constants available (or spectra when the constants are missing) in the spectral ranges relevant to current observations of TNOs, with particular emphasis on the visible and near-IR (0.3–5 μm).

3. ICES

Many laboratory spectra of ices have been recorded in the mid- and far-IR ranges ($\lambda > 2.5$ μm) at low temperatures, mostly for purely spectroscopic aims but also for the study of icy mantles condensed on dust grains in the interstellar medium (ISM). However, the ices of TNOs have compositions, physical states (phase, temperature, etc.), and thermal histories radically different from the ISM grains. Nevertheless, some of these data are useful in analyzing the fundamental bands of ices in the 3–5-μm region and will be briefly reviewed in this chapter. Note, however, that because of details of the radiative transfer, these strong bands are frequently saturated in reflectance spectra of surfaces, contrary to spectra of grains in the ISM.

The analysis of the near-IR spectra of the icy objects Triton and Pluto has already triggered several in-depth laboratory spectral studies of ices encompassing their specific physico-chemical conditions, and therefore also appropriate for TNOs (see sections 3.2–3.5). In particular, *Quirico and Schmitt* (1997a) studied the spectra of methane ice and methane diluted in nitrogen by growing crystals in a closed cell from the liquid phase, finding that these long pathlength samples grown under thermodynamical equilibrium are particularly appropriate for the surfaces of Triton and Pluto. These two similar bodies appear to have large surface exposures of polycrystalline N_2 in which small amounts of CH_4, CO, and perhaps other molecules are dissolved in vapor-pressure equilibrium with the tenuous atmospheres. Due to their complexity, these experiments are currently quite limited. However, thin films experiments with some volatile ices condensed from the gas phase can provide useful transmission spectra, particularly on the strongest near-IR bands, as long as the condensation conditions (and temperature) of the sample, and thus its physical state, are representative of the object studied. In any case, optical constants in both the visible and near-IR spectral ranges and at appropriate physical conditions for these cold objects are still seriously lacking for several important molecules or mixtures.

Many TNOs and Centaurs are covered, at least partly, by ices of low volatility, such as H_2O, that may be out of thermodynamic equilibrium (in terms of phase, for example) due to the low surface temperatures (~20–60 K) or to endogenic/exogenic processes (particle bombardment, impact gardening, etc.). Laboratory experiments should then consider these possibilities (see also chapter by Hudson et al.), as well as the effects of thermal history of the surface on the spectra. Most of these laboratory studies have been performed in the mid-IR with vapor-deposited samples.

The relevance of different types of laboratory measurements for ices (transmission through thin films and crystals, diffuse reflection in various geometries, etc.) to the study of solar system objects, as well as the methods for extracting the optical constants, have been discussed in some detail in a few books and papers (e.g., *Schmitt et al.,* 1998) and will not be repeated here, except when necessary. A detailed discussion of the effects of the various physical parameters that affect the spectra of ices and their mixtures is also available in *Schmitt et al.* (1998) and *Brown and Cruikshank* (1997).

In the following we will briefly review the available optical constants (or absorption coefficient, or spectra if only they are available) of the different (pure and mixed) ices of interest for TNO surface spectral analysis.

3.1. Water Ice

In addition to its presence on satellites of Jupiter, Saturn, and Uranus, water ice is found on Triton, Pluto (tentative), and seems to dominate the surface of Charon and many TNOs and Centaurs. Water ice shows about ten crystalline forms, with only two or three that are stable at low pressure (a cubic Ic, an hexagonal Ih, and possibly an orthorhombic ice XI), three amorphous forms (Ia), and a glassy phase.

Because water ice is one of the most widespread materials in the solar system, many authors have studied its optical properties through transmission and reflectance spectroscopy. Many of these studies were made with emphasis on the effects of temperature, phase, and thermal history on specific IR bands (e.g. *Schmitt et al.*, 1989; *Moore and Hudson,* 1992; *Smith et al.*, 1994). The most recent review, although partly updated, of the available spectra and optical constants of crystalline ice from the UV to the far-IR has been made by *Warren* (1984). All references prior to 1982 can be found in that paper. However, most of the data used for the calculation of the optical constants have been recorded at fairly high temperatures (between 80 and 263 K) and are thus not fully relevant for TNO studies.

More recently, a consistent set of absorption coefficients of crystalline water ice has been determined in the near IR (from 1 to 2.7 μm) at many different temperatures between 20 and 270 K (*Grundy and Schmitt,* 1998). In this dataset only the interband absorption near 1.1 μm has a substantial uncertainty. The spectral changes with temperatures were analyzed and the temperature-dependent spectrum was resolved as a set of 15 overlapping Gaussian curves with band parameters allowing synthesizing of the optical constants of crystalline water ice at any temperature.

Fink and Sill (1982) published the first near-IR transmission spectra of the amorphous phase of H_2O at different temperatures. *Schmitt et al.* (1998) also determined absorption coefficient spectra at 40 K and 140 K that have been used in several spectral models (e.g., *Cruikshank et al.,* 2000). Near-IR spectra at 10 K (*Gerakines et al.,* 2005) and between 40 and 125 K (*Mastrapa and Brown,* 2006) have also been recorded recently. However, due to the difficulty in obtaining thick amorphous samples these data are incomplete (no data below 1.4 μm).

In the mid-IR a dozen papers have concerned the calculation of the optical constants of water ice at low temperature (10–190 K) in its amorphous and crystalline forms (see *Warren,* 1984; *Hudgins et al.,* 1993; *Toon et al.,* 1994; *Trotta,* 1996; *Schmitt et al.,* 1998). There are significant differences in the indices calculated by various authors, mainly due to improved experimental techniques and optical constant extraction codes (for recent comparisons, see *Toon et al.,* 1994; *Trotta,* 1996).

Water ice may be mixed with some other ices on TNO surfaces. Many studies have been performed for the fundamental bands of water mixed or diluted in a large variety of other ices but only very limited data have been reported in the near-IR and only for the strongest water bands in N_2 ice (*Palumbo and Strazzulla,* 2003; *Satorre et al.,* 2001). The principal effect is that the broad water ice bands transform into narrow bands strongly shifted in wavelength.

3.2. Nitrogen Ice

Solid nitrogen exhibits two phases: below 35.6 K, the cubic α phase is stable; above this temperature it is the field of the hexagonal β phase. This last phase seems to be dominant on Pluto and Triton, and may be present on TNOs 90377 Sedna and 136199 Eris (formerly 2003 UB_{313}). On some TNOs with lower surface temperatures α-nitrogen may exist.

Very weak IR bands have been observed for both crystalline phases in the near-IR (first overtone region) (*Schmitt et al.,* 1990; *Green et al.,* 1991) and in the mid-IR (fundamental stretching vibration range) (*Tryka et al.,* 1995; *Quirico et al.,* 1996). Detailed studies and absorption coefficient spectra of both phases at various temperatures between 20 and 63 K were published (*Grundy et al.,* 1993; *Tryka et al.,* 1993, 1995; *Schmitt et al.,* 1998). The α phase displays very narrow absorptions, the main ones peaking at 2.148 μm and in the 4.15–4.30-μm range, requiring high spectral resolution for detection, while the β phase has broad bands around 2.15, 4.18, and 4.29 μm. The transition and possibly the spectral shape of the bands can be used as a surface thermometer (*Grundy et al.,* 1993).

When other molecules, especially CO_2 and H_2O, are dissolved in nitrogen ice, a very strong enhancement in the band strength of the N_2 absorption occurs. But this effect has been studied only for the fundamental vibration band of α-nitrogen (e.g., *Nelander,* 1976; *Bernstein and Sandford,* 1999).

3.3. Methane Ice

Methane ice has been found on Pluto and Triton either in pure form (on Pluto) or diluted in nitrogen (on both bodies). It is also observed on the TNOs 2005 FY_9 and Eris (2003 UB_{313}). Solid methane presents two phases: the cubic phase II below 20.4 K and the cubic phase I above.

Several near-IR reflectance (see, e.g., *Fink and Sill,* 1982) and transmission spectra (e.g., *Schmitt et al.,* 1992) have been published, but the first optical constants for CH_4 ice in both phases (at 10 K and 33 K) are from *Khare et al.* (1990a) and *Pearl et al.* (1991). *Quirico and Schmitt* (1997a) published a high-resolution near-IR absorption coefficient spectrum of phase I at 21 K. Complete absorption coefficient spectra covering the 0.7–5-μm range for various temperatures between 15 and 90 K were obtained for both phases of pure methane by *Grundy et al.* (2002). They also studied the use of some bands, especially in the region around 2.55–2.6 μm, as a thermometer for pure CH_4 frost. The spectrum at 40 K is shown in Fig. 1.

Several spectroscopic studies were performed in the mid-IR range (see review by *Quirico and Schmitt,* 1997a) and a few sets of optical constants of solid CH_4 are available for both phases between 10 and 30 K (*Pearl et al.,* 1991; *Hudgins et al.,* 1993; *Trotta,* 1996). Optical constants at higher temperatures (i.e., 38 K) have been published by *Schmitt et al.* (1998).

Spectra of methane diluted at low concentration in solid nitrogen have been studied by *Quirico and Schmitt* (1997a). They derived normalized absorption coefficients at different temperatures in both nitrogen ice phases that were used to simulate the spectra of Triton and Pluto (*Quirico et al.,* 1999; *Douté et al.,* 1999; *Cruikshank et al.,* 2000). The spectra were derived from transmission measurements through monocrystalline samples grown in closed cells. However, from the modeling of Triton and Pluto spectra it has been found that the existing spectral data for diluted methane are not yet sufficient to model the regions of weak absorptions between the strong methane bands. This problem may be even more crucial for some TNOs that have methane absorptions substantially stronger than those of Pluto, although if there is no N_2 the appropriate data are those for pure CH_4. The near-IR spectrum of methane-water ice mixtures has also been studied (*Bernstein et al.,* 2006); slight shifts in

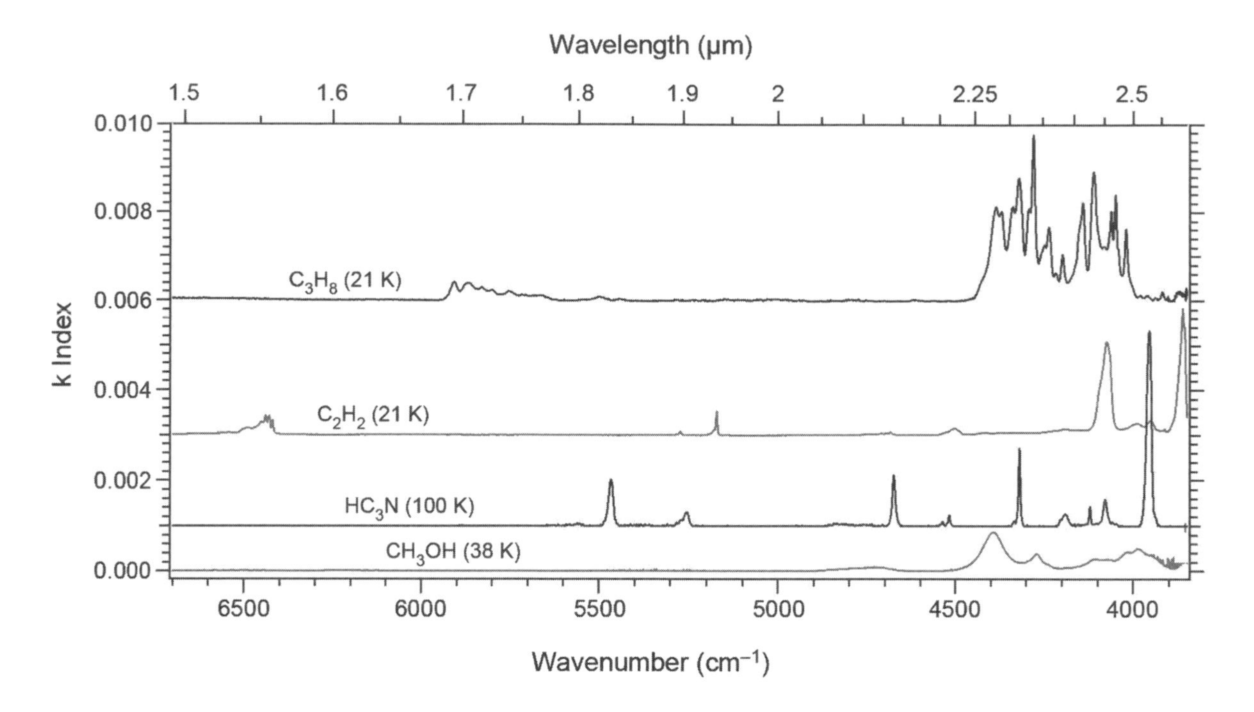

Fig. 2. Near-IR imaginary index (k) spectra of C_2H_2 (21 K), C_3H_8 (21K), HC_3N (100 K), and CH_3OH (38K) ices [derived from transmission spectra acquired at LPG and published in *Quirico et al.* (1999)]. The n index cannot be derived from these experiments.

band positions and significant band broadenings are observed.

3.4. Other Simple Solid Hydrocarbons

Light hydrocarbons can be efficiently produced by low-temperature irradiation of methane (see chapter by Hudson et al.). Although only ethane ice has been (tentatively) detected on the surface of Pluto so far, other simple, and possibly some large, hydrocarbons could be present as well, and are indeed predicted by photochemical models of Pluto's atmosphere (e.g., *Krasnopolsky and Cruikshank,* 1999).

3.4.1. Acetylene (C_2H_2). Acetylene presents a single solid-phase transition at 133 K. Only a few spectral studies have been devoted to solid acetylene, including a single set of mid-IR optical constants at 40 K (*Khanna et al.,* 1988) and a single near-IR spectrum at about 20 K (*Quirico et al.,* 1999). We show the corresponding imaginary index of the optical constants in Fig. 2.

3.4.2. Ethylene (C_2H_4). For ethylene, two phase transitions, one below 20 K and another near 50 K, are suspected. Near-IR measurements were performed by *Schmitt et al.* (1992), and *Quirico and Schmitt* (1997a) published absorption coefficients at 21 K. The effects of dilution at low concentration (1%) in nitrogen ice have also been studied by these authors. Although the mid-IR spectrum of ethylene has been extensively studied at many temperatures up to 93 K (see review by *Quirico and Schmitt,* 1997a), only a single set of optical constants (at 15 K) has been calculated so far (*Trotta,* 1996).

3.4.3. Ethane (C_2H_6). At low temperature ethane presents an amorphous phase below 25 K, a metastable phase between 25 and 60 K, and two crystalline phases (transition at 88.9 K). The optical constants of ethane at 30 K computed over part of the near-IR range by *Pearl et al.* (1991) have been completed by *Quirico and Schmitt* (1997a) from transmission spectra at 21 K (*Schmitt et al.,* 1992). These authors also recorded the spectrum of this molecule diluted at low concentration (1%) in nitrogen ice. The spectra of the different phases of solid C_2H_6 were recorded in the mid-IR range (see *Wisnosky et al.,* 1983) but the optical constants calculated at 15 and 30 K in this spectral range (*Trotta,* 1996; *Pearl et al.,* 1991) only concern the crystalline phase II.

3.4.4. Larger hydrocarbons. Hydrocarbons with three or more carbon atoms have been studied almost exclusively in the mid-IR where a few spectra are available at low temperature, especially for alkanes (e.g., *Goodman et al.,* 1983, *d'Hendecourt and Allamandola,* 1986). Optical constants have been published only for solid C_4H_2 at 70 K (*Khanna et al.,* 1988). *Bohn et al.* (1994) studied the mid-IR and part of the near-IR spectra (above 2 μm) of a series of large hydrocarbon molecules (up to C_8H_{18}) diluted in α-N_2 at very low temperature while *Quirico et al.* (1999) presented a first near-IR spectrum of solid propane (C_3H_8) at 21 K. We show the corresponding imaginary index of the optical constants in Fig. 2.

Curchin et al. (2006 and in preparation) have obtained reflectance spectra of a large number of hydrocarbons and other organic compounds in the solid phase, recorded at temperatures ~80–90 K, in the wavelength range 0.35–15.5 μm.

Their samples include alkanes, cycloalkanes, alkenes, alkynes, aromatics, nitriles, amines, and cyanides.

3.5. Carbon Monoxide and Carbon Dioxide Ices

Solid CO was observed on Triton and Pluto, while solid CO_2 was detected on Triton, Ariel, Umbriel, and Titania (for other outer planet satellites, the detected CO_2 is either trapped or complexed with other materials).

3.5.1. Carbon monoxide (CO). Solid carbon monoxide exibits a phase transition at 61.6 K; the low temperature α-phase is cubic and the high temperature β-phase is hexagonal.

The two overtone bands in the near-IR (1.577 and 2.404 μm) were first measured in transmission (see *Fink and Sill,* 1982; and review in *Quirico and Schmitt,* 1997a). The only published absorption coefficient spectrum of CO ice (at 20 K), with isotopic band assignments, can be found in *Quirico and Schmitt* (1997a) who also studied the spectra of CO diluted in the two phases of nitrogen ice at different temperatures (*Quirico and Schmitt,* 1997b). The fundamental region (around 4.68 μm) in the mid-IR has been studied by several authors but optical constants are only available at very low temperatures (10–15 K) (*Hudgins et al.,* 1993; *Trotta,* 1996).

3.5.2. Carbon dioxide (CO_2). Solid carbon dioxide has only one crystalline phase (cubic) at zero pressure, but amorphous and glassy CO_2 can form below 10–30 K. *Warren* (1986) reviewed all the available spectroscopic measurements on CO_2 ice and computed optical constants from the UV to the microwave region.

Reflectance spectra of this molecule in the near-IR have been recorded several times but only a few transmission spectra are available (*Schmitt et al.,* 1994; *Hansen,* 1996; *Bernstein et al.,* 2005). Absorption coefficients for the four strongest near-IR bands were estimated by *Warren* (1986) from data of *Fink and Sill* (1982) and completed by *Calvin* (1990) from reflectance measurements. High-spectral-resolution data for CO_2 ice have now been published (*Quirico and Schmitt,* 1997a). The weakest near-IR bands and the interband continuum absorption of CO_2 ice have been also measured in detail at 150 K by *Hansen* (1997, 2005) over a wide spectral range.

Several transmission studies of CO_2 ice have been performed in the mid-IR, and optical constants are given in those wavelength regions at a few temperatures between 10 and 160 K (*Warren,* 1986; *Hudgins et al.,* 1993; *Hansen,* 1997; *Trotta,* 1996; *Schmitt et al.,* 1998). Large differences exist among these different sets (see *Trotta,* 1996), mainly due to the extraction method.

The mid- and near-IR spectra of CO_2 diluted in various solids have been studied. Of relevance to TNOs are the studies in nitrogen ice (*Quirico and Schmitt,* 1997a) and in water and methanol ices (*Bernstein et al.,* 2005). Dilution permits the appearance of a new overtone band (inactive in pure CO_2) around 2.134 μm. This band is a potential indicator of the state of mixing of the CO_2.

3.6. Solid Ammonia and Ammonia-Hydrates

Ammonia ices have always been considered as a likely component of icy outer solar system satellites, helping to explain their complex geological histories. An absorption feature observed at 2.2 μm in the spectrum of Charon is probably a combination of ammonia and ammonia-hydrate.

Ammonia ice presents only one stable cubic crystalline phase, but an amorphous or glassy phase (below 60–80 K) and an intermediate, metastable phase (below 100–120 K) have been also observed.

After a first review of the spectroscopic measurements on NH_3 ice by *Taylor* (1973) and derivation of its optical constants in the IR, a more detailed study by *Martonchik et al.* (1984) covered the near-UV to the far-IR.

In the near-IR, a few reflectance spectra at low temperatures (77–150 K) have been published (e.g., *Fink and Sill,* 1982), as well as an absorption coefficient spectrum (down to 1.5 μm) of the cubic phase at 88 K (*Sill et al.,* 1980) used to calculate its optical constants (*Martonchick et al.,* 1984). Absorption coefficients over part of the near-IR for the metastable (40 K) and the cubic phases (70 K) of NH_3 ice between 1.8 and 2.5 μm were published by *Quirico et al.* (1999) and *Schmitt et al.* (1998). A partial spectrum of amorphous NH_3 at 10 K is also available (*Gerakines et al.,* 2005).

In addition to many spectroscopic studies of NH_3 ice in its various phases in the mid-IR, absorption coefficients of the crystalline phase are available in both ranges (*Sill et al.,* 1980) and optical constants have been calculated at various temperatures between 15 and 195 K (*Martonchik et al.,* 1984; *Mukai and Krätschmer,* 1986; *Trotta and Schmitt,* 1996).

There are at least three types of ammonia hydrates known to exist at low pressure: the hemihydrate ($2NH_3.H_2O$), the monohydrate ($NH_3.H_2O$, or NH_4OH), and the dihydrate ($NH_3.2H_2O$). Amorphous structures also exist at low temperature and NH_3 may also occur as a solid solution in water ice, in particular at low concentration. A series of studies has been conducted in the mid-IR to analyze spectra of these hydrates at 100 K (*Bertie and Shehata,* 1984, 1985; *Bertie and Devlin,* 1984). The hemihydrate, also studied at lower temperatures, shows two crystalline phases with a transition at 52 K (*Bertie and Devlin,* 1984).

In the near-IR a single set of reflectance spectra of ammonia-water mixtures at 77 K with four different concentrations (1–30%) is shown in one of the figures in *Brown et al.* (1988), and *Cruikshank et al.* (2005a) summarized their near-IR band positions. Clear spectral shifts (to shorter wavelengths for the 2.0- and 2.24-μm bands) and relative intensities changes occur upon hydration of ammonia (*Schmitt et al.,* 1998). Recently *Moore et al.* (2007) performed low-temperature (10–165 K) near- and mid-IR spectroscopic studies (1.8 to 20 μm) of ammonia ice and water-ammonia mixtures with special emphasis on features in the near-IR, but no absorption coefficients are derived. Spectral shifts of the bands toward higher wavenumbers occur upon dilution of NH_3, but in the near-IR no clear spectral differences are observed between the stoichiometric hydrates and icy mixtures with identical NH_3 concentrations.

3.7. Nitriles Ices

Simple nitriles, such as found in cometary comae, could be present as ices at the surface of TNOs (see chapter by Hudson et al.).

3.7.1. Hydrogen cyanide (HCN). Hydrogen cyanide has an orthorhombic low-temperature phase (<170 K). To the best of our knowledge, no spectra are available in the UV to near-IR range. Spectra of HCN polymer recorded at room temperature (*Cruikshank et al.,* 1991) are sometimes used in models instead.

Mid- and far-IR spectra have been recorded for the low-temperature phase of HCN between 35 and 95 K (see *Masterson and Khanna,* 1990; *Dello Russo and Khanna,* 1996) and optical constants were published in the mid-IR at 60 K (*Masterson and Khanna,* 1990). HCN polymers have also been studied from the UV to the mid-IR and optical constants have been calculated (see *Khare et al.,* 1994) (see also section 4.2).

3.7.2. Larger nitriles. The mid-IR transmission spectra of several larger solid nitriles have been studied at low temperatures (<100 K). *Dello Russo and Khanna* (1996) recorded the spectra at 35 K and 95 K of cyanogen (C_2N_2), dicyanoacetylene (C_4N_2), acetonitrile (CH_3CN), cyanoacetylene (HC_3N), cyanopropyne (CH_3C_3N), acrylonitrile (CH_2CHCN), and propionitrile (CH_3CH_2CN). They also systematically gave the integrated band intensities, and optical constants in the main bands above 12 μm. Mid-IR optical constants have been also published by other authors for C_2N_2 at 20 K (*Ospina et al.,* 1988), as well as for C_4N_2 and HC_3N at 60 K (*Masterson and Khanna,* 1990). The IR spectrum of CH_3CN has been studied by several authors (e.g., *d'Hendecourt and Allamandola,* 1986).

In the near-IR, absorption coefficients are available only for HC_3N (*Quirico et al.,* 1999). We present the corresponding imaginary index of the optical constants in Fig. 2.

Cruikshank et al. (1998) published a room-temperature reflectance spectrum of piperazine ($NHCH_2CH_2NHCH_2CH_2$), an amine similar in structure to trioxane [the formaldehyde trimer $(HCHO)_3$], 0.3–2.5 μm. They also published a reflectance spectrum in the same region of hexamethylenetetramine [$(CH_2)_6N_4$], a molecule of interest because it carries four atoms of N and its synthesis incorporates NH_3.

3.8. Methanol and Formaldehyde Ices

Methanol ice is probably present at the surface of Centaur Pholus and one TNO. Formaldehyde, detected in cometary comae, is another plausible icy surface component for TNOs.

Solid methanol has two hydrogen bonded crystalline phases with a transition at 157 K, and an amorphous phase below 78 K. A few spectra of methanol ice are available over part of the near-IR (*Cruikshank et al.,* 1998; *Quirico et al.,* 1999; *Brunetto et al.,* 2005, *Bernstein et al.,* 2005), but its spectrum is unknown below 2 μm. Figure 2 displays the imaginary index of the optical constants of CH_3OH at 38 K derived from *Quirico et al.* (1999).

On the other hand, there are numerous studies available for the mid- and far-IR transmission spectra of solid CH_3OH and its isotopic species (e.g., *d'Hendecourt and Allamandola,* 1986; *Hudson and Moore,* 1993; *Moore et al.,* 1994), and optical constants have been calculated between 10 and 120 K (*Hudgins et al.,* 1993; *Trotta,* 1996).

Formaldehyde (H_2CO) has been less studied, and essentially in the mid-IR range (e.g., *Harvey and Ogilvie,* 1962), but with no optical constants available as a pure solid. The only data published in the visible and near-IR (0.3–2.5 μm) are that of the paraformaldehyde polymer (*Cruikshank et al.,* 1998). Of relevance for TNOs, part of the near-IR spectrum (>2 μm) of formaldehyde molecules highly diluted in α-N_2 ice was also published by *Bohn et al.* (1994).

3.9. What is Missing

A number of laboratory studies have already recorded the spectra and derived the optical constants of several pure ices (H_2O, CH_4, N_2, CO, etc.) in many different physical conditions (temperature, phase, mixing). However, there are still several molecules that either have been studied only at temperatures irrelevant to TNOs or are available mostly as reflectance spectra. In other cases, the available transmission spectra incompletely cover the near-IR range. In particular, the molecules with incomplete data below 2 μm include NH_3, CH_3OH, C_2H_2, C_2H_4, C_3H_8, HC_3N, and numerous hydrocarbons. For a few others there are still no near-IR spectra available: HCN, H_2CO, all large hydrocarbons ($C \geq 3$, except C_3H_8), and most nitriles. Even when the transmission spectra have been measured, the absorption coefficients or the optical constants (needed for radiative transfer modeling) are not always derived. We note that, as mentioned in section 3.4.4, *Curchin et al.* (2006) have obtained reflectance spectra for a large suite of solid organics and nitriles that will soon become available.

When considering the potential mixtures that may occur on TNOs we can see that the near-IR spectra of many molecules have not yet been recorded in mixtures with the ices suspected to dominate at least part of their surfaces: H_2O, CO_2, β-N_2, CH_4. When data are available, in most cases only the strongest combination/overtone bands are recorded. In the case of the important CH_4-N_2 mixture, only few data have been recorded for mixtures with high concentrations of methane (*Quirico et al.,* 1996). Furthermore, wavelength regions where the absorption is very weak (defining the "continuum") in spectra of some CH_4 and N_2-rich TNOs need to be more accurately measured, as there are remaining and persistent difficulties in fitting these spectral regions with models.

Although significant progress has recently been achieved in understanding the spectra of different phases and types of mixtures of ammonia hydrate (*Moore et al.,* 2007), optical constants still must be measured for the amorphous phase, solid solutions, hemi-, mono-, dihydrates, etc.

Finally, several other species that may be formed by energetic processing of the TNO surface ices, such as H_2O_2, NO, NO_2, N_2O, C_2H_5OH, HCOOH, H_2CO_3, etc., or even cyanate (OCN^-) and ammonium ions (NH_4^+), clearly deserve further studies to extract their optical properties. Their formation and available spectral properties, not dealt with in this chapter, are described in the chapter by Hudson et al.

4. REFRACTORY CARBONACEOUS MATERIALS

Non-icy carbonaceous species of abiotic origin, ranging from very complex hydrocarbon macromolecular materials to pure elemental carbon solids (e.g., graphite, nanodiamonds), are present in primitive meteorites, IDPs, comets, asteroids, and interstellar and circumstellar dust (e.g., *Hayatsu and Anders,* 1981; *Kissel and Krueger,* 1987; *Kerridge,* 1999; *Sandford,* 1996; *Tielens,* 1997; *Pendleton and Allamandola,* 2002; *Huss et al.,* 2003; *Flynn et al.,* 2003; *Quirico et al.,* 2005; *Lisse et al.,* 2006; *Sandford et al.,* 2006). Such carbonaceous species usually have low visual albedos and a wide variety of colors in the visible and near-IR (VNIR) spectral region, ranging from red to neutral. In general, their colors become less red with increasing abundance and size of aromatic moieties in macromolecular material compared to the abundance of H-rich aliphatic chains and/or O-, N-functionalities. Complex hydrocarbon materials often show no absorption features in the VNIR spectral range. Many TNOs and Centaurs have low albedos and show a wide range of VNIR surface colors. This variety in surface optical properties may be (but is not necessarily) controlled by content, composition, and particle sizes of carbonaceous species. Therefore, it is important to acquire optical data on plausible carbonaceous analogs for use in spectral models of TNOs and Centaurs. Complex organics on TNOs and Centaurs may be indigenous, analogous to materials described in section 4.1, and/or may be products of irradiation of surface ices (see sections 4.2 and 4.3). In the absence of optical constants for other potentially useful organic analog materials, tholins are primarily used in models to reproduce red colors of TNOs and Centaurs, while elemental carbon of various kinds (ranging from structurally ordered graphite to amorphous C) is used to reproduce their low albedos.

4.1. Extraterrestrial and Terrestrial Carbonaceous Materials

Extraterrestrial organic materials available for laboratory studies are found in meteorites and IDPs. Over 650 individual organic molecules have been identified in primitive meteorites, but the largest portion ($\geq$90 wt.%) of meteoritic organic matter is represented by a complex macromolecular material insoluble in organic solvents and analogous to kerogen from terrestrial sedimentary rocks (*Durand,* 1980). Meteoritic "kerogens," often referred to as "IOM" (insoluble organic matter), are rich in polycyclic aromatic hydrocarbons.

Since primitive meteorites contain only up to a few weight percent of carbonaceous matter intimately mixed with minerals, a meteorite should be treated with acids to isolate organics for spectral/optical measurements. The acid treatment should be preceded by toluene/methanol extraction of soluble organics to avoid their reaction with acids, making it impossible to separate bulk meteoritic organic matter (insoluble plus soluble fractions) from minerals without its partial loss/damage (e.g., *Robert and Epstein,* 1982; *Halbout et al.,* 1990; *Wright et al.,* 1990; *Kerridge,* 1999). Therefore it must be borne in mind that the acid residues may be optically different from the bulk meteoritic organic matter in case of meteorites containing non-negligible soluble organic fraction, such as CI1 and CM2 chondrites. The careful use of appropriate soft demineralization procedures (e.g., *Cody et al.,* 2002) can minimize the damage to, at least, the insoluble (kerogen-like) fraction.

The optical constants (n, k) of insoluble organic residues from CM2 chondrite Murchison between 0.15 and 40 μm have been published by *Khare et al.* (1990b) and reflectance spectra from 0.2 to 6.5 μm by *Cruikshank et al.* (2001). *Hayatsu et al.* (1977) published transmittance spectra of Murchison IOM from 2.5 to 17 μm and *Flynn et al.* (2004) from 5 to 14 μm. Absorption spectra of the Murchison acid residue in the 3.4-μm spectral region have been published by *Pendleton* (1995) and *Flynn et al.* (2004), and by *Gardinier et al.* (2000) between 2.5 and 25 μm. Infrared transmittance spectra of acid residue from Orgueil (CI1) meteorite have been acquired by *Wdowiak et al.* (1988) (2.5–25 μm), *Ehrenfreund et al.* (1991) (2.5–25 μm), *Gardinier et al.* (2000) (2.5–25 μm), *Brownlee et al.* (2000) (3.2–3.7 μm), and *Flynn et al.* (2004) (5–14 μm). *Murae et al.* (1990) and *Murae* (1994) obtained IR spectra (2.5–25 μm) of carbonaceous insoluble residues from meteorites ALH 77307 and Y 791717 (CO3) and Allende (CV3).

Meteoritic IOM appears to be extremely diverse in composition (e.g., *Alexander et al.,* 1998; *Gardinier et al.,* 2000; *Cody and Alexander,* 2005). Composition and structure of IOM from meteorites of petrologic types higher than 2 (e.g., anhydrous CO3 and CV3 carbonaceous chondrites, ordinary chondrites) are controlled mostly by thermal metamorphism, although there is a debate regarding the heating of IOM on parent bodies vs. solar nebula (e.g., *Alexander et al.,* 1988; *Quirico et al.,* 2003; *Huss et al.,* 2003; *Bonal et al.,* 2006). Such IOM is highly carbonized compared to IOM from hydrated CR2, CI1, CM2 chondrites, and Tagish Lake (C2 ungrouped). Significant compositional and structural diversity exists also between organics from the hydrated meteorites, and it is unclear to what extent this diversity is affected by processing of IOM on the parent bodies. The fraction of aliphatic C compared to aromatic C decreases from CR2 through CI1, CM2 to Tagish Lake, while the content of O-bearing aliphatic functionalities increases (*Cody and Alexander,* 2005). However, *Gardinier at al.* (2000) reported higher aromaticity for the Orgueil (CI1) IOM compared to the Murchison (CM2) IOM.

These compositional and structural differences may also result in optical diversity of meteoritic IOM. The optical differences should be especially obvious between metamorphosed IOM from anhydrous meteorites and more primitive IOM from hydrated meteorites. The IOM from metamorphosed chondrites is highly carbonized, i.e., highly absorbing through the VNIR spectral range without any IR absorption features, while IOM from type 1 and 2 meteorites should be more transparent in the VNIR and shows a number of IR absorption bands (e.g., *Gardinier et al.,* 2000). However, optical variations may exist even between such more primitive IOM among different hydrated meteorites classified as types 1 and 2. It is obvious that the scarce available optical data on meteoritic organic residues cannot fully represent the optical diversity inherent to meteoritic organics.

The IOM content may be as high as 40 wt.% in some IDPs, but their tiny sizes make it very difficult to extract organic matter in quantities sufficient for optical measurements. Infrared transmittance spectra of organic acid residues from several IDPs have been published by *Brownlee et al.* (2000) (3.2–3.7 μm) and *Flynn et al.* (2004) (5–14 μm). Raman spectroscopy revealed the polyaromatic character of IOM from IDPs (*Wopenka,* 1988; *Quirico et al.,* 2005). Recent studies indicate that organic matter in both hydrated and anhydrous IDPs is very complex, resembling CI and CM IOM (*Flynn et al.,* 2003, 2004), while earlier studies suggested that most of the carbonaceous matter in anhydrous IDPs is represented by elemental amorphous carbon (*Keller et al.,* 1994). The complexity of organic matter in anhydrous IDPs is the direct evidence for the importance of indigenous complex organic matter for solar system objects formed beyond the "snow line," and not affected by aqueous alteration (*Flynn et al.,* 2003), suggesting that indigenous complex organics may also be important constituents of some TNOs and Centaurs. Dust particles captured from a Jupiter-family comet, 81P/Wild 2, by the Stardust spacecraft are probably the most relevant to TNOs compared to all other extraterrestrial samples available to date. Organic components of the returned samples resemble organic matter from primitive meteorites and IDPs but are relatively poor in IOM and rich in labile soluble components (*Sandford et al.,* 2006). In addition, the Stardust organics seem to be depleted in aromatics and enriched in aliphatics (probably long-chain ones) as well as in O- and N-functionalities compared to known meteoritic and IDP organic matter (*Sandford et al.,* 2006; *Keller et al.,* 2006). Infrared transmittance spectra of some organic-rich comet grains and aerogel tracks in the 3.4-μm spectral region have been published by *Sandford et al.* (2006) and *Keller et al.* (2006).

More accessible terrestrial organics have also been considered as analog materials. Natural kerogens from terrestrial sedimentary rocks may be useful analogs. *Hayatsu et al.* (1983), *Ehrenfreund et al.* (1991), and *Murae* (1994) noted compositional, structural, and spectral similarities between meteoritic IOM and relatively evolved terrestrial kerogens of type III. However, some differences have also been reported (*Ehrenfreund et al.,* 1991; *Binet et al.,* 2002, 2004; *Quirico et al.,* 2003). Kerogens show significant diversity in terms of composition and chemical structure, even within a single type-I, II, or III, depending on maturation grade and/or variations in the organic precursors. For example, the composition, chemical structure and optical properties of a young type II kerogen are significantly different from those of an evolved (mature) type II kerogen. Again, optical constants (0.15–40 μm) are available only for a single sample of the type II kerogen (*Khare et al.,* 1990b). However, for these last experiments, the demineralization procedure has been performed before the removal of the bitumen fraction, so that the latter could have reacted with acids (see above). In addition, the derived organic residue has been heated to 550°–750°C prior to optical measurements, which could have produced further alteration of the residue. Similar concerns exist regarding the Murchison organic residue from the same work.

Infrared absorbance spectra (2.5–25 μm) of terrestrial kerogens have been published, e.g., by *Espitalié et al.* (1973), *Robin and Rouxhet* (1976), and *Rouxhet et al.* (1980), and, with astronomical applications, by *Ehrenfreund et al.* (1991) and *Papoular* (2001). Analysis of relative intensities of several absorption bands in the IR absorbance spectra allows one to assess a kerogen type and maturation degree (*Ganz and Kalkreuth,* 1987). *Papoular* (2001) draws parallels between spectral changes induced by kerogen maturation and evolution of organic-rich interstellar dust. No VNIR reflectance spectra of pure terrestrial kerogens (types I, II, III) have been published, to our knowledge. We expect the VNIR spectral slopes to decrease and NIR overtone and combination features to disappear with increasing maturation degree. It is also possible that relatively immature type III kerogens show redder VNIR slopes and less-pronounced NIR absorption features compared to immature type I and II kerogens (see discussion of coal optical properties below).

As noted above, a kerogen-like fraction can be derived from any kind of partly insoluble organic matter. For example, the USGS spectral database (*Clark et al.,* 1993) contains the VNIR (0.2–3 μm) spectrum of a chemically uncharacterized kerogen-like material derived by D. P. Cruikshank from a coal tar sample.

The properties of kerogens are affected by the procedures used for their isolation, while some other natural analogs described below exist in concentrated form. The so-called solid oil bitumens — asphaltites, kerites, and anthraxolites — constitute another group of natural complex hydrocarbon solids suggested as analog materials for extraterrestrial macromolecular polymers (*Moroz et al.,* 1998, and references therein). The fraction of "kerogen" relative to the fraction of solvent-soluble bitumen, carbon aromaticity, and C/H ratio increase in this series from low-temperature asphaltites to thermally evolved high antraxolites. These changes are accompanied by a reduction of the spectral slope in the VNIR spectral range (Fig. 3). High kerites and low anthraxolites show compositional and structural similarities to organic matter from CM2 and CI1 chondrites, while high anthraxolites (shungites) resemble more carbonized organics of CV3 and CO3 chondrites. Reflectance spectra of solid oil bitumens between 0.5 and 17 μm have been published by *Moroz et al.* (1998), and transmittance spectra from 2.5 to 25 μm by *Moroz et al.* (1992). Optical constants (n and k) have been measured only for an asphaltite sample from 2.2 to 14 μm (unpublished data).

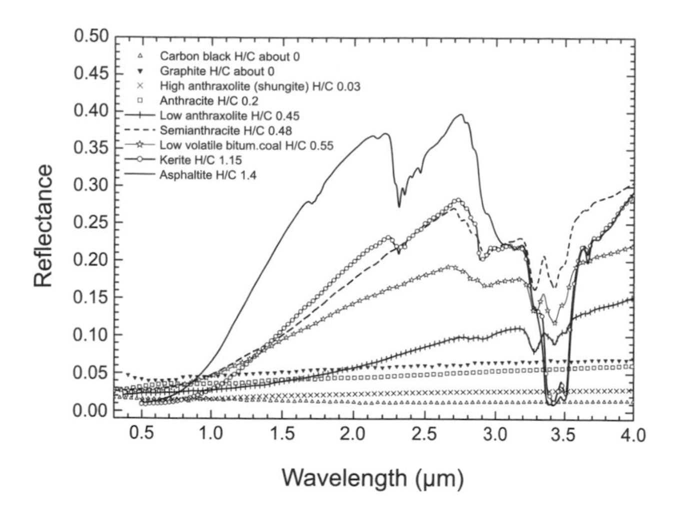

Fig. 3. Reflectance spectra of carbon black (90–125-μm powder); graphite (<45 μm), high rank coals (<100 μm) — low-volatile bituminous, semianthracite, and anthracite; solid oil bitumens (<25 μm) — asphaltite, kerite, low and high anthraxolites. The viewing geometry was biconical for solid bitumens (see *Moroz et al.,* 1998, for details); bidirectional (0.3–2.2 μm) and biconical (2.2–4 μm) for coals (spectra acquired at NASA RELAB facility); hemispherical for graphite (reproduced from the ASTER spectral library); hemispherical (0.3–2.7 μm) and biconical (2.1–4 μm) for carbon black (spectra acquired by R. Clark at the U.S. Geological Survey and J. Salisbury at Johns Hopkins University).

Natural coals of high ranks (low-volatility bituminous coals, semianthracites, anthracites) may also be useful analogs for extraterrestrial complex hydrocarbons (e.g., *Hayatsu et al.,* 1983; *Papoular et al.,* 1991, 1993; *Murae,* 1994; *Guillois et al.,* 1996). Low rank coals show additional NIR absorption features due to their high humidity and abundant mineral impurities (clays and carbonates), as shown in Fig. 4 for a lignite (brown coal) sample. Compared to solid oil bitumens with similar (H + O + N + S)/C ratios, coals are characterized by a higher degree of carbon aromaticity and are enriched in oxygen functionalities. As a result, NIR spectra of red bituminous coals and semianthracites lack absorptions at 2.3 μm and show much weaker absorptions at 3.4 μm compared to red bitumens, e.g., kerites (Fig. 3). Visible and near-IR spectral slopes of powdered carbonaceous materials with relatively high (H + O + N + S)/C ratios show significant dependence on particle size, as demonstrated in Fig. 4 for a lignite and a tholin sample, and in Fig. 2 from *Moroz et al.* (1998) for an asphaltite sample. This effect becomes less pronounced as the (H + O + N + S)/C ratios decrease. Reflectance spectra of some natural coals have been published by *Cloutis et al.* (1994) (0.3–2.6 μm) and *Cloutis* (2003) (0.3–26 μm). A number of authors published diffuse IR reflectance spectra of coal samples diluted in KBr or KCl. Many IR absorbance spectra of high rank natural coals have been published [e.g., *Painter et al.* (1985), 2.5–25 μm], some with astronomical applications [*Papoular et al.* (1991), 3–4 μm; *Guillois et al.* (1996), 3–15 μm; *Sourisseau et al.* (1992), 5–25 μm]. *Papoular et al.* (1991) published emissivity spectra of a semianthracite from 2 to 4.5 μm. Optical constants (n, k) of some coals have been obtained by *Van Krevelen* (1961) and *Papoular et al.* (1993) in the visible and by *Foster and Howarth* (1968) in the IR (1–10 μm).

Graphite — a very stable carbon allotrope composed of flat sheets of C atoms bonded into hexagonal structures — may form at the final stage of thermal evolution of natural organic solids such as coals, kerogens, solid oil bitumens, or meteoritic IOM. Visible and near-IR (0.3–2.6 μm) reflectance spectra of graphites (Fig. 3) have been published and discussed, e.g., by *Cloutis et al.* (1994). Fullerenes (large C molecules, e.g., C_{60}) are similar in structure to graphite but contain pentagonal or heptagonal rings. Fullerenes were detected in carbonaceous meteorites (*Becker et al.,* 1993, 1999; *Pizzarello et al.,* 2001) in low concentration of ~0.1 ppm. Fullerenes and buckyonions (multishell fullerenes) may be responsible for some UV and NIR bands in the ISM spectra (*Iglesias-Groth,* 2004, and references

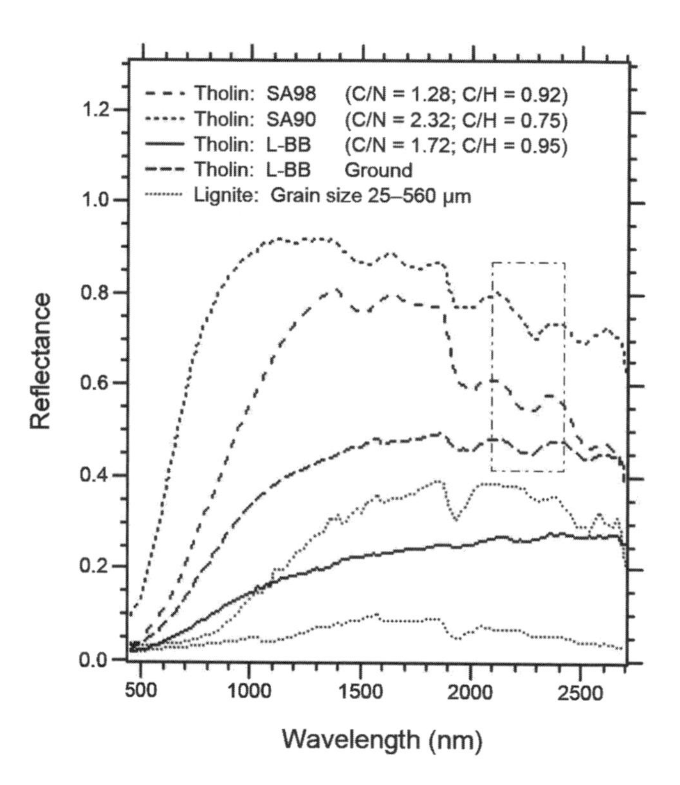

Fig. 4. Bidirectional reflectance spectra of Titan tholins and lignite (PSOC1532 Penn State Data Bank) coals (i = 0°, e = 30°). SA90 and SA98 tholins (two top curves) produced from the PAMPRE experiment (Service d'Aéronomie, Verrières-le-Buisson, France, courtesy of C. Szopa and G. Cernogora) have similar grain size distributions, but different chemical structures (*Szopa et al.,* 2006). The L-BB tholin is described in *Bernard et al.* (2006). This sample has also been ground in order to illustrate the effect of grain size. Note significant grain size effects and differences in the visible reflectance level and in the NIR vibrational bands. In particular, the –CN overtone near 2.2–2.3 μm (enclosed in the box) is useful for discriminating between the samples. Unlike tholins, lignites are black in the visible. Their NIR spectra show a band near 1.9 μm due to physically adsorbed and chemically bound water, and weaker bands between 2.1 and 2.6 μm due to combinations of metal–OH and OH bands in clays (see section 4.1) as well as combinations and overtones of C–H fundamentals.

therein). Optical constants of graphites and transmission spectra of fullerenes have been compiled in the Jena–St. Petersburg Database of Optical Constants (JPDOC) (*http://www.astro.uni-jena.de/Laboratory/Database/jpdoc*) (*Henning et al.,* 1999).

Thus, a review of the data available for various extraterrestrial and terrestrial carbonaceous materials shows a substantial lack of data on optical constants needed for modeling the spectra of TNOs and Centaurs.

4.2. Synthetic Carbonaceous Materials

The commonly used synthetic carbon materials as analogs of extraterrestrial carbonaceous materials are amorphous carbons, the so-called tholins, some disordered carbons (e.g., carbon blacks), and some irradiation products (described in section 4.3). These compounds indeed belong to the extensively wide family of amorphous and disordered carbon, which share numerous physical, electronic, and chemical properties. This section presents an overview of two classes of amorphous carbons of interest for analyzing TNOs spectral observations: pure and hydrogenated amorphous carbons and N_2:CH_4 (Titan and Triton) tholins. We particularly emphasize the physical and chemical control of their optical properties, and their potential use by astronomers.

4.2.1. Control of optical properties. Amorphous carbons and tholins can be considered as amorphous semiconductors. Their optical gap and electronic density of states, and therefore their optical properties in the visible and very-near-IR, are controlled by both the sp^2/sp^3 ratio and sp^2 structure. Note that sp^2 bonding is not restricted to aromatic species as in polyaromatic organics (coals, kerogens, etc.), and may include olefinic carbons in short chains, imides, etc.

The absorption in the visible range consists in a broad continuum, with the imaginary part of the refractive index related to the joint density of states (JDOS), which considers all types of π-π^* electronic transitions (*Knief and von Niessen,* 1999). Therefore, the parameters that control the spectral shape of this continuum are very complex and cannot be restricted to the Tauc or E_{04} gap, nor to the abundance and/or size of aromatic units. This explains why large spectral variations are reported from one compound to another (*Logothedis et al.,* 1995), thus frustrating attempts to derive physical information from the shape and/or slope of reflectance data in the visible.

In contrast, the spectral region from the near-IR to the far-IR is dominated by vibrational bands with no broad continuum. These are the bands that are the most interesting for identifying a specific compound, or to describe the chemical structure. The major limitation of using them comes from the spectral complexity due to the superimposition of numerous bands in complex macromolecular materials.

4.2.2. Amorphous carbons and hydrogenated carbons. Pure and hydrogenated amorphous carbons have been widely used as darkening agents in reflectance modeling of low-albedo solar system objects, and may be considered as analogs of irradiation products. Pure amorphous carbons constitute a wide range of carbon allotropes, which are classified according to their sp^2/sp^3 and their sp^2 structure, defined as the nature and extent of clustering and ordering of the sp^2 carbon bonds (*Ferrari and Robertson,* 2000).

Amorphous carbons are generally subdivided in two forms: (1) amorphous carbons (a-C), which are described as graphite-like sp^2 rich compounds; and (2) tetrahedral amorphous carbons (ta-C), described as sp^3-rich diamond-like carbons (DLC). A wide variety of compounds can be obtained in the laboratory, by using different deposition setups and/or experimental conditions, or by performing heating experiments on various precursor materials (*Beny-Bassez and Rouzaud,* 1985; *Logothetidis et al.,* 1995). Hydrogenated amorphous carbons contain up to 60% H, above which no solid can be formed, and are classified according to the same parameters as amorphous carbons, supplemented by the H/C ratio (*Ferrari and Robertson,* 2000). Introducing hydrogen within the carbon structure leads to the formation of sp^3 CH bonds. The general framework for interpreting their electronic structure and optical properties is similar to that applied to pure amorphous carbons. Note that tholins (see below) formed from CH_4 gas are indeed largely hydrogenated amorphous carbons (e.g., *Khare et al.,* 1987; *Mutsukura and Akita,* 1999), although aromatic and aliphatic structures, both with varying degrees of N substitution, are found in them (e.g., *Imanaka et al.,* 2004; *Bernard et al.,* 2006).

Only a few vibrational bands are present in the IR spectra. In the mid-IR, a broad and intense feature is reported at ~1250 cm^{-1} (~8 μm), associated with a weaker one at ~700 cm^{-1} (~14.3 μm) (*Rodil et al.,* 2001). The intensities of these features are correlated with the sp^2 content, therefore their IR activity has been interpreted by the presence of dynamic charges, allowed by the lack of local symmetry and the extent of conjugated π bonds (*Rodil et al.,* 2001, 2003). No overtones of these features have yet been reported in the near-IR, although they might be expected. No other features are detected in the far-IR for wavenumbers shorter than 400 cm^{-1} (>25 μm). Hydrogenated amorphous carbons furthermore present typical IR features of the CH_2 and CH_3 functional groups, as the well-known features around 2900 cm^{-1} (3.4 μm), plus the aromatic C–H stretching mode at 3040 cm^{-1} (3.29 μm) (e.g., *Schnaiter et al.,* 1999; *Pendleton and Allamandola,* 2002).

Many studies have been focused on the determination of optical properties of amorphous carbons (e.g., *Rouleau and Martin,* 1991; *Compagnini,* 1994; *Roush,* 1995), but few of them provide tabulated values of optical constants (*Arakawa et al.,* 1985; *Preibisch et al.,* 1993; *Zubko et al.,* 1996; *Jäger et al.,* 1998a). A major concern of astronomers is the selection of the most relevant sets of data. Indeed, large variations are reported among optical data from different studies and/or different kinds of carbons. For instance, the imaginary part (k) of the refractive index varies by more than two orders of magnitude within the data from *Arakawa et al.* (1985), *Preibisch et al.* (1993), and *Jäger et al.* (1998a).

Such variations reflect the wide diversity of amorphous carbons, and raise the problem of the selection of datasets.

The lack of systematic studies of the connection between optical properties and structural state of amorphous carbons also complicates the problem. Consequently, it is more prudent to select end members of various sets of data, in order to account for both absorption strength and spectral variations in the visible range. As noted above, no reliable information on surface composition can be derived from the slope of the visible reflectance curve, although the reddest surfaces cannot be explained by plausible minerals and thus require the presence of organic materials (*Cruikshank et al.,* 2001, 2005a,b).

4.2.3. N_2:CH_4 tholins. Tholins are complex organic compounds formed from initial gaseous mixtures exposed to electrical discharge or energetic photons. Various experiment configurations have been dedicated to studies of planetary relevance, including synthesis in cold plasma generated by a radio-frequency source (e.g., *Imanaka et al.,* 2004; *Szopa et al.,* 2006), spark discharge (*Khare et al.,* 1984), direct current discharge (e.g., *Coll et al.,* 1999), or purely ultraviolet light (*Ferris et al.,* 2005; *Tran et al.,* 2003). Various initial gaseous mixtures have been used, including numerous combinations of N_2, NH_3, CH_4, and H_2O. Ice tholins are compounds formed from irradiation of ice samples, and are described below along with other irradiation products. Note that tholins formed from gaseous mixtures are not strictly irradiation products, but are nevertheless used as analogs of irradiated ices present at the surface of TNOs or icy satellites. N_2:CH_4 has been by far the most extensively investigated mixture, because of its interest for simulating the atmospheric aerosols of Titan. Somewhat different are Triton tholins, which are formed from N_2:CH_4 mixtures with low CH_4 concentration (more relevant to Triton's atmosphere), but they are considered here in context with Titan tholins (*McDonald et al.,* 1994). We restrict the discussion in this section to N_2:CH_4 tholins due to the fact that they received much attention and that optical data are available.

N_2:CH_4 tholins are simply "polymer-like" amorphous hydrogenated carbon nitrides (e.g., *Mutsukura and Akita,* 1999; *Bernard et al.,* 2006). The term "tholin" was originally coined specifically to describe the materials synthesized in the context of organic materials in interstellar dust and planetary environments (*Sagan and Khare,* 1979), and is explicitly defined as organic solids produced by the irradiation of mixtures of cosmically abundant reducing gases (*Sagan et al.,* 1984). Compounds of this general type are rather common in materials science, and have been widely investigated using numerous techniques. Generally speaking, the classification scheme used for a-C and a-C:H compounds extends to any form of carbon nitrides, using the C/N ratio and N speciation as key factors controlling the chemical structure and sp^2 clustering. Nitrogen may present various sp^1, sp^2, and sp^3 bonding configurations with carbon, and also bonds with hydrogen as the amine functions $-NH_2$ and $-NH-$. As for other amorphous carbons, the system deposition and experimental parameters play a key role in the structure and composition of the materials. Four general classes of carbon nitrides have been defined for a current usage in materials science (*Rodil et al.,* 2001; *Ferrari et al.,* 2003). N_2:CH_4 tholins do not belong to any of these classes, but as noted above, can be described as a class of "polymer-like" hydrogenated carbon nitrides.

Numerous optical data are available for N_2:CH_4 tholins and in-depth systematic studies have revealed general trends in the control of samples' elemental composition, chemical structure, and optical properties with various experimental parameters: gas composition, pressure, discharge setup, injected power, and temperature (*McDonald et al.,* 1994; *McKay,* 1996; *Imanaka et al.,* 2004; *Bernard et al.,* 2006). The chemical structure of these compounds is not fully elucidated, however. Unlike other amorphous carbon nitrides, N_2:CH_4 tholins contain numerous functional chemical groups, as evidenced by their feature-rich IR spectra (*Imanaka et al.,* 2004; *Bernard et al.,* 2006; *Quirico et al.,* 2006) (see also Fig. 4). Therefore, optical properties in the IR range are controlled by the chemical composition, whereas the absorption in the visible and very near-IR is controlled by the sp^2 structure and the sp^2/sp^3 ratio. These latter are indeed correlated with the C/N ratio, as nitrogen favors sp^2 clustering. Finally, the so-called "HCN polymer" can be discussed with N_2:CH_4 tholins. This compound has been suggested as a widespread nitrogen-rich organic material in the solar system (e.g., *Minard et al.,* 1998). It is indeed an amorphous hydrogenated carbon nitride with a chemical structure similar to, but distinct from, that of N_2:CH_4 tholins (*Quirico et al.,* 2006).

We note that *Khare et al.* (2002) studied the time-dependent composition and structure of a N_2:CH_4 tholin, as well as chemical effects of exposure to oxygen and to laboratory air. Exposure to oxygen and moist air had no effect on the basic composition and structure of tholin, except to produce a weak absorption band attributed to CO_2 (2299–2398 cm^{-1}; 4.17–4.35 μm), and to remove (by oxidation) an aromatic C–H band around 3030 cm^{-1} (3.30 μm). These changes produced negligible differences in the pre- and post-exposure IR spectra, and changes in the optical constants would also therefore be negligible.

Among the sets of optical constants of N_2:CH_4 tholins available in the literature are those of *Khare et al.* (1984), *Ramirez et al.* (2002), *Tran et al.* (2003), and *Imanaka et al.* (2004). Within a series of tholins derived from similar experimental conditions by only varying the cell pressure, *Imanaka et al.* (2004) have shown that systematic variations of the optical constants arise. Indeed, tholins formed in different experimental configurations exhibit variations in their chemical structures, resulting in significant variations in their optical constants (e.g., *Ramirez et al.,* 2002; *Tran et al.,* 2003; *Imanaka et al.,* 2004). Therefore, the range of variations of optical properties of tholins should not be considered as a linear continuum, but rather as sets of data incoming from similar, but distinct, samples and experimental setups. For these reasons, as pointed out above for amorphous carbons, it appears relevant to select end members of available sets for modeling. This is of particular importance for fitting observational data in the visible spectral range. Additional focused efforts are needed to obtain measurements of optical constants along with in-depth structural and chemical characterization of samples of carbon nitrides in order to make significant advances in our knowledge of the chemical and structural control of the optical properties.

Reflectance spectra of tholins and HCN polymer have been published by *Cruikshank et al.* (1991, 1995), *Roush and Dalton* (1994), and *Bernard et al.* (2006) in which a wide variety of overtone and combination bands is observed in the near-IR (see, e.g., Fig. 4). These features, used along with the reflectance in the visible region, are valuable fingerprints of the chemical structure of N-rich organics. The 2.2-μm feature, which is the first overtone of the stretching modes of the –CN and –NC functional groups, may be particularly useful as a clue to the chemical structure. Indeed, the ~4.6-μm fundamental band has been proved as an interesting tracer of the C/N ratio within a series of tholins obtained from the same experimental configuration, and also helps to discriminate tholins from the so-called "HCN polymer" (*Imanaka et al.,* 2004; *Quirico et al.,* 2006). The visible reflectance, which consists mostly of a reddish monotonic curve, may also be useful. There is a rough correlation with the highest reflectance value and the spectral depth and position of the bands in the near-IR. This is evidence for a correlation between the optical gap controlling the visible absorption and the chemical structure and C/N ratio. Finally, the shape and position of the CN/NC 2.2-μm overtone, the absence or presence of CH_2/CH_3 and/or NH/NH_2 chemical function bands, together with the reflectance in the visible, should be considered as valuable fingerprints of "polymer-like," hydrogenated carbon nitrides.

4.3. Irradiation Products

Irradiating C-bearing ices, refractory hydrocarbon solids, and ice-refractory hydrocarbon mixtures with energetic particles or photons leads to their chemical, structural, and optical transformation to refractory carbonaceous materials. CH_4 ice may transform to a-C:H when irradiated by Lyman-α photons (*Dartois et al.,* 2005). Numerous studies have demonstrated the formation of amorphous or disordered carbons from ion irradiation experiments (see chapter by Hudson et al.). Most of these studies, reviewed by Hudson et al., demonstrate that amorphous carbons can be considered as valuable analogs of carbon-bearing irradiation products, although few optical constants are available. The so-called ice tholins are solids produced from icy samples exposed to plasma irradiation (*McDonald et al.,* 1991, 1996; *Khare et al.,* 1993). Different mixtures lead to solids with different chemical composition as evidenced by their IR spectra. These solids exhibit numerous oxygenated chemical functional groups, and they can be used as analogs of irradiation products of hydrocarbon ices containing water, or oxygen-bearing molecules like methanol in planetary conditions. Optical constants (measured at room temperature) are available for one ice tholin obtained from a C_2H_6/H_2O mixture irradiated at 77 K (*Khare et al.,* 1993).

5. MINERALS

At present, information on plausible mineral components of TNOs and Centaurs is based mostly on our knowledge of the composition of cometary dust and "cometary" (anhydrous "chondritic porous") IDPs. These data suggest Mg-rich anhydrous crystalline mafic silicates — forsterite and enstatite — as the most plausible major silicate components, and iron sulfides (e.g., troilite FeS, pyrrhotite $Fe_{1-x}S$) as the most plausible major non-silicate minerals (e.g., review by *Wooden et al.,* 2005, and references therein; *Zolensky et al.,* 2006; chapter by Gounelle et al.). Forsterite is the Mg-member of the olivine series (forsterite Mg_2SiO_4–fayalite Fe_2SiO_4), while enstatite is the Mg-member of low-Ca pyroxene series (enstatite $MgSiO_3$–ferrosilite $FeSiO_3$). Data on Fe^{2+}-bearing olivines and pyroxenes are also reviewed below, since such minerals, which are common components of many meteorites and asteroids, may also be present on TNOs and Centaurs; Fe^{2+}-bearing olivine significantly improved spectral models of Pholus around 1 μm (*Cruikshank et al.,* 1998), and *Lisse et al.* (2006) reported detection of Fe-rich pyroxene (ferrosilite) in Spitzer Space Telescope (SST) spectra of Comet 9P/Tempel 1 acquired during the Deep Impact encounter. In addition, a wide range of olivine and low-Ca pyroxene compositions has been detected in anhydrous IDPs and samples returned by the Stardust mission, although Mg-rich crystalline forsterites and enstatites were the most abundant silicates (*Zolensky et al.,* 2006). Anhydrous crystalline mafic silicates clearly dominate the silicate fraction in solar system materials and protoplanetary disks, while amorphous Fe-bearing silicates are abundant in the ISM, and some fraction of cometary silicates seems to be amorphous according to IR observations (*Wooden et al.,* 2005). Along with crystalline silicates, "cometary" IDPs contain glass with embedded metal and sulfides (GEMS) — spherules of uncertain origin, composed of amorphous Mg, Fe,Al,Si-silicate with embedded iron sulfides and Fe-metal grains (*Dai and Bradley,* 2005). Amorphous GEM-like particles with IR spectra resembling those of the ISM amorphous silicates (*Keller et al.,* 2006) are also present in Stardust mission samples, although it is not yet clear whether any fraction of the detected amorphous silicates existed before collection (*Brownlee et al.,* 2006; *Zolensky et al.,* 2006). In any case, data on amorphous anhydrous minerals are also discussed below.

No unambigous detections of hydrated minerals (e.g., serpentines, smectites) in comets have been reported (*Wooden et al.,* 2005). *Lisse et al.* (2006) reported emission features due to carbonates and hydrated silicates in the SST spectra of Comet 9P/Tempel 1, while none of these minerals have been detected in the samples returned by Stardust (*Brownlee et al.,* 2006; *Zolensky et al.,* 2006). However, such phases are abundant in primitive meteorites and some IDPs (see chapter by Gounelle et al.), and weak features possibly related to phyllosilicates have been reported for several TNOs and Centaurs (see chapter by Barucci et al.), although confirmation is needed. Therefore, we briefly mention some data on hydrated minerals as well.

Difficulties in detecting silicate features in the VNIR spectra of TNOs and Centaurs may be either due to the opacity of associated carbonaceous species and sulfides in this spectral region, or to the lack of Fe^{2+} in the silicates (see below). Recent SST detection of silicates on 8405 Asbolus (see chapter by Barucci et al.) and several Trojans (*Emery et al.,* 2006) shows that the thermal-IR region (TIR) ($\lambda >$ 10 μm) is potentially useful for silicate detection on TNOs and Centaurs, therefore we include TIR data on relevant minerals in this review.

5.1. Crystalline Olivines and Pyroxenes

Mid- and far-IR data on crystalline olivines and pyroxenes of various Fe/(Fe + Mg) contents have been compiled in JPDOC (*Henning et al.,* 1999). Mid-IR optical constants (at room temperature) have been published for crystalline olivines [e.g., *Mukai and Koike* (1990), 7–200 μm; *Fabian et al.* (2001), 8–130 μm], and for crystalline low-Ca pyroxenes [e.g., *Roush et al.* (1991), 5–25 μm; *Jäger et al.* (1998b), 5–100 μm]. *Henning and Mutschke* (1997) published optical constants of a Fe-bearing pyroxene (bronzite) sample between 6.7 and 500 μm as a function of temperature. Olivines show strong diagnostic absorption bands between 9 and 12 μm (Si-O stretches) as well as a number of other bands at longer wavelengths, which shift to shorter wavelengths with decreasing Fe/(Fe + Mg) content (e.g., *Koike et al.,* 2003) and temperature (*Koike et al.,* 2006). Pyroxene Si-O vibrations occur at somewhat shorter wavelengths and pyroxenes show more bands than olivines, especially Mg-rich pyroxenes, most of which shift to shorter wavelengths with decreasing Fe/(Fe + Mg) content, while the 10.5- and 11.5-μm bands shift to longer wavelengths (*Chihara et al.,* 2002).

Infrared reflectance spectra (2–25 μm) of crystalline olivines and pyroxenes of various compositions and several grain size ranges can be found in the ASTER spectral library (*http://speclib.jpl.nasa.gov*). The contrast of diagnostic IR features in reflectance and emittance spectra is strongly affected by grain size (*Lyon,* 1964).

Few laboratory mid-IR emissivity spectra exist for crystalline olivines and pyroxenes: spectra of coarse (710–1000 μm) olivine (forsterite and fayalite) and pyroxene separates from the Arizona State University (ASU) spectral library (*Christensen et al.,* 2000); spectra of forsterite fines (from <5 to 20–25 μm) from *Mustard and Hays* (1997); and spectra of forsterite, diopside, and Fe-bearing enstatite separates (from <25 to 125–250 μm) from the BED library (*Maturilli et al.,* 2007). Note that these measurements were made at ambient pressure, while emissivity spectra may be significantly different at low pressures (*Logan and Hunt,* 1970). Analysis of Spitzer observations of Trojans shows a need for laboratory IR emissivity data on hyperfine silicates (including crystalline and amorphous olivines and pyroxenes) and Fe-free silicates (*Emery et al.,* 2006).

Most observations of TNOs and Centaurs are made in the VNIR spectral range. Near-IR spectra of crystalline olivines are characterized by three overlapping absorptions around 1 μm due to crystal field transitions in the Fe^{2+} ion (*Burns,* 1970). Major Fe^{2+} crystal field absorptions occur in the spectra of pyroxenes at 0.9–1.15 μm (band I) and 1.8–2.3 μm (band II) (*Burns,* 1970; *Adams,* 1974). Wavelength positions and contrasts of Fe^{2+} bands increase with increasing Fe content (olivines and pyroxenes) and Ca content (pyroxenes), while the band widths and relative contrasts in VNIR reflectance spectra are affected not only by composition but also particle size (*Burns,* 1970; *Adams,* 1974; *King and Ridley,* 1987; *Cloutis and Gaffey,* 1991; *Sunshine and Pieters,* 1998). Room-temperature VNIR reflectance spectra (0.3–2.5 μm) of various crystalline olivines and pyroxenes are available in the USGS (*Clark et al.,* 1993) and RELAB (*Pieters and Hiroi,* 2004) spectral libraries and a number of publications. Very few data exist on VNIR optical constants of olivines and pyroxenes. *Lucey* (1998) derived k values from published olivine and pyroxene reflectance spectra. These data extend only to 2.3 or 2.5 μm. Terrestrial mafic minerals often contain some weathering products, fluid inclusions and adsorbed water, which affect the spectra beyond 2.2–2.3 μm and especially around 3 μm. However, modeling of observed TNO and Centaur spectra often requires optical constants covering wider spectral ranges (e.g., *Cruikshank et al.,* 1998). Optical constants from *Huffman and Stapp* (1973) (n: 0.08–8 μm), *Huffman* (1975) (k: 0.03–300 μm), and *Pollack et al.* (1994) (0.1–10^5 μm) are available for crystalline olivines, and only from *Pollack et al.* (1994) (0.1–10^6 μm) for crystalline orthopyroxene. No VNIR optical constants have been published for "featureless" pure Fe-free crystalline forsterites and enstatites, whose VNIR optical properties are dramatically different from those of crystalline Fe-poor forsterites and enstatites (e.g., *Adams,* 1975; *Cloutis et al.,* 1990). No VNIR optical constants have been measured at low T relevant to TNOs and Centaurs, while VNIR reflectance spectra of Fe-bearing crystalline olivines and pyroxenes are known to be affected by temperature (*Singer and Roush,* 1985; *Moroz et al.,* 2000; *Hinrichs and Lucey,* 2002).

5.2. Amorphous Olivines and Pyroxenes

Few data exist on amorphous pyroxenes and olivines, since natural samples are unavailable, and laboratory production is difficult. Amorphized olivines and pyroxenes for optical studies are usually produced by evaporation (*Koike and Tsuchiyama,* 1992), laser ablation (*Scott and Duley,* 1996; *Fabian et al.,* 2000; *Brucato et al.,* 1999, 2002), reactive sputtering (*Day,* 1979, 1981), or irradiation with low-energy ions (*Brucato et al.,* 2004). These publications show IR spectra of amorphous olivines and/or pyroxenes in transmittance or absorbance. Amorphous olivines and pyroxenes show a broad smooth absorption band centered near 10 μm and a weaker broader band around 20 μm. Optical constants of amorphous olivines and pyroxenes have been published by *Day* (1979) (forsterite and enstatite, 7–33 μm), *Day* (1981) (fayalite and ferrosilite, 7–300 μm), *Scott and Duley* (1996) (forsterite and enstatite, probably Fe-bearing, 0.12–17.5 μm), *Henning and Mutschke* (1997) (pyroxene glass, 6.7–500 μm). *Dorschner et al.* (1995) published optical constants of Mg-Fe-silicate glasses of olivine and pyroxene composition between 0.2 and 500 μm. Visible and near-IR optical properties of mafic glasses are strongly affected by cooling rate and oxygen fugacity of the ambient atmosphere. *Dorschner et al.* (1995) produced their glasses in air, so that a large fraction of the Fe^{2+} was oxidized to Fe^{3+}. Oxidation does not significantly affect the mid- and far-IR spectra (*Ossenkopf et al.,* 1992), but Fe^{3+} produces very strong absorption from the UV through the VNIR range, causing darkening and a steep red slope in the VNIR range. Basically, VNIR optical data of acceptable quality on pure amorphous olivines, which could be used in TNO and Centaur spectral models, are currently unavailable. For amorphous pyroxenes in the VNIR spectral range, only Fe-free pyroxene glass from *Dorschner et al.* (1995) is a reasonable simulation of a pure amorphous silicate.

5.3. Hydrated Minerals

The presence of hydrated minerals at the surface of TNOs has been suggested (see chapter by Barucci et al.). Although on Earth there is a vast range of hydrated minerals, not all terrestrial hydrated minerals are relevant to TNOs, which are thought to be largely primitive. Consequently, the most logical minerals to investigate through modeling are those found in pristine chondritic meteorites, IDPs, or the comet grains collected by the Stardust mission. However, no hydrated minerals have yet been detected in the comet samples returned by Stardust, and their absence cannot be explained by dehydration during capture (*Zolensky et al.,* 2006).

A number of hydrated minerals are plausible components of TNOs, such as phyllosilicates, salts, and hydrated oxides. The NIR spectral properties of hydrated minerals are dominated by features due to H_2O (~1.4, ~1.9, ~3 μm) and structural OH. The wavelengths of OH (~1.4, ~2.7 μm) and metal-OH (2.2–2.4 μm) bands in phyllosilicates are controlled by their cationic compositions (e.g., *Bishop et al.,* 2002a,b), resulting in a wide range of possibilities. Unlike ices, the identification of a specific mineral is often uncertain and cannot normally be accomplished with only one series of reference spectra. Reference spectra covering a wide range of compositions within a specific mineral type are essential.

The complicating effects are illustrated by a specific example: Fe^{3+}-bearing phyllosilicates may show absorption features near 0.43 μm and 0.6–0.8 μm, but these features are not diagnostic of hydrated silicates since bands at similar wavelengths are found in reflectance spectra of numerous *anhydrous* minerals. Such features may be assigned to hydrated minerals only if corresponding NIR absorption bands noted above are also present.

Visible and near-IR optical constants of relevant hydrated minerals are largely unavailable. *Bishop et al.* (2002a,b) have investigated the effect of cationic compositions on IR spectra of phyllosilicates, while reflectance spectra of hydrated oxides and sulfates have been published by *Bishop et al.* (1993) and *Cloutis et al.* (2006), respectively, and references for carbonates can be found in *Calvin et al.* (1994).

5.4. Irradiated Silicates

The possible contribution of irradiated silicates to VNIR optical properties of TNOs and Centaurs is an open question. Irradiation with low-energy solar wind plasma may darken and redden Fe-bearing silicates. Sputtering of Fe from silicate targets and its redeposition, causing darkening and reddening in experiments with high ion fluences (reviewed by *Hapke,* 2001), may be negligible for TNOs and Centaurs. Ion irradiation experiments on silicates, performed by the Catania group at low ion fluences (see chapter by Hudson et al.), where darkening and reddening are caused by radiation damage induced by elastic collisions in a thin (<1 μm) surface layer of silicates, seem to be more relevant. It is unclear whether optical effects of irradiation are negligible in the case of Fe-free silicates, since the possible role of Fe in the latter experiments remains to be clarified. Available data show that even very Fe-poor silicate samples darken and redden under irradiation, but no experiments have been performed on pure Fe-free silicate targets. Another question concerns the relative rates of irradiation-induced silicate darkening and reddening compared to the rates of competing regolith gardening on TNOs and Centaurs. Before we know the answers to these questions, we may consider irradiated silicates as possible end members in some TNOs/Centaur spectral models and simulate space-weathering effects by incorporation of fine Fe inclusions into silicate grains.

5.5. Iron Sulfides

Iron sulfides, sometimes enriched in Ni, such as pyrrhotite and troilite, are common opaque components of meteorites, IDPs, and cometary dust (*Jessberger et al.,* 1988; *Zolensky and Thomas,* 1995; *Schulze et al.,* 1997; *Zolensky et al.,* 2006). Infrared spectra of iron sulfides show a characteristic feature at ~23–24 μm (*Keller et al.,* 2002) and several weaker features at longer wavelengths (*Hony et al.,* 2002; *Kimura et al.,* 2005). Similar bands have been detected in IR spectra of planetary nebulae and C-rich stars (*Forrest et al.,* 1981) and assigned to sulfides (*Keller et al.,* 2002; *Hony et al.,* 2002). *Henning and Mutschke* (1997) published n and k of FeS from 10 to 500 μm. At wavelengths <23 μm iron sulfides are featureless, except for OH-bearing tochilinites, which are the most abundant sulfide phases in CM2 chondrites and show features near 2.7–2.8 μm due to O-H stretches (*Moroz et al.,* 2006). Visible and near-IR reflectance spectra of relevant sulfide powders have been published by *Britt et al.* (1992) (troilite), *Cloutis and Gaffey* (1993) (troilite, pyrrhotite), and *Moroz et al.* (2006) (tochilinite). Like other opaque semimetals, sulfide powders become darker with decreasing grain size. Troilite and pyrrhotite powders containing coarse grains show reddish slopes in reflectance spectra and may redden mixtures with other phases if present in high quantities. Fine (submicrometer) Fe sulfide grains do not redden but effectively darken a mixture with bright phases and suppress their VNIR absorption features. For example, in CM2 chondrites sulfide fines appear to be the most important darkening phases. *Egan and Hilgeman* (1977) published optical constants of troilite and pyrrhotite between 0.3 and 1.1 μm. *Pollack et al.* (1994) reported n and k of troilite from 0.1–10^5 μm, but the values between 1.1 and 3 μm and below 0.3 μm were estimated by extrapolation. New values of n and k of troilite between 0.1 and 10^5 μm, measured by the Jena group, can be found at *www.mpia-hd.mpg.de/homes/henning/Dust_opacities/Opacities/RI/troilitek.lnk.*

5.6. Iron-Nickel Metal

Metallic Fe or Fe,Ni alloys are usually only minor components of carbonaceous chondrites, IDPs, and comets, but tiny Fe metal particles may be products of space weathering of Fe-bearing silicates and therefore the data on iron metals are briefly summarized here. Similar to other metals, metallic iron is opaque and shows continuous featureless absorption from UV to FIR. Optical constants for various spectral ranges are summarized in the JPDOC database (*Henning et al.,* 1999). Visible and near-IR reflectance spectra of Fe,Ni from meteorites show an increase in reflectance as a function of wavelength (e.g., *Gaffey,* 1976; *Cloutis and Gaffey,* 1993). Fine metallic iron particles cause darkening and reddening of "space-weathered" Fe-bearing silicates and suppress absorption bands in their VNIR spectra (*Hapke,* 2001, and references therein).

6. CONCLUSION

We have reviewed the available laboratory data on detected or plausible TNOs and Centaurs surface materials. For each class of material (ices, refractory carbonaceous compounds, and minerals) we have indicated the laboratory data that are still missing. As we have seen, in many cases the materials of interest have been studied in some details only in the thermal-IR, while the near-IR and visible ranges are the most important ranges for TNOs and Centaurs. Furthermore, concerning the optical constants, which are required for a proper interpretation of the spectra, only very few are available for materials and conditions relevant to TNOs and Centaurs. In addition, very few studies of ice mixtures, and almost no study of mixtures of different classes of materials, have been made in the near-IR.

Concerning carbonaceous compounds, it is clear that the spectroscopic studies carried out so far do not cover the full range of possibilities. Other compounds should be studied in the VNIR, e.g., other carbon nitrides as well as other photolysis or radiolysis organic products (see also the chapter by Hudson et al.).

As we have noted, spectral slopes do not provide strong constraints on the surface composition because widely different classes of compounds can provide similar slopes (and shapes). However, investigators have been unable to find components other than complex refractory carbonaceous species to account for the very red colors of some TNOs. In particular, none of the plausible silicates that have been considered are red enough. Some classes of carbonaceous compounds with detectable features in their near-IR spectra, very often in the 2–2.5-μm region, have been noted. This is the case, for instance, of solid bitumens with relatively high H/C ratios, coals of not too high ranks, possibly relatively immature kerogens, and hydrogenated amorphous carbons with high H contents. In order to identify (or reject) these materials on TNOs, we therefore need spectra of higher signal precision (S/N), particularly in the spectral region 2–2.5 μm, than presently exist.

Another concern is the availability of the numerical files of the laboratory spectra and optical constants that have been obtained. Although a number of databases can be found on various websites (and these have been mentioned in the text), the existing databases are clearly insufficient. This is especially true for ices and organics. A major effort toward this aim, a website of spectroscopic data (transmission and reflectance spectra, optical constants, lists of absorption bands and attributions, etc.) of solid materials of planetary interest, including ices and organics, is currently under development at the Laboratoire de Planétologie de Grenoble (LPG) in France, and should begin providing data in 2008. Furthermore, recently acquired spectra of a number of hydrocarbons and other organic compounds will soon become available in digital form through the U.S. Geological Survey Spectroscopy Laboratory (R. N. Clark, personal communication). We encourage other people to do the same.

All these laboratory studies require important investments in laboratory equipment and the time of dedicated researchers. While new laboratory data are slow to materialize, thanks to new instrumentation, improved and more numerous data are becoming available for TNOs and Centaurs, showing their great diversity and the "richness" of this population. Furthermore, with the New Horizons mission, spatially resolved IR spectra of Pluto should become available in 2015. The spacecraft is expected to continue on to one or two TNOs, offering the possibility of spatially resolved spectra of these objects. It is therefore essential to carry out in parallel an ambitious program of laboratory spectroscopy.

REFERENCES

Adams J. B. (1974) Visible and near-infrared diffuse reflectance spectra of pyroxenes as applied to remote sensing of solid objects in the solar system. *J. Geophys. Res., 79,* 4829–4836.
Adams J. B. (1975) Interpretation of visible and near-infrared

diffuse reflectance spectra of pyroxenes and other rock-forming minerals. In *Infrared and Raman Spectroscopy of Lunar and Terrestrial Minerals* (C. Karr, ed.), pp. 91–116. Academic, New York.

Alexander C. M. O'D., Russell S. S., Arden J. W., Ash R. D., Grady M. M., and Pillinger C. T. (1998) The origin of chondritic macromolecular organic matter: A carbon and nitrogen isotope study. *Meteoritics & Planet. Sci., 33,* 603–622.

Arakawa E. T., Dolfini S. M., Ashley J. C., and Williams M. W. (1985) Arc-evaporated carbon films: Optical properties and electron properties and electron mean free paths. *Phys. Rev. B, 31,* 8097–8101.

Becker L., McDonald G. D., and Bada J. L. (1993) Carbon onions in meteorites. *Nature, 361,* 595.

Becker L., Bunch T. E., and Allamandola L. J. (1999) Higher fullerenes in the Allende meteorite. *Nature, 400,* 227–228.

Beny-Bassez C. and Rouzaud J-N. (1985) Characterization of carbonaceous materials by correlated electron and optical microscopy and Raman spectroscopy. *Scanning Electron Microscopy, 1,* 119–132.

Bernard J-M., Quirico E., Brissaud O., Montagnac G., Reynard B., McMillan P., Coll P., Nguyen M-J., Raulin F., and Schmitt B. (2006) Reflectance spectra and chemical structure of Titan's tholins. Application to the analysis of Cassini-Huygens observations. *Icarus, 185,* 301–307.

Bernstein M. P. and Sandford S. A. (1999) Variations in the strength of the infrared forbidden 2328.2 cm^{-1} fundamental of solid N_2 in binary mixtures. *Spectrochim. Acta, Part A, 55,* 2455–2466.

Bernstein M. P., Cruikshank D. P., and Sandford S. A. (2005) Near-infrared laboratory spectra of solid H_2O/CO_2 and CH_3OH/CO_2 ice mixtures. *Icarus, 179,* 527–534.

Bernstein M. P., Cruikshank D. P., and Sandford S. A. (2006) Near-infrared spectra of laboratory H_2O-CH_4 ice mixtures. *Icarus, 181,* 302–308.

Bertie J. E. and Devlin J. P. (1984) The infrared spectra and phase transitions of pure and isotopically impure $2ND_3.H_2O$, $2NH_3.D_2O$, $2NH_3.H_2O$, and $2ND_3.D_2O$ between 100 and 15 K. *J. Chem. Phys., 81,* 1559–1572.

Bertie J. E. and Shehata M. R. (1984) Ammonia dihydrate: Preparation, X-ray powder diffraction pattern and infrared spectrum of $NH_3.2H_2O$ at 100 K. *J. Chem. Phys., 81,* 27–30.

Bertie J. E. and Shehata M. R. (1985) The infrared spectra of $NH_3.H_2O$ and $ND_3.D_2O$ at 100 K. *J. Chem. Phys., 83,* 1449–1456.

Binet L., Gourier D., Derenne S., and Robert F. (2002) Heterogeneous distribution of paramagnetic radicals in insoluble organic matter from the Orgueil and Murchison meteorites. *Geochim. Cosmochim. Acta, 66,* 4177–4186.

Binet, L., Gourier D., Derenne S., Robert F., and Ciofini I. (2004) Occurence of abundant diradicaloid moieties in the insoluble organic matter from the Orgueil and Murchison meteorites: A fingerprint of its extraterrestrial origin? *Geochim. Cosmochim. Acta, 68,* 881–891.

Bishop J. L., Pieters C. M., and Burns R. G. (1993) Reflectance and Mössbauer spectroscopy of ferrihydrite-montmorillonite as Mars soil analog materials. *Geochim. Cosmochim. Acta, 57,* 4583–4595.

Bishop J., Madejova J., Komadel P., and Fröschl H. (2002a) The influence of structural Fe, Al and Mg on the infrared OH bands in spectra of dioctahedral smectites. *Clay Minerals, 37,* 607–616.

Bishop J., Murad E., and Dyar M. D. (2002b) The influence of octahedral and tetrahedral cation substitution on the structure of smectites and serpentines as observed through infrared spectroscopy. *Clay Minerals, 37,* 617–628.

Bohn R. B., Sandford S. A., Allamandola L. J., and Cruikshank D. P. (1994) Infrared spectroscopy of Triton and Pluto ice analogs: The case for saturated hydrocarbons. *Icarus, 111,* 151–173.

Bonal L., Quirico E., Bourot-Denise M., and Montagnac G. (2006) Determination of the petrologic type of CV3 chondrites by Raman spectroscopy of included organic matter. *Geochim. Cosmochim. Acta, 70,* 1849–1863.

Britt D. T., Bell J. F., Haack H., and Scott E. R. D. (1992) The reflectance spectrum of troilite (abstract). In *Lunar and Planetary Science XXIII,* pp. 167–168. Lunar and Planetary Institute, Houston.

Brown R. H. and Cruikshank D. P. (1997) Determination of the composition and state of icy surfaces in the outer solar system. *Annu. Rev. Earth Planet. Sci., 25,* 243–277.

Brown R. H., Cruikshank D. P., Tokunaga A. T., Smith R. G., and Clark R. N. (1988) Search for volatiles on icy satellites. I. — Europa. *Icarus, 74,* 262–271.

Brownlee D. E., Joswiak D. J., Bradley J. P., Gezo J. C., and Hill H. G. M. (2000) Spatially resolved acid dissolution of IDPs: The state of carbon and the abundance of diamonds in the dust (abstract). In *Lunar and Planetary Science XXXI,* Abstract #1921. Lunar and Planetary Institute, Houston (CD-ROM).

Brownlee D. E., Tsou P., Aléon J., Alexander C. M. O'D., Araki T., Bajt S., Baratta G. A., Bastien R., Bland P., Bleuet P., et al. (2006) Comet 81P/Wild 2 under a microscope. *Science, 314,* 1711–1716.

Brucato J. R., Colangeli L., Mennella V., Palumbo P., and Bussoletti E. (1999) Mid-infrared spectral evolution of thermally annealed amorphous pyroxene. *Astron. Astrophys., 348,* 1012–1019.

Brucato J. R., Mennella V., Colangeli L., Rotundi A., and Palumbo P. (2002) Production and processing of silicates in laboratory and in space. *Planet. Space. Sci., 50,* 829–837.

Brucato J. R., Strazzulla G., Baratta G., and Colangeli L. (2004) Forsterite amorphisation by ion irradiation: Monitoring by infrared spectroscopy. *Astron. Astrophys., 413,* 395–401.

Brunetto R., Baratta G. A., Domingo M., and Strazzulla G. (2005) Reflectance and transmittance spectra (2.2–2.4 μm) of ion irradiated frozen methanol. *Icarus, 175,* 226–232.

Burns R. G. (1970) *Mineralogical Applications to Crystal Field Theory.* Cambridge Univ., New York.

Calvin W. M. (1990) Additions and corrections to the absorption coefficients of CO_2 ice: Applications to the martian south polar cap. *J. Geophys. Res. (B), 95,* 14743–14750.

Calvin W. M., King T. V. V., and Clark R. N. (1994) Hydrous carbonates on Mars? Evidence from Mariner 6/7 infrared spectrometer and ground-based telescopic spectra. *J. Geophys. Res. (E), 99,* 14659–14675.

Chihara H., Koike C., Tsuchiyama A., Tachibana S., and Sakamoto D. (2002) Compositional dependence of infrared absorption spectra of crystalline silicates. I. Mg-Fe pyroxenes. *Astron. Astrophys., 391,* 267–273.

Christensen P. R., Bandfield J. L., Hamilton V. E., Howard D. A, Lane M. D., Piatek J. L., Ruff S. W., and Stefanov W. L. (2000) A thermal emission spectral library of rock-forming minerals. *J. Geophys. Res. (E), 105,* 9735–9740.

Clark R. N. and Roush T. L. (1984) Reflectance spectroscopy: Quantitative analysis techniques for remote sensing applications. *J. Geophys. Res., 89,* 6329–6340.

Clark R. N., Swayze G. A., Gallagher A. J., King T. V. V., and Calvin W. M (1993) *The U.S. Geological Survey, Digital Spectral Library: Version 1: 0.2 to 3.0 microns.* U.S. Geological Survey Open File Report 93–592, 1340 pp. Available online at *http://speclab.cr.usgs.gov*.

Cloutis E. A. (2003) Quantitative characterization of coal properties using bidirectional diffuse reflectance spectroscopy. *Fuel, 82,* 2239–2254.

Cloutis E. A. and Gaffey M. J. (1991) Pyroxene spectroscopy revisited: Spectral-compositional correlations and relationship to geothermometry. *J. Geophys. Res., 96,* 22809–22826.

Cloutis E. A. and Gaffey M. J. (1993) Accessory phases in aubrites: Spectral properties and implications for asteroid 44 Nysa. *Earth Moon Planets, 63,* 227–243.

Cloutis E. A., Gaffey M. J., Smith D. G. W., and Lambert R. St. J. (1990) Reflectance spectra of 'featureless' materials and the surface mineralogies of M- and E-class asteroids. *J. Geophys. Res., 95,* 281–293.

Cloutis E. A., Gaffey M. J., and Moslow T. F. (1994) Spectral reflectance properties of carbon-bearing materials. *Icarus, 107,* 276–287.

Cloutis E. A., Hawthorne F. C., Mertzman S. A., Krenn K., Craig M. A., Marcino D., Methot M., Strong J., Mustard J. F., Blaney D. L., Bell J. F. III, and Vilas F. (2006) Detection and discrimination of sulfate minerals using reflectance spectroscopy. *Icarus, 184,* 121–157.

Cody G. D. and Alexander C. M. O'D. (2005) NMR studies of chemical structural variation of insoluble organic matter from different carbonaceous chondrite groups. *Geochim. Cosmochim. Acta, 69,* 1085–1097.

Cody G. D., Alexander C. M. O'D., and Tera F. (2002) Solid-state (^{1}H and ^{13}C) nuclear magnetic resonance spectroscopy of insoluble organic residue in the Murchison meteorite: A self-consistent quantitative analysis. *Geochim. Cosmochim. Acta, 66,* 1851–1865.

Coll P., Coscia D., Smith N., Gazeau M.-C., Ramirez S. I., Cernogora G., Israel G., and Raulin F. (1999) Experimental laboratory simulation of Titan's atmosphere: Aerosols and gas phase. *Planet. Space Sci., 47,* 1331–1340.

Compagnini G. (1994) Optical constants of hydrogenated and unhydrogenated amorphous carbon in the 0.5–12 eV range. *Appl. Opt., 33,* 7377–7381.

Curchin J., Clark R. N., and Hoefen T. M. (2006) Cryogenic infrared reflectance spectra of organic ices and their relevance to the surface composition of Titan (abstract). *Bull. Am. Astron. Soc., 38,* 586–587.

Cruikshank D. P., Allamandola L. J., Hartmann W. K., Tholen D. J., Brown R. H., Matthews C. N., and Bell J. F. (1991) Solid CN bearing material on outer solar system bodies. *Icarus, 94,* 345–353.

Cruikshank D. P., Imanaka H., and Dalle Ore C. M. (1995) Tholins as coloring agents on outer solar system bodies. *Adv. Space Res., 36,* 178–183.

Cruikshank D. P., Roush T. L., Bartholomew M. J., Geballe T. R., Pendleton Y. J., White S. M., Bell J. F., Davies J. K., Owen T. C., de Bergh C., and 5 colleagues (1998) The composition of Centaur 5145 Pholus. *Icarus, 135,* 389–407.

Cruikshank D. P., Schmitt B., Roush T. L., Owen T. C., Quirico E., Geballe T. R., de Bergh C., Bartholomew M. J., Dalle Ore C., Douté S., and Meier R. (2000) Water ice on Triton. *Icarus, 147,* 309–316.

Cruikshank D. P., Dalle Ore C. M., Roush T. L., Geballe T. R., Owen T. C., de Bergh C., Cash M. D., Hartmann W. K., et al. (2001) Constraints on the composition of Trojan asteroid 624 Hektor. *Icarus, 153,* 348–360.

Cruikshank D. P., Owen T. C., Dalle Ore C., Geballe T. R., Roush T. L., de Bergh C., Sandford S. A., Poulet F., Benedix G. K., and Emery J. P. (2005a) A spectroscopic study of the surfaces of Saturn's large satellites: H_2O ice, tholins, and minor constituents. *Icarus, 175,* 268–283.

Cruikshank D. P., Imanaka H., and Dalle Ore C. M. (2005b) Tholins as coloring agents on outer solar system bodies. *Adv. Space Res., 36,* 178–183.

Dai Z. R. and Bradley J. P. (2005) Origin and properties of GEMS (glass with embedded metal and sulfides). In *Chondrites and the Protoplanetary Disk* (A. N. Krot et al., eds.), pp. 774–808. ASP Conf. Series 341, San Francisco.

Dartois E., Munoz Caro G. M., Deboffle D., Montagnac G., and d'Hendecourt L. (2005) Ultraviolet photoproduction of ISM dust. Laboratory characterisation and astrophysical relevance. *Astron. Astrophys., 432,* 895–908.

Day K. L. (1979) Mid-infrared optical properties of vapor-condensed magnesium silicates. *Astrophys. J., 234,* 158–161.

Day K. L. (1981) Infrared extinction of amorphous iron silicates. *Astrophys. J., 246,* 110–112.

Dello Russo N. and Khanna R. K. (1996) Laboratory infrared spectroscopic studies of crystalline nitriles with relevance to outer planetary systems. *Icarus, 123,* 366–395.

D'Hendecourt L. B.. and Allamandola L. J. (1986) Time dependent chemistry in dense molecular clouds. III. Infrared band cross sections of molecules in the solid state at 10 K. *Astron. Astrophys. Suppl. Ser., 64,* 453–467.

Dorschner J., Begemann B., Henning T., Jäger C., and Mutschke H. (1995) Steps toward interstellar silicate mineralogy. II. Study of Mg-Fe-silicate glasses of variable composition. *Astron. Astrophys., 300,* 503–520.

Douté S., Schmitt B., Quirico E., Owen T. C., Cruikshank D. P., de Bergh C., Geballe T. R., and Roush T. L. (1999) Evidence for methane segregation at the surface of Pluto. *Icarus, 142,* 421–444.

Douté S., Schmitt B., Langevin Y., Bibring J.-P., Altieri F., Bellucci G., Gondet B., Poulet F., and the MEX Team (2007) South pole of Mars: Nature and composition of the icy terrains from Mars Express OMEGA observations. *Planet. Space Sci., 55,* 113–133.

Durand B. (1980) Sedimentary organic matter and kerogen. Definition and qualitative importance of kerogen. In *Kerogen — Insoluble Organic Matter from Sedimentary Rocks* (B. Durand, ed.), pp. 13–34. Editions Technip, Paris.

Egan G. W. and Hilgeman T. (1977) The rings of Saturn — A frost-coated semiconductor. *Icarus, 30,* 413–421.

Ehrenfreund P., Robert F., D'Hendencourt L., and Behar F. (1991) Comparison of interstellar and meteoritic organic matter at 3.4 microns. *Astron. Astrophys., 252,* 712–717.

Emery J. P., Cruikshank D. P., and van Cleve J. (2006) Thermal emission spectroscopy (5.2–38 μm) of three Trojan asteroids with the Spitzer Space Telescope: Detection of fine-grained silicates. *Icarus, 182,* 496–512.

Espitalié J., Durand B., Roussel J. C., and Souron C. (1973) Etude de la matière organique insoluble (kérogène) des argiles du Toarcien du bassin de Paris. *Rev. Inst. Franç. du Pétrole, 28,* 37–66.

Fabian D., Jäger C., Henning Th., Dorschner J., and Mutschke H. (2000) Steps toward interstellar silicate mineralogy. V. Thermal evolution of amorphous magnesium silicates and silica. *Astron. Astrophys., 364,* 282–292.

Fabian D., Henning Th., Jäger C., Mutschke H., Dorschner J., and Wehrhan O. (2001) Steps toward interstellar silicate mineralogy. VI. Dependence of crystalline olivine IR spectra on iron content and particle shape. *Astron. Astrophys., 378,* 228–238.
Ferrari A. C. and Robertson J. (2000) Interpretation of Raman spectra of disordered and amorphous carbon. *Phys. Rev. B, 61,* 14095–14107.
Ferrari A. C., Rodil S. E., and Robertson J. (2003) Interpretation of infrared and Raman spectra of amorphous carbon nitrides. *Phys. Rev. B, 67,* 155306.
Ferris J., Tran B., Joseph J., Vuitton V., Briggs R., and Force M. (2005) The role of photochemistry in Titan's atmospheric chemistry. *Adv. Space Res., 36,* 251–257.
Fink U. and Sill G. T. (1982) The infrared spectral properties of frozen volatiles. In *Comets* (L. L. Wilkening, ed.), pp. 164–202. Univ. of Arizona, Tucson.
Flynn G. J., Keller L. P., Feser M., Wirick S., and Jacobsen C. (2003) The origin of organic matter in the solar system: Evidence from the interplanetary dust particles. *Geochim. Cosmochim. Acta, 67,* 4791–4806.
Flynn G. J., Keller L. P., Jacobsen C., and Wirick S. (2004) An assessment of the amount and types of organic matter contributed to the Earth by interplanetary dust. *Adv. Space Res., 33,* 57–66.
Forrest W. J., Houck J. R., and McCarthy J. F. (1981) A far-infrared emission feature in carbon-rich stars and planetary nebulae. *Astrophys. J., 248,* 195–200.
Foster P. J. and Howarth C. R. (1968) Optical constants of carbons and coals in the infrared. *Carbon, 6,* 719–729.
Gaffey M. J. (1976) Spectral reflectance characteristics of the meteorite classes. *J. Geophys. Res., 81,* 905–920.
Ganz H. and Kalkreuth W. (1987) Application of infrared spectroscopy to the classification of kerogen-types and the evaluation of source rock and oil shale potentials. *Fuel, 66,* 708–711.
Gardinier A., Derenne S., Robert F., Behar F., Largeau C., and Maquet J. (2000) Solid state CP/MAS ^{13}C NMR of the insoluble organic matter of the Orgueil and Murchison meteorites: Quantitative study. *Earth Planet. Sci. Lett., 184,* 9–21.
Gerakines P. A., Bray J. J., Davis A., and Richey C. R. (2005) The strengths of near-infrared absorption features relevant to interstellar and planetary ices. *Astrophys. J., 620,* 1140–1150.
Goodman M. A., Sweany R. L., and Flurry R. L. Jr. (1983) Infrared spectra of matrix-isolated, crystalline solid, and gas phase C_3-C_6 n-alkanes. *J. Phys. Chem., 87,* 1753–1757.
Green J. R., Brown R. H., Cruikshank D. P., and Anicich V. (1991) The absorption coefficient of nitrogen with application to Triton (abstract). *Bull. Am. Astron. Soc., 23,* 1208.
Grundy W. M. and Schmitt B. (1998) The temperature-dependent near-infrared absorption spectrum of hexagonal H_2O ice. *J. Geophys. Res. (E), 103,* 25809–25822.
Grundy W. M., Schmitt B., and Quirico E. (1993) The temperature dependent spectra of α and β nitrogen ice with application to Triton. *Icarus, 105,* 254–258.
Grundy W. M., Schmitt B., and Quirico E. (2002) The temperature-dependent spectrum of methane ice I between 0.7 and 5 μm and opportunities for near-infrared thermometry. *Icarus, 155,* 486–496.
Guillois O., Nenner I., Papoular R., and Reynaud C. (1996) Coal models for the infrared emission spectra of proto-planetary nebulae. *Astrophys. J., 464,* 810–817.
Halbout J., Robert F., and Javoy M. (1990) Hydrogen and oxygen isotope compositions in kerogen from the Orgueil meteorite — Clues to a solar origin. *Geochim. Cosmochim. Acta, 54,* 1453–1462.
Hansen G. B. (1996) The infrared absorption spectrum of carbon dioxide ice. Ph.D. thesis, Univ. of Washington.
Hansen G. B. (1997) The infrared absorption spectrum of carbon dioxide ice from 1.8 to 333 μm. *J. Geophys. Res. (E), 102,* 21569–21588.
Hansen G. B. (2005) Ultraviolet to near-infrared absorption spectrum of carbon dioxide ice from 0.174 to 1.8 μm. *J. Geophys. Res. (E), 110,* E11003.
Hapke B. (2001) Space weathering from Mercury to the asteroid belt. *J. Geophys. Res. (E), 106,* 10039–10074.
Hapke B. and Wells E. (1981) Bidirectional reflectance spectroscopy. 2. Experiments and observations. *J. Geophys. Res., 96,* 3055–3060.
Harvey K. B. and Ogilvie J. F. (1962) Infrared absorption of formaldehyde at low temperatures: Evidence for multiple trapping sites in an argon matrix. *Can. J. Chem., 40,* 85–91.
Hayatsu R. and Anders E. (1981) Organic compounds in meteorites and their origins. *Topics in Current Chemistry, 99,* 1–37.
Hayatsu R., Matsuoka S., Anders E., Scott R. G., and Studier M. H. (1977) Origin of organic matter in the early solar system. VII — The organic polymer in carbonaceous chondrites. *Geochim. Cosmochim. Acta, 41,* 1325–1339.
Hayatsu R., Scott R. G., and Winans R. E. (1983) Comparative structural study of meteoritic polymer with terrestrial geopolymers coal and kerogen. *Meteoritics, 18,* 310.
Henning Th. and Mutschke H. (1997) Low-temperature infrared properties of cosmic dust analogues. *Astron. Astrophys., 327,* 743–754.
Henning Th., Il'in V. B., Krivova N. A., Michel B., and Voshchinnikov N. V. (1999) WWW database of optical constants for astronomy. *Astron. Astrophys. Suppl. Ser., 136,* 405–406.
Hinrichs J. L. and Lucey P. G. (2002) Temperature-dependent near-infrared spectral properties of minerals, meteorites, and lunar soil. *Icarus, 155,* 169–180.
Hony S., Bouwman J., Keller L. P., and Waters L. B. F. M. (2002) The detection of iron sulfides in planetary nebulae. *Astron. Astrophys., 393,* L103–L106.
Hudgins D. M., Sandford S. A., Allamandola L. J., and Tielens A. G. G. M. (1993) Mid- and far-infrared spectroscopy of ices: Optical constants and integrated absorbances. *Astrophys. J. Suppl. Ser., 86,* 713–870.
Hudson R. L. and Moore M. H. (1993) Far-infrared investigations of a methanol clathrate hydrate — Implications for astronomical observations. *Astrophys. J., 404,* L29–L32.
Huffman D. R. (1975) Optical properties of particulates. *Astrophys. Space Sci., 34,* 175–184.
Huffman D. R. and Stapp J. L. (1973) Optical measurements on solids of possible interstellar importance. In *Interstellar Dust and Related Topics* (J. M. Greenberg and H. van der Hulst, eds.), pp. 297–302. IAU Symposium No. 52, Reidel, Dordrecht.
Huss G. R., Meshik A. P., Smith J. B., and Hohenberg C. M. (2003) Presolar diamond, silicon carbide, and graphite in carbonaceous chondrites: Implications for thermal processing in the solar nebula. *Geochim. Cosmochim. Acta, 67,* 4823–4848.
Iglesias-Groth S. (2004) Fullerenes and buckyonions in the interstellar medium. *Astrophys. J. Lett., 608,* L37–L40.
Imanaka H., Khare B. N., Elsila J. E., Bakes E. L. O., McKay C. P., Cruikshank D. P., Sugita S., Matsui T., and Zare R. N. (2004) Laboratory experiments of Titan tholin formed in cold plasma at various pressures: Implications for nitrogen-containing polycyclic aromatic compounds in Titan haze. *Icarus, 168,* 344–366.
Jäger C., Mutschke H., and Henning Th. (1998a) Optical proper-

ties of carbonaceous dust analogues. *Astron. Astrophys., 332,* 291–299.

Jäger C., Molster F. J., Dorschner J., Henning Th., Mutschke H., and Waters L. B. F. M. (1998b) Steps toward interstellar silicate mineralogy. IV. The crystalline revolution. *Astron. Astrophys., 339,* 904–916.

Jessberger E. K., Christoforidis A., and Kissel J. (1988) Aspects of the major element composition of Halley's dust. *Nature, 332,* 691–695.

Keller L. P., Thomas K. L., and McKay D. S. (1994) Carbon in primitive interplanetary dust particles. In *Analysis of Interplanetary Dust* (G. J. Flynn et al., eds.), pp. 159–164. AIP Conf. Proc. 310, American Institute of Physics, New York.

Keller L. P., Hony S., Bradley J. P., Molster F. J., Waters L. B. F. M., Bouwman J., de Koter A., Brownlee D. E., Flynn G. J., Henning Th., and Mutschke H. (2002) Identification of iron sulphide grains in protoplanetary disks. *Nature, 417,* 148–150.

Keller L. P., Bajt S., Baratta G., Borg J., Bradley J. P., Brownlee D. E., Busemann H., Brucato J. R., Burchell M., Colangeli L., et al. (2006) Infrared spectroscopy of Comet 81P/Wild 2 samples returned by Stardust. *Science, 314,* 1728–1731.

Kerridge J. F. (1999) Formation and processing of organics in the early solar system. *Space Sci. Rev., 90,* 275–288.

Khanna R. K., Ospina M. J., and Zhao G. (1988) Infrared band extinctions and complex refractive indices of crystalline C_2H_2 and C_4H_2. *Icarus, 74,* 527–535.

Khare B. N., Sagan C., Arakawa E. T., Suits F., Callcott T. A., and Williams M. W. (1984) Optical constants of organic tholins produced in a simulated titanian atmosphere: From soft X-ray to microwave frequencies. *Icarus, 60,* 127–137.

Khare B. N., Sagan C., Thompson W. R., Arakawa E. T., and Votaw P. (1987) Solid hydrocarbon aerosols produced in simulated Uranian and Neptunian stratospheres. *J. Geophys. Res., 92,* 15067–15082.

Khare B. N., Thompson W. R., Sagan C., Arakawa E. T., Bruel C., Judish J. P., Khanna R. K., and Pollack J. B. (1990a) Optical constants of solid methane. In *First International Conference on Laboratory Research of Planetary Atmospheres* (K. Fox et al., eds.), pp. 327–339. NASA CP-3077, Washington, DC.

Khare B. N., Thompson W. R., Sagan C., Arakawa E. T., Meisse C., and Gilmour I. (1990b) Optical constants of kerogen from 0.15 to 40 μm: Comparison with meteoritic organics. In *First International Conf. on Laboratory Research for Planetary Atmospheres* (K. Fox et al., eds.), p. 340. NASA CP-3077, Washington, DC.

Khare B. N., Thompson W. R., Cheng L., Chyba C., and Sagan C. (1993) Production and optical constants of ice tholin from charged particule irradiation of (1:6) C_2H_6/H_2O at 77 K. *Icarus, 103,* 290–300.

Khare B. N., Sagan C., Thompson W. R., Arakawa E. T., Meisse C., and Tuminello P. S. (1994) Optical properties of poly-HCN and their astronomical applications. *Can. J. Chem., 72,* 678–694.

Khare B. N., Bakes E. L. O., Imanaka H., McKay C. P., Cruikshank D. P., and Arakawa E. T. (2002) Analysis of the time-dependent chemical evolution of Titan haze tholin. *Icarus, 160,* 172–182.

Kimura Y., Tamura K., Koike C., Chihara H., and Kaito C. (2005) Laboratory production of monophase pyrrhotite grains using solid-solid reaction and their characteristic infrared spectra. *Icarus, 177,* 280–285.

King T. V. V. and Ridley W. I. (1987) Relation of the spectroscopic reflectance of olivine to mineral chemistry and some remote sensing implications. *J. Geophys. Res., 92,* 11457–11469.

Kissel J. and Krueger F. R. (1987) The organic component in dust from Comet Halley as measured by the PUMA mass spectrometer on board Vega 1. *Nature, 326,* 755–760.

Knief S. and von Niessen W. (1999) Disorder, defects, and optical absorption in a-Si and a-Si:H. *Phys. Rev. B, 59,* 12940–12946.

Koike C. and Tsuchiyama A. (1992) Simulation and alteration for amorphous silicates with very broad bands in infrared spectra. *Mon. Not. R. Astron. Soc., 255,* 248–254.

Koike C., Chihara H., Tsuchiyama A., Suto H., Sogawa H., and Okuda H. (2003) Compositional dependence of infrared absorption spectra of crystalline silicate. II. Natural and synthetic olivines. *Astron. Astrophys., 399,* 1101–1107.

Koike C., Mutschke H., Suto H., Naoi T., Chihara H., Henning Th., Jäger C., Tsuchiyama A., Dorschner J., and Okuda H. (2006) Temperature effects on the mid-and far-infrared spectra of olivine particles. *Astron. Astrophys., 449,* 583–596.

Krasnopolsky V. A. and Cruikshank D. P. (1999) Photochemistry of Pluto's atmosphere and ionosphere near perihelion. *J. Geophys. Res., 104,* 21979–21996.

Lisse C. M., VanCleve J., Adams A. C., A'Hearn M. F., Fernández Y. R., Farnham T. L., Armus L., Grillmair C. J., Ingalls J., Belton M. J. S., et al. (2006) Spitzer spectral observations of the Deep Impact ejecta. *Science, 313,* 635–640.

Logan L. and Hunt G. R. (1970) Emission spectra of particulate silicates under simulated lunar conditions. *J. Geophys. Res., 75,* 6539–6548.

Logothetidis S., Petalas J., and Ves S. (1995) The optical properties of a-C:H films between 1.5 and 10 eV and the effect of thermal annealing on the film character. *J. Appl. Phys., 79,* 1040–1050.

Lucey P. G. (1998) Model near-infrared optical constants of olivine and pyroxene as a function of iron content. *J. Geophys. Res. (E), 103,* 1703–1713.

Lyon R. J. P. (1964) Analysis of rocks by spectral infrared emission (8 to 25 microns). *Econ. Geol., 60,* 717–736.

Martonchik J. V., Orton G. S., and Appleby J. F. (1984) Optical properties of NH_3 ice from the far infrared to the near ultraviolet. *Appl. Opt., 23,* 541–547.

Masterson C. M. and Khanna R. K. (1990) Absorption intensities and complex refractive indices of crystalline HCN, HC_3N, and C_4N_2 in the infrared region. *Icarus, 83,* 83–92.

Mastrapa R. M. E. and Brown R. H. (2006) Ion irradiation of crystalline H_2O-ice: Effect on the 1.65-μm band. *Icarus, 183,* 207–214.

Maturilli A., Helbert J., and Moroz L. V. (2007) The Berlin Emissivity Database (BED). *Planet. Space Sci.,* in press.

McDonald G. D., Khare B. N., Thompson W. R., and Sagan C. (1991) $CH_4/NH_3/H_2O$ spark tholin: Chemical analysis and interaction with jovian aqueous clouds. *Icarus, 94,* 354–367.

McDonald G. D., Thompson W. R., Heinrich M., Khare B. N., and Sagan C. (1994) Chemical investigation of Titan and Triton tholins. *Icarus, 108,* 137–145.

McDonald G. D., Whited L. J., DeRuiter C., Khare B. N., Patnaik A., and Sagan C. (1996) Production and chemical analysis of cometary ice tholins. *Icarus, 122,* 107–117.

McKay C. O. (1996) Elemental composition, solubility, and optical properties of Titan's organic haze. *Planet. Space Sci., 44,* 741–747.

Minard R. D., Hatcher P. G., and Gourley R. C. (1998) Structural investigations of hydrogen cyanide polymers: New insights using TMAH thermochemolysis/GC-MS. *Origins Evol. Biosph., 28,* 461–473.

Moore M. H. and Hudson R. L. (1992) Far-infrared spectral studies of phase changes in water ice induced by proton irradiation. *Astrophys. J., 401,* 353–360.

Moore M. H., Ferrante R. F., Hudson R. L., Nuth J. A. III, and Donn B. (1994) Infrared spectra of crystalline phase ices condensed on silicate smokes at T < 20 K. *Astrophys. J. Lett., 428,* L81–L84.

Moore M. H., Ferrante R. F., Hudson R. L., and Stone J. N. (2007) Ammonia-water ice laboratory studies relevant to outer solar system surfaces. *Icarus, 190,* 260–273.

Moroz L. V., Pieters C. M., and Akhmanova M. V. (1992) Why the surfaces of outer belt asteroids are dark and red? (abstract). In *Lunar and Planetary Science XXIII,* pp. 931–932. Lunar and Planetary Institute, Houston.

Moroz L. V., Arnold G., Korochantsev A., and Wäsch R. (1998) Natural solid bitumens as possible analogs for cometary and asteroid organics: 1. Reflectance spectroscopy of pure bitumens. *Icarus, 134,* 253–268.

Moroz L. V., Schade U., and Wäsch R. (2000) Reflectance spectra of olivine-orthopyroxene-bearing assemblages at decreased temperatures: Implications for remote sensing of asteroids. *Icarus, 147,* 79–93.

Moroz L. V., Schmidt M., Schade U., Hiroi T., and Ivanova M. A. (2006) Synchrotron-based IR microspectroscopy as a useful tool to study hydration states of meteorite constituents. *Meteoritics & Planet. Sci., 41,* 1219–1230.

Mukai T. and Krätschmer W. (1986) Optical constants of the mixture of ices. *Earth Moon Planets, 36,* 145–155.

Mukai T. and Koike C. (1990) Optical constants of olivine particles between wavelengths of 7 and 200 microns. *Icarus, 87,* 180–187.

Murae T. (1994) FT-IR spectroscopic studies of major organic matter in carbonaceous chondrites using microscopic technique and comparison with terrestrial kerogen. *Proc. NIPR Symp. Antarct. Meteorites, 7,* 262–274.

Murae T., Masuda A., and Takahashi T. (1990) Spectroscopic studies of acid-resistant residues of carbonaceous chondrites. *Proc. NIPR Symp. Antarct. Meteorites, 3,* 211–219.

Mustard J. F. and Hays J. E. (1997) Effects of hyperfine particles on reflectance spectra from 0.3 to 25 μm. *Icarus, 125,* 145–163.

Mutsukura N. and Akita K-I. (1999) Infrared absorption spectroscopy measurements of amorphous CNx films prepared in CH_4/N_2 r.f. discharge. *Thin Solid Films, 349,* 115–119.

Nelander B. (1976) On the infrared spectrum of a carbon dioxide containing nitrogen matrix. *Chem. Phys. Lett., 42,* 187–189.

Ospina M., Zhao G., and Khanna R. K. (1988) Absolute intensities and optical constants of crystalline C_2N_2 in the infrared region. *Spectrochim. Acta A, 44,* 23–26.

Ossenkopf V., Henning Th., and Mathis J. S. (1992) Constraints on cosmic silicates. *Astron. Astrophys., 261,* 567–578.

Painter P., Starsinic M., and Coleman M. (1985) Determination of functional groups in coal by fourier transform interferometry. In *Fourier Transform Infrared Spectroscopy: Vol. 4, Application to Chemical Systems* (J. R. Ferraro and L. J.Basile, eds.), pp. 169–241. Academic, London.

Palumbo M. E. and Strazzulla G. (2003) Nitrogen condensation on water ice. *Can. J. Phys., 81,* 217–224.

Papoular R. (2001) The use of kerogen data in understanding the properties and evolution of interstellar carbonaceous dust. *Astron. Astrophys., 378,* 597–607.

Papoular R., Reynaud C., and Nenner I. (1991) The coal model for the unidentified infrared bands. II — The thermal emission mechanism. *Astron. Astrophys., 247,* 215–225.

Papoular R., Breton J., Gensterblum G., Nenner I., Papoular R. J., and Pireaux J.-J. (1993) The vis/UV spectrum of coals and the interstellar extinction curve. *Astron. Astrophys., 270,* L5–L8.

Pearl J., Ngoh M., Ospina M., and Khanna R. (1991) Optical constants of solid methane and ethane from 10000 to 450 cm^{-1}. *J. Geophys. Res., 96,* 17477–17482.

Pendleton Y. (1995) Laboratory comparisons of organic materials to interstellar dust and the Murchison meteorite. *Planet. Space Sci., 43,* 1359–1364.

Pendleton Y. J. and Allamandola L. J. (2002) The organic refractory material in the diffuse interstellar medium: Mid-infrared spectroscopic constraints. *Astrophys. J. Suppl. Ser., 138,* 75–98.

Pieters C. M. and Hiroi T. (2004) RELAB (Reflectance Experiment Laboratory): A NASA Multiuser Spectroscopy Facility (abstract). In *Lunar and Planetary Science XXXV,* Abstract #1720. Lunar and Planetary Institute, Houston (CD-ROM).

Pizzarello S., Huang Y., Becker L., Poreda R. J., Nieman R. A., Cooper G., and Williams M. (2001) The organic content of the Tagish Lake meteorite. *Science, 293,* 2236–2239.

Pollack J. B., Hollenbach D., Beckwith S., Simonelli D. P., Roush T. L., and Fong W. (1994) Composition and radiative properties of grains in molecular clouds and accretion disks. *Astrophys. J., 421,* 615–639.

Preibisch Th., Ossenkopf V., Yorke H.W., and Henning Th. (1993) The influence of ice-coated grains on protostellar spectra. *Astron. Astrophys., 279,* 577–588.

Quirico E. and Schmitt B. (1997a) Near-infrared spectroscopy of simple hydrocarbons and carbon oxides diluted in solid N_2 and as pure ices: Implications for Triton and Pluto. *Icarus, 127,* 354–378.

Quirico E. and Schmitt B. (1997b) A spectroscopic study of CO diluted in N_2 ice: Applications for Triton and Pluto. *Icarus, 128,* 181–188.

Quirico E., Schmitt B., Bini R., and Salvi P. R. (1996) Spectroscopy of some ices of astrophysical interest: SO_2, N_2 and N_2:CH_4 mixtures. *Planet. Space Sci., 44,* 973–986.

Quirico E., Douté S., Schmitt B., de Bergh C., Cruikshank D. P., Owen T. C., Geballe T. R., and Roush T.L. (1999) Composition, physical state and distribution of ices at the surface of Triton. *Icarus, 139,* 159–178.

Quirico E., Raynal P. I., and Bourot-Denise M. (2003) Metamorphic grade of organic matter in six unequilibrated ordinary chondrites. *Meteoritics & Planet. Sci., 38,* 795–811.

Quirico E., Borg J. Raynal P-Y., Montagnac G., and d'Hendecourt L. (2005) A micro-Raman survey of 10 IDPs and 6 carbonaceous chondrites. *Planet. Space Sci., 53,* 1443–1448.

Quirico E., Bernard J.-M., Montagnac G., Rouzaud J.-N., Szopa C., Cernogora G., Reynard B., McMillan P., Fray N., Schmitt B., Coll P., and Raulin F. (2006) Chemical structure and optical properties of Titan's tholins and HCN polymer. Implications for the analysis of Cassini-Huygens observations and refractory organics in cometary grains (abstract). In *Lunar and Planetary Science XXXVII,* Abstract #2105. Lunar and Planetary Institute, Houston (CD-ROM).

Ramirez S. I., Coll P., da Silva A., Navarro-Gonzalez R., Lafait J., and Raulin F. (2002) Complex refractive index of Titan's aerosol analogues in the 200–900 nm domain. *Icarus, 156,* 515–529.

Robert F. and Epstein S. (1982) The concentration and isotopic composition of hydrogen, carbon and nitrogen in carbonaceous meteorites. *Geochim. Cosmochim. Acta, 46,* 81–95.

Robin P. L. and Rouxhet P. G. (1976) Contribution des différentes fonctions chimiques dans les bandes d'absorption infrarouge des kérogènes situées à 1710, 1630 et 3430 cm^{-1}. *Rev. Inst. Franç. du Pétrole, 31,* 955–978.

Rodil S. E., Ferrari A. C., Robertson J., and Milne W. I. (2001) Raman and infrared modes of hydrogenated amorphous carbon nitride. *J. Appl. Phys., 89,* 5425–5430.

Rodil S. E., Ferrari A. C., Robertson J., and Muhl S. (2003) Infrared spectra of carbon nitride films. *Thin Solid Films, 420,* 122–131.

Rouleau F. and Martin P. (1991) Shape and clustering effects on the optical properties of amorphous carbon. *Astrophys. J., 377,* 526–540.

Roush T. L. (1995) Optical constants of amorphous carbon in the mid-IR (2.5–25 μm, 4000–400 cm^{-1}). *Planet. Space Sci., 43,* 1297–1301.

Roush T. L. (2003) Estimated optical constants of the Tagish Lake meteorite. *Meteoritics & Planet. Sci., 38,* 419–426.

Roush T. L. and Dalton J. B. (1994) Reflectance spectra of hydrated Titan tholins at cryogenic temperatures and implications for compositional interpretation of red objects in the outer solar system. *Icarus, 168,* 158–162.

Roush T., Pollack J., and Orenberg J. (1991) Derivation of mid-infrared (5–25 microns) optical constants of some silicates and palagonite. *Icarus, 94,* 191–208.

Rouxhet P. G., Robin P. L., and Nicaise G. (1980) Characterization of kerogen and their evolution by infrared spectroscopy. In *Kerogen — Insoluble Organic Matter from Sedimentary Rocks* (B. Durand, ed.), pp. 13–34. Editions Technip, Paris.

Sagan C. and Khare B. N. (1979) Tholins: Organic chemistry of interstellar grains and gas. *Nature, 277,* 102–107.

Sagan C., Khare B. N., and Lewis J. S. (1984) Organic matter in the solar system. In *Saturn* (T. Gehrels and M. S. Matthews, eds.), pp. 788–807. Univ. of Arizona, Tucson.

Sandford S. A. (1996) The inventory of interstellar materials available for the formation of the solar system. *Meteoritics & Planet. Sci., 31,* 449–476.

Sandford S. A., Aléon J., Alexander C. M. O'D., Araki T., Bajt S., Baratta G. A., Borg J., Bradley J. P., Brownlee D. E., Brucato J. R., et al. (2006) Organics captured from Comet 81P/Wild 2 by the Stardust spacecraft. *Science, 314,* 1720–1724.

Satorre M. A., Palumbo M. E., and Strazzulla G. (2001) Infrared spectra of N_2 rich ice mixtures. *J. Geophys. Res. (E), 106,* 33363–33370.

Schmitt B., Grim R. J. A., and Greenberg J. M (1989) Spectroscopy and physico-chemistry of $CO:H_2O$ and $CO_2:H_2O$ ices. In *22nd ESLAB Symposium: Infrared Spectroscopy in Astronomy,* pp. 213–219. ESA Spec. Publ. 290, Noordwijk, The Netherlands.

Schmitt B., Oehler A., Calvier R., Haschberger P., and Lindermeier E. (1990) The near infrared absorption features of solid nitrogen and methane on Triton (abstract). *Bull. Am. Astron. Soc., 22,* 1121.

Schmitt B., Quirico E., and Lellouch E. (1992) Near infrared spectra of potential solids at the surface of Titan. In *Proceedings of the Symposium on Titan,* pp. 383–388. ESA Spec. Publ. 338, Noordwijk, The Netherlands.

Schmitt B., de Bergh C., Lellouch E., Maillard J. P., Barbe A., and Douté S. (1994) Identification of three absorption bands in the two micron spectrum of Io. *Icarus, 111,* 79–105.

Schmitt B., Quirico E., Trotta F., and Grundy W. (1998) Optical properties of ices from UV to infrared. In *Solar System Ices* (B. Schmitt et al., eds.), pp. 199–240. Astrophys. Space Sci. Lib. Vol. 227, Kluwer, Dordrecht.

Schnaiter M., Henning Th., Mutschke H., Kohn B., Ehbrecht M., and Huisken F. (1999) Infrared spectroscopy of nano-sized carbon grains produced by laser pyrolysis of acetylene: Analog materials for interstellar grains. *Astrophys. J., 519,* 687–696.

Schulze H., Kissel J., and Jessberger E. K. (1997) Chemistry and mineralogy of comet Halley's dust. In *From Stardust to Planetesimals* (Y. J. Pendleton and A. G. G. M. Tielens, eds.), pp. 397–414. ASP Conf. Series 122, San Francisco.

Scott A. and Duley W. W. (1996) Ultraviolet and infrared refractive indices of amorphous silicates. *Astrophys. J. Suppl. Ser., 105,* 401–405.

Shkuratov Y., Starukhina L., Hoffmann H., and Arnold G. (1999) A model of spectral albedo of particulate surfaces: Implications for optical properties of the Moon. *Icarus, 137,* 235–246.

Sill G. S., Fink U., and Ferraro J. R. (1980) Absorption coefficients of solid NH_3 from 50 to 7000 cm^{-1}. *J. Opt. Soc. Am., 70,* 724–739.

Singer R. B. and Roush T. L. (1985) Effects of temperature on remotely sensed mineral absorption features. *J. Geophys. Res., 90,* 12434–12444.

Smith R. G., Robinson G., Hyland A. R., and Carpenter G. L. (1994) Molecular ices as temperature indicators for insterstellar dust: The 44 and 62 μm lattice features of H_2O ice. *Mon. Not. R. Astron. Soc., 271,* 481–489.

Sourisseau C., Coddens G., and Papoular R. (1992) On the 21-micron feature of pre-planetary nebulae. *Astron. Astrophys., 254,* L1–L4.

Sunshine J. M. and Pieters C. M. (1998) Determining the composition of olivine from reflectance spectroscopy. *J. Geophys. Res. (E), 103,* 13675–13688.

Szopa C., Cernogora G., Boufendi L., Correia J. J., and Coll P. (2006) PAMPRE: A dusty plasma experiment for Titan's tholins production and study. *Planet. Space Sci., 54,* 394–404.

Taylor F. W. (1973) Preliminary data on the optical properties of solid ammonia and scattering parameters for ammonia cloud particles. *J. Atmos. Sci., 30,* 677–683.

Tielens A. G. G. M. (1997) Circumstellar PAHs and carbon stardust. *Astrophys. Space Sci., 251,* 1–13.

Toon O. B., Tolbert M. A., Koehler B. G., Middlebrook A. M., and Jordan J. (1994) Infrared optical constants of H_2O ice, amorphous nitric acid solutions, and nitric acid hydrates. *J. Geophys. Res. (D), 99,* 25631–25654.

Tran B. N., Joseph J. C., Ferris J. P., Persans P. D., and Chera J. J. (2003) Simulation of Titan haze formation using a photochemical flow reactor. The optical constants of the polymer. *Icarus, 165,* 379–390.

Trotta F. (1996) Détermination des constantes optiques de glaces dans l'infrarouge moyen et lointain. Application aux grains du milieu interstellaire et des enveloppes circumstellaires. Thesis, LGGE — Université Joseph Fourier, Grenoble, France.

Trotta F. and Schmitt B. (1996) Determination of the optical constants of solids in the mid infrared. In *The Cosmic Dust Connection* (J. M. Greenberg, ed.), pp. 179–184. NATO ASI Series C 487, Kluwer, Dordrecht.

Tryka K. A., Brown R. H., Anicich V., Cruikshank D. P., and Owen T. C. (1993) Spectroscopic determination of the phase composition and temperature of nitrogen ice on Triton. *Science, 261,* 751–754.

Tryka K. A., Brown R. H., and Anicich V. (1995) Near-infrared absorption coefficients of solid nitrogen as a function of temperature. *Icarus, 116,* 409–414.

Van Krevelen D. W. (1961) Optical properties: Refractometric and spectrometric analysis of coal. In *Coal,* pp. 343–372. Elsevier, Amsterdam.

Warren S. G. (1984) Optical constants of ice from the ultraviolet to the microwave. *Appl. Opt., 23,* 1206–1223.

Warren S. G. (1986) Optical constants of carbon dioxide ice. *Appl. Opt., 25,* 2650–2674.

Wdowiak T. J., Flickinger G. C., and Cronin J. R. (1988) Insoluble organic material in Orgueil carbonaceous chondrite and the unidentified infrared bands. *Astrophys. J. Lett., 328,* L75–L79.

Wisnosky M. G., Eggers D. F., Fredrickson L. R., and Decius J. C. (1983) The vibrational spectra of solid II ethane and ethane-d_6. *J. Chem. Phys., 79,* 3505–3512.

Wooden D. H., Harker D. E., and Brearley A. J. (2005) Thermal processing and radial mixing of dust: Evidence from comets and primitive chondrites. In *Chondrites and the Protoplanetary Disk* (A. N. Krot et al., eds.), pp. 774–808. ASP Conf. Series 341, San Francisco.

Wopenka B. (1988) Raman observations on individual interplanetary dust particles. *Earth Planet. Sci. Lett., 88,* 221–231.

Wright I. P., McGarvie D. W., Grady M. M., and Pillinger C. T. (1990) The distribution of carbon in C1 to C6 carbonaceous chondrites. *Proc. NIPR. Symp. Antarct. Meteorites, 3,* 194–210.

Zolensky M. E. and Thomas K. L. (1995) Iron and iron-nickel sulfides in chondritic interplanetary dust particles. *Geochim. Cosmochim. Acta, 59,* 4707–4712.

Zolensky M. E., Zega T. J., Yano H.,Wirick S., Westphal A. J., Weisberg M. K., Weber I., Warren J. L., Velbel M. A., Tsuchiyama A., et al. (2006) Mineralogy and petrology of Comet 81P/Wild 2 nucleus samples. *Science, 314,* 1735–1739.

Zubko V., Mennella V., Colangeli L., and Bussoletti E. (1996) Optical constants of amorphous carbon grains. In *The Role of Dust in the Formation of Stars* (H. U. Käufl and R. Siebenmorgen, eds.), p. 333. ESO Astrophysics Symposia, European Southern Observatory/Springer-Verlag, Berlin.

Laboratory Studies of the Chemistry of Transneptunian Object Surface Materials

R. L. Hudson
Eckerd College and NASA Goddard Space Flight Center

M. E. Palumbo and G. Strazzulla
INAF-Osservatorio Astrofisico di Catania

M. H. Moore, J. F. Cooper, and S. J. Sturner
NASA Goddard Space Flight Center

Bombardment by cosmic-ray and solar wind ions alters the surfaces of transneptunian objects (TNOs) surfaces, and the influence of this weathering on candidate TNO materials has been extensively examined by laboratory scientists. Low-temperature radiation experiments with icy materials have demonstrated the existence of a rich TNO ice chemistry involving molecules such as H_2O, CH_4, N_2, and NH_3. These same experiments have provided insight into reaction mechanisms needed to predict yet-unseen chemical species. Near-IR and visible spectra of ion-irradiated candidate refractories have generated the data needed to understand TNO colors and spectral slopes. The planning, execution, and interpretation of these experiments have been influenced by new energetic particle measurements from Voyager and other heliospheric spacecraft and by models for TNO surface irradiation fluxes and dosages. Experiments and available surface irradiation models suggest specific timescales for reddening of TNO surfaces. Altogether, laboratory investigations and heliospheric radiation measurements contribute to the study of TNOs by aiding in the interpretation of astronomical observations, by suggesting new lines of investigation, and by providing the underlying knowledge needed to unravel the chemical and spectral evolution of objects in the outer solar system.

1. INTRODUCTION

The objects in the outer solar system can be organized into three different groups according to their observed surface IR spectra. First are objects with spectra dominated by H_2O-ice, such as some Centaurs, Charon, and several transneptunian objects (TNOs). A second group has spectra with prominent CH_4 features, and includes Triton, Pluto (both also have N_2-ice), and several TNOs. A third group includes objects having featureless spectra. This spectral diversity clearly indicates compositional differences in surface layers to a few millimeters or less in depth. More speculative are the compositions of the underlying layers and the processes by which they contribute material to TNO surfaces. For example, water ice may come from Enceladus-like outgassing (*Porco et al.,* 2006; *Waite et al.,* 2006), CH_4 could either be of internal primordial origin or be produced by surface irradiation, and more neutral featureless spectra could arise from long cumulative irradiation (*Moroz et al.,* 2003, 2004).

At present, six molecules (H_2O, CH_4, N_2, NH_3, CO, and CH_3OH) suffice to explain the spectral bands of icy TNO terrains. In addition, silicates and complex organics, presumably highly processed by cosmic radiation and/or micrometeorite bombardment, can explain spectral slopes and colors in the UV, visible, and near-IR regions. However, it is a challenge to reconcile these surface compositions with bulk compositions, as inferred from gas-phase observations of comets near the Sun. Note that the comets of the Jupiter family (i.e., coming from the Oort cloud) have compositions similar to those believed to have originated in the Kuiper belt. Within the current inventory of about 50 cometary molecules, nuclear ices are dominated by H_2O, CO, CO_2, and minor species such as CH_3OH, H_2CO, and CH_4. Some compositional differences among TNOs, comets, and other icy objects at the edge of the solar system can be understood in part by variations in formation and storage temperatures, which affect vapor pressure, and mass, which affects escape velocity. However, a complex evolutionary history for outer solar system objects is thought to include stochastic events as well as continuous exposure to ionizing radiation. The subject of this chapter is the use of laboratory data, theory, and spacecraft measurements to understand how long-term radiation exposure causes chemical changes in TNO ices and non-icy TNO surface materials.

Three observations are consistent with energetic processing of outer solar system objects. First, the visible and near-IR spectra of TNOs and ion-irradiated laboratory materials have similar slopes, which differ from those of ices and refractories that have not been irradiated (*Brunetto et al.,* 2006). Another observation is the detection of abundant C_2H_6 in comets C/1996 B2 Hyakutake (*Mumma et al.,* 1996) and C/1995 O1 Hale-Bopp (*Weaver et al.,* 1998). A C_2H_6 abundance comparable with that of CH_4 implies that these comets' ices did not originate in a thermochemically equilibrated region of the solar nebula, but were produced by processing of icy interstellar grain mantles. Both comets are thought to be Oort cloud objects with an origin somewhere in the Jupiter–Neptune region. Finally, the low surface reflectance and the neutral featureless color spectra of many objects are as expected from millions to billions of years of cosmic ray irradiation, if the most highly irradiated outer surface is not removed by plasma sputtering or meteoritic impacts (*Cooper et al.,* 2003, 2006a; *Strazzulla et al.,* 2003).

Over the past 30 years, laboratory research has shown that high-energy particles and photons cause irreversible physical and chemical changes in relevant solar system ices and analog surface materials. Since solar system surfaces have been exposed to radiation and have been altered over time, laboratory experiments can be used to investigate and predict radiation chemical changes.

In this chapter we summarize the results of many radiation chemistry experiments on relevant TNO ices and other surface materials. Based on past successes, a comprehensive picture of radiation processing is emerging, one that can be used to predict radiation products to be sought in upcoming missions and observing campaigns.

2. LABORATORY APPROACH

Assignments of TNO spectral features are based on comparisons to the spectra of materials available in laboratories (see chapter by de Bergh et al.). However, the alteration of TNO surfaces by energetic photons and ions means that specific experiments are needed to probe the resulting chemical and physical changes.

At present there are several laboratories where research is conducted to study processes, such as ion irradiation and UV photolysis, which can drive the evolution of TNO surface materials, such as ices, silicates, and carbonaceous solids. In the case of ices, experiments usually begin with the preparation of a sample by condensation of an appropriate gas, or gas-phase mixture, onto a 10–300 K substrate in a high or ultrahigh vacuum chamber (P ~ 10^{-7}–10^{-11} mbar). The ice's thickness can be measured by monitoring the interference pattern (intensity vs. time) from a laser beam reflected both by the vacuum-film and film-substrate interfaces (e.g., *Baratta and Palumbo,* 1998). The icy film produced can be processed by keV and MeV ions and electrons or by far-UV photons (e.g., Lyman-α, 10.2 eV). The resulting chemical and physical changes can be followed with visible, IR, and Raman spectroscopies before, during, and after processing. Experimental setups used to study refractory materials, such as silicates and carbonaceous compounds, are similar to those used to investigate ices.

Several different types of processing experiments have been performed. If the sample is thinner than the penetration depth of the impinging ions or photons, then they pass through the target. In some such cases the resulting spectrum shows only the more-intense IR absorptions. To enhance the weaker bands, irradiation can be done during sample deposition, building up a larger thickness of processed material. Finally, if the sample thickness is greater than the penetration depth of incident ions or photons, only the uppermost layers of the sample are altered. If the projectile is a reactive species, such as an H, C, N, O, or S ion, then it can be implanted into the ice to form new molecules that include the projectile.

Ions impinging on solids release energy mainly through elastic collisions with target nuclei, and inelastic interactions that cause the excitation and ionization of target species. For keV and MeV ion irradiations, doses can be estimated from a knowledge of the ion fluence (ions cm^{-2}), the energy of impinging ions (eV), the stopping power (eV/Å or eV cm^2 molecule^{-1}), and the penetration depth or range of the chosen projectiles (Å or molecules cm^{-2}). The ion energy and fluence are measured during laboratory irradiation, while the stopping power and range can be calculated, for example, with Ziegler's SRIM program (*www.srim.org*) (*Ziegler et al.,* 1985). In these experiments, low current densities, such as 0.001–1 µA cm^{-2}, are used to avoid macroscopic heating of the target. To facilitate comparisons between different samples and different energy sources, most workers use eV/16-amu-molecule as a standard unit of dose, even in UV-photolysis experiments. This unit is sometimes abbreviated, with occasional ambiguity, as eV/molecule. Note that 100 eV/16-amu at unit density corresponds to 60 gigarads, a dose that produces significant change in the bulk chemistry of irradiated materials.

Although laboratory experiments are commonly done with keV and MeV radiations, for theoretical models it is sometimes necessary to consider interactions at higher energies. For the MeV-to-GeV range, the GEANT radiation transport code, which includes secondary and higher-order interactions, is available at *wwwasd.web.cern.ch/wwwasd/geant* (*Sturner et al.,* 2003).

Irradiated samples can be analyzed by (1) near- and mid-IR transmission spectroscopy, in which case an IR-transparent substrate is used, such as KBr, CsI, or crystalline silicon; (2) transmission-reflection-transmission IR spectroscopy, in which case a substrate that reflects the IR beam, such as Al or Au, is used; (3) visible and near-IR diffuse reflectance spectroscopy, in which case an optically rough, diffusing Au substrate is used; and (4) Raman spectroscopy. Although laboratory Raman spectra cannot be directly compared with astronomical visible and IR observations, they provide valuable information on solid-phase radiation effects, in particular structural changes in carbonaceous materials and carbon-rich ice mixtures. Furthermore, because of the different selection rules that govern the interaction of light and matter in the IR and Raman techniques, these spectra give complementary information.

Additional details concerning experimental procedures can be found in *Baratta and Palumbo* (1998), *Palumbo et al.* (2004), *Moore and Hudson* (2003), *Gerakines et al.* (2005), and *Bernstein et al.* (2005).

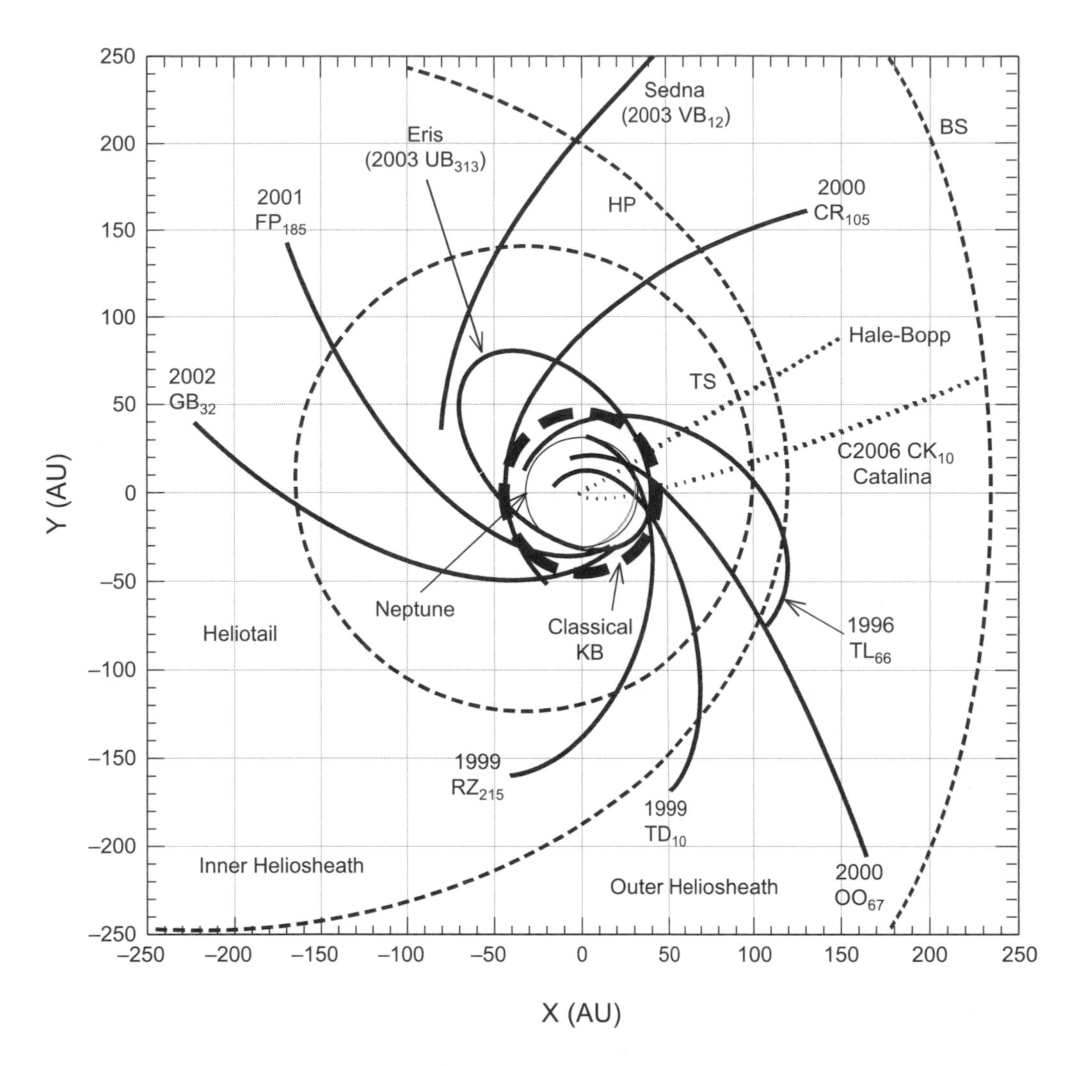

Fig. 1. Trajectories (solid curves) of selected TNOs and inner Oort cloud comets with highly eccentric orbits, many crossing the solar wind termination shock (TS). Other heliospheric boundaries (dashed curves) and regions include the inner heliosheath, heliopause (HP), outer heliosheath, and the bow shock (BS). See *Cooper et al.* (2006a) for more details on coordinates and boundaries.

3. RADIATION ENVIRONMENT

Since TNO surfaces are altered by exposure to high-energy particles and photons, it is important to understand the fluxes, energies, and distribution of radiations in the outer solar system. These have been studied with a combination of theoretical modeling and *in situ* measurements by spacecraft. Figure 1 shows the relevant parts of the heliosphere (region of the Sun's influence on plasma environment) and the solar system, with orbits of selected TNOs and other objects superimposed. Pioneers 10 and 11 combined to probe the 30–80-AU region, but ceased data return in 1995 and 2002, respectively. The solar wind plasma and energetic particle radiation environments beyond Neptune's 30-AU orbit were directly measured by Voyager 1 (only energetic particles due to failure of the plasma instrument in 1980), in 1987, and Voyager 2, in 1989 (*Cooper et al.,* 2003). Knowledge of the outer heliospheric environment dramatically advanced when Voyager 1 crossed the solar wind termination shock at 94 AU in December 2004 (*Burlaga et al.,* 2005; *Decker et al.,* 2005; *Stone et al.,* 2005). This boundary, labeled TS in Fig. 1, marks the transition from supersonic to subsonic flow of solar wind plasma.

Beyond the termination shock in the outer solar system is the heliopause, the theoretical boundary near minimum distance 120 AU between outward-flowing solar wind plasma and inflowing plasma of the local interstellar medium (LISM). Voyager 1 is currently passing through the intervening inner heliosheath region, while Voyager 2 is expected to enter it within the current decade. Some theoretical models suggest that interstellar plasma flowing inward across a theoretical bow shock (BS in Fig. 1) into the outer heliosheath region may also undergo a sonic transition.

The classical Kuiper belt, beyond the orbit of Neptune as shown in Fig. 1, now resides far sunward of the termination shock, although the latter may have occasionally have moved further inward in response to changes in the local interstellar environment of the Sun. More than 30 scattered disk objects have highly eccentric orbits traversing the termination shock in various directions (*Cooper et al.,* 2006a), and a few of these objects, in addition to more than 300 known comets, travel as far as the heliopause and the interstellar plasma environment beyond.

On going away from the Sun, the proton energy flux decreases and reaches a broad minimum in the middle of the supersonic heliosphere (upwind of TS), including the classical Kuiper belt region at 30–50 AU. This holds true over a large range of energy values. At even greater distances from the Sun, there is an increase in flux in going from the middle heliosphere to the inner heliosheath and then beyond into the local interstellar medium. *Cooper et al.* (2006a) have suggested that since the measured inner heliosheath spectrum is approaching that of a LISM model spectrum then the actual source may be in the interstellar environment beyond an ion-permeable heliopause. If so, a complete quantitative model for irradiation of TNOs in near-circular and highly eccentric orbits, as illustrated in Fig. 1, may be within reach.

As already stated, a common feature among TNOs is their exposure to energetic ions (mostly H^+, He^+, and O^+) and solar UV photons that slowly modify the chemistry of surface materials. Compositional abundances of solar wind plasma and energetic ions in the heliosphere are mostly similar to those of the solar photosphere (*Anders and Grevesse,* 1989), but there can be additional sources from solar, heliospheric, and galactic acceleration processes. Low-energy protons and heavier ions contribute to alteration of surface chemistry by direct implantation, while more energetic (keV–MeV) ions become important for inducing radiation-chemical reactions (see the following section) and sputtering of outer molecular layers (*Johnson,* 1990, 1995). Radiolytic chemical alteration of TNO surface ices at millimeter-to-meter depths is primarily driven by high-energy (keV–GeV) protons as the result of primary and secondary interactions (*Cooper et al.,* 2003, 2006a). Electrons are more penetrating than protons or ions of the same energy and may have significant radiolytic effects even at plasma energies. Figure 2 shows stopping range and differential energy loss rates for protons, selected heavier ions, and electrons as extended to lower energies from data above 10 keV (*Cooper et al.,* 2001) (note that the axis units in Figs. 12 and 13 of the latter reference were incorrectly labeled).

Vacuum-UV solar photons have a complex spectrum at 1 AU (*Hall et al.,* 1985; *Tobiska,* 2000; *Tobiska and Bouwer,* 2006). Their energy flux is generally higher than that from charged particles, except within the intense trapped radiation belt environment of the jovian magnetosphere (*Cooper et al.,* 2001). However, the UV penetration depth is only ~0.15 μm, as compared to the ~100-μm thickness of ice sampled by near-IR observations, so UV photolysis products can be removed by surface erosion or highly altered by radiation processing. Alternatively, UV photons can initiate gas-phase photochemistry on Pluto-sized TNOs with atmospheres (*Elliot and Kern,* 2003), and the reaction products can precipitate downward to potentially dominate surface compositions.

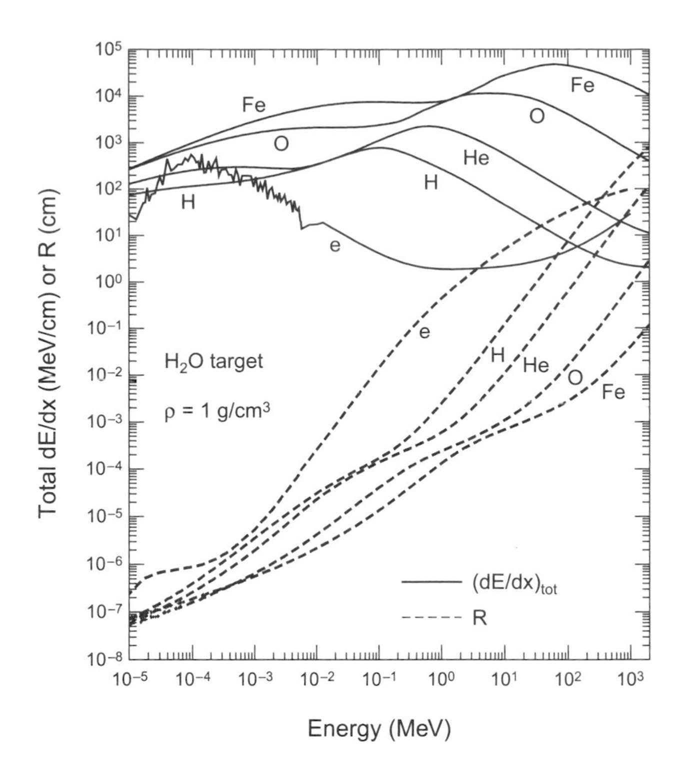

Fig. 2. Stopping ranges (R) and differential energy loss rates (dE/dx) in water at unit density of protons (H), the indicated heavier ions (He-Fe), and electrons (e). The loss rate is the sum of atomic ionization and nuclear collision components for protons and ions. For electrons, dE/dx is the sum of ionization and radiative energy losses. Proton and ion data are from the SRIM model (*Ziegler et al.,* 1985; *www.srim.org*). Electron data above 1 keV are from the ESTAR database (*physics.nist.gov/PhysRefData/Star/Text/ESTAR.html*) and at lower energies from published data for polystyrene at nearly (1.05 g/cm^3) unit density (*X-Ray Data Booklet,* Lawrence Berkeley National Laboratory, 2001).

Table 1 gives estimates of the doses accumulated in 4.6 G.y. by the outer 1 μm, 100 μm, and 1 m of an ice with an assumed density of 1 g cm^{-3} (*Cooper et al.,* 2003,2006a). Objects in a broad zone of the middle heliosphere, around 40 AU, experience moderate irradiation from galactic cosmic-ray ions at micrometer-to-meter depths (Pluto and dynamically cold TNOs). This region is bounded on the sunward side by increasing fluxes of solar energetic ions, resulting in increased surface doses for Centaurs. On the antisunward side there are rising fluxes of energetic ions diffusing inward from the heliosheath and perhaps also from the LISM. *Cooper et al.* (2003) have suggested that this radial separation of internal and outer heliospheric ion sources provides a potential explanation for predominately red colors of the more ancient objects in low-inclination, low-eccentricity orbits beyond 40 AU. That is, these red objects are less irradiated than those in other regions either closer to or further away from the Sun.

TABLE 1. Estimated radiation doses (eV/16-amu molecule) for ice-processing environments[*,†].

Object	Ices Detected	Distance (AU)	Dose at 1-μm Depth[†]	Dose at 100-μm Depth[†]	Dose at 1-m Depth[†]
Centaur	H_2O, CH-containing ices (CH_3OH?), silicates, organics ("tholin")	5–35	100[‡]–10,000[§]	100[‡]–200[§]	30[‡]
		48–1000	100[‡]–500,000[¶]	100[‡]–30,000[¶]	30[‡]–50[¶]
Triton	N_2, CH_4, CO, CO_2, H_2O				
Pluto	N_2, CH_4, CO (and H_2O?)	30–40	100[‡]	100[‡]	30[‡]
Charon	H_2O, NH_3, NH_3-hydrate				
TNO	H_2O, CH_4, NH_3, NH_3-hydrate?	<48	100[‡]	100[‡]	30[‡]
		~1000	500,000[¶]	30,000[¶]	50[¶]
Oort cloud comet	Gases**: H_2O, CO, CO_2, CH_3OH, CH_4, H_2CO, NH_3, OCS, HCOOH, HCN, C_2H_6, C_2H_2	~40,000–100,000	500,000[¶]	30,000[¶]	50[¶]

*Doses in eV (16-amu molecule)$^{-1}$ for 4.6 G.y., with an ice density of 1.0 g cm^{-3}.
†Solar minimum.
‡*Cooper et al.* (2003) extended with GEANT.
§J. F. Cooper et al. (unpublished data, 2006).
¶*Cooper et al.* (2006a).
**The assumed origin of these gases is the comet's nucleus.

It is important to note that the doses in Table 1 are based on direct observations of ion fluxes and on well-established theories of ion-matter interactions. Thus the processes that are studied in the laboratory must necessarily apply to the objects of Table 1, and will compete with resurfacing, collisional evolution, and micrometeoritic bombardment.

4. SPECTROSCOPY AND TRANSNEPTUNIAN OBJECT CHEMISTRY

Studies of TNO chemistry have been dominated by two interconnected approaches, astronomical observations and laboratory experiments. While polarization and photometry measurements have been valuable for understanding TNO density and porosity (e.g., *Bagnulo et al.,* 2006), spectroscopy has been the method of choice for probing TNO chemistry. Most TNO spectra have been measured at visible and near-IR wavelengths where the cold surfaces efficiently reflect solar radiation. [For an exception, see *Grundy et al.* (2002) for mid-IR data.] Visible-light measurements alone, and even some combined with near-IR data, usually give only sloping featureless spectra making unique chemical assignments difficult. These spectra will be examined more thoroughly in section 5.

The only firm assignments of TNO spectral features to specific molecules have come from near-IR data (see the chapter by Barucci et al.). As an example, the reflectance spectrum of Quaoar (*Jewitt and Luu,* 2004) exhibits absorptions near 1.5 and 2.0 μm that are characteristic of H_2O-ice, and a small dip at 1.65 μm shows that the ice is crystalline and at a temperature below 80 K. Evidence of an ammonia (NH_3) species comes from a small feature near 2.2 μm. The near-IR spectrum of Charon exhibits the same H_2O-ice bands (*Brown and Calvin,* 2000). Contrasting with these objects are Pluto and 2005 FY_9, whose near-IR spectra indicate the presence of CH_4-ice (*Licandro et al.,* 2006).

Other species considered in the analysis of TNO near-IR spectra include cyanides, both organic and inorganic (*Trujillo et al.,* 2007), and hydrocarbons (*Sasaki et al.,* 2005), although no firm spectral assignments to specific molecules have yet been published. The Centaur 5145 Pholus is thought to have originated in the Kuiper belt, and models show that solid CH_3OH (methanol) may be among its surface ices (*Cruikshank et al.,* 1998). A long series of near-IR observations of Pluto and Triton have revealed that they possess multicomponent ice surfaces. In addition to CH_4, both solid N_2 and CO have been detected on Pluto, with N_2 dominating in some regions and CH_4 in others (*Douté et al.,* 1999). For Triton, N_2, CH_4, and CO are also observed, but H_2O- and CO_2-ice features are suggested by the data as well (*Quirico et al.,* 1999).

All these TNO observations have motivated laboratory work whose goal is to record near-IR spectra of single- and multicomponent ices at temperatures relevant to the outer solar system. The work of Schmitt and colleagues has produced a collection of spectra and optical constants of single-component ices, and these data are treated elsewhere (see the chapter by de Bergh et al.). Near-IR band strengths have been published recently by *Gerakines et al.* (2005) for many TNO-relevant molecules. New spectra and band strengths of ice mixtures made from H_2O, CO_2, CH_4, and CH_3OH are also available (*Bernstein et al.,* 2005, 2006).

Here we consider the likely solid-phase chemistry of some known and suspected TNO molecules, with spectroscopy as the major investigative tool. The three types of ices to be considered are those made of a single component, those made of mixtures dominated by a very polar molecule (H_2O), and those made of mixtures dominated by a nonpolar material (N_2). The primary goal motivating the laboratory work is the discovery of efficient reactions leading from simple starting materials to more-complex species, to allow predictions of as-yet-unobserved TNO molecules.

4.1. Radiation Chemistry and Transneptunian Object Ices

Chemical reactions in TNO ices can be initiated both by solar far-UV photons, with energies in the 5–120-eV (250–5-nm) range, and cosmic-ray ion bombardment, with energies in the keV–MeV region and higher. Despite this great variation in energy, similar chemical products result from the two processes. The reason for this is that MeV radiation interacts with matter through a series of discrete steps, each involving energy loss until the eV level is reached (*Johnson,* 1990). Specifically, a single 1-MeV H^+ cosmic ray passing through an ice loses energy through the production of secondary electrons, which in turn lead to thousands of ionizations and excitations in the ice. These events result in the breakage of chemical bonds and the rearrangement of molecular fragments to give new molecules and ions. These eV-level ionizations and excitations resemble those of conventional UV photochemistry, so that the final products of photo- and radiation chemistry are quite similar, when comparable energy doses are involved. Differences do exist, however, as keV–MeV radiation is far more penetrating than UV photons, and can produce opaque surface materials that prevent penetration by UV or visible light (*Baratta et al.,* 2002). Note also that the chemical products from various MeV radiations (e.g., H^+, He^+, e^-, X-rays, γ-rays) acting on ices are essentially indistinguishable since it is the secondary electrons that cause the bulk of the chemical change, masking the identity of the original radiation. Finally, there are a few molecules, such as N_2, that are not dissociated by far-UV photons in a single step. In such cases the molecule can still be excited and react with other species, or vice versa, which may or may not result in dissociation. See *Moore and Hudson* (2003) for an example, the formation of DCN from both the ion irradiation and the UV photolysis of $N_2 + CD_4$ ices.

Table 1 of the previous section lists radiation doses thought to be typical for TNOs. Over several billion years every molecule within about 100 μm of a TNO's surface will receive 10^1–10^5 eV, depending on solar system position and heliospheric activity. A relatively constant 30–50 eV/molecule occurs everywhere at meter depths, due to very-high-energy galactic cosmic rays that are little affected by solar modulation. These doses are attainable in laboratories, and so experiments can be performed to explore radiation chemical reactions of TNO analog materials. We now survey some of the published literature on the low-temperature chemistry of TNO molecules.

4.2. Chemistry of Single-Component Ices

Spectroscopic detections of extraterrestrial H_2O-ice have been based on far-, mid-, and near-IR spectra. Features in the far-IR region arise from intermolecular transitions and are quite sensitive to the amorphous or crystalline nature of the ice. Mid-IR features of H_2O, and all other molecules, arise from characteristic intramolecular fundamental vibrations of groups of atoms. Near-IR bands, which have permitted identifications of specific ice molecules on TNO surfaces, arise from overtones and combinations of a molecule's fundamental vibrations. Near-IR bands are typically an order of magnitude weaker than those in the mid-IR.

Of these three spectral regions, laboratory workers studying reaction chemistry typically use the mid-IR as it is the most reliable for identification of product molecules. In contrast, TNO observers favor near-IR spectra. Here we consider both regions, but with an emphasis on results from the mid-IR to suggest with confidence the reaction products that one might expect in TNO ices.

The first ice reported on a classical TNO was H_2O (*Brown et al.,* 1999). Knowing that this ice is present, and that TNOs exist in a radiation environment, what other molecules might one reasonably expect to be present? Irradiated H_2O-ice has been studied for about a century, and still attracts attention. Early experiments showed that the molecular products of H_2O-ice decomposition, both by photo- and radiation chemistry, are H_2, O_2, and H_2O_2, as expected. In both photolysis and radiolysis, H_2 and H_2O_2 are thought to form by combination of radical-radical reactions, as indicated below.

$$H_2O \rightarrow H + OH \text{ (twice)}$$
$$OH + OH \rightarrow H_2O_2$$
$$H + H \rightarrow H_2$$

The mechanism for O_2 formation is still being studied, and recent work strongly suggests that trapped oxygen atoms are necessary precursors to O_2 formation (*Johnson et al.,* 2005).

Since H_2 and O_2 lack permanent dipole moments, their fundamental vibrational transitions are very weak, and do not lead to pronounced spectral features, leaving H_2O_2 as the most easily detectable product. Near-IR spectra of frozen H_2O_2 and H_2O_2-H_2O mixtures have been recorded down to ~9 K, and the H_2O_2 bands are found to strongly overlap those of H_2O-ice (*Hudson and Moore,* 2006). This suggests that while H_2O_2 is a well-known product of H_2O-ice irradiation (e.g., *Loeffler et al.,* 2006a; *Gomis et al.,* 2004a,b; *Moore and Hudson,* 2000), its near-IR detection on a TNO will be difficult. A more-promising approach might be to seek the mid-IR band near 3.5 μm that was used to identify H_2O_2 on Europa (*Carlson et al.,* 1999a).

There are cases for which characteristic IR features of CH_4-ice can be observed on TNO surfaces (*Grundy et al.,* 2002). This leads to questions about the products that might arise from the low-temperature irradiation of methane. Relevant experiments have been performed by several research groups (e.g., *Mulas et al.,* 1998; *Moore and Hudson,* 2003;

Baratta et al., 2003), and the consensus appears to be that radical-radical reactions can lead to more-complex hydrocarbons. For C_2H_6 (ethane) formation, the key path appears to be

$$CH_4 \rightarrow CH_3 + H \text{ (twice)}$$
$$CH_3 + CH_3 \rightarrow C_2H_6$$

although insertion reactions such as

$$CH_4 \rightarrow CH_2 + H_2 \text{ (or 2 H)}$$
$$CH_2 + CH_4 \rightarrow C_2H_6$$

probably play a role as well. In addition to C_2H_6, experiments have shown that the radiation products of frozen CH_4 include small hydrocarbons such as C_2H_4, C_2H_2, and C_3H_8 (e.g., *Mulas et al.,* 1998; *Moore and Hudson,* 2003; *Baratta et al.,* 2003). Overall, the safest prediction is that C_2H_6, a dominant product, will be present in CH_4-rich TNO ices.

The third molecule that has been reported to dominate a TNO surface is molecular nitrogen. Like H_2 and O_2, it has no permanent dipole moment, so its vibrational transitions are weak. Nevertheless, direct detection of N_2-ice has been made on Pluto (*Owen et al.,* 1993) and Triton (*Cruikshank et al.,* 1993) through the first overtone band of N_2. The radiation chemistry of N_2 ices is far more limited than that of either H_2O or CH_4, and only N_3 has been identified as a radiation product (*Hudson and Moore,* 2002). This radical was detected after ion bombardment of N_2-ice, and was identified through a single mid-IR feature, which rapidly decayed above 35 K.

To illustrate radiation-induced changes in an organic compound, we consider methanol (CH_3OH). *Brunetto et al.* (2005) presented both reflectance and transmission near-IR spectra (2.2–2.4 μm) of CH_3OH-ice at 16 K and 77 K, before and after irradiation with 30 keV He^+ ions and 200 keV H^+ ions. Their results confirmed the CO and CH_4 formation known from mid-IR studies. They also found evidence for a strong decrease in the intensity of the CH_3OH band at ~2.34 μm relative to the one at 2.27 μm. Figure 3 illustrates these near-IR results for H^+ irradiation at 16 K. In addition to the appearance of new spectral features, there is a change in the underlying spectral slope, indicating an alteration of the sample's color. We return to this observation in section 5, and to the radiation products of CH_3OH in section 4.3.

Table 2 summarizes radiation products for a wide variety of pure molecules, drawn from independent work in several laboratories. Temperatures for many of the experiments were well below those of TNOs, but in most cases this will make no difference in the radiation chemistry, which is not thermally driven.

4.3. Chemistry of H_2O-rich Ice Mixtures

Radiation chemistry experiments have been reported for essentially all common classes of organic molecules embedded in a H_2O-rich ice. Table 3 updates an earlier list (*Colangeli et al.,* 2005), starting with H_2O-hydrocarbon

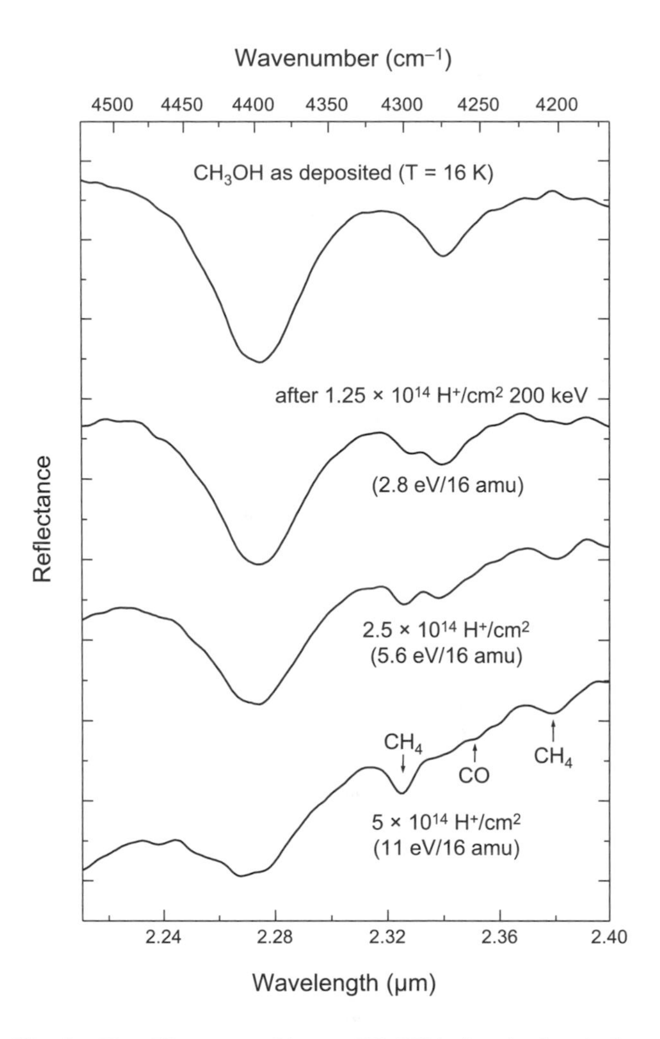

Fig. 3. Near-IR spectra of frozen CH_3OH before (top) and after various stages of irradiation, showing product formation. The four traces are offset for clarity (from *Brunetto et al.,* 2005).

mixtures. As with pure CH_4-ices, C_2H_6 is a radiation product in $H_2O + CH_4$ mixtures, which shows that H_2O does not block the reactions already given for methane. An important new product in these experiments is CH_3OH, thought to be made by radical-radical reactions as

$$CH_4 \rightarrow CH_3 + H$$
$$H_2O \rightarrow H + OH$$
$$CH_3 + OH \rightarrow CH_3OH$$

Recent work has shown that CH_2 insertion reactions such as

$$CH_4 \rightarrow CH_2 + H_2 \text{ (or 2 H)}$$
$$CH_2 + H_2O \rightarrow CH_3OH$$

also play a role in the CH_3OH formation (*Wada et al.,*2006). Spectra and selected band strengths for pure CH_3OH and $H_2O + CH_3OH$ mixtures are available to assist with analyzing TNO spectra (*Kerkhof et al.,* 1999), although optical constants are hard to locate (but see *Cruikshank et al.,* 1998).

Table 3 summarizes results for H_2O-rich mixtures containing the aforementioned radiation products, C_2H_6 and CH_3OH, of $H_2O + CH_4$ ices. If C_2H_6 is embedded in an

TABLE 2. Radiation products from one-component ices.

Ice	Reaction Products Identified in Ices	Least-Volatile Species	References
H_2O	H_2O_2	H_2O_2	[1,2,3,4]
CH_4	C_2H_2, C_2H_4, C_2H_6, C_3H_8, CH_3, C_2H_5	PAHs [4] and high molecular weight hydrocarbons	[5,6,7]
C_2H_6	CH_4, C_2H_2, C_2H_4	high molecular weight hydrocarbons	[8]
C_2H_4	CH_4 , C_2H_2, C_2H_6	high molecular weight hydrocarbons	[8]
C_2H_2	CH_4 [5], polyacetylene [8]	PAHs [4], polyacetylene [8]	[5,8,9]
CO	CO_2, C_3O_2, C_2O, C_4O, C_5O_2, C_7O_2	C_3O_2, C_5O_2, C_7O_2	[10,11,12]
CO_2	CO, O_3, CO_3	H_2CO_3 (from H^+ implantation) [11]	[10,13]
H_2CO	CO, CO_2, HCO, POM	polyoxymethylene (POM)	[14]
CH_3OH	CH_4, CO, CO_2, H_2CO, H_2O, $C_2H_4(OH)_2$, HCO, $HCOO^-$	$C_2H_4(OH)_2$	[15,16]
O_2	O_3		[9]
N_2	N_3		[17]
NH_3	NH_4^+	NH_4^+	[9,18]
HCN	HCN oligomers	HCN oligomers	[19]
CH_3CN	CH_4, H_2CCNH, CH_3NC, HCN	polymeric material	[20]
HCCCN	HCCNC (?)	polymeric material	[20]
HNCO	NH_4^+, OCN^-, CO, CO_2	NH_4OCN	[9]
SO_2	SO_3	S_8	[21,22]
H_2S	H_2S_2	H_2S_2	[22]
OCS	CO, CS_2	CS_2	[9]
$HC(O)CH_2OH$	CO, CO_2, CH_4, HCO, H_2CO, CH_3OH, $(CH_2OH)_2$	$(CH_2OH)_2$	[23]
$(CH_2OH)_2$	CH_4, H_2CO, CH_3OH?, CO, CO_2, $C(O)CH_2OH$	$HC(O)CH_2OH$	[23]

References: [1] *Gomis et al.* (2004a,b); [2] *Moore and Hudson* (2000); [3] *Zheng et al.* (2006); [4] *Loeffler et al.* (2006a); [5] *Kaiser and Roessler* (1998); [6] *Mulas et al.* (1998); [7] *Moore and Hudson* (2003); [8] *Strazzulla et al.* (2002); [9] Hudson and Moore (unpublished work); [10] *Gerakines and Moore* (2001); [11] *Trottier and Brooks* (2004); [12] *Loeffler et al.* (2005); [13] *Brucato et al.* (1997); [14] *Moore et al.* (2003); [15] *Hudson and Moore* (2000); [16] *Palumbo et al.* (1999); [17] *Hudson and Moore* (2002); [18] *Strazzulla and Palumbo* (1998); [19] *Gerakines et al.* (2004); [20] *Hudson and Moore* (2004); [21] *Moore* (1984); [22] *Moore et al.* (2007); [23] *Hudson et al.* (2005).

H_2O-rich TNO ice then experiments show that C_2H_5OH (ethanol) will also be present. The relevant chemical reactions are similar to those already presented for the formation of ethanol from $H_2O + CH_4$ mixtures. Mid-IR studies of irradiated $H_2O + CH_3OH$ ices show that $C_2H_4(OH)_2$, ethylene glycol is produced (*Hudson and Moore,* 2000), presumably by radical-radical coupling

$$CH_3OH \rightarrow H + CH_2OH \text{ (twice)}$$
$$CH_2OH + CH_2OH \rightarrow C_2H_4(OH)_2$$

Unfortunately, neither near-IR spectra nor optical constants have been published for H_2O-rich ices containing either C_2H_5OH or $C_2H_4(OH)_2$, although some mid-IR spectra are available (*Hudson et al.,* 2005). This is an all-too-common situation for most known or suspected TNO ices.

Continuing down Table 3, extensive work has been published on both H_2O + CO and $H_2O + CO_2$ ices. In the former, CO readily combines with H atoms to follow the sequence

$$CO \rightarrow HCO \rightarrow H_2CO \rightarrow CH_3O \text{ and/or } CH_2OH \rightarrow CH_3OH$$

leading to H_2CO (formaldehyde) and CH_3OH. Calculations of reaction yields are possible using intrinsic IR band strengths (*Hudson and Moore,* 1999). A similar sequence produces HCOOH (formic acid) by H and OH addition to CO (*Hudson and Moore,* 1999). In the case of $H_2O + CO_2$ ices, a major reaction product is H_2CO_3, carbonic acid, a molecule that long avoided direct laboratory detection. This molecule has been produced both photochemically and radiolytically, yields have been determined, its destruction rate has been measured, mid-IR band strengths are known, and isotopic variants have been examined. Enough is known about H_2CO_3 to safely predict that it will form in TNO ices containing H_2O and CO_2, and subjected to ionizing radiation (*Gerakines et al.,* 2000; *Brucato et al.,* 1997).

Mention already has been made of ammonia as a possible surface component of Quaoar (*Jewitt and Luu,* 2004), and there are reports of ammonia for Charon also (*Dumas et al.,* 2001; *Brown and Calvin,* 2000). In general, the radiation products of NH_3 ices have received little attention, although one would expect H_2 and N_2 to be formed and possibly N_2H_4 and NH_2OH as well. Radiolytic oxidation of NH_3 to form N_2 has been suggested (*Loeffler et al.,* 2006b) as a chemical energy source for the water ice plumes of Enceladus and to drive resurfacing on TNOs. Irradiation of $H_2O + NH_3$ mixtures demonstrates that ammonia is more easily lost than H_2O because of both a greater sputtering yield and a more effective chemical alteration. Thus the

TABLE 3. Radiation products from H_2O-dominated two-component ices.

Ice Mixture	Reaction Products Identified in Ices	References
$H_2O + CH_4$	CH_3OH, C_2H_5OH, C_2H_6, CO, CO_2	[1]
$H_2O + C_2H_6$	CH_4, C_2H_4, C_2H_5OH, CO, CO_2, CH_3OH	[2]
$H_2O + C_2H_2$	C_2H_5OH, CH_3OH, C_2H_6, C_2H_4, CO, CO_2, CH_4, C_3H_8, $HC(=O)CH_3$, $CH_2CH(OH)$	[1]
H_2O + CO	CO_2, HCO, H_2CO, CH_3OH, HCOOH, $HCOO^-$, H_2CO_3	[3]
$H_2O + CO_2$	H_2CO_3, CO, O_3, H_2O_2	[4,5]
$H_2O + H_2CO$	CO, CO_2, CH_3OH, HCO, HCOOH, CH_4	[3]
$H_2O + CH_3OH$	CO, CO_2, H_2CO, HCO, CH_4, $C_2H_4(OH)_2$, $HCOO^-$	[6,7]
$H_2O + O_2$	O_3, H_2O_2, HO_2, HO_3	[8,9]
$H_2O + N_2$	H_2O_2	[8]
$H_2O + NH_3$	NH_4^+	[2,10]
H_2O + HCN	CN^-, HNCO, OCN^-, $HC(=O)NH_2$, NH_4^+ (?), CO, CO_2	[11]
$H_2O + CH_3CN$	H_2CCNH, CH_4, OCN^-, HCN	[12]
H_2O + HCCCN	OCN^-	[12]
H_2O + HNCO	NH_4^+, OCN^-, CO, CO_2	[2]
$H_2O + SO_2$	H_3O^+, SO_4^{2-}, HSO_4^{2-}, HSO_3^{2-}	[13]
$H_2O + H_2S$	H_2S_2, SO_2	[13]
H_2O + OCS	CO, CO_2, SO_2, H_2CO (?), H_2O_2 (?)	[2]
$H_2O + HC(O)CH_2OH$	CO, CO_2	[14]
$H_2O + (CH_2OH)_2$	CO, CO_2, H_2CO, $HC(O)CH_2OH$	[14]

References: [1] *Moore and Hudson* (1998); [2] Hudson and Moore (unpublished data); [3] *Hudson and Moore* (1999); [4] *Brucato et al.* (1997); [5] *Gerakines et al.* (2000); [6] *Hudson and Moore* (2000); [7] *Palumbo et al.* (1999); [8] *Moore and Hudson* (2000); [9] *Cooper et al.* (2006b); [10] *Strazzulla and Palumbo* (1998); [11] *Gerakines et al.* (2004); [12] *Hudson and Moore* (2004); [13] *Moore et al.* (2007); [14] *Hudson et al.* (2005).

water/ammonia ratio progressively increases with radiation dose, and ammonia IR bands become less evident in transmission mid-IR spectra (*Strazzulla and Palumbo,* 1998). The observation of ammonia or an ammonia hydrate on TNO surfaces (e.g., on Quaoar) would then be evidence for a fresh exposed surface.

Ammonia also should play a role in the acid-base chemistry of TNO ices and experiments support this expectation. Irradiation of H_2O + CO ices produces HCOOH (formic acid), but similar experiments on $H_2O + CO + NH_3$ mixtures show no HCOOH but rather the $HCOO^-$ (formate) and NH_4^+ (ammonium) ions, as seen in Fig. 4 (*Hudson and Moore,* 2000). These ions are sufficiently stable, so as to accumulate on a TNO surface, or anywhere that sufficient energetic processing occurs. Beyond NH_4^+, other ions that have been studied in laboratory H_2O-rich ices include OCN^- from HNCO (*Hudson et al.,* 2001) and CN^- from HCN (*Moore and Hudson,* 2003). An earlier survey reported a near-IR band for NH_4^+, but nothing distinct for the other ions already mentioned (*Moore et al.,* 2003). Older laboratory work (*Maki and Decius,* 1958) shows that OCN^- may have near-IR bands suitable for searches in TNO spectra, but the experiments need to be repeated at more relevant temperatures and in the presence of H_2O-ice. H_3O^+ and OH^- also are likely in TNO ices, but difficult to detect by IR methods as they lack strong unobscured bands.

Of the organics remaining in Table 3 we mention only the nitriles, molecules containing the C≡N functional group. Ion-irradiated and UV-photolyzed nitrile-containing ices recently have been studied to understand their fate in H_2O-rich environments (*Hudson and Moore,* 2004). In all cases, nitriles were found to be unstable toward oxidation to OCN^-. This observation makes it unlikely that nitriles will be found in H_2O-rich TNO ices. Energetic processing also was found to give the H-atom transfer $CH_3CN \rightarrow H_2C{=}C{=}NH$ (keteni-

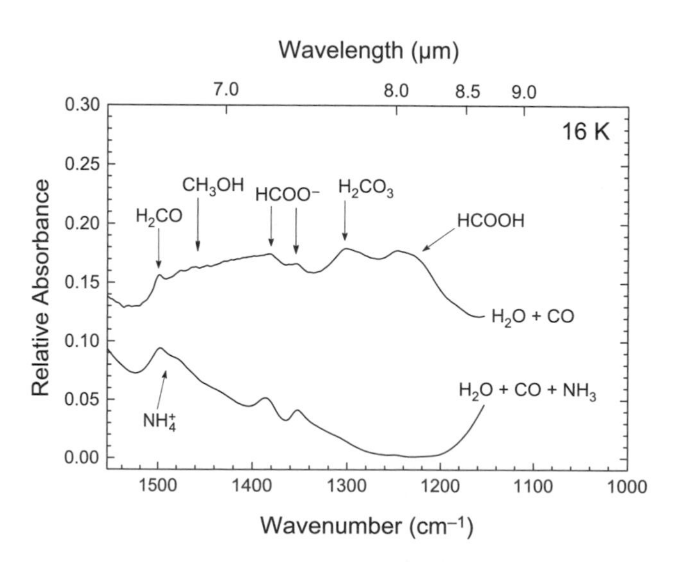

Fig. 4. IR spectra of two irradiated laboratory ices at 16 K, showing the influence of acid-base chemistry. The upper trace is an H_2O + CO (5:1) ice and the lower trace is an $H_2O + CO + NH_3$ (5:1:1) ice. Both were irradiated to about 22 eV 16-amu molecule^{-1} (after *Hudson et al.,* 2001).

mine). While CH_3CN (acetonitrile) has long been known in interstellar clouds, H_2CCNH has not, and so was predicted on the basis of the experiments (*Hudson and Moore,* 2004). Its recent discovery toward the star-forming region Sagittarius B2(N) (*Lovas et al.,* 2006) attests to the predictive power of the experimental approach.

Finally, Table 3 includes results from three sulfur-containing molecules, H_2S, OCS, and SO_2. All are known to be present both in comets and the interstellar medium, and OCS has been reported in the solid phase (*Palumbo et al.,* 1997). Given the oxidative power of ion-irradiated H_2O-ice, it is not surprising that the $H_2S \rightarrow SO_2 \rightarrow SO_4^{2-}$ sequence was found in recent laboratory measurements (*Moore et al.,* 2007). The implication for TNOs is that sulfur may well be present as SO_4^{2-}, similar to what is seen for Europa (*Carlson et al.,* 1999b).

4.4. Chemistry of N_2-rich Ice Mixtures

As already mentioned, models of Pluto and Triton spectra suggest that some TNO surfaces are dominated by N_2. Unfortunately, laboratory studies of N_2-rich ices are far fewer in number than those for H_2O-rich ices. Only a few examples of N_2-rich ice chemistry will be given here, involving CH_4 and CO (*Moore and Hudson,* 2003; *Palumbo et al.,* 2004).

Ion-irradiated $N_2 + CH_4$ ices near 12 K produce nitrogen-containing products HCN, HNC, and CH_2N_2, as well as NH_3 (*Moore and Hudson,* 2003). In addition, several hydrocarbons are identified, but their abundances depend on the initial N_2/CH_4 ratio. Figure 5 shows the 2.8–4-μm region of an irradiated $N_2 + CH_4$ mixture for three different initial N_2/CH_4 ratios, 100, 50, and 4, compared to pure irradiated CH_4. The abundances of aliphatic hydrocarbons, C_2H_6 and C_3H_8, are enhanced as the concentration of CH_4 increases. Therefore, TNO terrains rich in CH_4 are expected to have more C_2H_6 and C_3H_8 than those where CH_4 is diluted in N_2. When these ices are warmed to ~35 K, sharp features of HCN and HNC decrease as acid-based reactions produce both NH_4^+ and CN^- ions. Since these ions are stable under vacuum to about 150 K, they can accumulate on TNO surfaces. Diazomethane, CH_2N_2, also was seen in these experiments and was stable to at least ~35 K. The hydrocarbons C_2H_2, C_2H_6, and C_3H_8 are present to at least the 60–70 K range.

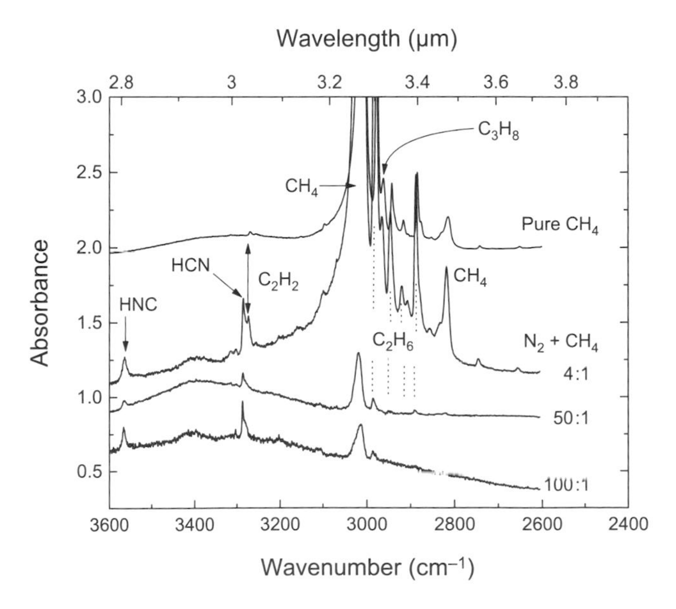

Fig. 5. Infrared spectra of 0.8 MeV irradiated pure CH_4 (top) and three $N_2 + CH_4$ mixtures. The formation of HCN, HNC, and hydrocarbons is indicated. The relative yields of HCN and HNC radiation products are greatest for the 100:1 mixture (*Moore and Hudson,* 2003).

Ion irradiations of N_2 + CO ices at 12 K produce the free radicals OCN, NO, NO_2, and N_3, as well as the C_3O_2 (carbon suboxide) and N_2O (nitrous oxide) molecules. The latter two are the species most likely to persist on TNO surfaces, the radicals being less stable (more reactive).

Mid-IR spectra of irradiated $N_2 + CO + CH_4$ ices are largely what would be expected from the binary mixtures just described. The main new product is HNCO, seen at 12 K. On warming to ~35 K, HNCO reacts with NH_3 to form OCN^- and NH_4^+, which are stable to about 200 K.

4.5. Residual Materials

It may be concluded from the foregoing that TNOs can possess surfaces made of both irradiated ices and residual species remaining after at least partial sublimation of more-volatile molecules such as N_2, CO, and CH_4. The sources of ions such as OCN^- CN^-, and NH_4^+ have already been discussed. Other molecules that might be residual materials include $C_2H_4(OH)_2$, C_3O_2, and H_2CO_3. Mid-IR band strengths have been published for all three, and near-IR values are available for the first two (*Moore et al.,* 2003; *Hudson et al.,* 2005).

In concluding this section we note that the doses used in most ice experiments are in the range of about 1–20 eV/16-amu-molecule. Doses of this size typically lead to a state of chemical equilibrium in which the ice's composition changes only very slowly. However, with increasing dose, and gradual loss of H_2, the carbon-to-hydrogen ratio of the original sample slowly rises (*Strazzulla et al.,* 1991). This process has important implications for understanding TNO surface colors and will be explored in the next section. Here we note that this carbonization is very different from the formation of specific long-chain molecules, such as polymeric HCN or polymeric H_2CO. Although the latter two polymers are sometimes invoked by astrochemists, the formation of unique polymeric materials in ice mixtures remains problematic.

5. SPECTROSCOPY AND TRANSNEPTUNIAN OBJECT COLORS

The measurements of irradiated ices just described were typically performed using transmission spectroscopy. We now turn to investigations of more-refractory materials, with spectra often recorded as diffuse reflectance measurements in the visible and near-IR regions (0.3–2.7 μm). These experiments investigate TNO chemistry through the study of spectral features, and reveal radiation-induced color changes in the underlying spectral continuum of TNO candidate materials. Diffuse reflectance spectroscopy also allows opaque refractories, such as silicates and various carbonaceous materials, to be studied more easily than by transmission methods. That such materials are important for understanding TNO surfaces is shown by the fit of the observed spectrum of the Centaur object 5145 Pholus to olivine (a silicate), tholins (a refractory organic material), H_2O-ice, frozen methanol, and carbon black (see *Cruikshank et al.,* 1998).

Among the materials studied by reflectance are terrestrial silicates (e.g., olivine and pyroxene) and carbons (e.g., natural bitumens such as asphaltite and kerite), meteorites (carbonaceous chondrites, ordinary chondrites, diogenites), and frozen ices such as methanol, methane, and benzene. Samples have been irradiated with different ions (H^+, He^+, Ar^+, Ar^{++}) having energies from 30 to 400 keV. All these materials show important spectral changes in their underlying visible-near-IR continuum after irradiation, usually reddening and darkening. Natural bitumens (asphaltite and kerite), however, are a noteworthy exception, being very dark in the visible region and possessing red-sloped spectra in the visible and near-IR (*Moroz et al.,* 1998). Ion irradiation experiments (*Moroz et al.,* 2004) showed that radiation-induced carbonization can gradually neutralize these spectral slopes.

The spectral slopes of ion-irradiated silicatic materials, namely olivine, pyroxene, and the olivine-rich meteorite Epinal, have been compared with observations of some S-type near-Earth asteroids. It has been found that the formation of solid-phase vacancies by solar wind ions can redden surfaces on a timescale of about 10^5 yr (*Strazzulla et al.,* 2005b). This means that radiation processing is the most efficient explanation for the observed color variety of both near-Earth and S-type main-belt asteroids (*Strazzulla et al.,* 2005b; *Brunetto and Strazzulla,* 2005; *Marchi et al.,* 2005). As an example of this work, Fig. 6 shows reflectance spectra (0.7–2.7 μm; normalized to 1 at 0.7 μm) of the meteorite Epinal before and after ion irradiation. The spectral reddening is quite evident. Another example comes from ion bombardment experiments with two carbonaceous chondrites, CV3 Allende and CO3 Frontier Mountain 95002. This work showed that those meteorites also are reddened by irradiation, regardless of whether the sample was a powder or a pressed pellet (*Lazzarin et al.,* 2006).

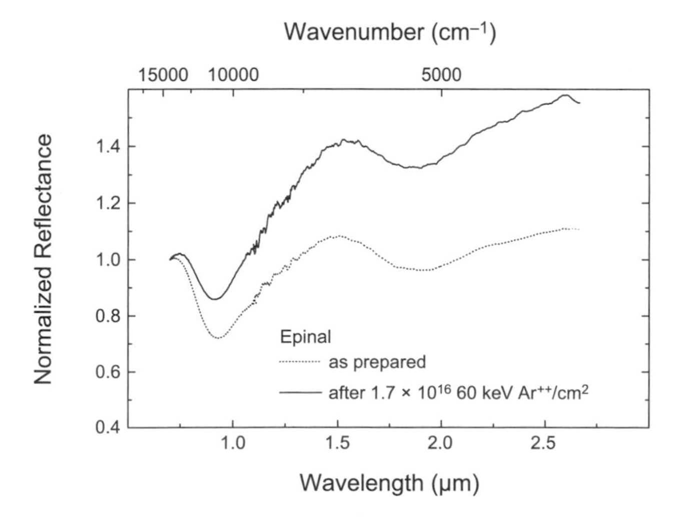

Fig. 6. Reflectance spectra (0.7–2.7 μm) of meteorite Epinal (H5) before and after ion irradiation. Spectra have been normalized to 1 at 0.7 μm (adapted from *Brunetto et al.,* 2005).

In addition to these results with complex starting materials, changes in colors also are found when fairly simple ices, such as CH_3OH, CH_4, C_6H_6, and $H_2O + CH_4 + N_2$ mixtures, are irradiated to doses of about 1000 eV/16-amu-molecule. Samples have been analyzed both by reflectance and Raman spectroscopies, and the formation of an organic refractory residue, and eventually the formation of amorphous carbon (*Ferini et al.,* 2004; *Palumbo et al.,* 2004), are seen. These materials cause a strong reddening and darkening of the visible and near-IR spectra (*Brunetto et al.,* 2006) of the original sample. Furthermore, *Brunetto et al.* (2006) showed that for these icy samples, it is the total dose (elastic plus inelastic contributions) that plays the main role in the reddening process. This is different from the case of silicates (*Brunetto and Strazzulla,* 2005) and bitumens (*Moroz et al.,* 2004), in which reddening effects are due solely to the elastic collisions between ions and target nuclei.

Figure 7 shows near-IR reflectance spectra of CH_4 and CH_3OH before and after irradiation at 16 K and 77 K, respectively. The spectra of the unprocessed ices are flat and bright, and show absorptions due to vibrational overtones and combinations. After irradiation, the original bands decrease and other features appear indicating the formation of new molecules discussed in section 4. Here we emphasize the change in the slope of the continuum after irradiation. It is clear from Fig. 7 that the spectrum of each compound becomes darker and redder with increasing dose.

Radiation-induced color variations have been compared with the observed spectra of some Centaurs and TNOs (after *Barucci and Peixinho,* 2005), and it has been shown (*Brunetto et al.,* 2006) that the observed TNO colors can be reproduced by ion irradiation experiments. This suggests that these objects possess a refractory organic crust developed after prolonged irradiation by cosmic ions, in analogy with what was previously suggested for Oort cloud comets (*Strazzulla et al.,* 1991). Many of the TNOs considered possess red colors that correspond to radiation doses between 10 and 100 eV/16-amu-molecule, but more-neutral-colored objects could have accumulated much higher dosages (*Moroz et al.,* 2003, 2004). It is estimated (*Strazzulla et al.,* 2003) that the surface layers (1–100 μm) of objects between 85 AU (solar wind termination shock) and the very local interstellar medium accumulate 100 eV/16-amu-molecule on timescales of 10^6–10^9 yr. This suggests that many icy objects in the outer solar system develop an irradiation mantle on

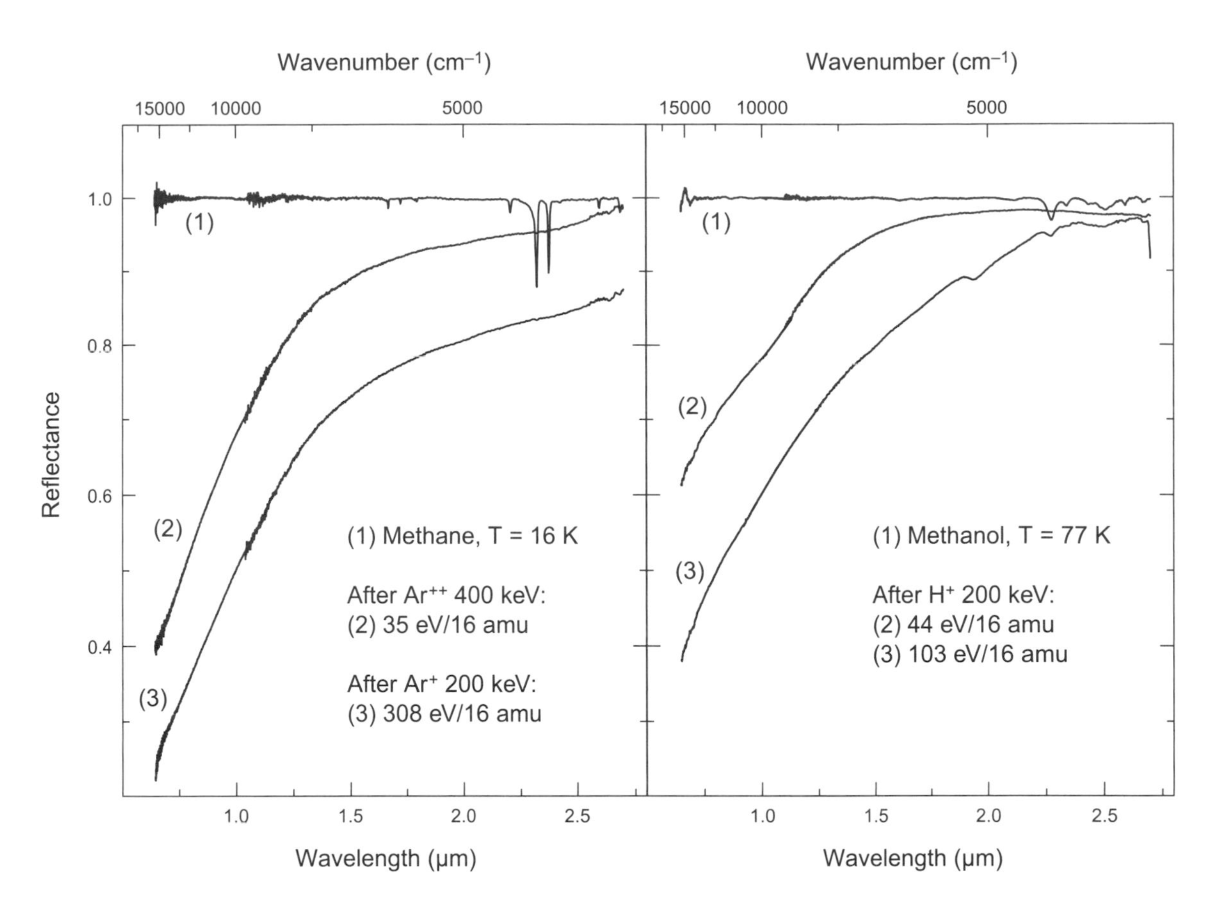

Fig. 7. Absolute visible-near-IR reflectance spectra of CH_4 (16 K) and CH_3OH (77 K) before and after irradiation with 200 keV Ar^+, 400 keV Ar^{++}, and 200 keV H^+ (adapted from *Brunetto et al.,* 2006).

timescales of 10^8 yr. From *Cooper et al.* (2003) and updates in Table 1, doses at 40–50 AU are lower, and volatile materials (e.g., N_2, CO, CH_4) can be better preserved at TNO surfaces.

The laboratory work summarized here and in section 4 implies that TNOs can possess a dark radiation mantle covering fresh subsurface ice hidden from observers. Transneptunian objects on which molecular ices (e.g., H_2O or CH_4) have been observed could either be poorly irradiated, have recently refreshed surfaces, and/or be lacking in carbon-bearing surface species to be converted into dark materials. In the first case, colored spectra are predicted depending on the surface portion recently refreshed. In the second case, spectra should be relatively flat with higher-than-average albedos.

Laboratory experiments also have shown that IR features of crystalline H_2O-ice are converted to those of amorphous material by irradiation at 10 K, that some crystalline IR features persist after irradiation at 50 K, and that at 70 K and higher the IR spectrum shows only slight changes (*Moore and Hudson,* 1992; *Strazzulla et al.,* 1992; *Mastrapa and Brown,* 2006). The observation of crystalline TNO H_2O-ice, in cases where it has been possible to distinguish between amorphous and crystalline material, is at present explained by freshly resurfaced layers (*Moore and Hudson,* 1992; *Strazzulla et al.,* 1992).

6. ASTROBIOLOGY

Important goals of astrobiology include understanding the origin of life on Earth and the possible forms of life present now, or in the past, on astronomical objects in and beyond the solar system. With this in mind, it must be admitted that TNO environments are very hostile to the origin or permanence of life as now exists on Earth. However, the possibility that terrestrial life, or the molecular ingredients from which such life originated, had an extraterrestrial source is actively debated. The "space vehicles" that could have delivered biologically relevant materials to the early Earth are thought to be meteorites, interplanetary dust particles, and comets from the Oort cloud and the Kuiper belt, supporting the relevance of TNOs to astrobiology.

A consideration of the chemistry reviewed in this chapter also supports the astrobiological importance of TNOs. First, the chemical processes in section 4, represented by products in Tables 2 and 3, are quite general for all irradiated ices. Reactions involving either bond breakage in a neutral molecule (e.g., $CH_3OH \rightarrow CO + 2\ H_2$) or the addition of a radical to a neutral (e.g., $H + CO \rightarrow HCO$) may require energy, but such can be provided by radiolysis. In contrast, little or no activation energy is encountered in most radical-radical combinations, proton transfers (acid-base reactions), or electron transfers (redox chemistry). Regardless of energetic considerations, all the aforementioned reactions are operative in irradiated ices, and do not depend on a reactant molecule's size. This implies that second- and third-generation products can form with accompanying increases in molecular complexity (e.g., chain length, molecular size, functional groups), albeit with lower yields. In this light, the reaction products seen to date are only the more-abundant ones in laboratory ices and, by extension, in TNO ices. This is important since biomolecules tend to be more complex than those found in either Table 2 or 3.

Many publications are available to support these points and to demonstrate biomolecular syntheses in ices, but only a few examples can be described here. Amino acids have been shown to form in processed ice mixtures after hydrolysis of the resulting radiation residues (*Kobayashi et al.,* 1995), while *Bernstein et al.* (2002) and *Muñoz Caro et al.* (2002) have found that UV-photolyzed ice mixtures contain precursors that lead to a suite of amino acids, some of which are found in meteorites. *Hudson et al.* (2005) reported the radiation synthesis of the simple sugar glycolaldehyde ($HOCH_2C(O)H$) in irradiated ethylene glycol, itself a product of CH_3OH radiolysis. Polycyclic aromatic hydrocarbon (PAH) molecules undergo sidegroup addition reactions to form alcohols, quinones, and ethers when irradiated in H_2O-ice (*Bernstein et al.,* 1999, 2003). *Tuleta et al.* (2001) studied the reaction products of simultaneously flowing H_2O vapor over 150-K anthracene (a PAH) and irradiating with 3.5-keV H_2^+ ions. Sidegroup addition to anthracene was found along with the oxidation product, anthraquinone.

In addition to experiments with icy mixtures, the growth of molecular chains, necessary for biomolecules to form, has been studied with model systems. *Strazzulla and Moroz* (2005) irradiated thin asphaltite films, both pure and covered with H_2O-ice layers. (Asphaltite was used as an analog of a complex carbonaceous material having both aliphatic and aromatic components.) After irradiation of both pure and H_2O-ice covered samples, carbon-carbon bonds characteristic of linear chains were found (carbynoids and cumulenes), as were new aromatic features. Low-temperature irradiations of solid acetylene (*Strazzulla et al.,* 2002), benzene, and cluster-assembled carbon thin films (*Strazzulla and Baratta,* 1991; *Strazzulla et al.,* 2005a) also showed evidence of carbon-chain extension, with the final products remaining present after warming the samples to ~300 K. The synthesis of such products appears to be typical of many irradiated hydrogen-bearing carbonaceous materials. The formation of carbon-carbon and carbon-nitrogen (section 4) triple bonds is particularly relevant to astrobiology. These materials, once delivered to the early Earth, could have been among the first ingredients to develop biochemical activity.

While the formation of larger structures from smaller ones is important, to understand equilibrium molecular abundances one must also understand destruction processes. As an example of how laboratory methods contribute to solving such problems, we end with the formamide molecule, $HCONH_2$. Formamide is made from the four major biogenic elements, has been observed in the interstellar medium (*Millar,* 2004), in the long-period comet C/1995 O1 Hale-Bopp (*Bockelée-Morvan et al.,* 2000), and tentatively in young stellar objects W33A (*Schutte et al.,* 1999) and NGC 7538 IRS9 (*Raunier et al.,* 2004). Formamide is made by room-temperature HCN hydrolysis, and it is the most-abundant pyrolysis product of HCN-polymer. Its role as a prebiotic precursor for the synthesis of nucleobases has been shown under a variety of conditions. In particular, various inorganic materials, such as cosmic-dust analogs, can catalyze formamide condensation to make many other compounds including purine and pyrimidine bases (*Brucato et al.,* 2006, and references therein). The radiation chemistry of frozen $HCONH_2$ has recently been studied at 20 K (*Brucato et al.,* 2006), and CO, CO_2, N_2O, isocyanic acid (HNCO), and ammonium cyanate ($NH_4^+OCN^-$) decomposition products were identified. Some of these species were stable even after warming to room temperature.

7. NEEDS AND CHALLENGES

Although extensive laboratory work has been done on solar system ice analogs, important and difficult tasks remain. High among these challenges is a clear demonstration of the extent and types of space weathering experienced by TNO surfaces. This chapter has documented some of the possible radiation-induced changes in spectra, colors, and chemical composition, but distinguishing these from other effects remains problematic.

Further quantification of previous work also needs to be done. Intrinsic strengths for many IR spectral features are known, but optical constants of many irradiated icy and refractory materials are needed. Only with laboratory optical constants of irradiated materials can such be included in quantitative models (e.g., Skuratov and Hapke) and fits of TNO spectra. However, the measurement of optical constants is difficult, and so many TNO candidate materials have not been studied.

The gaps in Table 2 and 3 also suggest avenues for future TNO ice experiments. Most entries in the tables come from publications covering only a few temperatures and but one radiation source. Additional work is needed to search for possible new product molecules, particularly those of astrobiological interest, and ions other than NH_4^+ and OCN^-.

We also note that there may be problems of scale to consider, since the behavior of deep-volume bulk ices irradiated at centimeter-to-meter depths by highly penetrating energetic particles and electromagnetic radiations is unknown. Input doses can be calculated with sophisticated radiation transport codes such as GEANT, but radiolytic yields can only be extrapolated from measurements on ices under ~1 mm in thickness. What are the long-range effects of penetrating particle ionization and mobile gas production in thicker ices? Are reaction rates enhanced by large internal surface areas in porous volume ice? Are chemical pathways altered by mobile electrons or H^+ in a bulk irradiated ice? Such questions require the transition of laboratory sample and irradiation system dimensions from microscopic to macroscopic scales.

Finally, there is a need for the chemistry represented in sections 4 (ices) and 5 (refractories) to be united in future laboratory experiments. For example, how will reactions to extend the length of carbon chains be altered when H_2O-ice is present? Conversely, Table 3 documents the radiation products expected in various H_2O-rich ices, but to what extent will product distributions be different in the presence of refractory materials?

Acknowledgments. The work in the Laboratory of Experimental Astrophysics in Catania has been financially supported by the National Institute for Astrophysics (INAF) in the framework of a national project. R.L.H. and M.H.M. acknowledge support through NASA's Outer Planets, Planetary Atmospheres, and Planetary Geology and Geochemistry programs. Recent support through the NASA Astrobiology Institute's Goddard Center for Astrobiology is also acknowledged. J.F.C. acknowledges past or current support from NASA's Heliophysics, Jovian System Data Analysis, Planetary Atmospheres, and Outer Planets programs.

REFERENCES

Anders E. and Grevesse N. (1989) Abundances of the elements — meteoritic and solar. *Geochim. Cosmochim. Acta, 53,* 197–214.

Bagnulo S., Boehnhardt H., Muinonen K., Kolokolova L., Belskaya I., and Barucci M. A. (2006) Exploring the surface properties of transneptunian objects and Centaurs with polarimetric FORS1/VLT observations. *Astron. Astrophys., 450,* 1239–1248.

Baratta G. A. and Palumbo M. E. (1998) Infrared optical constants of CO and CO_2 thin icy films. *J. Opt. Soc. Am. A, 15,* 3076–3085.

Baratta G. A., Leto G., and Palumbo M. E. (2002) A comparison of ion irradiation and UV photolysis of CH_4 and CH_3OH. *Astron. Astrophys., 384,* 343–349.

Baratta G. A., Domingo M., Ferini G., Leto G., Palumbo M. E., Satorre M. A., and Strazzulla G. (2003) Ion irradiation of CH_4-containing icy mixtures. *Nucl. Instr. Meth. B, 209,* 283–287.

Barucci M. A. and Peixinho N. (2005) Trans-Neptunian objects' surface properties. In *Asteroids, Comets, Meteors* (D. Lazzaro et al., eds.), pp. 171–190. IAU Symposium No. 229, Cambridge Univ., Cambridge.

Bernstein M. P., Sandford S. A., Allamandola L. J., Gillette J. S., Clemett S. J., and Zare R. N. (1999) UV irradiation of polycyclic aromatic hydrocarbons in ices: Production of alcohols, quinones, and ethers. *Science, 283,* 1135–1138.

Bernstein M. B., Dworkin J. P., Sandford S. A., Cooper G. W., and Allamandola L. J. (2002) Racemic amino acids from the ultraviolet photolysis of interstellar ice analogues. *Nature, 416,* 401–403.

Bernstein M. P., Moore M. H., Elsila J. E., Sandford S. A., Allamandola L. J., and Zare R. N. (2003) Side group addition to the polycyclic aromatic hydrocarbon coronene by proton irradiation in cosmic ice analogs. *Astrophys. J. Lett., 582,* L25–L29.

Bernstein M. B., Cruikshank D. P., and Sandford S. A. (2005) Near-infrared laboratory spectra of solid H_2O/CO_2 and CH_3OH/CO_2 ice mixtures. *Icarus, 179,* 527–534.

Bernstein M. B., Cruikshank D. P., and Sandford S. A. (2006) Near-infrared spectra of laboratory H_2O-CH_4 ice mixtures. *Icarus, 181,* 302–208.

Bockelée-Morvan D., Lis D. C., Wink J. E., Despois D., Crovisier J., Bachiller R., Benford D. J., Biver N., Colo P., Davies J. K., Gérard E., Germain B., Houde M., Mehringer D., Moreno R., Paubert G., Phillip T. G., and Raue H. (2000) New molecules found in comet C/1995 O1 (Hale-Bopp). Investigating the link between cometary and interstellar material. *Astron. Astrophys., 353,* 1101–1114.

Brown M. E. and Calvin W. M. (2000) Evidence for crystalline water and ammonia ices on Pluto's satellite Charon. *Science, 287,* 107–109.

Brown R. H., Cruikshank D. P., and Pendleton Y. (1999) Water ice on Kuiper belt object 1996 TO_{66}. *Astrophys. J. Lett., 519,* L101–L104.

Brucato J. R., Palumbo M. E., and Strazzulla G. (1997) Carbonic acid by ion implantation in water/carbon dioxide ice mixtures. *Icarus, 125,* 135–144.

Brucato J. R., Baratta G. A., and Strazzulla G. (2006) An infrared study of pure and ion irradiated frozen formamide. *Astron. Astrophys., 455,* 395–399.

Brunetto R. and Strazzulla G. (2005) Elastic collisions in ion irradiation experiments: A mechanism for space weathering of silicates. *Icarus, 179,* 265–273.

Brunetto R., Orofino V., and Strazzulla G. (2005) Space weathering on minor bodies induced by ion irradiation: Some experimental results. *Mem. S. A. Ital. Suppl., 6,* 45–50.

Brunetto R., Barucci M. A., Dotto E., and Strazzulla G. (2006) Ion irradiation of frozen methanol, methane, and benzene: Linking to the colors of Centaurs and trans-Neptunian objects. *Astrophys. J., 644,* 646–650.

Burlaga L. F., Ness N. F., Acuna M. H., Lepping R. P., Connerney J. E. P., Stone E. C., and McDonald F. B. (2005) Crossing the termination shock into the heliosheath: Magnetic fields. *Science, 309,* 2027–2029.

Carlson R. W., Anderson M. S., Johnson R. E., Smythe W. D., Hendrix A. R., Barth C. A., Soderblom L. A., Hansen G. B., McCord T. B., Dalton J. B., Clark R. N., Shirley J. H., Ocampo A. C., and Matson D. L. (1999a) Hydrogen peroxide on Europa. *Science, 283,* 2062–2064.

Carlson R. W., Johnson R. E., and Anderson M. S. (1999b) Sulfuric acid on Europa and the radiolytic sulfur cycle. *Science, 286,* 97–99.

Colangeli L., Brucato J. R., Bar-Nun A., Hudson R. L., and Moore M. H. (2005) Laboratory experiments on cometary materials. In *Comets II* (M. C. Festou et al., eds.), pp. 695–717. Univ. of Arizona, Tucson.

Cooper J. F., Johnson R. E., Mauk B. H., Garrett H. B., and Gehrels N. (2001) Energetic ion and electron irradiation of the icy Galilean satellites. *Icarus, 149,* 133–159.

Cooper J. F., Christian E. R., Richardson J. D., and Wang C. (2003) Proton irradiation of Centaur, Kuiper belt, and Oort cloud objects at plasma to cosmic ray energy. *Earth Moon Planets, 92,* 261–277.

Cooper J. F., Hill M. E., Richardson J. D., and Sturner S. J. (2006a) Proton irradiation environment of solar system objects in the heliospheric boundary regions. In *Physics of the Inner Heliosheath* (J. Heerikhuisen et al., eds.), pp. 372–379. AIP Conf. Proc. 858, American Institute of Physics, New York.

Cooper P. D., Moore M. H., and Hudson R. L. (2006b) Infrared detection of HO_2 and HO_3 radicals in water ice. *J. Phys. Chem. A, 110,* 7985–7988.

Cruikshank D. P., Roush T. L., Owen T. C., Geballe T. R., de Bergh C., Schmitt B., Brown R. H., and Bartholomew M. J. (1993) Ices on the surface of Triton. *Science, 261,* 742–745.

Cruikshank D. P., Roush T. L., Bartholomew M. J., Geballe T. R., Pendleton Y. J., White S. M., Bell J. F. III, Davies J. K., Owen T. C., de Bergh C., Tholen D. J., Bernstein M. P., Brown R. H., Tryka K. A., and Dalle Ore C. M. (1998) The composition of Centaur 5145 Pholus. *Icarus, 135,* 389–407.

Decker R. B., Krimigis S. M., Roelof E. C., Hill M. E., Armstrong T. P., Gloeckler G., Hamilton D. C., and Lanzerotti L. J. (2005) Voyager 1 in the foreshock, termination shock, and heliosheath. *Science, 309,* 2020–2024.

Douté S., Schmitt B., Quirico E., Owen T. C., Cruikshank D. P., de Bergh C., Geballe T. R., and Roush T. L. (1999) Evidence for methane segregation at the surface of Pluto. *Icarus, 142,* 421–444.

Dumas C., Terrile R. J., Brown R. H., Schneider G., and Smith B. A. (2001) Hubble Space Telescope NICMOS spectroscopy of Charon's leading and trailing hemispheres. *Astrophys. J., 121,* 1163–1170.

Elliot J. L. and Kern S. D. (2003) Pluto's atmosphere and a targeted-occultation search for other bound KBO atmospheres. *Earth Moon Planets, 92,* 375–393.

Ferini G., Baratta G. A., and Palumbo M. E. (2004) A Raman study of ion irradiated icy mixtures. *Astron. Astrophys., 414,* 757–766.

Gerakines P. A. and Moore M. H. (2001) Carbon suboxide in astrophysical ice analogs. *Icarus, 154,* 372–380.

Gerakines P. A., Moore M. H., and Hudson R. L. (2000) Carbonic acid production in $H_2O + CO_2$ ices: UV photolysis vs. proton bombardment. *Astron. Astrophys., 357,* 793–800.

Gerakines P. A., Moore M. H., and Hudson R. L. (2004) Ultraviolet photolysis and proton irradiation of astrophysical ice analogs containing hydrogen cyanide. *Icarus, 170,* 204–213.

Gerakines P. A., Bray J. J., Davis A., and Richey C. R. (2005) The strengths of near-infrared absorption features relevant to interstellar and planetary ices. *Astrophys. J., 620,* 1140–1150.

Gomis O., Satorre M. A., Strazzulla G., and Leto G. (2004a) Hydrogen peroxide formation by ion implantation in water ice and its relevance to the Galilean satellites. *Planet. Space Sci., 52,* 371–378.

Gomis O., Leto G., and Strazzulla G. (2004b) Hydrogen peroxide production by ion irradiation of thin water ice films. *Astron. Astrophys., 420,* 405–410.

Grundy W. M., Buie M. W., and Spencer J. R. (2002) Spectroscopy of Pluto and Triton at 3–4 microns: Possible evidence for wide distribution of nonvolatile solids. *Astron. J., 124,* 2273–2278.

Hall L. A., Heroux L. J., and Hinterregger H. E. (1985) Solar ultraviolet irradiance. In *Handbook of Geophysics and the Space Environment* (A. S. Jura, ed.), pp. 2-1 to 2-21. Air Force Geophysics Laboratory, Hanscom AFB, Massachusetts.

Hudson R. L. and Moore M. H. (1999) Laboratory studies of the formation of methanol and other organic molecules by water + carbon monoxide radiolysis: Relevance to comets, icy satellites, and interstellar ices. *Icarus, 140,* 451–461.

Hudson R. L. and Moore M. H. (2000) IR spectra of irradiated cometary ice analogues containing methanol: A new assignment, a reassignment, and a nonassignment. *Icarus, 145,* 661–663.

Hudson R. L. and Moore M. H. (2002) The N_3 radical as a discriminator between ion-irradiated and UV-photolyzed astronomical ices. *Astrophys. J., 568,* 1095–1099.

Hudson R. L. and Moore M. H. (2004) Reactions of nitriles in ices relevant to Titan, comets, and the interstellar medium: Formation of cyanate ion, ketenimines, and isonitriles. *Icarus, 172,* 466–478.

Hudson R. L. and Moore M. H. (2006) Infrared spectra and radiation stability of H_2O_2 ices relevant to Europa. *Astrobiology, 6,* 48–489.

Hudson R. L., Moore M. H., and Gerakines P. A. (2001) The formation of cyanate ion (OCN^-) in interstellar ice analogues. *Astrophys. J., 550,* 1140–1150.

Hudson R. L., Moore M. H., and Cook A. M. (2005) IR characterization and radiation chemistry of glycolaldehyde and ethylene glycol ices. *Adv. Space Res., 36,* 184–189.

Jewitt D. C. and Luu J. (2004) Crystalline water ice on the Kuiper belt object (50000) Quaoar. *Nature, 432,* 731–733.

Johnson R. E. (1990) *Energetic Charged Particle Interactions with Atmospheres and Surfaces.* Springer-Verlag, Heidelberg.

Johnson R. E. (1995) Sputtering of ices in the outer solar system. *Rev. Mod. Phys., 68,* 305–312.

Johnson R. E., Cooper P. D., Quickenden T. I., Grieves G. A., and Orlando T. M. (2005) Production of oxygen by electronically induced dissociations in ice. *J. Chem. Phys., 123,* 184715-1 to 184715-8.

Kaiser R. I. and Roessler K. (1998) Theoretical and laboratory studies on the interaction of cosmic-ray particles with interstellar ices. III. Suprathermal chemistry-induced formation of hydrocarbon molecules in solid methane, (CH_4), ethylene (C_2H_4), and acetylene (C_2H_2). *Astrophys. J., 503,* 959–975.

Kerkhof O., Schutte W. A., and Ehrenfreund P. (1999) The infrared band strengths of CH_3OH, NH_3, and CH_4 in laboratory simulations of astrophysical ice mixtures. *Astron. Astrophys., 346,* 990–994.

Kobayashi K., Kasamatsu T., Kaneko T., Koike J., Oshima T., Sait T., Yamamoto T., and Yanagawa H. (1995) Formation of amino acid precursors in cometary ice environments by cosmic radiation. *Adv. Space Res.,162,* 21–26.

Lazzarin M., Marchi S., Moroz L., Brunetto R., Magrin S., Paolicchi P., and Strazzulla G. (2006) Space weathering in the main asteroid belt: The big picture. *Astrophys. J. Lett., 647,* L179–L182.

Licandro J., Pinella-Alonso N., Pedani M., Oliva E., Tozzi G. P., and Grundy W. M. (2006) The methane ice rich surface of large TNO 2005 FY_9: A Pluto-twin in the trans-Neptunian belt? *Astron. Astrophys., 445,* L35–L38.

Loeffler M. J., Baratta G. A., Palumbo M. E., Strazzulla G., and Baragiola R. A. (2005) CO_2 synthesis in solid CO by Lyman-alpha photons and 200 keV protons. *Astron. Astrophys., 435,* 587–594.

Loeffler M. J., Raut U., Vidal R. A., Baragiola R. A., and Carlson R. W. (2006a) Synthesis of hydrogen peroxide in water ice by ion irradiation. *Icarus, 180,* 265–273.

Loeffler M. J., Raut U., and Baragiola R. A. (2006b) Enceladus: A source of nitrogen and an explanation for the water. *Astrophys. J. Lett., 649,* L133–L136.

Lovas F. J., Hollis J. M., Remijan A. J., and Jewell P. R. (2006) Detection of keteneimine (CH_2CNH) in SgrB2(N) hot cores. *Astrophys. J. Lett., 645,* L137–L140.

Maki A. and Decius J. C. (1958) Infrared spectrum of cyanate ion as a solid solution in a potassium iodide lattice. *J. Chem. Phys., 28,* 1003–1004.

Marchi S., Brunetto R., Magrin S., Lazzarin M., and Gandolfi D.

(2005) Space weathering of near-Earth and main belt silicate-rich asteroids: Observations and ion irradiation experiments. *Astron. Astrophys., 443,* 769–775.

Mastrapa R. M. E. and Brown R. H. (2006) Ion irradiation of crystalline H_2O ice: Effect on the 1.65-μm band. *Icarus, 183,* 207–214.

Millar T. J. (2004) Organic molecules in the interstellar medium. In *Astrobiology: Future Perspectives* (P. Ehrenfreund et al., eds.), p. 17. Astrophysics and Space Science Library, Vol. 305, Kluwer, Dordrecht.

Moore M. H. (1984) Infrared studies of proton irradiated SO_2 ices: Implications for Io. *Icarus, 59,* 114–128.

Moore M. H. and Hudson R. L. (1992) Far-infrared spectral studies of phase changes in water ice induced by proton irradiation. *Astrophys. J., 401,* 353–360.

Moore M. H. and Hudson R. L. (1998) Infrared study of ion irradiated water ice mixtures with organics relevant to comets. *Ica-rus, 135,* 518–527.

Moore M. H. and Hudson R. L. (2000) IR detection of H_2O_2 at 80 K in ion-irradiated ices relevant to Europa. *Icarus, 145,* 282–288.

Moore M. H. and Hudson R. L. (2003) Infrared study of ion-irradiated N_2-dominated ices relevant to Triton and Pluto: Formation of HCN and HNC. *Icarus, 161,* 486–500.

Moore M. H., Hudson R. L., and Ferrante R. F. (2003) Radiation products in processed ices relevant to Edgeworth-Kuiper-belt objects. *Earth Moon Planets, 92,* 291–306.

Moore M. H., Hudson R. L., and Carlson R. W. (2007) The radiolysis of SO_2 and H_2S in water ice: Implications for the icy jovian satellites. *Icarus, 189,* 409–423.

Moroz L. V., Arnold G., Korochantsev A. V., and Wasch R. (1998) Natural solid bitumens as possible analogs for cometary and asteroid organics. *Icarus, 134,* 253–268.

Moroz L. V., Baratta G., Distefano E., Strazzulla G., Dotto E., and Barucci M. A. (2003) Ion irradiation of asphalite: Optical effects and implications for trans-neptunian objects and Centaurs. *Earth Moon Planets, 92,* 279–289.

Moroz L., Baratta G. A., Strazzulla G., Starukhina L., Dotto E., Barucci M. A., Arnold G., and Distefano E. (2004) Optical alteration of complex organics induced by ion irradiation: 1. Laboratory experiments suggest unusual space weathering trend. *Icarus, 170,* 214–228.

Mulas G., Baratta G. A., Palumbo M. E., and Strazzulla G. A. (1998) Profile of CH_4 IR bands in ice mixtures. *Astron. Astrophys., 333,* 1025–1033.

Mumma M. J., DiSanti M. A., Dello Russo N., Fomenkova M., Magee-Sauer K., Kaminski C. D., and Xie D. X. (1996) Detection of abundant ethane and methane, along with carbon monoxide and water, in Comet C/1996 B2 Hyakutake: Evidence for interstellar origin. *Science, 272,* 1310–1314.

Muñoz Caro G. M., Meierhenrich U. J., Schutte W. A., Barbier B., Arcones Segovia A., Rosenbauer H., Thiemann W. H.-P, Brack A., and Greenberg J. M. (2002) Amino acids from ultraviolet irradiation of interstellar ice analogues. *Nature, 416,* 403–406.

Owen T. C., Roush T. L., Cruikshank D. P., Elliot J. L., Young L. A., de Bergh C., Schmitt B., Geballe T. R., Brown R. H., and Bartholomew M. J. (1993) Surface ices and the atmospheric composition of Pluto. *Science, 261,* 745–748.

Palumbo M. E., Geballe T. R., and Tielens A. G. G. M. (1997) Solid carbonyl sulfide (OCS) in dense molecular clouds. *Astrophys. J., 479,* 839–844.

Palumbo M. E., Castorina A. C., and Strazzulla G. (1999) Ion irradiation effects on frozen methanol (CH_3OH). *Astron. Astrophys., 342,* 551–562.

Palumbo M. E., Ferini G., and Baratta G. A. (2004) Infrared and Raman spectroscopies of refractory residues left over after ion irradiation of nitrogen-bearing icy mixtures. *Adv. Space Res., 33,* 49–56.

Porco C. C., et al. (2006) Cassini observes the active south pole of Enceladus. *Science, 311,* 1393–1401.

Quirico E., Douté S., Schmitt B., de Bergh C., Cruikshank D. P., Owen T. C., Geballe T., and Roush T. L. (1999) Composition, physical state, and distribution of ices at the surface of Triton. *Icarus, 139,* 159–178.

Raunier S., Chiavassa T., Duvernay F., Borget F., Aycard J. P., Dartois E., and d'Hendecourt L. (2004) Tentative identification of urea and formamide in ISO-SWS infrared spectra of interstellar ices. *Astron. Astrophys., 416,* 165–169.

Sasaki T., Kanno A., Ishiguro M., Kinoshita D., and Nakamura R. (2005) Search for nonmethane hydrocarbons on Pluto. *Astrophys. J. Lett., 618,* L57–L60.

Schutte W. A., Boogert A. C. A., Tielens A. G. G. M., Whittet D. C. B., Gerakines P. A., Chiar J. E., Ehrenfreund P., Greenberg J. M., van Dishoeck E. F., and de Graauw, Th. (1999) Weak ice absorption features at 7.24 and 7.41 μm in the spectrum of the obscured young stellar object W 33A. *Astron. Astrophys., 343,* 966–976.

Stone E. C., Cummings A. C., McDonald F. B., Heikkila B. C., Lal N., and Webber W. R. (2005) Voyager 1 explores the termination shock region and the heliosheath beyond. *Science, 309,* 2017–2020.

Strazzulla G. and Baratta G. A. (1991) Laboratory study of the IR spectrum of ion-irradiated frozen benzene. *Astron. Astrophys., 241,* 310–316.

Strazzulla G. and Moroz L. (2005) Ion irradiation of asphaltite as an analogue of solid hydrocarbons in the interstellar medium. *Astron. Astrophys., 434,* 593–598.

Strazzulla G. and Palumbo M. E. (1998) Evolution of icy surfaces: An experimental approach. *Planet. Space Sci., 46,* 1339–1348.

Strazzulla G., Baratta G. A., Johnson R. E., and Donn B. (1991) Primordial comet mantle — Irradiation production of a stable, organic crust. *Icarus, 91,* 101–104.

Strazzulla G., Baratta G. A., Leto G., and Foti G. (1992) Ion-beam-induced amorphization of crystalline water ice. *Europhys. Lett., 18,* 517–522.

Strazzulla G., Baratta G. A, Domingo M., and Satorre M. A. (2002) Ion irradiation of frozen C_2H_n (n = 2, 4, 6). *Nucl. Instr. Meth. B, 191,* 714–717.

Strazzulla G., Cooper J. F., Christian E. R., and Johnson R. E. (2003) Ion irradiation of TNOs: From the fluxes measured in space to the laboratory experiments. *Compt. Rend. Phys., 4,* 791–801.

Strazzulla G., Baratta G. A., Battiato S., and Compagnini G. (2005a) Ion irradiations of solid carbons. In *Polyynes: Synthesis, Properties, and Applications* (F. Cataldo, ed.), pp. 271–284. CRC, New York.

Strazzulla G., Dotto E., Binzel R., Brunetto R., Barucci M. A., Blanco A., and Orofino V. (2005b) Spectral alteration of the meteorite Epinal (H5) induced by heavy ion irradiation: A simulation of space weathering effects on near-Earth asteroids. *Icarus, 174,* 31–35.

Sturner S. J., Shrader C. R., Weidenspointner G., Teegarden B. J., Attié D., et al. (2003) Monte Carlo simulations and generation of the SPI response. *Astron. Astrophys., 411,* L81–L84.

Tobiska W. K. and Bouwer S. D. (2006) New developments in SOLAR2000 for space research and operations. *Adv. Space Res., 37,* 347–358.

Tobiska W. K., Woods T., Eparvier F., Viereck R., Floyd L., Bouwer D., Rottman G., and White O. R. (2000) The SOLAR2000 empirical solar irradiance model and forecast tool. *J. Atmos. Solar Terr. Phys., 62,* 1233–1250.

Trottier A. and Brooks R. L. (2004) Carbon-chain oxides in proton-irradiated CO ice films. *Astrophys. J., 612,* 1214–1221.

Trujillo C. A., Brown M. E., Barkume K. M., Schaller E. L., and Rabinowitz D. L. (2007) The surface of 2003 EL_{61} in the near infrared. *Astrophys. J., 655,* 1172–1178.

Tuleta M., Gabla L., and Madej J. (2001) Bioastrophysical aspects of low energy ion irradiation of frozen anthracene containing water. *Phys. Rev. Lett., 87,* id. 078103.

Wada A., Mochizuki N., and Hiraoka K. (2006) Methanol formation from electron-irradiated mixed H_2O/CH_4 ice at 10 K. *As-trophys. J., 644,* 300–306.

Waite J. H., et al. (2006) Cassini Ion and Neutral Mass Spectrometer: Enceladus plume composition and structure. *Science, 311,* 1419–1422.

Weaver H. A., Brooke T. Y., Chin G., Kim S. J., Bockelée-Morvan D., and Davies J. K. (1998) Infrared spectroscopy of Comet Hale-Bopp. *Earth Moon Planets, 78,* 71–80.

Zheng W., Jewitt D., and Kaiser R. I. (2006) Temperature dependence of the formation of hydrogen, oxygen, and hydrogen peroxide in electron-irradiated crystalline water ice. *Astrophys. J., 648,* 753–761.

Ziegler J. F., Biersack J. P., and Littmark U. (1985) *The Stopping and Range of Ions in Solids.* Pergamon, New York.

Meteorites from the Outer Solar System?

Matthieu Gounelle
Muséum National d'Histoire Naturelle, Paris,
and Natural History Museum, London

Alessandro Morbidelli
Observatoire de la Côte d'Azur

Philip A. Bland
Natural History Museum, London, and Imperial College London

Pavel Spurný
Astronomical Institute of the Academy of Sciences of the Czech Republic

Edward D. Young
University of California–Los Angeles

Mark Sephton
Imperial College London

We investigate the possibility that a small fraction of meteorites originate from the outer solar system, i.e., from the Kuiper belt, the Oort cloud, or from the Jupiter-family comet reservoir. Dynamical studies and meteor observations show that it is possible for cometary solid fragments to reach Earth with a velocity not unlike that of asteroidal meteorites. Cosmochemical data and orbital studies identify CI1 chondrites as the best candidates for being cometary meteorites. CI1 chondrites experienced hydrothermal alteration in the interior of their parent body. We explore the physical conditions leading to the presence of liquid water in comet interiors, and show that there are a range of plausible conditions for which comets could have had temperatures high enough to permit liquid water to interact with anhydrous minerals for a significant amount of time. Differences between CI1 chondrites and cometary nuclei can be ascribed to the diversity of cometary bodies revealed by recent space missions. If CI1 chondrites do indeed come from comets, it means that there is a continuum between dark asteroids and comets. We give some indications for the existence of such a continuum and conclude that there should be a small fraction of the ~30,000 meteorites that originate from comets.

1. INTRODUCTION

When Ernst Chladni first proposed that meteorites originated from outside the terrestrial atmosphere, he did not speculate much further on their origin. At that time, asteroids had not been discovered, and the only possible candidates for meteorite parent bodies were planets, satellites, and comets. At the time of the L'Aigle fall, when meteorites were recognized as extraterrestrial objects (*Gounelle,* 2006), Poisson quantitatively examined Laplace's hypothesis that meteorites originated from lunar volcanos (*Poisson,* 1803). With the discovery of Ceres in 1801, soon followed by the detection of several other small planets, asteroids became an additional possible source of meteorites, although they were not explicitly considered until the 1850s (*Marvin,* 2006).

Because most meteorites are primitive chondrites that originate from bodies too small to have differentiated, modern meteoriticists recognized asteroids and comets as the most likely source for meteorites. For some time, the two possibilities were considered as equally probable, mainly because of the difficulty in delivering meteorites onto Earth from the asteroid belt (e.g., *Wetherill,* 1971). The dynamics of meteorite delivery from the asteroid belt to Earth are now well understood (e.g., *Morbidelli and Gladman,* 1998; *Vokrouhlický and Farinella,* 2000). All meteorites for which a precise orbit has been calculated originate from the asteroid belt (e.g., *Gounelle et al.,* 2006, and references therein). There is at present a large consensus that asteroids are the source of all meteorites present in museum collections, except for a few tens of lunar or martian meteorites. It is still possible, however, that a *minor fraction* of meteorites originate from the outer solar system, i.e., from Jupiter-family comets (JFCs), long-period comets (LPCs), Halley-type comets (HTCs), or Kuiper belt objects (KBOs). As KBOs are the source for JFCs (*Levison and Duncan,* 1997), we will often use the generic term "cometary meteorites" when referring to meteorites coming from any outer solar system object.

The scope of this review is to examine critically the possibility that some of the meteorites present in museum collections originate from outer solar system objects. We will limit ourselves to objects larger than 1 mm, and will not discuss, except incidentally, the case of interplanetary dust particles (IDPs) and Antarctic micrometeorites. Thoughts about the origin of this submillimeter fraction of the extraterrestrial flux can be found in a diversity of recent reviews (e.g., *Engrand and Maurette,* 1998; *Rietmeijer,* 1998). On the topic of possible cometary meteorites, an excellent review was offered by *Campins and Swindle* (1998) some 10 years ago. Important developments both in our astronomical knowledge of the outer solar system (this volume) and in meteorite analyses (e.g., *Krot et al.,* 2005; *Lauretta and McSween,* 2005) were so numerous in the last 10 years that a reexamination of the question is both timely and necessary.

The importance of the question "Can meteorites originate from the outer solar system?" arises not only because asteroids and outer solar system small bodies, generally called comets, are perceived as radically different celestial objects both by laymen and astronomers, but because they sample different regions of the solar system. If we were certain that some meteorites originate from the outer solar system, this would give us the possibility of studying in the laboratory objects that spent most of their lifetime beyond the reach of modern analytical techniques. At present, only the dust brought back by the Stardust spacecraft from the JFC 81P/Wild 2 gives us the opportunity to study, with an unprecedented range of laboratory techniques, solid matter derived with certainty from an outer solar system body (*Brownlee et al.,* 2006).

Before entering detailed consideration of the possible cometary origin of some meteorites, it is important to realize that there are more than 30,000 meteorites present in the world's collections. If one takes into account the numerous desert meteorites not officially declared to the *Meteoritical Bulletin* because of their lack of scientific or commercial value, this number might be as high as 40,000. If there are no cometary meteorites, it means there are extremely powerful mechanisms that prevent them either being delivered to Earth, or surviving the effects of atmospheric entry. In other words, given the number of meteorites we have at hand, it would be surprising if *none* of them originated from comets.

Among the ~135 different meteorite groups (*Meibom and Clark,* 1999), what are the best candidates for representing cometary meteorites? It is unlikely that differentiated meteorites or metamorphosed chondrites originate from comets given the primitive nature of the latter. Volatile-rich, primitive carbonaceous chondrites are good candidates since they are chemically unfractionated relative to the Sun's composition (*Lodders,* 2003). Among these, the low-petrographic-type carbonaceous chondrites (types 1 and 2, representing respectively ~0.5% and ~1.3% of the meteorite falls) are the most likely candidates for being cometary meteorites because they are dark, rare, friable, and rich in organic matter, as expected for cometary meteorites (e.g., *McSween and Weissman,* 1989).

Because of the intricate nature of the subject addressed here, we will tackle the question raised in the title of this chapter in a diverse number of ways, sometimes apparently disconnected. All these approaches will be wrapped together into our conclusions. The paper is divided as follows. In section 2, we examine the possible dynamical routes between the outer solar system and Earth, and estimate the expected number of cometary meteorites relative to asteroidal meteorites. In section 3, we review the fireball evidence for hard, dense matter reaching Earth from cometary orbits. In section 4, we examine the similarities and differences between carbonaceous chondrites — our best bet for outer solar system meteorites — and comets. Section 5 is devoted to identifying the conditions under which hydrothermal alteration is possible within cometary interiors. Hydrothermal alteration in comets is a necessary, although not sufficient, condition for low petrographic carbonaceous chondrites being cometary. Section 6 discusses the diversity of comets revealed by recent space missions and emphasizes on the possible continuum between comets and asteroids. Its scope is to critically examine the implicit assumption stating that there is a clear distinction between comets and asteroids. In section 7, we summarize our main conclusions and give a tentative answer to the question enounced in the title.

2. DYNAMICAL PATHWAYS

It is well known that the Kuiper belt and scattered disk subpopulation (see chapter by Gladman et al.) is the source of JFCs (*Levison and Duncan,* 1997). The debiased orbital distribution of JFCs from the Kuiper belt has been computed through numerical simulations in *Levison and Duncan* (1997). These simulations followed the evolution of thousands of test particles, from their source region up to their ultimate dynamical elimination. The dynamical model included only the perturbations exerted by the giant planets, and neglected the effects of the inner planets and the effect of nongravitational forces. Consequently, the resulting JFCs had almost exclusively orbits with Tisserand parameter relative to Jupiter

$$T_J = \frac{a_J}{a} + 2\cos(i)\sqrt{(1-e^2)\frac{a}{a_J}}$$

smaller than 3. Actually, most of the simulated comets had $2 < T_J < 3$, in good agreement with the observed distribution. A follow-up of the Levison and Duncan work has been done by *Levison et al.* (2006), which included also the perturbations from the terrestrial planets. These perturbations allow some comets to acquire orbits with $T_J > 3$, although not much larger than this value. This is consistent with the existence of one active comet (P/Encke) with $T_J > 3$, and of a few other objects with sporadic activity (4015 Wilson-Harrington). Nevertheless, the ratio comets/asteroids should drop to a negligible value for $T_J > 3$ [see Fig. 5 of *Levison et al.* (2006)].

Bottke et al. (2002) developed a model of the orbital and absolute magnitude distributions of near-Earth objects (NEOs) that accounts for asteroids escaping from the main belt through various channels, as well as for active and dormant JFCs from *Levison and Duncan* (1997) simulations. Using the *Bottke et al.* (2002) model, and the albedo distribution model of *Morbidelli et al.* (2002) to convert absolute magnitudes to diameters, in addition to the orbit-dependent impact probabilities with Earth computed in the same work, we estimate that the impact rate of a JFC with our planet is only ~7% of that for an asteroid of the same size. Interestingly, even among objects with $T_J < 3$, comets account only for ~60% of the impacts, the rest being of asteroidal origin. This is important to keep in mind, when analyzing fireball data, as in section 3. The mean impact velocity of JFCs, weighted by collision probability, is ~21 km/s, similar to that of asteroids (20 km/s) (*Morbidelli and Gladman,* 1998), due to the predominance of low-inclination JFCs among cometary impact events. When considering these impact probability ratios, however, one should not forget that the *Bottke et al.* (2002) NEO model is calibrated for objects of about 1 km in size and it is not valid, *a priori*, for NEOs of size comparable to meteorite precursor bodies (typically a few meters). In fact, there is evidence that the size distribution of JFCs is shallower than that of asteroidal NEOs in the size range from a few thousand meters to a few kilometers (*Whitman et al.,* 2006). This is believed to be the consequence of the short physical lifetime of small comets. If this shallow size distribution can be extrapolated to meteorite sizes, the fraction of cometary vs. asteroidal meteoritic impacts would be much smaller than the 7% value quoted above. Taken to some extreme, one might speculate that meteorite-sized "comets" have such a short physical lifetime (a few days, consistent with observations of boulders released from comets) that they can be considered to be inexistent, in practice (*Boehnhardt,* 2004). On the other hand, active comets might continuously release a large number of meteorite-sized bodies into space, in particular close to perihelion. Thus, even if these fragments have a short physical lifetime, their population might be continuously regenerated in the inner solar system. Consequently, the JFC size distribution might become even steeper at small sizes than that estimated by *Whitman et al.* (2006) at intermediate sizes, increasing the fraction of meteorite-sized bodies of cometary origin relative to those of asteroidal origin. So, despite the intense modeling effort quoted above, the real ratio of cometary/asteroidal meteoritic impacts is still not well constrained. Impacts of HTCs and LPCs from the Oort cloud, conversely, should be negligible relative to JFC impacts (*Levison et al.,* 2002), and the much larger impact velocities should be prohibitive for allowing meteorites to reach the ground.

However, there is another dynamical path by which cometary bodies might impact Earth at the present time. Several cometary bodies, either from the Jupiter-Saturn region or from the transneptunian region, might have been trapped in the asteroid belt. Some of them might escape from the asteroid belt, along with normal asteroids, and contribute to the NEO population with $T_J > 3$, which, as we have seen, dominates the impact rate relative to the genuine JFC population.

There are three phases in solar system history when cometary objects might have been implanted in the main belt. The first phase occurs very early, prior to the formation of the giant planets. Icy bodies, accreted at the snow line, might have migrated inward due to gas drag (*Cyr et al.,* 1998). Because gas drag is important only for bodies smaller than a few kilometers in size, most of these icy bodies should not have survived to modern times. In fact, the collisional lifetime of bodies of this size in the main belt is smaller than the solar system lifetime (*Bottke et al.,* 2005). However, they or their fragments might have been incorporated into bigger bodies forming *in situ* (namely within the snow line, in the asteroid belt), delivering a significant amount of water to these objects (*Cyr et al.,* 1998). These water-enriched bodies could be transition objects between asteroids and comets (D-type asteroids?), and the meteorites released by them could be proxies for real cometary meteorites.

The second phase of implantation of comets in the asteroid belt occurred as soon as Jupiter formed. The scattering action of Jupiter dislodged the comets from their formation region in Jupiter's vicinity, making them evolve into JFCs. However, due to gas drag, the smallest members of this group could have had their orbital eccentricity reduced, reaching stable orbits in the asteroid belt. Again, due to the small sizes of these objects, only a tiny fraction of them should survive in the present-day asteroid belt. However, some bigger cometary objects might have been decoupled from Jupiter and inserted onto stable orbits by interaction with the numerous planetary embryos that should have existed at the time in the asteroid belt. Actually, the model postulating the existence of planetary embryos is the best one to explain the current properties of the main asteroid belt (*Petit et al.,* 2000). The possibility of capture of comets by embryos has never been explored in detail. This dynamical path has not been investigated yet, but should be analogous to that recently developed by *Bottke et al.* (2006) for the implantation of iron meteorite parent bodies from the terrestrial planet region into the asteroid belt. The cometary bodies implanted in the asteroid belt by this mechanism are not necessarily small, and therefore some might have survived up to the present time. These bodies would be genetically related with some of the Oort cloud objects (and therefore with some LPCs and HTCs), as a fraction of the Oort cloud was built from planetesimals originally in the Jupiter-Saturn zone (*Dones et al.,* 2004).

The third phase of implantation of comets should have occurred during the late heavy bombardment (LHB) of the inner solar system. Recent work by *Gomes et al.* (2005)

showed that the LHB may have been caused by a phase of dynamical instability of the giant planets, which destabilized the transneptunian planetesimal disk (see chapter by Morbidelli et al. for a detailed description of that model). *Morbidelli et al.* (2005) showed that during this phase of instability some of the originally transneptunian bodies should have been captured as permanent Trojans of Jupiter (see chapter by Dotto et al. for more details). In a similar way, transneptunian objects (TNOs) could also be captured in the outer main belt and in the 3:2 resonance with Jupiter (where the Hilda asteroids reside now). Figure 1 compares the semimajor axis and eccentricity distribution of the trapped bodies (Fig. 1b) with that of all known asteroids (Fig. 1a). The bodies trapped during the LHB should be genetically related to Kuiper belt/scattered disk bodies, and therefore with JFCs and Trojans.

D-type asteroids are the best candidates to be implanted cometary bodies, due to their spectral similarities with dormant cometary nuclei (see section 4.2). The objects captured during the LHB should have semimajor axes larger than 2.8 AU (Fig. 1). Those captured during an earlier phase might have, in principle, smaller semimajor axes. With the exception of one object with a ~ 2.25 AU, the currently known D-type asteroids in the main belt reside beyond 2.5 AU (see Fig. 1a). Five of them have a semimajor axis between 2.6 and 2.8 AU; the remainder (on the order of 20 objects) are beyond this threshold. According to the NEO model of *Bottke et al.* (2006), the outer main-belt asteroids (defined as objects with a > 2.8 AU) should contribute only ~10% of the asteroidal impacts on Earth. Because D-type asteroids are less than 50% of the outer belt population, even assuming that all D-type objects have a cometary origin, we should conclude that the putative cometary population trapped in the outer belt does not account for more than 5% of the total meteorite impacts.

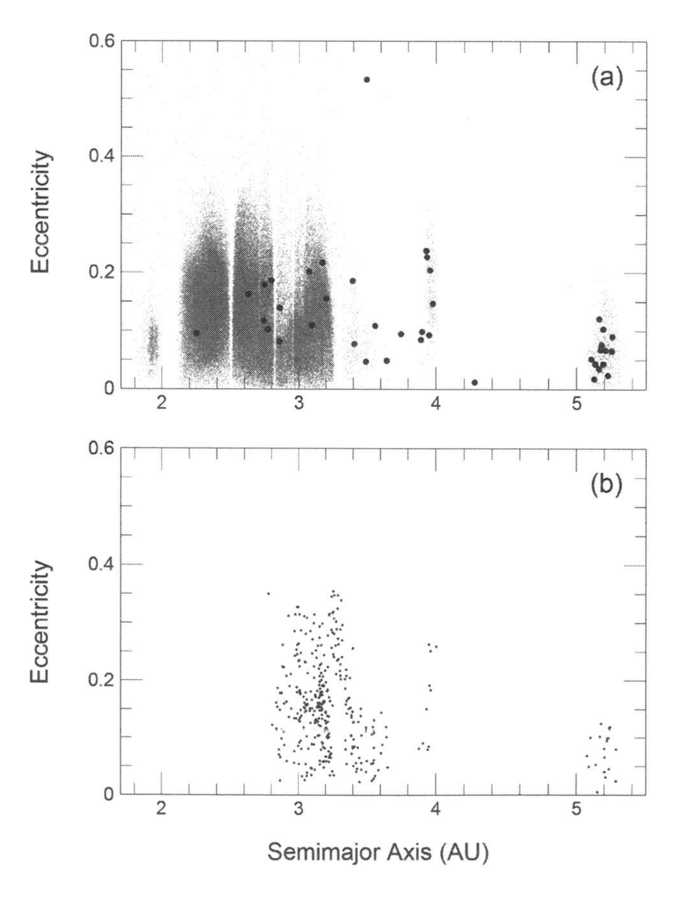

Fig. 1. **(a)** Semimajor axis vs. eccentricity distribution of main-belt asteroids (small gray dots) and of known D-type objects (large black dots). **(b)** Distribution of the particles of transneptunian origin captured in the asteroid belt during the late heavy bombardment (courtesy of H. F. Levison).

In conclusion, the indirect dynamical pathway that allows cometary bodies to impact Earth after a long period trapped in the main belt should not account for more than a few percent of objects striking the upper atmosphere. Therefore, from a dynamical point of view, it is not out of the question to have cometary samples in the current meteorite collections, but these samples should be rare.

3. EVIDENCE FROM METEORS AND FIREBALLS

In this section, we explore the possibility of cometary meteorites (i.e., objects larger than a millimeter) surviving atmospheric entry and reaching the ground. Meteor and fireball data help addressing that question in evaluating the fraction of solid materials originating from objects with a putative cometary orbit that reach Earth's surface.

Considering the meteor dataset, *Swindle and Campins* (2004) review the evidence for dense material in lightcurve data recorded from Leonid meteors whose cometary origin is secure. *Murray et al.* (1999) present several lightcurves that have a double-humped appearance, suggesting the presence of a harder portion that ablates later. Less than 10% of the lightcurves show this feature. Given that roughly half the mass of the double-humped meteor appears to be a high density solid, this would suggest that less than 5% of the total mass of cometary meteors is present as hard material. *Borovička and Jenniskens* (1998) also discuss a cometary fireball that had a hard portion continuing after the terminal burst. This harder component corresponds to 1 mg of a 1 kg meteor, approximately the size of a chondrule. This result clearly does not mean that chondrules are contained in comets, but it does suggest the presence of harder particles within them.

Fireball camera networks have the advantage that they record the entry of objects that survive the passage through the atmosphere and might be collected as meteorites. Camera network studies have hugely expanded our knowledge of meteoroid composition. The associated orbital information allows us to discriminate between cometary and asteroidal sources. Given the large dataset of camera network fireballs recorded over the last five decades, do we see evidence for the survival of cometary materials? In an early analysis of fireballs thought to be of cometary origin (based on orbital considerations), *Ceplecha and McCrosky* (1976) and *Wetherill and ReVelle* (1982) suggested that a small fraction of cometary material may survive atmospheric entry. *Wetherill and ReVelle* (1982) went so far as to suggest that the proportion may be high enough to account for all carbonaceous meteorites. Subsequent work does not support this level of cometary meteoroid survival, but suggests that a smaller portion could survive (see below).

In the Canadian network study, fireball events were categorized based on initial velocity into groups thought to contain (mostly) asteroidal and cometary material (*Halliday et al.,* 1996). There was no suggestion that the groups were exclusive — cometary events were included in the asteroidal group, and vice versa — but it was considered that the groups might have been dominated by one type of material. Although material in the cometary group had generally lower densities than material in the asteroidal group, 25% of the samples in the cometary category, for which a density was estimated, had densities >2 g cm^{-3} (*Halliday et al.,* 1996), raising the possibility that dense material may be contained within the set of cometary meteoroids.

It is important to note that high-density material can survive atmospheric entry even at large entry velocities. An interesting case involves fireball 892 from the MORP network dataset (*Halliday et al.,* 1996). This fireball had an initial velocity of 28.9 km/s, but produced a terminal mass of 62 g. Similarly, *Spurný* (1997) discussed a number of cases from the European Fireball Network. EN251095A "Tizsa" had an initial velocity of 29.2 km/s, and a probable terminal mass of 2.6 kg. Other (less spectacular) cases within this category include EN220405 "Kourim" (an entry velocity of 27.5 km/s, and terminal mass < 0.1 kg), and EN160196 "Ozd" (entry velocity of 25.6 km/s, and terminal mass <0.2 kg). Finally, *Wetherill and ReVelle* (1982) observed this type of event in the Prairie Network fireballs: PN39409 had a high initial velocity (31.7 km/s) and produced a terminal mass, as did PN40460B. All these fireballs are type I events, which represent (by definition) the densest, strongest part of the meteoroid complex. From the events described above it is clear that middle- to high-velocity material may survive to Earth's surface. But it is necessary to stress that entry velocity alone does not provide a robust discriminant between asteroidal and cometary material. As we have seen in section 2, the Tisserand parameter is a more robust criterion for distinguishing cometary and asteroidal bodies. All the events described above have $T_J > 3$, so they are likely to be asteroidal events despite their high entry velocity.

In the context of our current study, possibly the most interesting case comes from the European Fireball Network, discussed by *Spurný and Borovička* (1999). On June 1, 1997, three Czech camera stations recorded a fireball (EN 010697 "Karlštejn") that began its luminous trajectory at 93 km, with a very high initial velocity of 65 km/s. The object had an aphelion beyond the orbit of Jupiter, and was on a retrograde orbit. It has a Tisserand parameter of 0.22. This orbit is typical of HTCs. What makes the object unique is that it penetrated down to an altitude of 65 km above Earth's surface — about 25 km deeper than cometary meteoroids of similar velocity and mass (the European Network has observed ~500 cometary meteoroids on the basis of their Tisserand parameter). The atmospheric behavior of the Karlštejn object is consistent with it being a hard, strong, stony meteoroid. In addition, a spectral record taken at Ondřejov Observatory showed no trace of the sodium line in the Karlštejn fireball — a feature that is prominent in all other meteor spectra. It appears that the Karlštejn meteoroid had an approximately chondritic composition in terms of refractory elements (this does not imply that the object was a CI1 chondrite — many other meteorite groups also have approximately chondritic levels of refractory elements). What is curious, especially for an object on a cometary orbit, is that it was strongly depleted in volatile elements, to a degree not seen in other chondritic meteorites (*Spurný and Borovička,* 1999).

The Karlštejn event currently appears to be unique — we have not found any similar case among all other fireballs from photographic networks (>1000 events). In addition, there is no single event that shows a meteoroid on a clearly cometary orbit surviving atmospheric entry. Based on the meteor data, it therefore appears likely that the proportion of cometary meteorites is less than 1 in 1000. But the Karlštejn event suggests that dense, strong material is contained within comets. Therefore, at least a small portion of cometary material is capable of reaching the surface of Earth intact.

4. A BEST BET: CARBONACEOUS CHONDRITES

Among the meteorites present in our collections, hydrated or low petrographic type carbonaceous chondrites (petrographic type ≤2) are the best candidates for being cometary meteorites (e.g., *Campins and Swindle,* 1998). They indeed conform to the simplest criteria established for recognizing cometary meteorites: They should be dark, weak, porous, have nearly solar abundances of most elements, and have elevated contents of C, N, and H (*Mason,* 1963). C1 chondrites are mainly made of secondary minerals such as phyllosilicates, carbonates, and magnetite. Pyroxene and olivine are extremely rare among these meteorites. C2 chondrites are made of olivine and pyroxene coexisting with secondary minerals such as phyllosilicates and carbonates. Both C1 and C2 chondrites endured extensive hydrothermal alteration on their parent bodies, with the C1s being more altered than the C2s. Below, we develop the most important similarities and differences between carbonaceous chondrites and comets, emphasizing new data. Other relevant aspects of that discussion can be found in *Campins and Swindle* (1998).

4.1. Orbit of the Orgueil CI1 Chondrite

Nine precise meteorite orbits have been determined over the years using camera networks, satellite data, multiple video recordings, and visual observations (e.g., *Brown et al.,* 2004). Except for Tagish Lake (*Brown et al.,* 2000), all these meteorites are high petrographic type ordinary and enstatite chondrites. All originate from the asteroid belt.

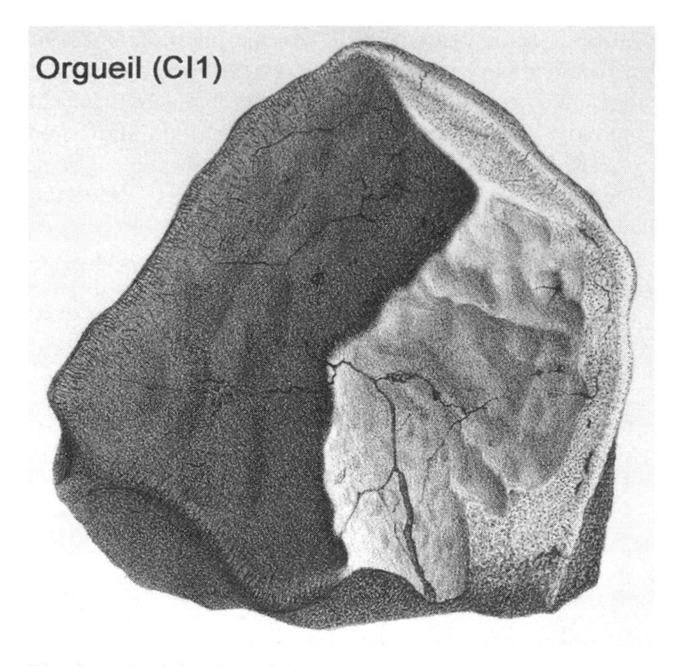

Fig. 2. Hand drawing of the Orgueil meteorite made by Daubrée at the time of the meteorite fall (1864). The orbit of the Orgueil meteorite was demonstrated to be compatible with that of a JFC (*Gounelle et al.,* 2006). If there are any cometary meteorites among our meteorites collections, CI1 chondrites such as Orgueil are the best candidates (see section 4).

Recently, the orbit of the Orgueil (CI1) meteorite was calculated using numerous visual observations communicated shortly after its fall (May 14, 1864) to the main mineralogist of the time, Auguste Daubrée (Fig. 2). Taking into consideration 13 visual observations, *Gounelle et al.* (2006) suggested that the orbit of the Orgueil meteorite was more compatible with that of a JFC than with that of an asteroid (aphelion $Q > 5.2$ AU and inclination $i \sim 0°$). Clearly, given the nature of the input data (150-year-old visual observations), this conclusion is more a best informed guess rather than a certainty. *Gounelle et al.* (2006) argue that if Orgueil has a cometary origin, it might also be the case for related carbonaceous chondrites such as other CI1 chondrites, Tagish Lake (ungrouped C2), or even CM2 chondrites. We note, however, that the orbit of Tagish Lake is asteroidal (*Brown et al.,* 2000), and that its cometary origin is possible only if it represents a comet that was trapped in the main asteroid belt (see section 2). This possibility is strengthened by the fact that Tagish Lake is similar to D-type asteroids that might be trapped outer solar system objects (see section 2).

4.2. Spectroscopy and Photometry Evidence

The direct comparison of carbonaceous chondrites' near-infrared spectra with that of KBOs or cometary nuclei might be meaningless for several reasons. First, cometary and KBO surfaces are subject to space weathering, i.e., modification of the surface optical properties due to micrometeorite impact or irradiation by the solar wind and galactic cosmic rays, that compromises a relevant comparison with the laboratory spectra of chondrites (e.g., *Jedicke et al.,* 2004). Second, it is possible that carbonaceous chondrites originate from the interior of a comet or a KBO, and therefore are different from their surfaces sampled by visible and infrared spectrometers. Third, it should not be forgotten that there is growing evidence that comets have a variety of compositions (see section 6.1.), suggesting that the limited number of near-infrared cometary nuclei spectra may not allow any definitive conclusion. Finally, it is now accepted that CI1 (and possibly other) chondrites widely interacted with the terrestrial atmosphere, resulting in a severe modification of their mineralogy (*Gounelle and Zolensky,* 2001), and consequently of their infrared spectrum. The prominent 3-μm feature seen in Orgueil and other hydrated carbonaceous chondrites (*Calvin and King,* 1997) might be enhanced by the ability of Orgueil and other carbonaceous chondrites to adsorb as much as 10 wt% of terrestrial water (e.g., *Gounelle and Zolensky,* 2001, and references therein).

Given the limitations exposed above, we will limit ourselves to mention a few explicit comparisons made in the literature. Tagish Lake near-infrared spectrum is similar to that of D asteroids (*Hiroi et al.,* 2001). Based on dynamical studies (*Morbidelli et al.,* 2005) and on the comparison of their spectra (*Licandro et al.,* 2003), there is a growing consensus that D asteroids and comets might have a common origin (see section 2). The Tagish Lake spectrum is also similar to the Oort cloud Comet C/2001 OG_{108} (*Abell et al.,* 2005). There is therefore a potential link between cometary nuclei and the Tagish Lake ungrouped C2 chondrite. On the other hand, the near-infrared spectrum of Tagish Lake is unlike that of Comet 19P/Borrelly (*Soderblom et al.,* 2004). We note that a relatively good match exists between the spectrum of Comet 162P/Siding Spring and that of the CI1 chondrite Alais (*Campins et al.,* 2006).

Carbonaceous chondrite mineralogy can be compared to the mineralogy of cometary dust as observed spectroscopically, bearing in mind the caveats discussed in the previous paragraph. Crystalline as well as amorphous olivine and pyroxene grains were detected in a variety of Oort cloud comets (*Wooden et al.,* 2005). Most of them are magnesium-rich, as are olivine and pyroxene grains in low petrographic type carbonaceous chondrites. Olivine and pyroxene are present but quite rare in CI1 chondrites (*Bland et al.,* 2004), and more abundant in type 2 chondrites such as Tagish Lake or CM2 chondrites (*Zolensky et al.,* 2002). While phyllosilicates are one of the main constituents of hydrated carbonaceous chondrites (*Bland et al.,* 2004), an upper limit of 1% has been proposed for the abundance of phyllosilicates in the Hale-Bopp dust, based on the montmorillonite 9.3-μm spectral feature (*Wooden et al.,* 1999). Recently, however, the analysis of the dust ejecta of the JFC 9P/Tempel 1 produced by the Deep Impact mission revealed the presence of phyllosilicates (*Lisse et al.,* 2006) (see section 4.5). The 0.7-μm spectral feature, attributed to the Fe^{2+}

to Fe^{3+} transition in phyllosilicates (*Rivkin et al.,* 2002), has also been tentatively identified in the TNO 2003 AZ_{84} (*Fornasier et al.,* 2004).

Comparing the albedo of carbonaceous chondrites to that of cometary nuclei or KBO objects is problematic because data are gathered under radically different experimental conditions (viewing geometries). The albedo of CI1 and CM2 chondrites ranges from 0.03 to 0.05 (*Johnson and Fanale,* 1973). Tagish Lake has an albedo of ~0.03 (*Hiroi et al.,* 2001). This compares well to the range of cometary nuclei geometric albedos, between 0.02 and 0.06 (*Campins and Fernández,* 2002). Centaur and Trojan asteroids have albedos varying from 0.04 to 0.17 (*Barucci et al.,* 2002) and 0.03 to 0.06 (e.g., *Emery et al.,* 2006) respectively. Small KBOs also have low albedos, while larger objects have higher albedos due to the presence of ices (see chapter from Barucci et al.). The hydrated carbonaceous chondrite albedos match those of comets, Trojan asteroids, and small KBOs.

4.3. Isotopic Data

The H-isotopic composition of comets, measured by radio spectroscopy, has often been considered as a key tool for identifying cometary components among meteorites. It is usually admitted that comets are enriched in D relative to the primitive solar system value, Earth, and chondritic meteorites (e.g., *Robert,* 2002). This led several authors to argue that IDPs, rather than carbonaceous chondrites, had a cometary origin based on large D excesses observed at the micrometer scale with secondary ion mass spectrometry (e.g., *Messenger et al.,* 2006).

This view relies on two misconceptions. First, when 2σ error bars (Fig. 3) and most recent measurement of comets are taken into account (*Bockelée-Morvan et al.,* 1998; *Crovisier et al.,* 2005; *Eberhardt et al.,* 1995; *Meier et al.,* 1998), only Comet 1P/Halley is clearly enriched in D relative to CI1 chondrites. We note also that only Oort cloud comets have known D/H ratios, while the D/H ratio of JFCs is unknown. To summarize, although *one* comet is clearly enriched in D relative to chondrites (1P/Halley), it is premature to conclude, as it is often the case, that *all* comets are enriched in D relative to chondrites. Second, it is now clear that the enrichment in D is not limited to IDPs. Recent carbonaceous chondrite studies revealed D enrichments larger than the ones found in IDPs (*Busemann et al.,* 2006; *Mostefaoui et al.,* 2006). This means either that D enrichments in extraterrestrial matter cannot be taken as a proof for a cometary origin, or that these carbonaceous chondrites originate from comets as well. The moderate enrichment in D of *bona fide* cometary samples from Comet 81P/Wild 2 supports the idea that D enrichments are not a definitive proof for a cometary origin (see section 4.5) (*McKeegan et al.,* 2006).

Besides H, the C- and N-isotopic composition of comets is also known (*Hutsemékers et al.,* 2005). The C-isotopic composition of Oort cloud comets and JFCs is identical within error bars to that of Earth (*Hutsemékers et al.,* 2005). This is compatible with bulk measurements of carbonaceous chondrites (e.g., *Robert,* 2002) and IDPs (*Floss et al.,* 2006). Thus, the C-isotopic composition of comets cannot be taken as a reliable indicator of an outer solar system origin. The N-isotopic composition of comets is terrestrial for HCN molecules and enriched in ^{15}N by a factor of 2 relative to Earth for CN radicals (*Hutsemékers et al.,* 2005). Most hydrous carbonaceous chondrites have bulk N-isotopic compositions similar to that of Earth (*Robert and Epstein,* 1982). Some IDPs as well as the anomalous CM2 chondrite Bells have bulk enrichment in ^{15}N relative to terrestrial as large as a few hundred permil (*Floss et al.,* 2006; *Kallemeyn et al.,* 1994). "Hotspots" with enrichments in ^{15}N as large as a factor of 2 (*Floss et al.,* 2006) and 3 were found in IDPs and in the anomalous CM2 chondrite Bells (*Busemann et al.,* 2006) respectively.

To summarize, the isotopic composition of comets is of no help at present to identify cometary meteorites for three fundamental reasons. First, the isotopic composition of comets is not well constrained because of the rarity of precise measurements (D/H ratios) and because it is unknown just how representative are its components (HCN molecules vs. CN radicals for the $^{15}N/^{14}N$ ratio). Second, within error bars, there is a variety of primitive materials (low petrographic type carbonaceous chondrites and IDPs) that have an isotopic composition compatible with that of comets. Third, it should be kept in mind that the spectroscopically measured isotopic composition of H, C, N in comets is that of the gas phase (i.e., ice), while that of putative cometary

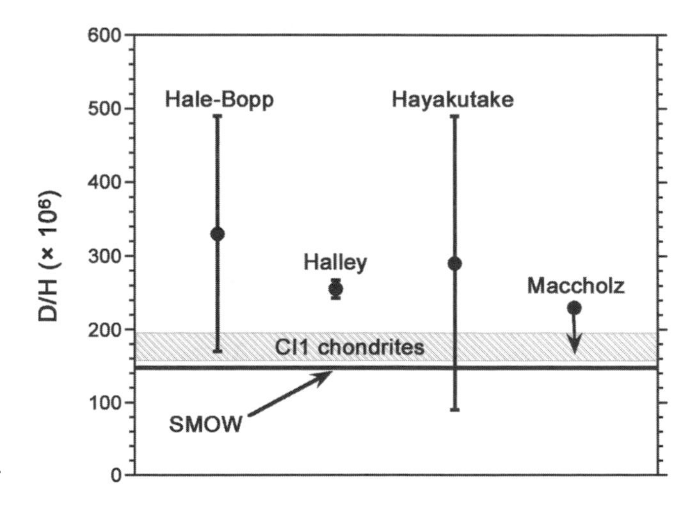

Fig. 3. Hydrogen-isotopic composition of Oort cloud comets (*Bockelée-Morvan et al.,* 1998; *Eberhardt et al.,* 1995; *Meier et al.,* 1998; *Crovisier et al.,* 2005) and CI1 chondrites (*Eiler and Kitchen,* 2004). Error bars are 2σ. Note that the D abundance measured in CI1 chondrites is probably a lower limit as these meteorites are notorious for having adsorbed terrestrial water whose D/H ratio is lower than that of the average value of CI1 chondrites (*Gounelle and Zolensky,* 2001).

samples is that of the solid phase (i.e., dust). Although little fractionation is expected between these phases, were they in equilibrium, these two components might, however, have formed in different loci.

4.4. Organics Record

The organics content of comets is poorly known. Carbon-rich grains that might contain pure C particles, polycyclic aromatic hydrocarbons (PAHs), branched aliphatic hydrocarbons, and more complex organic molecules were detected in the coma of Comet 1P/Halley (*Fomenkova,* 1997). Spectroscopic evidence indicates that amorphous carbon is an important component of cometary comae (*Ehrenfreund et al.,* 2004, and references therein). Indications of the presence of PAHs were seen in a diversity of comets (*Ehrenfreund et al.,* 2004, and references therein). These organic compounds are present in carbonaceous chondrites (e.g., *Gardinier et al.,* 2000; *Remusat et al.,* 2005). *Ehrenfreund et al.* (2001) suggested, on the basis of their amino acid peculiar abundances, that CI1 chondrites originate from comets. We show instead below that the distinctive amino acid population of CI1 chondrites as well as other important organic properties are the result of important parent-body processes unrelated to its asteroidal or cometary nature.

4.4.1. Amino acids. Amino acid compositions in CM2 meteorites display a wider variety in total abundance than in CI1s. However, the relative compositions (with respect to the glycine abundance) are similar. In contrast, the amino acid composition of the CI1 chondrites Orgueil and Ivuna is notably different from the CM2s, and very similar to each other (*Ehrenfreund et al.,* 2001). Glycine and β-alanine are the two most abundant amino acids in these meteorites, with much lower abundances of more complex amino acids. It was shown that relative amino acid abundances are a useful tool to investigate parent body processes (*Botta and Bada,* 2002). The observed differences between CM2s and CI1s prompted the suggestion that these meteorites originate from two completely different types of parent bodies, e.g., extinct comets could be the parent bodies of CI1s. However, it is interesting to note that the Nogoya meteorite, which is the most intensively altered CM2, displays a relative amino acid distribution that may show a trend from that of less-altered CM2s (*Cronin and Moore,* 1976), such as Murchison and Murray, toward that seen for the CI1s. This may indicate that progressive transformation of CM2 amino acids by aqueous alteration could generate the distributions observed in Nogoya and ultimately the CI1s.

4.4.2. Carboxylic acids. Aromatic acids detected in the CM2 Murchison include benzoic and methylbenzoic acids, phthalic and methylphthalic acids, as well as hydroxybenzoic acids (*Martins et al.,* 2006). The CI1 Orgueil, by contrast, displays a more simple distribution of carboxylic acids with abundant benzoic acid but few dicarboxylic acids, phthalic acids, or hydroxybenzoic acids. The distributions of carboxylic acids in the two meteorites can be attributed to the different levels of parent-body aqueous alteration affecting a common starting material. For instance, oxidation reactions occurring during the aqueous process could have selectively removed aliphatic carboxylic acids from the more extensively altered CI1 Orgueil (*Sephton et al.,* 2004), and a similar process may have removed the methyl and methoxy substituents of benzoic and phthalic acid units to leave behind the simple benzoic-acid-dominated distribution (*Martins et al.,* 2006).

4.4.3. Macromolecular materials. Macromolecular materials are relatively intractable organic entities and, as such, should provide a more robust record of aqueous transformations than is provided by free compounds. Hydrous pyrolysis of macromolecular material from Orgueil (CI1), Cold Bokkeveld (CM2), and Murchison (CM2) all contain volatile aromatic compounds with aliphatic side chains, hydroxyl groups, and thiophene rings attached (*Sephton et al.,* 2000). The macromolecular materials in these meteorites appear qualitatively similar. However, the pyrolysates show significant quantitative differences, with the pyrolysis products of ether linkages and condensed aromatic networks being less abundant in the more aqueously altered meteorites. In addition, the methylnaphthalene maturity parameter negatively correlates with aqueous alteration (*Sephton et al.,* 2000). These features are interpreted as the result of chemical reactions involving different amounts of water. Hence, the molecular architecture of the macromolecular materials can be explained by varying extents of aqueous alteration on a common organic progenitor.

4.4.4. Stable isotopes. Carbon- and N-isotopic variability within macromolecular materials was exploited by *Sephton et al.* (2003, 2004) to gain an insight into the parent-body alteration history of chondritic organic matter. The stable-isotopic data suggests that all carbonaceous chondrites accreted a common organic progenitor, which may have predated the formation of the solar system (*Alexander et al.,* 1998) or was formed in the presolar nebula (*Remusat et al.,* 2006). It is proposed that the organic starting material exhibited enrichments in the heavy isotopes of C and N and that these were progressively lost during increasing aqueous and thermal processing on the parent asteroid (*Sephton et al.,* 2003, 2004). If this hypothesis is correct, then the level of alteration endured by a carbonaceous chondrite can be assessed by establishing the preservation state of its stable-isotopic enrichments. Laboratory aqueous alteration experiments, performed on isolated macromolecular material from the Murchison CM2 meteorite, support the proposal that isotopic enrichments may be removed during parent-body alteration (*Sephton et al.,* 1998). Parent-body aqueous alteration, it seems, produces predictable and reproducible isotopic features consistent with the alteration of a common organic starting material for CM2 and CI1 meteorites.

With such compelling evidence of an alteration sequence of a common organic progenitor between the CI1 and CM2 chondrites, it follows that the organic content of low petrographic types is not diagnostic of its cometary or asteroidal origin.

4.5. Recent Insights from Space Missions

Stardust is a particularly important mission as it is the only solid sample return mission from a specific astronomical body other than the Moon (*Brownlee et al.,* 2006). Phyllosilicates typical of type 1 and type 2 carbonaceous chondrites were not found among the 25 well-characterized Stardust samples (*Zolensky et al.,* 2006). Magnetite, a typical mineral of CI1 chondrites, was also not identified. Although it is possible that phyllosilicates were destroyed during impact, it is more likely they were not present within the dust expelled by the 81P/Wild 2 cometary nucleus and harvested by the Stardust spacecraft (*Brownlee et al.,* 2006). The most common minerals found in Stardust samples are olivine, pyroxene, and iron sulfides. The wide composition range of olivine and the enrichment in some minor elements such as Mn are compatible with observations of IDPs, micrometeorites, and carbonaceous chondrites (*Zolensky et al.,* 2006). The mineralogy of abundant Ni-poor, Fe sulfides is, however, more compatible with anhydrous IDPs than with any other primitive material [with the exception of two pentlandite grains (*Zolensky et al.,* 2006)]. Some enrichments in D, ^{15}N, and ^{13}C, similar to those found in anhydrous IDPs (*Messenger et al.,* 2006) and in carbonaceous chondrites (*Busemann et al.,* 2006), were also found in 81P/Wild 2 dust (*McKeegan et al.,* 2006). Although we have only preliminary data, the organics record of Comet 81P/Wild 2 show similarities and differences with carbonaceous chondrites (*Sandford et al.,* 2006).

Results from the Deep Impact mission contrast with those of Stardust. On July 4, 2005, a 370-kg copper projectile impacted onto the Comet 9P/Tempel 1 nucleus surface (*A'Hearn et al.,* 2005). Near- and mid-IR observations made onboard the Spitzer Space Telescope revealed a large diversity of minerals (*Lisse et al.,* 2006). The signature of phyllosilicates and carbonates, in addition to the more classical cometary minerals, olivine and pyroxene, has been tentatively identified in the 9P/Tempel 1 dust from a high signal/noise ratio spectrum (*Lisse et al.,* 2006). Before we discuss the presence of phyllosilicates and carbonates in the 9P/Tempel 1 spectrum, we note that their identification does not rely on a search for individual spectral features as is usually the case (e.g., *Crovisier et al.,* 1996), but from the decrease of the residual χ^2 after a fit of the observed spectrum by a modeled spectrum. The weighted surface area of phyllosilicates and carbonates is estimated to be ~14% and 8% respectively. The mineralogy of Comet 9P/Tempel 1 dust (olivine, pyroxene, phyllosilicates, carbonates, sulfides) is not unlike that of CM2 chondrites or Tagish Lake (*Zolensky et al.,* 2002). It is, however, important to note that the magnetite signature was not detected in the dust ejecta of 9P/Tempel 1, and that the identified sulfide, niningerite, is not the typical Fe-Ni sulfide found in hydrated carbonaceous chondrites (e.g., *Bullock et al.,* 2005).Were the detection of phyllosilicates and carbonates confirmed in 9P/Tempel 1, this would provide a strong link between the nucleus of Comet 9P/Tempel 1 and hydrated carbonaceous chondrites.

The differences between the Stardust and Deep Impact results may be due to the sampling of different size fractions, to the different location of the dust within each comet, and to a strong diversity between cometary nuclei (see section 6.1). The difference between the dust gently released by the sublimation process (coming from the most outer layers of the comet) and the dust coming from the interior is demonstrated by the fact that the pre-impact spectrum of 9P/Tempel 1 dust is different from the post-impact spectrum (*Lisse et al.,* 2006). Only the post-impact spectrum revealed carbonates and phyllosilicates that might have formed in the comet interior.

5. HYDROTHERMAL ALTERATION IN COMETS

The possible cometary origin of CI1 chondrites raises the question of hydrothermal alteration in comets. It is not trivial that water-rich bodies formed in the outer solar system can have been hot enough to permit the circulation of liquid or vapor water. There have been numerous theoretical studies of the thermal history of comets (e.g., *Prialnik,* 1992, 2002; *Prialnik and Podolak,* 1999). The central issue in the context of a possible cometary origin of CI1 chondrites is the likelihood that comet interiors sustained liquid water for times sufficient to generate aqueous alteration of the type seen in CI meteorites (thousands to perhaps millions of years at ~273 K). Answering this specific question is complimentary to the chapter by Coradini et al., which addresses more generally the structure of Kuiper belt objects, and to that of de Bergh et al., who describe alteration minerals present in KBOs. We tackle this problem using a time-dependent energy balance equation for comets.

5.1. Model for the Thermal Evolution of a Comet

Here we will summarize the salient features of the thermal history of comets in the early stages of solar system evolution using a simple heat conduction model. Our basic state is a spherical body composed of rock/dust and water in all its various phases. In such a body there are two principle sources of heat: (1) radioactive decay of short-lived nuclides in the rock and dust and (2) exothermic crystallization of amorphous water ice to crystalline ice. As we are concerned primarily with the temperatures triggering the alteration, we will not include heat sources from reactions associated with alteration itself (reactions that form clay minerals, for example, are strongly exothermic). These two sources of heat are balanced by radiative cooling at the surface of the body and by endothermic phase changes in water such as sublimation and melting of water ice. Note that for the sake of simplicity, we do not take into account the cratering heat source, which might, however, play a significant role.

With these processes in mind, the temperature variations with time in the comet can be obtained using an energy conservation equation of the form

$$\frac{\partial T}{\partial t} = \kappa\left(\frac{\partial^2 T}{\partial r^2} + \frac{2}{r}\frac{\partial T}{\partial r}\right) + (1-\phi)\frac{Q}{c} \quad (1)$$

where T is temperature; t is time; κ is the bulk thermal diffusivity (m^2/s) at r; r is radial distance from the center of the spherical body (m); ϕ is the volume fraction of all H_2O phases (solid, liquid, gas) combined (a nominal value of 0.7 was adopted for the calculations here); Q is heat production due to decay of the radiogenic heat sources, primarily ^{26}Al and ^{60}Fe (W/kg); and c is the specific heat for rock modified to include the enthalpies of reaction. Equation (1) incorporates latent heats associated with phase changes into a variable specific heat (*Carslaw and Jaeger,* 1959). It is a suitable treatment of the problem where enthalpies of reaction are released over a finite temperature interval ΔT. Accordingly, values for c are specified by the relation

$$c = c_{rock} + \left(\frac{\Delta H_{xstl}}{\Delta T_{xstl}}\right)\left(\frac{\phi}{1-\phi}\right)\frac{\rho_{ice}}{\rho_{rock}}\xi_{xstl} + X_{sub}\left(\frac{\Delta H_{sub}}{\Delta T_{sub}}\right)\left(\frac{\phi}{1-\phi}\right)\frac{\rho_{ice}}{\rho_{rock}}\xi_{sub} + \left(\frac{\Delta H_{melt}}{\Delta T_{melt}}\right)\left(\frac{\phi}{1-\phi}\right)\frac{\rho_{w}}{\rho_{rock}}\xi_{melt} \quad (2)$$

where ΔH_{xstl} and ΔT_{xstl} are the heat of reaction and temperature interval for crystallization from amorphous to crystalline water ice, respectively; ΔH_{sub} and ΔT_{sub} are the analog values for sublimation of ice; ΔH_{melt} and ΔT_{melt} apply to melting to form liquid water; and ρ_w, ρ_{rock}, and ρ_{ice} are the mass densities of liquid water, rock, and ice. The ϕ and mass densities in equation (2) scale heats of reaction to a per kilogram rock basis and X_{sub} is the fraction of ice subjected to sublimation (see below). The ξ terms are reaction progress variables for each indicated phase change with a range of 0 to 1. For convenience in numerical calculation, these variables can be defined to provide a smooth transition over temperature ΔT using an error function formulation for process i

$$\xi_i = \mathrm{erf}\left(2\frac{T - T_{low}}{\Delta T_i}\right) \quad (3)$$

where T_{low} refers to the lower bound for the temperature interval for the phase transition.

Values for κ are obtained by combining the thermal diffusivities of the individual phases such that

$$\kappa = \phi\kappa_{H_2O} + (1-\phi)\kappa_{rock} \quad (4)$$

$$\kappa_{H2O} = \xi_{melt}\kappa_w + (1-\xi_{melt})\kappa_{ice} + X_{sub}(\xi_{sub}\kappa_{vap} + (1-\xi_{sub})\kappa_{ice})) \quad (5)$$

$$\kappa_{ice} = \xi_{xstl}\kappa_{xstl} + (1-\xi_{xstl})\kappa_{am} \quad (6)$$

In equations (4) through (6), $\xi_{sub} = 1$ when $\xi_{melt} = 0$ and vice versa. Values for Q in equation (1) are obtained using the expression $Q = 6.0 \times 10^{-3}\,(^{26}Al/^{27}Al)_o \exp(-\lambda_{26}t) + 1.2 \times 10^{-3}\,(^{60}Fe/^{56}Fe)_o \exp(-\lambda_{60}t)$, where λ_{26} (λ_{60}) is the decay constant for ^{26}Al (^{60}Fe), $(^{26}Al/^{27}Al)_o$ and $(^{60}Fe/^{56}Fe)_o$ are the initial radiogenic/stable isotope ratios at the time the body accretes, and the pre-exponentials include 6.4×10^{-13} J per decay of ^{26}Al and 3.8×10^{-14} J per decay for ^{60}Fe and typical chondritic concentrations of Al and Fe.

In the calculations that follow the fraction of sublimed ice, X_{sub}, represents the net loss of ice by sublimation from open pore walls in the comet. As a result, condensation of H_2O vapor is excluded as an explicit term in equation (2). The contribution of H_2O vapor to the thermal diffusivity is neglected ($\kappa_{vap} \sim 0$) because its contribution to the bulk diffusivity is minor, although advection of gas is an efficient agent for heat flow (*Prialnik et al.,* 2004).

For simplicity we adopt a fixed-temperature boundary condition across the body (the so-called fast-rotator approximation). The temperature at the outer boundary of the spherical body, T_s, is imposed by radiative heating from the star such that

$$T_s = [((1-A)L_\odot)/(4\pi R^2)/(\varepsilon\sigma)]^{1/4} \quad (7)$$

where R is the heliocentric distance from the star, ε is the surface emissivity (~1), σ is the Stefan-Boltzmann constant ($kg/(s^3\,K^4)$), $L_\odot$ is the solar luminosity (W), and A is the albedo of the comet surface.

Values for the various parameters in equations (1) through (7) are listed in Table 1. They are taken from the literature and are representative of the values used in most comet-related studies. Equations (1)–(6) with boundary condition (7) were solved using an explicit finite difference scheme in what follows.

5.2. Results Relevant to the Formation of Liquid Water

Previous work has shown that the most important control on the likelihood for liquid water early in the life of a comet is the degree of H_2O sublimation. This is because the enthalpy of sublimation is comparatively large and represents a substantial heat sink. For a given surface-to-volume ratio and temperature, the rate of sublimation of water ice is proportional to the difference between the vapor

TABLE 1. Parameters adopted for calculations shown in Fig. 4.

Description	Symbol	Value
Solar luminosity	$L_{\odot}$	3.83×10^{26} (W)
Surface albedo	A	0.05
Water volume fraction	ϕ	0.7
H_2O sublimation enthalpy	ΔH_{sub}	2.8×10^6 (J/kg)
H_2O crystallization enthalpy	ΔH_{xstl}	-9.0×10^4 (J/kg)
H_2O ice melting enthalpy	ΔH_{melt}	3.3×10^5 (J/kg)
Temperature interval for sublimation	ΔT_{sub}	273 K–180 K
Temperature interval for crystallization	ΔT_{xstl}	273 K–130 K
Temperature interval for melting	ΔT_{melt}	273 K–270 K
Thermal diffusivity of amorphous ice	κ_{am}	3.13×10^{-7} (m^2 s^{-1})
Thermal diffusivity of crystalline ice	κ_{xstl}	$[0.465 + 488.0/T(K)]/[\rho_{ice}$ (kg/m^3) 7.67 T(K)] (m^2 s^{-1})
Thermal diffusivity of liquid water	κ_{melt}	$[-0.581 + 0.00634$ T(K) $- 7.9 \times 10^{-6}$ T(K)]/ρ_w (kg/m^3) 4.186×10^3) (m^2 s^{-1})
Thermal diffusivity of rock/dust	κ_{rock}	$3.02 \times 10^{-7} + 2.78 \times 10^{-4}$/T(K) ($m^2$ s^{-1})
Mass density of ice	ρ_{ice}	917 (kg/m^3)
Mass density of liquid water	ρ_w	1000 (kg/m^3)
Mass density of rock/dust	ρ_{rock}	3000 (kg/m^3)
Initial $^{26}Al/^{27}Al$ of solar system	$(^{26}Al/^{27}Al)_o$	6×10^{-5}
Initial $^{60}Fe/^{56}Fe$ of solar system	$(^{60}Fe/^{56}Fe)_o$	1×10^{-6}

pressure of H_2O in the void space and the equilibrium vapor pressure. This difference in turn depends upon the details of the gas permeability within the comet. Models suggest that sublimation is limited in the deep interior while it is rampant nearer to the surface with the effect that the comet develops a weak zone composed of a fragile structure of sublimed ice (*Prialnik and Podolak,* 1999).

In order to estimate the amount of melting, one must specify the concentrations of ^{26}Al and ^{60}Fe in the dust/rock comprising the comet upon accretion. The concentration of ^{60}Fe turns out to be relatively unimportant as an agent for forming liquid water in comets as the heat contributed by ^{60}Fe is small in comparison to that provided by ^{26}Al. These initial concentrations depend on the initial $(^{26}Al/^{27}Al)_o$ and $(^{60}Fe/^{56}Fe)_o$ in the protoplanetary disk where the comet ultimately forms, and the time interval between formation of the solar protoplanetary disk and the accretion of the comet, i.e., the "free decay time" for each radionuclide. Some workers have considered the results of a protracted accretion interval (*Merk et al.,* 2002). The essential features of the thermal history can nonetheless be described by considering the free decay time followed by instantaneous accretion.

Weidenschilling (2000) suggested that the time τ required for accretion of planetesimals is expected to have varied across the disk according to the expression $\tau \sim 2000$ (yr) (R/1 AU$)^{3/2}$. We can therefore consider the melting of water ice in comets as a function of radial distance R from the Sun. The result is that even relatively small bodies on the order of 10 km in diameter (Fig. 4a) could have had liquid water in their interiors if the transport properties of the gas phase allowed for $\leq$5% of the mass of water ice to be lost to sublimation. We should expect that Oort cloud comets of even small sizes (e.g., 5 km), having formed within the planet-forming region of the early solar system ($\tau \sim$ 0.02 m.y.), will have had liquid water in their interiors for perhaps as long as 2 m.y. (Fig. 4a).

Liquid water could have been present in comets formed in or near the Kuiper belt (R > 40 AU), where the free decay time is longer ($\tau \geq 0.5$ m.y.), if their diameters approached 50 km or greater (Fig. 4b), and if the amount of ice sublimed in the interior was restricted to 30% or less by H_2O vapor buildup (Fig. 4b). The combined effects of sublimation and melting could serve to keep the maximum temperatures in the interiors of such comets to near the melting point of water ice several million years. This buffering capacity is overwhelmed for comets formed inside 35 AU (unshown calculations).

The simple models shown in Fig. 4 illustrate the fundamental point that liquid water could have persisted within comets for >1 m.y. Smaller bodies of ~5 km diameter formed in the region of planet formation (e.g., Oort cloud comets) will have contained liquid water in their first few million years if H_2O ice sublimation was limited by restricted gas transport to 5% of total solid water or less. Comets formed further out in the Kuiper belt will have contained liquid water for a million years or more if they formed as bodies with diameters approaching 50 km or more.

CI chondrites appear to have been altered at or very near to the melting temperature of water (*Young et al.,* 1999; *Young,* 2001). An argument can be made that the regulating effect on comet thermal histories by sublimation and the record of aqueous alteration in CI meteorites may be causally linked. Many model calculations, including those in

Fig. 4, result in maximum temperatures at or near the melting point of solid H_2O as a result of a balance between the buffering capacity of enthalpies of melting and the radiogenic heat remaining after sublimation. One can infer that many relatively ice-poor bodies (asteroid precursors) may have had too little water ice to prevent raising liquid water temperatures substantially above the melting point of H_2O. The result may have been aqueous alteration at higher temperatures.

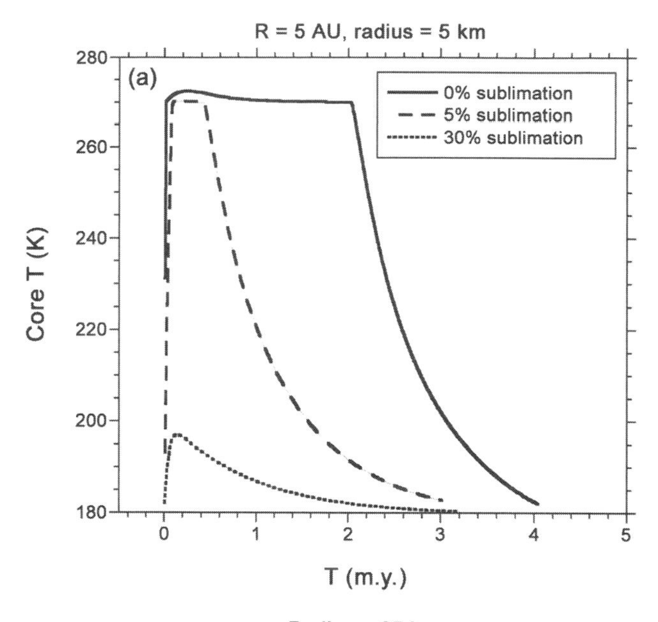

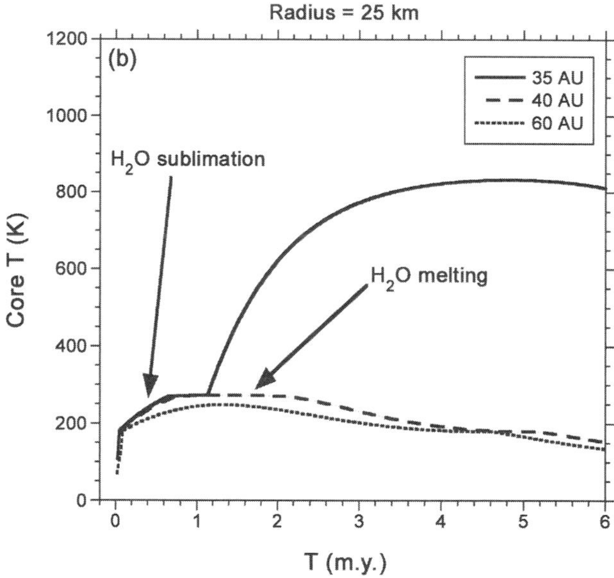

Fig. 4. **(a)** Calculated temperatures at the centre of a spherical comet formed in the planet-forming region at 5 AU as a function of time. The radius of the comet is 5 km. Different curves show the effects of different fractions of sublimed water ice. Results show that, with modest amounts of water ice sublimation, liquid water (occurring where T ≥ 273 K) is expected to have been present in small comets formed in the planet-accretion region of the solar protoplanetary disk. Calculations are based on equations described in the text and the parameters shown in Table 1. **(b)** Calculated temperature at the center of a 25-km-radius comet as a function of time. Plateaux correspond to the thermal buffer capacity of sublimation and melting of H_2O ice. Different curves refer to accretion of the comet at different distances from the Sun (R) based on the accretion rates of *Weidenschilling* (2000). At R = 35 AU liquid water also begins to heat up substantially.

6. THE ASTEROID-COMET CONTINUUM

In the previous sections, we implicitly assumed that comets were similar enough to each other to be distinguished, either on a dynamical or compositional basis, from asteroids. In the present section, we aim at criticizing that assumption. We will show that the closer one goes to a comet, the more different comets look (section 6.1), and that comets and water-rich asteroids might be more similar than once thought (section 6.2).

6.1. Comet Geological Diversity

Although all active comets undoubtedly share the common property of having a coma and look the same at a distance, it becomes increasingly clear that the term "comet" covers a large diversity of bodies. We already mentioned the fact that this diversity might explain the differences between the dust of 9P/Tempel 1 and 81P/Wild 2 (section 4.5).

The fact that comets are different from one another when looked at in detail is dramatically demonstrated by the different shapes and geological features exhibited by the three cometary nuclei explored by spacecraft (19P/Borrelly, 81P/Wild 2, and 9P/Tempel 1). 81P/Wild 2 and 9P/Tempel 1 are roughly spherical, while 19P/Borrelly and 1P/Halley are elongated (*A'Hearn et al.,* 2005; *Britt et al.,* 2004; *Brownlee et al.,* 2004). Comet 19P/Borrelly is characterized by the absence of impact craters and a diversity of complex geological units (smooth and mottled terrains, mesas) and features (*Britt et al.,* 2004). The dichotomy between mottled and smooth terrains in 19P/Borrelly is not observed in 81P/Wild 2 (*Britt et al.,* 2004; *Brownlee et al.,* 2004). The sculpting of the nucleus of 19P/Borrelly is attributed to sublimation processes (*Britt et al.,* 2004). In contrast, 81P/Wild 2 does not possess as many mesas as 19P/Borrelly (*Brownlee et al.,* 2004). Both 81P/Wild 2 and 9P/Tempel 1 have impact craters (*A'Hearn et al.,* 2005; *Brownlee et al.,* 2004). While the surface of 81P/Wild 2 is cohesive (*Brownlee et al.,* 2004), that of 9P/Tempel 1 is loosely bound (*A'Hearn et al.,* 2005). The geological diversity of comets is in agreement with the significant differences observed in the comets' molecular abundances (*Biver et al.,* 2006) and dust properties (*Lisse et al.,* 2004).

6.2. Continuum Between Asteroids and Comets

Although clear at first sight, the distinction between asteroids and comets might be more complex, as both objects are defined *observationally* and *dynamically*, as well as *compositionally*. Observationally, comets are defined by the presence of a coma of sublimated ice and dust. Dynamically, comets have a Tisserand parameter below 3 (see section 2). Compositionally, comets are ice-rich objects. All three definitions are linked, since ice-rich objects can develop a substantial coma only if they have orbits eccentric enough to be significantly heated by the sunlight. The limit between asteroids and comets is therefore blurred as, for example, an ice-rich object in a roughly circular orbit might be too far from the Sun to develop a coma and will appear observationally and dynamically as an asteroid, while it would be more of a comet compositionally.

Some objects having the dynamical properties of asteroids look indeed like comets as far as their composition is concerned. The Trojan binary asteroid 617 Patroclus has a density ($\rho = 0.8^{+0.2}_{-0.1}$ g/cm^3) compatible with that of an ice-rich body (*Marchis et al.,* 2006). Asteroid 1 Ceres is possibly made of a water-rich mantle (*Thomas et al.,* 2005). Water ice has been detected in the primitive asteroid 773 Irmintraud (*Kanno et al.,* 2003). The Centaur asteroid 2060 Chiron appears to have a cometary activity when close to perihelion (*Meech and Belton,* 1990). The dark B-type asteroids (3200) Phaeton and 2001 YB_5 are associated with meteor showers, suggesting some kind of so far undetected cometary activity (*Meng et al.,* 2004). Based on their albedos, there is a significant fraction of extinct comets in the near-Earth asteroids population (*Fernández et al.,* 2001). Recently, the existence of a new class of objects, main-belt comets, has been established by *Hsieh and Jewitt* (2006). 133P/Elst-Pizzaro, P/2005 U1 (Read), and 118401 (1999 RE_{70}) exhibit cometary activity, i.e., dust ejection driven by ice sublimation, while they are on asteroid-like orbits (*Hsieh and Jewitt,* 2006). This discovery of main-belt comets or *active asteroids* will probably be followed by others as the technological ability to detect faint comae will increase. The identification of such objects demonstrates the important point we were making above: Cometary bodies are hidden in the asteroid belt because their water emission is too faint to be detected.

It is also worth mentioning that recent results weakened the idea that comets are dirty snowballs. For example, it was recently realized that comets are richer in dust (dust/water ratios >1) (*Keller et al.,* 2005; *Lisse et al.,* 2006) than previously thought (*Lisse et al.,* 2004). It is also known that the active (i.e., water-rich) regions of comets cover a small surface of the nucleus (e.g., *Soderblom et al.,* 2004). If comets are more of an equal mixture of dust and ice rather than a water-rich body peppered with a small amount of dust, what differentiates dust-rich comets from water-rich asteroids? We note that *Clayton and Mayeda* (1999) estimate that water/dust ratios larger than 10 are needed to explain the O-isotopic composition of some carbonaceous chondrites conventionally considered to originate from asteroids. This water content is far larger than the expected water content of comets having water/dust ratios ~1 (*Keller et al.,* 2005; *Lisse et al.,* 2006).

A continuum between comets and asteroids is therefore expected. If we had the technical ability to tow a water-rich asteroid close enough to the Sun, it would undoubtedly display clear cometary activity. If we could observe a comet on sufficiently long timescales (~10^4 yr), we would observe the gradual disappearance of its cometary activity.

7. CONCLUSIONS

Although simple, the question asked at the beginning of this chapter has a complex answer. It is extremely puzzling that, given the ~30,000 samples present in museum collections, it is so difficult to positively identify a cometary meteorite. There can only be two solutions to that thought-provoking paradox. Either there are no cometary meteorites, or cometary meteorites are so similar to some of the asteroidal meteorites that there is no definitive way to identify them.

If no cometary meteorites exist, it means that there is an extremely powerful mechanism preventing them from reaching Earth. That possibility seems unlikely given the numerical simulations of the dynamical evolution of comets. Dynamical studies indeed show that JFCs can find direct and indirect dynamic routes to Earth. The existence of indirect routes (trapping of outer solar system objects in the asteroid belt) is experimentally confirmed by the presence of dormant comets in the near-Earth object population and by the recent discovery of active asteroids. Interestingly, the average entry velocity of these JFCs in Earth's atmosphere is similar to that of asteroids (21 vs. 20 km/s), relieving the caveat of destruction upon impact on the atmosphere (cometary fragments are usually thought to penetrate Earth's atmosphere at a high velocity, generating total destruction). When respective flux, collision probabilities, and entry velocities are taken into account, it appears that the proportion of cometary meteorites relative to their asteroidal counterparts is small but significantly *above zero*. Given that more than 30,000 meteorites are present in museum collections, dynamical observations clearly favor the existence of cometary meteorites. Although fireball observations have failed to positively identify a cometary meteorite so far, there is evidence for solid and dense material (i.e., meteorites) being contained in objects with cometary orbits.

Type 1 (and maybe type 2) carbonaceous chondrites are the best candidates for being cometary meteorites. They are rare, dark, have an unfractionated chemical composition, are rich in volatile elements, and are rich in the light elements H, C, N, and O. The orbit of the Orgueil CI1 chondrite is compatible with that of a JFC. The infrared spectrum of the ungrouped C2 chondrite Tagish Lake is similar to that of D asteroids, which have been linked to comets, while the infrared spectrum of Comet 162P/Siding Spring is not unlike that of the CI1 chondrite Alais. Some differences between comets and low petrographic type carbonaceous chondrites exist, however. The orbit of the Tagish Lake meteorite is asteroidal rather than cometary. The D/H ratio of CI1 chondrites is lower than that of comets, although this discrepancy is based on the *single* measurement of Comet 1P/Halley. The D/H ratio of JFCs is unknown. We also show that the distinctive organics record of CI1 chondrites could be the result of parent-body processes unrelated to its asteroidal or cometary nature. The 81P/Wild 2 cometary samples delivered to Earth by the Stardust spacecraft are unlike CI1 chondrites. On the other hand, the ejecta of the 9P/Tempel 1 exposed by the Deep Impact mission might contain abundant phyllosilicates and carbonates, minerals typical of CI1 chondrites and Tagish Lake. The differences between the two comets illustrates that there might be huge differences from one comet to the other, complicating the task of identifying cometary meteorites.

If CI1 (and some C2) chondrites originate from comets, this has strong implications for the evolution of cometary nuclei. It means that they endured vigorous hydrothermal alteration as recorded by CI1 chondrites. We show that the history of sublimation of water ice is the crucial parameter determining the prospects for liquid water in a comet. Importantly, this alteration would take place in the interior of the comet, meaning that the spectroscopic record, which samples the outer surface of celestial bodies, should be taken with caution. It might also explain why the results of Stardust are different from that of Deep Impact. If comets endured hydrothermal activity and were, to some extent, heated, it also implies that we should keep in mind the provocative possibility that other carbonaceous chondrites [such as CBs, which are enriched in ^{15}N and contain osbornite (TiN), a rare refractory mineral found among Stardust samples] also have a cometary origin.

Although it is possible that some carbonaceous chondrites originate from comets, it is probably not the case for all of them. Dark asteroids such as C types are still a significant source for carbonaceous chondrites. We would argue, however, that there is a continuum between comets and asteroids rather than a sharp distinction. This continuum is easy to understand as the cometary (ice-rich) or asteroidal (ice-poor) nature of a given object depends on its position at formation relative to the snow line. There is no reason for the snow line to have always occupied the same location in the protoplanetary accretion disk, nor to define an abrupt transition between water-poor and water-rich bodies. Type 1 and 2 carbonaceous chondrites might sample the continuum between asteroids and comets. The answer to our question is therefore *yes*.

Acknowledgments. We thank F. Robert, M. Zolensky, D. Bockelée-Morvan, and P. Abell for stimulating discussions. H. Levison kindly provided the unpublished Fig. 1. Two anonymous reviewers as well as the associate editor, H. Boehnhardt, provided significant comments that helped improving the paper. The organizers of the TNO workshop in Catania are warmly thanked for putting together an extremely stimulating meeting. The Programme National de Planétologie and the CNRS France-Etats-Unis program partly funded this project. This is IARC publication 2006-0946.

REFERENCES

Abell P. A., Fernández Y. R., Pravec P., French L. M., Farnham T. L., et al. (2005) Physical characteristics of Comet Nucleus C/2001 OG_{108} (LONEOS). *Icarus, 179,* 174–194.

A'Hearn M. F., Belton M. J. S., Delamere W. A., Kissel J., Klaasen K. P., et al. (2005) Deep Impact: Excavating comet Tempel 1. *Science, 310,* 258–264.

Alexander C. M. O. D., Russell S. S., Arden J. W., Ash R. D., Grady M. M., and Pillinger C. T. (1998) The origin of chondritic macromolecular organic matter: A carbon and nitrogen isotope study. *Meteoritics & Planet. Sci., 33,* 603–622.

Barucci M. A., Cruikshank D. P., Mottola S., and Lazzarin M. (2002) Physical properties of Trojan and Centaur asteroids. In *Asteroids III* (W. F. Bottke Jr. et al., eds.), pp. 273–287. Univ. of Arizona, Tucson.

Biver N., Bockelée-Morvan D., Crovisier J., Lis D. C., Moreno R., et al. (2006) Radio wavelength molecular observations of comets C1999 T1 (McNaught-Hartley), C/2001 A2 (LINEAR), C/2000 WM_1 (LINEAR) and 153P/Ikeya-Zhang. *Astron. Astrophys., 449,* 1255–1270.

Bland P. A., Cressey G., and Menzies M. A. (2004) Modal mineralogy of carbonaceous chondrites by X-ray diffraction and Mössbauer spectroscopy. *Meteoritics & Planet. Sci., 39,* 3–16.

Bockelée-Morvan D., Gautier D., Lis D. C., Young K., Keene J., et al. (1998) Deuterated water in Comet C/1996 B2 (Hyakutake) and its implications for the origin of comets. *Icarus, 133,* 147–162.

Boehnhardt H. (2004) Split comets. In *Comets II* (M. C. Festou et al., eds.), pp. 301–316. Univ. of Arizona, Tucson.

Borovička J. and Jenniskens P. (1998) Time resolved spectroscopy of a Leonid fireball afterglow. *Earth Moon Planets, 82-83,* 399–428.

Botta O. and Bada J. L. (2002) Extraterrestrial organic compounds in meteorites. *Surv. Geophys., 23,* 411–467.

Bottke W. F., Morbidelli A., Jedicke R., Petit J.-M., Levison H. F., Michel P., and Metcalfe T. S. (2002) Debiased orbital and absolute magnitude distribution of the near-Earth objects. *Icarus, 156,* 399–433.

Bottke W. F., Durda D. D., Nesvorný D., Jedicke R., Morbidelli A., Vokrouhlický D., and Levison H. F. (2005) Linking the collisional history of the main asteroid belt to its dynamical depletion and excitation. *Icarus, 179,* 63–94.

Bottke W. F., Nesvorný D., Grimm R. E., Morbidelli A., and O'Brien D. P. (2006) Iron meteorites are remnants of planetesimals from the terrestrial planet region. *Nature, 439,* 821–824.

Britt D. T., Boice D. C., Buratti B. J., Campins H., Nelson R. M., et al. (2004) The morphology and surface processes of Comet 19/P Borrelly. *Icarus, 167,* 45–53.

Brown P. G., Hildebrand A. R., Zolensky M. E., Grady M., Clayton R. N., et al. (2000) The fall, recovery, orbit, and composition of the Tagish Lake meteorite: A new type of carbonaceous chondrite. *Science, 290,* 320–325.

Brown P., Pack D., Edwards W. N., Revelle D. O., and Yoo B. B. (2004) The orbit, atmospheric dynamics, and initial mass of the Park Forest meteorite. *Meteoritics & Planet. Sci., 39,* 1781–1796.

Brownlee D. E., Horz F., Newburn R. L., Zolensky M., Duxbury T. C., et al. (2004) Surface of young Jupiter family Comet 81P/Wild 2: View from the Stardust spacecraft. *Science, 304,* 1764–1769.

Brownlee D. E., Tsou P., Aléon J., Alexander C. M. O.'D., Araki T., et al. (2006) Comet 81P/Wild 2 under a microscope. *Science, 314,* 1711–1716.

Bullock E. S., Gounelle M., Lauretta D. S., Grady M. M., and Russell S. S. (2005) Mineralogy and texture of Fe-Ni sulphides in CI1 chondrites: Clues to the extent of aqueous alteration on the CI1 parent-body. *Geochim. Cosmochim. Acta, 69,* 2687–2700.

Busemann H., Young A. F., Alexander C. M. O. D., Hoppe P., Mukhopadhyay S., and Nittler L. R. (2006) Interstellar chemistry recorded in organic matter from primitive meteorites. *Science, 312,* 727–730.

Calvin W. M. and King T. V. (1997) Spectral characteristics of iron-bearing phyllosilicates: Comparison to Orgueil (CI1), Murchison and Murray (CM2). *Meteoritics & Planet. Sci., 32,* 693–701.

Campins H. and Fernández Y. R. (2002) Observational constraints on surface characteristics of cometary nuclei. *Earth Moon Planets, 59,* 117–134.

Campins H. and Swindle T. D. (1998) Expected characteristics of cometary meteorites. *Meteoritics & Planet. Sci., 33,* 1201–1211.

Campins H., Ziffer J., Licandro J., Pinilla-Alonso N., Fernández Y. R., De León J., Mothé-Diniz T., and Binzel R. P. (2006) Nuclear spectra of Comet 162P/Siding Spring (2004 TU12). *Astrophys. J., 132,* 1346–1353.

Carslaw H. S. and Jaeger J. C. (1959) *Conduction of Heat in Solids.* Oxford Univ., Oxford. 510 pp.

Ceplecha Z. and McCrosky R. E. (1976) Fireball end heights — A diagnostic for the structure of meteoric material. *J. Geophys. Res., 81,* 6257–6275.

Clayton R. N. and Mayeda T. K. (1999) Oxygen isotope studies of carbonaceous chondrites. *Geochim. Cosmochim. Acta, 63,* 2089–2104.

Cronin J. R. and Moore C. B. (1976) Amino acids of the Nogoya and Mokoia carbonaceous chondrites. *Geochim. Cosmochim. Acta, 40,* 853–857.

Crovisier J., Brooke T. Y., Hanner M. S., Keller H. U., Lamy P. L., et al. (1996) The infrared spectrum of Comet C/1995 O1 (Hale-Bopp) at 4.6 AU from the Sun. *Astron. Astrophys., 315,* L385–L388.

Crovisier J., et al. (2005) Chemical diversity of comets observed at radio wavelengths in 2003–2005. *Bull. Am. Astron. Soc., 37,* 646.

Cyr K. E., Sears W. D., and Lunine J. I. (1998) Distribution and evolution of water ice in the solar nebula: Implications for solar system body formation. *Icarus, 135,* 537–548.

Dones L., Weissman P. R., Levison H. F., and Duncan M. J. (2004) Oort cloud formation and dynamics. In *Comets II* (M. C. Festou et al., eds.), pp. 153–174. Univ. of Arizona, Tucson.

Eberhardt P., Reber M., Krankowsky D., and Hodges R. R. (1995) The D/H and $^{18}O/^{16}O$ ratios in water from Comet P/Halley. *Astron. Astrophys., 302,* 301–316.

Ehrenfreund P., Glavin D. P., Botta O., Cooper G., and Bada J. L. (2001) Extraterrestrial amino acids in Orgueil and Ivuna: Tracing the parent body of CI type carbonaceous chondrites. *Proc. Natl. Acad. Sci. USA, 98,* 2138–2141.

Ehrenfreund P., Charnley S. B., and Wooden D. H. (2004) From interstellar material to cometary particles and molecules. In *Comets II* (M. C. Festou et al., eds.), pp. 115–133. Univ. of Arizona, Tucson.

Eiler J. M. and Kitchen N. (2004) Hydrogen isotopic evidence for the origin and evolution of the carbonaceous chondrites. *Geochim. Cosmochim. Acta, 68,* 1395–1411.

Emery J.-P., Cruikshank D. P., and Van Cleve J. (2006) Thermal emission spectroscopy (5.2–38 μm) of three Trojan asteroids with the Spitzer Space Telescope: Detection of fine-grained silicates. *Icarus, 182,* 496–512.

Engrand C. and Maurette M. (1998) Carbonaceous micrometeorites from Antarctica. *Meteoritics & Planet. Sci., 33,* 565–580.

Fernández Y. R., Jewitt D. C., and Sheppard S. S. (2001) Low albedos among extinct comets candidates. *Astrophys. J. Lett., 553,* L197–L200.

Floss C., Stadermann F. J., Bradley J. P., Dai Z. R., Bajt S., Graham G., and Lea A. S. (2006) Identification of isotopically primitive interplanetary dust particles: A nanoSIMS isotopic imaging study. *Geochim. Cosmochim. Acta, 70,* 2371–2399.

Fomenkova M. N. (1997) Organic components of cometary dust. In *From Stardust to Planetesimals* (Y. J. Pendelton and A. G. G. M. Tielens), pp. 415–421. ASP Conf. Series 122, San Francisco.

Fornasier S., Doressoundiram A., Tozzi G. P., Barucci M. A., Boehnhardt H., de Bergh C., Delsanti A., Davis J., and Dotto E. (2004) ESO Large Program on physical studies of transneptunian objects and Centaurs: Final results of the visible spectrophotometric observations. *Astron. Astrophys., 421,* 353–363.

Gardinier A., Derenne S., Robert F., Behar F., Largeau F., and Maquet J. (2000) Solid state CP/MAS ^{13}C NMR of the insoluble organic matter of the Orgueil and Murchison meteorites: Quantitative study. *Earth Planet. Sci. Lett., 184,* 9–21.

Gomes R., Levison H. F., Tsiganis K., and Morbidelli A. (2005) Origin of the cataclysmic late heavy bombardment period of the terrestrial planets. *Nature, 435,* 466–469.

Gounelle M. (2006) The meteorite fall at l'Aigle and the Biot report: Exploring the cradle of Meteoritics. In *The History of Meteoritics and Key Meteorite Collections: Fireballs, Finds and Falls* (G. J. H. McCall et al., eds.), pp. 73–89. Geological Society Special Publications, London.

Gounelle M. and Zolensky M. E. (2001) A terrestrial origin for sulfate veins in CI1 chondrites. *Meteoritics & Planet. Sci., 36,* 1321–1329.

Gounelle M., Spurný P., and Bland P. A. (2006) The atmospheric trajectory and orbit of the Orgueil meteorite. *Meteoritics & Planet. Sci., 41,* 135–150.

Halliday I. A., Griffin A. A., and Blackwell A. T. (1996) Detailed data for 259 fireballs from the Canadian camera network and inferences concerning the influx of large meteoroids. *Meteoritics & Planet. Sci., 31,* 185–217.

Hiroi T., Zolensky M. E., and Pieters C. M. (2001) The Tagish Lake meteorite: A possible sample from a D-type asteroid. *Science, 293,* 2234–2236.

Hsieh H. J. and Jewitt D. (2006) A population of comets in the main asteroid belt. *Nature, 312,* 561–563.

Hutsemékers D., Manfroid J., Jehin E., Arpigny C., Cochran A., Schulz R., Stüwe J. A., and Zucconi J.-M. (2005) Isotopic abundances of carbon and nitrogen in Jupiter-family and Oort cloud comets. *Astron. Astrophys., 440,* L21–L24.

Jedicke R., Nesvorný D., Whiteley R., Ivezić Ž., and Juric M. (2004) An age-colour relationship for main-belt S-complex asteroid. *Nature, 429,* 275–277.

Johnson T. V. and Fanale F. P. (1973) Optical properties of carbonaceous chondrites and their relationship to asteroids. *J. Geophys. Res., 78,* 8507–8518.

Kallemeyn G. W., Rubin A. E., and Wasson J. T. (1994) The compositional classification of chondrites: The CR carbonaceous chondrite group. *Geochim. Cosmochim. Acta, 58,* 2873–2888.

Kanno A., Hiroi T., Nakamura R., Abe M., Ishiguro M., et al. (2003) The first detection of water absorption on a D type asteroid. *Geoph. Res. Lett., 30(17),* 1909–1913.

Keller H. U., Jorda L., Küppers M., Gutierrez P. J., Hviid S. F., et al. (2005) Deep Impact observations by Osiris onboard the Rosetta spacecraft. *Science, 310,* 281–283.

Krot A. N., Scott E. R. D., and Reipurth B., eds. (2005) *Chondrites and the Protoplanetary Disk.* ASP Conf. Series 341, San Francisco. 1029 pp.

Lauretta D. S. and McSween H. Y. Jr., eds. (2005) *Meteorites and the Early Solar System II.* Univ. of Arizona, Tucson. 943 pp.

Levison H. F. and Duncan M. J. (1997) From the Kuiper belt to Jupiter-family comets: The spatial distribution of ecliptic comets. *Icarus, 127,* 13–32.

Levison H. F., Morbidelli A., Dones L., Jedicke R., Wiegert P., and Bottke W. F. Jr. (2002) The mass distribution of the Oort cloud comets. *Science, 296,* 2212–2215.

Levison H. F., Terrell D., Wiegert P., Dones L., and Duncan M. J. (2006) On the origin of the unusual orbit of Comet 2P/Encke. *Icarus, 182,* 161–168.

Licandro J., Campins H., Hergenrother C., and Lara L. M. (2003) Near-infrared spectroscopy of the nucleus of Comet 124P/Mrkos. *Astron. Astrophys., 398,* L45–L48.

Lisse C. M., Fernández Y. R., A'Hearn M. F., Grün E., Käufl H. U., et al. (2004) A tale of two very different comets: ISO and MSK measurements of dust emission from 126P/IRAS (1996) and 2P/Encke (1997). *Icarus, 171,* 444–462.

Lisse C. M., VanCleve J., Adams A. C., A'Hearn M. F., Fernández Y. R., et al. (2006) Spitzer spectral observations of the Deep Impact ejecta. *Science, 313,* 635–640.

Lodders K. (2003) Solar system abundances and condensation temperatures of the elements. *Astrophys. J., 591,* 1220–1247.

Marchis F., Hestroffer D., Descamps P., Berthier J., Bouchez A. H., et al. (2006) A low density of 0.8 g cm^{-3} for the Trojan binary asteroid 617 Patroclus. *Nature, 439,* 565–567.

Martins Z., Watson J., Sephton M. A., Botta O., Ehrenfreund P., and Gilmour I. (2006) Free carboxylic acids in the carbonaceous chondrites Murchison and Orgueil. *Meteoritics & Planet. Sci., 41,* 1073–1080.

Marvin U. B. (2006) Meteorites in history: An overview from the Renaissance to the 20th century. In *The History of Meteoritics and Key Meteorite Collections: Fireballs, Finds and Falls* (G. J. H. McCall et al., eds.), pp. 15–71. Geological Society Special Publications, London.

Mason B. (1963) The carbonaceous chondrites. *Space Sci. Rev., 1,* 621–646.

McKeegan K. D., Aléon J., Bradley J., Brownlee D., Busemann H., et al. (2006) Isotopic compositions of cometary matter returned by Stardust. *Science, 314,* 1724–1728.

McSween H. Y. and Weissman P. R. (1989) Cosmochemical implications of the physical processing of cometary nuclei. *Geochim. Cosmochim. Acta, 53,* 3263–3271.

Meech K. J. and Belton M. J. S. (1990) The atmosphere of 2060 Chiron. *Astron. J., 100,* 1323–1338.

Meibom A. and Clark B. E. (1999) Evidence for the insignificance of ordinary chondritic material in the asteroid belt. *Meteoritics & Planet. Sci., 34,* 7–24.

Meier R., Owen T. C., Matthews H. E., Jewitt D. C., Bockelée-Morvan D., Biver N., Crovisier J., and Gautier D. (1998) A determination of the HDO/H_2O ratio in Comet C/1995 O1 (Hale-Bopp). *Science, 279,* 842–844.

Meng H., Zhu J., Gong X., Li Y., Yang B., Gao J., Guan M., Fan Y., and Xia D. (2004) A new asteroid-associated meteor shower and notes on comet-asteroid connection. *Icarus, 169,* 385–389.

Merk R., Breuer D., and Spohn T. (2002) Numerical modeling of ^{26}Al-induced radioactive melting of asteroids considering accre-tion. *Icarus, 159,* 183–191.

Messenger S., Sandford S., and Brownlee D. E. (2006) The starting materials. In *Meteorites and the Early Solar System II* (D. S. Lauretta and H. Y. McSween Jr., eds.), pp. 187–208. Univ. of Arizona, Tucson.

Morbidelli A. and Gladman B. (1998) Orbital and temporal distributions of meteorites originating in the asteroid belt. *Meteoritics & Planet. Sci., 33,* 999–1016.

Morbidelli A., Jedicke R., Bottke W. F. Jr., Michel P., and Tedesco E. F. (2002) From magnitudes to diameters: The albedo distribution of near Earth objects and the Earth collision hazard. *Icarus, 158,* 329–342.

Morbidelli A., Levison H. F., Tsiganis K., and Gomes R. (2005) Chaotic capture of Jupiter's Trojan asteroids in the early solar system. *Nature, 435,* 462–465.

Mostefaoui S., Robert F., Derenne S., and Meibom A. (2006) Spatial distribution of deuterium hot-spots in the insoluble organic matter: A nanoSIMS study (abstract). *Meteoritics & Planet. Sci., 41 (Suppl.),* Abstract #5255.

Murray I. S., Hawkes R. L., and Jenniskens P. (1999) Airborne intensified charge-coupled device observations of the 1998 Leonid shower. *Meteoritics & Planet. Sci., 34,* 949–958.

Petit J.-M., Morbidelli A., and Chambers J. E. (2000) The primordial excitation and clearing of the asteroid belt. *Icarus, 153,* 338–347.

Poisson S.-D. (1803) Sur les substances minérales que l'on suppose tombées du ciel sur la terre. *Bull. Soc. Philomatique, 3,* 180–182.

Prialnik D. (1992) Crystallization, sublimation, and gas release in the interior of a porous comet nucleus. *Astrophys. J., 388,* 196–202.

Prialnik D. (2002) Modeling the comet nucleus interior — Applications to Comet C/1995 Hale-Bopp. *Earth Moon Planets, 89,* 27–52.

Prialnik D. and Podolak M. (1999) Changes in the structure of comet nuclei due to radioactive heating. *Space Sci. Rev., 90,* 169–178.

Prialnik D., Benkhoff J., and Podolak M. (2004) Modeling the structure and activity of comet nuclei. In *Comets II* (M. Festou et al., eds.), pp. 359–387. Univ. of Arizona, Tucson.

Remusat L., Derenne S., and Robert F. (2005) New insights on aliphatic linkages in the macromolecular organic fraction of Orgueil and Murchison meteorites through ruthenium tetroxide oxidation. *Geochim. Cosmochim. Acta, 69,* 4377–4386.

Remusat L., Palhol F., Robert F., Derenne S., and France-Lanord C. (2006) Enrichment of deuterium in insoluble organic matter from primitive meteorites: A solar system origin? *Earth Planet. Sci. Lett., 243,* 15–25.

Rietmeijer F. J. M. (1998) Interplanetary dust particles. In *Planetary Materials* (J. J. Papike, ed.), pp. 2.1 to 2.94. Reviews in Mineralogy, Vol. 36, Mineralogical Society of America.

Rivkin A. S., Howell E. S., Vilas F., and Lebofsky L. A. (2002) Hydrated minerals on asteroids: The astronomical record. In *Asteroids III* (W. F. Bottke Jr. et al., eds.), pp. 235–253. Univ. of Arizona, Tucson.

Robert F. (2002) Water and organic matter D/H ratios in the solar system: A record of an early irradiation of the nebula? *Planet. Space Sci., 50,* 1227–1234.

Robert F. and Epstein S. (1982) The concentration and isotopic composition of hydrogen, carbon and nitrogen in carbonaceous chondrites. *Geochim. Cosmochim. Acta, 46,* 81–95.

Sandford S. A., Aléon J., Alexander C. M. O'D., Araki T., Bajt S., et al. (2006) Organics captured from Comet 81P/Wild 2 by the Stardust spacecraft. *Science, 314,* 1720–1724.

Sephton M. A., Pillinger C. T., and Gilmour I. (1998) $\delta^{13}C$ of free and macromolecular aromatic structures in the Murchison meteorite. *Geochim. Cosmochim. Acta, 62,* 1821–1828.

Sephton M. A., Pillinger C. T., and Gilmour I. (2000) Aromatic moieties in meteorite macromolecular materials: Analyses by hydrous pyrolysis and $\delta^{13}C$ of individual molecules. *Geochim. Cosmochim. Acta, 64,* 321–328.

Sephton M. A., Verchovsky A. B., Bland P. A., Gilmour I., Grady M. M., and Wright I. P. (2003) Investigating the variations in carbon and nitrogen isotopes in carbonaceous chondrites. *Geochim. Cosmochim. Acta, 67,* 2093–2108.

Sephton M. A., Verchovsky A. B., and Wright I. P. (2004) Carbon and nitrogen isotope ratios in meteoritic organic matter: Indicators of alteration processes on the parent asteroid. *Intl. J. Astrobiol., 3,* 221–227.

Soderblom L. A., Britt D. T., Brown R. H., Buratti B. J., Kirk R. L., Owen T. C., and Yelle R. V. (2004) Short-wavelength infrared (1.3–2.6 μm) observations of the nucleus of Comet 19/P Borrelly. *Icarus, 167,* 100–112.

Spurný P. (1997) Exceptional fireballs photographed in Central Europe during the period 1993–1996. *Planet. Space Sci., 45,* 541–555.

Spurný P. and Borovička J. (1999) Detection of a high density meteoroid on cometary orbit. In *Evolution and Source Regions of Asteroids and Comets* (J. SvoreH et al., eds.), pp. 163–168. Astronomical Institute of the Slovak Academy of Sciences, Vol. 173, Praha.

Swindle T. D. and Campins H. (2004) Do comets have chondrules and CAIs? Evidence from the Leonid meteors? *Meteoritics & Planet. Sci., 39,* 1733–1740.

Thomas P. C., Parker J. W., McFadden L. A., Russell C. T., Stern S. A., Sykes M. V., and Young E. F. (2005) Differentiation of the asteroid Ceres as revealed by its shape. *Nature, 437,* 224–226.

Vokrouhlický D. and Farinella P. (2000) Efficient delivery of meteorites to the Earth from a wide range of asteroid parent bodies. *Nature, 407,* 606–608.

Weidenschilling S. J. (2000) Formation of planetesimals and accretion of the terrestrial planets. *Space Sci. Rev., 92,* 295–310.

Wetherill G. W. (1971) Cometary versus asteroidal origin of chondritic meteorites. In *Physical Studies of Minor Planets* (T. Gehrels, ed.), pp. 447–460. NASA SP-267, Washington, DC.

Wetherill G. W. and Revelle D. O. (1982) Relationship between comets, large meteors and meteorites. In *Comets* (L. L. Wilkening, ed.), pp. 297–319. Univ. of Arizona, Tucson.

Whitman K., Morbidelli A., and Jedicke R. (2006) The size frequency distribution of dormant Jupiter family comets. *Icarus, 183,* 101–114.

Wooden D. H., Harker D. E., Woodward C. E., Butner H. M., Koike C., Witteborn F. C., and McMurty C. W. (1999) Silicate mineralogy of the dust in the inner coma of Comet C/1995 O1 (Hale-Bopp) pre- and post-perihelion. *Astrophys. J., 517,* 1034–1058.

Wooden D. H., Harker D. E., and Brearley A. J. (2005) Thermal processing and radial mixing of dust: Evidence from comets and primitive chondrites. In *Chondrites and the Protoplanetary Disk* (A. N. Krot et al., eds.), pp. 774–810. ASP Conf. Series 341, San Francisco.

Young E. D. (2001) The hydrology of carbonaceous chondrite parent bodies and the evolution of planet progenitors. *Philos. Trans. R. Soc. Lond. A, 359,* 2095–2110.

Young E. D., Ash R. D., England P., and Rumble D. I. (1999) Fluid flow in chondritic parent bodies: Deciphering the compositions of planetesimals. *Science, 286,* 1331–1335.

Zolensky M. E., Nakamura K., Gounelle M., Mikouchi T., Kasama T., Tachikawa O., and Tonui E. (2002) Mineralogy of Tagish Lake: An ungrouped type 2 carbonaceous chondrite. *Meteoritics & Planet. Sci., 37,* 737–762.

Zolensky M. E., Zega T. J., Yano H., Wirick S., Westphal A. J., et al. (2006) Mineralogy and petrology of Comet 81P/Wild 2 nucleus samples. *Science, 314,* 1735–1739.

Part X: Perspectives

The Kuiper Belt Explored by Serendipitous Stellar Occultations

F. Roques
Paris Observatory

G. Georgevits
University of New South Wales

A. Doressoundiram
Paris Observatory

The possibility of exploring the Kuiper belt by searching for fortuitous stellar occultations has been under development for several years. This technique has the potential to permit exploration of the Kuiper belt down to objects of subkilometer radius. High-speed photometric observations provide lightcurves in which the occultation signatures appear as very brief dips. Depending on the target stars, the occultation waveform may exhibit Fresnel diffraction effects. The star's size and the geometry of the occultation must be carefully taken into account to interpret the occultation data in terms of Kuiper belt object (KBO) size and distance. The observation programs dedicated to this type of research are described in this chapter. Three such programs have recently announced positive detections. These first results are described briefly, along with the information the data may provide. Potential results expected from these campaigns could help answer many questions concerning the Kuiper belt, its nature, and its relation with the Oort cloud: Is there an extended "cold" Kuiper belt? What is the size distribution for the Kuiper disk population? Is the maximum size of the KBO population decreasing with distance? What is the radial extent of the Kuiper belt? Is there a connection with the Oort cloud? What is the mass of the Kuiper belt? What fraction of KBOs are binaries? Are the smaller KBOs regular or elongated? These programs should also provide valuable data for theorists attempting to model solar system formation.

1. INTRODUCTION

For centuries, occultations of stars by the Moon have been used successfully for studying the binarity and angular size of stars. Nowadays, the stellar occultation method is commonly used to study the "dark matter" in the solar system, i.e., small objects or material invisible by direct imaging. The method consists of recording the flux of a star. Short-duration dips in the stellar flux indicate the passage of an object in front of the star.

A distinction must be made between occultations by known objects, for which we can predict the occultation conditions (timing, geometry, etc.), and occultations by unknown objects. The latter are random events. This chapter focuses on the serendipitous stellar occultation method, i.e., the search for random occultations by unknown/small objects belonging to the Kuiper belt.

The study of predictable occultation events has been applied to observe various solar system bodies. Such observations have been conducted using both large telescopes and smaller mobile instrumentation. The complementary nature of large and smaller instruments has proved crucial in producing the results obtained so far.

Stellar occultations have the potential to provide precise measurement of *asteroid* sizes and shapes. The applicability of this technique to the asteroids was understood as early as 1952 (*Taylor,* 1952). However, because of poor organization and equipment, as well as inaccurate predictions, the first successful occultation of a star by the asteroid Pallas was not observed until 1978 (*Wasserman et al.,* 1979). This observation permitted the determination of the dimensions of Pallas with an uncertainty of less than 2%, a precision never before achieved. The size determination, in turn, led us to derive the asteroid's density. To date, stellar occultation has proved to be the most accurate method (except for *in situ* exploration) for determining asteroid sizes and shapes, with the measurements of tens of asteroids as well as other bodies such as planetary satellites.

Stellar occultations can also be used to probe *planetary atmospheres*. This technique has proved to be a powerful tool for the detection of very tenuous atmospheres at microbar levels. Pluto's tenuous atmosphere was first detected by this means in 1985 (*Brosch,* 1995). The principle of the technique is to study the dimming of the starlight as it passes through the planet's atmosphere. The starlight interacts with the atmosphere through refraction and extinction.

Consequently, the stellar occultation method can provide the temperature, pressure, and density profiles of a planetary atmosphere with a typical vertical resolution of a few kilometers. Furthermore, depending on the quality and completeness of the dataset, this technique has sometimes been applied to determine local density variations, atmospheric composition, the presence of aerosol content, zonal wind speed, and temporal and spatial variations of the whole atmosphere. For example, the Pluto occultation of 2002 showed that Pluto's atmospheric pressure had, very surprisingly, increased by a factor of 2 since 1988, a phenomenon probably caused by frost-migration effects on the surface (see chapter by Stern and Trafton). Stellar-occultation studies of planetary atmospheres have included all solar system planets and many satellites like Triton, Titan, and Charon (see the review by *Elliot and Olkin,* 1996).

Finally, stellar occultations can also detect invisible material like *planetary rings and arcs* (see, e.g., *Sicardy et al.,* 1991). Thus, stellar occultations have led to the discovery of planetary rings and arcs around Uranus (1977) and Neptune (1984).

Stellar occultations by known members of the *Kuiper belt* are a new challenge, since they generally involve the use of fainter stars, and above all, bodies that subtend very small angles. Typically, a Kuiper belt object (KBO) subtends less than 30 milliarcsec on the sky. This makes occultations difficult to predict (*Denissenko,* 2004). The limitation on astrometric predictions of the shadow path on Earth severely complicates the logistics of the observations. Occultations by the largest KBOs have yet to be observed. The potential reward to be gained from detecting an occultation by a KBO is tremendous, however (*Elliot and Kern,* 2003).

First, this is a unique way to measure sizes at *kilometric* accuracy, using a number of observing stations on the ground and careful timing of the event. To measure the size, one does not need high photometric precision, hence small telescopes can be used. The size can be used, in turn, to derive the albedos of these bodies, and for those KBOs with satellites, their density.

Second, for the largest of these bodies, the sizes of which lie between the radii of Charon and Pluto, this is the only way to detect a possible tenuous atmosphere, down to a surface pressure of *30 nbar* for typical nitrogen, methane, or carbon monoxide atmospheres (see chapter by Stern and Trafton).

2. THE METHOD OF SERENDIPITOUS OCCULTATIONS

Stellar occultation is a powerful tool for detecting otherwise invisible objects in the outer solar system, using the method of serendipitous stellar occultation (*Bailey,* 1976). Using this method, one searches for random occultations of the background stars by passing objects. This permits us to study the invisible part of the KBO size distribution, for which the object's radius is less than a few kilometers, as well as the outer regions of the Kuiper belt. This method could also be used to obtain information about the larger members of the Oort cloud population.

2.1. Justification

The large distance from Earth to the Kuiper belt and the small size of KBOs limits the possibility of direct detection to the biggest objects. Current estimates suggest that there are a total of 10^5 KBOs larger than 100 km. The multikilometer radius objects have a differential size distribution $N(r) = N_0 r^{-q}$, r being the object radius, with q = 4, after *Luu and Jewitt* (2002; see chapter by Petit et al.). One expects a shallower size distribution for smaller objects (*Bernstein et al.,* 2004). The position of the expected turnover radius is a critical clue to understanding the first stages of planetary formation in the outer solar system. As a hectometer-sized KBO at 40 AU is expected to have a magnitude of ≈40, reliable statistics on such small objects are unlikely to be obtainable by direct observation. Thus, exploration of the Kuiper belt by stellar occultations is the only way to obtain information about this population. Moreover, stellar occultation is less limited by the distance of the occultor, hence this technique can be used to explore the outer frontier of the Kuiper belt at ~10^2–10^5 AU, inaccessible by direct detection (Fig. 1).

A simple estimate of the power of the method can be made using geometrical optics: An object of radius r, passing in front of a star of angular radius α, creates a signal

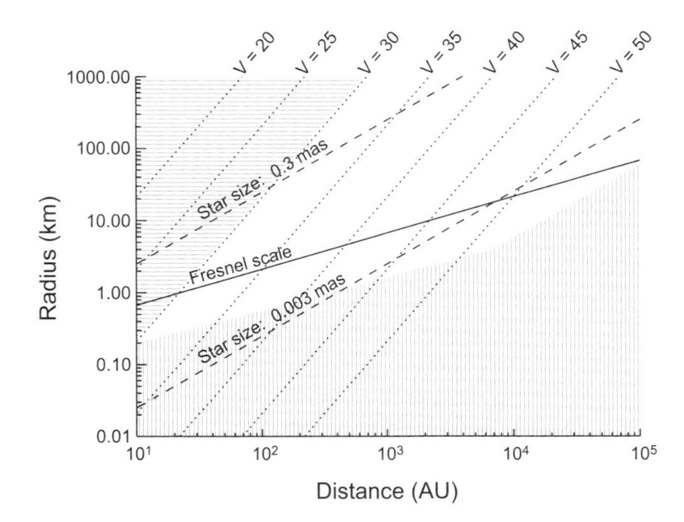

Fig. 1. Comparison of the direct detection method with the occultation method: The dotted lines show the visual magnitude. The dashed lines correspond to the projected stellar radius; the upper one corresponds to an M5 star with V = 12 and the lower one to an O5 star with V = 12. Occultations of stars larger than the Fresnel scale are geometric and the limit of detection is a fraction of the stellar radius. When the star is smaller than the Fresnel scale, the occultation is a diffracting phenomenon and the limit of detection is a fraction of the Fresnel scale. Objects in the white zone are invisible (V > 30) and can only be detected by occultation.

decay $\delta F = [r/(\alpha D)]^2$, where D is the object-to-Earth distance (for an impact parameter smaller than $\alpha D - r$). $\alpha D = R_*$ is the radius of the star projected at the distance of the occulting object. Note that if the star's apparent size is known, the occultation profile provides some information on the object's size without any hypothesis about its albedo. The angular sizes of the stars depend on their spectral and luminosity class and on their distance (see section 2.5). Figure 1 shows that, for a large proportion of stars, there would be full extinction for KBOs under 1 km in radius if the occultation were purely geometrical in nature. In reality, diffraction plays an important role for such occultations. And as a result, the occultations do not achieve full extinction. The diffraction shadow has specific properties that allow us to validate the detection of such events and also provide information about the size and distance of the occulting objects (see below).

The limitation of the serendipitous occultations method is that the objects detected are otherwise unknown, insofar as it is not possible to see them due to their very faint magnitude. Also, the detection of an object by occultation is not a reproducible observation. It is not a "discovery" because it is not possible to deduce a precise trajectory from a single observation, and hence observe it again later. Information about the physical properties of the Kuiper belt must be derived from statistical analysis of many occultation profiles or by the analysis of the very specific signature of single occultation profiles containing diffraction fringes. In addition, it is possible under some observing conditions to misinterpret occultation events caused by asteroids at much closer distances. Estimating the distance to the occultor by analysis of the diffraction fringes then becomes an important issue.

2.2. A Diffracting Phenomenon

Figure 1 shows that occultation of most stars is a simple geometric phenomenon. However, for some well-chosen stars (small angular size), occultations are a diffracting phenomenon up to a distance of 10^4 AU, i.e., for the outer Kuiper belt and the inner part of the Oort cloud.

Let us now focus on the properties of occultations involving diffraction. The light emitted by a point source (assumed to be at infinity so as to yield planar waves), incident on a sharp-edged obstacle (such as a KBO), is diffracted. Because of the Huygens-Fresnel principle of wave propagation, each point on a wave front may be considered as the center of a secondary disturbance giving rise to spherical wavelets, which mutually interfere. If part of the original wave front is blocked by an obstacle, the system of secondary waves is incomplete, so that diffraction fringes are generated. When observed at a finite distance D from the obstacle, this effect is known as "Fresnel diffraction," and falls within the scope of the Kirchhoff diffraction theory, which remains valid as long as the dimensions of the diffracting obstacles are large compared to the observed wave length λ and small compared to D (cf. *Born and Wolf,* 1980). The characteristic scale of the Fresnel diffraction effect (i.e., roughly speaking, the broadening of the object shadow) is the so-called Fresnel scale $F_s = \sqrt{(\lambda D/2)}$ (*Warner,* 1988). [Note that some authors give different definitions for the Fresnel scale: $\sqrt{(\lambda D/2\pi)}$, $\sqrt{(\lambda D)}$, or $\sqrt{(\lambda D)/2}$.] The Fresnel scale at 40 AU, for a wavelength of 0.4 μm (blue light), is 1.1 km and therefore diffraction must be taken into account when analyzing occultations by (under) kilometer-sized KBOs. Similarly, the Fresnel scale at 40 AU is 42 m for X-rays (0.6 nm). In fact, the Fresnel scale is a scaling factor of the profile. The shadows of occulting objects on Earth, *expressed in Fresnel scale,* are the same if the objects have the same size, *expressed in Fresnel scale.* This degeneracy can be broken if we have a measure of the relative KBO velocity in the sky plane and have a good knowledge of the star's apparent size. And finally, for non-monochromatic observations, the Fresnel scale must be averaged over the detector bandwidth and weighted by the detector response as a function of wavelength.

2.3. The Occultation Signature

The computation of the profile of a KBO occultation requires several steps, described below.

2.3.1. Computation of the diffraction pattern with a monochromatic point source. The modeling of the occultation of a point source by a disk is relatively easy and makes for a good approximation, even though most of the small KBOs are probably not spherical. Let us now consider the case of a monochromatic point source occulted by an opaque spherical object of radius ρ. If ρ denotes the distance between the line of sight (the star's direction) and the center of the object, and if the lengths r and ρ are expressed in Fresnel scale units, the normalized light intensity $I_r(\rho)$ is given by the following (see appendix B of *Roques et al.,* 1987):

Outside the geometric shadow ($\rho > r$)

$$I_r(\rho) = 1 + U_1^2(r,\rho) + U_2^2(r,\rho) - 2U_1(r,\rho)\sin\frac{\pi}{2}(r^2 + \rho^2) + 2U_2(r,\rho)\cos\frac{\pi}{2}(r^2 + \rho^2) \quad (1)$$

Inside the geometric shadow ($\rho < r$)

$$I_r(\rho) = U_0^2(\rho,r) + U_1^2(\rho,r) \quad (2)$$

where U_0, U_1, and U_2 are the Lommel functions defined by (for $x < y$)

$$U_n(x,y) = \sum_{k=0}^{\infty} - 1^k (x/y)^{n+2k} J_{n+2k}(\pi xy) \quad (3)$$

where J_n is the Bessel function of order n.

Figure 2 shows the diffraction pattern of a circular object occulting a point star. It shows that objects a fraction

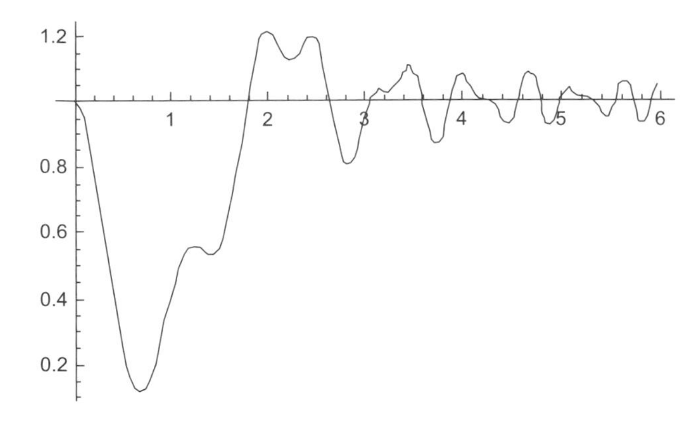

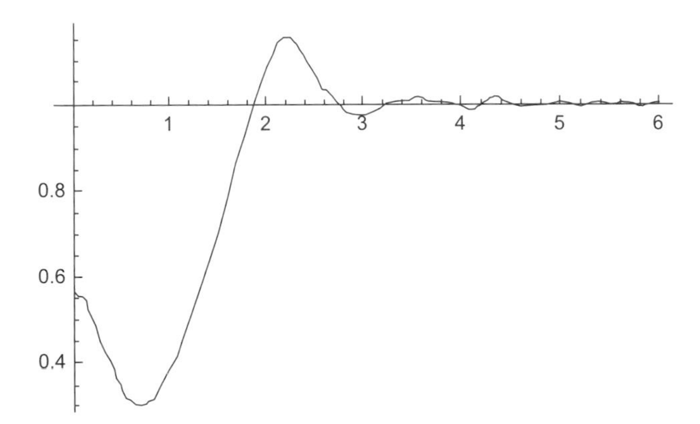

Fig. 2. Synthetic profiles for a 1-F_s radius occultor. The x axis unit is in Fresnel scale units. The upper diagram is for a projected stellar radius of 0.01 F_s, and the lower one for 0.5 F_s.

of a Fresnel scale in size produce a significant decrease in starlight. Furthermore, the size of the shadow is larger than the geometric shadow and overall, the diffraction fringes are visible at a large distance from the object. The computation of the diffraction pattern of an elliptical, or an irregular object is more complex (*Roques et al.,* 1987; *Roques and Moncuquet,* 2000). However, the diffraction fringes are significant for objects the size of which is on the order of, or smaller than, the Fresnel scale.

2.3.2. Wavelength dependency of the profile. As the profile depends on wavelength, averaging over the bandwidth of the detector smoothes the diffraction fringes. If the observation is done simultaneously at more than one wavelength (say, red and blue), the profiles of an occultation event are different in the different channels. The profile is larger, but shallower for red light. In fact, if expressed in Fresnel scales, the same object is smaller in the red than in the blue. This property can be used to verify the authenticity of the event.

2.3.3. Convolution of the profile with the projected disk of the target star. If we consider the light source to be a distributed rather than a point source, we must convolve the occultation profile with the source shape. In the case of a star, the source shape is really a disk. To perform the convolution, one needs to know the star's angular size with good precision. The normalized light intensity measured during the occultation of a stellar disk of apparent radius R_* (expressed in Fresnel scale) is

$$I_r^*(\rho) = \frac{2}{\pi R_*^2} \int_0^{R_*} s\,ds \int_0^{\pi} I_r\left(\sqrt{\rho^2 + s^2 + 2\rho s \cos\theta}\right) d\theta \qquad (4)$$

The diffraction fringes are strongly smoothed when the star's apparent size is larger than roughly half the Fresnel scale (Fig. 2). When the star's apparent size is larger than the Fresnel scale, the occultation shadow is closely approximated by the geometric occultation case, and the decrease of light intensity is simply the ratio of the surface area of the occultor to the surface area of the star disk as projected at the occultor's distance.

2.3.4. Dependency on integration time. The profile must also be averaged over the integration time (i.e., the time taken for one data sample). The occultation dips are very brief events, so the sampling rate must be fast enough to provide good resolution of the occultation profile (see below).

2.4. The Geometry of the Observation

If the KBO has a circular orbit in the ecliptic plane around the Sun, its velocity v in the sky plane is given by

$$v = v_E\left(\cos\omega - \sqrt{\frac{1}{D_{AU}}}\right) \qquad (5)$$

where D_{AU} is the heliocentric distance of the KBOs in AU, v_E is the Earth's orbital speed ≈30 km/s, and ω is the angle between the KBO and the antisolar direction (called "opposition" hereinafter). ω is called the observation angle (Fig. 3).

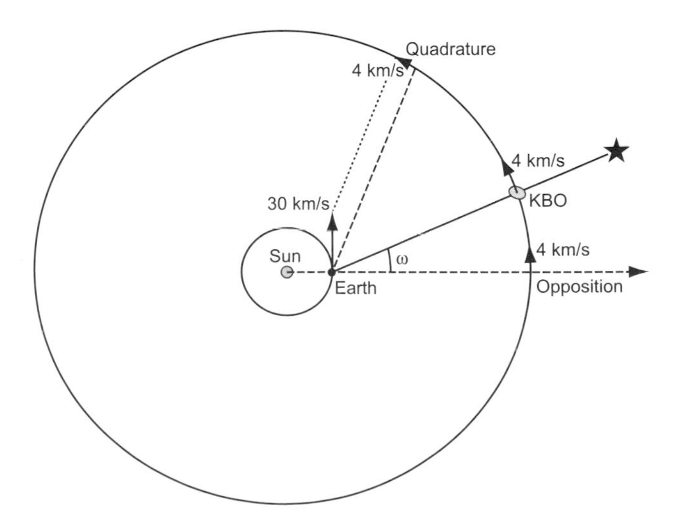

Fig. 3. Geometry of an occultation by a KBO, showing the observation angle ω.

Because v depends on the distance of the occulting objects, so do both the duration of the occultation ($dt \alpha 1/v$) and the number of occultations ($N_{occ} \alpha v$). The velocity is maximum toward the opposition and minimum toward a direction, called the "quadrature" hereinafter, for which cos $\omega = 1\sqrt{D_{AU}}$. This property allows us to discriminate between occultations by a KBO population and an asteroid population, because quadrature for KBOs and asteroids are at different positions (see Table 4 in *Roques and Moncuquet,* 2000).

At 40 AU, we get v ≈ 30(cosw–0.16) km/s ≈(cosw–0.16) mas/s, the velocity is ≈25 km/s toward the opposition and the quadrature at ≈81°. An occultation by a small KBO could last from about 0.1 s to a few seconds, as ω goes from the opposition to the quadrature. If the data acquisition rate is not fast enough, the occultation profile is degraded. For good resolution of the events, the acquisition rate must be larger than v/F_s. To satisfactorily resolve events toward opposition, this rate must be greater than 600 Hz at 0.6 nm and 20 Hz at 0.4 μm.

When the acquisition rate is not fast enough, observations toward quadrature may still detect bodies that are invisible toward opposition (because their occultation dips are too brief to be detected). Observing toward quadrature makes this technique less restrictive in terms of the acquisition rate. Thus, *Brown and Webster* (1997) proposed to search for KBO occultations toward quadrature with the MACHO experiment.

2.5. The Target Stars

Selecting appropriate target stars is critical to the success of any occultation observation program. The smaller the occulted star, the smaller the detectable KBO, and because small KBOs are more numerous, the higher the occultation rate. It is essential that the selected target stars project a disk onto the Kuiper belt plane that is of the same size or preferably smaller than the Fresnel scale.

The radius of the star, projected at the distance D, is $R_* = \alpha D$, so that

$$\log R_* = -\frac{V}{5} = \frac{M}{5} + \log\left(\frac{R}{R_S}\right) + \log\left(\frac{DR_S}{d_{pc}}\right) \quad (6)$$

where V is the apparent magnitude of the star, M its absolute magnitude, and R its radius; R_S is the radius of the Sun, and d_{pc} = 10 pc. One way to favor a small projected stellar radius is to reduce the apparent magnitude. Unfortunately, this degrades the signal-to-noise ratio of the lightcurve. Another method is to consider hot stars, which, for a given apparent magnitude, project a smaller angular size. An O5 star of V = 12 has a projected radius of 130 m at 40 AU. With the same magnitude, an M5 star has a projected radius of 10 km (Fig. 1). The above equation does not take into account interstellar extinction or reddening. If extinction is significant, the apparent magnitude will be underestimated and so will the star's apparent size. However, if the reddening is not taken into account, the star is hotter than estimated, and the apparent size is smaller than estimated. An estimate for interstellar extinction and reddening can be derived from studies such as that done by *Kharchenko et al.* (1996) toward the center of the galaxy. However, the errors associated with any such estimates are likely to be large. The stellar size distribution is considered by *Cooray* (2003) in a geometric framework.

The magnitude and spectral class of the target stars need to be carefully chosen. The effective telescope aperture sets the lower limit criterion for a target star's observed magnitude. Below this limit, scintillation noise and photon noise degrade the photometry to the point where it is no longer usable for occultation work. Unfortunately, the above selection criteria dictate that for any given range of magnitudes, the vast majority of stars are unsuitable as occultation target candidates. The hottest stars have the highest light output per unit of surface area. It so happens that the very hottest stars (types O, B, and A) provide the best candidates for occultation work. As an example, for a 1.2-m telescope fitted with a high-efficiency detector and no filter, the limiting magnitude for KBO detection work is V ~ 12. Stars of type O, B, and A with 10 < V < 11.5 are suitable for detecting KBOs down to a radius of ~250 m when observing at opposition. Field crowding can be a problem, particularly toward the center of the galaxy. Long exposure plates in the vicinity of each star need to be checked to make certain that confusion is not a problem.

Finally, the distribution of KBOs dictates that all observations be carried out on target stars located close to the ecliptic. The galactic plane crosses the ecliptic at two locations on the sky. The large number of stars at these two locations makes finding suitable target star candidates easier for any observation program. Furthermore, one crossing occurs very close to the direction of the galactic center. This particular location provides the densest possible star fields anywhere on the sky. However, in the long term, exploring the Kuiper belt by the occultation method will make it necessary to find appropriate fields all along the ecliptic.

3. TELESCOPES AND INSTRUMENTATION

Choosing a telescope and high-speed instrumentation for KBO occultation work involves a judicious selection of various conflicting parameters, all of which affect the probability of detecting occultation events. This section describes instruments used in the various programs dedicated to the detection of KBOs by occultation. It does not cover the X-ray satellite used for this subject (see below).

3.1. Telescope

Using a wide-field telescope and appropriate detection instrumentation, it is possible to monitor many suitable target stars in the same field of view. This greatly increases the probability of detecting occultation events. However, the signal-to-noise ratio will be relatively poor due to the effects of scintillation — typically a 5% noise level for a 1-m-aperture telescope (see *Young,* 1967; *Warner,* 1988). If one employs a larger telescope (e.g., 4-m-aperture or larger), the field of view will be much smaller. However, due to the larger telescope aperture, it will be possible to monitor fainter stars. With a suitable detector, a 1% noise level can be achieved, and this should be sufficient to permit diffraction fringes to be observed.

The issue of whether to attempt observing diffraction events needs to be assessed in terms of system requirements. Any system targeting diffraction-based occultation events must be capable of high photometric precision (ideally ~1% or better), and must use a narrowband blue filter and target stars whose projected disk size is preferably less than one-tenth that of the Fresnel scale. These criteria imply that one must use the largest possible telescope, along with a detector capable of fast, high-precision photometry. This latter constraint usually implies the use of ultra-rapid camera, which immediately limits the number of target stars that can be observed simultaneously to a few at most. Thus the probability of detecting any event will be low, thereby necessitating many hours of monitoring.

If one uses a wide-field telescope to monitor 100 or more suitable target stars simultaneously, the chances of detection are greatly improved, albeit under somewhat noisy photometric conditions. To obtain the best possible photometry, such observations should be carried out using a high-efficiency detector without any intervening filter. The demands on medium-sized telescopes are not as great, so longer observing runs are feasible.

3.2. Detectors Suitable for Multi-Object Occultation Work

Two types of detector suitable for multi-object occultation work are available. A multi-photometer instrument can simultaneously record the flux from several stars with a high sampling rate and good signal-to-noise ratio (*Roques et al.,* 2003). Another approach consists of using a CCD camera to image the target star field directly. Unfortunately, the readout time for large CCD images is too slow for satisfactory KBO occultation work. To compound the problem, CCDs tend to be noisy when working in fast readout mode. By judicious selection of target field, CCD camera type and special programming techniques — such as pixel binning and window readout — it is possible to acquire data at a sufficiently high rate (e.g., 50 ms integration time) and with a satisfactory signal-to-noise ratio to have a chance of detecting KBO occultations. An observation of a typical KBO occultation using the above-mentioned approach is likely to yield two to three data point detections. A fiber-based multi-object spectrograph/CCD camera, combined with a suitable setup and CCD control software, allows acquisition rates as fast as 10 ms. Using such a scheme, a typical occultation event consists of 10 to 15 data points. Thus, each event can be nicely resolved. As an additional bonus, some information about the shape of the occulting object can be gleaned from the occultation waveform. Finally, new CCD technology now permits the use of very fast response cameras with a 20-ms integration time (*Dhillon and Marsh,* 2001) (see also *www.shef.ac.uk/physics/people/vdhillon/ultracam/*).

4. DATA REDUCTION

Detection of occultations relies on monitoring the light received from one or more stars. The acquired data will vary in format, noise content, and event characteristics, depending on the nature of the instruments used for the acquisition. For CCD camera images, the target stars represent signal peaks in two-dimensional signal space. To compute the lightcurves, one needs to choose an aperture size. The choice is done in a manner that minimizes the r.m.s. fluctuations of the lightcurve. For CCD images derived from a multi-object spectrograph, the light is collected from each target star in a separate fiber. The fibers are then physically aligned along a row in front of the CCD camera. This permits fast acquisition, as only the first few rows of the CCD image need to be read.

The aim of data reduction is to find signal changes similar to those expected for a distant KBO passing in front of a target star. Event-detection algorithms may be based on simple level change detection, σ threshold detection, or cross-correlation with one or more suitable event mask(s). This latter technique is a powerful means of detecting all types of events, including Fresnel diffraction waveforms, in the presence of noise. Such waveforms may be mathematically synthesized, and the parameters tuned for the relevant detection conditions.

The small KBOs we are seeking to detect by occultation are effectively invisible from Earth with all other present-day observing techniques. We can never be 100% certain that a dip in the lightcurve of a star is due to an occulting KBO. Observational artifacts can be due to instrumental effects, near-Earth phenomena and unknown effects. Simple tests can be performed to eliminate most of the following artifacts:

1. *Correlated events.* Systems that simultaneously monitor several nearby stars possess a powerful means of discriminating against false detections. An event that occurs simultaneously on more than one star (hereafter called a correlated event) *must* be an observational artifact! Kuiper belt objects are far too rare, small, and distant to occult more than one star at a time. As an example, a bird or a plane flying through the telescope field of view will cause a brief reduction in light intensity. This reduction is proportional to the percentage of the aperture that an object obstructs as it flies past. This reduction in intensity occurs for all objects in the field of view, not just for a single star. Thus, a correlated event is generated. Other examples of artifacts that can cause correlated events include clouds passing through the field of view, electronic glitches, and telescope tracking errors.

2. *Nearby telescopes.* Simultaneous observation using two nearby telescopes permits us to confirm the reality of an occultation by an interplanetary object. The distance between the telescopes must be smaller than one Fresnel scale unit. The simultaneity of an event serves to confirm the reality of the occultation. The observation with several telescopes on an east-west line, could, in principle, permit us to retrieve the event with a time shift, which, in turn, gives us information about the KBO shadow velocity.

3. *Multi-wavelength observations.* The profiles of an occultation event observed at different wavelengths are different: The dips are larger and shallower in red than in blue. This difference between the profiles will confirm that it is a diffracting event. The comparison between the two profiles gives information about the size and the distance of the occulting KBO.

4. *Single data point events.* In order to assist with differentiating against artifacts, an observation scheme must be capable of resolving KBO-related occultation events into more than just one data point. For example, single-data-point events may be caused by Earth-orbiting objects or cosmic rays. An artificial satellite will completely extinguish a star during a millisecond.

To check for instrument-related artifacts linked to detection electronics, it should be possible to run a check by observing a suitably illuminated white screen for an extended period of time during the day. Any artifacts generated by the instrumentation will be immediately evident.

5. *Profile fitting.* Occulation profiles have special signatures that depend on the observing conditions, the size of the target star's projected disk and the shape, and size of the occultor(s). For example, Fresnel diffraction simulations show that occultations by small KBOs have a characteristic shape and should never reach 100% depth. This is a powerful sanity check. Also, correlating observed events with synthetic profiles permits us to further probe the reality of the occultation.

6. *Event statistics.* To further differentiate between real events and artifacts, one can make use of event statistics. The event statistics should change with the angle of observation ω (see section 2.4). On the one hand, the event rate should drop as ω decreases. For example, toward $\omega = 60°$, the event rate for any KBO size should drop by a little over one-half with respect to the events rate at $\omega = 30°$ (equation (5)). However, as we observe closer toward quadrature, the observing geometry permits us to detect progressively smaller objects due to the longer event duration. As smaller objects are thought to be more numerous, this effect may partially counteract the reduction due to the decreasing probability of detection.

Similarly, there are other statistics-based checks that can be performed for the purpose of event verification, e.g., the comparison of dips with stars of different sizes, such as O/A stars and K/M stars with the same visual magnitude: In the latter, the stellar disks projected onto the Kuiper belt plane will be too large to produce any deep occultation events.

Another very robust test is the monitoring of stars well away from the ecliptic. Here, we should detect very few events, because we are no longer looking through the main part of the Kuiper belt.

5. ANALYSIS OF THE OCCULTATION EVENTS

5.1. Population Parameters

Occultation data provides an instantaneous view of the position of the KBOs in the sky plane, but does not give access to the orbit parameters. This is because it is not possible to derive the orbit parameters of any object from a single observation. Note also that the density in the sky plane of KBOs detected by occultation cannot be compared to that of objects observed directly because the sensitivity of the two methods to the distance parameter is different. If the observations provide enough events in a given configuration (star size, position in the sky, value of the angle of observation ω), and taking into account the geometry of the observation, it may be possible to use the event rate to estimate the density of KBOs in the sky plane. Unfortunately, in occultation work, there exists a degeneracy between the size of a KBO, its distance, and the impact parameter of the occultation. This is because there are three unknowns and only two measured variables (the event duration and its depth). Under certain conditions, this degeneracy can be lifted. For example, a mean value for the distance to the Kuiper belt can be derived from observations done at two (or more) widely separated observing angles, ω.

If a fixed distance is assumed, the density in the sky plane gives an indication of the density of objects. This, in turn, can be compared with the large KBO size distribution. Fitting diffraction profiles with synthetic profiles gives access to the velocity of the occultor in the sky plane. This permits us to estimate the distance of the object, if we make some assumptions about the orbit (e.g., circular orbit). On the other hand, occultations will give a direct measure of the positions of the occultors with respect to the ecliptic plane, and this provides information about the thickness and the potential azimuthal variations of the density of the Kuiper belt. If the occultation profiles are well defined, comparison with synthetic profiles could give some information about the shape of the objects, and on their roundness. The proportion of binaries can also be constrained by occultation data.

5.2. Lower Limit of Detection

Noise in the measured lightcurve imposes a lower limit on the detectability of dips due to occultation events. The level of scintillation noise, in turn, is primarily governed by the telescope aperture (see section 3.1). The level of photon noise is governed by the star magnitude. On the other hand, the smallest detectable size also depends on whether diffraction fringes are present. If d_{min} is the limit in the detectable event depth, the radius of the detectable object is roughly $\sqrt{d_{min}/3} \cdot F_s$ for diffracting occultations and $R_* \sqrt{d_{min}}$ for geometrical occultations. F_s is the Fresnel scale (section 2.2) and R_* is the stellar radius projected at the KBO distance (section 2.5). For a 1-m (4-m) telescope, these limits are ≈0.3 F_s (0.15 F_s) and ≈0.4 R_* (0.2 R_*). Figure 1 shows that if the projected stellar disks are smaller than the Fresnel scale, objects of few hundred meters in radius are detectable.

The integration time can also limit the shortest detectable events. If the acquisition frequency is smaller than v/F_s, the smaller objects may not be detected because the occultation duration is shorter than the integration time. Information about smaller objects can best be obtained at larger observation angles ω, since v decreases with increasing observing angle (see section 2.4).

6. THE RESEARCH PROGRAMS

Several research programs are in progress using different approaches. The expected number of detections with each program is very difficult to estimate because it varies with hypotheses on the size and spatial distribution of KBOs and on the efficiency of the method. Two programs have very recently announced positive detections: three detections for the Paris program (*Roques et al.,* 2006), and hundreds of events from the University of New South Wales (UNSW) program (*Georgevits,* 2006).

6.1. The Taiwan-America Occultation Survey Program

The Taiwan-America Occultation Survey (TAOS) project is the first project to conduct a survey of the Kuiper belt using stellar occultations (*Alcock et al.,* 2003). This collaboration, involving the Lawrence Livermore National Laboratory (USA), Academia Sinica and National Central University (both of Taiwan), and Yonsei University (South Korea), uses the occultation technique in conjunction with an array of four wide-field robotic telescopes. Their aim is to estimate the number of KBOs of size greater than a few kilometers radius. The 50-cm f/1.9-aperture telescopes simultaneously point to a 3 deg^2 area of the sky and record light from the same ~1000 stars (V < 14). The array is located in the Yu Shan (Jade Mountain) area of central Taiwan (longitude 120° 50' 28"; latitude 23° 30' N). The detection scheme is operating in real time, to be able to alert more powerful telescopes that could then follow the motion of a potential occultor. A large amount of data is generated on a nightly basis, yielding about 10,000 GB of data and 10^{10}–10^{12} occultation tests per year. The expected number of occultations ranges from a few to a few hundreds.

Each telescope is equipped with an SI800 2 K camera, and the data acquisition system runs on a shutterless mode. The chip integrates ("pause") and then reads out a block of pixels ("shift") once at a time, instead of the whole frame. This "pause-and- shift" operating mode permits it to sample stellar lightcurves at up to ~5 Hz.

Three telescopes have been operational since early 2005. Some 10^9 stellar photometric measurements have been collected. So far, no events have been detected. A few predicted asteroid occultations (e.g., 51 Nemausa and 1723 Klemola) were recorded to demonstrate the capability of the system.

6.2. The Paris Program

After a theoretical analysis of the serendipitous occultation method (*Roques and Moncuquet,* 2000), the team from Observatoire de Paris lead by F. Roques conducted several observation campaigns. Some of the results obtained so far are:

1. The first observations were carried out at Pic du Midi with the 2-m Bernard-Lyot Telescope with a multi-object photometer. These observations showed the relevance of this approach, brought a first constraint on the size distribution of KBOs, and delivered a possible profile of detection of KBOs (*Roques et al.,* 2003).

2. Observations have been conducted on larger telescopes: the La Palma 4.2-m William Hershel Telescope and the European Southern Observatory, Paranal, 8.2-m, Very Large Telescope with the ultrarapid CCD Ultracam camera. *Roques et al.* (2006) reported the first and unambiguous detection of hectometer-sized KBOs at a few hundred AU.

The approach of the Paris team is to monitor few well-chosen single stars toward opposition with the ultrarapid triple-beam CCD Ultracam (*Dhillon and Marsh,* 2001) (see also *www.shef.ac.uk/physics/people/vdhillon/ultracam/*). This instrument allows a very fast acquisition rate and observations at three different wavelengths. Observations are made in two windows, each 34 × 34 $arcsec^2$, with an acquisition frequency of 42 Hz. Two (or more) stars are observed simultaneously. Comparison of the lightcurves from the different stars eliminates "false events" due to observing mishaps (of technical or human origin) and arising from near-Earth artifacts (clouds, birds, satellite, etc). If an event is observed simultaneously or quasi-simultaneously on more than one star, it can be assumed to be an Earth-connected artifact.

The detection algorithm used by the Paris team computes the standard deviation of the stellar flux and filters out all fluctuations below a 5σ standard deviation level. High-frequency fluctuations, either due to cosmic rays, clouds, or electronic glitches, have a clear signature and are eliminated by visual examination of the time series.

A further, more robust statistical technique is used to isolate sources of noise such as scintillation. This technique is known as the vectorial "Variability Index" (VI). For a given interval, one can define VI(int) with coordinates equal to the standard deviations of the normalized stellar flux in two different wavelength bands. Each standard deviation is computed relatively to the mean standard deviation of the dataset and expressed in units of standard deviation: sig = stddev(int) is the standard deviation for the interval int; meansig is the mean standard deviation of the data (for the whole night), then VI(int) = (sig(int)–meansig)/stddev(sig).

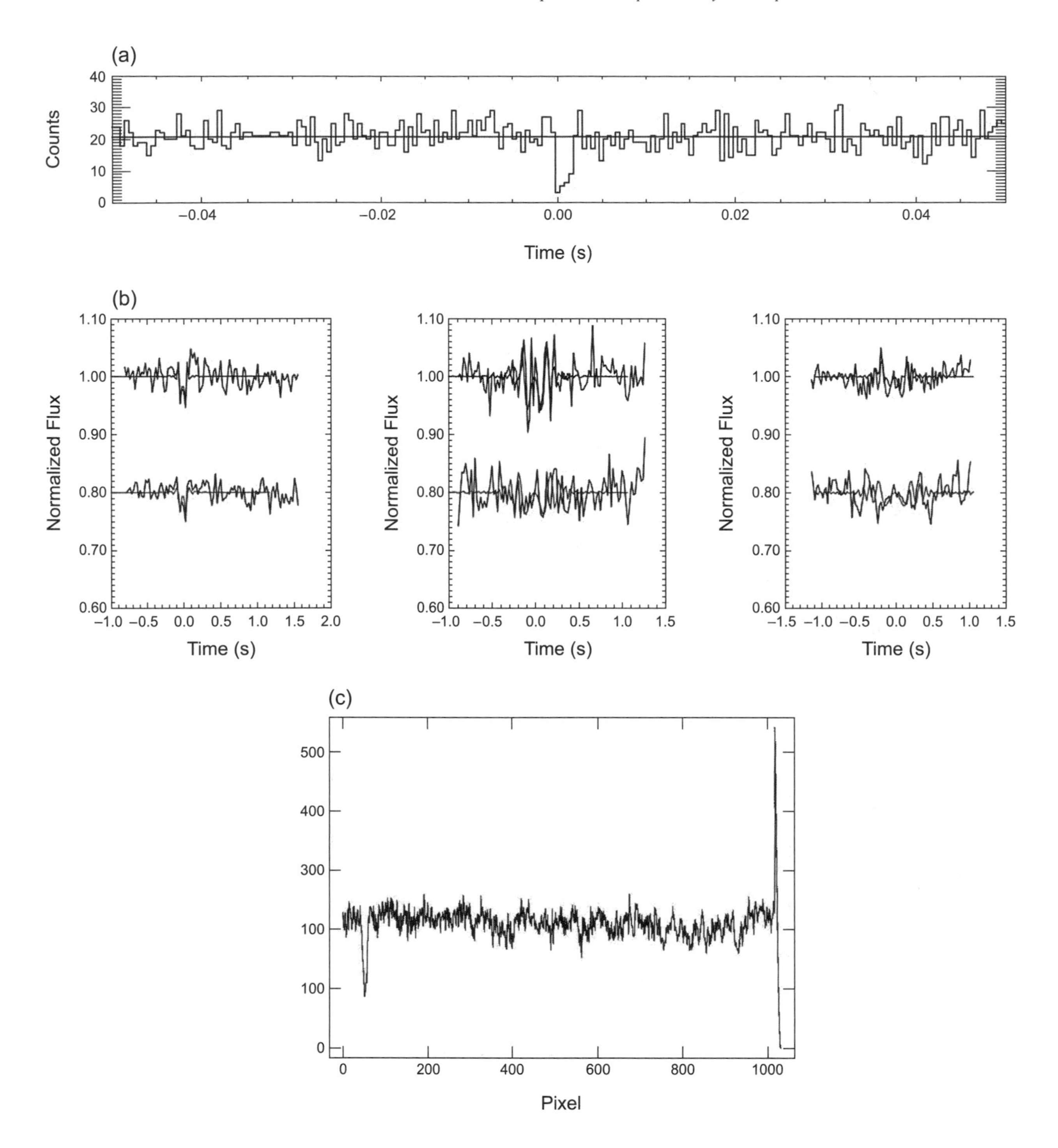

Fig. 4. Examples of events: **(a)** X-ray event (*Chang et al.,* 2006); **(b)** Paris team events (*Roques et al.,* 2006); **(c)** UNSW event. The 1000 pixels in the x-axis correspond to 10 seconds of acquisition.

For a purely random dataset, a two-dimensional plot of VI appears as an isotropic cloud of vectors centered on the origin. The distribution is Gaussian, therefore concentrated around the mean. Consequently, any deviant point should represent the detection of a nonrandom event.

The Paris team observations reported the detection of three objects (*Roques et al.,* 2006) using this technique. Cumulative observations from the William Hershel Telescope and from the Very Large Telescope yielded a first result on the density of objects on the sky plane. The number of occultors with a profile depth of ~5% (a 0.1-F_s object) is $2 \pm 1 \times 10^{10}$ deg^{-2}. The comparison of the dips with synthetic profiles suggests that one object is 100 m in radius and located at 10–20 AU and the two others are 300-m objects at 140 and 210 AU (Fig. 4). Other observations will be conducted to add to these results.

An additional program with the satellite Corot, a high-precision photometry mission, will be dedicated to the search of occultation by KBOs. The main objectives of this satellite are stellar seismology and the search of exoplanetary transits. Such a program will provide information on the population of decameter-sized objects.

6.3. The University of New South Wales (UNSW)/ Anglo Australian Observatory (AA) Program

A pilot set of occultation observation runs were conducted on the UNSW 0.5-m Automated Patrol Telescope in 2004 by G. Georgevits' team. The target field was the O–B association cluster NGC6611, located at a distance of 2.2 Kpc. Using the CCD techniques outlined above, a 50-ms integration time was achieved. Some two-pixel events were detected. The results appeared sufficiently favorable to warrant further observational work using a larger telescope. To follow up on this work, the 1.2-m UK Schmidt Telescope (UKST)/6df multi-object spectrograph (6° field) was used to undertake a more ambitious occultation program. The UKST is a survey telescope with a very wide field of view. It was originally designed to photograph 6.6° × 6.6° areas of the sky. It is now been fitted with a fiber-based multi-object spectrograph.

A new "through" mode of operation was developed for the multi-object spectrograph to support the UNSW/AAO occultation program. This involved using the spectrograph without an intervening diffraction grating. In addition, the acquisition software was modified to permit a very fast 10-ms integration time. The CCD camera has peak efficiency of around 65%, and better than 40% efficiency over the range 400–800 nm. The UNSW/AAO Occultation Observation Program was conducted over a period of four months in 2005. Data was acquired at observation angles of 30° before opposition, and 30° and 60° past opposition. The spectrograph equipment configuration permitted simultaneous monitoring of ~90 specially selected target stars. A total of nearly 7000 h of stellar lightcurve data was acquired.

The target stars consisted mostly of O-, B-, and A-class stars. They were selected from 4° × 4° fields located close to the center of the galaxy, centered on the point where the ecliptic crosses the galactic equator. Target star magnitudes ranged from V = 10.5 to 11.5, with most stars having B–V < 0. It is difficult to get an accurate value for reddening or extinction when observing toward the center of the galaxy, due to the copious amounts of gas and dust in this region. "Best estimate" based on *Kharchenko et al.* (1996) allows us to say that most target stars used for this work have a disk radius in the range 300–600 m when projected at 40 AU.

Most of the artifact identification/elimination techniques and statistical checks described above have been employed as part of the data verification process. At this stage, it appears that a large number of occultation events have been detected (*Georgevits,* 2006). An example is given Fig. 4. The analysis of the data is still in progress and the implications relating to the KBO population have yet to be derived.

6.4. The Canadian Project

Another project is led by JJ Kavelaars of the Queen University of Kingston, Canada. Their approach consists of correlating the observed photometric time series with synthetic diffraction patterns. They have analyzed 6.5 star-hours of 40-Hz photometry to search for occultations but found no viable candidate event. The search algorithm uses the cross-correlation of template occultation events with a time series. Several different templates are used to cover the parameter space of KBO size and distance, and each one is also run at 20 different impact-parameter distances. The detection algorithm flags any point in the time series that produces a peak in the cross-correlation greater than six times the standard deviation of the cross-correlation series. Then the candidate events are subsequently evaluated by eye.

Artificial occultation events were added to the data before the detection process was run to determine the efficiency of the algorithm. With their data, they were able to recover the artificial events corresponding to objects as small as 250 m in radius, and having impact-parameters up to 1 km. This implies a cumulative surface density less than 2.7×10^{10} deg^{-2}.

They also developed a way to simulate scintillation noise and produce artificial time series. By running their detection algorithm on these artificial data, they were able to find candidate events with fairly high significance (6.4 times the standard deviation of the cross-correlation series was the highest). Unfortunately, this means that the noise can mimic occultation events quite convincingly. What remains to be done is to determine the rates at which false-positive events will occur due to scintillation noise. To do this, they are running their detection code on artificial time series made at different noise levels, and with different slopes in the power spectrum.

6.5. Proposed Space-Based Project

Whipple is a space-based occultation survey of the outer solar system proposed by the Smithonian Astrophysical Observatory, Ball Aerospace, and the Jet Propulsion Laboratory. The aim of this mission is to measure the size distribution of KBOs down to 300 m, to map out the spatial distribution of the Kuiper belt, to determine if a Sedna-like population exists, and to determine the KBO population and distribution in the inner and outer Oort cloud. The proposed spacecraft is based on a 95-cm Schmidt optical system with a 100-deg^2 field and a hybrid CMOS focal plane array (Rockwelle HyViSI/Hawaii-2RG). The flux of 140,000 stars will be recorded at a frequency of 40 Hz. Whipple will go into an Earth-trailing orbit and will scan fields at all ecliptic longitudes and latitude. The raw data rate of 1 Gbps requires significant onboard processing. The expected event rate is thousands to tens of thousands per year for KBOs. Several tens of Oort cloud objects could also be detected per year. *Nihei et al.* (2007) describes this project and compares it with the TAOS observations and observations with a 6.5-m telescope.

6.6. X-Ray Occultation Data

X-ray occultations may stand a good chance of revealing the existence of small KBOs as long as the background X-ray source is bright enough to allow statistically meaningful determination at short timescales. Scorpius X-1 is the brightest and first-discovered X-ray source outside the solar system. Moreover, it is only 6° north of the ecliptic. Analysis of archival data sets of this object by the satellite RXTE has been conducted by the Taiwanese team of H. Chang. The Proportional Counter Array (PCA) instrument (2–60 keV) onboard RXTE has registered a raw count rate of about 10^5 counts per second. This count rate enables them to perform an examination of the lightcurve at timescales as short as 1 ms to search for possible occultations. The first results (*Chang et al.,* 2006) show events compatible with occultation by KBOs of a few tens of meters (Fig. 4). Unfortunately, these reported dip events seem to be contaminated by those due to high-energy particles (*Jones et al.,* 2006). A definite identification of X-ray occultation events probably cannot be achieved until adequately configured observations are conducted in the future (*Chang et al.,* 2007).

7. CONCLUSIONS

The exploration of the Kuiper belt using the method of serendipitous stellar occultations has been under development for several years. The very first results appeared in 2006. The occultation method permits us to detect subkilometer-radius objects throughout the solar system and in particular in the region of the giant planets and beyond. This method also provides a potential way of detecting cometary nuclei in the Oort cloud. The main limitation of the occultation technique is that most of the detected objects cannot be observed again. Hence, information about the Kuiper belt population can only come from statistical analysis.

The method does not allow determining the distance of the occulting object from the analysis of a single profile. An indirect method is to map the spatial distribution of several events, because this spatial and temporal distribution depends on the distance of the KBO population. This difficulty could be bypassed by using very small angular size target stars. The diffraction profiles would then provide information on the distance and size of the occulting object. The size distribution of the population can be retrieved if the distance to the occultors is known. Three different research programs based on this technique have recently announced positive detections. These promising results show that there exists a large population of small objects in the outer solar system and that the occultation method is a powerful tool to explore this population. The analysis of these results is likely to bring very significant progress in a short time frame. The Paris program announced the possible existence of an extended cold disk. More detections and comparisons with other programs will further test these results and bring answers to decisive questions regarding the size and spatial distribution of the Kuiper belt, its connection with the Oort cloud, the shape distribution of KBOs, and the percentage of binary objects in the population.

A connected research field, not treated in this chapter, is occultations by the known KBOs. It will soon give important results about the size, shape, and density of the largest KBOs. Indeed, only the largest KBOs allow sufficiently precise astrometric predictions. This improved knowledge of the structure of the Kuiper belt, and in particular of the unperturbed part of the disk, will provide an important input for a more complete model of solar system formation.

Acknowledgments. It is a pleasure to thank J. Elliot, an anonymous referee, R. Courtin, and the editor for very useful comments that greatly improved this chapter.

REFERENCES

Alcock C., Dave R., Giammarco J., Goldader J., Mehner M., et al. (2003) TAOS: The Taiwanese-American Occultation Survey. *Earth Moon Planets, 92,* 459–464.

Bailey M. E. (1976) Can invisible bodies be observed in the solar system. *Nature, 259,* 290–291.

Bernstein G. M., Trilling D. E., Allen R. L., Brown M. E., Holman M., et al. (2004) The size distribution of trans-neptunian bodies. *Astron. J., 128,* 1364–1390.

Born M. and Wolf E. (1980) Elements of the theory of diffraction. In *Principles of Optics, Sixth Edition*, pp. 370–458. Pergamon, New York.

Brosch N. (1995) The 1985 stellar occultation by Pluto. *Mon. Not. R. Astron. Soc., 276,* 571–578.

Brown M. J. I. and Webster R. L. (1997) Occultations by Kuiper belt objects. *Mon. Not. R. Astron. Soc., 289,* 783–786.

Chang H., King S., Liang J., Wu P., Lin L., et al. (2006) Occultation of X-rays from Scorpius X-1 by small trans-neptunian objects. *Nature, 442,* 660–663.

Chang H.-K., Liang J.-S., Liu C.-Y., and King S.-K. (2007) Millisecond dips in the RXTE/PCA light curve of Sco X-1 and trans-Neptunian object occultation. *Mon. Not. R. Astron. Soc., 378,* 1287–1297.

Cooray A. (2003) Occultation searches for Kuiper belt objects. *Astrophys. J. Lett., 587,* L125–L128.

Denissenko D. V. (2004) Occultations of stars brighter than 15 mag by the largest trans-neptunian objects in 2004–2014. *Astron. Lett., 30,* 630–633.

Dhillon V. S. and Marsh T. (2001) ULTRACAM studying astrophysics on the fastest timescales. *New Astron. Rev., 45,* 91–95.

Elliot J. L. and Kern S. D. (2003) Pluto's atmosphere and a targeted-occultation search for other bound 2b6 atmospheres. *Earth Moon Planets, 92,* 375–393.

Elliot J. L. and Olkin C. B. (1996) Probing planetary atmospheres with stellar occultations. *Annu. Rev. Earth. Planet. Sci., 24,* 89–123.

Georgevits G. (2006) Detection of small Kuiper belt objects by stellar occultation. *Bull. Am. Astron. Soc., 38,* 551.

Jones T. A., Levine A. M., Morgan E. H., and Rappaport S. (2006)

Millisecond dips in Sco X-1 are likely the result of high-energy particle events. *The Astronomer's Telegram, 949.*

Kharchenko et Schilbach E. (1996) Schmidt plate toward the galactic centre. 2. Stellar statistic in the direction of the Sagittarius-Carina arm. *Astron. Nachr., 317(2),* 117–126.

Luu J. X. and Jewitt D. C. (2002) Kuiper belt objects: Relics from the accretion disk of the sun. *Annu. Rev. Astron. Astrophys., 40,* 63–101.

Nihei T. C., Lehner M. J., Bianco F. B., King S.-K., Giammarco J. M., and Alcock C. (2007) Detectability of occultations of stars by objects in the Kuiper belt and Oort cloud. *Astron. J., 134,* 1596–1612.

Roques F. and Moncuquet M. (2000) A detection method for small Kuiper belt objects: The search for stellar occultations. *Icarus, 147,* 530–544.

Roques F., Moncuquet M., and Sicardy B. (1987) Stellar occultations by small bodies: Diffraction effects. *Astron. J., 93,* 1549–1558.

Roques F., Moncuquet M., Lavilloniere N., Auvergne M., Chevreton M., et al. (2003) A search for small Kuiper belt objects by stellar occultations. *Astron. J., 594,* L63–L66.

Roques F., Doressoundiram A., Dhillon V., Marsh T., Bickerton S., et al. (2006) Exploration of the Kuiper belt by high precision photometric stellar occultations: First results. *Astron. J., 132,* 819–822.

Sicardy B., Roques F., and Brahic A. (1991) Neptune's rings, 1983–1989 ground-based stellar occultation observations. *Icarus, 89,* 220–243.

Taylor G. E. (1952) An occultation by a minor planet. *Mon. Not. Astron Soc. South Africa, 11,* 33.

Warner B. (1988) *High Speed Astronomical Photometry.* Cambridge Univ., Cambridge.

Wasserman L. H., Millis R. L., Franz O. G., Bowell E., White N. M., et al. (1979). The diameter of Pallas from its occultation of SAO 85009. *Astron. J., 84,* 259–268.

Young A. T. (1967) Photometric error analysis VI. Confirmation of Reiger's theory of scintillation. *Astron. J., 72,* 747–753.

New Horizons: NASA's Pluto-Kuiper Belt Mission

Harold A. Weaver
Johns Hopkins University Applied Physics Laboratory

S. Alan Stern
NASA Headquarters

The New Horizons (NH) mission was selected by NASA in November 2001 to conduct the first *in situ* reconnaissance of Pluto and the Kuiper belt. The NH spacecraft was launched on January 19, 2006, received a gravity assist from Jupiter during closest approach on February 28, 2007, and is currently heading for a flyby encounter with the Pluto system. NH will study the Pluto system for nearly seven months beginning in early 2015, with closest approach currently planned for mid-July 2015 at an altitude of ~12,500 km above Pluto's surface. If NASA approves an extended mission phase, the NH spacecraft will be targeted toward a flyby encounter with one or more small (~50 km diameter) Kuiper belt objects (KBOs) after the Pluto flyby. The NH spacecraft has a total dry mass of only 400 kg and was launched with 76.8 kg of hydrazine propellant to provide in-flight trajectory correction and spacecraft attitude control. The launch performance was virtually flawless, so less fuel was used for trajectory correction than originally budgeted, which means that more fuel should be available for targeting KBOs beyond the Pluto system. NH carries a sophisticated suite of seven scientific instruments, altogether weighing approximately 30 kg and drawing less than 30 W of power, which includes panchromatic and color imagers, ultraviolet and infrared spectral imagers, a radio science package, plasma and charged particle sensors, and a dust counting experiment. These instruments enable the first detailed exploration of a new class of solar system objects, the dwarf planets, which have exotic volatiles on their surfaces, escaping atmospheres, and satellite systems. NH will also provide the first dust density measurements beyond 18 AU and cratering records that document both the ancient and present-day collisional environment in the outer solar system down to sizes of tens of meters. In addition, NH is the first principal-investigator-led mission to be launched to the outer solar system, potentially opening the door to other nontraditional exploration of the outer solar system in the future.

1. HISTORICAL BACKGROUND

New Horizons (NH) is a flyby reconnaissance mission that will conduct the first *in situ* exploration of the Pluto system and other Kuiper belt objects (KBOs). NH is also the first mission in NASA's New Frontiers series of medium class, robotic, planetary exploration missions, and the first mission to the outer solar system led by a principal investigator (PI) rather than a space agency or laboratory. In this section, we provide an overview of the long, and sometimes torturous, path that finally led to the successful launch of NH in January 2006. An extensive discussion of the history of the NH mission is given by *Stern* (2008), who also discusses previously proposed Pluto-KBO missions, including international collaborations. Other sources for historical background on Pluto-KBO missions include *Terrile et al.* (1997) and *Stern and Mitton* (2005). We focus below on the NASA initiatives, which led to the only successful Pluto-KBO mission to date.

1.1. Early Pluto Mission Concepts

The genesis of NH can be traced back to at least 1989–1990, when a study of a Pluto flyby mission (now referred to as "Pluto-350") was carried out under the auspices of NASA's Discovery Program Science Working Group (DPSWG). The idea was to explore Pluto and Charon with a "minimalist" scientific payload; at that time the Kuiper belt and Pluto's smaller moons Nix and Hydra had not yet been discovered, N_2 had not yet been observed on Pluto's surface and its temperature was thought to be significantly higher than it actually is, the variable nature of Pluto's atmosphere was largely unknown, and Charon was even less well-characterized than Pluto. The resulting spacecraft (*Farquhar and Stern,* 1990) was a 350-kg vehicle, powered by a radioisotope thermoelectric generator (RTG), and carrying four instruments: a visible light imager, an ultraviolet (UV) spectrometer, a radio science experiment, and a plasma package. At that time, such a modest spacecraft, weighing only half as much as the Voyager spacecraft that flew by the outer planets during the 1980s, was considered controversial, both in terms of its small scope and its perceived high risk.

Shortly after the Pluto-350 study, NASA studied a much larger, Cassini-class Mariner Mark II mission to Pluto. This mission, although much more costly, was thought to have lower risk and broader scientific potential. The design included a short-lived, deployable second flyby spacecraft

TABLE 1. Major milestones of New Horizons mission.

Date	Milestone
January 2001	NASA Announcement of Opportunity for Pluto-KBO mission
November 2001	New Horizons selected by NASA
May 2002	Systems requirements review
October 2002	Mission preliminary design review
July 2002	Selection of the Boeing STAR-48 upper stage
March 2003	Nonadvocate review and authorization for phase C/D
July 2003	Selection of the Lockheed-Martin Atlas V 551 launch vehicle
October 2003	Mission critical design review
May 2004	Spacecraft structure complete
September 2004	First instrument delivery
March 2005	Final instrument delivery
April 2005	Spacecraft integration complete
May 2005	Start of spacecraft environmental testing
September 2005	Spacecraft shipment to the launch site in Florida
December 2005	Spacecraft mating with its launch vehicle
January 2006	Launch
February 2007	Closest approach to Jupiter
July 2015	Closest approach to Pluto (planned)
2016–2020	Other KBO encounters (if extended mission approved)

TABLE 2. New Horizons scientific objectives.

Group 1 (Primary Objectives)
Characterize the global geology and morphology of Pluto and Charon
Map surface composition of Pluto and Charon
Characterize the neutral atmosphere of Pluto and its escape rate
Group 2 (Secondary Objectives)
Characterize the time variability of Pluto's surface and atmosphere
Image Pluto and Charon in stereo to measure surface topography
Map the terminators of Pluto and Charon with high resolution
Map the surface composition of selected areas of Pluto and Charon with high resolution
Characterize Pluto's ionosphere and solar wind interaction
Search for neutral species including H, H_2, HCN, C_xH_y, and other hydrocarbons and nitriles in Pluto's upper atmosphere, and obtain isotopic discrimination where possible
Search for an atmosphere around Charon
Determine bolometric Bond albedos for Pluto and Charon
Map the surface temperatures of Pluto and Charon
Group 3 (Tertiary Objectives)
Characterize the energetic particle environment of Pluto and Charon
Refine bulk parameters (radii, masses, densities) and orbits of Pluto and Charon
Search for additional satellites and rings

designed to fly over Pluto's far hemisphere some 3.2 days (one Pluto half-rotation) before or after the mother ship. This mission was adopted as a high priority in the Solar System Exploration Subcommittee (SSES) 1990s planetary exploration plan derived in a "community shoot out" meeting in February 1991. Following this, NASA's Solar System Exploration Division formed the Outer Planets Science Working Group (OPSWG) (S. A. Stern, Chair) to shape the mission's scientific objectives, document its rationale, and prepare for an instrument selection process by the mid-1990s. By 1992, OPSWG had completed most of its assigned mission study support tasks. Owing to tight budgets at NASA, OPSWG was also asked to debate the large Mariner Mark II vs. the much smaller Pluto-350 mission concepts. In early 1992, OPSWG selected Pluto-350 as the more pragmatic choice.

However, in the late spring of 1992, a new, more radical mission concept called Pluto Fast Flyby (PFF) was introduced by the Jet Propulsion Laboratory (JPL) as a "faster, better, cheaper" alternative to the Mariner Mark II and Pluto-350 Pluto mission concepts. As initially conceived, PFF was to weigh just 35–50 kg and carry only 7 kg of highly miniaturized (then nonexistent) instruments, and fly two spacecraft to Pluto for <$500M. PFF found a ready ally in then NASA Administrator D. Goldin, who directed all Pluto-350 and Mariner Mark II work to cease in favor of PFF. PFF would have launched its two flyby spacecraft on Titan IV-Centaur launchers; these low-mass spacecraft would have shaved the Pluto-350 and Mariner Mark II flight times from 12–15 years down to 7 or 8 years. Like Mariner Mark II and Pluto-350, PFF involved RTG power and Jupiter gravity assists (JGAs). The heavier missions also involved Earth and Venus gravity assists on the way to Jupiter. All these mission concepts were developed by JPL mission study teams.

Shortly after PFF was introduced, however, it ran into problems. One was mass growth, which quickly escalated the flight system to ~140 kg with no increase in science payload mass. A second issue involved cost increases, largely due to a broad move within NASA to include launch vehicle costs in mission cost estimates; since two Titan IV launchers alone cost over $800M, this pushed PFF to well over $1B. A third issue was the turmoil introduced into NASA's planetary program by the loss of the Mars Observer in 1993. These events caused PFF to lose favor at NASA, and the concept never made it into the development phase. Nevertheless, during 1994–1995 PFF did solicit, select, and fund the breadboard/brassboard development of a suite of miniaturized imagers, spectrometers, and radio science and plasma instruments, whose successors would ultimately become the science payload on NH.

Owing to the rapidly expanding interest in the Kuiper belt by the mid-1990s, NASA directed JPL to re-invent PFF as Pluto Express, later named, and more commonly known as, Pluto-Kuiper Express (PKE). PKE was a single spacecraft PFF mission with a 175-kg spacecraft and a 9-kg science payload. It would have launched in the 2001–2006 JGA launch window. A science definition team (SDT) (J. I. Lunine, Chair) was formed in 1995 and delivered its report in 1996 for an anticipated instrument selection in 1996–1997. However, in late 1996, PKE mission studies were drastically cut by Administrator Goldin and no instrument selection was initiated. By 1999, however, NASA did release a solicitation for PKE instruments with proposals due in March 2000. These proposals were evaluated and ranked, but never selected. By September 2000, NASA cancelled PKE, still in Phase A, owing to mission cost increases that pushed the projected mission cost over the $1B mark. Following this cancellation, intense scientific and public pressure spurred then NASA Associate Administrator for Space Science E. Weiler to solicit mission proposals in 2001 for a Pluto Kuiper belt (PKB) flyby reconnaissance mission, which we discuss next.

1.2. Pluto Kuiper Belt Mission Announcement of Opportunity and Selection of New Horizons

NASA's decision to solicit PKB mission proposals was announced at a press conference on December 20, 2000, and the formal PKB Announcement of Opportunity (AO) was released on January 19, 2001. The AO (NASA 01-OSS-10) mandated a two-step selection process with initial proposals due March 20, 2001, later extended to April 6, 2001. Following a down-select to two teams, Phase A studies would be performed with due dates in the August–September timeframe. Since no PI-led mission to the outer planets, nor any PI-led mission involving RTGs, had ever been selected in NASA's history, the AO was termed "experimental" by NASA, which made it clear that the agency might not select *any* of the proposals.

The PKB AO required responders to propose an entire PKB mission, to meet at least the basic (i.e., "Group 1") scientific objectives specified in the 1996 PKE SDT report, to complete a Pluto flyby by 2020, to launch onboard a U.S. Atlas V or Delta IV launch vehicle, and to do so within a complete mission cost cap of $506M FY2001 dollars. Launch vehicle selection between the Atlas V and Delta IV was planned for 2002. Two spare Cassini-Galileo RTGs were made available for use to proposal teams, with associated costs of $50M and $90M (the latter with higher power).

Shortly after the PKB AO release, on February 6, 2001, the newly elected Bush Administration released its first budget, which canceled PKB by not funding it in FY02 and future years. Within days, NASA announced the suspension of the PKB AO as well. However, within another week, following intensive work on Capitol Hill by the science community, the U.S. Senate directed NASA to proceed with the AO so as not to limit Congressional authority to override the PKB cancellation decision.

Five PKB proposals were submitted to NASA by the April 6, 2001, deadline. S. A. Stern partnered with the Johns Hopkins University Applied Physics Laboratory (APL) on a proposal called New Horizons, whose name was meant to symbolize both the new scientific horizons of exploring the Pluto system and the Kuiper belt, as well as the programmatic new horizons of PI-led outer planet missions. *Stern and Cheng* (2002) and *Stern* (2002) summarized the NH mission as proposed.

After a two-month technical and programmatic review process, on June 6, 2001, NASA announced the selection of JPL's Pluto Outer Solar System Explorer (POSSE) (L. Esposito, PI) and APL's NH (S. A. Stern, PI) for Phase A studies and further competition. Both teams were given $500K to refine their mission concepts and prepare revised proposals by September 18, 2001. The deadline was pushed back to September 25 owing to the interruption of U.S. government activities by the September 11 terrorist attacks. Formal oral briefings on the two proposals to a NASA Concept Study Evaluation Review Board were held for NH and POSSE on October 17 and 19, respectively. NASA announced the selection of NH on November 29, 2001.

The selection of NH did *not* mean smooth sailing from that point on. NASA explained that many hurdles had to be overcome before the mission could enter full development, including lack of funding, lack of a nuclear qualified launch vehicle, and lack of sufficient fuel to power an RTG. Furthermore, NASA postponed the earliest launch date from December 2004 to January 2006, incurring a three-year delay in the Pluto arrival time, from 2012 to

2015. In the summer of 2002, after hard-fought battles on Capitol Hill and the key endorsement of a Pluto-KBO mission as NASA's highest priority new start for solar system exploration by the National Research Council's Decadal Report in Planetary Sciences [the *Decadal Survey* (*Belton et al.,* 2002)], NASA finally became committed to NH and strongly supported its development to launch and beyond. The most important milestones in the NH mission are listed in Table 1; the only milestones not yet completed are the Pluto encounter in July 2015 and any subsequent KBO encounters, assuming that NASA approves an extended mission phase.

2. SCIENTIFIC OBJECTIVES

As previously discussed, the scientific objectives of a PKB mission were developed by NASA's OPSWG in 1992 and slightly refined and then re-ratified by the PKE SDT in 1996. The specific measurements needed to achieve the scientific objectives of the mission were also described in detail in the reports prepared by the OPSWG and PKE SDT. These scientific and measurement objectives were adopted by NASA for the PKB mission AO that led to the selection of NH.

The NH scientific objectives are ranked in three categories, called Group 1, Group 2, and Group 3 (Table 2). Group 1 objectives represent an irreducible floor for the mission science goals at the Pluto system. Group 2 goals add depth and breadth to the Group 1 objectives and are termed highly desirable. The Group 3 objectives add further depth and are termed desirable, but they have a distinctly lower priority than the Group 2 objectives.

2.1. Primary Objectives (Group 1)

The Group 1 objectives address the most basic questions about Pluto and Charon: What do they look like, what are they made of, and what is the nature of Pluto's atmosphere? As discussed further in section 4, the NH science payload is capable of mapping the entire sunlit surfaces of Pluto and Charon at a best resolution of ~0.5 km pixel^{-1} in panchromatic visible light images, and of making four-color visible light maps of Pluto and Charon at a best resolution of ~5 km pixel^{-1}. These images will be used to determine the global geology and the surface morphology of Pluto and Charon. New Horizon's infrared (IR) spectral imager will map the distributions of N_2, CO, CH_4, and H_2O, as well as other species yet to be discovered, on the sunlit surfaces of Pluto and Charon at a best resolution of ~10 km pixel^{-1}.

Radio uplink and ultraviolet (UV) solar occultation observations of Pluto's atmosphere will be used to measure its pressure and temperature as a function of height above the surface, its composition, and its escape rate. Ultraviolet airglow measurements will be used to search for atomic and molecular emissions excited by charged particles or solar fluorescence. The atmospheric escape rate will be probed by NH's charged particle instruments, which can measure the solar wind stand-off distance and detect energetic ions produced when escaping neutral molecules charge-exchange with the solar wind. Both the imaging observations at large solar phase angles and the occultation measurements will be used to search for hazes in Pluto's atmosphere, which may be important in determining the atmosphere's thermal structure.

2.2. Secondary Objectives (Group 2)

The NH science payload will also be used to address *all* the SDT Group 2 objectives. During the Pluto encounter approach phase, NH's visible light imagers will make global maps over at least a dozen rotational periods of Pluto-Charon to search for temporal variability on their surfaces, and NH's UV spectral imager will monitor airglow emissions to search for variability in Pluto's atmosphere. Visible light images taken at slightly different times will be used to create stereo views of the surfaces of Pluto and Charon, which provide information on the surface topography. Small areas near the terminator will be mapped at a resolution of ~50 m pixel^{-1} with panchromatic imaging, which will provide a sensitive probe of surface features having unusual morphologies. Visible light imaging observations at multiple phase angles will be used to determine the Bond albedos of Pluto and Charon. Using the widths and positions of N_2, CO_2, and H_2O bands as thermometers, NH's IR spectral imager will map Pluto's surface temperature wherever there is ice, and NH's radio science package will be used in radiometer mode to measure the average global temperatures of Pluto and Charon, on both the daytime and nighttime hemispheres.

The SDT objectives do not include any discussion of Pluto's small moons Nix and Hydra because they had not yet been discovered. Fortunately, there are more than eight years available to plan how best to investigate Nix and Hydra, and the NH Science Team is treating the compositional and geological mapping of these bodies as an additional Group 2 objective to be addressed by the mission.

2.3. TERTIARY OBJECTIVES (GROUP 3)

The NH science payload will also be used to address all the SDT Group 3 objectives, *except* the objective of measuring magnetic fields. Pluto's magnetic field is probably extremely weak, if it exists at all, and a large boom would be needed to make sensitive magnetic field measurements. Since indirect information on Pluto's magnetic field can be gleaned from the two plasma instruments onboard NH (SWAP and PEPSSI), the NH Science Team decided during the proposal phase not to complicate the spacecraft design by adding a boom with a magnetometer.

The best full-disk observations of Pluto and Charon will be used to refine the radii, masses, densities, and orbits of both objects. Investigations of craters observed in these images will also enable a determination of the KBO size frequency distribution down to the meter-class scale. Visible light images before and after Pluto closest approach will be used to search for new satellites and for dust rings. We note that the presence of Nix and Hydra raises the prospects for discovering ephemeral dust rings in the Pluto system because the surface gravities on those small satellites are too small to capture material excavated from their surfaces by impacts with boulder-sized debris passing through the system (*Stern et al.,* 2006).

Additional discussion of NH's objectives is provided in section 4, where the individual NH instruments are described. However, *Young et al.* (2008) should be consulted for a much more extensive and detailed discussion of the scientific and measurement objectives of the NH mission.

3. MISSION DESIGN

The NH mission design was driven by the desire to reach Pluto by at least 2020, preferably earlier. Pluto reached perihelion in 1989, at a heliocentric distance of 30 AU, and is currently heading toward aphelion at 49 AU in 2114. Besides the obvious advantage of trying to reach Pluto before it gets even farther away, the earlier the spacecraft arrival date the better the chance that Pluto's tenuous atmosphere will still be observable; the expected decrease in surface temperature with increasing heliocentric distance may eventually cause all the atmosphere's constituents to freeze out on the surface. Although the exact time when atmospheric freeze-out will occur is unknown, models suggest that Pluto's atmospheric density should start steadily decreasing around 2020 (*Hansen and Paige,* 1996). In addition, Pluto's obliquity is 120°, and the fraction of Pluto's surface receiving sunlight is continually decreasing between now and the next solstice in 2029, when the subsolar latitude reaches –57.5°. Thus, the earlier the arrival date, the more of Pluto's surface is available for imaging. Finally, the longer the mission duration, the higher the risk of failure before the mission is completed, which is yet another reason to design the mission to reach Pluto as quickly as possible.

Several constraints imposed by the geometry required for the Pluto encounter also affected the NH mission design, and those are discussed below. But here we mention one other important mission design consideration, which was to make the launch window as wide as possible. Restricting the range of possible launch dates would make the mission more vulnerable to a wide variety of circumstances that can delay launches (e.g., bad weather, technical problems, etc.). NH had a relatively long, 35-day launch window (January 11 through February 14, 2006); launches during the first 23 days employed a Jupiter gravity assist, while later launch dates were Pluto-direct trajectories. NH also had a 14-day backup launch window in January 2007, but all those were Pluto-direct trajectories with Pluto arrivals in 2019–2021. Fortunately, NH launched on January 19, 2006, on a Jupiter-assist trajectory, with a Pluto arrival in mid-2015. A summer arrival date is preferred because Pluto will be nearly at opposition when viewed from Earth, which is the best geometry for the radio uplink occultation experiment.

A launch energy (C3) of nearly 170 $km^2 s^{-2}$ was needed to propel the NH spacecraft to Pluto, which was accomplished using the powerful Lockheed-Martin Atlas 551 launch vehicle in tandem with its Centaur second stage and a Boeing Star 48 third stage. The launch performance was virtually flawless, and the change in velocity needed for trajectory correction (called "ΔV") was only ~18 m s^{-1}, whereas 100 m s^{-1} was budgeted pre-flight. NH departed Earth faster than any other spacecraft (~16 km s^{-1}), passing the orbit of the Moon in only 9 hours and reaching Jupiter in a record time of only 13 months. Plate 13 displays NH's trajectory.

We briefly discuss below the circumstances for the four encounters that NH will experience during its journey; further details on NH's mission design can be found in *Guo and Farquhar* (2008).

3.1. Asteroid Encounter

NH serendipitously flew past the small (~3 km diameter) S-type (*Tubiana et al.,* 2007) asteroid APL (2002 JF56 = 132524) on June 13, 2006, at a distance of 102,000 km. The door on NH's highest-resolution camera (LORRI) was still closed at that time, but panchromatic and color visible light images, and infrared spectral images, were obtained by the Ralph instrument. Observations were obtained over a range of solar phases angles not possible from the ground (4°–90°), and the barely resolved asteroid displayed an unusual morphology, indicative of either a highly irregular shape or possibly an orbiting companion (*Olkin et al.,* 2006). But the main utility of the asteroid encounter was the opportunity to test NH's ability to track a relatively fast-moving object. The apparent motion of APL near closest approach was ~270 μrad s^{-1}, which was much larger than the highest rate encountered during the Jupiter flyby (~9.7 μrad s^{-1}). Although Pluto's apparent motion at closest approach (~1275 μrad s^{-1}) will be approximately five times faster than the asteroid's motion, the observations of asteroid APL still provided an important opportunity for the NH operations team to perform an encounter sequence on a real planetary target, a test that NH passed with flying colors.

3.2. Jupiter Encounter

NH made its closest approach to Jupiter on February 28, 2007. The primary objective of the Jupiter encounter was to send the NH spacecraft through an aim point located ~32 jovian radii (~2.3 million km) from the center of the planet, where a gravity assist increased its speed by ~20% and reduced the travel time to Pluto by approximately three years. A secondary objective was to perform instrument calibration observations in the jovian system that could not

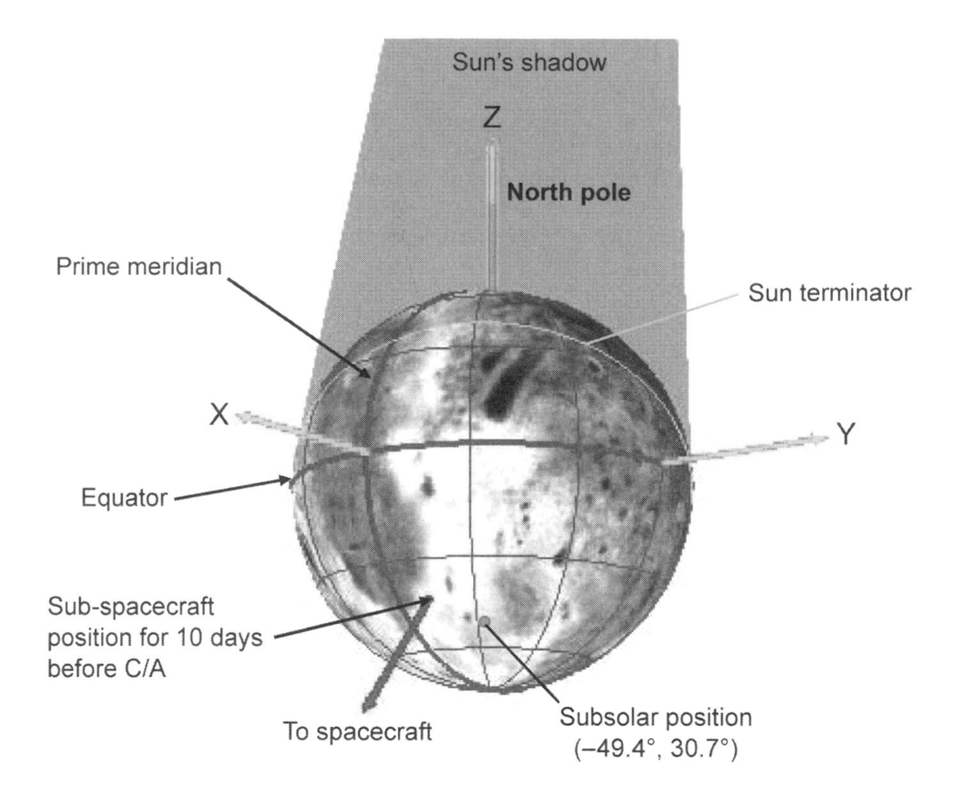

Fig. 1. The geometry during NH observations of Pluto 10 days before closest approach is depicted, showing the subspacecraft and subsolar points, and the day-night terminator, in a plutocentric frame. From *Guo and Farquhar* (2008).

be obtained any other time during the mission (e.g., on large, extended targets). Another important objective was to put NH through an encounter that would root out any flaws in the encounter strategy, so that any mistakes during the Jupiter system observations would not be repeated during the Pluto system observations, when the scientific success of the mission is at stake. To accomplish this latter objective, which essentially amounted to a "stress test" for the encounter strategy, an ambitious set of observations of the jovian system was executed. Jovian science was the beneficiary as ~700 observations were planned for the Jupiter encounter compared to ~350 for the baseline Pluto encounter. The NH science team had spent several years designing a program of observations that could return unique and exciting scientific results from the Jupiter encounter and was extremely gratified to have most of its "wish list" of observations actually performed.

During approach, NH monitored the jovian meteorology and found a surprisingly clear mid-latitude atmosphere with very little turbulence around the Great Red Spot, in striking contrast to what was seen by Voyager, Galileo, and Cassini. Also during approach, UV spectral observations monitored changes in Jupiter's auroral activity and emissions from the Io plasma torus. An intense and complex observational sequence was executed during a ±4-day period centered on the closest approach date (February 28, 2007) that involved imaging and spectroscopic observations of Jupiter, its satellites, and its rings. Throughout the approach phase and for ~100 days post-encounter, the NH particle and plasma instruments investigated Jupiter's magnetosphere and performed the first *in situ* measurements down the magnetotail of a giant planet. The initial results from the NH encounter with Jupiter are discussed in detail in a series of papers published in the journal *Science*.

3.3. Pluto Encounter

NH will have its closest approach to Pluto in mid-July 2015. The following constraints for the Pluto encounter were drivers for the mission's design:

1. The closest approach distance to Pluto must be ~12,500 km, so that the spatial resolution measurement objectives can be achieved.

2. The spacecraft must fly through Pluto's shadow, so that both solar and Earth occultations occur when at least two NASA DSN stations can track the spacecraft at elevation angles >15°.

3. The spacecraft must fly through Charon's shadow to achieve a solar occulation, and preferably an Earth occultation too, when at least two NASA DSN stations are tracking the spacecraft at elevation angles >15°.

4. The Pluto closest approach must occur first, and Charon must be placed so that the Pluto nightside can be imaged in reflected light from Charon. NH will approach Pluto from its southern hemisphere at a solar phase angle of ~15° (Fig. 1). The spacecraft's heliocentric velocity at encounter time is ~14 km s^{-1}. The occultation geometry for the Pluto encounter is illustrated in Fig. 2.

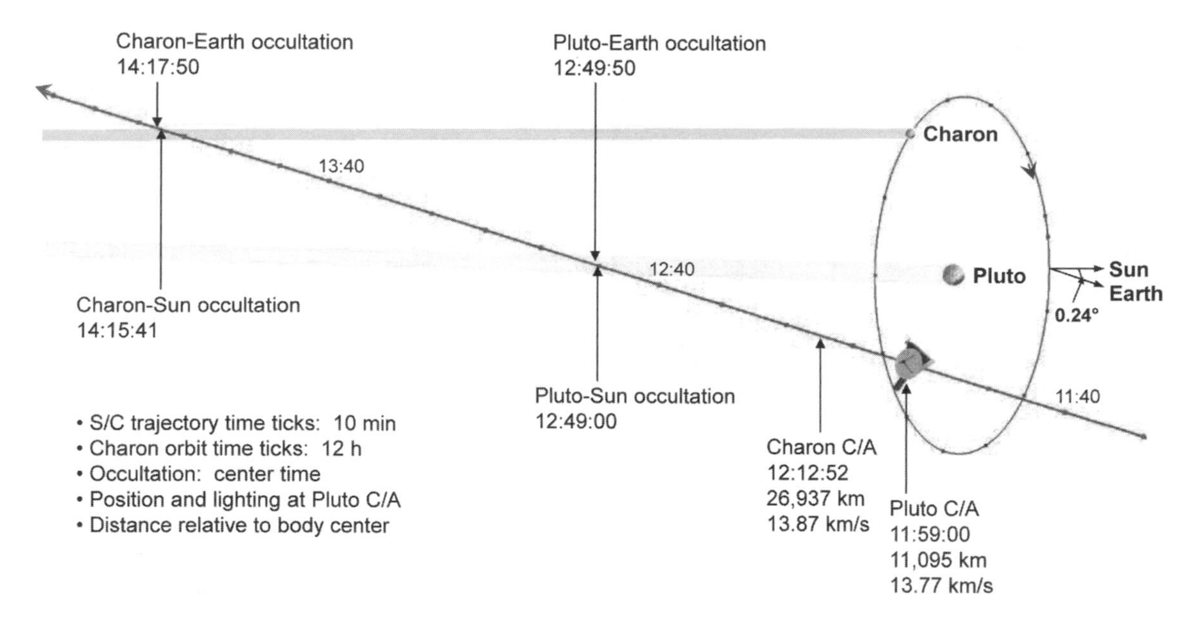

Fig. 2. This figure illustrates the timing and geometry for the NH occultation measurements to be performed shortly after the closest approach to Pluto on July 14, 2015. A test trajectory case is depicted; the absolute times are likely to change for the final design, but the relative timing of the various events should be well-represented. From *Guo and Farquhar* (2008).

3.4. Kuiper Belt Object Encounters

Pending NASA approval of the extended mission phase, NH will perform a target trajectory maneuver approximately two weeks after Pluto closest approach to send the spacecraft to an encounter with another, yet to be identified, KBO. Depending on the availability of fuel for maneuvers, the baseline strategy is to pass within ~25,000 km of the targeted KBO.

At launch, 76.8 kg of hydrazine propellant was loaded onto the NH spacecraft. The propellant is used for inflight trajectory correction maneuvers and for spacecraft attitude control. Owing to the excellent launch performance, the propellant usage for the trajectory corrections has been well below the pre-flight estimates. As of mid-April 2007 (i.e., after the Jupiter encounter) only ~12 kg of propellant had been consumed, including both trajectory corrections and all spacecraft-pointing maneuvers. The best current estimate is that ~47 kg should be available after the Pluto flyby, which corresponds to a ΔV capacity of ~235 m s^{-1}. With this amount of fuel available for trajectory course correction, the Monte Carlo model of *Spencer et al.* (2003), which incorporates recent estimates on the size distribution and dynamical structure of the Kuiper belt, indicates that NH has more than a 95% probability of reaching one ~50 km diameter KBO. The probability of reaching two KBOs larger than ~40 km is ~50%. The first KBO encountered is most likely to be at a heliocentric distance of ~42 AU, which NH will reach in 2018. NH is expected to retain operational readiness through ~2020, at which time the spacecraft will be at a heliocentric distance of ~50 AU.

A campaign has been mounted to search for potential KBO targets for the NH mission using large, groundbased telescopes. However, the region of the sky where potential targets are located is currently near the galactic plane, and the high density of stars makes it difficult to find KBOs, especially the small, faint ones, which require a limiting magnitude of $V \approx 27$ to detect objects ~50 km in diameter. Of course, it is possible to be lucky and find a large KBO in the region accessible to NH, but so far none have been found. The detectability of candidate KBOs should become much easier after 2010 because the sky region of interest should be much farther from the galactic plane, and the sensitivity of the telescopic facilities should improve as well. The NH Project expects to have selected its KBO target by ~2012, well before the Pluto encounter period.

4. SCIENTIFIC INSTRUMENTS

The challenges associated with sending a spacecraft to Pluto in less than 10 years and performing an ambitious suite of scientific investigations at such large heliocentric distances (>32 AU) are formidable and required the development of lightweight, low-power, and highly sensitive instruments. Fortunately, all the NH instruments successfully met these daunting technical challenges without compromising any of the mission's original scientific objectives.

All the fundamental (Group 1) scientific objectives for the NH mission can be achieved with the *core* payload, comprising the Alice ultraviolet (UV) imaging spectroscopy remote sensing package, the Ralph visible and infrared imaging and spectroscopy remote sensing package, and the Radio Science Experiment (REX) radio science package.

The *supplemental* payload is not required to achieve minimum mission success, but it both deepens and broadens the scientific objectives and provides functional redundancy across scientific objectives. The supplemental payload comprises the Long Range Reconnaissance Imager

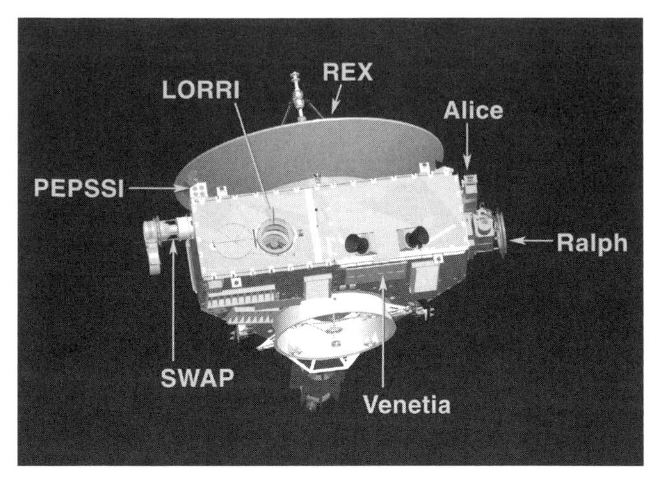

Fig. 3. This drawing shows the locations of the instruments on the New Horizons spacecraft. The antenna diameter is 2.1 m, which provides a scale for the figure. The boresights of LORRI, Ralph, and the Alice airglow channel are approximately coaligned. The boresights of the Alice solar occultation channel and the antenna are approximately coaligned and are approximately orthogonal to the LORRI, Ralph, and Alice airglow channel boresights.

(LORRI), which is a long-focal-length, panchromatic visible light imager; the Solar Wind Around Pluto (SWAP) charged particle detector; the Pluto Energetic Particle Spectrometer Science Investigation (PEPSSI),which detects charged particles at higher energies than those detected by SWAP; and the Venetia Burney Student Dust Counter (VB-SDC), which is an interplanetary dust detection and mass characterization experiment. VB-SDC was a late addition to the supplemental payload approved by NASA as an Education and Public Outreach (EPO) initiative.

The locations of the instruments on the NH spacecraft are shown in Fig. 3. Pictures of all seven instruments are displayed in Fig. 4, which also lists the mass and power consumption of each instrument. The primary measurement objectives and the principal characteristics of each instrument are summarized in Table 3. We provide below a high-level overview of all the NH instruments; a more detailed description of the NH science payload is given by *Weaver et al.* (2008) and references therein, while *Young et al.* (2008) provide more discussion on the scientific objectives of each instrument.

4.1. Alice

Alice, a UV imaging spectrometer with a dual-delay line microchannel plate detector, is sometimes called Pluto-Alice (P-Alice) to distinguish it from its predecessor, Rosetta-Alice (R-Alice), which is a similar instrument being flown on the European Space Agency (ESA) Rosetta mission to Comet 67P/Churyumov-Gerasimenko. Compared to R-Alice, P-Alice has a somewhat different bandpass and various enhancements to improve reliability. P-Alice also includes a separate solar occultation channel, which is not available on R-Alice, to enable sensitive measurements of Pluto's upper atmosphere by observing the sun as the NH spacecraft enters Pluto's shadow. Both P-Alice and R-Alice are significantly improved versions of the Pluto mission

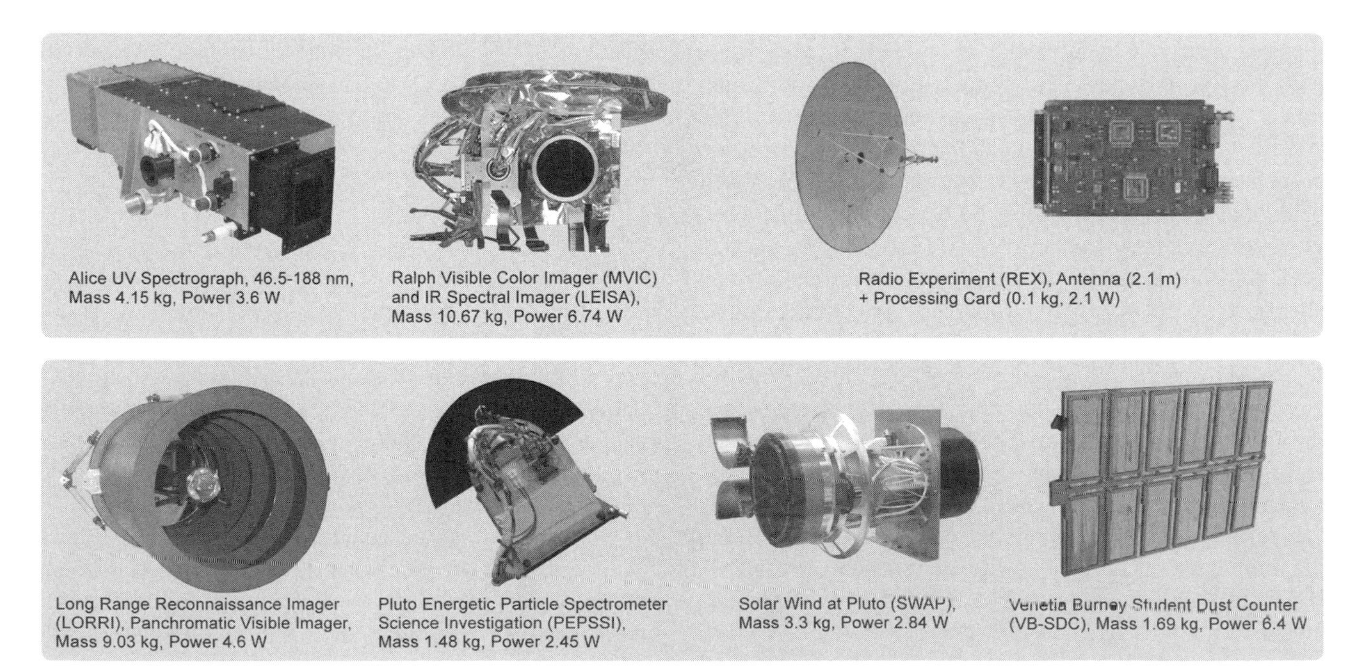

Fig. 4. The three instruments comprising the New Horizons core payload are shown along the top row, and the instruments comprising the supplemental payload are displayed along the bottom row. All these instruments are included in the as-launched spacecraft. The total mass of the entire science payload is 30.4 kg, and the total power drawn by all the instruments is 28.7 W.

TABLE 3. New Horizons science payload.

Instrument	Measurement Objectives	Characteristics
Alice	• **Upper atmospheric temperature and pressure profiles of Pluto** • **Temperature and vertical temperature gradient measured to ~10% at a vertical resolution of ~100 km for atmospheric densities ≥10^9 cm^{-3}** • **Search for atmospheric haze at a vertical resolution <5 km** • **Mole fractions of N_2, CO, CH_4, and Ar in Pluto's upper atmosphere** • **Atmospheric escape rate from Pluto** • Minor atmospheric species at Pluto • Search for an atmosphere of Charon • Constrain escape rate from upper atmospheric structure	UV spectral imaging: 465–1880 Å; FOV 0.1° × 4° and 2° × 2°; dispersion 1.8 Å $pixel^{-1}$; spatial resolution 5 mrad $pixel^{-1}$; airglow and solar occultation channels; modes: time-tag, histogram; A_{eff} ≈ 0.3 cm^2 (peak)
Ralph-MVIC	• **Hemispheric panchromatic maps of Pluto and Charon at best resolution exceeding 0.5 km $pixel^{-1}$** • **Hemispheric four-color maps of Pluto and Charon at best resolution exceeding 5 km $pixel^{-1}$** • **Search for/map atmospheric hazes at a vertical resolution <5 km** • High-resolution panchromatic maps of the terminator region • Panchromatic, wide-phase-angle coverage of Pluto, Charon, Nix, and Hydra • Panchromatic stereo images of Pluto and Charon, Nix, and Hydra • Orbital parameters, bulk parameters of Pluto, Charon, Nix, and Hydra • Search for rings • Search for additional satellites	Visible CCD imaging: 400–975 nm (panchromatic), four color filters (blue, red, methane, near-IR); FOV 0.15° × 5.7° (stare, pan), FOV 5.7° × arbitrary (scan, color + pan); IFOV 20 µrad (4")
Ralph-LEISA	• **Hemispheric near-IR spectral maps of Pluto and Charon at best resolution exceeding 10 km $pixel^{-1}$** • **Map hemispheric distributions of N_2, CO, and CH_4 on Pluto at a best resolution exceeding 10 km $pixel^{-1}$** • Surface temperature mapping of Pluto and Charon • Phase angle dependent spectral maps of Pluto and Charon	IR spectral imaging: 1.25–2.5 µm; $\lambda/\delta\lambda \approx 240$ (1.25–2.50 µm); $\lambda/\delta\lambda \approx 550$ (2.10–2.25 µm); FOV 0.9° × 0.9°; IFOV 62 µrad (12")
REX	• **Temperature and pressure profiles of Pluto's atmosphere to the surface** • **Surface number density to ±1.5%, surface temperature to ±2.2 K, and surface pressure to ±0.3 µbar** • Surface brightness temperatures on Pluto and Charon • Masses and chords of Pluto and Charon; detect or constrain J_2 • Detect, or place limits on, an ionosphere for Pluto	X-band (7.182 GHz uplink, 8.438 GHz downlink); radiometry T_{Noise} < 150 K; Ultra-Stable Oscillator (USO) frequency stability: $\delta f/f = 3 \times 10^{-13}$ over 1 s
LORRI	• **Hemispheric panchromatic maps of Pluto and Charon at best resolution exceeding 0.5 km $pixel^{-1}$** • **Search for/map atmospheric hazes at a vertical resolution <5 km** • High-resolution panchromatic maps of the terminator region • Panchromatic, wide-phase-angle coverage of Pluto, Charon, Nix, and Hydra • Panchromatic stereo images of Pluto and Charon, Nix, and Hydra • Orbital parameters, bulk parameters of Pluto, Charon, Nix, and Hydra • Search for rings • Search for additional satellites	Visible CCD panchromatic images: 350–850 nm; FOV 0.29° × 0.29°; IFOV 5 µrad (1"); optical navigation
SWAP	• **Atmospheric escape rate from Pluto** • Solar wind velocity and density, low-energy plasma fluxes and angular distributions, and energetic particle fluxes at Pluto system • Solar wind interaction of Pluto and Charon	Solar wind detector; FOV 10° × 200°; energy range 0.25–7.5 keV; energy resolution: RPA: 0.5 V (<1.5 keV), ESA: 0.4 δE/E (>1.4 keV)
PEPSSI	• Composition and density of pick-up ions from Pluto, which indirectly addresses the atmospheric escape rate • Solar wind velocity and density, high-energy plasma fluxes and angular distributions, and energetic particle fluxes in the Pluto system	Energetic particle detector; energy range 1 keV–1MeV; FOV 12° × 160°; IFOV 12° × 25°; 12 energy channels; mass resolution: <5 amu (>1.7 keV $nucleon^{-1}$), <2 amu (>5 keV $nucleon^{-1}$)
VB-SDC	• Trace the density of dust in the solar system along the NH trajectory from Earth to Pluto and beyond	12 PVF panels to detect dust impacts and 2 control panels shielded from impacts

Boldface text specifies measurement objectives that achieve Group 1 scientific objectives. A_{eff} is the effective area of the Alice instrument and is a measure of its throughput or sensitivity. IFOV gives the individual pixel field of view for the MVIC, LEISA, and LORRI instruments, in both microradians (µrad) and arcseconds ("). RPA and EPA refer to the retarding potential analyzer and the electrostatic analyzer, respectively, for SWAP.

"HIPPS" UV spectrograph (HIPPS/UVSC) (*Stern et al.,* 1995), which was developed at Southwest Research Institute (SwRI) in the mid-1990s with funds from NASA, JPL, and SwRI. Further details on Alice and its performance can be found in *Stern et al.* (2008).

Alice was designed to measure Pluto's upper atmospheric composition and temperature, which is a Group 1 scientific objective. Alice will also obtain model-dependent escape rate measurements from Pluto's atmosphere and will provide some limited surface mapping and surface composition capabilities in the UV. Alice's spectral bandpass includes lines of CO, atomic H, Ar, and Ne, which may be detectable as airglow, and the electronic bands of N_2, CH_4, and other hydrocarbons and nitriles, which are detectable during solar and stellar occultation observations.

4.2. Ralph

Ralph is essentially two instruments rolled into a single package. The Multispectral Visible Imaging Camera (MVIC) is a visible light panchromatic and color imager, and the Linear Etalon Imaging Spectral Array (LEISA) is an infrared imaging spectrometer. The common telescope assembly for Ralph has a three-mirror, off-axis anastigmat design with a 7.5-cm primary mirror. A dichroic reflects the visible light to the MVIC focal plane and transmits the IR light to the LEISA focal plane. Only one focal plane is active at a time, with a relay used to select either MVIC or LEISA. The Ralph boresight is approximately co-aligned with the LORRI and Alice airglow boresights.

The MVIC focal plane has seven independent CCD arrays mounted on a single substrate. Six of the arrays have 5000 (columns) × 32 (rows) photosensitive pixels and operate in time-delay integration (TDI) mode, in which the spacecraft is scanned in cadence with the transfer of charge from one row to the next until the desired scene is covered. Two of the TDI arrays provide panchromatic (400–975-nm) images, and the other four TDI arrays provide, respectively, color images in blue (400–550 nm), red (540–700 nm), near-IR (780–975 nm), and narrowband methane (860–910 nm) channels. The nominal scan rate for MVIC is 1000 μrad s^{-1} (200" s^{-1}) for its color images, and 1600 μrad s^{-1} (320" s^{-1}) for its panchromatic images. The frame transfer array has 5000 × 128 pixels and provides panchromatic images (400–975 nm) that can be used for either science or optical navigation. Each panchromatic array can be operated independently, for redundancy. The four color arrays are operated in tandem. Further details on MVIC and its performance can be found in *Reuter et al.* (2008).

MVIC images in the three broadband colors will provide information on spectral slopes of Pluto's surface and on its atmospheric properties. The narrowband filter permits mapping of the surface methane abundance, as the well-known 890-nm absorption band is the strongest methane feature available at visible light wavelengths. The 700–780-nm gap between the red and near-IR bandpasses overlaps another methane band at 740 nm; combining data from the panchromatic, blue, red, and near-IR filters can provide some information about band depth in this "virtual" filter.

LEISA's dispersive capability is provided by its wedged etalon (a linear variable filter, or LVF), which is mounted ~100 μm above its 256 × 256 pixel HgCdTe PICNIC array. The etalon covers 1.25–2.5 μm, a spectral region populated with many absorption features of N_2, CH_4, H_2O, NH_3, CO, and other molecules, at a resolving power of ~250. A higher-resolution subsegment, covering 2.10–2.25 μm at a resolving power of ~550, will be used to discern grain sizes, mixing states, and pure vs. solid-solution abundances (*Quirico et al.,* 1999). The higher-resolution segment is also critical for taking advantage of the temperature sensitive N_2 bands (*Grundy et al.,* 1993, 1999), and the symmetric, doubled $\nu_2 + \nu_3$ CH_4 band that is diagnostic of pure vs. diluted CH_4 abundances (*Quirico and Schmitt,* 1997).

As was the case for MVIC, LEISA images are obtained by scanning its field of view across the target with the frame transfer rate synchronized with the spacecraft scan rate. The LVF is oriented so that wavelength varies along the scan direction. Thus, scanning LEISA over a target produces images at different wavelengths, unlike the case for MVIC where the scanning simply increases the signal 32-fold (i.e., by the number of rows in the TDI array). LEISA builds up a conventional spatial-spectral data cube (256 monochromatic images) by scanning the FOV across all portions of the target at a nominal scan rate of 120 μrad s^{-1} (24" s^{-1}). A nominal framing rate of 2 Hz, corresponding to a single-frame exposure time of 0.5 s, will be used to maintain less than 1 pixel attitude smear and provide good signal-to-noise ratio during observations in the Pluto system. Further details on LEISA and its performance can be found in *Reuter et al.* (2008).

4.3. Radio Science Experiment (REX)

REX is unique among the NH instruments in that it is physically and functionally incorporated within the spacecraft telecommunications subsystem. Because this subsystem is entirely redundant, so is REX. The two copies of REX can be used simultaneously to increase the SNR, as well as provide dual polarization capability.

NASA's Deep Space Network (DSN) transmits radio signals to NH, which are received by REX via the 2.1-m High Gain Antenna (HGA). The carrier frequency signal at 7.182 GHz is down-converted to an intermediate frequency (IF) by an ultrastable oscillator (USO), which operates at 30 MHz. During a radio occultation event, REX measures phase delays in the received signal as a function of time, which can be inverted into a temperature, number density profile of the intervening atmosphere. REX can also operate in a passive radiometry mode to measure radio brightness temperatures at its carrier frequency. Further details on REX and its performance can be found in *Tyler et al.* (2008).

REX addresses the Group 1 scientific objective of obtaining Pluto's atmospheric temperature and pressure profiles down to the surface using the unique uplink radio occultation technique described above. REX will also address Group 2 and Group 3 scientific objectives by probing Pluto's ionospheric density, searching for Charon's atmosphere, refining bulk parameters like mass and radius, and measuring the surface emission brightness at a wavelength of 4.2 cm, which permits the determination of both the dayside and nightside brightness temperatures with an angular resolution of ~1.2° (full-width between the 3 dB points).

4.4. Long Range Reconnaissance Imager (LORRI)

The Long Range Reconnaissance Imager (LORRI) is a narrow-angle, high-resolution, panchromatic imaging system that was added to NH to augment and provide redundancy for Ralph's panchromatic imaging capabilities. LORRI's input aperture is 20.8 cm in diameter, making it one of the largest telescopes flown on an interplanetary spacecraft. The large aperture, in combination with a high throughput ($QE_{peak} \approx 60\%$) and wide bandpass, will allow LORRI to achieve SNR > 100 during disk-resolved observations of Pluto, even though exposure times must be kept below 100 ms to prevent smearing from pointing drift. A frame transfer 1024 × 1024 pixel (optically active region), thinned, backside-illuminated charge-coupled device (CCD) detector records the image in the telescope focal plane. LORRI image exposure times can be varied from 0 ms to 29,967 ms in 1-ms steps, and images can be accumulated at a maximum rate of 1 image per second. LORRI's large dynamic range allowed it to be an imaging workhorse during the Jupiter encounter, when saturation limited MVIC observations to relatively large solar phase angles. LORRI operates in an extreme thermal environment, mounted inside the warm spacecraft and viewing cold space, but the telescope's monolithic, silicon carbide construction allows the focus to be maintained over a large temperature range (–120°C to 50°C) without any focus adjustment mechanisms. Indeed, LORRI has no moving parts, making it a relatively simple, reliable instrument that is easy to operate.

LORRI is also serving as the prime optical navigation instrument on NH. During a typical 100 ms exposure using the full format (1024 × 1024) mode, LORRI can achieve SNR ≈ 5 on V = 13 stars. On-chip 4 × 4 binning, used in conjunction with a special pointing control mode that permits exposing up to 10 s while keeping the target within a single rebinned pixel, allows imaging of point sources as faint as V ≈ 18, which will permit LORRI to detect a 50-km-diameter KBO approximately seven weeks prior to encounter, thereby enabling accurate targeting to the KBO. Further details on LORRI and its performance can be found in *Cheng et al.* (2008).

LORRI first detected Pluto on September 21, 2006, at a distance of 28 AU. LORRI's resolution at Pluto will start exceeding that available from the Hubble Space Telescope approximately three months prior to closest approach. Enroute to Pluto, LORRI will obtain rotationally resolved phase curves of Pluto and later Charon, once the two can be separately resolved. LORRI will obtain panchromatic maps over at least 10 Pluto rotations during approach, with the final complete map of the sunlit hemisphere exceeding a resolution of 0.5 km pixel^{-1}. LORRI will map small regions near Pluto's terminator with a resolution of ~50 m pixel^{-1}, depending on the actual closest approach distance. LORRI will also be heavily used for studies requiring high geometrical fidelity, such as determining the shapes of Pluto, Charon, Nix, and Hydra and refining the orbits of all these objects relative to the system barycenter. LORRI observations at high phase angles will provide a sensitive search for any particulate hazes in Pluto's atmosphere.

4.5. Solar Wind Around Pluto (SWAP)

The Solar Wind Around Pluto (SWAP) instrument is one of two particle detection *in situ* instruments onboard NH. SWAP is comprised of a retarding potential analyzer (RPA), a deflector (DFL), and an electrostatic analyzer (ESA). Collectively, these elements are used to select the angles and energies of solar wind ions entering the instrument. The selected ions are directed through a thin foil into a coincidence detection system: The ions themselves are detected by one channel electron multiplier (CEM), and secondary electrons produced from the foil are detected by another CEM. SWAP can measure solar wind particles in the energy range from 25 eV up to 7.5 keV with a resolution of $\delta E/E < 0.4$. SWAP has a fan-shaped field of view that extends ~200° in the XY-plane of the spacecraft by ~10° out of that plane. Further details on SWAP and its performance can be found in *McComas et al.* (2008).

SWAP was designed to measure the interaction of the solar wind with Pluto, which addresses the Group 1 scientific objective of measuring Pluto's atmospheric escape rate. Additionally, SWAP has a specific goal of characterizing the solar wind interaction with Pluto as a Group 2 objective. SWAP also addresses the Group 3 objectives of characterizing the energetic particle environment of Pluto and searching for magnetic fields, which it does indirectly.

4.6. Pluto Energetic Particle Spectrometer Science Investigation (PEPSSI)

The Pluto Energetic Particle Spectrometer Science Investigation (PEPSSI) is the other *in situ* particle instrument onboard NH, and it provides measurements of both the energy and the composition of the plasma near the spacecraft. The PEPSSI design is derived from that of the Energetic Particle Spectrometer (EPS), which is flying on the MESSENGER mission to Mercury. PEPSSI has thinner foils than EPS, which enables measurements down to smaller energy ranges. PEPSSI also has a slightly increased geometric factor and draws less power than EPS. Both EPS and PEPSSI trace their heritage to a NASA PIDDP program in the 1990s to develop a particle instrument for use on a Pluto flyby mission.

PEPSSI is a compact, radiation-hardened instrument comprised of a time-of-flight (TOF) section feeding a solid-state silicon detector (SSD) array. Each SSD has four pixels, two dedicated to ions, and two for electrons. PEPSSI's field of view (FOV) is fan-like and measures 160° × 12°, divided into six angular sectors of 25° × 12° each. Ions entering the PEPSSI FOV generate secondary electrons as they pass through entrance and exit foils in the TOF section, providing "start" and "stop" signals detected by a microchannel plate (MCP). Particle energy information, measured by the SSD, is combined with TOF information to identify the particle's composition. Each particle's direction is determined by the particular 25° sector in which it is detected. Event classification electronics determine incident mass and energy, with 12 channels of energy resolution. Protons can be detected in the energy range 40–1000 keV, electrons in the range 25–500 keV, and CNO ions in the range 150–1000 keV. TOF-only measurements extend to <1 keV for protons, to 15 eV for CNO ions, and to 30 keV for N_2^+. TOF measurements are possible in the range 1–250 ns to an accuracy of ±1 ns. The mass resolution of PEPSSI varies with energy: For CNO ions, it is <5 amu for >1.7 keV nucleon^{-1}, and <2 amu for >5 keV nucleon^{-1}. Further details on PEPSSI and its performance can be found in *McNutt et al.* (2008).

By measuring energetic pickup ions from Pluto's atmosphere, PEPSSI provides information related to the atmospheric escape rate on Pluto, which is a Group 1 scientific objective. PEPSSI's primary role, however, is to address the Group 3 objective of characterizing the energetic particle environment in the Pluto system. Fluxes of energetic pickup ions may be measured as far as several million kilometers from Pluto (see *Bagenal et al.,* 1997), and PEPSSI observations will be used to determine the mass, energy spectra, and directional distributions of these energetic particles (*Bagenal and McNutt,* 1989). Secondarily, PEPSSI will also provide low resolution, supporting measurements of the solar wind flux, complementing SWAP.

4.7. Venetia Burney Student Dust Counter (VB-SDC)

The Venetia Burney Student Dust Counter (SDC), which was named in 2006 in honor of the student who named Pluto in 1930, is an impact dust detector that will be used to map the spatial and size distribution of interplanetary dust along the trajectory of the NH spacecraft from the inner solar system to and through the Kuiper belt. Unlike all the other instruments, the VB-SDC was not part of the original NH proposal and was added by NASA as an Education and Public Outreach (EPO) experiment. For the first time ever, students were given the opportunity to design, build, and operate an instrument for an interplanetary mission, under the supervision of NASA-certified personnel. Approximately 20 undergraduate physics and engineering students at the University of Colorado worked on the VB-SDC, which was the first instrument delivered to the NH spacecraft.

The VB-SDCs sensors are thin, permanently polarized polyvinylidene fluoride (PVDF) plastic films that generate an electrical signal when dust particles penetrate their surface. The VB-SDC has a total sensitive surface area of ~0.1 m^2, comprising 12 separate film patches, each 14.2 cm × 6.5 cm, mounted onto the top surface of a support panel. In addition, there are two reference sensor patches mounted on the backside of the detector support panel, protected from any dust impacts. These reference sensors, identical to the top surface sensors, are used to monitor the various background noise levels, from mechanical vibrations or cosmic ray hits. The entire support panel is mounted on the exterior of the NH spacecraft, outside the spacecraft multilayer insulating (MLI) blanket, facing the ram (–Y) direction. The VB-SDC observations are most useful during the cruise phases of the mission, when the spacecraft is spinning and the other instruments are turned off. Thruster firings during three-axis operations generate large VB-SDC background signals, which make it very difficult to detect true interplanetary dust particle (IDP) impacts, thus ruling out dust measurements during the Pluto encounter itself. The VB-SDC was designed to resolve, to within a factor of ~2, the masses of IDPs in the range of $10^{-12} < m < 10^{-9}$ g, which corresponds roughly to a size range of 1–10 μm in particle radius. Bigger grains are also recorded, but their masses cannot be resolved. With the characteristic spacecraft speed during cruise of ~13 km s^{-1}, current models of the dust density in the solar system (*Divine,* 1993) suggest that the VB-SDC should record approximately one IDP hit per week. Further details on VB-SDC and its performance can be found in *Horanyi et al.* (2008).

5. SPACECRAFT SUBSYSTEMS AND MISSION OPERATIONS

Details on the NH spacecraft and its subsystems are discussed by *Fountain et al.* (2008). Here we provide a high-level summary of the most important subsystem components.

The NH spacecraft mechanical design is similar to that of the Ulysses spacecraft. The principal moment of inertia, which is the axis about which the spacecraft spins during much of the mission, is aligned with the antenna boresight. The instruments and propulsion system thrusters are placed around the main structure so that they do not interfere with each other. The RTG is cantilevered off the main structure to place it as far as possible from the electronics and instruments, thereby minimizing their exposure to radiation.

The power for the NH spacecraft is supplied by a General Purpose Heat Source Radioisotope Thermoelectric Generator (GPHS-RTG). The "F8" GPHS-RTG supplied by the U.S. Department of Energy is the latest in a series of RTGs of the same design supplied for NASA missions since the late 1980s. The unit converts heat generated by the radioactive decay of 72 plutonium dioxide pellets into electricity using silicon-germanium (SiGe) thermocouples. The RTG provided 244 W of power at launch, has performed flawlessly in flight, and is expected to deliver ~199 W during the Pluto encounter in 2015; at least 180 W is needed to accomplish the mission's scientific objectives.

Only a few elements of the NH subsystems do not have redundancy (e.g., the RTG, the propulsion system tank, the radio subsystem hybrid coupler, and the high gain antenna), but in all cases those components have a robust design and a history of failure-free service over time periods longer than the expected NH mission duration.

The average temperature of the spacecraft structure was slightly under 50°C during early operations with the lower deck facing the Sun at 1 AU and will be maintained close to room temperature (~20°–30°C) for most of the mission; the spacecraft structure must always be somewhat above 0°C to ensure that the hydrazine propellant does not freeze. The propulsion system components are thermally tied to the spacecraft bus and are kept warm through thermal contact with the structure.

The NH spacecraft spends much of its time spinning at ~5 rpm around the antenna boresight. In this mode, useful data can be obtained by REX, SWAP, PEPSSI, and the VBSDC, but typically not by any of the other instruments. For virtually all observations made by the imaging instruments, three-axis pointing control mode is required. In three-axis mode, the spacecraft can be slewed to a targeted location to an accuracy of ±1024 μrad (±200", 3σ) and controlled to that location within a typical "deadband" of ±500 μrad (±100"). For some Alice observations, when the target must be kept near the center of its narrow slit, the deadband can be reduced to ±250 μrad (±50"). The drift rate is controlled to within ±34 μrad s^{-1} (±7" s^{-1}, 3σ) for both fixed and scanning observations. The post-processing knowledge of the attitude and drift rate derived from the star tracker and gyro data are ±350 μrad (±70", 3σ) and ±7.5 μrad s^{-1} (±1.5" s^{-1}, 3σ), respectively. Ralph observations usually require the spacecraft to scan about its Z axis, which can be performed accurately enough to keep smearing below 1 pixel for both the MVIC or LEISA images. Further details about the NH guidance and control system can be found in *Rogers et al.* (2006).

The NH Mission Operations Center (MOC) is located at JHU/APL, which also runs several other space missions. NH mission operations is divided into several phases: launch and early operations, commissioning, cruise, and encounter.

As soon as communications with NH were established ~45 min after launch, the MOC took over complete control of the spacecraft. The first month of operations was primarily devoted to checking the performance of the spacecraft subsystems. By late February 2006, the health and safety of the spacecraft had been verified, and the operations team had demonstrated that the observatory was ready for business.

Instrument commissioning proceeded systematically during the following months, during which the performance of each instrument was investigated in detail and calibration observations obtained. The "first light" observations, when an astronomical or interplanetary target is viewed for the first time, were staggered throughout the May to September 2006 period. By mid-2007, only a handful of commissioning activities remained, all of which will be executed during the late summer or early fall of 2007.

NH is a long-duration mission, with the primary objective not being completed until nearly 10 years after launch. For most of the eight years between the Jupiter and Pluto encounter phases (2007–2014, inclusive), the spacecraft will be placed into "hibernation" mode, with all nonessential subsystems, including the science payload, powered off to preserve the life of the components. During the hibernation period, beacon radio tones are sent periodically from the spacecraft to Earth that allow flight controllers to verify the basic health and safety of the spacecraft. Additionally, monthly telemetry passes are scheduled to collect engineering trending data.

Although the spacecraft is kept in hibernation to reduce component use prior to the Pluto encounter, it is important to verify periodically the performance of the spacecraft subsystems and instruments and to keep the mission operations team well trained and prepared for the Pluto encounter activities. Therefore, the spacecraft will be brought out of hibernation each year for roughly 60 days, called "annual checkouts" (ACOs), during which the performance of the spacecraft subsystems and instruments can be verified. ACOs also provide an opportunity for *cruise science* data to be collected, such as interplanetary charged particle measurements, studies of the hydrogen distribution in the interplanetary medium, and extensive phase curve studies of Pluto, Charon, Uranus, Neptune, Centaurs, and KBOs, none of which can be obtained from spacecraft near Earth.

Three full rehearsals of the Pluto encounter are being planned (during January 2009, July 2014, and January 2015) that will serve both to verify that the Pluto encounter sequence will work and to provide essential training for the mission operations team in preparation for the actual encounter.

6. SCIENCE OPERATIONS AND DATA ARCHIVING

The Science Operations Center (SOC) for NH is located at the SwRI office in Boulder, Colorado. The SOC is responsible for designing and implementing the science activity plans (SAPs) and for processing and archiving all science data.

The so-called "low-speed" data (e.g., all data from the particle instruments and engineering "housekeeping" data from the other instruments) are packetized by a telemetry processor and then downlinked to the NASA Deep Space Network (DSN). The "high-speed" data (e.g., all science data from the remote sensing instruments) are normally losslessly compressed (reducing the data volume by at least a factor of 2) before telemetry processing and downlinking. Lossy compression of high-speed data is also possible, typically reducing the data volume by a factor of ~5–10 with only moderate loss of information. During the Pluto encounter, lossy compressed data will be downlinked for a critical subset of observations to provide relatively quick feedback on the quality of the data and to mitigate any potential problems that develop during the approximately six-month period following closest approach when the losslessly compressed data will be downlinked. There is also the capability to select portions of the high-speed data ("windows") for downlink, which should be useful in cases when much of the image is not expected to be useful (e.g., when observing sparse star fields, or when the target only fills a small fraction of the field of view). The SOC combines the low-speed and high-speed (when relevant) data for each instrument, along with information describing the spacecraft's trajectory and orientation ("SPICE" files), into "Level 1" FITS files, with one Level 1 FITS file for each observation.

The Level 1 data files are essentially raw data and contain instrumental signatures that need to be removed before scientific analysis can proceed. A *calibration pipeline* has been developed for each instrument that accepts the Level 1 file as input and processes those data into a "Level 2" file that has instrumental signatures removed and allows conversion of the data from engineering units into physical (e.g., scientific) units. The Level 2 files also provide error estimates and quality flags marking data that may be compromised (e.g., "hot," "dead," or saturated pixels).

All Level 1 and Level 2 files produced by the NH mission will be archived at the Small Bodies Node (SBN) of NASA's Planetary Data System (PDS), where they will be publicly accessible. There is no proprietary period for NH data. However, there will typically be an approximately nine-month lag between the production of Level 1 files on the ground and the archiving of Level 1 and Level 2 files at the SBN, so that the NH Project can verify that the Level 2 files have been properly calibrated and that the data formats meet the PDS requirements.

NASA has funded the NH Science Team to reduce and analyze the data returned by NH, and to publish the scientific results in the refereed literature. In addition, NASA expects to fund other scientific investigators who want to make use of NH data through special data analysis programs (DAPs), which will be announced through a call for proposals in the relevant year. The NH Jupiter DAP is expected to be announced in 2007 with proposals due in February 2008 and selections made in May–June 2008. A similar DAP is expected for the data returned from the Pluto portion of the mission, and from the KBO portion should an extended mission be approved by NASA.

7. MISSION STATUS

As of mid-2007, the NH spacecraft is heading to Pluto on its nominal trajectory having completed a successful launch and 1.5 years of operations, including encounters with an asteroid and Jupiter. Inflight performance tests have verified that the science payload meets its measurement objectives and, thus, can fulfill all the scientific objectives of the mission. Assuming that no serious problems develop during the remaining eight-year cruise to Pluto, we can all look forward in the summer of 2015 to a wealth of exciting scientific results on a new class of solar system objects, the dwarf planets, which have exotic volatiles on their surfaces, escaping atmospheres, and giant-impact-derived satellite systems. And if the Pluto encounter is successful, we can hope to extend the *in situ* exploration deeper into the transneptunian region and conduct the first reconnaissance of a small KBO beyond the orbit of Pluto. The path to the Kuiper belt is a long and difficult one, but the knowledge unveiled will certainly justify our patience and our resolve to explore this primitive region of the solar system.

Acknowledgments. This chapter is dedicated to the pioneering work of Kenneth Edgeworth, Gerard Kuiper, and Clyde Tombaugh, who paved the way for the New Horizons mission. We thank all the New Horizons team members, and numerous contractors, for their extraordinary efforts in designing, developing, testing, and delivering a highly capable spacecraft that promises to revolutionize our understanding of the Pluto system and the Kuiper belt. Partial financial support for this work was provided by NASA contract NAS5-97271 to the Johns Hopkins University Applied Physics Laboratory.

REFERENCES

Bagenal F. and McNutt R. L. Jr. (1989) Pluto's interaction with the solar wind. *Geophys. Res. Lett., 16,* 1229–1232.

Bagenal F., Cravens T. E., Luhmann J. G., McNutt R. L. Jr., and Cheng A. F. (1997) Pluto's interaction with the solar wind. In *Pluto and Charon* (S. A. Stern and D. J. Tholen, eds.), pp. 523–555. Univ. of Arizona, Tucson.

Belton M. J., et al. (2002) *New Frontiers in the Solar System. An Integrated Exploration Strategy.* National Research Council.

Cheng A. F., Weaver H. A., Conard S. J., Morgan M. F., Barnouin-Jha O., et al. (2008) Long Range Reconnaissance Imager on New Horizons. *Space Sci. Rev.,* in press.

Divine N. (1993) Five populations of interplanetary meteoroids. *J. Geophys. Res., 98,* 17029–17051.

Farquhar R. and Stern S. A. (1990) Pushing back the frontier — A mission to the Pluto-Charon system. *Planetary Report, 10,* 18–23.

Fountain G. F., Kusnierkiewicz D. Y., Hersman C. B., Herder T. S., Coughlin T. B., et al. (2008) The New Horizons Spacecraft. *Space Sci. Rev.,* in press.

Grundy W. M., Schmitt B., and Quirico E. (1993) The temperature dependent spectra of alpha and beta nitrogen ice with application to Triton. *Icarus, 105,* 254–258.

Grundy W. M., Buie M. W., Stansberry J. A., Spencer J. R., and Schmitt B. (1999) Near-infrared spectra of icy outer solar system surfaces: Remote determination of H_2O ice temperatures. *Icarus, 142,* 536–549.

Guo Y. and Farquhar R. (2008) New Horizons mission design. *Space Sci. Rev.,* in press.

Hansen C. H. and Paige D. A. (1996) Seasonal nitrogen cycles on Pluto. *Icarus, 120,* 247–265.

Horanyi M., Hoxie V., James D., Poppe A., Bryant C., et al. (2008) The Student Dust Counter on the New Horizons mission. *Space Sci. Rev.,* in press.

McComas D., Allegrini F., Bagenal F., Casey P., Delamere P., et al. (2008) The Solar Wind Around Pluto (SWAP) instrument aboard New Horizons. *Space Sci. Rev.,* in press.

McNutt R. E., Livi S. A., Gurnee R. S., Hill M. E., Cooper K. A., et al. (2008) The Pluto Energetic Particle Spectrometer Science Investigation (PEPSSI) on New Horizons. *Space Sci. Rev.,* in press.

Olkin C. B., Reuter D., Lunsford A., Binzel R. P., and Stern S. A. (2006) The New Horizons distant flyby of asteroid 2002 JF56. *Bull. Am. Astron. Soc., 38,* abstract #9.22.

Quirico E. and Schmitt B. (1997) A spectroscopic study of CO diluted in N_2 ice: Applications for Triton and Pluto. *Icarus, 128,* 181–188.

Quirico E., Doute S., Schmitt B., de Bergh C., Cruikshank D. P., Owen T. C., Geballe T. R., and Roush T. L. (1999) Composition, physical state, and distribution of ices at the surface of Triton. *Icarus, 139,* 159–178.

Reuter D., Stern S. A., Scherrer J., Jennings D. E., Baer J., et al. (2008) Ralph: A visible/infrared imager for the New Horizons mission. *Space Sci. Rev.,* in press.

Rogers G. D., Schwinger M. R., Kaidy J. T., Strikwerda T. E., Casini R., Landi A., Bettarini R., and Lorenzini S. (2006) Autonomous star tracker performance. In *Proc. 57th IAC Congress,* Valenica, Spain.

Spencer J., Buie M., Young L., Guo Y., and Stern A. (2003) Finding KBO flyby targets for New Horizons. *Earth Moon Planets, 92,* 483–491.

Stern S. A. (2002) Journey to the farthest planet. *Sci. Am., 286,* 56–59.

Stern S. A. (2008) The New Horizons Pluto Kuiper belt mission: An overview with historical context. *Space Sci. Rev.,* in press.

Stern S. A. and Cheng A. F. (2002) NASA plans Pluto-Kuiper belt mission. *Eos Trans. AGU, 83,* 101–106.

Stern S. A. and Mitton J. (2005) *Pluto and Charon: Ice Worlds on the Ragged Edge of the Solar System.* Wiley-VCH, New York.

Stern S. A., et al. (1995) The Highly Integrated Pluto Payload System (HIPPS): A sciencecraft instrument for the Pluto mission. In *EUV, X-Ray, and Gamma-Ray Instrumentation for Astronomy VI* (O. H. W. Siegmund and J. Vallerga, eds.), pp. 39–58. Proc. SPIE, Vol. 2518.

Stern S. A., Weaver H. A., Steffl A. J., Mutchler M. J., Merline W. J., Buie M. W., Young E. F., Young L. A., and Spencer J. R. (2006) A giant impact origin for Plutos small moons and satellite multiplicity in the Kuiper belt. *Nature, 439,* 946–948.

Stern S. A., Slater D. C., Scherrer, Stone J., Dirks G., et al. (2008) Alice: The ultraviolet imaging spectrometer aboard the New Horizons Pluto-Kuiper belt mission. *Space Sci. Rev.,* in press.

Terrile R. J., Stern S. A., Staehle R. L., Brewster S. C., Carraway J. B. Henry P. K., Price H., and Weinstein S. S. (1997) Spacecraft missions to the Pluto and Charon system. In *Pluto and Charon* (S. A. Stern and D. J. Tholen, eds.), pp. 103–124. Univ. of Arizona, Tucson.

Tubiana C., Duffard R., Barrera L., and Boehnhardt H. (2007) Photometric and spectroscopic observations of (132524) 2002 JF56: Fly-by target of the New Horizons mission. *Astron. Astrophys., 463,* 1197–1199.

Tyler G. L., Linscott I. R., Bird M. K., Hinson D. P., Strobel D. F., Pätzold M, Summers M. E., and Sivaramakrishnan K. (2008) The New Horizons Radio Science Experiment (REX). *Space Sci. Rev.,* in press.

Weaver H. A., Gibson W. C., Tapley M. B., Young L. A., and Stern S. A. (2008) Overview of the New Horizons science payload. *Space Sci. Rev.,* in press (preprint available at *arxiv.org/abs/0709.4261*).

Young L. A., Stern S. A., Weaver H. A., Bagenal F., Binzel R. P., et al. (2008) New Horizons: Anticipated scientific investigations at the Pluto system. *Space Sci. Rev.,* in press.

Future Surveys of the Kuiper Belt

Chadwick A. Trujillo
Gemini North Observatory

The next decade of Kuiper belt object (KBO) science will be completely dominated by the output of two surveys. These two surveys, the Panoramic Survey Telescope and Rapid Response System (Pan-STARRS) and the Large Synoptic Survey Telescope (LSST), will increase the total number of known KBOs by factors of 25 to 150 over the next decade. This discovery rate will not be uniform — it will come as a large flood of information during the first year of operation. If software development allows, additional depth may be gained in successive years if data can be combined across years to find bodies that are too faint to be detected in an individual visit. Not only will the surveys increase the number of objects dramatically, but they will also be sensitive to heliocentric distances far beyond 100 AU due to their multiyear survey methodology. In this work, we outline a basic timeline of operation of these two powerful surveys and the science that will be enabled by their data output, as well as other significant advances in KBO surveys expected in the next decade.

1. INTRODUCTION

The coming decade should see a dramatic increase in our knowledge of the Kuiper belt object (KBO) population due entirely to new telescopic survey experiments. This advance in our knowledge of the Kuiper belt population statistics is hard to underestimate, as even the most conservative estimates suggest increases of factors of 25 in the total number of known KBOs as well as the possibility of finding bodies in solar orbit well beyond 100 AU.

Methods employed in Kuiper belt surveys can be very crudely divided into two categories: (1) all-sky shallow surveys and (2) targeted deep surveys. In each of these categories, past surveys have made a large scientific impact. It is difficult to quantify the impact of a survey, but by examining refereed publications and citations directly related to a given survey, one can get an indication of how important these two types of surveys are to the larger scientific community. No survey has covered the entire sky in the search for solar system bodies, but the closest modern digital survey (cf. *Tombaugh,* 1946) is the Caltech wide-area survey, which has covered most of the sky available from Palomar Mountain (roughly 25,000 deg^2) to about 21st mag in a custom filter encompassing approximately the ri bandpasses, roughly equivalent to $m_R \sim 21.5$ (*Trujillo and Brown,* 2003). This single survey has discovered many candidate dwarf planets, including (50000) Quaoar, (55565) 2002 AW_{197}, (90377) Sedna, (90482) Orcus, (136108) 2003 EL_{61}, (136199) Eris, and (136472) 2005 FY_9, which have led to over 20 refereed publications by a variety of authors, most of which are about the physical surfaces of these relatively bright bodies (*Hughes,* 2003; *Marchi et al.,* 2003; *Trujillo and Brown,* 2003; *Brown and Trujillo,* 2004; *Morbidelli and Levison,* 2004; *Stevenson,* 2004; *Jewitt and Luu,* 2004; *Wickramasinghe et al.,* 2004; *Barucci et al.,* 2005; *de Bergh et al.,* 2005; *Cruikshank et al.,* 2005; *Stern,* 2005; *Trujillo et al.,* 2005; *Gaudi et al.,* 2005; *Brown et al.,* 2005; *Matese et al.,* 2005; *Doressoundiram et al.,* 2005; *Licandro et al.,* 2006; *Rabinowitz et al.,* 2006; *Barkume et al.,* 2006; *Brown et al.,* 2006; *Bertoldi et al.,* 2006; *Trujillo et al.,* 2007). These 20 papers have about 60 citations combined, which is likely an underestimate of the impact of the survey as all of the publications are less than a few years old. The targeted deep surveys are responsible for most of the dynamical information known about the Kuiper belt population. The impact of these surveys is also difficult to estimate; however, the 12 published surveys reviewed in the chapter by Petit et al. have a combined citation count of about 400 (*Irwin et al.,* 1995; *Jewitt et al.,* 1998; *Gladman et al.,* 1998; *Chiang and Brown,* 1999; *Sheppard et al.,* 2000; *Larsen et al.,* 2001; *Trujillo et al.,* 2001a,b; *Gladman et al.,* 2001; *Bernstein et al.,* 2004; *Elliot et al.,* 2005; *Petit et al.,* 2006). At roughly 30 citations per paper, this is well above the astronomy average of 4 to 10 citations per paper, depending on the journal (*Henneken et al.,* 2006). Even more of an example of the scientific utility of survey work is the pivotal work of *Jewitt and Luu* (1993), leading to the discovery of the first KBO, (15760) 1992 QB_1, a single paper that has been cited more than 160 times. Thus, the impact of both all-sky surveys as well as targeted surveys has been very high for the Kuiper belt. The next decade will see the most significant impact from all-sky surveys, although some additional impact can be made in targeted deep surveys.

The two largest survey projects of the next decade will be the Panoramic Survey Telescope and Rapid Response System (Pan-STARRS) and the Large Synoptic Survey Telescope (LSST). Both Pan-STARRS and LSST will cover the whole sky to red magnitude $m_R = 24$ or better, at least 10 times fainter than the current state of the art in all-sky

TABLE 1. Current and future survey power.

Instrument	Site	D (m)	Ω (deg²)	θ (arcsec)	SP*	All Sky?[†]	Science Start
LSST	Cerro Pachón	6.7[‡]	10.0	0.7	720	yes	2014
Pan-STARRS PS4	Mauna Kea	3.2[§]	7.0	0.5	225	yes	2010
VISTA (Visible)	Cerro Paranal	4.0	3.0	0.6	105	yes	2010?[¶]
Pan-STARRS PS1	Haleakalā	1.6[§]	7.0	0.6	39	yes	2008
DCT	Happy Jack, AZ	4.2	2.0	0.8	43	no?[**]	2009
VLT Survey Tel.	Cerro Paranal	2.6	1.0	0.6	15	no[††]	2008?
Caltech QUEST	Palomar	1.2	10.0	2.0	3	yes	2001
SkyMapper	Siding Spring	1.3	8.0	2.0	3	yes	2008?
Subaru Suprime	Mauna Kea	8.0	0.3	0.5	51	no	2000
CFHT Megacam	Mauna Kea	3.6	1.0	0.7	21	no	2004
Magellan IMACS	Las Campanas	6.5	0.2	0.6	18	no	2003
MMT Megacam	Mt. Hopkins	6.5	0.2	0.8	8	no	2004
UH 2.2 m 8 k	Mauna Kea	2.2	0.3	0.7	2	no	1995

Current and future surveys divided into dedicated survey telescopes (top) and other instruments (bottom). The table is ordered by survey power (SP) in each of the sections. Future instruments for nonsurvey telescopes were omitted, as were surveys with SP <1.

[*] Defined in equation (1), units are $m^2 deg^2 arcsec^{-2}$.

[†] Given reasonable assumptions, can the telescope survey the entire sky for KBOs? Telescopes with large fractions of other commitments are marked "no."

[‡] The LSST will have an unobstructed aperture similar to a 6.7-m-diameter mirror, although the actual outer mirror diameter will be 8.4 m.

[§] The unobstructed aperture of each of the Pan-STARRS telescopes is equivalent to 1.6 m, although the outer diameter is 1.8 m.

[¶] Note that VISTA was recently shipped with an infrared camera, which is low efficiency for KBO discovery due to limited field size and high telluric background, but may have a visible camera added at an unspecified future date.

[**] It is currently unknown if the Discovery Channel Telescope (DCT) will operate in survey mode.

[††] The VLT Survey Telescope will likely cover only a small amount (~300 deg²) of the ecliptic in a mode conducive to KBO detection due to the high demand for the telescope.

surveys, the Caltech survey (*Trujillo and Brown,* 2003). In terms of raw numbers, we should expect the total number of KBOs to increase dramatically, from the current small fraction (~1%) of all KBOs larger than 100 km in diameter cataloged to *all KBOs* larger than 100 km cataloged. In addition, the deepest surveys will be able to image ~10 times the area of the current deepest surveys, allowing the detection of significant number of the smallest (~25 km diameter) KBOs. This chapter will deal only with the impact of the various surveys on the Kuiper belt in the next decade. Each of these surveys has many other programs of research beyond the scope of this work, from the most distant bodies in the universe to the near-Earth asteroid population. For more detailed accounts of such survey products, consult the websites for Pan-STARRS (*pan-starrs.ifa.hawaii.edu*) and LSST (*www.lsst.org*).

Targeted deep surveys are also a possibility and require a large aperture and large field, as they study a small patch of sky to very faint limits by imaging the same field many times. There are likely to be several very deep surveys conducted by independent researchers using the largest telescopes in search of the very faintest KBOs, including the use of Pan-STARRS and LSST themselves, both of which have deep surveys on selected sky regions planned. However, most of these deep sky regions will not be much deeper than surveys that have already been conducted, although they will cover significantly more area. The targeted deep surveys are discussed in more detail in section 2.5.

2. OVERVIEW OF SURVEYS

We rank the relative strengths of existing and future surveys in Table 1, which outlines the basic parameters of various facilities. Etendue is the traditional method of measuring a survey's strength, consisting of mirror area [approximated as $\pi(D/2)^2$, where D is telescope diameter] multiplied by the field of view Ω. Although widely used, etendue tells only part of the story. Image quality is critical to any survey, so site selection is very important. Thus the median seeing θ of a survey's site must also be included in estimates of survey power. Total survey power (SP) can be estimated by the formula

$$SP = \pi\left(\frac{D/2}{m}\right)^2\left(\frac{\Omega}{deg^2}\right)\left(\frac{arcsec}{\theta}\right)^2 \quad (1)$$

following the definition in *Jewitt* (2003). Using either the etendue or SP calculation, one finds that the two main surveys of interest in coming years are Pan-STARRS and LSST. In this section, we briefly introduce some of the major survey products, with a more detailed look at both Pan-STARRS and LSST in the next section.

It must be noted that the survey power only estimates the ability of a survey to discover KBOs on an instrumental level. An even more important component is the fraction of time a survey will spend looking for KBOs. A comparison of operational models is very difficult to make for future surveys as much of the instrumental capabilities are still in flux. Nonetheless, all the dedicated surveys described in Table 1 have proposed placing significant resources into survey methodology conducive to KBO discovery. In most, KBO discovery is a cornerstone of the proposed survey products. Thus, factors of a few differences in SP can be overcome by adjusting fractions of time allocated to KBO surveys, but differences beyond this are unlikely to occur.

2.1. Panoramic Survey Telescope and Rapid Response System (Pan-STARRS)

The final Pan-STARRS project configuration (called PS4) will consist of four 1.8-m-diameter telescopes either in a common mount or in four independent mounts. Each telescope will have a 7-deg^2 field of view with a camera system independent of the other telescopes. The major innovations of Pan-STARRS over current survey telescopes are (1) cost reduction by combining four smaller telescopes into a single system; (2) the use of Orthogonal Transfer CCDs (OTCCDs), which allow for solid state tip/tilt image compensation across the focal plane; (3) site selection to take advantage of the excellent seeing on Mauna Kea, Hawai'i, or Haleakalā, Maui; and (4) use of a wide gri filter for solar system targets. The gri filter allows a gain of about 0.4 mag over a simple r or R filter as it covers roughly three times the bandwidth with a ~ $\sqrt{4}$ increase in background noise. The final system will thus have the equivalent of a magnitude depth of $m_R = 24.0$ mag (5σ detection) in about 60 s. The performance of the OTCCDs is quite critical to the system's effectiveness, as the power SP of any survey is inversely proportional to the square of the seeing. On-sky performance for the entire array of OTCCDs is not currently available, but if they performed as expected, it seems that 0.5 arcsec median seeing is a reasonable performance for purely tip/tilt correction on Mauna Kea. The plate scale of Pan-STARRS will be 0.3 arcsec per pixel, meaning that the field may be slightly undersampled in the best conditions, and adequately sampled in poor conditions.

The Pan-STARRS prototype telescope (PS1) achieved first light on Haleakalā, Maui, on June 30, 2006, and is currently funded under a grant from the U.S. Air Force. The telescope mount and enclosure are pictured in Fig. 1. The camera is still under development with the first OTCCDs being tested in the Pan-STARRS laboratory. The timescale for moving the project from the single telescope PS1 prototype to the final four-telescope system, PS4, is currently unclear; however, routine operations of PS1 are expected in 2008 and PS4 will likely be on the order of two years after this. For the sake of this article, we will assume that PS4 will be fully operational beginning in 2010.

The final PS4 system will be able to observe the entire visible sky from Hawai'i, about 30,000 deg^2 once per week, with the solar system survey focusing on the opposition area. Thus, in a single year, the PS4 system should be able to find all KBOs brighter than red magnitude m_R ~24 visible from Hawai'i in terms of single visit depth. The prototype PS1 system will be able to perform similarly, although with somewhat reduced depth. Final performance will greatly depend on the observing strategies employed, in particular how much time is allocated to the solar system survey compared to more distant science goals. The major source of construction funds were granted through the U.S. Air Force to find near-Earth asteroids (NEAs). However, science goals parallel to the NEA task such as KBO detection and other time-variable science products have been instrumental in granting funding to the project. Although NEAs have much larger rates of motion (~100 arcsec/h) than the KBOs (~3 arcsec/h), the two can be found simultaneously if the field of view is large enough. For example, a 1-h cadence between visits would easily allow KBO detection without sacrificing NEA detection capability for the PS1 7-deg^2 field of view. Thus, it is very likely that KBO detection will be one of the primary activities of PS1 and PS4.

Fig. 1. The Pan-STARRS PS1 prototype telescope located atop Haleakalā, Maui. The final Pan-STARRS PS4 will consist of four PS1 type telescopes, likely in a single enclosure and possibly in a single mount on either Haleakalā, Maui, or Mauna Kea, Hawai'i.

2.2. Large Synoptic Survey Telescope (LSST)

The Large Synoptic Survey Telescope (LSST) is in the planning stages and represents the deepest all-sky survey that will be undertaken in the next decade. One potential mirror and mount design is pictured in Fig. 2. The LSST

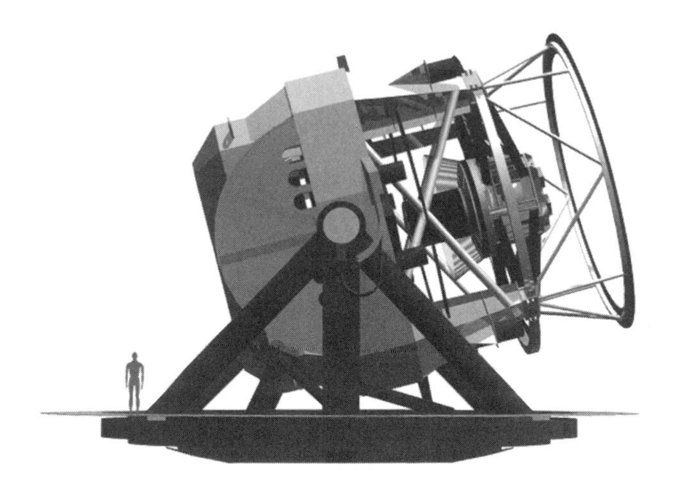

Fig. 2. The LSST design, a single 8.4-m telescope to be located at El Peñón, Cerro Pachón, Chile.

will consist of a 8.4-m-diameter telescope with a 10-deg^2 field of view. Due to the three-mirror design of the LSST, a 5-m-diameter section of the primary is obscured, yielding a clear aperture equivalent to a 6.7-m-diameter unobscured mirror. The main camera of interest to wide-field surveys will be a 3 gigapixel focal plane array. The largest technical obstacle to overcome in the design of LSST is image quality. Since the telescope is a very fast design, it has a very narrow depth of field, requiring strict tolerance on the placement of the arrays in the focal plane, to approximately 10 μm, or less than the size of a single detector pixel. The plate scale of the LSST will be 0.2 arcsec per 50 μm pixel, so therefore should be Nyquist-sampled in all conditions.

First light is scheduled for the end of 2012 with science operations beginning in 2014. The LSST will be located on Cerro Pachón (near the Gemini South and SOAR telescopes), at the El Peñón summit in Chile. The current plans for the LSST estimate that it will achieve a 5σ detection for a $m_R = 24.5$ AB magnitude star (equivalent to an $m_R = 24.25$ KBO for average KBO colors in the Vega magnitude system) when observing in the r band for 30 s per visit. Thus, while the single-visit depth of LSST is only slightly deeper than Pan-STARRS, the total number of visits available to LSST is a factor of ~3 larger, roughly the ratio between the two survey SP values. The final depth achieved by the survey will depend on the total number of visits to a single point, which is highly dependent on survey methodology, as discussed later.

2.3. The Case for Ultradeep Multiyear Detection Software

The software effort for finding moving objects is daunting for both Pan-STARRS and LSST. Each image will require 2–6 GB of disk space for the surveys. With one or two visits every minute, several terabytes of data are expected every operational night. Finding moving objects in this vast amount of data is a soluble problem only due to rapidly increasing computing power [i.e., Moore's law (from *Moore,* 1965)]. Each survey will achieve roughly similar single-visit depth, as the smaller Pan-STARRS will spend longer at each location and use a wider filter than the LSST. Thus, the total depth available to each survey will depend very critically on the ability to find moving objects beyond the single-visit depth. The software methods to detect objects beyond the single-visit depth by combining data collected over months or years are very computationally intensive, but are quite important if the full potential for each survey is to be realized.

Such an ultradeep multiyear map will certainly be produced for stationary objects as both Pan-STARRS and LSST consider this a primary science goal. However, producing a similar product for moving objects is many orders of magnitude more difficult than the original single-image identification process. Such methods have already been developed for groundbased and spaceborne KBO surveys with time bases of nights to months (*Gladman et al.,* 1998; *Bernstein et al.,* 2004). However, searching for objects over multiple years is a considerably more difficult proposition, as nonlinear orbital motion must be estimated. The computational requirements for analyzing an entire multiyear all-sky survey for all possible KBOs are not practical. However, a more limited search is computationally possible, even for the LSST, which will generate more data than any of the other surveys considered.

We first consider analyzing a single years' data confined to within 10° of the ecliptic for the LSST, the most data-intensive of the future surveys. Since the LSST proposes to cover the whole sky several times each lunation, a given ecliptic longitude will be visible to the LSST for about 6 months of the year with about 20 visits to a given field per year. Thus, on a monthly basis, roughly 600 deg^2 of sky imaged 20 times each need to be searched for bodies beyond the single-visit detection limit. Storage requirements are easily satisfied for this project, as with a 0.2 arcsec/pixel plate scale roughly ~10 TB of data per month will be analyzed, well within current (2008) technology as 1 TB disks are now available for less than $250.

Processing requirements are much more difficult to overcome. The closest work to the LSST is the HST survey of *Bernstein et al.* (2004), who found the number of required grid points for image shifting and combining follows $P^{-5}\Delta T^3$, where P is roughly equivalent to the image quality and T is the timebase. The HST search required several CPU years for a 2.4-GHz Pentium 4 processor to search $\Delta T = 1$ d at $P = 0.03$ arcsec for 5×10^9 pixels. A deep LSST ecliptic survey would require days, $\Delta T \sim 150$ arcsec and roughly 4×10^{12} pixels, increasing the processing requirement by a factor of ~200 over the *Bernstein et al.* (2004) work. This requires roughly 10^3 CPU years for a single 2.4-GHz Pentium 4 processor. Since we require a monthly cycle time for the processing, this requires a total of ~10^4 CPUs, or ~10 Tflops. Such a CPU is available for ~$100 street price as of this writing, but by 2014, the first year of LSST operation, Moore's law predicts that prices should fall by a factor of ~10, allowing such a search to be feasible for ~$200K, likely a small fraction of the entire LSST construction budget. As the LSST already plans to have about 60 Tflops of computing power, a deep multiyear search would require a 15% increase in the total computing power of the LSST project, while increasing KBO discovery depth by 1.6 mag per year.

A limited multiple-year deep survey is also computationally feasible. To allow multiple years, each year would have to be processed independently using the method described above to produce a list of candidate objects. Objects above the formal ~5σ significance level for a single year's worth of observations could be immediately reported as real with preliminary orbits published. Objects below the ~5σ significance level would be saved until the second year's data was collected and analyzed. If two ~3σ objects were found on successive years with similar orbits, then they could be considered a real object. Recomputation of the candidate object's orbit would be only a marginal computational requirement as once an object's orbit is roughly known, only a small amount of phase space must be searched for orbital refinement.

The extra computational effort is well worth the cost due to the steep number density of KBOs, especially for the steeper size distribution of the $22 < m_R < 23.5$ KBOs. *Bernstein et al.* (2004) describe a double power-law model where the medium faintness $m_R \sim 22$ follows a power law of $\alpha_1 = 0.88$ and the very faintest KBOs may be described by a shallower power law with $\alpha_2 = 0.32$. By combining data collected during adjacent years, a magnitude gain can be realized in the background limited case. Such a gain would mean a factor $10^{0.4\alpha_1} = 2.25$ increase in the number of $m_R \sim 23$ KBOs found and a factor $10^{0.4\alpha_2} = 1.34$ for the faintest KBOs. The true utility of an ultradeep multiyear moving object survey will depend on the actual luminosity distributions measured for the KBOs, as the luminosity distributions for faint bodies are poorly known.

2.4. Other Surveys

2.4.1. The Discovery Channel Telescope (DCT). The Discovery Channel Telescope (DCT) with a 2-deg² field of view will be located atop Happy Jack mountain in Arizona, a dark site with median seeing of 0.8 arcsec. With a 4.3-m-diameter primary mirror, the DCT has the ability to beat PS1 in terms of performance. However, if PS4 is delivered on time, it will likely be able to outperform the DCT. With likely DCT science beginning in 2010, about the same year PS4 science will begin, the role of DCT as a primary survey tool is somewhat diminished. If the PS4 experiment is delayed for any reason, then the DCT would be a good candidate for covering the whole sky to a fainter depth than PS1. Regardless of operation timetable, the DCT will be a useful facility for follow-up science such as the science products discussed in section 4. In particular, color measurements of the PS1 or PS4 KBOs could be a niche that the DCT could fill.

2.4.2. Visible and Infrared Survey Telescope for Astronomy (VISTA). The Visible and Infrared Survey Telescope for Astronomy (VISTA) is a 4-m telescope as well, slightly smaller than the DCT in diameter but with a larger field of view of 3 deg². The VISTA project is located at Cerro Paranal, and the primary mirror has already been installed, giving it the potential for being a very powerful survey telescope as it is between PS1 and PS4 in power and will be operational by 2008 or earlier. Unfortunately, the initial camera is infrared only, which is not useful for KBO discovery due to the extremely high telluric background in the infrared as well as the limited field of view (0.6 deg²) that is typical of near-infrared cameras. There are tentative plans for a visible camera, but no firm date of delivery or funding at the present time.

For analysis of the impact of VISTA, we have split the project into two possible modes, the current near-infrared mode and a possible future visible mode upgrade. The current near-infrared mode can reach KBOs down to AB ~ 21.5 using the Y filter, the shortest wavelength (and lowest background) filter currently in the camera. Thus, the current VISTA is roughly equivalent to a 1-m telescope in the visible in terms of depth, but with only a 0.6-deg² field of view. Thus, the value of SP is quite close to 1 for the current VISTA project and is not considered further in this work. If a visible VISTA camera were produced, the telescope would be very powerful, with SP ~ 10. Since the schedule for such a camera is uncertain, it seems likely that the much more powerful PS4 will be available prior to a visible VISTA camera. Thus, we do not further consider VISTA in this work, although as with the DCT, a survey of KBO colors could be a useful project for VISTA.

2.5. Targeted Deep Surveys

Neither Pan-STARRS nor LSST will be able to beat existing telescopes in survey depth by large factors. Most of the deepest surveys with existing technology have already been attempted, with detections of KBOs fainter than $m_R > 27$ reported from Keck in a survey of 0.01 deg² by *Chiang and Brown* (1999) and from the Hubble Space Telescope (HST) in an survey of 0.02 deg² (*Bernstein et al.,* 2004). Both surveys found objects consistent with sky densities of ~100 KBOs per square degree near the ecliptic. A similar project could be done at Subaru using the existing 0.3-deg² Suprime-Cam, which could also reach $m_R \sim 27$ during a full night of excellent seeing. Thus, it seems reasonable that a project covering ~1 deg² to $m_R = 27$ is likely to be completed prior to the arrival of either PS4 or the LSST.

Although technology in the next decade may increase the survey area by large factors, it will not significantly increase the threshold of faintness because such research is currently aperture limited. So although deep discovery surveys have been pivotal to studying the smallest KBOs in the previous decade, in the next decade, KBO surveys are not expected to go much deeper, only wider. The LSST does propose a very deep survey (although the specific area of which is not certain at this point), which is critical to understanding the maximum depth that could be achieved. With a ~10-deg^2 field of view, the LSST could improve on the existing sky area surveyed by a factor of ~50 or more, resulting in 10,000 very faint (m_R ~ 28) KBOs discovered. This would be a very useful determination of the faint end of the size distribution, but not impacting the entire KBO field as much as the main survey itself, which could find ~100,000 KBOs. Thus, targeted deep surveys will likely be much overshadowed by the all-sky surveys of Pan-STARRS and LSST in the coming decade (reviewed in section 4) until the advent of completely new facilities capable of breaking the m_R ~ 29 barrier, such as the James Webb Space Telescope (JWST), the Thirty-Meter Telescope (TMT), and the Extremely Large Telescope (ELT) as discussed in section 5.

3. THE NEXT DECADE OF KUIPER BELT OBJECT DISCOVERY

In this section, we outline a rough timeline of probable survey discoveries in the next decade. As each instrument comes to science production — PS1, PS4, and then the LSST — there will be two phases of discovery. The initial discovery phase will be single-lunation discovery where KBOs visible in a single month are identified. Then the very deep, processing-intensive survey can be produced after multiple years of data have been collected *only* if suitable software effort is applied. This ultradeep multiyear work would allow significant gains if the processing difficulties can be overcome. The fourth year of any of the surveys listed is the point where diminishing returns is reached since signal-to noise ratios only increase with the square root of the number of visits for the background-limited KBO surveys. We summarize the basic numbers and depth of KBO detections in Fig. 3.

3.1. Pan-STARRS PS1: 2007 to 2010

The Pan-STARRS PS1 prototype telescope is the first large-survey telescope that will come on line. Although its mirror is relatively small (1.8 m diameter, compared to 2.4 m for Sloan and 8.4 m for the LSST), the use of OTCCDs will boost Pan-STARRS' sensitivity beyond surveys with smaller telescopes such as the current generation Caltech survey (*Trujillo and Brown,* 2003). The Caltech survey uses 150-s exposures, a 1.2 m-diameter mirror, and has quite poor image quality (around 2.5 arcsec) due to the large focal plane and poor natural seeing. As discussed above, the PS1 OTCCDs will allow Pan-STARRS to perform tip/tilt correction on each chip. Actual performance should be around 0.5 arcsec on Mauna Kea and maybe around 0.6 arcsec on Haleakalā. Thus, by comparing the Caltech survey aperture (1.8 m/1.2 m), seeing (2.5 arcsec/0.6 arcsec) and typical exposure times (60 s/150 s), one can determine that the PS1 should be able to go about 1.5 mag deeper than the existing Caltech survey, achieving depths of m_R ~ 23 in about 60 s. With a field of view of 7 deg^2, the survey will cover the entire sky. In operation, the survey will likely take three 60-s exposures of each location per night. Assuming 50% of the nights are photometric, 70% of the photometric nights have seeing acceptable enough to conduct the survey, and roughly 66% efficiency due to telescope slew and image readout (30 s per location) and no observations within three days of the full Moon (75% of a lunation used), the entire survey efficiency will be about 18%. For 10-h nights, the survey could cover

$$90000\frac{\text{deg}^2}{\text{yr}} \approx 0.18 \times \frac{365\text{ nts}}{\text{yr}}\frac{10\text{ h}}{\text{nt}}\frac{20\text{ triplets}}{\text{h}}\frac{7\text{ deg}^2}{\text{triplet}} \quad (2)$$

or the entire visible sky three times over per year (in triplets). The solar system survey will not be the only survey conducted, but if at least one-third of the allocated time is devoted to the solar system survey (a reasonable number given that solar system science is a primary goal of the experiment), it should be able to discover all the KBOs within its survey depth. For simplicity, we assume that the PS1 system will be able to survey all KBOs to m_R ~ 23 in a single year. Follow-up will occur naturally as part of the data processing pipeline as candidate objects will all have three triplets per year. In general, if subsequent visits to the same location are separated by a month, nearly all object links can be made from month to month. Multiple years will be able to add significant depth to revisits given suitable software tools, as quantified in Fig. 3 as an increase in depth with time. Note that the total object numbers computed within this document are scaled from existing surveys. Thus, when the next generation of surveys comes online and factors of ~100 to ~1000 increases in sky area surveyed are realized, population statistics are very likely to differ from expected values.

Initially, the PS1 experiment should find ~3000 KBOs, tripling the currently known sample after a year of operation, nominally by mid-2008. However, as the experiment progresses to the multiyear stage, ultradeep multiyear analysis would enable a much larger number of bodies to be found, about 7500 KBOs in total by the start of the PS4 experiment in 2010. All bodies will be automatically recovered, so the 3000 first-year KBOs will have multiyear arcs starting in mid-2008. By 2010, all the 7500 discovered KBOs will have three-year arcs. For comparison to the current KBOs population, as of this writing in the spring of 2007, 1200 KBOs are known, only 700 of which have multiyear arcs and 600 of which have three-year arcs. In terms of the KBO size distribution (for which no orbits are needed), accuracy will be increased by a factor of ~10 more objects by 2008 over the largest survey to date in terms of discovery statistics, the Deep Ecliptic Survey (DES) (*Elliot et al.,* 2005), and by a factor of ~20 by 2010. For dynamical studies, the number of "test particles" (i.e., total number of known KBOs with reliable orbits) will be increased by a factor of 2 to 10 from 2008 to 2010. It should be noted that the full impact for dynamical studies will only be made at the third year since orbits must be well known, but limited information such as heliocentric distance and inclination data will be available for objects during the very first year of observations.

3.2. Pan-STARRS PS4: 2010 to 2014

The development and operation of PS4 will effectively mean the end of the PS1 project. Utilizing four 1.8-m mirrors on Mauna Kea instead of one on Haleakalā, the Pan-STARRS project will be able to produce a significantly deeper survey. With tip/tilt corrected seeing of 0.5 arcsec vs. 0.6 arcsec and the additional mirror area, the PS4 experiment should be able to reach about a magnitude deeper than the original PS1 experiment in the same exposure time. From Fig. 3, we see that at outset, the PS4 system will be sensitive to $m_R \sim 24$ KBOs, corresponding to 10,000 KBO discoveries from its first year of operation alone. By the time the LSST becomes operational in 2014, the survey depth will reach $m_R \sim 24.75$, resulting in 30,000 KBO detections if multiyear ultradeep detections can be made. Thus, PS4 will result in a size distribution and dynamical works using 5–10 times more objects than PS1.

3.3. Large Survey Synoptic Telescope: 2014 to 2017

Although the PS4 will likely stay operational for many years after inception for hazardous asteroid defense purposes, the advent of the LSST will take a large amount of the science interest away from PS4, at least for KBO discovery purposes. The LSST will be a significant step up from the PS4 in terms of survey power mainly due to its larger collecting area (8.4 m/four 1.8 m) and wider field of view (10 deg^2/7 deg^2). However, the LSST will suffer from somewhat poorer seeing compared to the PS4 experiment due to its use of traditional guiding (0.7 arcsec/0.5 arcsec).

Operational efficiency for the LSST is likely to be a factor of ~3 higher than Pan-STARRS. The most critically important overhead with the LSST will be the use of six filters, which will reduce sky time spent in the r filter, the most sensitive to KBOs. The current specifications for the LSST suggest that the r filter will be the most used, with 40% of visits budgeted for the r filter. The i filter, which has only slightly less sensitivity to KBOs than the r filter, will be used for 30% of visits, so together, these two filters should limit LSST efficiency to be 30% lower than PS4 after considering bandwidth differences. Due to site differences, there may be on the order of 10% more photometric nights for LSST compared to PS1 and PS4. Slew and readout overheads will also be somewhat lower for the LSST, as the main part of the survey proposes to spend roughly 9 s for telescope slew and image readout for every 30 s of on-sky exposure, yielding 77% efficiency on the sky as opposed to that of Pan-STARRS, ~66%. Thus, overall operational efficiency of the LSST will be similar to that of Pan-STARRS PS4 after accounting for filter bandwidth, readout,

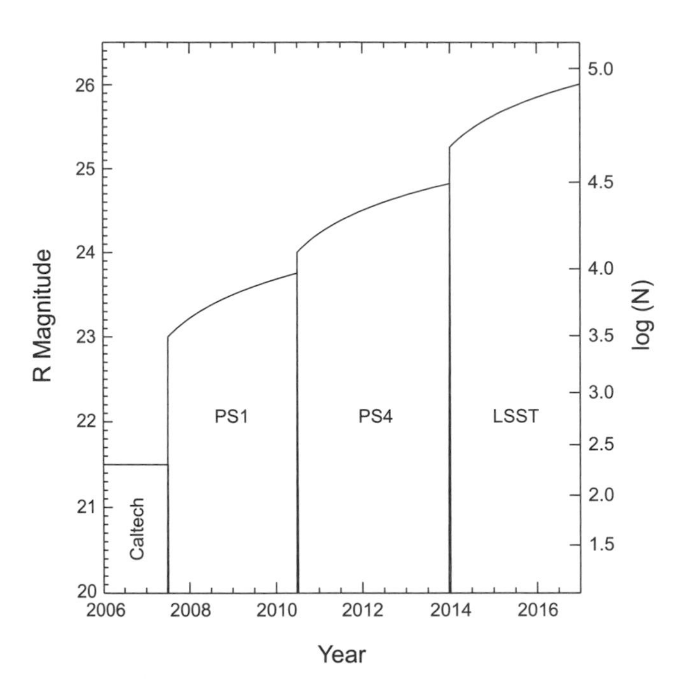

Fig. 3. Summary of the next decade of KBO all-sky survey work. Lines represent the likely survey sensitivity (left vertical axis) and the log number N of KBOs found [*Bernstein et al.* (2004), double power law model, right vertical axis] as a function of year. The three primary surveys are the Pan-STARRS PS1, the Pan-STARRS PS4, and the LSST. Also depicted is the current generation Caltech survey, the only modern all-sky survey to date with SP >1. The curved portion of each survey's coverage corresponds to increasing depth with multiple visits, and is only applicable if the survey can perform multiyear ultradeep searches for KBOs too faint for detection in a single pass.

and site differences. One large advantage that LSST has over Pan-STARRS is that nearly all the survey time can be used to find moving objects, where for Pan-STARRS, only the fraction allocated to the solar system search will be used (approximately one-third).

Combining the basic instrument advantage of the LSST (in terms of SP) with the operational differences means that the LSST will be slightly deeper than the PS4 in single-visit depth, but will have a factor of ~10 more visits available, ultimately beating the PS4 experiment by about 1.25 mag in depth over similar timescale, or about $m_R \sim 25.25$ per year. Note that this depth assumes (as do the depths calculated for Pan-STARRS) that data collected throughout the year can be combined to find moving objects, which as described in section 2.3, is a nontrivial task. Although the LSST formally has a decade long lifecycle, we only consider the impact of the first few years in this work in our decade outlook.

The LSST may also spend ~10% of telescope time performing a very deep survey of a small portion of the ecliptic, beyond $m_R \sim 27$. This very deep survey will provide an additional ~10,000 very faint KBOs. This represents a few percent of the KBOs in the sky that are beyond the depth of the primary all-sky LSST survey. The number of KBOs found in the very deep survey is very difficult to estimate as the number distribution at the faint end is currently very uncertain. As previously discussed in section 2.5, this very deep survey will provide information on some of the faintest KBOs available, but this impact will be much less than the main survey for KBO research.

The LSST's location in the southern hemisphere will allow it to identify some KBOs that cannot be seen from Hawai'i, but not a significantly large fraction. Mauna Kea is about 20° north of the equator while Cerro Pachón is about 30° south of the equator. The practical limit for reasonable survey seeing is around 1.7 airmasses, which occurs near 54° zenith angle. Thus, the PS4 survey should be able to observe as far south as –34° latitude, which will include all KBOs within about 11° of the ecliptic even near the summer opposition point. Rough estimates from the KBO latitude distribution measured by *Elliot et al.* (2005) indicate that roughly 85% of all KBOs should be northward of this limit. Thus, the only major impact that the LSST will make over PS4 due to its southern latitude will be if any particularly interesting KBOs happen to be present south of the PS4's range, such as unusually large objects.

In its first year, the LSST will detect about 50,000 KBOs, increasing the sample collected by the PS4 experiment by about a factor of 5. After a few years of operation, the LSST will be probing all-sky depths near $m_R \sim 26$. Again, if suitable moving object detection software is in place to take advantage of combining multiple years of observations, about 100,000 KBOs will be found by 2017, tripling the number of KBOs found by the PS4 experiment. Although the use of several filters lowers the LSST detection efficiency, it does allow a huge amount of color information to be gathered in the main r and i filters, as described further in section 4.7.

Overall, by the end of 2017, the additional breadth of information will be astonishing by today's standards. The total known KBO population will have increased by a factor of 80. The total number of KBOs with multiyear orbits will have increased even more dramatically, by a factor of about 150.

4. SCIENCE ENABLED BY KUIPER BELT OBJECT DISCOVERIES

The increase in knowledge over the next decade will come in sudden bursts every few years as each of the upcoming surveys are brought into science production. Between these bursts, the total number of bodies will still be growing as depth and size of the surveys increase if multiyear depth increases can be made. It is impossible to predict the science gleaned by the factor of ~100 increase in number of KBOs by 2017 compared to present-day values. However, what is clear is that certain areas of KBO research that have been stymied by lack of sample size will have many bodies available.

4.1. Luminosity Function

Measurements of the luminosity function of the KBOs will likely be the first science products to emerge from the next generation of KBO surveys. While much work has been done on the subject in the past decade, increasing the total population of KBOs by factors of ~5 every few years will dramatically reduce the random and systematic uncertainties in current estimates of the luminosity function and the associated implications for size distribution. As it now seems clear that the size distribution of bodies is not constant over the factor of ~100 range of sizes of currently observed KBOs, specific breakpoints (if they exist) can be identified and may be associated with physical or dynamical causes.

Of particular interest is the number of bodies with $m_R \sim 22$, corresponding to diameters D ~ 200 km at 45 AU. To date, a large variety of albedos have been seen in the Kuiper belt (see chapter by Stansberry et al.). What is not clear is why the very largest bodies appear to have uniformly high albedos, such as (50000) Quaoar, (136199) Eris, and (136108) 2003 EL_{61} (*Brown and Trujillo,* 2004; *Rabinowitz et al.,* 2006; *Brown et al.,* 2006). Finding a bend in the luminosity function of the bright KBOs would certainly suggest that the bright (large) KBOs could have a different albedo distribution than the medium-sized KBOs, although it is possible that this could be explained by a change in the size distribution itself as well. Such a study requires all-sky coverage to maximize counting statistics.

Also of particular interest is comparing the luminosity distribution of distinct dynamical populations. This can shed light on the possibility of unique origins for disparate classes of objects. Such a study could improve theories of separate origins for the high- and low-inclination bodies, for example.

4.2. Inclination Distribution

The inclination distribution and the ecliptic latitude distribution of bodies will be measured with extreme accuracy compared to our current state of knowledge. Although the ecliptic latitude distribution of the KBOs is known in enough detail to make rough estimates of the KBO population, this knowledge will be further refined. More interesting yet are aspects of the inclination/latitude distribution for which we have no data. The brightest bodies appear to be significantly more inclined than the fainter objects, as seen from population-averaged work as well as in initial results from the Caltech survey (*Levison and Stern,* 2001; *Trujillo and Brown,* 2003). At what point this break occurs in terms of brightness and latitude is unclear as only ecliptic surveys have the ability to find substantial numbers of faint objects and only full-sky surveys have the ability to find bright objects due to the rarity of objects. An intermediate survey that is wide enough to find significant numbers of $21 < m_R < 22$ objects is needed to explore the brightness/

inclination connection without bias. The PS1 survey fulfills just these criteria, and it is unlikely that any smaller survey will do so unless it can cover a few thousand square degrees to depths of $m_R \sim 22$.

4.3. Highly Eccentric and Distant Objects

Highly eccentric objects have extreme observational biases. In particular, Kepler's law requires objects of high eccentricity to spend most of their time near aphelion, where they are difficult to observe. Thus objects like (90377) Sedna, with an eccentricity of 0.85, are very difficult to discover since flux $f \propto R^{-4}$, where R is heliocentric distance. Thus, the ~4-mag increase from the present-day Caltech all-sky survey to the 2017 LSST all-sky survey will allow objects of similar size to be discovered at distances six times farther from the Sun than current works. Thus, bodies such as (90377) Sedna, which was found at R = 89.6 AU, and (136199) Eris, which was found at R = 97.0 AU, could be found at over R > 200 AU.

Parallactic motion at such extreme distances is very low, about 0.3 arcsec per hour, so under most observing strategies any potential objects would appear as point sources except in multiyear surveys where the ~1-arcmin-per-year orbital motion can be detected or in multinight surveys where the ~10-arcsec-per-day motion could be seen. There is no doubt that such objects are present and in significant numbers. For instance, (90377) Sedna was only detectable in the Caltech survey for about 1% of its orbit. By extrapolation, there should be many hundreds of such high-eccentricity objects that could be detected by surveys such as Pan-STARRS and LSST. There are no sound observational constraints on even more distant objects to date, so the radial distribution of objects in the 100-AU to 1000-AU distance range will be completely determined by the output of Pan-STARRS and LSST.

4.4. Small and Large Objects

Although much has been made of the recent discovery of (136199) Eris and other dwarf planets similar in size to Pluto, such discoveries could very easily become commonplace when each of the new surveys begins. Even a Mars-sized body is quite easy to hide from present-day surveys. Assuming a Mars-like albedo of 25%, and a diameter of 7000 km, a Mars-like body would be $m_R = 23$ at 300 AU, outside the range of any all-sky surveys sensitive to such objects. And of course there could be many Pluto-sized bodies at 200 AU that would be $m_R = 23$, and have remained undetected because surveys with such depths have only covered tiny fractions of the available sky, a few percent at best.

Small bodies in the Kuiper belt, those near the detection limit in terms of faintness, will make up the majority of any survey's discoveries. Thus, the greatest impact that these surveys will make, at least in number, is the vast amount of smaller KBOs. Typical sizes of interest for the surveys vary depending on the heliocentric distance at which they are found and their albedo. However, if we assume 25% albedo, and that most bodies are found at 45 AU, in the middle of the classical Kuiper belt, then PS1 will be most sensitive to diameter D ~ 100 km bodies, PS4 will be most sensitive to D ~ 50 km bodies, and the LSST will be most sensitive to D ~ 35 km bodies. The true albedo of small KBOs is unknown and is very difficult to measure. However, the approximate sizes of bodies to be discovered by the various surveys are outlined in Fig. 4 for a 25% albedo. Since body diameter is inversely proportional to the square root of albedo for a given apparent magnitude, assuming 10% albedo would result in about a 60% increase in diameter for Fig. 4.

4.5. Multiple Systems

One of the most interesting physical phenomena in the Kuiper belt is the large number of binary systems, with the possibility of increasing fractions of binaries for the larger bodies. Since most of the objects found will be near the faint limit of an individual survey, any secondaries found will likely be of similar brightness to the primary. To date, only a few binaries are known with similar brightness fractions, including the first discovered KBO binary system after Pluto/Charon, 1998 WW_{31} and its companion (*Veillet et al.,* 2002). To date, the size distribution of secondaries has not been well measured, but adding data to the equal-mass bi-

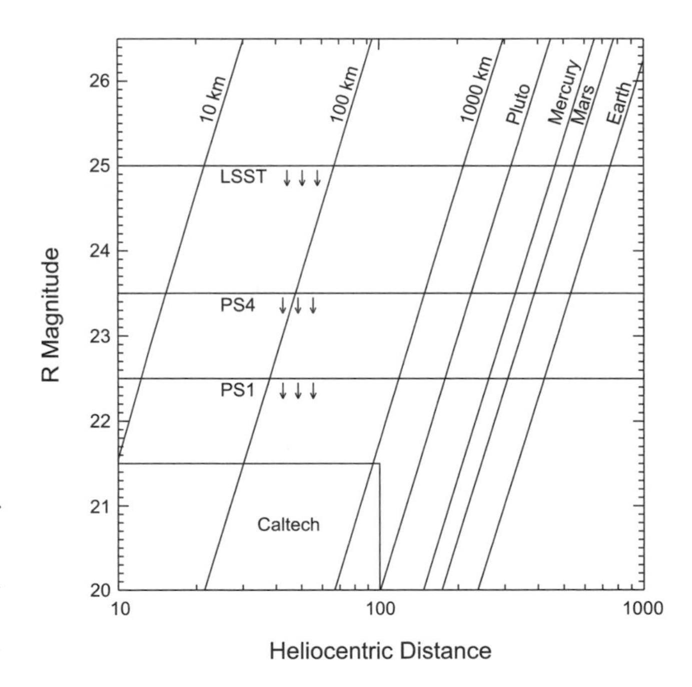

Fig. 4. Diagram of the sizes of bodies that can be detected by the current Caltech survey (rectangle in lower left) and future surveys, which enclose all observable regions below their respective horizontal lines. Due to their multiyear coverage, the future surveys have the ability to find objects at very great distances spanning a tremendous size range.

naries will likely be helpful in any future dynamical efforts studying binary production mechanisms.

4.6. Trace Populations

Some of the most interesting bodies in the Kuiper belt are the ones that are unusual in dynamical terms. For instance, the presence of high-perihelion bodies such as (90377) Sedna and others represents a population whose origins are currently being debated (*Morbidelli and Levison,* 2004; *Stern,* 2005; *Matese et al.,* 2005). Increasing the number of high-perihelion bodies by factors of 100 will add a wealth of data for comparison to dynamical models.

Other trace populations of interest are likely to be the bodies on the edge of strong Neptune resonances, which will help map the relative strength and width of the strong resonances such as the 2:3 (Pluto-Neptune) resonance. Resonances that today have only a few bodies and are not well defined will have a dramatic increase in numbers. For instance, there are about 200 bodies in 2:3 resonance, while resonances such as the 1:2, which currently only have ~20 known members, will have ~1000 or more members by 2017. Unusual objects such as 2004 XR_{190}, which has a high inclination and low eccentricity, will become much more prevalent (*Allen et al.,* 2006).

4.7. Surface Information

Very little detailed surface information will be available for most bodies due to faintness. However, once the LSST is in production with multiple color filters integrated into the survey, no additional color campaigns will be required for study. Thus the major questions found in the color surveys to date will be fed factors of 100 more data points. One potential project for either the DCT or a VISTA visible camera could be a large color survey conducted prior to the LSST. The impact of such a survey is probably best estimated after the PS1 KBOs are found, when sample sizes are known and can be weighed against telescope subscription rates and timescales for construction.

Having color information for the 50,000 LSST KBOs with $m_R < 25$ will add an unmatched amount of information to the discussion about KBO surfaces. To date, there are roughly 200 KBOs with some amount of published color information, which is a combination of many researchers' work and over 50 publications [for a compilation, see the MBOSS website (*Hainaut and Delsanti,* 2002)]. Neither PS1 nor PS4 will provide color information as only one wide filter will be used. It is likely that researchers will continue measuring colors of found bodies, but as it takes about a night of 8-m to 10-m telescope time to collect 10–20 basic colors for $m_R > 24$-mag objects, it is likely by the time the LSST arrives that no more than 1000 KBO colors will be known. Thus, during the first year of operation, the LSST will increase the number of known colors by a factor of ~50 (or a factor $\sqrt{50} \approx 7$ in signal to-noise ratio assuming Gaussian statistics) using a uniform telescope and method. This should allow the detection of many 5σ color trends that would otherwise be completely unobservable prior to the LSST. Such a large amount of color information could allow discernment of basic KBO surface types even in what are now considered trace dynamical populations, such as the weak Neptune resonances.

Such a huge amount of color information has been collected by the Sloan survey for the asteroids, which is a hint of what may be seen for the KBOs (*Ivezić et al.,* 2002). In addition, bodies with peculiarly large lightcurves, such as smaller versions of (136108) 2003 EL_{61} and (20000) Varuna, may be identified in survey data. It is completely uncertain at this point if such unusual rotators are limited to the largest KBOs. However, the PS4 and in particular the LSST with its short ~3-d visit cadence, will increase the total number of KBOs with measured photometric variability by a factor of ~500 or more over the few dozen that have currently been studied.

4.8. Fundamental Plane

Questions regarding the fundamental plane of the Kuiper belt will be easily answered with the large numbers of bodies expected as survey products. Of particular interest in this area of research will be a comparison of Kuiper belt planes for various dynamical subpopulations. It would be interesting to compare the high-perihelion bodies, for example, with the classical KBOs, to see if their different heliocentric distances have any effect on the axis about which they orbit.

5. SPECIAL SCIENCE 2015 AND BEYOND

The key areas of research a decade from now will be very different from today's. Scientists as a whole will have three very large bodies of data, collected in largely different manners. First, the basic survey data from PS1, PS4, and the LSST will provide sheer numbers of objects to study. Second, follow-up science performed at other observatories including the next generation of giant telescopes will be able to study particularly interesting objects in great detail. Third, the results of the New Horizons mission, with its Pluto-Charon encounter in 2015 and possible KBO encounters until 2020, will provide a wealth of physical data on a few selected bodies beyond 30 AU.

5.1. James Webb Space Telescope (JSWT)

The James Webb Space Telescope (JWST) will be an infrared-optimized observatory with a 6.6-m primary mirror, scheduled for launch in 2013 (*Gardner et al.,* 2006). In terms of object discovery, the JWST will be able to probe interesting areas of the Kuiper belt that will not be able to be seen from the ground. Its greatest science impact for KBO research is likely to be low-resolution spectroscopy of the faintest KBOs using the NIRSpec camera. Initial requirements, which are subject to change, are that the camera will be able to collect spectra with signal-to noise ratios of S/N ~ 10 for an object with m_{AB} ~ 26 in 10,000 s around 2 μm (K band). For useful KBO spectral information, one needs approximately S/N ~10 or better, requiring m_{AB} ~ 24.5, which for KBOs with mean ($m_V - m_K$) ~ 1.75 (Vega scale) requires objects brighter than m_K ~ 20 for reasonable science return. Currently, KBOs with m_K ~ 20 are very difficult to study from the largest telescopes on the ground, so this will be a factor of ~100 increase over the number of KBOs for which spectra can be collected. For the first time, spectral studies of KBOs may be more limited by telescope oversubscription rates than being limited by the small number of bright targets.

The imaging instrument NIRCam can also be used for survey work. One could target a single field for several hours and detect point sources as faint as m_{AB} ~ 28.8 in the K band, which is equivalent to about m_K ~ 25. The field of view of the detector is quite small, with only 0.0025 deg^2 covered. However, at these very faint magnitudes, which approach m_R ~29, the sky density of KBOs may be very high, a few hundred per square degree, thus one KBO will be detected for every few hours of telescope time. Such a search would not be very productive in numbers, but it would provide basic data approaching the ~10-km size regime, nearly the size of the typical cometary nucleus. A far better instrument could be the TMT, however.

5.2. Thirty Meter Telescope (TMT) and Extremely Large Telescope (ELT)

The Thirty Meter Telescope (TMT) is planned to be available for science starting around the year 2016. There are two obvious uses for the TMT in KBO surveys. The first is in the visible, where the giant collecting area can increase the depth of observation over existing telescopes in the visible, if instrumentation allows. The TMT will have a 20-arcmin field of view, covering about 0.1 deg^2 to depths of m_R ~ 29 in several hours, probing the same size range as the JWST but with factors of ~40 more objects discovered. Thus, the telescope could find some of the faintest KBOs in the visible regime. However, it is not clear yet whether the TMT will have a wide-field optical imager. It is possible that the TMT will be only equipped with near-infrared instruments due to the great performance increases that adaptive optics (AO) can provide for the near-infrared. For certain, the TMT will have a wide-field near-infrared imager, since the TMT's large size (and thus small diffraction limit of ~15 milliarcsec at 2 μm) will greatly benefit from AO correction, which generally delivers light at wavelengths longer than 1 μm. Compared to current 8-m to 10-m seeing-limited groundbased telescopes, the AO-corrected TMT will provide great advances. Sky background noise will be reduced by factors of ~4 even with the larger collecting area, since plate scales will be very fine (around 7 milliarcsec compared to 0.1 arcsec). Signal will be increased by factors of ~3.5 over current groundbased infrared instruments due to the larger collecting area. Overall, magnitudes of m_K ~ 26 (equivalent to m_R ~ 27.75) should be achievable with the TMT using AO for point sources with a few hours on source. Thus, the TMT will probably achieve sensitivities somewhat less than the JWST, but with factors of ~50 more sky area. Together, the JWST and TMT will complement one another in terms of number and size of KBOs discovered.

The largest telescope likely to be built in the next decade is the Extremely Large Telescope (ELT), under feasibility studies by the European Southern Observatory (ESO). Construction is to begin in 2010, with completion of construction in 2017. The primary mirror will be 42 m in diameter, in a segmented design. The possible instrumentation suite is still unclear, but if a visible capability is added, it should be able to break m_R ~ 29 in a few hours in seeing-limited mode. In the near-infrared, it should be able to image even fainter KBOs than the TMT, with the ability to approach m_K ~ 26.5 objects with AO correction, or m_R ~28 given typical KBO colors. The ELT will thus have access to the smallest KBOs, probably some with diameters of 5 km, those bodies that could be future cometary nuclei.

5.3. New Horizons

The science produced by New Horizons (see chapter by Weaver and Stern), which will have closest approach with Pluto in July 2015, is probably the biggest unknown for KBO science in the next decade. Although the capabilities of the instruments are quite certain, what is very poorly known is what geologic features the surface of Pluto harbors. No doubt, it will be much, much more complex than imagined from the several resolution elements across the surface that we now have. It will be very challenging to draw conclusions about KBOs as a whole from the study of Pluto/Charon and the few KBOs that may be visited in the years after the Pluto/Charon flyby. After 2015, for the first time, we will have a very deep knowledge of the largest of KBOs, and a very rudimentary knowledge of the vast population of much smaller KBOs. Connecting such depth and breadth of these disparate datasets will be the largest challenge of the next decade of KBO research.

Acknowledgments. Special thanks to S. Sheppard (Carnegie Institute of Washington), who contributed an outline of current survey power with future estimates of survey power as well as other useful discussions about the future of survey work. Thanks are also in order to G. M. Bernstein (University of Pennsylvania) and Ž. Ivezić (University of Washington) who contributed useful information about the LSST survey depth, coverage, and timescale of development. Also appreciated was communication with T. Grav (University of Hawai'i), who contributed survey parameters and telescope updates for Pan-STARRS. R. L. Allen contributed helpful information about future CFHT survey capabilities. Both reviewers of this work, J-M. Petit and an anonymous referee, helped significantly with their constructive comments. This work was supported by the Gemini Observatory, which is operated by the Association of Universities for Research in Astronomy, Inc., on behalf of the international Gemini partnership of Argentina, Australia, Brazil, Canada, Chile, the United Kingdom, and the United States.

REFERENCES

Allen R. L., Gladman B., Kavelaars J. J., Petit J.-M., Parker J. W., and Nicholson P. (2006) Discovery of a low-eccentricity, high-inclination Kuiper belt object at 58 AU. *Astrophys. J. Lett., 640,* L83–L86.

Barkume K. M., Brown M. E., and Schaller E. L. (2006) Water ice on the satellite of Kuiper belt object 2003 EL_{61}. *Astrophys. J. Lett., 640,* L87–L89.

Barucci M. A., Cruikshank D. P., Dotto E., Merlin F., Poulet F., Dalle Ore C., Fornasier S., and de Bergh C. (2005) Is Sedna another Triton? *Astron. Astrophys., 439,* L1–L4.

Bernstein G. M., Trilling D. E., Allen R. L., Brown M. E., Holman M., and Malhotra R. (2004) The size distribution of trans-neptunian bodies. *Astron. J., 128,* 1364–1390.

Bertoldi F., Altenhoff W., Weiss A., Menten K. M., and Thum C. (2006) The trans-neptunian object UB_{313} is larger than Pluto. *Nature, 439,* 563–564.

Brown M. E. and Trujillo C. A. (2004) Direct measurement of the size of the large Kuiper belt object (50000) Quaoar. *Astron. J., 127,* 2413–2417.

Brown M. E., Bouchez A. H., Rabinowitz D., Sari R., Trujillo C. A., van Dam M., Campbell R., Chin J., Hartman S., Johansson E., Lafon R., Le Mignant D., Stomski P., Summers D., and Wizinowich P. (2005) Keck Observatory laser guide star adaptive optics discovery and characterization of a satellite to the large Kuiper belt object 2003 EL_{61}. *Astrophys. J. Lett., 632,* L45–L48.

Brown M. E., Schaller E. L., Roe H. G., Rabinowitz D. L., and Trujillo C. A. (2006) Direct measurement of the size of 2003 UB_{313} from the Hubble Space Telescope. *Astrophys. J. Lett., 643,* L61–L63.

Chiang E. I. and Brown M. E. (1999) Keck pencil-beam survey for faint Kuiper belt objects. *Astron. J., 118,* 1411–1422.

Cruikshank D. P., Stansberry J. A., Emery J. P., Fernández Y. R., Werner M. W., Trilling D. E. and Rieke G. H. (2005) The high-albedo Kuiper belt object (55565) 2002 AW_{197}. *Astrophys. J. Lett., 624,* L53–L56.

de Bergh C., Delsanti A., Tozzi G. P., Dotto E., Doressoundiram A., and Barucci M. A. (2005) The surface of the transneptunian object 90482 Orcus. *Astron. Astrophys., 437,* 1115–1120.

Doressoundiram A., Barucci M. A., Tozzi G. P., Poulet F., Boehnhardt H., de Bergh C., and Peixinho N. (2005) Spectral characteristics and modeling of the transneptunian object (55565) 2002 AW_{197} and the Centaurs (55576) 2002 GB_{10} and (83982) 2002 GO_9: ESO Large Program on TNOs and Centaurs. *Planet. Space Sci., 53,* 1501–1509.

Elliot J. L., Kern S. D., Clancy K. B., Gulbis A. A. S., Millis R. L., Buie M. W., Wasserman L. H., Chiang E. I., Jordan A. B., Trilling D. E., and Meech K. J. (2005) The Deep Ecliptic Survey: A search for Kuiper belt objects and Centaurs. II. Dynamical classification, the Kuiper belt plane, and the core population. *Astron. J., 129,* 1117–1162.

Gardner J. P., Mather J. C., Clampin M., Doyon R., Greenhouse M. A., Hammel H. B., Hutchings J. B., Jakobsen P., Lilly S. J., Long K. S., Lunine J. I., Mc-Caughrean M. J., Mountain M., Nella J., Rieke G. H., Rieke M. J., Rix H.-W., Smith E. P., Sonneborn G., Stiavelli M., Stockman H. S., Windhorst R. A., and Wright G. S. (2006) The James Webb Space Telescope. *Space Sci. Rev., 123,* 485–606.

Gaudi B. S., Stanek K. Z., Hartman J. D., Holman M. J., and McLeod B. A. (2005) On the rotation period of (90377) Sedna. *Astrophys. J., Lett., 629,* L49–L52.

Gladman B., Kavelaars J. J., Nicholson P. D., Loredo T. J., and Burns J. A. (1998) Pencil-beam surveys for faint trans-neptunian objects. *Astron. J., 116,* 2042–2054.

Gladman B., Kavelaars J. J., Petit J.-M., Morbidelli A., Holman M. J., and Loredo T. (2001) The structure of the Kuiper belt: Size distribution and radial extent. *Astron. J., 122,* 1051–1066.

Hainaut O. R. and Delsanti A. C. (2002) Colors of minor bodies in the outer solar system. A statistical analysis. *Astron. Astrophys., 389,* 641–664.

Henneken E. A., Kurtz M. J., Eichhorn G., Accomazzi A., Grant C., Thompson D., and Murray S. S. (2006) Effect of e-printing on citation rates in astronomy and physics. *J. Electronic Publishing, 9,* 2.

Hughes D. W. (2003) Planets: Quaoar and the Edgeworth-Kuiper belt. *Astron. Geophys., 44,* 21–3.

Irwin M., Tremaine S., and Zytkow A. N. (1995) A search for slow-moving objects and the luminosity function of the Kuiper belt. *Astron. J., 110,* 3082.

Ivezić Ž., Lupton R. H., Jurić M., Tabachnik S., Quinn T., Gunn J. E., Knapp G. R., Rockosi C. M., and Brinkmann J. (2002) Color confirmation of asteroid families. *Astron. J., 124,* 2943–2948.

Jewitt D. (2003) Project Pan-STARRS and the outer solar system. *Earth Moon Planets, 92,* 465–476.

Jewitt D. and Luu J. (1993) Discovery of the candidate Kuiper belt object 1992 QB_1. *Nature, 362,* 730–732.

Jewitt D. C. and Luu J. (2004) Crystalline water ice on the Kuiper belt object (50000) Quaoar. *Nature, 432,* 731–733.

Jewitt D., Luu J., and Trujillo C. (1998) Large Kuiper belt objects: The Mauna Kea 8K CCD survey. *Astron. J., 115,* 2125–2135.

Larsen J. A., Gleason A. E., Danzl N. M., Descour A. S., McMillan R. S., Gehrels T., Jedicke R., Montani J. L., and Scotti J. V. (2001) The Spacewatch Wide-Area Survey for bright Centaurs and trans-neptunian objects. *Astron. J., 121,* 562–579.

Levison H. F. and Stern S. A. (2001) On the size dependence of the inclination distribution of the main Kuiper belt. *Astron. J., 121,* 1730–1735.

Licandro J., Grundy W. M., Pinilla-Alonso N., and Leisy P. (2006) Visible spectroscopy of 2003 UB_{313}: Evidence for N_2 ice on the surface of the largest TNO? *Astron. Astrophys., 458,* L5–L8.

Marchi S., Lazzarin M., Magrin S., and Barbieri C. (2003) Visible spectroscopy of the two largest known trans-neptunian objects: Ixion and Quaoar. *Astron. Astrophys., 408,* L17–L19.

Matese J. J., Whitmire D. P., and Lissauer J. J. (2005) A widebinary solar companion as a possible origin of Sedna-like objects. *Earth Moon Planets, 97,* 459–470.

Moore G. E. (1965) Cramming more components onto integrated circuits. *Electronics, 38.*

Morbidelli A. and Levison H. F. (2004) Scenarios for the origin of the orbits of the trans-neptunian objects 2000 CR_{105} and 2003 VB_{12} (Sedna). *Astron. J., 128,* 2564–2576.

Petit J.-M., Holman M. J., Gladman B. J., Kavelaars J. J., Scholl H., and Loredo T. J. (2006) The Kuiper belt luminosity function from $m_R = 22$ to 25. *Mon. Not. R. Astron. Soc., 365,* 429–438.

Rabinowitz D. L., Barkume K., Brown M. E., Roe H., Schwartz M., Tourtellotte S., and Trujillo C. (2006) Photometric observations constraining the size, shape, and albedo of 2003 EL_{61}, a rapidly rotating, Pluto-sized object in the Kuiper belt. *Astrophys. J., 639,* 1238–1251.

Sheppard S. S., Jewitt D. C., Trujillo C. A., Brown M. J. I., and Ashley M. C. B. (2000) A wide-field CCD survey for Centaurs and Kuiper belt objects. *Astron. J., 120,* 2687–2694.

Stern S. A. (2005) Regarding the accretion of 2003 VB_{12} (Sedna) and like bodies in distant heliocentric orbits. *Astron. J., 129,* 526–529.

Stevenson D. J. (2004) Planetary science: Volcanoes on Quaoar? *Nature, 432,* 681–682.

Tombaugh C. W. (1946) The search for the ninth planet, Pluto. *Leaflet Astron. Soc. Pac., 5,* 73.

Trujillo C. A. and Brown M. E. (2003) The Caltech Wide Area Sky Survey. *Earth Moon Planets, 92,* 99–112.

Trujillo C. A., Jewitt D. C., and Luu J. X. (2001a) Properties of the trans-neptunian belt: Statistics from the Canada-France-Hawaii Telescope Survey. *Astron. J., 122,* 457–473.

Trujillo C. A., Luu J. X., Bosh A. S., and Elliot J. L. (2001b) Large bodies in the Kuiper belt. *Astron. J., 122,* 2740–2748.

Trujillo C. A., Brown M. E., Rabinowitz D. L., and Geballe T. R. (2005) Near-infrared surface properties of the two intrinsically brightest minor planets: (90377) Sedna and (90482) Orcus. *Astrophys. J., 627,* 1057–1065.

Trujillo C. A., Brown M. E., Barkume K. M., Schaller E. L., and Rabinowitz D. L. (2007) The surface of 2003 EL_{61} in the near-infrared. *Astrophys. J., 655,* 1172–1178.

Veillet C., Parker J. W., Griffin I., Marsden B., Doressoundiram A., Buie M., Tholen D. J., Connelley M., and Holman M. J. (2002) The binary Kuiper-belt object 1998 WW_{31}. *Nature, 416,* 711–713.

Wickramasinghe J. T., Wickramasinghe N. C., and Napier W. M. (2004) Sedna's missing moon. *The Observatory, 124,* 300–302.

Color Section

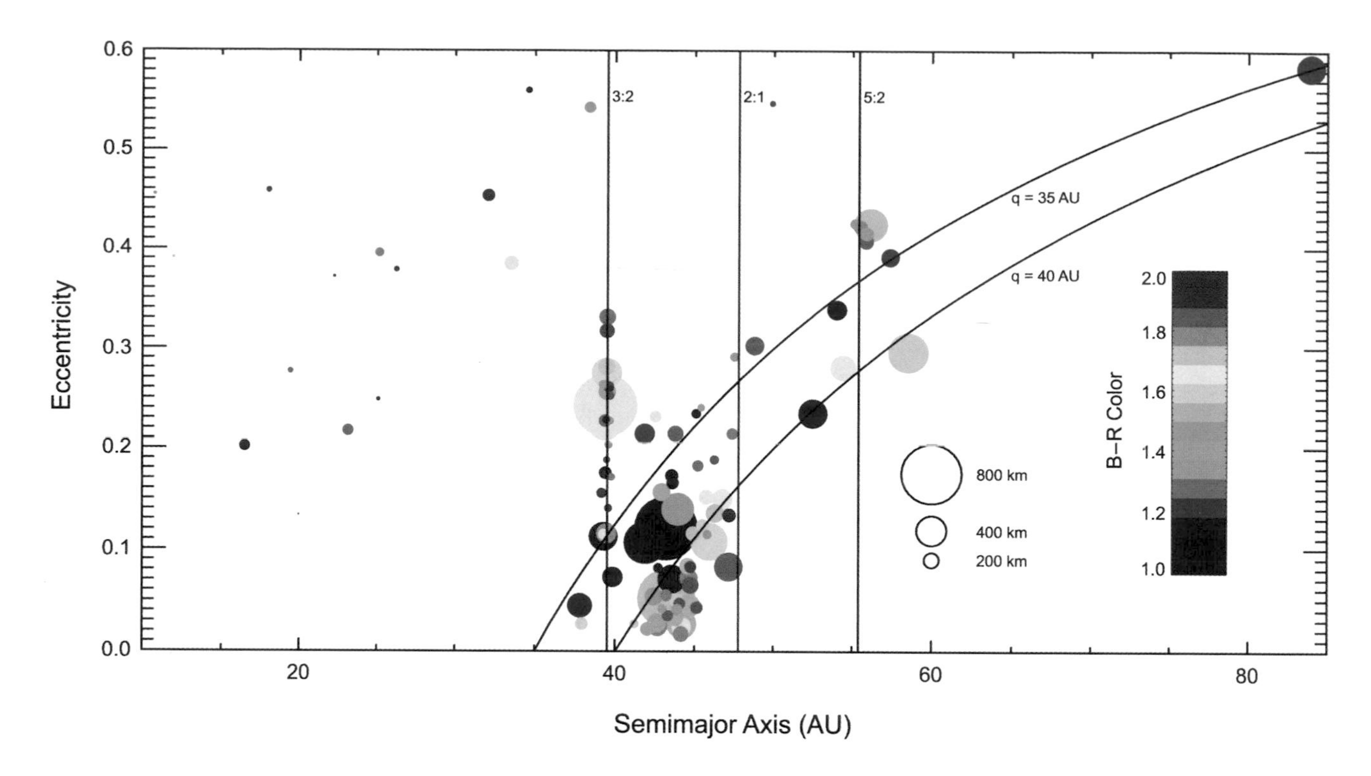

Plate 1. Colors of TNOs and Centaurs (more than 100 objects) in the orbital eccentricity vs. semimajor axis plane. The advantage of this representation is that it offers to the eye the global color distribution of the TNOs. The sizes of the symbols are proportional to the corresponding object's diameter (an average R geometric albedo of 0.09 has been assumed). A color palette has been adopted to scale the color spread from B–R = 1.0 (coded as dark blue) to B–R = 2.0 (coded as red). In comparison, B–R = 1.03 for the Sun and about 2 for the Centaur 5145 Pholus (one of the reddest known objects in the solar system). 3:2 (a ~ 39.5 AU), 2:1 (a ~ 48 AU), and 5:2 (a ~ 55.4 AU) resonances with Neptune are marked, as well as the q = 40 AU perihelion curve. A wide color diversity characterizes the outer solar system objects. Interesting patterns clearly emerge from this color map. For instance, objects with perihelion distances around and beyond 40 AU are *mostly* very red. Classical objects (mostly between the 2:3 and 1:2 resonances) with high eccentricity (and also inclination) are preferentially neutral/slightly red. In contrast, no clear trend is obvious for SDOs (a > 50 AU), nor for the Plutinos, which appear to lack any trends in their surface colors. Updated from *Doressoundiram et al.* (2005).

Accompanies chapter by Doressoundiram et al. (pp. 91–104).

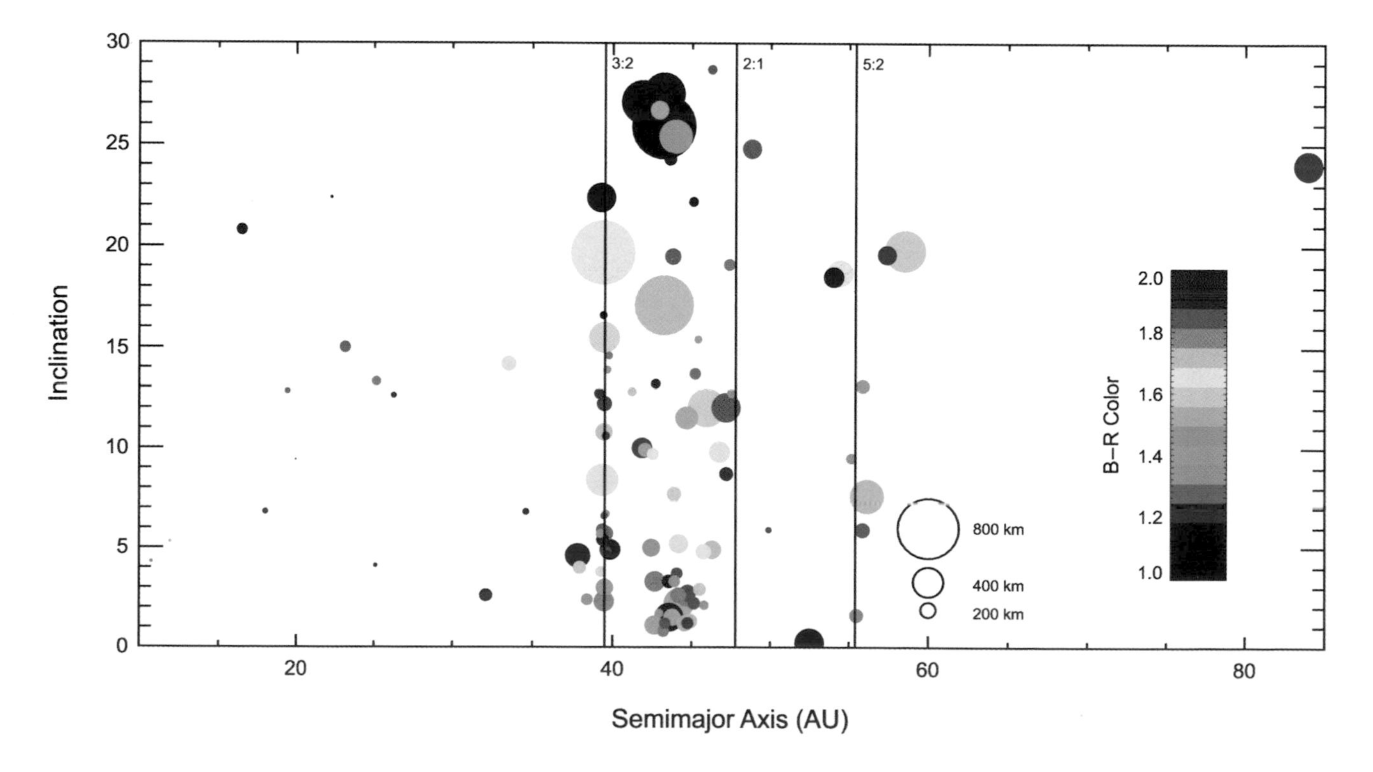

Plate 2. Same as Plate 1 in the orbital inclination vs. semimajor axis plane.

Accompanies chapter by Doressoundiram et al. (pp. 91–104).

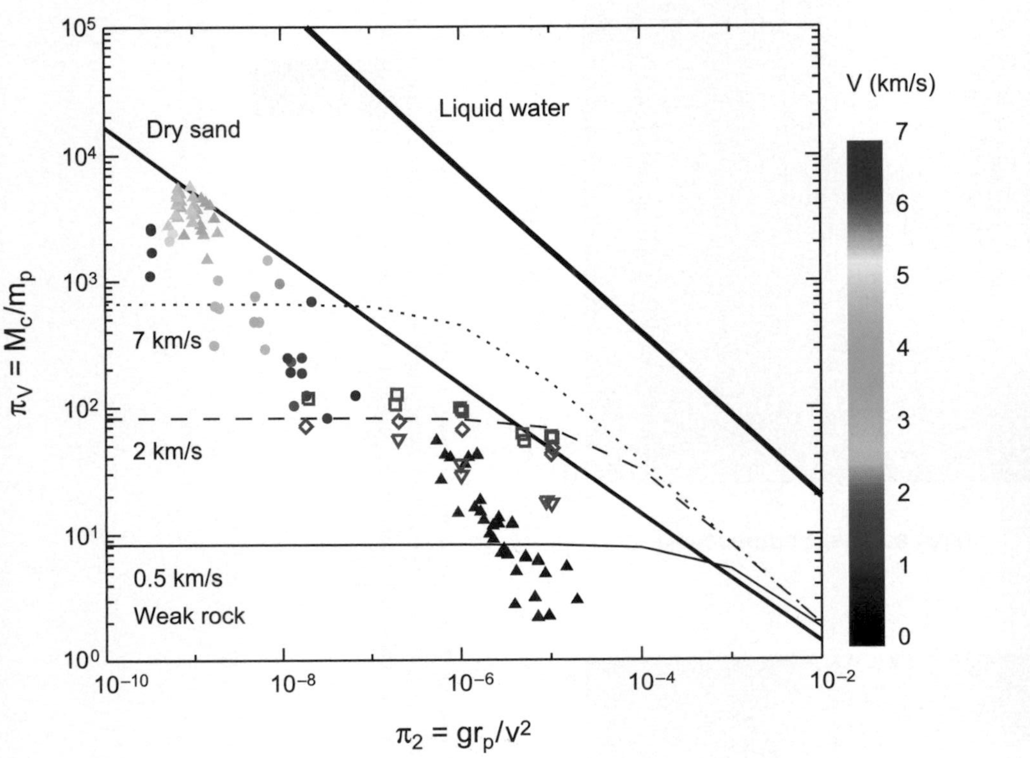

Plate 3. Cratering efficiencies (ratio of ejected and displaced mass to projectile mass, p_V) as a function of the ratio of gravitational to inertial forces (inverse Froude number, p_2) for different target materials at different velocities. Lines denote fitted cratering efficiencies in liquid water, dry sand (35% porosity), and weak rock using equation (10). Experimental data (see text): nonporous ice, ▲; 50% porous ice, ●; various crushable nonicy mixtures with porosities of about 40%, □; 70%, ◇; and 96%, ▽. Colors denote impact velocity.

Accompanies chapter by Leinhardt et al. (pp. 195–211).

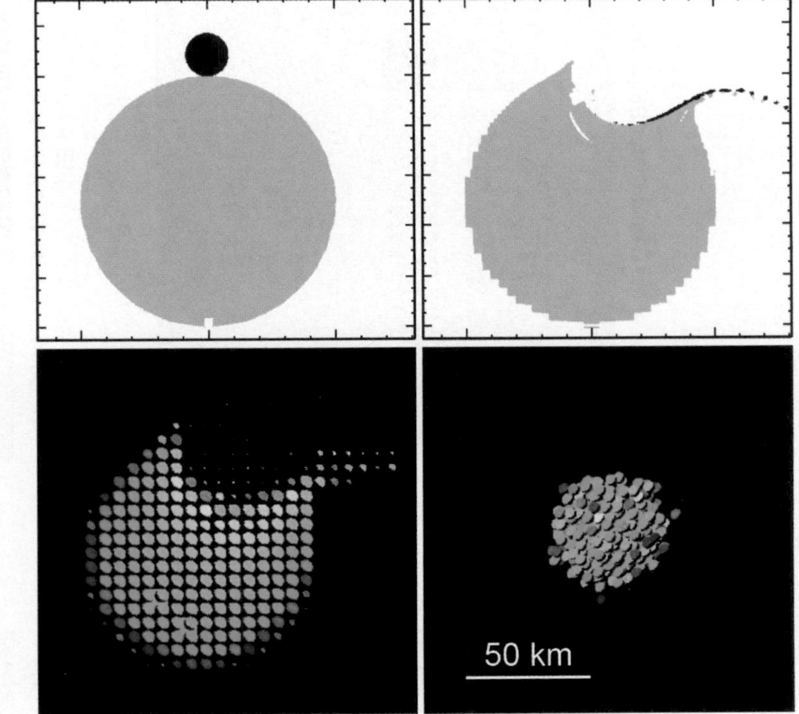

Plate 4. An example of a hybrid hydrocode to N-body numerical simulation of the catastrophic disruption of a solid ice target. The first row shows the positions of the target (gray) and projectile (black) at time 0 and 30 s. The second row shows the target after handoff from the hydrocode to the N-body code at 60 s and the largest remnant (r = 35 km) at 140 h. The projectile (r = 8.4 km) hit the target (r = 50 km) at 45° and 3 km s^{-1}. The first three frames show a slice through the three-dimensional target and projectile along the y = 0 plane. The final frame shows the surface of the largest postcollision remnant. The color coding in the N-body frames show the peak pressure attained by each mass element. Blue is the lowest peak pressure (2×10^6 dynes cm^{-2}), and red is the highest (1×10^{11} dynes cm^{-2}). The peak pressure is stored in Lagrangian tracer particles during the hydrocode component of the simulation. The few noncolored (gray or white) particles in the N-body images are blocks from the hydrocode grid that did not contain any tracer particles. The surface of the largest remnant shows a mixture of high and low levels of shock deformation.

Accompanies chapter by Leinhardt et al. (pp. 195–211).

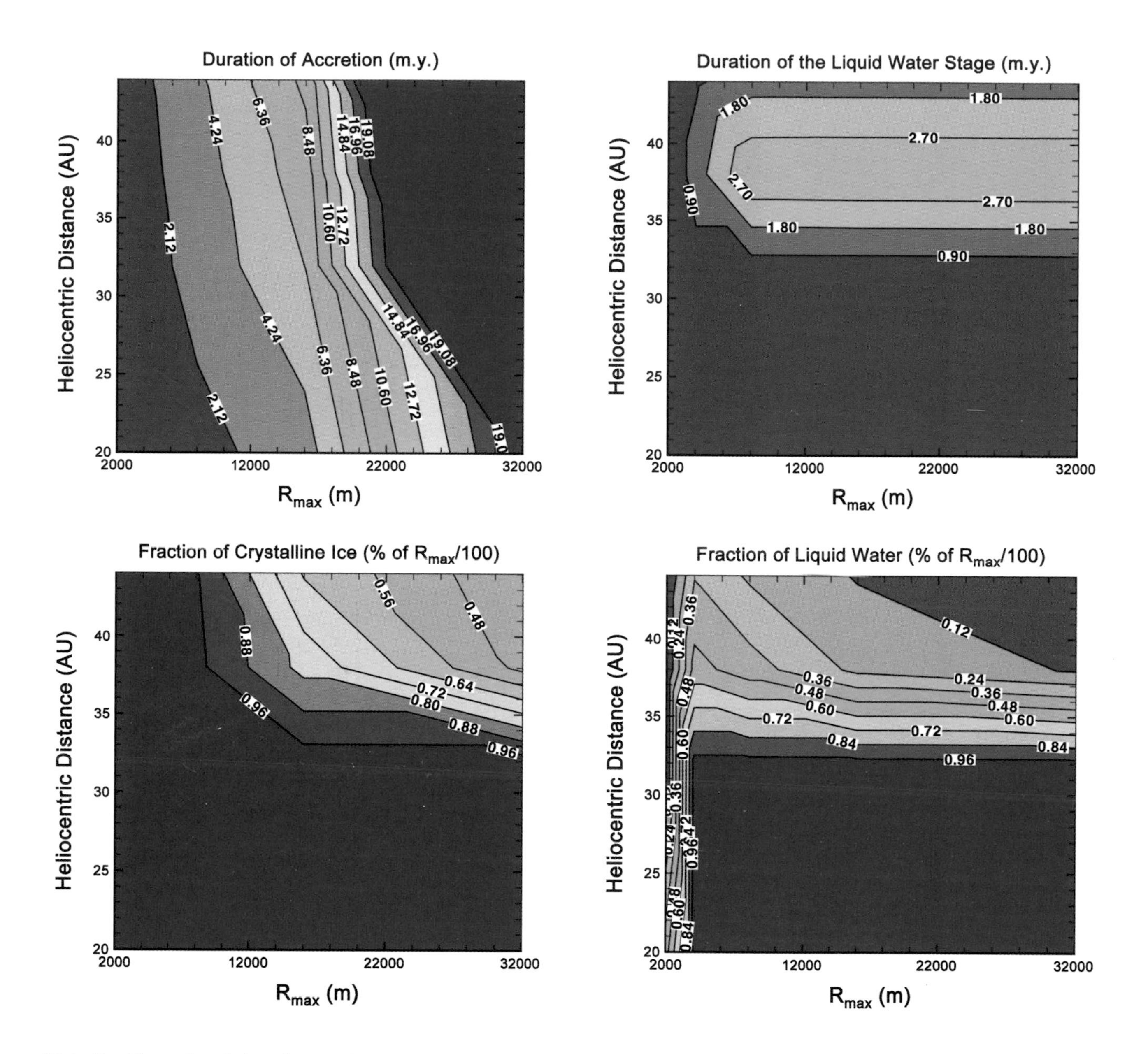

Plate 5. Thermal evolution characteristics of growing bodies with a composition of 0.775/0.225 dust-to-ice by mass, porosity of 0.5, and negligible permeability: top left, accretion time; top right, duration of liquid-water stage; bottom left, effective radial (not volume) extent of crystalline ice zone; bottom right, effective maximal radial extent of liquid water. Modified from *Merk and Prialnik* (2006).

Accompanies chapter by McKinnon et al. (pp. 213–241).

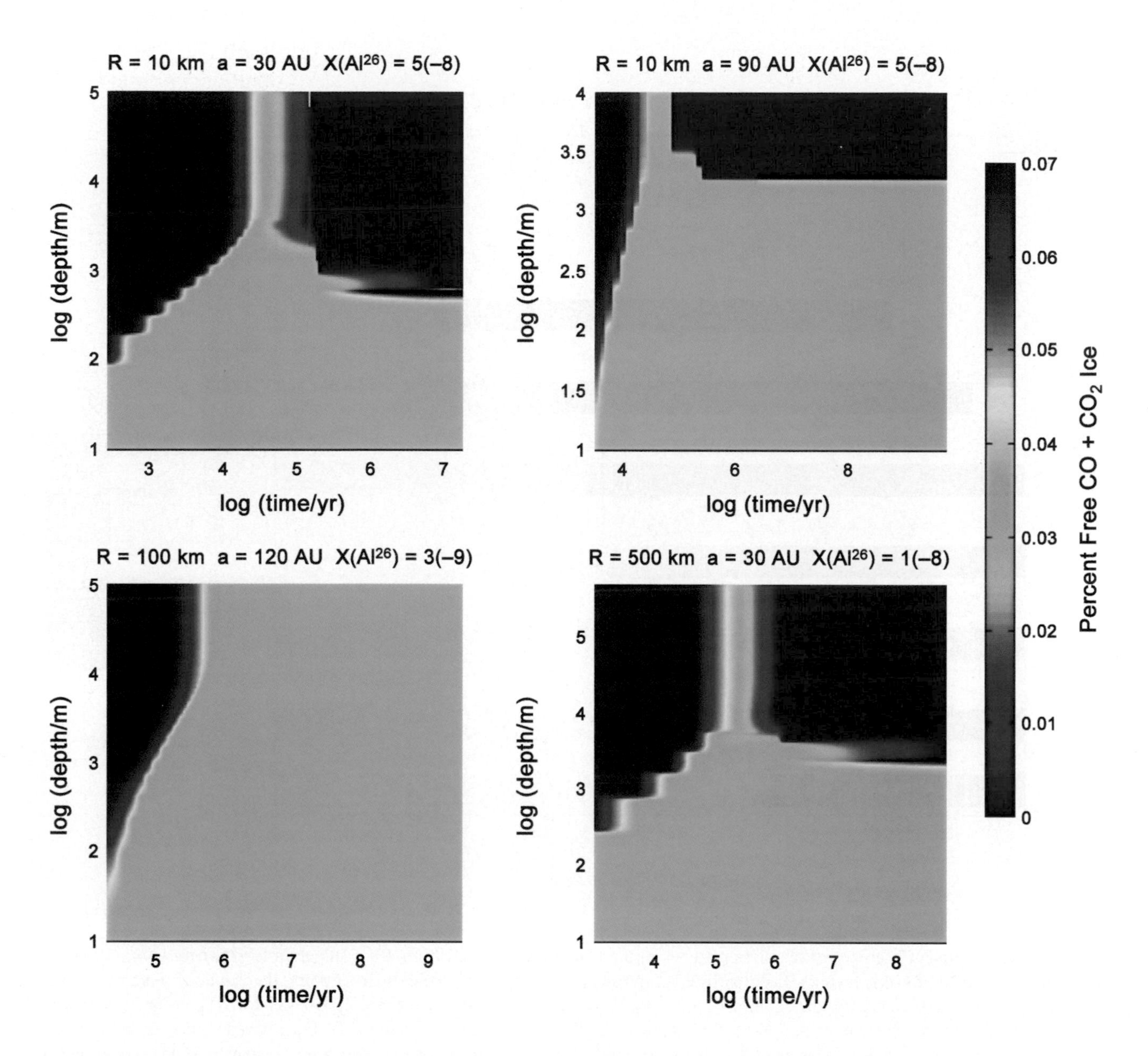

Plate 6. Map of the total mass density of free CO and CO_2 ices (initially 0.035 g cm^{-3} each, corresponding to CO/H_2O and CO_2/H_2O abundances of 0.08 and 0.05, respectively), as it changes with time throughout the body, for different combinations of body radius, heliocentric distance (i.e., surface temperature), and initial ^{26}Al content, as indicated. CO sublimates and escapes first, followed by the CO_2; in some cases CO_2 gas from the interior refreezes at the base of the CO_2-retaining outer layer, or rind. Initial temperature T_0 = 10 K, rock mass fraction is 0.5, and porosity is 53%. Modified from *Choi et al.* (2002).

Accompanies chapter by McKinnon et al. (pp. 213–241).

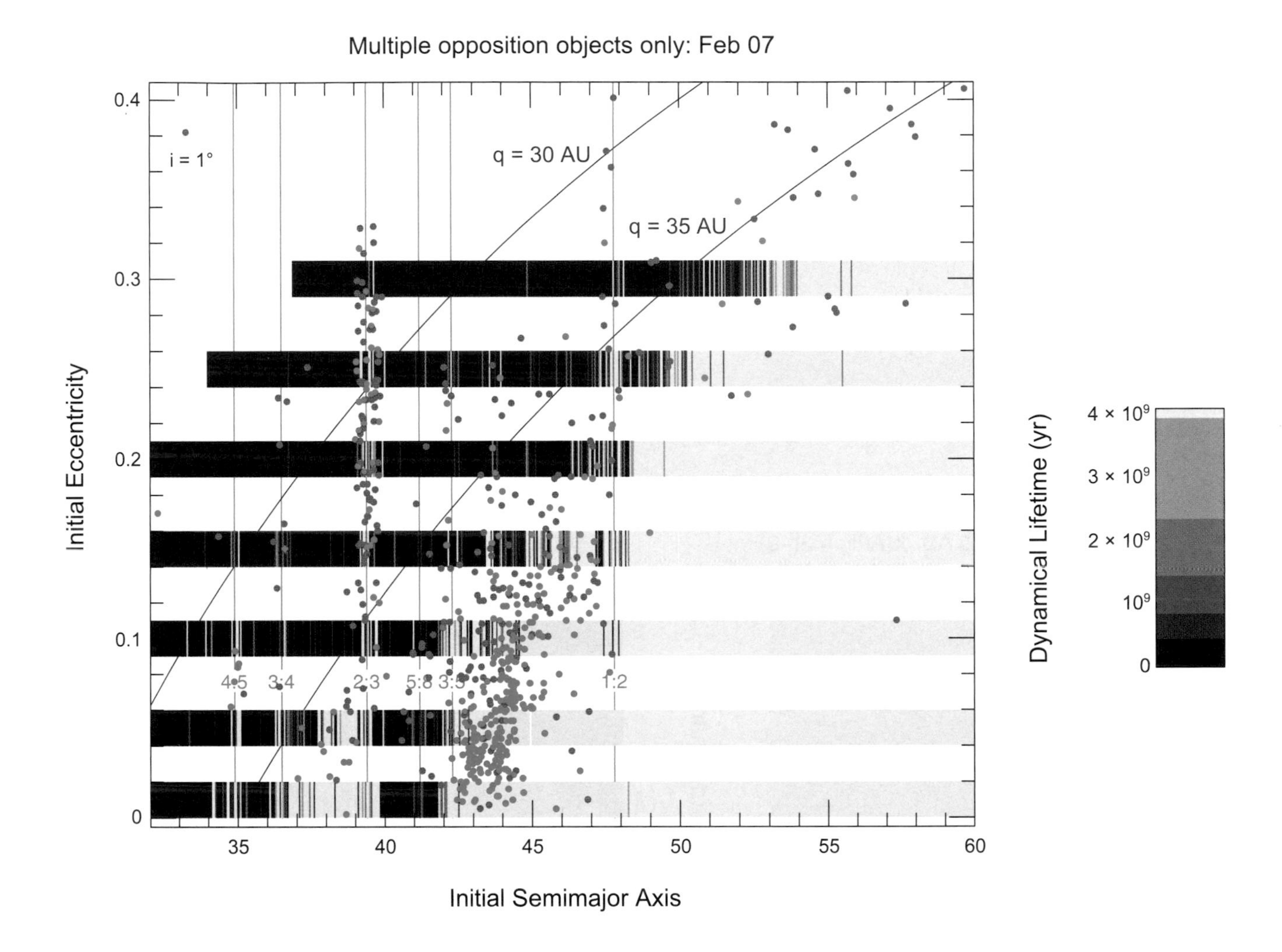

Plate 7. The dynamical lifetime for massless particles in the Kuiper belt derived from 4-b.y. integrations (*Duncan et al.,* 1995, but extended to a = 60 AU for this review). Each particle is represented by a narrow vertical strip of color, the center of which is located at the particle's initial eccentricity and semimajor axis (the initial orbital inclination for all objects was 1°). The color of each strip represents the dynamical lifetime of the particle, as reported on the scale on the righthand side. For reference, the locations of the important Neptune mean-motion resonances are shown in blue and two curves of constant perihelion distance, q, are shown in red. The (a, e) elements of the Kuiper belt objects with orbits determined over three oppositions are also shown. Green dots are for i < 4°, magenta dots otherwise.

Accompanies chapter by Morbidelli et al. (pp. 275–292).

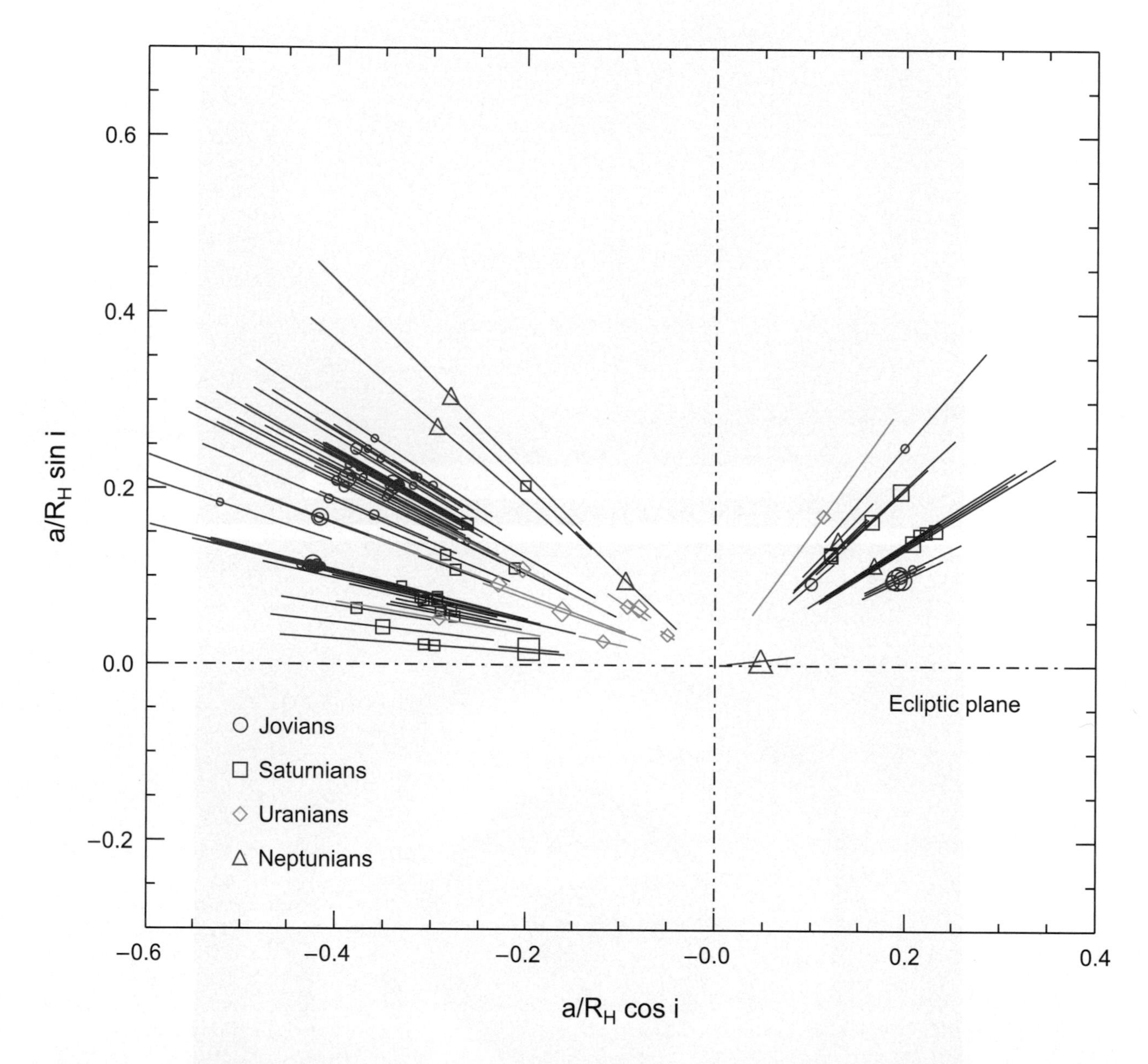

Plate 8. A comparison of the mean orbital elements of the irregular satellites, updated from Fig. 1 of *Gladman et al.* (2001a). The distance from the origin represents the semimajor axis, scaled by the planet's Hill's sphere radius (see Table 1), while the polar angle from the X-axis represents the inclination relative to the ecliptic. A radial line through each data point indicates the periapse and apoapse radii, while the size of the symbol is proportional to the logarithm of the object's estimated size. This plot excludes Triton and the nine saturnian satellites discovered in 2006, whose orbits are still relatively uncertain.

Accompanies chapter by Nicholson et al. (pp. 411–424).

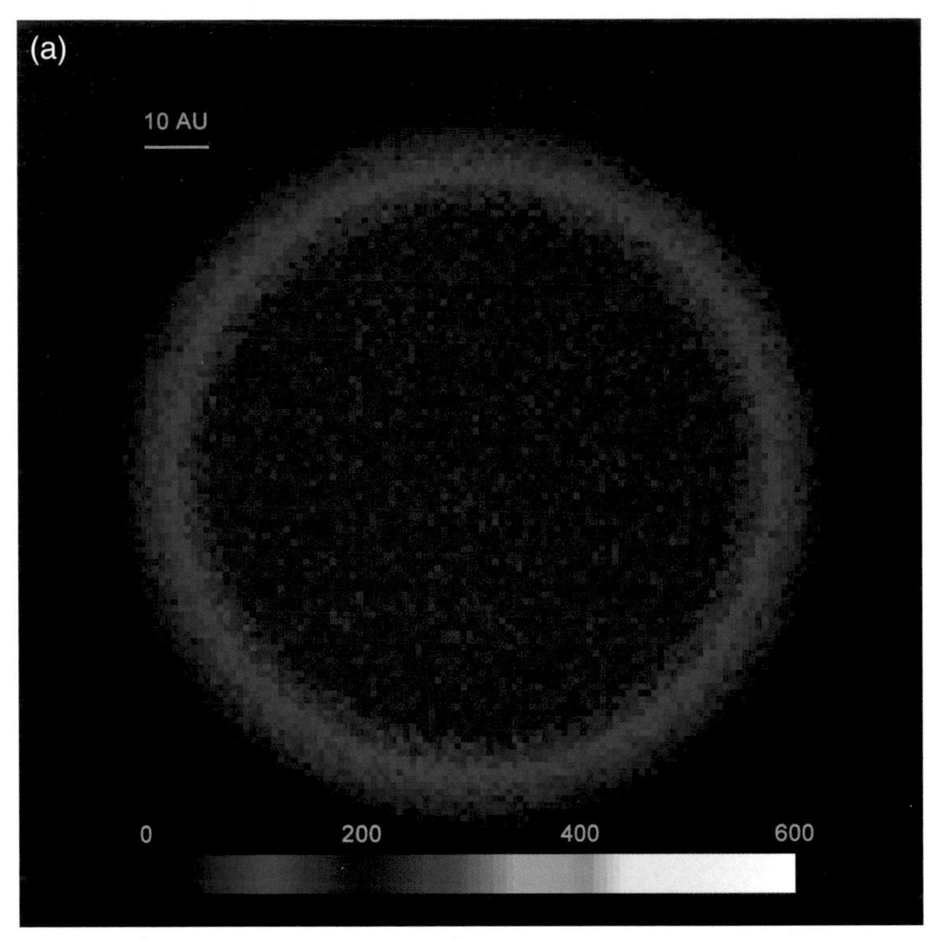

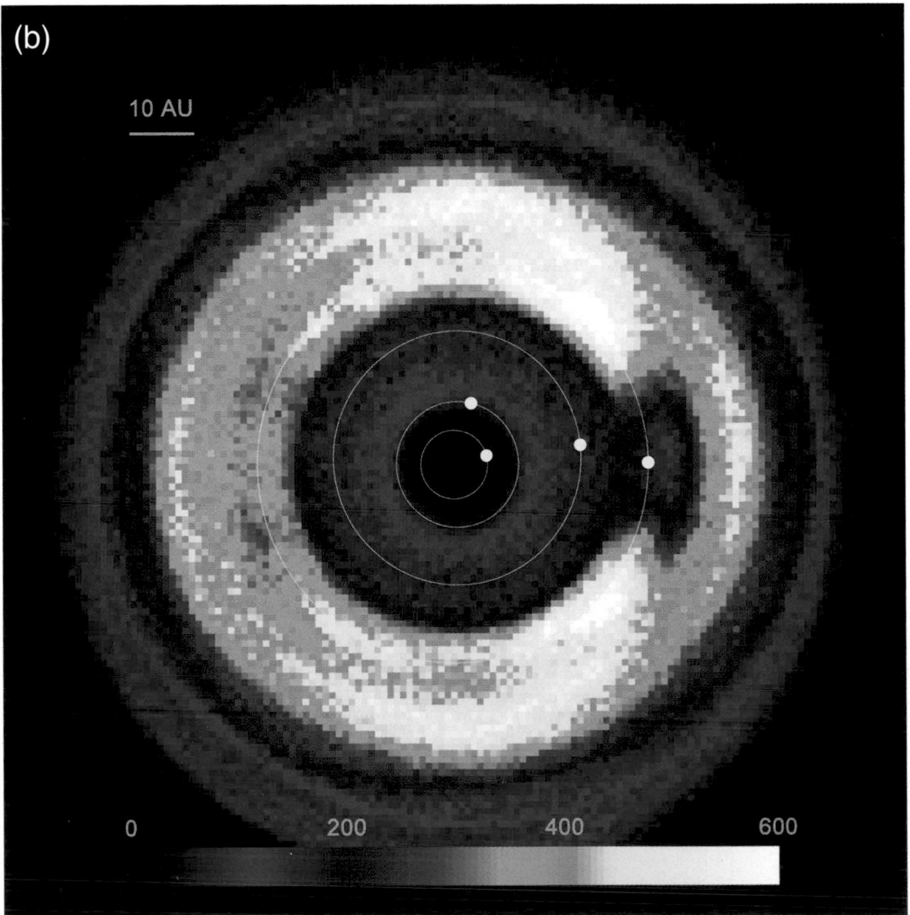

Plate 9. Column density maps of the simulated 23-μm KB dust particles: **(a)** without planets and **(b)** with seven planets included in the numerical simulations.

Accompanies chapter by Liou and Kaufmann (pp. 425–439).

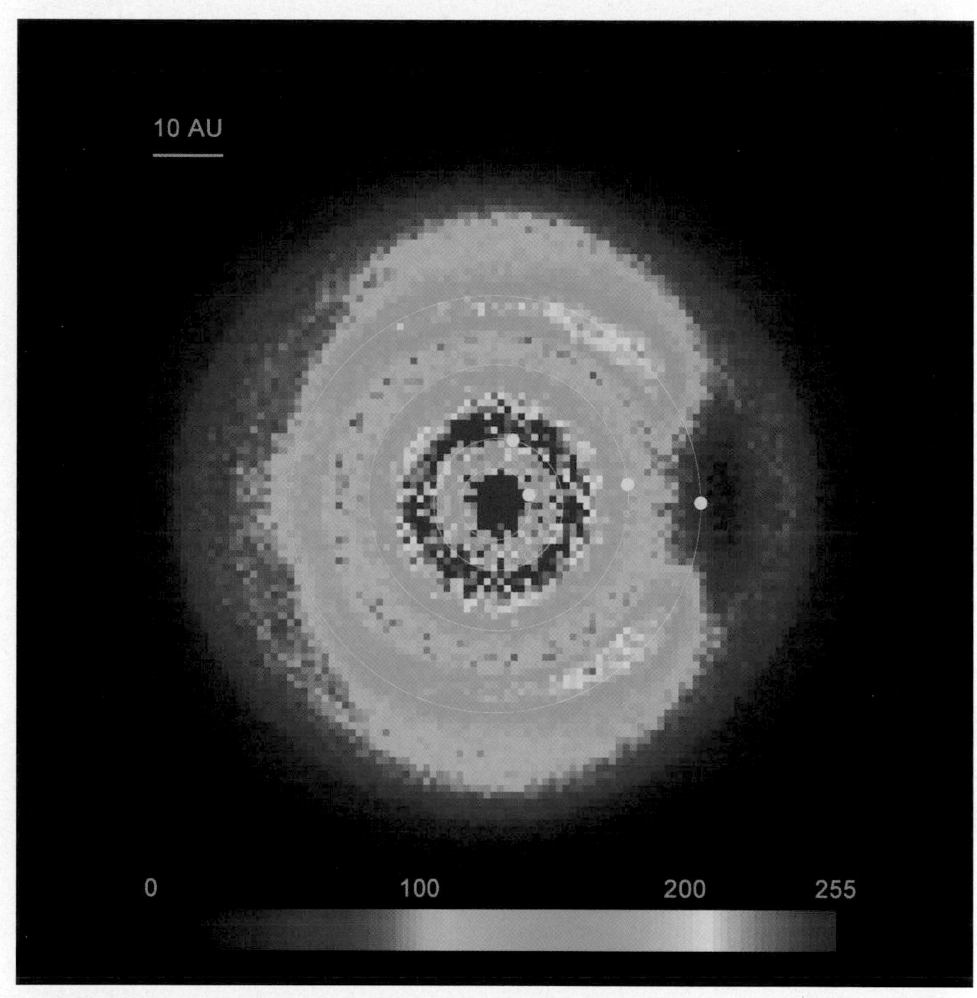

Plate 10. False-color images of the calculated brightness of a simulated KB dust disk. The orbits and positions of the four giant planets are also labeled for reference.

Accompanies chapter by Liou and Kaufmann (pp. 425–439).

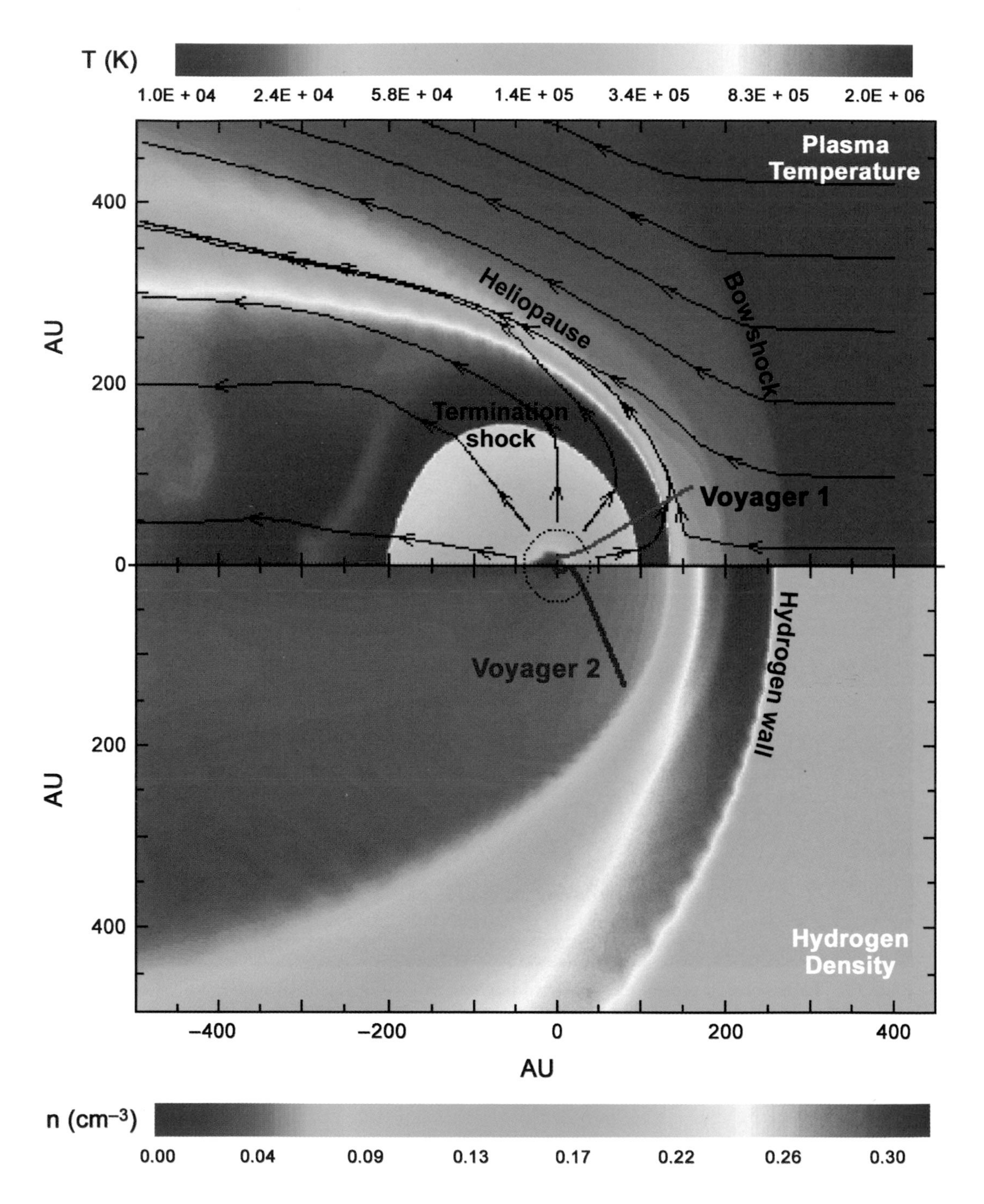

Plate 11. A plot of the heliosphere from a plasma (top) and neutral (bottom) perspective. The figure shows the equatorial plane from a model of *Müller et al.* (2006). The color bar on the top panel shows the plasma temperature. The lines show the plasma flow. The main boundaries, termination shock, heliopause, and bow shock are labeled. The color bar on the bottom panel shows the H density; the hydrogen wall in front of the heliopause is labeled and the trajectories of the Voyager spacecraft are shown. Figure courtesy of H. Müller.

Accompanies chapter by Richardson and Schwadron (pp. 443–463).

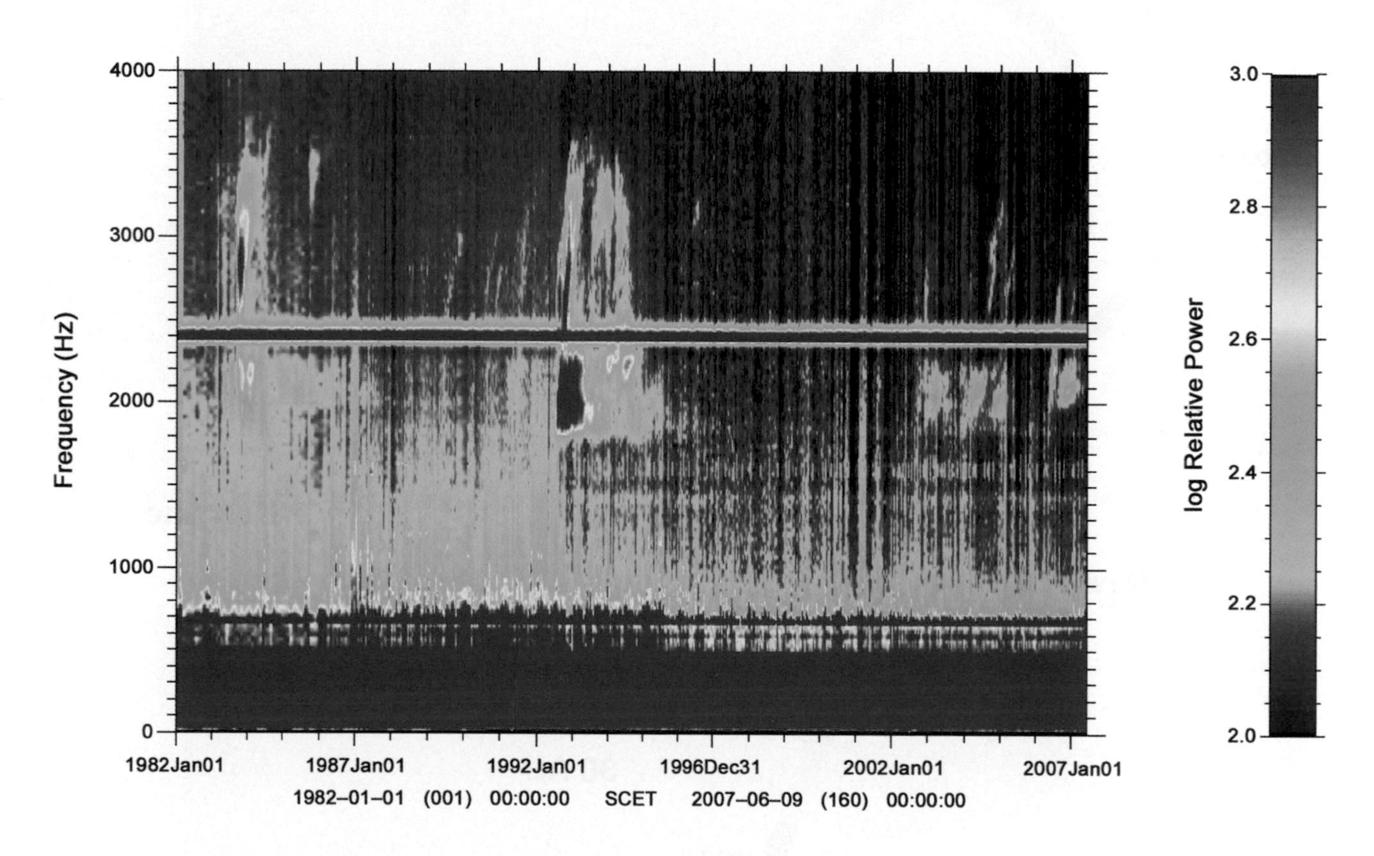

Plate 12. Plasma wave spectra showing heliospheric radio emissions. The horizontal red bar and colors at low frequencies are noise. The radio emissions are observed at 1.8–3.6 kHz in 1983–1985, 1992–1995, and 2003–2007. Figure courtesy of D. Gurnett.

Accompanies chapter by Richardson and Schwadron (pp. 443–463).

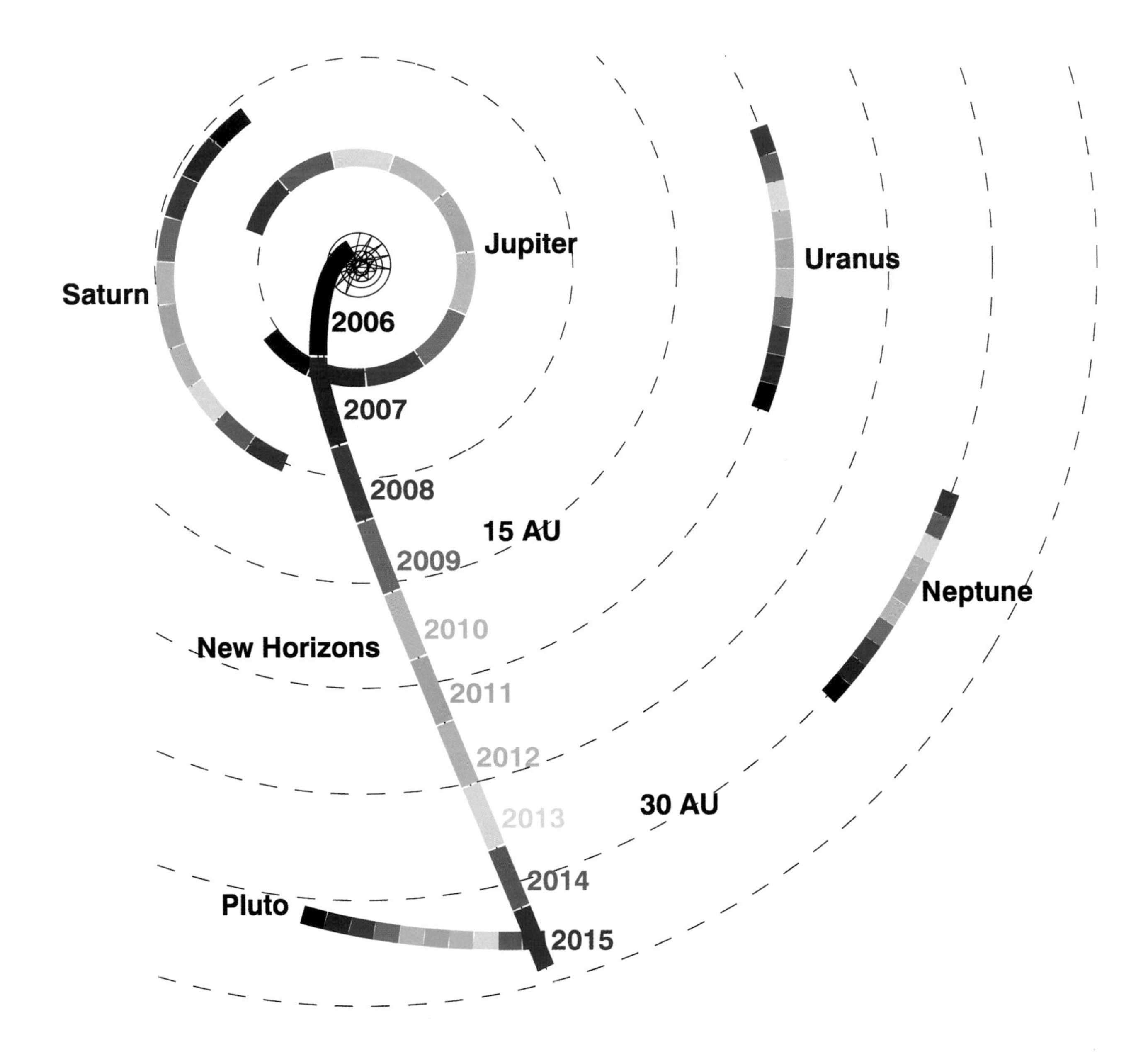

Plate 13. The trajectory of New Horizons (NH) and the locations of the planets, as projected on the ecliptic plane, are displayed. Different colors are used for different years. NH was launched on January 19, 2006, had its closest approach to Jupiter on February 28, 2007, and will encounter Pluto in July 2015. From *Young et al.* (2008).

Accompanies chapter by Weaver and Stern (pp. 557–571).

Index

Page numbers refer to specific pages on which an index term or concept is discussed. "ff" indicates that the term is also discussed on the following pages.